王文兴院士（1927—　）

一、中学时期——抗日战争期间（1941~1946）

国立第二十一中学初二，家乡沦陷，萧县中学学生跟游击队来此，学校提供衣食住。安徽太和（1942）

萧县部分同乡欢送初三毕业同学，二排右三王文兴，安徽太和（**1944.7**）。日军侵占京汉铁路后太和危机，教育部命令学校西迁到安全地带。**1944** 年 **10** 月开始西迁，经河南到陕西，**1945** 年 **7** 月到蓝田。

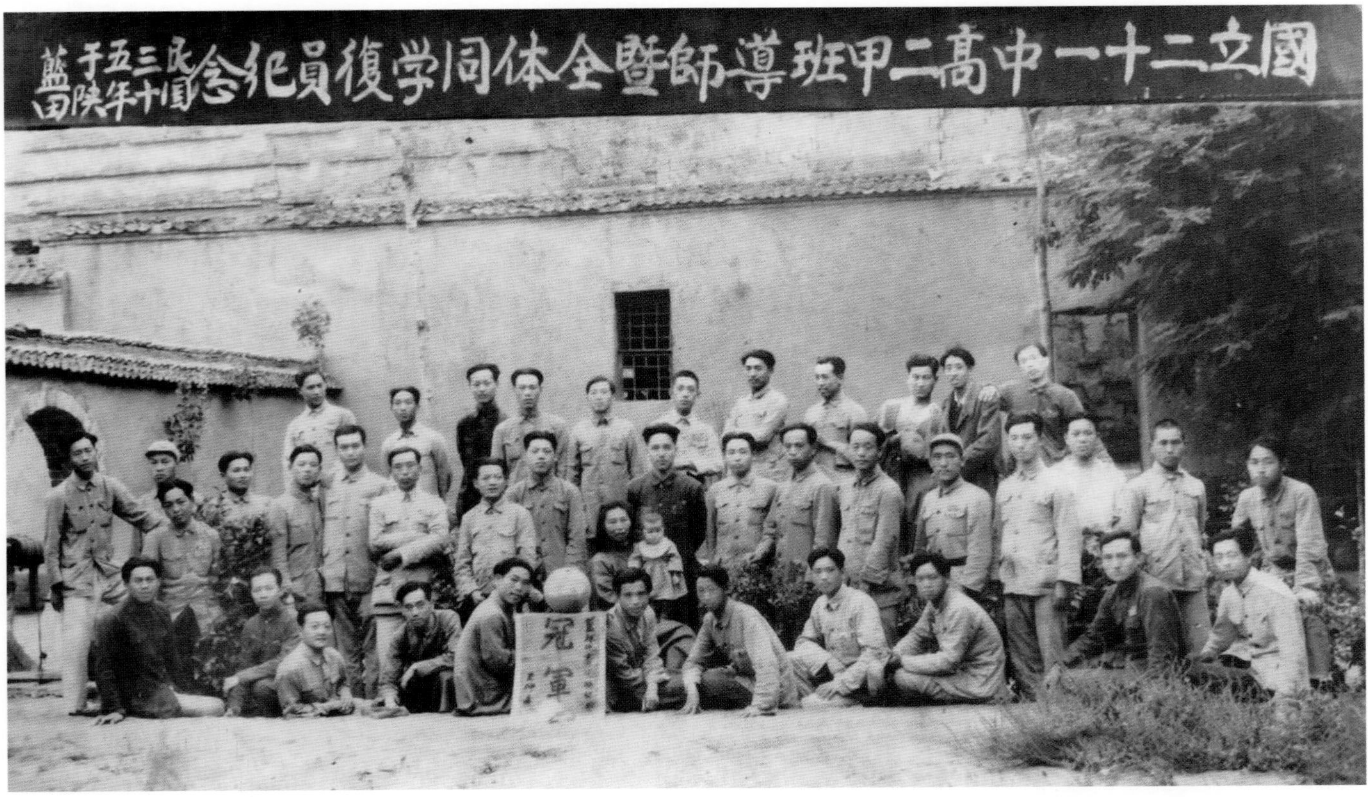

国立第二十一中学高二甲班导师及全体同学抗日战争胜利复员纪念，前排右四王文兴、中排右九代数老师马菊圃及其夫人（抱小孩者），陕西蓝田（**1946.6**）。国立中学随着抗日战争胜利完成了历史使命，教育部决定教职工、学生结束战时生活，复员回原籍。**1946** 年 **7** 月江苏籍学生回到在徐州新建的江苏省立连云中学，学生的待遇不变，**1947** 年 **7** 月王文兴在此高中毕业。后来，连云中学合并到江苏省立新海高级中学。

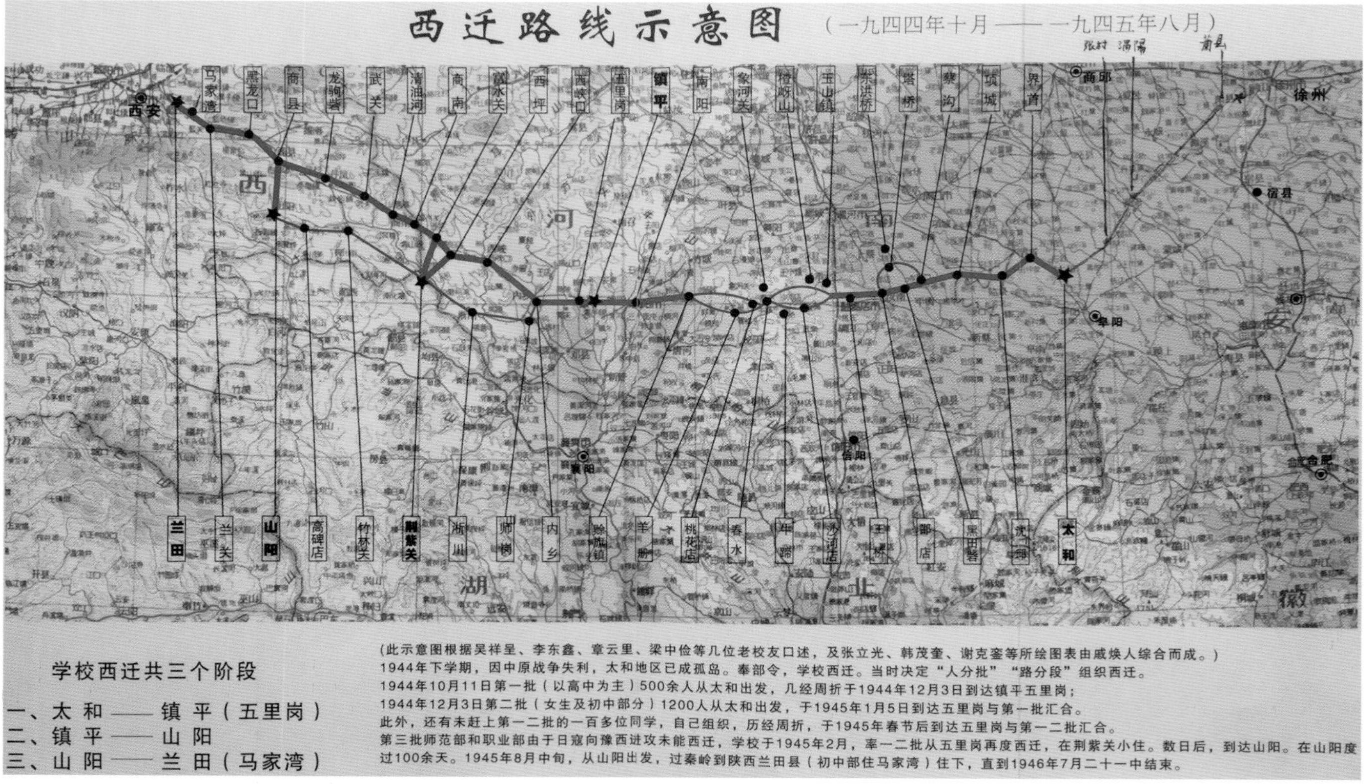
西迁路线示意图
（一九四四年十月——一九四五年八月）
学校西迁共三个阶段
一、太和——镇平（五里岗）
二、镇平——山阳
三、山阳——兰田（马家湾）
(此示意图根据吴祥呈、李东鑫、章云里、梁中俭等几位老校友口述，及张立光、韩茂奎、谢克銮等所绘图表由戚焕人综合而成。)
1944年下学期，因中原战争失利，太和地区已成孤岛。奉部令，学校西迁。当时决定“人分批”“路分段”组织西迁。
1944年10月11日第一批（以高中为主）500余人从太和出发，几经周折于1944年12月3日到达镇平五里岗；
1944年12月3日第二批（女生及初中部分）1200人从太和出发，于1945年1月5日到达五里岗与第一批汇合。
此外，还有未赶上第一二批的一百多位同学，自己组织，历经周折，于1945年春节后到达五里岗与第一二批汇合。
第三批师范部和职业部由于日寇向豫西进攻未能西迁，学校于1945年2月，率一二批从五里岗再度西迁，在荆紫关小住。数日后，到达山阳。在山阳度过100余天。1945年8月中旬，从山阳出发，过秦岭到陕西兰田县（初中部住马家湾）住下，直到1946年7月二十一中结束。
西安
马家湾
黑龙口
商县
龙驹砦
武关
清油河
商南
富水关
西坪
西峡口
五里岗
镇平
南阳
象河关
橙岈山
玉山镇
东洪桥
塔桥
蔡沟
项城
界首
商邱
徐州
宿县
阜阳
兰田
兰关
山阳
高碑店
竹林关
荆紫关
淅川
内乡
春水
太和

国立第二十一中学及其前身江苏萧县中学抗日战争中学校迁移，被喻为“小长征”

二、大学与进修期间（1947~1957）

国立安徽大学数学系一年级，安徽安庆，公费，免食宿费。（1947.10）

院書敷敬

国立安徽大学旧址。敬敷书院，原名培原书院，后又改称安徽大学堂、安徽高等学堂，现为国家重点文物保护单位。这里曾是国立安徽大学的学生宿舍，1947年开始王文兴在此居住一年多，安徽安庆。

山东大学化学系，青岛（1949 年夏）

山东大学化学系二年级全体同学，青岛（1950）

山东大学化学系欢送一九五一届毕业生纪念，三排左七物理化学教授刘遵宪、右九有机化学教授刘椽、四排右八王文兴，青岛（1951.9.12）

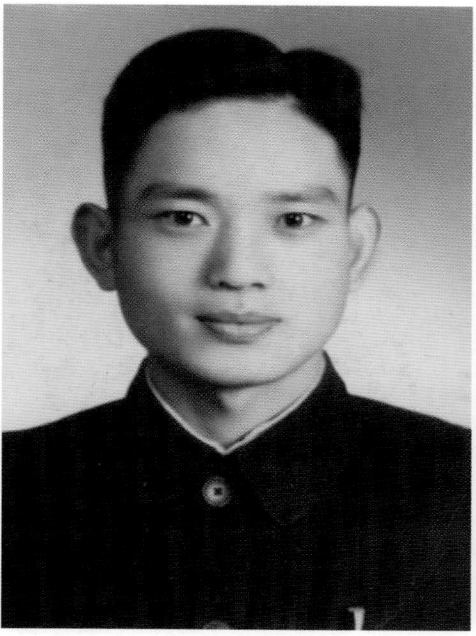

东北人民大学（吉林大学前身）化学系研究生班进修，长春（1956）

东北人民大学化学系物化班，左起傅献彩、王文兴、戴树珊，长春（1956.7）

在前苏联卡尔波夫物理化学研究所进修，莫斯科（1959）

周末与物化研究所同事在莫斯科郊外，后排左一王文兴（1959）

左孙天赋、中王文兴、右主任吴冰颜，莫斯科（1962.2）

在克里姆林宫大剧院，左李鑫、中王文兴、右孙天赋（1962）

在住处，中王文兴，莫斯科（1962）

三、重工业部与化工部属科研单位（1957~1975）

1956 年 5 月化工部成立。1957 年化工部沈阳化工综合研究所一分为四，分别建立沈阳、北京、天津和上海四个化工研究院。王文兴先后在沈阳、北京和天津化工研究院工作二十多年。

化工部沈阳化工研究院物化室催化组，前排中王文兴（1956）

重工业部沈阳化工综合研究所所部，催化组在此建立（1954），这是化工部最早的催化实验室。1958 年随物化室迁到北京。

化工部北京化工研究院物化室欢送兰州大学学生完成实验及论文合影，前排左三王文兴（**1964**）

化工部北京化工研究院实验大楼，王文兴在此楼物化室工作近九年 （**1958-1966**）

化工部天津化工研究院部同事回访合影，后排左五王文兴（**1991**）

化工部天津化工研究院实验大楼，王文兴在此工作近 **10** 年。（**1966.6-1976.1**）

20 世纪 70 年代中国工业已发展到一定规模，环境污染已经显现，为借鉴发达国家环境治理经验，1973 年 6 月中国环境科学技术考察团赴工业革命发源地英国考察工业对环境的影响。成员由政府部门领导、科研和工业界专家 11 人组成，在英国进行一个多月的考察。访问了伦敦、曼彻斯特、谢菲尔德等城市，考察了钢铁、化工等工业，以及泰晤士河、川特河水质污染与控制等。部分代表列席了当年八月在北京召开的全国第一次环保工作大会。

中国环境科学技术考察团赴英国考察工业对环境的影响，前排左起三李澄、四丁秀、五赵士修，后排左三刘东生、六王文兴，伦敦（**1973.6.2-1973.7.3**）

中国环境科学技术考察团在英国考察期间，休息日三人同行，在马克思墓前合影，左起王文兴、刘东生、刘永清，伦敦（**1973**）

四、天津市环保办、环保局（1976~1983，1980.3 借调到中国环科院）

全国第一次环境科学技术情报会议，左王文兴、右戴树桂，黄山（1976）

黄渤海污染调查课题，在金星号考察船上，二排左二王文兴、三郭方、四徐柏麟、五周静（1977）

五、中国环境科学研究院（1980~）

中国环境科学研究院于1978年12月31日正式成立，1980初边建设、边科研，大气环境研究先行建设。研究院全部新建，科研人员面向全国招聘，很快集中一批年富力强的科研人员。

中国环境科学研究院建院初期，右边平房是当时仅有的几栋房子之一。在后边的临建房里开展科研。左一朱铨鈞、三黄新民、四王文兴（1980）

中国环境科学研究院建院初期在板房内召开研讨会，右排左起一缪天成、二王文兴、三黄新民、四唐孝炎、左排左一朱铨鈞（1980）

中国环境科学研究院建院初期开展工作和住宿的板房，冬天冷，夏天热（1980）

中国环境科学研究院和中国监测总站建设初期，建设中的实验大楼（20 世纪 80 年代初）

中国环境科学研究院建院初期在板房内召开研讨会，中间黄新民、右排后王文兴（1981）

建院初期在板房内与日本学者秋元肇（右前）讨论光化学研究，左前黄新民、右后王文兴（1981）

联合国资助项目大气化学和大气物理学科建设涉及建设先进的光化学烟雾箱和大型环境风洞。为此成立 10 人考察组，先后赴日本和美国各考察一个月。在日本考察了日本国立公害研究所、资源和环境研究所，东京大学、东京工业大学、千叶大学等高校。在美国考察了美国环保署（EPA）三角科技园大气环境相关研究所、北卡罗来纳大学、乔治亚理工学院、纽约州立大学奥尔巴尼分校、美国 NOAA、NCAR、加州理工学院、加州大学河滨分校等单位。

在日本国立公害研究所与所长交谈，左起王文兴、近藤次郎所长、全浩，日本筑波（1980.12）

参观日本霞浦湖水质保护，前排左三王文兴（1980.12）

考察组在美国环保署总部了解大气环境保护和科学研究，前排左二王文兴、三吴博士，华盛顿（1981.1）

在北卡莱罗纳大学参观室外烟雾箱，左一王文兴、二 **Kamens**，北卡 **Chapel Hill**（**1981**）

考察组在美国乔治亚理工学院考察，左王文兴、右江家驷（地球物理学院院长），亚特兰大（**1981.1**）

在加州理工学院参观，左一唐孝炎、二王文兴、三田炳申，加州 **Pasadena**（**1981.1**）

全国大气环境质量标准审议会，**1981** 年 **12** 月在武汉召开，这是我国第一个环境空气质量标准，国家环境保护办公室委托王文兴主持本次审议会。

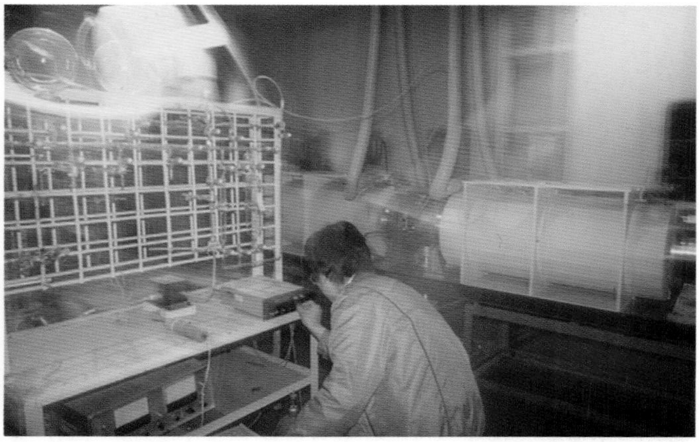

中国环境科学研究院建设的可抽真空、长光程 **FTIR** 烟雾箱，左配气系统，右反应器系统（**1985**）

兰州西固地区光化学污染研究课题，观测现场，左起唐孝炎、王文兴、黄建国、任阵海、陈长河、彭贤安（1981）

太原地区大气综合观测项目，飞机航测（双水獭飞机，除双水獭外还有米 -8 和 BO-105），前排左五山西省副省长兼太原市委书记王茂林（1982）

左二王文兴、三太原市环保局副局长钮骏岭等在飞机航测现场（1982.12）

右一王文兴等在太原飞机航测现场（BO-105 飞机）（1982.2）

我国酸雨研究项目，峨眉山金顶酸雨观测（1985）

在峨眉山森林考察酸雨污染情况（1985）

在乐山考察酸雨对乐山大佛的影响（1985）

国环办组织在海南考察生态环境，前排左一王文兴、二梁思翠、后排尹改（1985）

国家“七五”科技攻关项目：华南酸雨研究，南宁，前排左二丁永福、左三班玲，后排左一郝吉明、二张婉华、三王文兴（1987）

国家“八五”科技攻关项目浙江酸雨课题鉴定会后合影，左二何纪力、三王文兴、四李柱国（1996）

1998

2005

2016

王文兴于 1998（左图）和 2005 年（中图）到美国 Mitchell 山，考察酸雨破坏森林的恢复情况，与 2016 年森林现状（右图为美国朋友拍摄）的对比，查看近二十年来该地区森林生长状况

王文兴在广州酸雨监测培训班上讲解酸雨监测质量控制和质量保证（**1988.1**）

王文兴在国家“八五”科技攻关验收评审会上汇报工作，科技部组织验收（**1998**）

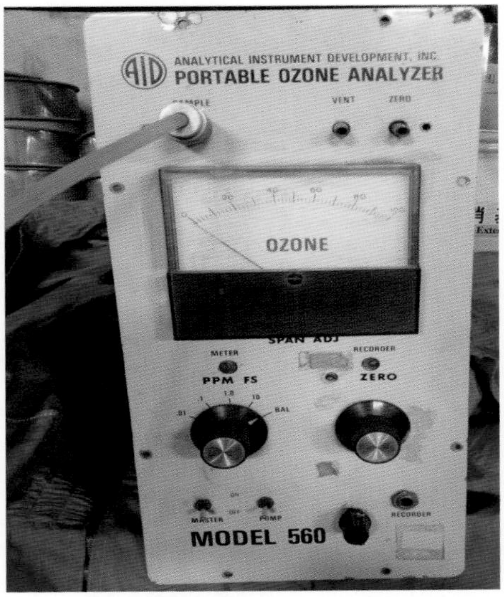

这台从美国进口的臭氧仪，自 **1981** 年起王文兴携带它参加兰州光化学烟雾观测、太原和沈阳煤烟型大气污染观测、峨眉山臭氧立体观测，在中国环科院早期大气环境研究中起到重要作用

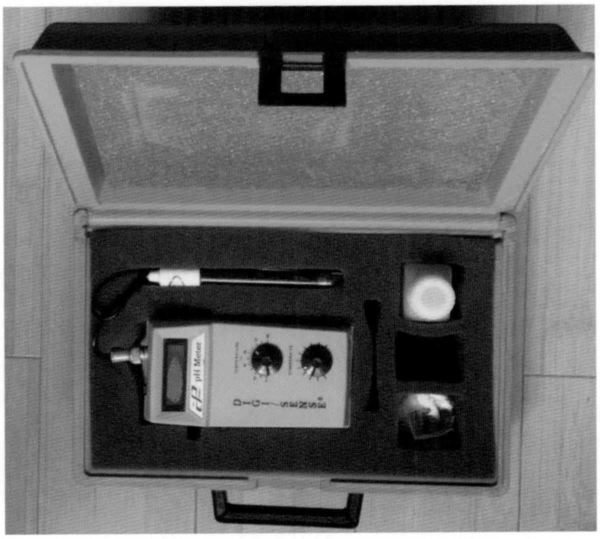

这台从美国进口的便携式 **pH** 计，自 **1986** 年起王文兴携带它走遍我国大地，在吉林图们和辽宁凤凰山的夏季、舟山群岛及东南沿海的冬季发现强酸雨及其来源，利用它核对当地监测数据，在酸雨国家科技攻关项目中起到重要作用

美国北卡莱罗纳大学 **Kamens** 教授访问中国环境科学研究院，他为中国培养了大批人才。左起王文兴、**Kamens**、孟伟院长（**2006**）

王文兴与 **W. E.Wilson** 夫妇合影（**2005**）。他们三十多年来致力于中美大气环境合作研究

六、山东大学 济南（2002—　）

山东大学环境研究院成立大会，常务副校长王琪珑（左一）主持，宣布聘任王文兴（左二）任院长，展涛（左三）校长讲话，山东省环保厅厅长张凯（左四）致辞，左六娄红祥副校长，济南（2003.11）

访问江西省环科院，与黄云院长等讨论降水化学问题，商讨在庐山进行大气观测研究事宜，南昌（2004.10）

山东东部海滨考察，后排左起王文兴、张婉华、刘春博（2004）

2007 年设立“王文兴院士环境科学奖学金”并为首届获奖研究生颁发证书，前排左起五胡敬田、六王文兴、七王琪珑副校长、八张庆竹，济南（2007）

王文兴在山东大学中心校区理科综合楼给研究生讲授《大气环境化学》课程，济南（2010）

2010 年 4 月王文兴受聘为山东大学终身教授，与校领导等合影，前排左二王文兴、四校党委书记朱正昌、五校长徐显明，济南（2010）

2014 年 4 月山东大学环境研究院建院十周年与山东大学张荣校长（前排左五）、院士专家合影，济南（2014）

山东大学与中国科学院生态环境研究中心合办的首届“环境与健康菁英班”毕业典礼，左四山东大学校党委书记李守信、五王文兴、八江桂斌，济南（2015.5）

考察庐山气象站并指导酸雨监测工作（2006）

山东泰山酸雨观测实验，前排左起张庆竹、郭佳、任宇，中排左起周杨、王艳、张佳玟，后排左起王哲、薛丽坤、王进、王文兴、王韬、潘振南（2007）

湖南衡山酸雨观测实验，前排左起王琳琳、王新明、张庆竹、王文兴、高晓梅、范绍佳、马强，后排左起李彭辉、薛丽坤、王韬、雷国强、徐政（2009）

中国环境科学研究院大气观测站，左起高健、陈义珍、柴发合、张晖、王文兴、张婉华、王淑兰、张新民（2007）

北京昌平奥运会大气观测实验，左起聂玮、高健、王静、王文兴、张庆竹、高锐、王新锋（2008）

在山东大学实验室指导研究生开展量子化学与分子模拟研究，前排王文兴，后排左起张庆竹、李善青、王晓粉，济南（2005）

山东大学环境研究院量子化学研究团队召开国家自然科学基金重点项目启动会，前排左起胡敬田、王文兴、张庆竹，济南（2013）

山东大学与清华大学联合举办第 11 届国际大气科学与空气质量（**ASAAQ**）会议，左起贺克斌、王文兴、郝吉明、张庆竹，济南（**2009**）

在山东大学 **110** 周年校庆之际举办山东大学环境科学高层论坛，前排左起江桂斌、任阵海、孙铁珩、王文兴、梁文平、朱正昌、张波、金鉴明、蔡道基、侯立安、娄红祥（**2011**）

美国北卡罗来纳大学 **R. Kamens** 教授及夫人来访。二排左起王海宁、张庆竹、王文兴、**R. Kamens**、**Kamens** 夫人、高健，济南（**2006**）

美国德克萨斯 **A&M** 大学张人一教授来访。前排左起刘建、张庆竹、张人一、王文兴、叶兴南、孙廷利、王艳、戴九兰，济南（**2007**）

美国哈佛大学 **Scott Martin** 教授来访。前排左四王文兴、五 **Scott Martin**、六张庆竹，济南（**2009**）

王文兴在山东大学环境研究院建院十周年环境科学高端学术论坛上作“中国雾霾污染控制若干科学问题”学术报告，济南（**2014**）

参加第 22 届大气环境科学与技术大会，与参会的山东大学环境研究院的师生及院友合影，上海（**2016.11**）

举办大气环境高峰论坛暨“中国大气环境问题的演变和未来”主题研讨会，山大省委常委济南市委书记王文涛（前排十二）、山东大学校长张荣（前排十）与专家合影，济南（**2016.12.25**）

七、社会活动与学术兼职

参加世界工程师大会。左张维、右王文兴，内罗毕（1983.10）

在世界工程师大会会上，王文兴在环境分会场作了中国大气污染及其控制报告，内罗毕（1983.10）

访问美国乔治亚理工学院与 **D. Davis** 教授商讨在中国举办首次国际大气环境化学会议（**1989**）

国际学术会议后王文兴（中）与日本大喜多教授（左）应邀到韩国 **YS Chung** 教授（右）清州家中做客（**1992**）

应邀参加日本第 **34** 届大气污染学会，与参会专家学者合影，左起吕世宗、王文兴、盐泽、铃木（**1993**）

访问加拿大约克大学化学系，左一 **Niki**、二王文兴、三化学系主任、四 **YE**（**2005**）

参加第 10 届 ASAAQ 会议，左一高赞明、二郝吉明、三潘宗光、四王文兴、五王韬，香港（2007）

参加国家“973”酸雨项目第三课题学术研讨会合影，前排左三王韬、四王文兴，北京（2010）

《中国环境科学》第二届编委会合影，北京（1984），左六王文兴，当选主编

清华大学朱天乐（左四）博士学位论文答辩，左三李国鼎、五王文兴、六郝吉明等教授

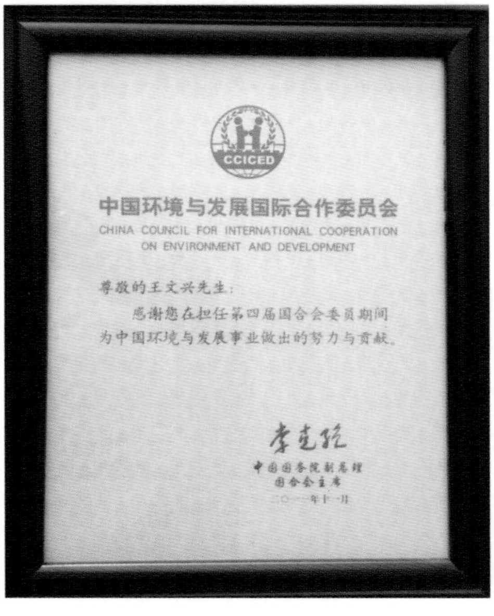
CCICED
中国环境与发展国际合作委员会
CHINA COUNCIL FOR INTERNATIONAL COOPERATION ON ENVIRONMENT AND DEVELOPMENT

尊敬的王文兴先生：

感谢您在担任第四届国合会委员期间为中国环境与发展事业做出的努力与贡献。

李克强
中国国务院副总理
国合会主席
二〇一一年十一月

担任中国环境与发展国际合作委员会委员（2011）

聘书

兹聘 王文兴 同志任国家环境咨询委员会委员。

二〇一二年三月

国家环境咨询委员会委员聘书（2012）

八、亲情、友情、记忆

全家福（1962）

全家福（1973）

王文兴与父亲，北京（1962）

全家福（2017）

王文兴与大姐、姐夫合影（1980）

王文兴陪同夫人张婉华等回访战时江西保育院旧址

与大哥（左一）、大嫂合影，北京（1985）

与大哥、弟弟及侄子合影，安徽萧县（2016.10）

与大哥、妹妹、弟弟等合影，安徽萧县（2016.10）

王文兴儿女们及其家人旅游聚会

王文兴与南京大学傅献彩教授合影，南京（2004）

王文兴夫妇等南京酸雨会议后参观孙中山临时大总统办公室，南京（1990）

王文兴夫妇在黄花岗七十二烈士陵园前，广州（1990）

王文兴与中学和大学同学张立光参观包公墓，合肥（1991.9）

九、奖励和荣誉

王文兴常说，他在科教工作中取得的任何奖励和荣誉都离不开合作者的共同努力、组织领导和家人的支持，这些永远铭记在心。

王文兴共获得省部级及以上奖励 17 项，其中包括国家科技进步奖五项：

1. 中国大气酸沉降及其生态环境影响研究：国家科技进步一等奖
2. 我国酸雨来源和影响及其控制对策：国家科技进步二等奖
3. 大气环境容量研究：国家科技进步二等奖
4. 兰州西固大气光化学污染规律和防治：国家科技进步二等奖，
5. 太原地区大气环境综合观测研究：国家科技进步三 等奖。

国家科技进步奖奖章，其中一等奖 1 个、二等奖 3 个、三等奖 1 个

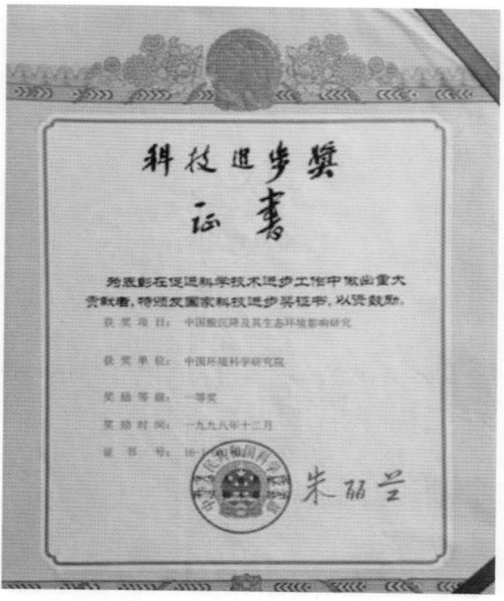

国家科技进步一等奖证书，获奖单位：中国环境科学研究院

国家科技进步一等奖证书，获奖者：王文兴

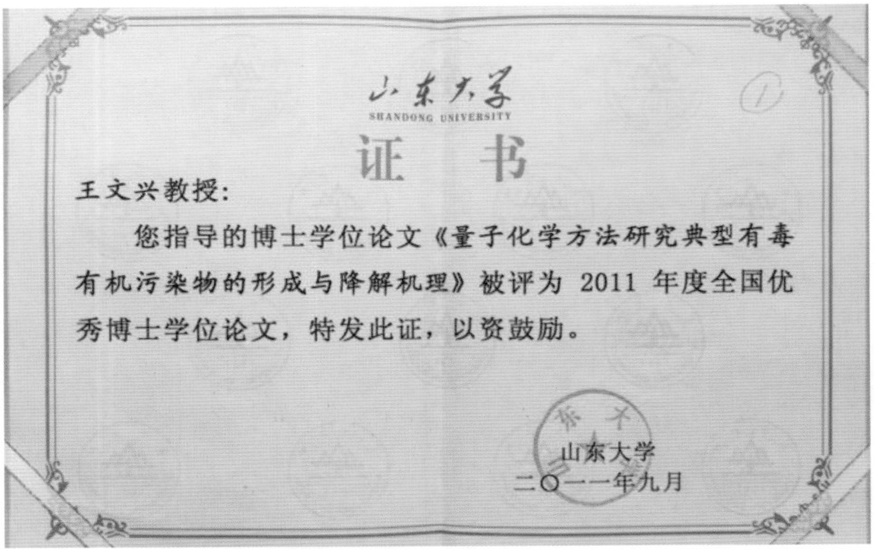

王文兴指导的博士学位论文被评为 **2011** 年度全国优秀博士学位论文（**2011**）

荣获山东大学育才功勋（**2009**）

王文兴在院士大会上荣获第十届“光华科技工程奖”，北京（**2014**）

王文兴在第 **18** 届中国大气环境科学与技术大会上荣获“环境科学终身成就奖”，杭州（**2011**）

第十届光华工程科技奖（**2014**）

国际大气科学与空气质量应用会议 **ASAAQ** 特别贡献奖（**2015**）

王文兴文集

（第二卷）

——环境科学前沿学术论文集

《王文兴文集》（第二卷）编辑组　编

科学出版社

北　京

内 容 简 介

《王文兴文集》（第二卷）是一本环境科学研究论文集，它是2008年出版的《王文兴文集》的续集之一，内容包括王文兴院士及其指导的研究生近十年发表的具有代表性的论文。这部分论文反映了作者在大气环境、环境化学和环境生态学等研究领域取得的重要学术成果和最新研究进展。除此之外，本书还附有王文兴院士求学及工作各时期的珍贵照片，这些照片见证了我国环境科学研究事业的发展历程，具有珍贵的史料价值。

全书正文部分共分为3篇，即大气环境化学篇、环境量子化学篇、环境生态学篇。

本书可供从事大气环境科学、环境化学、环境生态学领域的科学研究人员，环境保护和环境管理部门的工程技术人员以及高等院校相关专业的师生参考。

图书在版编目（CIP）数据

王文兴文集．第二卷，环境科学前沿学术论文集 /《王文兴文集》（第二卷）编辑组编．—北京：科学出版社，2017.3

ISBN 978-7-03-052230-6

Ⅰ.①王…　Ⅱ.①王…　Ⅲ.①王文兴－文集②环境科学－文集　Ⅳ.①X-53

中国版本图书馆CIP数据核字（2017）第054789号

责任编辑：朱　丽 / 责任校对：张怡君
责任印制：肖　兴 / 封面设计：耕者设计工作室

科学出版社 出版
北京东黄城根北街16号
邮政编码：100717
http://www.sciencep.com
北京通州皇家印刷厂 印刷
科学出版社发行　各地新华书店经销
*
2017年3月第　一　版　开本：889×1194 1/16
2017年3月第一次印刷　印张：32 1/2　插页：12
字数：1 020 000

定价：298.00元

（如有印装质量问题，我社负责调换）

王文兴院士简介

王文兴，男，1927年10月出生，安徽萧县人，祖籍山东临沂。中国工程院院士，山东大学终身教授、博导、环境研究院院长，中国环境科学研究院学术顾问。目前主要从事大气环境化学、大气污染与控制、环境量子化学等方向的研究工作。

1952年毕业于山东大学（青岛）化学系物理化学专业，1955年在吉林大学化学系研究生班进修物理化学，1959年赴苏联卡尔波夫物理化学研究所进修催化动力学。大学毕业后至七十年代中期，先后任重工业部沈阳化工研究所工程师、化工部北京化工研究院研究室主任、天津化工研究院副院长。早期主要从事烃类催化氧化研究，应用放射性同位素示踪技术、电磁泵流动循环法，研究了烃类催化氧化反应机理与动力学，其研究结果在烃类氧化工业催化剂的研制方面有重要参考价值。编著了我国第一本《工业催化》（1978）。

1973年随中国科学技术代表团赴英国考察一个月，工业发展对环境的影响给他留下了深刻印象。1976年被任命为天津市环境保护办公室副主任，分管科技。组建了中国最早的环境学会——天津市环境科学学会，并任理事长。针对天津大气污染，提出并实施了烟气治理、工厂搬迁、集中供暖、居民炉灶气化等重要举措，对改善天津市空气质量起到重要作用。

1980年初受邀参与筹建中国环境科学研究院。先后担任中国环境科学研究院副院长、学术委员会主任、学术顾问。连续承担国家“六五”至“九五”科技攻关项目，在大气光化学污染规律和防治、煤烟型大气污染与控制、大气环境容量、酸沉降化学与污染控制等方面，组织开展了大量的现场观测和实验室模拟工作。建立了当时国内最先进的、可抽真空的、具有FTIR长光路的光化学反应装置，与合作者共同发现我国兰州光化学烟雾和煤烟型污染的形成机理及长距离传输规律。在酸沉降的观测和实验研究方面取得了重要成就，首次计算了全国二氧化硫和氨的排放量与排放强度，查清了全国酸雨现状及分布规律和沉降通量，发现我国酸雨面积已达三百万平方公里、成为世界第三大酸雨区，创建了我国第一套材料暴露自动试验装置并建立了材料二元损伤函数式，为我国酸雨控制对策和相关环保法规的制定起到了关键作用。

2002年受邀到山东大学任教。2003年组建了山东大学环境研究院，担任院长。在环渤海地区较早开展区域性PM2.5与雾霾研究，利用地面、高山和航测证实了高浓度的PM2.5是该地区雾霾频发的根本原因，并结合野外观测、实验室烟雾箱模拟和化学传输模式模拟基本查明了区域性雾霾形成的化学机制和气象成因。近十年来创建了环境量子化学计算团队和研究新领域，指导研究了高温条件下二噁英的气相形成机理、探索了POPs等环境有机污染物的大气转化和降解机理，模拟了典型环境污染物在生物酶及表面催化剂作用下的催化降解机制，建立了新的理论模型和研究方法，取得了一系列重要成果。

王文兴院士积极参与国家环保事务和环境科学学术活动。担任国家环境咨询委员会委员，曾任中国国家环境与发展国际合作委员会中方委员、国家环境保护部科学技术顾问委员会委员、前国家环境保护总局科学技术委员会委员。曾任中国环境科学学会第一至四届常务理事、五至六届理事会顾问、大气分会理事长，中华环境奖评委会第一、第二届主任。担任《中国环境科学》杂志主编，曾任国际 ASAAQ 组织委员以及美国 *Environment Research* 杂志副主编。

王文兴院士成果丰硕，成就斐然。先后承担了国家“六五”至“九五”科技攻关重大项目，曾主持科技部项目 6 项、国家自然科学基金委员会项目 4 项、省部级项目多项。共发表论文 350 余篇，出版著作 6 部。获国家科技进步奖一等奖 1 项、二等奖 3 项、三等奖 1 项及省部级奖多项。荣获首届环境科学终身成就奖、第十届光华科技工程奖、国际大气科学及其空气质量应用会议（ASAAQ）特别贡献奖。

王文兴院士培养了一批优秀的研究生。近十年来在山东大学培养研究生 40 人，其中博士 21 人、硕士 19 人，博士毕业生中 1 人获得全国百篇优秀博士学位论文奖、2 人获得山东省优秀博士学位论文奖、3 人获得山东大学优秀博士学位论文奖。目前在读研究生 18 人，其中博士研究生 7 人、硕士研究生 11 人。先后指导博士后 12 人。2013 年被评为山东省优秀研究生指导教师。

王文兴院士是我国著名的环境化学家，为我国环境科学理论技术的发展和环境保护事业的进步做出了突出贡献。他淡泊名利、心胸豁达、厚德载物、乐于奉献，为人谦虚谨慎，做事兢兢业业，深受环保部门、科研单位、高等学校等社会各界人士的尊敬和爱戴。

序　一

王文兴院士是我国著名的环境化学家。他从事科学研究及教育工作六十余载，勤勤恳恳，兢兢业业，为我国环境科学理论的积淀、环境科技工作者的培养、环境保护事业的发展，做出了杰出贡献。

王文兴院士化学专业出身，早期从事工业催化研究。他 1952 年毕业于山东大学化学系，1955 年被推荐到吉林大学化学系物化研究生班进修，1959 年又被派往苏联莫斯科卡尔波夫物理化学研究所进修，期间得到多位著名化学家的指导，为之后开展科学研究积累了坚实的理论基础。回国之后，他在北京化工研究院物化研究室筹建了当时国内先进的催化实验室，并先后担任物化研究室副主任和主任。他应用放射性同位素跟踪技术和电磁泵流动循环法，专注研究烃类催化氧化反应机理与动力学，编著了我国第一本《工业催化》（1978 年），对我国烃类氧化工业催化剂的研制和催化化学的发展有重要意义。

王文兴院士是我国最早从事大气环境科学研究的工作者之一。早在 20 世纪 70 年代初，他随中国科学技术代表团赴英国考察了工业对环境的影响，深刻感受到工业发展对环境的污染和破坏，认识到发达国家对环境科研、环保技术和环境管理的高度重视和巨大投入，一个多月的考察给他留下深刻的印象。1976 年他被调到天津市环境保护办公室，任副主任，分管科技，工作内容也随之转入环境保护。他组建了中国最早的环境学会——天津市环境科学学会，并任理事长。针对大气污染问题提出并实施了烟气治理、工厂搬迁、集中供暖、居民炉灶气化等一系列重要举措，树立了我国早期大气污染防治工作的典范。在中国环境科学研究院从 80 年代初到 90 年代末的近二十年间，他先后承担了多项国家科技攻关项目、国家级和省部级重要科研项目，针对兰州光化学污染、太原煤烟型污染、我国酸雨污染的来源、分布、影响及控制对策等，领导并参与开展了深入系统的科学研究，相关研究成果发表学术论文 100 余篇，出版专著多部，获国家科技进步一等奖 1 项、二等奖 3 项及省部级奖励多项，对我国大气环境立法、大气污染和酸雨控制起到了重要作用。

王文兴院士是中国环境科学研究院的主要创建者和领导人之一。1980 年初他受邀参与筹建中国环境科学研究院，是中国环境科学研究院领导班子的重要成员，分管科技。中国环境科学研究院很快成为当时国内大气环境化学领域先进的科研平台，他也被任命为中国环境科学研究院副院长、首届学术委员会主任。在中国环境科学研究院二十余年间，他连续承担国家“六五”至“九五”科技攻关项目，在大气光化学、煤烟型大气污染和大气酸沉降领域进行长期研究。特别在酸沉降的观测和实验研究方面取得了重大成果，首次计算了全国二氧化硫和氨的排放量与排放强度，查清了全国酸雨现状及分布规律和沉降通量，发现并证明我国已成为继北美、欧洲之后的世界第三大酸雨区，为我国酸雨控制对策和相关环保法规的制定起到了关键作用，使我国免受欧美地区曾经出现的大面积土壤酸化、森林死亡、鱼虾绝迹的严重灾害。凭借在酸雨研究方面的突出贡献，王文兴院士及其研究团队先后获得国家科学技术进步奖二等奖和一等奖各一项，后者是我国环境科学技术领域国家三大奖中的第一个，也是迄今为止唯一的一项国家一等奖。1993 年从中国环境科学研究院离休之后，王文兴院士一直担任学术顾问，对中国环境科学研究院的长期发展做出了积极贡献。

王文兴院士指导、培养了一批优秀的环保科技和环境科研工作者。他领导承担的国家“六五”至“九五”科技攻关项目，涉及众多科研机构、大学、地方单位，上千人参与项目的研究工作。在项目和课题的组织与实施过程中，他对专业研究人员和地方业务人员都给予了热情的指导，对我国环保工作者的大批涌现和科研水平的快速提升起到了重要的促进作用。2002 年，王文兴院士受邀前往母校山东大学，组建了山东大学环境研究院，从事研究生培养、科研和教学工作，瞄准国家需要和国际前沿，在大气环境化学、环境量子化学及环境生态领域开展了大量的应用基础研究，发表学术论文 200 余篇，培养了四十余名博士和硕士，为我国高校、科研机构和环保部门注入了新生力量。

王文兴院士积极参与国家和国际环保事务与活动，推动了我国环境保护事业的发展和进步。他是国家环境咨询委员会委员，曾任中国国家环境与发展国际合作委员会中方委员、国家环境保护部科学技术顾问委员会委员、前国家环境保护总局科学技术委员会委员。他曾任中国环境科学学会第一至四届常务理事、大气分会理事长，中华环境奖评委会第一、第二届主任。他是《中国环境科学》杂志主编，曾任国际 ASAAQ 组织委员以及美国 *Environment Research* 杂志副主编。由于他对我国环境科学技术的发展和环境保护工作的突出贡献，获得了国家科技进步奖五项、光华工程科技奖、国际 ASAAQ 会议特别奖等多项荣誉。

《王文兴文集》（第二卷）即将出版之际，恰逢王文兴院士九十华诞，我与王老一起工作三十余载，目睹了他在环境科学领域的科研活动和为人做事、杰出的科研成就和对我国环保事业的突出贡献，王老淡泊名利、心胸豁达、厚德载物，为人谦虚谨慎，做事兢兢业业，深受同事称赞。在此，我向王老表示衷心的祝贺，祝他健康长寿，继续为我国的环境科学和环境保护事业做出新的贡献。

全国人民代表大会环境与资源保护委员会副主任委员

中国工程院院士

2016 年 10 月

序　二

王文兴先生是我国著名环境化学家，中国工程院院士，山东大学终身教授。他从事环境科学研究、管理及人才培养工作六十余载，为我国环境保护和教育事业做出了突出贡献。今年适逢王老九十华诞，他的弟子们商议将其近年所著之论文编辑成册，以期付梓，请我作序，我深感荣幸！

山东大学自1901年创建以来，始终秉承“为天下储人才，为国家图富强”的办学宗旨，为国家和社会输送了40余万各类人才，王文兴院士是其中的杰出代表。1952年，刚刚毕业走出山东大学校门，他便怀着满腔的报国热情，投入到新中国火热的建设事业，倾注了自己的全部心血。他早期从事工业催化研究并取得骄人成就，撰写的《工业催化》是我国该领域的第一部专著。20世纪70年代初，随着国家对环境问题的日益重视，他开始投身环境科学研究，成为我国最早进入这一科学领域的研究人员。作为我国大气环境科学的早期研究者之一，他见证了这一领域发展的全过程，在大气光化学污染规律和防治、煤烟型大气污染与控制、大气酸沉降化学等方面取得了一系列重大成果。他领导科研团队砥砺二十载，跑遍了中华大地的山川海岛，查清了全国致酸物质的排放、酸雨现状与分布规律及其成因，为我国大气环境立法和酸雨防控提供了坚实的理论与技术支撑，并因此获得我国环境科技领域的第一个国家科技进步一等奖。

王老求学山东大学时期，适逢山大办学史上“第二次辉煌”，校内名家云集，学术氛围浓厚，由此在他心中深深地种下一生挥之不去的山大情结。毕业以后，他虽辗转工作于沈阳、天津、兰州、北京多地，却丝毫未曾淡忘母校的培育之情，时刻关心着山大的建设发展。2001年，在接到母校百年校庆的邀约后，王老激动不已，欣然赴命；在校庆典礼上收到回母校任教的邀请时，他毅然放弃了本应含饴弄孙的退休生活，只身来到山大，执起教鞭。2003年，山东大学环境研究院正式成立，王老出任院长。其时学校环境学科建设不久，学科方向以水污染治理为主，其他领域甚为薄弱。王老审时度势，决定创建大气环境化学和环境量子化学两个学科方向，同时考虑服务地方的需要，发展环境生态学方向，富有前瞻性地拓展和丰富了山东大学环境学科的研究方向与领域。十三年来，在王老及各位同仁的努力下，山东大学环境研究院由一株稚嫩的幼苗逐步成长为参天大树，在队伍建设、学科发展、人才培养、科学研究、社会服务等各个方面都取得了长足进步。2012年，山东大学“环境科学与工程”一级学科在教育部学科评估中名列第12位；在国家自然科学基金委发布的学科发展态势评估研究报告《化学十年：中国与世界》中，山东大学在量子化学密度泛函领域的国际排名位列第四。2015年，山东大学的“环境与生态学”学科排名进入ESI世界前1%。这些令人瞩目的成就，其中无不凝结着王老及其团队的辛劳。

在山东大学一提起王老，无论教师还是学生都钦敬不已。这不仅因为他开拓创新的科学精神、严谨求实的治学态度、诲人不倦的师者风范，更因他有一种强大的人格力量。王老德高望重，生活简朴，平易近人。

他经常拿出自己的工资积蓄，慷慨资助身边家庭经济困难的优秀学生，还捐献个人积蓄40万元设立了环境科学奖学金。他强调，做科研就像进庙修行一样，不要想着发财，因为这不是发财的工作。他常说："国家和人民培养了我，我只是在回报，用我的智慧和汗水，竭尽全力，鞠躬尽瘁。"诚如梅贻琦先生所言，"所谓大学者，非谓有大楼之谓也，有大师之谓也"。王老的这种高风亮节，是山东大学"学无止境，气有浩然"校训精神的生动诠释和集中体现。

对于王老而言，工作就是人生乐趣，就是意义所在，就是生命的存在方式。他日复一日地奔波在祖国大地，坚守在实验室，为我国科技事业的进步，不断地贡献智慧和汗水，从青年到壮年，从壮年到老年。王老治学一生，耄耋之年仍笔耕不辍，著作颇丰，今次编纂之文集更是反映了其近十多年来的工作成绩之精要，必可为当前年轻一代的环境工作者提供宝贵借鉴。值此机会，我谨代表山东大学，对王老表示衷心祝贺，祝愿他身体健康、工作顺利、学术常青，继续为我国的环境保护和教育事业做出新的贡献！

是为序。

山东大学　校长

2016年12月

目　录

第一部分　大气环境化学

一、大气细颗粒物化学、来源及生成

Airborne fine particulate pollution in Jinan, China: concentrations, chemical compositions and influence on visibility impairment (1)
Lingxiao Yang, Xuehua Zhou, Zhe Wang, Yang Zhou, Shuhui Cheng, Pengju Xu, Xiaomei Gao, Wei Nie, Xinfeng Wang, Wenxing Wang

Asian dust storm observed at a rural mountain site in southern China: chemical evolution and heterogeneous hotochemistry (10)
Wei Nie, Tao Wang, Likun Xue, Aijun Ding, Wenxing Wang, Xiaomei Gao, Zheng Xu, Yangchun Yu, Chao Yuan, Shengzhen Zhou, RuiGao, Xiaohan Liu, Yan Wang, Shaojia Fan, Steven C.N. Poon, Qingzhu Zhang., Wenxing Wang

Characterization of aerosol acidity at a high mountain site in central eastern China (21)
Yang Zhou, Likun Xue, Tao Wang, Xiaomei Gao, Zhe Wang, Xinfeng Wang, Jiamin Zhang, Qingzhu Zhang, Wenxing Wang

Enhanced formation of fine particulate nitrate at a rural site on the North China Plain in summer: the important roles of ammonia and ozone (31)
Liang Wen, Jianmin Chen, Lingxiao Yang, Xinfeng Wang, Caihong Xu, Xiao Sui, Lan Yao, Yanhong Zhu, Junmei Zhang, Tong Zhu, Wenxing Wang

Formation of secondary organic carbon and cloud impact on carbonaceous aerosols at Mount Tai, north China (40)
Zhe Wang, Tao Wang, Jia Guo, Rui Gao, Likun Xue, Jiamin Zhang, Yang Zhou, Xuehua Zhou, Qingzhu Zhang, Wenxing Wang

HONO and its potential source particulate nitrite at an urban site in North China during the cold season (52)
Liwei Wang, Liang Wen, Caihong Xu, Jianmin Chen, Xinfeng Wang, Lingxiao Yang, Wenxing Wang, Xue Yang, Xiao Sui, Lan Yao, Qingzhu Zhang

Impacts of firecracker burning on aerosol chemical characteristics and human health risk levels during the Chinese New Year Celebration in Jinan, China (61)
Lingxiao Yang, Xiaomei Gao, Xinfeng Wang, Wei Nie, Jing Wang, Rui Gao, Pengju Xu, Youping Shou, Qingzhu Zhang, Wenxing Wang

Photochemical evolution of organic aerosols observed in urban plumes from Hong Kong and thePearl River Delta of China (69)
Shengzhen Zhou, Tao Wang, Zhe Wang, Weijun Li, Zheng Xu, Xinfeng Wang, Chao Yuan, C. N. Poon, Peter K. K. Louie, Connie W. Y. Luk, Wenxing Wang

Secondary organic aerosol formation from xylenes and mixtures of toluene and xylenes in an atmospheric urban hydrocarbon mixture (80)
Yang Zhou, Haofei Zhang, Harshal M. Parikh, Eric H. Chen, Weruka Rattanavaraha, Elias P. Rosen, Wenxing Wang, Richard M. Kamens

Semi-continuous measurement of water-soluble ions in $PM_{2.5}$ in Jinan, China: temporal variations and source apportionments (89)
Xiaomei Gao, Lingxiao Yang, Shuhui Cheng, Rui Gao, Yang Zhou, Likun Xue, Youping Shou, Jing Wang, Xinfeng Wang, Wei Nie, Pengju Xu, Wenxing Wang

Severe haze episodes and seriously polluted fog water in Ji'nan, China (98)
Xinfeng Wang, Jianmin Chen, Jianfeng Sun, Weijun Li, Lingxiao Yang, Liang Wen, Wenxing Wang, Xinming Wang, Jeffrey L. Collett Jr. Yang Shi, Qingzhu Zhang, Jingtian Hu, Lan Yao, Yanhong Zhu, Xiao Sui, Xiaomin Sun, Abdelwahid Mellouki

Sources apportionment of $PM_{2.5}$ in a background site in the North China Plain (103)
Lan Yao, Lingxiao Yang, Qi Yuan, Chao Yan, Can Dong, Chuanping Meng, Xiao Sui, Fei Yang, Yaling Lu, Wenxing Wang

Study on ambient air quality in Beijing for the summer 2008 Olympic Games (112)
Wenxing Wang, Fahe Chai, Kai Zhang, Shulan Wang, Yizhen Chen, Xuezhong Wang, Yaqin Yang

Temporal variations, acidity, and transport patterns of $PM_{2.5}$ ionic components at a background site in the Yellow River Delta, China (118)
Qi Yuan, Lingxiao Yang, Can Dong, Chao Yan, Chuanping Meng, Xiao Sui, Wenxing Wang

The secondary formation of inorganic aerosols in the droplet mode through heterogeneous aqueous reactions under haze conditions (129)
Xinfeng Wang, Wenxing Wang, Lingxiao Yang, Xiaomei Gao, Wei Nie, Yangchun Yu, PengjuXu, Yang Zhou, Zhe Wang

Understanding of regional air pollution over China using CMAQ, part I performance evaluation and seasonal variation (138)
Xiaohuan Liu, Yang Zhang, Shuhui Cheng, Jia Xing, Qiang Zhang, David G. Streets, Carey Jang, Wenxing Wang, Jiming Hao

Understanding of regional air pollution over China using CMAQ, part II. Process analysis and sensitivity of ozone and particulate matter to precursor emissions (150)
Xiaohuan Liu, Yang Zhang, Jia Xing, Qiang Zhang, Kai Wang, David G. Streets, Carey Jang, Wenxing Wang, JimingHao

二、大气颗粒物的粒径分布与光学性质

Aerosol size distributions in urban Jinan: Seasonal characteristics and variations between weekdays and weekends in a heavily polluted atmosphere (159)
Pengju Xu, Wenxing Wang, Lingxiao Yang, Qingzhu Zhang, Rui Gao, Xinfeng Wang, Wei Nie, Xiaomei Gao

Airborne particulate polycyclic aromatic hydrocarbon (PAH) pollution in a background site inthe North China Plain: concentration, size distribution, toxicity and sources (173)
Yanhong Zhu, Lingxiao Yang, Qi Yuan, Chao Yan, Can Dong, Chuanping Meng, Xiao Sui, Lan Yao, Fei Yang, Yaling Lu, Wenxing Wang

Aircraft measurements of the vertical distribution of sulfur dioxide and aerosol scattering coefficient in China (185)
Likun Xue, Aijun Ding, Jian Gao, Tao Wang, Wenxing Wang, Xuezhong Wang, Hengchi Lei, Dezhen Jin, Yanbin Qi

Measurement of aerosol number size distributions in the Yangtze River delta in China: formation and growth of particles under polluted conditions (190)
Jian Gao, Tao Wang, Xuehua Zhou, Waishing Wu, Wenxing Wang

Size-fractionated water-soluble ions, situ pH and water content in aerosol on hazy days and the influences on visibility impairment in Jinan, China (198)
Shuhui Cheng, Lingxiao Yang, Xuehua Zhou, Likun Xue, Xiaomei Gao, Yang Zhou, Wenxing Wang

三、大气酸沉降与光化学污染

中国大气降水化学研究进展 (208)
王文兴，许鹏举

Air quality during the 2008 Beijing Olympics: secondary pollutants and regional impact (224)
Tao Wang, Wei Nie, Jian Gao, Likun Xue, Xiaomei Gao, Xinfeng Wang, Jun Qiu, StevenC.N. Poon, Simone Meinardi, Donald R. Blake, Shulan Wang, Aijun Ding, Fahe Chai, Qingzhu Zhang, Wenxing Wang

Development of a chlorine chemistry module for the Master Chemical Mechanism (237)
Likun Xue, Sandra M. Saunders, Tao Wang, Rui Gao, Xinfeng Wang, Qingzhu Zhang, Wenxing Wang

Oxidative capacity and radical chemistry in the polluted atmosphere of Hong Kong and Pearl River Delta region: analysis of a severe photochemical smog episode (249)
Likun Xue, Rongrong Gu, Tao Wang, Xinfeng Wang, Sandra Saunders, Donald Blake, Peter K. K. Louie, Connie W. Y. Luk, Isobel Simpson, Zheng Xu, Zhe Wang, Yuan Gao, Shuncheng Lee, Abdelwahid Mellouki, Wenxing Wang

Significant increase of summertime ozone at Mount Tai in central eastern China (262)
Lei Sun, Likun Xue, Tao Wang, Jian Gao, Aijun Ding, Owen R. Cooper, Meiyun Lin, Pengju Xu, Zhe Wang, Xinfeng Wang, Liang Wen, Yanhong Zhu, Tianshu Chen, Lingxiao Yang, Yan Wang, Jianmin Chen, Wenxing Wang

Source of surface ozone and reactive nitrogen speciation at Mount Waliguan in western China: new insights from the 2006 summer study (276)
Likun Xue, Tao Wang, Jiamin Zhang, Xiaochun Zhang, Deliger, Steven C.N. Poon, Aijun Ding, Xuehua Zhou, Wai Shing Wu, Jie Tang, Qingzhu Zhang, Wenxing Wang

第二部分　环境量子化学

一、二噁英的形成机理

A quantum mechanical study on the formation of PCDD/Fs from 2-chlorophenol as precursor (288)
Qingzhu Zhang, Shanqing Li, Xiaohui Qu, Xiangyan Shi, Wenxing Wang

Dioxin formations from the radical/radical cross-condensation of phenoxy radicals with 2-chlorophenoxy radicals and 2,4,6-trichlorophenoxy radicals (296)
Fei Xu, Wanni Yu, Rui Gao, Qin Zhou, Qingzhu Zhang, Wenxing Wang

Formation of bromophenoxy radicals from complete series reactions of bromophenols with H and OH radicals (303)
Rui Gao, Fei Xu, Shanqing Li, Jingtian Hu, Qingzhu Zhang, Wenxing Wang

Kinetic properties for the complete series reactions of chlorophenols with OH radicals—relevance for dioxin formation (312)
Fei Xu, Hui Wang, Qingzhu Zhang, Ruixue Zhang, Xiaohui Qu, Wenxing Wang

Mechanism and direct kinetic study of the polychlorinated dibenzo-p-dioxin and dibenzofuran formations from the radical/radical cross-condensation of 2,4-dichlorophenoxy with 2-chlorophenoxy and 2,4,6-trichlorophenoxy (318)
Fei Xu, Wanni Yu, Qin Zhou, Rui Gao, Xiaoyan Sun, Qingzhu Zhang, Wenxing Wang

Mechanistic and kinetic studies on the homogeneous gas-phase formation of PCDD/Fs from 2,4,5-trichlorophenol (326)
Xiaohui Qu, Hui Wang, Qingzhu Zhang, Xiangyan Shi, Fei Xu, Wenxing Wang

Mechanism and direct kinetics study on the homogeneous gas-phase formation of PBDD/Fs from 2-BP, 2,4-DBP, and 2,4,6-TBP as precursors (334)
Wanni Yu, Jingtian Hu, Fei Xu, Xiaoyan Sun, Rui Gao, Qingzhu Zhang, Wenxing Wang

Mechanism and thermal rate constants for the complete series reactions of chlorophenols with H (343)
Qingzhu Zhang, Xiaohui Qu, Hui Wang, Fei Xu, Xiangyan Shi, Wenxing Wang

PBCDD/F formation from radical/radical cross-condensation of 2-chlorophenoxy with 2-bromophenoxy, 2,4-Dichlorophenoxy with 2,4-Dibromophenoxy, and 2,4,6-Trichlorophenoxy with 2,4,6-Tribromophenoxy (351)
Xiangli Shi, Wanni Yu, Fei Xu, Qingzhu Zhang, Jingtian Hu, Wenxing Wang

二、环境有机污染物大气降解机理

Mechanism and kinetic studies for OH radical-initiated atmospheric oxidation of methyl propionate (359)
Xiaoyan Sun, Yueming Hu, Fei Xu, Qingzhu Zhang, Wenxing Wang

Mechanistic and kinetic studies on the OH-initiated atmospheric oxidation of fluoranthene (367)
Juan Dang, Xiangli Shi, Qingzhu Zhang, Jingtian Hu, Jianmin Chen, Wenxing Wang

Mechanical and kinetic study on gas-phase formation of dinitronaphthalene from 1-and 2-nitronaphthalene (375)
Zixiao Huang, Qingzhu Zhang, Wenxing Wang

Mechanism and kinetic study on the gas-phase reactions of OH radical with carbamate insecticide isoprocarb (385)
Chenxi Zhang, Wenbo Yang, Jing Bai, Yuyang Zhao, Chen Gong, Xiaomin Sun, Qingzhu Zhang, Wenxing Wang

Mechanism of OH-initiated atmospheric photooxidation of the organophosphorus insecticide $(C_2H_5O)_3PS$ (392)
Qin Zhou, Xiangyan Shi, FeiXu, Qingzhu Zhang, Maoxia He, Wenxing Wang

OH-initiated oxidation mechanisms and kinetics of 2,4,4′-tribrominated diphenyl ether (400)
Haijie Cao, Maoxia He, Dandan Han, Jing Li, Mingyue Li, Wenxing Wang, Side Yao

OH radical-initiated oxidation degradation and atmospheric lifetime of N-ethylperfluorobutyramide in the presence of O_2/NO_x (410)
Yanhui Sun, Qingzhu Zhang, Hui Wang, Wenxing Wang

Role of water molecule in the gas-phase formation process of nitrated polycyclic aromatic hydrocarbons in the atmosphere: a computational study (419)
Qingzhu Zhang, Rui Gao, Fei Xu, Qin Zhou, Guibin Jiang, Tao Wang, Jianmin Chen, Jingtian Hu, Wei Jiang, Wenxing Wang

Theoretical investigation on mechanistic and kinetic transformation of 2,2′,4,4′,5- pentabromodiphenyl ether (426)
Haijie Cao, Dandan Han, Xin Li, Mingyue Li, Maoxia He, Wenxing Wang

三、表面与酶催化反应机理

Adsorption and transformation mechanism of NO_2 on NaCl (100) surface: A density functional theory study (434)
Chenxi Zhang, Xue Zhang, Lingyan Kang, Ning Wang, Mandi Wang, Xiaomin Sun, Wenxing Wang

Catalytic mechanism of C-F bond cleavage: insights from QM/MM analysis of fluoroacetatedehalogenase (440)
Yanwei Li, Ruiming Zhang, Likai Du, Qingzhu Zhang, Wenxing Wang

Computational evidence for the detoxifying mechanism of epsilon class glutathione transferase toward the insecticide DDT (448)
Yanwei Li, Xiangli Shi, Qingzhu Zhang, Jingtian Hu, Jianmin Chen, Wenxing Wang

Dehydrochlorination mechanism of gamma-hexachlorocyclohexane degraded by dehydrochlorinase LinA from Sphingomonas paucimobilis UT26 (457)
Xiaowen Tang, Ruiming Zhang, Qingzhu Zhang, Wenxing Wang

Heterogeneous reaction mechanism of gaseous HNO_3 with solid NaCl: a density functional theory study (467)
Nan Zhao, Qingzhu Zhang, Wenxing Wang

第三部分 环 境 生 态

Adsorption and desorption of divalent mercury (Hg^{2+}) on humic acids and fulvic acids extracted from typical soils in China (477)
Jie Zhang, Jiulan Dai, Renqing Wang, Fasheng Li, Wenxing Wang

Distribution and sources of petroleum-hydrocarbon in soil profiles of the Hunpu wastewater-irrigated area, China's northeast (485)
Juan Zhang, Jiulan Dai, Xiaoming Du, Fasheng Li, Wenxing Wang, Renqing Wang

Heavy Metal Bioaccumulation and Health Hazard Assessment for Three Fish Species from Nansi Lake, China (494)
Pengfei Li, Jian Zhang, Huijun Xie, Cui Liu, Shuang Liang, Yangang Ren, Wenxing Wang

Atmospheric Environment 55 (2012) 506–514

Contents lists available at SciVerse ScienceDirect

Atmospheric Environment

journal homepage: www.elsevier.com/locate/atmosenv

Airborne fine particulate pollution in Jinan, China: Concentrations, chemical compositions and influence on visibility impairment

Lingxiao Yang [a,b], Xuehua Zhou [a], Zhe Wang [a], Yang Zhou [a], Shuhui Cheng [a], Pengju Xu [a], Xiaomei Gao [a], Wei Nie [a], Xinfeng Wang [a], Wenxing Wang [a,*]

[a] *Environment Research Institute, Shandong University, Jinan 250100, China*
[b] *School of Environmental Science and Engineering, Shandong University, Jinan 250100, China*

ARTICLE INFO

Article history:
Received 1 December 2011
Received in revised form
30 January 2012
Accepted 7 February 2012

Keywords:
$PM_{2.5}$
Chemical composition
Mass closure
Visibility impairment

ABSTRACT

Daily $PM_{2.5}$ samples were collected simultaneously at an urban site (SD) and a rural site (MP) in Jinan, China from March 2006 to February 2007. The samples were analyzed for major inorganic and organic water-soluble ions, 24 elements and carbonaceous species to determine the spatial and temporal variations of $PM_{2.5}$ mass concentrations and chemical compositions and evaluate their contributions to visibility impairment. The annual average concentrations of $PM_{2.5}$ were 148.71 μg m^{-3} and 97.59 μg m^{-3} at SD and MP, respectively. The predominant component of $PM_{2.5}$ was $(NH_4)_2SO_4$ at SD and organic mass at MP, which accounted for 28.71% and 37.25% of the total mass, respectively. The higher SOR (sulfur oxidation ratio) and ratios of OC/EC at SD indicated that the formation of secondary inorganic ions and secondary organic aerosols (SOA) could be accelerated in the urban area. Large size $(NH_4)_2SO_4$ and large size organic mass were the most important contributors to visibility impairment at SD and MP, accounting for 43.80% and 41.02% of the light extinction coefficient, respectively.

1. Introduction

China is a world leader as both a producer and consumer of coal, steel and cement. Furthermore, due to its rapid economic development, the number of vehicles in China has increased by 15% annually (National Bureau of Statistics of China, 2008). As a result, China faces serious atmospheric pollution. Fine particulate matter is the principal pollutant in most urban areas (Zhang et al., 2009b), drawing worldwide attention for its adverse impact on visibility (Ghim et al., 2005) and public health (Hong et al., 2002).

Shandong Province (15.36×10^4 km^2 and 1.6% of the total area in China) is located in Northern China. Coal-fired power plants, which account for 10.8% of total electricity production, are distributed densely throughout the province to meet the rapidly increasing demand for electric power of industrialization and urbanization. Accordingly, the emission intensities of SO_2, NO_x and PM in Shandong Province rank first in China and in the world (Zhang et al., 2009b). The high intensities of air pollutant emissions in Shandong Province have caused deterioration in the air quality and decreased visibility. Shandong Province has become a highly recognized hazy region near the Bohai Sea, which is one of the four major hazy areas in China (Cheng et al., 2011a). In addition to local air quality problems, there are regional concerns. Using the CMAQ modeling system, Streets et al. (2007) found that Shandong Province might exert a significant influence on the concentration of $PM_{2.5}$ in Beijing during the Olympic Games. Trajectory clustering has shown that Shandong Province is the most important contributor to particulate matter pollution in Tianjin (Kong et al., 2010), and Kim et al. (2009) reported that the Korean Peninsula may be heavily influenced by air masses originating in Shandong.

Jinan, the capital of Shandong Province, is a semi-enclosed area surrounded by the 1000 m peaks of the Taishan and Lushan Mountain and the Yellow River. This region is characterized by poor atmospheric diffusion, which is most notable in winter when the predominant wind direction is NE. It ranked 6 out of the 10 most heavily air-polluted cities in the world in 1999 (Sheng et al., 2000). Despite intensifying atmospheric pollution control measures in recent decades, the air quality in Jinan has not substantially improved. The annual average concentration of $PM_{2.5}$ at the urban site in Jinan was as high as 148.71 μg m^{-3} in this study, which is among the highest levels reported in the world (Cheng et al., 2011a). Obviously, such a high concentration of airborne fine particles severely reduces visibility; predictably, over the past four

* Corresponding author. Environment Research Institute, Shandong University, Jinan 250100, China. Tel./fax: +86 531 88366072.
E-mail address: wxwang@sdu.edu.cn (W. Wang).

1352-2310/$ – see front matter © 2012 Elsevier Ltd. All rights reserved.
doi:10.1016/j.atmosenv.2012.02.029

decades, visibility has been decreasing in Jinan. In recent years, Jinan has suffered from a heavy haze, and the percentage of seriously hazy days (with visibility below 5 km) reached 8.5% in 2006 (Yang et al., 2007).

Significant progress has been made in our understanding of water-soluble ions in $PM_{2.5}$ (Cheng et al., 2011a,b; Gao et al., 2011) and size-distribution of particles in Jinan (Xu et al., 2011). However, our understanding of the systematic, long-term variations of fine particle concentration levels and their chemical constituents and impact on visibility impairment, remains incomplete. In this paper, samples of $PM_{2.5}$ collected on a continuous basis from March 2006 to February 2007 simultaneously at an urban site and at a rural site in Jinan were analyzed to obtain detailed information about the chemical characteristics of $PM_{2.5}$ in Jinan, which included 24 inorganic elements and major organic and inorganic water-soluble ions together with organic carbon and elemental carbon. Using these observation data, we analyzed the seasonal and spatial variations of $PM_{2.5}$ mass concentrations, their chemical constituents and the influence exerted on them by meteorological conditions. The revised IMPROVE formula was then applied to estimate the contributions of $PM_{2.5}$ chemical components to the aerosol light extinction coefficient.

2. Methodology

2.1. Site description

The city of Jinan (36°40′04″N,117°02′01″E), with a population of approximately 6,000,000, is located in north central Shandong Province, between Mount Tai in the south and the Yellow River in the north. $PM_{2.5}$ samples were simultaneously collected at an urban site and a rural site (Fig. 1). The urban sampling site is 20 m above ground, on the rooftop of the Information Science and Engineering Building at Shandong University. Because the site is in a prosperous business district with convenient traffic and high population density, it provides information about the exposure of the population to $PM_{2.5}$ in Jinan. The rural sampling site, on the other hand, is located at the Miao Pu National Forest Park Monitoring Station of the Jinan Environmental Protection Bureau, a part of the Jinan Air Quality Monitoring Network, which lies approximately 10 km from the urban area. There are few large factories or main communication lines near the site; therefore, it could be considered a clean monitoring station in Jinan. The sampling site is on the roof of a two-story building in Miao Pu National Forest Park Monitoring Station, approximately 7 m above ground.

2.2. Instrumentation and sampling

The Thermo Anderson Chemical Speciation Monitor (Thermo Electron Corporation, Model RAAS2.5-400) was used to collect $PM_{2.5}$ samples at the urban site at Shandong University every 6 days from March 2006 to February 2007. Two airstreams flying at the rate of 16.7 L min^{-1} were connected to a Teflon filter (1 μm pore size and 47 mm diameter, Pall Gelman Inc.) for the $PM_{2.5}$ mass, while water-soluble ions and element analysis and a quartz filter (1 μm pore size and 47 mm diameter, Pall Gelman Inc.) for the determination of organic and elemental carbon, respectively. A dichotomous sampler (Graseby Anderson G241) was used to collect samples of $PM_{2.5}$ and $PM_{2.5-10}$ at the MP site. The main flow at the rate of 15.03 L min^{-1} was connected to a Teflon filter (1 μm pore size and 37 mm diameter, Pall Gelman Inc.) for the $PM_{2.5}$ mass, while water soluble ions and element analysis or a quartz filter (1 μm pore size and 37 mm diameter, Pall Gelman Inc.) for the determination of OC and EC every 6 days, alternately. A minor flow at the rate of 1.67 L min^{-1} was connected to glass fiber filters (Ø37 mm) to measure the mass concentrations of $PM_{2.5-10}$. The mass concentration of PM_{10} at MP could subsequently be obtained ($PM_{2.5} + PM_{2.5-10}$).

Each sample was collected for 23.5 hours, beginning at 8:00 am. The glass fiber filters and quartz filters were heated at 600 °C to eliminate all organic species prior to sampling. The Teflon filters were heated at 60 °C and equilibrated in a clean chamber with temperature and humidity automatically controlled at 20 ± 1 °C and 50 ± 5%, respectively, for at least 24 h, then weighed on an analytical balance (Sartorius, detection limit 0.001 mg) before and after sampling to obtain the $PM_{2.5}$ mass. After collection, the samples were sealed in clean plastic bags, transported to the laboratory and stored in a freezer at −4 °C before analysis.

The data of SO_2 at the MP site was provided by the Miao Pu National Forest Park Monitoring Station of the Jinan Environmental Protection Bureau and was co-located with the $PM_{2.5}$ monitor. The data of the SO_2 at the SD site was provided by the Jinan Shi Zhan Monitoring Station of the Jinan Environmental Protection Bureau, which was about 1000 m from the $PM_{2.5}$ monitor.

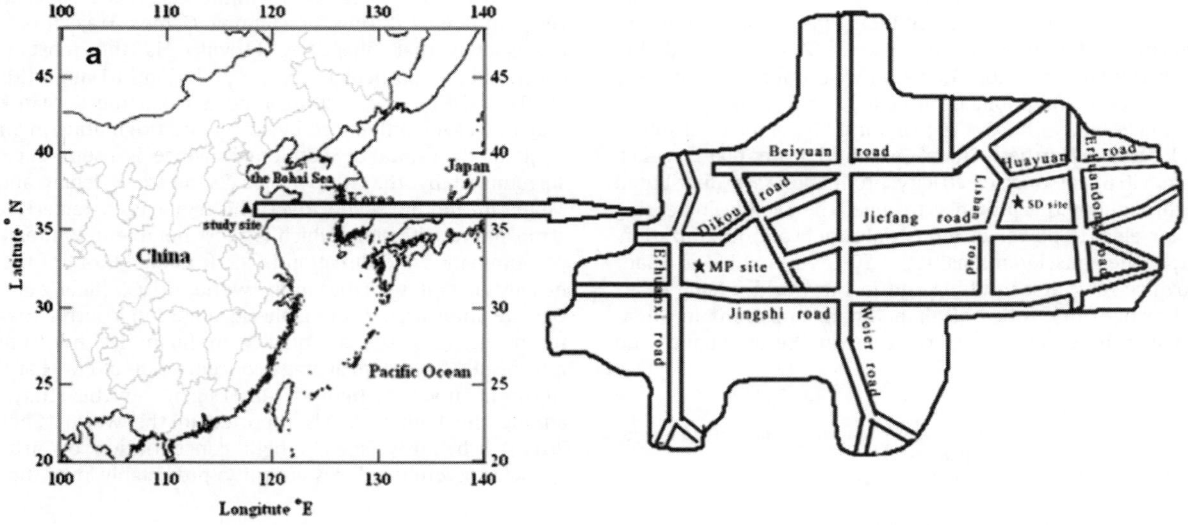

Fig. 1. Aerosol sampling at an urban site and a rural site in Jinan.

2.3. Aerosol chemical analyses

Three inorganic ions (NO_3^-, SO_4^{2-} and NH_4^+) and three organic acids (formic, oxalic and acetic acids) were analyzed using ion chromatography (IC; Dionex 2500, USA). A separation column (Dionex Ionpac AS11), a guard column (Dionex Ionpac AG 11) and a self-regenerating suppressor (ASRS-ULTR AΠ4-mm) were used for anion analysis; a separation column (Dionex Ionpac CS14), a guard column (Dionex Ionpac CG 14) and a self-regenerating suppressor (CSRS ULTRA Π 4-mm) were used for cation analysis. The details of this procedure were described by Yang et al. (2007). The Teflon filters were analyzed for 24 elements (Al, Na, Cl, Mg, Si, K, Ca, Sc, Ti, V, Cr, Mn, Fe, Co, Ni, Cu, Zn, As, Se, Br, Sr, Cd, Ba and Pb) using X-ray fluorescence (XRF). The carbon contents were determined using thermal/optical transmittance (TOT) protocol (Zhi et al., 2011). The number of filters for the analysis of mass, water-soluble ions, OC/EC and elements is 52, 52, 53, 54 and 109, 51, 53, 51 at SD and MP respectively.

3. Results and discussion

3.1. Ambient concentrations of $PM_{2.5}$

The variations of $PM_{2.5}$ concentration during the sampling period are shown in Table 1 and Fig. 2; these range from 30.16–392.55 μg m^{-3} at SD and 15.05–294.83 μg m^{-3} at MP, with an average of 148.71 μg m^{-3} at SD and 97.59 μg m^{-3} at MP. These measurements are 9.91 and 6.51 times the annual US NAAQS standards of $PM_{2.5}$ (15 μg m^{-3}), respectively. The results clearly show that fine particulate pollution is particularly serious in Jinan, especially in the urban area.

The concentrations of $PM_{2.5}$ at MP and SD showed a high correlation ($r = 0.75$), implying that $PM_{2.5}$ may have similar sources at MP and SD. Jinan is surrounded by mountains on 3 sides, and the anticyclone system under the height of 600 m in Shandong Province could hinder the diffusion of pollutants (Cheng et al., 2011a). The prevailing wind direction is northeast in winter and southwest in the other seasons. Because SD is located northeast of MP, these factors are likely to influence the concentrations of $PM_{2.5}$ at MP and SD.

As illustrated in Fig. 2, the $PM_{2.5}$ mass concentration showed an apparent seasonal variation, in which high values were common in winter and autumn and lower values predominated in spring and summer. Wind speed multiplies the mixing height, which is defined as atmospheric dilution rate per unit length; this can be used to evaluate the diffusion capability of the atmosphere (Chua et al., 2004). The meteorological parameters of the four seasons are shown in Table 2. The dilution rate of the atmosphere in winter was noticeably lower, 54.47% of the dilution rate in spring. A combination of low atmospheric diffusion capacity and an increase of particulate matter emission related to heating in winter led to serious fine particle pollution. During the dust storm period, the concentration of $PM_{2.5}$ was 0.84–1.35 times the annual average, while the concentration of PM_{10} was 1.35–2.19 times the annual average, indicating that dust storms have a greater impact on PM_{10} than $PM_{2.5}$. After a sandstorm, the concentration of PM_{10} decreased rapidly, demonstrating that the impact is short-term and does not result in an accumulation of particulate matter in the air. The averages of $PM_{2.5}/PM_{10}$ at SD and MP were 0.52 and 0.43, respectively, implying that fine particulate pollution at the urban site is more serious than at the rural site.

Table 1
Seasonal average concentrations of chemical species in $PM_{2.5}$ in Jinan (unit: μg m^{-3}).

	SD					MP				
	Spring	Summer	Autumn	Winter	Average	Spring	Summer	Autumn	Winter	Average
Mass (ug m^{-3})	143.25	129.04	134.89	204.89	148.71	93.46	69.56	93.21	146.80	97.59
Acetic	0.05	0.08	0.08	0.10	0.08	0.12	0.11	0.08	0.09	0.10
Formic	0.07	0.11	0.09	0.11	0.09	0.05	0.05	0.08	0.09	0.07
Oxalic	0.32	0.50	0.46	0.28	0.40	0.25	0.27	0.20	0.16	0.22
Nitrate	9.06	6.71	9.15	20.12	10.58	7.27	1.83	4.32	9.73	5.49
Sulfate	21.91	37.36	27.80	37.70	30.92	15.40	24.63	14.73	18.34	18.43
Ammonium	9.69	14.67	12.57	20.47	13.99	6.66	8.17	6.11	11.15	7.97
OC	15.76	14.99	23.96	35.75	22.19	13.44	9.30	22.50	35.24	20.20
OC_{sec}	8.20	7.80	13.50	22.30	12.10	7.50	3.00	11.20	23.10	10.90
EC	3.25	3.08	4.49	5.74	4.10	2.80	3.19	5.27	5.67	4.33
Al	1.64	0.41	0.69	0.78	0.86	0.63	0.30	0.24	0.27	0.34
Na	0.48	0.42	0.57	0.63	0.52	0.23	0.21	0.26	0.28	0.24
Cl	1.66	0.63	2.96	6.77	2.70	1.46	0.09	1.68	3.81	1.70
Mg	0.37	0.10	0.19	0.18	0.21	0.15	0.06	0.06	0.07	0.08
Si	4.00	0.91	1.57	1.81	2.02	1.49	0.68	0.60	0.68	0.81
K	3.47	4.32	4.73	4.58	4.30	2.55	2.22	2.53	2.74	2.51
Ca	1.59	0.61	1.25	1.09	1.13	0.73	0.63	0.45	0.62	0.59
Sc	0.01	0.01	0.01	0.02	0.01	–	–	–	–	–
Ti	0.15	0.04	0.06	0.09	0.08	0.05	0.03	0.03	0.03	0.03
V	0.01	0.01	–	–	–	–	–	–	0.01	–
Cr	0.02	0.02	0.02	0.03	0.02	0.01	0.02	0.01	0.01	0.01
Mn	0.09	0.05	0.12	0.16	0.10	0.06	0.03	0.06	0.07	0.06
Fe	1.94	0.99	1.61	2.02	1.59	0.83	0.50	0.62	0.61	0.63
Co	0.01	0.01	0.01	0.02	0.01	0.01	0.01	0.01	0.01	0.01
Ni	0.01	0.01	0.01	0.01	0.01	0.01	–	0.01	–	0.01
Cu	0.02	0.02	0.03	0.05	0.03	0.01	0.02	0.02	0.02	0.02
Zn	0.38	0.56	0.55	0.99	0.59	0.29	0.22	0.29	0.51	0.32
As	0.02	0.03	0.02	0.03	0.02	0.01	0.01	0.01	0.01	0.01
Se	0.01	0.01	0.02	0.04	0.02	0.01	0.01	0.01	0.02	0.01
Br	0.06	0.04	0.08	0.15	0.08	0.03	0.02	0.05	0.09	0.05
Sr	0.02	0.01	0.01	0.02	0.02	0.01	0.01	0.01	0.01	0.01
Cd	0.01	0.01	–	–	0.01	0.01	0.01	0.01	0.01	0.01
Ba	0.07	0.04	0.07	0.07	0.06	0.04	0.04	0.03	0.03	0.03
Pb	0.20	0.28	0.31	0.43	0.30	0.24	0.29	0.19	0.26	0.25

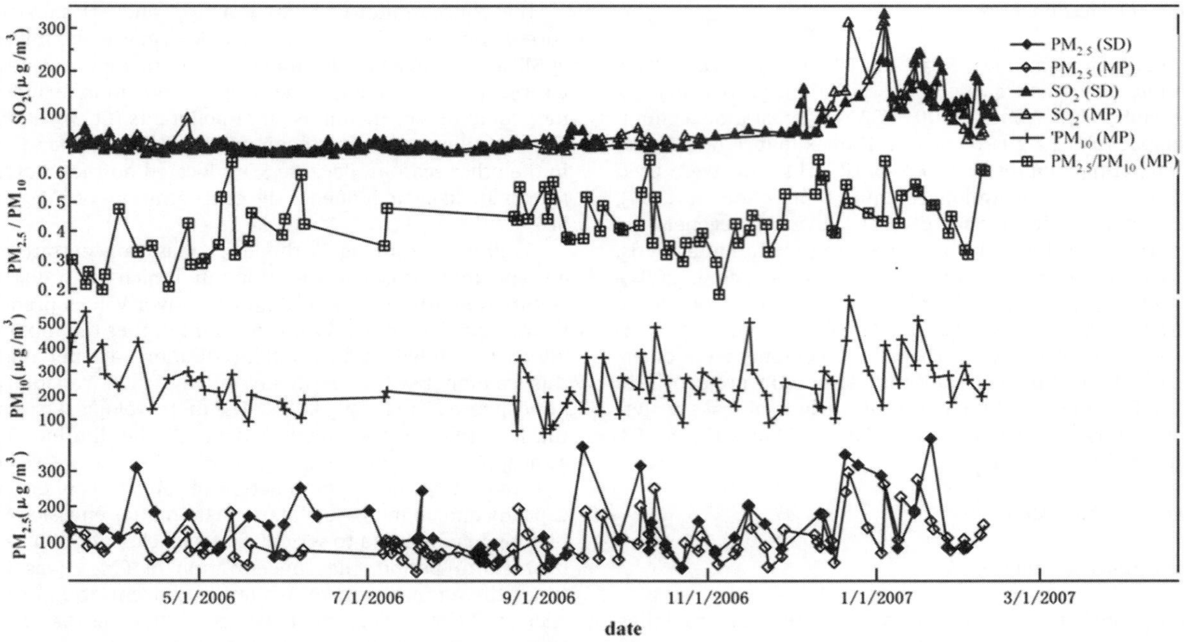

Fig. 2. Temporal variations of $PM_{2.5}$ mass, PM_{10} mass and gaseous pollutant concentrations in Jinan from March 2006 to February 2007.

3.2. Chemical compositions

3.2.1. Water-soluble ions

The concentrations of major chemical components in $PM_{2.5}$ are provided in Table 1. The sum of SO_4^{2-}, NO_3^- and NH_4^+ contributed 38.1% and 37.5% to $PM_{2.5}$ mass concentrations at SD and MP, respectively, indicating that SO_4^{2-}, NO_3^- and $NH_4^+$are the dominant ions in $PM_{2.5}$ in Jinan.

SO_4^{2-} and NH_4^+ exhibit a similar seasonal trend, which is high in the summer and winter and low in spring. However, the maximum and minimum concentrations of NO_3^- appeared in winter and summer, respectively. The SOR may be used to indicate the reaction process and the extent of transformation from SO_2 to SO_4^{2-}. At SD, the SOR was as follows: summer (0.62) > autumn (0.41) > spring (0.38) > winter (0.18). At MP, on the other hand, the SOR was as follows: summer (0.49) > autumn (0.24) > spring (0.23) > winter (0.10). The formation of SO_4^{2-} from SO_2 mainly includes the gas-phase reaction of SO_2 and OH radical and heterogeneous reaction (metal catalyzed oxidation or H_2O_2/O_3 oxidation) (Ziegler, 1979). The gas-phase oxidation of SO_2 to SO_4^{2-} by an OH radical is positively related to the temperature (Seinfeld, 1986). The average RH in winter is lower than the average for the entire year, which also does not favor an aqueous heterogeneous reaction. The lowest SOR was observed in winter, as expected. The high concentration of SO_4^{2-} in winter may be due to the high concentration of SO_2, but the strong solar radiation (high temperature) and high RH and O_3 concentrations in summer may accelerate the secondary conversion in the atmosphere and lead to the high SOR and SO_4^{2-} concentrations. The SOR at SD is higher than at MP, which implies that the highly polluted atmosphere is favorable for the transformation of SO_2.

The total concentrations of formic, acetic and oxalic acids accounted for 0.39% and 0.45% of the mass of $PM_{2.5}$ at SD and MP, respectively. The concentrations of these organic acids in Jinan were comparable with those in Beijing (Wang et al., 2007) and New York City (Khwaja, 1995), indicating that the pollution condition of these organic acid among the big cities was similar. Acetic acid is mainly produced from primary emissions, while formic acid comes mainly from secondary transformations. The ratio of acetic to formic acid (A/F) is used to evaluate the relative importance of primary emissions and secondary emissions. The A/F ratios were relatively low at SD (<1), implying that secondary formation is important. In comparison, the higher annual A/F ratio at MP (>1) indicates that primary emissions are the dominant source of carboxylic acids. The much higher A/F ratios in the spring and summer at MP (>2) are likely due to rapid leaf expansion (Wang et al., 2007).

Oxalic acid showed apparent correlations with formic acid at SD $r = 0.45$) and at MP ($r = 0.67$) and was principally observed in droplet form (Yao et al., 2002). Therefore, we may infer that secondary formation is the main source of oxalic acid and that an in-cloud or heterogeneous formation process was expected.

3.2.2. Carbonaceous species

Variations of the concentrations of OC and EC at SD and MP are presented in Fig. 3. There was no apparent spatial difference between the concentrations of OC and EC at SD and MP. Furthermore, similar seasonal variations of the concentrations of OC and EC at SD and MP were observed, with the lowest in spring and summer; the concentrations continued to increase steadily, reaching their maximum concentrations in winter. The concentrations of OC and EC often reached their peak value at the same

Table 2
Seasonal average meteorological conditions during the sampling period in Jinan.

	Concentration of $PM_{2.5}$ ($\mu g\ m^{-3}$)	Temperature (°C)	Mixing height (m)	Wind speed ($m\ s^{-1}$)	Relative humidity (%)	Atmospheric dilution rate
Spring	143.3	15.4	975.8	3.8	45.0	4423.9
Summer	129.0	26.9	722.1	2.8	66.2	2599.4
Autumn	134.9	17.2	680.0	2.9	62.8	2508.9
Winter	204.9	2.7	586.3	3.1	52.6	2409.5

time, implying the inner correlation of OC and EC and their common sources.

The correlation of OC and EC may reflect the common sources of carbonaceous aerosol; in other words, the positive correlation between OC and EC may indicate that they have a common source (Turpin and Huntzicker, 1991). The regressive relationship between OC and EC is shown in Fig. 4. The correlations between OC and EC at SD and MP were both higher in autumn and winter, suggesting that OC and EC may have common sources in autumn and winter, possibly biomass burning and coal combustion (Zhang et al., 2009a). The lower correlation between OC and EC that appears in summer at SD and spring at MP implies that OC and EC may have different sources in spring and summer, which might be related to the influence of dust storms coming from the northwest of China in spring and the high temperatures in summer that favor the formation of SOA (Turpin and Huntzicker, 1995). It has been suggested that if the ratio of OC/EC > 2.0, SOC will be formed (Turpin et al., 1990). The ratios of OC/EC at SD and MP were in the range of 2.34–15.41 and 2.14–11.30, with averages of 5.4 and 4.7, respectively, indicating that SOC may be formed both at SD and MP and that the SOC pollution at SD was more serious than that at MP. The concentration of SOC was estimated through the following equation: $OC_{sec} = (OC)_{tot} - (EC)^{*}(OC/EC)_{pri}$. The least square method was used to estimate the value of $(OC/EC)_{pri}$ (2.4); the detail of this method is described in the references (Strader et al., 1999). SOC accounted for 48.3% and 47.5% of the OC at SD and MP, respectively, suggesting that SOC is an important organic component of $PM_{2.5}$. The extremely high values of OCsec /OC in winter may be related to the combination of the increasing emissions of the precursors of SOC, including VOC and SVOC, the poor conditions for diffusion and the scant amount of rainfall in winter, which may favor the active photochemical reaction of VOC and SVOC to form SOC (Lin, 2002).

3.2.3. *Inorganic elements*

In this study, the elements were divided into two groups, one related to anthropogenic activity and the other to typical crustal elements. The crustal elements (Ca, Al, Si, Fe, Mg and Ti) showed a similar seasonal variation, with high values in spring and lower values in summer. The concentrations of the indicator elements of coal burning (Cl, Br and Se) in winter are much higher than in other seasons, and no apparent seasonal variations were observed in the elements that come primarily from industrial sources (Sc, V, Zn, Cr, Mn, Co, Ni, Cu, Ba, Pb, Cd, As and Sr). This indicates that the contribution of burning coal to elemental pollution increased in winter, while the contributions of industrial sources were consistent.

The enrichment factor, defined as $EF = (C_i/C_n)$ environment/(C_i/C_n) background is useful in judging the contributions of natural and anthropogenic sources (Gao and Anderson, 2001). In this study, Al is chosen as the reference element. The seasonal variations of EF at SD and MP are shown in Fig. 5. The EF of Al, Mg, Fe, Na and Ca are all lower than 10, indicating that these elements are mostly from the local and adjacent soil. The EF values of Cl, K, Mn, Ni, Cu, Zn, Br and Pb are much higher than 10, demonstrating that these elements had close relationships with anthropogenic activity and are less influenced by soil dust (Duan et al., 2006). Except for Cl at MP, the EF of all the elements reached their minimum levels in spring, implying an increase of the contribution of soil sources, which may be related to frequent dust storms. The EF of Cl and Br was higher in winter, which may be related to the rapidly increase in burning coal for heat in the winter. Apparent variations of the EF of Mn, Ni, Zn, Cu and Pb were not observed, except that they are lower in the spring, indicating that contribution from industrial sources is stable throughout the four seasons.

3.2.4. *Mass closure*

The mass closure of $PM_{2.5}$ was made under three assumptions. First, in addition to carbon, organic mass contains other elements, such as hydrogen, oxygen and nitrogen. The concentration of organic mass is calculated as 1.8 times the concentration of OC, according to the revised IMPROVE formula. Second, we assume that crustal elements exist as oxides and the soil dust concentration is the sum of the oxides of the main crustal elements (Al_2O_3, SiO_2, CaO, FeO, Fe_2O_3, TiO_2). The formula as follows was used: $[Soil] = 2.20[Al] + 2.49[Si] + 1.63[Ca] + 2.42[Fe] + 1.94[Ti]$ (Kim et al., 2001). Third, the concentration of trace elements is also assumed to be the sum of oxides of the corresponding elements, except the Al, Si, Ca, Fe and Ti.

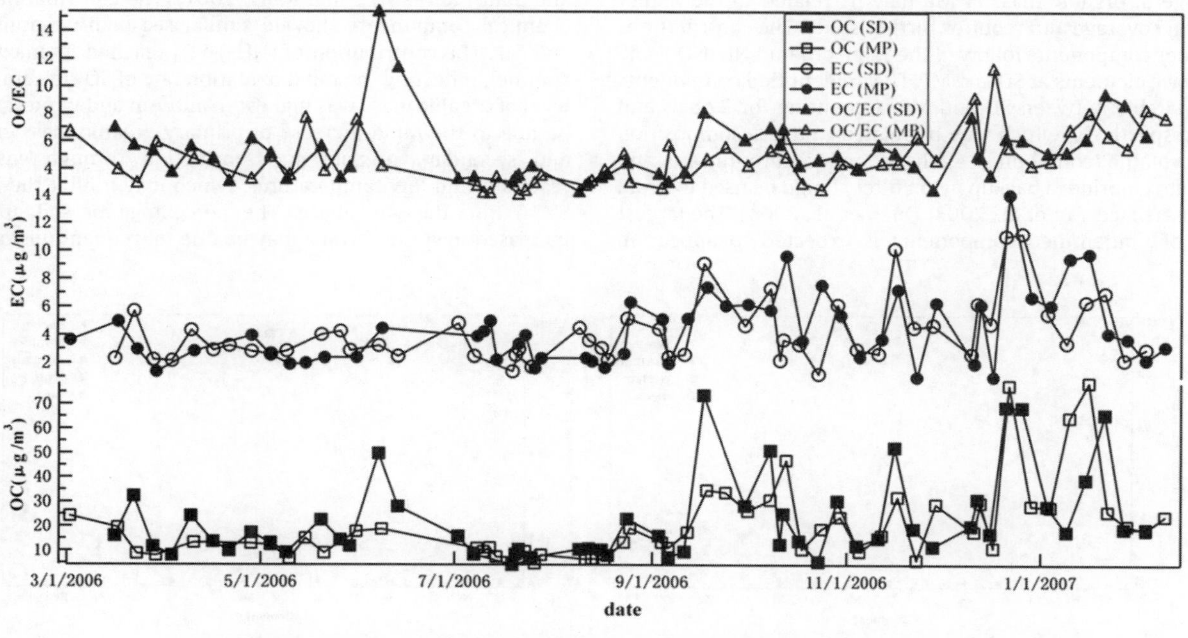

Fig. 3. OC and EC concentrations and OC/EC ratio in Jinan during the sampling period.

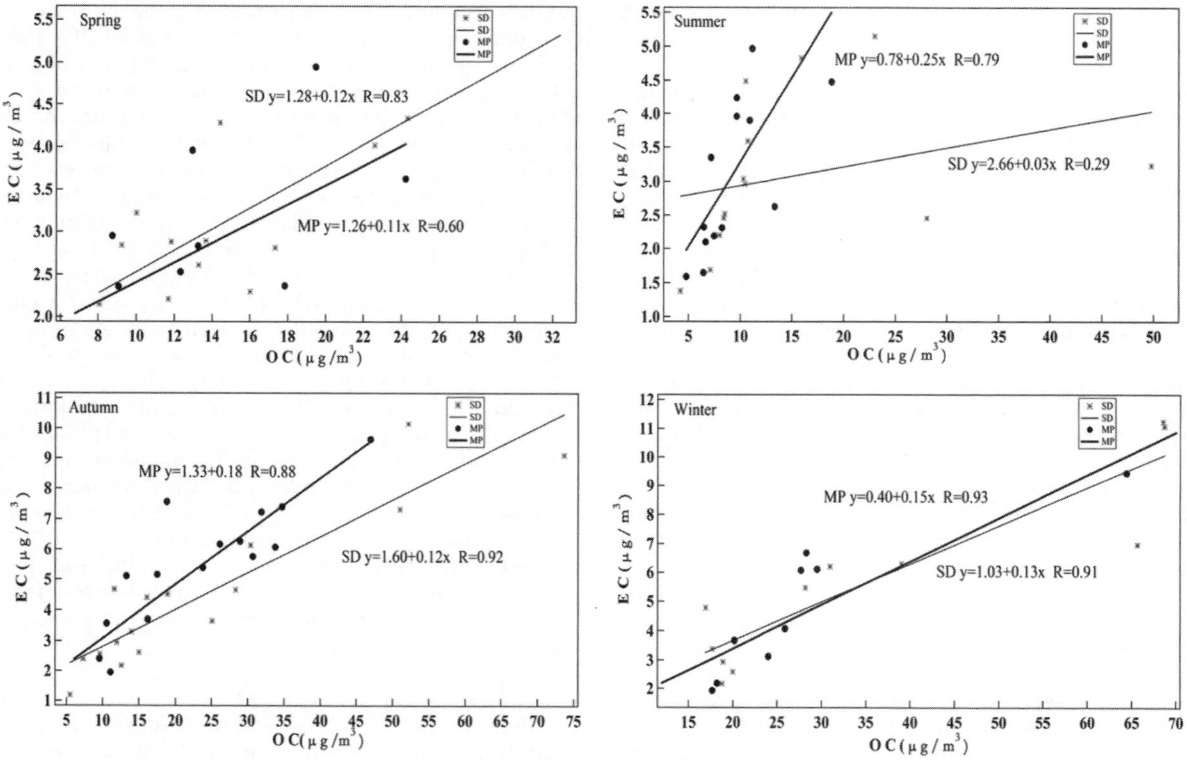

Fig. 4. Regression relationship between OC and EC in four seasons in Jinan.

The mass closures of $PM_{2.5}$ at SD and MP in four seasons are presented in Fig. 6. The chemical component percentages in $PM_{2.5}$ at SD and MP are different that $(NH_4)_2SO_4$, which accounts for 28.71% of the mass of $PM_{2.5}$ and is the primary component of $PM_{2.5}$ at SD; organic mass ranked second, accounting for 26.98% of the $PM_{2.5}$ mass. However, the contribution of organic mass ranked first in $PM_{2.5}$ at MP and $(NH_4)_2SO_4$ ranked second, accounting for 37.25% and 25.96% of the $PM_{2.5}$ mass, respectively. The higher percentage of organic mass at MP may be related to the higher vegetation coverage and requires further study. The contributions of the other components followed the same order of NH_4NO_3, soil, EC and trace elements at SD and MP. The unidentified components are comparable between SD and MP, accounting for 22.81% and 18.82%, respectively, which may be related to the decomposition of semi-volatile compounds such as ammonium nitrate and ammonium chloride; a backup nylon filter should be used to avoid sampling artifacts (Ye et al., 2003; Duan et al., 2006). The largest amount of unidentified components is expected to appear in summer, when the high temperature can promote the decomposition of semi-volatile compounds. However, the largest unidentified fraction appeared in spring at both SD and MP, which indicates that the temperature did not have a substantial impact on the unidentified percentage; the unexplained category might result from the contributions of some nonanalyzed crustal and trace elements species coming primarily from the soil. Similar results were observed in the study of $PM_{2.5}$ in Taiwan (Lin, 2002) and South Korea (Lee and Kang, 2001). The contributions of the chemical components showed similar seasonal variations at SD and MP. The contribution of $(NH_4)_2SO_4$ reached its maximum in summer, reflecting the rapid oxidation rate of SO_2 in summer. The level of organic mass was highest in autumn and winter; this may be due to the rapid increase of primary organic mass emissions and secondary organic mass formation through gas-aerosol reactions and low temperatures, which may inhibit the VOC and SVOC into the gas phase. The percentage of soil apparently increased in spring, which may be due to frequent dust storms.

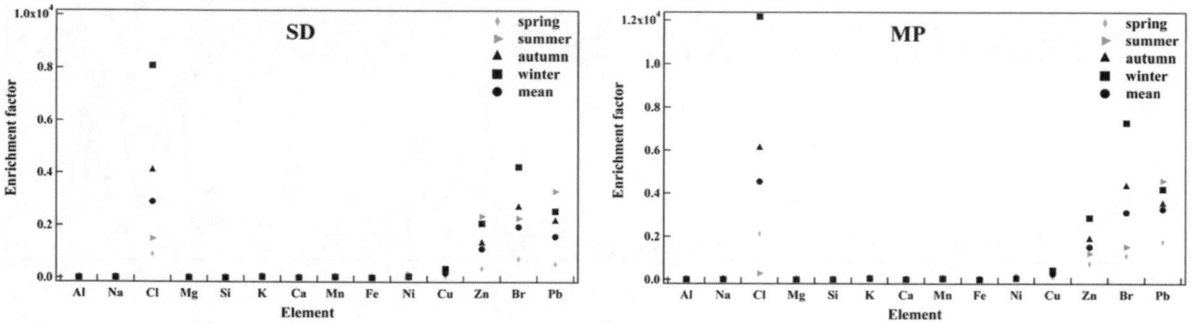

Fig. 5. Seasonal variations of enrichment factors of the elements in $PM_{2.5}$ in Jinan.

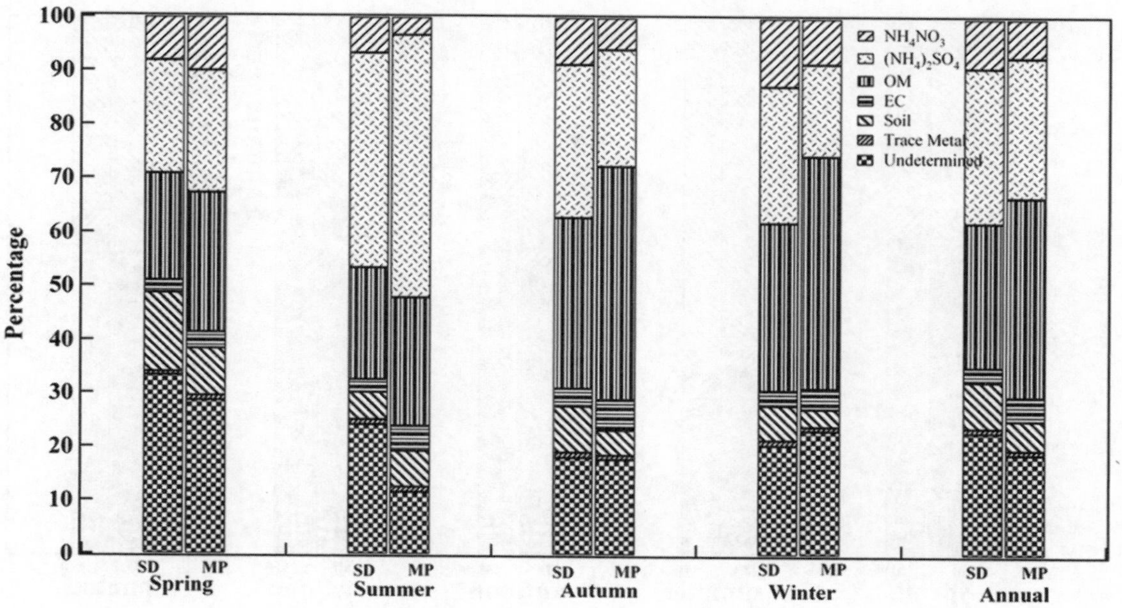

Fig. 6. Annual and seasonal contributions of chemical species to $PM_{2.5}$ mass concentrations in Jinan.

3.3. Contributions of the species in $PM_{2.5}$ to the b_{ext}

The visual range can be calculated as follows (Koschmieder, 1925):

$$V = 3.91/b_{ext}$$

where b_{ext} is the light extinction coefficient. Visibility is inversely proportional to the b_{ext}, so the higher the b_{ext}, the lower the visibility.

The revised algorithm proposed by the IMPROVE program for use in estimating light extinction from aerosol chemical compositions was applied in this study. $b_{ext} \approx 2.2 \times f_S(RH) \times$ [Small Sulfate] + $4.8 \times f_L(RH) \times$ [Large Sulfate] + $2.4 \times f_S(RH) \times$ [Small Nitrate] + $5.1 \times f_L(RH) \times$ [Large Nitrate] + $2.8 \times$ [Small Organic Mass] + $6.1 \times$ [Large Organic Mass] + $10 \times$ [Elemental Carbon] + $1 \times$ [Fine Soil] + $1.7 \times f_{SS}(RH) \times$ [Sea Salt] + $0.6 \times$ [Coarse Mass] + Rayleigh Scattering + $0.33 \times [NO_2(ppb)]$.

The major novelties of using the revised IMPROVE algorithm from the equation in our study (Yang et al., 2007) include the following: splitting the total sulfate, nitrate and organic carbon into two fractions, large and small sizes; changing the coefficient (from 1.4 to 1.8) to calculate the organic mass from organic carbon; and the new addition of sea salt and NO_2.The computational process of the revised algorithm was obtained from http://vista.cira.colostate.edu/improve/Publications/GrayLit/019_RevisedIMPROVEeq. Because the soil and coarse mass contribute only small fractions to the b_{ext} (Cheung et al., 2005), soil and coarse mass were excluded from this study. Given that the size of the hygroscopic species may increase with relative humidity and result in the variation of the b_{ext}, the hygroscopic species growth function f(RH) was introduced to the formula to indicate the influence of the relative humidity. Three water growth adjustment terms (f_S(RH), f_L(RH) and f_{SS}(RH), respectively) were used for the small-size and large-size distribution sulfate and nitrate compounds and for sea salt, respectively. The value of the Rayleigh scattering is approximately 12 Mm^{-1} for sites near sea level.

The result of the b_{ext} calculated from the data set from March 2006 to February 2007 was 778.16 Mm^{-1}at SD and 528.48 Mm^{-1} at MP, higher than in our previous study (Yang et al., 2007), indicating that the visibility in Jinan continued to deteriorate. The b_{ext} similarly showed clear seasonal variations at both SD and MP. The highest measurements were obtained in winter, which may be due to the high concentration of fine particles caused by the combination of increasing pollutant emissions and inverse atmospheric diffusion conditions. The contributions of chemical components in $PM_{2.5}$ to b_{ext} at SD and MP across four seasons are detailed in Fig. 7. The relative average contributions of the species in $PM_{2.5}$ at SD and MP are different in that large size ammonium sulfate is the largest contributor at SD (43.80%), similar to our previous study at SD and studies at Hongkong (Cheung et al., 2005) and Brisban (Chan et al., 1999). The contribution of large size organic mass (41.02%) at MP indicates that the rural area of Jinan experiences more serious organic pollution. The contribution of large size ammonium sulfate reached its maximum in the summer, possibly due to the large fraction of sulfate in $PM_{2.5}$ and its ability to absorb water vapor, which can enhance light scattering under the high relative humidity of summer. Large size organic mass is the largest contributor to b_{ext} in autumn and winter, which is closely related to the relatively high concentration of organic mass in the autumn and winter.

4. Conclusion

The daily $PM_{2.5}$ mass concentrations ranged from 30.16–392.55 μg m^{-3} and 15.05–294.83 μg m^{-3} at SD and MP, respectively, with an average of 148.71 μg m^{-3} and 97.59 μg m^{-3} and were 9.91 and 6.51 times the U.S.EPA's NAAQS. This indicates that fine particulate pollution was very serious in Jinan and worst in the urban area. The most abundant species contributing to the $PM_{2.5}$ were ammonium sulfates and organic mass, accounting for a total of 55.69% and 63.21% at SD and MP, respectively. $(NH_4)_2SO_4$ was the primary component of $PM_{2.5}$ at SD, while the content of organic mass ranked first in $PM_{2.5}$ at MP. The higher SOR at SD showed that the secondary inorganic ion pollution was more serious in the urban area. The correlation coefficients between OC and EC showed that OC and EC may have common sources in winter and autumn. The high ratio of OC/EC in Jinan implied that the formation of secondary organic aerosol was active in Jinan and stronger at the urban location. The results of the revised IMPROVE formula showed that the

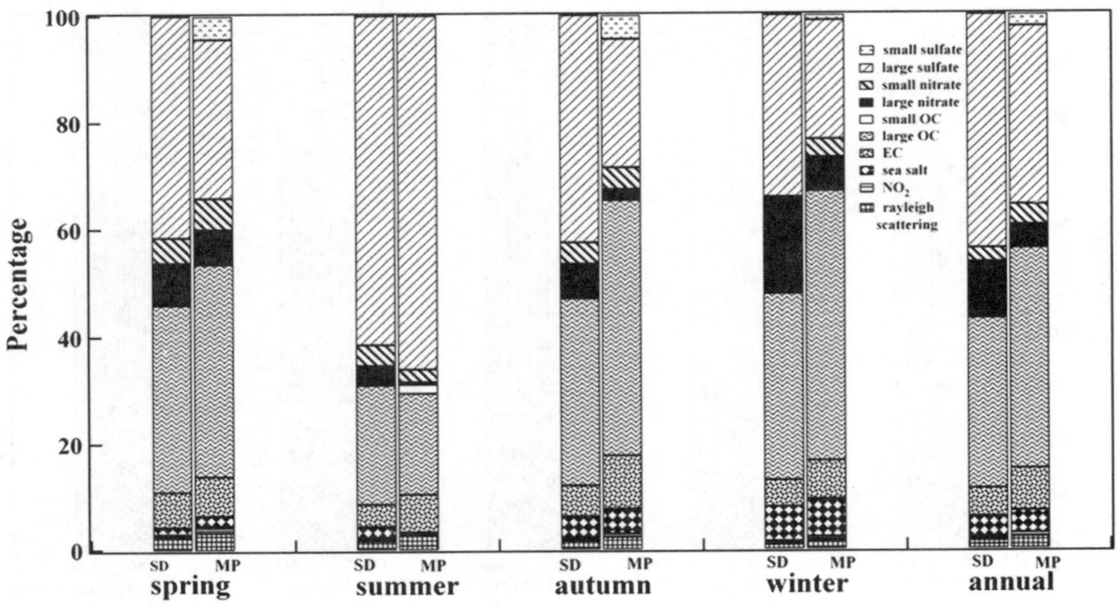

Fig. 7. Annual and seasonal contributions of chemical components in $PM_{2.5}$ to the light extinction coefficient in Jinan.

contributions of large size sulfate and large size organic mass in fine particle form to light extinction were significant at SD and MP, respectively.

Acknowledgements

This work was supported by the National Basic Research Program (973 Program) of China (2005CB422203), Key Project of Shandong Provincial Environmental Agency (2006045), Promotive Research Fund for Young and Middle-aged Scientists of Shandong Province (BS2010HZ010), Independent Innovation Foundation of Shandong University (2009TS024) and Special Research for Public-Beneficial Environment Protection (201009001-1).

References

Chan, Y.C., Simpson, R.W., Mctainsh, G.H., Vowles, P.D., Cohen, D.D., Bailey, G.M., 1999. Source apportionment of visibility degradation problems in Brisbane (Australia) using the multiple linear regression techniques. Atmospheric Environment 33, 3237–3250.

Cheng, S., Yang, L., Zhou, X., Wang, Z., Zhou, Y., Gao, X., Nie, W., Wang, X., Xu, P., Wang, W., 2011a. Evaluating $PM_{2.5}$ ionic components and source apportionment in Jinan, China from 2004 to 2008 using trajectory statistical methods. Journal of Environmental Monitoring 13, 1662–1671.

Cheng, S., Yang, L., Zhou, X., Xue, L., Gao, X., Zhou, Y., Wang, W., 2011b. Size-fractionated water-soluble ions, situ pH and water content in aerosol on hazy days and the influences on visibility impairment in Jinan, China. Atmospheric Environment 45, 4631–4640.

Cheung, H.C., Wang, T., Baumann, K., Guo, H., 2005. Influence of regional pollution outflow on the concentrations of fine particulate matter and visibility in the coastal area of southern China. Atmospheric Environment 39, 6463–6474.

Chua, S.H., Paisiea, J.W., Jang, B.W., 2004. PM data analysis—a comparison of two urban areas: Fresno and Atlanta. Atmospheric Environment 38, 3155–3164.

Duan, F.K., He, K.B., Ma, Y.L., Yang, F.M., Yu, X.C., Cadle, S.H., Chan, T., Mulawa, P.A., 2006. Concentration and chemical characteristics of $PM_{2.5}$ in Beijing, China: 2001–2002. Science of the Total Environment 355, 264–275.

Gao, X., Yang, L., Cheng, S., Gao, R., Zhou, Y., Xue, L., Shou, Y., Wang, J., Wang, X., Nie, W., Xu, P., Wang, W., 2011. Semi-continuous measurement of water-soluble ions in $PM_{2.5}$ in Jinan, China: temporal variations and source apportionments. Atmospheric Environment 45, 6048–6056.

Gao, Y., Anderson, J.R., 2001. Characterization of Chinese aerosols determined by individual-particle analyses. Journal of Geophysical Research 106, 18037–18045.

Ghim, Y.S., Moon, K.C., Lee, S., Kim, Y.P., 2005. Visibility trends in Korea during the past two decades. Journal of the Air & Waste Management Association 55, 73–82.

Hong, Y.C., Lee, J.T., Kim, H., Ha, E.H., Schwartz, J., Christiani, D.C., 2002. Effects of air pollutions on acute stroke mortality. Environmental Health Perspectives 110, 187–191.

Kim, K.W., Kim, Y.J., Oh, S.J., 2001. Visibility impairment during Yellow Sand periods in the urban atmosphere of Kwangju, Korea. Atmospheric Environment 35, 5157–5167.

Kim, Y.J., Woo, J.H., Ma, Y., Kim, S., Nam, J.S., Sung, H., Choi, K.C., Seo, J., Kim, J.S., Kang, C.H., Lee, G., Ro, C.U., Chang, D., Sun, Y., 2009. Chemical characteristics of long-range transport aerosol at background sites in Korea. Atmospheric Environment 43, 5556–5566.

Khwaja, H.A., 1995. Atmospheric concentration of carboxylic acids and related compounds at a semiurban site. Atmospheric Environment 29, 127–139.

Kong, S., Han, B., Bai, Z., Chen, L., Shi, J., Xu, Z., 2010. Receptor modeling of $PM_{2.5}$, PM_{10} and TSP in different seasons and long-range transport analysis at a coastal site of Tianjin, China. Science of the Total Environment 408, 4681–4694.

Koschmieder, H., 1925. Theorie der horizontalen sichtweite :kontrast und sichtweite beitrage zur physik der freien. Atmosphere 12, 171–181.

Lee, H.S., Kang, B.W., 2001. Chemical characteristics of principal $PM_{2.5}$ species in Chongju, South Korea. Atmospheric Environment 35, 739–746.

Lin, J.J., 2002. Characterization of the major chemical species in $PM_{2.5}$ in the Kaohsiung City, Taiwan. Atmospheric Environment 36, 1911–1920.

National Bureau of Statistics of China, 2008. China Statistical Yearbook. China Statistics Press, Beijing.

Seinfeld, 1986. Atmospheric Chemistry and Physics of Air Pollution. Wiley, New York, NY.

Sheng, L., An, J., Dong, L., 2000. Jinan–facing the pollution. Walking to the World 3, 24 (in Chinese).

Strader, R., Lurmann, F., Pandis, S.N., 1999. Evaluation of secondary organic aerosol formation in winter. Atmospheric Environment 33, 4849–4863.

Streets, D.G., Fu, J.S., Jang, C.J., Hao, J., He, K., Tang, X., Zhang, Y., Wang, Z., Li, Z., Zhang, Q., Wang, L., Wang, B., Yu, C., 2007. Air quality during the 2008 Beijing Olympic games. Atmospheric Environment 41, 480–492.

Turpin, B.J., Cary, R.A., Huntzicker, J.J., 1990. An in-situ, time-resolved analyzed for aerosol organic and elemental carbon. Aerosol Science and Technology 12, 161–171.

Turpin, B.J., Huntzicker, J.J., 1991. Secondary formation of organic aerosol in the Los Angeles Basin: a descriptive analysis of organic and elemental carbon concentrations. Atmospheric Environment 25A, 207–215.

Turpin, B.J., Huntzicker, J.J., 1995. Identification of secondary organic aerosol episodes and quantitacation of primary and secondary organic aerosol concentration during SCAQS. Atmospheric Environment 29, 3527–3544.

Wang, Y., Zhuang, G., Chen, S., An, Z., Zheng, A., 2007. Characteristics and sources of formic, acetic and oxalic acids in $PM_{2.5}$ and PM_{10} aerosols in Beijing, China. Atmospheric Research 84, 169–181.

Xu, P., Wang, W., Yang, L., Zhang, Q., Gao, R., Wang, X., Nie, W., Gao, X., 2011. Aerosol size distributions in urban Jinan: Seasonal characteristics and variations between weekdays and weekends in a heavily polluted atmosphere. Environmental Monitoring and Assessment 179, 443–456.

Yang, L., Wang, D., Cheng, S., Wang, Z., Zhou, Y., Zhou, X., Wang, W., 2007. Influence of meteorological conditions and particulate matter on visual range impairment in Jinan, China. Science of the Total Environment 383, 164–173.

Yao, X., Chan, C.K., Fang, M., Cadle, S., Chan, T., Mulawa, P., He, K., Ye, B., 2002. The water-soluble ionic composition of $PM_{2.5}$ in Shanghai and Beijing, China. Atmospheric Environment 36, 4223–4234.
Ye, B., Ji, X., Yang, H., Yao, X., Chan, C.K., Cadle, S.H., Chan, T., Mulawa, P.A., 2003. Concentration and chemical composition of $PM_{2.5}$ in Shanghai for a 1-year period. Atmospheric Environment 37, 499–510.
Zhang, R., Ho, K.F., Cao, J., Han, Z., Zhang, M., Cheng, Y., Lee, S.C., 2009a. Organic carbon and elemental carbon associated with PM_{10} in Beijing during spring time. Journal of Hazardous Materials 172, 970–977.
Zhang, Q., Streets, D., Carmichael, G., He, K., Huo, H., Kannari, A., Klimont, Z., Park, I., Reddy, S., Fu, J., 2009b. Asian emissions in 2006 for the NASA INTEX-B mission. Atmospheric Chemistry and Physics 9, 5131–5153.
Zhi, G., Chen, Y., Sun, J., Chen, L., Tian, W., Duan, J., Zhang, G., Chai, F., Sheng, G., Fu, J., 2011. Harmonizing aerosol carbon measurements between two conventional thermal/optical analysis methods. Environmental Science & Technology 45, 2902–2908.
Ziegler, E.N., 1979. Sulfate-formation mechanism: theoretical and laboratory studies. Advances in Environmental Science and Engineering 1, 184–194.

Atmos. Chem. Phys., 12, 11985–11995, 2012
www.atmos-chem-phys.net/12/11985/2012/
doi:10.5194/acp-12-11985-2012

Atmospheric Chemistry and Physics

Asian dust storm observed at a rural mountain site in southern China: chemical evolution and heterogeneous photochemistry

W. Nie[1,2], T. Wang[2,1], L. K. Xue[2], A. J. Ding[3], X. F. Wang[1,2], X. M. Gao[1,2], Z. Xu[1,2], Y. C. Yu[1], C. Yuan[1,2], Z. S. Zhou[1,2], R. Gao[1], X. H. Liu[6], Y. Wang[4], S. J. Fan[5], S. Poon[2], Q. Z. Zhang[2], and W. X. Wang[2]

[1]Environment Research Institute, Shandong University, Jinan, China
[2]Department of Civil and Environmental Engineering, The Hong Kong Polytechnic University, Hong Kong, China
[3]Institute for Climate and Global Change Research, Nanjing University, Nanjing, China
[4]School of Environmental Science and Engineering, Shandong University, Jinan, China
[5]School of Environmental Science and Engineering, Sun Yat-Sen University, Guangzhou, China
[6]College of Environmental Science and Engineering, Ocean University of China, Qingdao, China

Correspondence to: T. Wang (cetwang@polyu.edu.hk)

Received: 12 July 2012 – Published in Atmos. Chem. Phys. Discuss.: 2 August 2012
Revised: 20 November 2012 – Accepted: 30 November 2012 – Published: 17 December 2012

Abstract. Heterogeneous processes on dust particles are important for understanding the chemistry and radiative balance of the atmosphere. This paper investigates an intense Asian dust storm episode observed at Mount Heng (1269 m a.s.l.) in southern China on 24–26 April 2009. A set of aerosol and trace gas data collected during the study was analyzed to investigate their chemical evolution and heterogeneous photochemistry as the dust traveled to southern China. Results show that the mineral dust arriving at Mt. Heng experienced significant modifications during transport, with large enrichments in secondary species (sulfate, nitrate, and ammonium) compared with the dust composition collected at an upwind mountain top site (Mount Hua). A photochemical age "clock" ($-\mathrm{Log}_{10}(NO_x/NO_y)$) was employed to quantify the atmospheric processing time. The result indicates an obvious increase in the abundance of secondary water-soluble ions in dust particles with the air mass atmospheric processing time. Based on the observations, a 4-stage evolution process is proposed for carbonate-containing Asian dust, starting from fresh dust to particles coated with hydrophilic and acidic materials. Daytime-enhanced nitrite formation on the dust particles was also observed, which indicates the recent laboratory result of the TiO_2 photocatalysis of NO_2 as a potential source of nitrite and nitrous acid.

1 Introduction

Mineral dust is injected into the atmosphere under specific meteorological conditions in deserts or semiarid areas, generating dust storms. In East Asia, the Taklimakan Desert in western China and the Gobi Desert in Mongolia and northern China are the two major source regions of dust, with annual emissions between 100 Mt yr^{-1} and 460 Mt yr^{-1} (Laurent et al., 2006). Upon entering the atmosphere, especially when transported over heavily polluted regions, the chemical and surface nature of mineral dust particles undergo significant changes, in turn influencing a number of atmospheric processes (Zhuang et al., 1992; Usher et al., 2003; Formenti et al., 2011, and the references therein).

In the past few decades, several large field campaigns (e.g., TRACP, ACE-Asia and INTEX-B) have studied the atmospheric processes of Asian dust (Jocob et al., 2003; Arimoto et al., 2006; Singh et al., 2009). The evolutionary processes of dust particles during their eastward transport to the western Pacific have been already investigated in northeast Asia. These studies have significantly improved understanding of the heterogeneous reactions between Asian pollution and mineral dust. Sullivan et al. (2007a) found that nitrate tended to accumulate on calcium-rich dust, while sulfate tended to accumulate on aluminosilicate-rich dust. Li and Shao (2009) demonstrated nitrate being coated onto dust particles. The direct uptake of chlorine on dust particles by

Published by Copernicus Publications on behalf of the European Geosciences Union.

heterogeneous reaction with HCl was clearly verified by Sullivan et al. (2007b). Sullivan and Prather (2007) observed internally mixed organic dicarboxylic acid (DCA) with dust particles and proposed a mechanism of heterogeneous oxidation of DCA on the surface of Asian dust. To our knowledge, there have been very few studies of the chemical evolution of Asian dust during its transport to southern China, where the dust is subject to higher temperature, higher humidity, and possibly different emissions compared with eastward transport.

Heterogeneous reactions on dust particles have been extensively investigated, and most previous studies have focused on nighttime chemistry (Usher et al., 2003). Photo-induced/enhanced reactions of some reactive gases on dust particles have recently been demonstrated by laboratory studies (Cwiertny et al., 2008 and the references therein). Nicolas et al. (2009) observed photo-enhanced O_3 decomposition on mimic mineral dust particles of TiO_2/SiO_2 mixtures; Ndour et al. (2008) detected the photo-enhanced uptake of NO_2 on both TiO_2/SiO_2 mixed particles and real Saharan and Arizona dust particles. The latter reaction received great attention because it may be a missing source of daytime nitrous acid (HONO). However, the extent of that reaction has not been observed in the real atmosphere.

As part of China's National Basic Research Project on acid rain, a comprehensive field campaign was conducted from March to May 2009 at Mt. Heng in Hunan province, southern China (Fig. 1), during which an intense dust storm was observed on 24–26 April. The aim of the specific work is to investigate the impact of this Asian dust storm on the atmospheric chemistry at rural and remote mountain regions in southern China. Special attention is given to the chemical evolution of the dust by comparison with the aerosol composition concurrently observed at another mountain site near the dust source region (Wang et al., 2011), as well as the role of heterogeneous photochemistry in the production of nitrite. The general features and transport pathway of this dust storm are described first. The chemical evolution of the dust particles during transport from Mt. Hua to Mt. Heng is then investigated. A photochemical age "clock" ($-Log_{10}(NO_x/NO_y)$) is employed to examine the changes in the major secondary water-soluble ions with the atmospheric processing time. The evidence of the daytime production of nitrite on dust particles is also presented, and its possible mechanism and potential contribution to the gas-phase HONO is discussed.

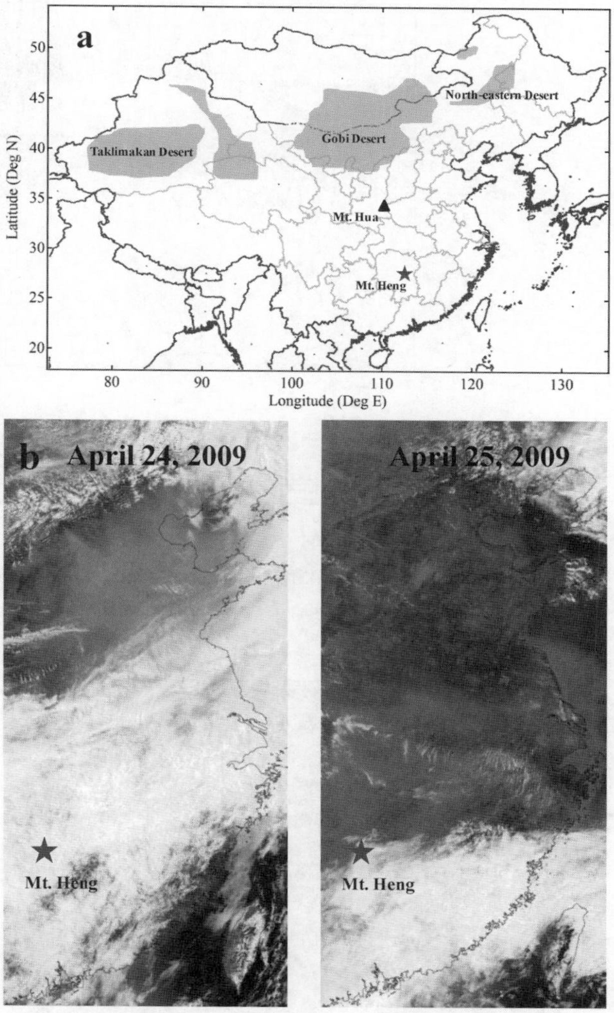

Fig. 1. (a) Map showing Mt. Heng and Mt. Hua. The major deserts in East Asia are marked by the shadows in the map. **(b)** MODIS true-color imagery of eastern China on 24 and 25 April 2009.

2 Experimental methodologies

2.1 Site description

Mt. Heng is situated in Hunan Province, southern China, and is approximately 500 km and 900 km from the South China Sea and the East China Sea, respectively (Figs. 1a and 2). The field campaign was conducted at a meteorological station at the summit of Mt. Heng (27°18′ N, 112°42′ E, 1269 m a.s.l.). There are a few temples and a road around the mountain top of Mt. Heng. Emissions from incense burning and traffic may occasionally affect the trace gas and aerosol measurements at the site. There are few other local emission sources around the measurement site. The closest cities are Hengyang and Xiangtan, which are approximately 50 km to the south and 70 km to the north, respectively.

2.2 Field campaign, sample collection and measurement techniques

The field campaign covered a 3-month period from March to May 2009. A comprehensive suite of trace gases, aerosols, and rain and cloud water samples was measured/collected

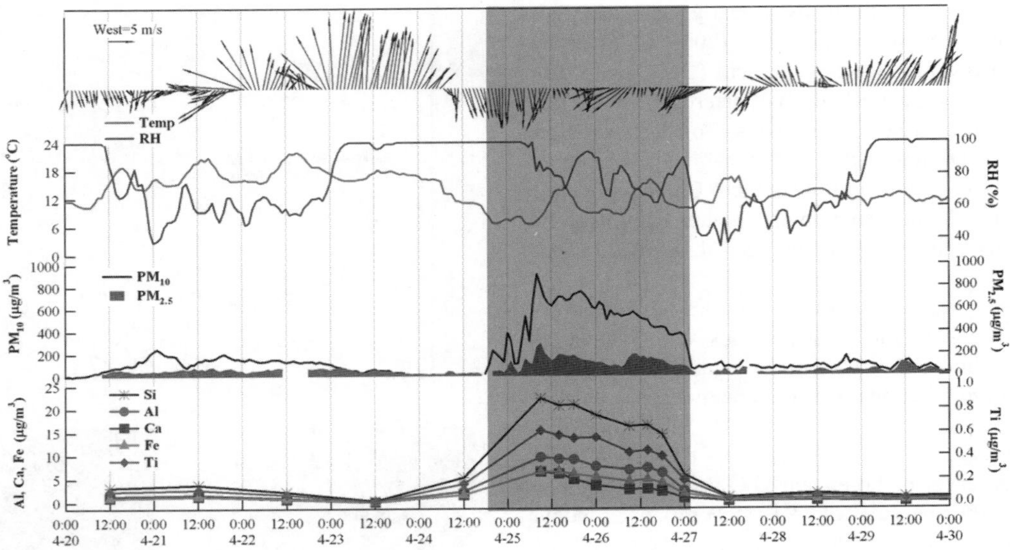

Fig. 2. Temporal variations of PM_{10}, $PM_{2.5}$, mineral elements (Si, Al, Ca, Fe, Ti) in $PM_{2.5}$ and meteorological parameters at Mt. Heng during 20 to 29 April 2009.

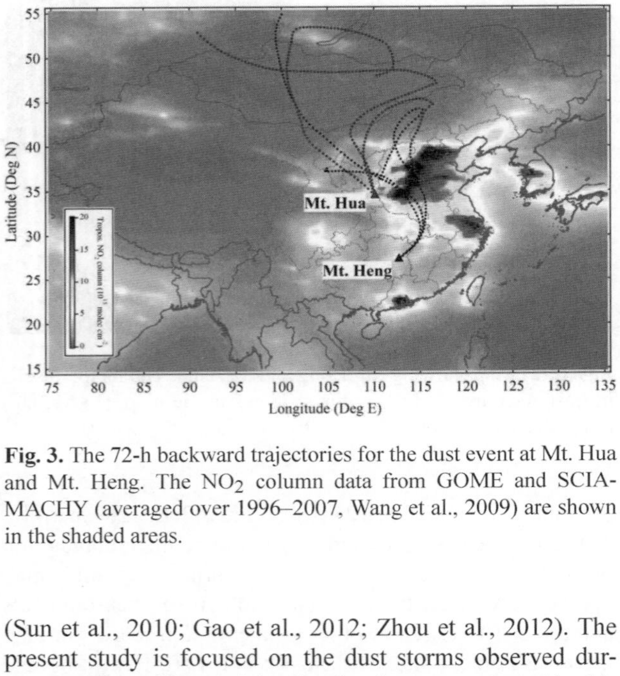

Fig. 3. The 72-h backward trajectories for the dust event at Mt. Hua and Mt. Heng. The NO_2 column data from GOME and SCIAMACHY (averaged over 1996–2007, Wang et al., 2009) are shown in the shaded areas.

(Sun et al., 2010; Gao et al., 2012; Zhou et al., 2012). The present study is focused on the dust storms observed during 20–29 April 2009 (Fig. 3). The dataset employed in this paper includes continuous data of SO_2, NO_x, NO_y, $PM_{2.5}$, PM_{10} and black carbon (BC); filter-based element data of Si, Al, Fe, Ca and Ti in $PM_{2.5}$ and Al in TSP; and size-resolved water-soluble ions (SO_4^{2-}, NO_3^-, NO_2^-, NH_4^+, Ca^{2+}).

Measurements of trace gases were done as follows: SO_2 was measured with a pulsed UV fluorescence analyzer (TEI model 43C); NO_y (defined as NO_x and its oxidation products and intermediates) was converted to NO with heated molybdenum oxide (MoO) and measured with a chemiluminescence analyzer (TEI model 42CY); NO_2 was converted to NO with a blue light converter (Meteorologieconsult GmbH) and measured with another NO analyzer (TEI model 42I). The NO_2 conversion efficiencies, determined by the method of gas-phase titration, were approximately 40 % on average during the whole sampling period. The calibration methods followed those reported in Wang et al. (2001). In this study, hourly-averaged data were employed for all the gaseous species and the uncertainties were estimated to be within ± 10 % (Wang et al., 2001, 2003).

Sampling and analysis of aerosols were done as follows: $PM_{2.5}$ and PM_{10} mass concentrations were continuously measured with a tapered element oscillating microbalance (TEOM, Thermo Electron Corporation, East Greenbush, NY, USA). BC was measured using an Aethalometer (Magee Scientific, Berkeley, California, USA, Model AE-21) at a wavelength of 880 nm. $PM_{2.5}$ filter samples were collected using a four-channel particle sampler (Thermo Andersen Chemical Speciation Monitor, RAAS2.5-400, Thermo Electron Corporation) with Teflon filters (Teflon™, 2 μm pore size and 47 mm diameter, Pall Inc.) at a flow rate of 16.7 liters per minute (LPM) (Wu and Wang, 2007). Total Suspended Particulates (TSP) samples were collected using medium-volume air samplers (TH-105C, Wuhan Tianhong Intelligent Instrument Corp. Co., Ltd., China) at flow rate of 100 LPM on glass fiber filters (GFFs, 90 mm in diameter). The size-resolved particles were collected with a Micro-Orifice Uniform Deposit Impactor (MOUDI, MSP Company) with aluminum substrates (MSP Company) at a flow rate of 30 LPM (Nie et al., 2010). The impactor collected

Table 1. Comparison of chemical species in non-dust and dust storm periods during the Mt. Heng study.

Species	Dust		Non-dust		PM size	Sampling instruments
	Concentrations	Sample numbers	Concentrations	Sample numbers		
PM_{10} ($\mu g\,m^{-3}$)	502 ± 196	55	45.6 ± 32.4	1195	PM_{10}	TEOM
$PM_{2.5}$ ($\mu g\,m^{-3}$)	125 ± 55	54	32.3 ± 22.6	1273	$PM_{2.5}$	TEOM
$PM_{2.5}/PM_{10}$	0.25 ± 0.08	54	0.63 ± 0.23	621		TEOM
Si ($\mu g\,m^{-3}$)	18.5 ± 2.9	7	1.1 ± 0.6	7	$PM_{2.5}$	RAAS
Al ($\mu g\,m^{-3}$)	8.1 ± 1.3	7	0.47 ± 0.25	7	$PM_{2.5}$	RAAS
Fe ($\mu g\,m^{-3}$)	5.4 ± 1.0	7	0.39 ± 0.17	7	$PM_{2.5}$	RAAS
Ca ($\mu g\,m^{-3}$)	4.2 ± 1.7	7	0.20 ± 0.10	7	$PM_{2.5}$	RAAS
Ti ($\mu g\,m^{-3}$)	0.5 ± 0.1	7	0.03 ± 0.02	7	$PM_{2.5}$	RAAS
Sulfate ($\mu g\,m^{-3}$)	20.0 ± 3.9	4	13.8 ± 6.1	21	TSP	MOUDI
Nitrate ($\mu g\,m^{-3}$)	16.9 ± 3.4	4	8.2 ± 5.4	21	TSP	MOUDI
NO_y (ppbv)	6.5 ± 2.1	55	3.7 ± 5.4	1505		42CY
SO_2 (ppbv)	4.2 ± 2.2	55	2.1 ± 2.4	1396		43C
NO_x (ppbv)	2.1 ± 0.7	55	1.0 ± 0.8	671		42I

particles in nine size ranges: > 18 µm (inlet), 10–18 µm, 5.6–10 µm, 3.2–5.6 µm, 1.8–3.2 µm, 1–1.8 µm, 0.56–1 µm, 0.32–0.56 µm, and 0.18–0.32 µm. Submicron particle ($PM_{<1}$) concentrations are calculated by the sum of last three stages. Supermicron particle ($PM_{>1}$) concentrations are calculated by the sum of first five stages. The sampled filters were stored at −5 °C in order to minimize artifacts. In this paper, the discussions on the chemical evolution of mineral dust particles focus on the supermicron dust particles ($PM_{>1}$).

The elements in $PM_{2.5}$ and TSP were measured using X-ray fluorescence (XRF) and inductively coupled plasma mass spectrometry (ICP-MS), respectively. The water-soluble ions were analyzed using ion chromatography (IC) (Dionex 90) (Wu and Wang, 2007). The measurement uncertainties of water-soluble ions and elements were estimated to be ± 10 % (Vecchi et al., 2008; Zhou et al., 2009). The evaporation loss of ammonium nitrate can be ignored due to the low ambient temperature (< 20 °C, Fig. 2) (Nie et al., 2010). During the dust storm of 24–26 April, we collected four sets of size-resolved aerosol samples using MOUDI: sample-I, sample-II, sample-III and sample-IV, with sample-I and sample-III collected during daytime (08:00–20:00 LT) and sample-II and sample-IV at night (20:00–08:00 of next day, LT) (Fig. 5).

3 Results and discussion

3.1 General description of the dust episodes

Figure 2 shows the temporal variations of meteorological parameters, particle mass, and chemical species observed at Mt. Heng during 20–29 April 2009, including wind speed, wind direction, temperature, RH, PM_{10}, $PM_{2.5}$, and major mineral elements (Si, Al, Ca, Fe, Ti) in $PM_{2.5}$. During this period, a moderate dust storm was observed on 21–22 April, with hourly averaged PM_{10} and $PM_{2.5}$ concentrations of 153 $\mu g\,m^{-3}$ and 57 $\mu g\,m^{-3}$, respectively. A more intense dust storm attacked the sampling site on 24–26 April, as highlighted by the shadow in Fig. 2, with the hourly averaged PM_{10} and $PM_{2.5}$ concentrations reaching 502 $\mu g\,m^{-3}$ and 125 $\mu g\,m^{-3}$, respectively. Prevailing northerly winds occurred during the dust storm periods (Fig. 2). The second dust storm on 24–26 April was also observed at three other sites at Mt. Hua in the Shaanxi province of central China (Fig. 1a), Mt. Tai in the Shandong province of central-eastern China (Wang et al., 2011) and the island of Taiwan. This dust event was also clearly shown in the MODIS true-color image for 24 and 25 April (Fig. 1b). These observations indicate a large geographical region of southern and eastern China being influenced by this dust storm. Due to the significant particle loading and more extensive transport, the specific analysis is focused on the second dust storm event (24 to 26 April).

Table 1 compares a set of species/parameters measured in non-dust periods and dust storm periods (24–26 April), including PM_{10}, $PM_{2.5}$, $PM_{2.5}/PM_{10}$ ratio, mineral elements (Si, Al, Ca, Fe and Ti) in $PM_{2.5}$, sulfate and nitrate in TSP, NO_y, NO_x and SO_2. Compared to non-dust periods, all the species in the dust storm event showed higher concentrations: PM_{10} and mineral elements in $PM_{2.5}$ increased by 10–20 fold; $PM_{2.5}/PM_{10}$ ratio decreased from 0.63 to 0.25; anthropogenic pollutants (sulfate, nitrate, NO_y, SO_2 and NO_x) increased only moderately, by 0.4–1.2 fold. These results reveal that the dust storm of 24–26 April brought to Mt. Heng high loading mineral dusts that had been mixed with moderate anthropogenic pollutants. Gao et al. (2012) have examined the size distributions of major water-soluble ions in non-dust periods, which showed the presence of sulfate and ammonium as a single peak in the fine particle mode and nitrate as dual peaks in the fine and coarse mode. As to be shown

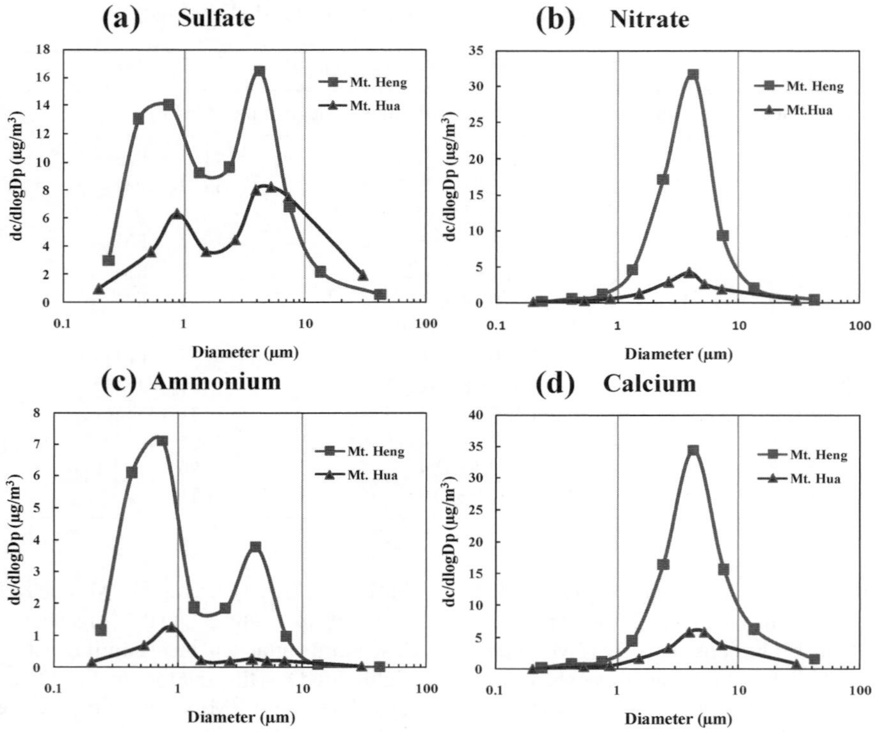

Fig. 4. Size distributions of **(a)** sulfate, **(b)** nitrate, **(c)** ammonium and **(d)** calcium at Mt. Hua and Mt. Heng during the dust event on 24–26 April.

in Sect. 3.2.1, during the dust storm of 24–26 April, the size distributions of these ions were significantly modified.

In order to characterize the origin and transport pathway of the air masses, a NOAA HYSPLIT4 model (Hybrid Single-Particle Lagrangian Integrated Trajectory) was employed to calculate 72-h backward trajectories arriving at Mt. Hua and Mt. Heng during the dust event of 24–26 April. As illustrated in Fig. 3a, the dust originating in the Gobi Desert moved southward and passed over central China (Shaanxi province), the North China Plain (Hebei province, Shanxi province and Henan province) and southern China (Hubei province and Hunan province) and is expected to travel farther to the South China Sea.

3.2 Chemical evolution of mineral dust particles during transport

3.2.1 Comparison with samples collected near the dust source region

The concurrent campaigns during the dust storm of 24–26 April at Mt. Hua (Wang et al., 2010) and Mt. Heng (this study) provide an opportunity to directly investigate the evolution of mineral dust particles from a near source region in the north to a southern mountainous area. Figure 4 illustrates the size distributions of four major water-soluble ions, including sulfate, nitrate, ammonium and calcium, at Mt. Hua (sample of 24 April) and Mt. Heng (sample of 25 April). The size distribution patterns of sulfate, nitrate and calcium at both mountain sites were similar: sulfate had a bimodal size distribution, with one peak in the submicron range and another larger peak in the supermicron range; nitrate and calcium showed single peaks in the coarse mode. For ammonium, the size distribution exhibited different patterns at the two sites, being unimodal at Mt. Hua, but bimodal at Mt. Heng. Compared with Mt. Hua, all four ions in the supermicron particles at Mt. Heng showed significant enhancement (Fig. 4): sulfate increased by approximately 56 %, but the other three ions were enhanced up to 5–7 times. The enrichments of water-soluble ions on dust particles at Mt. Heng can be generally attributed to the injection of anthropogenic emissions as the dust plumes passed over the polluted North China Plains (see Fig. 3).

The different degrees of enrichment for the different ions at Mt. Heng may be influenced by the relative abundance of SO_2, NO_x and NH_3 in the atmosphere and the factors affecting their uptake/conversion on dust particles. Due to a higher hygroscopicity of $Ca(NO_3)_2$ than $CaSO_4$, the formation of $Ca(NO_3)_2$ would significantly enhance the uptake of water (and other soluble gases), resulting in a positive feedback until all the calcium transformed to calcium nitrate (Formenti et al., 2011). This process may offer an explanation for the much larger increase in nitrate than that of sulfate on the mineral particles during the transport from Mt. Hua to Mt. Heng (Fig. 4). The enrichment of ammonium will be discussed in Sect. 3.2.3.

3.2.2 Changes of mineral dust characteristics with atmospheric processing time

In order to further investigate the evolution of the mineral dust, the atmospheric processing time of the air masses was calculated by the photochemical age ratio $-\mathrm{Log}_{10}(NO_x/NO_y)$ (Kleinman et al., 2008), and then was related to the changes in water-soluble ions on the mineral dust particles. As illustrated in Fig. 5a, the photochemical age generally showed an upward trend during the dust event (from 19:00, 24 April to 08:00, 27 April, LT). The sudden decrease in the values of $-\mathrm{Log}_{10}(NO_x/NO_y)$ was related to the changes of northerly wind to easterly wind (Fig. 2), which brought freshly emitted NO_x from the incense burning in the nearby temples. The photochemical ages of the first three samples increased gradually, while the values of the sample-III and sample-IV were similar (approximately 0.7).

The ratios of sulfate, nitrate and ammonium to the PM mass in the supermicron range are shown in Figs. 5b. These results indicate overall increases in the relative abundances of all three ions with the atmospheric processing time of the air parcels. During the time scale of the four samples, sulfate increased from 1.7 to 2.4 %, nitrate from 2.5 to 4.3 %, and ammonium from 0.2 to 1.2 %. The increased water-soluble ions could significantly enhance the solubility of mineral particles, thus affecting their lifetimes and cloud condensation nuclei (CCN) or ice nuclei (IC) activity. It is worth noting that the large increases in the abundance of the secondary ions in sample III and sample IV with similar photochemical age indicate the contribution of additional factor(s) to the formation of these ions.

Acidity is another important factor that affects the chemical and physical properties of aerosols (Jang et al., 2002). For Asian dust particles, calcium is the major cation. Therefore, the ratios of the sum of major anions to calcium in equivalence in the supermicron range particles were calculated to investigate the changes in acidity of the mineral particles with the atmospheric processing time of the air mass. As illustrated in Fig. 5c, evident acidification occurred with the aging of the air mass. The ratio increased from 0.48 in the first sample to 0.96 in the third sample and to 1.20 in the fourth sample.

3.2.3 Ammonium enrichment and a conceptual model for dust evolution

In our study, large enrichments in ammonium on supermicron range particles were observed at Mt. Heng. The results show an absence of coarse-mode ammonium at Mt. Hua

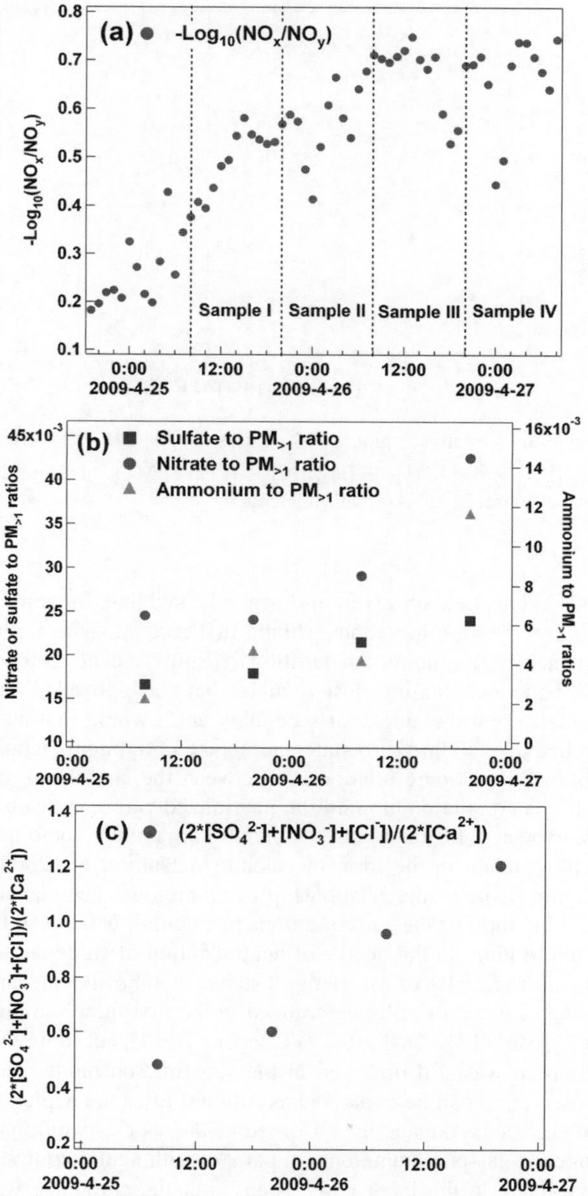

Fig. 5. **(a)** Photochemical age (defined as $-\mathrm{Log}_{10}(NO_x/NO_y)$), **(b)** concentration ratios of sulfate, nitrate and ammonium to PM in the supermicron particles, **(c)** sum of major anions to calcium in equivalence in the supermicron particles, during the dust event on 24–26 April at Mt. Heng.

(Fig. 4c), an increase to a content of 2–5 % in sample I to sample III, and a further increase to ~ 12 % in sample IV at Mt. Heng (Fig. 5c). These observations are in contrast to previous that which suggest suppression of the formation of ammonium aerosol by the invasion of a dust storm due to dilution of the precursor ammonia and the uptake of acidic gases on dust (Song and Carmichael, 1999; Huang et al., 2010).

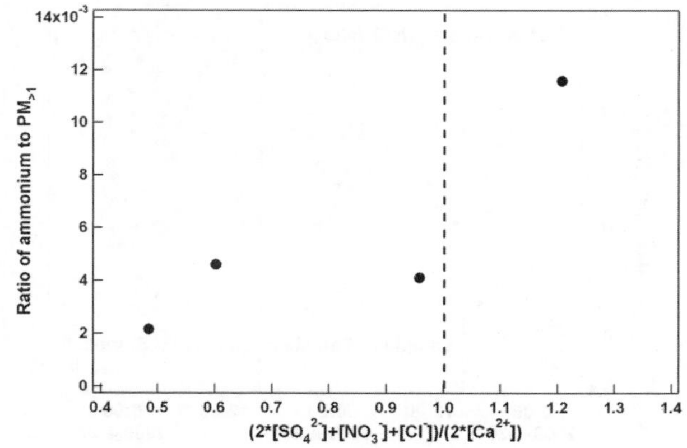

Fig. 6. Scatter plots of ratios of NH_4^+/PM and $(2\times[SO_4^{2-}]+[NO_3^-]+[Cl^-])/(2\times[Ca^{2+}])$ in $PM_{>1}$ during the dust event on 24–26 April at Mt. Heng.

Previous field observations have indicated that, for ammonium to form, mineral dust should first take up some acidic species that ammonia can partition to (Sullivan et al., 2007a). Carbonate-containing dust particles have a high alkalinity initially, and the uptake of secondary acids would first neutralize the alkaline carbonate and uptake of ammonia would follow. Thus some relationship between the amount of accumulated ammonium and the neutralized carbonate would be expected. In Asian dust, carbonates are mainly combined with calcium in the form of calcite (Matsuki et al., 2005). Figure 6 shows the relationship of ammonium abundances and the ratio of the sum of anions to calcium, which can be used to indicate the degree of neutralization of carbonate in the dust particles of Mt. Heng. It shows that the alkaline carbonates were not fully neutralized in the first three samples $((2\times[SO_4^{2-}]+[NO_3^-]+[Cl^-])/(2\times[Ca_2^+]) < 1)$, but some ammonium was still observed in the supermicron mode. This observation can be explained as follows: after the depletion of surface carbonate, the absorbed acidic species would uptake the gas-phase ammonia in parallel with neutralizing the carbonate in the inner core. Then, after the carbonate was fully consumed, ammonium was enhanced significantly in the fourth sample (Fig. 6).

Based on the above discussions, a conceptual model is proposed to describe dust evolution into a four-stage process (Fig. 7): freshly emitted dust (stage 1), the uptake of secondary acidic species with little ammonium (stage 2, as represented by the sample of Mt. Hua; see Fig. 4c), dust with ammonium accumulated on but without the depletion of carbonate (stage 3, represented by sample-I to III of Mt. Heng; see Figs. 5c and 6), and dust with all carbonate neutralized and significantly enhanced ammonium (stage 4, represented by sample-IV of Mt. Heng). This 4-stage evolution has important implications for atmospheric chemistry and the impact of dust on climate. When the Asian dust particles have reached the third stage (i.e., with accumulated ammonium), the surface properties of the dust have changed from hydrophobic and alkaline to hydrophilic and acidic, thus significantly enhancing their CCN/IN capacity. At the fourth stage, the dust particles would be no different from the secondarily formed particles and could serve as favorable media to promote further reactions. It should be noted that the conceptual model can only apply to the carbonate-containing dust particles and that ambient ammonia concentrations will influence the development of the four stages.

3.3 Photo-enhanced nitrite formation

High nitrite concentrations on dust particles were observed during the dust event of 24–26 April. Figure 8 shows the nitrite concentrations in TSP in the four samples following the temporal order during the dust episode of 24–26 April. The average nitrite concentration during this dust event was 2.5 $\mu g\,m^{-3}$, much higher than that during the non-dust days (the mean value = 0.3 $\mu g\,m^{-3}$), indicating an enhancement of nitrite formation on the dust particles. Furthermore, the enhanced nitrite formation during the dust event occurred only in the daytime samples (with a concentration up to 6 $\mu g\,m^{-3}$ in sample-III; see Fig. 8), suggesting a photochemical pathway yielding the observed nitrite.

Daytime HONO (as well as nitrite) can be produced from gas phase sources and photo-related heterogeneous sources (Kleffmann, 2007). The reaction of NO and OH radicals is the dominant gas phase process yielding HONO. However, due to the fast photolysis of HONO, it has been demonstrated that this reaction is not a net source and cannot explain the frequently observed high daytime HONO concentrations (e.g., Kleffmann, 2007; Sörgel et al., 2011). The insignificance of this gas-phase source in our study is supported by the low NO concentrations (< 0.5 ppbv) observed during the dust event. Therefore, there must be other photo-related heterogeneous processes contributing to the observed daytime nitrite.

Heterogeneous photochemical reactions producing HONO are generally thought to occur via four different processes: the surface photolysis of nitric acid (HNO_3) (Zhou et al., 2003; Ramazan et al., 2004); photo-induced NO_2 conversion on soot surfaces (Aubin and Abbatt, 2007); the heterogeneous photochemistry of NO_2 on some organic surfaces, such as aromatic compounds and humic acids (George et al., 2005; Stemmler et al., 2006); and the surface TiO_2 photocatalysis of NO_2 (Gustafsson et al., 2006; Ndour et al., 2008). In order to determine the key process contributing to the observed nitrite formation, the concentrations of nitrite, nitrate and BC (representing soot) during the dust storm of 24–26 April were compared with those during a non-dust pollution episode (denoted as PE) occurring on 8–9 May (Table 2). Compared to those during PE, the nitrate and BC concentrations during the dust storm were lower, but the nitrite concentrations were much higher.

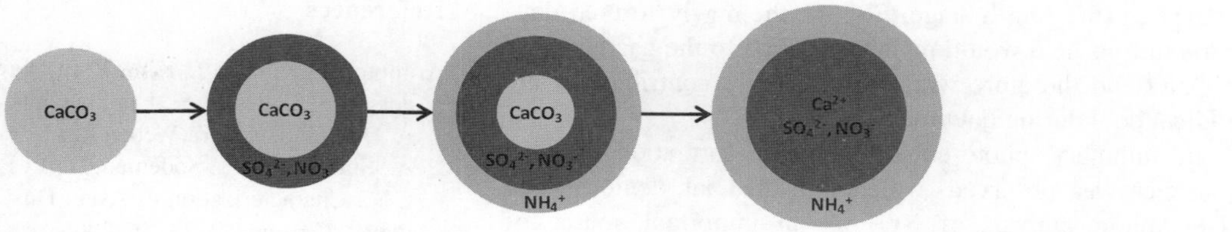

Fig. 7. Conceptual model showing the four aging stages for the carbonate-containing dust particles.

Table 2. Concentrations of nitrite in TSP, nitrate in TSP and BC in $PM_{2.5}$ measured during the dust storm of 24–26 April and a non-dust pollution episode that occurred on 8–9 May at Mt. Heng.

	DS-II on 25–26 April	Pollution episode on 8–9 May
Nitrite in TSP ($\mu g\,m^{-3}$)	2.5	0.6
Nitrate in TSP ($\mu g\,m^{-3}$)	15.6	20.1
BC in $PM_{2.5}$ ($\mu g\,m^{-3}$)	3.2	3.5

This result suggests that the first two reaction pathways were not the major contributors to the observed nitrite formation. Despite the fact that aromatic compounds and humic acids were not measured, their concentrations were not presumed to increase much during the dust storm because mineral dust particles are not their sources. Considering that the daytime enhancement of nitrite production was only observed in the dust storm event, the third process should also be excluded as the key contributor.

The fourth process of TiO_2 photocatalysis of NO_2 yielding HONO was recently demonstrated in the laboratory (Gustafsson et al., 2006; Ndour et al., 2008). The reaction mechanism can be simply described as nitrogen dioxides being reduced to nitrite ions by photo-produced electrons (the detailed mechanism was proposed in Ndour's study, 2008). In the present study, the titanium concentrations (in the form of TiO_2, Chen et al., 2012) indeed increased significantly during the dust storm of 24–26 April (see Fig. 2), implying that the surface TiO_2 photocatalysis may be responsible for the daytime nitrite formation. To further support this hypothesis, the correlations of daytime nitrite with titanium and with nitrate were also examined. Noting that the titanium concentration in the coarse mode particles ($PM_{>2.5}$) was unavailable, their values were calculated via the formula $Ti_{coarse} = Ti_{fine} \times (Al_{coarse}/Al_{fine})$ under the assumption that aluminum and titanium have the same size distributions (Zhang et al., 2003). As tabulated in Table 3, for sample-I, the ratio of $NO_2^-{}_{(coarse)}/NO_2^-{}_{(fine)}$ was approximately 3.0, which was similar to that of titanium (approximately 2.8), but largely differed from that of nitrate (approximately 9.1). The result for sample-III was similar. These results further indicate the role of TiO_2 photocatalysis on the observed daytime nitrite formation.

Table 3. Concentrations of titanium and nitrite in both fine particle mode ($PM_{2.5}$) and coarse particle mode ($PM_{>2.5}$) in the two daytime samples during the dust storm of 24–26 April at Mt. Heng.

Sample	Fine particles ($\mu g\,m^{-3}$)			Coarse particles ($\mu g\,m^{-3}$)		
	NO_2^-	Ti	NO_3^-	NO_2^-	Ti*	NO_3^-
04:25 Daytime	0.68	0.58	2.0	2.02	1.61	18.2
04:26 Daytime	1.5	0.42	1.0	4.49	1.12	10.7

* Ti in coarse particles was calculated by $Ti_{coarse} = Ti_{fine} \times (Al_{coarse}/Al_{fine})$.

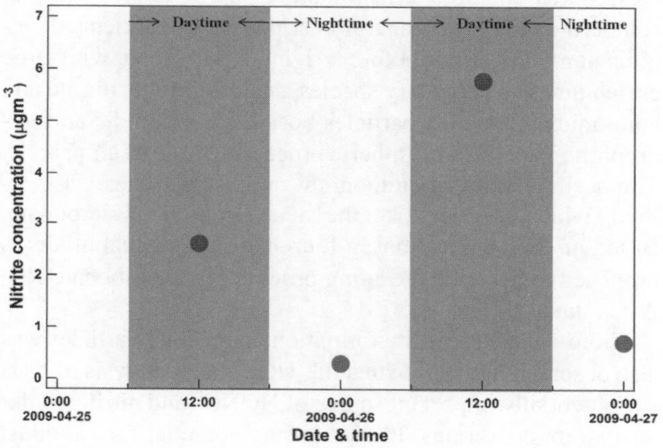

Fig. 8. Nitrite concentrations in TSP collected during the dust event on 24–26 April at Mt. Heng.

The partition between gas-phase HONO and aerosol nitrite is affected by acidity of dust particles. In the two daytime samples (sample-I and sample-III), the supermicron particles showed strong alkalinity (Fig. 5d). Under such a condition, the nitrous acid is normally thought to favor partitioning to the aerosol phase. However, Ndour et al. (2008) observed 2.3 ppbv of gas-phase HONO produced from 3 ppbv of NO_2 on irradiated real Sahara dust particles, suggesting the potential source of the gas-phase HONO from the heterogeneous photochemical reactions of dust particles even under the condition of strong alkalinity. In addition, the presence of coarse mode ammonium (samples I to III, the third stage, Fig. 7) may suggest the surface acidification of the dust particles, which would promote the partitioning of nitrous acid to the gas phase in samples I–III. For sample IV, strong acidity was indicated (the fourth-stage, Fig. 7); the newly formed daytime nitrous acid would mainly partition to the gas phase as HONO and therefore would significantly contribute to the radical pool during daytime.

In summary, photo-enhanced nitrite formation on dust particles was observed, which indicated lab-demonstrated TiO_2 photocatalysis of NO_2 as an important source of daytime HONO under dust storm conditions. Given that there is increasing evidence for anthropogenic sources of photocatalyst-containing (including TiO_2) nanodusts emitted from industrial nanomaterials (Vicki, 2009; Chen et al., 2012), this process may play a potentially important role in atmospheric chemistry on a global scale. We therefore recommend more comprehensive studies to better evaluate the contribution of photoactive particles to gas-phase HONO and, in turn, OH radicals.

4 Summary and conclusions

An intense Asian dust storm that originated from the Gobi Desert was captured at a mountain site (Mount Heng) in southern China. The mineral dust particles experienced significant modifications before arriving at Mt. Heng, with large enrichments in secondary species, such as sulfate, nitrate and ammonium. The dust particles became more acidic and hydrophilic with the atmospheric processing time of air parcels. The accumulation of ammonium on dust particles was verified, which is related to the neutralization of carbonate. Based on the observations, a four-stage conceptual model is proposed to describe the aging process of the carbonate-rich Asian dust.

Photo-enhanced nitrite formation on the dust particles was also observed, indicating that the TiO_2 photocatalysis of NO_2 is a potentially important source of HONO (and nitrite) under dust storm conditions. In view of the potential for nanodust to become a global source of photocatalysts (e.g., TiO_2), this process may be an important daytime HONO source. More studies are needed in order to examine further the extent of photo-enhanced nitrite formation in the atmosphere and its implications for atmospheric chemistry.

Acknowledgements. We thank Jia Guo, Weijun Li, Yuhua Li, Jie Zhou, Penghui Li and Minghu Sun for their contribution to the field work and data processing, and Kin Fai Ho and Hongmei Xu for their help on the elements analysis. We are grateful to the NOAA Air Resources Laboratory for providing the HYSPLIT model. This study was funded by the National Basic Research Program of China (2005CB422203) and the Niche Area Development Program of the Hong Kong Polytechnic University (1-BB94).

Edited by: Y.-S. Chung

References

Arimoto, R., Kim, Y. J., Kim, Y. P., Quinn, P. K., Bates, T. S., Anderson, T. L., Gong, S., Uno, I., Chin, M., Huebert, B. J., Clarke, A. D., Shinozuka, Y., Weber, R. J., Anderson, J. R., Guazzotti, S. A., Sullivan, R. C., Sodeman, D. A., Prather, K. A., and Sokolik, I. N.: Characterization of Asian Dust during ACE-Asia, Global Planet. Change, 52, 23–56, 2006.

Aubin, D. G. and Abbatt, J. P. D.: Interaction of NO_2 with Hydrocarbon Soot:? Focus on HONO Yield, Surface Modification, and Mechanism, J. Phys. Chem. A, 111, 6263–6273, 2007.

Chen, H., Nanayakkara, C. E., and Grassian, V. H.: Titanium Dioxide Photocatalysis in Atmospheric Chemistry, Chem. Rev., 112, 5919–5948, doi:10.1021/cr3002092, 2012.

Cwiertny, D. M., Young, M. A., and Grassian, V. H.: Chemistry and photochemistry of mineral dust aerosol, Annu. Rev. Phys. Chem., 59, 27–51, 2008.

Formenti, P., Schütz, L., Balkanski, Y., Desboeufs, K., Ebert, M., Kandler, K., Petzold, A., Scheuvens, D., Weinbruch, S., and Zhang, D.: Recent progress in understanding physical and chemical properties of African and Asian mineral dust, Atmos. Chem. Phys., 11, 8231–8256, doi:10.5194/acp-11-8231-2011, 2011.

Gao, X., Xue, L., Wang, X., Wang, T., Yuan, C., Gao, R., Zhou, Y., Nie, W., Zhang, Q., and Wang, W.: Aerosol ionic components at Mt. Heng in central southern China: Abundances, size distribution, and impacts of long-range transport, Sci. Total Environ., 433, 498–506, doi:10.1016/j.scitotenv.2012.06.095, 2012.

George, C., Strekowski, R. S., Kleffmann, J., Stemmler, K., and Ammann, M.: Photoenhanced uptake of gaseous NO_2 on solid organic compounds: a photochemical source of HONO?, Faraday Discuss., 130, 195–210, 2005.

Gustafsson, R. J., Orlov, A., Griffiths, P. T., Cox, R. A., and Lambert, R. M.: Reduction of NO_2 to nitrous acid on illuminated titanium dioxide aerosol surfaces: implications for photocatalysis and atmospheric chemistry, Chem. Commun., 3936–3938, 2006.

Huang, K., Zhuang, G., Li, J., Wang, Q., Sun, Y., Lin, Y., and Fu, J. S.: Mixing of Asian dust with pollution aerosol and the transformation of aerosol components during the dust storm over China in spring 2007, J. Geophys. Res., 115, D00K13, doi:10.1029/2009jd013145, 2010.

Jacob, D. J., Crawford, J. H., Kleb, M. M., Connors, V. S., Bendura, R. J., Raper, J. L., Sachse, G. W., Gille, J. C., Emmons, L., and Heald, C. L.: Transport and Chemical Evolution over the Pacific (TRACE-P) aircraft mission: Design, execution, and first results, J. Geophys. Res., 108, 9000, doi:10.1029/2002jd003276, 2003.

Jang, M., Czoschke, N. M., Lee, S., and Kamens, R. M.: Heterogeneous Atmospheric Aerosol Production by Acid-Catalyzed Particle-Phase Reactions, Science, 298, 814–817, 2002.

Kleffmann, J.: Daytime Sources of Nitrous Acid (HONO) in the Atmospheric Boundary Layer, Chem. Phys. Chem., 8, 1137–1144, 2007.

Kleinman, L. I., Springston, S. R., Daum, P. H., Lee, Y.-N., Nunnermacker, L. J., Senum, G. I., Wang, J., Weinstein-Lloyd, J., Alexander, M. L., Hubbe, J., Ortega, J., Canagaratna, M. R., and Jayne, J.: The time evolution of aerosol composition over the Mexico City plateau, Atmos. Chem. Phys., 8, 1559–1575, doi:10.5194/acp-8-1559-2008, 2008.

Laurent, B., Marticorena, B., Bergametti, G., and Mei, F.: Modeling mineral dust emissions from Chinese and Mongolian deserts, Global Planet. Change, 52, 121–141, 2006.

Li, W. J. and Shao, L. Y.: Observation of nitrate coatings on atmospheric mineral dust particles, Atmos. Chem. Phys., 9, 1863–1871, doi:10.5194/acp-9-1863-2009, 2009.

Matsuki, A., Iwasaka, Y., Shi, G. Y., Chen, H. B., Osada, K., Zhang, D., Kido, M., Inomata, Y., Kim, Y. S., Trochkine, D., Nishita, C., Yamada, M., Nagatani, T., Nagatani, M., and Nakata, H.: Heterogeneous sulfate formation on dust surface and its dependence on mineralogy: Balloon-borne observations from ballon-borne measurements in the surface of Beijing, China, Water Air Soil Poll., 5, 101–132, 2005.

Ndour, M., D'Anna, B., George, C., Ka, O., Balkanski, Y., Kleffmann, J., Stemmler, K., and Ammann, M.: Photoenhanced uptake of NO_2 on mineral dust: Laboratory experiments and model simulations, Geophys. Res. Lett., 35, L05812, doi:10.1029/2007gl032006, 2008.

Nicolas, M. L., Ndour, M., Ka, O., D'Anna, B., and George, C.: Photochemistry of Atmospheric Dust: Ozone Decomposition on Illuminated Titanium Dioxide, Environ. Sci. Technol., 43, 7437–7442, 2009.

Nie, W., Wang, T., Gao, X., Pathak, R. K., Wang, X., Gao, R., Zhang, Q., Yang, L., and Wang, W.: Comparison among filter-based, impactor-based and continuous techniques for measuring atmospheric fine sulfate and nitrate, Atmos. Environ., 44, 4396–4403, 2010.

Ramazan, K. A., Syomin, D., and Finlayson-Pitts, B. J.: The photochemical production of HONO during the heterogeneous hydrolysis of NO_2, Phys. Chem. Chem. Phys, 6, 3836–3843, 2004.

Singh, H. B., Brune, W. H., Crawford, J. H., Flocke, F., and Jacob, D. J.: Chemistry and transport of pollution over the Gulf of Mexico and the Pacific: spring 2006 INTEX-B campaign overview and first results, Atmos. Chem. Phys., 9, 2301–2318, doi:10.5194/acp-9-2301-2009, 2009.

Song, C. H. and Carmichael, G. R.: The aging process of naturally emitted aerosol (sea-salt and mineral aerosol) during long range transport, Atmos. Environ., 33, 2203–2218, 1999.

Sörgel, M., Regelin, E., Bozem, H., Diesch, J.-M., Drewnick, F., Fischer, H., Harder, H., Held, A., Hosaynali-Beygi, Z., Martinez, M., and Zetzsch, C.: Quantification of the unknown HONO daytime source and its relation to NO_2, Atmos. Chem. Phys., 11, 10433–10447, doi:10.5194/acp-11-10433-2011, 2011.

Stemmler, K., Ammann, M., Donders, C., Kleffmann, J., and George, C.: Photosensitized reduction of nitrogen dioxide on humic acid as a source of nitrous acid, Nature, 440, 195–198, 2006.

Sullivan, R. C. and Prather, K. A.: Investigations of the diurnal cycle and mixing state of oxalic acid in individual particles in Asian aerosol outflow, Environ. Sci. Technol., 41, 8062–8069, 2007.

Sullivan, R. C., Guazzotti, S. A., Sodeman, D. A., and Prather, K. A.: Direct observations of the atmospheric processing of Asian mineral dust, Atmos. Chem. Phys., 7, 1213–1236, doi:10.5194/acp-7-1213-2007, 2007a.

Sullivan, R. C., Guazzotti, S. A., Sodeman, D. A., Tang, Y., Carmichael, G. R., and Prather, K. A.: Mineral dust is a sink for chlorine in the marine boundary layer, Atmos. Environ., 41, 7166–7179, 2007b.

Sun, M., Wang, Y., Wang, T., Fan, S., Wang, W., Li, P., Guo, J., and Li, Y.: Cloud and the corresponding precipitation chemistry in south China: Water-soluble components and pollution transport, J. Geophys. Res., 115, D22303, doi:10.1029/2010jd014315, 2010.

Usher, C. R., Michel, A. E., and Grassian, V. H.: Reactions on mineral dust, Chem. Rev., 103, 4883–4939, 2003.

Vecchi, R., Chiari, M., D'Alessandro, A., Fermo, P., Lucarelli, F., Mazzei, F., Nava, S., Piazzalunga, A., Prati, P., Silvani, F., and Valli, G.: A mass closure and PMF source apportionment study on the sub-micron sized aerosol fraction at urban sites in Italy, Atmos. Environ., 42, 2240–2253, doi:10.1016/j.atmosenv.2007.11.039, 2008.

Vicki, H. G.: New Directions: Nanodust – A source of metals in the atmospheric environment?, Atmos. Environ., 43, 4666–4667, 2009.

Wang, G., Li, J., Cheng, C., Hu, S., Xie, M., Gao, S., Zhou, B., Dai, W., Cao, J., and An, Z.: Observation of atmospheric aerosols at Mt. Hua and Mt. Tai in central and east China during spring 2009 – Part 1: EC, OC and inorganic ions, Atmos. Chem. Phys., 11, 4221–4235, doi:10.5194/acp-11-4221-2011, 2011.

Wang, T., Cheung, V. T. F., Anson, M., and Li, Y. S.: Ozone and related gaseous pollutants in the boundary layer of eastern China: Overview of the recent measurements at a rural site, Geophys. Res. Lett., 28, 2373–2376, 2001.

Wang, T., Ding, A. J., Blake, D. R., Zahorowski, W., Poon, C. N., and Li, Y. S.: Chemical characterization of the boundary layer outflow of air pollution to Hong Kong during February–April 2001, J. Geophys. Res., 108, 8787, doi:10.1029/2002jd003272, 2003.

Wang, T., Wei, X. L., Ding, A. J., Poon, C. N., Lam, K. S., Li, Y. S., Chan, L. Y., and Anson, M.: Increasing surface ozone concentrations in the background atmosphere of Southern China, 1994–2007, Atmos. Chem. Phys., 9, 6217–6227, doi:10.5194/acp-9-6217-2009, 2009.

Wang, T., Nie, W., Gao, J., Xue, L. K., Gao, X. M., Wang, X. F., Qiu, J., Poon, C. N., Meinardi, S., Blake, D., Wang, S. L., Ding, A. J., Chai, F. H., Zhang, Q. Z., and Wang, W. X.: Air quality during the 2008 Beijing Olympics: secondary pollutants and regional impact, Atmos. Chem. Phys., 10, 7603–7615, doi:10.5194/acp-10-7603-2010, 2010.

Wang, Y., Sun, M., Li, P., Li, Y., Xue, L., and Wang, W.: Variation of low molecular weight organic acids in precipitation and cloudwater at high elevation in South China, Atmos. Environ., 45, 6518–6525, 2011.

Wu, W. S. and Wang, T.: On the performance of a semi-continuous $PM_{2.5}$ sulphate and nitrate instrument under high loadings of particulate and sulphur dioxide, Atmos. Environ., 41, 5442–5451, 2007.

Zhang, X. Y., Gong, S. L., Arimoto, R., Shen, Z. X., Mei, F. M., Wang, D., and Cheng, Y.: Characterization and Temporal Variation of Asian Dust Aerosol from a Site in the Northern Chinese Deserts, J. Atmos. Chem., 44, 241–257, doi:10.1023/a:1022900220357, 2003.

Zhou, S., Wang, Z., Gao, R., Xue, L., Yuan, C., Wang, T., Gao, X., Wang, X., Nie, W., Xu, Z., Zhang, Q., and Wang, W.: Formation of secondary organic carbon and long-range transport of carbonaceous aerosols at Mount Heng in South China, Atmos. Environ., 63, 203–212, doi:10.1016/j.atmosenv.2012.09.021, 2012.

Zhou, X., Gao, H., He, Y., Huang, G., Bertman, S. B., Civerolo, K., and Schwab, J.: Nitric acid photolysis on surfaces in low-NO_x environments: Significant atmospheric implications, Geophys. Res. Lett., 30, 2217, doi:10.1029/2003gl018620, 2003.

Zhou, Y., Wang, T., Gao, X., Xue, L., Wang, X., Wang, Z., Gao, J., Zhang, Q., and Wang, W.: Continuous observations of water-soluble ions in $PM_{2.5}$ at Mount Tai (1534 m.a.s.l.) in central-eastern China, J. Atmos. Chem., 64, 107–127, doi:10.1007/s10874-010-9172-z, 2009.

Zhuang, G., Yi, Z., Duce, R. A., and Brown, P. R.: Link between iron and sulphur cycles suggested by detection of Fe(n) in remote marine aerosols, Nature, 355, 537–539, 1992.

Atmospheric Environment 51 (2012) 11–20

Contents lists available at SciVerse ScienceDirect

Atmospheric Environment

journal homepage: www.elsevier.com/locate/atmosenv

Characterization of aerosol acidity at a high mountain site in central eastern China

Yang Zhou [a], Likun Xue [a], Tao Wang [a,b,c], Xiaomei Gao [a], Zhe Wang [a], Xinfeng Wang [a,b], Jiamin Zhang [b], Qingzhu Zhang [a], Wenxing Wang [a,c,*]

[a] Environment Research Institute, Shandong University, Ji'nan, Shandong 250100, China
[b] Department of Civil and Structural Engineering, The Hong Kong Polytechnic University, Hong Kong, China
[c] Chinese Research Academy of Environmental Sciences, Beijing 100012, China

ARTICLE INFO

Article history:
Received 18 June 2011
Received in revised form
16 December 2011
Accepted 24 January 2012

Keywords:
Semi-continuous measurements
Aerosol acidity
AIM-II
Mt. Tai
$PM_{2.5}$
Secondary organic aerosol formation

ABSTRACT

Aerosol acidity plays an important role in the formation of secondary organic aerosols. In the present study, strong and aerosol acidity properties of $PM_{2.5}$ were evaluated based on the highly time-resolved measurements of $PM_{2.5}$ ionic compositions obtained at the highest mountain site in central eastern China in spring and summer of 2007. Overall, $PM_{2.5}$ was weakly acidic at Mt. Tai with 57.2% and 81.3% of the observations being acidic aerosols in spring and summer, respectively. Strong and aerosol acidities showed higher levels in summer (mean ± stand deviation: 142.65 ± 115.23 and 35.27 ± 30.88 nmol m^{-3}) and lower concentrations in spring (64.82 ± 75.07 and 25.25 ± 32.23 nmol m^{-3}). Aerosol pH exhibited an opposite seasonal trend with less acidic aerosols in summer compared to the aerosols in spring due to high water content in the particles in summer. Strong acidity showed a well-defined diurnal profile with a broad peak during the daytime, while aerosol acidity was at relatively low level in the daytime. The effects of ambient RH and atmospheric aging on the acidities of $PM_{2.5}$ were examined. Aerosol water content facilitated the release of free H^+ in the aerosol droplet via hydrolysis processes of bisulphate and acidic aerosols were often associated with more processed air masses. Several cases with formation and accumulation of secondary organic aerosols occurring were investigated. The results indicated that the increase of secondary organic aerosols was probably due to effects of acidity promotion and aqueous phase formation. This is the first attempt to investigate aerosol acidity based on high resolution measurements of aerosol ions in central eastern China.

1. Introduction

Fine particles (i.e., $PM_{2.5}$) are well known to be associated with human health, air quality deterioration, and even global climate change (Charlson et al., 1992; Nel, 2005). Aerosol acidity is an important property of particles. It can influence aerosol phase reactions by altering the uptake of precursors and the partitioning of volatile and semi-volatile compounds between the gas phase and particle surfaces (Grassian, 2001; Usher et al., 2002; Hatch and Grassian, 2008). It can also catalyze heterogeneous reactions to enhance the formation of secondary inorganic aerosols (Underwood et al., 2001; Ullerstam et al., 2002; Manktelow et al., 2010) and secondary organic aerosols (SOA) (Jang et al., 2002; Gao et al., 2004; Cao and Jang, 2010). Therefore, understanding of aerosol acidity and its impact is a fundamental issue in aerosol sciences.

Acidity of fine aerosols is mainly determined by the balance of acidic ionic components with basic ones, namely, sulfate (SO_4^{2-}), nitrate (NO_3^-) and ammonium (NH_4^+). There are two indicators of particle acidity: strong acidity (H^+_{strong}) and actual free acidity of aerosol (H^+_{air}). H^+_{strong}, is the total amount of acid contributed by the strong acids, such as sulfuric and/or nitric acid, in the aqueous extract of the aerosols collected on the filter or from semi-continuous sampler. H^+_{air}, defined as the moles of free hydrogen ions in the aqueous phase of aerosols per unit of air (nmol m^{-3}), is the actual acidity in the droplets of the aerosol. However direct measurement of H^+_{air} is not possible since the aqueous water of the aerosol is very small. Indirect methods, such as thermodynamic models, including Aerosol Inorganic Model (AIM-II) (Clegg et al., 1998), ISORROPIA (Nenes et al., 1998), SCAPE2 (Meng et al., 1995) and GFEMN (Ansari and Pandis, 1999), can supply suitable approaches for predicting H^+_{air} and pH can also be calculated based on the output H^+_{air} and liquid volume of the particles. H^+_{strong} not only contains the actual free hydrogen ion in the aqueous phase of aerosols (H^+_{air}), also includes other hydrogen ion released from either undissociated sulfuric acid or bisulfate in the presence of

* Corresponding author. Environment Research Institute, Shandong University, Ji'nan, Shandong 250100, China.
E-mail address: wenxwang@hotmail.com (W. Wang).

1352-2310/$ – see front matter © 2012 Elsevier Ltd. All rights reserved.
doi:10.1016/j.atmosenv.2012.01.061

large excesses of water. Thus in amount $[H^+]_{strong}$ is the sum of $[H^+]_{air}$, $[HSO_4^-]$ and any other $[H^+]$ in the solid phase with sulfate and/or nitrate at equilibrium (Pathak et al., 2009; Yao et al., 2006). Thus H^+_{air} is a more accurate indicator of aerosol acidic nature (Pathak et al., 2009) and more relevant than strong acidity in understanding the chemical behavior and subsequent environmental impacts of atmospheric aerosols.

Several applications have been reported the performance of AIM-II (http://www.aim.env.uea.ac.uk/aim/aim.htm) is the most accurate in evaluating the acidity nature of particles (Ansari and Pandis, 1999; Pathak et al., 2003; Yao et al., 2006). So AIM-II was selected to calculate H^+_{air} in this study. On the other hand, most of pervious studies were based on integrated measurements (i.e., 24-h filter measurements) of ion concentrations, which cannot provide useful information on the time evolution of aerosol acidity and its impacts on aerosol phase reaction processes. Furthermore, long-time integrated sampling of particulate matter also leads to many artifacts (Nie et al., 2010), which in turn interfere with AIM-II's calculation of aerosol acidity. Highly time-resolved data of ion concentrations are valuable for investigating temporal variation of aerosol properties and for detailed process analysis of aerosol pollution; however only a few studies are based on the real time measurement (Zhang et al., 2007; Takahama et al., 2006; Tanner et al., 2009).

Central eastern China is home to three megacities, i.e., Beijing, Tianjin, and Shanghai, and parts or the entirety of Hebei, Shandong, Henan, Hubei, and Jiangsu provinces. It is the largest emitter in China for many chemically and radioactively important pollutants such as SO_2 and particles (Zhang et al., 2009), and thus is of particular interest with regard to air quality studies. High concentrations of carbonaceous aerosols have also been reported at Mt. Tai (the highest mountain in this area), and it was also reported that substantial SOA was formed through both photochemical process and cloud processing (Wang et al., 2011, 2012). This abundant SOA may be attributed to the high concentration of precursors (e.g., VOC) and atmospheric oxidants (e.g., O_3) (Gao et al., 2004; Suthawaree et al., 2010). Intensive field observations were conducted at Mt. Tai in spring and summer of 2007. In this study hourly measurements of SO_4^{2-}, NO_3^- and NH_4^+ were used to investigate the acidity of $PM_{2.5}$ including strong acidity, aerosol acidity and pH. This is the first attempt to investigate aerosol acidity based on such high resolution measurements of aerosol ions at a mountain site in China. Several factors that influence the acidity were discussed. To date conflicting conclusions have been obtained by different measurements about the influence of acidity on the SOA formation (Takahama et al., 2006; Zhang et al., 2007; Tanner et al., 2009; Rengarajan et al., 2011). It is thus worthwhile to investigate the relationship between acidity and SOA formation in this mountain site.

2. Experiment and methods

2.1. Experiment

Two phases of field experiments were conducted at the peak of Mount Tai (36°16′ N, 117°6′ E, 1534 m a.s.l.) from March 21 to April 23 (spring campaign) and from June 15 to July 15 (summer campaign) in 2007. Mount Tai is the highest mountain on the central eastern plain of China. It is located 15 km north of Tai'an city (population: ~500,000) and about 230 km from the Bohai and Yellow Seas. Previous study indicated that Mount Tai is relatively isolated from local emissions and most air masses arriving at Mount Tai have undergone long-range transport (Zhou et al., 2010), thus it is an ideal location to study the regional-scale air pollution and atmospheric processes in central eastern China. The details of the measurement station had been described (Zhou et al., 2010).

A large suite of air pollutants including trace gases and aerosol parameters were measured during the campaigns (Zhou et al., 2010). In this paper, only a brief description is presented. Ionic compositions in $PM_{2.5}$ were measured on an hourly basis by using a semi-continuous ambient ion monitor (Model URG 9000B), which had been described in Zhou et al. (2010). Ten inorganic ions, i.e., F^-, Cl^-, NO_2^-, NO_3^-, SO_4^{2-}, Na^+, NH_4^+, K^+, Mg^{2+} and Ca^{2+} were determined by two ion chromatographs. The detection limits (at the 99% confidence level) are 0.054, 0.010, and 0.045 $\mu g\, m^{-3}$ for SO_4^{2-}, NO_3^-, and NH_4^+ which were used in this study, with measurement uncertainties of approximate ±10% (Zhou et al., 2010). On other species relevant to this study (e.g., OC, EC, NO_x, O_3, NH_3 and CO), detailed descriptions of measurement techniques, accuracy and precision, and quality assurance/quality control methods were described elsewhere (Wang et al., 2003, 2011).

2.2. Calculation of aerosol acidities

2.2.1. Strong acidity

In the present work, hourly concentrations of SO_4^{2-}, NO_3^- and NH_4^+ were used to evaluate the acidic characteristics of $PM_{2.5}$. These three ions were selected due to the fact that they contributed more than 90% of the total ionic compounds in $PM_{2.5}$ and thus controlled aerosol acidity (Zhou et al., 2010). And water soluble organic ions only contributed less than 1% to the total ions. This method is not adapted for the area with large fraction of coarse or organic aerosol. Neutralization degree (denoted by F in this paper), defined as the extent to which acidic aerosol is neutralized, was derived from the mole concentration ratio of NH_4^+ to $(2\times SO_4^{2-}+NO_3^-)$.

$$F = \left[NH_4^+\right] / \left(2 \times \left[SO_4^{2-}\right] + \left[NO_3^-\right]\right) \tag{1}$$

where $[NH_4^+]$, $[SO_4^{2-}]$ and $[NO_3^-]$ denote the mole concentrations ($nmol\, m^{-3}$) of each species in air. F ranges from 0 (no neutralization has occurred) to $1 \pm \sigma$ (anions have been neutralized fully by NH_4^+), in which σ stands for the analytical error based on error propagation of corresponding ion's measurement uncertainty. On average, $\sigma = 10\%$ was estimated for the present method.

For acidic aerosols with $F < 1$, strong acidity was estimated from the difference between mole concentrations of $(2 \times SO_4^{2-} + NO_3^-)$ and NH_4^+.

$$\left[H^+\right]_{strong} = \left[H^+\right] + \left[HSO_4^-\right] \approx \left(2 \times \left[SO_4^{2-}\right] + \left[NO_3^-\right] - \left[NH_4^+\right]\right) \tag{2}$$

It is worth noting that a small portion of data was excluded from the acidity analysis in this study. These data were mainly obtained from several dust storm cases in spring when mineral ions and carbonic acid accounted for a larger fraction of the aerosol mass, and from cloud/fog events in summer when the concentrations of ions were very low. Finally, a total of 430 and 443 sets of hourly data were available for the calculation of strong acidity, accounting for ~71% and ~63% of the total measurements in spring and summer, respectively.

2.2.2. Aerosol acidity and pH

Aerosol acidity was calculated by the online version of Aerosol Inorganic Model II (AIM-II) with gas-aerosol interaction disabled (Clegg et al., 1998). Hourly measurements of SO_4^{2-}, NO_3^-, NH_4^+, ambient temperature (T), relative humidity (RH), and the calculated H^+_{strong} were entered into the model as inputs. After calculation, the model outputted the aqueous phase concentrations of free ions including H^+, SO_4^{2-}, NO_3^-, NH_4^+ and HSO_4^-, and water content in the aerosol droplets per m^3 air. The pH was predicted by the mole concentrations of free H^+ in the aqueous phase of particle droplets,

$$pH = -\log\left[\gamma \times \left[H^+\right]_{air} / (V_{aq}/1000)\right] \quad (3)$$

where γ and V_{aq} denote the activity coefficient of H^+_{air} (mol m^{-3}) and the volume of particle aqueous phase in air (cm^3 m^{-3}).

The H^+_{air} predicted by AIM-II highly depends on the presence of an aqueous phase in the thermodynamic equilibrium. When ambient RH is lower than the deliquescence point (DRH), the particle is considered by AIM-II to exist as a pure solid phase and thus there was no output for the H^+_{air}. In addition, the model cannot predict the H^+_{air} for fully neutralized aerosols (i.e., $F \geq 1$). That means only the aerosol acidity of acidic ones were considered. In the present study, a total of 214 and 339 sets of hourly data had valid output to evaluate the aerosol acidity of $PM_{2.5}$ in spring and summer, respectively.

3. Results and discussion

3.1. General results

3.1.1. Overview of aerosol acidity

Fig. 1 shows the frequency distributions of neutralization degrees of $PM_{2.5}$ obtained in the spring and summer phases at Mt. Tai. In the present study, acidic aerosol is defined as the sample with a neutralization degree lower than 0.9. $PM_{2.5}$ samples collected at Mt. Tai showed a general weakly acidic nature. In spring, 57.2% of the valid samples were acidic aerosols with an average F (±standard deviation) of 0.88 (±0.23), while 81.3% of the valid observations were acidic with a mean F of 0.78 (±0.17) in summer. Furthermore, about 28.1% of the observations in spring were more acidic with F smaller than 0.75, compared to an approximate fraction of 48.5% in summer. These results indicate higher acidities of particles in summer than those in spring.

Strong acidity, aerosol acidity, and pH of $PM_{2.5}$ at Mt. Tai and other compositions modeled by AIM-II are summarized in Table 1. Fig. 2 shows the time series of $[H^+]_{air}$ and pH together with other acidity indicators ($[H^+]_{strong}$ and F) for the study periods. The average concentration of H^+_{strong} (±standard deviation) was 64.82 (±75.07) nmol m^{-3} in spring, which was much lower than that of 142.65 (±115.23) nmol m^{-3} in summer. This is in line with the higher frequency of acidic particles in summer. The mean concentration of total water-soluble ions in spring was about 25% lower than that in summer (Zhou et al., 2010) with most accounted for sulfate. Thus higher acidities were observed in summer.

Similar to strong acidity, aerosol acidities of $PM_{2.5}$ in summer (mean = 35.27 nmol m^{-3}) were also higher than those in spring (mean = 25.25 nmol m^{-3}). The amount of H^+_{air} accounted for about 20% of the strong acidity in summer, which was slightly lower than that of 24% in spring. In contrast, aerosol pH was much higher with a mean value of −0.04(±1.01) in summer than that in spring (pH = −0.32 ± 1.38), indicating lower acidity in summer than that in spring. This is because the aerosol water content was almost two times higher in summer (mean = 78.57 μg m^{-3}) than in spring (mean = 47.89 μg m^{-3}), which led to less ionic strength (19.39 ± 11.09 mol kg^{-1} water in summer compare to 21.48 ± 12.86 mol kg^{-1} water in spring) and less H^+ in the aqueous phase. The impact of aerosol water content on the aerosol acidity of $PM_{2.5}$ will be discussed in section 3.3.1.

3.1.2. Comparison with other studies

We compared the acidities of $PM_{2.5}$ at Mt. Tai with the data obtained from other areas in the world (Table 2). The strong acidity of $PM_{2.5}$ at Mt. Tai was much higher than those reported in US, Korea, Japan, India, Hong Kong Guangzhou and Lanzhou, but lower than those in Beijing and Shanghai (Table 2). The seasonal variation at Mt. Tai was opposite to that in Durham in New England (Ziemba et al., 2007) and Hong Kong (Pathak et al., 2003), which may be caused by different air mass transportation. According to the limited reports, we can find the aerosol acidity of aerosol in Mt. Tai was higher than those from Indian, Pittsburgh, Guangzhou and Lanzhou, but lower than Beijing and Shanghai, which was similar to the case of strong acidity.

The overall acidity of aerosols is balanced by both acidic and basic components. NH_3 plays an important role in neutralizing the acidic components in the atmosphere. Atmospheric NH_3 emission is often positively associated with human activities (Galloway et al., 2004). Thus more acidic aerosols are often observed in rural areas than in urban areas (Table 2), due to lower NH_3 emissions caused by less human activities in rural areas (Liu et al., 1996). However Mt. Tai, a place far from human activities, had lower aerosol acidity than a rural site in Beijing, indicating that relatively abundant NH_3 may exist in this area. This is evidenced by our measurement that the average NH_3 concentration at this mountain site was 10.06 and 13.75 ppb in spring and summer respectively, higher than the concentration reported in the rural site in Beijing in 2007 (annual average: 4.5 ± 4.6 ppb) (Meng et al., 2011). In this study the fractions of acidic sulfate and nitrate ($[H^+]_{strong}/[2\times SO_4^{2-} + NO_3^-]$) with

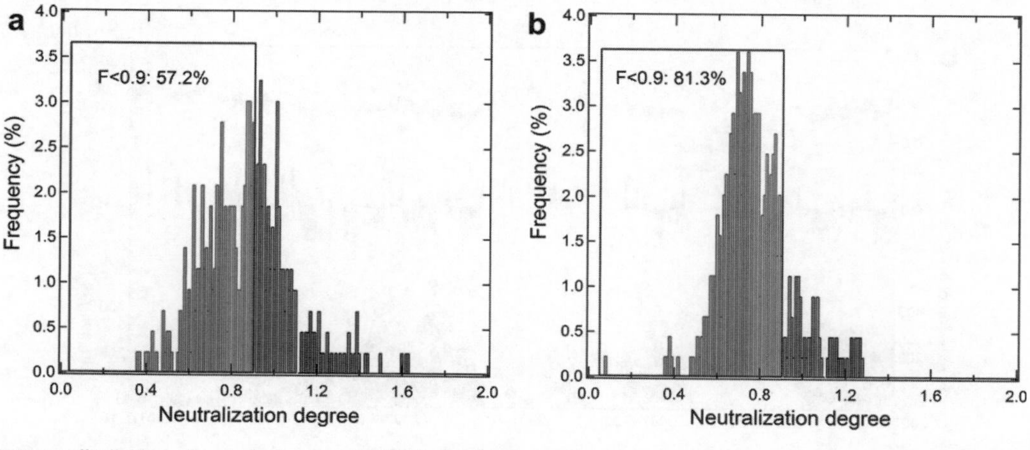

Fig. 1. Frequency distributions of neutralization degrees of $PM_{2.5}$ in (a) spring and (b) summer campaigns at Mt. Tai with acidic frequency in the box.

Table 1
Statistics of strong acidity ($[H^+]_{strong}$), neutralization degree (F) and aerosol acidity ($[H^+]_{air}$) modeled by AIM-II in $PM_{2.5}$ in spring and summer campaigns at Mt. Tai.

	Spring			Summer		
Strong acidity	All valid data	$F<0.75$	$0.75<F<0.9$	All valid data	$F<0.75$	$0.75<F<0.9$
Number	430	121 (28.1%)	125 (29.1%)	443	215 (48.5%)	145 (32.7%)
F	0.88 ± 0.23	0.64 ± 0.09	0.83 ± 0.05	0.78 ± 0.17	0.65 ± 0.08	0.82 ± 0.04
H^+_{strong} (nmol m^{-3})	64.82 ± 75.07	134.38 ± 76.01	72.82 ± 53.44	142.65 ± 115.23	209.42 ± 110.29	118.42 ± 72.54
Aerosol acidity	All valid data			All valid data		
Number	214			339		
H^+_{air} (nmol m^{-3})	25.25 ± 32.23			35.27 ± 30.88		
pH	-0.32 ± 1.38			-0.04 ± 1.01		
HSO_4^- (nmol m^{-3})	67.66 ± 57.65			121.34 ± 79.05		
H^+_{strong}[a] (nmol m^{-3})	111.52 ± 82.64			180.02 ± 104.4		
HSO_4^-/H^+_{air}	3.8 ± 2.51			4.74 ± 2.99		
H^+_{air}/H^+_{strong}[a]	0.24 ± 0.14			0.2 ± 0.11		
Water content (μg m^{-3})	47.89 ± 77.14			78.57 ± 136.99		
RH (%)	65 ± 22			70 ± 16		
T (K)	277.98 ± 4.32			291.35 ± 2.24		

[a] Stands for the strong acidity for the aerosol with valid H^+_{air}.

these two anions concentrations were also compared (Fig. 3). The results showed that high $[H^+]_{strong}/[2\times SO_4^{2-}+NO_3^-]$ ratios were accompanied with low to moderate anion concentrations in both spring and summer. This trend was opposite to that in many other locations where a limited amount of ammonia was available (Liu et al., 1996). It was reported that short or long range transport of air mass from other areas dominated the source of the aerosol at Mt. Tai during the study (Zhou et al., 2010). Thus the NH_3 concentration in this mountain site was probably affected by upslope transport of low-land emissions or by transport from areas with high NH_3 emissions.

The observations at Mt. Tai showed the existence of high ammonia emission which to a large extent neutralized the aerosol acidity. However the aerosol acidity at Mt. Tai was still at a high level compared with that in many other sites in the world, which suggests extremely serious acidity pollution in this area.

3.2. Diurnal variations

Fig. 4 shows the diurnal profiles of H^+_{strong}, H^+_{air} and pH, together with RH and water content in aerosols obtained at Mt. Tai. Overall, H^+_{strong} exhibited a well-defined diurnal pattern in both

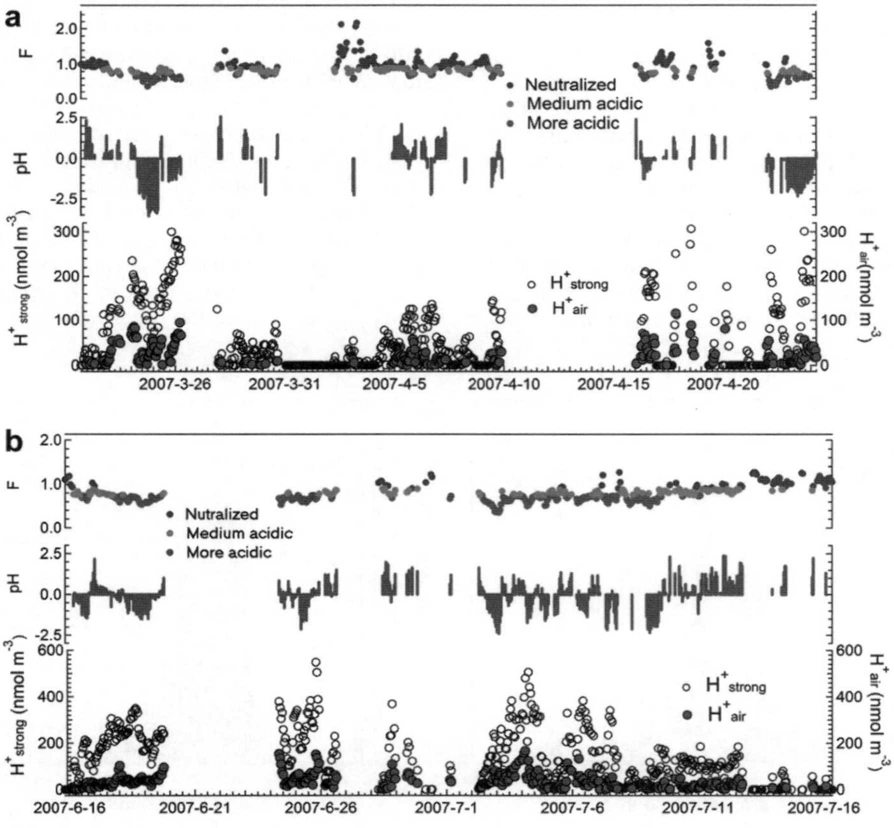

Fig. 2. Temporal variations of H^+_{strong}, H^+_{air}, pH and F of $PM_{2.5}$ in (a) spring and (b) summer campaigns at Mt. Tai.

Table 2
Comparison of aerosol acidity of $PM_{2.5}$ at Mt. Tai and other sites in the world.

	Location	Study periods	H^+_{strong}	H^+_{air}	Reference[a]
Mt. Tai, China (1532 m)	36°16′ N, 117°6′ E	2007 Mar.–Apr.	64.82 ± 75.07	25.25 ± 32.23	This study
		2007 Jun.–Jul.	142.65 ± 115.23	35.27 ± 30.88	
Camp Dodge, US (452 m)	41° 42′ N, 93° 42′ W	1991–1997	25.9 ± 44.4		Murray et al., 2009
		1998–2007	24.0 ± 27.9		
Lakes of the Clouds, US (1540 m)	46°48′ N, 89°45′ W	1991–1997	18.2 ± 31.8		
		1998–2007	25.4 ± 34.0		
Pittsburgh, US (urban site)	40°26′ N, 80°0′ W	2002 Sep.	28		Zhang et al., 2007
		1990 summer	38.2 ± 5.1		Liu et al., 1996
Uniontown, US (semi-rural site)	39°54′ N 79°43′ W	1990 summer	128.0 ± 2.6		
State College, US (semi-rural site)	40°47′ N, 77°51′ W	1990 summer	68.6 ± 3.6		
New England,[b] US (rural site)	43°6′ N, 70°56′ W	2000–2004 spring	7.46		Ziemba et al., 2007
		2000–2004 summer	6.58		
Seoul, Korea (urban site)	37°34′ N, 126°58′ E	1996 fall	5.9		Lee et al., 1999
Goto, Japan (rural site)	32°50′ N, 129°0′ E	1997 Jan. and Dec.	24.2		Shimohara et al., 2001
Dazaifu, Japan (urban site)	33°31′ N, 130°31′ E	1997 Jan. and Dec.	11.4		
Ahmedabad, India (semi-arid region)	23°02′ N, 72°32′ E	2006 Dec. ~ 2007 Jan.	11–250		Rengarajan et al., 2011
Beijing, China (280 m rural site)	40°21′ N, 116°18′ E	2005 Jun.–Aug.	390 ± 545	228 ± 344	Pathak et al., 2009
Shanghai, China (urban site)	31°27′ N, 121°06′ E	2005 May–Jun.	220 ± 225	96 ± 136	
Lanzhou, China (suburban site)	36° N, 104° E	2006 Jun.–Jul.	65 ± 44	7 ± 6	
Guangzhou, China (suburban site)	22°42′ N, 113°30′ E	2004 May	70 ± 58	25 ± 29	
Hong Kong, China (rural and urban sites)	22° N, 114° E	2000 spring	49		Pathak et al., 2003
		2000 summer	27		

[a] This study used the online ion determination method, Zhang et al. (2007) used the aerosol mass spectrometer method and other studies used the filter-based method.
[b] Medium data.

seasons. In spring, H^+_{strong} had high concentrations ($[H^+]_{strong}$: 55.45 ~ 81.67 nmol m^{-3}) in the afternoon and evening, and low levels ($[H^+]_{strong}$: 37.36 ~ 54.06 nmol m^{-3}) at night and in the early morning. While in summer, it showed a broader daytime peak with a maximum value of 173.5 nmol m^{-3} compared to the nighttime with a minimum value of 78.99 nmol m^{-3}. These patterns were consistent with those of major secondary inorganic ions, namely, SO_4^{2-}, NO_3^- and NH_4^+ (Zhou et al., 2010), and to some extent coincided with those of Semi-volatile organic carbon (SVOC) and Non-volatile organic carbon (NVOC) (Wang et al., 2011). Such diurnal variations were a result of the upslope transport of polluted air from the lowland areas due to the mountain-valley breeze and enhanced convective mixing during the daytime (Ren et al., 2009; Zhou et al., 2010). Much newly-formed sulfate may lead to enhanced acidity in the aerosol in the summer campaign.

Compared to H^+_{strong}, diurnal profiles of H^+_{air} and pH were less various. This is in large part due to the fact that the level of H^+_{air} is a complex function of not only the concentrations of SO_4^{2-}, NO_3^- and NH_4^+, also a function of water content. In spring, H^+_{air} generally showed low concentrations (~8.25 nmol m^{-3}) in the afternoon. Low RH and less particle water (H_2Op) in the afternoon in spring may lead to more solid-phase transitions in aerosols, resulting in low concentrations of free H^+ in the aqueous phase of aerosol. Aerosol pH showed the highest acidity concentration in the aqueous phase in the afternoon, which may be attributed to the low aerosol water content in the afternoon. In summer, H^+_{air} generally showed low concentrations during the daytime, consistent with the broad peak of pH. An interesting phenomenon is that pH showed similar diurnal trends with RH. The correlation of pH and RH for all the hourly measurements was examined, and excellent correlations in both seasons ($R = 0.96$ in spring and $R = 0.91$ in summer) were derived, suggesting RH plays a significant role in the acidity characteristics of aerosols. The specific impact of ambient RH on aerosol acidity will be discussed in detail in the following section.

3.3. Factors affecting acidity

3.3.1. Aerosol water content

As stated above, water content in aerosol droplets plays a key role in the acidity of $PM_{2.5}$. One of the effects is the release of free

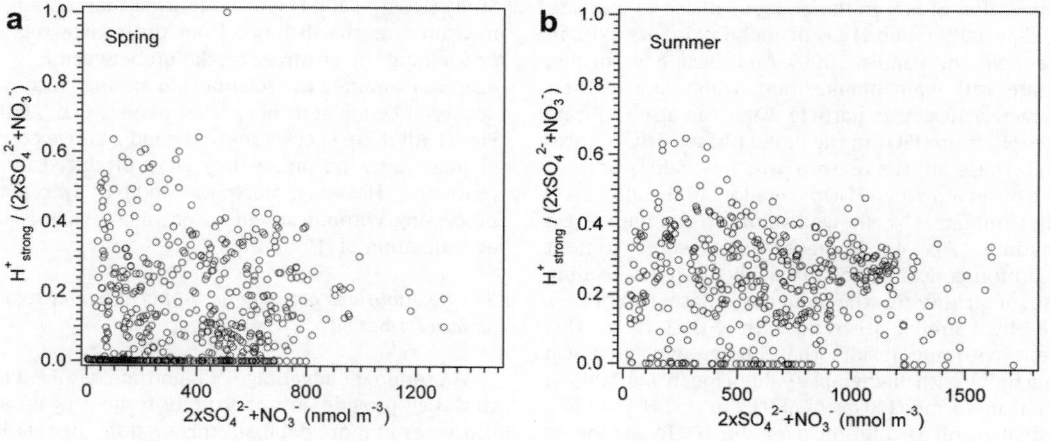

Fig. 3. Mole ratios of acid to sulfate and nitrate vs. sulfate and nitrate concentrations in $PM_{2.5}$ at Mt.Tai for (a) spring and (b) summer campaigns.

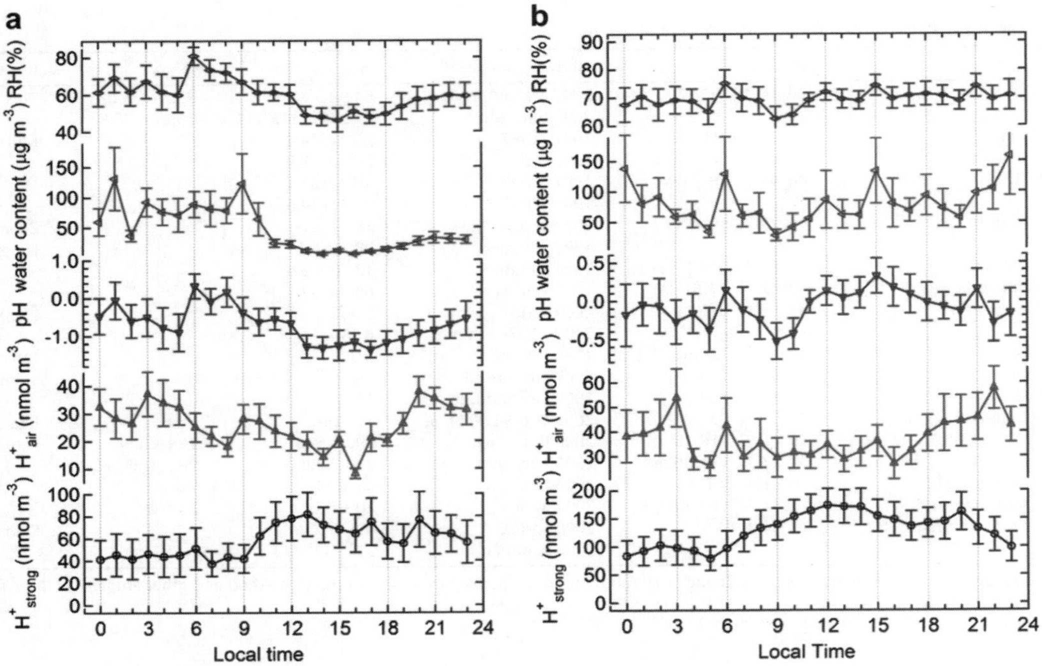

Fig. 4. Average diurnal variations of H^+_{strong}, H^+_{air}, pH, RH and water content in $PM_{2.5}$ at Mt. Tai during (a) spring and (b) summer campaigns. Note that the diurnal variations of H^+_{air}, pH, RH and water content are only for periods when AIM-II gave valid outputs as shown in Fig. 2, and whisker stands for 1/4 standard deviation.

H^+ from HSO_4^- in the liquid phase of the aerosol (Yao et al., 2007). To evaluate the impact of the HSO_4^- hydrolysis process on the aerosol acidity, the relationship of $[HSO_4^-]/[H^+_{air}]$ against RH was examined (Fig. 5). Under conditions with RH > 65%, the mole ratio of $[HSO_4^-]/[H^+_{air}]$ decreased with increasing RH, indicating that increasing aerosol water led to the dissociation of HSO_4^- to form free H^+. According to the outputs of AIM-II, the DRH for the aerosols in the present study were estimated at a range of 60%∼65%, which further evidences that dissociation of HSO_4^- in aerosols is dependent on the difference between ambient RH and DRH (Pathak et al., 2003).

The interaction between water content and aerosol acidity is complicated. Some water content measurements found that when the aerosol was acidic, it could retain water at low RH, preventing the aerosol from drying in summer (Khlystov et al., 2005). On the other hand, elevated water content can supply a larger water surface to uptake more SO_2, H_2SO_4 and HNO_3, and accelerate the oxidation of SO_2 in the aqueous phase to increase aerosol acidity, especially when H_2O_2 or metal oxidants exist in the particles (Seinfeld and Pandis, 2006). And these new formed sulfate and nitrate will again uptake more water due to their hygroscopic characteristics. Also, particle water can also facilitate the release of free H^+ from HSO_4^- in the liquid phase of the aerosol (Yao et al., 2007). These are the reasons why low values of H^+_{air} (nmol m^{-3}) were observed when H_2Op were low in the afternoon in spring, and in summer, H^+_{air} showed similar trend with H_2Op during the daytime (Fig. 4). However high water content can also dilute proton concentrations in acidic aerosols, leading to the decrease of acidity in the aqueous phase (nmol L^{-1}). This is evidenced by a similar pH trend with H_2Op (Fig. 4). This is also consistent with the results that higher average H_2Op (mean = 78.57 μg m^{-3}) with higher pH value (mean = −0.04) in summer than that in spring (H_2Op = 47.89 μg m^{-3}; pH = −0.32), indicating the final result is dilution over the H^+ formation in the aqueous phase.

3.3.2. *Atmospheric processing*

In this section, the relationship between aerosol acidities and atmospheric processing of air masses is explored. The ratio of NO_x to NO_y is usually used to evaluate the extent of atmospheric processing of air masses (Mao and Talbot, 2004). In the present study, $-\mathrm{Log}(NO_x/NO_y)$ was used as a proxy of photochemical age of air masses arriving at Mt. Tai, with larger values indicating higher degree of atmospheric aging (Kleinman et al., 2008). Fig. 6 shows the correlation between H^+_{strong} and air mass age for both seasons. From the figure, H^+_{strong} correlated very well with the air mass age with *R* of 0.94 and 0.88 in spring and summer, respectively; indicating aerosols were more acidic in more processed air masses. The RMA slopes in Fig. 6 indicated increases of ∼9.20 and ∼18.25 nmol m^{-3} in H^+_{strong} for each increment of 0.01 in photochemical age in spring and summer respectively. This positive correlation was similar to the result observed by Quinn et al. (2006) in a cruise measurement during second New England Air Quality Study (NEAQS, 2004) which reported that more acidic aerosol was measured as the distance from the source region increased. To understand the positive correlation between H^+_{strong} and air mass age, we examined the relationship among air mass age with main acidity-affecting aerosol species, namely SO_4^{2-}, NO_3^- and NH_4^+ (see Fig. 7). All these species showed good positive correlation with the air mass age, indicating they were produced during the aerosol processing. However, more SO_4^{2-} and NO_3^- were formed during the processing without an equivalent increase in NH_4^+, leading to an accumulation of H^+.

3.4. *Relationships between aerosol acidity and secondary organic aerosol formation*

A prominent advantage of high time-resolved measurements is that they provide an opportunity to investigate aerosol chemical processes in more detail. In this section, the relationship between acidity and secondary organic aerosol for summer campaign were

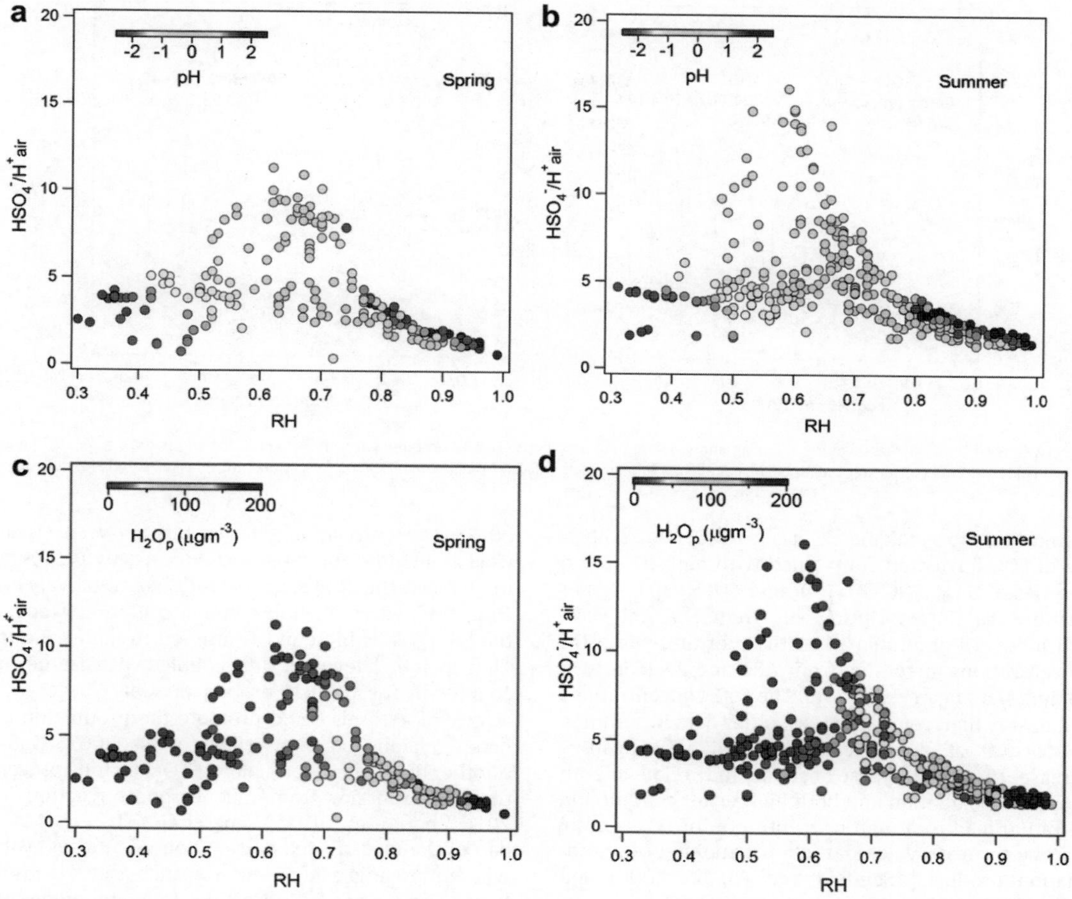

Fig. 5. Variations of $[HSO_4^-]/[H^+_{air}]$ ratios with RH for spring and summer campaigns (a and b colored by pH, c and d by aerosol water content). (For interpretation of the references to color in this figure legend, the reader is referred to the web version of this article.)

investigated. 18 out of 29 days of validate data were examined (11 days influenced by fog and rain were excluded). Because changes of wind direction were often observed in the morning and evening due to the influence of valley wind at this mountain site, the time series between 8:00 to18:00 were investigated. Air parcel with similar wind direction are assumed underwent similar aerosol formation processes. The concentration of secondary organic carbon (SOC) was estimated by EC-tracer method which had been already described (Wang et al., 2012).

3.4.1. Case of SOC increase with high acidity

If acidity enhanced SOA formation exists, possible evidence would be illustrated by an increase of the SOC concentration accompanying or following a period of high H^+_{air} concentrations

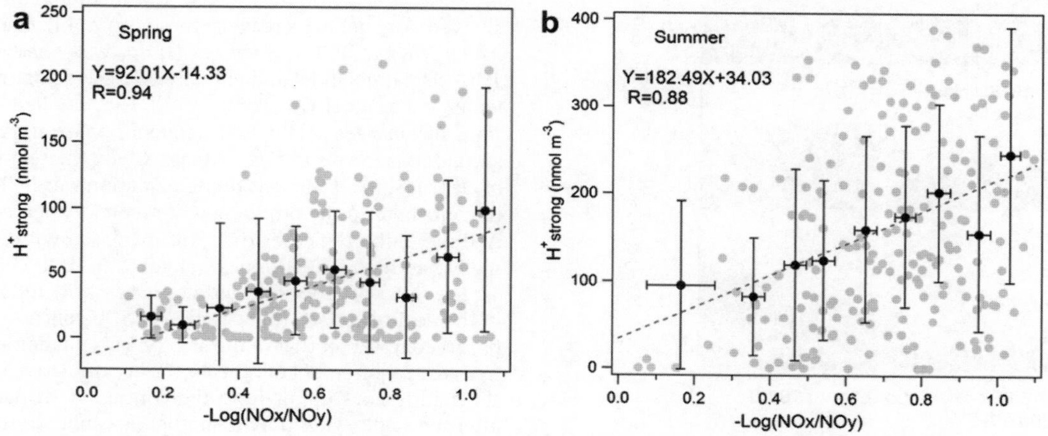

Fig. 6. Correlation between H^+_{strong} and air mass age in (a) spring and (b) summer campaigns (Grey data points are the H^+_{strong} during the campaigns, dark data points are the average values of H^+_{strong} in each specific age bin, and whisker stands for standard deviation).

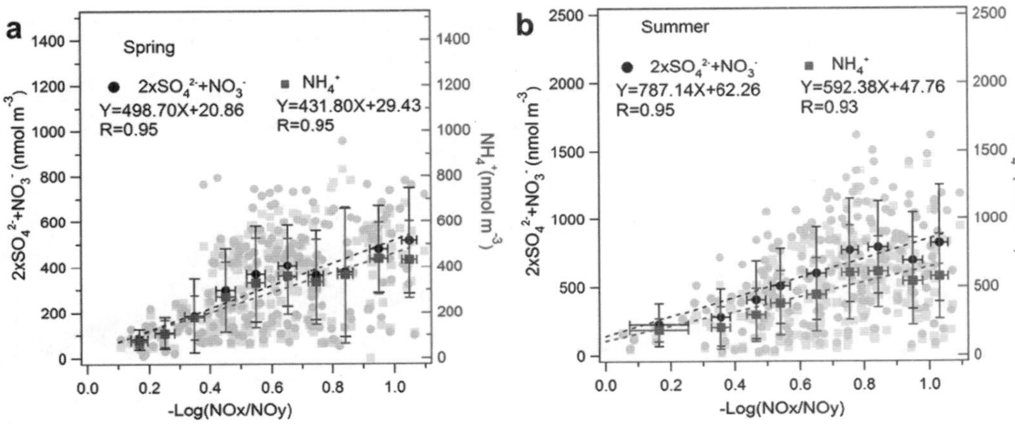

Fig. 7. Correlation between ions and air mass age in (a) spring and (b) summer campaigns (Light data points are the ions concentrations during the campaigns, solid clear data points are the average values of ions concentrations in each specific age bin, circle stands for anion, square stands for cation, and whisker stands for standard deviation).

with similar wind direction (Takahama et al., 2006). Fig. 8 illustrates an event of SOC formation companied with high $[H^+]_{air}$ on July 3rd. Time series of H^+_{air}, SOC, SO_4^{2-}, O_3 and CO, H_2Op together with surface wind were plotted. During this event, SOC and SO_4^{2-} were quickly formed and accumulated in the morning with SOC and sulfate concentrations increasing from 2.82 and 24.25 $\mu g\ m^{-3}$ at 8:00 to 5.63 and 47.53 $\mu g\ m^{-3}$ at 14:00. The SOC concentrations were continued to stay high compared with other data in summer. The aerosol water was at a high level (H_2Op: 57 ~ 136 $\mu g\ m^{-3}$) during this increase. Thus the increase of sulfate may relate to both gas and aqueous phase oxidation and both high water content and sulfate concentration lead to a high concentration of H^+_{air} (nmol m^{-3}). SOA could be formed by gas-particle partitioning of photochemical oxidation products (Seinfeld and Pandis, 2006) and aqueous phase reactions (Blando and Turpin, 2000; Wang et al., 2012). The correlation relationship between O_3 and CO can be used as an indicator of photochemical activity (Cooper et al., 2002). In this case, the RMA slope of O_3/CO was relatively low (slope: 0.05) compared with the whole summer condition (slope: 0.09), whereas the $[H^+]_{air}$ was high and continued to increase during the event. Thus such high levels of H^+_{air} coupled with a large amount of water content in the particle droplet possibly suggested that the high acidity of aerosols could promote the production of SOA through heterogeneous reactions (Jang et al., 2002; Gao et al., 2004). Another possibility is enhanced H_2Op or cloud processing increased the SOA formation from aqueous phase reactions (Kamens et al., 2011; Zhou et al., 2011; Wang et al., 2012). At Mt. Tai, more than half of the investigated days (10 out of 18 days) were of this type, where inorganic acidity and SOC increased together under conditions of constant wind direction. This phenomenon at Mt. Tai was probably due to the effects of acidity promotion and aqueous phase formation. To be noted, the SOC may sometimes lag off the increase of H^+_{air}.

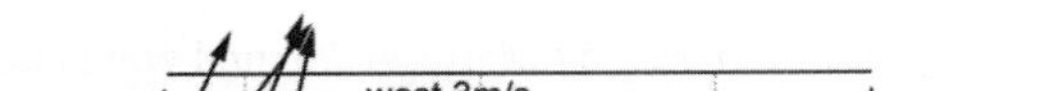

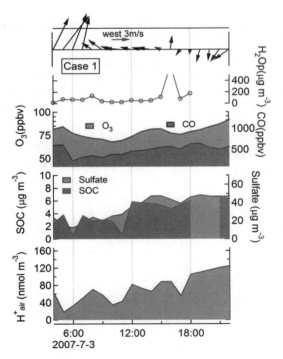

Fig. 8. Time series for the concentrations of H^+_{air}, SOC, O_3, CO, water content and related surface wind direction on July 3rd.

3.4.2. Case in no clear relationship was observed

In some cases no obvious relationship was observed between SOC concentration and acidity (8 cases), suggesting possibilities of other processes dominating the change in organic aerosol. Fig. 9 shows an example on July 10th. SO_4^{2-} was quickly formed and accumulated during the daytime with its concentrations increased from 17.67 $\mu g\ m^{-3}$ at 8:00 to 33.29 $\mu g\ m^{-3}$ at 12:00. But it began to decrease after 12:00 and stayed at a medium concentration. SOC showed a general increasing trend from 1.17 $\mu g\ m^{-3}$ at 8:00 to 3.43 $\mu g\ m^{-3}$ at 15:00. However, $[H^+]_{air}$ was relatively low (range: 1.36 ~ 11.5 nmol m^{-3}) and exhibited a decreasing trend due to low water content ($H_2Op < 20\ \mu g\ m^{-3}$). This implied no direct relationship between acidity in the aqueous phase and the formation of secondary aerosols in this air mass. Given the relatively high slope of O_3/CO (slope: 0.10), gas phase oxidation was believed important to contribute to the production of secondary aerosol in this event. And we found the cases that SOC increased with low acidity normally had high O_3/CO slopes (range: 0.10 ~ 0.16).

The SOC concentration did not necessarily follow the variation of the acidity. In fact, the variation of SOC concentration was often observed to lag off that of the acidity. This phenomenon may result from the requirement of time for the accumulation of the SOC when the acidity goes up, or from the continued formation of the SOC after the acidity goes down. Another possibility is that the acidity is under the threshold of affecting SOC formation. The wind direction of these two cases is very different, which may suggest different air

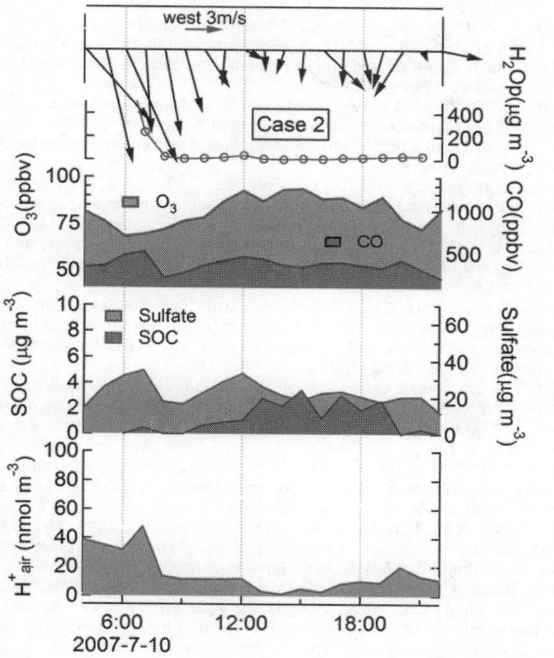

Fig. 9. Time series for the concentrations of H^+_{air}, SOC, O_3, CO, water content and related surface wind direction on July 10th.

mass sources. All the air sources were checked each hour during the summer campaign and generally two categories were found. The acidities of aerosols from east (51.3% of the campaign) and south (7.0% of the campaign) were high with $[H^+]_{strong}$ at 161.32 ± 116.12 nmol m^{-3} and $[H^+]_{air}$ at 41.13 ± 32.43 nmol m^{-3} respectively. While when the air blew from the north direction (8.1% of the campaign), and west or northwest direction (33.62% of the campaign), the acidities were relatively low ($[H^+]_{strong}$: 115.87 ± 108.77 nmol m^{-3}; $[H^+]_{air}$: 27.40 ± 27.16 nmol m^{-3}). This indicated the air mass from east and south brought highly polluted aerosol in the summer campaign at Mt. Tai. This is consistent with the results that low pH of wet deposition was observed at Mt. Tai when the air mass came from the south and east in previous study (Wang et al., 2008).

4. Summary

Near continuous measurements of water-soluble ions in $PM_{2.5}$ were made in the spring and summer of 2007 at Mt. Tai, the highest mountain in central eastern China. The hourly data were used to estimate the strong and aerosol acidities of $PM_{2.5}$. Aerosol acidities characteristics and their relationship with SOC formation were investigated.

$PM_{2.5}$ was on average weakly acidic with approximate 57.2% and 81.3% of the samples being acidic particles in spring and summer, respectively. Strong acidity had higher concentrations in summer with an average of 142.65 nmol m^{-3} and lower levels in spring with a mean value of 64.82 nmol m^{-3}. Similarly, aerosol acidity also showed elevated concentrations in summer (mean = 35.27 nmol m^{-3}) compared to lower levels in spring (mean = 25.25 nmol m^{-3}). In contrast, aerosol pH exhibited an opposite seasonal trend with less acidic aerosol in summer (mean = −0.04) and more acidic ones in spring (mean = −0.32). Aerosol acidities of $PM_{2.5}$ at Mt. Tai were much higher than those reported for other locations in US, Korea, and Japan, but reasonably lower than those obtained from Beijing and Shanghai, in central eastern China. Strong acidity showed broad maximum levels during the daytime in both seasons. Aerosol acidity generally showed relatively low concentrations during the day. Several factors influencing aerosol acidities, such as ambient RH, existing state of sulfate, and atmospheric aging of air masses were investigated. Water content in aerosol droplets can enhance the release of free H^+ through the hydrolysis process of HSO_4^-, but also dilute the proton concentrations in the liquid phase. Acidic aerosols were often associated with more processed air masses. Time evolution of sulfate, SOC and aerosol acidity in several events of secondary organic aerosols production was examined. SOC formation with high acidities and high water content possibly suggested that aerosol acidity and aqueous phase reaction may play significant parts in the secondary organic aerosol formation in central eastern China.

Acknowledgement

The authors would like to acknowledge Ding Aijun, Steven Poon, Wang Jin, Zhou Xuehua, Sun Tingli, Gao Jian, Nie Wei, and Xu Pengju for their help in organizing the field study and in the laboratory work. We are also grateful to the Mt. Tai Meteorological Observatory for providing meteorological data. And we also thank Ge Xinlei and Anthony S. Wexler for their help in running AIM-II model. This research was funded by the National Basic Research Program of China (973 Project No. 2005CB422203), Shandong Provincial Environmental Protection Department (2006045), and the Hong Kong Polytechnic University's Niche Area Development Scheme (1-BB94). Thanks to Dr. Edward C. Mignot, for linguistic advice.

References

Ansari, A.S., Pandis, S.N., 1999. An analysis of four models predicting the partitioning of semivolatile inorganic aerosol components. Aerosol Science & Technology 31, 129–153.

Blando, J.D., Turpin, B.J., 2000. Secondary organic aerosol formation in cloud and fog droplets: a literature evaluation of plausibility. Atmospheric Environment 34, 1623–1632.

Cao, G., Jang, M., 2010. An SOA model for toluene oxidation in the presence of inorganic aerosols. Environmental Science & Technology 44, 727–733.

Charlson, R., Schwartz, S., Hales, J., Cess, R., Coakley Jr., J., Hansen, J., Hofmann, D., 1992. Climate forcing by anthropogenic aerosols. Science 255, 423–433.

Clegg, S.L., Brimblecombe, P., Wexler, A.S., 1998. Thermodynamic model of the system $H + -NH_4 + -Na + -SO_4^{2-}-NO^{3-}-Cl^{-}-H_2O$ at 298.15 K. The Journal of Physical Chemistry A 102, 2155–2171.

Cooper, O., Moody, J., Parrish, D., Trainer, M., Holloway, J., Hübler, G., Fehsenfeld, F., Stohl, A., 2002. Trace gas composition of midlatitude cyclones over the western North Atlantic Ocean: a seasonal comparison of O3 and CO. Journal of Geophysical Research 107. doi:10.1029/2001JD000902.

Gao, S., Keywood, M., Ng, N.L., Surratt, J., Varutbangkul, V., Bahreini, R., Flagan, R.C., Seinfeld, J.H., 2004. Low-molecular-weight and oligomeric components in secondary organic aerosol from the ozonolysis of cycloalkenes and alpha-pinene. Journal of Physical Chemistry A 108, 10147–10164.

Galloway, J.N., Dentener, F.J., Capone, D.G., Boyer, E.W., Howarth, R.W., Seitzinger, S.P., Asner, G.P., Cleveland, C., Green, P., Holland, E., 2004. Nitrogen cycles: past, present, and future. Biogeochemistry 70, 153–226.

Grassian, V.H., 2001. Heterogeneous uptake and reaction of nitrogen oxides and volatile organic compounds on the surface of atmospheric particles including oxides, carbonates, soot and mineral dust: implications for the chemical balance of the troposphere. International Reviews in Physical Chemistry 20, 467–548.

Hatch, C.D., Grassian, V.H., 2008. 10th Anniversary review: applications of analytical techniques in laboratory studies of the chemical and climatic impacts of mineral dust aerosol in the Earth's atmosphere. Journal of Environmental Monitoring 10, 919–934.

Jang, M., Czoschke, N., Lee, S., Kamens, R., 2002. Heterogeneous atmospheric aerosol production by acid-catalyzed particle-phase reactions. Science 298, 814.

Kamens, R.M., Zhang, H.F., Chen, E.H., Zhou, Y., Parikh, H.M., Wilson, R.L., Galloway, K.E., Rosen, E.P., 2011. Secondary organic aerosol formation from toluene in an atmospheric hydrocarbon mixture: water and particle seed effects. Atmospheric Environment 45, 2324–2334.

Khlystov, A., Stanier, C.O., Takahama, S., Pandis, S.N., 2005. Water content of ambient aerosol during the Pittsburgh Air Quality Study. Journal of Geophysical Research 110, D07S10.
Kleinman, L.I., Springston, S.R., Daum, P.H., Lee, Y.N., Nunnermacker, L.J., Senum, G.I., Wang, J., Weinstein-Lloyd, J., Alexander, M.L., Hubbe, J., Ortega, J., Canagaratna, M.R., Jayne, J., 2008. The time evolution of aerosol composition over the Mexico City plateau. Atmospheric Chemistry and Physics 8, 1559–1575.
Lee, H.S., Kang, C.M., Kang, B.W., Kim, H.K., 1999. Seasonal variations of acidic air pollutants in Seoul, South Korea. Atmospheric Environment 33, 3143–3152.
Liu, L., Burton, R., Wilson, W., Koutrakis, P., 1996. Comparison of aerosol acidity in urban and semi-rural environments. Atmospheric Environment 30, 1237–1245.
Manktelow, P.T., Carslaw, K.S., Mann, G.W., Spracklen, D.V., 2010. The impact of dust on sulfate aerosol, CN and CCN during an East Asian dust storm. Atmospheric Chemistry and Physics 10, 365–382.
Mao, H., Talbot, R., 2004. O3 and CO in New England: temporal variations and relationships. Journal of Geophysical Research 109, 1304.
Meng, Z., Seinfeld, J.H., Saxena, P., Kim, Y.P., 1995. Atmospheric gas–aerosol equilibrium: IV. Thermodynamics of carbonates. Aerosol Science and Technology 23, 131–154.
Meng, Z., Lin, W., Jiang, X., Yan, P., Wang, Y., Zhang, Y., Jia, X., Yu, X., 2011. Characteristics of atmospheric ammonia over Beijing, China. Atmospheric Chemistry and Physics 11, 6139–6151.
Murray, G.L.D., Kimball, K., Bruce Hill, L., Allen, G.A., Wolfson, J.M., Pszenny, A., Seidel, T., Doddridge, B.G., Boris, A., 2009. A comparison of fine particle and aerosol strong acidity at the interface zone (1540 m) and within (452 m) the planetary boundary layer of the Great Gulf and Presidential-Dry River Class I Wildernesses on the Presidential Range, New Hampshire USA. Atmospheric Environment 43, 3605–3613.
Nel, A., 2005. Air pollution-related illness: effects of particles. Science 308, 804–806.
Nenes, A., Pandis, S.N., Pilinis, C., 1998. ISORROPIA: a new thermodynamic equilibrium model for multiphase multicomponent inorganic aerosols. Aquatic Geochemistry 1, 123–152.
Nie, W., Wang, T., Gao, X., Pathak, R.K., Wang, X., Gao, R., Zhang, Q., Yang, L., Wang, W., 2010. Comparison among filter-based, impactor-based and continuous techniques for measuring atmospheric fine sulfate and nitrate. Atmospheric Environment 44, 4396–4403.
Pathak, R.K., Yao, X., Lau, A.K.H., Chan, C.K., 2003. Acidity and concentrations of ionic species of PM2.5 in Hong Kong. Atmospheric Environment 37, 1113–1124.
Pathak, R.K., Wu, W.S., Wang, T., 2009. Summertime PM2.5 ionic species in four major cities of China: nitrate formation in an ammonia-deficient atmosphere. Atmospheric Chemistry and Physics 9, 1711–1722.
Quinn, P.K., Bates, T.S., Coffman, D., Onasch, T.B., Worsnop, D., Baynard, T., de Gouw, J.A., Goldan, P.D., Kuster, W.C., Williams, E., Roberts, J.M., Lerner, B., Stohl, A., Pettersson, A., Lovejoy, E.R., 2006. Impacts of sources and aging on submicrometer aerosol properties in the marine boundary layer across the Gulf of Maine. Journal of Geophysical Research-Atmospheres 111. doi:10.1029/2006JD007582.
Ren, Y., Ding, A.J., Wang, T., Shen, X.H., Guo, J., Zhang, J.M., Wang, Y., Xu, P.J., Wang, X.F., Gao, J., Collett, J.L., 2009. Measurement of gas-phase total peroxides at the summit of Mount Tai in China. Atmospheric Environment 43, 1702–1711.
Rengarajan, R., Sudheer, A.K., Sarin, M.M., 2011. Aerosol acidity and secondary organic aerosol formation during wintertime over urban environment in western India. Atmospheric Environment 45, 1940–1945.
Seinfeld, J., Pandis, S., 2006. From Air Pollution to Climate Change. Atmospheric Chemistry and Physics. John Wiley & Sons, New York.
Shimohara, T., Oishi, O., Utsunomiya, A., Mukai, H., Hatakeyama, S., Eun-Suk, J., Uno, I., Murano, K., 2001. Characterization of atmospheric air pollutants at two sites in northern Kyushu, Japan-chemical form, and chemical reaction. Atmospheric Environment 35, 667–681.
Suthawaree, J., Kato, S., Okuzawa, K., Kanaya, Y., Pochanart, P., Akimoto, H., Wang, Z., Kajii, Y., 2010. Measurements of volatile organic compounds in the middle of Central East China during Mount Tai Experiment 2006 (MTX2006): observation of regional background and impact of biomass burning. Atmospheric Chemistry and Physics 10, 1269–1285.
Takahama, S., Davidson, C.I., Pandis, S.N., 2006. Semicontinuous measurements of organic carbon and acidity during the Pittsburgh air quality study: implications for acid-catalyzed organic aerosol formation. Environmental Science & Technology 40, 2191–2199.
Tanner, R.L., Olszyna, K.J., Edgerton, E.S., Knipping, E., Shaw, S.L., 2009. Searching for evidence of acid-catalyzed enhancement of secondary organic aerosol formation using ambient aerosol data. Atmospheric Environment 43, 3440–3444.
Ullerstam, M., Vogt, R., Langer, S., Ljungström, E., 2002. The kinetics and mechanism of SO2 oxidation by O3 on mineral dust. Physical Chemistry Chemical Physics 4, 4694–4699.
Underwood, G.M., Song, C.H., Phadnis, M., Carmichael, G.R., Grassian, V.H., 2001. Heterogeneous reactions of NO2 and HNO3 on oxides and mineral dust: a combined laboratory and modeling study. Journal of Geophysical Research-Atmospheres 106, 18055–18066.
Usher, C.R., Al-Hosney, H., Carlos-Cuellar, S., Grassian, V.H., 2002. A laboratory study of the heterogeneous uptake and oxidation of sulfur dioxide on mineral dust particles. Journal of Geophysical Research-Atmospheres 107. doi:10.1029/2002JD002051.
Wang, T., Poon, C.N., Kwok, Y.H., Li, Y.S., 2003. Characterizing the temporal variability and emission patterns of pollution plumes in the Pearl River Delta of China. Atmospheric Environment 37, 3539–3550.
Wang, Y., Wai, K.M., Gao, J., Liu, X., Wang, T., Wang, W., 2008. The impacts of anthropogenic emissions on the precipitation chemistry at an elevated site in North-eastern China. Atmospheric Environment 42, 2959–2970.
Wang, Z., Wang, T., Gao, R., Xue, L.K., Guo, J., Zhou, Y., Nie, W., Wang, X.F., Xu, P.J., Gao, J.A., Zhou, X.H., Wang, W.X., Zhang, Q.Z., 2011. Source and variation of carbonaceous aerosols at Mount Tai, North China: results from a semi-continuous instrument. Atmospheric Environment 45, 1655–1667.
Wang, Z., Wang, T., Guo, J., Gao, R., Xue, L., Zhang, J., Zhou, Y., Zhou, X., Zhang, Q., Wang, W., 2012. Formation of secondary organic carbon and cloud impact on carbonaceous aerosols at Mount Tai, North China. Atmospheric Environment 46, 516–527.
Yao, X., Ling, T.Y., Fang, M., Chan, C.K., 2006. Comparison of thermodynamic predictions for in situ pH in PM2.5. Atmospheric Environment 40, 2835–2844.
Yao, X., Ling, T.Y., Fang, M., Chan, C.K., 2007. Size dependence of in situ pH in submicron atmospheric particles in Hong Kong. Atmospheric Environment 41, 382–393.
Zhang, Q., Jimenez, J.L., Worsnop, D.R., Canagaratna, M., 2007. A case study of urban particle acidity and its influence on secondary organic aerosol. Environmental Science & Technology 41, 3213–3219.
Zhang, Q., Streets, D.G., Carmichael, G.R., He, K.B., Huo, H., Kannari, A., Klimont, Z., Park, I.S., Reddy, S., Fu, J.S., Chen, D., Duan, L., Lei, Y., Wang, I.T., Yao, Z.L., 2009. Asian emissions in 2006 for the NASA INTEX-B mission. Atmospheric Chemistry and Physics 9, 5131–5153.
Zhou, Y., Wang, T., Gao, X.M., Xue, L.K., Wang, X.F., Wang, Z., Gao, J.A., Zhang, Q.Z., Wang, W.X., 2010. Continuous observations of water-soluble ions in PM2.5 at Mount Tai (1534 ma.s.l.) in central-eastern China. Journal of Atmospheric Chemistry 64, 107–127.
Zhou, Y., Zhang, H., Parikh, H.M., Chen, E.H., Rattanavaraha, W., Rosen, E.P., Wang, W., Kamens, R.M., 2011. Secondary organic aerosol formation from xylenes and mixtures of toluene and xylenes in an atmospheric urban hydrocarbon mixture: water and particle seed effects (II). Atmospheric Environment 45, 3882–3890.
Ziemba, L.D., Fischer, E., Griffin, R.J., Talbot, R.W., 2007. Aerosol acidity in rural New England: temporal trends and source region analysis. Journal of Geophysical Research 112, D10S22.

Atmospheric Environment 101 (2015) 294–302

Contents lists available at ScienceDirect

Atmospheric Environment

journal homepage: www.elsevier.com/locate/atmosenv

Enhanced formation of fine particulate nitrate at a rural site on the North China Plain in summer: The important roles of ammonia and ozone

Liang Wen [a], Jianmin Chen [a, b, c], Lingxiao Yang [a, b, *], Xinfeng Wang [a], Caihong Xu [a], Xiao Sui [a], Lan Yao [a], Yanhong Zhu [a], Junmei Zhang [a], Tong Zhu [d], Wenxing Wang [a]

[a] Environment Research Institute, Shandong University, Jinan 250100, China
[b] School of Environmental Science and Engineering, Shandong University, Jinan 250100, China
[c] Shanghai Key Laboratory of Atmospheric Particle Pollution and Prevention (LAP3), Fudan Tyndall Centre, Department of Environmental Science and Engineering, Fudan University, Shanghai 200433, China
[d] State Key Lab for Environment Simulation and Pollution Control, College of Environmental Sciences and Engineering, Peking University, Beijing 100871, China

HIGHLIGHTS

- High concentration of nitrate in $PM_{2.5}$ was observed at a rural site in North China.
- High ammonia in the early morning accelerated the formation of fine nitrates.
- The formation of nitrates at night was mainly attributed to the hydrolysis of N_2O_5.

ARTICLE INFO

Article history:
Received 28 August 2014
Received in revised form
14 November 2014
Accepted 16 November 2014
Available online 18 November 2014

Keywords:
Fine particulate nitrates
Secondary formation
Ammonia
Ozone
North China Plain

ABSTRACT

Severe $PM_{2.5}$ pollution was observed frequently on the North China Plain, and nitrate contributed a large fraction of the elevated $PM_{2.5}$ concentrations. To obtain a comprehensive understanding of the formation pathways of these fine particulate nitrate and the key factors that affect these pathways, field measurements of fine particulate nitrate and related air pollutants were made at a rural site on the North China Plain in the summer of 2013. Extremely high concentrations of fine particulate nitrate were frequently observed at night and in the early morning. The maximum hourly concentration of fine particulate nitrate reached 87.2 μg m^{-3}. This concentration accounted for 29.9% of the $PM_{2.5}$. The very high NH_3 concentration in the early morning significantly accelerated the formation of fine particulate nitrate, as indicated by the concurrent appearance of NH_3 and NO_3^- concentration peaks and a rising neutralization ratio (the equivalent ratio of NH_4^+ to the sum of SO_4^{2-} and NO_3^-). On a number of other episode days, strong photochemical activity during daytime led to high concentrations of O_3 at night. The fast secondary formation of fine particulate nitrate was mainly attributed to the hydrolysis of N_2O_5, which was produced from O_3 and NO_2. Considering the important roles of NH_3 and O_3 in fine particulate nitrate formation, we suggest the control of NH_3 emissions and photochemical pollution to address the high levels of fine particulate nitrate and the severe $PM_{2.5}$ pollution on the North China Plain.

1. Introduction

Particulate nitrate account for a large fraction of the $PM_{2.5}$ in rural, suburban, urban and industrial areas, usually averagely ranging from 4.5 to 25 % (Ye et al., 2003; Wang et al., 2005; Pathak et al., 2009; Du et al., 2011; Squizzato et al., 2012). Nitrate aerosols can have a stronger impact on visibility than sulphate aerosols in the same concentrations (Lei and Wuebbles, 2013) because the

* Corresponding author. Environment Research Institute, Shandong University, Jinan 250100, China.
E-mail address: yanglingxiao@sdu.edu.cn (L. Yang).

http://dx.doi.org/10.1016/j.atmosenv.2014.11.037
1352-2310/

scattering albedo of nitrate aerosols is larger than that of sulphate aerosols in low RH conditions (Zhang et al., 2012). High concentrations of secondary nitrate, sulphate, and organic aerosols have led to frequent haze episodes in North China (Guo et al., 2010; Gao et al., 2011; Wang et al., 2014a,b).

Fine particulate nitrate can be formed through the homogeneous reaction of gaseous HNO_3 and NH_3 (Feng and Penner, 2007). HNO_3 is primarily produced from the reaction between NO_2 and OH radicals during the daytime (Calvert and Stockwell, 1983) and later combines with NH_3 to produce fine particles of NH_4NO_3. Field observations in urban Shanghai show that NH_3 can neutralize HNO_3 in the gas phase and liquid phase, playing a vital role in the increase in particulate nitrate during haze episodes (Ye et al., 2011). NH_4NO_3 aerosols are unstable under conditions of high temperature and low humidity due to the reversible phase equilibrium with HNO_3 and NH_3 (Mozurkewich, 1993). Laboratory studies have shown that low temperature and high RH favoured the formation of nitrate aerosols (Hu et al., 2011; Shi et al., 2014).

In addition, a large fraction of fine particulate nitrate can be produced via the heterogeneous hydrolysis of N_2O_5 on the wet surface of aerosols in the dark (Ravishankara, 1997). N_2O_5 primarily accumulates at night-time via the reversible reaction between the NO_2 and NO_3 radicals which are produced from the reaction of NO_2 with O_3 (Mentel et al., 1996). Several finding shave indicated that strong photochemical activity and high levels of NO_2 promoted the formation of fine particulate nitrate during the night-time (Wang et al., 2009). At a site downwind of Beijing in summertime, the contribution of N_2O_5 hydrolysis to the enhancement of fine particulate nitrate was estimated at up to 50 %–100 % by using a thermodynamic model (Pathak et al., 2009, 2011).

North China, one of the major agricultural and industrial bases, is densely populated and suffered from serious particulate matter pollution and photochemical pollution in the last decade (Wang et al., 2006, 2010; Guo et al., 2010; Luo et al., 2013). Large amounts of NO_x and NH_3 were emitted from industry, vehicles, agriculture, and daily activities on the North China Plain (Meng et al., 2010; Shen et al., 2011; Carslaw and Rhys-Tyler, 2013). In recent years, the NO_x emissions have kept rising because of the continuous increase in the number of motor vehicles (Richter et al., 2005; Liu et al., 2013). Consequently, both the concentration of fine particulate nitrate and the fraction of fine particles containing nitrate show an increasing trend (Lei and Wuebbles, 2013).

In this study, intensive field measurements, including $PM_{2.5}$, fine particulate nitrate and other water-soluble ions, aerosol surface area, trace gases and meteorological parameters, were conducted at a rural site in the middle of the North China Plain from June 18 to June 30, 2013. Surprisingly, extremely high concentrations of fine particulate nitrate and fast nitrate formation were observed. The dominant formation pathways of the elevated fine particulate nitrate were analysed in detail based on real-time observations in combination with numerical calculations.

2. Experiments and methods

2.1. Sampling site

The measurement site of this study was located in rural Yucheng, Dezhou, Shandong Province, China (36.87°N, 116.57°E, ~23 m a.s.l.), which is almost in the centre of the North China Plain (as shown in Fig. 1). The selected site was in an open field surrounded by farmland. Yucheng and the surrounding areas are famous for their agriculture (e.g., wheat and corn) and grazing land (e.g., donkeys and chickens). In addition, the site near 20 to 30 Km radius located several factories in the production of inorganic and organic fertilizers. These, in addition to the application of fertilisers to farmland emitted a great deal of NH_3 (Zhao et al., 2012). A previous study has shown that the NH_3 concentrations in this region were very high, with annual average concentrations of 21.4, 31.9, and 27.7 ppbv in Quzhou, Shouguang, and Wuqiao, respectively (Shen et al., 2011, with locations shown in Fig. 1b). Both the emissions and the concentration of NH_3 in the summer were highest of any of the four seasons of the year (Meng et al., 2010).

Online instruments were installed in a steel container, with sampling inlets crossing the roof vertically (~4 m a.g.l.). The field campaign was conducted from June 1 to 30, June, 2013. In this study, only data during June 18 to June 30 were used to research the atmospheric chemical processes of fine nitrate formation because the data of fine nitrate in the first half of the month were significantly affected by biomass burning.

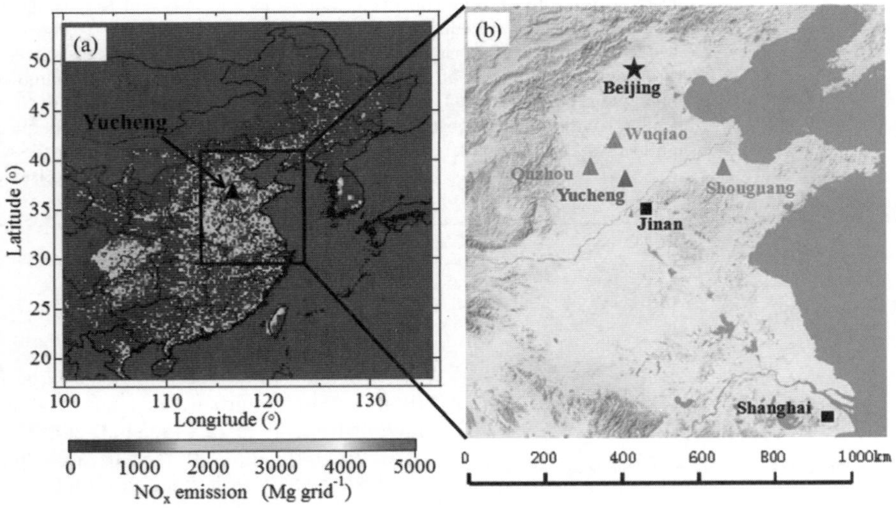

Fig. 1. (a) Location in of the measurement site of Yucheng with colour showing the emission intensity of NO_x in 2006 (Zhang et al., 2009), (b) topographic map of the study region which shows the locations of Yucheng site in this study (marked in red) and some other rural sites in literature (marked in blue). (For interpretation of the references to colour in this figure legend, the reader is referred to the web version of this article.)

2.2. Instruments

In this study, a Monitor for Aerosols and Gases (MARGA, ADI20801, Applikon-ECN, Netherlands) was deployed to continuously measure the concentrations of acid and alkaline gases and inorganic water-soluble ions in $PM_{2.5}$ with a time resolution of 1 h. The flow rate was 16.7 L min^{-1}. The water-soluble gases, including HCl, HNO_2, HNO_3, SO_2, and NH_3, were collected by a WRD (Wet Rotating Denuder), and the inorganic water-soluble ions in $PM_{2.5}$, including Cl^-, NO_3^-, SO_4^{2-}, Na^+, NH_4^+, K^+, Mg^{2+} and Ca^{2+}, were collected by a SJAC (Steam Jet Aerosol Collector). These substances were dissolved in the absorption solution (DI water with 15 ppmv H_2O_2) and were analysed by ion chromatography. The cationic eluent and the anionic eluent were prepared by methanesulphonic acid (308 mg L^{-1}) and $NaHCO_3$ (672 mg L^{-1}) – Na_2CO_3 (742 mg L^{-1}), respectively. A solution of 4.0 mg L^{-1} LiBr as an internal standard was periodically injected and mixed with the sample solutions for subsequent detection. The detection limits of HCl, HNO_2, HNO_3, SO_2, NH_3, Cl^-, NO_3^-, SO_4^{2-}, Na^+, NH_4^+, K^+, Mg^{2+} and Ca^{2+} are 0.00614, 0.0238, 0.00711, 0.0105, 0.0658 ppbv, 0.01, 0.05, 0.04, 0.05, 0.05, 0.09, 0.06, and 0.1 μg m^{-3}, respectively. Multi-point calibrations of nine water-soluble ions were conducted immediately before and after the campaign.

$PM_{2.5}$ was measured with a particle monitor (Model 5030 SHARP Monitor, Thermo Fisher Scientific, USA), and the hourly average concentrations were quantified by the scattering coefficient of 880 nm light and the absorption coefficient of beta rays. The surface area concentration of aerosols in the range of 5 nm to 1 μm was measured with a Wide-Range Particles Spectrometer (WPS, Model 1000XP, MSP Corporation, USA) with the assumption of all aerosols being spherical. The concentration of NO_x (NO and NO_2) was detected by the chemiluminescence method (42C, Thermo Electron Corporation, USA) with a molybdenum oxide catalytic converter. The concentration of O_3 was measured by the ultraviolet absorption method (49C, Thermo Electron Corporation, USA). In addition, the meteorological data were measured with an automatic meteorological station (MILOS520, Vaisala, Finland).

2.3. Steady-state predictions

Due to their short lifetimes, the concentrations of the NO_3 radical and N_2O_5 can be predicted by steady-state calculations in the situation that measurement data are lacking (Osthoff et al., 2006). The formation and loss of NO_3 and N_2O_5 are dominated by a series of chemical reactions listed in Table 1. Consequently, the NO_3 concentration can be calculated by Eq. (1) (Wang et al., 2014a,b),

$$[NO_3]_{cal.} = \frac{k_1 \cdot [NO_2] \cdot [O_3]}{k_3 \cdot [NO] + j_4 + j_5 + k_6 + k_7 + (k_8 + k_9) \cdot k_{eq} \cdot [NO_2]} \tag{1}$$

A fast equilibrium exists between N_2O_5 and its decomposition products, NO_3 and NO_2. Similarly, the N_2O_5 concentration can be calculated by Eq. (2) (Wang et al., 2014a,b),

$$[N_2O_5]_{cal.} = k_{eq}\,[NO_2] \cdot [NO_3]_{cal.} \tag{2}$$

In this study, to help elucidate the contribution of N_2O_5 hydrolysis to nitrate formation, Eq. (1) and Eq. (2) were used to calculate the steady-state concentrations of NO_3 and N_2O_5. The involved reactions and rate constants together with their provenances are included in Table 1. Due to the lack of VOC measurement data, the total reaction rates of NO_3 with VOCs were assumed to be equal to the NO_3 loss rate caused by the heterogeneous hydrolysis of N_2O_5 (Aldener et al., 2006). For the heterogeneous processes on wet aerosol surfaces, the uptake coefficients of the NO_3 radical and N_2O_5 (γ_{NO_3} and $\gamma_{N_2O_5}$) used in this study were 0.004 and 0.03, respectively (Aldener et al., 2006).

Table 1
Major chemical reactions involved in NO_3 and N_2O_5 and the reaction rate constants.

Reaction	Reaction rate constant	Provenance
(R1)$NO_2 + O_3 \rightarrow NO_3 + O_2$	k_1	a, b
(R2)$NO_3 + NO_2 \leftrightarrow N_2O_5$	K_{eq}	a, b
(R3)$NO_3 + NO \rightarrow NO_2 + NO_2$	k_3	a, b
(R4)$NO_3 \rightarrow NO_2 + O$	j_4	c
(R5)$NO_3 \rightarrow NO + O_2$	j_5	c
(R6)$NO_3 + \xrightarrow{VOC}$ products	$k_6 = \sum(k_{VOC,i} \cdot [VOC]_i)$	b
(R7)$NO_3 \xrightarrow{Heterogeneous}$ products	$k_7 = \frac{1}{4} \cdot C_{NO_3} \cdot \gamma_{NO_3} \cdot S_{aerosol}$	d, e
(R8)$N_2O_5 \xrightarrow{Heterogeneous}$ products	$k_8 = \frac{1}{4} \cdot C_{N_2O_5} \cdot \gamma_{N_2O_5} \cdot S_{aerosol}$	e, f
(R9)$N_2O_5 \xrightarrow{Homogeneous} 2HNO_3$	$k_9 = k_I \cdot [H_2O] + k_{II} \cdot [H_2O]^2$	g

[a] Master Chemical Mechanism (MCM version 3.1http://mcm.leeds.ac.uk/MCM/) (Saunder et al., 2006).
[b] Atkinson and Arey, 2003.
[c] Calculated as functions of solar zenith angles (MCM).
[d] Evans and Jacob, 2005.
[e] Aldener et al., 2006.
[f] Osthoff et al., 2006.
[g] Wahner et al., 1998.

3. Results and discussion

3.1. High levels of fine particulate nitrates

3.1.1. Statistic results and comparison with other locations

The concentrations of major water-soluble ions in $PM_{2.5}$ at the Yucheng site and other locations in the summertime are listed in Table 2. At the Yucheng site, the average concentrations (±standard deviations) of $PM_{2.5}$, NO_3^-, SO_4^{2-}, and NH_4^+ were 155.9 (±88.4), 22.5 (±18.6), 28.7 (±17.7), and 18.2 (±11.9) μg m^{-3}, respectively. The fine particulate nitrate concentration exhibited a large variation, ranging from 1.5 to 87.2 μg m^{-3}. The average ratios of $NO_3^-/PM_{2.5}$ and NO_3^-/SO_4^{2-} were 0.14 (range, 0.014–0.45) and 0.79 (range, 0.06–2.9), respectively. When compared to other locations, the fine particulate nitrate concentration at the Yucheng site is found to be extremely high. The average concentration of fine particulate nitrate at Yucheng was close to those in urban Jinan, four times higher than those in urban Fuzhou and one to six times higher than those observed at rural or suburban sites in Beijing, Shanghai, Lanzhou, and Guangzhou (Pathak et al., 2009; Gao et al., 2011; Zhang et al., 2013). Furthermore, the fine particulate nitrate concentration at Yucheng was dozens of times higher than those measured at urban or rural-coastal sites in Japan and Europe (Khan et al., 2010; Squizzato et al., 2012). The $NO_3^-/PM_{2.5}$ ratio at Yucheng was close to that in Beijing and much higher than those at other sites both domestic and abroad. The NO_3^-/SO_4^{2-} ratio was much higher than those at all sites listed in Table 2. The results of the elevated fine particulate nitrate concentrations and the very high ratios of $NO_3^-/PM_{2.5}$ and NO_3^-/SO_4^{2-} at the Yucheng site demonstrate severe fine particulate nitrate pollution in this region.

3.1.2. Time series of fine particulate nitrate concentration

The time dependence of the concentrations of $PM_{2.5}$, NO_3^-, together with meteorological parameters and NR (NR = $[NH_4^+]/(2[SO_4^{2-}] + [NO_3^-])$) from June 18 to June 30 are shown in Fig. 2. The four highest concentration peaks of NO_3^- in $PM_{2.5}$ were 73.7, 64.9, 82.9, and 87.2 μg m^{-3}, which appeared in the early morning of June 19, 20, 24, and 25, respectively (marked as the red arrows in Fig. 2). The variation pattern of fine particulate nitrate concentration

Table 2
Concentrations of $PM_{2.5}$ and secondary water-soluble inorganic ions at Yucheng and other sites during summer in domestic and abroad in recent years.

Site	Types of study sites	Measurement periods	$PM_{2.5}$ ($\mu g\ m^{-3}$)	NO_3^- ($\mu g\ m^{-3}$)	SO_4^{2-} ($\mu g\ m^{-3}$)	NH_4^+ ($\mu g\ m^{-3}$)	$NO_3^-/PM_{2.5}$	NO_3^-/SO_4^{2-}	Provenance
Yucheng, China (Min.)	Rural	Jun. 2013	48.9	1.5	3.7	4.2	0.014	0.06	This study
Yucheng, China (ave.)			155.9	22.5	28.7	18.2	0.14	0.79	
Yucheng, China (max.)			356.2	87.2	99.7	54.9	0.45	2.9	
Beijing, China	Rural	Jun.–Aug. 2005	68	9.9	22.6	4.7	0.15	0.44	Pathak et al., 2009
Shanghai, China	Suburban	May–Jun. 2005	67	7.1	15.8	4.1	0.11	0.45	
Lanzhou, China	Suburban	Jun.–Jul. 2006	65	3.2	9.8	4.1	0.049	0.33	
Guangzhou, China	Suburban	May 2006	55	5.2	13.1	4.8	0.1	0.4	
Fuzhou, China	Urban	Sept. 2007	69.5	5.6	14.4	5.9	0.080	0.39	Zhang et al., 2013
Jinan, China	urban	Jun. 2008	173.2	19.2	64.3	28.0	0.11	0.3	Gao et al., 2011
Yokohama, Japan	urban	Jun.–Aug. 2008	20.8	0.2	4.6	4.4	0.008	0.037	Khan et al., 2010
Punta Sabbioni, Europe	rural-coastal	Jun.–Jul. 2009	11	0.5	2.4	1	0.045	0.21	Squizzato et al., 2012

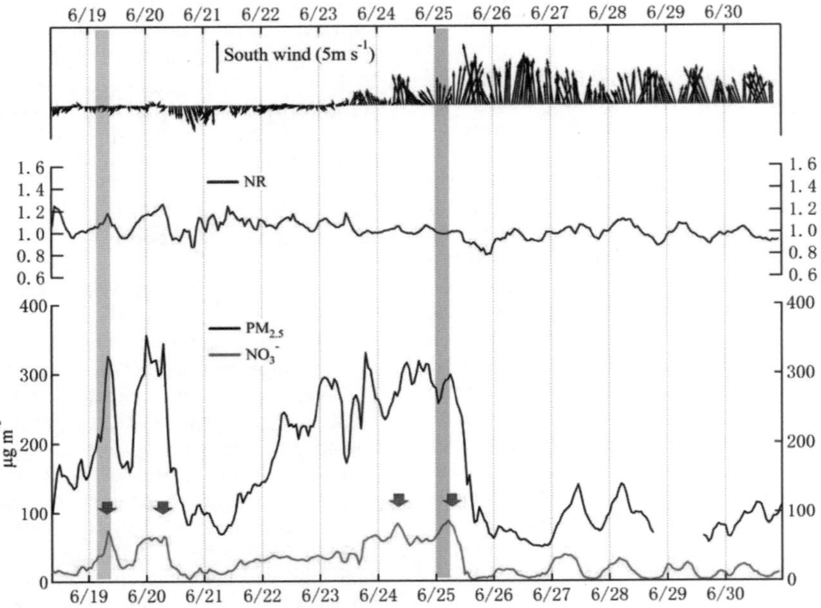

Fig. 2. Time series of concentrations of NO_3^-, $PM_{2.5}$, wind direction, wind speed, and NR (NR = $[NH_4^+]/(2[SO_4^{2-}] + [NO_3^-])$) at Yucheng site from June 18 to June 30, 2013. Four serious polluted cases of fine particulate nitrate were noted by red arrows. The green and blue shadows show two typical cases and will be discussed in 3.2 and 3.3, respectively. (For interpretation of the references to colour in this figure legend, the reader is referred to the web version of this article.)

tracked well with that of $PM_{2.5}$ (slope = 0.19 and R^2 = 0.72 for the data during periods from 20:00 to 7:00 the next morning).

It is well-known that the formation of fine particulate nitrate largely depends on meteorological conditions and the mixing ratios of precursors and oxidants. During the measurement period, the wind direction and speed were relatively stable, with a low northerly wind speed (1.5 m s^{-1} on average) before June 23 and high southerly wind speed (4.2 m s^{-1} on average) after June 24.

At the Yucheng site, the NH_3 concentration was notably high. The highest, average, and night-time average (20:00–7:00 the next morning) concentrations of NH_3 at the Yucheng site were 120.7, 34.9, and 34.3 ppbv, respectively, much higher than the summer mean value (13.9 ppbv) which was detected at a rural site nearby several swine production facilities in eastern North Carolina, USA (Robargea et al., 2002). During the night-time, more than 59.5% of the NH_3 concentrations were above 30 ppbv and the average value of NR reached 1.09. In addition, the O_3 was also observed at high levels. The maximum, average and night-time average concentrations of O_3 were 175.5, 65.4, and 41.9 ppbv, respectively. During the night-time, more than 65.9% of the ozone concentrations exceeded 40 ppbv. The detailed effects of NH_3 and O_3 on night-time fine particulate nitrate formation will be discussed later.

3.1.3. Diurnal variation of fine particulate nitrates

The average diurnal variations of concentrations of fine particulate nitrate, HNO_3, NH_3, NO_2, O_3, and the $NO_3^-/PM_{2.5}$ ratio from June 18 to June 30, 2013 are shown in Fig. 3. The fine particulate NO_3^- concentration and $NO_3^-/PM_{2.5}$ ratio generally increased gradually during the night-time and decreased during the daytime. The average concentration of fine particulate NO_3^- showed a peak of 34.7 $\mu g\ m^{-3}$ at 7:00 in the early morning, accounting for 22.3% of $PM_{2.5}$. The valley concentration of 12.3 $\mu g\ m^{-3}$ appeared at 16:00, representing a fraction of 9.1% of $PM_{2.5}$.

NO_2, the major primary precursor of nitrate, showed a peak concentration of 23.9 ppbv at 9:00 in the morning. During the night-time, the NO_2 concentration remained high, only a little lower than that during the daytime. The valley concentration of NO_2 reached 14.2 ppbv, appearing at 23:00. Gaseous HNO_3 is readily converted to particulate nitrate via reaction with NH_3 and/or heterogeneous uptake on surfaces of alkaline aerosols or small

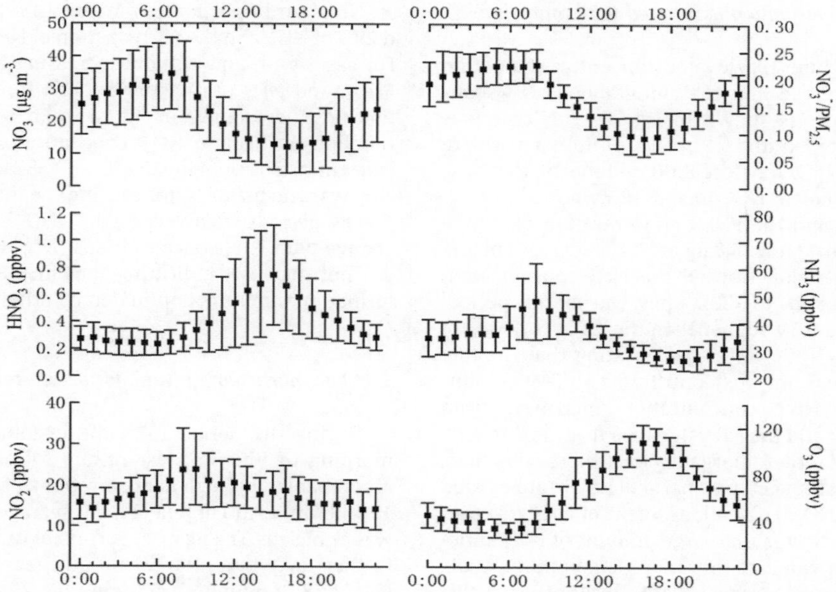

Fig. 3. Average diurnal variations of concentrations of fine particulate nitrates, HNO_3, NH_3, NO_2, O_3, and $NO_3^-/PM_{2.5}$ ratio at Yucheng for the period from June 18 to June 30, 2013. The standard deviations are presented as the error bars.

droplets. The HNO_3 concentration was relatively low, showing a maximum average concentration of 0.75 ppbv at 15:00. Surprisingly, the NH_3 concentration was quite high at the Yucheng site, exhibiting a peak of 47.9 ppbv appearing at 8:00 in the morning. O_3, one of the most important oxidants, exhibited a large concentration peak in the afternoon with maximum value of 107.8 ppbv at 16:00. This very high O_3 concentration demonstrates that the study region experienced serious photochemical pollution. The night-time O_3 concentration remained high (e.g., 44.6 ppbv at 0:00), indicating a relatively strong oxidation capacity, even at night.

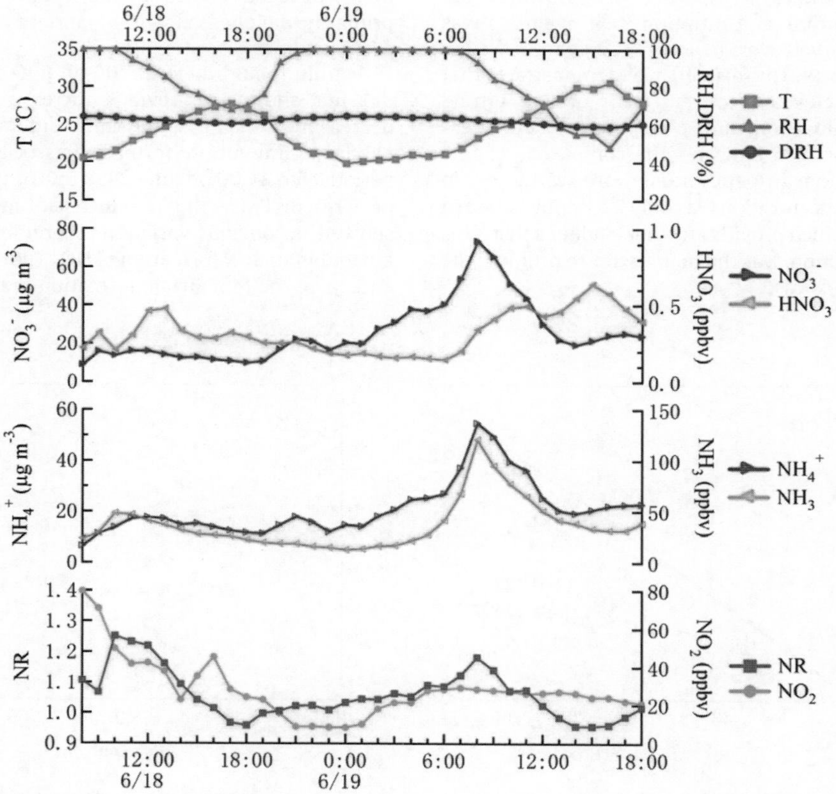

Fig. 4. Time series of meteorological data and the concentrations of air pollutants related to nitrate formation on June 18 and June 19.

3.2. Fine particulate nitrate formation associated with ammonia

In the early morning of June 19, elevated concentrations of fine particulate NO_3^- were observed and the simultaneous NH_3 concentrations were very high (as shown in Fig. 4). This case was selected to analyse the detailed impact of NH_3 on fine particulate nitrate formation. Within the 4 h before 8:00 on June 19, the mass concentration of fine particulate NO_3^- increased by a factor of 2.0 from 37.0 to 73.7 μg m^{-3} and the $PM_{2.5}$ concentration rose by a factor of 1.6 from 203.8 μg m^{-3} to 325.8 μg m^{-3}. The increase in fine particulate nitrate concentration tracked the NH_3 concentration well, which rose from 27.4 ppbv to 120.7 ppbv. During this period, the $\Delta NO_3^-/\Delta PM_{2.5}$ ratio was 0.30, higher than the $\Delta SO_4^{2-}/\Delta PM_{2.5}$ of 0.27 and the $\Delta NH_4^+/\Delta PM_{2.5}$ ratio of 0.24, suggesting that the fine particulate nitrate formation obvious contributed to $PM_{2.5}$ pollution in this case. The NH_4^+ concentration increased from 25.3 μg m^{-3} to 54.9 μg m^{-3} and the NR value rose from 1.09 to 1.18, but the concentrations of other inorganic ions were very low, indicating that the newly produced fine particulate nitrate existed mainly in the form of NH_4NO_3. The enhancement of NH_4^+ together with the increase in NR indicates that large amount of NH_3 anticipated the fine nitrate formation and thus played an important role in the rapid accumulation of fine particulate nitrate. During this period, the ambient RH was substantially higher than the DRH (deliquescence relative humidity of NH_4NO_3, DRH $= e^{((723.7/T)+1.6954)}$, Lin et al., 2007) during this period. Therefore, the fine particulate nitrate mainly dissolved in small droplets or in water on aerosol surfaces.

Assuming that solid or aqueous NH_4NO_3 in aerosols was in equilibrium with its gaseous precursors of HNO_3 and NH_3, the fine nitrate quantitative relationship between concentrations of them can be expressed by a series of equations (Lin and Cheng, 2007). In this study, the concentration of NH_4NO_3 was also calculated based on the observed concentrations of HNO_3 and NH_3. However, the average calculated fine nitrate concentration (0.58 μg m^{-3}) was much lower than that observed average value (51.2 μg m^{-3}) within the 4 h before 8:00 in June 19. This discrepancy demonstrates that the ammonia-associated elevated concentration of fine nitrate possibly formed in other locations and transported to our measurement site. This finding also indicates the complexity of fine nitrate formation, and further in-depth studies are warranted. In addition, the average O_3 concentration was only 22.1 ppbv between 5:00 and 8:00 in June 19, which provides further evidence that NH_3, not the photochemical reaction, was the main factor to promote the fine nitrate formation in this case.

NR is an important parameter to represent the relative abundance of NH_4^+ and the participation of NH_3 in fine nitrate formation. For cases with and without increasing NR, scatter plots of NO_3^- in $PM_{2.5}$ and NH_3 during the periods from 20:00 to 7:00 the next morning were shown in Fig. 5a and b, respectively. When NR rose, the fine particulate NO_3^- concentration tracked well with NH_3 concentration (slope = 0.82, $R^2 = 0.582$, six nights), suggesting that NH_3 was the major factor causing the fine nitrate formation during these periods. However, the NO_3^- concentrations did not change with NH_3 concentrations ($R^2 = 0.003$, six nights) if the NR did not increase, indicating that in these nights the fine nitrate formation might be dominated by other factors.

3.3. Fine particulate nitrate formation related to ozone

During the period from the evening of June 24 to the early morning of June 25, fast nitrate formation was observed to be accompanied by high concentrations of O_3 (see Fig. 6). Within the 10 h before 21:00 in June 24, the average concentration of fine NO_3^- was as high as 57.4 μg m^{-3}. Surprisingly, the fine NO_3^- concentration further increased after 21:00 and reached to an extremely high level of 87.2 μg m^{-3} at 5:00 in June 25. When the maximum NO_3^- concentration appeared, the NO_3^- accounted for a very large fraction of 29.9% in $PM_{2.5}$, ~50% higher than SO_4^{2-}. The fine NO_3^- concentration increased at a fast rate. The average and the maximum rates of increase of fine NO_3^- within the 8 h before 5:00 were 3.7 μg m^{-3} h^{-1} and 7.1 μg m^{-3} h^{-1} (appeared at 2:00 in June 25), respectively. At the same time, high concentrations of night-time O_3 were observed. The O_3 concentration was in a high level of 106.5 ppbv at 18:00 in June 24 and was gradually dropped to 41.8 ppbv at 6:00 in June 25, demonstrating strong oxidation capacity in the whole night. Moreover, the ambient RH kept obviously higher than the DRH during the whole night, indicating that nitrate primarily deliquesced on the surfaces of wet aerosols or small droplets.

Within the 8 h before 5:00 on June 25, the NH_3 concentration did not show any obvious increase and the NR value kept decreasing, which is significantly different from the other cases of enhanced fine nitrate formation associated with NH_3. During the period from 21:00 on June 24 to 5:00 on June 25, the concentration of SO_4^{2-} in $PM_{2.5}$ did not increase and the NH_4^+ concentration showed no obvious variation. Therefore, the decrease of NR was attributed to new formation of NO_3^- (not SO_4^{2-}) and the NH_3 did not anticipate the fine nitrate formation in this case.

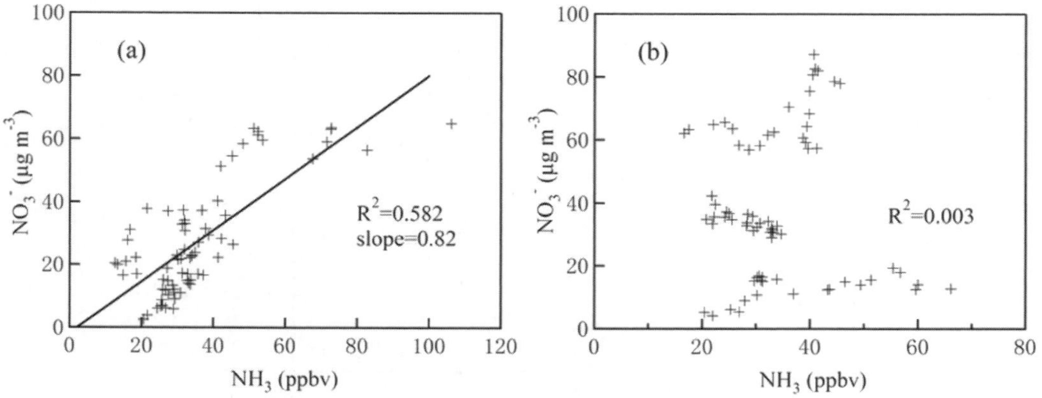

Fig. 5. Scatter plots of hourly concentrations of NO_3^- and NH_3 during periods from 20:00 to 7:00 in the next morning for cases (a) with and (b) without increasing NR. The data were selected from June 18 to June 20 and from June 26 to June 30 for the left plot and from June 20 to 26 for the right plot.

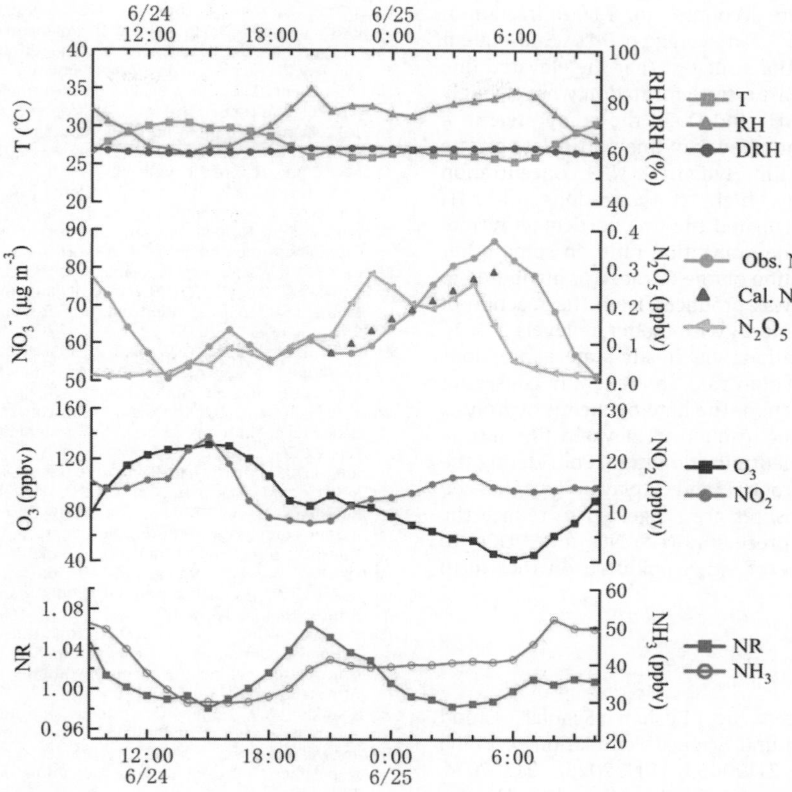

Fig. 6. Time series of meteorological data and the concentrations of air pollutants related to nitrate formation from June 24 to June 25.

In this study, the N_2O_5 concentrations were calculated by using steady-state expressions from 10:00 on June 24 to 9:00 on June 25 and are shown in Fig. 6. The calculated N_2O_5 concentrations also exhibited a large peak from 21:00 on June 24 to 5:00 on June 25 with a maximum hourly concentration of 0.28 ppbv. The fine nitrate concentration of 57.4 μg m^{-3} at 21:00 on June 24 was selected as the initial concentration of the calculated concentration within 8 h after 21:00 on June 24. Hence the calculated NO_3^- concentration in the 8 h period (as shown in Fig. 6) can be calculated by adding the increase in fine nitrate concentration from N_2O_5 hydrolysis during each hour. The calculated NO_3^- concentration of 79.0 μg m^{-3} at 5:00 on June 25 was very close to the observed NO_3^- peak concentration of 87.2 μg m^{-3}. Furthermore, the average increase rate of the calculated NO_3^- from the N_2O_5 hydrolysis was 2.7 μg m^{-3} h^{-1}, close to that of the observed NO_3^- rate (3.7 μg m^{-3} h^{-1}). These results suggest that the heterogeneous hydrolysis of N_2O_5 on wet aerosol surfaces was the major formation pathway for the fast fine nitrate formation in this case.

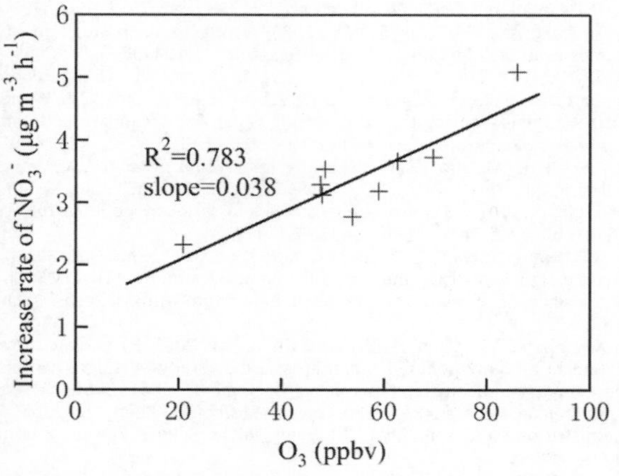

Fig. 7. Scatter plot of increase rate of fine NO_3^- and the average night-time O_3 concentration during the periods with fine nitrate formation from June 18 to June 20 and from June 23 to June 30.

To further understand the impact of O_3 on night-time fine nitrate formation, the data with high night-time NO_3^- concentration peaks (excluding days when it was raining) were selected to perform a correlation analysis. Fig. 7 shows the scatter plot of the average increase rate of fine NO_3^- versus the average O_3 concentration during periods with obvious night-time fine nitrate formation. The increase rate of NO_3^- strongly correlated and rose positively with the average O_3 concentration (R^2 = 0.783, slope = 0.038). This suggests that hydrolysis of N_2O_5 is one of the major formation pathways of fine nitrate and significantly contributed to the elevated fine nitrate concentrations at night-time during summer in rural areas in the North China Plain.

4. Summary and conclusions

Online measurements of fine particulate nitrate, related air pollutants and meteorological parameters were conducted from June 18 to June 30, 2013 at a rural site on the North China Plain, where a large amount of ammonia was emitted, and severe photochemical pollution frequently appeared during the summer. During the measurement period, notably high levels of fine nitrate were usually observed from the evening to the next morning, with a maximum hourly concentration of 87.2 μg m^{-3} appearing at 5:00

on June 25. The observed nitrate accounted for a large fraction of $PM_{2.5}$, ranging from 1.4 % to 45 % (an average of 14%). Subsequent case analyses and data calculations suggest that the elevated fine nitrate concentration and the fast nitrate formation were strongly related to the high levels of NH_3 and O_3 in the study area. In a number of cases, the concentration peak of fine particulate nitrate appeared exactly at the same time when the NH_3 concentration exhibited a peak. Extremely high concentrations of NH_3 (120.7 ppbv in maximum) anticipated the fine particulate nitrate formation and led to a rising neutralization ratio. In some other cases, the fast secondary formation of fine nitrate was attributed to the hydrolysis of N_2O_5 which was produced from the reaction of high concentrations of night-time O_3 with moderate levels of NO_2. The calculated N_2O_5 concentrations via steady-state expressions and the resulting nitrate formation rate are generally consistent with the observed nitrate formation. The heterogeneous hydrolysis of N_2O_5 on wet aerosol surfaces resulted in elevated fine nitrate concentrations and a reduced neutralization ratio. Considering the important roles that high levels of NH_3 and O_3 play in the enhanced nitrate formation, control measures are suggested to reduce the emissions of NH_3 and ozone precursors (i.e., NO_x and VOCs) to address the serious occurrence of $PM_{2.5}$ pollution on the North China Plain.

Acknowledgements

This work was supported by Taishan Scholar Grand (ts20120552), the National Natural Science Foundation of China (Nos. 41375126, 41275123, 21190053, 21177025, 21307074, 21177073), the Fundamental Research Funds of Shandong University (No. 2014GN010), the Shanghai Science and Technology Commission of Shanghai Municipality (Nos. 13XD1400700, 12DJ1400100), the Strategic Priority Research Program of the Chinese Academy of Sciences (No. XDB05010200), Jiangsu Collaborative Innovation Center for Climate Change, Key Project of Shandong Provincial Environmental Agency (2006045) and Special Research for Public-Beneficial Environment Protection (201009001-1).

References

Aldener, M., Brown, S.S., Stark, H., Williams, E.J., Lerner, B.M., Kuster, W.C., Goldan, P.D., Quinn, P.K., Bates, T.S., Fehsenfeld, F.C., Ravishankara, A.R., 2006. Reactivity and loss mechanisms of NO_3 and N_2O_5 in a polluted marine environment: result from in situ measurements during New England Air Quality Study 2002. J. Geophys. Res. 111, D23S73.

Atkinson, R., Arey, J., 2003. Atmospheric degradation of volatile organic compounds. Chem. Revel. 103, 4605–4638.

Calvert, J.G., Stockwell, W.R., 1983. Acidic generation in the troposphere by gas phase chemistry. Environ. Sci. Technol. 17, 428–443.

Carslaw, D.C., Rhys-Tyler, G., 2013. New insights from comprehensive on-road measurements of NO_x, NO_2 and NH_3 from vehicle emission remote sensing in London, UK. Atmos. Environ. 81, 339–347.

Du, H.H., Kong, L.D., Cheng, T.T., Chen, J.M., Du, J.F., Li, L., Xia, X.G., Leng, C.P., Huang, G.H., 2011. Insights into summertime haze pollution events over Shanghai based on online water-soluble ionic composition of aerosols. Atmos. Environ. 45, 5131–5137.

Evans, M.J., Jacob, D.J., 2005. Impact of new laboratory studies of N_2O_5 hydrolysis on global model budgets of tropospheric nitrogen oxides, ozone, and OH. Geophys. Res. Lett. 32, 1–4.

Feng, Y., Penner, J.E., 2007. Global modeling of nitrate and ammonium: interaction of aerosols and tropospheric chemistry. J. Geophys. Res. 112, D1.

Gao, X.M., Yang, L.X., Cheng, S.H., Gao, R., Zhou, Y., Xue, L.K., Shou, Y.P., Wang, J., Wang, X.F., Nie, W., Xu, P.J., Wang, W.X., 2011. Semi-continuous measurement of water-soluble ions in $PM_{2.5}$ in Jinan, China: temporal variations and source apportionments. Atmos. Environ. 45, 6048–6056.

Guo, S., Hu, M., Wang, Z.B., Slanina, J., Zhao, Y.L., 2010. Size-resolved aerosol water-soluble ionic compositions in the summer of Beijing: implication of regional secondary formation. Atmos. Chem. Phys. 10, 947–959.

Hu, D.W., Chen, J.M., Ye, X.N., Li, L., Yang, X., 2011. Hygroscopicity and evaporation of ammonium chloride and ammonium nitrate: relative humidity and size effects on the growth factor. Atmos. Environ. 45, 2349–2355.

Khan, M.F., Shirasuna, Y., Hirano, K., Masunaga, S., 2010. Characterization of $PM_{2.5}$, $PM_{2.5-10}$ and PMN_{10} in ambient air, Yokohama, Japan. Atmos. Res. 96, 159–172.

Lei, H., Wuebbles, D.J., 2013. Chemical competition in nitrate and sulfate formations and its effect on air quality. Atmos. Environ. 80, 472–477.

Lin, Y.C., Cheng, M.T., 2007. Evaluation of formation rates of NO_2 to gaseous and particulate nitrate in the urban atmosphere. Atmos. Environ. 41, 1903–1910.

Liu, X.J., Zhang, Y., Han, W.X., Tang, A.H., Shen, J.L., Cui, Z.L., Peter, V., Jan, W.E., Keith, G., Peter, C., Andreas, F., Zhang, F.S., 2013. Enhanced nitrogen deposition over China. Nature 28, 459–463.

Luo, X.S., Liu, P., Tang, A.H., Liu, J.Y., Zong, X.Y., Zhang, Q., Kou, C.L., Zhang, L.J., Fowler, D., Fangmeier, A., Christie, P., Zhang, F.S., Liu, X.J., 2013. An evaluation of atmospheric Nr pollution and deposition in North China after the Beijing Olympics. Atmos. Environ. 74, 209–216.

Meng, Z.Y., Xu, X.B., Wang, T., Zhang, X.Y., Yu, X.L., Wang, S.F., Lin, W.L., Chen, Y.Z., Jiang, Y.A., An, X.Q., 2010. Ambient sulphur dioxide, nitrogen dioxide, and ammonia at ten background and rural sites in China during 2007–2008. Atmos. Environ. 44, 625–2631.

Mentel, T.F., Bleilebens, D., Wahner, A., 1996. A study of nighttime nitrogen oxide oxidation in a large reaction chamber — the fate of NO_2, N_2O_5, HNO_3, and O_3 at different humidities. Atmos. Environ. 30, 4007–4020.

Mozurkewich, M., 1993. The dissociation constant of ammonium nitrate and its dependence on temperature, relative humidity and particle size. Atmos. Environ. 27, 261–270.

Osthoff, H.D., Sommariva, R., Baynard, T., Pettersson, A., Williams, E.J., Lerner, B.M., Roberts, J.M., Stark, H., Goldan, P.D., Kuster, W.C., Bates, T.S., Coffman, D., Ravishankara, A.R., Brown, S.D., 2006. Observation of daytime N_2O_5 in the marine boundary layer during New England Air Quality Study–Intercontinental transport and chemical transformation 2004. J. Geophys. Res. 111, D23S14.

Pathak, R.K., Wu, W.S., Wang, T., 2009. Summertime $PM_{2.5}$ ionic species in four major cities of China: nitrate formation in an ammonia-deficient atmosphere. Atmos. Chem. Phys. 9, 1711–1722.

Pathak, R.K., Wang, T., Wu, W.S., 2011. Nighttime enhancement of $PM_{2.5}$ nitrate in ammonia-poor atmospheric conditions in Beijing and Shanghai: plausible contributions of heterogeneous hydrolysis of N_2O_5 and HNO_3 partitioning. Atmos. Environ. 45, 1183–1191.

Ravishankara, A.R., 1997. Heterogeneous and multiphase chemistry in the troposphere. Science 276, 1058–1065.

Richter, A., Burrows, J.P., Nuß, H., Granier, C., Niemeier, U., 2005. Increase in tropospheric nitrogen dioxide over China observed from space. Nature 437, 129–132.

Robargea, W.P., Walkerb, J.T., McCullochc, R.B., Murrayd, G., 2002. Atmospheric concentrations of ammonia and ammonium at an agricultural site in the southeast United States. Atmos. Environ. 36, 1661–1674.

Saunder, S.P., Golden, D., Kurylo, M., Moortgat, G., Wine, P., Ravishankara, A., Kolb, C., Molina, M., Finlayson-Pitts, B., Huie, R., 2006. Chemical Kinetics and Photochemical Data for Use in Atmospheric Studies. Evaluation Number 17. Jet Propulsion Laboratory, National Aeronautics and Space Administration, Pasadena, CA.

Shen, J.L., Liu, X.J., Zhang, Y., Fangmeier, A., Goulding, K., Zhang, F.S., 2011. Atmospheric ammonia and particulate ammonium from agricultural sources in the North China Plain. Atmos. Environ. 45, 5033–5041.

Shi, Y., Chen, J.M., Hu, D.W., Wang, L., Yang, X., Wang, X.M., 2014. Airborne submicron particulate (PM_1) pollution in Shanghai, China: chemical variability, formation/dissociation of associated semi-volatile components and the impacts on visibility. Sci. Total Environ. 473–474, 199–206.

Squizzato, S., Masiol, M., Innocente, E., Pecorari, E., Rampazzo, G., Pavoni, B., 2012. A procedure to assess localand long-range transport contributions to $PM_{2.5}$ and secondary inorganic aerosol. J. Aerosol Sci. 46, 64–76.

Wahner, A., Mentel, T.F., Sohn, M., 1998. Gas phase reaction of N_2O_5 with water vapor: Importance of heterogeneous hydrolysis of N_2O_5 and surface desorption of HNO_3 in a large Teflon chamber. Geophys. Res. Lett. 25, 2169–2172.

Wang, T., Ding, A.J., Gao, J., Wu, W.S., 2006. Strong ozone production in urban plumes from Beijing, China. Geophys. Res. Lett. 33, L21806.

Wang, T., Nie, W., Gao, J., Xue, L.K., Gao, X.M., Wang, X.F., Qiu, J., Poon, C.N., Meinardi, S., Blake, D., Wang, S.L., Ding, A.J., Chai, F.H., Zhang, Q.Z., Wang, W.X., 2010. Air quality during the 2008 Beijing Olympics: secondary pollutants and regional impact. Atmos. Chem. Phys. 10, 7603–7615.

Wang, X.F., Chen, J.M., Sun, J.F., Li, W.J., Yang, L.X., Wen, L., Wang, W.X., Wang, X.M., Collett Jr., J.L., Shi, Y., Zhang, Q.Z., Hu, J.T., Yao, L., Zhu, Y.H., Sui, X., Sun, X.M., Mellouki, A., 2014a. Severe haze episodes and seriously polluted fog water in Ji'nan, China. Sci. Total Environ. 493, 133–137.

Wang, X.F., Wang, T., Yan, C., Tham, Y.J., Xue, L.K., Xu, Z., Zha, Q.Z., 2014b. Large daytime signals of N_2O_5 and NO_3 inferred at 62 amu in a TD-CIMS: chemical interference or a real atmospheric phenomenon? Atmos. Meas. Technol. 7, 1–12.

Wang, X.F., Zhang, Y.P., Chen, H., Yang, X., Chen, J.M., 2009. Particulate nitrate formation in a highly polluted urban area: a case study by single-particle mass spectrometry in Shanghai. Environ. Sci. Technol. 43, 3061–3066.

Wang, Y., Zhuang, G.S., Tang, A., Yuan, H., Sun, Y., Chen, S., Zheng, A., 2005. The ion chemistry and the source of $PM_{2.5}$ aerosol in Beijing. Atmos. Environ. 39, 3771–3784.

Ye, B., Ji, X., Yang, H., Yao, X., Chan, C.K., Cadle, S.H., Chan, T., Mulawa, P.A., 2003. Concentration and chemical compositions of $PM_{2.5}$ in Shanghai for 1-year period. Atmos. Environ. 37, 499–510.

Ye, X.N., Ma, Z., Zhan, J.C., Du, H.H., Chen, J.M., Chen, H., Yang, X., Gao, W., Geng, F.H., 2011. Important role of ammonia on haze formation in Shanghai. Environ. Res. Lett. 6, 24019–24023.

Zhang, F.W., Xu, L.L., Chen, J.S., Chen, X.Q., Niu, Z.C., Lei, T., Li, C.M., Zhao, J.P., 2013. Chemical characteristics of $PM_{2.5}$ during haze episodes in the urban of Fuzhou, China. Particuology 11, 264–272.

Zhang, H., Shen, Z., Wei, X., Zhang, M., Li, Z., 2012. Comparison of optical properties of nitrate and sulfate aerosol and the direct radiative forcing due to nitrate in China. Atmos. Res. 113, 113–125.

Zhang, Q., Streets, D.G., Carmichael, G.R., He, K.B., Huo, H., Kannari, A., Klimont, Z., Park, I.S., Reddy, S., Fu, J.S., Chen, D., Duan, L., Lei, Y., Wang, L.T., Yao, Z.L., 2009. Asian emissions in 2006 for the NASA INTEX-B mission. Atmos. Chem. Phys. 9, 5131–5153.

Zhao, B., Wang, P., Ma, J.Z., Zhu, S., Pozzer, A., Li, W., 2012. A high-resolution emission inventory of primary pollutants for the Huabei region, China. Atmos. Chem. Phys. 12, 481–501.

Atmospheric Environment 46 (2012) 516–527

Contents lists available at ScienceDirect

Atmospheric Environment

journal homepage: www.elsevier.com/locate/atmosenv

Formation of secondary organic carbon and cloud impact on carbonaceous aerosols at Mount Tai, North China

Zhe Wang [a], Tao Wang [a,b,c,*], Jia Guo [b], Rui Gao [a], Likun Xue [a,b], Jiamin Zhang [b], Yang Zhou [a], Xuehua Zhou [a], Qingzhu Zhang [a,**], Wenxing Wang [a,c]

[a] Environment Research Institute, Shandong University, Ji'nan, Shandong 250100, PR China
[b] Department of Civil and Structural Engineering, The Hong Kong Polytechnic University, Hong Kong, PR China
[c] Chinese Research Academy of Environmental Sciences, Beijing 100012, PR China

ARTICLE INFO

Article history:
Received 9 May 2011
Received in revised form
18 July 2011
Accepted 8 August 2011

Keywords:
Secondary organic aerosol (SOA)
In-cloud SOA formation
Cloud scavenging
Carbonaceous aerosol
Mount Tai (Mt. Tai)
Multiple linear regression

ABSTRACT

Carbonaceous aerosols measured at Mount Tai in north China in 2007 were further examined to study the formation of secondary organic carbon (SOC) and the impact of clouds on carbonaceous species. A constrained EC-tracer method and a multiple regression model showed excellent agreement in estimating SOC concentration. The average concentrations of non-volatile and semi-volatile SOC (SOC_{NV} and SOC_{SV}) were 2.61, 5.58 $\mu g\,m^{-3}$ in spring and 2.81, 10.44 $\mu g\,m^{-3}$ in summer. The total SOC accounted for 57.3% and 71.2% of total organic carbon in spring and summer, respectively, indicating the presence of high loading of secondary organic aerosol (SOA) in the North China Plain. The fraction of SOC_{NV} increased with photochemical age (as indicated by NO_x/NO_y ratios) of air mass, whereas SOC_{SV} was also influenced by the dynamic equilibrium between formation and sink. Significant scavenging by clouds of non-volatile organic carbon (OC_{NV}) and elemental carbon (EC) was observed, whereas semi-volatile organic carbon (OC_{SV}) concentrations increased during clouds, suggesting substantial SOA formation through aqueous-phase reactions in clouds. A mass balance model was proposed to quantify the scavenging coefficients for OC_{NV}, EC and formation rates of OC_{SV} in clouds. The scavenging coefficient constant of EC (K_{EC}) varied from 0.11 to 0.90 h^{-1}, and was higher than that of OC_{NV} ($K_{NV\text{-}OC}$: 0.07–0.55 h^{-1}), implying internal mixing of EC with more hygroscopic species. The formation rate constant ($J_{SV\text{-}OC}$) and sink constant ($S_{SV\text{-}OC}$) of OC_{SV} ranged from 0.09 to 1.39 h^{-1} and 0.001 to 1.07 h^{-1}, respectively. These field derived parameters could be incorporated into atmospheric models to help close the gap between predicted and observed SOA loadings in the atmosphere.

1. Introduction

Carbonaceous aerosols, including organic carbon (OC) and elemental carbon (EC), account for a substantial fraction (20–80%) of fine particulate matter, affecting human health, visibility, and global climate (Seinfeld and Pandis, 1998). EC is believed to be exclusively from primary sources, involving combustion in transport, industrial processes, and biomass burning. OC includes primary emission from these same sources (i.e., primary organic carbon, POC), and also could be formed through atmospheric oxidation of reactive organic gases with subsequent gas-to-particle conversion or aqueous-phase reactions (i.e., secondary organic carbon, SOC) (Seinfeld and Pandis, 1998; Blando and Turpin, 2000).

The apportionment of atmospheric OC concentrations to primary or secondary source has been problematic (e.g., Turpin and Huntzicker, 1991; Pachon et al., 2010), and there is no direct measurement approach that can definitively differentiate POC and SOC. The difficulties arise from large spatial and temporal variations of combination sources, incomplete information on precursors and source profiles, the large number of organic species which have not been totally identified or quantified, and the complexity of formation mechanisms for SOC (Blanchard et al., 2008; Chou et al., 2010).

Nonetheless, several indirect approaches have been developed to estimate the amount of SOC. One technique that relies on the use of tracer species is the EC-tracer method (Turpin and Huntzicker, 1991). EC is a good tracer of primary combustion-generated carbon and can be used to estimate POC co-emitted from these

* Corresponding author. Department of Civil and Structural Engineering, The Hong Kong Polytechnic University, Hong Kong, PR China. Tel.: +852 2766 6059; fax: +852 2334 6389.
** Corresponding author. Tel.: +86 531 88364435; fax: +86 531 88369788.
E-mail addresses: cetwang@polyu.edu.hk (T. Wang), zqz@sdu.edu.cn (Q. Zhang).

1352-2310/$ – see front matter © 2011 Elsevier Ltd. All rights reserved.
doi:10.1016/j.atmosenv.2011.08.019

sources. A second approach is the multiple regression method, involving tracers of both primary emission and photochemical activity (Blanchard et al., 2008). The third approach is receptor models, such as Chemical Mass Balance (CMB) and Positive Matrix Factorization (PMF) methods (Hu et al., 2010; Pachon et al., 2010). Additional methods include chemical transport models and non-reactive transport models (Strader et al., 1999; Yu et al., 2004). Some recent studies have also used organic molecular markers and specific compounds to estimate the SOC formation, such as biogenic and anthropogenic SOA tracer, OC/CO ratio and carbon isotopic method (Blanchard et al., 2008; Hu et al., 2008; Fu et al., 2010).

There has been rich literature on the interaction between cloud/fog and inorganic species, but studies on the processing of carbonaceous aerosols by cloud/fog are still limited (Herckes et al., 2007; Collett et al., 2008). The uncertainties in the parameterization of the in-cloud scavenging process can lead to significant differences in predicted aerosol concentrations by models. For example, the simulation by the aerosol-climate model ECHAM5-HAM underestimated BC concentrations observed by aircraft by up to two orders of magnitude (Croft et al., 2010). Besides photochemical oxidation and oligomer/polymer formation processes, aqueous-phase reactions in cloud/fog processing, which leads to substantial sulfate formation, could also be important contributors to SOA formation (Blando and Turpin, 2000). This heterogeneous process has recently emerged as an important SOA pathway and drawn increasing interest.

Several laboratory and modeling studies have demonstrated the formation of low-volatility products and potential SOA through aqueous-phase processing of organic compounds (e.g., glyoxal, methylglyoxal) in clouds (Hallquist et al., 2009; De Haan et al., 2011). Model calculations by Fu et al. (2008) have suggested that in-cloud formed SOA is comparable in magnitude to the SOA formed through the traditional pathway. However, the uncertainties of these estimates are quite large, and the mechanism of aqueous-phase reactions in SOA formation is not well understood. Clouds can also remove atmospheric aerosols at rates dependant on chemical composition, but there have been few observational studies on the scavenging of carbonaceous aerosols and SOA formation in clouds.

As part of a China's National Basic Research project on acid rain, two intensive studies were conducted in spring and summer of 2007 at the summit of Mount Tai (Mt. Tai) in the North China Plain, which is one of the greatest anthropogenic emission regions in the country. Mt. Tai, as the highest mountain in this area, provides a unique place to study photochemistry, cloud process, gas–particle–cloud interactions and their impacts on regional scales. High concentrations of non-volatile organic carbon (OC_{NV}) and semi-volatile organic carbon (OC_{SV}) were observed during the campaigns, and a principal component analysis has indicated substantial SOA formation through the photochemical process and cloud processing (Wang et al., 2011b).

The present study further analyzes carbonaceous and other relevant data to quantify the formation and cloud scavenging of carbonaceous aerosols. Both constrained EC-tracer and multiple regression methods were used to estimate SOC formation at Mt. Tai, SOC with photochemical age was characterized, cloud scavenging of carbonaceous aerosols was examined, and OC_{SV} formation through aqueous-phase reactions in clouds was investigated. We also propose models to determine the cloud scavenging coefficients and OC_{SV} formation rate in clouds.

2. Methodology

2.1. Measurement site and instrumentation

The measurement site was located in a meteorological observatory on the summit of Mt. Tai (36.25 N, 117.10E, 1532.7 m asl) in Shandong Province of China. Intensive campaigns were performed from March 22 to April 24 in spring and from June 16 to July 20 in summer. More detailed description of the site and the weather conditions during campaigns were provided in Wang et al. (2011b).

Hourly carbonaceous concentrations in $PM_{2.5}$ were measured by a semi-continuous OC/EC analyzer (Dual-oven model, Sunset Laboratory Inc.) equipped with an upstream parallel-plate organic denuder (Sunset Laboratory Inc.) and a backup carbon-impregnated filter (CIF). The principle and operation of this instrument has been previously described (Wang et al., 2011b). In brief, OC_{NV} and EC were sampled on a quartz filter in the first oven, and analyzed by the thermal-optical transmittance method (TOT) with a two-stage thermal procedure (250–450–620 °C in a He atmosphere and 500–700–830 °C in an oxidizing atmosphere (98% He + 2% O_2)). OC_{SV} was collected by CIF in the second oven and analyzed in an oxidizing atmosphere (98% He + 2% O_2) with a thermal program of 170–310 °C.

Tracer gases (including O_3, CO, SO_2, NO, NO_x and NO_y), gas-phase total peroxides, and real-time water soluble ions in $PM_{2.5}$ were also measured during the campaigns. The setup, precision and accuracies of these instruments have been described previously (Ren et al., 2009; Wang et al., 2010; Zhou et al., 2010). For details the reader is referred to the previous publications.

2.2. Empirical estimation methods for SOC concentration

2.2.1. Constrained EC-tracer method

The EC-tracer method estimates the SOC concentration from the relationship of POC and EC, which is assumed to have a fixed ratio in the study region. The POC is obtained by

$$(POC) = (OC/EC)_{pri}\times(EC) + (OC)_{nc} \tag{1}$$

where $(OC)_{nc}$ is the OC from non-combustion source, and $(OC/EC)_{pri}$ is the primary ratio of OC to EC corresponding to combustion sources. Then SOC can be estimated by

$$(SOC) = (OC)_{total}-(OC/EC)_{pri}\times(EC) - (OC)_{nc} \tag{2}$$

The non-volatile SOC (SOC_{NV}) and semi-volatile SOC (SOC_{SV}) can be separately determined by

$$(SOC_{NV}) = (OC_{NV})_{total}-(OC_{NV}/EC)_{pri}\times(EC) - (OC_{NV})_{nc} \tag{3}$$

$$(SOC_{SV}) = (OC_{SV})_{total}-(OC_{SV}/EC)_{pri}\times(EC) - (OC_{SV})_{nc} \tag{4}$$

The advantage of this method is that it is simple and straightforward requiring only the observed ambient concentrations of OC and EC, and thus it has received the widest application (e.g., Raman et al., 2008; Chou et al., 2010). However, there are also some potential problems with this approach, including the uncertainty of the measurement technique, variation of OC/EC emission ratio between sources, and the influence of other factors (e.g., meteorology condition) (Strader et al., 1999; Chu, 2005). The selection of the primary OC/EC ratio is crucial, which is typically determined from linear regression of observed OC and EC concentrations during the period when SOC is expected to be negligible.

Some researchers have used minimum or the lowest 10–20% OC/EC ratios as the primary OC/EC ratio (Cao et al., 2007; Ram et al., 2008; Chou et al., 2010). In the present study, advantage was taken of highly time resolved data obtained from various continuous instruments. The primary ratios were selected from the period when primary sources dominated and no cloud or rain was present. The constraints that disaggregate data characterized by low photochemical activity were defined as carbonaceous data corresponding to O_3 concentration, OC_{NV}/CO and OC_{SV}/CO ratios being below their respective 25th percentile in each season (specifically, 55 ppb, 0.012 and 0.007 in spring; 61 ppb, 0.008 and 0.017 in summer). The extracted data points during daytime and nighttime for both campaigns are shown in Fig. 1, and $(OC_{NV}/EC)_{pri}$, $(OC_{SV}/EC)_{pri}$, $(OC_{NV})_{nc}$ and $(OC_{SV})_{nc}$ were determined from these datasets through Deming regression, which takes into account errors in both coordinates (Chu, 2005).

2.2.2. Multiple regression method

The multiple regression method includes both tracers of primary emission and secondary activity to determine POC and SOC. EC and CO are good tracers for primary emissions, K^+ is a marker for biomass burning, O_3 is an indicator for photochemical production, and SO_4^{2-} represents additional oxidation activity and aqueous-phase oxidation (Blanchard et al., 2008; Pachon et al., 2010). In the present study, their approach was modified by substituting NO_3^- with NO_y (total odd reactive nitrogen; defined as the sum of NO_x and its oxidation products, including aerosol nitrate) as a tracer of SOC, because NO_y contains a large fraction of oxidation products and better represents the degree of photochemical air pollution together with ozone (Wang et al., 2010). Thus the regression equation for OC_{NV} and OC_{SV} can be expressed as follows:

$$(OC_{NV}\text{ or }OC_{SV}) = a + b \times (EC) + c \times (CO) + d \times (O_3) + e \times (NO_y) + f \times \left(SO_4^{2-}\right) + g \times \left(K^+\right) \quad (5)$$

where '*a*' through '*g*' are regression coefficients, and are independently derived from stepwise multiple regressions for OC_{NV} or OC_{SV}.

Primary and secondary OC_{NV} and OC_{SV} can be estimated respectively by

$$(POC_{NV}\text{ or }POC_{SV}) = a + b \times (EC) + c \times (CO) + g \times \left(K^+\right) \quad (6)$$

$$(SOC_{NV}\text{ or }SOC_{SV}) = d \times (O_3) + e \times (NO_y) + f \times \left(SO_4^{2-}\right) \quad (7)$$

To ensure that the sum of predicted non-volatile (or semi-volatile) POC and SOC does not exceed the measured total OC_{NV} (or OC_{SV}), the POC_{NV} and SOC_{NV} (or POC_{SV} and SOC_{SV}) are rescaled separately as (Pachon et al., 2010)

$$(POC_r) = \left(\frac{POC}{POC + SOC}\right) \times (OC) \quad (8)$$

$$(SOC_r) = \left(\frac{SOC}{POC + SOC}\right) \times (OC) \quad (9)$$

2.3. Photochemical age

NO_x/NO_y ratio is a good indicator of photochemical processing, and has been used to study the time evolution of organic aerosols during photochemical aging (Kleinman et al., 2008; Slowik et al., 2011). The photochemical age, defined as $-\log_{10}(NO_x/NO_y)$, has

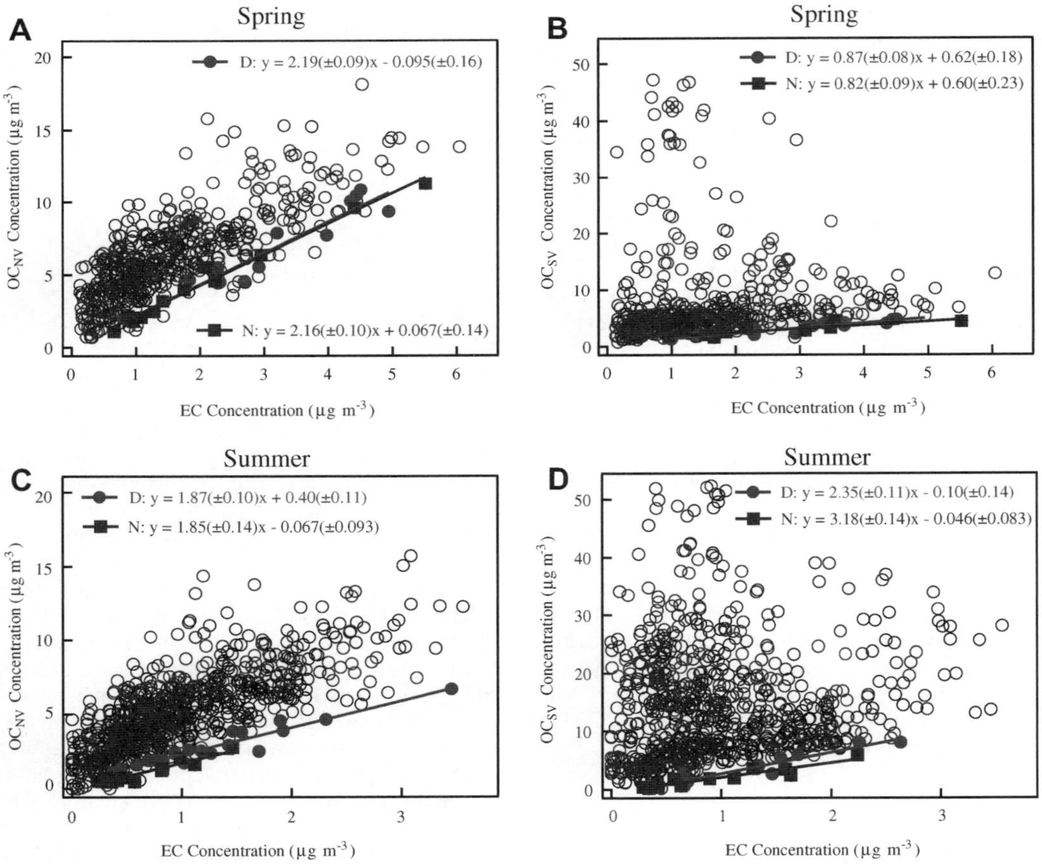

Fig. 1. Scatter plot of (A) OC_{NV} versus EC and (B) OC_{SV} versus EC in spring, and (C) OC_{NV} versus EC and (D) OC_{SV} versus EC in summer of 2007 at Mt. Tai, China. Open circles are the measured data during campaigns. Solid circles and cubes represent the selected dataset characterized as primary condition (see text) in daytime and nighttime, respectively. The solid lines are the Deming regression fits.

a value of 0 for fresh emissions ($NO_y = NO_x$) and a value of 1 when 90% of NO_x has been oxidized into NO_y (Slowik et al., 2011). Based on the oxidation mechanism of NO_x, a photochemical age of 0–1 corresponds to an estimated time scale of about 24 h (Kleinman et al., 2008).

2.4. Cloud processing model for aerosols

Cloud scavenging is an important removal mechanism for carbonaceous aerosols, and scavenging efficiency represents a first order approximation of the particle transfer rate into cloud droplets (Bae et al., 2009). The governing equation of the scavenging process of particles is

$$\frac{dC_i(t)}{dt} = -K_i \cdot C_i(t) \tag{10}$$

in which $C_i(t)$ is the concentration ($\mu g\,m^{-3}$) of aerosol-phase species i in atmosphere at time t, and K_i is the cloud scavenging coefficient constant (h^{-1}) for species i, which is a function of character of the clouds. Thus the aerosol-phase concentration of species i during a cloud event is related to the initial concentration ($C_{i,0}$) and decreases exponentially with time based on K_i by integrating Eq. (10),

$$C_i(t) = C_{i,0} \cdot \exp(-K_i \cdot t) \tag{11}$$

and the decay factor (DF), defined as $DF_i = C_i(t)/C_{i,0}$ ($C_i(t) < C_{i,0}$), is given by

$$DF_i(t) = \exp(-K_i \cdot t) \tag{12}$$

The above equations consider only the removal effect of clouds, not the formation process in the clouds for OC_{SV}, which has been observed at Mt. Tai (Wang et al., 2011b). A more general model can be expressed as

$$\frac{dC_i(t)}{dt} = Q_i - S_i \cdot C_i(t) \tag{13}$$

where Q_i is the formation rate ($\mu g\,m^{-3}\,h^{-1}$), and S_i is the sink constant (h^{-1}), which represents all the processes leading to decrease in species concentration, including scavenging, condensation, reaction, and evaporation. The time-dependent formation rate of OC_{SV} is hard to determine due to the complexity of and lack of organic precursors and relationship between precursors and particulate yield. Therefore, an open system that treats gas precursor concentrations as constant, which is a convenient assumption in aqueous-phase sulfur chemistry (Seinfeld and Pandis, 1998), is employed here and assumes constant formation rates. Then the species concentration can be obtained by integrating Eq. (13),

$$C_i(t) = \frac{Q_i}{S_i} - \left(\frac{Q_i}{S_i} - C_{i,0}\right) \cdot \exp(-S_i \cdot t) \tag{14}$$

and the growth factor (i.e., $GF_i = C_i(t)/C_{i,0}$, $C_i(t) > C_{i,0}$) is expressed as,

$$GF_i(t) = \frac{J_i}{S_i} - \left(\frac{J_i}{S_i} - 1\right) \cdot \exp(-S_i \cdot t) \tag{15}$$

in which J_i is the formation rate constant (h^{-1}), which is equal to Q_i divided by the initial species concentration. In cases with no or negligible formation source, the general equations (Eqs. (13)–(15)) will reduce to the specific scavenging equations (i.e., Eqs. (10)–(12)). The constants J_i and S_i are determined by fitting the Eq. (15) with the observed GF as a function of time of a cloud event.

3. Results and discussions

3.1. Estimation of SOA formation

3.1.1. Comparison of constrained EC-tracer and multiple regression methods

The relationships between OC_{NV}, OC_{SV} and EC concentrations observed at Mt. Tai during the spring and summer campaigns in 2007 are presented in Fig. 1. The Deming regression was performed on the extracted 'primary' dataset (cf. Section 2.2.1), and derived parameters are summarized in Table 1. The $(OC_{NV}/EC)_{pri}$ was 2.19 (2.16) and 1.87 (1.85) for daytime (nighttime) in spring and summer, respectively, and these values are consistent with that from many urban cities in southern and northern China (1.29–1.99 for summer, 2.13–2.81 for winter) (Cao et al., 2007), suggesting the regional representativeness of aerosols measured at Mt. Tai. The $(OC_{NV}/EC)_{pri}$ here is also comparable to most rural sites in Asia and US, as summarized in Table 2. The $(OC_{SV}/EC)_{pri}$ was estimated for the first time, being 0.87 (0.82) in daytime (nighttime) for spring and 3.18 (2.35) for summer (Table 1). The $(OC_{NV}/EC)_{pri}$ was stable from daytime to nighttime in both seasons, while $(OC_{SV}/EC)_{pri}$ showed a diurnal difference in summer, possibly suggesting the different combustion sources for OC_{NV} and OC_{SV} and that semi-volatile species would receive more influence from local and upward transport of emissions during summer daytime. Most of the intercepts were close to zero, implying low background concentrations of OC from non-combustion sources at Mt. Tai.

Multiple regressions were performed on hourly data of variables (cf. Eq. (5)) for OC_{NV} and OC_{SV} in each season, and the determined regression coefficients are listed in Table 3. The intercept term in Eq. (5) was not statistically significant at probability (p) < 0.05 for either season; therefore the regressions were performed with zero-intercept. Additional regressions were conducted separately for daytime and nighttime in each season, and the results showed no significant difference to that from entire season but displayed larger uncertainties due to fewer data in each subgroup (Table S1). Therefore the results from entire season are employed in the following analysis. The regression coefficients of EC, CO and O_3 were all statistically significant for OC_{NV} and OC_{SV} in both seasons. O_3 coefficients were higher for OC_{SV} (0.887–0.907) than that for OC_{NV} (0.260–0.306), and are apparently higher than those for daily

Table 1
EC-tracer parameters determined from Deming regression of selected data shown in Fig. 1 for both OC_{NV} and OC_{SV} in spring and summer of 2007.

	$(OC_{NV}/EC)_{pri}$	$(OC_{NV})_{nc}$	Number	R	$(OC_{SV}/EC)_{pri}$	$(OC_{SV})_{nc}$	Number	R
Spring								
Daytime	2.19	0[a]	16	0.95	0.87	0.60	15	0.92
Nighttime	2.16	0	13	0.99	0.82	0.62	9	0.86
Summer								
Daytime	1.87	0.4	21	0.94	3.18	0	10	0.92
Nighttime	1.85	0	15	0.95	2.35	0	12	0.93

[a] Zero non-combustion source is employed when the intercept is not significantly different from zero.

Table 2
Comparison of estimation method, primary OC/EC ratio and SOC concentration at Mt. Tai with other high-altitude sites, rural background sites and urban sites reported in the literature.

Location	Elevation (km)	Sampling Period	Size	Method	$(OC/EC)_{pri}$	SOC ($\mu g\ m^{-3}$)	SOC/TOC (%)	Reference
Manora Peak, India	1.95	Feb–Mar 2005	TSP	EC-tracer	5.7	1.6	13.8	Ram et al., 2008
		Apr–Jun 2005	TSP		4.8	4.3	51.6	
		Dec–Mar 2006	TSP		5.7	2.7	25.7	
		Apr–Jun 2006	TSP		4.8	3.2	36.5	
Mt. Abu, India	1.7	Dec–Mar 2005	TSP	EC-tracer	3.4	1.3	27.7	Ram et al., 2008
Mt. Tai, China	1.545	May–Jun 2006	TSP	Biogenic tracer	–	1.6–1.7[a]	10	Fu et al., 2010
Dunhuang, China	1.139	2006	PM_{10}	EC-tracer	2.2	19.6	67	Zhang et al., 2008
Cape Fuguei, Taiwan	Rural	2003–2007	PM_{10}	EC-tracer	1.62	2.3	47.9	Chou et al., 2010
Penghu, Taiwan	Rural	2003–2007	PM_{10}	EC-tracer	1.75	1.5	45.5	
Cape Fuguei, Taiwan	Rural	2003–2007	$PM_{2.5}$		1.73	1.6	42.1	
Penghu, Taiwan	Rural	2003–2007	$PM_{2.5}$		1.51	1.0	45.5	
Potsdam, New York, US	Rural background	2002–2005	$PM_{2.5}$	EC-tracer	1.35–3.97	–	66–72	Raman et al., 2008
Stockton, New York, US	Rural background	2002–2005	$PM_{2.5}$		2.24–5.70	–	47–64	
Cities in northern China	Urban	Summer 2003	$PM_{2.5}$	EC-tracer	1.99	–	44.8	Cao et al., 2007
		Winter 2003	$PM_{2.5}$		2.81	–	29.6	
Cities in Southern China	Urban	Summer 2003	$PM_{2.5}$		1.29	–	53.1	
		Winter 2003	$PM_{2.5}$		2.13	–	32.8	
Mt. Tai, China	1.533	Mar–Apr 2007	$PM_{2.5}$	EC-tracer	2.16–2.19[b] 0.82–0.87[c]	8.54	60.2	This study
		Jun–Jul 2007	$PM_{2.5}$	EC-tracer	1.85–1.87[b] 2.35–3.18[c]	13.78	73.4	
		Mar–Apr 2007	$PM_{2.5}$	Multiple regression	–	8.19	57.3	
		Jun–Jul 2007	$PM_{2.5}$		–	13.26	71.2	

[a] Biogenic SOC from oxidation of isoprene, α/β-pinene and β-caryophyllene.
[b] Primary ratio for OC_{NV}.
[c] Primary ratio for OC_{SV}.

Table 3
Coefficients of multiple regressions for OC_{NV} and OC_{SV} in each season. The regression equation is described as Eq. (5), and the regression was performed with zero-intercept (see text). Regression coefficients for daytime and nighttime in each season are summarized in Table S1.

	Number	b (EC)/$\mu g\ m^{-3}$	c CO/10 ppb	d O_3/10 ppb	e (NO_y)/10 ppb	f (SO_4^{2-})/$\mu g\ m^{-3}$	g (K^+)/$\mu g\ m^{-3}$	R
Non-volatile OC								
Spring	554	0.651 ± 0.068	0.052 ± 0.005	0.306 ± 0.032	0.097 ± 0.015	0.041 ± 0.007	0.055 ± 0.014	0.95
Summer	706	2.082 ± 0.161	0.005 ± 0.002	0.260 ± 0.028	0.535 ± 0.088	NS[a]	0.190 ± 0.035	0.96
Semi-volatile OC								
Spring	543	1.523 ± 0.170	0.007 ± 0.002	0.907 ± 0.103	NS	0.012 ± 0.006	NS	0.83
Summer	703	2.424 ± 0.368	0.012 ± 0.003	0.887 ± 0.120	2.030 ± 0.392	NS	NS	0.88

[a] NS, not significant at $p < 0.05$

filter-based OC (mostly OC_{NV}) at urban and rural sites in the US (0.11–0.24) (Blanchard et al., 2008), suggesting a higher SOC production efficiency from photochemistry in this region, with an increase of OC (approximately 1.2 $\mu g\ m^{-3}$) for each increment of O_3 (10 ppbv). The higher K^+ coefficient in summer was consistent with the more influence of biomass burning observed at Mt. Tai in summer of 2007 (Wang et al., 2011b).

The estimated POC and SOC concentrations by the constrained EC-tracer and multiple regression methods are compared in Table 4. As shown, the two methods gave similar average concentrations for species in both seasons, with a difference of less than 10% (except for POC_{SV}). Fig. 2 depicts the temporal variations of SOC estimated by these two methods, and the trends show excellent agreement. Larger fluctuations with more extreme peaks and valleys were observed for SOC concentrations from EC-tracer method. Air masses strongly impacted by biomass burning at Mt. Tai in April and June (Zhou et al., 2010; Wang et al., 2011b) could partially explain the higher values of SOC from EC-tracer method during late April and June (Fig. 2). As the multiple regression method involves several primary and secondary tracers, it is thought to better represent the temporal evolution of the secondary formation process. Pachon et al. (2010) compared several different methods with multiple regression giving the lowest uncertainty. In the present study, the very good overall agreement between the two methods demonstrates their validity and applicability at Mt. Tai during two study seasons. The following

Table 4
Comparison of average concentrations ($\mu g\ m^{-3}$) of SOC and POC estimated by the constrained EC-tracer method and multiple regression method during spring and summer campaigns at Mt. Tai.

	POC_{NV}		SOC_{NV}		POC_{SV}		SOC_{SV}	
	Mean (±SD)	Range	Mean (±SD)	Range	Mean (±SD)	Range	Mean (±SD)	Range
EC-tracer								
Spring	3.54 ± 2.44	0.29–13.19	2.65 ± 1.86	0–11.17	2.11 ± 1.25	0.74–12.33	5.89 ± 5.53	0–46.00
Summer	2.11 ± 1.38	0–8.31	3.02 ± 1.85	0–11.72	2.91 ± 2.19	0–11.33	10.78 ± 8.66	0–47.78
Multiple regression								
Spring	3.46 ± 2.09	0–11.39	2.61 ± 1.12	0.08–8.48	2.67 ± 2.07	0–25.33	5.58 ± 5.41	0.75–39.46
Summer	2.29 ± 1.62	0–9.45	2.81 ± 1.19	0–6.78	3.06 ± 2.58	0–19.89	10.44 ± 6.49	0.16–41.32

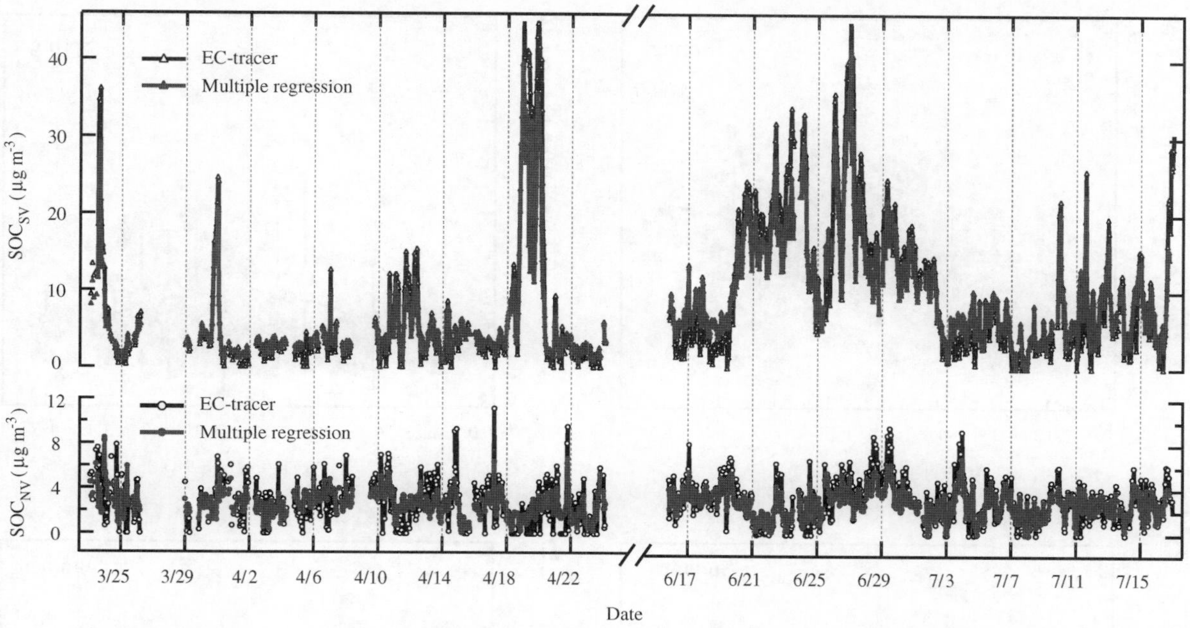

Fig. 2. Time series of concentrations of SOC_{NV} and SOC_{SV} estimated by the constrained EC-tracer method and multiple regression method in spring and summer of 2007 at Mt. Tai.

analysis is based on the multiple regressions results unless otherwise stated.

3.1.2. Characterization of SOC formation

The average concentration of non-volatile and semi-volatile SOC were 2.61, 5.58 $\mu g\ m^{-3}$ in spring and 2.81, 10.44 $\mu g\ m^{-3}$ in summer (Table 4), which are much higher than the biogenic SOC (1.6–1.7 $\mu g\ m^{-3}$) at Mt. Tai estimated by Fu et al. (2010), indicating that the anthropogenic sources are more important to SOC at Mt. Tai. The average total SOC concentrations at Mt. Tai were 8.19 and 13.26 $\mu g\ m^{-3}$ in spring and summer, respectively, and are much higher than most reported results from worldwide mountain and rural sites summarized in Table 2, suggesting strong SOA formation at Mt. Tai during campaigns and revealing the high SOA loading in the planetary boundary layer (PBL) of the North China Plain.

The contributions of different species to TOC are illustrated in Fig. 3. As shown, SOC_{NV} and SOC_{SV} accounted for 18.2% and 39.1% of TOC in spring, and the latter increased to 56.1% in summer, indicating that more SOA formed in summer is semi-volatile. The increased SOC concentrations and contributions to TOC (71.2%) in summer reflect enhanced SOA formation from more active photochemical processes. POC contributed 42.7% and 28.8% of TOC in spring and summer, respectively, with a higher non-volatile concentration in spring and a higher semi-volatile concentration in summer. The different seasonal variation of POC species is likely due to the temperature dependence of phase partitioning of POC emissions (Shrivastava et al., 2006). Reduced concentrations at night were observed for all species except SOC_{SV} (Fig. 3C). The decreased concentrations of photochemical oxidants at night would slow down the aging process and the conversion of semi-volatile to non-volatile phase, and the lower temperature and frequent cloud/fog process at night would favor the condensation and formation of semi-volatile species.

The dependence of SOC_{NV} and SOC_{SV} concentrations on the photochemical age at Mt. Tai is illustrated in Fig. 4. The data have been divided into several subsets, spanning 0.1 or more units of age according to the number of datasets, and the average concentration of each subset was subsequently determined. In spring, SOC_{NV} concentration exhibited a clear positive correlation with photochemical age, and the linear regression indicates an increase of 0.78 $\mu g\ m^{-3}$ in SOC_{NV} for each increase of 0.1 in photochemical age. This is similar to the increasing oxygenated organic aerosol (OOA) with photochemical age observed in Mexico City by Kleinman et al. (2008) and in Ontario by Slowik et al. (2011). SOC_{SV} concentration

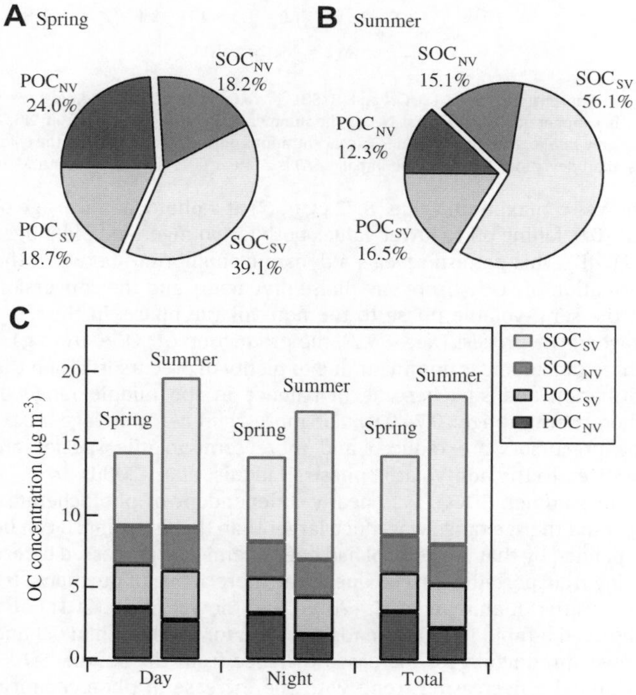

Fig. 3. Percentage contribution of POC and SOC to TOC during (A) spring and (b) summer of 2007, and (C) average concentrations of different fractions in daytime and nighttime in both seasons estimated by using coefficients in Table 3.

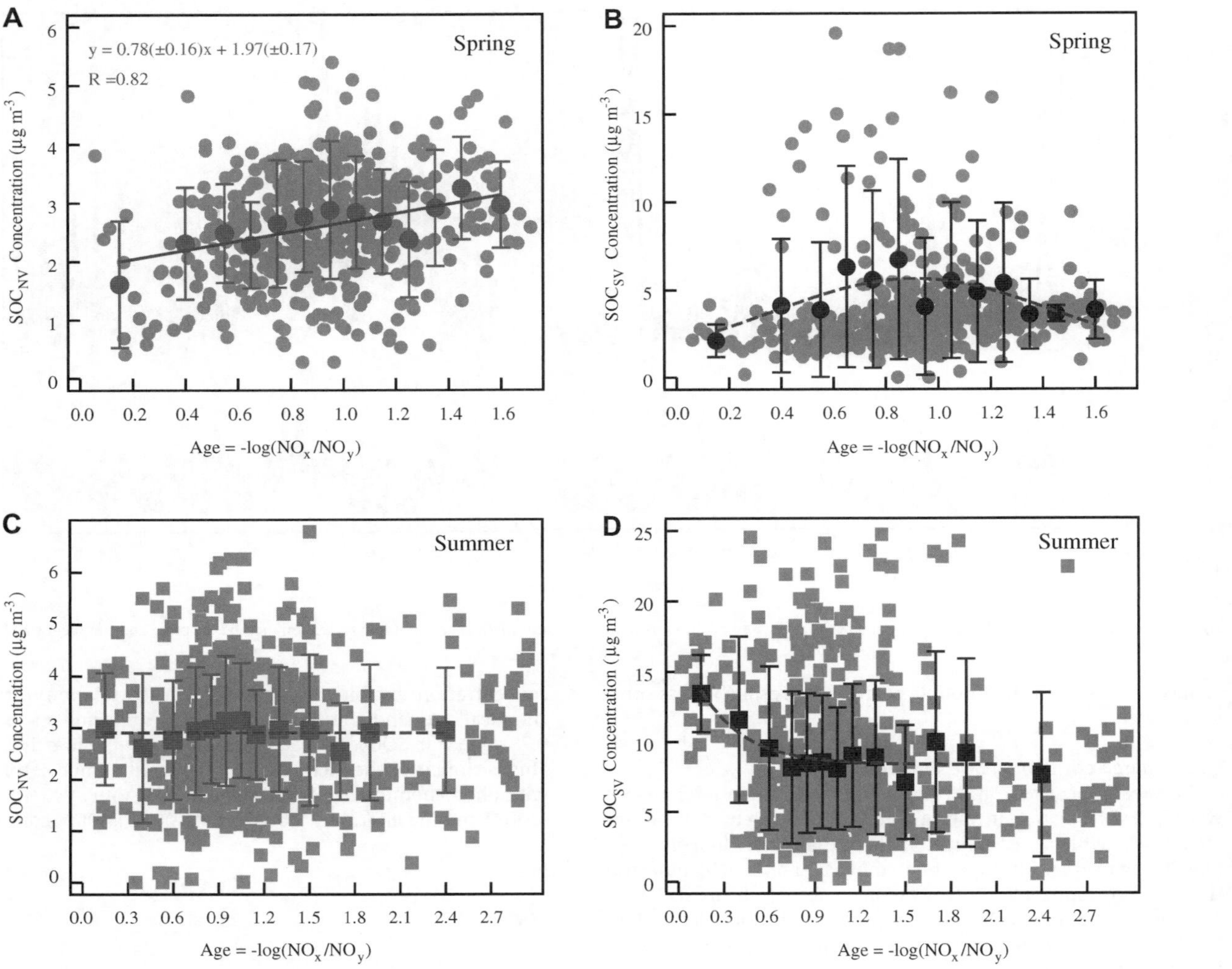

Fig. 4. Concentrations of (A) SOC_{NV} and (B) SOC_{SV} as a function of photochemical age (see text) in spring, and (C) SOC_{NV} and (D) SOC_{SV} concentrations as a function of photochemical age in summer of 2007 at Mt. Tai. NO_x in the summer period was calculated from NO_y − PAN − nitric acid − aerosol nitrate, as it was not directly measured in the summer campaign. The grey markers are the derived SOC concentrations during campaigns, and the color markers represent the averaged concentration for each age bin, and the error bars show the standard deviation. The solid line in panel (A) is a linear fit to average concentrations, and the dash lines in other panels are not model fits but are drawn to aid the eye.

showed a maximum value (8.72 μg m^{-3}) at a photochemical age of 0.8–0.9, falling off to lower values both for younger and older ages (Fig. 4B). This pattern reveals a dynamic equilibrium between the formation of SOC_{SV} from gas-phase precursors and the conversion of the semi-volatile phase to the non-volatile phase. In the early stage of the process (age < 0.7), the production of SOC_{SV} from gas-phase precursors is dominant due to plenty of precursors. Then the formation and sink become in balance in the middle range of photochemical age (0.7–0.9). For more aged aerosols (age < 0.9), the precursors are reduced and more semi-volatile species are oxidized to the non-volatile phase (Hallquist et al., 2009).

In summer, SOC_{NV} was nearly independent of photochemical age, and the age range was much larger than that in spring. It can be explained by that the aerosol had been chemically processed before being transported to Mt. Tai due to the more active photochemistry in summer. Kleinman et al. (2008) and Slowik et al. (2011) also observed a rapid initial formation of OOA for age less than 0.3 and slower production for increased distance from the source. SOC_{SV} exhibited a decreasing trend with the increase of photochemical age in summer. It represents the third step of the dynamic equilibrium and is controlled by the continued oxidation of semi-volatile species to the non-volatile phase.

3.2. Impact of clouds on carbonaceous aerosols

Clouds frequently cover the summit of Mt. Tai. According to a 30-years observation record, the average annual number of foggy/cloudy days is 176 (Wang et al., 2011a). In the present study, nine cloud events lasting longer than 6-h are selected to examine their impact on the carbonaceous aerosols.

3.2.1. Cloud scavenging effect on non-volatile species

Fig. 5 illustrates the temporal variation of carbonaceous and related species during two typical cloud events observed on April 17 and June 26. As shown, OC_{NV} and EC concentration decreased remarkably in both cloud events, indicating significant cloud scavenging effect and that a large fraction of organic and EC particles act as cloud condensation nuclei (CCN) in clouds. The concentrations of gas species (particularly CO) during the clouds showed only minor fluctuations, indicating no significant changes of air mass throughout the events. The decay factors (DF) of OC_{NV} and EC during these two cases, as well as other cloud events, are depicted in Figs. 6 and 7. Eq. (12) was applied to model the scavenging process for OC_{NV} and EC, and the scavenging coefficient constants ($K_{NV\text{-}OC}$ and K_{EC}) for each event were determined and

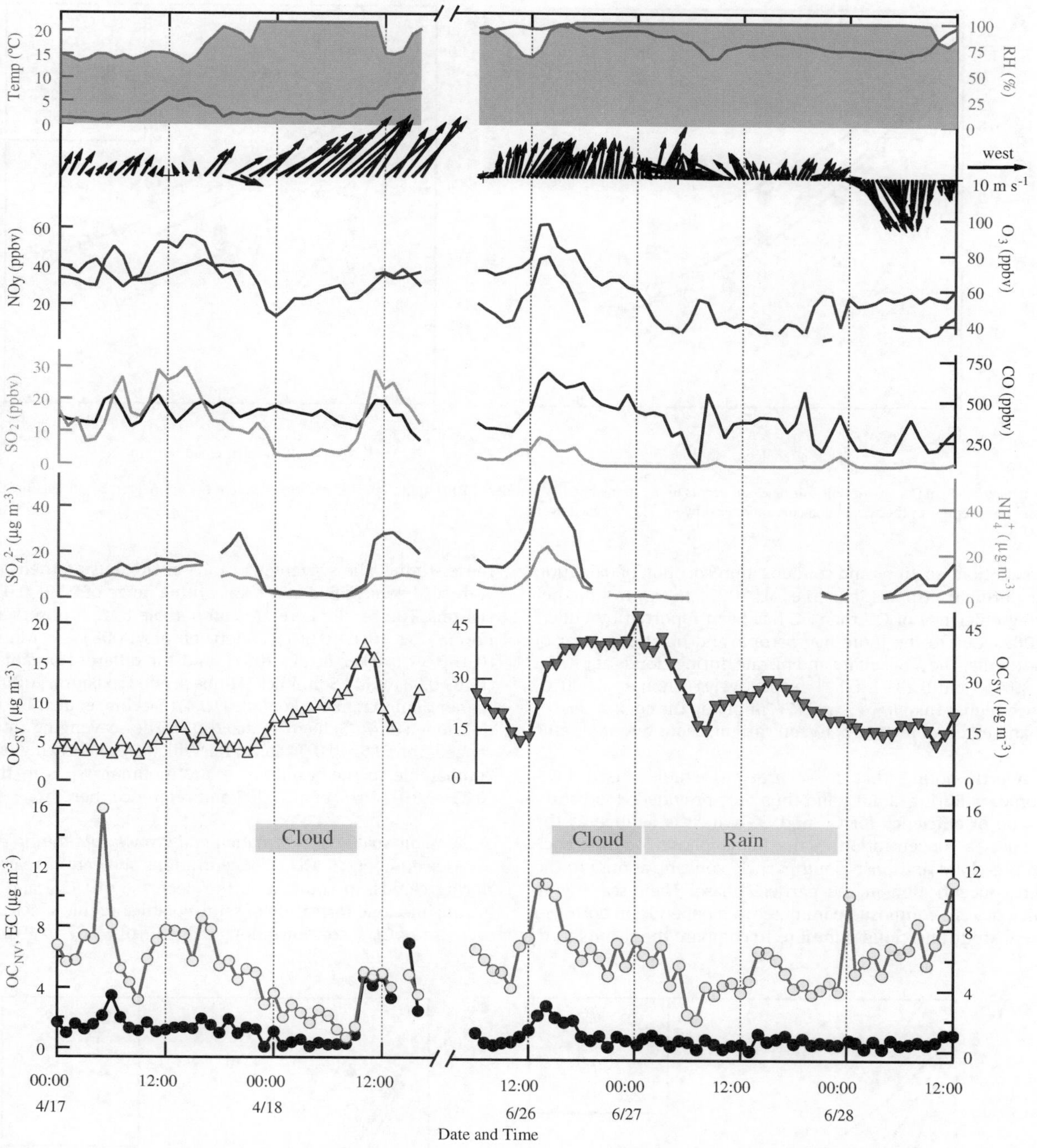

Fig. 5. Time series of concentrations of OC_{NV}, EC, OC_{SV}, trace gases, soluble ions and meteorological parameters during cloud events observed on April 17 and June 26 of 2007 at Mt. Tai.

summarized in Table 5, as well as the average *K* estimated from Eq. (12) by using the total dataset for each season.

The derived $K_{NV\text{-}OC}$ varied from 0.094 to 0.55 h^{-1} with an average of 0.16 h^{-1} in spring, and ranged from 0.070 to 0.30 h^{-1} with an average of 0.11 h^{-1} in summer. The significant difference between events could be attributed to the different organic chemical composition, volatility and solubility of species, and characteristic of the clouds (Limbeck and Puxbaum, 2000; Hitzenberger et al., 2001; Sellegri et al., 2003; Collett et al., 2008). Seasonal variation was also found for $K_{NV\text{-}OC}$, with lower values in summer than in spring, implying that OC_{NV} is mixed with more inorganic hydrophilic species or contains more water soluble organic species (e.g., dicarboxylic acids) in spring.

K_{EC} ranged from 0.11 to 0.90 h^{-1} in spring and from 0.18 to 0.28 h^{-1} in summer, with similar seasonal variation with $K_{NV\text{-}OC}$. The highest and lowest K_{EC} were observed concurrently with $K_{NV\text{-}OC}$, except the event on July 12. Moreover, most of K_{EC} were higher than $K_{NV\text{-}OC}$, which is consistent with higher EC scavenging efficiency than OC observed at some remote mountain sites (Hitzenberger et al., 2001; Sellegri et al., 2003). It indicates internal mixing of EC with more hygroscopic species (i.e., sulfate and nitrate), and that hydrophobic particles could become more hydrophilic through

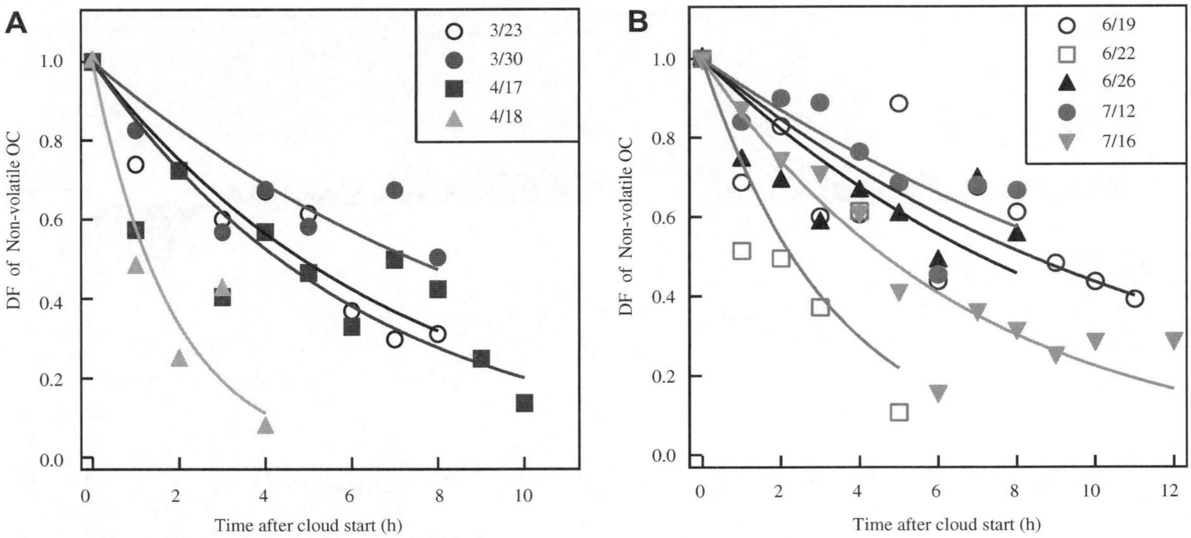

Fig. 6. Decay factor (DF) of OC_{NV} with time during cloud events in (A) spring and (B) summer of 2007 at Mt. Tai. The markers represent the normalized DF (cf. Section 2.3) of OC_{NV}, and the solid lines represent the scavenging curve modeled by Eq. (12) for each event.

surface oxidation reactions and condensation/sorption of oxidation products during transport (Sellegri et al., 2003). In contrast, higher scavenging efficiency of OC than EC has been reported by Collett et al. (2008), due to the more hygroscopic and polar character of OC constituents. The K of sulfate and nitrate during clouds at Mt. Tai (0.33–1.39 h^{-1} and 0.23–1.18 h^{-1}, respectively) (Zhou et al., 2010) were much higher than $K_{NV\text{-}OC}$ and K_{EC}, implying the coexistence of internal and external mixing between carbonaceous and inorganic species.

It is worth noting that most previous studies used low-resolution data during clouds, and thus only provided cloud scavenging ratio or efficiency for OC and EC, which is defined as the ratio of species concentration in the liquid phase (or difference between pre-cloud and post-cloud particle concentrations) to the pre-cloud concentration in the particle phase. Their scavenging ratio/efficiency is a composite parameter that depends on both K in the present study and cloud duration. To compare the results from these studies, the scavenging ratios of OC_{NV} using their definition were also determined at Mt. Tai, with a range of 0.33–0.93 in both seasons. The results here are comparable to that observed in California fog (0.41–0.90) (Collett et al., 2008), at Mt. Brocken (0.4–0.65) (Acker et al., 2002), and for different organic species (0.16–0.98) at Mt. Sonnblick (Limbeck and Puxbaum, 2000), but are higher than that in the Po Valley (0.4) (Facchini et al., 1999) and Puy de Dome (0.14) (Sellegri et al., 2003). The scavenging ratios of EC ranged from 0.62 to 0.94 at Mt. Tai, which are slightly higher than or comparable to the results at many mountain sites in the world (0.33–0.80) (Cozic et al., 2007 and reference therein).

3.2.2. *Semi-volatile OC formation and growth rate during clouds*

As seen in Fig. 5, OC_{SV} concentrations showed increasing trend during clouds, in contrast to the decrease of OC_{NV} and EC, suggesting in-cloud formation of semi-volatile organic species. The net increase of OC_{SV} concentrations ranged from 3.67 to 19.04 $\mu g\ m^{-3}$

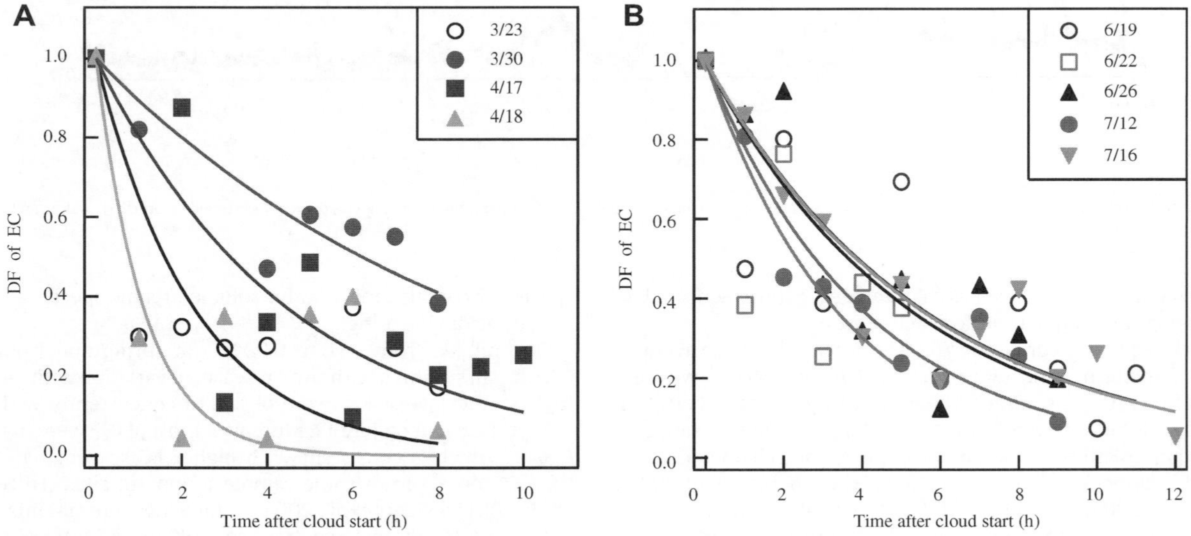

Fig. 7. Decay factor (DF) of EC with time during cloud events in (A) spring and (B) summer of 2007 at Mt. Tai. Notations are the same as in Fig. 6.

Table 5
Summary of cloud processing parameters for OC_{NV}, EC and OC_{SV} during clouds at Mt. Tai in spring and summer of 2007. The model and equations are described in Section 2.4.

Event	Start time (LT)	Parameters for OC_{NV} and EC		Parameters for OC_{SV}		
		$K_{NV\text{-}OC}$ (h^{-1})	K_{EC} (h^{-1})	Formation rate constant J (h^{-1})	Formation rate Q ($\mu g\ m^{-3}\ h^{-1}$)	Sink constant S (h^{-1})
Spring						
3–23	12:00	0.14	0.43	–	–	–
3–30	15:00	0.094	0.11	0.65	5.10	0.19
4–17	22:00	0.16	0.22	0.43	1.89	0.16
4–18	17:00	0.55	0.90	1.17	19.7	0.49
Total average		0.16	0.25	0.88	–	0.36
Summer						
6–19	19:00	0.08	0.18	0.17	2.38	0.097
6–22	20:00	0.30	0.28	0.09	1.68	0.001
6–26	15:00	0.098	0.19	1.26	43.5	1.07
7–12	18:00	0.070	0.25	0.10	1.40	0.028
7–16	19:00	0.15	0.18	1.39	29.3	1.00
Total average		0.11	0.20	0.45	–	0.33

for all cloud events in both seasons, which are much higher than the estimated in-cloud SOA formation ($0.28\ \mu g\ m^{-3}$ for 24-h average and $0.60\ \mu g\ m^{-3}$ for 1-h average) by Chen et al. (2007), and are of the same magnitude as the aerosol mass increase ($0.03–27.8\ \mu g\ m^{-3}$ within 22-h) from aqueous-phase reactions of methacrolein with OH radicals observed by El Haddad et al. (2009). The enhanced OC_{SV} formation at Mt. Tai can be attributed to the high concentration of precursors and abundant scavenged OC_{NV} in clouds, which could increase the uptake of organic vapors into droplets (Raja and Valsaraj, 2004). The SOA formation through cloud processing may involve uptake of organic gases or gas-phase oxidation products by cloud droplets, formation of potential low-volatility products through aqueous-phase reactions inside or on the surface of droplets, followed by subsequent partitioning to particle phase post evaporation of the hydrometeor (Blando and Turpin, 2000; Chen et al., 2007).

The dependence of OC_{SV} and SOC_{SV} concentrations on ambient relative humidity (RH) during two campaigns is investigated in Fig. 8. Apparently, OC_{SV} and SOC_{SV} concentrations increased with the increasing RH, and significant enhancement was found when RH was above 60%, suggesting that water vapor can influence organic partitioning and that high RH would accelerate the formation of SOC_{SV}. It is consistent with the increased water soluble OC when RH is higher than 70% reported by Hennigan et al. (2008). Volkamer et al. (2009) have also observed a linear correlation between SOA yields and aerosol liquid water concentrations.

The growth factor (GF) of OC_{SV} is illustrated as a function of time after cloud starts in Fig. 9. The GF was much higher in spring than that in summer, indicating the variation of growth and formation rate between seasons. The growth curve for each cloud event was predicted by Eq. (15) and is plotted as a solid line in Fig. 9. The formation rate constant ($J_{SV\text{-}OC}$) and sink constant ($S_{SV\text{-}OC}$) were consequently determined and are presented in Table 5, as well as the formation rate ($Q_{SV\text{-}OC}$) estimated from Eq. (14) with concentration data. The determined $J_{SV\text{-}OC}$ ranged from 0.43 to 1.17 h^{-1} in spring, with an average of 0.88 h^{-1} obtained by fitting the Eq. (15) with the total dataset in spring. In summer, $J_{SV\text{-}OC}$ presented higher fluctuations, with a range of 0.09–1.39 h^{-1} and an average of 0.45 h^{-1}. A similar trend was found for $S_{SV\text{-}OC}$, which varied between 0.16 and 0.49 h^{-1} in spring and 0.001 and 1.07 h^{-1} in summer. $Q_{SV\text{-}OC}$ also showed obvious differences between events, with $1.89–19.7\ \mu g\ m^{-3}\ h^{-1}$ in spring and $1.40–43.5\ \mu g\ m^{-3}\ h^{-1}$ in summer, which are higher than the mean in-cloud oxidation fluxes ($0.12–0.5\ \mu g\ m^{-3}\ h^{-1}$) of C2–C4 organic gas precursors estimated by Tilgner and Herrmann (2010).

It should be noted that some high values of $J_{SV\text{-}OC}$ (>1.0) were observed for cloud events on April 18, June 26 and July 16 (Table 5). Correspondingly, $S_{SV\text{-}OC}$ in those events was also higher than other

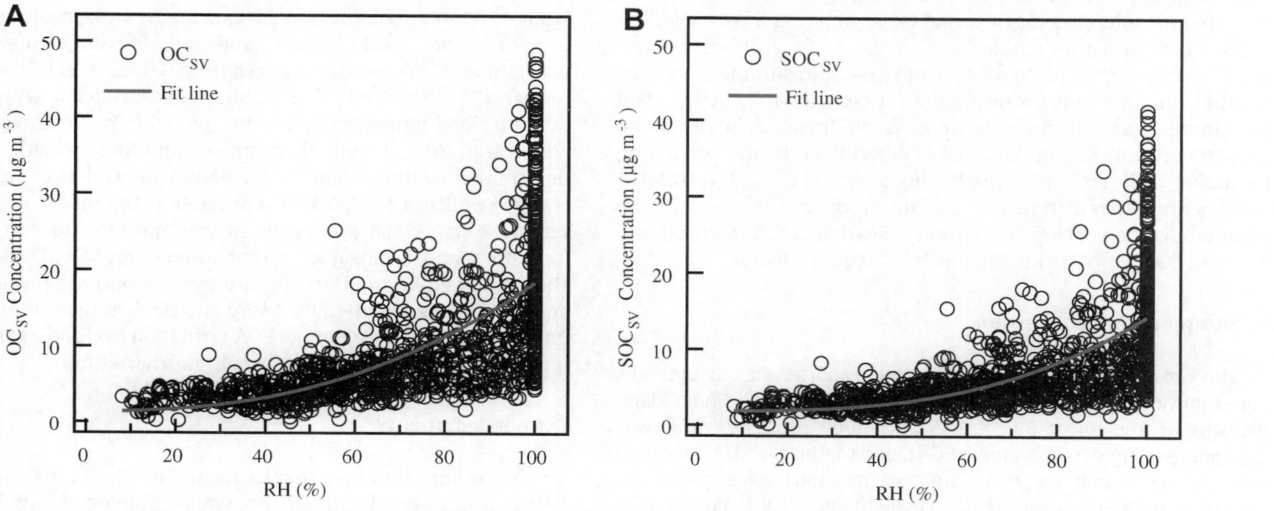

Fig. 8. Correlations between (A) OC_{SV} concentration and RH, (B) SOC_{SV} concentrations and RH in both campaigns in 2007 at Mt. Tai.

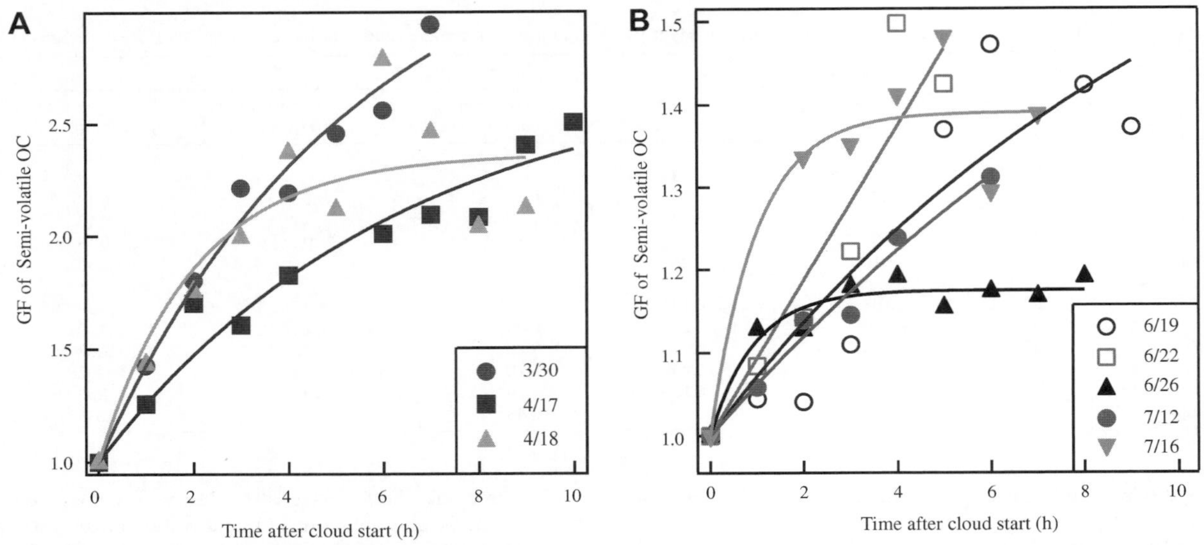

Fig. 9. Growth factor (GF) of OC_{SV} with time during cloud events in (A) spring and (B) summer of 2007 at Mt. Tai. The markers represent the normalized GF (cf. Section 2.3) of OC_{SV}, and the solid lines represent the growth curve modeled by Eq. (15) for each event.

cases. The GF of these events increased sharply in the first 2–3 h of the cloud, but slowly grew after 3 h (Fig. 9). The indication is that a high growth rate of OC_{SV} is present in the early stage of the cloud, but slows down afterward due to the domination by some sink mechanisms, and that the formation and sink of OC_{SV} balance in the late stage of the cloud. De Haan et al. (2011) have reported that nitrogen-containing oligomers formed from aqueous reactions of methylglyoxal and amines could extensively take place within minutes. Anttila and Kerminen (2003) have also suggested that the formation of low-volatility organic compounds in aqueous-phase process would contribute significantly to particle growth in the early stages of droplet formation. In addition, the growth rate is also sensitive to the concentration of pre-existing aerosols, precursors and oxidants (Hallquist et al., 2009). In comparison, the growth rates in other events were more stable through the cloud duration, and concentrations showed continuous increase with low sink constants (e.g., April 17 in Figs. 5 and 9).

$S_{SV\text{-}OC}$ exhibited a similar trend with $K_{NV\text{-}OC}$ in spring, indicating that the cloud scavenging process is the most important sink for OC_{SV} in spring. The difference between $S_{SV\text{-}OC}$ and $K_{NV\text{-}OC}$ suggests the different solubility for OC_{SV} and OC_{NV}, and that lower volatility species have a higher solubility (Sellegri et al., 2003). However, $S_{SV\text{-}OC}$ showed more discrepancy from $K_{NV\text{-}OC}$ in summer, suggesting that sink mechanisms rather than scavenging may be involved in summer and contribute to depress the increase in OC_{SV}. The growth curve of OC_{SV} in the cloud is determined by the competing formation and sink processes. The scavenging and formation coefficients derived from the present study can be used in the parameterization of cloud scavenging of carbonaceous aerosols and in-cloud SOA formation in chemistry transport models.

4. Summary and conclusion

The present work extends the results reported in our previous paper on carbonaceous aerosols at Mt. Tai in the North China Plain, focusing on the formation of SOA and cloud processes. SOC was estimated using a constrained EC-tracer method and a modified multiple regression method. The two methods gave consistent results for both SOC_{NV} and SOC_{SV}. The estimated SOC accounted for 57.3% and 71.2% of TOC in spring and summer, respectively, indicating a strong SOA formation in the PBL above the North China Plain. An examination of SOC with photochemical age reveals different relationships of non-volatile and semi-volatile SOC with age and also a different seasonal dependence. SOC_{NV} increased with photochemical age in spring, while SOC_{SV} exhibited a mono-peak pattern. These results indicate continuing formation of SOC_{NV} and SOC_{SV} being determined by the dynamic equilibrium between formation from gas-phase precursors and conversion to the non-volatile phase in spring. This dependence was not significant in more aged air masses observed in summer, revealing that the aerosol had been fully processed before being transported to Mt. Tai.

The significant scavenging of OC_{NV} and EC by clouds was observed at Mt. Tai. The scavenging coefficients constant (K) derived from the observations ranged from 0.070 to 0.55 h^{-1} for OC_{NV}, and 0.11 to 0.90 h^{-1} for EC. The relatively high K_{EC} suggests internal mixing of EC with more hygroscopic species. In contrast, OC_{SV} concentrations increased during clouds, indicating the semi-volatile SOA formation through aqueous-phase reactions in cloud processing. The observed OC_{SV} growth in clouds is much higher than the in-cloud SOA formation in other laboratory and model studies. A mass balance model was applied to predict the OC_{SV} growth curve during clouds, and the derived formation rate constant and sink constant ranged from 0.09 to 1.39 h^{-1} and 0.001 to 1.07 h^{-1}, respectively. The results suggest that the OC_{SV} growth rate in cloud processing depends on both pre-existing aerosol concentration and sink mechanism, and the growth curve is determined by the competing formation and sink processes.

The results of the present work add to the increasing body of literature that cloud processing plays important roles in the SOA formation and removal of carbonaceous aerosols. These results should be helpful in efforts to improve the performance of atmospheric models. Further laboratory and field studies are needed to unravel the mechanism of the SOA formation in clouds, particularly under polluted conditions such as in northern China.

Acknowledgments

The authors gratefully thank Dr. Jian Gao, Dr. Aijun Ding, Dr. Yan Wang, Mr. Steven Poon, Dr. Jin Wang, Xinfeng Wang, Wei Nie, Xiaomei Gao and Pengju Xu for their contributions to the field studies and laboratory analysis. The Mt. Tai Meteorological Observatory is gratefully acknowledged for its support of the field study and for providing the meteorological data. We also thank two anonymous referees for their suggestions, and Dr. Edward C. Mignot in Shandong University for linguistic advice. This work is funded by the National Basic Research Project of China (973 Project No. 2005CB422203).

Appendix. Supplementary material

Supplementary data related to this article can be found online at doi:10.1016/j.atmosenv.2011.08.019.

References

Acker, K., Mertes, S., Moller, D., Wieprecht, W., Auel, R., Kalass, D., 2002. Case study of cloud physical and chemical processes in low clouds at Mt. Brocken. Atmospheric Research 64, 41–51.

Anttila, T., Kerminen, V.M., 2003. Aerosol formation via aqueous-phase chemical reactions. Journal of Geophysical Research-Atmospheres 108. doi:10.1029/2002jd002764.

Bae, S.Y., Jung, C.H., Kim, Y.P., 2009. Relative contributions of individual phoretic effect in the below-cloud scavenging process. Journal of Aerosol Science 40, 621–632.

Blanchard, C.L., Hidy, G.M., Tanenbaum, S., Edgerton, E., Hartsell, B., Jansen, J., 2008. Carbon in southeastern US aerosol particles: empirical estimates of secondary organic aerosol formation. Atmospheric Environment 42, 6710–6720.

Blando, J.D., Turpin, B.J., 2000. Secondary organic aerosol formation in cloud and fog droplets: a literature evaluation of plausibility. Atmospheric Environment 34, 1623–1632.

Cao, J.J., Lee, S.C., Chow, J.C., Watson, J.G., Ho, K.F., Zhang, R.J., Jin, Z.D., Shen, Z.X., Chen, G.C., Kang, Y.M., Zou, S.C., Zhang, L.Z., Qi, S.H., Dai, M.H., Cheng, Y., Hu, K., 2007. Spatial and seasonal distributions of carbonaceous aerosols over China. Journal of Geophysical Research-Atmospheres 112. doi:10.1029/2006jd008205.

Chen, J., Griffin, R.J., Grini, A., Tulet, P., 2007. Modeling secondary organic aerosol formation through cloud processing of organic compounds. Atmospheric Chemistry and Physics 7, 5343–5355.

Chou, C.C.K., Lee, C.T., Cheng, M.T., Yuan, C.S., Chen, S.J., Wu, Y.L., Hsu, W.C., Lung, S.C., Hsu, S.C., Lin, C.Y., Liu, S.C., 2010. Seasonal variation and spatial distribution of carbonaceous aerosols in Taiwan. Atmospheric Chemistry and Physics 10, 9563–9578.

Chu, S.H., 2005. Stable estimate of primary OC/EC ratios in the EC tracer method. Atmospheric Environment 39, 1383–1392.

Collett, J.L., Herckes, P., Youngster, S., Lee, T., 2008. Processing of atmospheric organic matter by California radiation fogs. Atmospheric Research 87, 232–241.

Cozic, J., Verheggen, B., Mertes, S., Connolly, P., Bower, K., Petzold, A., Baltensperger, U., Weingartner, E., 2007. Scavenging of black carbon in mixed phase clouds at the high alpine site Jungfraujoch. Atmospheric Chemistry and Physics 7, 1797–1807.

Croft, B., Lohmann, U., Martin, R.V., Stier, P., Wurzler, S., Feichter, J., Hoose, C., Heikkila, U., van Donkelaar, A., Ferrachat, S., 2010. Influences of in-cloud aerosol scavenging parameterizations on aerosol concentrations and wet deposition in ECHAM5-HAM. Atmospheric Chemistry and Physics 10, 1511–1543.

De Haan, D.O., Hawkins, L.N., Kononenko, J.A., Turley, J.J., Corrigan, A.L., Tolbert, M.A., Jimenez, J.L., 2011. Formation of nitrogen-containing oligomers by methylglyoxal and amines in simulated evaporating cloud droplets. Environmental Science & Technology 45, 984–991.

El Haddad, I., Yao, L., Nieto-Gligorovski, L., Michaud, V., Temime-Roussel, B., Quivet, E., Marchand, N., Sellegri, K., Monod, A., 2009. In-cloud processes of methacrolein under simulated conditions. Part 2. Formation of secondary organic aerosol. Atmospheric Chemistry and Physics 9, 5107–5117.

Facchini, M.C., Fuzzi, S., Zappoli, S., Andracchio, A., Gelencser, A., Kiss, G., Krivacsy, Z., Meszaros, E., Hansson, H.C., Alsberg, T., Zebuhr, Y., 1999. Partitioning of the organic aerosol component between fog droplets and interstitial air. Journal of Geophysical Research-Atmospheres 104, 26821–26832.

Fu, T.M., Jacob, D.J., Wittrock, F., Burrows, J.P., Vrekoussis, M., Henze, D.K., 2008. Global budgets of atmospheric glyoxal and methylglyoxal, and implications for formation of secondary organic aerosols. Journal of Geophysical Research-Atmospheres 113. doi:10.1029/2007jd009505.

Fu, P.Q., Kawamura, K., Kanaya, Y., Wang, Z.F., 2010. Contributions of biogenic volatile organic compounds to the formation of secondary organic aerosols over Mt Tai, Central East China. Atmospheric Environment 44, 4817–4826.

Hallquist, M., Wenger, J.C., Baltensperger, U., Rudich, Y., Simpson, D., Claeys, M., Dommen, J., Donahue, N.M., George, C., Goldstein, A.H., Hamilton, J.F., Herrmann, H., Hoffmann, T., Iinuma, Y., Jang, M., Jenkin, M.E., Jimenez, J.L., Kiendler-Scharr, A., Maenhaut, W., McFiggans, G., Mentel, T.F., Monod, A., Prevot, A.S.H., Seinfeld, J.H., Surratt, J.D., Szmigielski, R., Wildt, J., 2009. The formation, properties and impact of secondary organic aerosol: current and emerging issues. Atmospheric Chemistry and Physics 9, 5155–5236.

Hennigan, C.J., Bergin, M.H., Weber, R.J., 2008. Correlations between water-soluble organic aerosol and water vapor: a synergistic effect from biogenic emissions? Environmental Science & Technology 42, 9079–9085.

Herckes, P., Chang, H., Lee, T., Collett, J.L., 2007. Air pollution processing by radiation fogs. Water, Air, and Soil Pollution 181, 65–75.

Hitzenberger, R., Berner, A., Glebl, H., Drobesch, K., Kasper-Giebl, A., Loeflund, M., Urban, H., Puxbaum, H., 2001. Black carbon (BC) in alpine aerosols and cloud water–concentrations and scavenging efficiencies. Atmospheric Environment 35, 5135–5141.

Hu, D., Bian, Q., Li, T.W.Y., Lau, A.K.H., Yu, J.Z., 2008. Contributions of isoprene, monoterpenes, beta-caryophyllene, and toluene to secondary organic aerosols in Hong Kong during the summer of 2006. Journal of Geophysical Research-Atmospheres 113. doi:10.1029/2008jd010437.

Hu, D., Bian, Q.J., Lau, A.K.H., Yu, J.Z., 2010. Source apportioning of primary and secondary organic carbon in summer $PM_{2.5}$ in Hong Kong using positive matrix factorization of secondary and primary organic tracer data. Journal of Geophysical Research-Atmospheres 115. doi:10.1029/2009jd012498.

Kleinman, L.I., Springston, S.R., Daum, P.H., Lee, Y.N., Nunnermacker, L.J., Senum, G.I., Wang, J., Weinstein-Lloyd, J., Alexander, M.L., Hubbe, J., Ortega, J., Canagaratna, M.R., Jayne, J., 2008. The time evolution of aerosol composition over the Mexico City plateau. Atmospheric Chemistry and Physics 8, 1559–1575.

Limbeck, A., Puxbaum, H., 2000. Dependence of in-cloud scavenging of polar organic aerosol compounds on the water solubility. Journal of Geophysical Research-Atmospheres 105, 19857–19867.

Pachon, J.E., Balachandran, S., Hu, Y.T., Weber, R.J., Mulholland, J.A., Russell, A.G., 2010. Comparison of SOC estimates and uncertainties from aerosol chemical composition and gas phase data in Atlanta. Atmospheric Environment 44, 3907–3914.

Raja, S., Valsaraj, K.T., 2004. Uptake of aromatic hydrocarbon vapors (benzene and phenanthrene) at the air-water interface of micron-size water droplets. Journal of the Air & Waste Management Association 54, 1550–1559.

Ram, K., Sarin, M.M., Hegde, P., 2008. Atmospheric abundances of primary and secondary carbonaceous species at two high-altitude sites in India: Sources and temporal variability. Atmospheric Environment 42, 6785–6796.

Raman, R.S., Hopke, P.K., Holsen, T.M., 2008. Carbonaceous aerosol at two rural locations in New York State: characterization and behavior. Journal of Geophysical Research-Atmospheres 113. doi:10.1029/2007jd009281.

Ren, Y., Ding, A., Wang, T., Shen, X., Guo, J., Zhang, J., Wang, Y., Xu, P., Wang, X., Gao, J., Collett Jr., J.L., 2009. Measurement of gas-phase total peroxides at the summit of Mount Tai in China. Atmospheric Environment 43, 1702–1711.

Seinfeld, J.H., Pandis, S.N., 1998. Atmospheric Chemistry and Physics: From Air Pollution to Climate Change. John Wiley, New York.

Sellegri, K., Laj, P., Dupuy, R., Legrand, M., Preunkert, S., Putaud, J.P., 2003. Size-dependent scavenging efficiencies of multicomponent atmospheric aerosols in clouds. Journal of Geophysical Research-Atmospheres 108. doi:10.1029/2002jd002749.

Shrivastava, M.K., Lipsky, E.M., Stanier, C.O., Robinson, A.L., 2006. Modeling semi-volatile organic aerosol mass emissions from combustion systems. Environmental Science & Technology 40, 2671–2677.

Slowik, J.G., Brook, J., Chang, R.Y.W., Evans, G.J., Hayden, K., Jeong, C.H., Li, S.M., Liggio, J., Liu, P.S.K., McGuire, M., Mihele, C., Sjostedt, S., Vlasenko, A., Abbatt, J.P.D., 2011. Photochemical processing of organic aerosol at nearby continental sites: contrast between urban plumes and regional aerosol. Atmospheric Chemistry and Physics 11, 2991–3006.

Strader, R., Lurmann, F., Pandis, S.N., 1999. Evaluation of secondary organic aerosol formation in winter. Atmospheric Environment 33, 4849–4863.

Tilgner, A., Herrmann, H., 2010. Radical-driven carbonyl-to-acid conversion and acid degradation in tropospheric aqueous systems studied by CAPRAM. Atmospheric Environment 44, 5415–5422.

Turpin, B.J., Huntzicker, J.J., 1991. Secondary formation of organic aerosol in the Los Angeles Basin: a descriptive analysis of organic and elemental carbon concentrations. Atmospheric Environment 25, 207–215.

Volkamer, R., Ziemann, P.J., Molina, M.J., 2009. Secondary organic aerosol formation from acetylene (C_2H_2): seed effect on SOA yields due to organic photochemistry in the aerosol aqueous phase. Atmospheric Chemistry and Physics 9, 1907–1928.

Wang, T., Nie, W., Gao, J., Xue, L.K., Gao, X.M., Wang, X.F., Qiu, J., Poon, C.N., Meinardi, S., Blake, D., Wang, S.L., Ding, A.J., Chai, F.H., Zhang, Q.Z., Wang, W.X., 2010. Air quality during the 2008 Beijing Olympics: secondary pollutants and regional impact. Atmospheric Chemistry and Physics 10, 7603–7615.

Wang, Y., Guo, J., Wang, T., Ding, A.J., Gao, J.A., Zhou, Y., Collett, J.L., Wang, W.X., 2011a. Influence of regional pollution and sandstorms on the chemical composition of cloud/fog at the summit of Mt. Taishan in northern China. Atmospheric Research 99, 434–442.

Wang, Z., Wang, T., Gao, R., Xue, L., Guo, J., Zhou, Y., Nie, W., Wang, X., Xu, P., Gao, J., Zhou, X., Wang, W., Zhang, Q., 2011b. Source and variation of carbonaceous aerosols at Mount Tai, North China: results from a semi-continuous instrument. Atmospheric Environment 45, 1655–1667.

Yu, S.C., Dennis, R.L., Bhave, P.V., Eder, B.K., 2004. Primary and secondary organic aerosols over the United States: estimates on the basis of observed organic carbon (OC) and elemental carbon (EC), and air quality modeled primary OC/EC ratios. Atmospheric Environment 38, 5257–5268.

Zhang, X.Y., Wang, Y.Q., Zhang, X.C., Guo, W., Gong, S.L., 2008. Carbonaceous aerosol composition over various regions of China during 2006. Journal of Geophysical Research-Atmospheres 113. doi:10.1029/2007jd009525.

Zhou, Y., Wang, T., Gao, X., Xue, L., Wang, X., Wang, Z., Gao, J., Zhang, Q., Wang, W., 2010. Continuous observations of water-soluble ions in $PM_{2.5}$ at Mount Tai (1534 m a.s.l.) in central-eastern China. Journal of Atmospheric Chemistry 64, 107–127.

Science of the Total Environment 538 (2015) 93–101

Contents lists available at ScienceDirect

Science of the Total Environment

journal homepage: www.elsevier.com/locate/scitotenv

HONO and its potential source particulate nitrite at an urban site in North China during the cold season

Liwei Wang [a], Liang Wen [a], Caihong Xu [a], Jianmin Chen [a,b], Xinfeng Wang [a,*], Lingxiao Yang [a,b], Wenxing Wang [a], Xue Yang [a], Xiao Sui [a], Lan Yao [a], Qingzhu Zhang [a]

[a] Environment Research Institute, Shandong University, Ji'nan 250100, China
[b] School of Environmental Science and Engineering, Shandong University, Ji'nan 250100, China

HIGHLIGHTS

- HONO and particulate nitrite were simultaneously measured in North China.
- Vehicle exhausts and NO_2 reactions are identified as major sources of HONO.
- Heterogeneous reactions of NO_2 produced a large amount of particulate nitrite.
- Particulate nitrite acted as a potential source of HONO especially in the daytime.

GRAPHICAL ABSTRACT

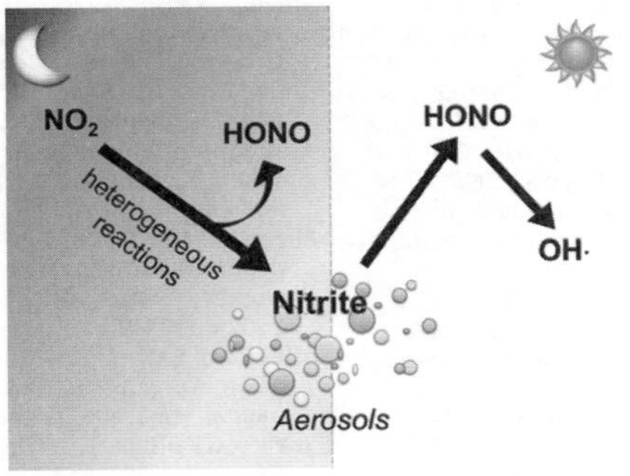

ARTICLE INFO

Article history:
Received 26 June 2015
Received in revised form 5 August 2015
Accepted 7 August 2015
Available online 22 August 2015

Editor: D. Barcelo

Keywords:
HONO
Particulate nitrite
Source identification
Heterogeneous conversion
Transformation

ABSTRACT

Characteristics and transformation of nitrous acid (HONO) and particulate nitrite were investigated with high time-resolution field measurements at an urban site in Ji'nan, China from Nov. 2013 to Jan. 2014. During the sampling period, averages of 0.35 ppbv HONO and 2.08 μg m^{-3} fine particulate nitrite were observed. HONO and particulate nitrite exhibited similar diurnal variation patterns but differed in the time at which concentration peaks and valleys occurred. Elevated nocturnal HONO concentration peaks were mainly associated with primary emissions from vehicle exhaust and secondary formation via heterogeneous reactions of NO_2. In fresh air masses dominated by vehicle emissions, the average $HONO/NO_x$ ratio was 0.58%. The nocturnal heterogeneous reactions of NO_2 contributed to about half of the elevated HONO concentration peaks, with the conversion rates in the range of 0.05% to 0.96% h^{-1}. Meanwhile, a large amount of particulate nitrite, which greatly exceeded the concentration of the gas-phase HONO, was also produced through the heterogeneous reactions of NO_2. The large yields of particulate nitrite were facilitated by abundant ammonia and particulate cations in urban Ji'nan. Notably, in the daytime, particulate nitrite acted as a potential source of HONO, especially in conditions of low humidity and acidic aerosols, which possibly has subsequent effects on photochemistry in the boundary layer.

* Corresponding author.
E-mail address: xinfengwang@sdu.edu.cn (X. Wang).

http://dx.doi.org/10.1016/j.scitotenv.2015.08.032
0048-9697/

1. Introduction

Nitrous acid (HONO) normally accumulates at nighttime and photodissociates during the daytime. The photolysis of HONO in the sunlight enhances the OH radicals (R1) present in the troposphere, especially in the early morning when other sources are weak (Zhou et al., 2001; Alicke et al., 2002; Acker et al., 2006; Kleffmann, 2007). In addition to the high levels of nocturnal HONO, a considerable amount of HONO is also frequently observed at daytime, suggesting the complexity of HONO sources and the existence of unknown ones (Kleffmann et al., 2003; Su et al., 2008a; Sörgel et al., 2011a; Li et al., 2012, 2014; Ma et al., 2013). Therefore, identification and evaluation of the various sources of HONO have become crucial to the recognition of current HONO chemistry (VandenBoer et al., 2014a).

$$HONO \overset{h\nu}{\leftrightarrow} NO + OH \quad (R1)$$

Ambient HONO originates not only from primary emission but also secondary formation. The predominant HONO source varies with locations and periods. In urban cities, a large fraction of HONO is emitted from vehicle exhaust especially during rush hours. In farmlands or forests, soil nitrite and bacteria act as an important emission sources (Su et al., 2011; Oswald et al., 2013). In most of other cases, however, chemical reactions of nitrogen oxides and/or other nitrogen-containing compounds produce more HONO. Among them, heterogeneous process of NO_2 on available surfaces is considered as the dominant formation pathway (R2). Both laboratory and field studies have reported that a significant amount of ambient HONO is converted from NO_2 on wet surfaces such as atmospheric aerosols (mineral dusts, soot, sulfates, organics, etc.), ground, plants, sea, and snow with different mechanisms and conversion rates (Zhou et al., 2001; Herrmann et al., 2010; Khalizov et al., 2010; Bedjanian and El Zein, 2012; Zha et al., 2014). The conversion coefficient mainly depends on the abundance of surface to volume ratios, the surface properties, and the ambient relative humidity (Finlayson-Pitts et al., 2003; Su et al., 2008b; Zha et al., 2014). Besides, other homogeneous and heterogeneous reactions and surface photolysis involving NO, NO_2, and HNO_3 also contribute to the ambient HONO in some specific conditions (Li et al., 2012; Liu et al., 2014). Due to the various constituents of atmospheric trace pollutants and surfaces properties, researchers are far from having a full understanding of the sources of HONO, particularly in a polluted environment with intensive NO_x emissions and high aerosol loading such as occurs in North China.

$$NO_2 + NO_2 + H_2O_{ads} \xrightarrow{surf} HONO + HNO_3 \quad (R2)$$

Particulate nitrite (NO_2^-) has the same oxidation state of the nitrogen atom as HONO. Owing to its very low abundance and chemical instability (Lammel and Cape;, 1996), little attention has been paid to particulate nitrite in past decades. In recent years, however, high levels of particulate nitrite in the troposphere have been observed, and it was revealed that particulate nitrite may serve as a potential reservoir of HONO (Song et al., 2009; VandenBoer et al., 2014b). The production of particulate nitrite from gas-phase HONO may occur on solution layers exposed to the atmosphere, such as wet aerosols, fog droplets, and surface water (Moore et al., 2004; He et al., 2006; Sörgel et al., 2011b; VandenBoer et al., 2014b). R3 demonstrates the reversible reaction and gas-aerosol partition between particulate nitrite and HONO, which depends on such factors as concentrations, acidity of the active surfaces, ambient temperature, and relative humidity. In acidic conditions, particulate nitrite reversibly associates with H^+ ions to yield HONO, such that high levels of particulate nitrite may also contribute to the sources of ambient HONO in the regions with severe particulate pollution. Therefore, recognition of particulate nitrite and its linkages to HONO is essential to provide complete insights into HONO chemistry in such regions.

$$HONO_{(g)} \longleftrightarrow HNO_{2(aq)} \longleftrightarrow NO_{2(aq)}^- + H_{(aq)}^+ \quad (R3)$$

Ji'nan, the capital city of Shandong province, is located almost in the center of North China (a fast-growing region in past decades). The city has a temperate, semi-humid continental monsoon climate, which is rainless in spring, rainy and hot in summer, and dry and cold in autumn and winter. It covers an area of 8177.21 km^2 and has a population of approximately 7.0 million (SPBS, 2014). By the end of 2013, the number of vehicles in Ji'nan had surged to about 1.4 million, a rate of increase of about 14%. Due to the intensive emissions of air pollutants from vehicles and other sources such as industries, urban Ji'nan has been suffering from severe particulate matter pollution in the past decade, and elevated concentrations of particulate nitrite were observed there in our previous studies (e.g., Gao et al., 2011; Wang et al., 2012, 2014). In recent years, the city has devoted itself to reducing pollutant emissions by speeding up the elimination of "yellow label" vehicles and implementing stricter emission standards.

In this study, simultaneous on-line measurements of HONO and nitrite in $PM_{2.5}$ were made in urban Ji'nan during the cold season. Characteristics, sources, and transformation of HONO and particulate nitrite were analyzed in detail with the aid of related air pollutants, with the expectations of obtaining a comprehensive understanding of gas-phase HONO and particulate nitrite in the polluted urban boundary layer in North China and providing new insights into the chemistry of HONO.

2. Experiment and methods

2.1. Site description

Field measurements were conducted at the Atmospheric Environment Observation Station of Shandong University (AEOS-SDU) in urban Ji'nan (N36°40′, E117°03′). It is located on the rooftop of a six-story teaching building on the central campus of the university, approximately 20 m above the ground. The inlets of the online instruments were installed about 1.5 m above the rooftop of the observation station. The sampling period lasted 41 days, from Nov. 26, 2013 to Jan. 5, 2014.

The AEOS-SDU site is surrounded by educational, residential, and commercial districts. It is located to the north of Shanda S Road and to the east of Shanda Road. Shanda N Road lies to the north, and Hongjialou S Road lies to the east (Fig. 1). There are several busy roads nearby: Jiefang Road (about 750 m to the south), Lishan Road (about 1200 m to the west), 2nd Ring Road E (about 1300 m to the east), and Huayuan Road (about 1100 m to the north). In addition to vehicle emissions, large-scale industries in suburban areas are important sources of emissions in Ji'nan. These industries are mainly distributed in the northeast (coal-combustion power plants, steel plant, and chemical plants) and southwest (cement plants).

2.2. Instruments

Data collected in this study included the concentrations of $PM_{2.5}$ water-soluble ions such as nitrite, nitrate, ammonia, and sulfate; $PM_{2.5}$ mass; trace gases of HONO, HNO_3, NO_x, and O_3; and meteorological parameters. The involved monitors, analyzers, and sensors are described in detail below.

Water-soluble gases and ions in $PM_{2.5}$ were measured using the online Monitor for AeRosols and GAses in ambient air (MARGA, ADI20801, Applikon-ECN, Netherlands). A WRD (Wet Rotating Denuder) was used to capture the acidic and alkaline gases including HCl, HNO_2, HNO_3, and NH_3, and a SJAC (Steam Jet Aerosol Collector) was equipped to collect the water-soluble ions in $PM_{2.5}$. The sample solutions were then analyzed by two ion chromatographs (Makkonen et al., 2012; Khezri et al., 2013). Note that the measured data of HONO using the WRD

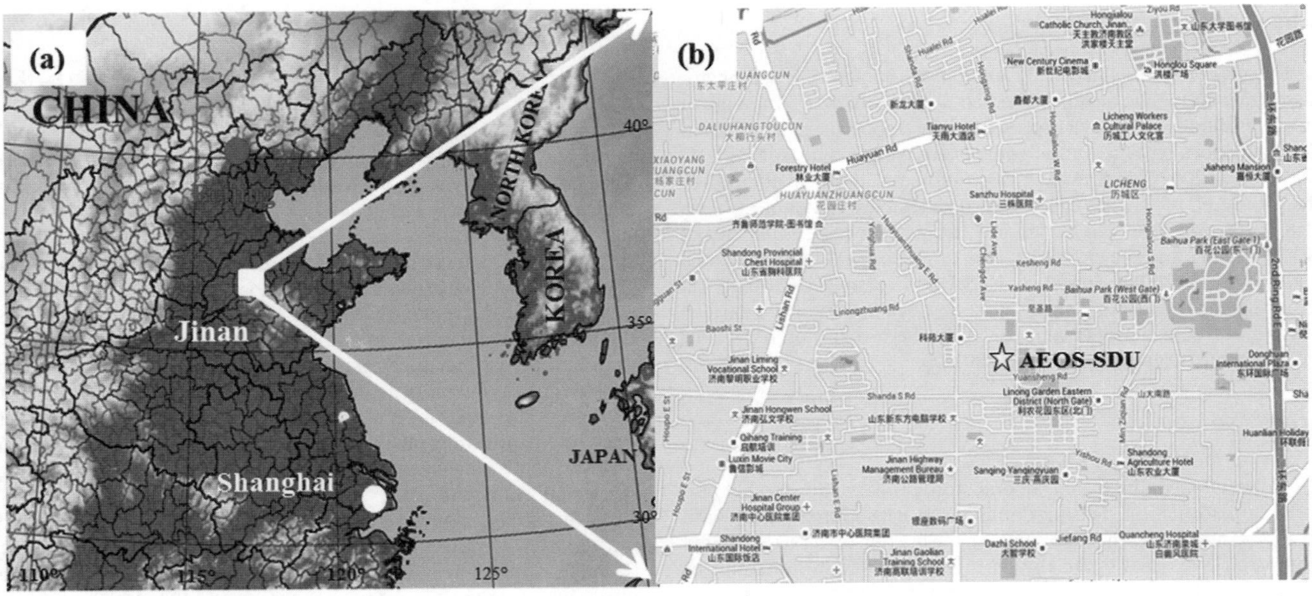

Fig. 1. (a) Topographic map of the study region which shows the locations of Ji'nan (marked in square) and some other major cities. (b) Specific location of the measurement site and roads nearby.

with absorption liquid of 25 mg L^{-1} Na_2CO_3 was previously discovered to be overestimated (Genfa et al., 2003). Therefore, 15 ppmv H_2O_2 in water instead of Na_2CO_3 was applied as absorption solution, which could efficiently oxidize S (IV) and avoid its reaction with NO_2. With the H_2SO_4 formed, the absorption solution is acidified, and the heterogeneous reactions of NO_2 in the WRD are impeded. Thus, any artifact present was the result of the conversion of NO_2 on the surface of the sampling tube, and the overestimation was relatively small. Recent studies have also shown that the overestimation of HONO using WRD mainly occurs in the daytime (Khezri et al., 2013; Nie et al., 2015). Thus, we used only nocturnal HONO data except as described in Section 3.4. During the field study, LiBr solution (4.0 mg L^{-1}) was added to each sample as an internal standard. The aerosol and gas samples were collected by the MARGA at a flow rate of 16.7 L min^{-1} and analyzed at a time resolution of 1 h. During the collection period, 897 hourly samples were taken and analyzed.

$PM_{2.5}$ mass concentration was continuously measured by a Synchronized Hybrid Ambient Real-Time Particulate Monitor (SHARP Monitor, Model 5030; Thermo Fisher Scientific, USA) with a flow rate of 16.7 L min^{-1}. The absorption coefficient of beta rays and the scattering coefficient of 880 nm light were used to quantify the $PM_{2.5}$ concentrations (Wen et al., 2015).

The concentrations of NO and NO_2 were measured by chemiluminescence method (Model 42C, TEC, USA), which was equipped with a molybdenum oxide catalytic converter to convert NO_2 into NO. An ultraviolet absorption method was applied to detect the O_3 concentration (Model 49C, TEC, USA) (Wang et al., 2012; Wen et al., 2015). An automatic meteorological station (PC-4, JZYG, China) was used to collect meteorological data including temperature, relative humidity, and wind direction and speed. In the following analysis, 798 sets of hourly average data of these trace gases and meteorological parameters were collected and used.

3. Results and discussion

3.1. General results of the observation

3.1.1. Characteristics and concentrations

The average concentrations of HONO, NO, NO_2, O_3, $PM_{2.5}$, fine particulate nitrite and nitrate (NO_3^-), and the ratios among them in urban Ji'nan are listed in Table 1. During the measurement period, the average mixing ratios of HONO, NO, NO_2, and O_3 were 0.35, 58.5, 46.9, and 16 ppbv, respectively. The average concentration of fine particulate nitrite was up to 2.08 μg m^{-3}. The average molar or mass ratios of HONO/NO_x, HONO/NO_2, NO_2^-/NO_2, and NO_3^-/$PM_{2.5}$ were 0.006, 0.009, 0.021, and 0.17, respectively. The average HONO concentration observed in Ji'nan exceeded that measured at a background site in Hong Kong (0.126 ppbv, Zha et al., 2014), was comparable to that measured in urban Seoul (0.36 ppbv, Song et al., 2009), but was lower than those measured in most other megacities including Beijing (1.04 ppbv, Spataro et al., 2013), Shanghai (0.92 ppbv, Wang et al., 2013), Guangzhou (2.8 ppbv, Qin et al., 2009), and Santiago (1.5 ppbv, minimum of daily average, Elshorbany et al., 2009). Of note, the HONO concentration in Guangzhou (measured in the summer of 2006) exceeded that in Beijing (measured in the winter of 2006), probably due to the differences in emissions, season, and climate. Guangzhou has a subtropical monsoon climate (humid), whereas Beijing has a temperate monsoon climate. The HONO concentration in Shanghai (measured from 2010 to 2012) was lower than those in both Guangzhou and Beijing, partly because of the implementation of strict vehicle exhaust emission standards in recent years. Compared with that in other cities in China, the HONO concentration in Ji'nan in the cold season was very low,

Table 1
Statistics of major gaseous pollutants and water-soluble ions in $PM_{2.5}$.

Species	Concentration	Sampling instrument
$PM_{2.5}$ (μg m^{-3})	134.9 ± 114.5	SHARP
Nitrate (μg m^{-3})	23.43 ± 19.93	MARGA
Sulfate (μg m^{-3})	24.04 ± 22.46	MARGA
Nitrite (μg m^{-3})	2.08 ± 2.31	MARGA
HONO (ppbv)	0.35 ± 0.5	MARGA
HNO_3 (ppbv)	1.18 ± 1.86	MARGA
NO (ppbv)	58.5 ± 76.9	42C
NO_2 (ppbv)	46.9 ± 21.7	42C
NO_x (ppbv)	105.5 ± 93.6	42C
SO_2 (ppbv)	43 ± 33	43C
O_3 (ppbv)	16 ± 13	49C
HONO/NO	0.081 ± 0.483	
HONO/NO_x	0.0055 ± 0.0149	
HONO/NO_2	0.0093 ± 0.018	
NO_2^-/NO_2	0.021 ± 0.018	
NO_2^-/$\sum$N(III)	0.81 ± 0.17	
NO_3^-/$PM_{2.5}$	0.17 ± 0.07	

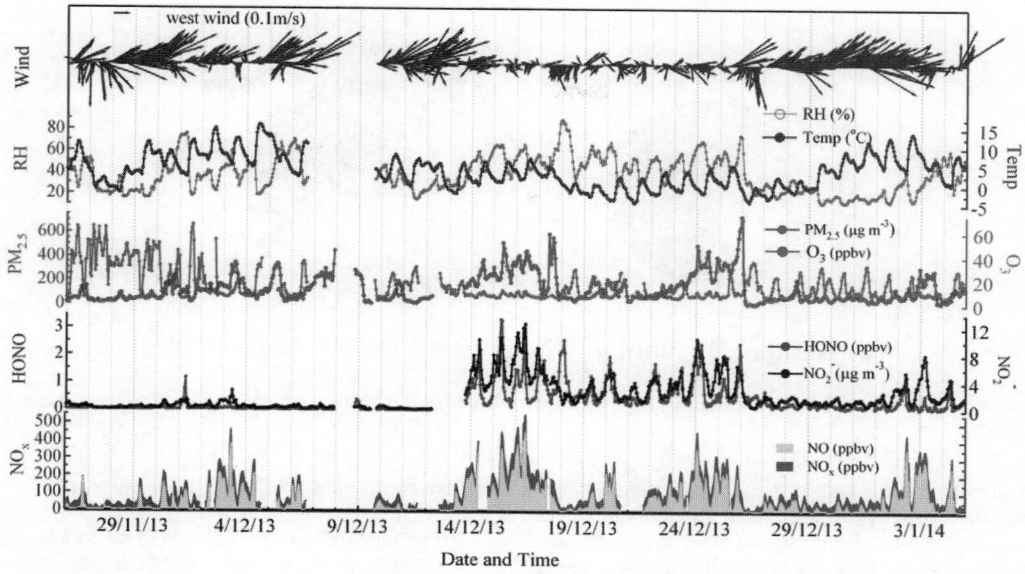

Fig. 2. Time series of concentrations of NO, NO_2, HONO, O_3, particulate nitrite, and $PM_{2.5}$, and meteorological parameters of wind, temperature, and relative humidity during the sampling periods.

owing to the emissions, weather conditions, and the chemical processes occurring in this region, which are discussed in the following sections.

Despite the low concentrations of gas-phase HONO, a high level of fine particulate nitrite was observed in Ji'nan (2.08 μg m^{-3} on average, with an average $PM_{2.5}$ concentration of 134.9 μg m^{-3}), which was much higher than those recorded in Bakersfield, California (0.15 μg m^{-3}, VandenBoer et al., 2014a) and Beijing (0.12 μg m^{-3}, Wang et al., 2005) and was comparable to the concentration in Seoul (1.41 μg m^{-3}, Song et al., 2009). The average molar ratio of fine particulate nitrite to total nitrite (sum of HONO and fine particulate nitrite) was 0.81, and 97% of the NO_2^- data exceeded the HONO concentrations. On average, approximately 1.37 ppbv (corresponding to 2.80 μg m^{-3}) of total nitrite was observed in urban Ji'nan, which is much higher than that in Gwangju (0.677 ppbv, Chang et al., 2008) and Seoul (1.05 ppbv, Song et al., 2009), indicating high levels of N (III) pollutants and possibly important roles of particulate nitrite in the atmospheric chemistry of this area.

Temporal variations of concentrations of aerosols and trace gases and meteorological parameters in Ji'nan are depicted in Fig. 2. Haze episodes occurred frequently, particularly from Dec. 13 to Dec. 26 when the wind speed was low and humidity was high. Throughout this period, high concentrations of HONO and particulate nitrite were observed along with high levels of NO_x and $PM_{2.5}$. The maximum hourly concentration of HONO was 3.39 ppbv, which occurred at 8:00 LT (local time) on Dec. 15, with simultaneous concentrations of NO, NO_2, and $PM_{2.5}$ being 288.4 ppbv, 66.6 ppbv, and 270.0 μg m^{-3}, respectively. The highest particulate nitrite concentration measured was 12.62 μg m^{-3}, which occurred at 10:00 LT on Dec. 16, and simultaneously, concentrations of 452.8 ppbv for NO, 133.2 ppbv for NO_2 (the maximum value recorded during the measurement period), and 443.3 μg m^{-3} for $PM_{2.5}$ were observed. The concurrent appearance of concentration peaks of these air pollutants suggests that high levels of HONO and particulate nitrite in urban Ji'nan are combined results of primary emissions, secondary formation via heterogeneous reactions of NO_2 on aerosol surfaces, and/or accumulation in adverse weather conditions. However, during the period from Nov. 27 to Dec. 12, very low concentrations of both HONO and nitrite were observed in conjunction with high levels of NO_x and $PM_{2.5}$. This was possibly associated with a special meteorological condition (e.g., high-speed wind from the southwest), intense solar radiation (as indicated by good visibility and low humidity), and relatively high concentrations of atmospheric oxidants (e.g., ozone), which accelerated the oxidation of N (III).

3.1.2. Diurnal variation

Concentrations of HONO and particulate nitrite exhibited apparent diurnal variations (as shown in Fig. 3). From late afternoon, HONO concentrations started to increase and reached a peak (maximum average value of 0.58 ppbv) at 7:00 LT in the early morning. After sunrise, decomposition started in the presence of sunlight, and thus the concentration decreased to a minimum average of 0.16 ppbv at 13:00 LT in the early afternoon. Although the diurnal pattern of particulate nitrite was similar to that of HONO, the appearance of a concentration peak and valley occurred 1 or 2 h later than that of HONO. The maximum average concentration of particulate nitrite (3.00 μg m^{-3}) appeared at 8:00 LT, and the valley value was 1.17 μg m^{-3} at 15:00 LT. These results indicate

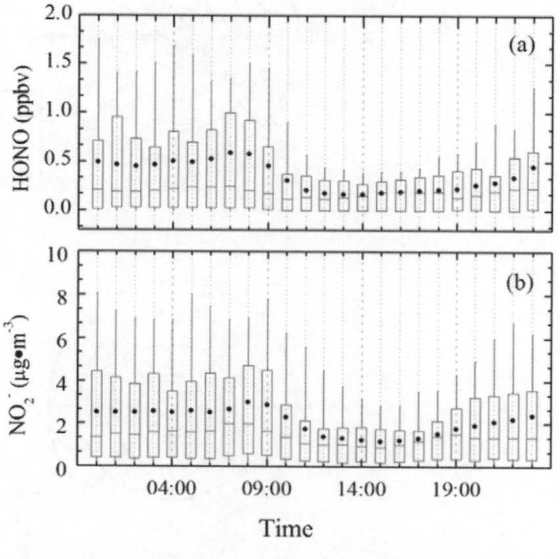

Fig. 3. Diurnal variations of (a) HONO and (b) fine particulate nitrite. The dot and the line in the boxes refer to the mean and median values, respectively. The boxes represent 50% (25% to 75%) and the lengths of whiskers donate 90% of the data.

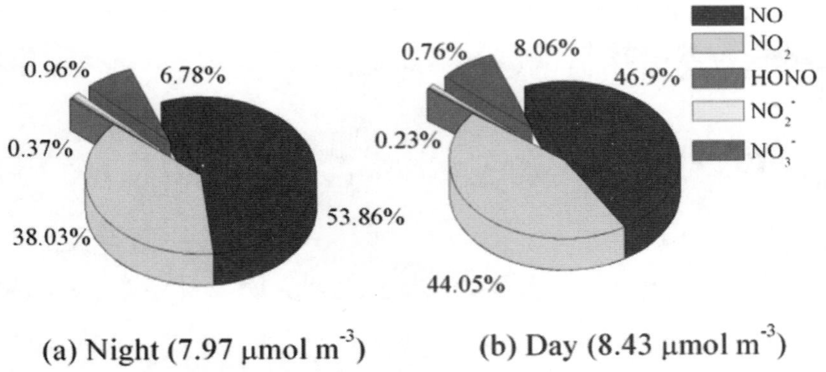

Fig. 4. Proportions of major nitrogen-containing components at (a) nighttime and (b) daytime (data of HNO_3 are not included because the concentrations were very low).

striking resemblances in sources and formation pathways between HONO and particulate nitrite, with the time lag attributable to the chemical transformation and equilibrium among them and related substances. The HONO concentration was the highest in autumn, followed by winter, summer, and spring in Beijing (Hendrick et al., 2014) and Hong Kong (Xu et al., 2015). The concentration of particulate nitrite in Ji'nan was high in winter and summer and relatively low in autumn and spring according to our previous study (Gao et al., 2011).

In general, the concentrations of HONO and particulate nitrite at nighttime were much higher than those at daytime. Moreover, the nocturnal proportions of HONO and nitrite in reactive nitrogen oxides were also higher than those during daytime: 0.37% and 0.96% versus 0.23% and 0.76% (Fig. 4). In contrast, the proportions of oxidized nitrogen-containing compounds such as NO_2 and particulate nitrate were remarkably higher at daytime than those at night, which was attributed to the strong oxidation capacity of photochemical oxidants in the presence of sunlight (Ma et al., 2013).

3.2. Direct emissions of HONO

To understand the contribution of direct emissions to the HONO concentration peaks, nocturnal HONO mixing ratios were analyzed in combination with wind directions and speeds. As shown in Fig. 5, elevated HONO concentrations (>0.5 ppbv) mostly occurred during periods of low wind speed (<1 m s^{-1}). When the wind speed was above 1.5 m s^{-1}, the HONO concentration was substantially below 0.3 ppbv. However, there was no significant linear correlation between the HONO concentration and wind speed. These results suggest that local emissions contribute significantly to the frequent nocturnal concentration peaks of HONO. The distribution characteristics of HONO with winds were quite similar to those of NO, with the majority of the increased concentrations emerging at low wind speeds (<1 m s^{-1}) and lower concentrations emerging at high wind speeds, but were significantly different from those of NO_2 and SO_2, indicating that traffic emissions had a strong influence on HONO peaks recorded at the sampling site.

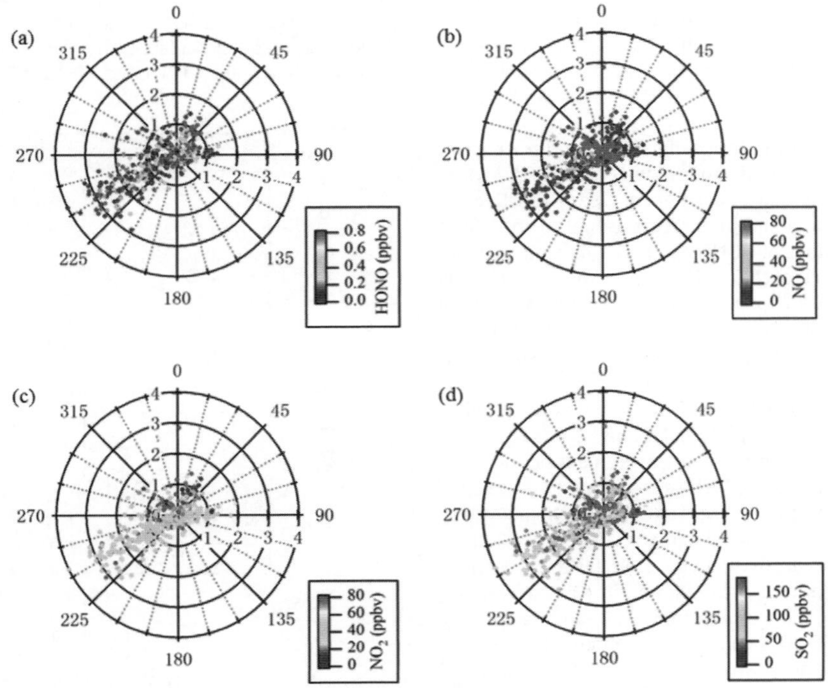

Fig. 5. Nighttime frequencies of wind directions and speeds for (a) HONO, (b) NO, (c) NO_2, and (d) SO_2 during the measurement periods.

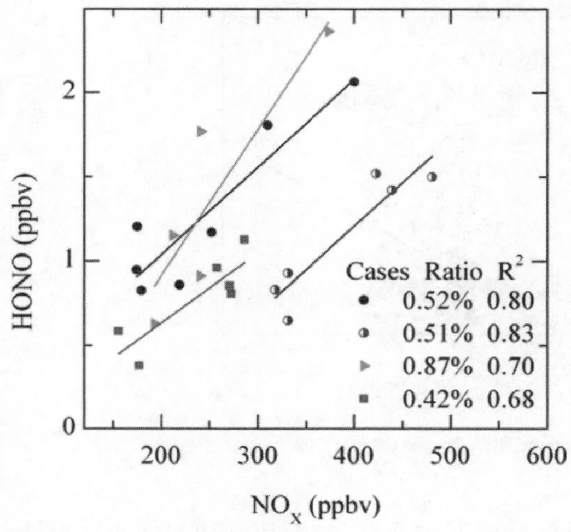

Fig. 6. Scatter plots of HONO and NO_x for four nocturnal cases.

The primary emission of HONO in urban Ji'nan is mainly attributed to vehicle exhaust. To further clarify the contributions of traffic emissions to the HONO concentration peaks, pollution cases were selected according to the following criteria: (a) only nocturnal data (to avoid the effects of photolysis in the daytime); (b) NO/NO_x ratio > 0.7; and (c) NO > 88 ppbv (the highest 25% of NO data). Totally four cases were chosen here and analyzed in detail. As shown in Fig. 6, HONO correlated well with NO_x ($0.68 \le R^2 \le 0.83$), with the $HONO/NO_x$ ratio ranging from 0.42% to 0.87%. These results indicate that traffic emissions are a principal source of the HONO peaks. Under conditions dominated by traffic emissions, an average $HONO/NO_x$ ratio of 0.58% was obtained at our sampling site. The ratios of $HONO/NO_x$ have been reported to range from 0.3% to 0.8% in freshly emitted vehicle exhaust based on tunnel studies (e.g., Kleffmann et al., 2003; Kurtenbach et al., 2001). Therefore, 0.58% was reasonably chosen as the emission factor at this site and used to adjust the HONO data to exclude the contribution from traffic emissions in Section 3.3.

3.3. Heterogeneous formation of HONO and particulate nitrite

3.3.1. Heterogeneous formation of HONO

Apart from direct emissions, a large fraction of HONO was produced from the heterogeneous reaction of NO_2 on wet surfaces. To understand the contribution of the heterogeneous uptake of NO_2 to the elevated HONO concentrations and the conversion efficiency, seven nighttime cases were selected and analyzed here. Instead of the original HONO concentration, an adjusted HONO concentration, HONO*, was used to remove (or reduce) the influence from traffic emissions. The HONO* concentration was calculated by subtracting the NO_x concentration timing factor of 0.58% obtained in the previous section (Eq. (1)). Considering all of the nocturnal data with HONO concentration peaks greater than 1 ppbv, approximately 42% of the HONO was contributed by traffic emissions, and about half (~58%) was attributed to heterogeneous reactions of NO_2.

$$[\mathrm{HONO}*] = [\mathrm{HONO}] - 0.58\% \times [\mathrm{NO}] \quad (1)$$

Subsequently, the NO_2-to-HONO conversion efficiency was calculated following Eq. (2) (Alicke et al., 2002; Li et al., 2012).

$$C_{\mathrm{HONO}} = \frac{[\mathrm{HONO}*]_{t_2} - [\mathrm{HONO}*]_{t_1}}{\overline{[\mathrm{NO_2}]} \times (t_2 - t_1)} \quad (2)$$

Here, $\overline{[NO_2]}$ represents the average NO_2 concentration over the time interval t_1 to t_2.

C_{HONO} and HONO mixing ratios in these cases are summarized in Table 2. The adjusted HONO concentrations showed large variability (less than 0.1 to 2.5 ppbv). As depicted, HONO* concentrations correlated well with the NO_2 mixing ratios in most cases ($0.54 \le R^2 \le 0.89$), suggesting that the heterogeneous reactions of NO_2 were the major formation pathways of the enhanced HONO during these periods. Similar to HONO*, C_{HONO} also showed great variability, from 0.05% to 0.96% h^{-1} with an average of 0.29% h^{-1}. The conversion efficiency in urban Ji'nan was lower than those observed in Guangzhou (1.6% h^{-1}, Su et al., 2008a; Li et al., 2012), Shanghai (0.7% h^{-1}, Wang et al., 2013), and Hong Kong (0.52% h^{-1}, Xu et al., 2015). Apart from case 3 on Dec. 17, the C_{HONO} values correlated well with $PM_{2.5}$ ($R^2 = 0.84$), suggesting that the aerosol surface is of vital importance to the NO_2-to-HONO conversion efficiency under severe haze conditions. In addition, very good correlation was found between relative humidity and NO_2-to-HONO conversion efficiency ($R^2 = 0.92$). It is possible that the conversion rate is strongly influenced by the liquid water content, which correlates positively with relative humidity.

3.3.2. Heterogeneous formation of particulate nitrite

Besides HONO, particulate nitrite is another important product of the heterogeneous reactions of NO_2 in the troposphere. Moderately good correlations were found between nocturnal nitrite and NO_2 (see Fig. 7). In conditions of moderate and high humidity (RH > 50%), the concentration of particulate nitrite increased as the concentration of NO_2 rose with an average slope of 0.10 (μg m^{-3} $ppbv^{-1}$), indicating the formation of particulate nitrite from the heterogeneous reaction of NO_2. When the relative humidity was low (<50%), the level of particulate nitrite was relatively low and changed little with NO_2 concentration in that the heterogeneous reactions of NO_2 were inhibited.

Six nocturnal cases were further analyzed to explore the formation of elevated concentrations of particulate nitrite (Fig. 8). These cases lasted from 6 to 14 h, during which the concentration of nitrite rose positively with NO_2 concentration. Strong correlations ($0.68 \le R^2 \le 0.92$) between the concentrations indicated that the heterogeneous process is one of the major formation pathways of particulate nitrite and makes a significant contribution to the elevated concentrations of nocturnal particulate nitrite during the cold season in urban areas of Ji'nan. In the six pollution cases, the slope of fine NO_2^- versus NO_2 ranged from 0.045 to 0.250 (μg m^{-3} $ppbv^{-1}$), with the maximum slope of 0.250 occurring during the night of Dec. 15.

Table 2
C_{HONO}, R^2 and other supportive information for seven nocturnal cases.

Case	Date	Time	R^2	C_{HONO} (% h^{-1})	HONO* (ppbv)	RH[a] (%)	$PM_{2.5}$ [a](μg m^{-3})
1	Dec. 15, 2013	01:00–07:00	0.58	0.34	0.88–1.53	65	227.4
2	Dec. 17, 2013	00:00–05:00	0.52	0.13	0.45–0.82	51	162.9
3	Dec. 17, 2013	19:00–03:00	0.90	0.96	0.41–2.50	81	91.3
4	Dec. 18, 2013	19:00–22:00	0.90	0.10	0.26–0.39	48	69.1
5	Dec. 19, 2013	19:00–23:00	0.82	0.05	0.29–0.43	51	124.1
6	Dec. 24, 2013	03:00–07:00	0.92	0.42	0.20–1.10	69	342.5
7	Jan. 04, 2014	19:00–23:00	0.93	0.12	0.11–0.29	50	144.3

[a] Mean value of the selected case period.

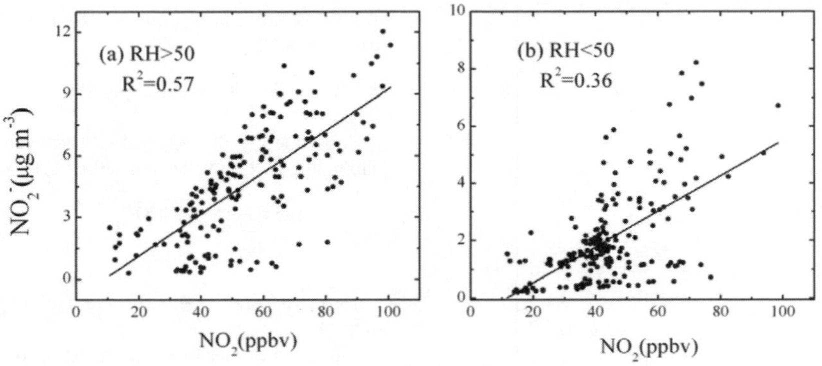

Fig. 7. Scatter plots of fine particulate nitrite versus NO_2 with (a) RH > 50% and (b) RH < 50%.

Generally, the NO_2-to-nitrite conversion efficiency is promoted with a larger active water layer surface (e.g., higher relative humidity and $PM_{2.5}$ mass concentration), favorable surface chemical environment (e.g., alkaline aerosols and less oxidants), and steady meteorological conditions. On Dec. 15, high relative humidity (65% on average), elevated $PM_{2.5}$ mass concentration, sufficient ammonium cations, low levels of ozone, and low wind speeds all facilitated the heterogeneous reactions and led to an extremely high conversion efficiency.

The high ratios of particulate nitrite to total nitrites and the good correlations between particulate nitrite and NO_2 suggest that a large fraction of heterogeneous products is presented in particulate form instead of gas-phase HONO. Here, the concentration of free cation (equivalent concentration) was calculated to evaluate the existence of nitrite with consideration of all of the major ions in fine particles (Eq. (3)).

$$\text{Cation}_{(p,\,free)} = \left(NH_{4\,(p)}^{+} + Na_{(p)}^{+} + K_{(p)}^{+} + \left(2 \times Ca_{(p)}^{2+}\right) + \left(2 \times Mg_{(p)}^{2+}\right)\right) - \left(\left(2 \times SO_{4\,(p)}^{2-}\right) + NO_{3\,(p)}^{-} + Cl_{(p)}^{-}\right) \tag{3}$$

It is worth noting that ammonium contributed the most among the five major cations and was produced from the rich ammonia (9.1 ± 6.3 ppbv on average) present in urban Ji'nan.

Throughout the nocturnal elevations, the majority of the free cation concentrations were higher than the peak concentrations of particulate nitrite, i.e., the cations were in excess. Good correlation ($R^2 = 0.68$) was found between them, and particulate nitrite concentrations increased positively with the free cations in most of the periods. Therefore, the formation of particulate nitrite in urban Ji'nan is enhanced by the presence of rich ammonia, and there are abundant cations left to neutralize the newly produced nitrite in wet aerosols or small droplets. As a result, particulate nitrite became the dominant product of the heterogeneous processes of NO_2, and the nocturnal HONO presented relatively low concentration in North China during the cold season.

3.4. Release of HONO from particulate nitrite during the daytime

Gas-phase HONO could be produced from high levels of particulate nitrite in suitable conditions. The release of HONO from particulate nitrite is promoted when the ambient temperature is high, humidity is low, and in particular, the aerosols are acidic.

Neutralization degree (*F*) is used here to represent the acidity of the aerosols and is calculated using the equivalent concentrations of the major ions: sulfate, nitrate, and ammonium (Eq. (4)) (Zhou et al., 2012).

$$F = [NH_4^+]/\left(2 \times [SO_4^{2-}] + [NO_3^-]\right) \tag{4}$$

As illustrated in Fig. 9, gas-phase HONO formation was observed in urban Ji'nan with the concurrent decrease in particulate nitrite concentration in the daytime. From the late morning to the early afternoon (11:00–15:00 LT) on Dec. 18, there was a sharp decrease in the relative humidity (from 52% to 35%), leading to a very dry environment (little liquid water available on aerosol surfaces). Moreover, the aerosols were slightly acidic (*F* value <1 at most times during the period). The HONO concentration exhibited a continuous increase from 0.25 to 0.53 ppbv. Within the 4 h, the concentration of fine nitrite decreased from 2.00 to 1.04 μg m^{-3}. The ratio of fine particulate nitrite to the total nitrites also decreased from 0.77 to 0.50. The rate of decrease in particulate nitrite concentration was approximately 0.24 μg m^{-3} h^{-1}. Simultaneously, the rate of increase of the HONO concentration was 0.06 ppbv · h^{-1} (~0.13 μg m^{-3} h^{-1}), suggesting the release of HONO from nitrite. On Dec. 26, from 10:00 LT in the morning to 15:00 LT in the afternoon, the relative humidity was very low (ranging from 22% to 30%), and the aerosols were acidic with the *F* value decreased to ~0.8. During this period, although the particulate nitrite concentration was low (substantially below 0.90 μg m^{-3}), a significant increase in the HONO concentration (from 0.39 to 0.60 ppbv) was observed with a simultaneous decline in the fine nitrite concentration. The average rate of increase of the HONO concentration was 0.03 ppbv · h^{-1} (~0.06 μg m^{-3} h^{-1}), and the rate of decrease of the fine nitrite was 0.08 μg m^{-3} h^{-1} – close to the rate of increase of the HONO. In this case, the proportion of fine particulate nitrite in total nitrites decreased from 0.52 to 0.30, i.e., almost half of the fine particulate nitrite was converted into gas-phase HONO. Moreover, the NO_x concentrations during these two periods were substantially low (less than 15 and 9 ppbv, respectively), indicating limited contributions to HONO from other sources such as traffic emissions and heterogeneous reactions of NO_2.

It is well known that photolysis of HONO is an important source of OH radicals in the daytime. The high levels of particulate nitrite in

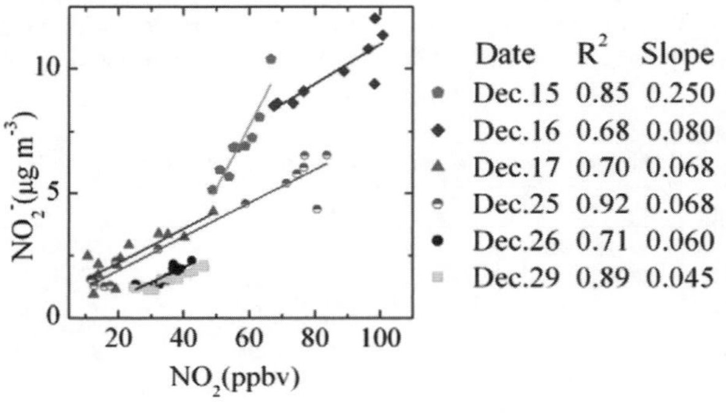

Fig. 8. Correlations between fine particulate nitrite and NO_2 for six nocturnal cases.

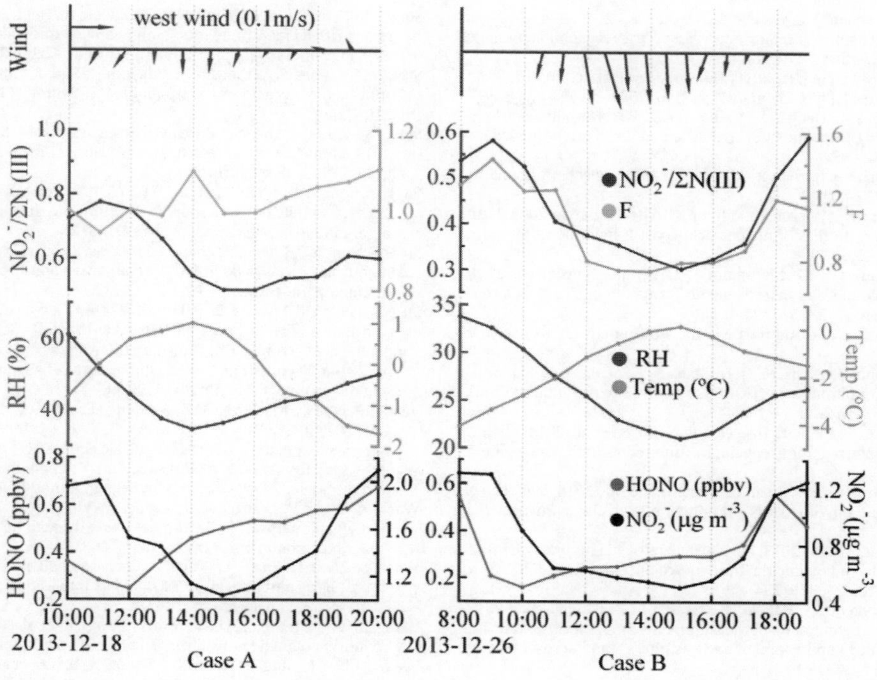

Fig. 9. Temporal variations of HONO, fine particulate nitrite, ambient temperature, RH, $NO_2^-/\sum N(III)$ ratio and *F* value for two selected cases. Case A lasted from 10:00 to 20:00 LT on Dec. 18 and case B started on 8:00 LT and ended at 19:00 LT on Dec. 26. The NO_2^- ratio and *F* value were estimated based on the measured data (see main text).

urban Ji'nan and the enhanced aerosol-gas conversion in conditions of low humidity and acidic aerosols made the particulate nitrite an important source of HONO, especially during the daytime, and produced subsequent effects on the photochemistry in the boundary layer.

4. Summary and conclusions

High time-resolution field measurements of HONO, particulate nitrite in $PM_{2.5}$, related air pollutants, and meteorological parameters were conducted in urban Ji'nan in North China from Nov. 26, 2013 to Jan. 4, 2014. During the measurement period, the mean concentrations of HONO and fine particulate nitrite were 0.35 ppbv and 2.08 μg m^{-3}, with the maximum hourly concentrations recorded of 3.39 ppbv and 12.62 μg m^{-3}, respectively. The concentrations of both HONO and fine particulate nitrite exhibited distinct diurnal cycles with peaks in the early morning and valleys in the early afternoon. Elevated nocturnal HONO concentration peaks were associated with primary emissions from vehicle exhaust and secondary formation via heterogeneous reactions of NO_2. The average $HONO/NO_x$ ratio during the periods with fresh vehicle emissions was 0.58%. The efficiency of the heterogeneous conversion of NO_2 to HONO, ranging from 0.05% to 0.96% h^{-1}, strongly depended on the humidity and the aerosol loading. Overall, secondary formation contributed to about half of the elevated HONO concentration peaks. In addition, the heterogeneous reactions of NO_2 on aerosol surfaces also produced significant amounts of particulate nitrite. Abundant ammonia and alkaline substance in fine particles enhanced the heterogeneous formation of particulate nitrite and thus made it the major product instead of gas-phase HONO in urban Ji'nan. However, during the daytime, particulate nitrite acted as a potential source of HONO. Gas-phase HONO was released especially in conditions of low humidity and acidic aerosols. The enhanced aerosol-gas conversion and high levels of particulate nitrites make non-negligible contributions to the production of HONO and thus have subsequent effects on the photochemistry in the boundary layer in North China. With the implementation of stricter vehicle emission standards in China, the emissions and productions of HONO and particulate nitrite are expected to decrease in the future. However, the concurrent control of ammonia emissions may lead to enhanced release of gas-phase HONO from particulate nitrite in the acid aerosol condition.

Acknowledgments

This work was supported by Taishan Scholar Grand (ts20120552), National Natural Science Foundation of China (Nos. 41275123, 41375126, 21407094, 21307074), Natural Science Foundation of Shandong Province (No. ZR2014BQ031), and Strategic Priority Research Program of the Chinese Academy of Sciences (No. XDB05010200).

References

Acker, K., Möller, D., Wieprecht, W., Meixner, F.X., Bohn, B., Gilge, S., et al., 2006. Strong daytime production of OH from HNO_2 at a rural mountain site. Geophys. Res. Lett. 33.
Alicke, B., Platt, U., Stutz, J., 2002. Impact of nitrous acid photolysis on the total hydroxyl radical budget during the Limitation of Oxidant Production/Pianura Padana Produzione di Ozono study in Milan. J. Geophys. Res. 107.
Bedjanian, Y., El Zein, A., 2012. Interaction of NO_2 with TiO_2 surface under UV irradiation: products study. J. Phys. Chem. A 116, 1758–1764.
Chang, W., Choi, J., Hong, S., Lee, J.H., 2008. Simultaneous measurements of gaseous nitrous acid and particulate nitrite using diffusion scrubber/steam chamber/luminol chemiluminescence. Bull. Kor. Chem. Soc. 29, 1525–1532.
Elshorbany, Y., Kurtenbach, R., Wiesen, P., Lissi, E., Rubio, M., Villena, G., et al., 2009. Oxidation capacity of the city air of Santiago, Chile. Atmos. Chem. Phys. 9, 2257–2273.
Finlayson-Pitts, B.J., Wingen, L.M., Sumner, A.L., Syomin, D., Ramazan, K.A., 2003. The heterogeneous hydrolysis of NO_2 in laboratory systems and in outdoor and indoor atmospheres: an integrated mechanism. Phys. Chem. Chem. Phys. 5, 223–242.
Gao, X., Yang, L., Cheng, S., Gao, R., Zhou, Y., Xue, L., et al., 2011. Semi-continuous measurement of water-soluble ions in $PM_{2.5}$ in Jinan, China: temporal variations and source apportionments. Atmos. Environ. 45, 6048–6056.
Genfa, Z., Slanina, S., Boring, C.B., Jongejan, P.A., Dasgupta, P.K., 2003. Continuous wet denuder measurements of atmospheric nitric and nitrous acids during the 1999 Atlanta Supersite. Atmos. Environ. 37, 1351–1364.
He, Y., Zhou, X., Hou, J., Gao, H., Bertman, S.B., 2006. Importance of dew in controlling the air — surface exchange of HONO in rural forested environments. Geophys. Res. Lett. 33.
Hendrick, F., Müller, J.F., Clémer, K., Wang, P., De Mazière, M., Fayt, C., et al., 2014. Four years of ground-based MAX-DOAS observations of HONO and NO_2 in the Beijing area. Atmos. Chem. Phys. 14, 765–781.

Herrmann, H., Hoffmann, D., Schaefer, T., Bräuer, P., Tilgner, A., 2010. Tropospheric aqueous-phase free-radical chemistry: radical sources, spectra, reaction kinetics and prediction tools. ChemPhysChem 11, 3796–3822.

Khalizov, A.F., Cruz-Quinones, M., Zhang, R., 2010. Heterogeneous reaction of NO_2 on fresh and coated soot surfaces. J. Phys. Chem. A 114, 7516–7524.

Khezri, B., Mo, H., Yan, Z., Chong, S.-L., Heng, A.K., Webster, R.D., 2013. Simultaneous on-line monitoring of inorganic compounds in aerosols and gases in an industrialized area. Atmos. Environ. 80, 352–360.

Kleffmann, J., 2007. Daytime sources of nitrous acid (HONO) in the atmospheric boundary layer. ChemPhysChem 8, 1137–1144.

Kleffmann, J., Kurtenbach, R., Lörzer, J., Wiesen, P., Kalthoff, N., Vogel, B., et al., 2003. Measured and simulated vertical profiles of nitrous acid—part I: field measurements. Atmos. Environ. 37, 2949–2955.

Kurtenbach, R., Becker, K.H., Gomes, J.A.G., Kleffmann, J., Lörzer, J.C., Spittler, M., et al., 2001. Investigations of emissions and heterogeneous formation of HONO in a road traffic tunnel. Atmos. Environ. 35, 3385–3394.

Lammel, G., Cape, J.N., 1996. Nitrous acid and nitrite in the atmosphere. Chem. Soc. Rev. 25, 361–369.

Li, X., Brauers, T., Häseler, R., Bohn, B., Fuchs, H., Hofzumahaus, A., et al., 2012. Exploring the atmospheric chemistry of nitrous acid (HONO) at a rural site in Southern China. Atmos. Chem. Phys. 12, 1497–1513.

Li, X., Rohrer, F., Hofzumahaus, A., Brauers, T., Haeseler, R., Bohn, B., et al., 2014. Missing gas-phase source of HONO inferred from Zeppelin measurements in the troposphere. Science 344, 292–296.

Liu, Z., Wang, Y., Costabile, F., Amoroso, A., Zhao, C., Huey, L.G., et al., 2014. Evidence of aerosols as a media for rapid daytime HONO production over China. Environ. Sci. Technol. 48, 14386–14391.

Ma, J., Liu, Y., Han, C., Ma, Q., Liu, C., He, H., 2013. Review of heterogeneous photochemical reactions of NO_y on aerosol — a possible daytime source of nitrous acid (HONO) in the atmosphere. J. Environ. Sci. 25, 326–334.

Makkonen, U., Virkkula, A., Mäntykenttä, J., Hakola, H., Keronen, P., Vakkari, V., et al., 2012. Semi-continuous gas and inorganic aerosol measurements at a Finnish urban site: comparisons with filters, nitrogen in aerosol and gas phases, and aerosol acidity. Atmos. Chem. Phys. 12, 5617–5631.

Moore, K.F., Eli Sherman, D., Reilly, J.E., Hannigan, M.P., Lee, T., Collett Jr., J.L., 2004. Drop size-dependent chemical composition of clouds and fogs. Part II: relevance to interpreting the aerosol/trace gas/fog system. Atmos. Environ. 38, 1403–1415.

Nie, W., Ding, A.J., Xie, Y.N., Xu, Z., Mao, H., Kerminen, V.M., et al., 2015. Influence of biomass burning plumes on HONO chemistry in eastern China. Atmos. Chem. Phys. 15, 1147–1159.

Oswald, R., Behrendt, T., Ermel, M., Wu, D., Su, H., Cheng, Y., et al., 2013. HONO emissions from soil bacteria as a major source of atmospheric reactive nitrogen. Science 341, 1233–1235.

Qin, M., Xie, P., Su, H., Gu, J., Peng, F., Li, S., et al., 2009. An observational study of the HONO–NO_2 coupling at an urban site in Guangzhou City, South China. Atmos. Environ. 43, 5731–5742.

Song, C.H., Park, M.E., Lee, E.J., Lee, J.H., Lee, B.K., Lee, D.S., et al., 2009. Possible particulate nitrite formation and its atmospheric implications inferred from the observations in Seoul, Korea. Atmos. Environ. 43, 2168–2173.

Sörgel, M., Regelin, E., Bozem, H., Diesch, J.-M., Drewnick, F., Fischer, H., et al., 2011a. Quantification of the unknown HONO daytime source and its relation to NO_2. Atmos. Chem. Phys. 11, 10433–10447.

Sörgel, M., Trebs, I., Serafimovich, A., Moravek, A., Held, A., Zetzsch, C., 2011b. Simultaneous HONO measurements in and above a forest canopy: influence of turbulent exchange on mixing ratio differences. Atmos. Chem. Phys. 11, 841–855.

Spataro, F., Ianniello, A., Esposito, G., Allegrini, I., Zhu, T., Hu, M., 2013. Occurrence of atmospheric nitrous acid in the urban area of Beijing (China). Sci. Total Environ. 447, 210–224.

SPBS (Shandong Provincial Bureau of Statistics), 2014. Shandong Statistical Yearbook 2014. China Statistics Press, Beijing (Available online at http://www.stats-sd.gov.cn/tjnj/nj2014/indexch.htm).

Su, H., Cheng, Y.F., Shao, M., Gao, D.F., Yu, Z.Y., Zeng, L.M., et al., 2008a. Nitrous acid (HONO) and its daytime sources at a rural site during the 2004 PRIDE-PRD experiment in China. J. Geophys. Res. 113, D14312.

Su, H., Cheng, Y.F., Cheng, P., Zhang, Y.H., Dong, S., Zeng, L.M., et al., 2008b. Observation of nighttime nitrous acid (HONO) formation at a non-urban site during PRIDE-PRD2004 in China. Atmos. Environ. 42, 6219–6232.

Su, H., Cheng, Y.F., Oswald, R., Behrendt, T., Trebs, I., Meixner, F.X., et al., 2011. Soil nitrite as a source of atmospheric HONO and OH radicals. Science 333, 1616–1618.

VandenBoer, T.C., Young, C.J., Talukdar, R.K., Markovic, M.Z., Brown, S.S., Roberts, J.M., et al., 2014a. Nocturnal loss and daytime source of nitrous acid through reactive uptake and displacement. Nat. Geosci. 8, 55–60.

VandenBoer, T.C., Markovic, M.Z., Sanders, J.E., Ren, X., Pusede, S.E., Browne, E.C., et al., 2014b. Evidence for a nitrous acid (HONO) reservoir at the ground surface in Bakersfield, CA, during CalNex 2010. J. Geophys. Res. 119, 9093–9106.

Wang, Y., Zhuang, G., Tang, A., Yuan, H., Sun, Y., Chen, S., et al., 2005. The ion chemistry and the source of $PM_{2.5}$ aerosol in Beijing. Atmos. Environ. 39, 3771–3784.

Wang, X., Wang, W., Yang, L., Gao, X., Nie, W., Yu, Y., et al., 2012. The secondary formation of inorganic aerosols in the droplet mode through heterogeneous aqueous reactions under haze conditions. Atmos. Environ. 63, 68–76.

Wang, S., Zhou, R., Zhao, H., Wang, Z., Chen, L., Zhou, B., 2013. Long-term observation of atmospheric nitrous acid (HONO) and its implication to local NO_2 levels in Shanghai, China. Atmos. Environ. 77, 718–724.

Wang, X., Chen, J., Sun, J., Li, W., Yang, L., Wen, L., et al., 2014. Severe haze episodes and seriously polluted fog water in Ji'nan, China. Sci. Total Environ. 493, 133–137.

Wen, L., Chen, J., Yang, L., Wang, X., Caihong, X., Sui, X., et al., 2015. Enhanced formation of fine particulate nitrate at a rural site on the North China Plain in summer: the important roles of ammonia and ozone. Atmos. Environ. 101, 294–302.

Xu, Z., Wang, T., Wu, J., Xue, L., Chan, J., Zha, Q., et al., 2015. Nitrous acid (HONO) in a polluted subtropical atmosphere: seasonal variability, direct vehicle emissions and heterogeneous production at ground surface. Atmos. Environ. 106, 100–109.

Zha, Q., Xue, L., Wang, T., Xu, Z., Yeung, C., Louie, P.K.K., et al., 2014. Large conversion rates of NO_2 to HNO_2 observed in air masses from the South China Sea: evidence of strong production at sea surface? Geophys. Res. Lett. 41, 7710–7715.

Zhou, X., Beine, H.J., Honrath, R.E., Fuentes, J.D., Simpson, W., Shepson, P.B., et al., 2001. Snowpack photochemical production of HONO: a major source of OH in the Arctic boundary layer in springtime. Geophys. Res. Lett. 28, 4087–4090.

Zhou, Y., Xue, L., Wang, T., Gao, X., Wang, Z., Wang, X., et al., 2012. Characterization of aerosol acidity at a high mountain site in central eastern China. Atmos. Environ. 51, 11–20.

Science of the Total Environment 476–477 (2014) 57–64

Contents lists available at ScienceDirect

Science of the Total Environment

journal homepage: www.elsevier.com/locate/scitotenv

Impacts of firecracker burning on aerosol chemical characteristics and human health risk levels during the Chinese New Year Celebration in Jinan, China

Lingxiao Yang [a,b,*], Xiaomei Gao [a,d], Xinfeng Wang [a], Wei Nie [a], Jing Wang [a], Rui Gao [a], Pengju Xu [a], Youping Shou [a], Qingzhu Zhang [a], Wenxing Wang [a,c]

[a] Environment Research Institute, Shandong University, Jinan 250100, China
[b] School of Environmental Science and Engineering, Shandong University, Jinan 250100, China
[c] Chinese Research Academy of Environmental Sciences, Beijing 100012, China
[d] School of Resources and Environment, University of Jinan, Jinan 250022, China

HIGHLIGHTS

- The effect of firecracker burning on aerosol characteristics and human health was assessed.
- The burning of firecrackers elevated the concentrations of particles and water-soluble ions.
- The burning of firecrackers varied the chemical composition of $PM_{2.5}$ and the number size distribution of particles.
- The burning of firecrackers did not alter the mass size distributions of the water-soluble ions.
- Pollutants emitted from the firecracker burning caused high non-carcinogenic risks to human health.

ARTICLE INFO

Article history:
Received 2 September 2013
Received in revised form 23 December 2013
Accepted 23 December 2013
Available online 21 January 2014

Keywords:
Chemical component
Risk assessment
Firecrackers
Chinese New Year
Jinan

ABSTRACT

Measurements for size distribution and chemical components (including water-soluble ions, OC/EC and trace elements) of particles were taken in Jinan, China, during the 2008 Chinese New Year (CNY) to assess the impacts of firecracker burning on aerosol chemical characteristics and human health risk levels. On the eve of the CNY, the widespread burning of firecrackers had a clear contribution to the number concentration of small accumulation mode particles (100–500 nm) and $PM_{2.5}$ mass concentration, with a maximum $PM_{2.5}$ concentration of 464.02 μg/m^3. The firecracker activities altered the number size distribution of particles, but had no influence on the mass size distribution of major water-soluble ions. The concentrations of aerosol and most ions peaked in the rush hour of firecracker burning, whereas the peaks of NO_3^- and NH_4^+ presented on the day following the burning of firecrackers. K^+, SO_4^{2-} and Cl^- composed approximately 62% of the $PM_{2.5}$ mass, and they existed as KCl and K_2SO_4 during the firecracker period. However, during the non-firecracker period, organic matter (OM), SO_4^{2-}, NO_3^- and NH_4^+ were the major chemical components of the $PM_{2.5}$, and major ions were primarily observed as $(NH_4)_2SO_4$ and NH_4NO_3. Estimates of non-carcinogenic risk levels to human health showed that the elemental risk levels during the firecracker period were substantially higher than those observed during the non-firecracker period. The total elemental risk levels in Jinan for the three groups (aged 2–6 years, 6–12 years and ≥70 years) were higher than 2 during the firecracker period, indicating that increased pollutant levels emitted from the burning of firecrackers over short periods of time may cause non-carcinogenic human health risks.

1. Introduction

Festivals worldwide, such as Independence Day in the US, France's Commemoration of the French Revolution, the Las Fallas in Spain, the Lantern Festival and Spring Festival in China, Diwali Festival during October/November in India, and New Year's Eve celebrations throughout the world, are often celebrated with the extensive burning of firecrackers. The burning of firecrackers is responsible for elevated levels of pollutants, including gaseous pollutants (e.g., SO_2, NO_x and O_3) (Attri et al., 2001; Ravindra et al., 2003; Moreno et al., 2007; Barman et al., 2008; Godri et al., 2010; Singh et al., 2010; Nishanth et al., 2012) and particles (e.g., TSP, PM_{10} and $PM_{2.5}$) with water-soluble ions and

* Corresponding author at: Environment Research Institute, Shandong University, Jinan 250100, China. Tel.: +86 531 88366072.
E-mail address: yanglingxiao@sdu.edu.cn (L. Yang).

0048-9697/$ – see front matter © 2014 Elsevier B.V. All rights reserved.
http://dx.doi.org/10.1016/j.scitotenv.2013.12.110

trace metals (Kulshrestha et al., 2004; Drewnick et al., 2006; Moreno et al., 2007; Vecchi et al., 2008; Camilleri and Vella, 2010; Moreno et al., 2010; Perrino et al., 2011). In addition, the burning of firecrackers often causes degradation in air quality (Clark, 1997; Vecchi et al., 2008) and health hazards (e.g., chronic lung diseases, cancer, neurological and haematological diseases) (Becker et al., 2000; Kamp et al., 2005; Godri et al., 2010; Moreno et al., 2010). As a result, the pollution caused by the burning of firecrackers has recently received serious attention in the scientific community. However, most of the above studies were located abroad in areas with relatively low air pollutant levels, but data is still limited for China.

The Spring Festival and the Lantern Festival are two important celebrations with intensive burning of various firecrackers in China, a country that already has suffered serious air pollution. In China, several studies have been conducted to characterise the impacts of firecracker burning on air quality. These studies have indicated significant increases in the levels of $PM_{2.5}$ and PM_{10} with elements and water-soluble ions (Wang et al., 2007; Chang et al., 2011; Huang et al., 2012), and in the number concentration of particles in the size range of 100–500 nm (Zhang et al., 2010) due to the extensive burning of firecrackers. Li et al. (2013) also found that the emissions from the firecracker burning significantly changed the morphology and chemical composition of individual airborne particles and the transformation pathway from SO_2 to SO_4^{2-}. The above studies provided limited information regarding the aerosols, water-soluble ions and metal components emitted from the burning of firecrackers. However, the impacts of firecracker burning on aerosol chemical characteristics, especially for $PM_{2.5}$ and its chemical components, and human health risk levels have not been systematically studied in China.

To investigate the impacts of firecracker burning on aerosol chemical characteristics and human health risk levels, a campaign was conducted from February 3rd to 26th 2008 in Jinan, China, which spans the Chinese New Year. Jinan is the capital of Shandong Province, the hometown of Confucius, and is often the site of Chinese New Year celebrations that include the extensive burning of firecrackers. In addition, Jinan suffers from serious air pollution, especially particulate matter pollution (Baldasano et al., 2003; Yang et al., 2007, 2012; Gao et al., 2011). Therefore, it is important to understand whether the extensive burning of firecrackers has significant impacts on aerosol chemical characteristics and human health risk levels in this highly polluted region. In this manuscript, we discuss the impacts of firecracker burning on the mass size distribution of water-soluble ions, number concentration and size distribution of particles, and the chemical compositions of $PM_{2.5}$ during the firecracker period. And then we choose three highly sensitive groups, including children aged 2 to 6 years, children aged 6 to 12 years and older adults (≥70 years) to assess the potential health impact of $PM_{2.5}$ from the firecracker burning.

2. Methodology

2.1. Sampling site

The study was conducted at two urban sites in Jinan, the capital of Shandong Province (36°69′ N, 117°06′ E), from February 3rd to February 26th 2008. The filters for $PM_{2.5}$ and size-segregated aerosols were collected on the rooftop of a six-storied teaching building on the Centre Campus of Shandong University, approximately 20 m above ground level. Online instruments for particle number concentration and water-soluble ions in $PM_{2.5}$ were located at the rooftop (15 m above ground level) of public teaching building on the Hongjialou Campus of Shandong University, 1 km away from the Centre Campus. The inlets for aerosols were 1.5 m above the laboratory rooftop. These two sampling sites were surrounded by densely populated residential and commercial areas. The specific event of this study was the Chinese New Year and is characterised by the extensive burning of firecrackers from the night of February 6th to the morning of the following day when the city was shrouded in fume and smoke, particularly in the densely populated residential areas. In this study, we defined the day of February 6th as the firecracker period, while the other days as the non-firecracker period.

2.2. Instruments

2.2.1. Filter-based instruments

$PM_{2.5}$ samples were collected manually by using a Reference Ambient Air Sampler (Model RAAS 2.5-400, Thermo Andersen) with Teflon filters (Teflo™, 2 μm pore size and 47 mm diameter, Pall Inc.) at a flow rate of 16.7 L/min. Size-resolved aerosol samples were collected on aluminium substrates (MSP) by using the MOUDI (Micro-Orifice Uniform Deposit Impactor 110 with rotator, MSP) at a flow rate of 30 L/min. The MOUDI has eight stages with the size ranges of ≥18 μm, 10–18 μm, 5.6–10 μm, 3.2–5.6 μm, 1.8–3.2 μm, 1.0–1.8 μm, 0.56–1.0 μm, 0.32–0.56 μm and 0.18–0.32 μm. The sampling time for both $PM_{2.5}$ and size-resolved aerosol samples was approximately 24 h, normally from 9:00 a.m. to 8:45 a.m. the following day from February 3rd to February 26th 2008. A total of 20 samples for $PM_{2.5}$, and 13 sets of size-resolved aerosol samples with each set comprising of nine samples, were collected during the whole campaign. The flow rates of $PM_{2.5}$ sampler and MOUDI were calibrated before the field campaign, and field blanks were collected at the start and end of the field campaign. After sampling, all the filters were kept in plastic Petri dishes and then stored in a refrigerator at −4 °C for subsequent analysis in laboratory. In the laboratory, the concentrations of water-soluble ions were determined by ion chromatography (ICs, model Dionex 90) (Zhou et al., 2010), OC and EC in $PM_{2.5}$ were analysed by a semi-continuous OC/EC analyser (Sunset-DOSCOCEC, Sunset Lab, Portland, OR) (Wang et al., 2011), and trace metals in $PM_{2.5}$ were determined by using X-ray fluorescence (XRF) (Yang et al., 2013).

2.2.2. Real-time instruments

Particle number concentration at the range of 10 nm–10 μm was measured by a wide-range Particle Spectrometer™ (WPS model 1000XP, MSP Co., USA). This instrument combines the principles of differential mobility analysis (DMA), condensation particle counting (CPC), and laser light scattering (LPS). The detailed information for the principles of these parts can be found in Xu et al. (2011). The time of WPS measurements was from February 4th to February 9th 2008. Before and after the measurement, PSL spheres with sizes of (0.269 μm and 0.1007 μm mean diameter) and (0.701 μm, 1.36 μm, 1.6 μm, and 4.0 μm mean diameter) were used to calibrate DMA and LPS respectively. The DMA and CPC can measure particle number size distribution at the range of 10–500 nm in up to 96 channels. The LPS covers the 350–10,000 nm range in 24 additional channels. In this study we chose the sample mode with 60 channels in DMA and 24 channels in LPS. It took about 8 min for one complete scanning of the entire size range.

An ambient ion monitor (AIM; Model URG-9000B, URG Co.) was deployed to measure the hourly concentrations of water-soluble inorganic ions in $PM_{2.5}$. The AIM measurements started from February 3rd and ended on February 9th. Multi-point calibrations were performed every four days after changing the eluent solutions. The uncertainties were approximately 10%, and the estimated detection limits ranged from 0.010 to 0.084 μg/m^3 for all ions. The data from AIM had been compared to the filter samples, and they perfectly matched (Gao et al., 2011).

2.3. Elemental risk level calculations in $PM_{2.5}$

In this study, representative elemental components of $PM_{2.5}$ were applied to calculate elemental risk levels to assess the possible impacts on human health.

Based on the US EPA (2001) and the experimental data in this study, the following equation was used to obtain one-sided (1-a) UCL on the mean:

$$UCL_{1-a} = \overline{X} + t_{a,n-1}S/\sqrt{n}, \quad (1)$$

where $\overline{X}$ is the mean concentration, S is the standard deviation, t is the Student's t value, which can be found in Gilbert (1987), and n is the sample size. To evaluate the long-term or chronic impact of pollutant exposure, the average amount of pollutant exposure per an individual's body weight over a given time span for the three sensitive groups (aged 2–6 years, 6–12 years and ≥70 years) was calculated as (USEPA, 1989):

$$DE = \frac{C \times I \times F \times D}{t \times W} \quad (2)$$

where the terms are DE: dose of exposure (mg/kg-day); C: mean concentrations (mg/m^3); I: inhalation rate (m^3/day); F: exposure frequency (days/year); D: exposure duration (years); t: average time; and W: body weight (kg). The elemental risk (R) was calculated by Eq. (3). Here RD is the reference dose (USEPA, 1989).

$$R = {}^{DE}\!/_{RD} \quad (3)$$

3. Results and discussion

3.1. Overview of measurement data

Fig. 1 depicts the daily concentrations of $PM_{2.5}$, water-soluble ions and OC/EC measured in Jinan from February 3rd to 26th 2008. From this figure, it can be observed that the $PM_{2.5}$ concentration in Jinan varied over a large range during the measurement period. Heavy aerosol pollution occurred on February 6th when the firecrackers were extensively displayed, with the daily $PM_{2.5}$ concentration of 464.02 μg/m^3. Light aerosol pollution was observed with the daily $PM_{2.5}$ concentration of 33.08 μg/m^3, slightly lower than the 24 h US National Ambient Air Quality Standard (35 μg/m^3).

Table 1 summarises the statistics of the $PM_{2.5}$, water-soluble ions and OC/EC for the study period, including subset intervals for periods with and without the firecracker burning. Overall, serious aerosol pollution was illustrated, especially during the firecracker period. During the study period, the average concentration (±standard deviation) of $PM_{2.5}$ was 134.51 (±97.40) μg/m^3, which is approximately nine times the annual US National Ambient Air Quality Standard of $PM_{2.5}$ (15 μg/m^3) and is substantially higher than the annual standard in China (35 μg/m^3). Water-soluble ions and carbonaceous species (OC + EC) contributed to 57% and 14%, respectively, of the $PM_{2.5}$ mass. Water-soluble ions were the most abundant species; among all ions observed, SO_4^{2-} was the most abundant composition, with a mean value of 26.41 (±20.25) μg/m^3, followed by NO_3^- and NH_4^+ with average concentrations of 18.07 (±11.99) and 12.21 (±6.93) μg/m^3, respectively. These three species together composed approximately 76% of all water-soluble ions. Cl^- and K^+ also had relatively large concentrations of 10.11 (±16.88) and 10.78 (±29.20) μg/m^3, respectively. Other ions (F^-, Na^+, Mg^{2+}, Ca^{2+}) normally were observed at very low concentrations, and accounted for a minor fraction of the total $PM_{2.5}$ water-soluble ions.

Table 1
Average concentration of chemical species in $PM_{2.5}$ in Jinan, China (unit: μg/m^3).

Species	$PM_{2.5}$	SO_4^{2-}	NO_3^-	Cl^-	F^-	Na^+
All data	134.51	26.41	18.07	10.11	0.43	0.69
Firecrackers	464.02	86.85	14.63	74.54	1.08	1.81
Non-firecrackers	113.92	22.64	18.29	6.08	0.39	0.62
Species	NH_4^+	K^+	Mg^{2+}	Ca^{2+}	OC	EC
All data	12.21	10.78	0.69	0.76	14.77	1.70
Firecrackers	7.25	123.69	6.66	1.05	19.56	3.63
Non-firecrackers	12.52	3.72	0.32	0.74	14.47	1.58

3.2. Air pollution caused by the burning of firecrackers

3.2.1. Variation in $PM_{2.5}$ and water-soluble ion concentrations

The holiday for celebrating the 2008 Chinese New Year lasted seven days from February 6th to 12th, including the lunar New Year's Eve on February 6th, when firecrackers were displayed through the following morning. The mass concentration of $PM_{2.5}$ was high with daily value of 464.02 μg/m^3 during the firecracker period (Fig. 1 and Table 1), three times higher than the average values (113.92 μg/m^3) obtained during the non-firecracker period. Similarly, the total concentrations of water-soluble ions were also elevated, with hourly values (Fig. 2) at their highest concentrations of 1620.09 μg/m^3 at 1:00 a.m. on February 7th when firecrackers were extensively displayed. Because industrial activity and traffic are drastically reduced during the lunar New Year's Eve, anthropogenic emissions should only minimally contribute to air pollution during this period (Li et al., 2013). Thus, increased aerosol pollution can be attributed to firecracker emissions.

As shown in Fig. 2, water-soluble ions can be classified into two groups based on the variations in their concentrations: (1) K^+, Cl^-, SO_4^{2-}, Mg^{2+}, Ca^{2+}, F^- and Na^+, and (2) NO_3^- and NH_4^+. The ions K^+, Cl^-, SO_4^{2-}, Mg^{2+}, Ca^{2+}, F^- and Na^+ exhibited a sharp concentration

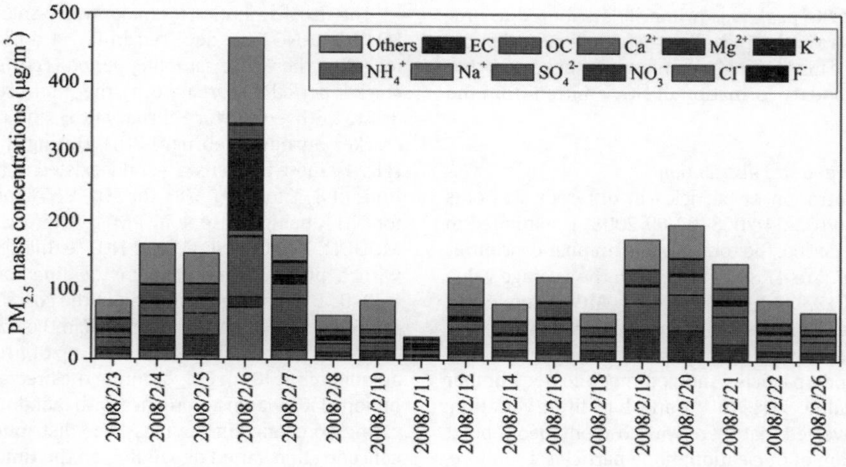

Fig. 1. Time series of daily concentrations of $PM_{2.5}$ and its chemical components in Jinan, China.

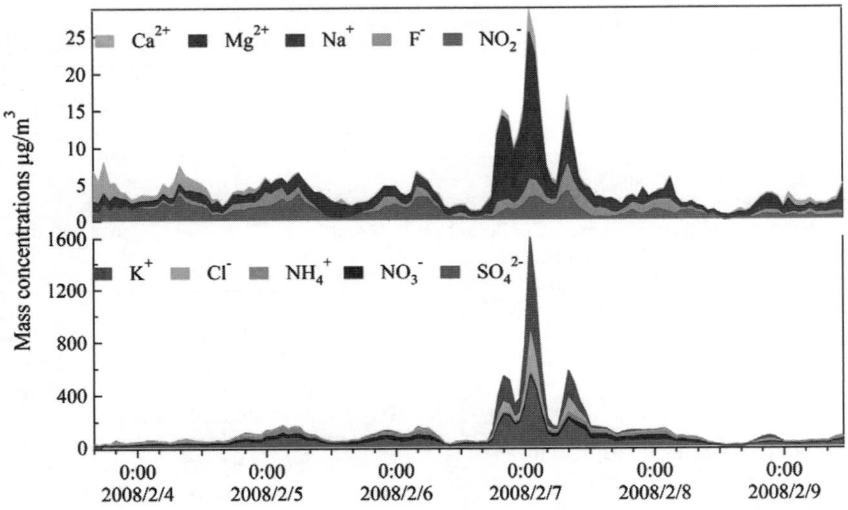

Fig. 2. Hourly concentrations of water-soluble ions during sampling period in Jinan, China.

peak on the night when firecrackers were extensively displayed, followed by a rapid decrease in their concentrations the following day. Thus, these ions were most likely released from the burning of firecrackers, with weak contribution from other local sources. Figs. 1 and 2 showed that NO_3^- and NH_4^+ had lower concentrations during the extensive firecracker burning period, and their concentrations gradually increased the following day. The concentrations of K^+, Mg^{2+}, Cl^-, SO_4^{2-}, F^- and Na^+ in $PM_{2.5}$ were approximately 33, 21, 12, 4, 3 and 3 times greater, respectively, during the firecracker period compared to the period when the firecrackers were not displayed (Table 1). K^+ was found in large quantities and showed the largest increase during the firecracker burning. The daily and hourly K^+ concentrations reached values of 123.69 and 717.49 μg/m³, respectively, due to the extensive firecracker burning, which indicated that potassium salts might be one of the major compounds used in the firecrackers. KNO_3, $KClO_3$ and $KClO_4$ are often widely used as oxidiser that sustain the burning of firecrackers and are restored into KNO_2 and KCl (Wang et al., 2007). In this study, NO_2^- varied little during the firecracker burning (Fig. 2), whereas Cl^- was highly elevated, suggesting that $KClO_3$ and $KClO_4$ were used as the major oxygen sources in the firecrackers. However, the concentrations of NO_3^- and NH_4^+ peaked the day following the firecracker burning. The concentration peaks for NO_3^- and NH_4^+ occurred at 8:00 and 12:00, respectively, on the day following the extensive firecracker burning. In contrast, their concentration peaks occurred before dawn on days without firecracker use. The moderate elevation in NO_3^- just after dawn suggested that the peak was primarily attributed to firecracker emissions, and less to anthropogenic activities (e.g., coal burning and traffic exhausts). This "tailing" phenomena suggested that NO_3^- may be from the secondary formation of NO_x emitted from the burning of firecrackers.

3.2.2. Number concentration and size distribution

Hourly number concentration of particles in different size bins during the sampling period (02/04/2008–02/09/2008) is exhibited in Fig. 3. During the sampling period, the total particle number concentration ranged from 5377 cm^{-3} to 47,888 cm^{-3}, with the average value (± standard deviation) of 19,928 (±8590) cm^{-3}. Aitken mode and accumulation mode particles were the dominant size fractions, accounting for 57% and 42%, respectively, of the total particle number concentration. The nucleation mode particles showed much lower number concentration (average value: 204 cm^{-3}) and had little variation. Examination of the data revealed that there was no spontaneous burst in the number concentration of nucleation mode particles, therefore, no new particle formation events were observed during the sampling period. However, one particle growth process was clearly observed in the afternoon on February 6th. The growth rate (GR) was calculated to be 6.7 nm h^{-1} following the methods by Kulmala et al. (2004), which was within the typical urban particle GR range (1–20 nm/h) in previous study (Kulmala et al., 2004; Gao et al., 2012).

The episode of extensive firecracker burning started at ~18:00 LT on February 6th and ended with the high point after midnight (1:00 LT) on February 7th. Fig. 3 showed that a significant increase in the number concentration of large particles, which also lead to an elevation of geometrical mean diameter (GMD). During the firecracker period the average total number concentration was 24,783 cm^{-3}, which increased 30.3% compared to the non-firecracker period. Moreover, the number concentration of accumulation mode particles increased to 15,023 cm^{-3} (60.8% of the total) during the firecracker period, much higher than 7007 cm^{-3} (36.9% of the total) during the non-firecracker period, which was in accordance with some previous studies (Agus et al., 2008; Mönkkönen et al., 2004; Zhang et al., 2010). In contrast, the number concentration of nucleation and Aitken mode particles reduced by 15% and 19%, respectively, during the firecracker period compared to the non-firecracker period, which was due to the large coagulation sink (Zhang et al., 2010). Therefore, we concluded that the intensive firecracker burning released a large amount of accumulation mode particles, and changed the particle number size distribution.

3.2.3. Size distribution of major water-soluble ions for different periods

The size distribution for major water-soluble ions (i.e., NH_4^+, K^+, Cl^-, SO_4^{2-} and NO_3^-) is depicted in Fig. 4 during four different periods, including the whole sampling period (February 3rd to 26th), the "firecracker period" (February 6th), the "non-firecracker period" (February 3rd to 26th except for February 6th), and the day following the firecracker burning (February 7th). During the whole sampling period (Fig. 4), these five ions generally existed in the fine particles with fractions of 87% for NO_3^-, 78% for SO_4^{2-}, 85% for Cl^-, 85% for K^+ and 98% for NH_4^+ being present in $PM_{1.8}$ (there is no 2.5 μm cut point for MOUDI). K^+, Cl^-, SO_4^{2-} and NO_3^- exhibited a bimodal distribution, with a predominant peak in the fine mode in the size range of 0.32–0.56 μm and a small peak in the coarse mode in the size range of 3.2–5.6 μm. NH_4^+ had a single peak in the size range of 0.32–0.56 μm.

As mentioned above, the burning of firecrackers can emit a large amount of the K^+ in $PM_{2.5}$. During the firecracker period, the size distribution of K^+ was in accordance with that during the other periods, indicating no change in the mass size distribution for K^+. However, the concentration varied depending on the time period. For each size-bin, the K^+ concentration during the firecracker period was 20–80 times

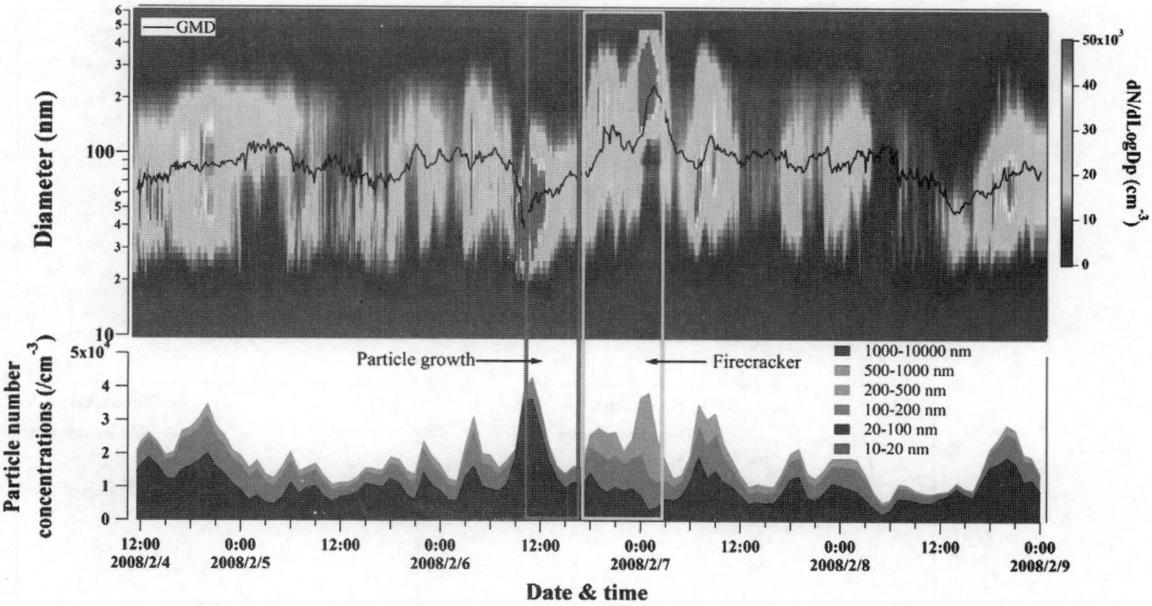

Fig. 3. Number size distribution and concentration of different size bins from February 4th to 9th, 2008.

that of the non-firecracker period. The K^+ concentration then decreased rapidly to the average value for each size-bin on the day following firecracker burning, which suggested that the firecracker burning made a strong contribution while other local sources made a weak contribution to the K^+ concentration. Variations in Cl^- size distribution and concentration were in accordance with those in K^+, and Cl^- showed strong correlation with K^+ for all size-bins ($R = 0.97$, $p \ll 0.01$). Thus Cl^- likely originated from the same source (i.e., firecracker burning) as K^+. The variation in SO_4^{2-} was similar to that of K^+; the only difference was the concentration variation on the day following the firecracker burning (February 7th). Similarly, the SO_4^{2-} concentration declined after the burning of firecrackers. However, the SO_4^{2-} concentration in the fine particles was still higher than the average value in all samples. This result indicated that primary particles emitted directly from the firecracker burning could be quickly removed by dry deposition, whereas pollution gases from the burning activities were likely to be oxidised to secondary inorganic components and retained in the particles. NO_3^- and NH_4^+ showed similar characteristics in size distribution variation, with no change over all the periods. However, their concentration peaks appeared on the day following firecracker burning (February 7th), which was in contrast with the peak in K^+ concentration during the firecracker period (February 6th). These results indicated that NO_3^- and NH_4^+ were not directly produced by the burning of firecrackers, unlike K^+, and that they may have been formed from the gases emitted by the burning of firecrackers.

3.2.4. Chemical composition of $PM_{2.5}$

Particles emitted from firecracker burning are likely to have different chemical compositions compared with the normal aerosols. To identify the variation in chemical composition, the chemical composition of $PM_{2.5}$ was determined under four assumptions. First, the concentration of organic matter (OM) was calculated as 1.8 times the concentration of OC, according to the revised IMPROVE formula. Second, the soil dust concentration was assumed to be the sum of the oxides of the main crustal elements (Kim et al., 2001): $[Soil] = 2.20 * [Al] + 2.49 * [Si] + 1.63 * [Ca] + 2.42 * [Fe] + 1.94 * [Ti]$. Third, the concentration of trace metal was assumed to be the sum of the oxides of the corresponding elements, except for Al, Si, Ca, Fe and Ti. Fourth, the firework matter was calculated as the sum of K^+ and Cl^-, as their concentrations were highly elevated during the night of firecracker burning and could serve as firecracker indicators (Wang et al., 2007). The contributions of different components of $PM_{2.5}$ during the firecracker and non-firecracker periods are depicted in Fig. 5.

The results showed that, for the samples collected during the non-firecracker period, the pattern was dominated by secondary inorganic matter (the sum of SO_4^{2-}, NH_4^+ and NO_3^-; 47%) and organic matter (23%) in $PM_{2.5}$. The contribution of firework matter (the sum of K^+ and Cl^-) was relatively lower (9%). However, the samples collected during the firecracker period showed a different pattern, with greater amounts of firework matter (43%) and less secondary inorganic (24%) and organic matter (7%). This finding suggested that the chemical components of $PM_{2.5}$ were significantly influenced by the burning of firecrackers. The reduced fraction of secondary inorganic matter associated with the burning of firecrackers was due to the lower amounts of NH_4^+ and NO_3^-. The fraction of SO_4^{2-} in $PM_{2.5}$ was 19% during the firecracker period and 20% during the non-firecracker period, whereas NH_4^+ and NO_3^- declined from 11% and 16%, respectively, during the non-firecracker period to 1% and 3%, respectively, during the firecracker period.

To further identify the chemical forms of the major ions (i.e., SO_4^{2-}, NH_4^+, NO_3^-, K^+ and Cl^-), the correlation among ion concentrations was analysed using the hourly data for the firecracker and non-firecracker periods, respectively, and the results are shown in Table 2. During the non-firecracker period, SO_4^{2-} showed stronger correlations with NH_4^+ ($R = 0.96$, $p \ll 0.01$). The RMA slope of the regression between equivalent concentrations of NH_4^+ and SO_4^{2-} was 1.34, which indicated the complete neutralisation of SO_4^{2-} by NH_4^+. The RMA slope of the regression between equivalent concentrations of NH_4^+ and ($SO_4^{2-} + NO_3^-$) was 1.00 ($R = 0.96$, $p \ll 0.01$), which implied complete neutralisation of SO_4^{2-} and NO_3^- by NH_4^+. K^+ showed little correlation with SO_4^{2-}, NO_3^- or Cl^-. These results indicated that $(NH_4)_2SO_4$ and NH_4NO_3 were the major chemical species in $PM_{2.5}$ during the non-firecracker period. However, during the firecracker period, different results were obtained. SO_4^{2-} showed stronger correlations with K^+, Na^+, Mg^{2+}, and Ca^{2+} ($R = 0.98$, 0.91, 0.91 and 0.89, respectively, $p \ll 0.01$) rather than NH_4^+ ($R = 0.07$, $p = 0.78$). K^+ can serve as a tracer for firecracker emissions. This finding indicated that SO_4^{2-} was largely produced from the firecracker burning and existed as K_2SO_4, and Na_2SO_4, $MgSO_4$ and $CaSO_4$ were also other major existing forms of SO_4^{2-}. Cl^- was highly correlated with K^+, Na^+, Mg^{2+}, and Ca^{2+}, with

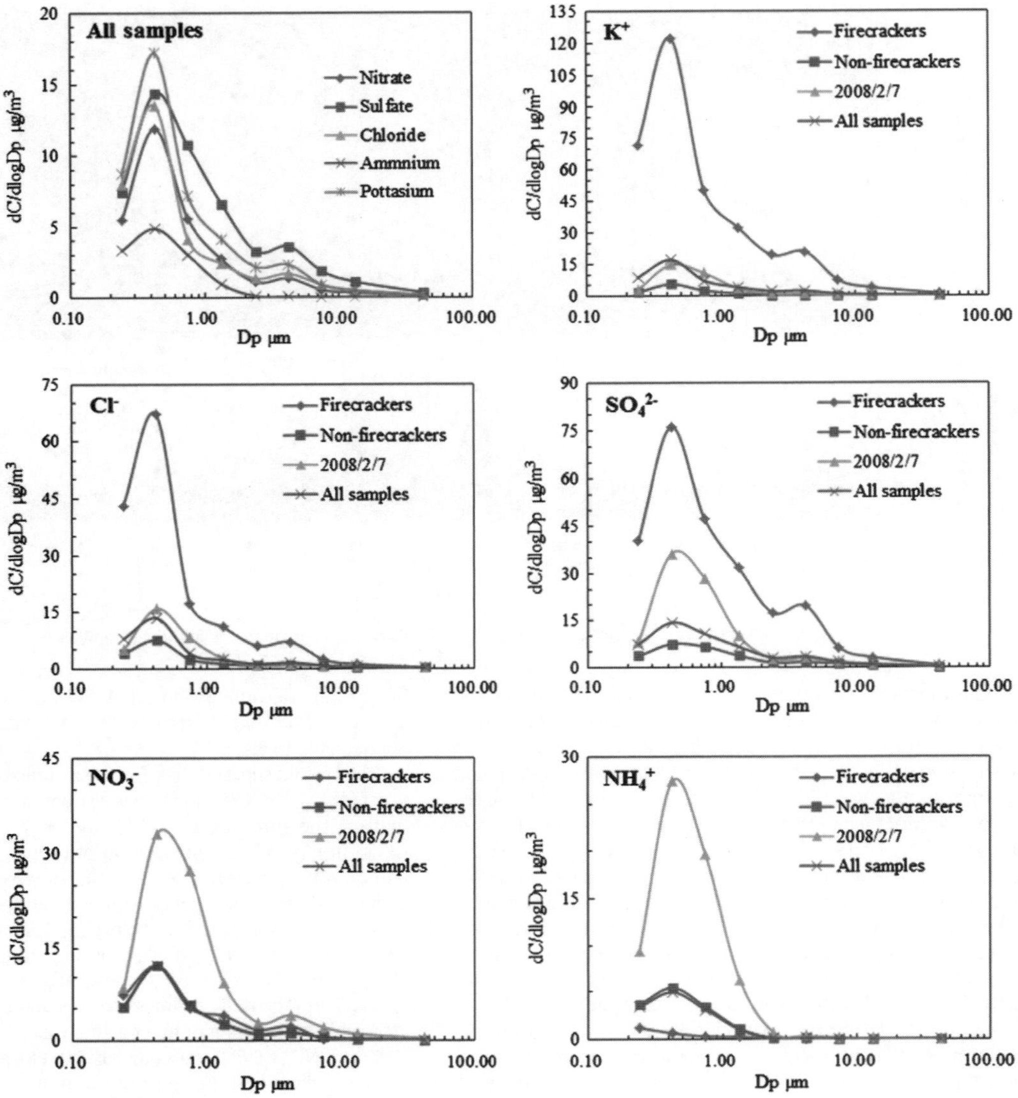

Fig. 4. Mass size distribution of major water-soluble ions during the four sampling periods.

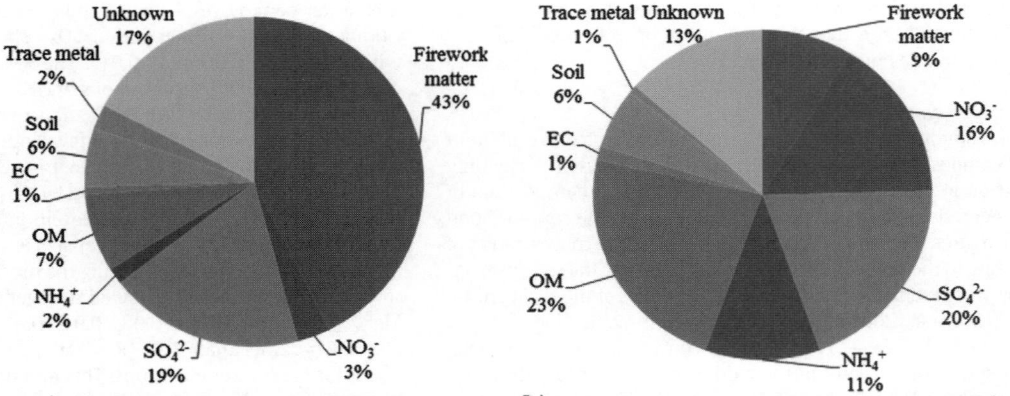

Fig. 5. Chemical composition of $PM_{2.5}$ in Jinan during (a) the firecracker period and (b) the non-firecracker period.

Table 2
Correlation between water-soluble ions using the hourly data for (a) the non-firecracker period and (b) the firecracker period.

(a)	Cl^-	NO_2^-	NO_3^-	SO_4^{2-}	Na^+	NH_4^+	K^+	Mg^{2+}	Ca^{2+}
Cl^-	1								
NO_2^-	0.54**	1							
NO_3^-	0.45**	0.37**	1						
SO_4^{2-}	0.48**	0.49**	**0.86****	1					
Na^+	0.01	0.17	0.28**	0.42**	1				
NH_4^+	0.46**	0.60**	**0.84****	**0.96****	0.40**	1			
K^+	0.42**	−0.16	0.26**	0.26**	−0.06	0.1	1		
Mg^{2+}	0.25**	−0.16	−0.13	−0.06	−0.18**	−0.18*	0.78**	1	
Ca^{2+}	0.11	0.28**	−0.31**	−0.35**	−0.05	−0.28**	−0.21*	0.04	1
(b)	Cl^-	NO_2^-	NO_3^-	SO_4^{2-}	Na^+	NH_4^+	K^+	Mg^{2+}	Ca^{2+}
Cl^-	1								
NO_2^-	0.54*	1							
NO_3^-	−0.29	−0.01	1						
SO_4^{2-}	**0.96****	0.50*	−0.35	1					
Na^+	**0.92****	0.45	−0.4	**0.91****	1				
NH_4^+	0.18	0.61**	**0.71****	0.07	0.03	1			
K^+	**0.99****	0.51*	−0.32	**0.98****	**0.92****	0.11	1		
Mg^{2+}	**0.84****	0.33	−0.58**	**0.91****	**0.84****	−0.26	**0.87****	1	
Ca^{2+}	**0.94****	**0.73****	−0.33	**0.89****	**0.84****	0.26	**0.92****	**0.77****	1

The bold and underlined values indicate strong correlation.
* Correlation is significant at the 0.05 level (2-tailed).
** Correlation is significant at the 0.01 level (2-tailed).

correlation coefficients of 0.99, 0.92, 0.84 and 0.94, respectively. These correlation patterns suggested that chloride salt might be the main chemical form for these metals in the firecrackers. Furthermore, NH_4^+ and NO_3^- showed moderate correlation, but they showed a smaller or negative correlation with other ions (e.g., K^+, Na^+, Mg^{2+}, and Ca^{2+}). This result reinforced that NH_4^+ and NO_3^- were not directly emitted from the burning of firecrackers.

3.3. Assessment of health risk levels from metals in $PM_{2.5}$

The burning of firecrackers can release a large amount of heavy metals into the atmosphere (Gao et al., 2002; Wang et al., 2007). Particles of these heavy metals can penetrate into the human body through direct inhalation, ingestion and dermal contact and then cause short- and long-term health problems, especially for children and elderly (Kong et al., 2012). Therefore, it is essential to evaluate the human health risks related to heavy metal exposure. In this study, nine major metals (Cu, Pb, Cr, Co, S, Mn, Zn, As and Ni) were used to assess the possible non-carcinogenic human health risks via direct inhalation during the firecracker burning and non-firecracker burning scenarios. The non-carcinogenic risk assessment results for the firecracker period and the non-firecracker period are listed in Table 3.

De Miguel et al. (2007) suggested that an elemental risk level larger than 0.1 had adverse health effects on children. As shown in Table 3, the metals in $PM_{2.5}$ with risk values higher than 0.1 for the study populations in Jinan were Mn, Co and S for the non-firecracker period and Mn, Co, S, Cr and Pb for the firecracker period. The risk values for Cu, Pb, Cr, Co, S, Mn, Zn, As and Ni were approximately 16, 9, 6, 4, 3, 3, 3, 2 and 2 times greater, respectively, during the firecracker period compared to the non-firecracker period. Considering the cumulative effect of the non-carcinogenic risk levels of heavy metals, the total risk values were summed for the nine metals. The total risk values were 0.83 for children aged 2–6 years, 0.85 for children aged 6–12 years, and 0.63 for elderly adults during the non-firecracker period, and these values exceeded 2.0 during the firecracker period. These results indicated that exposure to metals found in $PM_{2.5}$ may pose a serious public health risk in this study area and that the non-carcinogenic elemental inhalation risk is greatly increased due to the burning of firecrackers. Cu, used to make blue colours in firecrackers, and Pb, used to achieve a steady and reproducible burning rate, increased the most (Conkling, 1985). During the burning of firecrackers, a large amount of Cu and Pb was emitted into the atmosphere, causing increases in their concentrations (Gao et al., 2002; Wang et al., 2007). In addition, a comparison was conducted between children and adults, and the results showed that children were the most sensitive group to non-carcinogenic effects and should avoid possible exposure to these contaminants.

4. Summary

To assess the impacts of firecracker burning on aerosol chemical characteristics and human health risk levels, chemical components of $PM_{2.5}$, number concentration and size distribution of particles, and

Table 3
Health risks based on the chemical elemental components of $PM_{2.5}$ for the three groups during periods with and without firecracker detonation.

Firecrackers period				Non-firecrackers period		
Elements	Children (2–6)	Children (6–12)	Adult	Children (2–6)	Children (6–12)	Adult
S	0.49	0.50	0.38	0.16	0.16	0.12
Cr	0.43	0.44	0.33	0.08	0.08	0.06
Mn	1.11	1.12	0.84	0.39	0.39	0.29
Ni	1.61E−04	1.63E−04	1.22E−04	9.113E−05	9.251E−05	6.916E−05
Cu	2.45E−03	2.49E−03	1.86E−03	1.55E−04	1.57E−04	1.18E−04
Zn	6.29E−04	6.39E−04	4.78E−04	2.23E−04	2.27E−04	1.69E−04
Pb	0.11	0.11	0.09	1.25E−02	1.26E−02	9.45E−03
As	0.02	0.02	0.02	9.27E−03	9.41E−03	7.04E−03
Co	0.69	0.70	0.52	0.19	0.20	0.15
Total	2.86	2.90	2.17	0.83	0.85	0.63

mass size distribution of water-soluble ions were measured in Jinan during the 2008 Chinese New Year. The firecrackers displayed to celebrate the Chinese New Year elevated number concentration of particles and mass concentration of $PM_{2.5}$ and water-soluble ions, and varied the number size distribution and chemical components of $PM_{2.5}$; however, the firecracker burning only minimally altered the mass size distribution of major water-soluble ions. The results of elemental risk assessment suggested that the pollutants emitted from the burning of firecrackers even over a short duration may cause high non-carcinogenic risk levels to human health.

Conflict of interest

We declare that we have no conflict of interest.

Acknowledgements

The authors are grateful to the Jinan Monitoring Station for providing the field platform and Elsevier Language Editing Service for polishing the English writing. This work was supported by the National Basic Research Program (973 Program) of China (2005CB422203), the Key Project of Shandong Provincial Environmental Agency (2006045), the Promotive Research Fund for Young and Middle-aged Scientists of Shandong Province (BS2010HZ010), the Special Research for Public-Beneficial Environment Protection (201009001-1) and Jinan university scientific research fund (youth projectXKY1326).

References

Agus EL, Lingard JJ, Tomlin AS. Suppression of nucleation mode particles by biomass burning in an urban environment: a case study. J Environ Monit 2008;10:979–88.
Attri AK, Kumar U, Jain VK. Microclimate-formation of ozone by fireworks. Nature 2001;411:1015.
Baldasano JM, Valera E, Jiménez P. Air quality data from large cities. Sci Total Environ 2003;307:141–65.
Barman SC, Singh R, Negi MPS, Bhargava SK. Ambient air quality of Lucknow City (India) during use of fireworks on Diwali Festival. Environ Monit Assess 2008;137:495–504.
Becker JM, Iskandrian S, Conkling J. Fatal and near-fatal asthma in children exposed to fireworks. Ann Allergy Asthma Immunol 2000;85:512–3.
Camilleri R, Vella AJ. Effect of fireworks on ambient air quality in Malta. Atmos Environ 2010;44:4521–7.
Chang SC, Lin TH, Young CY, Lee CT. The impact of ground-level fireworks (13 km long) display on the air quality during the traditional Yanshui. Environ Monit Assess 2011;172:463–79.
Clark H. New directions—light blue touch paper and retire. Atmos Environ 1997;31:2893–4.
Conkling JA. Chemistry of pyrotechnics: basic principles and theory. New York, NY: Marcel Dekker, Inc.; 1985.
De Miguel E, Iribarren I, Chacón E, Ordoñez A, Charlesworth S. Risk based evaluation of the exposure of children to trace elements in playgrounds in Madrid (Spain). Chemosphere 2007;66:505–13.
Drewnick F, Hings SS, Curtius J, Eerdekens G, Williams J. Measurement of fine particulate and gas-phase species during the New Year's fireworks 2005 in Mainz, Germany. Atmos Environ 2006;40:4316–27.
Gao Y, Nelson ED, Field MP, Ding Q, Li H, Sherrell RM, et al. Characterization of atmospheric trace elements on $PM_{2.5}$ particulate matter over the New York–New Jersey harbor estuary. Atmos Environ 2002;36:1077–86.
Gao XM, Yang LX, Cheng SH, Gao R, Zhou Y, Xue LX, et al. Semi-continuous measurement of water-soluble ions in $PM_{2.5}$ in Jinan, China: temporal variations and source apportionments. Atmos Environ 2011;45:6048–56.
Gao J, Chai FH, Wang T, Wang SX, Wang WX. Particle number size distribution and new particle formation: new characteristics during the special pollution control period in Beijing. J Environ Sci 2012;24:14–21.
Gilbert. Statistical methods for environmental pollution monitoring. New York, NY, USA: Van Nostrand Reinhold; 1987.
Godri KJ, Green DC, Fuller GW, Dall'osto M, Beddows DC, Kelly FJ, et al. Particulate oxidative burden associated with firework activity. Environ Sci Tech 2010;44:8295–301.
Huang K, Zhuang G, Lin Y, Wang Q, Fu JS, Zhang R, et al. Impact of anthropogenic emission on air quality over a megacity — revealed from an intensive atmospheric campaign during the Chinese Spring Festival. Atmos Chem Phys 2012;12:11631–45.
Kamp IV, Van Der Velden PG, Stellat RK, Roorda J, Loon JV, Kleber RJ, et al. Physical and mental health shortly after a disaster: first results from the Enschede firework disaster study. Eur J Public Health 2005;16:252–8.
Kim KW, Kim YJ, Oh SJ. Visibility impairment during Yellow Sand periods in the urban atmosphere of Kwangju, Korea. Atmos Environ 2001;35:5157–67.
Kong SF, Lu B, Ji YQ, Zhao XY, Bai ZP, Xu YH, et al. Risk assessment of heavy metals in road and soil dusts within $PM_{2.5}$, PM_{10} and PM_{100} fractions in Dongying City, Shandong Province, China. J Environ Monit 2012;14:791–803.
Kulmala M, Vehkamaki H, Petajda T, Dal Maso M, Lauri A, Kerminen V-M, et al. Formation and growth rates of ultrafine atmospheric particles: a review of observations. J Aerosol Sci 2004;35:143–76.
Kulshrestha UC, Nageswara Rao T, Azhaguvel S, Kulshrestha MJ. Emissions and accumulation of metals in the atmosphere due to crackers and sparkles during Diwali festival in India. Atmos Environ 2004;38:4421–5.
Li WJ, Shi ZB, Yan C, Yang LX, Dong C, Wang WX. Individual metal-bearing particles in a regional haze caused by firecracker and firework emissions. Sci Total Environ 2013;443:464–9.
Mönkkönen P, Koponen IK, Lehtinen KEJ, Uma R, Srinivasan D, Hämeri K, et al. Death of nucleation and Aitken mode particles: observations at extreme atmospheric conditions and their theoretical explanation. J Aerosol Sci 2004;35:781–7.
Moreno T, Querol X, Alastuey A, Minguillon MC, Pey J, Rodriguez S, et al. Recreational atmospheric pollution episodes: inhalable metalliferous particles from firework displays. Atmos Environ 2007;41:913–22.
Moreno T, Querol X, Alastuey A, Amato F, Pey J, Pandolfi M, et al. Effect of fireworks events on urban background trace metal aerosol concentrations: is the cocktail worth the show? J Hazard Mater 2010;183:945–9.
Nishanth T, Praseed KM, Rathnakaran K, Satheesh Kumar MK, Krishna RR, Valsaraj KT. Atmospheric pollution in a semi-urban, coastal region in India following festival seasons. Atmos Environ 2012;47:295–306.
Perrino C, Tiwari S, Catrambone M, Torre SD, Rantica E, Canepari S. Chemical characterization of atmospheric PM in Delhi, India, during different periods of the year including Diwali festival. Atmos Pollut Res 2011;2:418–27.
Ravindra K, Mor S, Kaushik CP. Short-term variation in air quality associated with firework events: a case study. J Environ Monit 2003;5:260–4.
Singh DP, Gadi R, Mandal TK, Dixit CK, Singh K, Saud T, et al. Study of temporal variation in ambient air quality during Diwali festival in India. Environ Monit Assess 2010;169:1–13.
USEPA (United States Environmental Protection Agency). Human health evaluation manual. EPA/540/1-89/002Risk assessment guidance for superfund, vol. I. Office of Solid Waste and Emergency Response; 1989.
USEPA (United States Environmental Protection Agency). Supplemental guidance for developing soil screening levels for superfund sites. OSWER 9355.4-24. Office of Solid Waste and Emergency Response; 2001.
Vecchi R, Bernardoni V, Cricchio D, D'Alessandro A, Fermo P, Lucarelli F, et al. The impact of fireworks on airborne particles. Atmos Environ 2008;42:1121–32.
Wang Y, Zhuang GS, Xu C, An ZS. The air pollution caused by the burning of fireworks during the lantern festival in Beijing. Atmos Environ 2007;41:417–31.
Wang Z, Wang T, Gao R, Xue LK, Guo J, Zhou Y, et al. Source and variation of carbonaceous aerosols at Mount Tai, North China: results from a semi-continuous instrument. Atmos Environ 2011;45:1655–67.
Xu PJ, Wang WC, Yang LC, Zhang QZ, Gao R, Wang XF, et al. Aerosol size distributions in urban Jinan: seasonal characteristics and variations between weekdays and weekends in a heavily polluted atmosphere. Environ Monit Assess 2011;179:443–56.
Yang LX, Wang DC, Cheng SH, Wang Z, Zhou Y, Zhou XH, et al. Influence of meteorological conditions and particulate matter on visual range impairment in Jinan, China. Sci Total Environ 2007;383:164–73.
Yang LX, Zhou XH, Wang Z, Zhou Y, Cheng SH, Xu PJ, et al. Airborne fine particulate pollution in Jinan, China: concentrations, chemical compositions and influence on visibility impairment. Atmos Environ 2012;55:506–14.
Yang LX, Cheng SH, Wang XF, Nie W, Xu PJ, Gao XM, et al. Source identification and health impact of PM2.5 in a heavily polluted urban atmosphere in China. Atmos Environ 2013;75:265–9.
Zhang M, Wang XM, Chen JM, Cheng TT, Wang T, Yang X, et al. Physical characterization of aerosol particles during the Chinese New Year's firework events. Atmos Environ 2010;44:5191–8.
Zhou Y, Wang T, Gao XM, Xue LK, Wang XF, Wang Z, et al. Continuous observations of water-soluble ions in $PM_{2.5}$ at Mount Tai (1534 m a.s.l.) in central-eastern China. J Atmos Chem 2010;64:107–27.

Atmospheric Environment 88 (2014) 219–229

Contents lists available at ScienceDirect

Atmospheric Environment

journal homepage: www.elsevier.com/locate/atmosenv

Photochemical evolution of organic aerosols observed in urban plumes from Hong Kong and the Pearl River Delta of China

Shengzhen Zhou [a,b], Tao Wang [a,b,*], Zhe Wang [b], Weijun Li [a], Zheng Xu [a,b], Xinfeng Wang [a], Chao Yuan [a], C.N. Poon [b], Peter K.K. Louie [c], Connie W.Y. Luk [c], Wenxing Wang [a,d]

[a] Environment Research Institute, Shandong University, Ji'nan, Shandong 250100, PR China
[b] Department of Civil and Environmental Engineering, The Hong Kong Polytechnic University, Hong Kong, PR China
[c] Environmental Protection Department, Government of the Hong Kong Special Administrative Region, Hong Kong, PR China
[d] Chinese Research Academy of Environmental Sciences, Beijing 100012, PR China

HIGHLIGHTS

- Hourly measurements of carbonaceous aerosols in Hong Kong are presented in this paper.
- Photochemical evolutions of organic aerosols were examined during the photochemical episodes.
- The SOC production rates ranged from 1.31 to 3.86 μg m^{-3} $ppmv^{-1}$ h^{-1}.
- Thick organic coatings were internally mixed with inorganic sulfate/nitrate in the aged plumes.

ARTICLE INFO

Article history:
Received 10 October 2013
Received in revised form
9 January 2014
Accepted 15 January 2014
Available online 25 January 2014

Keywords:
Carbonaceous aerosols
Secondary organic aerosol
Photochemical evolution
Individual aerosol particles
$PM_{2.5}$

ABSTRACT

Organic aerosols influence human health and global radiative forcing. However, their sources and evolution processes in the atmosphere are not completely understood. To study the aging and production of organic aerosols in a subtropical environment, we measured hourly resolved organic carbon (OC) and element carbon (EC) in $PM_{2.5}$ at a receptor site (Tung Chung, TC) in Hong Kong from August 2011 to May 2012. The average OC concentrations exhibited the highest values in late autumn and were higher during the daytime than at night. The secondary organic carbon (SOC) concentrations, which were estimated using an EC-tracer method, comprised approximately half of the total OC on average. The SOC showed good correlation with odd oxygen ($O_x = O_3 + NO_2$) in the summer and autumn seasons, suggestive of contribution of photochemical activities to the formation of secondary organic aerosols (SOA). We calculated production rates of SOA using the photochemical age (defined as $-Log_{10}(NO_x/NO_y)$) in urban plumes from the Pearl River Delta (PRD) region and Hong Kong during pollution episodes in summer and autumn. The CO-normalized SOC increased with the photochemical age, with production rates ranging from 1.31 to 1.82 μg m^{-3} $ppmv^{-1}$ h^{-1} in autumn and with a larger rate in summer (3.86 μg m^{-3} $ppmv^{-1}$ h^{-1}). The rates are in the range of the rates observed in the outflow from Mexico City, the eastern U.S. and Los Angeles. Microscopic analyses of the individual aerosol particles revealed large contrasts of aerosol physico-chemical properties on clean and smoggy days, with thick organic coatings internally mixed with inorganic sulfate for all particle sizes in the aged plumes from the PRD region.

1. Introduction

Carbonaceous aerosols, which are composed of organic carbon (OC) and element carbon (EC, also termed black carbon or soot), have significant effects on global climate and human health (Nel, 2005; Bond et al., 2013). OC can be either directly emitted from primary combustion sources together with EC (primary OC, POC) or formed through the gas-particle conversion of reactive organic gas

* Corresponding author. Department of Civil and Environmental Engineering, The Hong Kong Polytechnic University, Hong Kong, PR China. Tel.: +852 2766 6059; fax: +852 2330 9071.
E-mail address: cetwang@polyu.edu.hk (T. Wang).

http://dx.doi.org/10.1016/j.atmosenv.2014.01.032
1352-2310/© 2014 Elsevier Ltd. All rights reserved.

oxidation and atmospheric heterogeneous reactions (secondary OC, SOC) (Seinfeld and Pandis, 2006). EC is emitted from incomplete combustions, such as biomass burning and fossil fuel combustion.

Recent studies have indicated that secondary organic aerosols (SOA) account for 63–95% of the organic aerosols (OA) in urban, rural, and remote sites (Zhang et al., 2007b), and the POC can be further oxidized to SOC after emission (Robinson et al., 2007). Although substantial efforts have been made in understanding SOA formation, their sources and atmospheric evolution processes remain inadequately characterized (Jimenez et al., 2009). The current air quality models, which are mainly based on the parameters defined by laboratory experiments, cannot represent the magnitudes and evolution of organic aerosols in the real atmosphere (Jimenez et al., 2009), and the mass of modeled and measured organic aerosols were not always agreed (Johnson et al., 2006; Dzepina et al., 2010).

Urban areas act as the sources of aerosols and aerosol precursors (e.g., SO_2, NO_X, volatile organic compounds (VOCs)). Secondary aerosols are produced from the oxidation of such precursors, and both primary aerosols and secondary aerosols undergo atmospheric processes within and downwind of the source regions. Field studies have been conducted to investigate the processing of aerosols in urban centers and their surrounding areas, such as in Mexico City (Kleinman et al., 2008; Jimenez et al., 2009; DeCarlo et al., 2010), the northeastern U.S. (De Gouw et al., 2008), the southern and southeastern U.S. (Weber et al., 2007; Bahreini et al., 2009), Tokyo (Miyakawa et al., 2008), Ontario (Slowik et al., 2011), and California (Hayes et al., 2013). These studies have shown the amount of SOA to be several times that of the initial primary organic aerosol concentration after a few hours of photochemical aging.

The Pearl River Delta region (PRD region, including the cities of Guangzhou, Shenzhen, Dongguan, Foshan, Jiangmen, Zhongshan and Zhuhai, and the urban areas of Huizhou and Zhaoqing) is situated in the central southern part of Guangdong Province alongside the Pearl River Estuary (Fig. 1). It is one of the most urbanized and industrialized areas in China. Although the PRD region covers only 0.5% of China's total land area and holds about 4% of its total population, its gross domestic product comprised about 14% of China's total in 2010 (Zhong et al., 2013). Along with rapid economic and population growth, the PRD region and adjacent Hong Kong (HK) metropolitan area suffer from severe photochemical smog pollution and high concentrations of fine particles (Chan and Yao, 2008; Zhang et al., 2008b; Wang et al., 2009). As one of the major components of submicron aerosols, carbonaceous aerosols have been studied intensively in HK-PRD region. Some topics of carbonaceous aerosols have been well investigated, such as the spatio-temporal distributions of mass concentrations (Cao et al., 2004; Yu et al., 2004; Hagler et al., 2006), size distributions (Huang et al., 2006; Gnauk et al., 2008; Yu et al., 2010), optical properties (Cheng et al., 2006; Andreae et al., 2008), source identifications (Zheng et al., 2011) and apportionment of primary and secondary organic aerosols (Yuan et al., 2006; Hu et al., 2010; Ding et al., 2012). These studies were mostly based on an integrated filter sampling method with a low time resolution (hours or days) and thus could not quantify the chemical processing of carbonaceous aerosols occurring on a short time scale. Recently, several field studies have reported the measurement results of carbonaceous aerosols using high-resolution instruments in the HK-PRD region. With a quadrupole aerosol mass spectrometer, Xiao et al. (2009) observed good correlation between a SOA tracer (i.e., m/z 44) and sulfate in the condensation mode in July 2006 in the PRD outflow and suggested that SOA was formed from the gas-phase oxidation of VOCs. Hu et al. (2012) studied the temporal variation of carbonaceous aerosols at a rural site of the PRD (Back Garden) in July 2006 and quantified the POC and SOC using a modified EC-tracer method with a semi-continuous thermal-optical OC/EC analyzer. Lee et al. (2013) reported size-resolved chemical compositions of the non-refractory submicron aerosol species (including organics) using a HR-ToF-AMS in Hong Kong in May 2011.

During August 2011–May 2012, we conducted a comprehensive field study in select months at a suburban Hong Kong site that frequently received plumes from the urban areas of Hong Kong and the PRD. This paper presents its findings on hourly resolved carbonaceous aerosols together with other relevant chemical data and aerosol morphology. Compared with the previous work on carbonaceous aerosols, our study was unique in that (1) it made

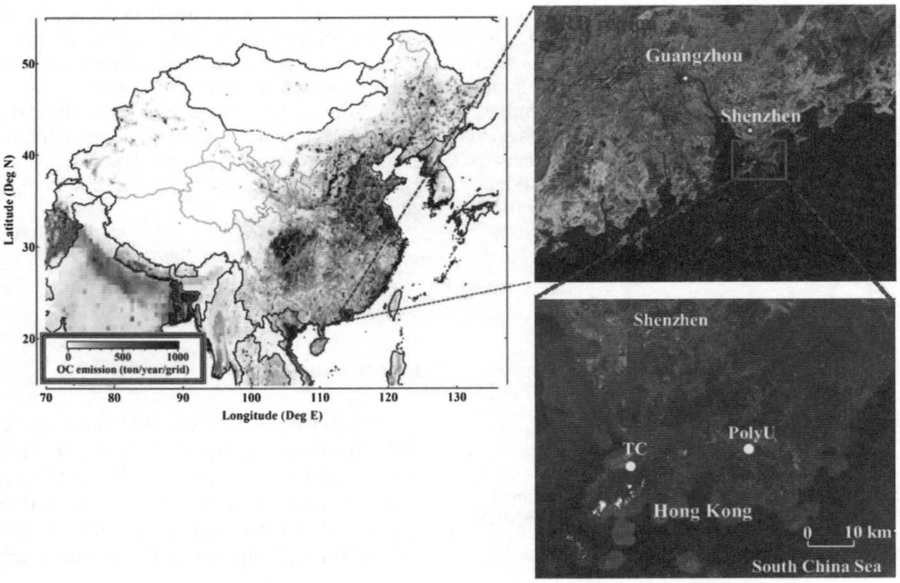

Fig. 1. Map of China and locations of the sampling sites in Hong Kong: a receptor site in Tung Chung (TC, 22.289°N, 113.943°E) and an urban site on the campus of the Hong Kong Polytechnic University (PolyU, 22.303°N, 114.179°E). The girded Asia OC emissions in 2006 are also shown (http://mic.greenresource.cn/intex-b2006).

hourly resolved trace gases and aerosol compositions concurrently available, enabling us to quantify photochemical age and thus the production rates of SOC in a polluted subtropical environment, and (2) aerosol morphology information was examined together with bulk aerosol chemical data. We began by introducing abundances and seasonal/diurnal variations of carbonaceous aerosols, and estimated the fraction of SOC using an EC-tracer method with the aid of other information on photochemical age and biomass burning tracers. We then focused on examining the production rates of SOC in individual urban plumes from Hong Kong and the PRD region, and finally investigated the mixing state of organic aerosol using transmission electron microscope (TEM). This multi-technique approach led to an improved understanding of the sources and formation of carbonaceous aerosols in the outflow from sub-tropical urban and industrial regions.

2. Experimental

2.1. Sampling location and periods

Hong Kong is located on the eastern side of the Pearl River Estuary on the South China coast. Its climate is governed by the Asian monsoons. The prevailing synoptic winds are the northerlies and northeasterlies in winter, the easterlies in spring and autumn and the southwesterlies in summer (Wang et al., 2001). Our measurement site was located at Tung Chung (TC, 22.289°N, 113.943°E) on Lantau Island southwest of Hong Kong. TC is a new town with a population of approximately 80,000, and Che Lap Kok International Airport is located 3 km to the north of the site. The sampling inlet of carbonaceous aerosols was placed on the rooftop of a health center in the town. The site is unique in that it is surrounded by the region's major urban areas (Wang et al., 2001). The measurements were conducted in 4 non-consecutive months: August 3–September 7, 2011 in late summer; November 1–December 3, 2011 in late autumn; February 18–March 9, 2012 in late winter and May 1–31, 2012 in late spring.

2.2. Measurement instruments

The OC and EC concentrations in $PM_{2.5}$ were measured hourly using an in-situ semi-continuous OC and EC analyzer (Dual-Oven Model, Sunset Laboratory Inc., USA). Wang et al. (2011) presented a detailed description of the instrument's principle and operation. In this study, a 1-h measurement cycle was used, with 43 min devoted to sampling and 17 min to analysis during the summer, autumn and winter measurements. A 2-h cycle with 103 min devoted to sampling and the remaining time to analysis was used for the spring measurement due to the low aerosol mass concentrations. The detection limit for OC and EC was 0.3 and 0.1 $\mu g\ m^{-3}$, respectively (Wang et al., 2011).

A suite of trace gases was measured continuously during the field study, including $NO/NO_2/NO_x$ (TEI, Model 42i with a blue light converter), NO_y (TEI, Model 42i-Y), CO (TEI, Model 48C), O_3 (Teledyne, API 400), and SO_2 (TEI, Model 43A). Wang et al. (2003) and Xu et al. (2013) provided detailed information on these trace-gas analyzers. An ambient ion monitor (AIM, Model URG-9000B, USA) was applied to measure the hourly concentrations of water soluble inorganic ions, including SO_4^{2-}, NO_3^-, Cl^-, NH_4^+, Na^+, K^+, Mg^{2+} and Ca^{2+} (Gao et al., 2012). In addition, hourly mass concentration of $PM_{2.5}$ was measured using a tapered element oscillating microbalance (TEOM 1405-DF, Thermo Scientific) with a Filter Dynamic Measurement System (FDMS).

Individual aerosol particles were also collected onto the carbon-film-coated copper TEM grids (carbon type-B, 300-mesh copper, Tianld Co., China) using a single-stage cascade impactor with a jet nozzle 0.5 mm in diameter at a flow rate of 1.0 l min^{-1} (Li et al., 2013). The collected individual aerosol samples were analyzed using a high-resolution transmission electron microscope (TEM, JEOL JEM-2100) operated at 200 kV to determine the morphology, size and the mixing state of the individual aerosol particles. An energy-dispersive X-ray spectrometer (EDS) was used to obtain the compositions of the targeted particles.

2.3. EC-tracer method and photochemical age

2.3.1. EC-tracer method

The EC-tracer method has widely been used to estimate SOC concentrations due to its simplicity (Turpin and Huntzicker, 1995). The key step of this method is to determine the primary OC/EC ratio in the study period and region. In previous studies, the minimum OC/EC ratio or the lowest 5–10% OC/EC ratios have been applied as the primary OC/EC ratio (Lim and Turpin, 2002; Zhang et al., 2008a). In this present study, we made use of highly time-resolved measurements of the OC and EC and simultaneous, continuous measurements of other gas-phase and aerosol pollutants to determine the primary OC/EC ratio. We selected a subset of the data to represent primary OC and EC: that is, the data corresponding to the lowest 25% OC/EC ratios and the lowest 25% of O_3 concentrations in each season. The above selected data were subject to additional checks. We used the NO_x/NO_y ratio, which is an indicator of photochemical aging, to verify the primary OC and EC data. The results showed that the NO_x/NO_y ratios corresponding to the primary OC/EC data in each season were much higher (>0.92) than the respective monthly average ratio (summer: 0.78, autumn: 0.79, winter: 0.89, and spring: 0.84), indicating that the emissions during the selected periods were fresher. The EC-tracer method has an intrinsic drawback when a constant primary OC/EC ratio is adopted. Ding et al. (2012) reported that the EC-tracer method might have overestimated the SOC concentrations in the PRD region during the fall-winter biomass burning events, when high primary OC/EC ratios (>7) were observed. Therefore, we excluded the OC and EC data that could have been affected by biomass burning according to certain criteria (i.e., OC/EC ratios over 7 and obvious enhancement of potassium ion, a commonly used biomass burning tracer). In this study, 2% and 1% of the OC and EC data were excluded in summer and spring respectively, and no data were excluded during autumn and winter based on the criteria. This suggests that biomass burning had a minor influence on the TC site during our observation periods. The 'final' primary OC/EC ratio was derived through a linear regression of the selected sub data points. Since the source emission intensities and meteorological factors were different between the daytime (7:00–18:00) and nighttime (19:00–6:00), we derived the primary OC/EC ratios separately for daytime and nighttime.

2.3.2. Photochemical age

The NO_x/NO_y ratio has often been used as a photochemical 'clock' to help quantify the atmospheric evolution of organic aerosols (Kleinman et al., 2008; Miyazaki et al., 2009; Slowik et al., 2011; Wang et al., 2012). NO_x is emitted mostly as NO and then converted to NO_2 by rapidly reacting with O_3. Further oxidation of NO_2 leads to the formation of higher oxidation state compounds, and NO_y is the sum of NO_x and its oxidation products. As suggested by Kleinman et al. (2008), the photochemical age is defined as $-\log_{10}(NO_x/NO_y)$:

$$\int \kappa[OH]dt = -2.303\log_{10}(NO_x/NO_y) = -\ln(NO_x/NO_y) \quad (1)$$

where κ is the rate constant of NO_2 reacting with OH, with a value about 7.9×10^{-12} $molecules^{-1}\ cm^3\ s^{-1}$ at 1 atm and 300 K (Slowik et al., 2011). [OH] is the average daytime OH radical

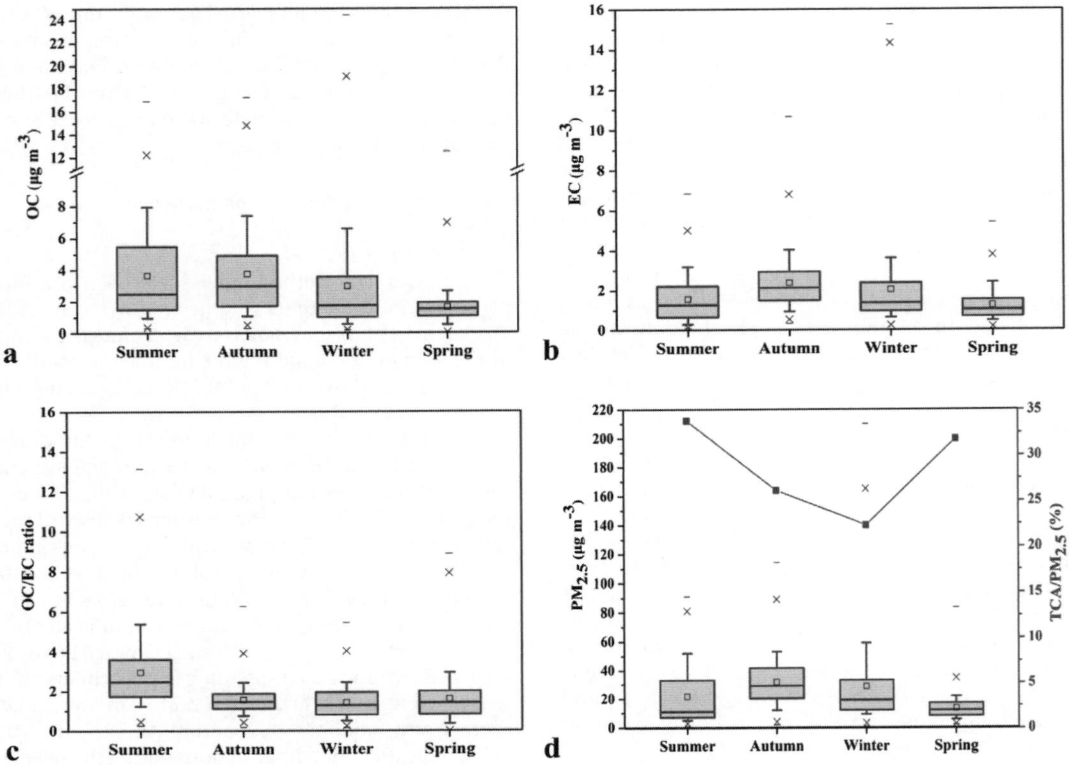

Fig. 2. Seasonal variations of **(a)** OC, **(b)** EC, **(c)** OC/EC ratio and **(d)** $PM_{2.5}$ and TCA/$PM_{2.5}$ ratio at TC. The squares denote the mean concentrations. The whiskers denote the 10th and 90th percentiles. "×" and "–" represent the 99th percentile and maximum and 1st percentile and minimum, respectively.

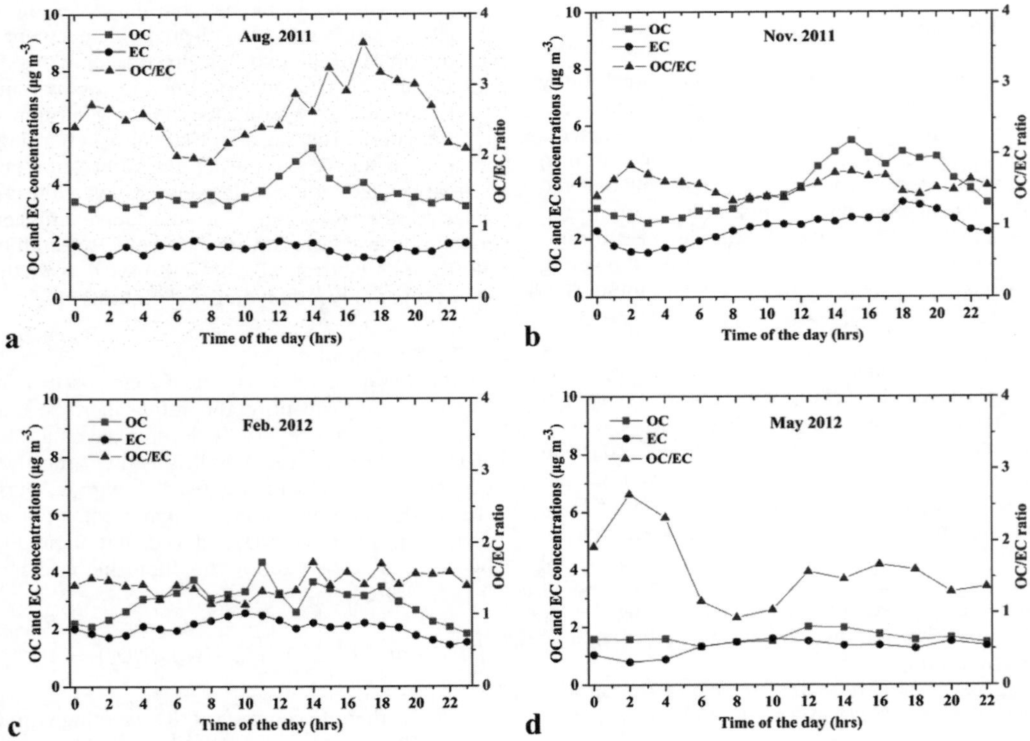

Fig. 3. Mean diurnal variations of OC, EC and OC/EC at TC in **(a)** late summer, **(b)** late autumn, **(c)** late winter and **(d)** late spring.

Table 1
Statistics of the estimated SOC concentrations during the four seasons in Tung Chung (TC), Hong Kong. σ stands for standard deviation.

Season	$(OC/EC)_{pri}$		SOC (μg m^{-3})		SOC/OC (%)	
	Day	Night	Day (Mean ± σ)	Night (Mean ± σ)	Day (Mean ± σ)	Night (Mean ± σ)
Summer	1.35	1.72	2.77 ± 2.35	1.77 ± 1.30	46.9 ± 19.4	32.8 ± 18.8
Autumn	0.90	1.11	2.28 ± 2.50	1.41 ± 1.47	41.0 ± 17.5	32.5 ± 18.0
Winter	0.65	0.73	2.23 ± 2.73	1.58 ± 1.69	51.2 ± 20.2	48.9 ± 20.7
Spring	0.55	0.56	1.17 ± 1.12	1.01 ± 0.49	58.3 ± 18.4	62.1 ± 20.4

concentration, which is about 5.2×10^6 molecules cm^{-3} and 1.6×10^6 molecules cm^{-3} modeled at the TC site in summer and autumn, respectively (Wu et al., manuscript under preparation). The summer average daytime OH concentration at TC is comparable with the summer values of 4.0×10^6 molecules cm^{-3} in Tokyo (Miyakawa et al., 2008) and 3.6×10^6 molecules cm^{-3} in Ontario (Slowik et al., 2011); the autumn OH concentration is comparable with the 1.3×10^6 molecules cm^{-3} in California during the 2010 CalNex Campaign (Hayes et al., 2013) and the 1.6×10^6 molecules cm^{-3} in Mexico City (DeCarlo et al., 2010). At TC, the photochemical ages (i.e. $-Log_{10}(NO_x/NO_y)$) of 0.1, 0.4 and 0.8 were approximately 1.6, 6.2 and 12.5 h in summer and 5, 20 and 40 h in autumn, respectively. The longer times in the autumn season are due to smaller OH concentrations but comparable NO_x/NO_y ratios in summer and autumn. The uncertainty in using the above photochemical age has been investigated previously (Kleinman et al., 2008; Hayes et al., 2013). They point out that the NO_x emission sources are spatially distributed, which may revert the NO_x/NO_y ratio during the transport. As a result, this method may underestimate the photochemical ages, especially for the more aged plumes.

3. Results and discussion

3.1. Seasonal and diurnal variations of carbonaceous aerosols

Fig. 2 presents the seasonal variations of the mean mass concentrations of OC, EC and $PM_{2.5}$ and the OC/EC ratios. The average OC concentrations were 3.66, 3.76, 2.94 and 1.62 μg m^{-3} in summer, autumn, winter and spring, respectively, with an annual mean concentration of 3.00 μg m^{-3}. The EC and $PM_{2.5}$ showed similar seasonal patterns, with the highest concentrations in autumn, followed by summer, winter and spring. The OC/EC ratio reached its highest value in summer, and the ratios in the other 3 months did not vary greatly. The total carbonaceous aerosol mass concentrations (TCA, TCA = 1.6 × OC + EC) accounted for about one third of the $PM_{2.5}$ mass at TC (Fig. 2d).

The diurnal variations of the OC and EC concentrations and the OC/EC ratios in $PM_{2.5}$ in the four seasons are illustrated in Fig. 3. The EC showed high concentrations during the morning and evening rush hours, and the concentrations either dropped slightly (in summer, winter and spring) or kept a constant high level (in autumn) in the afternoon due to a deeper planetary boundary layer. The OC peaked in the noon or early afternoon, similar to the diurnal pattern of O_3 (Wang et al., 2001). This result suggests that the formation of O_3 and organic aerosols may have experienced the same photochemical process. Fig. 3 shows the afternoon peaks of the OC/EC ratio suggesting that SOC was formed through the daytime photochemical processes.

3.2. Estimation of SOC concentrations

Table 1 shows the calculated primary OC/EC ratios, the estimated mean SOC concentrations, and the contributions of SOC to the OC in the four seasons. The estimated primary OC/EC ratio in our study is 1.35, 0.9, 0.65, and 0.55 in daytime and 1.72, 1.11, 0.73, and 0.56 in nighttime in summer, autumn, winter, and spring, respectively. These ratios are similar to the value (0.6) observed in a tunnel in Hong Kong (Cheng et al., 2010), whereas our summer ratio is somewhat larger which may be due to stronger evaporative and biogenic emissions at high temperature. The SOC concentration was the highest in summer, followed by that in winter, autumn and spring. The daytime SOC concentrations were normally higher than those in the nighttime in all four seasons. Strong correlations between the SOC and odd oxygen ($O_x = O_3 + NO_2$) were observed in the summer and autumn seasons (Fig. 4), indicating that profound photochemical production of SOC in these two seasons. The SOC concentration in winter was comparable with that in autumn, a similar result to that of Yuan et al. (2006). Overall, the SOC concentrations accounted for about 47% of the total measured OC concentrations at TC, which is generally consistent with the previous work performed in Hong Kong (46%) for 1998–2002 (Yuan et al., 2006) and at a rural site (Back Garden) in the PRD region (47%) during the summer 2006 (Hu et al., 2012). The SOC to OC ratio was the largest in spring, followed by that in winter, summer and autumn.

3.3. Photochemical evolution and SOC formation in summer and autumn

3.3.1. Case study periods and plume identification

Fig. 5 illustrates the time series of the OC, EC, inorganic ions and gas-phase pollutants during the study periods in summer and autumn when several photochemical episodes were observed with

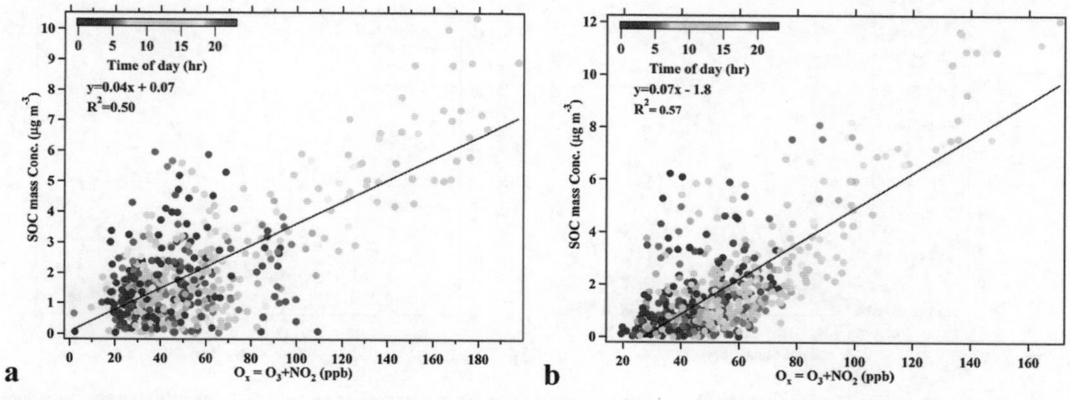

Fig. 4. Correlations between the SOC and odd oxygen ($O_x = O_3 + NO_2$) colored by time of day in **(a)** summer (August, 2011) and **(b)** autumn (November, 2011) in TC site.

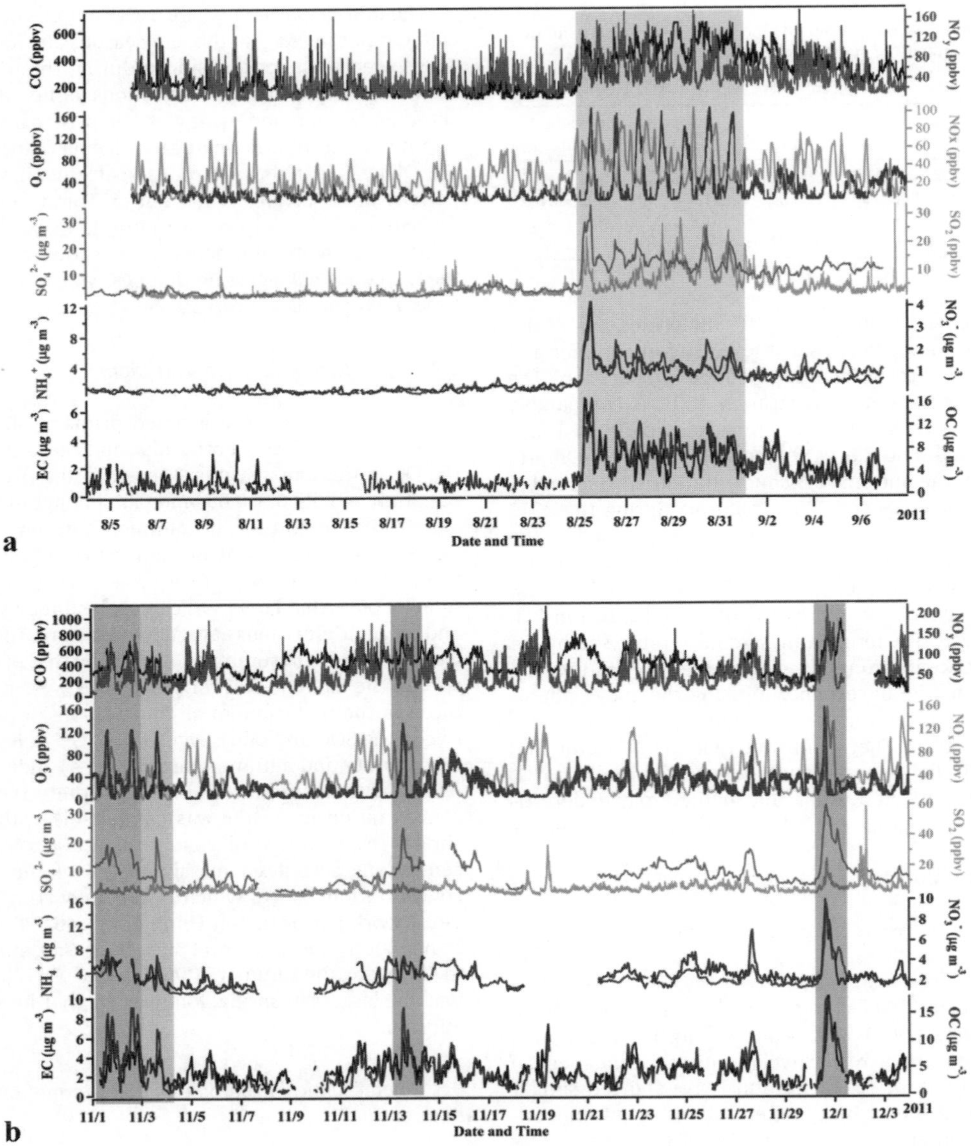

Fig. 5. Time series of OC, EC, inorganic aerosol composition and gas-phase pollutants in **(a)** August 2011 and **(b)** November 2011 at TC. The shaded regions represent the case-study periods in the text.

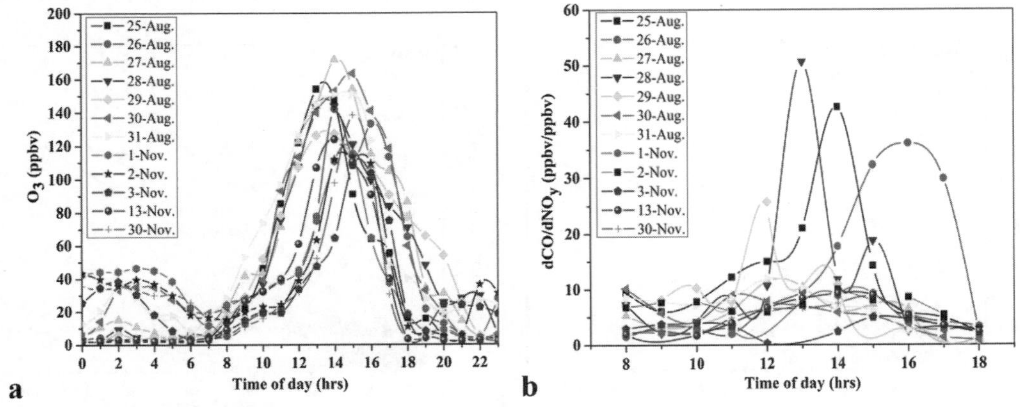

Fig. 6. **(a)** Diurnal variations of O_3 observed at TC during the smoggy days with hourly averaged O_3 concentrations exceeding 100 ppbv. **(b)** Daytime variations of the "24 hourly" dCO/dNO_y ratios at TC during the smoggy days.

the hourly averaged maximum O_3 concentrations of over 100 ppbv (Fig. 6a). Elevated OC, EC and inorganic ion concentrations were also observed during these episodes, shown as the shaded regions in Fig. 5.

To identify the origins of the plumes at TC during the photochemical episodes, two methods (i.e., chemical tracer and back trajectory analysis) were adopted, which are similar to those described by Zhang et al. (2007a). Wang et al. (2003) showed large differences in CO/NO_y emission ratios between the PRD cities and Hong Kong (i.e., CO/NO_y ratio>20 ppbv $ppbv^{-1}$ in the PRD plumes but about 3.9–6 ppbv $ppbv^{-1}$ in the Hong Kong plumes). Following the approach of Zhang et al. (2007a), we calculated "24 hourly" dCO/dNO_y ratio (dCO = the measured 1-h averaged CO concentrations subtracted the minimum CO concentration that day; dNO_y = the measured 1-h averaged NO_y concentrations subtracted the minimum NO_y concentration that day). Fig. 6b shows the daytime variations of the dCO/dNO_y ratios during the aforementioned photochemical episodes. Back trajectories were used to check the validity of the chemical tracer method (dCO/dNO_y ratios). The back trajectories were calculated using the NOAA/ARL HYSPLIT 4.9 model driven by the GDAS meteorological data (Draxler and Rolph, 2003). Fig. 7 shows the representative types of air masses on August 28, November 3 and November 30, 2011. Despite the coarse resolution of the GDAS data (i.e., 1° × 1°), the back trajectories generally verified the results from the dCO/dNO_y ratios, that is, the cases of high CO/NO_y ratios corresponded to air masses from the inland PRD region. Using the above methods, we classified the photochemical events in summer and autumn as HK, the PRD, and mixed HK and the PRD (see Table 2), and these events are subject to further analysis.

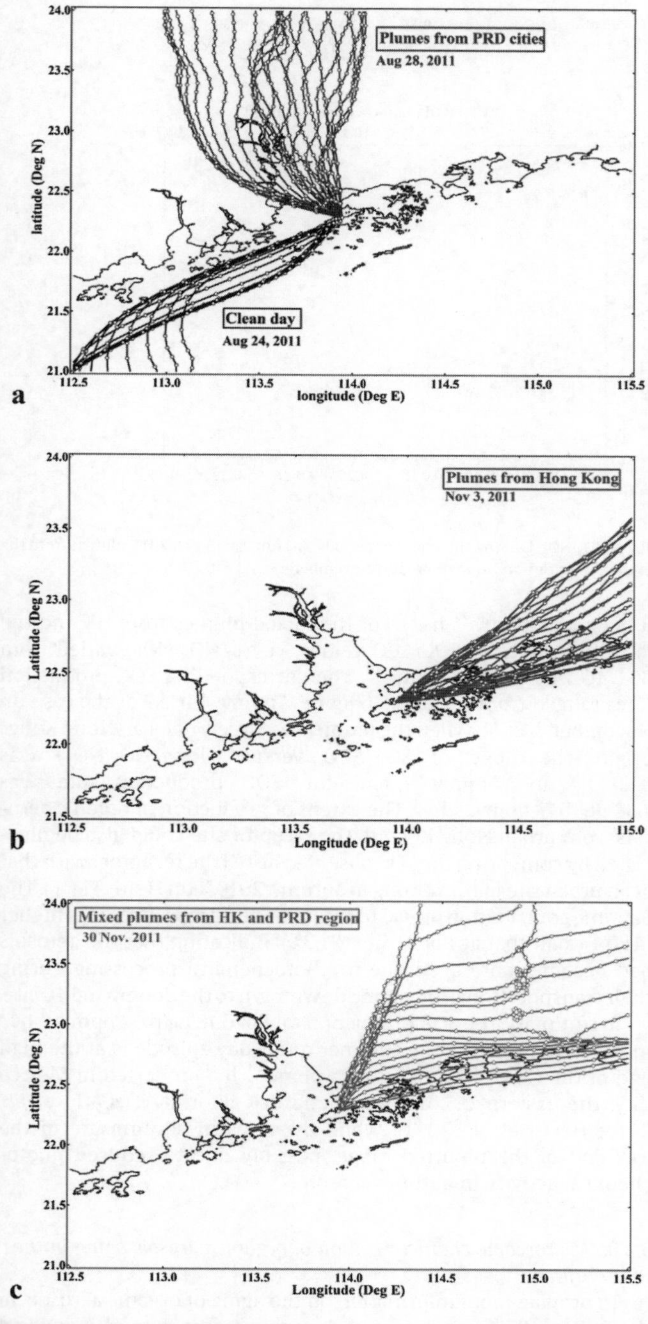

Fig. 7. Back trajectories of the air masses arriving at TC on **(a)** August 28, 2011, **(b)** November 3, 2011 and **(c)** November 30, 2011.

3.3.2. Photochemical evolution of OA in urban plumes

We next calculate the SOC production rates in the identified photochemical episodes. The SOC concentrations were normalized by the excess CO (the measured CO concentrations minus the background CO concentration) in order to reduce the influence of dilution (Kleinman et al., 2008). In this study, the background CO concentrations were obtained from a background Hok Tsui (HT) monitoring station, which is located on the southeastern tip of Hong Kong Island and is considered as the background site of Hong Kong. An examination of the CO data at HT in August and November 2011 indicated absence of a typical double-peak pattern for an urban site but only slightly higher daytime concentrations, confirming only small influence from urban sources on HT (figures not shown). The background CO was the average concentration in the "Marine" air group in August and "Aged continental" air mass in November, as these two air masses arriving at HT were from long distant sources in the respective months (Wang et al., 2009). Using this approach, the background CO concentration was 92 and 186 ppbv in August and November 2011, respectively.

Fig. 8 shows the ΔCO-normalized SOC as a function of the photochemical age on the episode days. The normalized SOC concentrations increased with the increase in photochemical age, indicating that SOC was produced as the plume advected to the TC site. The estimated aging time in late summer was less than 6 h, much shorter than the times of urban plumes observed in other downwind sites: ~24 h in Mexico City in spring (Kleinman et al., 2008), 15 h in Ontario in summer (Slowik et al., 2011), and 12 h in Tokyo in summer (Miyakawa et al., 2008). This can be partly explained by a more approximation of the TC site to urban areas as

Table 2
Time period, air mass categorization, and SOC production of pollution episodes in summer and autumn.

Date	Daily maximum 1-h O_3 concentration (ppbv)	Air mass categorization	Aging time (h)	SOC production rate (μg m^{-3} $ppmv^{-1}$ h^{-1})
Aug. 25–31	127.2–171.7	The PRD	5.5	3.86
Nov. 1–2	115.4	Mixed HK and the PRD	13.9	1.31
Nov. 3	108.4	HK	15.9	1.82
Nov. 13	123.9	Mixed HK and the PRD	16.4	1.32
Nov. 30	138.5	Mixed HK and the PRD	14.7	1.46

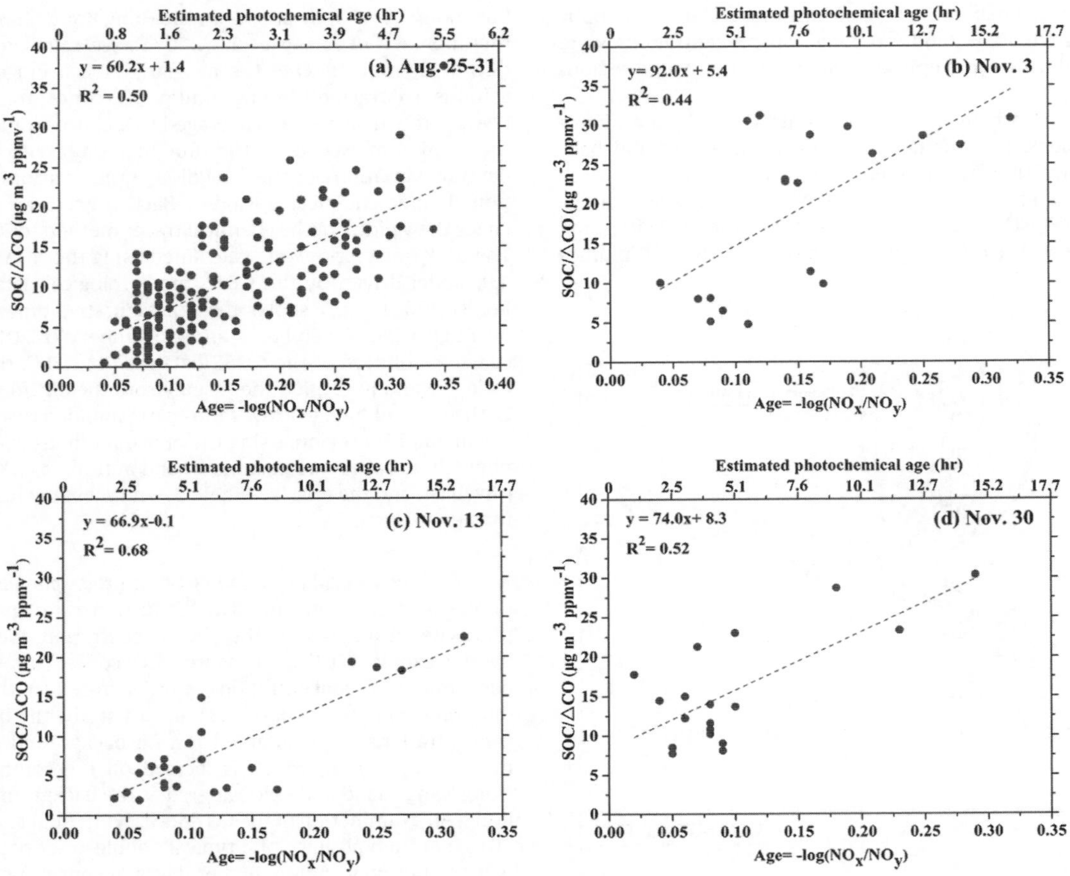

Fig. 8. ΔCO-normalized SOC concentrations (SOC/ΔCO) as a function of photochemical age ($-\log_{10}(NO_x/NO_y)$) during the polluted periods: **(a)** August 25–31, 2011, plumes from the PRD cities; **(b)** November 3, 2011, plumes from Hong Kong and **(c)** and **(d)** plumes from both the PRD cities and Hong Kong on November 13 and 30, 2011.

compared to other downwind sites. The slopes of SOC/ΔCO versus estimated photochemical age (hours) are the SOC production rates, which varied slightly among the three types of air masses at TC (Fig. 8 and Table 2).

We first considered the urban plumes from the PRD cities observed during the multi-day episode (August 25–31). The slope of the SOC/ΔCO versus $-\log_{10}(NO_x/NO_y)$ was 60.2 μg m^{-3} ppmv^{-1}. Assuming a factor of 1.6 to convert SOC to SOA (Turpin and Lim, 2001), the slope is 96.3 μg m^{-3} ppmv^{-1}, which is comparable with that observed in the Mexico City outflow (71.4 μg m^{-3} ppmv^{-1}) (Kleinman et al., 2008), but much lower than that measured downwind Detroit/Winsor in Canada (189.2 μg m^{-3} ppmv^{-1}) (Slowik et al., 2011). Using the more conventional time scale (hour), the average SOC production rate was 3.86 μg m^{-3} ppmv^{-1} h^{-1} for PRD plumes (with a SOA production rate of about 6.18 μg m^{-3} ppmv^{-1} h^{-1}). For the mixed plumes from HK and the PRD, the slopes of SOC/ΔCO versus $-\log_{10}(NO_x/NO_y)$ varied from 66.2 to 74.0 μg m^{-3} ppmv^{-1}. The corresponding SOC production rates ranged from 1.31 to 1.46 μg m^{-3} ppmv^{-1} h^{-1}. For the case on November 3, 2011, when the plume was largely of local (Hong Kong) origin, the slope of SOC/ΔCO versus $-\log_{10}(NO_x/NO_y)$ was 92.0 μg m^{-3} ppmv^{-1} and the SOC production rate was 1.82 μg m^{-3} ppmv^{-1} h^{-1}. The extent of production of organic aerosols from urban Hong Kong to the receptor site could also be illustrated by comparing the SOC observed at TC (the receptor) with that at an urban site in Hong Kong in autumn 2010 (PolyU site, Fig. 1). The daytime contribution of SOC to the OC at the receptor site was higher (41.0%) than that at PolyU site (31.3%), indicating that the aerosols had become more aged due to photochemical processing during their transport from Hong Kong downtown to the downwind TC site.

In summary, the SOA production rate of 6.18 μg m^{-3} ppmv^{-1} h^{-1} in Hong Kong during the summer multi-day episode is at the high end of the range (~2–5 μg m^{-3} ppmv^{-1} h^{-1}) reported in Mexico City, the eastern U.S. and Pasadena in California (DeCarlo et al., 2010; Hayes et al., 2013), while the rates in autumn are in the low end of the reported range, possibly due to reduced photochemical activity in autumn season.

Table 3
Sampling time and atmospheric conditions of individual aerosol particle samples at TC.

Date	Start time	End time	RH (%)	Temp. (°C)	WD	WS (m s^{-1})	$PM_{2.5}$ (μg m^{-3})	O_3 (ppbv)
Clean day								
2011/8/24	13:56	14:01	51%	36.5	E	1.6	11.0	35.1
2011/8/24	17:27	17:32	60%	35.0	W	1.5	7.8	23.3
Smoggy days								
2011/8/25	10:13	10:15	71%	33.8	NNW	1.5	72.0	37.0
2011/8/25	18:56	18:59	68%	31.9	N	0.6	33.5	4.0
2011/8/27	12:15	12:16	57%	34.9	NW	2.6	95.0	104.0
2011/8/31	17:47	17:48	56%	34.2	NW	2.4	78.4	55.0

3.3.3. *Microscopic characterization of organic aerosols aging in the PRD outflow*

To provide more information on the aging of organic aerosols in the PRD outflow, we analyzed the microscopic data of individual aerosol particles obtained from the high-resolution TEM. The

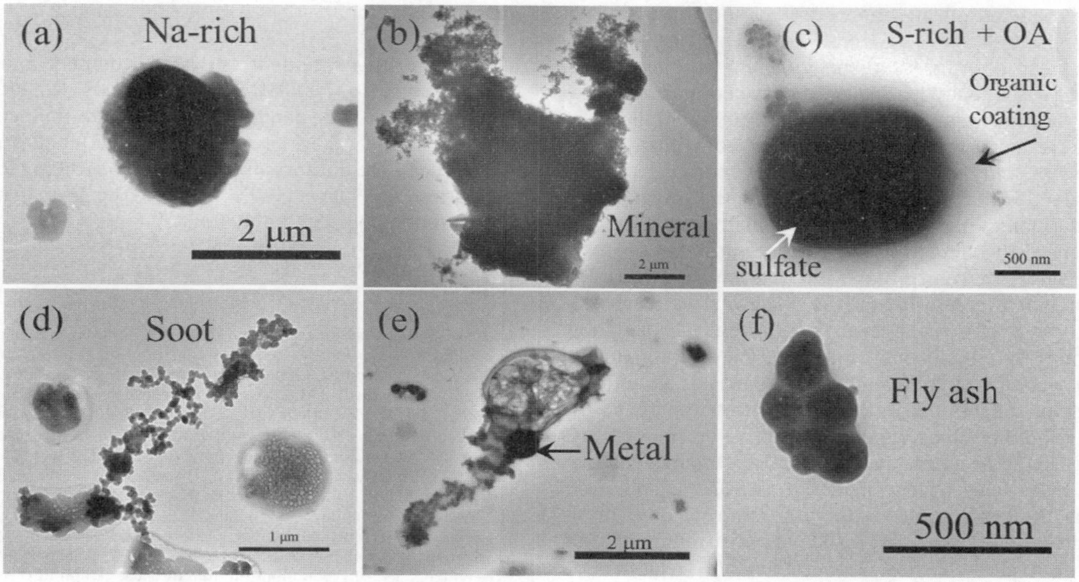

Fig. 9. TEM images of six types of individual particles collected on August 24–31, 2011: **(a)** Na-rich, **(b)** mineral, **(c)** sulfate and organic aerosols, **(d)** soot, **(e)** metal and **(f)** fly ash. The compositions of the individual aerosol particles were determined by TEM/EDS.

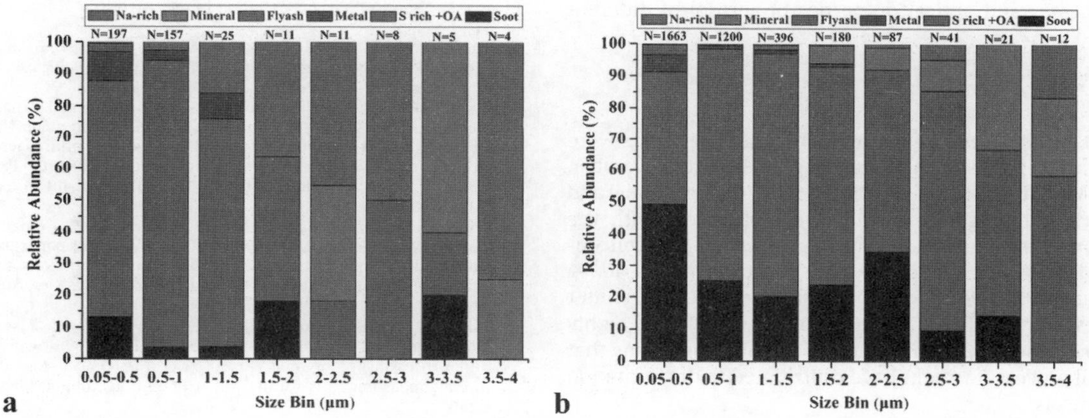

Fig. 10. The proportions of different aerosol particle types (Na-rich, mineral, fly ash, metal, sulfate and organic aerosols (S rich + OA), and soot) in different size ranges for **(a)** the clean day and **(b)** the smoggy days in summer (see in Table 3).

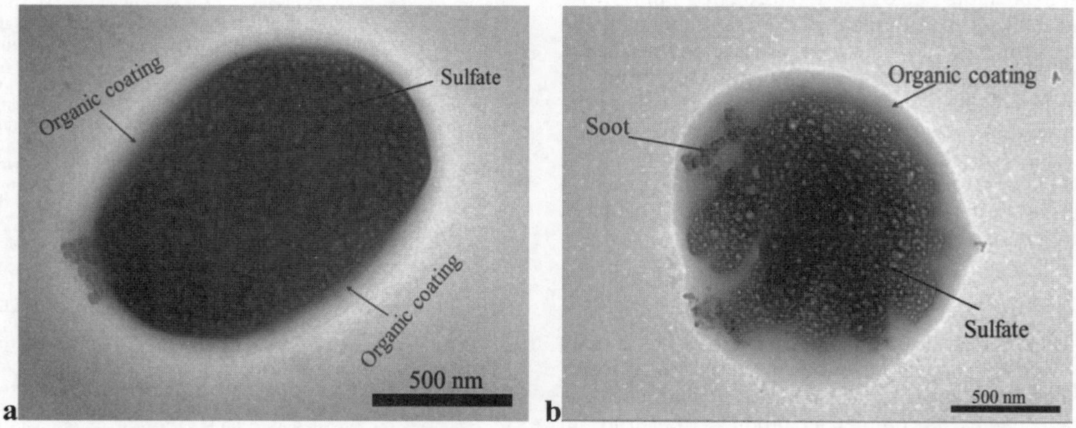

Fig. 11. **(a)** Thin organic coating on the clean day (Aug. 24, 2011). **(b)** Thick organic coating on the smoggy day (Aug. 25, 2011).

particles were collected during one clean day and three smoggy days in August 2011 (Table 3). A total of 4018 particles from 6 samples (2 on clean day and 4 on smoggy days) were examined, and six types of individual particles were obtained, including Na-rich, mineral, fly ash, metal, sulfate/organics, and soot particles (Fig. 9). On the clean day, the particle diameters were mostly lower than 1.5 μm in size, and the sulfates and organics comprised about 80% of the total individual particles (Fig. 10a). On the smoggy days, the sulfate and organics had a wider range of diameters compared with those on the clean day, and sulfate and organics were the main components even in the coarse mode (up to 4 μm) (Fig. 10b). The results suggest that the aerosols on the smoggy days were subjected to intensive processing (i.e., photochemical oxidation, coagulation and heterogeneous nucleation), resulting in an enrichment of secondary materials in coarse particles. In addition to the presence of sulfate and organics in larger-size particle, the organic coatings on the sulfate were much thinker on smoggy days (e.g., Fig. 11b) than on the clean day (e.g., Fig. 11a). Organics coatings on sulfate particles have been considered as SOA (Li and Shao, 2010), and the thinker organic coatings on smoggy days imply more SOA than on the clean day, which is consistent with SOC concentration from the real-time measurements (1.32 μg m^{-3} on clean days versus 3.58 μg m^{-3} on the smoggy days in August 2011). The morphology data indicate that organics were totally internally mixed with sulfate particles in chemically processed air, highlighting the complexity in understanding and predicting the chemical processes and optical properties of aerosol of different ageing.

4. Summary

OC, EC, $PM_{2.5}$ and other gaseous and particulate pollutants were measured at a receptor site in Hong Kong in August, November, February and May during the year of 2011–2012. Clear seasonal and diurnal variations of organic aerosols were observed, with the highest OC concentrations in autumn. The estimated SOC concentrations were 2.14, 1.85, 1.91 and 1.09 μg m^{-3}, accounting for 40.7%, 36.8%, 50.1% and 60.2% of the total OC in summer, autumn, winter and spring, respectively. The SOC and odd oxygen (O_x) were highly correlated during the summer and autumn seasons, suggesting that photochemical processing might have contributed to the formation of organic aerosols.

The photochemical aging of organic aerosol was quantitatively studied during the smog episodes in summer and autumn. The results showed that the PRD cities and Hong Kong were the source regions and that aerosol particles became more aged as they were advected toward the receptor TC site. The SOC production rates were estimated to be in the range of 1.31–1.82 μg m^{-3} $ppmv^{-1}$ h^{-1} in autumn and 3.86 μg m^{-3} $ppmv^{-1}$ h^{-1} in summer, with the summer rate in the upper end of the range observed in the U.S. and Mexico City. The reduced rates in the autumn episodes may be due to decreased level of photochemistry. TEM images revealed organic coatings on the sulfates and soot in all particle sizes (0.05–4.0 μm) during the episodes, indicating complex physico-chemical processing of organic aerosols. The results of this study can help improve modeling the formation of organic aerosols in polluted sub-tropical environments.

Disclaimer

The opinions expressed in this paper are those of the authors and do not necessarily reflect the views or policies of the Government of the Hong Kong Special Administrative Region, nor does mention of trade names or commercial products constitute an endorsement or recommendation of their use.

Acknowledgements

This work was funded by the Environment and Conservation Fund (Project No. 7/2009). We thank Dr. K.F. Ho, Prof. S.C. Lee, Dr. K.S. Lam, Dr. Wei Nie and Mr. Qiaozhi Zha for their contributions to the field study and Dr. Likun Xue and Miss Jueqi Wu for providing the OH data. The authors also thank Dr. Robert Cary for his technical support on the Sunset OC/EC analyzer. The participation of Dr. W.J. Li was funded by NSFC (41105088).

References

Andreae, M.O., Schmid, O., Yang, H., Chand, D., Zhen Yu, J., Zeng, L.M., Zhang, Y.H., 2008. Optical properties and chemical composition of the atmospheric aerosol in urban Guangzhou, China. Atmos. Environ. 42 (25), 6335–6350.

Bahreini, R., Ervens, B., Middlebrook, A., Warneke, C., de Gouw, J., DeCarlo, P., Jimenez, J., Brock, C., Neuman, J., Ryerson, T., 2009. Organic aerosol formation in urban and industrial plumes near Houston and Dallas, Texas. J. Geophys. Res. Atmos. 114, D00F16. http://dx.doi.org/10.1029/2008JD011493.

Bond, T.C., Doherty, S.J., Fahey, D.W., Forster, P.M., Berntsen, T., DeAngelo, B.J., Flanner, M.G., Ghan, S., Kärcher, B., Koch, D., Kinne, S., Kondo, Y., Quinn, P.K., Sarofim, M.C., Schultz, M.G., Schulz, M., Venkataraman, C., Zhang, H., Zhang, S., Bellouin, N., Guttikunda, S.K., Hopke, P.K., Jacobson, M.Z., Kaiser, J.W., Klimont, Z., Lohmann, U., Schwarz, J.P., Shindell, D., Storelvmo, T., Warren, S.G., Zender, C.S., 2013. Bounding the role of black carbon in the climate system: a scientific assessment. J. Geophys. Res. Atmos. 118. http://dx.doi.org/10.1002/jgrd.50171.

Cao, J., Lee, S., Ho, K., Zou, S., Fung, K., Li, Y., Watson, J., Chow, J., 2004. Spatial and seasonal variations of atmospheric organic carbon and elemental carbon in Pearl River Delta Region, China. Atmos. Environ. 38 (27), 4447–4456.

Chan, C.K., Yao, X., 2008. Air pollution in mega cities in China. Atmos. Environ. 42 (1), 1–42.

Cheng, Y., Lee, S.C., Ho, K.F., Chow, J.C., Watson, J.G., Louie, P.K.K., Cao, J.J., Hai, X., 2010. Chemically-speciated on-road PM2.5 motor vehicle emission factors in Hong Kong. Sci. Total Environ. 408 (7), 1621–1627.

Cheng, Y.F., Eichler, H., Wiedensohler, A., Heintzenberg, J., Zhang, Y.H., Hu, M., Herrmann, H., Zeng, L.M., Liu, S., Gnauk, T., Brüggemann, E., He, L.Y., 2006. Mixing state of elemental carbon and non-light-absorbing aerosol components derived from in situ particle optical properties at Xinken in Pearl River Delta of China. J. Geophys. Res. Atmos. 111, D20204. http://dx.doi.org/10.1029/22005JD006929.

De Gouw, J., Brock, C., Atlas, E., Bates, T., Fehsenfeld, F., Goldan, P., Holloway, J., Kuster, W., Lerner, B., Matthew, B., 2008. Sources of particulate matter in the northeastern United States in summer: 1. Direct emissions and secondary formation of organic matter in urban plumes. J. Geophys. Res. Atmos. 113, D08301. http://dx.doi.org/10.1029/02007JD009243.

DeCarlo, P., Ulbrich, I., Crounse, J., de Foy, B., Dunlea, E., Aiken, A., Knapp, D., Weinheimer, A., Campos, T., Wennberg, P., 2010. Investigation of the sources and processing of organic aerosol over the Central Mexican Plateau from aircraft measurements during MILAGRO. Atmos. Chem. Phys. 10 (12), 5257–5280.

Ding, X., Wang, X.-M., Gao, B., Fu, X.-X., He, Q.-F., Zhao, X.-Y., Yu, J.-Z., Zheng, M., 2012. Tracer-based estimation of secondary organic carbon in the Pearl River Delta, south China. J. Geophys. Res. Atmos. 117, D05313. http://dx.doi.org/10.1029/02011JD016596.

Draxler, R., Rolph, G., 2003. HYSPLIT (HYbrid Single-particle Lagrangian Integrated Trajectory) Model. NOAA Air Resources Laboratory, Silver Spring, MD. Access via NOAA ARL READY Website. http://www.arl.noaa.gov/ready/hysplit4.html.

Dzepina, K., Cappa, C.D., Volkamer, R.M., Madronich, S., DeCarlo, P.F., Zaveri, R.A., Jimenez, J.L., 2010. Modeling the multiday evolution and aging of secondary organic aerosol during MILAGRO 2006. Environ. Sci. Technol. 45 (8), 3496–3503.

Gao, X., Xue, L., Wang, X., Wang, T., Yuan, C., Gao, R., Zhou, Y., Nie, W., Zhang, Q., Wang, W., 2012. Aerosol ionic components at Mt. Heng in central southern China: abundances, size distribution, and impacts of long-range transport. Sci. Total Environ. 433, 498–506.

Gnauk, T., Muller, K., van Pinxteren, D., He, L.Y., Niu, Y., Hu, M., Herrmann, H., 2008. Size-segregated particulate chemical composition in Xinken, Pearl River Delta, China: OC/EC and organic compounds. Atmos. Environ. 42 (25), 6296–6309.

Hagler, G., Bergin, M., Salmon, L., Yu, J., Wan, E., Zheng, M., Zeng, L., Kiang, C., Zhang, Y., Lau, A., 2006. Source areas and chemical composition of fine particulate matter in the Pearl River Delta region of China. Atmos. Environ. 40 (20), 3802–3815.

Hayes, P.L., Ortega, A.M., Cubison, M.J., Froyd, K.D., Zhao, Y., Cliff, S.S., Hu, W.W., Toohey, D.W., Flynn, J.H., Lefer, B.L., Grossberg, N., Alvarez, S., Rappenglück, B., Taylor, J.W., Allan, J.D., Holloway, J.S., Gilman, J.B., Kuster, W.C., de Gouw, J.A., Massoli, P., Zhang, X., Liu, J., Weber, R.J., Corrigan, A.L., Russell, L.M., Isaacman, G., Worton, D.R., Kreisberg, N.M., Goldstein, A.H., Thalman, R., Waxman, E.M., Volkamer, R., Lin, Y.H., Surratt, J.D., Kleindienst, T.E., Offenberg, J.H., Dusanter, S., Griffith, S., Stevens, P.S., Brioude, J., Angevine, W.M., Jimenez, J.L., 2013. Organic aerosol composition and sources in Pasadena,

California during the 2010 CalNex campaign. J. Geophys. Res. Atmos. 118. http://dx.doi.org/10.1002/jgrd.50530.
Hu, D., Bian, Q., Lau, A.K.H., Yu, J.Z., 2010. Source apportioning of primary and secondary organic carbon in summer PM2. 5 in Hong Kong using positive matrix factorization of secondary and primary organic tracer data. J. Geophys. Res. Atmos. 115, D16204. http://dx.doi.org/10.1029/12009JD012498.
Hu, W.W., Hu, M., Deng, Z.Q., Xiao, R., Kondo, Y., Takegawa, N., Zhao, Y.J., Guo, S., Zhang, Y.H., 2012. The characteristics and origins of carbonaceous aerosol at a rural site of PRD in summer of 2006. Atmos. Chem. Phys. 12 (4), 1811–1822.
Huang, X.F., Yu, J., He, L.Y., Yuan, Z., 2006. Water-soluble organic carbon and oxalate in aerosols at a coastal urban site in China: size distribution characteristics, sources and formation mechanisms. J. Geophys. Res. Atmos. 111, D22212. http://dx.doi.org/10.1029/22006JD007408.
Jimenez, J., Canagaratna, M., Donahue, N., Prevot, A., Zhang, Q., Kroll, J.H., DeCarlo, P.F., Allan, J.D., Coe, H., Ng, N., 2009. Evolution of organic aerosols in the atmosphere. Science 326 (5959), 1525–1529.
Johnson, D., Utembe, S., Jenkin, M., Derwent, R., Hayman, G., Alfarra, M., Coe, H., McFiggans, G., 2006. Simulating regional scale secondary organic aerosol formation during the TORCH 2003 campaign in the southern UK. Atmos. Chem. Phys. 6 (2), 403–418.
Kleinman, L., Springston, S., Daum, P., Lee, Y.-N., Nunnermacker, L., Senum, G., Wang, J., Weinstein-Lloyd, J., Alexander, M., Hubbe, J., 2008. The time evolution of aerosol composition over the Mexico City plateau. Atmos. Chem. Phys. 8, 1559–1575.
Lee, B.P., Li, Y.J., Yu, J.Z., Louie, P.K.K., Chan, C.K., 2013. Physical and chemical characterization of ambient aerosol by HR-ToF-AMS at a suburban site in Hong Kong during springtime 2011. J. Geophys. Res. Atmos. 118 (15), 8625–8639.
Li, W., Shao, L., 2010. Mixing and water-soluble characteristics of particulate organic compounds in individual urban aerosol particles. J. Geophys. Res. Atmos. 115 (D2). http://dx.doi.org/10.1029/2009JD012575.
Li, W., Wang, T., Zhou, S., Lee, S., Huang, Y., Gao, Y., Wang, W., 2013. Microscopic observation of metal-containing particles from Chinese Continental outflow observed from a non-industrial site. Environ. Sci. Technol. 47 (16), 9124–9131.
Lim, H.J., Turpin, B.J., 2002. Origins of primary and secondary organic aerosol in Atlanta: results of time-resolved measurements during the Atlanta supersite experiment. Environ. Sci. Technol. 36 (21), 4489–4496.
Miyakawa, T., Takegawa, N., Kondo, Y., 2008. Photochemical evolution of submicron aerosol chemical composition in the Tokyo megacity region in summer. J. Geophys. Res. Atmos. 113, D14304. http://dx.doi.org/10.1029/12007JD009493.
Miyazaki, Y., Kondo, Y., Shiraiwa, M., Takegawa, N., Miyakawa, T., Han, S., Kita, K., Hu, M., Deng, Z., Zhao, Y., 2009. Chemical characterization of water-soluble organic carbon aerosols at a rural site in the Pearl River Delta, China, in the summer of 2006. J. Geophys. Res. Atmos. 114, D14208. http://dx.doi.org/10.1029/12009JD011736.
Nel, A., 2005. Air pollution-related illness: effects of particles. Science 308 (5723), 804–806.
Robinson, A.L., Donahue, N.M., Shrivastava, M.K., Weitkamp, E.A., Sage, A.M., Grieshop, A.P., Lane, T.E., Pierce, J.R., Pandis, S.N., 2007. Rethinking organic aerosols: semivolatile emissions and photochemical aging. Science 315 (5816), 1259–1262.
Seinfeld, J., Pandis, S., 2006. Atmospheric Chemistry and Physics:From Air Pollution to Climate Change. John Wiley & Sons, New York.
Slowik, J.G., Brook, J., Chang, R.Y.W., Evans, G.J., Hayden, K., Jeong, C.H., Li, S.M., Liggio, J., Liu, P.S.K., McGuire, M., Mihele, C., Sjostedt, S., Vlasenko, A., Abbatt, J.P.D., 2011. Photochemical processing of organic aerosol at nearby continental sites: contrast between urban plumes and regional aerosol. Atmos. Chem. Phys. 11 (6), 2991–3006.
Turpin, B., Huntzicker, J., 1995. Identification of secondary organic aerosol episodes and quantitation of primary and secondary organic aerosol concentrations during SCAQS. Atmos. Environ. 29 (23), 3527–3544.
Turpin, B., Lim, H., 2001. Species contributions to $PM_{2.5}$ mass concentrations: revisiting common assumptions for estimating organic mass. Aerosol Sci. Technol. 35 (1), 602–610.
Wang, T., Wu, Y., Cheung, T., Lam, K., 2001. A study of surface ozone and the relation to complex wind flow in Hong Kong. Atmos. Environ. 35 (18), 3203–3215.
Wang, T., Poon, C., Kwok, Y., Li, Y., 2003. Characterizing the temporal variability and emission patterns of pollution plumes in the Pearl River Delta of China. Atmos. Environ. 37 (25), 3539–3550.
Wang, T., Wei, X.L., Ding, A.J., Poon, C.N., Lam, K.S., Li, Y.S., Chan, L.Y., Anson, M., 2009. Increasing surface ozone concentrations in the background atmosphere of Southern China, 1994–2007. Atmos. Chem. Phys. 9 (16), 6217–6227.
Wang, Z., Wang, T., Gao, R., Xue, L., Guo, J., Zhou, Y., Nie, W., Wang, X., Xu, P., Gao, J., 2011. Source and variation of carbonaceous aerosols at Mount Tai, North China: results from a semi-continuous instrument. Atmos. Environ. 45 (9), 1655–1667.
Wang, Z., Wang, T., Guo, J., Gao, R., Xue, L., Zhang, J., Zhou, Y., Zhou, X., Zhang, Q., Wang, W., 2012. Formation of secondary organic carbon and cloud impact on carbonaceous aerosols at Mount Tai, North China. Atmos. Environ. 46, 516–527.
Weber, R., Sullivan, A., Peltier, R., Russell, A., Yan, B., Zheng, M., De Gouw, J., Warneke, C., Brock, C., Holloway, J., 2007. A study of secondary organic aerosol formation in the anthropogenic-influenced southeastern United States. J. Geophys. Res. Atmos. 112, D13302. http://dx.doi.org/10.1029/12007JD008408.
Xiao, R., Takegawa, N., Kondo, Y., Miyazaki, Y., Miyakawa, T., Hu, M., Shao, M., Zeng, L.M., Hofzumahaus, A., Holland, F., Lu, K., Sugimoto, N., Zhao, Y., Zhang, Y.H., 2009. Formation of submicron sulfate and organic aerosols in the outflow from the urban region of the Pearl River Delta in China. Atmos. Environ. 43 (24), 3754–3763.
Xu, Z., Wang, T., Xue, L.K., Louie, P.K.K., Luk, C.W.Y., Gao, J., Wang, S.L., Chai, F.H., Wang, W.X., 2013. Evaluating the uncertainties of thermal catalytic conversion in measuring atmospheric nitrogen dioxide at four differently polluted sites in China. Atmos. Environ. 76, 221–226.
Yu, H., Wu, C., Wu, D., Yu, J., 2010. Size distributions of elemental carbon and its contribution to light extinction in urban and rural locations in the Pearl River delta region, China. Atmos. Chem. Phys. 10 (11), 5107–5119.
Yu, J., Tung, J., Wu, A., Lau, A., Louie, P., Fung, J., 2004. Abundance and seasonal characteristics of elemental and organic carbon in Hong Kong PM10. Atmos. Environ. 38 (10), 1511–1521.
Yuan, Z.B., Yu, J.Z., Lau, A.K.H., Louie, P.K.K., Fung, J.C.H., 2006. Application of positive matrix factorization in estimating aerosol secondary organic carbon in Hong Kong and its relationship with secondary sulfate. Atmos. Chem. Phys. 6 (1), 25–34.
Zhang, J., Wang, T., Chameides, W., Cardelino, C., Kwok, J., Blake, D., Ding, A., So, K., 2007a. Ozone production and hydrocarbon reactivity in Hong Kong, Southern China. Atmos. Chem. Phys. 7, 557–573.
Zhang, Q., Jimenez, J., Canagaratna, M., Allan, J., Coe, H., Ulbrich, I., Alfarra, M., Takami, A., Middlebrook, A., Sun, Y., 2007b. Ubiquity and dominance of oxygenated species in organic aerosols in anthropogenically-influenced Northern Hemisphere midlatitudes. Geophys. Res. Lett. 34, L13801. http://dx.doi.org/10.1029/12007GL029979.
Zhang, X., Wang, Y., Zhang, X., Guo, W., Gong, S., 2008a. Carbonaceous aerosol composition over various regions of China during 2006. J. Geophys. Res. Atmos. 113, D14111. http://dx.doi.org/10.1029/12007JD009525.
Zhang, Y., Hu, M., Zhong, L., Wiedensohler, A., Liu, S., Andreae, M., Wang, W., Fan, S., 2008b. Regional integrated experiments on air quality over Pearl River Delta 2004 (PRIDE-PRD2004): overview. Atmos. Environ. 42 (25), 6157–6173.
Zheng, M., Wang, F., Hagler, G., Hou, X., Bergin, M., Cheng, Y., Salmon, L., Schauer, J.J., Louie, P.K.K., Zeng, L., 2011. Sources of excess urban carbonaceous aerosol in the Pearl River Delta Region, China. Atmos. Environ. 45 (5), 1175–1182.
Zhong, L., Louie, P.K.K., Zheng, J., Yuan, Z., Yue, D., Ho, J.W.K., Lau, A.K.H., 2013. Science-policy interplay: air quality management in the Pearl River Delta region and Hong Kong. Atmos. Environ. 76, 3–10.

Atmospheric Environment 45 (2011) 3882–3890

Contents lists available at ScienceDirect

Atmospheric Environment

journal homepage: www.elsevier.com/locate/atmosenv

Secondary organic aerosol formation from xylenes and mixtures of toluene and xylenes in an atmospheric urban hydrocarbon mixture: Water and particle seed effects (II)

Yang Zhou [a,b], Haofei Zhang [b], Harshal M. Parikh [b], Eric H. Chen [b], Weruka Rattanavaraha [b], Elias P. Rosen [b], Wenxing Wang [a,c], Richard M. Kamens [b,*]

[a] Environment Research Institute, Shandong University, Jinan, Shandong 250100, PR China
[b] Department of Environmental Sciences and Engineering, University of North Carolina, Chapel Hill, NC 27599, USA
[c] Chinese Research Academy of Environmental Sciences, Beijing 100012, PR China

ARTICLE INFO

Article history:
Received 3 October 2010
Received in revised form
14 December 2010
Accepted 17 December 2010

Keywords:
Xylene SOA
Secondary organic aerosol modeling
SOA yields
Glyoxal
Aqueous phase
Particle water

ABSTRACT

Secondary organic aerosol (SOA) formation from the photooxidation of *o*-, *p*-xylene, and toluene with xylene mixtures was investigated in the UNC dual outdoor smog chambers. Experiments were performed with different initial background aerosol concentrations and levels of relative humidity (RH) in the environment of an eleven component mixture of non-SOA-forming dilute urban hydrocarbon mixture, oxides of nitrogen and sunlight. Post-nucleation was observed in most of the experiments in the 14–20 nm range except under the conditions with high background aerosol (>5 μg m^{-3}) and with low *o*-xylene concentrations (<0.092 ppmv). The SOA yields of *o*-xylene varied from 0.8% to 6.5% depending on the RH and initial seed concentrations. *p*-Xylene had a lower SOA yield compared with *o*-xylene and the yields in experiments with toluene and xylene mixtures ranged from 1.1% to 10.3%. SOA yield was found to be positively correlated with the particle water (H_2Op) content. A new condensed aromatic kinetic mechanism employing uptake of organics in H_2Op as a key parameter was applied to all the experiments and the simulations showed reasonable fits to the observed data.

1. Introduction

Aromatic hydrocarbons are an important class of volatile organic compounds present in the atmosphere (Calvert et al., 2002), which can have direct and indirect negative impacts on public health because they can act as precursors to harmful pollutants such as ozone and secondary organic aerosols (SOA) (Song et al., 2007a). Xylenes (*m*-, *p*-, and *o*-xylene) are the second most important ambient aromatic hydrocarbons after toluene and have hydroxyl radical (OH) rate coefficients that are approximately two to four times that of toluene.

Experimentally determined SOA yields based on the ratio of observed SOA (μg m^{-3}) formation to the reacted aromatic concentration (μg m^{-3}) are typically used to estimate SOA formation in air quality models (Henry et al., 2008; Carlton et al., 2010). Odum et al. (1996) expressed chamber observations of increasing SOA yields with increasing organic aerosol concentrations by the following equation:

$$Y = \frac{\Delta M}{\Delta \mathrm{HC}} = M \sum_i \frac{\alpha_i K_{om,i}}{1 + MK_{om,i}} \qquad (1)$$

where α_i and $K_{om,i}$ are the mass-based stoichiometric coefficient and gas-particle partitioning coefficient of the reaction product i and M is the mass concentration of absorbing organic aerosol present. One or two products are typically employed in Equation (1) ($i = 1$ or 2). Current air quality models using the Odum parameters based on experimentally determined aerosol yields, predict lower SOA level in general than the ambient measurements (De Gouw et al., 2005; Heald et al., 2005; Volkamer et al., 2006; Appel et al., 2008). Potential reasons for this under prediction may involve several factors: lack of representation of all SOA precursors, misrepresentation of some SOA species with intermediate volatilities (Robinson et al., 2007), incorrect temperature dependence on SOA yields (Odum et al., 1996), idealized seed backgrounds vs. real aerosol seed backgrounds (Cao and Jang, 2010), as well as variable ambient relatively humidities (Cocker et al., 2001; Volkamer et al., 2009; Kamens et al., 2011).

The chemistry and partitioning processes involved in SOA formation are complex and dependent on a variety of atmospherically

* Corresponding author.
E-mail address: kamens@UNC.edu (R.M. Kamens).

1352-2310/$ – see front matter © 2010 Elsevier Ltd. All rights reserved.
doi:10.1016/j.atmosenv.2010.12.048

relevant conditions. A dependence of SOA formation on hydrocarbon to nitrogen oxide ratios has been observed (Cocker et al., 2001; Kroll et al., 2007; Ng et al., 2007; Song et al., 2007b), which are normally carried out under dry conditions. This is due to the formation of more volatile products under low HC_0/NO_x ratios versus products that are formed under high HC_0/NO_x ratios. Temperature is another major factor that influences SOA yields. The yield is generally higher at lower temperatures (Odum et al., 1996; Takekawa et al., 2003; Hildebrandt et al., 2009). Further, positive temperature dependence of SOA yields is observed under humid conditions, compared with expected temperature dependences under dry conditions, indicating chemical reactions in particle aqueous-phase are influenced by temperature (Von Hessberg et al., 2009).

Particle seed characteristics can also play an important role in SOA formation. Kleindienst et al. (1999), Edney et al. (2001) and Cocker et al. (2001) found that $(NH_4)_2SO_4$ background seed had no significant effect on SOA formation in aromatic photooxidation systems. However, Kroll et al. (2007) and Lu et al. (2009) recently observed that SOA yield was enhanced in the presence of $(NH_4)_2SO_4$ seeds with relative humidity (RH) at 4%~7% and 56%. Kroll et al. (2007) suggested this enhancement may be attributed to particle-phase reactions which form products with high-molecular and low volatility. These processes deplete the semivolatile reactants in the particle phase and enhance SOA formation by shifting the gas-particle equilibrium towards the particle liquid phase. Nevertheless, previous observations of no enhancement in SOA formation due to the presence of $(NH_4)_2SO_4$ seed in experiments of Kleindienst et al. (1999), Edney et al. (2001) and Cocker et al. (2001) cannot be completely explained. Lu et al. (2009) proposed that high concentrations of the seed aerosol and thin organic layers are the major factors in the particle-phase heterogeneous reactions. The fact that only effects of $(NH_4)_2SO_4$ seed impact SOA yields while $CaSO_4$ and $Ca(NO_3)_2$ do not, suggest that the unique acidic ammonium ion may induce and catalyze heterogeneous reactions. In addition, most studies show acidic aerosol enhances SOA yields (Jang et al., 2002; Liggio et al., 2005; Cao and Jang, 2010).

Ambient conditions have a varying range of relative humidities (RH). Therefore, it is critical to study the effect of RH on SOA formation for different SOA precursors. Cocker et al. (2001) and Edney et al. (2001) reported that increased RH does not increase the aerosol mass produced from aromatics-oxides of nitrogen (NO_x) systems. The chamber RH in their experiments ranged from 2%~50% and 52%~70%, respectively. Seinfeld et al. (2001) predict increases of both the amount of condensed organic mass and the amount of liquid water in the SOA phase with increasing RH in the oxidation systems for a variety of SOA precursors. Further, recent research, in both chamber and field studies, have shown major SOA formation due to uptake of glyoxal which is a major aromatic oxidation product, to the particle aqueous-phase (Liggio et al., 2005; Volkamer et al., 2007; Corrigan et al., 2008; Ervens and Volkamer, 2010). Volkamer et al. (2009) confirmed that water soluble organic carbon (WSOC) photochemistry of glyoxal in the particle water (H_2Op) is responsible for the high SOA formation in the ethyne (C_2H_2) and glyoxal oxidation system. The implication is that for certain systems, humidity may be associated with SOA formation due to its strong influence on seed acidity and particle liquid water concentration. Therefore, there is a need to include RH in the pathways and kinetics of the particle water related chemical reactions in the next generation of SOA models (Donahue et al., 2009; Lu et al., 2009; Von Hessberg et al., 2009).

Specifically for *m*-xylene, SOA formation has been widely investigated in indoor smog chambers with different levels of RH, different levels of (NO_x), and different dry nonacid or acidic background aerosols (Cocker et al., 2001; Ng et al., 2007; Song et al., 2007a,b; Huang et al., 2008; Lu et al., 2009). All these studies employed inorganic salts, but none have used natural background aerosol as initial seed. Also most studies did not evaluate SOA aromatic precursor systems with natural sunlight, over a large range of RH, or in the environment of a non-SOA-forming urban hydrocarbon mixture. Further, only a few studies have addressed the SOA potential of other two xylene isomers, *p*-xylene and *o*-xylene, under different NOx conditions (Izumi and Fukuyama, 1990; Odum et al., 1997; Song et al., 2007b).

Recently Kamens et al. (2011) developed a condensed toluene SOA kinetic mechanism that calculates particle-phase water when $(NH_4)_2SO_4$ is present in initial seed particles. A similar approach is applied to new experiments with *o*-xylene, *p*-xylene and toluene with xylene mixtures conducted in the UNC outdoor smog chamber. In particular, the goal of this study is to evaluate the SOA kinetic mechanism and investigate the impacts of different humidities and initial rural seed aerosol on SOA formation in a gas-phase environment truly representative of ambient air conditions.

2. Experimental section

Experiments were performed in the UNC 274 m^3 dual outdoor aerosol smog chamber (Leungsakul et al., 2005; Hu and Kamens, 2007; Kamens et al., 2011). Simultaneous experiments were performed in both halves of the 274 m^3 chamber. One side of the chamber will be designated as North and the other as South. The chambers were vented with rural North Carolina background air. After venting, drying of the chamber was performed with a 250 L min^{-1} Aadco clean air generator. Depending on the level of drying and outdoor conditions, individual experiments began with different levels of rural background particles (called background seed) and humidities. Oxides of nitrogen and an eleven component hydrocarbon mixture (HC mix) (Kamens et al., 2011) that ranges in carbon number from ethylene to trimethylpentane, were injected from high-pressure gas cylinders before injecting toluene and xylenes. Xylenes and toluene were individually introduced into the chamber by vaporizing liquids of the pure liquid toluene (98%, Aldrich, Milwaukee, WI), *o*-xylene (98%, Aldrich, Milwaukee, WI), and *p*-xylene (98%, Aldrich, Milwaukee, WI) into a heated, flowing dry nitrogen stream. In selected experiments, a mixture of toluene and either *o*-xylene or *p*-xylene was also used. The measurement of O_3, NOx and aromatics from the chambers are described in detail elsewhere (Hu et al., 2007; Li et al., 2007). Aerosol size measurements were made with a Scanning Mobility Particle Sizer (SMPS, TSI 3080 with CPC TSI 3022 A, Shoreview, MN).

3. Condensed aromatic kinetics mechanism

A 37 step gas phase kinetic mechanism for simulating O_3, NO, NO_2 from the oxidative reactions of toluene (not including radical-radical reactions) was recently developed (Kamens et al., 2011) and substituted for the toluene reactions in a Carbon Bond mechanism framework (Gery et al., 1989; Yarwood et al., 2005). Gas phase oxidation of toluene by hydroxyl radicals (OH) or nitrate radicals led to four generalized categories that can be accommodated by the particle phase. These include highly oxygenated molecules that have carbonyl and alcohol groups (*polycarb*), aliphatic hydroxy nitrates (*OHCarbNO*$_3$), polymerized products (*GLYpoly*) from glyoxal (*GLY*) and methylglyoxal (*MGLY*), and highly oxygenated carboxylic acids (*carbacid*). *Polycarb*, *OHCarbNO*$_3$ and *carbacid* were permitted to dimerize (*oligomer*) in the particle phase at a rate of 2% per hour as per Kamens et al. (2011). Particle phase interactions are represented by particle water uptake as well as gas-particle partitioning in the Kamens' condensed mechanism. A complete listing of the kinetic mechanism is given in Kamens et al. (2011). The oxidation of *o*- and *p*-xylene was added to the above mechanism by having the

OH and NO_3 adducts of these compounds shunt directly into the toluene mechanism, where OH attack on toluene, *o*- and *p*-xylene is illustrated as:

$$OH + TOL \rightarrow 0.76*TOLO2 + 0.06*benzald + 0.06*XO2 +0.24*HO2 + 0.18*CRES$$
$$k = 2660*\exp(338/T)\ ppm^{-1}\ min^{-1}$$

$$OH + o - Xylene \rightarrow 0.56*TOLO2 + 0.3*MGLY + 1.1*PAR +0.06*benzald + 0.06*XO2 + 0.26*HO2 + 0.10*CRES +0.07*OH + 0.07*MeHexdial$$
$$k = 19,900\ ppm^{-1}\ min^{-1}$$

$$OH + p - Xylene \rightarrow 0.23*TOLO2 + 0.75*MGLY + 1.1*PAR +0.06*benzald + 0.06*XO2 + 0.26*HO2 + 0.10*CRES +0.07*OH + 0.07*MeHexdial$$
$$k = 20,900\ ppm^{-1}\ min^{-1}$$

In the above reactions TOLO2 is an aromatic peroxy radical, benzald represents aromatic aldehydes; CRES represents aromatic phenols; MeHexdial represents a C7 or C8 dialkene diketones; PAR is the CB4 and CB05 representation for one paraffinic carbon, and XO2 is a Carbon Bond virtual operator radical (Gery et al., 1989). The original CB4 toluene/xylene chemistry was compared with CB4 plus the new UNC toluene/xylene chemistry for the behavior of temporal profiles for HO_2, OH and XO_2. Similar behavior was observed by these two mechanisms. The CB4-UNC mechanism simulated ozone much better than the original CB4 mechanism for experiments with aromatics and the non-SOA forming HC mixture.

4. Results and discussion

4.1. Description of the basic gas phase system

To develop and evaluate the new condensed mechanism with *o*- and *p*-xylene, experiments with different initial background levels (seed) of aerosol that are typically present in rural North Carolina air were conducted. Experiments were run under clear sunlight and different diurnal temperatures and humidities. Tables 1 and 2 summarize the initial experimental conditions and SOA model results. To be consistent with previous toluene experiments (Kamens et al., 2011), 1 ppmC aromatics, 0.2 ppm NOx with 3 ppmC urban mix were typically employed in this study. For the *o*-xylene system, initial aromatic to NOx ratios were in the 0.6 range. Experiments could be separated into three categories: "wet" (%RH = 90 to 40), "medium wet" (%RH = 70 to 20) and "dry" (%RH = 20 to 10). All of *p*-xylene experiments were conducted under wet conditions, but they had different initial seed aerosol levels; the initial aromatic to NOx ratio varied from 0.3 to 0.6. Experiments with toluene and xylene mixtures were also conducted to develop and test the mechanism for simple aromatic mixtures (Table 1, experiments Toa, b and TPa, b, c). We chose a ratio of 2:1 (ppmC/ppmC) of toluene to either *o*-xylene or *p*-xylene to approximate ambient concentrations (Calvert et al., 2002). Initial chamber temperatures ranged from −3 °C to 25 °C and the highest temperatures ranged from 20 °C to 38 °C. The performance of the condensed mechanism in the gas-phase is illustrated in Fig. 1. For the lower NOx case (Fig. 1d, experiment TPb) the model tends to perform well for simulating O_3, NO and NO_2. For the higher NOx case (Fig. 1b, experiment P2), we tend to under-predict O_3. This is primarily due to the removal of nitrogen in the form of $OHCarbNO_3$. Under experimental conditions there is a possibility of some conversion back to NO_2. This is not currently represented in this mechanism and likely leads to the slight under prediction in simulating O_3.

4.2. General SOA observations

The xylene experiments were conducted at different combinations with varying ranges of initial seed and relative humidities. SOA mass concentration is obtained based on the volume concentration measured by SMPS. The calculation of SOA and the particle phase water concentration (H_2Op) is described below and in detail by Kamens et al. (2011). Low chamber background aerosol concentrations (<2 $\mu g\ m^{-3}$) were achieved by injecting clean air (usually overnight) before the experiment began. As mentioned in

Table 1
Initial conditions for *o*-xylene, *p*-xylene and with toluene mix experiments.

ID	Date	Parent VOC	HC mix	Initial $[NO]_0$	Initial $[NO_2]_0$	Temp Range	RH range	Initial seed	Initial H_2Op
	MMDDYY	ppmV	ppmC	ppm	ppm	K	%	$\mu g\ m^{-3}$	$\mu g\ m^{-3}$
o-xylene									
O1[a]	07Jun17S	0.124 *o*-xylene	3	0.179	0.028	293–311	89–45	5.8	6.3
O2a[a]	08Oct04N	0.117 *o*-xylene	3	0.179	0.023	279–304	95–38	3.0	3.5
O2b[a]	08Oct04S	0.106 *o*-xylene	1.5	0.080	0.023	279–304	95–37	3.7	4.3
O3	08Oct31N	0.125 *o*-xylene	3	0.183	0.025	270–296	82–40	0.1	0.1
O4	07Jul12S	0.123 *o*-xylene	3	0.158	0.052	294–310	60–30	0.7	0.2
O5a	08Oct02N	0.125 *o*-xylene	3	0.173	0.014	279–299	74–25	0.2	0.1
O5b[a]	08Oct02S	0.060 *o*-xylene	1.5	0.092	0.014	280–299	61–24	0.2	0.1
O6	07Aug14S	0.125 *o*-xylene	3	0.185	0.028	298–313	20–12	0.3	0.01
p-xylene									
P1	08Oct31S	0.125 *p*-xylene	3	0.177	0.031	270–296	90–44	0.3	0.3
P2	08Oct26N	0.125 *p*-xylene	3	0.375	0.011	278–297	82–40	0.2	0.1
P3	08Nov01N	0.233 *p*-xylene	4	0.395	0.017	273–301	94–58	2.5	3.0
Mix[b]									
TOa	07Sep17N	0.093T + 0.048O	3	0.173	0.008	285–303	20–9	0.8	0.0
TOb	07Sep17S	0.089T + 0.049O	3	0.170	0.009	285–303	43–18	0.9	0.1
TPa[a]	08Nov02N	0.094T + 0.041P	3	0.199	0.001	275–301	93–58	7.1	8.4
TPb	08Nov10N	0.060T + 0.031P	2	0.104	0.003	271–293	98–68	1.1	1.3
TPc	08Nov10S	0.094T + 0.041P	3	0.203	0.007	271–293	96–66	1.5	1.7

[a] No post nucleation (14–20 nm) was observed due to scavenging of high initial background seed (O1, O2a, b and TPa) or low initial aromatic concentration (O5b).
[b] T is toluene, O is *o*-xylene, P is *p*-xylene, TO is toluene with *o*-xlene, TP is toluene with *p*-xylene.

Table 2
Measured and model results for *o*-xylene, *p*-xylene and with toluene mix experiments.

ID	Date	Parent VOC	Max TSP		SOA		Mo[a]	ΔHC[b]	SOA yield	H_2Op^c in chamber
			Measured	Simulated	Measured	Simulated				
	MMDDYY	ppmV	$\mu g\ m^{-3}$	$\mu g\ m^{-3}$	$\mu g\ m^{-3}$	$\mu g\ m^{-3}$	$\mu g\ m^{-3}$	$\mu g\ m^{-3}$	%	$\mu g\ m^{-3}$
o-xylene										
O1	07Jun17S	0.124 *o*-xylene	26.8	28.3	20.6	23.1	29.8	457	6.5	1.56
O2a	08Oct04N	0.117 *o*-xylene	10.2	13.8	7.9	11.1	10.1	284	3.6	0.66
O2b	08Oct04S	0.106 *o*-xylene	12.1	15.3	9.2	11.6	12.3	278	4.4	0.9
O3	08Oct31N	0.125 *o*-xylene	6.5	5.9	6.6	5.3	8.3	377	2.4	0.03
O4	07Jul12S	0.123 *o*-xylene	5.2	5.4	4.4	4.9	6.2	376	1.7	0.06
O5a	08Oct02N	0.125 *o*-xylene	4.7	3.7	4.6	3.4	6.4	347	1.8	0.04
O5b	08Oct02S	0.060 *o*-xylene	1.8	1.6	1.6	1.3	2.0	168	1.1	0.02
O6	07Aug14S	0.125 *o*-xylene	3.7	4.2	3.5	4.1	4.6	416	0.8	0.00
p-xylene										
P1	08Oct31S	0.125 *p*-xylene	2.9	2.6	2.6	2.4	2.4	248	1.2	0.12
P2	08Oct26N	0.125 *p*-xylene	3.3	2.7	3.0	2.4	3.7	300	1.2	0.07
P3	08Nov01N	0.233 *p*-xylene	20.2	20.7	17.0	18.3	21.3	475	4.5	1.25
Mix[d]										
TOa	07Sep17N	0.093T + 0.048O	2.8	2.6	2.3	2.2	2.4	227	1.1	0.01
TOb	07Sep17S	0.089T + 0.049O	2.7	2.8	2.1	2.4	3.2	244	1.3	0.04
TPa	08Nov02N	0.094T + 0.041P	26.7	22.0	17.1	14.5	19.8	175	11.3	3.46
TPb	08Nov10N	0.060T + 0.031P	3.3	3.3	2.1	2.2	2.2	101	2.3	0.59
TPc	08Nov10S	0.094T + 0.041P	5.2	5.4	2.7	4.1	3.3	185	1.8	0.80

[a] Mo is the maximum SOA mass corrected for particle wall losses over time.
[b] ΔHC is corrected by dilution.
[c] H_2Op is the average during each experiment.
[d] T is toluene, O is *o*-xylene, P is *p*-xylene, TO is toluene with *o*-xlene, TP is toluene with *p*-xylene.

the experimental section, the high background seed experiments employed rural North Carolina background air with an aerosol size mode that ranged from 100 to 175 nm.

Experiments that began with less than 5.5 $\mu g\ m^{-3}$ of initial aerosol showed evidence of "post-nucleation" in the SMPS data (14–20 nm). Nucleation began below the sizes measured by the SMPS and we could only see the "post-nucleated" particles as they grew out of the nucleation range. Experiments that began with initial concentrations equal to or greater than 6.5 $\mu g\ m^{-3}$ exhibited particle growth, but did not exhibit particle nucleation. Experiment O1 in Table 1 began with a high aerosol background (12 $\mu g\ m^{-3}$) and wet conditions (RH = 89 to 45%); it did not show any post-nucleation, but rather growth into the larger particle sizes (100 nm to 730 nm). This resulted in an increase in particle mass from 12 to

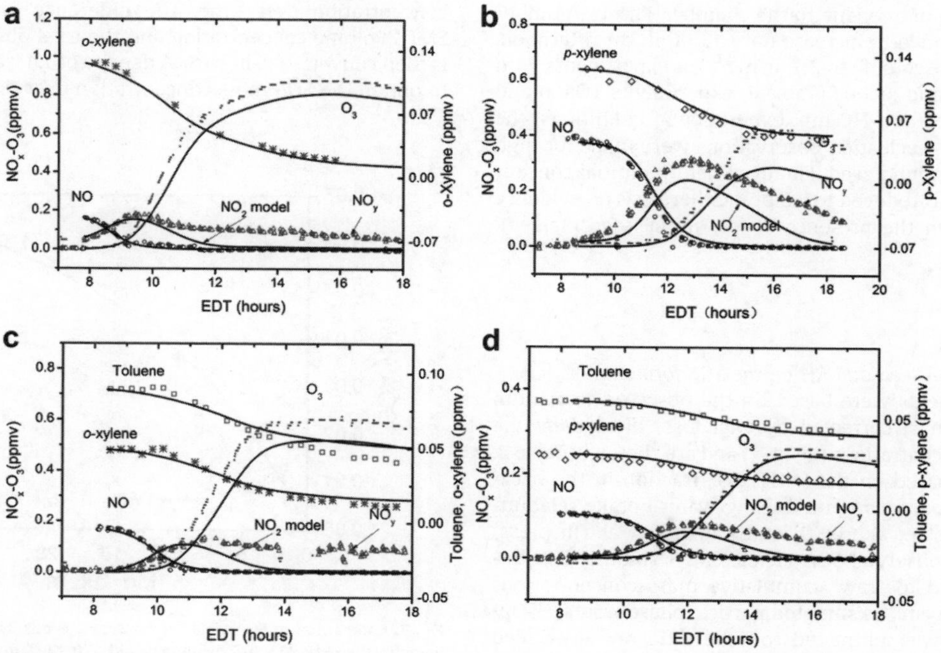

Fig. 1. Comparison of model simulated and measured concentrations of *o*-, *p*-xylene, toluene, NOx and ozone for (a) O4-2007July12S, (b) P2-2008Oct26N, (c) TOb-2007Sep17S and (d) TPb-2008Nov10N. The solid lines represent model simulations. The following species were experimentally determined: dashed (–) O_3; triangles (△) NO_y; circles (○) NO; asterisks (*) *o*-xylene; diamonds (◇) *p*-xylene; and squares (□) toluene.

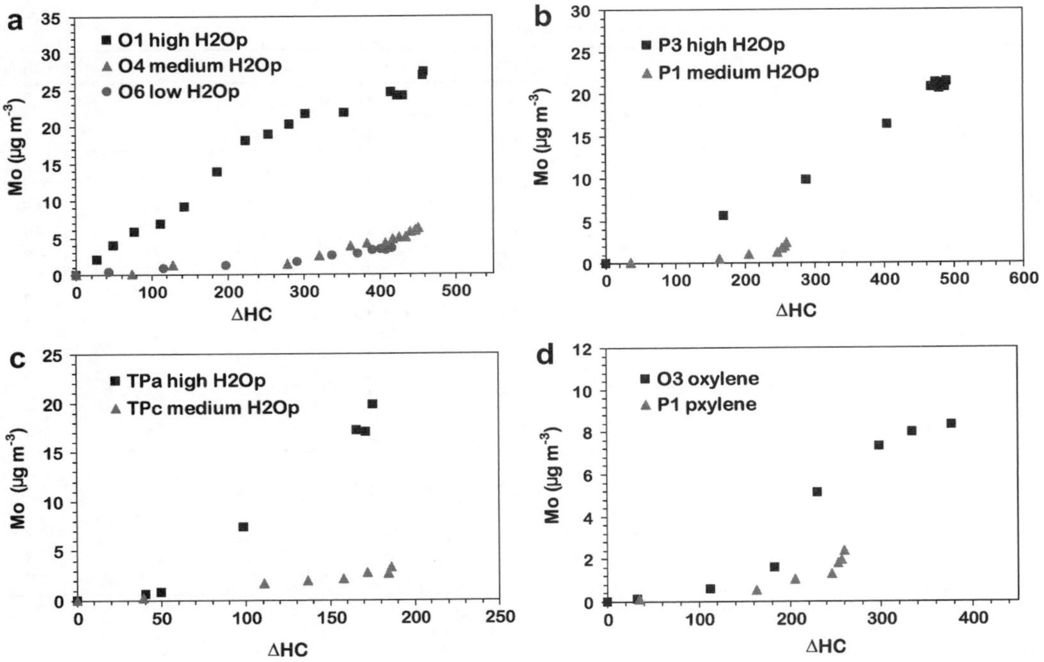

Fig. 2. Concentration of secondary organic aerosol mass formed as a function of reacted hydrocarbon concentration and particle phase water (H_2Op): (a) *o*-xylene, (b) *p*-xylene, (c) toluene + *p*-xylene mixture and (d) the comparison between *o*-xylene and *p*-xylene under similar conditions. Initial conditions were listed in Table 1.

27 μg m^{-3}. Newly nucleated particles or the molecules that have the potential to nucleate appear to be scavenged by larger particles. A similar observation was made with the toluene system (Kamens et al., 2011).

By contrast, experiment O6 in Table 1, had very similar gas-phase conditions to experiment O1, except initial background seed particles were much lower (0.3 μg m^{-3}) and the air was significantly dryer (RH = 20–12%). Nucleation occurred approximately 3 h after the addition of *o*-xylene to the chamber. Particle number concentrations continued to increase until 13:00 in the afternoon, while the mass increased only to 3.7 μg m^{-3}. For experiments with mixtures of toluene and xylene (Table 1, experiments TOa, b and TPa) and similar NOx, and HC mix experimental conditions, the same nucleation/non-nucleation observations were made for high and low initial background seed. Finally, when the initial concentration of *o*-xylene was reduced to 0.5 ppmC there was no evidence of nucleation, even in the presence of low initial seed (Table 1, experiment O5b).

4.3. SOA yields

4.3.1. Effects of aerosol seed and RH on the SOA formation

Calculated SOA yields were based on the observed maximum SOA formation (μg m^{-3}, corrected by wall loss) divided by the reacted aromatic concentrations (μg m^{-3}) and are shown in Table 2. Average H_2Op was based on the $(NH_4)_2SO_4$ fraction in the background seed and RH and computed using water-uptake relationship from Kleindienst et al. (1999) and Kamens et al. (in press) (Table 2). To derive observed SOA yields, SMPS volume distributions were converted to "raw" cumulative mass concentrations using a density of 1 g cm^{-3}. Estimated particle phase water (H_2Op) (Kamens et al., 2011) was subtracted from the initial raw SMPS seed mass concentration. According to sulfate, nitrate, and ammonium measurements of the background aerosol at the chamber facility, ~35% of the mass was assumed to be inorganic compounds (equilibrium RH at 50%), which was primarily composed of ammonium sulfate and nitrate with an average density of 1.76 g cm^{-3}. The remainder was assumed to be composed of 50% aged organic aerosol and 50% fresh organic emissions with and average density of 1.1 g cm^{-3}. From the initial ammonium sulfate an organics corrected for its density and wall loss decay, and the water content calculated based on the corrected ammonium sulfate and chamber RH, it is possible to compute an "adjusted" SMPS volume concentration over time. The difference between the adjusted SMPS volume concentration and the total observed SMPS volume, is then converted using a SOA density of 1.4 g cm^{-3} (Ng et al., 2007) to obtain the SOA mass concentration over time.

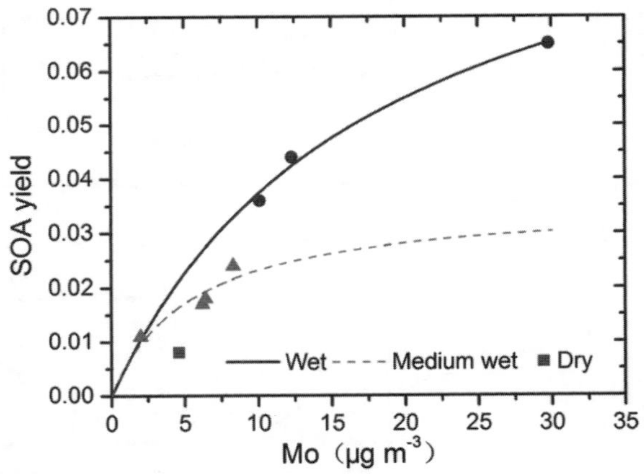

Fig. 3. Comparison of SOA yield for *o*-xylene HC mix system varying with relative humidity (wet RH: 90~40%, medium wet RH: 70~40%, dry RH: 20~10%). Also shown are yield curves generated from one-product model (eq. (1), model lines are $\alpha = 0.10485$, $K_{om} = 0.05515$ for wet condition and $\alpha = 0.0357$, $K_{om} = 0.1846$ for medium wet condition).

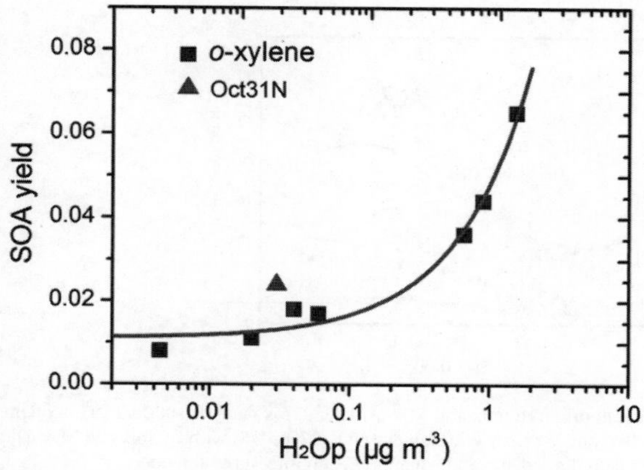

Fig. 4. SOA yields for the *o*-xylene with NOx-HC Mix system as a function of the average H_2Op (data from Table 2) concentration.

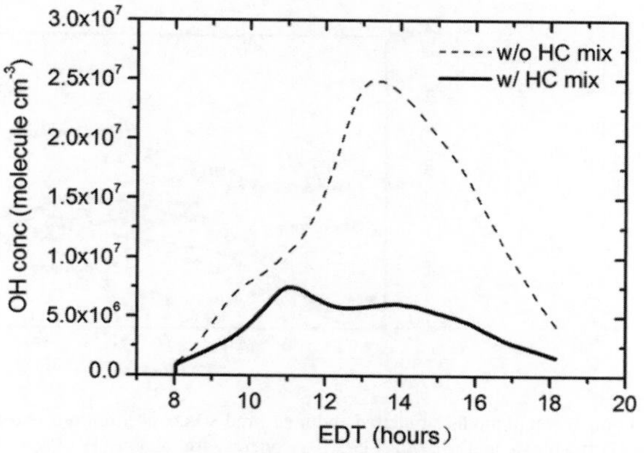

Fig. 6. Comparison of OH radical modeling results from w/and w/o UNC mix system (but with *o*-xylene) based on experiment O4-July12S.

SOA formation is a function not only of semivolatile compounds, but also the aerosol mass into which they can partition (Kroll et al., 2007), and the particle water that can uptake glyoxal and other water soluble compounds (Volkamer et al., 2009; Ervens and Volkamer, 2010). Kamens et al. (2011) used H_2Op as one of the major parameters to simulate the SOA formation from toluene-NOx-HC mix sunlight systems. Fig. 2a–c illustrates the association of H_2Op with SOA accumulation for different aromatic systems. For each of the xylene systems the greatest SOA formation was associated with the highest H_2Op experiments. This is further confirmed by looking at the yield for wet experiment O_3 (Table 2, 08Oct31N) where the yield was three times that of the dry experiment O6 (Table 2. 07Aug14S) under almost similar gas phase and initial seed conditions. For the base case, comparison of SOA forming potential between *o*- and *p*-xylene (Fig. 2d and Table 2, experiments O3 and P1) showed approximately twice the SOA yield for *o*-xylene (0.024) than the *p*-xylene (0.012).

SOA yield as a function of SOA mass concentration in the *o*-xylene system is shown in Fig. 3. Also shown are the yield curves generated from a one-product model (eq. (1), estimated $\alpha = 0.10485$, $K_{om} = 0.05515$ for wet condition and $\alpha = 0.0357$, $K_{om} = 0.1846$ for medium wet condition) under different humidity conditions. As with toluene (Kamens et al., 2011), high initial seed and its associated particle phase water strongly influence SOA yield. In Fig. 3, the yield curve for high humidity experiments is higher compared to medium-wet and dry experiments. If, however, there is low initial seed with high humidity (experiments O3, P1, P2), the overall availability of particle phase water (H_2Op) is still very low, which results in limited uptake of compounds such as glyoxal and other water-soluble organics into particle water phase. Further, in outdoor smog chamber experiments using toluene and the HC mixture under conditions of high seed but low RH, we did not see a direct effect of initial seed aerosol concentrations on SOA (Kamens et al., 2011). A plot of SOA yields and average H_2Op over the entire experiment for xylene is illustrated in Fig. 4 and suggests an exponential relationship between SOA and particle phase water. The increase in SOA yields is particularly dramatic above 1 $\mu g\,m^{-3}$ of particle-phase water; a similar relationship can be observed for the toluene system.

4.3.2. Comparison of the SOA yields with previous studies

As shown in Fig. 5a, the *o*-xylene SOA yields are generally in the range of previous studies. Odum et al. (1997) suggest similar yield from *m*-, *o*-, *p*-xylene under similar conditions (temperature ~37 °C, RH < 2%, HC/NOx < 1, no seed). Our yield from dry *o*-xylene system falls on the low part of the Odum et al. (1997) curve. A higher aerosol yield curve than Odum et al. (1997) was observed by Song et al. (2007b), which is possibly due to the fact that their experimental temperature was about 10 °C lower than that of our study, and they did not employ propene (C_3H_6) as an initial reactant as in the Odum experiments. Our simulations suggest that significantly lower OH concentrations occur in experiments initiated with the HC Mix (Fig. 6), and this will correspondingly reduce the SOA formation potential.

p-Xylene SOA yields shown in Fig. 5b from this study are in the range observed by other investigators (Odum et al., 1997; Song et al., 2007b). Toluene with *o*-and *p*-xylene mixture experiments

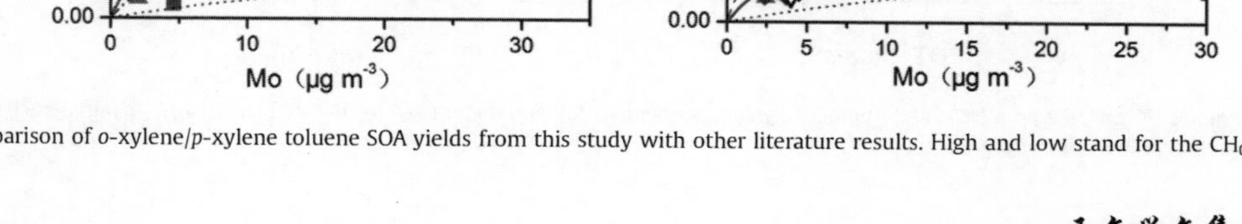

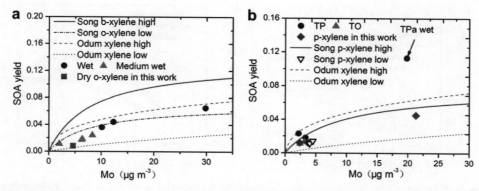

Fig. 5. Comparison of *o*-xylene/*p*-xylene toluene SOA yields from this study with other literature results. High and low stand for the CH_0:NOx ratio.

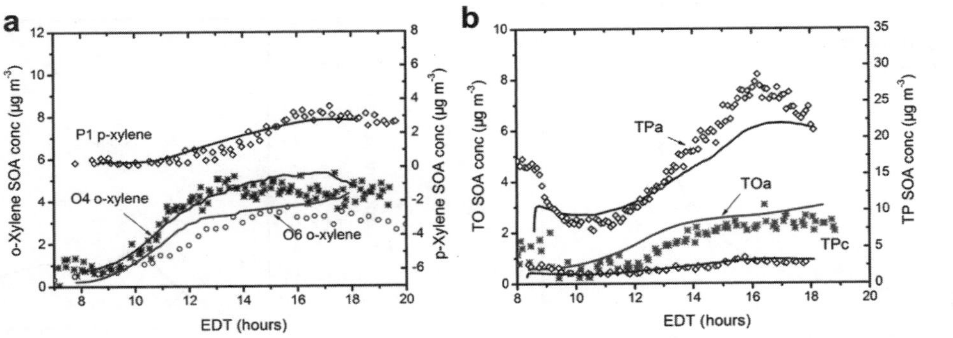

Fig. 7. Comparison of model simulated and measured SOA concentrations of *o*-xylene, *p*-xylene with an urban HC mix and NOx (a) 1ppmC *p*-xylene (P1-08Oct26N), *o*-xylene (O4-07July12S) and *o*-xylene (O6-07Aug14S); (b) *p*-xylene with toluene (TPa-08Nov02N and TPc-08Nov10S) and *o*-xylene with toluene (TOa-07Sep17S) with HC mix and NOx. The solid lines (—) represent model simulations; asterisks (*) are measured *o*-xylene; diamonds (◇) are measured *p*-xylene; and squares (□) are measured toluene.

were carried out under very different conditions, and one experiment (TPa) was out of the range of the Odum plots of Fig. 5 because it took place under a very high RH regime with temperatures that were on average 10 °C cooler than all the other experiments. Our modeling results suggest that at these cooler temperatures the gas-particle partitioning channel dramatically influences SOA accumulation (Sheehan and Bowman, 2001).

4.4. SOA modeling and comparisons of different aromatic systems

Comparisons between observed SMPS mass concentration and simulated particle mass concentrations profiles are illustrated for representative experiments in Fig. 7. Averaged H_2Op treatment was used in the model (Kamens et al., 2011). The initial average H_2Op was the initial model generated H_2Op at the start of the experiment. In general for all experiments, the trends exhibited by the data are followed by the simulations. Comparisons of maximum measured and simulated particle concentrations for all the experiments are shown in Table 2. The maximum SOA derived from the SMPS data vs. the maximum simulated SOA are in good agreement. Simulations suggest that half of the SOA for wet experiments were associated with GLYpoly formation; for dry o-xylene system, oligomer is the major SOA product and GLYpoly is the next dominant product. GLYpoly, polycarb and OHCarbNO3 are simulated in equal amounts in the *p*-xylene experiments with wet conditions; under dry condition simulation results are similar with *o*-xylene with the exception of lower carbacid concentration.

For the toluene and *o*-xylene mixture experiments (the xylene was ~ half of the toluene concentration), the model showed a little more SOA produced by *o*-xylene (54%, and 53% for TOa and TOb respectively) than toluene (45% and 47% for TOa and TOb respectively). A similar SOA distribution with model results was estimated by calculating the SOA formation from the reacted *o*-xylene concentration and the yield predicted from the medium wet curve in Fig. 3. For the toluene with *p*-xylene mixture experiments, model results show similar SOA was produced by these two aromatics in experiment TPb (50% vs 50% for *p*-xylene and toluene), however in experiment TPa and TPc, toluene produced more SOA than the *p*-xylene. The contribution from toluene was 55 and 58% for TPa and TPc respectively. The results suggest that xylenes, although lower in concentration than toluene contribute about the same amount of SOA as toluene.

To compare the SOA formation potentials of *o*-xylene, *p*-xylene, and toluene in the model, the gas phase injection conditions of experiment O4 were selected (1 ppmC *o*-xylene/*p*-xylene/toluene, 3 ppmC HC mix, 0.2 NOx) and simulated with the mechanism developed for *o*-xylene, *p*-xylene and toluene in this study. Initial seed levels of 1 and 10 $\mu g\,m^{-3}$ for background aerosol were used (H_2Op content is included in this seed concentration). The "dry" simulations had a dew point of −10 °C, and started at an 11% RH and decreased to 5% RH during the day. The humid simulations used a dew point that started at 20 °C, with an equivalent RH of 94%. Over the course of this simulation temperature increased from 294 K to 311 K and %RH decreased from 94% to 46%. Fig. 8 demonstrates that the potential trends of SOA yield for 1 and 10 $\mu g\,m^{-3}$ conditions. Wet conditions result in higher SOA yields than dry conditions. *o*-Xylene had the highest aromatic reactivity, followed by toluene, and then *p*-xylene. We also estimated the SOA as a function of reacted aromatics of these three systems under the assumed conditions (Fig. 9). For 10 $\mu g\,m^{-3}$ background seed

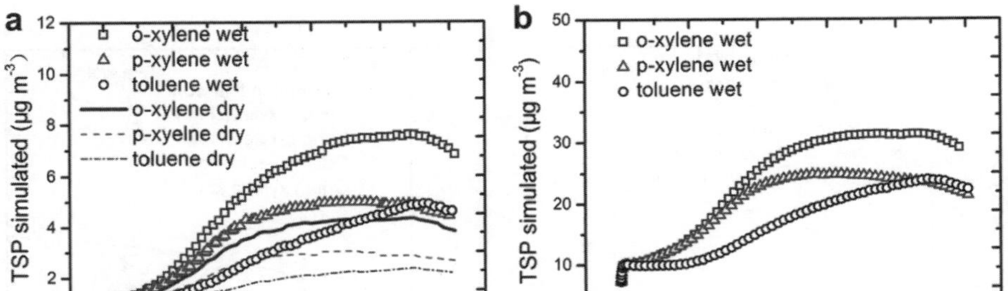

Fig. 8. Simulations with 3 ppmC HC mix, 0.2 ppm NOx, and 1ppmC *o*-xylene, *p*-xylene or toluene under (a) 1 $\mu g\,m^{-3}$ and (b) 10 $\mu g\,m^{-3}$ initial background particles in dry and humid conditions. Particle concentrations include particle phase water.

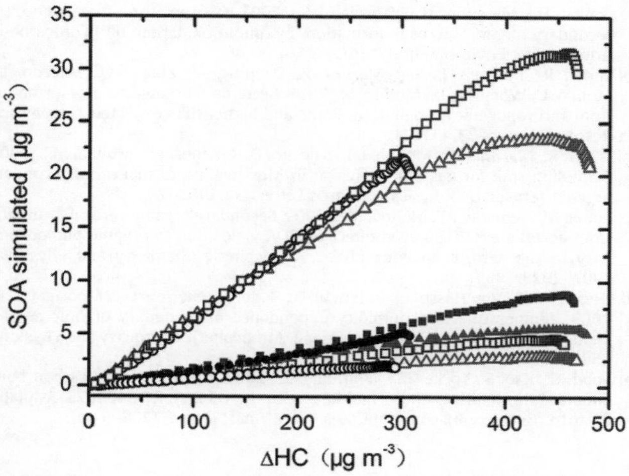

Fig. 9. SOA formation ($\mu g\ m^{-3}$) as a function of reacted aromatics in *o*-xylene (square), *p*-xylene (triangle) and toluene (circle) system under different conditions (lines with symbols stand for 10 $\mu g\ m^{-3}$ initial seed and wet conditions, solid symbols stand for 1 $\mu g\ m^{-3}$ initial seed wet conditions, open symbols stand for 1 $\mu g\ m^{-3}$ initial seed under dry conditions).

condition, much more SOA was simulated in the wet condition than in the dry condition.

5. Summary and conclusions

Based on our experiments and model simulations, xylenes can have a significant impact on SOA formation due to their high reactivity and relative abundance in the atmosphere in comparison to other aromatic hydrocarbons. The magnitude of xylene's contribution to SOA formation can depend on the relative humidity (RH), temperature, along with the composition and concentration of the background seed aerosol. In an urban hydrocarbon and NOx environment, SOA yields with ~0.125 ppmv xylene increase by a factor of 2 to 3 under high compared to low RH. This behavior is similar to that observed in experiments with toluene in an urban hydrocarbon mixture and NOx system. In addition, the impact of initial background seed on SOA yield was observed in these experiments where increased initial seed resulted higher SOA yield. A comparison of *o*-, *p*-xylene and toluene with NOx, HC mix systems under natural sunlight showed the highest yield in the *o*-xylene system compared to toluene and *p*-xylene. Similar yields are obtained from *p*-xylene and toluene. This study is the first investigation on the SOA formation from these three aromatics in an environment of a non-SOA-forming urban type HC mixture. The results clearly indicate the importance of studying SOA formation over a comprehensive range of environmental conditions. Only such evaluations can lead to meaningful SOA mechanisms to be implemented in air quality models.

Acknowledgements

This research was supported almost entirely by a USA NSF grant (ATM-0711097) to the University of North Carolina. A Postgraduate fellowship for Yang Zhou was provided by the China Scholarship Council, PRC.

References

Appel, K.W., Bhave, P.V., Gilliland, A.B., Sarwar, G., Roselle, S.J., 2008. Evaluation of the community multiscale air quality (CMAQ) model version 4.5: Sensitivities impacting model performance; Part II—particulate matter. Atmospheric Environment 42, 6057–6066.

Calvert, J.G., Atkinson, R., Becker, K.H., Kamens, R.M., Seinfeld, J.H., Wallington, T.J., Yarwood, G., 2002. The Mechanisms of Atmospheric Oxidation of Aromatic Hydrocarbons. Oxford University Press, Inc., New York.

Cao, G., Jang, M., 2010. An SOA model for toluene oxidation in the presence of inorganic aerosols. Environmental Science & Technology 44, 727–733.

Carlton, A.G., Bhave, P.V., Napelenok, S.L., Edney, E.D., Sarwar, G., Pinder, R.W., Pouliot, G.A., Houyoux, M., 2010. Model representation of secondary organic aerosol in CMAQv4.7. Environmental Science & Technology 44, 8553–8560.

Cocker III, D.R., Mader, B.T., Kalberer, M., Flagan, R.C., Seinfeld, J.H., 2001. The effect of water on gas–particle partitioning of secondary organic aerosol: II. m-xylene and 1,3,5-trimethylbenzene photooxidation systems. Atmospheric Environment 35, 6073–6085.

Corrigan, A.L., Hanley, S.W., Haan, D.O., 2008. Uptake of glyoxal by organic and inorganic aerosol. Environmental Science & Technology 42, 4428–4433.

De Gouw, J., Middlebrook, A., Warneke, C., Goldan, P., Kuster, W., Roberts, J., Fehsenfeld, F., Worsnop, D., Canagaratna, M., Pszenny, A., 2005. Budget of organic carbon in a polluted atmosphere: results from the New England Air Quality Study in 2002. Journal of Geophysical Research 110, D16305.

Donahue, N.M., Robinson, A.L., Pandis, S.N., 2009. Atmospheric organic particulate matter: from smoke to secondary organic aerosol. Atmospheric Environment 43, 94–106.

Edney, E.O., Driscoll, D.J., Weathers, W.S., Kleindienst, T.E., Conver, T.S., McIver, C.D., Li, W., 2001. Formation of polyketones in irradiated toluene/propylene/NOx/air mixtures. Aerosol Science and Technology 35, 998–1008.

Ervens, B., Volkamer, R., 2010. Glyoxal processing by aerosol multiphase chemistry: towards a kinetic modeling framework of secondary organic aerosol formation in aqueous particles. Atmospheric Chemistry and Physics 10, 8219–8244.

Gery, M.W., Whitten, G.Z., Killus, J.P., Dodge, M.C., 1989. A photochemical kinetics mechanism for urban and regional scale computer modeling. Journal of Geophysical Research 94, 12925–12956.

Heald, C., Jacob, D., Park, R., Russell, L., Huebert, B., Seinfeld, J., Liao, H., Weber, R., 2005. A large organic aerosol source in the free troposphere missing from current models. Geophysical Research Letters 32, L18809.

Henry, F., Coeur-Tourneur, C., Ledoux, F., Tomas, A., Menu, D., 2008. Secondary organic aerosol formation from the gas phase reaction of hydroxyl radicals with m-, o- and p-cresol. Atmospheric Environment 42, 3035–3045.

Hildebrandt, L., Donahue, N.M., Pandis, S.N., 2009. High formation of secondary organic aerosol from the photo-oxidation of toluene. Atmospheric Chemistry and Physics 9, 2973–2986.

Hu, D., Kamens, R.M., 2007. Evaluation of the UNC toluene-SOA mechanism with respect to other chamber studies and key model parameters. Atmospheric Environment 41, 6465–6477.

Hu, D., Tolocka, M., Li, Q., Kamens, R.M., 2007. A kinetic mechanism for predicting secondary organic aerosol formation from toluene oxidation in the presence of NOx and natural sunlight. Atmospheric Environment 41, 6478–6496.

Huang, M.Q., Zhang, W.J., Hao, L.Q., Wang, Z.Y., Zhao, W.W., Gu, X.J., Fang, L., 2008. Low-molecular weight and oligomeric components in secondary organic aerosol from the photooxidation of p-xylene. Journal of the Chinese Chemical Society 55, 456–463.

Izumi, K., Fukuyama, T., 1990. Photochemical aerosol formation from aromatic hydrocarbons in the presence of NOx. Atmospheric Environment. Part A. General Topics 24, 1433–1441.

Jang, M., Czoschke, N., Lee, S., Kamens, R., 2002. Heterogeneous atmospheric aerosol production by acid-catalyzed particle-phase reactions. Science 298, 814.

Kamens, R.M., Zhang, H., Chen, E.M., Zhou, Y., Parikh, H.M., Wilson, R., Galloway, K., Rosen, E.P., 2011. Secondary organic aerosol formation from toluene in an atmospheric hydrocarbon mixture: water and particle seed effects. Atmospheric Environment 45 (11), 2324–2334.

Kleindienst, T.E., Smith, D.F., Li, W., Edney, E.O., Driscoll, D.J., Speer, R.E., Weathers, W.S., 1999. Secondary organic aerosol formation from the oxidation of aromatic hydrocarbons in the presence of dry submicron ammonium sulfate aerosol. Atmospheric Environment 33, 3669–3681.

Kroll, J.H., Chan, A.W.H., Ng, N.L., Flagan, R.C., Seinfeld, J.H., 2007. Reactions of semivolatile organics and their effects on secondary organic aerosol formation. Environmental Science & Technology 41, 3545–3550.

Leungsakul, S., Jaoui, M., Kamens, R.M., 2005. Kinetic mechanism for predicting secondary organic aerosol formation from the reaction of d-Limonene with ozone. Environmental Science &Technology 39, 9583–9594.

Li, Q.F., Hu, D., Leungsakul, S., Kamens, R.M., 2007. Large outdoor chamber experiments and computer simulations: (I) Secondary organic aerosol formation from the oxidation of a mixture of d-limonene and alpha-pinene. Atmospheric Environment 41, 9341–9352.

Liggio, J., Li, S.M., McLaren, R., 2005. Reactive uptake of glyoxal by particulate matter. Journal of Geophysical Research 110, D10304.

Lu, Z.F., Hao, J.M., Takekawa, H., Hu, L.H., Li, J.H., 2009. Effect of high concentrations of inorganic seed aerosols on secondary organic aerosol formation in the m-xylene/NOx photooxidation system. Atmospheric Environment 43, 897–904.

Ng, N.L., Kroll, J.H., Chan, A.W.H., Chhabra, P.S., Flagan, R.C., Seinfeld, J.H., 2007. Secondary organic aerosol formation from m-xylene, toluene, and benzene. Atmospheric Chemistry and Physics 7, 3909–3922.

Odum, J.R., Hoffmann, T., Bowman, F., Collins, D., Flagan, R.C., Seinfeld, J.H., 1996. Gas/particle partitioning and secondary organic aerosol yields. Environmental Science & Technology 30, 2580–2585.

Odum, J.R., Jungkamp, T.P.W., Griffin, R.J., Forstner, H.J.L., Flagan, R.C., Seinfeld, J.H., 1997. Aromatics, reformulated gasoline, and atmospheric organic aerosol formation. Environmental Science & Technology 31, 1890–1897.
Robinson, A.L., Donahue, N.M., Shrivastava, M.K., Weitkamp, E.A., Sage, A.M., Grieshop, A.P., Lane, T.E., Pierce, J.R., Pandis, S.N., 2007. Rethinking organic aerosols: semivolatile emissions and photochemical aging. Science 315, 1259–1262.
Seinfeld, J.H., Erdakos, G.B., Asher, W.E., Pankow, J.F., 2001. Modeling the formation of secondary organic aerosol (SOA). 2. The predicted effects of relative humidity on aerosol formation in the alpha-pinene-, beta-pinene-, sabinene-, Delta(3)-Carene-, and cyclohexene-ozone systems. Environmental Science & Technology 35, 1806–1817.
Sheehan, P.E., Bowman, F.M., 2001. Estimated effects of temperature on secondary organic aerosol concentrations. Environmental Science & Technology 35, 2129–2135.
Song, C., Na, K., Warren, B., Malloy, Q., Cocker, D.R., 2007a. Impact of propene on secondary organic aerosol formation from m-xylene. Environmental Science & Technology 41, 6990–6995.
Song, C., Na, K., Warren, B., Malloy, Q., Cocker, D.R., 2007b. Secondary organic aerosol formation from the photooxidation of p- and o-xylene. Environmental Science & Technology 41, 7403–7408.
Takekawa, H., Minoura, H., Yamazaki, S., 2003. Temperature dependence of secondary organic aerosol formation by photo-oxidation of hydrocarbons. Atmospheric Environment 37, 3413–3424.
Volkamer, R., Jimenez, J., San Martini, F., Dzepina, K., Zhang, Q., Salcedo, D., Molina, L., Worsnop, D., Molina, M., 2006. Secondary organic aerosol formation from anthropogenic air pollution: Rapid and higher than expected. Geophysical Research Letters 33, L17811.
Volkamer, R., Martini, F.S., Molina, L.T., Salcedo, D., Jimenez, J.L., Molina, M.J., 2007. A missing sink for gas-phase glyoxal in Mexico City: formation of secondary organic aerosol. Geophysical Research Letters 34, L19807.
Volkamer, R., Ziemann, P.J., Molina, M.J., 2009. Secondary organic aerosol formation from acetylene (C2H2): seed effect on SOA yields due to organic photochemistry in the aerosol aqueous phase. Atmospheric Chemistry and Physics 9, 1907–1928.
Von Hessberg, C., Von Hessberg, P., Poschl, U., Bilde, M., Nielsen, O.J., Moortgat, G.K., 2009. Temperature and humidity dependence of secondary organic aerosol yield from the ozonolysis of beta-pinene. Atmospheric Chemistry and Physics 9, 3583–3599.
Yarwood, G., Rao, S., Yocke, M., Whitten, G.Z., 2005. Updates to the carbon bond chemical mechanism: CB05 Final Report to the US EPA, RT-0400675. Available at: http://www.camx.com/publ/pdfs/CB05_Final_Report_120805.pdf.

Atmospheric Environment 45 (2011) 6048–6056

Contents lists available at ScienceDirect

Atmospheric Environment

journal homepage: www.elsevier.com/locate/atmosenv

Semi-continuous measurement of water-soluble ions in $PM_{2.5}$ in Jinan, China: Temporal variations and source apportionments

Xiaomei Gao [a], Lingxiao Yang [a,b,*], Shuhui Cheng [a], Rui Gao [a], Yang Zhou [a], Likun Xue [a], Youping Shou [a], Jing Wang [a], Xinfeng Wang [a], Wei Nie [a], Pengju Xu [a], Wenxing Wang [a,c]

[a] Environment Research Institute, Shandong University, Jinan 250100, China
[b] School of Environmental Science and Engineering, Shandong University, Jinan 250100, China
[c] Chinese Research Academy of Environmental Sciences, Beijing 100012, China

ARTICLE INFO

Article history:
Received 18 April 2011
Received in revised form
21 July 2011
Accepted 24 July 2011

Keywords:
semi-continuous
Water-soluble ions
$PM_{2.5}$
Seasonal and diurnal variations
Transport patterns
Sources
Jinan

ABSTRACT

To better understand secondary aerosol pollution and potential source regions, semi-continuous measurement of water-soluble ions in $PM_{2.5}$ was performed from December 2007 to October 2008 in Jinan, the capital of Shandong Province. The data was analyzed with the aid of backward trajectory cluster analysis in conjunction with redistributed concentration field (RCF) model and principal component analysis (PCA). SO_4^{2-}, NO_3^- and NH_4^+ were the most abundant ionic species with annual mean concentrations (±standard deviations) of 38.33 (±26.20), 15.77 (±12.06) and 21.26 (±16.28) μg m^{-3}, respectively, which are among the highest levels reported in the literatures in the world. Well-defined seasonal and diurnal patterns of SO_4^{2-}, NO_3^- and NH_4^+ were observed. The fine sulfate and nitrate oxidation ratios (SOR and NOR) were much higher in summer (SOR: 0.47 ± 0.13; NOR: 0.28 ± 0.03) than those in other seasons (SOR: 0.17–0.30; NOR: 0.12–0.14), indicating more extensive formations of SO_4^{2-} and NO_3^- in summer. The most frequent air masses connected with high concentrations of SO_4^{2-}, NO_3^- and NH_4^+ originated from Shandong Province in spring, autumn and winter, while from the Yellow Sea in summer, and then slowly traveled in Shandong Province to Jinan. RCF model indicated that Shandong Province was the main potential source region for SO_4^{2-} and NO_3^- and other potential source regions were also identified including the provinces of Hebei, Henan, Anhui and Jiangsu and the Yellow Sea. Principal component analysis indicated that the major sources contributing to $PM_{2.5}$ pollution were secondary aerosols, coal/biomass burnings and traffic emissions.

1. Introduction

The rapid industrialization and urbanization in China have inevitably led to remarkable increase of air pollutants emissions in the past two decades. Coal combustions and automobile exhausts are mainly responsible for the severe air pollution in large cities in China (Chen et al., 2004). Primary air pollutants such as SO_2 and dust have been successfully reduced due to the enforcement of control measurements in recent years (Chan and Yao, 2008), while $PM_{2.5}$ has emerged as the biggest concern. $PM_{2.5}$ is believed to be a predominant factor to scatter and absorb solar radiation and reduce visibility (Sloane et al., 1991), and it can easily penetrate into lungs and lead to the respiratory and mutagenic diseases (Hughes et al., 1998). Water-soluble ions account for about half of the $PM_{2.5}$ mass (Zhang et al., 2007b; Chan and Yao, 2008). The major water-soluble ions such as sulfate, nitrate and ammonium have effects on the hydroscopic nature and acidity of aerosols (Ocskay et al., 2006), while their characteristics vary significantly with seasons and geographic locations.

Shandong Province is located on the central coast of China and adjacent to Korea and Japan. The area of Shandong constitutes only 1.6% of total China area while anthropogenic emissions from Shandong contributed approximately 10% for SO_2, 8% for NO_x, and 9% for $PM_{2.5}$ to China emissions in 2006 (National Bureau of Statistics of China, 2009; Zhang et al., 2009). Regional transport of air pollutants from Shandong was found to contribute to the aerosol pollution in Beijing under prevailing south and southeast winds (Streets et al., 2007). Besides, Shandong was identified as a potential source region for secondary inorganic aerosol in Seoul, Korea (Heo et al., 2009). As the capital of Shandong Province, Jinan was listed in the group of large cities with the highest concentrations of SO_2, NO_x and TSP in the world (Baldasano et al., 2003). Previous study showed that Jinan suffered serious $PM_{2.5}$ pollution

* Corresponding author. Environment Research Institute, Shandong University, Jinan 250100, China. Tel./fax: +86 531 88366072.
E-mail address: yanglingxiao@sdu.edu.cn (L. Yang).

1352-2310/$ – see front matter Crown Copyright © 2011 Published by Elsevier Ltd. All rights reserved.
doi:10.1016/j.atmosenv.2011.07.041

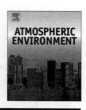

and SO_4^{2-} and NO_3^- were major contributors to the visibility reduction (Yang et al., 2007). However, temporal variations (especially diurnal variation) and source apportionments of water-soluble ions in $PM_{2.5}$ in Jinan are still unclear.

In order to better understand secondary aerosol pollution and potential source regions in Jinan, semi-continuous measurement of water-soluble ions in $PM_{2.5}$ was performed, in conjunction with trace gases and meteorological parameters in 2008. This paper presents the overall results of water-soluble ions. We first show the seasonal and diurnal variations of major water-soluble ions, and then deploy backward trajectory cluster analysis and redistributed concentration field (RCF) model to allocate the potential source regions for secondary ions in Jinan. Finally, principal component analysis (PCA) is used to uncover the underlying factors contributing to the $PM_{2.5}$ pollution in Jinan.

2. Experiments and methodologies

2.1. Sampling sites

Four intensive measurements were conducted from December 2007 to October 2008. In winter (Dec 1 2007–Jan 3 2008) and spring (Apr 1–18 2008), the observation site was chosen at the rooftop of public teaching building in Hongjialou Campus of Shandong University (in brief "HJLC"; 36°69′N, 117°06′E), and the detailed information about this site was given by Xu et al. (2010). In summer (Jun 5–17 2008) and autumn (Sep 12–Oct 15 2008), the study site was set up on the fourth floor at the building of Environmental Science and Engineering in Central Campus of Shandong University (in brief "CC"; 36°40′N, 117°03′E) (Shou et al., 2010), 1 km away from the HJLC. The sampling inlet was ~15 m above the ground level at the two sites. These two sites are both located in the urban area in Jinan, being surrounded by residential or commercial districts (Xu et al., 2010).

2.2. Instruments

An ambient ion monitor (AIM; Model URG 9000B, URG Corporation) was used to measure hourly concentrations of water-soluble ions in $PM_{2.5}$, including F^-, Cl^-, NO_2^-, NO_3^-, SO_4^{2-}, Na^+, NH_4^+, K^+, Mg^{2+} and Ca^{2+}. The instrument has been used in several field campaigns, and the details can refer to Zhou et al. (2010). To avoid positive interference from SO_2 to the SO_4^{2-} measurement (Wu and Wang, 2007; Zhou et al., 2010), a NaOH solution (5 mmol L^{-1}) was substituted for the original ultra-pure water as the denuder liquid to enhance the absorption of SO_2.

$PM_{2.5}$ samples were collected on Teflon membranes using a commercially available filter-based sampler (Reference Ambient Air Sampler, Model RAAS 2.5–400, Thermo Andersen) and 101 sets of samples were obtained during our observation. The collection of samples and analysis of water-soluble ions have been described elsewhere (Wu and Wang, 2007; Zhou et al., 2010). In this study, we compared the results obtained from AIM and traditional filter-based measurements in Section 3.1.1.

Other instruments for measuring SO_2 (TEI, Model 43C), NO_x (TEI Model 42i-TL), O_3 (TEI, Model 49C), CO (API Model 300E) and BC (Magee Scientific, Berkeley, California, USA, Model AE-21) have been described in our previous studies (Zhou et al., 2009; Wang et al., 2010). And the meteorological data were directly obtained from an automatic meteorological station (Xu et al., 2010).

2.3. Trajectories calculation and cluster analysis

Three-day backward trajectories, terminated at 50 m a.s.l., were computed every hour by the Hybrid Single-Particle Lagrangian Integrated Trajectory model (HYSPLIT, version 4.9) with the Global Data Assimilation system (GDAS) meteorological data (Draxler and Rolph, 2003). A total of 2278 backward trajectories with 72 hourly trajectory endpoints in four seasons were used as input for further analysis. A K-means cluster approach was then used to classify the trajectories into several different clusters (Salvador et al., 2010) and five suitable clusters were chosen in four seasons.

2.4. Redistributed concentration field (RCF) model

In this study, C_k is the concentration measured at the receptor site for trajectory k. If C_{ik} is the mean concentration of the grid cells which are hit by segment i ($i = 1, N_k$) of trajectory k (Salvador et al., 2010), then the distribution of air pollutants for trajectory k is

$$C_{ik} = C_k \frac{C_{ik}N_k}{\sum_{i=1}^{N_k} C_{ik}},\ i = 1, N_k \tag{1}$$

After the redistribution of all individual trajectories, the new concentration field $\overline{C}_{mn}$ is calculated by the redistributed concentration C_{ik}:

$$\log\overline{C}_{mn} = \frac{1}{\sum_{k=1}^{M}\sum_{i=1}^{N_k}\tau_{mnik}} \sum_{k=1}^{M}\sum_{i=1}^{N_k}\log(C_{ik})\tau_{mnik} \tag{2}$$

In eq. (2), τ_{mnil} is the residence time of segment i for trajectory k in grid cell (m, n). The new concentration filed is repeated until the average difference for the concentration fields of two successive iterations is below a threshold value of 0.5%. The geophysical regions passed by the trajectories were divided into 1.0° × 1.0° grids.

3. Results and discussions

3.1. Overall statistics of water-soluble ions in $PM_{2.5}$

3.1.1. Comparison of results from AIM and traditional filter-based measurements

To evaluate the performance of modified AIM, hourly data from AIM were averaged to match the collection time of filter samples for comparison. The results from AIM and traditional filter-based measurements are plotted in Fig. 1. Excellent correlations were found for major ionic species, namely, SO_4^{2-}, NO_3^-, NH_4^+, Cl^- and K^+ ($R^2 = 0.84$–0.95, RMA slope = 0.83–1.08), and good correlations ($R^2 = 0.46$–0.90, RMA slope = 0.83–1.08) were obtained for F^-, Na^+, Mg^{2+} and Ca^{2+} with relatively low concentrations. NO_2^- showed no correlation and far higher concentration measured by AIM than that by traditional filter-based measurement (RMA slope = 42.90; $R^2 = 0.08$), and this difference could be due to the loss of NO_2^- from the filters (Chang et al., 2007; Zhang et al., 2007a). Overall, AIM worked well for measuring major water-soluble ions in $PM_{2.5}$.

3.1.2. Mass concentrations

The hourly mean concentrations and standard deviations of water-soluble ions in $PM_{2.5}$ are summarized in Table 1. The concentrations of water-soluble ions followed the order of $SO_4^{2-} > NH_4^+ > NO_3^- > Cl^- \sim NO_2^- \sim K^+ > Na^+ > Ca^{2+} > F^- > Mg^{2+}$ and this order changed slightly with seasons. SO_4^{2-}, NH_4^+ and NO_3^- were the dominant ions, and contributed more than 80% to the total measured water-soluble ions. SO_4^{2-} was the most abundant water-soluble ion in Jinan and its annual mean concentration was 38.33 ± 26.20 μg m^{-3}, accounting for 44.65 ± 11.30% of the total measured water-soluble ions. It is worth noting that the concentration of SO_4^{2-} alone was more than twice the annual US National

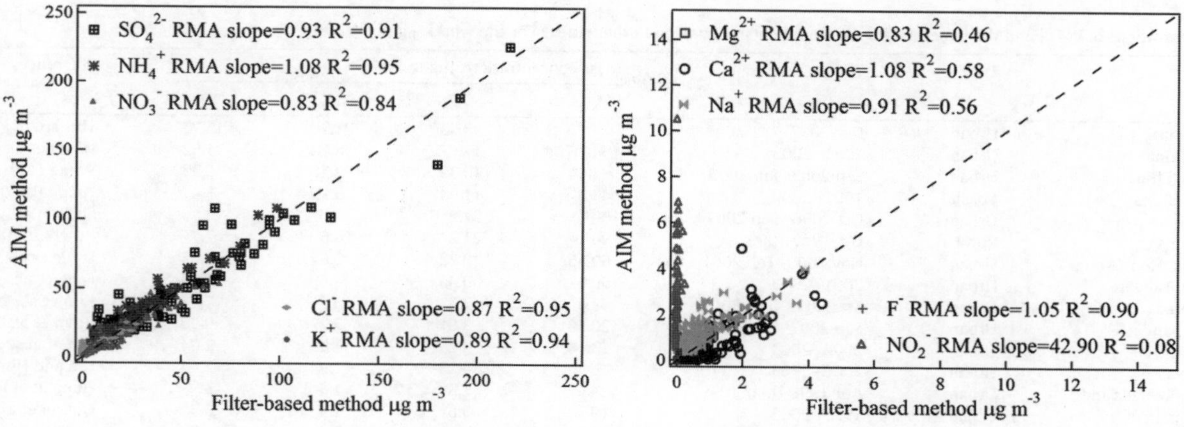

Fig. 1. Scatter plots of major water-soluble ions from AIM and Filter-based methods.

Ambient Air Quality Standards of $PM_{2.5}$ (15 μg m^{-3}) and the hourly maximum concentration of SO_4^{2-} could be up to 227 μg m^{-3}. NH_4^+ (21.16 ± 16.28 μg m^{-3}) and NO_3^- (15.77 ± 12.06 μg m^{-3}) were another major components, accounting for 17.63 ± 7.61% and 23.07 ± 5.85% of the total water-soluble ions respectively.

Table 2 compares the concentrations of SO_4^{2-}, NH_4^+ and NO_3^- in $PM_{2.5}$ in Jinan with those measured in other cities over the world. Obviously, the levels of these compounds in Jinan were substantially (5–10 times) higher than those in cities of USA, Europe, Japan and Korea. Moreover, they were also higher than those of other Chinese cities (e.g. Beijing, Shanghai), which are well-known to suffer serious aerosol pollution. These results demonstrated the severity of secondary inorganic aerosol pollution in Jinan.

3.2. *Temporal variations of water-soluble ions in $PM_{2.5}$*

3.2.1. *Seasonal variations of major water-soluble ions*

Different seasonal variations were observed for individual ion due to their differences in emission sources and formation mechanisms (Table 1). Secondary ions, namely SO_4^{2-}, NO_3^-, NH_4^+ and NO_2^- showed higher values in summer (64.27 ± 31.00, 19.22 ± 11.84, 28.01 ± 16.27 and 2.93 ± 2.45 μg m^{-3}) and winter (42.84 ± 31.72, 21.77 ± 15.05, 29.19 ± 20.72 and 2.48 ± 1.94 μg m^{-3}), and lower levels in spring (27.11 ± 11.41, 10.19 ± 6.00, 13.28 ± 8.75 and 1.50 ± 1.27 μg m^{-3}) and autumn (30.99 ± 14.15, 11.69 ± 7.33, 15.13 ± 7.36 and 1.75 ± 1.61 μg m^{-3}). The summertime peak could be attributed to more active photochemistry process which can facilitate formation of secondary species, while the higher levels in winter may associate with huge emissions of primary pollutants (such as SO_2 and NO_x) from coal combustion for heating and worsen atmospheric dispersion.

The fine sulfate and nitrate oxidation ratios (SOR and NOR) are defined as SOR = $nSO_4^{2-}/(nSO_4^{2-} + nSO_2)$ and NOR = $nNO_3^-/(nNO_3^- + nNO_x)$ to indicate the process and extent of formations from SO_2 to SO_4^{2-} and NO_x to NO_3^- (Wang et al., 2005). SOR and NOR are represented in Fig. 2 and their average values were both larger than 0.10, reflecting occurrence of secondary formation in Jinan (Wang et al., 2005). SOR in summer was 0.47 ± 0.13, much larger than that in spring (0.22 ± 0.05), autumn (0.30 ± 0.04) and winter (0.17 ± 0.02), indicating stronger oxidation of SO_2 to SO_4^{2-} in summer leading to the highest concentration of SO_4^{2-} in spite of relatively lower SO_2 concentrations among four seasons (SO_2: Summer = 26.25 ± 28.73 ppb; Spring = 32.14 ± 26.86 ppb; Autumn = 22.31 ± 16.41 ppb; Winter = 58.59 ± 32.98 ppb). The formation of SO_4^{2-} from SO_2 mainly includes gas-phase reaction of SO_2 and OH radical affected by temperature and solar radiation (Seinfeld, 1986), and heterogeneous reaction which is a function of RH (metal catalyzed oxidation or H_2O_2/O_3 oxidation) (Dlugi et al., 1981). The seasonal variation of SOR was consistent with temperature in Fig. 2, indicating that gas-phase oxidation of SO_2 played a major role in the formation of SO_4^{2-} in the whole year (Wang et al., 2005). NOR showed the highest level in summer (0.28 ± 0.03), the lowest level in winter (0.12 ± 0.01), and comparable level in spring (0.14 ± 0.01) and autumn (0.14 ± 0.01), indicating that high temperature and high RH promoted the faster formation of NO_3^- in spite of more dissociation of NH_4NO_3 at high temperature in summer.

The mass ratio of NO_3^-/SO_4^{2-} has been used as an indicator of relative importance of mobile (e.g. vehicles) vs. stationary sources (e.g. power plant) in the air pollution (Yao et al., 2002; Wang et al., 2006). High NO_3^-/SO_4^{2-} mass ratios have been measured in southern California, with 2 in downtown Los Angeles and 5 in

Table 1
Concentrations of water-soluble ions (mean concentrations ± standard deviation (SD)) in four seasons in Jinan (μg m^{-3}).

Species	Mean ± SD				
	Annual (N = 2282)	Spring (N = 448)	Summer (N = 276)	Autumn (N = 773)	Winter (N = 785)
F^-	0.33 ± 0.42	0.22 ± 0.31	0.21 ± 0.28	0.15 ± 0.15	0.61 ± 0.53
Cl^-	4.19 ± 5.36	1.59 ± 2.34	3.18 ± 3.74	1.44 ± 1.56	8.75 ± 6.37
NO_2^-	2.10 ± 0.86	1.50 ± 1.27	2.93 ± 2.45	1.75 ± 1.61	2.48 ± 1.94
NO_3^-	15.77 ± 12.06	10.19 ± 6.00	19.22 ± 11.84	11.69 ± 7.33	21.77 ± 15.05
SO_4^{2-}	38.33 ± 26.20	27.11 ± 11.41	64.27 ± 31.00	30.99 ± 14.15	42.84 ± 31.72
Na^+	1.22 ± 0.75	0.85 ± 0.38	2.12 ± 1.04	1.04 ± 0.62	1.34 ± 0.65
NH_4^+	21.26 ± 16.28	13.28 ± 8.75	28.01 ± 16.27	15.13 ± 7.361	29.19 ± 20.72
K^+	2.36 ± 2.32	1.32 ± 1.14	4.62 ± 3.08	1.44 ± 1.17	3.07 ± 2.54
Mg^{2+}	0.11 ± 0.15	0.12 ± 0.10	0.03 ± 0.05	0.01 ± 0.07	0.22 ± 0.17
Ca^{2+}	0.76 ± 1.18	0.83 ± 0.17	0.29 ± 0.46	0.23 ± 0.59	1.41 ± 1.43

Table 2
Mass concentrations of $PM_{2.5}$ and the major chemical components in Jinan and other cities over the world ($\mu g\ m^{-3}$).

Site	Type	Time	Mass concentrations ($\mu g\ m^{-3}$)				References
			$PM_{2.5}$	SO_4^{2-}	NH_4^+	NO_3^-	
Jinan, China	Urban	Dec 2007–Oct 2008		38.33	21.26	15.77	This study
Beijing, China	Urban	2001–2003	154.26	17.07	8.72	11.52	Wang et al., 2005
Shanghai, China	Urban	Sep 2003–Jan 2005	94.64	10.39	3.78	6.23	Wang et al., 2006
Qingdao, China	Coastal	1997–2000	43.6	11.94	5.79	3.4	Hu et al., 2002
Xi'an, China	Urban	Oct 2006–Sep 2007	130	27.9	7.6	12	Shen et al., 2009
Linan, China	Rural	Oct–Nov 1999	90	21.2	8.6	7.7	Xu et al., 2002
Mong Kok, Hong Kong	Urban	Nov 2000–Feb 2001	69.15	10.32	3.84	2.65	Louie et al., 2005
Taichung, Taiwan	Urban	2001–2003	59.8	9.45	4.49	1.93	Fang et al., 2002
Seoul, Korea	Urban	Mar 2003–Feb 2005	42.8	7.5	5.5	7.1	Kim et al., 2007
Tokyo, Japan	Urban	Sep 2007–Aug 2008	20.58	3.8	2.27	0.96	Khan et al., 2010
New York, US	Urban	2002–2003	13.16	4.29	1.93	2.04	Qin et al., 2006
St. Louis, US	Urban	2000–2003	16.4	4.23	1.94	2.48	Lee and Hopke, 2006
Kerbside, Switzerland	Urban	Apr 1998–Mar 1999	24.6	2.8	1.6	3	Hueglin et al., 2005
Huelva, Spain	Urban	1999–2005	19	3.6	1.4	0.5	Querol et al., 2008
Milan, Italy	Urban	Aug 2002–Nov 2003	20.2	4	2.2	4.6	Lonati et al., 2005
Milan, Italy	Urban	Aug 2002–Nov 2003	53.7	5.8	5.2	20.2	Lonati et al., 2005

Rubidoux, which was due to less use of coal (Kim et al., 2000); However in Chinese cities (e.g. Beijing, Shanghai), lower ratios had been reported as a result of the wide use of sulfur-containing coal (Yao et al., 2002). In our study, the NO_3^-/SO_4^{2-} mass ratio ranged from 0.03 to 1.52, with the annual mean of 0.44. These results indicated that like other cities in China, stationary sources were more important compared with vehicle emissions in Jinan. The NO_3^-/SO_4^{2-} mass ratios showed clear seasonal variations with the highest ratio in winter (0.53 ± 0.05), the lowest ratio in summer (0.34 ± 0.03), and comparable ratio in spring (0.40 ± 0.06) and autumn (0.42 ± 0.06) (see Fig. 2). The reason is that in summer high temperature and high RH are more favorable for formation of SO_4^{2-} compared to NO_3^- due to enhanced evaporation of NH_4NO_3 at high temperature.

Cl^- showed higher concentrations in winter (8.75 ± 6.37 $\mu g\ m^{-3}$) than that in other seasons (Table 1). Cl^- is a major component of sea-salt particle, and is also released from coal combustion (Sun et al., 2006). In winter, the air generally originated from the northwest (Xu et al., 2010) and thus the influence of sea-salt would be minor. The mass ratio of Cl^-/Na^+ was calculated as 6.18 ± 2.96 in winter, which is much higher than that detected for sea water (1.797) (Moller, 1990). Therefore, the elevated concentration of Cl^- in winter was due to the enhanced coal combustion. K^+ exhibited higher levels in summer (4.62 ± 3.08 $\mu g\ m^{-3}$) and winter (3.07 ± 2.54 $\mu g\ m^{-3}$) than that in spring and autumn (1.32 ± 1.14 and 1.44 ± 1.17 $\mu g\ m^{-3}$). In summer, extensive activities of biomass burning around Shandong was the main factor contributing to the elevated concentration of K^+ (http://maps.geog.

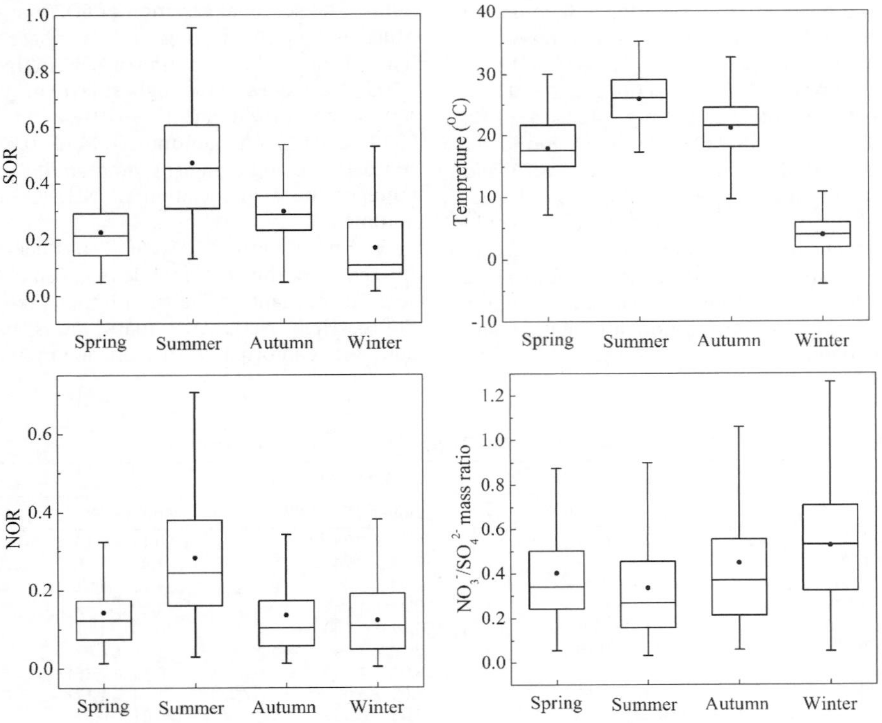

Fig. 2. Seasonal variations of SOR, NOR, NO_3^-/SO_4^{2-} mass ratio and temperature.

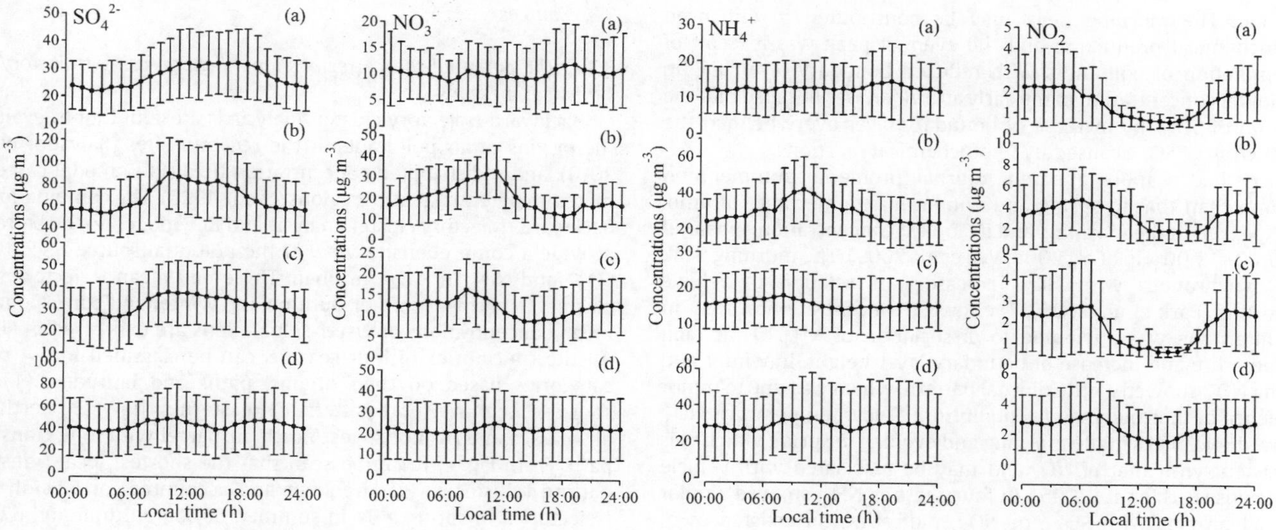

Fig. 3. The diurnal profiles of SO_4^{2-}, NO_3^-, NH_4^+ and NO_2^- in four seasons (a) Spring; (b) Summer; (c) Autumn; (d) Winter.

umd.edu/firms/). High concentrations of K^+ in winter may be associated with coal combustion as implied from the strong correlation ($r = 0.82$) between K^+ with Cl^- (Westberg et al., 2003). Other water-soluble ions were not relevant for seasonal variations due to their low concentrations.

3.2.2. Diurnal variations of secondary water-soluble ions

The diurnal variations of secondary water-soluble ions in $PM_{2.5}$ in four seasons are shown in Fig. 3. In general, SO_4^{2-} exhibited similar diurnal profiles in spring, summer and autumn, with an evident increase as sun rising and a broad daytime maximum, which was consistent with those reported in other cities (e.g. Beijing, PRD) (Hu et al., 2008; Wu et al., 2009). This typical pattern can be explained by the fact that photochemical production is more extensive during the daytime with stronger solar radiation. Compared with other seasons, in winter SO_4^{2-} had a little diurnal variation and showed two peaks in the morning and evening, which may be related with boundary layer height and photochemical

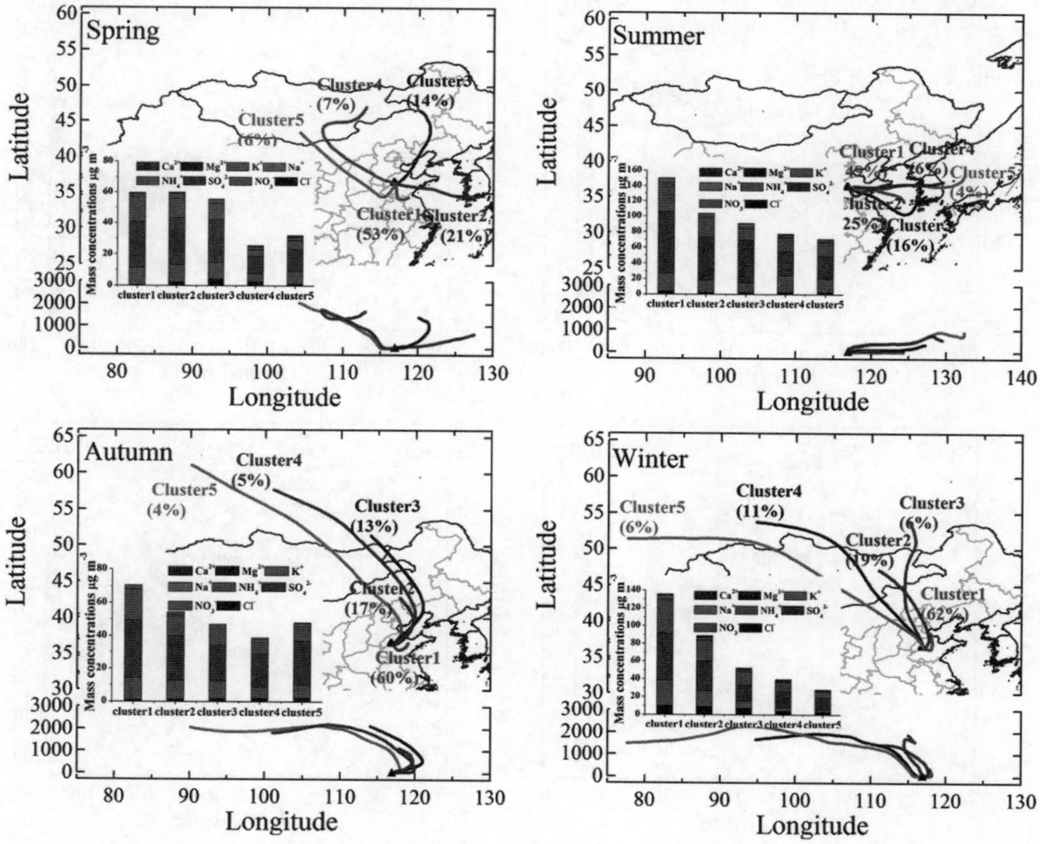

Fig. 4. Mean clusters and the corresponding mean ions concentrations in four seasons.

reaction. The morning peak may be contributed to enhanced photochemical production and the evening peak was a result of accumulation of pollutants with reduced boundary layer height. The lower concentration in the early afternoon was because that the dilution of SO_4^{2-} by increase of boundary layer overwhelmed the production of SO_4^{2-} caused by photochemical reactions.

NO_3^- had a more apparent diurnal profile in summer and autumn than that in spring and winter. In summer and autumn, NO_3^- peaked in the morning, and its lowest concentration appeared at 16:00–18:00 (Hu et al., 2008; Wu et al., 2009). The morning peak was synchronous with NO_x, indicating its relation to vehicle emissions (Park et al., 2005). The lowest concentration of NO_3^- in the afternoon was attributed to dissociation of NH_4NO_3 at high temperature and increase of boundary layer height. In winter and spring NO_3^- showed a little diurnal pattern, which was due to minor influence of thermodynamic equilibrium at low temperature. NO_3^- showed two peaks in the morning and evening in spring which was consistent with that of NO_x and may be associated with vehicle emissions (Park et al., 2005). In four seasons NH_4^+ showed similar diurnal profiles with SO_4^{2-} or NO_3^-, indicating the existences of $(NH_4)_2SO_4$ and NH_4NO_3.

NO_2^- showed similar diurnal profiles in four seasons with higher concentrations at night and lower concentrations during daytime (Fig. 3), and similar profile was also observed in Beijing (Zhang et al., 2007a). Higher NO_2^- concentration at night than that during daytime probably was related with lower oxidized extent due to weak solar radiation and nighttime accumulation of NO_2^-.

3.3. Sources

3.3.1. The potential source regions identification using trajectory statistical methods

Backward trajectory cluster analysis is a useful tool to evaluate the origins of air pollutants at the receptor sites (Salvador et al., 2010), and redistributed concentration field (RCF) model can estimate the potential source regions (Stohl, 1996). The combination of backward trajectory cluster analysis and RCF model can be better to provide a comprehensive view of the potential source regions for SO_4^{2-} and NO_3^- in $PM_{2.5}$ in Jinan. Three-day mean trajectories for clusters in spring, summer, autumn and winter and corresponding mean concentrations of water-soluble ions are expressed in Fig. 4. All the trajectories in four seasons can be classified into 4 main categories based on their origins, paths and latitudes: (1) the shortest/local transport pattern, (2) eastern airflow, (3) northeast air masses and (4) northwest/north air parcel with long transport path. From Fig. 4, it can be seen that the shortest/local transport pattern (cluster 1) was frequent and accounted for 53% of total trajectories in spring, 49% in summer, 60% in autumn and 62% in winter. Eastern airflow was dominant in summer and contributed for 21% of total trajectories in spring (cluster 2). Northeast/north air masses were observed in spring (cluster 3, 14%) and in autumn (cluster 2, 17%). Other clusters generally originated from northwest/north of China and traveled fast at the highest altitude.

The highest concentrations of SO_4^{2-}, NO_3^- and NH_4^+ were observed in the shortest cluster (cluster 1) in four seasons, indicating that secondary ions in $PM_{2.5}$ were easy to be enriched in the

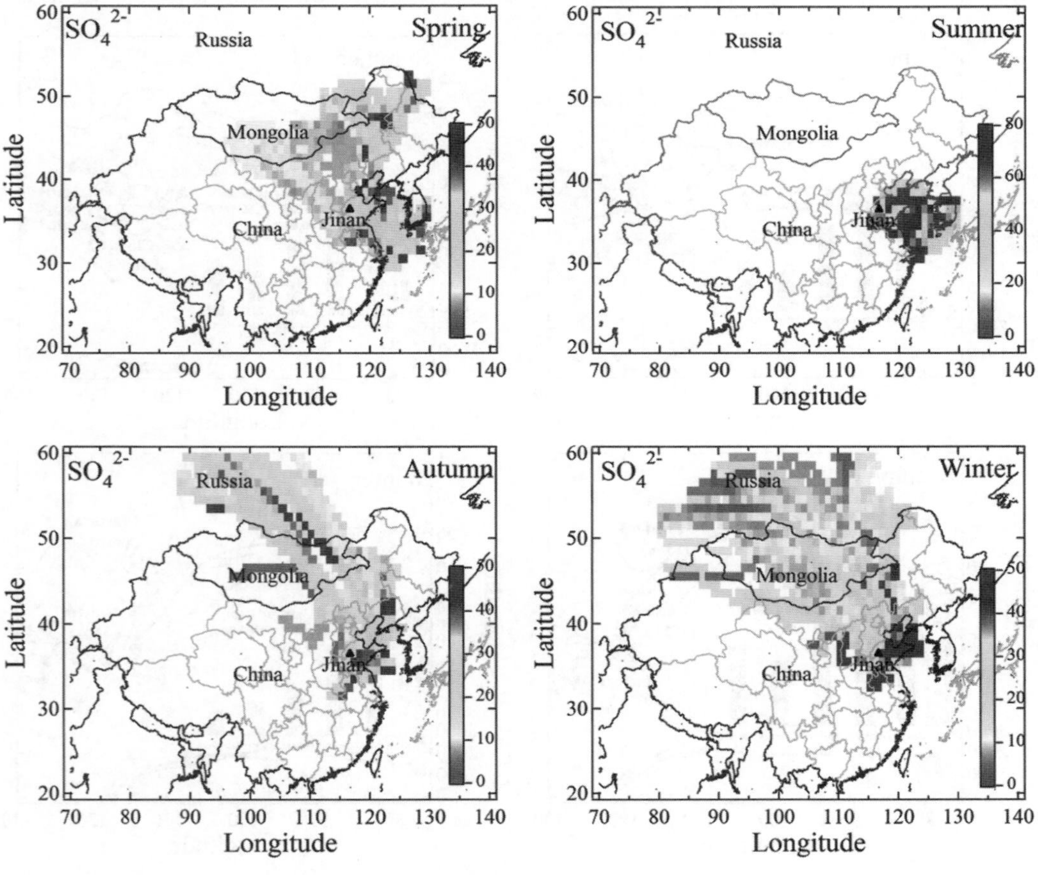

Fig. 5. RCF distribution for sulfate in Jinan in four seasons.

short trajectories from upwind regional and local emission sources (Karaca and Camci, 2010). Cluster 1 generally originated from the middle of Shandong Province, moved southerly and finally turned westerly to Jinan in spring, autumn and winter, while in summer it derived from the Yellow Sea, moved northwesterly to Jinan. A major big petro-chemical corporation, power plants and cement production base are located in the middle (Zibo city) and southwest (Jining city and Zaozhuang city) of Shandong Province. Cluster 1 spent much time on passing over industrial zones with high emissions of primary pollutants (e.g. SO_2, NO_x), leading to the high concentrations of SO_4^{2-}, NO_3^- and NH_4^+.

Much higher concentrations of SO_4^{2-}, NO_3^- and NH_4^+ were associated with air masses from northeast or northwest, including cluster 3 in spring, clusters 2 in autumn and cluster 2 in winter. The flow patterns of cluster 3 in spring and cluster 2 in autumn were both typically originated from Inner Mongolia, flowed over Liaoning Province and Bohai Gulf before arriving at Jinan. While cluster 2 in winter derived from Outer Mongolia, passed through Inner Mongolia, Hebei Province and then to Jinan. These trajectories all passed over the Bohai economic zone, which is one of the most populated and industrial zones in China and has the highest SO_2 and NO_x emissions (Zhang et al., 2009).

Low concentrations of SO_4^{2-}, NO_3^- and NH_4^+ occurred in cluster 4 and 5 in spring, autumn and winter. These clusters derived from northwest of China and moved faster at the higher latitudes compared to other clusters.

Compared to other seasons, in summer higher concentrations of SO_4^{2-}, NO_3^- and NH_4^+ could be explained by the combination of Dimethyl Sulfide (DMS) oxidation (Salvador et al., 2010) and long residence time over industrial zones with large amount emissions of primary pollutants. It was worth noting that there was also a significant fraction of the flow from Korea in summer (cluster 4 and 5, 10%), which passed westerly over the Yellow Sea and Jiaodong Peninsula before arriving at Jinan.

Figs. 5 and 6 show the results of RCF analysis for SO_4^{2-} and NO_3^-, respectively. The high potential source region of SO_4^{2-} and NO_3^- was Shandong, and Hebei, Henan, Anhui, Jiangsu, Liaoning, and Inner Mongolia in spring, autumn and winter and eastern Jiangsu Province, South Korea and the Yellow Sea in summer were also identified as the potential source regions.

3.3.2. The sources identification by principal component analysis (PCA)

In order to identify the sources of water-soluble ions in $PM_{2.5}$, principal component analysis (PCA) was applied. PCA is a widely used statistical technique to quantitatively identify a smaller number of independent factors among the compound concentrations, which can explain the variance of the data, by using the eigenvector decomposition of a matrix of pair-wise correlations (Johnson and Wichern, 1998; Miller et al., 2002). PCA is conducted using a commercially available software package (SPSS). Hourly values of Nss-SO_4^{2-}, NO_3^-, NH_4^+, Cl^-, NO_2^-, Na^+, K^+, Mg^{2+}, Ca^{2+}, NO_x, SO_2, CO, BC and O_3 were used for PCA and the results are shown in Table 3.

In spring, four principal components were obtained and accounted for 82% of the total variance. The principal component 1 accounted for 45% of the total variance, and comprised NO_x, NO_2^-, BC and CO, while anti-correlated with O_3, indicating its relation to

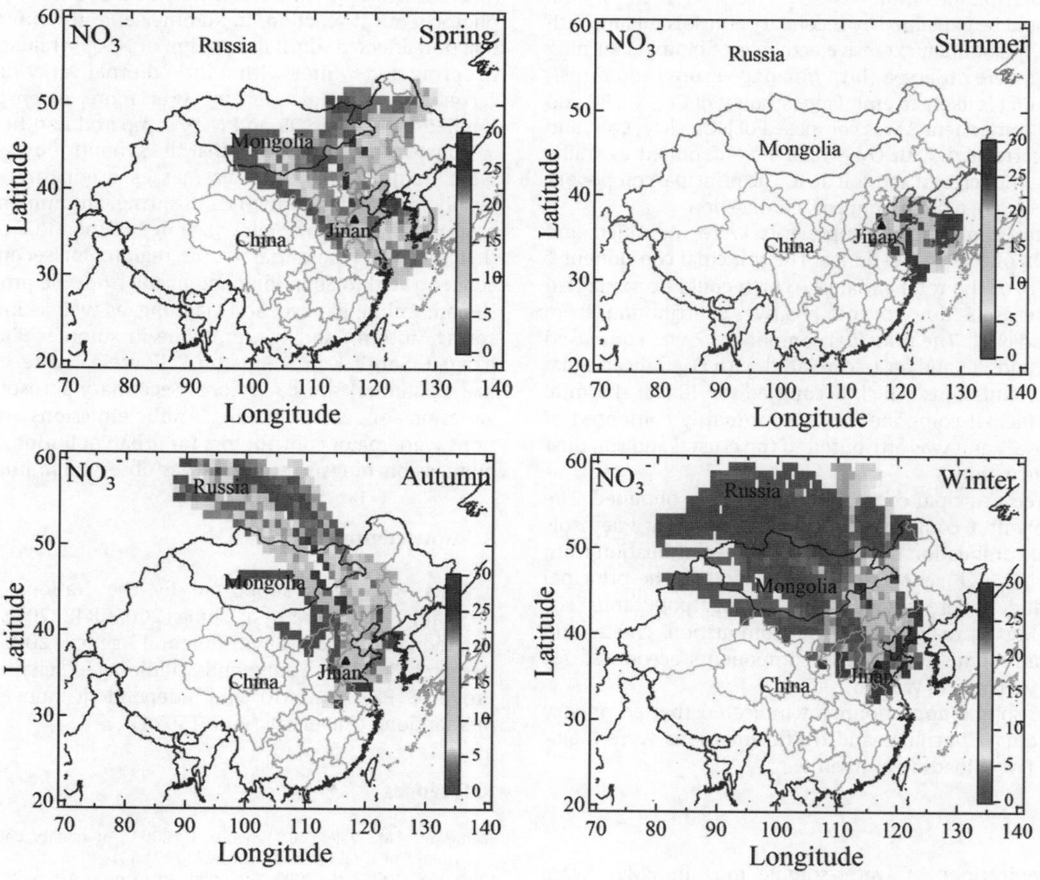

Fig. 6. RCF distribution for nitrate in Jinan in four seasons.

Table 3
Factor loadings from PCA in the four seasons.

	Spring				Summer			Autumn			Winter		
	1	2	3	4	1	2	3	1	2	3	1	2	3
Nss-SO_4^{2-}	−0.08	**0.76**	**0.51**	0.11	**0.73**	−0.06	0.56	**0.94**	0.03	0.04	**0.93**	−0.11	0.10
NO_3^-	0.20	**0.91**	0.02	0.03	**0.91**	0.18	−0.07	**0.63**	0.49	−0.24	**0.78**	−0.18	0.26
NH_4^+	0.21	**0.84**	0.28	0.14	**0.90**	0.05	0.35	**0.91**	0.44	−0.19	**0.95**	−0.03	0.12
Cl^-	0.46	−0.08	**0.78**	−0.12	**0.75**	0.41	−0.07	0.40	0.03	0.23	**0.85**	0.40	0.05
NO_2^-	**0.85**	0.27	0.17	0.09	0.38	**0.76**	−0.01	−0.05	**0.77**	0.06	**0.83**	0.15	−0.03
Na^+	0.01	0.27	0.68	0.48	0.12	0.23	0.45	−0.01	0.04	**0.67**	**0.84**	0.15	−0.01
K^+	0.39	0.26	0.55	0.32	**0.87**	−0.02	0.09	**0.69**	0.20	0.50	**0.91**	0.06	0.06
Mg^{2+}	0.12	0.17	0.05	**0.95**	0.30	**0.61**	0.02	0.08	0.17	**0.72**	0.28	**0.74**	0.14
Ca^{2+}	0.23	−0.02	−0.03	**0.89**	−0.10	**0.84**	0.02	0.05	0.45	**0.78**	0.20	**0.86**	−0.11
NOx	**0.83**	0.13	0.22	0.30	0.18	**0.84**	0.23	0.00	**0.73**	0.53	−0.19	0.36	**0.69**
SO_2	0.11	0.25	**0.77**	−0.18	−0.02	−0.11	**0.85**	**0.69**	−0.22	0.37	−0.14	**0.74**	0.25
CO	**0.57**	0.31	**0.57**	0.22	**0.71**	0.60	0.21	0.31	0.73	0.34	**0.81**	0.41	−0.08
O_3	**−0.88**	0.09	−0.17	−0.01	−0.23	**−0.75**	0.36	0.26	**−0.42**	−0.07	−0.32	0.03	**−0.64**
BC	**0.70**	−0.08	0.00	0.42	**0.72**	0.41	−0.10	0.29	**0.83**	0.25	**0.85**	0.33	−0.05
Variance	45%	15%	13%	9%	45%	18%	10%	39%	18%	12%	52%	16%	7%

Note: Values in bold indicate loading factors discussed in this paper.

traffic emissions. The principal component 2 could be explained by secondary aerosols due to the positive contribution from Nss-SO_4^{2-}, NO_3^- and NH_4^+, and this factor accounted for 15% of the total variance. The principal component 3 contained Cl^- and SO_2 with 13% of the total variance, which was related with stationary source emissions such as coal combustion. The principal component 4 was primarily composed of Ca^{2+}, Mg^{2+} and Na^+, and was attributed to the crustal and soil dust from urban constructions.

In summer, three principal components were identified and accounted for 73% of the total variance. The principal component 1 had highly positive contributions from Nss-SO_4^{2-}, NO_3^-, NH_4^+, Cl^-, K^+, BC and CO, which could be explained by secondary aerosols mixing with biomass burning. In summer, extensive activities of biomass burning around Shandong were observed (http://maps.geog.umd.edu/firms/) and biomass burning is likely to emit large amount of Cl^-, K^+, BC and CO. The principal component 2 was composed of NO_x, NO_2^-, Ca^{2+} and Mg^{2+}, and anti-correlated with O_3, which was identified as traffic emissions mixing with crustal and soil dust. The principal component 3 was composed of SO_2, mainly from coal combustion.

In autumn, three principal components were identified and accounted for 69% of the total variance. The principal component 1 accounted for 39% of the total variance, which could be explained by secondary aerosols due to the positive contribution from Nss-$SO_4^{2-}$$NO_3^-$ and NH_4^+. The principal component 2 was composed of NO_x, NO_2^-, CO and BC, and anti-correlated with O_3, indicating its relation to traffic emissions which accounted for 18% of the total variance. The principal component 3 was primarily composed of Ca^{2+}, Mg^{2+} and Na^+, and was attributed to the crustal and soil dust from urban constructions.

In winter, three principal components also were obtained. The principal component 1 could be identified as secondary aerosols mixing with coal combustion due to the positive combustion from Nss-SO_4^{2-}, NO_3^-, NH_4^+, Cl^-, NO_2^-, Na^+, K^+, BC and CO. The principal components 2 and 3 were composed of primary pollutants (e.g. SO_2, NO_x, Ca^{2+}, Mg^{2+}), mainly from coal combustion, crustal and soil dust and traffic emissions. These components accounted for 75% of the total variance in winter.

All in all, the source apportionment indicated that secondary aerosols, coal/biomass burnings and traffic emissions were major contributors for $PM_{2.5}$ loading in Jinan.

4. Summary

Hourly concentrations of water-soluble ions in $PM_{2.5}$ were measured to investigate secondary aerosol pollution and potential source regions in Jinan, Shandong Province from Dec 2007 to Oct 2008.

The results verified that Jinan was suffering more serious fine particle pollution compared with other cities in the world. Accelerated photochemistry reaction under high temperature, O_3 concentrations and strong solar radiation in summer and high emissions of SO_2 and NO_x from coal combustion in winter led to higher concentrations of SO_4^{2-}, NO_3^- and NH_4^+ in these two seasons than those in spring and autumn. The diurnal variations of SO_4^{2-} in spring, autumn and winter were dominated by photochemical process, while in winter controlled by boundary layer height and photochemical reaction. In summer and autumn thermodynamic reaction affected diurnal variation of NO_3^-, while it rarely occurred in spring and winter with a little diurnal variation. Production of secondary inorganic aerosol was more extensive in summer implied by higher SOR and NOR compared to other seasons.

Cluster analysis showed that the synoptic flows arriving at Jinan were dominated by the air masses originating and circulating locally in Shandong Province in spring, autumn and winter, while originating from the Yellow Sea in summer. RCF results indicated that the major potential source regions for secondary ions were concentrated in Shandong and partly from the provinces of Hebei, Henan, Anhui, Jiangsu, and Liaoning, as well as Inner Mongolia in spring, autumn and winter, while in summer Shandong, eastern Jiangsu, South Korea and the Yellow Sea were identified as the main potential source regions. Secondary aerosol dominated the variations of aerosol loading. Traffic emissions and coal combustions were major contributors for urban pollution. The influence of biomass burnings in summer was observed in Jinan.

Acknowledgments

This work was supported by the National Basic Research Program (973 Program) of China (2005CB422203), a key project of Shandong Provincial Environmental Agency (2006045), Promotive Research Fund for Young and Middle-aged Scientists of Shandong Province (BS2010HZ010) and Independent Innovation Foundation of Shandong University (2009TS024).

References

Baldasano, J.M., Valera, E., Jiménez, P., 2003. Air quality data from large cities. Science of Total Environment 307 (1–3), 141–165.
Chan, C.K., Yao, X.H., 2008. Air pollution in mega cities in China. Atmospheric Environment 42 (1), 1–42.

Chang, S.Y., Lee, C.T., Chou, C.C.K., Liu, S.C., Wen, T.X., 2007. The continuous field measurements of soluble aerosol compositions at the Taipei Aerosol Supersite, Taiwan. Atmospheric Environment 41 (9), 1936–1949.
Chen, B.H., Hong, C.J., Kan, H.D., 2004. Exposures and health outcomes from outdoor air pollutants in China. Toxicology 198 (1–3), 291–300.
Dlugi, R., Jordan, S., Lindemann, E., 1981. The heterogeneous formation of sulfate aerosols in the atmosphere. Journal of Aerosol Science 12, 185–197.
Draxler, R.R., Rolph, G.D., 2003. HYSPLIT (Hybrid Single-Particle Lagrangian Integrated Trajectory) Model Access via NOAA ARL READY Websiter. NOAA Air Resources Laboratory, Silver Spring, MD. Available at: http://www.arl.noaa.gov/HYSPLIT.php (accessed 2009).
Fang, G.C., Chang, C.N., Wu, Y.S., Fu, P.P.C., Yang, C.J., Chen, C.D., Chang, S.C., 2002. Ambient suspended particulate matters and related chemical species study in central Taiwan, Taichung during 1998–2001. Atmospheric Environment 36 (12), 1921–1928.
Heo, J.B., Hopke, P.K., Yi, S.M., 2009. Source apportionment of $PM_{2.5}$ in Seoul, Korea. Atmospheric Chemistry and Physics 9, 4957–4971.
Hu, M., Wu, Z.J., Slanina, J., Lin, P., Liu, S., Zeng, L.M., 2008. Acidic gases, ammonia and water-soluble ions in $PM_{2.5}$ at a coastal site in the Pearl River Delta, China. Atmospheric Environment 42 (25), 6310–6320.
Hueglin, C., Gehrig, R., Baltensperger, U., Gysel, M., Monn, C., Vonmont, H., 2005. Chemical characterisation of $PM_{2.5}$, PM_{10} and coarse particles at urban, near-city and rural sites in Switzerland. Atmospheric Environment 39 (4), 637–651.
Hughes, L.S., Cass, G.G., Gone, J., Ames, M., Olmez, I., 1998. Physical and chemical characterization of atmospheric ultrafine particles in the Los Angeles area. Environmental Science & Technology 32 (9), 1153–1161.
Hu, M., He, L.Y., Zhang, Y.H., Wang, M., Kim, Y.P., Moon, K.C., 2002. Seasonal variation of ionic species in fine particles at Qingdao, China. Atmospheric Environment 36 (38), 5853–5859.
Johnson, R.A., Wichern, D., 1998. Applied Multivariate Statistical Analysis. Prentice Hall, Englewood Cliffs, NJ.
Karaca, F., Camci, F., 2010. Distant source contributions to PM_{10} profile evaluated by SOM based cluster analysis of air mass trajectory sets. Atmospheric Environment 44 (7), 892–899.
Khan, M.F., Shirasuna, Y., Hirano, K., Masunaga, S., 2010. Characterization of $PM_{2.5}$, $PM_{2.5-10}$ and $PM_{>10}$ in ambient air, Yokohama, Japan. Atmospheric Research 96 (1), 159–172.
Kim, B.M., Teffera, S., Zeldin, M.D., 2000. Characterization of $PM_{2.5}$ and PM_{10} in the South Coast Air Basin of Southern California: part 1 – spatial variations. Journal of the Air and Waste Management Association 50 (12), 2034–2044.
Kim, H.S., Huh, J.B., Hopke, P.K., Holsen, T.M., Yi, S.M., 2007. Characteristics of the major chemical constituents of $PM_{2.5}$ and smog events in Seoul, Korea in 2003 and 2004. Atmospheric Environment 41 (32), 6762–6770.
Lee, J.H., Hopke, P.K., 2006. Apportioning sources of $PM_{2.5}$ in St. Louis, MO using speciation trends network data. Atmospheric Environment 40 (2), 360–377.
Lonati, G., Giugliano, M., Butelli, P., Romele, L., Tardivo, R., 2005. Major chemical components of $PM_{2.5}$ in Milan (Italy). Atmospheric Environment 39 (10), 1925–1934.
Louie, P.K.K., Watson, J.G., Chow, J.C., Chen, A., Sin, D.W.M., Lau, A.K.H., 2005. Seasonal characteristics and regional transport of $PM_{2.5}$ in Hong Kong. Atmospheric Environment 39 (9), 1695–1710.
Moller, D., 1990. The Na/Cl ration in rainwater and the seasalt chloride cycle. Tellus B 423, 254–262.
Miller, S.L., Anderson, M.J., Daly, E.P., Milford, J.B., 2002. Source apportionment of exposures to volatile organic compounds. I. Evaluation of receptor models using simulated exposure data. Atmospheric Environment 36 (22), 3629–3641.
National Bureau of Statistics of China, 2009. China Statistical Yearbook. China Statistics Press, Beijing (in Chinese).
Ocskay, R., Salma, I., Wang, W., Maenhaut, W., 2006. Characterization and diurnal variation of size-resolved inorganic water-soluble ions at a rural background site. Journal of Environmental Monitoring 8 (2), 300–306.
Park, S.S., Ondov, J.M., Harrison, D., Nair, N.P., 2005. Seasonal and shorter-term variations in particulate atmospheric nitrate in Baltimore. Atmospheric Environment 39 (11), 2011–2020.
Qin, Y.J., Kim, E., Hopke, P.K., 2006. The concentrations and sources of $PM_{2.5}$ in metropolitan New York City. Atmospheric Environment 40 (2), 312–332.
Querol, X., Alastuey, A., Moreno, T., Viana, M.M., Castillo, S., Pey, J., Rodríguez, S., Artíñano, B., Salvador, P., Sánchez, M., Garcia Dos Santos, S., Herce Garraleta, M.D., Fernandez-Patier, R., Moreno-Grau, S., Negral, L., Minguillón, M.C., Monfort, E., Sanz, M.J., Palomo-Marín, R., Pinilla-Gil, E., Cuevas, E., de la Rosa, J., Sánchez de la Campa, A., 2008. Spatial and temporal variations in airborne particulate matter (PM_{10} and $PM_{2.5}$) across Spain 1999–2005. Atmospheric Environment 42 (17), 3964–3979.
Salvador, P., Artíñano, B., Pio, C., Afonso, J., Legrand, M., Puxbaum, H., Hammer, S., 2010. Evaluation of aerosol sources at European high altitude background sites with trajectory statistical methods. Atmospheric Environment 44 (19), 2316–2329.
Seinfeld, J.H., 1986. Atmospheric Chemistry and Physics of Air Pollution. Wiley, New York.
Shen, Z.X., Cao, J.J., Arimoto, R., Han, Z.W., Zhang, R.J., Han, Y.M., Liu, S.X., Okuda, T., Nakao, S., Tanaka, S., 2009. Ionic composition of TSP and $PM_{2.5}$ during dust storms and air pollution episodes at Xi'an, China. Atmospheric Environment 43 (18), 2911–2918.
Shou, Y.P., Gao, X.M., Wang, J., Yang, L.X., Wang, W.X., 2010. Online measurement of water-soluble ions in fine particles from the atmosphere in autumn in Jinan. Research of Environmental Sciences 23 (1), 41–47.
Sloane, C.S., Watson, J., Chow, J., Pritchett, L., Richards, L.W., 1991. Size-segregated fine particle measurements by chemical species and their impact on visibility impairment in Denver. Atmospheric Environment. Part A. General Topics 25 (5–6), 1013–1024.
Stohl, A., 1996. Trajectory statistics-a new method to establish source-receptor relationships of air pollutants and its application to the transport of particulate sulfate in Europe. Atmospheric Environment 30 (4), 579–587.
Streets, D.G., Fu, J.S., Jang, C.J., Hao, J.M., He, K.B., Tang, X.Y., Zhang, Y.H., Wang, Z.F., Li, Z.P., Zhang, Q., Wang, L.T., Wang, B.Y., Yu, C., 2007. Air quality during the 2008 Beijing Olympic Games. Atmospheric Environment 41 (3), 480–492.
Sun, Y.L., Zhuang, G.S., Tang, A.H., Wang, Y., An, Z.S., 2006. Chemical characteristics of $PM_{2.5}$ and PM_{10} in haze-fog episodes in Beijing. Environmental Science & Technology 40 (10), 3148–3155.
Wang, T., Nie, W., Gao, J., Xue, L.K., Gao, X.M., Wang, X.F., Qiu, J., Poon, C.N., Meinardi, S., Blake, D., Wang, S.L., Ding, A.J., Chai, F.H., Zhang, Q.Z., Wang, W.X., 2010. Air quality during the 2008 Beijing Olympics: secondary pollutants and regional impact. Atmospheric Chemistry and Physics 10, 7603–7615.
Wang, Y., Zhuang, G.S., Tang, A.H., Yuan, H., Sun, Y.L., Chen, S., Zheng, A.H., 2005. The ion chemistry and the source of $PM_{2.5}$ aerosol in Beijing. Atmospheric Environment 39 (21), 3771–3784.
Wang, Y., Zhuang, G.S., Zhang, X.Y., Huang, K., Xu, C., Tang, A.H., Chen, J.M., An, Z.S., 2006. The ion chemistry, seasonal cycle, and sources of $PM_{2.5}$ and TSP aerosol in Shanghai. Atmospheric Environment 40 (16), 2935–2952.
Westberg, H.M., Byström, M., Leckner, B., 2003. Distribution of potassium, chlorine, and sulfur between solid and vapor phases during combustion of wood chips and coal. Energy & Fuels 17 (1), 18–28.
Wu, W.S., Wang, T., 2007. On the performance of a semi-continuous $PM_{2.5}$ sulphate and nitrate instrument under high loadings of particulate and sulphur dioxide. Atmospheric Environment 41 (26), 5442–5451.
Wu, Z.J., Hu, M., Shao, K.S., Slanina, J., 2009. Acidic gases, NH_3 and secondary inorganic ions in PM_{10} during summertime in Beijing, China and their relation to air mass history. Chemosphere 76 (8), 1028–1035.
Xu, J., Bergin, M.H., Yu, X., Liu, G., Zhao, J., Carrico, C.M., Baumann, K., 2002. Measurement of aerosol chemical, physical and radiative properties in the Yangtze delta region of China. Atmospheric Environment 36 (2), 161–173.
Xu, P.J., Wang, W.X., Yang, L.X., Zhang, Q.Z., Gao, R., Wang, X.F., Nie, W., Gao, X.M., 2010. Aerosol size distributions in urban Jinan: seasonal characteristics and variations between weekdays and weekends in a heavily polluted atmosphere. Environmental Monitoring and Assessment. doi:10.1007/s10661-010-1747-2.
Yang, L.X., Wang, D.C., Cheng, S.H., Wang, Z., Zhou, Y., Zhou, X.H., Wang, W.X., 2007. Influence of meteorological conditions and particulate matter on visual range impairment in Jinan, China. Science of Total Environment 383 (1–3), 164–173.
Yao, X.H., Chan, C.K., Fang, M., Cadle, S., Chan, T., Mulawa, P., He, K.B., Ye, B., 2002. The water-soluble ionic composition of $PM_{2.5}$ in Shanghai and Beijing, China. Atmospheric Environment 36 (26), 4223–4234.
Zhang, K., Wang, Y.S., Wen, T.X., Liu, G.R., Xu, H.H., 2007a. On-line analysis the water-soluble chemical of $PM_{2.5}$ in late summer and early autumn in Beijing. Acta Scientiae Circumstantiae 27 (3), 459–465.
Zhang, Q., Jimenez, J.L., Canagaratna, M.R., Allan, J.D., Coe, H., Ulbrich, I., Alfarra, M.R., Takami, A., Middlebrook, A.M., Sun, Y.L., Dzepina, K., Dunlea, E., Docherty, K., DeCarlo, P.F., Salcedo, D., Onasch, T., Jayne, J.T., Miyoshi, T., Shimono, A., Hatakeyama, S., Takegawa, N., Kondo, Y., Schneider, J., Drewnick, F., Borrmann, S., Weimer, S., Demerjian, K., Williams, P., Bower, K., Bahreini, R., Cottrell, L., Griffin, R.J., Rautiainen, J., Sun, J.Y., Zhang, Y.M., Worsnop, D.R., 2007b. Ubiquity and dominance of oxygenated species in organic aerosols in anthropogenically-influenced Northern Hemisphere midlatitudes. Geophysical Research Letters 34, L13801. doi:10.1029/2007GL029979.
Zhang, Q., Streets, D.G., Carmichael, G.R., He, K.B., Huo, H., Kannari, A., Klimont, Z., Park, I.S., Reddy, S., Fu, J.S., Chen, D., Duan, L., Lei, Y., Wang, L.T., Yao, Z.L., 2009. Asian emissions in 2006 for the NASA INTEX-B mission. Atmospheric Chemistry and Physics 9, 5131–5153.
Zhou, X.H., Cao, J., Wang, T., Wu, W.S., Wang, W.X., 2009. Measurement of black carbon aerosols near two Chinese megacities and the implications for improving emission inventories. Atmospheric Environment 43 (25), 3918–3924.
Zhou, Y., Wang, T., Gao, X.M., Xue, L.K., Wang, X.F., Wang, Z., Gao, J., Zhang, Q.Z., Wang, W.X., 2010. Continuous observations of water-soluble ions in $PM_{2.5}$ at Mount Tai (1534 ma.s.l.) in central-eastern China. Journal of Atmospheric Chemistry. doi:10.1007/s10874-010-9172-z.

Science of the Total Environment 493 (2014) 133–137

Contents lists available at ScienceDirect

Science of the Total Environment

journal homepage: www.elsevier.com/locate/scitotenv

Short Communication

Severe haze episodes and seriously polluted fog water in Ji'nan, China

Xinfeng Wang [a], Jianmin Chen [a,b,*], Jianfeng Sun [a], Weijun Li [a], Lingxiao Yang [a], Liang Wen [a], Wenxing Wang [a,*], Xinming Wang [c], Jeffrey L. Collett Jr. [d], Yang Shi [b], Qingzhu Zhang [a], Jingtian Hu [a], Lan Yao [a], Yanhong Zhu [a], Xiao Sui [a], Xiaomin Sun [a], Abdelwahid Mellouki [a,e]

[a] *Environment Research Institute, School of Environmental Science and Engineering, Shandong University, Ji'nan 250100, China*
[b] *Shanghai Key Laboratory of Atmospheric Particle Pollution and Prevention (LAP³), Fudan Tyndall Centre, Department of Environmental Science & Engineering, Fudan University, Shanghai 200433, China*
[c] *State Key Laboratory of Organic Geochemistry, Guangzhou Institute of Geochemistry, Chinese Academy of Sciences, Guangzhou 510640, China*
[d] *Department of Atmospheric Science, Colorado State University, Fort Collins, CO 80523, USA*
[e] *Institut de Combustion, Aérothermique, Réactivité et Environnement, CNRS, 45071 Orléans cedex 02, France*

HIGHLIGHTS

- Spatial distributions of $PM_{2.5}$ over China during server haze episodes were given.
- Fog water was seriously polluted due to longtime and large-scale haze pollution.
- Fog events reduced ambient air pollutants but difficult to cleanse the air.

ARTICLE INFO

Article history:
Received 17 January 2014
Received in revised form 14 April 2014
Accepted 29 May 2014
Available online 15 June 2014

Editor: Xuexi Tie

Keywords:
Regional haze
$PM_{2.5}$
Fog
Carbonaceous materials
Water-soluble ions
Eastern China

ABSTRACT

Haze episodes often hit urban cities in China recently. Here, we present several continuous haze episodes with extremely high $PM_{2.5}$ levels that occurred over several weeks in early 2013 and extended across most parts of the northern and eastern China—far exceeding the Beijing–Tianjin–Hebei region. Particularly, the haze episode covered ~1 million km^2 on January 14, 2013 and the daily averaged $PM_{2.5}$ concentration exceeded 360 μg m^{-3} in Ji'nan. The observed maximum hourly $PM_{2.5}$ concentration in urban Ji'nan reached 701 μg m^{-3} at 7:00 am (local time) in January 30. During these haze episodes, several fog events happened and the concurrent fog water was found to be seriously polluted. For the fog water collected in Ji'nan from 10:00 pm in January 14 to 11:00 am in January 15, sulfate, nitrate, and ammonium were the major ions with concentrations of 1.54×10^6, 8.98×10^5, and 1.75×10^6 μeq L^{-1}, respectively, leading to a low in-situ pH of 3.30. The sulfate content in the fog sample was more than 544 times as high as those observed in other areas. With examination of the simultaneously observed data on $PM_{2.5}$ and its chemical composition, the fog played a role in scavenging and removing fine particles from the atmosphere during haze episodes and thus was seriously contaminated. However, the effect was not sufficient to obviously cleanse air pollution and block haze episodes.

1. Introduction

Large quantities of pollutants have been emitted into the atmosphere along with rapid economic expansion in China (Tie et al., 2006). The annual average $PM_{2.5}$ concentration crept upward from approximately 60 μg m^{-3} in 1999–2000 to 90 μg m^{-3} in 2005–2006 in Shanghai (Ye et al., 2003; Feng et al., 2009). Carbonaceous materials and SO_4^{2-}, NO_3^-, and NH_4^+ are main components of $PM_{2.5}$ in haze episodes (Du et al., 2011; P. Li et al., 2011; W. Li et al., 2011; Yang et al., 2012; Sun et al., 2013). The concentrations of water-soluble ions in $PM_{2.5}$ during haze episodes can be 3.5 times higher than those observed in clear days (M. Zhang et al., 2013; Y. Zhang et al., 2013).

Carbonaceous materials and water-soluble ions in fog water are particularly interesting as they represent the result of hydrometeor interactions with ambient pollution (Herckes et al., 2006; P. Li et al., 2011; W. Li et al., 2011). From the 1980s to 1990s, fog events happened frequently in the Yangtze River Delta, the Sichuan basin, and the Gansu and Shanxi region (Wang et al., 2005; Niu et al., 2010). However, a significant decrease of fog days has been observed in the Gansu and Shanxi region (Wang et al., 2005) and precipitation at Mt. Hua in Shanxi province

* Corresponding authors. Tel.: +86 53188363711; fax: +86 531 88361990.
E-mail addresses: jmchen@sdu.edu.cn (J. Chen), wxwang@sdu.edu.cn (W. Wang).

http://dx.doi.org/10.1016/j.scitotenv.2014.05.135
0048-9697/

has decreased by 30% to 50% during hazy conditions (Rosenfeld et al., 2007). They suggest that the high level of fine particles in haze episodes plays an important role in effecting wet precipitation (Rosenfeld et al., 2008). While the content of pollutants in fog water is related to air quality due to the scavenging of fine particles and soluble gasses (P. Li et al., 2011; W. Li et al., 2011), little work has been reported on the fog chemistry and the capacity of fog for fine particle removal during severe haze episodes.

In January, 2013, extremely severe haze episodes appeared in northern and eastern China, which attracted broad attentions from both domestic and international communities, partly due to the serious public health damage (M. Zhang et al., 2013; Y. Zhang et al., 2013). To understand the pollution characteristics, meteorological conditions, sources, and formation mechanism of the extraordinary haze episodes, several studies have been made and published recently. From ground-based measurements, it was found that the most polluted areas were Beijing–Tianjin–Hebei region, west Shandong, and north Henan provinces, and atmospheric aerosols mostly concentrated in boundary layer below 1500 m (H. Wang et al., 2014). Secondary sulfate, nitrate and organic aerosols were the major components of $PM_{2.5}$ during the haze episodes, accounting for 65.7% of the $PM_{2.5}$ in urban Shanghai (Zhou et al., 2013). Secondary aerosols mostly formed via heterogeneous reactions on particle surfaces, which changed the particle size, hygroscopicity and optical properties, causing large negative aerosol radiative forcing efficiency at surface and accelerating the formation of haze episodes (Che et al., 2014; Y. Wang et al., 2014). Modeling studies indicate the important role of pollutant transport to the regional haze (Z. Wang et al., 2014), and identify the major emission sources from industry, domesticity, and agriculture (Wang et al., 2013). Besides emissions and transformations of air pollutants, the unusual meteorological conditions, i.e., weak southerly winds in the middle and low troposphere, high pressure at 500 hPa, and inversion in near surface, were also responsible to the severe haze episodes (Y. Wang et al., 2014; Zhang et al., 2014).

During the severe haze episodes in January, 2013, several heavy fog events simultaneously happened within this region, which provide a unique opportunity to investigate the pollution characteristics of the fog and the capacity to remove the very high concentrations of fine particles.

In this study, we use the Geographic Information System (GIS) to reconstruct the severe haze episodes that occurred over a large region and an extended time period in January, 2013 in northern and eastern China. In the urban area of one of the most polluted cities, Ji'nan in Shandong province, on-line measurements of trace gases, $PM_{2.5}$ and the aerosol components were conducted during this period and the fog samples were collected. The pollutant levels in the seriously polluted fog water were analyzed and compared with those in other locations. The influence of fog on cleansing the atmosphere during the severer haze episodes was also investigated.

2. Materials and methods

2.1. Inversion of $PM_{2.5}$ concentrations

The $PM_{2.5}$ levels in provinces or cities in China shown in this study were mostly derived from the air pollution index (API) records from air quality daily reports published by the China National Environmental Monitoring Center (a small fraction from local air quality daily reports). Firstly, we calculated the daily PM_{10} concentrations from the API records by using the following equation: $C = [(I - I_{low})/(I_{high} - I_{low})] \times (C_{high} - C_{low}) + C_{low}$ based on the classification of API described by Bian et al. (2011). Here, C is the PM_{10} concentration; I is the API. I_{low} and I_{high} stand for API grading limits that are lower and larger than I, respectively; C_{high} and C_{low} represent the concentrations of PM_{10} corresponding to I_{high} and I_{low}, respectively. This derivation is a common method to obtain PM_{10} concentrations for a large number of cities in China (Wang et al., 2006; Qu et al., 2010). The daily $PM_{2.5}$ levels were then estimated from the obtained PM_{10} data with assumption of a $PM_{2.5}/PM_{10}$ ratio of 0.6 based on previous studies in China (Cao et al., 2003; Sun et al., 2006; Fu et al., 2008; Gu et al., 2010; Kong et al., 2010). Note that the $PM_{2.5}/PM_{10}$ ratio of 0.6 used here was almost the lowest limit. The $PM_{2.5}$ levels in the provincial capital or the neighboring city were used to represent the status of fine particles pollution over the whole province. Although uncertainty existed in the representativeness of $PM_{2.5}$ levels by API data due to the difference in $PM_{2.5}/PM_{10}$ ratio among cities, the derived $PM_{2.5}$ data were a good indication of the pollution degree and the concentration variation patterns on a large temporal or spatial scale.

2.2. Collection and chemical analyses of fog samples

Fog samples were collected using a CASCC2 fog/cloud collector (Demoz et al., 1996) with a lower droplet size cut of 3.5 μm on the rooftop of a three-story building on the campus of Shandong University in Ji'nan (36.67° N, 117.05° E). The pH and electrical conductivity were determined immediately after the sampling stopped. OC and EC concentrations in the fog water were measured using an OC/EC analyzer (Sunset Lab) after a 1:100 dilution, by applying 10 μL diluted sample on the quartz filter suspended in the quartz insert of the analyzer. The water-soluble ions were analyzed using ion chromatograph (Dionex, ICS-90) after a 1:100,000 dilution and subsequent filtration.

2.3. On-line measurements of air pollutants and meteorological parameters

Ambient concentrations of air pollutants and meteorological parameters were concurrently measured in real-time from January 18 to February 1, 2013 at the site on the campus. $PM_{2.5}$ levels were measured using a Beta attenuation and optical analyzer (Thermo Scientific, model 5030 SHARP monitor). Nine inorganic water-soluble ions in $PM_{2.5}$ were analyzed using two on-line ion chromatographs coupling with a wet rotating denuder and a steam jet aerosol collector (Applikon-ECN, MARGA ADI 2080). OC and EC in $PM_{2.5}$ were analyzed using a semi-continuous thermo-optical OC/EC analyzer (Sunset Lab). SO_2 was measured using a pulsed UV fluorescence analyzer (Thermo Scientific, model 43C), and O_3 was measured using a UV photometric analyzer (Thermo Scientific, model 49C). NO and NO_2 were measured by a commercial chemiluminescence analyzer equipped with a molybdenum oxide converter (Thermo Scientific, model 42C). The relative humidity was measured using an automatic meteorological station (JZYG, PC-4), and visibility was measured using a forward-scattering visibility sensor (Vaisala, PWD22).

3. Results and discussion

3.1. $PM_{2.5}$ distribution in eastern China

Based on the API records from air quality daily reports of 24 provinces or cities, the derived 24-h average concentration of $PM_{2.5}$ in Ji'nan exceeded 360 $\mu g\ m^{-3}$ on January 14, 2013 (see Supplementary Material Table 1), 4.8 times higher than the 24-h average concentration limit of $PM_{2.5}$ (75 $\mu g\ m^{-3}$) of the Ambient Air Quality Standards of Class II of China. As shown in Fig. 1, the $PM_{2.5}$ concentration above 200 $\mu g\ m^{-3}$ covered a large region in north and middle eastern China with an area of ~1 million km^2 on January 14. The super-regional fine particle pollution was also indicated by the Moderate Resolution Imaging Spectroradiometer true color images (see Supplementary Material Fig. 1), extending to East China Sea in the east, Mt. Hua in the west, Yangzte River in the south, and Heilungkiang River in the north. Based on the spatial distribution of $PM_{2.5}$, the area suffering from haze episode far exceeds the Beijing–Tianjin–Hebei region and the North China Plain reported by Che et al. (2014) and Wang et al. (2014a), also including the Northeast Plain, the Yangtze River

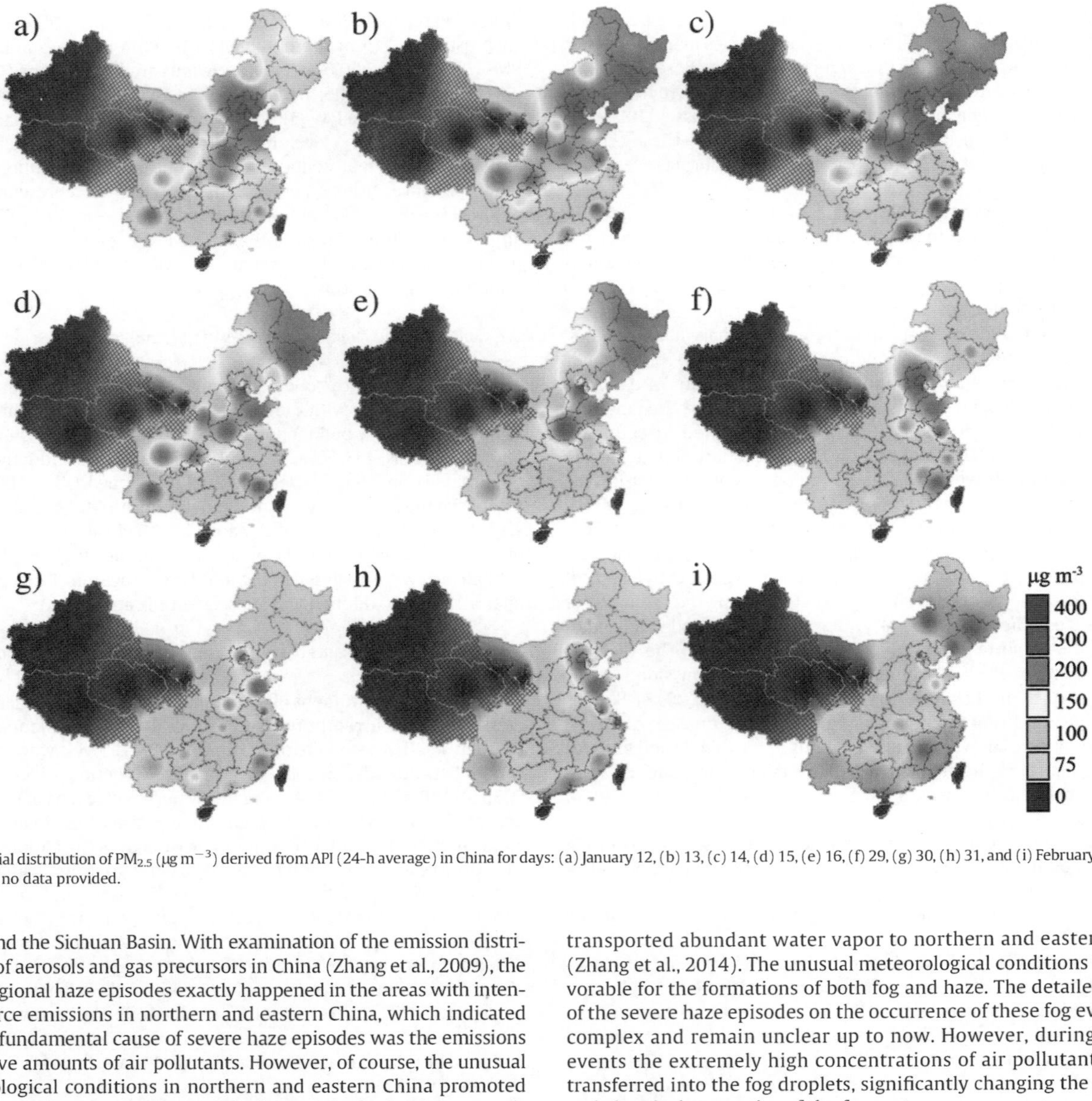

Fig. 1. Spatial distribution of $PM_{2.5}$ (μg m^{-3}) derived from API (24-h average) in China for days: (a) January 12, (b) 13, (c) 14, (d) 15, (e) 16, (f) 29, (g) 30, (h) 31, and (i) February 1. Grids in blue mean no data provided.

valley, and the Sichuan Basin. With examination of the emission distributions of aerosols and gas precursors in China (Zhang et al., 2009), the super-regional haze episodes exactly happened in the areas with intensive source emissions in northern and eastern China, which indicated that the fundamental cause of severe haze episodes was the emissions of massive amounts of air pollutants. However, of course, the unusual meteorological conditions in northern and eastern China promoted the accumulation of pollutions and the formation of secondary aerosols (Y. Wang et al., 2014; Zhang et al., 2014).

During the severe haze episodes in northern and eastern China in January, 2013, Ji'nan was among the top ten most polluted cities. The 24-h average concentration of $PM_{2.5}$ in Ji'nan was never less than 75 μg m^{-3} for a period of 28 days from January 5 to February 1, much longer than the average duration of 5 days for the "sawtooth circle" of fine particle concentration described for Beijing (Jia et al., 2008). The most polluted city was Shijiazhuang, the capital of Hebei province. The number of days with 24-h average $PM_{2.5}$ concentrations no less than 360 μg m^{-3} was 10 between January 5 and 29.

3.2. Carbonaceous materials and water-soluble ions in fog water

During the haze episodes, dense fog events occurred in Ji'nan and neighboring areas in the early morning of January 15, from the midday of January 29 to the morning of January 30, and from the night of January 30 to the morning of February 1. The occurrence of the fog events was mainly attributed to the weakened southerly surface winds and the reduced vertical shear of horizontal winds. The abnormally warm and humid air flow moved slow from the south and transported abundant water vapor to northern and eastern China (Zhang et al., 2014). The unusual meteorological conditions were favorable for the formations of both fog and haze. The detailed effects of the severe haze episodes on the occurrence of these fog events ere complex and remain unclear up to now. However, during the fog events the extremely high concentrations of air pollutants partly transferred into the fog droplets, significantly changing the physical and chemical properties of the fog water.

Fog water samples collected in Ji'nan were charcoal grey in color, indicating very high concentrations of carbonaceous materials. As shown in Fig. 2, elemental carbon (EC) concentrations were in the range of 0.8×10^3–1.25×10^4 mg L^{-1}. Organic carbon (OC) concentrations varied from 3.5×10^3 to 3.25×10^4 mg L^{-1}. The fog samples were also acidic, with in-situ pH values ranging from an extremely acidic value of 2.62 up to 4.20. The fog samples contained extremely high levels of water-soluble ions, with electrical conductivity values all above 2000 μS cm^{-1} (the upper limit of the Conductivity Meter). SO_4^{2-}, NO_3^-, and NH_4^+ were the most abundant water-soluble ions, with concentrations of 0.6×10^5–1.54×10^6, 0.4×10^5–8.98×10^5, and 0.9×10^5–1.75×10^6 μeq L^{-1} (μN), respectively. The contents of Cl^- and Ca^{2+} were also very rich, which exhibited maximum concentrations of 4.35×10^5 and 4.28×10^5 μN, respectively. Compared to fog collected at other sites in the world (see Supplementary Material Table 2), the major water-soluble ions in the Ji'nan fog samples are much higher, more than 22–544 times for SO_4^{2-}, 16–371 for NO_3^-, 23–437 for NH_4^+, 18–369 for Cl^-, and 10–207 for Ca^{2+}. The Cl^-/Na^+ ratios of collected fog samples were in the range of 3.4–11.8, suggesting that the pollutants in the fog are strongly influenced by

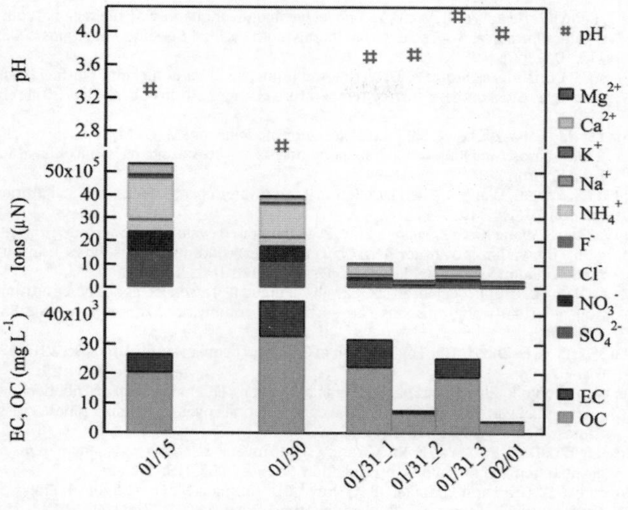

Fig. 2. The pH, water-soluble ions, EC and OC concentrations of fog water samples.

coal combustion in the local area and across the North China Plain (Ye et al., 2003; Mira-Salama et al., 2008).

3.3. Pollution tide during fog events

Pollutant timelines during the Ji'nan fog events (see Fig. 3) illustrated the positive effects that fog can have on air quality. The most impressive effect appeared during a fog event at the end of January. Beginning at 7:00 pm in the night of January 30 and extending until 8:00 am in the morning of February 1, this extended fog event was accompanied by substantial reductions, as shown in Fig. 3, in SO_2 and NO_x, and fine particles. CO mixing ratios also decreased obviously during the event; however, the reduction lagged when compared with those of SO_2, NO_x, and $PM_{2.5}$. The hourly $PM_{2.5}$ level (the $PM_{2.5}$ data recorded at the site of Shandong University in Ji'nan) was as high as 701 μg m^{-3} immediately before the fog event on January 31, and dropped to 115 μg m^{-3} in 36 h after the fog event. During the Ji'nan fog event, the percentage reduction of fine particle SO_4^{2-} was 90%. Reductions of NO_3^-, OC, and EC were 81%, 76%, and 71%, respectively. The overall $PM_{2.5}$ reduction was 84%, significantly higher than that of CO (~70%, as a benchmark to a certain degree considering the changes in air mass and mixing layer). This reduction indicated that during severe haze episodes fog had capacity to scavenge fine particles followed by wet deposition to the surface. While fog deposition is not characterized here, others (Collett et al., 2001) have demonstrated the efficient scavenging and deposition of inorganic pollutants in radiation fog in California. During the fog event from the midday of January 29 to the morning of January 30, apparent reductions in fine sulfate and SO_2 concentrations were also observed. We believe that the extremely high concentrations of carbonaceous materials and water-soluble ions in fog water came from the fog scavenging of fine particles and possibly some gas precursors. The role of fog in removing fine particles can also be seen in our previous studies in North China (Zhou et al., 2009; Wang et al., 2012).

Despite the apparent cleaning effect on the atmosphere by fog scavenging and deposition, the $PM_{2.5}$ level never fell to 75 μg m^{-3} in the fog events in this study. $PM_{2.5}$ concentration sharply increased from 115 to 195 μg m^{-3} within 4 h after the fog event in the morning of February 1. These results showed that air pollutants were efficiently removed by fog, but it could hardly clean up air pollutants down to good air quality during such severe haze episodes.

4. Summary and conclusions

In summary, from January 1 to February 1, 2013, north and middle areas of eastern China experienced continuous severe fine particle pollution. The super-regional haze episodes covered an area of ~1 million km^2,

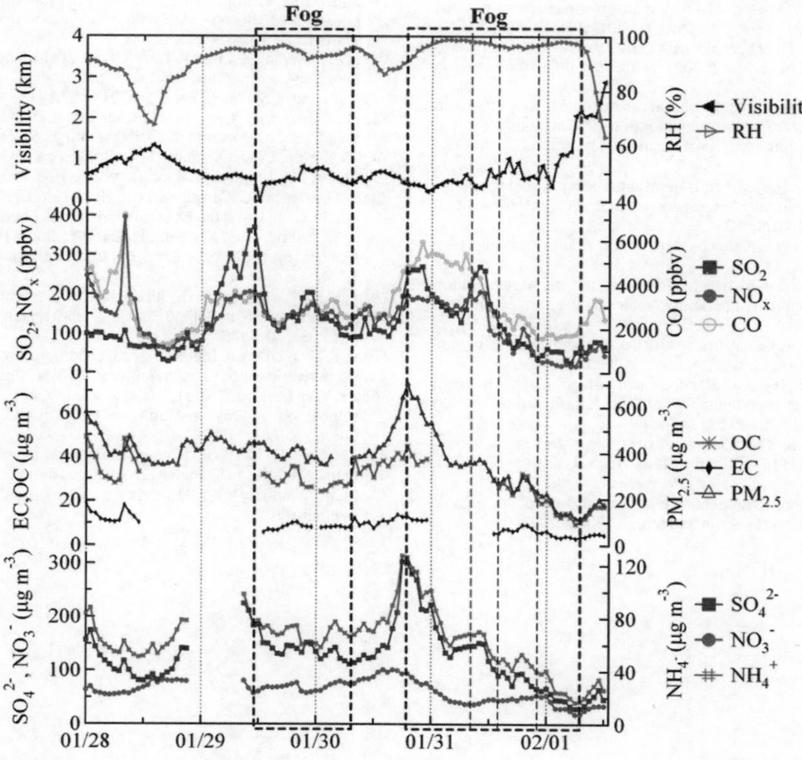

Fig. 3. Hourly average concentrations of $PM_{2.5}$, OC, EC, SO_4^{2-}, NO_3^-, NH_4^+, SO_2, NO_x, CO, visibility, and RH during fog events.

far beyond the Beijing–Tianjin–Hebei region. In particular on January 14, the daily average concentrations of $PM_{2.5}$ in most of cities in this region were above 200 $\mu g\ m^{-3}$. The severe haze episodes happened in China in early 2013 sound the alarm again to the government and researchers. They emphasize the urgent need to find approaches to control $PM_{2.5}$. During the severe haze episodes, several fog events occurred. The fog scavenged a large proportion of the fine particles in atmosphere and thus contained extremely high levels of carbonaceous materials and water-soluble ions. Nevertheless, it was difficult for fog to clean up fine particles to eliminate the regional haze episodes. The incorporation of pollutants in fog is also a concern as the deposited pollutants may damage ecosystems and the capture of light absorbing black carbon in the regional fog could significantly alter the optical properties and thus impact climate change (Jones et al., 2011; Allen et al., 2012). Further studies are required to give a comprehensive evaluation.

Acknowledgments

This work was supported by Taishan Scholar Grant (ts20120552), the National Natural Science Foundation of China (Nos. 21190053, 41375126, 21177025), the Shanghai Science and Technology Commission of Shanghai Municipality (Nos. 13XD1400700, 12DJ1400100), the Priority fields for Ph.D. Programs Foundation of Ministry of Education of China (No. 20110071130003), the Strategic Priority Research Program of the Chinese Academy of Sciences (No. XDB05010200), and FP7 project (AMIS, no. 069720).

Appendix A. Supplementary data

Supplementary data to this article can be found online at http://dx.doi.org/10.1016/j.scitotenv.2014.05.135.

References

Allen RJ, Sherwood SC, Norris JR, Zender CS. Recent Northern Hemisphere tropical expansion primarily driven by black carbon and tropospheric ozone. Nature 2012;485:350–4.

Bian H, Tie X, Cao J, Ying Z, Han S, Xue Y. Analysis of a severe dust storm event over China: application of the WRF-Dust Model. Aerosol Air Qual Res 2011;11:419–28.

Cao JJ, Lee SC, Ho KF, Zhang XY, Zou SC, Fung K, et al. Characteristics of carbonaceous aerosol in Pearl River Delta Region, China during 2001 winter period. Atmos Environ 2003;37:1451–60.

Che H, Xia X, Zhu J, Li Z, Dubovik O, Holben B, et al. Column aerosol optical properties and aerosol radiative forcing during a serious haze-fog month over North China Plain in 2013 based on ground-based sunphotometer measurements. Atmos Chem Phys 2014;14:2125–38.

Collett Jr J, Sherman DE, Moore K, Hannigan M, Lee T. Aerosol particle processing and removal by fogs: observations in chemically heterogeneous central California radiation fogs. Water Air Soil Pollut Focus 2001;1:303–12.

Demoz BB, Collett Jr JL, Daube Jr BC. On the Caltech active strand cloudwater collectors. Atmos Res 1996;41:47–62.

Du H, Kong L, Cheng T, Chen J, Du J, Li L, et al. Insights into summertime haze pollution events over Shanghai based on online water-soluble ionic composition of aerosols. Atmos Environ 2011;45:5131–7.

Feng Y, Chen Y, Guo H, Zhi G, Xiong S, Li J, et al. Characteristics of organic and elemental carbon in $PM_{2.5}$ samples in Shanghai, China. Atmos Res 2009;92:434–42.

Fu Q, Zhuang G, Wang J, Xu C, Huang K, Li J, et al. Mechanism of formation of the heaviest pollution episode ever recorded in the Yangtze River Delta, China. Atmos Environ 2008;42:2023–36.

Gu J, Bai Z, Liu A, Wu L, Xie Y, Li W, et al. Characterization of atmospheric organic carbon and element carbon of $PM_{2.5}$ and PM_{10} at Tianjin, China. Aerosol Air Qual Res 2010; 10:167–76.

Herckes P, Leenheer JA, Collett JL. Comprehensive characterization of atmospheric organic matter in Fresno, California fog water. Environ Sci Technol 2006;41:393–9.

Jia Y, Rahn KA, He K, Wen T, Wang Y. A novel technique for quantifying the regional component of urban aerosol solely from its sawtooth cycles. J Geophys Res Atmos 2008; 113. [D21309].

Jones GS, Christidis N, Stott PA. Detecting the influence of fossil fuel and bio-fuel black carbon aerosols on near surface temperature changes. Atmos Chem Phys 2011;11: 799–816.

Kong S, Han B, Bai Z, Chen L, Shi J, Xu Z. Receptor modeling of $PM_{2.5}$, PM_{10} and TSP in different seasons and long-range transport analysis at a coastal site of Tianjin, China. Sci Total Environ 2010;408:4681–94.

Li P, Li X, Yang C, Wang X, Chen J, Collett Jr JL. Fog water chemistry in Shanghai. Atmos Environ 2011;45:4034–41.

Li W, Zhou S, Wang X, Xu Z, Yuan C, Yu Y, et al. Integrated evaluation of aerosols from regional brown hazes over northern China in winter: concentrations, sources, transformation, and mixing states. J Geophys Res 2011;116. [D09301].

Mira-Salama D, Grüning C, Jensen NR, Cavalli P, Putaud JP, Larsen BR, et al. Source attribution of urban smog episodes caused by coal combustion. Atmos Res 2008;88: 294–304.

Niu S, Lu C, Yu H, Zhao L, Lü J. Fog research in China: an overview. Adv Atmos Sci 2010;27: 639–62.

Qu WJ, Arimoto R, Zhang XY, Zhao CH, Wang YQ, Sheng LF, et al. Spatial distribution and interannual variation of surface PM_{10} concentrations over eighty-six Chinese cities. Atmos Chem Phys 2010;10:5641–62.

Rosenfeld D, Dai J, Yu X, Yao Z, Xu X, Yang X, et al. Inverse relations between amounts of air pollution and orographic precipitation. Science 2007;315:1396–8.

Rosenfeld D, Lohmann U, Raga GB, O'Dowd CD, Kulmala M, Fuzzi S, et al. Flood or drought: how do aerosols affect precipitation? Science 2008;321:1309–13.

Sun Y, Zhuang G, Tang A, Wang Y, An Z. Chemical characteristics of $PM_{2.5}$ and PM_{10} in haze–fog episodes in Beijing. Environ Sci Technol 2006;40:3148–55.

Sun Z, Mu Y, Liu Y, Shao L. A comparison study on airborne particles during haze days and non-haze days in Beijing. Sci Total Environ 2013;456–457:1–8.

Tie X, Brasseur GP, Zhao C, Granier C, Massie S, Qin Y, et al. Chemical characterization of air pollution in Eastern China and the Eastern United States. Atmos Environ 2006; 40:2607–25.

Wang L, Chen S, Dong A. The distribution and seasonal variations of fog in China. Acta Geograph Sin 2005;60:134–9.

Wang S, Yuan W, Shang K. The impacts of different kinds of dust events on PM10 pollution in northern China. Atmos Environ 2006;40:7975–82.

Wang Z, Wang T, Guo J, Gao R, Xue L, Zhang J, et al. Formation of secondary organic carbon and cloud impact on carbonaceous aerosols at Mount Tai, north China. Atmos Environ 2012;46:516–27.

Wang LT, Wei Z, Yang J, Zhang Y, Zhang FF, Su J, et al. The 2013 severe haze over the southern Hebei, China: model evaluation, source apportionment, and policy implications. Atmos Chem Phys Discuss 2013;13:28395–451.

Wang H, Tan S-C, Wang Y, Jiang C, Shi G-Y, Zhang M-X, et al. A multisource observation study of the severe prolonged regional haze episode over Eastern China in January 2013. Atmos Environ 2014a;89:807–15.

Wang Y, Yao L, Wang L, Liu Z, Ji D, Tang G, et al. Mechanism for the formation of the January 2013 heavy haze pollution episode over central and eastern China. Sci China Earth Sci 2014b;57:14–25.

Wang Z, Li J, Wang Z, Yang W, Tang X, Ge B, et al. Modeling study of regional severe hazes over mid-eastern China in January 2013 and its implications on pollution prevention and control. Sci China Earth Sci 2014c;57:3–13.

Yang F, Chen H, Du J, Yang X, Gao S, Chen J, et al. Evolution of the mixing state of fine aerosols during haze events in Shanghai. Atmos Res 2012;104–105:193–201.

Ye B, Ji X, Yang H, Yao X, Chan CK, Cadle SH, et al. Concentration and chemical composition of $PM_{2.5}$ in Shanghai for a 1-year period. Atmos Environ 2003;37:499–510.

Zhang Q, Streets DG, Carmichael GR, He K, Huo H, Kannari A, et al. Asian emissions in 2006 for the NASA INTEX-B mission. Atmos Chem Phys 2009;9:5131–53.

Zhang M, Chen J, Chen X, Cheng T, Zhang Y, Zhang H, et al. Urban aerosol characteristics during the World Expo 2010 in Shanghai. Aerosol Air Qual Res 2013;13: 36–48.

Zhang Y, Ma G, Yu F, Cao D. Health damage assessment due to $PM_{2.5}$ exposure during haze pollution events in Beijing–Tianjin–Hebei region in January 2013. Natl Med J China 2013;93:2707–10.

Zhang R, Li Q, Zhang R. Meteorological conditions for the persistent severe fog and haze event over eastern China in January 2013. Sci China Earth Sci 2014;57:26–35.

Zhou Y, Wang T, Gao X, Xue L, Wang X, Wang Z, et al. Continuous observations of water-soluble ions in $PM_{2.5}$ at Mount Tai (1534 m a.s.l.) in central-eastern China. J Atmos Chem 2009;64:107–27.

Zhou M, Chen C, Qiao L, Lou S, Wang H, Huang H, et al. The chemical characteristics of particulate matters in Shanghai during heavy air pollution episode in Central and Eastern China in January 2013. Acta Sci Circumst 2013;33:3118–26. [in Chinese with Abstract in English].

Science of the Total Environment 541 (2016) 590–598

Contents lists available at ScienceDirect

Science of the Total Environment

journal homepage: www.elsevier.com/locate/scitotenv

Sources apportionment of $PM_{2.5}$ in a background site in the North China Plain

Lan Yao [a], Lingxiao Yang [a,b,*], Qi Yuan [a], Chao Yan [a], Can Dong [a], Chuanping Meng [a], Xiao Sui [a], Fei Yang [a], Yaling Lu [a], Wenxing Wang [a,c]

[a] *Environment Research Institute, Shandong University, Jinan 250100, China*
[b] *School of Environmental Science and Engineering, Shandong University, Jinan 250100, China*
[c] *Chinese Research Academy of Environmental Sciences, Beijing 100012, China*

HIGHLIGHTS

- The Yellow River Delta National Nature Reserve (YRDNNR) suffered serious air pollution in 2011.
- $PM_{2.5}$ at YRDNNR was characterized by inorganic and organic fraction when air mass originated from the south and north direction, respectively.
- Sources and likely source areas of $PM_{2.5}$ at YRDNNR were first identified.

GRAPHICAL ABSTRACT

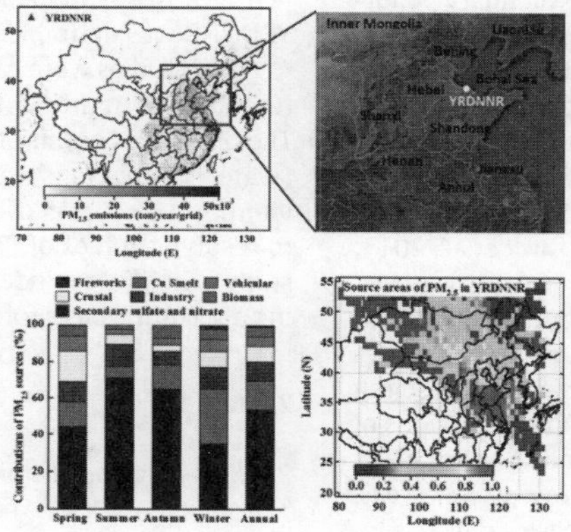

ARTICLE INFO

Article history:
Received 1 July 2015
Received in revised form 22 September 2015
Accepted 24 September 2015
Available online 1 October 2015

Editor: D. Barcelo

Keywords:
$PM_{2.5}$
Source apportionment
PMF
PSCF

ABSTRACT

To better understand the sources and potential source regions of $PM_{2.5}$, a field study was conducted from January 2011 to November 2011 at a background site, the Yellow River Delta National Nature Reserve (YRDNNR) in the North China Plain. Positive matrix factorisation (PMF) analysis and a potential source contribution function (PSCF) model were used to assess the data, which showed that YRDNNR experienced serious air pollution. Concentrations of $PM_{2.5}$ at YRDNNR were 71.2, 92.7, 97.1 and 62.5 μg m^{-3} in spring, summer, autumn and winter, respectively, with 66.0% of the daily samples exhibiting higher concentrations of $PM_{2.5}$ than the national air quality standard. $PM_{2.5}$ mass closure showed remarkable seasonal variations. Sulphate, nitrate and ammonium were the dominant fractions of $PM_{2.5}$ in summer (58.0%), whereas $PM_{2.5}$ was characterized by a high load of organic aerosols (40.2%) in winter. PMF analysis indicated that secondary sulphate and nitrate (54.3%), biomass burning (15.8%), industry (10.7%), crustal matter (8.3%), vehicles (5.2%) and copper smelting (4.9%) were important sources of $PM_{2.5}$ at YRDNNR on an annual average. The source of secondary sulphate and nitrate was probably industrial coal combustion. PSCF analysis indicated a significant regional impact on $PM_{2.5}$ at YRDNNR all year

* Corresponding author.
E-mail address: yanglingxiao@sdu.edu.cn (L. Yang).

http://dx.doi.org/10.1016/j.scitotenv.2015.09.123
0048-9697/

round. Local emission may be non-negligible at YRDNNR in summer. The results of the present study provide a scientific basis for the development of $PM_{2.5}$ control strategies on a regional scale.

1. Introduction

Regional haze has frequently occurred over China and attracted worldwide attention in recent years (Guo et al., 2014; Han et al., 2014; Huang et al., 2014; Wang et al., 2014a; Zhao et al., 2013). Haze pollution has a serious impact on health and leads to huge economic loss. For example, it was reported that substantial health effects, including 201 cases of premature death, 10,132 cases of acute bronchitis and 7643 cases of asthma, accompanied by health-related economic losses of 489 million China Yuan, were caused by a six-day haze pollution in Beijing (Xie et al., 2014). The direct cost to health and transportation attributed to the heavy haze in China during January 2013 was at least 23 billion China Yuan (Mu and Zhang, 2013). Previous studies have found that $PM_{2.5}$ is the primary pollutant causing haze, especially secondary constituents of $PM_{2.5}$, which account for 30–77% of $PM_{2.5}$ (Huang et al., 2014). $PM_{2.5}$ control strategies to mitigate haze pollution are urgently needed. Establishing the sources of $PM_{2.5}$ is the first step. The composition of $PM_{2.5}$ is complex, and includes components from various sources including industry, vehicle emissions, residences, biomass burning and other human activities. The formation of secondary pollutants by the atmospheric oxidation of primary pollutants (SO_2, NOx and VOCs) is an important source of $PM_{2.5}$. Positive matrix factorisation (PMF) is a powerful tool for source apportionment and potential source contribution function (PSCF) analysis is able to identify possible source areas by combining aerosol data with air mass backward trajectories. PMF and PSCF have successfully identified the sources of aerosols in many areas (Aldabe et al., 2011; Cesari et al., 2014; Cusack et al., 2013; Heo et al., 2009).

The North China Plain, with a regional GDP contribution of 23.2% in 2013 (China Statistical Yearbook), has experienced rapid economic development, population expansion and urbanization in the past ten years, and suffered serious air pollution at the same time. Emissions of $PM_{2.5}$ in the North China Plain are much higher than in other areas (Fig. 1), which may aggravate the air pollution of downwind areas. The Yellow River Delta National Nature Reserve (YRDNNR), which is located in the estuary of the Yellow River into the Bohai Sea (Fig. 1), occupies an important geographical location. As a typical wetland, YRDNNR was established to protect rare species, and with an area of 153,000 ha, it has become a significant over-wintering and breeding station for migrant birds of the Northeast Asian Inland and Western Pacific Rim (Cui et al., 2009). Despite its environmental and ecological functions, YRDNNR is in the downwind area of the North China Plain and may suffer serious air pollution. The prevailing winds at YRDNNR are south-southeast winds all year round. In summer, south-southeast winds are dominated winds, whereas north-northwest winds are prevailing at YRDNNR in winter (Yuan and Peng, 1998). Its particular geographic location makes it an ideal background site for the study of the regional transport of atmospheric pollutants, and local emissions. However, few studies concerning aerosol pollution have been conducted in this area until recently. We previously investigated inorganic and organic components of $PM_{2.5}$ at YRDNNR, including water soluble ions, polycyclic aromatic hydrocarbons (PAH), new particle formation and carbonaceous aerosols (Sui et al., 2015; Yuan et al., 2014a, 2014b; Zhu et al., 2014). Despite these valuable reports, the sources of $PM_{2.5}$ and their contributions were poorly understood in this area. Thus, a comprehensive investigation focussing on the $PM_{2.5}$ components and sources at YRDNNR was desirable.

Field studies were carried out at the Yellow River Delta National Nature Reserve in the North China Plain from January to November 2011. The objectives of this study were to determine the $PM_{2.5}$ mass closure (water soluble ions, carbonaceous fraction and trace elements) and identify the sources of $PM_{2.5}$ at YRDNNR. PMF and PSCF were applied to resolve sources of $PM_{2.5}$ and their likely source areas. The current study is helpful to understand the sources of $PM_{2.5}$ at YRDNNR, providing a scientific basis for $PM_{2.5}$ control policies.

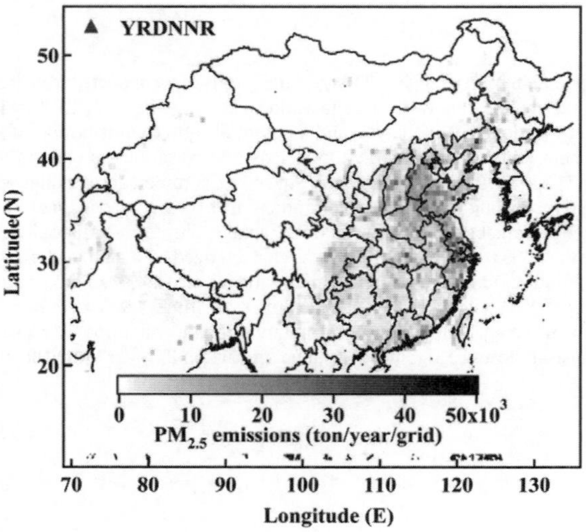

Fig. 1. Emissions of $PM_{2.5}$ over China (Zhang et al., 2009). The study site is marked in blue.

2. Method

2.1. Sampling and chemical analysis

Field research was conducted at YRDNNR (38°03′N, 118°44′E, Fig. 1) in Dongying, Shandong Province, from 14 January to 14 February (winter), April 1 to 3 May (spring), 4 July to 29 July (summer) and 11 October to 7 November (autumn), 2011. There are few nearby stationary sources of $PM_{2.5}$. The sampling site is ~20 m above the ground, on the roof of the manager's station building. $PM_{2.5}$ sampler (Thermo Electron Corporation, Model RAAS2.5-400) with four parallel channels was employed to collect daily $PM_{2.5}$ samples (from 8:30 a.m. to 8:00 a.m.) at the flow rate of 16.7 L min^{-1}. One channel equipped with quartz filter (1 μm pore size and 47 mm diameter, Pall Gelman Inc.) was used to collect $PM_{2.5}$ samples for analysis of water soluble ions and the carbonaceous fraction (organic carbon and elemental carbon). One channel installed with Teflon filter (1 μm pore size and 47 mm diameter, Pall Gelman Inc.) was used to collect trace elements for analysis. One set of field blank samples were taken in each season. Prior to sampling, quartz filters were heated at 600 °C for 2 h to remove organic species, and Teflon filters were heated at 60 °C. Teflon filters were weighted (Sartorius ME-5F, readability: 1 μg) before and after sampling at the temperature of 20 ± 1 °C and relative humidity of 50% ± 2% to determine $PM_{2.5}$ mass concentrations. The detailed site description, samplers used and filter pre-treatment method can be found in our previous studies (Yang et al., 2012; Yuan et al., 2014b). A total of 106 sets of valid samples were taken.

Chemical compositions were analysed with multiple techniques. Water soluble ions (Cl^-, SO_4^{2-}, NO_3^-, Na^+, NH_4^+, K^+, Mg^{2+}, and Ca^{2+}) were analysed by ion chromatography (Dionex-2500, USA). Organic carbon (OC) and elemental carbon (EC) were measured using IMPROVE thermal optical reflectance (TOR) (DRI, USA). Twenty-four trace elements were analysed with X-ray fluorescence (XRF). Chemical constitutes measurements of $PM_{2.5}$ samples were corrected for their field blanks. These instruments have been widely used and the principles have been previously described (Sui et al., 2015; Yang et al., 2012).

2.2. Positive matrix factorisation (PMF)

PMF is a convenient and helpful multivariate factor analysis tool. It has advantages over principal component analysis (PCA) and chemical mass balance (CMB). Negative factor loadings are usually obtained with PCA, and with CMB, the user must provide source profiles that the model uses to apportion mass. PMF is based on non-negative constraints to obtain more realistic factors. It is able to deal with missing and below-detection-limit data, quantifying the contribution of sources to samples based on the composition or 'fingerprints' of the sources. PMF 5.0 was used to identify sources of $PM_{2.5}$ due to our large database (at least 100 samples is suggested by the PMF 5.0 User Guide, US EPA, 2014) and wide-ranging chemical compositions. Two data files are needed to initialise PMF: one containing concentration values and one containing uncertainty values for each species. Uncertainty was determined as follows (Polissar et al., 1998):

$$\text{If MDL} < C_i \leq 3\text{MDL}, \text{Unc} = \frac{1}{3} \times \text{MDL} + 0.2 \times C_i \tag{1}$$

$$\text{If } C_i \geq 3\text{MDL}, \quad \text{Unc} = \frac{1}{3} \times \text{MDL} + 0.1 \times C_i. \tag{2}$$

Values below the method detection limit (MDL) were replaced by half of the MDL and the corresponding uncertainty is calculated as 5/6 of the MDL. Missing values were replaced by the geometric mean of the observed values, and associated uncertainties were set at four times the geometric mean (Heo et al., 2009). The principals of PMF have been described in the PMF 5.0 User Guide and in our previous study (Yang et al., 2013). In this study, V, Cr, Co, Br and Cd were set as weak variables. Arsenic (As) was set as a bad variable because 90% of the samples of $PM_{2.5}$-associated As were below MDL. Strong correlation was observed between SO_4^{2-} and S ($r^2 = 0.70$) and only SO_4^{2-} was included in the analysis. Ultimately, 29 chemical species were used for PMF analysis and seven factors were resolved at YRDNNR. The Q_{Robust} (2468) was close to the theoretical value of Q (2129), which was calculated by 106 × 29–7 × (106 + 29) in this study.

2.3. Potential source contribution function (PSCF)

PSCF calculates the probability that a source is located at latitude *i* and longitude *j*. The PSCF values for the grid cells are calculated by counting the trajectory segment endpoints that terminate within each cell by the following equation:

$$PSCF_{ij} = \frac{m_{ij}}{n_{ij}} \tag{3}$$

where n_{ij} is the number of times that the trajectories passed through the cell (*i*, *j*) and m_{ij} is the number of times that a source concentration was higher than a criterion value when the trajectories passed through the cell (*i*, *j*). In this study, the criterion value was set to the 75th percentile concentration of a source of $PM_{2.5}$. The Hybrid Single-Particle Lagrangian Integrated Trajectory (HYSPLIT) model was used to calculate air mass back trajectory. Three-day back trajectories were calculated every 6 h each day, with a starting height of 50 m. The trajectory domain extended from 80° E to 130° E and from 15° N to 65° N, comprising 2500 cells 1 × 1° in latitude and longitude, with 12 endpoints per cell on average. To reduce the effect of small values of n_{ij}, the PSCF values were multiplied by a weighting function W_{ij}:

$$W_{ij} = \begin{Bmatrix} 1.0 & 36 < n_{ij} \\ 0.7 & 12 < n_{ij} \leq 36 \\ 0.4 & 8 < n_{ij} \leq 12 \\ 0.2 & 0 < n_{ij} \leq 8 \end{Bmatrix} \tag{4}$$

3. Results and discussion

3.1. Seasonal variation of $PM_{2.5}$ concentrations

According to the ambient air quality standards in China (GB 3096-2012), 35 μg m^{-3} and 15 μg m^{-3} are the respective recommended limits of daily and annual $PM_{2.5}$ concentration for ambient air quality function zones Class I (e.g., nature reserves). The annual $PM_{2.5}$ concentration at YRDNNR was 78.9 μg m^{-3}, 4.3 times higher than that required by the ambient air quality standards. And 66.0% of the daily samples at YRDNNR had $PM_{2.5}$ concentrations higher than 35 μg m^{-3}. $PM_{2.5}$ at YRDNNR exhibited clear seasonal variations, with concentrations of 71.2, 92.7, 97.1 and 62.5 μg m^{-3} during spring, summer, autumn and winter, respectively. $PM_{2.5}$ concentrations at YRDNNR were clearly higher in summer and lower in winter, the opposite of findings from urban areas in China, such as urban Beijing and Jinan (Yang et al., 2012; Zhao et al., 2009). Unlike urban areas, YRDNNR does not have many strong local emission sources. Thus, the transport of pollutants seems highlighted factors in influencing $PM_{2.5}$ concentrations at such a background site. Mean 72-hour backward trajectory clusters at YRDNNR in summer and winter was shown in Fig. 2. Obviously, the seasonal air mass patterns entirely differed in summer and winter. In summer, a large proportion of air masses sourced from Shandong and Liaoning and air masses backward trajectories were short, indicating air masses move slowly and pollutants were readily to accumulate. However, most air masses originated from Mongolia and Russian and moved faster in winter. Thus, the strong flow was helpful to dilute $PM_{2.5}$ concentrations in winter. Similar seasonal variation of $PM_{2.5}$ was observed in Shangdianzi, a regional Global Atmosphere Watch (GAW) station which is located 100 km away from Beijing (Zhao et al., 2009). Zhao et al. reported $PM_{2.5}$ concentrations of 59.8, 58.5, 45.1 and 44.3 μg m^{-3} during spring, summer, autumn and winter, respectively, at Shangdianzi in 2007. Concentrations of $PM_{2.5}$ at YRDNNR were 1.2–2.2 times those in Shangdianzi. Zibo and Weifang (both ~130 km away from YRDNNR), located in the south of YRDNNR, are heavy polluted cities in Shandong. For example, SO_2 emissions of Zibo and Weifang were 164,000 and 116,000 t, which were even higher than that in Beijing (115,000 t) in 2010 (Shandong Statistical Yellowbook, 2011; China Statistical Yellowbook, 2011). As south-southeast wind dominated at YRDNNR throughout the year, air masses from south-southeast winds brought massive pollutants to YRDNNR. YRDNNR is facing a serious air pollution problem and measures should be taken to protect its ecosystems by controlling $PM_{2.5}$.

3.2. $PM_{2.5}$ mass closure

$PM_{2.5}$ is composed of inorganic and organic fractions, including the secondary inorganic aerosols (SIA: sulphate, nitrate and ammonium), dust, organic aerosol (OA) and EC. Si and Ca, along with Al, Fe and Ti, are major crustal elements. These elements can be used to estimate dust mass (Pettijohn, 1975), which can be expressed as: [Dust] = 2.20 * [Al] + 2.49 * [Si] + 1.63 * [Ca] + 1.42 * [Fe] + 1.94 * [Ti]. Organic aerosols (OA) are divided into primary organic aerosols (POA) and secondary organic aerosols (SOA). Generally, for rural areas OA is estimated as organic carbon (OC) multiplied by 2.1 (Turpin and Lim, 2001). In this study SOA and POA were represented by secondary organic carbon (SOC) and primary organic carbon (POC = OC − SOC), each multiplied

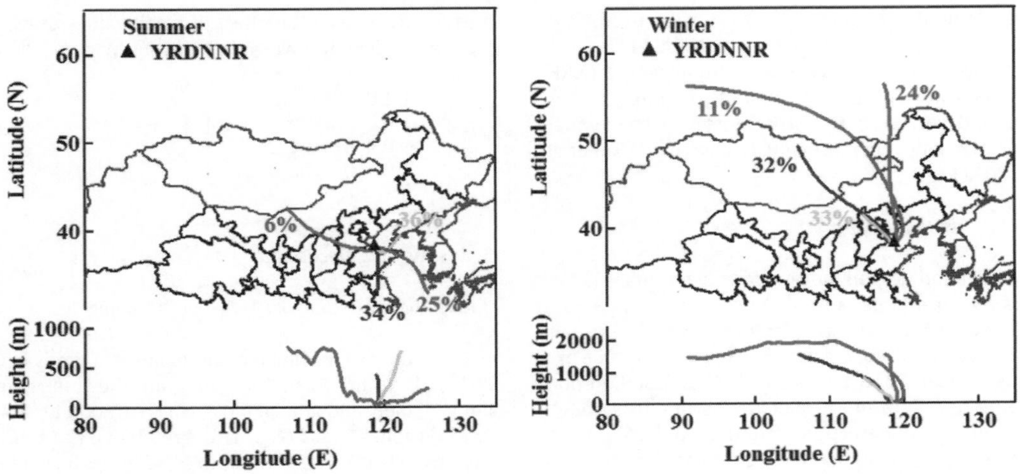

Fig. 2. Mean 72-hour backward trajectory clusters at YRDNNR in summer and winter.

by a factor of 2.1. SOC was estimated by the EC-tracer method. For details refer to our previous study (Sui et al., 2015).

$PM_{2.5}$ mass closures in YRDNNR during different seasons are depicted in Fig. 3. In spring, $PM_{2.5}$ was mainly composed of SIA (38.7%), OA (26.2%; 16.2% for SOA and 10.0% for POA) and dust (13.2%). Compared with other seasons, the high dust fraction characteristic of $PM_{2.5}$ during spring, probably resulted from higher wind speeds, a dry climate and dust storms. In summer, SIA was the dominant fraction of $PM_{2.5}$ (58.0%), especially SO_4^{2-} (33.3%), whereas OA only accounted for 13.0% (9.8% POA and 3.2% SOA). Summertime POA preponderated over SOA may indicate local emission contribution to $PM_{2.5}$ at YRDNNR was non-negligible in summer. The unidentified fraction was higher in summer (20.2%), which might be associated with water bound to aerosols due to high relative humidity during summer. Models have suggested that water content in aerosols ranges from 20 to 35% at relative humidity of 50% (Tsyro, 2005). Thus, water content is a non-negligible fraction in $PM_{2.5}$ during summer. In autumn, SIA was still the main components of $PM_{2.5}$ (55.2%). However, OA increased to 26.5% (15.2% POA and 11.3% SOA), which twice the summer level. In winter, $PM_{2.5}$ was characterized by high loads of OA (40.2%) and SIA (45.8%). SOA (25.5%) was the highest individual fraction during winter, which was likely associated with the combined effect of lower

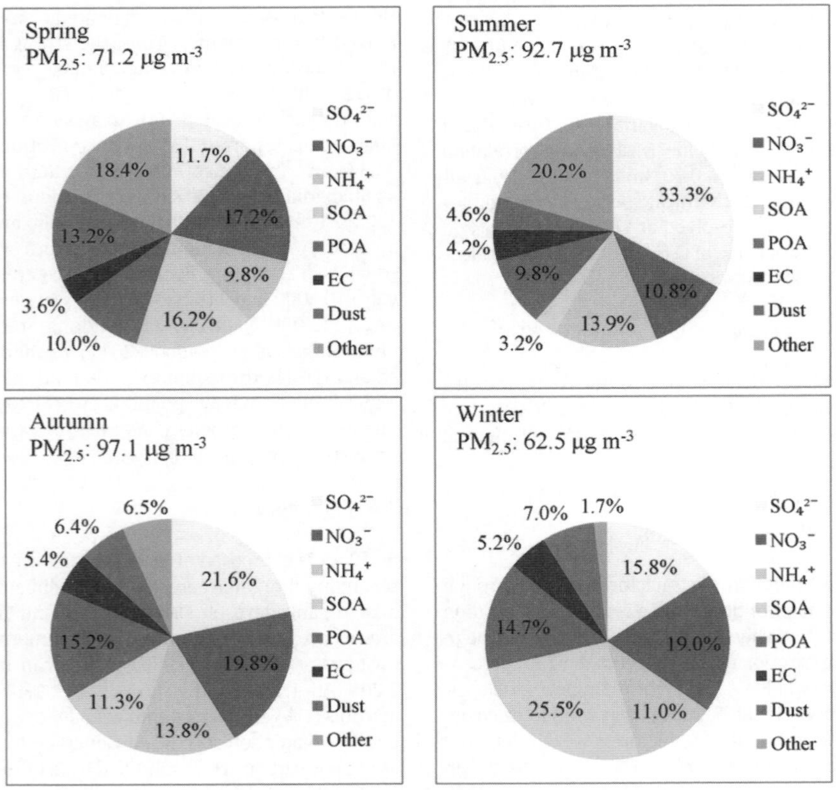

Fig. 3. Mass closure of $PM_{2.5}$ at YRDNNR during different seasons.

temperature (−1.6 °C) and the relatively high POA fraction (14.7%) in winter. Lower temperature favours gas–particle equilibrium to move towards the particle phase and leads to the adsorption and condensation of semi-volatile compounds onto particles, resulting in SOA formation. EC was minor component of $PM_{2.5}$ throughout the year. The proportion of EC ranged from 3.6–5.4%, with an average of 4.6%. Despite its small fraction in $PM_{2.5}$, EC has become the second greatest contributor to global warming after CO_2 in the troposphere, playing an important role in climate change (Ramanathan and Carmichael, 2008). The identified components accounted for 79.8–98.3% of $PM_{2.5}$, explaining 88.3% of $PM_{2.5}$ on average. Thus, in this study, sources of $PM_{2.5}$ can be almost fully resolved using these identified components.

3.3. Source identification

Fig. 4 shows a comparison of the measured and simulated $PM_{2.5}$ mass concentrations deduced from PMF at YRDNNR. The reconstructed $PM_{2.5}$ mass concentrations agreed well with the observed $PM_{2.5}$ concentrations (slope: 0.98 ± 0.02, $r^2 = 0.98$). The resolved source profiles of $PM_{2.5}$ in YRDNNR are shown in Fig. 5. Seasonal and annual contributions of each source are listed in Table 1. Fig. 6 shows the results of PSCF analysis to indicate the likely source areas of each source resolved by PMF.

The first factor, identified as biomass burning, is represented by a high load of OC (40.8%), EC (31.8%), K (36.7%) and NO_3^- (34.6%) (Heo et al., 2009). Simultaneous high concentrations of OC, EC and K^+ are generally indicators of biomass burning. PSCF results for biomass burning show two source areas of highest probability (Fig. 6). One is the southern source area, including Jiangsu and the area south of Shandong near Henan. The other is the northern source area, covering Mongolia, Inner Mongolia, Hebei and Beijing. Fire data (http://hjj.mep.gov.cn/stjc/index_20.htm) shows areas of concentrated fire dots north of Hebei, south of Shandong and in Jiangsu. Biomass burning contributions to $PM_{2.5}$ at YRDNNR were significant throughout the year, especially in winter (Table 1). Biomass burning is the largest source of OC on a global scale and is strongly associated with SOA (Cheng et al., 2013). From $PM_{2.5}$ mass closure, SOA proportion had a large increase in winter (Fig. 3). The increased SOA was probably associated with the significant biomass burning contribution in winter. Similar result of wintertime maxima of anhydrosugars (an indicator of biomass burning) was observed in Okinawa, Japan, which may be associated with open burning and domestic

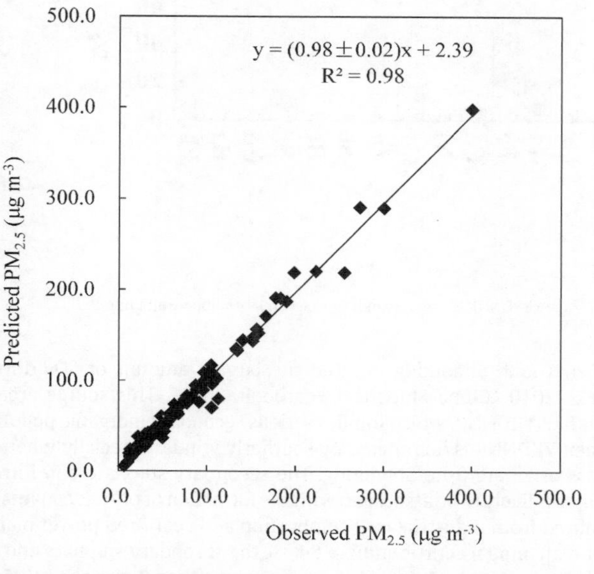

Fig. 4. Predicted versus observed $PM_{2.5}$ mass concentrations (μg m^{-3}) at YRDNNR (95% confidence interval).

heating and cooking in northern and northeastern China, Mongolia and Russia (Zhu and Kawamura, 2015). Cheng et al. investigated biomass burning contribution to Beijing aerosol in 2011 (Cheng et al., 2013). In his study, average concentrations of levoglucosan (a unique biomass burning indicator) in Beijing were 0.23 μg m^{-3} in summer and 0.59 μg m^{-3} in winter. And levoglucosan/OC mass ratio was 0.019 in summer and 0.024 in winter, which probably indicated that biomass burning in Beijing was more frequent in winter than that in summer. Zhang et al. emphasized that levoglucosan exhibited a background concentration all year round (Zhang et al., 2008), which may be explained by biofuel combustion in the countryside of suburban Beijing and surrounding provinces (e.g. Hebei). In winter, air parcels at YRDNNR exclusively originated from the north and passed through Inner Mongolia, Hebei and Beijing (Fig. 2). Thus, the higher biomass burning contribution to $PM_{2.5}$ at YRDNNR in winter results from regional impact.

The second factor was characterized by the highest load of Ba (85.8%) and median load of K (15.6%). It should be noted that the sampling period during winter (16 January–13 February) included the Spring Festival (2–8 Feb), when fireworks are traditional. Ba has been shown to exhibit a sharp increase (264 fold) during fireworks displays, as has K (18 fold) (Sarkar et al., 2010). This factor made a higher contribution (2.2–8.4%) during Spring Festival and the cause was identified as fireworks. Fireworks accounted for 2.2% $PM_{2.5}$ in winter and its contribution was negligible in other seasons (Table 1). The PSCF analysis of fireworks is not shown due to the negligible contribution to the annual average.

The third factor was identified as vehicular emissions, with a relatively high load of OC (32.7%), Zn (61.1%), Pb (35.3%), Mn (47.7%), Fe (31.8%) and a median load of NO_3^- (12.7%) and EC (14.2%). OC and EC are the main pollutants from gasoline and diesel combustion. Zn is used as an additive in lubricating oil in two-stroke engines (Begum et al., 2004). Zn and Pb are linked to metal brake wear (Sternbeck et al., 2002). Fe and Mn typically exist in brake wear dust (Yang et al., 2013). The relatively higher load of NO_3^- than SO_4^{2-} also suggested vehicular sources. PSCF analysis of vehicular emission indicated Hebei Province and Shandong Province were the likely source areas. Possessions of private vehicles in Shandong and Hebei ranked second and fifth, respectively, over China (China Statistical Yellowbook, 2011). Vehicular emissions contributed 2.8–8.1% to $PM_{2.5}$, with an annual contribution of 5.2%. The vehicular factor contribution was highest in spring, which was likely related to higher wind speed and dry climate in spring, leading to easy re-suspension of road dust related to traffic activity. Comparable result (7.3%) was obtained at the Gosan background site in East Asia (Han et al., 2006), but the result was much lower than that previously recorded in urban areas, such as Seoul (25.3%) (Heo et al., 2009).

The fourth factor was characterized by the highest load of Cu (62.4%), with median loads of Zn and Pb (23.1% for both Zn and Pb). This factor was identified as a copper smelting source. There is a Fangyuan copper smelting company, with annual Cu production exceeding 100,000 t in 2005, located in Dongying, which is 50 km from the YRDNNR. Abundant Zn and Pb are also emitted from copper smelters (Kabala and Singh, 2001). Thus, local emission in Dongying is likely to contribute to this factor. From the PSCF results, other possible regional source areas were also identified, including Inner Mongolia, the north of Hebei, Anhui and Liaoning. Copper mine may be possible Cu sources in these areas. Inner Mongolia possesses the second largest amount of copper ore and copper resources in Anhui ranks sixth in China (China Statistical Yellowbook, 2011). The copper smelting factor contributed 2.1–6.6% to $PM_{2.5}$, with an average contribution of 4.9%.

The fifth factor was secondary sulphate and nitrate, represented by the highest load of SO_4^{2-} (66.6%), NO_3^- (42.3%), NH_4^+ (66.0%) and Se (51.7%), with a median load of Br (25.4%). Se is a trace element typically associated with coal combustion and a good correlation (r = 0.74) between S and Se was obtained. Coal combustion in the industrial sector is the largest source of Se, accounting for 56.2% on a national scale and

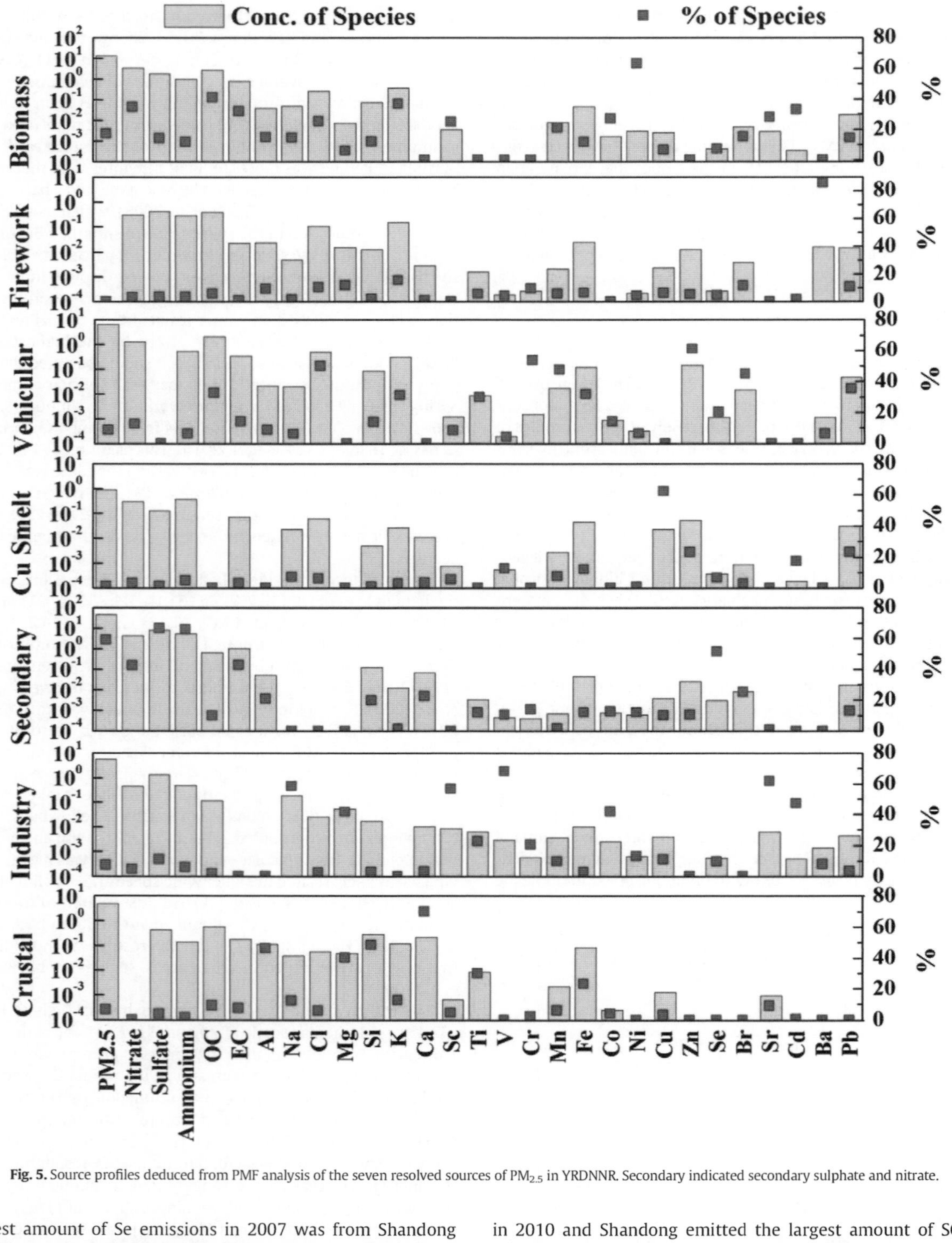

Fig. 5. Source profiles deduced from PMF analysis of the seven resolved sources of $PM_{2.5}$ in YRDNNR. Secondary indicated secondary sulphate and nitrate.

the greatest amount of Se emissions in 2007 was from Shandong (289.11 t, Tian et al., 2010). The ratio of S/Se in $PM_{2.5}$ at YRDNNR was 281, close to that from coal-fired power plant emissions in Vermont (214, Polissar et al., 2001). Br also has a high load in coal combustion sources of $PM_{2.5}$ (Yang et al., 2013) and was significantly correlated with Se ($P < 0.01$, $r = 0.51$). PSCF results of the secondary sulphate and nitrate sources indicated that Shandong Province and its adjacent areas, including the north of Henan and Anhui, and Jiangsu, were the probable source areas (Fig. 6). The emission of SO_2 in Shandong, Henan and Jiangsu ranked first, third and ninth, respectively, in China in 2010 and Shandong emitted the largest amount of SO_2 during 2002–2010 (China Statistical Yearbook, 2011). This source area is south of YRDNNR, which implies serious secondary inorganic pollution when YRDNNR is dominated by southerly winds, especially when air mass originate from Shandong. The secondary sulphate and nitrate source is likely to be associated with the formation of primary pollutants emitted from industrial coal combustion and coal-fired power plants. With an annual contribution of 54.3%, the secondary sulphate and nitrate source contributions show a large seasonal variation (35.3–71.8%). In summer, the factor accounted for up to 71.8%, which is

Table 1
Seasonal and annual source contributions to $PM_{2.5}$ mass concentrations in YRDNNR (%).

Season	Biomass	Fireworks	Vehicular	Cu smelt	Secondary sulphate and nitrate	Industry	Crustal
Spring	13.8	0.2	8.1	5.6	44.5	11.4	16.4
Summer	6.0	0.2	2.8	2.1	71.8	12.2	5.0
Autumn	13.3	0.1	3.6	6.6	65.6	7.2	3.6
Winter	30.3	2.2	6.5	5.4	35.3	12.1	8.4
Annual	15.8	0.7	5.2	4.9	54.3	10.7	8.3

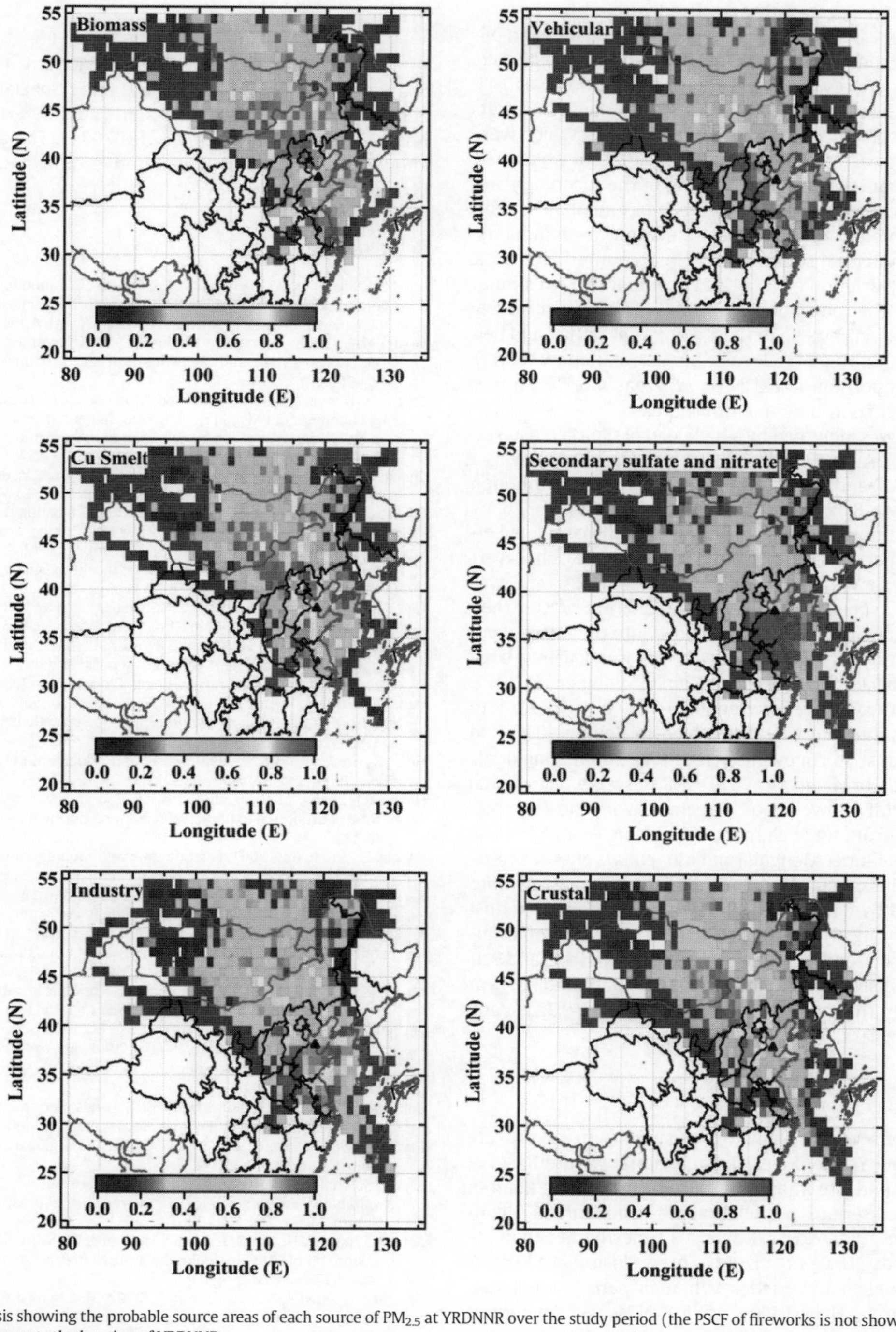

Fig. 6. Results of PSCF analysis showing the probable source areas of each source of $PM_{2.5}$ at YRDNNR over the study period (the PSCF of fireworks is not shown due to its small contribution). The black triangle represents the location of YRDNNR.

consistent with the higher conversion rate from SO_2 to SO_4^{2-} and NO_X to NO_3^- (SOR = 0.65, NOR = 0.30) in summer (Yuan et al., 2014b). Additionally, a large proportion of air masses at YRDNNR passed through Shandong during summer (Fig. 2), and could have brought in high concentrations of sulphate. Compared with the contribution of secondary sulphate and nitrate in other seasons, the secondary sulphate and nitrate contributed the lowest proportion of $PM_{2.5}$ during winter, probably as a result of a lower conversion rate of the primary pollutants (SO_2 and NO_2) in winter (SOR = 0.27, NOR = 0.10, Yuan et al., 2014b).

The sixth resolved factor, characterized by a high load of V (68.3%), Co (41.9%), Cd (47.3%) and Sc (56.9%) with a median load of Cr (20.5%), was identified as coming from industry. As a typical trace element of oil combustion, V is generally associated with oil-fired power plants and steam boilers (Karnae and John, 2011). Dongying is an oil-rich city in which the second largest oil field (Shengli oil field) is located. Oil is one of the major energy sources and oil consumption was 14.2 million tons in Dongying in 2008 (Kong et al., 2012). The PSCF results in Fig. 6 show the most likely industrial source locations, which were Dongying, the other areas of Shandong and its adjacent areas. The PSCF plot of industry source was similar to that of the secondary sulphate and nitrate, which suggested that the secondary sulphate and nitrate were mainly formed from the primary emissions from industry. The contribution of industrial sources to $PM_{2.5}$ accounted for 10.7%, with a small seasonal variation of 7.2–12.2%. Compared with similar studies in urban areas, the contribution from this factor was relatively higher than that in cities, such as Jinan (2.9%, Yang et al., 2013) and Beijing (6%, Song et al., 2006). This is likely to be because stationary sources are relatively more important at YRDNNR, whereas mobile sources make a relatively higher contribution in urban areas.

The seventh factor was identified by a high load of typical crustal elements like Ca (69.8%), Al (46.3%), Si (48.5%) and Mg (40.2%), with a median load of Ti (30.0%) and Fe (23.1%). The crustal factor contribution to $PM_{2.5}$ in YRDNNR was 8.3%, with a seasonal variation of 3.6–16.4%. This factor was highest during spring, resulting from the combined effect of dry climate, higher wind speeds and dust storms. Seasonal average relative humidity at YRDNNR varied from 57.7–78.8%, and was lowest in spring. Wind speed in spring was 1.7–2.7 times faster than in other seasons, meaning that dust re-suspended into the atmosphere more easily. Dust storm events usually occur in spring and affect a large area (e.g. urban and background sites) in China (Wang et al., 2011, 2014b). In our study, the Ca/Si ratios were 0.16–0.29 from 29 April to 3 May, which were similar to those observed in Beijing (0.2) and Seoul (0.18–0.29) in dust storm events (Heo et al., 2009; Song et al., 2006), indicating dust storm impact at YRDNNR in spring. The crustal factor resolved from PMF showed good agreement with the dust fraction (7.8%) estimated from trace elements (Al, Si, Ca, Fe and Ti). The PSCF analysis found that Inner Mongolia and Mongolia were the probable source areas of the crustal factor (Fig. 6). Large areas of Inner Mongolia are covered by desert, which is a potential source of dust storms. The PSCF plot (Fig. 6) suggests that the crustal fraction originates from Inner Mongolia and Mongolia, passes through Hebei and Beijing, and arrives at YRDNNR. The crustal factor contribution to $PM_{2.5}$ at YRDNNR was similar to that in urban areas such as Jinan (9.3%) (Yang et al., 2013) and Beijing (9%) (Song et al., 2006).

4. Conclusions

Field measurements were taken in four seasons to investigate sources of $PM_{2.5}$ at the Yellow River Delta National Nature Reserve (YRDNNR), a background site in the east of the North China Plain in 2011. YRDNNR suffered serious air pollution as the annual $PM_{2.5}$ concentration (78.9 μg m^{-3}) at YRDNNR was 4.3 times higher than required by the standards (15 μg m^{-3}). $PM_{2.5}$ mass closure at YRDNNR exhibited clear seasonal characteristics. When air parcels originated from the north (Beijing and Hebei), $PM_{2.5}$ at YRDNNR was characterized by organic aerosols, whereas inorganic aerosols were dominated in $PM_{2.5}$ at YRDNNR when air masses derived from the south (Shandong). Secondary sulphate and nitrate (54.3%), biomass burning (15.8%), industry (10.7%), crustal (8.3%), vehicle (5.2%) and Cu smelting (4.9%) were important sources throughout the year. The PSCF results implied the secondary source and industry source had similar source areas (Shandong Province), confirming that secondary pollutants were mainly formed from primary pollutants emitted by industry. This study provides an intuitional understanding of $PM_{2.5}$ sources at YRDNNR and gives us clues to mitigate air pollution in this area. Enlarging degree of pollution prevention and treatment in Shandong is helpful to improve air quality at YRDNNR.

Acknowledgments

The study was supported by the Fundamental Research Funds of Shandong University (2014QY001), the Special Research for Public-Beneficial Environment Protection (201009001-1) and the National Natural Science Foundation (21307074). The authors gratefully acknowledge the Jiangsu Collaborative Innovation Center for Climate Change.

References

Aldabe, J., Elustondo, D., Santamaría, C., Lasheras, E., Pandolfi, M., Alastuey, A., et al., 2011. Chemical characterisation and source apportionment of $PM_{2.5}$ and PM_{10} at rural, urban and traffic sites in Navarra (north of Spain). Atmos. Res. 102, 191–205.
Begum, B.A., Kim, E., Biswas, S.K., Hopke, P.K., 2004. Investigation of sources of atmospheric aerosol at urban and semi-urban areas in Bangladesh. Atmos. Environ. 38, 3025–3038.
Cesari, D., Genga, A., Ielpo, P., Siciliano, M., Mascolo, G., Grasso, F.M., et al., 2014. Source apportionment of $PM_{2.5}$ in the harbour–industrial area of Brindisi (Italy): identification and estimation of the contribution of in-port ship emissions. Sci. Total Environ. 497–498, 392–400.
Cheng, Y., Engling, G., He, K.B., Duan, F.K., Ma, Y.L., Du, Z.Y., et al., 2013. Biomass burning contribution to Beijing aerosol. Atmos. Chem. Phys. 13, 7765–7781.
Cui, B.S., Yang, Q.C., Yang, Z.F., Zhang, K.J., 2009. Evaluating the ecological performance of wetland restoration in the Yellow River Delta, China. Ecol. Eng. 35, 1090–1103.
Cusack, M., Pérez, N., Pey, J., Alastuey, A., Querol, X., 2013. Source apportionment of fine PM and sub-micron particle number concentrations at a regional background site in the western Mediterranean: a 2.5 year study. Atmos. Chem. Phys. 13, 5173–5187.
Guo, S., Hu, M., Zamora, M.L., Peng, J.F., Shang, D.J., Zheng, J., et al., 2014. Elucidating severe urban haze formation in China. Proc. Natl. Acad. Sci. U. S. A. 111, 17373–17378.
Han, J.S., Moon, K.J., Lee, S.J., Kim, Y.J., Ryu, S.Y., Cliff, S.S., et al., 2006. Size-resolved source apportionment of ambient particles by positive matrix factorization at Gosan background site in east Asia. Atmos. Chem. Phys. 6, 211–223.
Han, S.Q., Wu, J.H., Zhang, Y.F., Cai, Z.Y., Feng, Y.C., Yao, Q., et al., 2014. Characteristics and formation mechanism of a winter haze–fog episode in Tianjin, China. Atmos. Environ. 98, 323–330.
Heo, J.B., Hopke, P., Yi, S.M., 2009. Source apportionment of $PM_{2.5}$ in Seoul, Korea. Atmos. Chem. Phys. 9, 4957–4971.
Huang, R.J., Zhang, Y.L., Bozzetti, C., Ho, K.F., Cao, J.J., Han, Y.M., et al., 2014. High secondary aerosol contribution to particulate pollution during haze events in China. Nature 514, 218–222.
Kabala, C., Singh, B.R., 2001. Fractionation and mobility of copper, lead, and zinc in soil profiles in the vicinity of a copper smelter. J. Environ. Qual. 30, 485–492.
Karnae, S., John, K., 2011. Source apportionment of fine particulate matter measured in an industrialized coastal urban area of south Texas. Atmos. Environ. 45, 3769–3776.
Kong, S.F., Lu, B., Ji, Y.Q., Zhao, X.Y., Bai, Z.P., Xu, Y.H., et al., 2012. Risk assessment of heavy metals in road and soil dusts within $PM_{2.5}$, PM_{10} and PM_{100} fractions in Dongying city, Shandong Province, China. J. Environ. Monit. 14, 791–803.
Mu, Q., Zhang, S., 2013. An evaluation of the economic loss due to the heavy haze during January 2013 in China. China Environ. Sci. 11, 034.
Pettijohn, F.J., 1975. Sedimentary rocks. Earth Data. Harper and Row, New York, NY, USA.
Polissar, A.V., Hopke, P.K., Paatero, P., et al., 1998. Atmospheric aerosol over Alaska: 2. Elemental composition and sources. J. Geophys. Res. Atmos. (1984–2012) 103 (D15), 19045–19057.
Polissar, A.V., Hopke, P.K., Poirot, R.L., 2001. Atmospheric aerosol over Vermont: chemical composition and sources. Environ. Sci. Technol. 35, 4604–4621.
Ramanathan, V., Carmichael, G., 2008. Global and regional climate changes due to black carbon. Nat. Geosci. 1, 221–227.
Sarkar, S., Khillare, P.S., Jyethi, D.S., Hasan, A., Parween, M., 2010. Chemical speciation of respirable suspended particulate matter during a major firework festival in India. J. Hazard. Mater. 184, 321–330.
Song, Y., Zhang, Y.H., Xie, S.D., Zeng, L.M., Zheng, M., Salmon, L.G., et al., 2006. Source apportionment of $PM_{2.5}$ in Beijing by positive matrix factorization. Atmos. Environ. 40, 1526–1537.
Sternbeck, J., Sjödin, Å., Andréasson, K., 2002. Metal emissions from road traffic and the influence of resuspension–results from two tunnel studies. Atmos. Environ. 36, 4735–4744.

Sui, X., Yang, L.X., Yi, H.Y., Yuan, Q., Yan, C., Dong, C., et al., 2015. Influence of seasonal variation and long-range transport of carbonaceous aerosols on haze formation at a seaside background site, China. Aerosol Air Qual. Res. 15, 1251–1260.
Tian, H.Z., Wang, Y., Xue, Z.G., Cheng, K., Qu, Y.P., Chai, F.H., et al., 2010. Trend and characteristics of atmospheric emissions of Hg, As, and Se from coal combustion in China, 1980–2007. Atmos. Chem. Phys. 10, 11905–11919.
Tsyro, S., 2005. To what extent can aerosol water explain the discrepancy between model calculated and gravimetric PM_{10} and $PM_{2.5}$? Atmos. Chem. Phys. 5, 515–532.
Turpin, B.J., Lim, H.J., 2001. Species contributions to $PM_{2.5}$ mass concentrations: revisiting common assumptions for estimating organic mass. Aerosol Sci. Technol. 35, 602–610.
USEPA, 2014. Positive Matrix Factorization (PMF) 5.0 Fundamentals & User Guide. Office of Research and Development.
Wang, H.L., An, J.L., Shen, L.J., Zhu, B., Pan, C., Liu, Z.R., et al., 2014a. Mechanism for the formation and microphysical characteristics of submicron aerosol during heavy haze pollution episode in the Yangtze River Delta, China. Sci. Total Environ. 490, 501–508.
Wang, G.H., Cheng, C.L., Huang, Y., Tao, J., Ren, Y.Q., Wu, F., et al., 2014b. Evolution of aerosol chemistry in Xi'an, inland China, during the dust storm period of 2013—part 1: sources, chemical forms and formation mechanisms of nitrate and sulfate. Atmos. Chem. Phys. 14 (21), 11571–11585.
Wang, G.H., Li, J.J., Cheng, C.L., Hu, S., Xie, M., Gao, S., et al., 2011. Observation of atmospheric aerosols at Mt. Hua and Mt. Tai in central and east China during spring 2009—part 1: EC, OC and inorganic ions. Atmos. Chem. Phys. 11, 4221–4235.
Xie, Y.B., Chen, J., Li, W., 2014. An assessment of $PM_{2.5}$ related health risks and impaired values of Beijing residents in a consecutive high-level exposure during heavy haze days. Huan Jing Ke Xue 35, 1–8.
Yang, L.X., Cheng, S.H., Wang, X.F., Nie, W., Xu, P.J., Gao, X.M., et al., 2013. Source identification and health impact of $PM_{2.5}$ in a heavily polluted urban atmosphere in China. Atmos. Environ. 75, 265–269.
Yang, L.X., Zhou, X.H., Wang, Z., Zhou, Y., Cheng, S.H., Xu, P.J., et al., 2012. Airborne fine particulate pollution in Jinan, China: concentrations, chemical compositions and influence on visibility impairment. Atmos. Environ. 55, 506–514.
Yuan, J.L., Peng, S.F., 1998. Dongying Traffic Record. China Communication Press, Beijing.
Yuan, Q., Yang, L.X., Dong, C., Yan, C., Meng, C.P., Sui, X., et al., 2014a. Particle physical characterisation in the Yellow River Delta of eastern China: number size distribution and new particle formation. Air Qual. Atmos. Health 1–12.
Yuan, Q., Yang, L.X., Dong, C., Yan, C., Meng, C.P., Sui, X., et al., 2014b. Temporal variations, acidity, and transport patterns of $PM_{2.5}$ ionic components at a background site in the Yellow River Delta, China. Air Qual. Atmos. Health 7, 143–153.
Zhang, Q., Streets, D.G., Carmichael, G.R., He, K., Huo, H., Kannari, A., et al., 2009. Asian emissions in 2006 for the NASA INTEX-B mission. Atmos. Chem. Phys. 9, 5131–5153.
Zhang, T., Claeys, M., Cachier, H., Dong, S.P., Wang, W., Maenhaut, W., et al., 2008. Identification and estimation of the biomass burning contribution to Beijing aerosol using levoglucosan as a molecular marker. Atmos. Environ. 42, 7013–7021.
Zhao, X.J., Zhang, X.L., Xu, X.F., Xu, J., Meng, W., Pu, W.W., 2009. Seasonal and diurnal variations of ambient $PM_{2.5}$ concentration in urban and rural environments in Beijing. Atmos. Environ. 43, 2893–2900.
Zhao, X.J., Zhao, P.S., Xu, J., Meng, W., Pu, W.W., Dong, F., et al., 2013. Analysis of a winter regional haze event and its formation mechanism in the North China Plain. Atmos. Chem. Phys. 13, 5685–5696.
Zhu, C., Kawamura, K., 2015. Effect of biomass burning over the western north Pacific rim: wintertime maxima of anhydrosugars in ambient aerosols from Okinawa. Atmos. Chem. Phys. 15, 1959–1973.
Zhu, Y.H., Yang, L.X., Yuan, Q., Yan, C., Dong, C., Meng, C.P., et al., 2014. Airborne particulate polycyclic aromatic hydrocarbon (PAH) pollution in a background site in the North China Plain: concentration, size distribution, toxicity and sources. Sci. Total Environ. 466–467, 357–368.

Air Qual Atmos Health (2008) 1:31–36
DOI 10.1007/s11869-008-0003-1

REVIEW ARTICLE

Study on ambient air quality in Beijing for the summer 2008 Olympic Games

Wen-xing Wang · Fa-he Chai · Kai Zhang · Shu-lan Wang · Yi-zhen Chen · Xue-zhong Wang · Ya-qin Yang

Received: 22 January 2008 / Accepted: 26 February 2008 / Published online: 30 April 2008

Abstract With the coming/approaching of the Olympic Games in 2008, air pollution in Beijing attracts the attention of government and people. The objective of this study is to define the air quality during the Olympic Games; we conducted the observation of SO_2, NO, CO, NO_2, O_3, and PM_{10} from August 7 to September 30 in 2007 in Beijing. The results showed that the average daily concentrations of SO_2, NO_2, CO, and PM_{10} during observation were 0.024, 0.072, 2.25, and 0.19 mg m^{-3}, respectively. Compared with the National Ambient Air Quality Standard II, the concentrations of SO_2 and CO in the observation were low, the concentration of NO_2 basically satisfied the National Ambient Air Quality Standard II, and the concentrations of O_3 and PM_{10} were much higher than the values of the standard. The characteristics of diurnal variation of NO, NO_2, CO, and PM_{10} were similar, and the lower concentrations of these pollutants were observed by day and the higher concentrations at night. The concentration of SO_2 in the daytime was a little higher than that at night. The highest 1-h concentration of O_3 occurred at 14:00 local time.

Keywords Beijing · Air quality · Pollution characteristic

W.-x. Wang (✉) · F.-h. Chai · K. Zhang · S.-l. Wang · Y.-z. Chen · X.-z. Wang
Chinese Research Academy of Environmental Sciences,
Beijing 100012, China
e-mail: wxwang99@hotmail.com

Y.-q. Yang
Key Lab of Marine Environmental Science and Ecology Ministry of Education, Ocean University of China,
Qingdao 266100, China

Introduction

Beijing (39.13–41.08° N, 115.22–117.50° E), the capital and a major metropolis of the People's Republic of China, is located in the north of Huabei (North China) Plain and is a part of the North Temperate Zone. Since the 1980s, the rapid industrial development and urbanization and rapid increasing traffic have resulted in severe air pollution in Beijing, especially particulate pollution (Zhang et al. 2000; He et al. 2001; Yao et al. 2003). Primary particles from coal combustion and vehicle emission with the secondary fine particle resulted in a complex particulate pollutant mixture (Lun et al. 2003; Shrestha et al. 2005; Sun et al. 2004). Pollutants come not only from local emissions but also from surrounding regional/background sources (An et al. 2007). For example, sandstorms often occurred, which intensified the particulate pollution in Beijing during the springtime (Song et al. 2006). Power plants are also significant sources of sulfur dioxide (SO_2) and nitrogen oxides (NO_x), which are harmful at high concentrations and contribute to the formation of atmospheric fine particles. SO_2 and NO_x can be oxidized in the atmosphere to SO_4^{2-} and NO_3^- and remain in the air for a long time (Vesilind 1982). The ambient concentrations of NO_x and particles were very high due to the rapid growth of the vehicles in Beijing (Jraiw 2002; Shrestha et al. 2005). It has been estimated that the transport sector was responsible for about 74% of NO_x emission in Beijing in 2007 (http://www.bjee.org.cn/news/index.php?ID=1483).

The ambient air quality for particle matter in National Ambient Air Quality Standard includes only PM_{10}, not $PM_{2.5}$. The annual average concentrations of PM_{10} ranged between 0.14 and 0.165 mg m^{-3} and the annual average concentrations of $PM_{2.5}$ about 0.1 mg m^{-3}, which accounted for about 60% of PM_{10} mass concentrations in

Beijing (He et al. 2001). The scientific conditions' research on particulates especially fine particulates has been conducted including sources (Sun et al. 2004), size distribution (John et al. 1990; Zhang et al. 2000), chemical compositions (Lun et al. 2003; Yao et al. 2002, 2003), visibility (He et al. 2001; Okuda et al. 2004), and their impacts on human health (Zhang et al. 2000). Okuda et al. (2004) and He et al. (2001) found that organic components, secondary aerosol, soil-derived particles, and coal combustion were primary contributors to PM_{10} in Beijing. Zhang et al. (2007) analyzed nitrate, sulfate, and ammonium in the polluted atmosphere of Beijing and found that the mean concentration of water-soluble ions increased during heavily polluted atmospheric conditions.

There have been few studies analyzing the air quality in summer in Beijing at present. To discuss the potential air quality during the Olympic Games in Beijing in 2008, selected atmospheric pollutants were measured continuously during this period in 2007 when the Para-Olympic Games were held. Thus, analysis of air pollution characteristics during this time period can provide valuable guidance in developing control plans on air quality for the 2008 Summer Olympic Games.

Materials and methods

Experimental site and period

The sampling instruments were installed on the roof of a 15-m building of the Chinese Research Academy of Environmental Sciences. The building is located near a residential area in the north part of Beijing (40°02′ N, 116°24′ E) outside the fifth ring roads of the city (Fig. 1). The distance from the sampling site to the 2008 Olympic Games' stadiums is only 5.7 km and to the Olympic village is less than 3 km. The sampling results potentially represent air quality levels in the vicinity of the Olympic venues as there are no industrial air pollution sources nearby. Continuous measurements of the gaseous pollutants were conducted during the period August 7 to September 30, 2007 inclusive, whereas that of continuous PM_{10} measurements were collected for the period August 17 to September 30, 2007.

Monitoring methods

The concentration of SO_2 was measured using the high-density impulse fluorescence reaction (Thermo Fisher Model 43i Trace Level), $NO–NO_2–NO_x$ concentrations were measured using the chemiluminescent reaction (Thermo Fisher Model 42i Trace Level), CO using the gas filter correlation reaction (Thermo Fisher Model 48i Trace Level), O_3 using the UV photometric reaction, and PM_{10} using β-ray method (Thermo Fisher Model FH62C14) made by Thermo Fisher Environmental Instrument, USA. The sampling frequency was 1 min.

Experiment quality assurance and quality control

Quality assurance and quality control were strictly adhered to during the measurement period. Constant temperature and humidity were maintained. Calibrations of equipments were done before and after the study period with automatic zeroing at midnight, and every 2 weeks, manual calibrations were performed. Invalid data were eliminated.

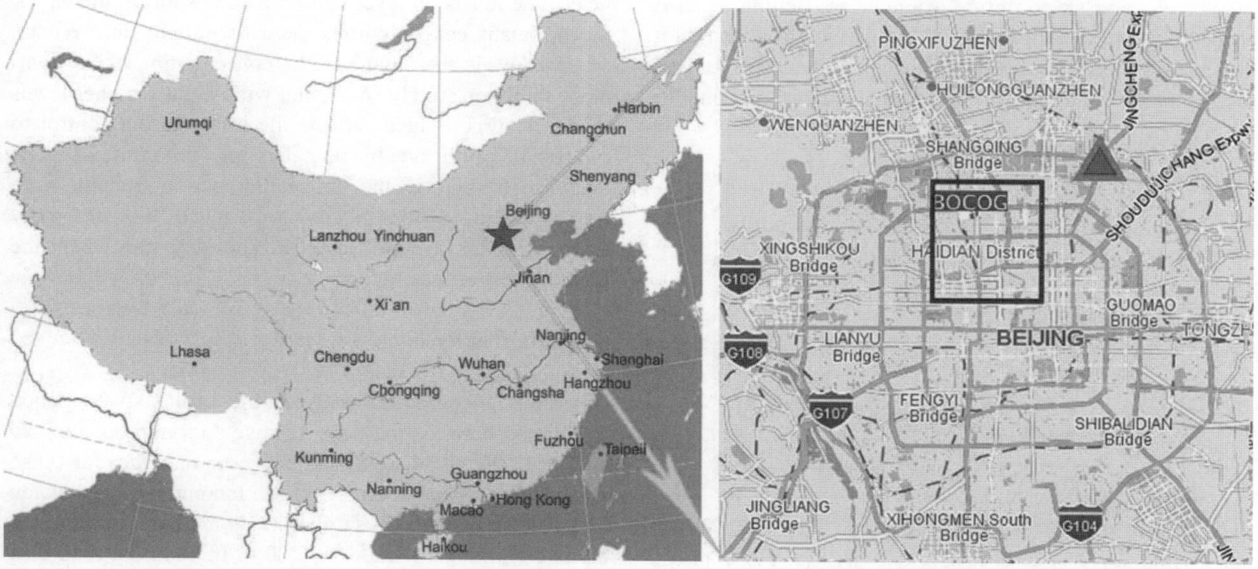

Fig. 1 Map showing Beijing city and the sampling site. (*triangles*, the sampling site; *area boxed*, The Olympic Games' stadiums)

Results and discussions

Characteristics of pollutant concentrations

The daily average concentrations of SO_2, NO_x, CO, O_3, and PM_{10} were measured during the summer period August 7 to September 30, 2007. The daily average concentration of SO_2 was 0.024 mg m^{-3} with the range of 0.01×0.07mg m^{-3} (Fig. 2), lower than the National Ambient Air Quality Standard II (Table 1), and close to the observation of Zhang et al. (2004) in October 1999 but higher than that in July 1999. The concentrations of SO_2 were strongly affected by weather conditions (Yan and Huang 2002) and varied distinctively with seasons. The concentrations of SO_2 were higher in winter (heating period) because of the burning of high-sulfur coal (Zhang et al. 2004) and lower in summer than other seasons. The daily average concentration of NO_2 was 0.072 mg m^{-3} with the range of 0.03×0.12 mg m^{-3}, lower than the National Ambient Air Quality Standard II except on 1 day. The daily average concentration of CO was 2.25 mg m^{-3} with the range of 0.72×3.46 mg m^{-3}, which was lower than the National Ambient Air Quality Standard II and close to the observation of Xue et al. (2006) during the nonheating period. The hourly average concentration exceeding the National Ambient Air Quality Standard II of O_3 happened on 30 days, and the unattainment rate was 54.5%. The daily average concentration of PM_{10} was high (0.19 mg m^{-3}) with a wide range of 0.04×0.34 mg m^{-3}, and the exceeding of the National Ambient Air Quality Standard II happened on 34 days (over the standard rate was 73.3%); on 8 days (17.8%), the National Ambient Air Quality Standard III was exceeded.

Recent research showed that the pollution of NO_x caused by vehicle emissions during summer in Beijing is also attracting more attention. The rising of NO_x concentration not only affects the air quality directly but also causes secondary air pollution, an important precursor of ozone. In recent years, the number of automobiles especially private cars in Beijing has increased sharply at an annual growth rate of 10–20% over the period (Yao et al. 2003). The total number of vehicles in Beijing was nearly 1.7 million at the end of 2001 (Shrestha et al. 2005) and broke through 3 million by the end of May 2007. Pollution from vehicle emission is becoming more and more serious in Beijing. In contrast, as a result of industrial pollution controls, in recent years, the concentration of SO_2 during summer in Beijing has been cut down obviously.

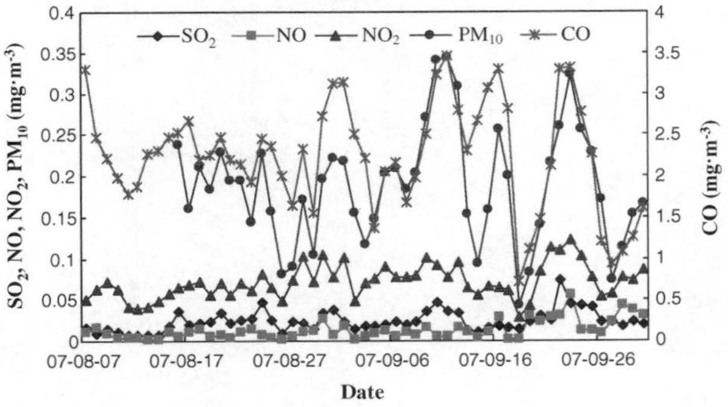

Fig. 2 Daily average concentrations of SO_2, NO, NO_2, CO, and PM_{10} in August and September, 2007

From Fig. 2, the variations of daily average concentrations of SO_2, NO_x, CO, and PM_{10} are similar, and periodical changes appear during the period of observation. August 28 to September 1, September 9 to 12, September 15 to 16, and September 20 to 25(high-pollution periods) were cloudy and rainy days.

Diurnal variation

In general, the concentrations of PM_{10}, NO, NO_2, and CO were higher at night than during the daytime (Fig. 3). The diurnal variation of PM_{10} shows double peaks at 7:00–8:00 and 18:00 local time, synchronized with the time of peak traffic. The photochemical reaction activity leads to an obvious decrease of NO_x (NO and NO_2) in the daytime. In contrast, the peak value of O_3 appears between 11:00 and 18:00 local time (maximum appears at 14:00 local time) when the solar radiation is extremely high in the daytime. Meanwhile, the concentrations of PM_{10}, CO, and NO_2 are affected by meteorological conditions. When the shallow inversion layer appears frequently during urban summer night and the atmosphere boundary layer was stable, air pollutants tend to accumulate in the lower atmospheric layer. The inversion layer is destroyed at sunrise; thereafter, the pollutant concentrations decrease rapidly and remain relative low in the daytime. The concentration of SO_2 was more stable, relatively small, and with slightly higher levels in the daytime, which reflects the effort of the control to industrial pollution by the Beijing government. The emission of SO_2 in the nonheating period (mainly from few coal-fired power plants) was much less and only impacted near ground level in the daytime when the boundary layer was higher.

Correlation analysis

Based on the correlation analyses, the result showed a good correlation between the daily average concentrations of CO and PM_{10} (r=0.741), which suggests that they are also related to the same source—traffic. Incomplete combustion of vehicles would cause the large amount of CO and particle emissions (Xue et al. 2006). In addition, road dust and the particle from vehicle exhaust are two important

Table 1 National Ambient Air Quality Standard Levels I, II, and III (GB 3095-1996)

Pollutants	Time	Concentration (mg m^{-3})		
		Level I	Level II	Level III
SO_2	Annual average	0.02	0.06	0.1
	Daily average	0.05	0.15	0.25
	1 h average	0.15	0.5	0.7
NO_2	Annual average	0.04	0.08	0.08
	Daily average	0.08	0.12[a]	0.12
	1 h average	0.12	0.24[a]	0.24
CO	Daily average	4	4	6
	1 h average	10	10	20
O_3	1 h average	0.16[a]	0.2[a]	0.2
PM_{10}	Annual average	0.04	0.1	0.15
	Daily average	0.05	0.15	0.25
TSP	Annual average	0.08	0.2	0.3
	Daily average	0.12	0.3	0.5

[a] This standard was modified in 2001. The index of NO_x was cancelled. The level II of daily average concentration of NO_2 changed from 0.08 to 0.12 mg m^{-3}; 1 h average concentration level changed from 0.12 to 0.24 mg m^{-3}. 3. One hour average concentration level I of O_3 changed from 0.12 to 0.16 mg m^{-3}; level II changed from 0.16 to 0.20 mg m^{-3}.

sources of PM_{10}. Vehicle growth is much faster than the adding of roads, which results in the increase in CO emission from incomplete combustion of gas during congested traffic and low-speed driving conditions. The correlation between NO_x and PM_{10} and CO is not obvious, although NO_x is another vehicle exhaust pollutant. The reason is that the strong activity of photochemical reaction of NO_x makes it easily transform into NO_y or secondary particles by the reactions with OH, oxides, or NH_3.

The concentrations of O_3 are negatively correlated with NO_x. NO_x is an important precursor of O_3 in the daytime, which constantly consumed by the photochemical reactions and produced lots of O_3, and the reacting intensity changes with the solar radiation. The peak value of O_3 appears at about 14:00 local time. At night, O_3 is consumed by the reducing chemical such as NO_x and without any accumulated process, so the concentration of O_3 decreases rapidly.

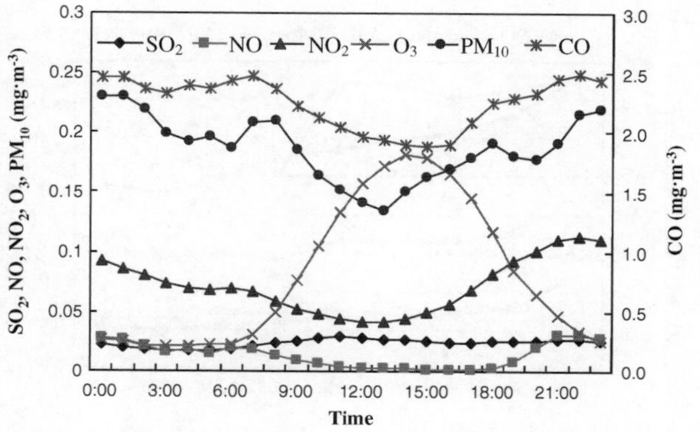

Fig. 3 Diurnal variations of SO_2, NO, NO_2, CO, O_3, and PM_{10} in August and September, 2007

The observed ratios of $[O_3]/([NO_2]/[NO])$ are plotted in Fig. 4 (the concentrations are at gas volume fraction $\times10^{-9}$). The ratio remains at low level (below 5) at night, when there is more fresh NO in air. After sunrise, the ratio of $[NO_2]/[NO]$ rises up rapidly. Most parts of NO are oxidized into NO_2; some are transformed into NO_y, and the NO_x in the air aged gradually. The peak value of the $[NO_2]/[NO]$ ratio (higher than 25) appear at 18:00–19:00 local time, when NO_x is fully aged. Thereafter, the photochemical reaction becomes weaker as the night falls, and the ratio of $[NO_2]/[NO]$ decreases rapidly below 5 again as fresh NO is accumulated in the absence of solar radiation.

The model study results of MM5 and RADM by Yang and Li (1999) in China show the relationship between $[O_3]$ and $[NO_2]/[NO]$ was linearly connected with the ratio of 15:1 during stable photochemical conditions. However, this condition rarely occurs in the atmosphere, and the actual observation results show that the ratio is always below 15. For example, the ratio calculated from the observed data in

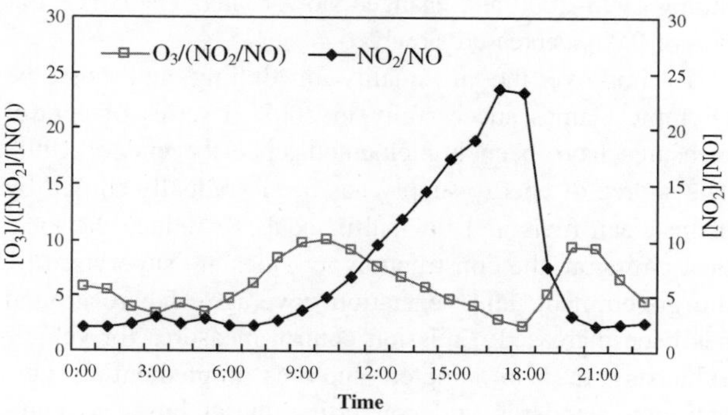

Fig. 4 Ratio of $O_3/(NO_2/NO)$ and NO_2/NO

Table 2 Emissions of SO_2, soot, and industrial dust in Beijing, 2000–2005 (unit: kt)

Year	SO_2			Soot			Particle
	Total	Life	Industry	Total	Life	Industry	Industry
2000	223.9	77.5	146.4	100.3	48.5	51.8	93.7
2001	200.7	74.4	126.3	90.3	46.5	43.8	62.7
2002	192	71.4	120.6	80.9	47.8	33.1	46.5
2003	182.8	68.8	114	70.8	41.6	29.2	32.1
2004	191.2	65.8	125.4	70.1	41.2	28.9	35.5
2005	190.6	85.1	105.5	57.6	39.9	17.7	32.5

the lower atmospheric layer in summer, 2000, by Liu et al. (2002) in Beijing was 12.5 on average. The variations of the ratio of $[O_3]/([NO_2]/[NO])$ calculated from the data in this paper are plotted in Fig. 4. The ratio is lower at night and higher in the daytime; the peak value appears at about 10:00 local time, and the daily average value is about 6 because of the high value of $[NO_2]/[NO]$ during the highly aged NO in the summer daytime's strong solar radiation.

Emissions and concentration characteristics of pollutions of Beijing in recent years

In mountainous areas surrounding Beijing in all directions except the south, dry air and lack of rain are not favorable for the diffusion and wet deposition of pollutants. Many factories and companies surrounding the city produce significant air emissions. Meanwhile, the growth of car quantity and vehicle emission and lack of adequate control measures make the air pollution more serious. The Beijing municipal government has paid much attention to air pollution, and measures have been taken to address such issues. After several years' efforts by taking measures and policy, the emissions of air pollutants in Beijing have been reduced significantly (Table 2). Figure 5 shows that the concentrations of pollutants have also decreased accordingly, especially the decrease in CO, whose concentration was decreased by 36.4%, from 3.3 (1998) to 2.1 mg m^{-3} (2006). The concentrations of SO_2 and NO_2 showed a clear decrease during 1998–2000 and changed slowly later. The concentration of PM_{10} decreased slowly.

To improve the air quality in Beijing and host the Olympic Games successfully in 2008, a series of control measures have been implemented since the end of 1998. The source of energy supply has been gradually shifted by using clean fuels and low-sulfur coal. To reduce the local dust emission, the construction activities are supervised by the government, and vegetation coverage of bare ground has been increased. Emission control measures for vehicle exhausts are also adopted such as implementing new emission standards and converting diesel buses to compressed natural gas. Through the effort of environmental protection and the action of "Green Olympics," the Beijing government obtained the "air pollution governance extra prize" and "extension clean automobile award" in "oxygen series award" in France in November 9, 2007. However, compared with some other Olympic cities abroad, such as Helsinki, Los Angeles, Barcelona, and Sydney (Li et al., 2004), Beijing's air quality needs to improve. The Beijing government plans to reduce the emissions and the concentrations of air pollutions by a series of measures, so Beijing will offer good air quality for the Olympic Games in 2008.

Conclusions

The following are the conclusions from this study:

1. The daily average concentrations of SO_2 and CO during the study period were lower than that of the National Ambient Air Quality Standard II, and the NO_2 concentrations basically met the National Ambient Air Quality Standard II. However, O_3 and PM_{10} concentrations were higher than that of the National Ambient Air Quality Standard II.
2. The variations of daily average concentrations of air pollutants showed basically consistency, which might be related closely with source emissions and meteorological conditions.

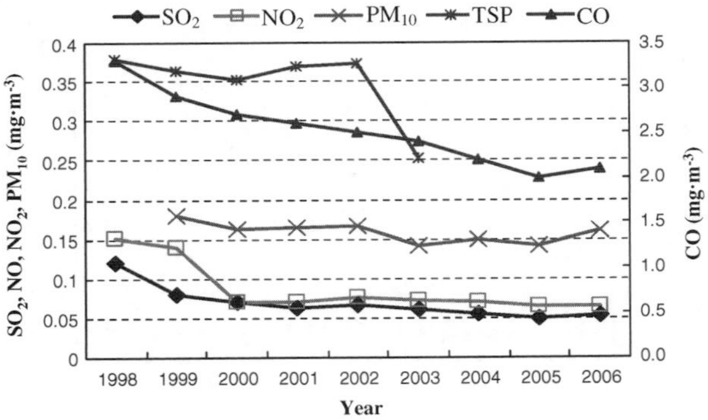

Fig. 5 Variations of annual average concentrations of pollutants in Beijing during 1998–2006

3. The lowest concentrations of PM_{10}, CO, NO, and NO_2 appeared at 13:00–14:00; at this time, the atmospheric diffusions are better. The daily variation of NO, NO_2, and O_3 agreed with the general roles of photochemical reactions.

Acknowledgments This work was financially supported by Beijing "Air Pollutant transportation and transformation in Beijing and Its Ambient Areas and Its Regulation-Controlling Principles" project (Grant no. HB200504-4).

References

An X, Zhu T, Wang Z, Li C, Wang Y (2007) A modeling analysis of a heavy air pollution epiosde occured in Beijing. Atmos Chem Phys 7:3103–3114

He K, Yang F, Ma Y, Zhang Q, Yao X, Chan CK, Cadle S, Chan T, Mulawa P (2001) The characteristics of $PM_{2.5}$ in Beijing, China. Atmos Environ 38:4959–4970

John W, Wall SM, Ondo JL, Winklmayr W (1990) Models in the size distributions of atmospheric inorganic aerosol. Atmos Environ 24A:2349–2359

Jraiw K (2002) Cleaning the air: Vehicular emission in the People Republic of China (PRC). ADB 35th Annual Meeting, 12 May, Shanghai International Convention Center, PRC. Available at: http://www.adb.org/AnnualMeeting/2002/media/vehicle_emissions.asp

Li J, Xiao Z, Yang S, Shao L (2004) Analysis of pollution characteristics of inhalable particulate matter in Beijing and a comparison with some other Olympic Games cities. Environ Sci Trends 3:26–28 (in Chinese)

Liu F, Chen H, Liu Y (2002) Study on the vertical observations of NOx and O_3 in the low level of ABL in Beijing area. J Ocean Univ Qingdao 32(2):179–185 (in Chinese)

Lun X, Zhang X, Mu Y, Nang A, Jiang G (2003) Size fractionated speciation of sulfate and nitrate in airborne particulates in Beijing, China. Atmos Environ 37:2581–2588

Okuda T, Kato J, Mori J, Tenmoku M, Suda Y, Tanaka S, He K, Ma Y, Yang F, Yu X, Duan F, Lei Y (2004) Daily concentrations of trace metals in aerosols in Beijing, China, determined by using inductively coupled plasma mass spectrometry equipped with laser ablation analysis, and source identification of aerosols. Sci Total Environ 330(1/3):145–158

Shrestha RM, Anandarajah G, Adhikari S, Jiang K, Zhu S (2005) Energy and environmental implications of NOx emission reduction from the transport sector of Beijing: a least-cost planning analysis. Transp Res Part D Transp Environ 10:1–11

Song Y, Zhang Y, Xie S et al (2006) Source apportionment of $PM_{2.5}$ in Beijing by positive matrix factorization. Atmos Environ 40:1526–1537

Sun Y, Zhuang G, Wang Y, Han L, Guo J, Dan M, Zhang W, Wang Z, Hao Z (2004) The air-borne particulate pollution in Beijing-concentration, composition, distribution and sources. Atmos Environ 38:5991–6004

Vesilind PA (1982) Environmental pollution and control. Ann Arbor Science, Ann Arbor, MI

Xue M, Wang Y, Sun Y, Hu B, Wang M (2006) Measurement on the atmospheric CO concentration in Beijing. Environ Sci 27 (2):200–206 (in Chinese)

Yan P, Huang J (2002) A preliminary study of effects of surrounding sources on surface SO_2 in Beijing. J Appl Meteorol Sci 13 (Suppl):144–152 (in Chinese)

Yang X, Li X (1999) A numerical study of photochemical reaction mechanism of ozone variation in surface layer. Chin J Atmos Sci 23(4):427–438 (in Chinese)

Yao XH, Chan CK, Fang M, Cadle S, Chan T, Mulawa P, He K, Ye B (2002) The water-soluble ionic composition of $PM_{2.5}$ in Shanghai and Beijing, China. Atmos Environ 36:4223–4234

Yao XH, Lau AP, Fang M, Chan CK, Hu M (2003) Size distributions and formation of ionic species in atmospheric particulate pollutants in Beijing, China: inorganic ions. Atmos Environ 37:2991–3000

Zhang J, Song H, Tong S, Li L, Liu B, Wang L (2000) Ambient sulfate concentration and chronic disease mortality in Beijing. Sci Total Environ 262:63–71

Zhang R, Cai X, Song Y (2004) Spatial–temporal variation and accumulation effect of air pollutants over Beijing area. Acta Sci Nat Univ Pekinensis 40(6):930–938 (in Chinese)

Zhang K, Wang Y, Wen T, Meslmani Y, Murray F (2007) Properties of nitrate, sulfate and ammonium in typical polluted atmospheric aerosols (PM_{10}) in Beijing. Atmos Res 84:67–77

Air Qual Atmos Health (2014) 7:143–153
DOI 10.1007/s11869-014-0236-0

Temporal variations, acidity, and transport patterns of $PM_{2.5}$ ionic components at a background site in the Yellow River Delta, China

Qi Yuan · Lingxiao Yang · Can Dong · Chao Yan · Chuanping Meng · Xiao Sui · Wenxing Wang

Received: 4 November 2013 / Accepted: 6 January 2014 / Published online: 21 January 2014

Abstract To better understand the pollution characteristics and potential sources of $PM_{2.5}$ ionic components at the Yellow River Delta (YRD), a semicontinuous measurement was conducted to observe water-soluble ions in $PM_{2.5}$ at a nature reserve in Dongying of Shandong province, China, in 2011. The results showed that SO_4^{2-}, NO_3^-, and NH_4^+ were the dominant ionic species (constituting 93 % of the total ionic mass) with their annual average concentrations of 22.48, 12.77, and 11.21 μg/m^3, respectively. These three ion concentrations were generally lower than those observed in major cities in China but higher than those in other rural and nature reserve sites. Ion concentrations exhibited large seasonal variations, and maximum values were observed in summer. SO_4^{2-} concentration presented a daytime peak in summer, autumn, and winter, while in spring, a relative flat diurnal cycle was observed. NO_3^- concentration changed with that of SO_4^{2-} during most of measurement period. Transport from surrounding areas contributed to the diurnal cycle of secondary ions. In addition, photochemical reaction and thermodynamic equilibrium played important roles on the diurnal variation of SO_4^{2-} and NO_3^-, respectively. The aerosol at the YRD was weakly acidic, and it was most acidic in winter. A cluster analysis showed that fine particle pollution at the YRD was mainly affected by southwest local emissions and northern middle- to long-distance transport.

Q. Yuan · L. Yang (✉) · C. Dong · C. Yan · C. Meng · X. Sui · W. Wang
Environment Research Institute, Shandong University, Jinan 250100, China
e-mail: yanglingxiao@sdu.edu.cn

L. Yang
School of Environmental Science and Engineering, Shandong University, Jinan 250100, China

W. Wang
Chinese Research Academy of Environmental Sciences, Beijing 100012, China

Keywords Water-soluble ions · High time resolution · Acidity · Transport · Yellow River Delta

Introduction

Atmospheric fine particulate matter ($PM_{2.5}$, with the aerodynamic diameters ≤2.5 μm) is one of the most complex and harmful pollutants in urban atmospheres, and it is the cause of significant public concern (Borja-Aburto et al. 1998; Schwartz et al. 1996). Water-soluble inorganic ions, particularly the secondary ions (e.g., NO_3^-, SO_4^{2-}, and NH_4^+), are the major components of $PM_{2.5}$. They play important roles in reducing atmospheric visibility, altering radiation forcing, changing surface temperatures, and influencing cloud formation and wet deposition (Fu et al. 2008; Jung et al. 2009; Kim et al. 2008; Ocskay et al. 2006). Ionic composition determines the acidity of fine particles, which has potential influences on the ecosystem and environmental materials through deposition (Zhou et al. 2012). Thus, the seasonal and diurnal variations in the water-soluble ion concentrations of $PM_{2.5}$ are important in studying aerosol characteristics.

China is one of the largest coal, iron, and steel producing countries in the world. The consumption of coal and the rapid increase in motor vehicles in recent decades have led to large amounts of SO_2 and NO_x emissions (Zhang et al. 2009). The Bohai Sea rim region generates the largest amount of pollutants, especially for particulate matters in China. However, most previous studies in this area have focused on the aerosol chemical properties of large cities, such as Beijing and Jinan (Gao et al. 2011; Sun et al. 2006; Wang et al. 2006). The Yellow River Delta Nature Reserve is an ecological protection zone near the Bohai Sea and features a unique environment. Because of the restriction of human activity in this reserve, the environment and the ecosystem remain intact. Some industrial cities near the Yellow River Delta (YRD), such as Zibo and

Jinan, have significant local emissions and suffer serious air pollution, which may affect the air quality of the YRD. Because of its geographic positional specificity and environment sensitivity, the YRD Nature Reserve has received a significant amount of attention. Most studies on pollution have focused on the organic contaminants of soil and water, such as PAHs and nitrobenzene (He et al. 2006; Zhang et al. 2011), but rarely on air pollution, especially on chemical composition of $PM_{2.5}$.

The filter-based measurement and online monitor are the two primary methods for measuring chemical particle composition. Traditional filter-based measurement is time-consuming and is unable to provide finer spatial and temporal variations. Online measurement can provide highly time-resolved data for investigating more detailed information, such as temporal variation, process analysis, and source of water-soluble ions (Oms et al. 1996; Stolzenburg and Hering 2000; Trebs et al. 2004). In this study, a semicontinuous ion monitor (URG 9000B) was used for high time-resolution measurement of $PM_{2.5}$ water-soluble ions at a background site in the YRD. This study presents the overall results for $PM_{2.5}$ water-soluble ions, including their seasonal and diurnal variations and the relationships between water-soluble ions and gaseous precursors and meteorological parameters. Additionally, this study investigates the aerosol acidity and source of pollutants using a backward-trajectory cluster analysis.

Experimental

Sampling site

The sampling site was chosen at the YRD Nature Reserve Yi Qian Er Management Station in Dongying, Shandong province (38°03′N, 118°44′E, 5 m a.s.l.). This regional background site is located in the east of North China Plain, 10 km away from the Bohai Sea (Fig. 1) and far away from the major cities in Shandong and Beijing-Tianjin-Hebei region. There are rare coal-fired power plants and industrial activities nearby, and the nearest plant is located in the southwest Dongying district with 58 km away from the sampling site. Measuring instruments including an automatic meteorological station were set up on the fourth floor of the station building, which is 15 m above the ground level. Four series of measurements were collected intensively in 2011, in spring (April 26–May 3), summer (July 8–July 31), autumn (October 16–November 7), and winter (December 30 of 2011 to January 9 of 2012).

Instruments

An ambient ion monitor (AIM, model URG 9000B, URG Corporation) was used to measure hourly concentration of water-soluble ions in $PM_{2.5}$. Some previous research (Gao et al. 2011) has showed excellent correlations (R=0.92–0.97) between AIM and the traditional filter-based measurement for measuring the major water-soluble ions (SO_4^{2-}, NO_3^-, NH_4^+, Cl^-, and K^+). Ambient air samples were drawn into the instrument at a flow rate of 3 L min^{-1}. The particles first entered a denuder that removed acidic and alkaline interfering gases before the particles were collected. An H_2O_2 solution (6 mmol/L) was substituted for the original ultrapure water as the denuder liquid to enhance the absorption of SO_2 and to avoid interference with the SO_4^{2-} measurement from SO_2. The particles were then mixed with supersaturated steam to form water droplets. The droplets passed through a condensing tube, and the condensed solutions were collected into two syringes every hour. The solutions were injected into two ion chromatographs (Dionex, ICs90) to detect the major inorganic ions, which included F^-, Cl^-, NO_2^-, NO_3^-, SO_4^{2-}, Na^+, NH_4^+, K^+, Mg^{2+}, and Ca^{2+}. Multipoint calibration was performed with standard ion solutions ordered from Dionex Corporation. Standard reference materials produced by the National Research Center for Certified Reference Materials (Beijing, China) were analyzed for quality control and assurance purposes.

Other instruments for measuring SO_2 (TEI model 43C), NO_y (TEI model 42CY), and O_3 (TEI model 49C) have been described previously (Gao et al. 2005; Wang et al. 2003). The 43C model is based on the principle that SO_2 molecules absorb ultraviolet (UV) light, become excited at one wavelength, and then decay to a lower energy state, emitting UV light at a different wavelength. NO_y was detected with a modified commercial MoO/chemiluminescence analyzer. Meteorological data (including temperature, relative humidity (RH), wind speed, wind direction, etc.) were obtained from the meteorological station.

In situ acidity

Aerosol acidity plays a particular role in secondary aerosol formation and can promote the formation of secondary inorganic aerosols by catalyzing heterogeneous reactions (Underwood et al. 2001). In this study, we adopted the online version of the aerosol inorganic model (AIM-II) (Clegg et al. 1998) to investigate the acidic characteristics of $PM_{2.5}$ with hourly concentrations of SO_4^{2-}, NO_3^-, and NH_4^+. Hourly measurements of SO_4^{2-}, NO_3^-, NH_4^+, ambient temperature (T), RH, and H^+_{strong} ($H^+_{strong}=2\times[SO_4^{2-}]+[NO_3^-]-[NH_4^+]$) were inputted into the model to obtain the aqueous phase concentrations of free ions including in situ acidity (H^+_{air}), SO_4^{2-}, NO_3^-, NH_4^+, and HSO_4^-, and water content in the aerosol droplets. The pH was predicted by the equation of $pH=-Log[\gamma\times H^+_{air}/(V_{eq}/1000)]$, in which γ and V_{eq} denote the activity coefficient for H^+_{air} and the volume of particle aqueous phase in air (cm^3/m^3).

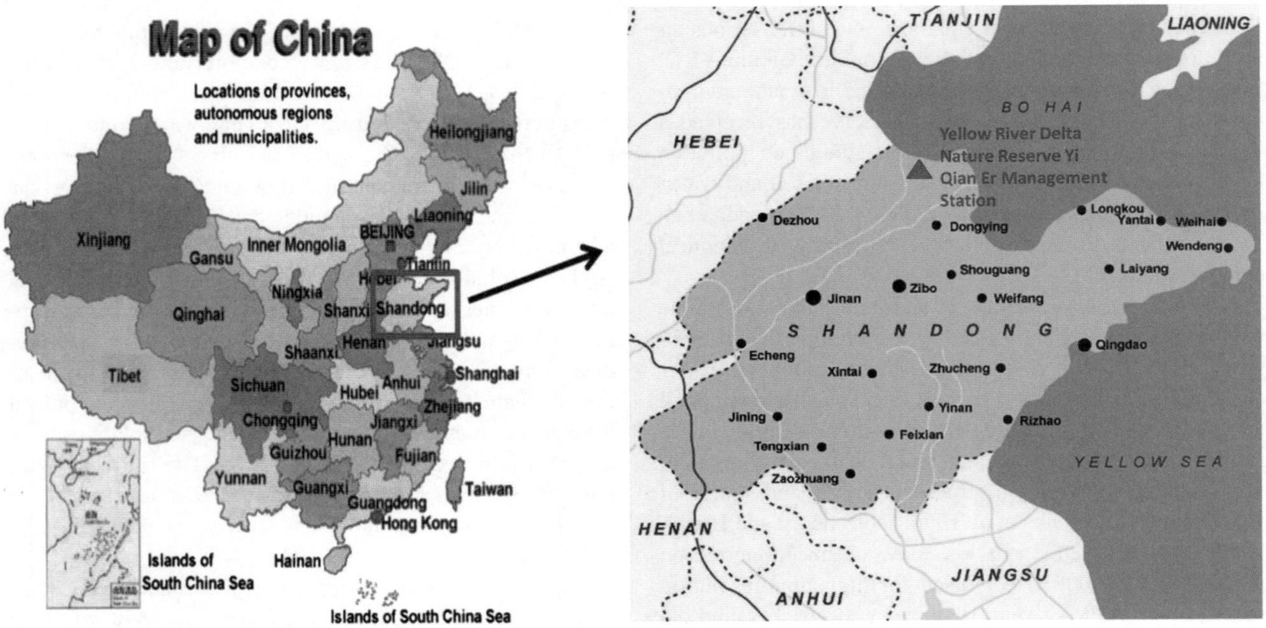

Fig. 1 Location of the sampling site in the Yellow River Delta, Shandong province, China

Backward-trajectory cluster analysis

For a more accurate analysis of the major ion sources, the backward trajectories and cluster analysis were applied. In this study, the backward trajectories were run using the Hybrid Single-Particle Lagrange Integrated Trajectory (HYSPLIT4) model (Draxler and Rolph 2003) that was developed by the National Oceanic and Atmospheric Administration (NOAA). First, 72-h backward trajectories beginning at 50 m above ground level in YRD (38°03′N, 118°44′E) were calculated every 1 h during the sampling period. The trajectories were then clustered according to their similarity in spatial distribution using the HYSPLIT 4 software. The clustering principles and processes are described in the user's guide of the software (Draxler et al. 1999).

Weather conditions

Weather conditions during the sampling period are shown in Table 1. The ambient average annual temperature is 13.0 °C, with the highest monthly average in July at 26.8 °C and the lowest monthly average in January at −1.6 °C. The RH in YRD during the period was somewhat high (at an annual average of 64.4 %), although the largest variations occurred in summer (78.8 %) and lowest in spring (57.7 %). The average annual wind speed was 1.3 m/s, with higher speeds in spring (2.1 m/s) and lower speeds in summer and autumn (0.80 and 0.78 m/s, respectively). The prevailing wind direction in winter was northern, from northwest to northeast; in summer, it was southern and eastern, from southeast to southwest. Dust storm invaded the sampling site on April 30th and May 1st during the spring measurement, when the wind speed increased to 4 m/s.

Results and discussions

Statistics for major water-soluble ions

Concentrations of water-soluble ions over the four seasons are shown in Table 2 and Fig. 2. The annual average TWSI (total water-soluble ions) was 49.72 μg/m^3, with the highest value in summer (59.91 μg/m^3) and the lowest value in spring (24.25 μg/m^3). In autumn and winter, average TWSI concentrations were 58.20 and 38.26 μg/m^3, respectively. SO_4^{2-}, NO_3^-, and NH_4^+ were the dominant ionic species and accounted for approximately 93 % of the total ions. Cl^-, NO_2^-, Na^+, K^+, Mg^{2+}, and Ca^{2+} contributed approximately 7 % of the TWSI on the annual average. The concentration of F^- was always below the limits of detection.

Table 1 Meteorological data from different seasons in the Yellow River Delta during the sampling period

	Temperature (°C)	RH (%)	Wind speed (m/s)
Annual mean	13.0	64.4	1.28
Spring	11.8	57.7	2.11
Summer	26.8	78.8	0.80
Autumn	14.1	68.5	0.78
Winter	−1.6	59.3	0.62

Table 2 Concentrations of water-soluble ions during four seasons in the Yellow River Delta

	Cl^-	NO_2^-	NO_3^-	SO_4^{2-}	Na^+	NH_4^+	K^+	Mg^{2+}	Ca^{2+}	TWSI
Spring	1.12	0.40	7.25	8.19	0.75	5.31	0.63	0.14	0.46	24.25
Summer	0.91	0.23	12.10	30.32	0.57	15.06	0.54	0.09	0.09	59.91
Autumn	1.34	0.34	18.72	23.95	0.71	12.03	0.85	0.13	0.13	58.20
Winter	2.26	0.42	7.28	19.62	0.70	7.13	0.65	0.05	0.05	38.26
Annual mean	1.31	0.33	12.77	22.48	0.66	11.21	0.65	0.10	0.19	49.72

TWSI total water-soluble ions, μg/m^3

SO_4^{2-} was the most abundant ions, accounting for 50.6 % of the total ion mass. It showed clear seasonal variation with the highest value in summer (30.32 μg/m^3), lowest concentration in spring (8.19 μg/m^3). High concentration in summer might be the result of the high RH, abundant photochemical oxidants, and strong solar radiation, while low emissions and good atmospheric diffusion conditions caused low levels in spring. The average NO_3^- concentration in autumn (18.72 μg/m^3) was much higher than that in the other seasons. The variation of NH_4^+ coincided with that of SO_4^{2-}, indicating that NH_4^+ largely originated from ammonia neutralizing acidic species.

The average concentrations of major ions (NO_3^-, SO_4^{2-}, and NH_4^+) in this research were compared with observations from other world cities (shown in Table 3). Compared with the other cities, these three ion concentrations at YRD were lower than those measured in Jinan, Xi'an, and Raipur, but they were still higher than those in Beijing, Shanghai, and Guangzhou in China and the other major world cities. As a regional background site, the major ion concentrations at YRD were substantially higher than those in other rural sites, indicating serious secondary inorganic aerosol pollution at the YRD. This may be the result of intensive emissions of SO_2, NO_x, NH_3, and particulate matters in the Bohai Sea rim region (Zhang et al. 2009). Effective emission reduction measures are required to address the severe $PM_{2.5}$ pollution in this region.

The mass ratio of NO_3^-/SO_4^{2-} has been used as an indicator to show the relative importance of mobile (e.g., vehicles) versus stationary sources (e.g., power plant) in air pollution (Yao et al. 2002). In this study, the value of NO_3^-/SO_4^{2-} was 0.89 in spring, 0.40 in summer, 0.78 in autumn, and 0.67 in winter, with an annual mean of 0.57. It was substantially higher than 0.44 in Jinan (Gao et al. 2011), 0.46 in Xi'an (Zhang et al. 2011), and 0.41 in Fuzhou (Xu et al. 2012). The high NO_3^-/SO_4^{2-} values were most likely related to less usage of coal because petroleum products and natural gas are the main energy sources in local area, where there are many large oil and gas fields. However, they were significantly lower than that in the NO_3^-/SO_4^{2-} mass ratio of 2 in downtown Los Angeles (Kim et al. 2000), suggesting that stationary source emissions are also very important in the study area. The NO_3^-/SO_4^{2-} exhibited the lowest value in summer, which might hypothetically be the result of higher SO_4^{2-} levels and the evaporative loss of NO_3^- with a higher temperature.

Given that the sampling site is a coastal city, sea salt aerosol is the major source of Cl^- and Na^+. This was verified by the Cl^-/Na^+ molar ratio. The Cl^-/Na^+ ratios were 1.01, 1.06, and

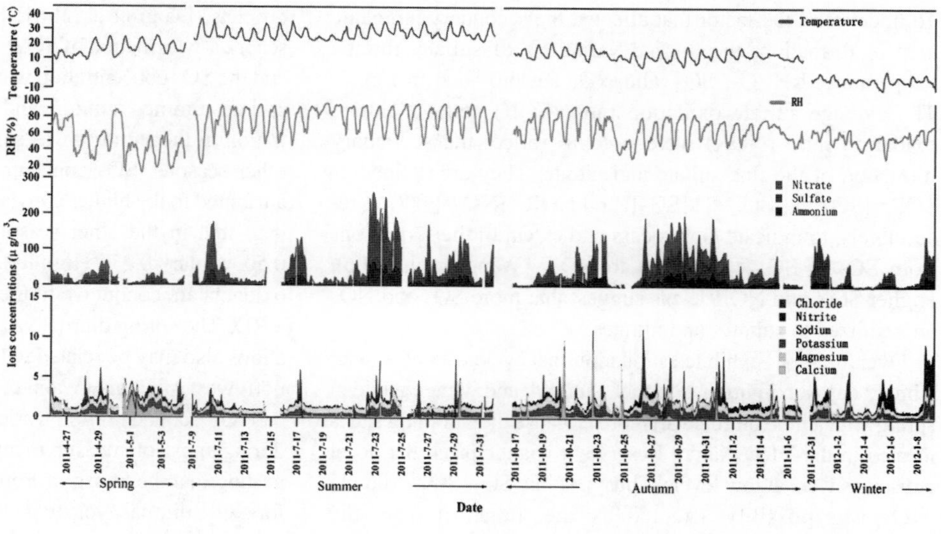

Fig. 2 Temporal variation of SO_4^{2-}, NO_3^-, NH_4^+, and meteorological conditions during the sampling period

Table 3 Mass concentrations of major chemical components in the YRD and other cities in the remainder of the world (μg/m^3)

Site	Type	Time	Major ion concentrations (μg/m^3)			References
			NO_3^-	SO_4^{2-}	NH_4^+	
Yellow River Delta	Rural/coastal	2011–2012	12.77	22.48	11.21	
Jinan, China	Urban	Dec. 2007–Oct. 2008	15.77	38.33	21.26	(Gao et al. 2011)
Xi'an, China	Urban	Mar. 2006–Mar. 2007	16.4	35.6	11.4	(Zhang et al. 2011)
Guangzhou, China	Urban	Apr. 2007	9.5	21.6	7.3	(Tao et al. 2009)
Beijing, China	Urban	June–August 2011	12.4	9	8	(Sun et al. 2012)
Beijing, China	Urban	June–August 2005	5.5	16.2		(Wu and Wang 2007)
Shanghai, China	Urban	May–June 2005	5.8	23.1		
Qingdao, China	Coastal	1997–2000	3.4	11.94	5.79	(Hu et al. 2002)
Xinken, China	Rural/coastal	Oct.–Nov. 2004	7.2	24.1	9.2	(Hu et al. 2008)
Chongming Island, China	Rural/coastal	Jun. 2006	10.89	23.14	10.28	(Li et al. 2010)
Jianfengling Nature Reserve, China	Rural/coastal	Nov. 2007	0.13	2.17	0.56	
Taibei summer	Urban	Aug. 2003	1.71	3.47	2.04	(Chang et al. 2007)
Taibei winter	Urban	Dec. 2003–Jan. 2004	3.66	12.01	5.67	
Gwangju, Korea	Rural	Oct.–Nov. 2003	2.89	3.86	2.62	(Hong et al. 2008)
Saitama City, Japan	Urban	Jan. 2010	5.67	2.26	2.22	(Kim et al. 2011)
Cario, Egypt	Urban	Nov. 2004–Mar. 2005	6.1	14.2	2.5	(Favez et al. 2008)
Raipur, India	Urban	Apr. 2005–Mar. 2006	8.16	46.5	8.76	(Verma et al. 2010)
Dearborn, USA	Urban	July–Aug. 2007	0.81	3.69		(Pancras et al. 2013)
Brigantine, USA	Rural/coastal	Nov. 2003	1.26	2.27	1.32	(Lee et al. 2008)

1.12 in spring, summer, and autumn, respectively, and 1.96 in winter, while the Cl^-/Na^+ molar ratio is 1.17 in seawater. The ratio was highest in winter, indicating that more part of Cl^- was related to nonmarine sources such as coal-fired emission. Cl^- levels decreased during the summer months as a consequence of its volatilization as HCl during the formation of $NaNO_3$ from gaseous HNO_3 and marine NaCl.

Secondary formation of SO_4^{2-} and NO_3^-

To understand the factor that affected the secondary inorganic aerosol formation, the diurnal variations of sulfate, nitrate, temperature, RH, O_3, SO_2, and NO_y are provided in Fig. 3. The average sulfate oxidation ratio (SOR) and the nitrate oxidation ratio (NOR) were used to reflect the secondary formation of the fine sulfate and nitrate. They are defined as $SOR=[SO_4^{2-}]/([SO_4^{2-}]+[SO_2])$ and $NOR=[NO_3^-]/[NO_y]$, respectively, to indicate the process and extent of the formations from SO_2 to SO_4^{2-} and NO_x to NO_3^- (Wang et al. 2006). Higher SOR and NOR levels suggest that more SO_2 and NO_x are oxidized to sulfates and nitrates.

Overall, SO_4^{2-} exhibited similar diurnal cycle with SO_2, with a broad day maximum in summer, autumn, and winter, while in spring with a little diurnal variation. Because the sampling site is located in the YRD Nature Reserve, it is anticipated that local emissions were at low levels. Thus, just like SO_2, SO_4^{2-} diurnal cycle was possibly affected by the transport from the surrounding areas concentrated with coal-fired power plants to this sampling site. In addition, secondary formation may be another factor in affecting diurnal variation of SO_4^{2-}. In summer, autumn, and winter, SO_4^{2-} concentration began to increase with the rising of sun when SO_2 and O_3 concentrations were rapidly elevated. The enhancing of solar radiation and O_3 concentrations promoted photochemical production of SO_4^{2-}. Generally, sulfates were produced from homogeneous oxidation of SO_2 by OH radicals by following subsequent condensation or from heterogeneous reactions of SO_2 on surfaces of aerosols or droplets (Dlugi et al. 1981; Meng and Seinfeld 1994). Thus, secondary formation of SO_4^{2-} depended on SO_2 concentration, and the SO_2 concentration was 12.1, 7.9, 12.7, and 20.0 ppbv in spring, summer, autumn, and winter. Although SO_4^{2-} concentration in the summer was significantly higher than that in the other seasons, SO_2 concentration was lowest. This could be attributed to the higher conversion rate (SOR=0.65) in summer than that in the other seasons (SOR spring=0.31, autumn=0.50, winter=0.27). High RH, temperature, and photochemical oxidants are conducive to the second generation in summer in YRD. The spring diurnal variation of SO_4^{2-} with small fluctuations also may be related with invasion of dust storm from the northwest of China (Wang et al. 2005).

NO_3^- concentration varied with that of SO_4^{2-} and NO_y during most of measurement period except in summer, suggesting that the transport from source areas may be one reason for NO_3^- diurnal cycle (e.g., the urban cities as Jinan and Zibo).

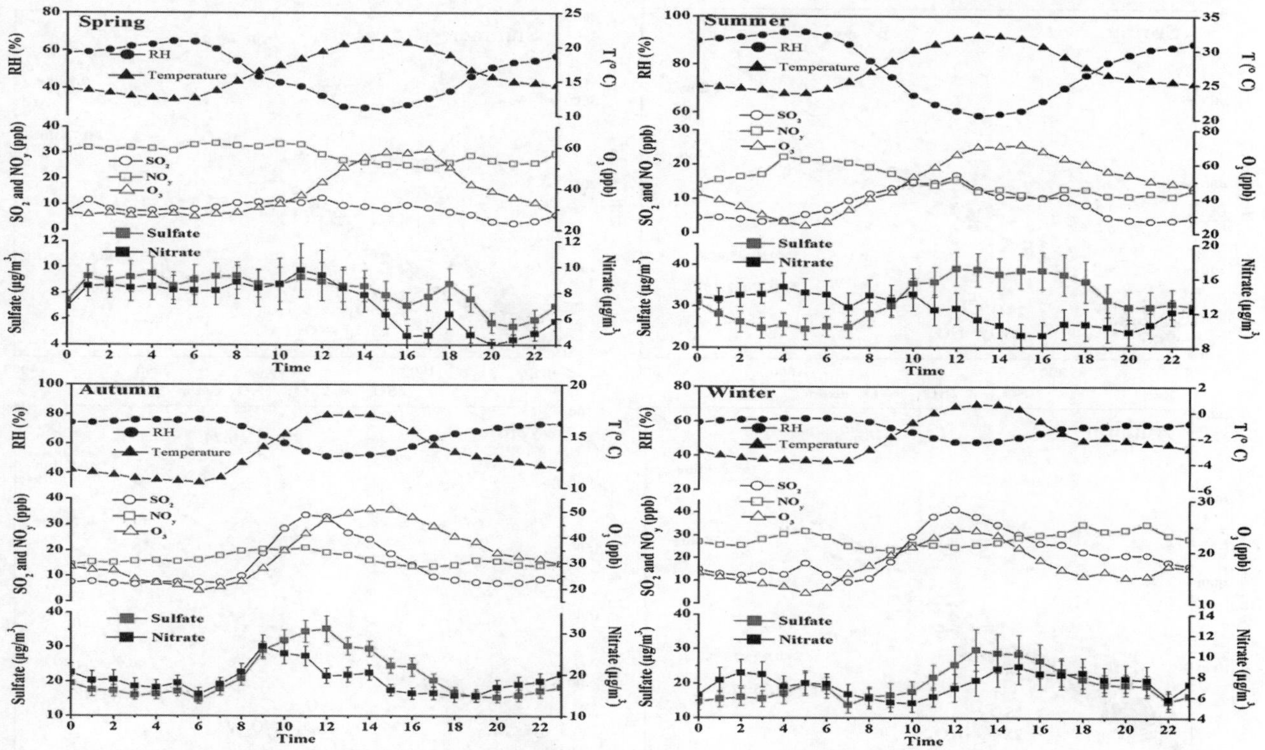

Fig. 3 Diurnal variation of SO_4^{2-} and NO_3^- and related parameters at the Yellow River Delta during different seasons

Besides, in spring, summer, and autumn, NO_3^- showed lowest concentration in the afternoon when the temperature was highest in a day, which was driven by the thermodynamic equilibrium of NO_3^-. It is well known that NH_4NO_3 tends to decompose into gaseous HNO_3 and NH_3 in hot and dry conditions. In winter, due to low temperature, thermodynamic equilibrium had small influence on NO_3^- concentration variation, while NO_3^- was mainly affected by transport in the afternoon and accumulation during the night. Thus, there were two peaks, which appeared in the afternoon and early in the morning. In addition, the nitrate concentrations varied with NO_y during most of measurement period except in spring with high correlation coefficient (average of 0.83 in summer, autumn, and winter, 0.45 in spring), suggesting that the NO_3^- could be mainly formed by the oxidation from NO_x. In our study, NOR values were obviously higher in summer (NOR=0.30) and autumn (NOR=0.29) than those in spring (NOR=0.07) and winter (NOR=0.10). Higher NOR value in summer and autumn led to high concentrations of NO_3^-.

Aerosol acidity

Aerosol acidity is an important parameter for investigating the aerosol phase reactions, acidity-dependent heterogeneous formation of secondary aerosols (Jang et al. 2002; Surratt et al. 2007). The neutralization degree ($F=NH_4^+/(2SO_4^{2-}+NO_3^-)$) was defined as the extent to which acidic aerosol is neutralized. In this study, the samples with F lower than 0.9 and 0.75 are regarded as acidic and more acidic aerosols (Zhou et al. 2012). It is noted that this method was not adapted while mineral ions (e.g., Ca^{2+}, Mg^{2+}, Na^+) contributed to a large proportion of $PM_{2.5}$, and these data were excluded from this analysis. Finally, a total of 1,450 sets of hourly data were used to calculate the neutralization degree, which accounted for approximately 95 % of the total measurements. Figure 4 shows the neutralization degree ($F=NH_4^+/(2SO_4^{2-}+NO_3^-)$). In this study, the aerosol at YRD presented as weakly acidic (annual average F of being 0.89). In spring and summer, acidic aerosols (F<0.9) accounted for 35.4 and 45.6 % of all samples, respectively, and only 11.1 and 8.8 %, respectively, of $PM_{2.5}$ samples presented strongly acidic (F<0.75). In autumn and winter, more than 50 % of $PM_{2.5}$ samples showed acidic (F<0.9), with 22.5 and 32.1 % being strongly acidic (F<0.75). These results indicated higher acidities in autumn and winter than those in spring and summer. Strong acidity (H^+_{strong}) and in situ acidity (H^+_{air}) were further modeled for acidic aerosols with F<1, and a total of 850 sets of hourly data were used to model the aerosol acidity. The pH, strong acidity (H^+_{strong}), in situ acidity (H^+_{air}), and water content of $PM_{2.5}$ obtained from AIM-II model at YRD and the other sites of China are provided in Table 4. The average concentrations of H^+_{strong} and H^+_{air} in winter were 251.24 and 76.03 nmol/m^3,

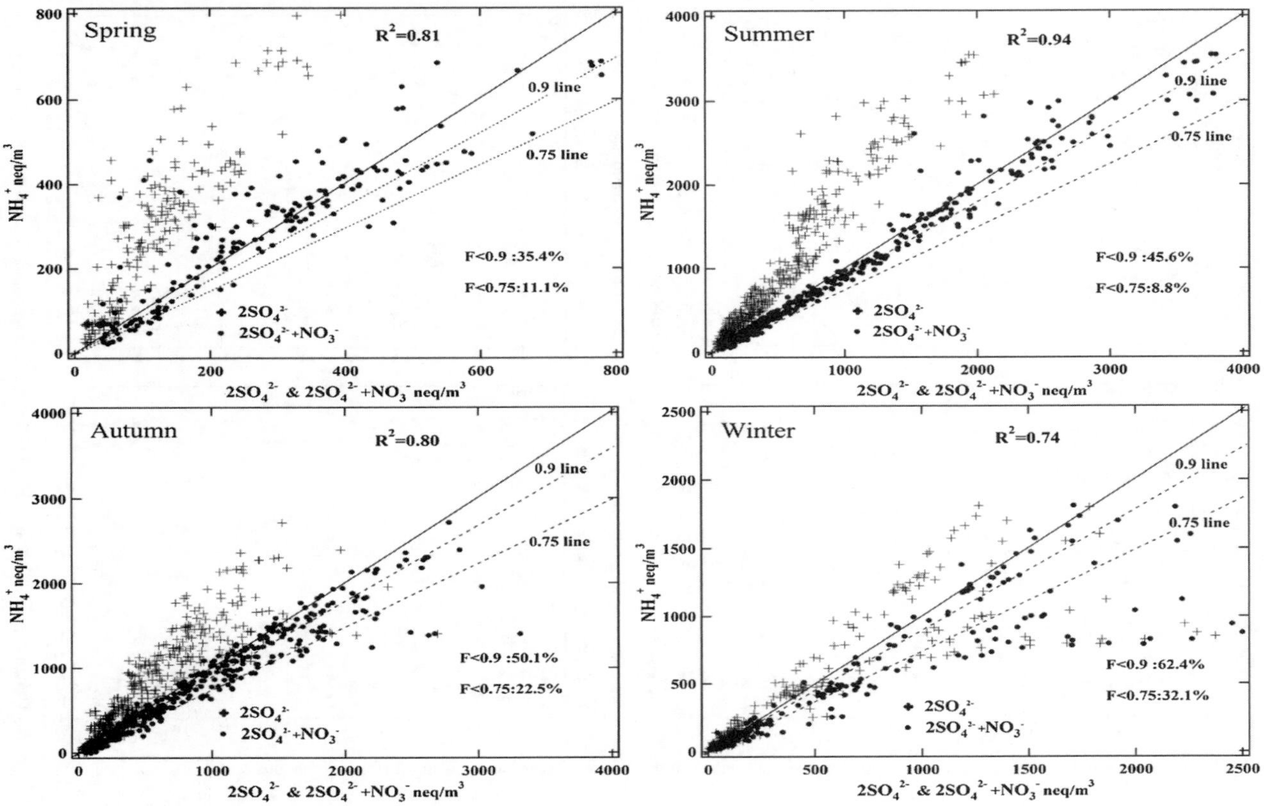

Fig. 4 The extent of neutralization of SO_4^{2-} and NO_3^- by NH_4^+ in $PM_{2.5}$ in different seasons

which were much higher than those in the other seasons. Compared with the other sites in China, the aerosol acidity at YRD was medium level, i.e., lower than that of Mt. Tai, Beijing, and Shanghai and higher than that of Mt. Heng, Lanzhou, and Guangzhou.

In addition, the molar ratio of $[NH_4^+]/[SO_4^{2-}]$ was also used to identify the ammonium sulfate and the reaction between nitrate and ammonia (Pathak et al. 2009). When the ratio is >1.5, the homogenous gas-phase formation of nitrate was significant by forming nitrate or the nitrate–sulfate salts of ammonium. After statistical analysis, the molar ratios of $[NH_4^+]/[SO_4^{2-}]$ in most samples were larger than 1.5 in spring, summer, and autumn, which suggested that NO_3^- was mainly formed by homogeneous reaction between the

Table 4 Aerosol acidity of $PM_{2.5}$ modeled by AIM-II at the YRD and other sites in China

Location	Season	pH	H_{air}^+ (nmol/m³)	H_{strong}^+ (nmol/m³)	Water content (μg/m³)	Reference
Yellow River Delta	Spring	−0.41±1.14	6.60±6.37	60.10±50.72	7.70±8.45	This study
	Summer	1.01±0.87	11.37±11.36	81.48±96.83	91.79±128.92	
	Autumn	0.23±1.15	24.31±49.86	132.93±183.07	48.50±58.88	
	Winter	−0.98±0.60	76.03±149.37	251.24±358.83	24.69±34.39	
	Annual	0.34±1.19	25.55±68.34	125.28±197.14	60.80±95.42	
Mt. Tai	Spring	−0.32±1.38	25.25±32.23	64.82±75.07	47.89±77.14	(Zhou et al. 2012)
Mt. Tai	Summer	−0.04±1.01	35.27±30.38	142.65±115.23	78.57±136.99	
Mt. Heng	Spring	0.64±0.96	13.3±15.4	53.4±38.7	67.7±143.3	(Gao et al. 2012)
Beijing	Summer	–	–	326±421	66±89	(Pathak et al. 2011)
Shanghai	Summer	–	–	196±201	27±24	
Lanzhou	Summer	–	–	59±49	9±12	
Guangzhou	Summer	–	–	90±60	60±80	

"–" indicates no data

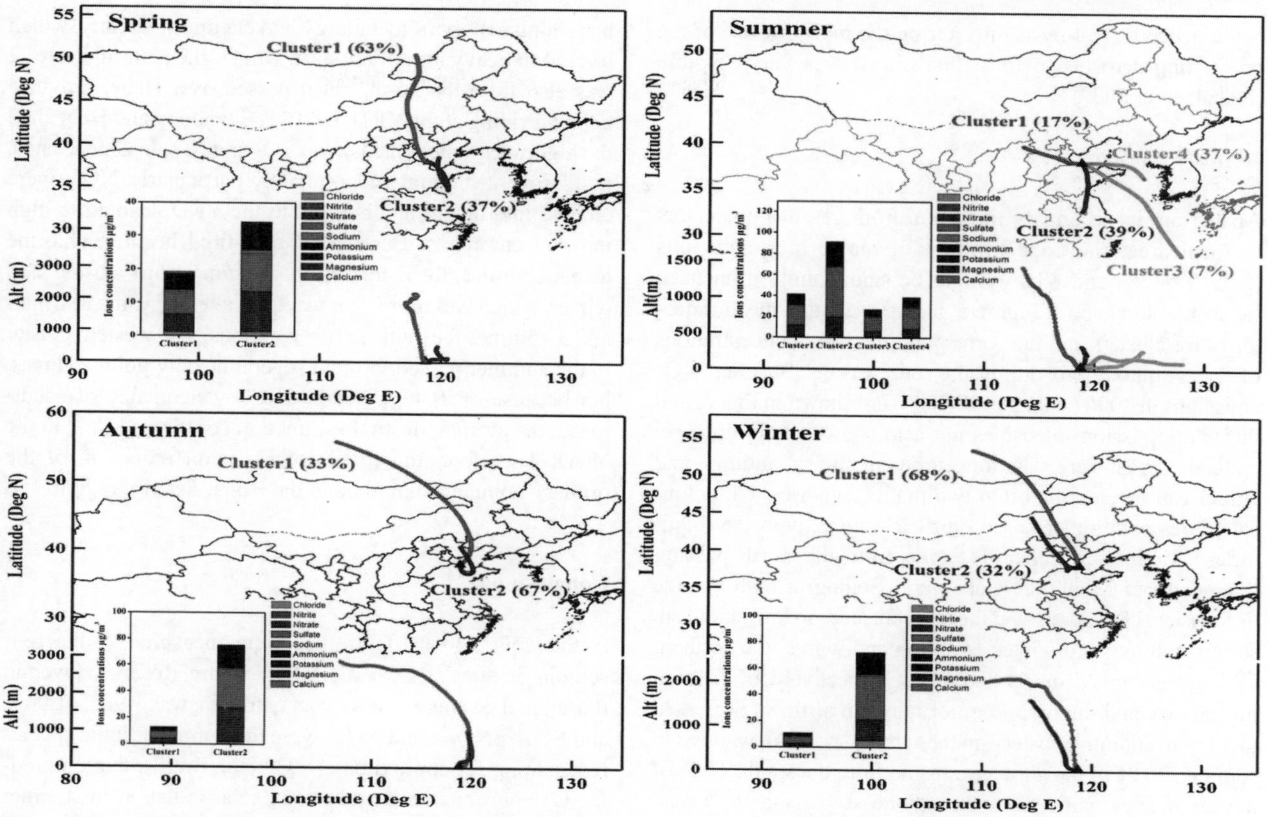

Fig. 5 Three-day backward trajectories of air mass and corresponding ion concentrations of the clusters at the Yellow River Delta during different seasons

ambient ammonia and nitric acid. While about 30 % of samples were ammonium poor in winter with their $[NH_4^+]/[SO_4^{2-}]$ molar ratios of below 1.5. In winter, the concentration of NH_4^+ was 7.13 μg/m^3, which is much lower than that in summer (15.06 μg/m^3). The low level of NH_4^+ was attributed to low emissions of NH_3 (Stelson and Seinfeld 1982), and NH_3 played an important role in the fine aerosol acidity in the Bohai Sea rim region. Therefore, under poor-ammonium conditions,

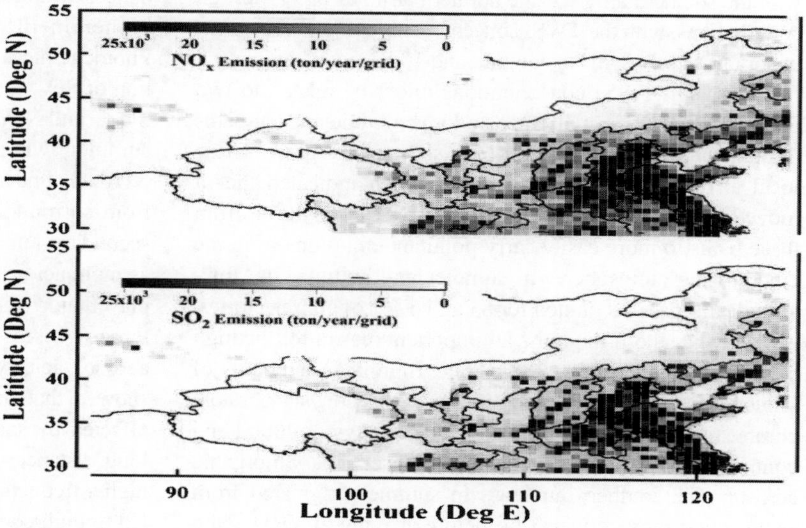

Fig. 6 SO_2 and NO_x emissions in China in 2006 (Zhang et al. 2009)

heterogeneous hydrolysis of N_2O_5 on the moist surface of the preexisting aerosols to form fine nitrate was very possible (Pathak et al. 2009).

Air mass transport

Apart from its secondary formation from gaseous precursors in local area, the concentration of major inorganic ions (SO_4^{2-}, NO_3^-, and NH_4^+) might be significantly influenced by air mass transport. Figure 5 shows the 3-day mean trajectories for clusters and the corresponding mean concentrations of water-soluble ions during the four seasons. SO_2 and NO_x emissions in 2006 (Zhang et al. 2009) are shown in Fig. 6, and different emission intensities are aslo shown along different paths of the clusters. The trajectories in spring, autumn, and winter can be classified into two main categories, while four categories in summer based on their origins, paths, and latitudes. Air masses generally came from the north passing through Inner Mongolia, Hebei, and Beijing at high altitude or from local Shandong and the adjacent areas in low altitude in spring, autumn, and winter. In spring and winter, the northern airflows accounted for 63 and 68 %, respectively, of the total trajectories, and the proportion of northern airflows decreased to 33 % in autumn (cluster 1 in these three seasons), whereas in summer, 39 % of the air masses came from the south, 44 % of the air masses came from the southeast passing the ocean (clusters 3 and 4), and only 17 % came from the northwest passing through Shanxi and Hebei province (cluster 1).

Generally, the air masses from local areas contained the highest concentrations of TWSI with concentrations being 39.22 μg/m^3 in spring (cluster 2), 75.19 μg/m^3 in autumn (cluster 2), and 91.25 μg/m^3 in summer (cluster 2). SO_4^{2-} levels from the southern air masses were highest in summer (47.53 μg/m^3), and NO_3^- of the southern airflows had the highest concentrations in autumn (25.63 μg/m^3). Compared with the southern airflows, the northern airflows had lower ion contributions, with the TWSI concentrations being 19.22 μg/m^3 in spring, 14.30 μg/m^3 in autumn, and 10.52 μg/m^3 in winter.

Local high TWSI concentrations might be related to two main causes: (1) short distance and low altitude and (2) paths through heavily polluted areas (as shown in Fig. 6). These local airflows were short and low, which indicated that it moved more slowly at lower altitude. This caused the air from these paths to more easily carry pollutant emissions from the cities to the atmosphere. In summer and autumn, the low-altitude air mass accounted for 83 and 67 % of all trajectories, respectively, and it might be an important reason for the high concentrations of sulfate and nitrate. Significant amounts of industrial activities and surface primary emission, such as coal combustion and vehicle emission, led to heavy pollution and contributed to elevated concentrations of secondary inorganic aerosol. The southern air mass in summer originated from Jiangsu and moved over Zibo before arriving at YRD. Zibo has a long history of metallurgy and chemical industry, which has led to heavy air pollution. In winter, the path of cluster 2 was also from the north, but it passed over Hebei province before arriving at the YRD. In 2011, Shandong and Hebei had the highest emission intensity of SO_2 and NO_x in China. Thus, a significant amount of pollutants, particularly NO_x, were emitted into the air and brought to the YRD, leading to high ion concentrations. Because of coal-fired heating, chlorine levels significantly increased to 5 μg/m^3 from cluster 2 in winter, which was more than twice the average value in winter and 3~5 times the value in other seasons. The western cluster in the summer passed through several heavily polluted areas, but because of its high height and low frequency of occurrence, the air mass from this direction contributed little to the YRD. Therefore, the focus for emission reduction of the primary pollutant should be in the Bohai Sea rim region.

Conclusions

A total of 1530 h of samples were measured in the four seasons in the YRD at the YRD Nature Reserve. Overall, the annual average TWSI concentration was 49.72 μg/m^3, and SO_4^{2-}, NO_3^-, and NH_4^+ were the dominant ionic species (accounting for approximately 93 % of TWSI). High humidity, O_3 concentrations, and strong solar radiation in summer and autumn, and high emissions of SO_2 and NO_x from coal combustion in winter led to higher ion concentrations in these three seasons compared with spring. These three major ion concentrations at the YRD were lower compared with the cities in China and abroad; however, as a reserve, the YRD showed a high pollution level compared with the rural sites. Stationary source emissions were more important than vehicle emissions in the source areas, as implied by low ratios of NO_3^-/SO_4^{2-}. SO_4^{2-} concentration varied with SO_2 abundance, exhibited a peak near the noon, and it was significantly higher in the summer than that in the other seasons. Photochemical reaction played important roles on the production of SO_4^{2-}. The nitrate concentration changed with that of SO_4^{2-} and NO_y during most of measurement period except in summer, which is attributed to thermodecomposition of NH_4NO_3 under the condition of high temperature. Transport from surrounding areas contributed to the diurnal cycle of secondary ions. The aerosol in autumn and winter presented as more acidic, although the high NO_3^- and low NH_4^+ levels play major roles in affecting aerosol acidity at the YRD in these two seasons. Compared with other sites in China, the aerosol acidity at YRD was medium. A cluster analysis showed that fine particle pollution in the YRD was mainly affected by southwest local emissions and northern middle-long distance transmission. In the summer and autumn, the higher frequency of southern local and low-altitude air masses led to high concentrations of sulfate and nitrate.

Acknowledgments The authors would like to thank Dong Can, Yan Chao, Meng Chuanping, Sui Xiao, and Yu Yangchun for their help in setting up the instruments. The HYSPLIT mode was provided by the NOAA Air Resources Laboratory. This work was supported by the National Basic Research Program (973 Program) of China (2005CB422203), Key Project of Shandong Provincial Environmental Agency (2006045), Promotive Research Fund for Young and Middle-aged Scientists of Shandong Province (BS2010HZ010), Independent Innovation Foundation of Shandong University (2009TS024), National Natural Science Foundation (21307074), and Special Research for Public-Beneficial Environment Protection (201009001-1).

References

Borja-Aburto VH, Castillejos M, Gold DR, Bierzwinski S, Loomis D (1998) Mortality and ambient fine particles in southwest Mexico City, 1993–1995. Environ Health Perspect 106:849–855

Chang SY, Lee CT, Chou CCK, Liu SC, Wen TX (2007) The continuous field measurements of soluble aerosol compositions at the Taipei Aerosol Supersite, Taiwan. Atmos Environ 41:1936–1949

Clegg SL, Brimblecombe P, Wexler AS (1998) Thermodynamic model of the system H^+-NH_4^+-SO_4^{2-}-NO_3^--H_2O at tropospheric temperatures. J Phys Chem A 102:2137–2154

Dlugi R, Jordan S, Lindemann E (1981) The heterogeneous formation of sulfate aerosols in the atmosphere. J Aerosol Sci 12:185–197

Draxler R, Rolph G (2003) HYSPLIT (HYbrid Single-Particle Lagrangian Integrated Trajectory) model access via NOAA ARL READY website (http://www.arl.noaa.gov/ready/hysplit4.html). NOAA Air Resources Laboratory, Silver Spring

Draxler R, Stunder B, Rolph G, Stein A, Taylor A (1999) Hysplit4 User Guide. National Oceanic and Atmospheric Administration Technical Memorandum ERL ARL-230

Favez O, Cachier H, Sciare J, Alfaro SC, El-Araby TM, Harhash MA et al (2008) Seasonality of major aerosol species and their transformations in Cairo megacity. Atmos Environ 42:1503–1516

Fu Q, Zhuang G, Wang J, Xu C, Huang K, Li J et al (2008) Mechanism of formation of the heaviest pollution episode ever recorded in the Yangtze River Delta, China. Atmos Environ 42:2023–2036

Gao J, Wang T, Ding A, Liu C (2005) Observational study of ozone and carbon monoxide at the summit of mount Tai (1534 m asl) in central-eastern China. Atmos Environ 39:4779–4791

Gao X, Yang L, Cheng S, Gao R, Zhou Y, Xue L et al (2011) Semi-continuous measurement of water-soluble ions in $PM_{2.5}$ in Jinan, China: temporal variations and source apportionments. Atmos Environ 45:6048–6056

Gao X, Xue L, Wang X, Wang T, Yuan C, Gao R et al (2012) Aerosol ionic components at Mt. Heng in central southern China: abundances, size distribution, and impacts of long-range transport. Sci Total Environ 433:498–506

He M, Sun Y, Li X, Yang Z (2006) Distribution patterns of nitrobenzenes and polychlorinated biphenyls in water, suspended particulate matter and sediment from mid-and down-stream of the Yellow River (China). Chemosphere 65:365–374

Hong S, Kim D, Ryu S, Kim Y, Lee J (2008) Chemical characteristics of $PM_{2.5}$ ions measured by a semicontinuous measurement system during the fall season at a suburban site, Gwangju, Korea. Atmos Res 89:62–75

Hu M, He LY, Zhang YH, Wang M, Pyo Kim Y, Moon K (2002) Seasonal variation of ionic species in fine particles at Qingdao, China. Atmos Environ 36:5853–5859

Hu M, Wu Z, Slanina J, Lin P, Liu S, Zeng L (2008) Acidic gases, ammonia and water-soluble ions in PM2.5 at a coastal site in the Pearl River Delta, China. Atmos Environ 42:6310–6320

Jang M, Czoschke NM, Lee S, Kamens RM (2002) Heterogeneous atmospheric aerosol production by acid-catalyzed particle-phase reactions. Science 298:814–817

Jung J, Lee H, Kim YJ, Liu X, Zhang Y, Gu J et al (2009) Aerosol chemistry and the effect of aerosol water content on visibility impairment and radiative forcing in Guangzhou during the 2006 Pearl River Delta campaign. J Environ Manage 90:3231–3244

Kim BM, Teffera S, Zeldin MD (2000) Characterization of PM2. 5 and PM10 in the south coast air basin of southern California: Part 1-Spatial variations. J Air Waste Manag Assoc 50:2034–2044

Kim KW, Kim YJ, Bang SY (2008) Summer time haze characteristics of the urban atmosphere of Gwangju and the rural atmosphere of Anmyon, Korea. Environ Monit Assess 141:189–199

Kim KH, Sekiguchi K, Furuuchi M, Sakamoto K (2011) Seasonal variation of carbonaceous and ionic components in ultrafine and fine particles in an urban area of Japan. Atmos Environ 45:1581–1590

Lee T, Yu XY, Kreidenweis SM, Malm WC, Collett JL (2008) Semi-continuous measurement of PM2. 5 ionic composition at several rural locations in the United States. Atmos Environ 42:6655–6669

Li L, Wang W, Feng J, Zhang D, Li H, Gu Z et al (2010) Composition, source, mass closure of $PM_{2.5}$ aerosols for four forests in eastern China. J Environ Sci 22:405–412

Meng Z, Seinfeld JH (1994) On the source of the submicrometer droplet mode of urban and regional aerosols. Aerosol Sci Technol 20:253–265

Ocskay R, Salma I, Wang W, Maenhaut W (2006) Characterization and diurnal variation of size-resolved inorganic water-soluble ions at a rural background site. J Environ Monit 8:300–306

Oms M, Jongejan P, Veltkamp A, Wyers G, Slanina J (1996) Continuous monitoring of atmospheric HCl, HNO2, HNO3 and SO2 by wet-annular denuder air sampling with on-line chromatographic analysis. Int J Environ Anal Chem 62:207–218

Pancras JP, Landis MS, Norris GA, Vedantham R, Dvonch JT (2013) Source apportionment of ambient fine particulate matter in Dearborn, Michigan, using hourly resolved PM chemical composition data. Sci Total Environ 448:2–13

Pathak RK, Wu W, Wang T (2009) Summertime PM 2.5 ionic species in four major cities of China: nitrate formation in an ammonia-deficient atmosphere. Atmos Chem Phys 9:1711–1722

Pathak RK, Wang T, Ho K, Lee S (2011) Characteristics of summertime $PM_{2.5}$ organic and elemental carbon in four major Chinese cities: Implications of high acidity for water-soluble organic carbon (WSOC). Atmos Environ 45:318–325

Schwartz J, Dockery DW, Neas LM (1996) Is daily mortality associated specifically with fine particles? J Air Waste Manage Assoc 46:927–939

Stelson A, Seinfeld J (1982) Relative humidity and temperature dependence of the ammonium nitrate dissociation constant. Atmos Environ 16:983–992, 1967

Stolzenburg MR, Hering SV (2000) Method for the automated measurement of fine particle nitrate in the atmosphere. Environ Sci Technol 34:907–914

Sun Y, Zhuang G, Tang A, Wang Y, An Z (2006) Chemical characteristics of PM2.5 and PM10 in haze-fog episodes in Beijing. Environ Sci Technol 40:3148–3155

Sun Y, Wang Z, Dong H, Yang T, Li J, Pan X et al (2012) Characterization of summer organic and inorganic aerosols in Beijing, China with an Aerosol Chemical Speciation Monitor. Atmos Environ 51:250–259

Surratt JD, Kroll JH, Kleindienst TE, Edney EO, Claeys M, Sorooshian A, Ng NL, Offenberg JH, Lewandowski M, Jaoui M (2007) Evidence for organosulfates in secondary organic aerosol. Environ Sci Technol 41:517–527

Tao J, Ho KF, Chen L, Zhu L, Han J, Xu Z (2009) Effect of chemical composition of PM2.5 on visibility in Guangzhou, China, 2007 spring. Particuology 7:68–75

Trebs I, Meixner F, Slanina J, Otjes R, Jongejan P, Andreae M (2004) Real-time measurements of ammonia, acidic trace gases and water-soluble inorganic aerosol species at a rural site in the Amazon Basin. Atmos Chem Phys 4:967–987

Underwood G, Song C, Phadnis M, Carmichael G, Grassian V (2001) Heterogeneous reactions of NO2 and HNO3 on oxides and mineral dust: A combined laboratory and modeling study. J Geophys Res 106:18055–18066

Verma SK, Deb MK, Suzuki Y, Tsai YI (2010) Ion chemistry and source identification of coarse and fine aerosols in an urban area of eastern central India. Atmos Res 95:65–76

Wang T, Poon C, Kwok Y, Li Y (2003) Characterizing the temporal variability and emission patterns of pollution plumes in the Pearl River Delta of China. Atmos Environ 37:3539–3550

Wang Y, Zhuang G, Tang A, Yuan H, Sun Y, Chen S, Zheng A (2005) The ion chemistry and the source of $PM_{2.5}$ aerosol in Beijing. Atmos Environ 39:3771–3784

Wang Y, Zhuang G, Zhang X, Huang K, Xu C, Tang A et al (2006) The ion chemistry, seasonal cycle, and sources of $PM_{2.5}$ and TSP aerosol in Shanghai. Atmos Environ 40:2935–2952

Wu WS, Wang T (2007) On the performance of a semi-continuous $PM_{2.5}$ sulphate and nitrate instrument under high loadings of particulate and sulphur dioxide. Atmos Environ 41:5442–5451

Xu L, Chen X, Chen J, Zhang F, He C, Zhao J et al (2012) Seasonal variations and chemical compositions of $PM_{2.5}$ aerosol in the urban area of Fuzhou, China. Atmos Res 104:264–272

Yao X, Chan CK, Fang M, Cadle S, Chan T, Mulawa P et al (2002) The water-soluble ionic composition of PM2.5 in Shanghai and Beijing, China. Atmos Environ 36:4223–4234

Zhang Q, Streets DG, Carmichael GR, He K, Huo H, Kannari A et al (2009) Asian emissions in 2006 for the NASA INTEX-B mission. Atmos Chem Phys 9:5131–5153

Zhang T, Cao J, Tie X, Shen Z, Liu S, Ding H et al (2011) Water-soluble ions in atmospheric aerosols measured in Xi'an, China: Seasonal variations and sources. Atmos Res 102:110–119

Zhou Y, Xue L, Wang T, Gao X, Wang Z, Wang X et al (2012) Characterization of aerosol acidity at a high mountain site in central eastern China. Atmos Environ 51:11–20

Atmospheric Environment 63 (2012) 68–76

Contents lists available at SciVerse ScienceDirect

Atmospheric Environment

journal homepage: www.elsevier.com/locate/atmosenv

The secondary formation of inorganic aerosols in the droplet mode through heterogeneous aqueous reactions under haze conditions

Xinfeng Wang, Wenxing Wang*, Lingxiao Yang, Xiaomei Gao, Wei Nie, Yangchun Yu, Pengju Xu, Yang Zhou, Zhe Wang

Environment Research Institute, Shandong University, Ji'nan, Shandong 250100, China

HIGHLIGHTS

► Size-resolved inorganic aerosol compositions were analyzed for a highly polluted area.
► Humidity-dependent heterogeneous formation of inorganic aerosols was investigated.
► Relationships were given between the SO_4^{2-}, NO_3^-, NH_4^+ in the droplet mode and the RH.
► Dominant formation pathways of sulfates and nitrates were analyzed for haze events.

ARTICLE INFO

Article history:
Received 6 June 2012
Received in revised form
3 September 2012
Accepted 10 September 2012

Keywords:
Secondary inorganic aerosols
Mass size distribution
Droplet mode
Heterogeneous aqueous reaction
Haze pollution
Jinan

ABSTRACT

Secondary inorganic aerosols play important roles in visibility reduction and in regional haze pollution. To investigate the characteristics of size distributions of secondary sulfates and nitrates as well as their formation mechanisms under hazes, size-resolved aerosols were collected using a Micro-Orifice Uniform Deposit Impactor (MOUDI) at an urban site in Jinan, China, in all four seasons (December 2007–October 2008). In haze episodes, the secondary sulfates and nitrates primarily formed in fine particles, with elevated concentration peaks in the droplet mode (0.56–1.8 μm). The fine sulfates and nitrates were completely neutralized by ammonia and existed in the forms of $(NH_4)_2SO_4$ and NH_4NO_3, respectively. The secondary formation of sulfates, nitrates and ammonium (SNA) was found to be related to heterogeneous aqueous reactions and was largely dependent on the ambient humidity. With rising relative humidity, the droplet-mode SNA concentration, the ratio of droplet-mode SNA to the total SNA, the fraction of SNA in droplet-mode particles and the mass median aerodynamic diameter of SNA presented an exponential, logarithmic or linear increase. Two heavily polluted multi-day haze episodes in winter and summer were analyzed in detail. The secondary sulfates were linked to heterogeneous uptake of SO_2 followed by the subsequent catalytic oxidation by oxygen together with iron and manganese in winter. The fine nitrate formation was strongly associated with the thermodynamic equilibrium among NH_4NO_3, gaseous HNO_3 and NH_3, and showed different temperature-dependences in winter and summer.

1. Introduction

The size distribution of aerosols is crucial in understanding the particle emissions, *in-situ* formation and the subsequent conversion processes of secondary aerosols and is also important in assessing the effects of aerosols on human health and the global radiation budget (Salma et al., 2002; Mather et al., 2003; Liu et al., 2008; Haywood et al., 2008). Typically, the mass distribution is dominated by three modes (or sub-modes): the condensation mode (~0.1–0.5 μm), the droplet mode (~0.5–2 μm) and the coarse mode (~2–50 μm) (John et al., 1990; Seinfeld and Pandis, 2006; Guo et al., 2010).

Haze is defined as the weather phenomenon featuring a high concentration of fine particles that leads to horizontal visibility below 10 km at a relative humidity (RH) less than 90% (Wu, 2006). Haze pollution is characterized by the elevated levels and high fractions of the secondary components of sulfates, nitrates and ammonium (SNA) in fine particles (Sun et al., 2006; Wang et al., 2006; Tan et al., 2009). Generally, the sulfates are primarily produced through the gas-phase oxidation of SO_2 by the OH radical followed by nucleation and condensational growth, or are produced by the heterogeneous uptake of SO_2 on pre-existing particles or

* Corresponding author. Tel.: +86 531 88369788; fax: +86 531 88361990.
E-mail address: wenxwang@hotmail.com (W. Wang).

1352-2310/$ – see front matter © 2012 Elsevier Ltd. All rights reserved.
http://dx.doi.org/10.1016/j.atmosenv.2012.09.029

cloud droplets followed by being oxidized by O_3, H_2O_2, or O_2 catalyzed by Fe(III) and Mn(II) (Berresheim and Jaeschke, 1986; Mather et al., 2003; Seinfeld and Pandis, 2006; Yao and Zhang, 2011). The fine nitrate formation is dominated by the reactions of gaseous HNO_3 or nitric acid in droplets with NH_3 under ammonia-rich conditions or by heterogeneous hydrolysis of N_2O_5 on aerosol surfaces in ammonia-poor environments (Schryer, 1982; Seinfeld and Pandis, 2006; Pathak et al., 2009). Ammonium in fine mode particles is mostly combined with sulfates and/or nitrates (Feng and Penner, 2007). Several recent studies have emphasized the important roles of heterogeneous reactions in secondary aerosol formation during polluted scenarios. Field measurements during a wood smoke episode and the subsequent modeling simulation indicate the evidence of secondary sulfate formation via heterogeneous surface reactions on smoke particles (Buzcu et al., 2006). Both laboratory experiments and *in-situ* observations also indicate the importance of the aqueous oxidation of SO_2 catalyzed by metals in sulfate formation under humid conditions (Turšič et al., 2003; Li et al., 2011). During particulate matter (PM) pollution episodes, the dominant formation pathway of fine nitrates is found to vary with the location as well as with the sampling period and is strongly dependent on the ambient abundance of ammonia (e.g., Pathak et al., 2009; Wang et al., 2009; Ye et al., 2011). In polluted environments of high particle loading and high levels of gas precursors in China, the aerosol mass size distributions usually exhibit a sharp peak within the size range of 0.56–1.8 μm when fine particle pollution occurred (Hu et al., 2005; Huang et al., 2006; Guo et al., 2010; Cheng et al., 2011b). In spite of a large contribution of droplet-mode aerosols to the PM, field studies have rarely focused on the secondary formation of droplet-mode aerosols in such environments.

Jinan, the capital of the Shandong Province, has a population of 6.8 million and an area of 8177 square kilometers. It is located near the center of the North China Plain. The city is surrounded by mountains to the south and hills to the east (Fig. 1), which is disadvantageous for air pollutants to disperse. With the rapid urbanization and industrial development in the past several decades, Jinan has suffered from deteriorated air quality in particular, from heavy PM pollution (Yang et al., 2007; Cheng et al., 2011a). In 2008, the average aerosol optical depth over Jinan was very high – exceeding 0.6 (see Fig. 1a). The annual average $PM_{2.5}$ in urban Jinan was at a level of 124–156 μg m^{-3} (Cheng et al., 2011a). Sulfates, nitrates and ammonium, the dominant water-soluble ions, contributed approximately 47% of the fine particles. The elevated aerosol loading is responsible for reducing atmospheric visibility. Jinan experiences haze pollution on nearly one-third of the days in a year (Cheng et al., 2011b). Furthermore, the MODIS (moderate-resolution imaging spectroradiometer) true-color images of haze episodes over this area always show a regional-scale brown haze, covering most of the North China Plain (e.g., Li et al., 2011).

To obtain a comprehensive understanding of the droplet-mode aerosol formation and the involved chemical mechanisms, a large number of size-segregated aerosol samples were collected in an urban area in Jinan and underwent chemical analysis in the laboratory. We first compare the size-resolved aerosol compositions in haze episodes with those found on clear days to identify the characteristics of aerosol size distributions and chemical compositions for the haze pollution. Then, the humidity-dependent heterogeneous formation of SNA is analyzed, and the relationships between the droplet-mode SNA and the ambient humidity are given based on the field measurements. Two severe, multi-day haze episodes in winter and summer are also discussed in detail to investigate the dominant formation pathways of sulfates and nitrates in the droplet mode and to investigate the factors affecting the secondary aerosol formation.

2. Experiment and methods

2.1. Sampling site

Size-segregated aerosols were collected from an urban area of Jinan, China. The sampling site is located on the rooftop of a seven-story teaching building in the center campus of Shandong University (36.67° N, 117.05° E and 50 m asl) and is surrounded by a residential area. There are several large-scale industries in Jinan, including steel plants, thermal power plants, cement plants, oil refineries and chemical plants, which are considered the major industrial emission sources of the local air pollution. Size-resolved aerosol samples were collected discontinuously in all four seasons from November 2007 to October 2008. Normally, atmospheric aerosols were collected over 24-h periods every two or three days. When pollution episodes occurred, sample collection was increased to take place once or twice per day. During the sampling periods, the relative humidity in Jinan covered a large range of 8%–91%, and the temperatures varied from −2 to 32 °C.

2.2. Sample collection and analysis

In this study, a MOUDI (Micro-Orifice Uniform Deposit Impactor, Model 110 with rotator, MSP) was deployed to collect the aerosol samples at a flow rate of 30 L min^{-1}. The MOUDI has eight stages with size ranges of >18 μm (inlet), 10–18 μm, 5.6–10 μm, 3.2–5.6 μm, 1.8–3.2 μm, 1.0–1.8 μm, 0.56–1 μm, 0.32–0.56 μm, 0.18–

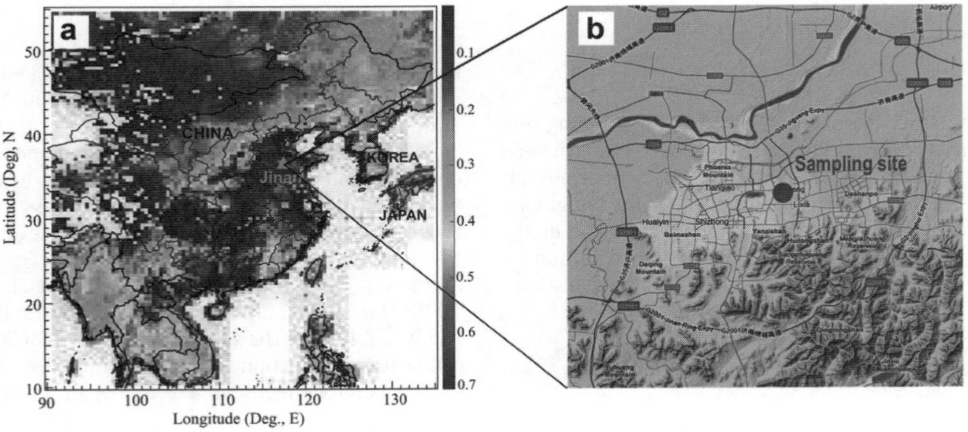

Fig. 1. Map showing the study area, (a) the aerosol optical depth in 2008 (http://aatsraerosol.fmi.fi/DwnLdPage_Globe.html) and (b) the location of our sampling site.

0.32 μm. Aluminum substrates (MSP) were used to collect the aerosol samples. Before sampling, the substrates were pre-heated to 500 °C using a muffle furnace to remove residual organics. After sampling, the aluminum substrates were placed in plastic Petri dishes and stored in a refrigerator at a temperature below −5 °C for subsequent gravimetric and chemical analysis in the laboratory. The substrates were then weighed using a microbalance (ME5, Sartorius). Before weighing, the substrates were balanced for 48 h with constant temperature (20 ± 0.5 °C) and relative humidity (50 ± 2%). As found in our previous study by Nie et al. (2010) during the summer in Beijing, the use of aluminum substrates in MOUDI sampler is subject to apparent evaporation loss for ammonium nitrates in high-temperature conditions. For the present study in Jinan, the temperature was moderate, and the sampling artifacts were generally less than 20%.

In the laboratory, the atmospheric aerosol samples collected on aluminum substrates were completely dissolved in deionized water by ultrasonication. Inorganic water-soluble ions in the sample solutions were detected using an ion chromatograph (ICS-90, Dionex). The anions, including F^-, Cl^-, NO_3^- and SO_4^{2-}, were analyzed using an AS14A Column with an AMMS 300 Suppresser and were eluted with 3.5 mmol L^{-1} Na_2CO_3–1.0 mmol L^{-1} $NaHCO_3$. The cations Na^+, NH_4^+, K^+, Mg^{2+} and Ca^{2+} were analyzed using a CS12A Column with a CSRS Ultra II Suppresser and were eluted with 20 mmol L^{-1} methanesulfonic acid.

Concurrently, a number of other air pollutants and parameters were measured during this study. SO_2 was measured using a pulsed UV fluorescence analyzer (Model 43C, TEC – Thermo Electron Corporation), and O_3 was measured using a UV photometric analyzer (Model 49C, TEC). Nitric oxide (NO) and NO_2 were measured using a commercial chemiluminescence analyzer equipped with a molybdenum oxide catalytic converter. Fine particulate matter was also collected using a four-channel $PM_{2.5}$ sampler (RAAS 2.5-400, Thermo Andersen) and was followed by a chemical analysis of the inorganic water-soluble ions (ICS-90, Dionex), OC and EC (Sunset-DOSCOCEC, Sunset Lab), and trace elements (X-ray fluorescence). In addition, the ambient temperature and relative humidity were measured using a portable automatic meteorological station. The visibility data used in this study were obtained from the Weather Underground, with a resolution of 3 h. These data agreed well with those obtained from visual inspection during the sampling periods.

2.3. Data processing

The cut point of the MOUDI is defined as a diameter with aerosol collection efficiency of 50% (Marple et al., 1991), which means that the collected particles in each size bin exhibit an approximately normal distribution against the particle diameter on a logarithmic scale. In this study, the geometric mean value of the two cut-point diameters was used as the average diameter of particles within the size range, and $dC/d\log_{10}Dp$ was calculated to represent the size-resolved concentrations in figures with an *X* axis of particle diameter on a logarithmic scale. To simplify the calculation, the particle modes were divided directly by the cut points. The condensation-mode particles were designated within the size range of 0.18–0.56 μm, droplet-mode particles were within 0.56–1.8 μm and coarse-mode particles were within 1.8–100 μm.

3. Results and discussions

3.1. Size-resolved aerosol composition

The size-segregated aerosol mass concentrations during the sampling periods are depicted in Fig. 2. As shown, the atmospheric

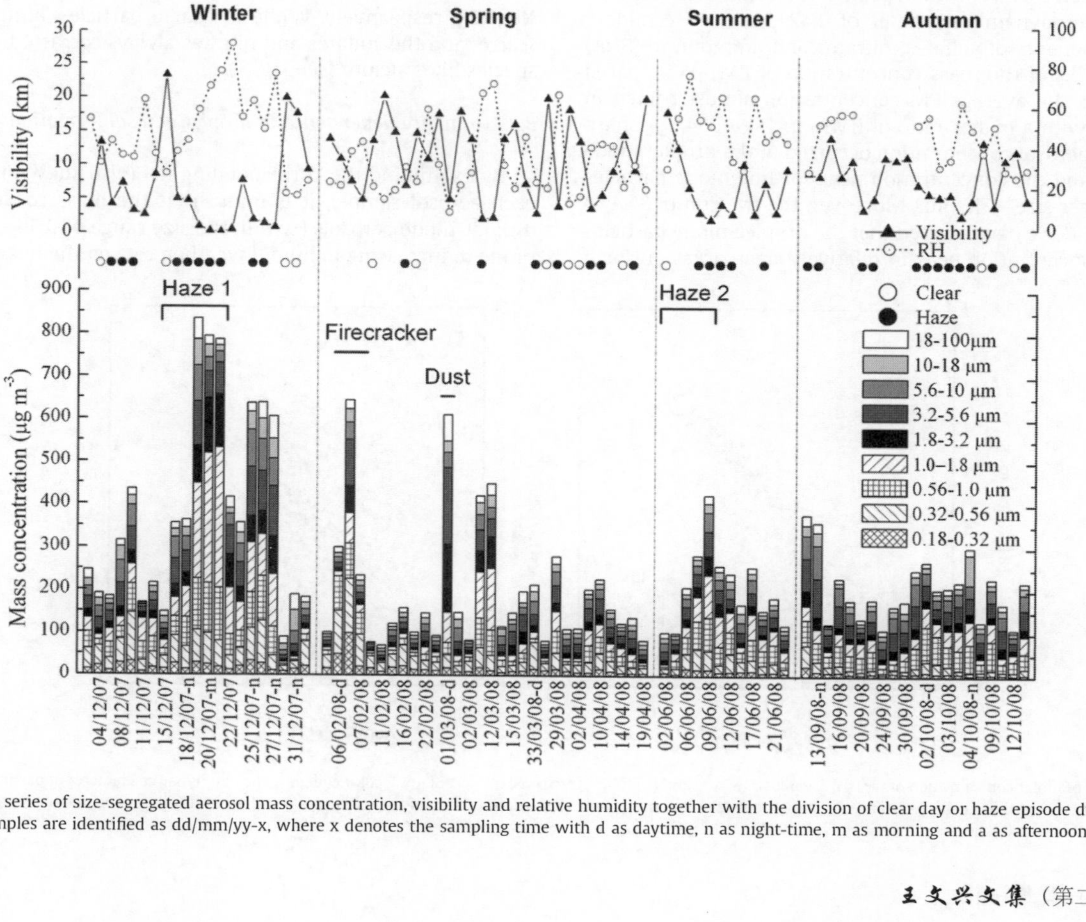

Fig. 2. The time series of size-segregated aerosol mass concentration, visibility and relative humidity together with the division of clear day or haze episode during the sampling periods. The samples are identified as dd/mm/yy-x, where x denotes the sampling time with d as daytime, n as night-time, m as morning and a as afternoon.

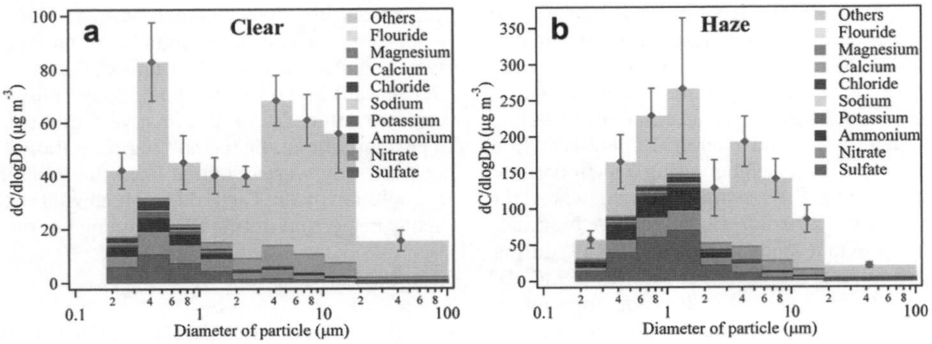

Fig. 3. Size-resolved aerosol compositions for (a) clear days ($n = 13$) and (b) haze episodes ($n = 34$). The marker shows the average mass concentration in each size bin with the error bar representing 1/3 of the standard deviation.

concentrations of PM in Jinan possessed a large variance. Both light and heavy particulate pollution episodes were observed in all four seasons. With a definition of a "clear" day as having all 3-h visibility above 10 km and "haze" as having all 3-h visibility below 10 km without any extreme weather event of fog, rain or snow, 13 sets of size-resolved aerosol samples were collected during clear days and 34 sets were collected during haze episodes (marked in Fig. 2). Most of the remaining sampling days featured haze pollution during part of the day and clear weather in the remaining time. In addition to the haze episodes, heavy PM pollution also occurred during the firecracker episodes (on the Chinese New Year's Eve and on the New Year's Day) and during a dust event (on 1 March 2008).

Fig. 3 compares the averaged aerosol compositions in different size bins for clear days and haze episodes. Generally, particulate matter was found in moderately low concentrations on clear days and was distributed evenly in the fine and coarse modes. The fine particles exhibited a concentration peak in the condensation mode with a mean aerodynamic diameter of 0.42 μm. The secondary inorganic components of sulfates, nitrates and ammonium (SNA) accounted for 32% of the mass concentration of $PM_{1.8}$. Compared with clear days, the average PM concentration almost tripled in haze episodes, with a higher ratio of $PM_{1.8}$ to TSP – 54%. A sharp increase in the aerosol concentration occurred in the droplet mode (0.56–1.8 μm) and the concentration peak for fine mode particles shifted to a larger size – 1.3 μm. Moreover, the fraction of SNA in $PM_{1.8}$ rose to 48%. We have noticed that the droplet-mode particles exhibit a flat concentration peak in relatively clean areas in North America and that the mean aerodynamic diameters are always below 1 μm for the MOUDI data (e.g., Mather et al., 2003; Kleeman et al., 2008; Yao and Zhang, 2011). While in the polluted environment in China, the concentration peaks for droplet-mode particles are sharp, and they always cover a broad size range from 0.56 to 1.8 μm (Hu et al., 2005; Huang et al., 2006; Guo et al., 2010). Compared to North America, the larger size of droplet-mode aerosols in polluted areas of China is believed to be associated with much higher levels of precursory air pollutants including both gases and particles.

Both on clear and hazy days in Jinan, the fine ammonium was found to be well correlated with the fine sulfates and nitrates (Fig. 4). The molar concentrations of ammonium in fine particles were higher than the molar concentrations of sulfates, and the excess ammonium ($NH_4^+ - 2*SO_4^{2-}$) almost equaled to the amount of nitrates, indicating the fine ammonium primarily combined with sulfates and nitrates in the forms of $(NH_4)_2SO_4$ and NH_4NO_3, respectively. While in coarse particles, ammonium was scarce, and the sulfates and nitrates always coexisted with crustal species like calcium (see Fig. 3).

3.2. Humidity dependence of droplet aerosol formation

By examining the corresponding RH (also shown in Fig. 2) for each aerosol sample, it is found that the mass concentration of droplet-mode aerosols (within the size range of 0.56–1.8 μm) was prone to increasing in humid weather, e.g., on the heavy hazy days

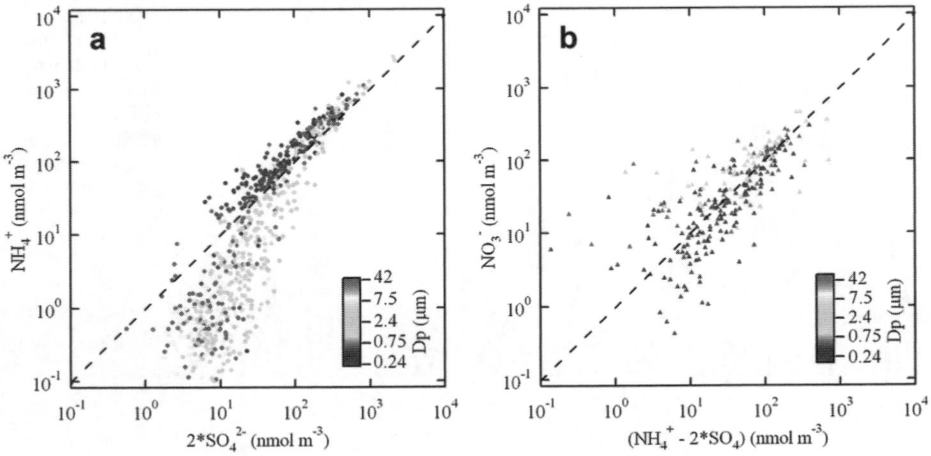

Fig. 4. Scatter plots of molar concentration of (a) NH_4^+ versus $2*SO_4^{2-}$ and (b) NO_3^- versus ($NH_4^+ - 2*SO_4^{2-}$), color coded by the aerodynamic diameter of particles. Note that all samples are included except those collected in the firecracker episodes and dust event. (For interpretation of the references to colour in this figure legend, the reader is referred to the web version of this article.)

on 19 and 20 December 2007. During all sampling periods, the ambient RH was generally lower than 80% and there was no fog. In the droplet-mode aerosols, abundant highly hygroscopic NH_4NO_3 significantly reduced the deliquescence RH to below 60% (Wexler and Seinfeld, 1991; Tang and Munkelwitz, 1993). Subsequent calculation using an aerosol thermodynamic model of E-AIM II (Extended Aerosol Inorganic Model, http://www.aim.env.uea.ac.uk/aim/aim.php) (Clegg et al., 1998) shows that liquid water can exist in droplet-mode aerosols even at an ambient RH of 20% (Fig. 5a). Generally, the mass ratio of liquid water to the droplet-mode particles was lower than 30% in this study; therefore, the moisture absorption had only a minor influence on the aerosol size. However, the aqueous phase on the aerosol surfaces provided the possibility of the heterogeneous gas–liquid conversion of gaseous precursors such as SO_2, HNO_3, N_2O_5, NO_x and NH_3 to produce secondary inorganic aerosols in the droplet mode at a fast rate.

To investigate the effects of humidity on the droplet-mode aerosol formation, scatter plots were shown of the droplet-mode SNA concentration, the ratio of droplet-mode SNA to total SNA, the fraction of SNA in droplet-mode particles and the MMAD (mass median aerodynamic diameter) of SNA versus the RH in Fig. 5b, c, d and e, respectively. The relationships between them were also given in the figures, which provide potential parameterized formulas for modeling studies of the secondary formation and size distribution of inorganic aerosols in this area. There was light rain for 4 h on the day with highest RH of 91% (22 December 2007); thus, the data point for that day was excluded from Fig. 5b & c. As observed in the figure, the droplet-mode SNA concentration was enhanced with rising RH. If the average RH was below 50%, the droplet-mode SNA concentration was generally lower than 40 μg m^{-3}. Once the RH was above 50%, the concentration of droplet-mode SNA increased sharply, with a maximum of 251 μg m^{-3} at the RH of 77%. Under very dry conditions, e.g., RH < 20%, only a small part (<30%) of SNA were produced in the droplet mode (see Fig. 5c). When the RH was above 50%, more than half of the SNA formed in the droplet mode. The ratio of the droplet-mode SNA to the total SNA (SNA-DL/SNA-total) tended to rise logarithmically with ascending RH. A logarithmic relationship is also found between the fraction of SNA in droplet-mode particles and the RH, and the fraction increased to over 45% when the RH

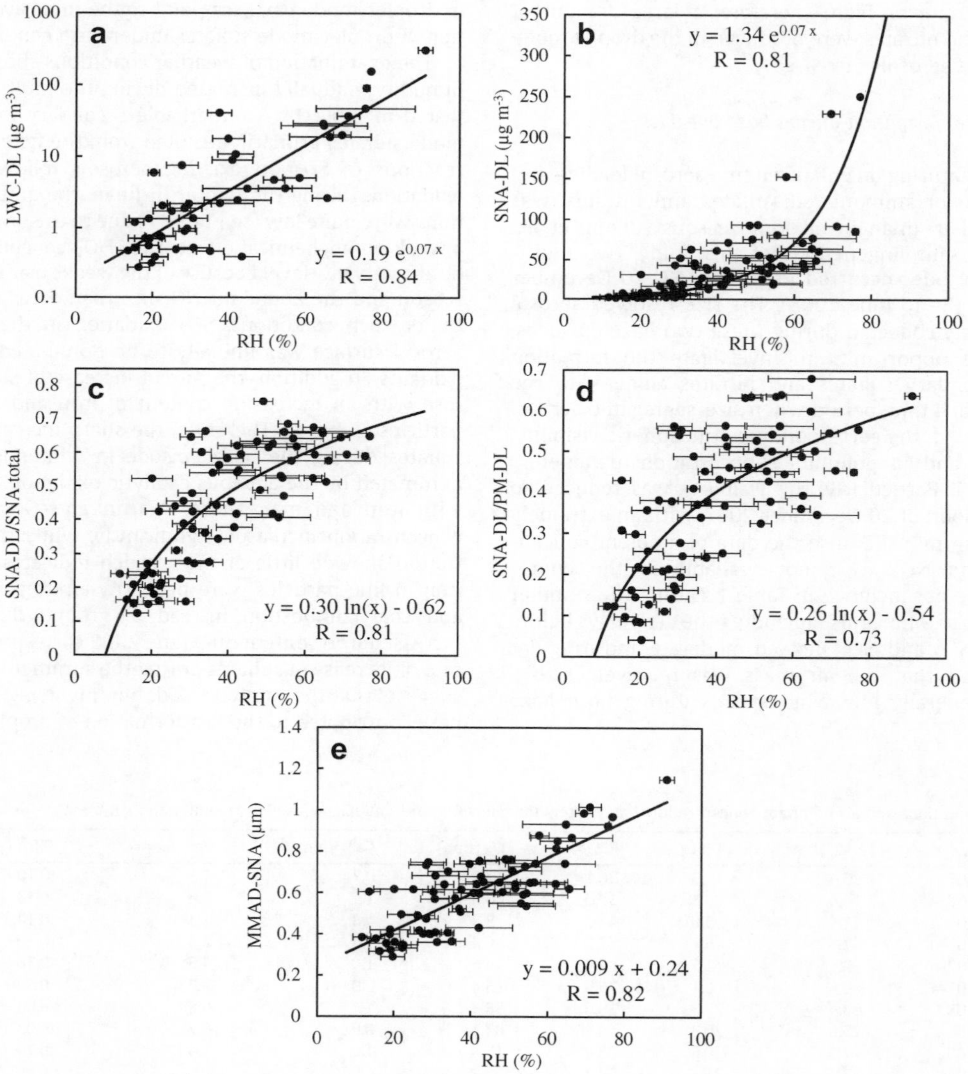

Fig. 5. Scatter plots of (a) liquid water content (LWC) in droplet-mode particles versus RH, (b) droplet-mode SNA concentration versus RH, (c) ratio of droplet-mode SNA to total SNA versus RH, (d) fraction of SNA in droplet-mode particles versus RH and (e) MMAD of SNA versus RH. Note that all data are included here except those collected during the dust event and that the error bar stands for 1/2 standard deviation.

was above 50% (Fig. 5d). Apparently, the elevated SNA-DL/SNA-total ratio in humid conditions would lead to the growth of particle size in the fine mode. As shown in Fig. 5e, the MMAD of SNA grew linearly with the RH. When the RH was 20%, 50% and 80%, the MMAD of SNA was approximately 0.4, 0.7 and 1.0 μm, respectively. In summary, the humidity-related heterogeneous aqueous reactions significantly enhanced the SNA concentrations and their fractions in droplet-mode aerosols under haze conditions and thus played an important role in the secondary formation of droplet-mode aerosols.

The heterogeneous formation of secondary aerosols is associated with the abundances of gas precursors, the aerosol surface area, the aerosol composition and the liquid water content in aerosols. In the urban area of Jinan, intensive source emissions lead to high levels of SO_2, NO_x and ammonia as well as to elevated aerosol loading. The gas precursors are relatively abundant and the aerosol surface area is relatively large, and thus the aerosol water content acts as the limiting factor for the occurrence of the heterogeneous uptake of gas precursors on the aerosol surface. Therefore, the heterogeneous formation of secondary aerosols enhances as the RH increases. During haze episodes in Jinan, the humidity was moderate or high; therefore, a large amount of secondary sulfates and nitrates were produced in the droplet mode and their fractions in aerosols increased.

3.3. *Secondary aerosol formation during haze episodes*

Haze pollution, featuring an enhanced fine aerosol loading and an elevated fraction of ammoniated sulfates and nitrates, was frequently observed in Jinan in all four seasons (Cheng et al., 2011b). During the sampling periods of this study, two multi-day, severe-haze episodes occurred in winter (17–20 December 2007) and summer (8–10 June 2008). The size-resolved aerosol samples were densely collected during these two haze episodes, which provided an opportunity to investigate the formation mechanism of secondary sulfates and nitrates along with the causal factors. For each time period when size-segregated aerosol samples were collected, the corresponding atmospheric visibility, weather conditions and air pollutant concentrations/parameters are listed in Table 1. Particularly, the visibility was reduced to 0.6 km in the afternoon of 20 December 2007, with an extremely high $PM_{1.8}$ of 532 μg m^{-3}. Because the data of ambient concentrations of SO_2 and NO_x were not available in the winter campaign, they were not included in Table 1. During the summer haze episode from 2 to 9 June (also including the clear days before the haze episode), SO_2 and NO_x showed moderate and irregular variance. It is noted that the air mass was relatively stable, characterized by generally low wind speeds during both haze episodes.

3.3.1. *Secondary formation of sulfates*

Fig. 6a & b clearly demonstrate the continuous increase of sulfate concentrations during the multi-day haze episodes. At the early stage, sulfates exhibited a flat concentration peak in the condensation mode with a size range of 0.32–0.56 μm (i.e., on 17 December 2007 and 2 June 2008). Following with aggravated particle pollution, droplet-mode sulfates increased remarkably, whereas condensation-mode sulfates showed only a slight increase. In the last stage of the haze episodes, the sulfate concentration peak became sharper and shifted to a larger size – droplet mode – compared with clear days in the beginning. During the haze episodes, sulfates exhibited an obvious positive increase in the droplet mode (0.56–1.8 μm) (see Fig. 6c & d). The integrated rate of increase of sulfates in $PM_{1.8}$ varied from 0.3 to 4.1 μg m^{-3} h^{-1} in the winter haze episode and from 0.2 to 1.0 μg m^{-3} h^{-1} in the summer haze episode. Furthermore, the fraction of sulfates in droplet-mode particles consistently became higher – from approximately 10% and gradually increasing to 30%. The particle size of sulfates became progressively larger, with the MMAD increasing from approximately 0.4 to 1.0 μm (Fig. 6e & f). The elevated sulfate concentration and the elevated fraction of sulfates in droplet-mode particles point to the intensive secondary formation of droplet-mode sulfates under haze conditions.

The examination of weather conditions shows that the relative humidity gradually increased during the multi-day haze episodes (listed in Table 1). A remarkable secondary formation of droplet mode sulfates primarily resulted from the heterogeneous aqueous reactions of SO_2 on existing aerosols followed by subsequent oxidation. During the winter in Jinan, the ambient O_3 concentrations were quite low (see Table 1) due to the low light intensity and a weak photochemical effect. The H_2O_2 concentration should also be at a very low level because of the weak solar radiation in the cold season and the abundant NO_x in urban areas (Kang et al., 2002). Under such conditions, SO_2 oxidation in the aqueous layer on aerosol surface was unlikely to be dominated by photochemical oxidants. In addition, the rate of increase of sulfate concentration rose with an increasing content of iron and magnesium in fine particles (Table 1). Therefore, the sharp increase of droplet-mode sulfates during the haze episode in winter is considered to be dominated by the aqueous catalytic oxidation by oxygen together with iron and manganese (Martin and Good, 1991). The iron concentration increased significantly, while the calcium concentration showed little change, which indicates that the abundant iron in fine particles were primarily emitted from steel smelting and coal combustion instead of ground dusts. Moreover, the increased iron content on 9 June 2008 corresponded to the highest rate of increase of sulfates during the summer haze episode, which suggests that the catalytic oxidation involving in iron and manganese contributed to the fast formation of droplet-mode sulfates in

Table 1
The visibility, weather conditions and air pollutant concentrations/parameters for selected multi-day haze episodes in winter and summer.

Season	Sample ID[a]	Visi. (km)	T (°C)	RH (%)	O_3 (ppb)	Ca[b] (μg m^{-3})	Fe[b] (μg m^{-3})	Mn[b] (μg m^{-3})	OC/EC[b]
Winter	15/12/07	25.4	3.5	26.1	7.8	0.7	1.3	0.10	5.0
	17/12/07	14.4	6.3	37.0	6.9	1.7	1.6	0.13	5.8
	18/12/07-n	11.8	5.6	42.7	8.3	0.8	1.3	0.14	7.8
	19/12/07-d	2.4	5.1	57.7	6.4	1.4	2.6	0.20	7.6
	20/12/07-m	2.4	1.5	69.4	4.6	1.0	5.5	0.36	8.3
	20/12/07-a	0.6	0.1	76.9	5.4	0.9	3.2	0.34	8.4
Summer	02/06/08	23.4	22.3	29.0	38.3	1.1	0.8	0.04	
	04/06/08	15.0	20.3	54.5	48.1	0.6	0.7	0.03	4.0
	06/06/08	7.3	20.6	75.4	15.4	1.0	1.4	0.10	5.1
	08/06/08	7.1	26.3	53.7	58.2				6.4
	09/06/08	7.2	27.2	50.5	38.2	1.5	2.4	0.13	6.7

[a] The "n" stands for night-time, "d" for daytime, "m" for morning and "a" for afternoon.
[b] The calcium, iron, magnesium concentrations and the OC/EC ratio were for $PM_{2.5}$.

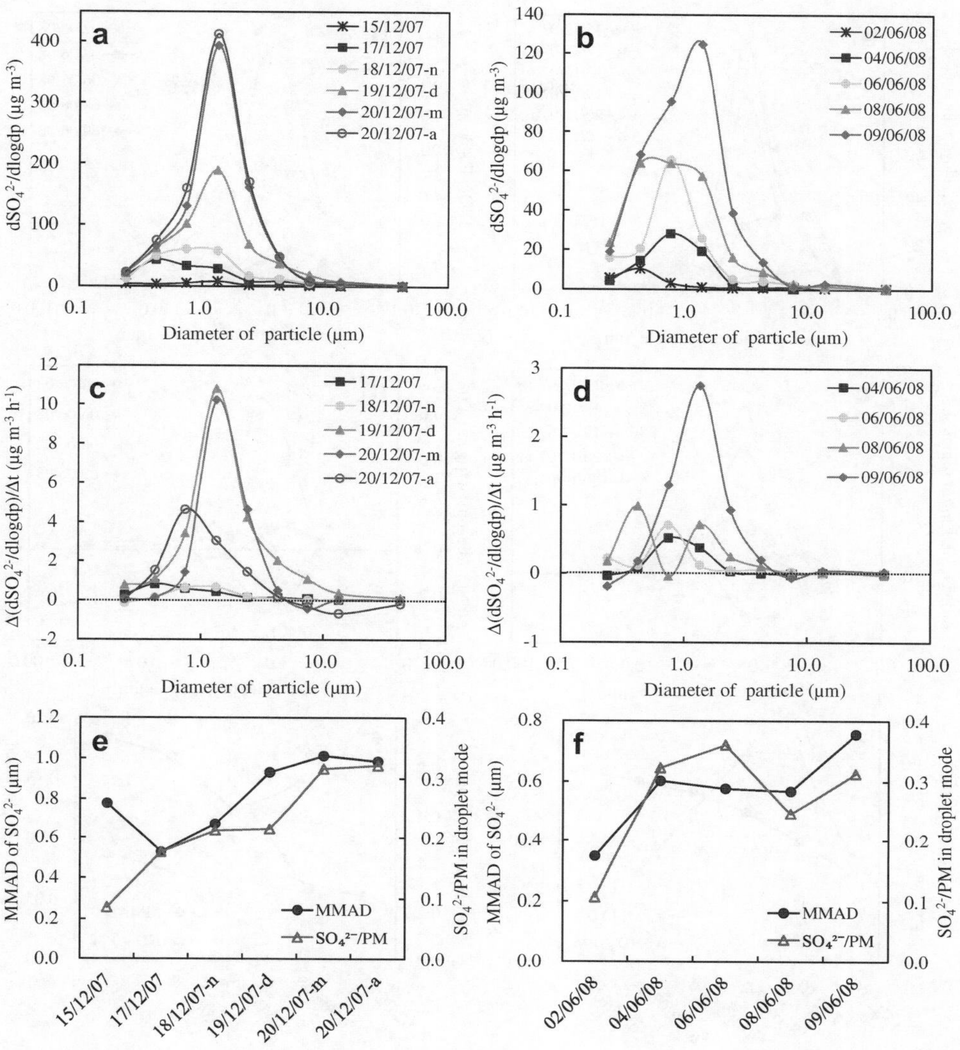

Fig. 6. Size distributions of the sulfate concentration and their rate of increase, variations of the sulfate MMAD and the sulfate fraction in droplet-mode particles for the selected multi-day haze episodes in (a, c & e) winter and (b, d & f) summer. Note: *increase rate* = $(C_n - C_{n-1})/(t_n - t_{n-1})$, with n standing for sample ID, C for concentration and t for the starting time of the sampling.

the hot season. The sulfates primarily formed in the droplet mode, and thus the concentration peak (and the MMAD) shifted to the droplet mode. During summer in Jinan, photochemical oxidant concentrations were relatively high because of the intense solar radiation together with the abundant precursors. In this condition, the homogeneous reaction of SO_2 with the OH radical followed by condensational growth also contributed to secondary sulfate formation; e.g., the daily average O_3 was 58.2 ppb on 8 June 2008 and an apparent increase was observed for the condensation-mode sulfates.

Additionally, the OC/EC ratio in the fine particles increased on heavily polluted hazy days (also shown in Table 1), which indicates relatively aged air masses from the regional-scale (Pio et al., 2011). The long-lived air mass during the haze episode was also favorable for the aerosols to grow to a larger size through coagulation and coalescence.

3.3.2. Secondary formation of nitrates

As shown in Fig. 7a & b, a remarkable increase of nitrates was also observed during the multi-day haze episodes. On clear days in the beginning, the fine nitrates exhibited a concentration peak in the condensation mode, whereas the coarse nitrates exhibited a moderately small peak in size range of 3.2–5.6 μm. With the aggravation of the haze episode, droplet-mode nitrates apparently increased and the concentration peak of fine nitrates shifted to a larger size – the droplet mode. The coarse nitrates were also enhanced by a small degree, possibly due to more abundant precursors and the accumulation of pollutants on hazy days. Compared to the sulfates, the temporal variation of fine nitrates was less consistent with the rising RH, which indicates that some other factors were likely to influence the formation processes of secondary nitrates. The rate of increase of $PM_{1.8}$ nitrates was generally low, with a maximum of 1.5 μg m^{-3} h^{-1} observed on 19 December 2007 (Fig. 7c & d). During the haze episode, the MMAD of nitrates rose gradually, from 0.44 to 0.96 μm in winter and from 0.62 to 0.93 μm in summer (see Fig. 7e & f). However, the fraction of nitrates in the droplet-mode particles did not undergo a continuous increase and always dropped back to below 10% on the heavily polluted hazy days. These suggest that the secondary formation of fine nitrates was suppressed under certain conditions.

As mentioned earlier, fine nitrates in Jinan always combined with ammonium in the form of NH_4NO_3; therefore, under haze

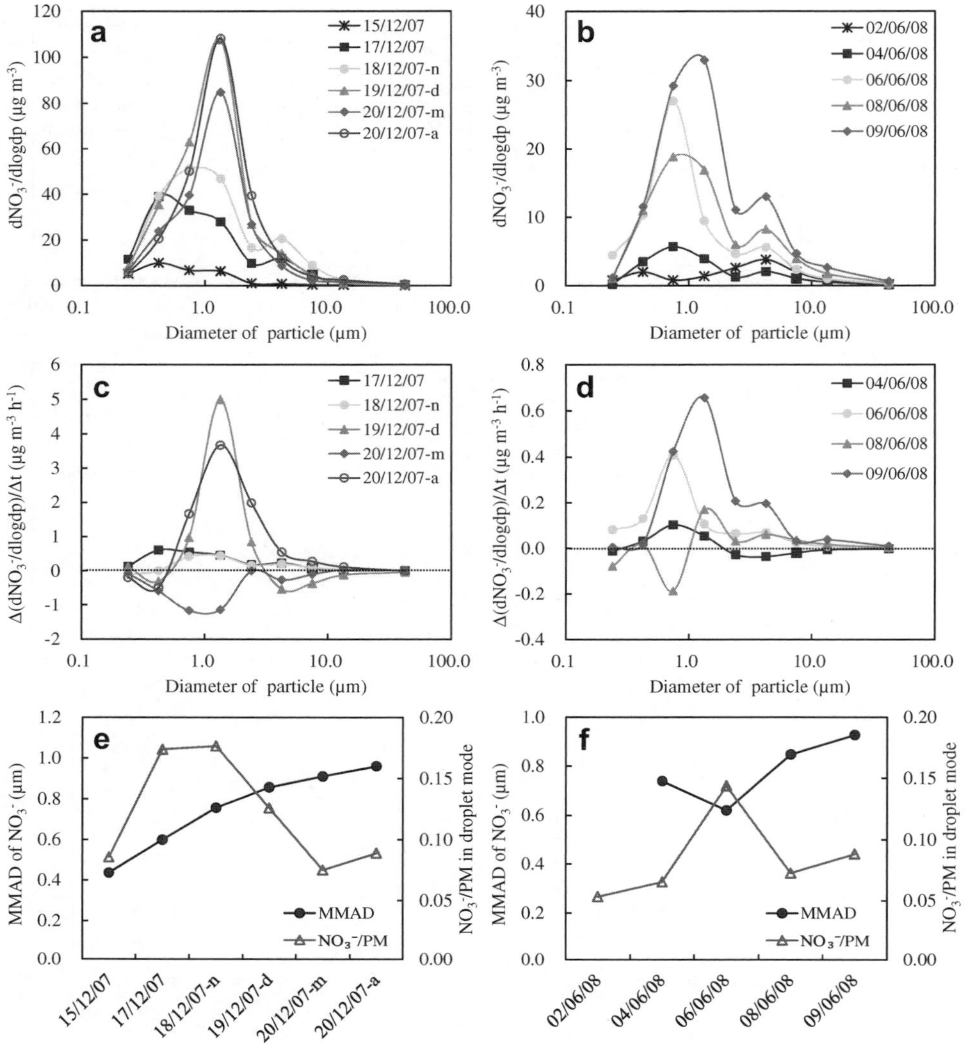

Fig. 7. Size distributions of the nitrate concentration and their rate of increase, variations of the nitrate MMAD and the nitrate fraction in droplet-mode particles for the selected multi-day haze episodes in (a, c & e) winter and (b, d & f) summer.

conditions, the droplet-mode nitrates are believed to form primarily via the heterogeneous aqueous reaction of gaseous nitric acid (HNO_3) and ammonia (NH_3) on the wet surfaces of pre-existing aerosols. As we know, the NH_4NO_3 in fine particles readily establishes a temperature- and humidity-dependent equilibrium with gaseous HNO_3 and NH_3 (Stelson and Seinfeld, 2007). During the summer haze episode, the ambient temperature was high and varied from 20.3 to 27.2 °C (shown in Table 1). Under this condition, the aerosol NH_4NO_3 readily dissociated into HNO_3 and NH_3, and the dissociation amplified with rising temperature and descending RH (Seinfeld and Pandis, 2006). The highest fraction of nitrates in droplet-mode particles (14%) occurred on 6 June 2008, with nearly the lowest temperature and highest RH occurring during that day. This result indicates that low temperature and high humidity were favorable to the secondary formation of fine nitrates in the droplet mode, whereas the secondary formation would be suppressed at high ambient temperatures during summer. In contrast, the temperature during the winter haze episode was rather low, ranging from 0.1 to 6.3 °C. In such conditions, the particulate NH_4NO_3 hardly decomposed, and the HNO_3 tended to react with NH_3 to produce NH_4NO_3. During winter in Jinan, the weak solar radiation would reduce the photochemical production of HNO_3. Furthermore, the ammonia emission was generally low in the cold season and was prone to decrease at lower temperature (Shen et al., 2011). Therefore, the temperature was possibly the dominating factor that limited the fine nitrate formation in winter. For the aerosol samples on the days of 17–19 December 2007, the elevated concentration of droplet nitrates and their rising contribution to particulate matter demonstrate the increased secondary formation of droplet-mode nitrates at high temperature in the winter haze episode. The low temperature suppressed the further production of fine nitrate (e.g., the samples on 20 December 2007) due to a lack of nitric acid and ammonia.

4. Summary and conclusion

Size-segregated aerosols were collected using a MOUDI at the urban area of Jinan, China, from December 2007 to October 2008, and the water-soluble components were subsequently analyzed by ion chromatography. The size-resolved aerosol ionic composition in haze episodes was characterized by sharply increased SNA in the droplet mode (0.56–1.8 μm), which was substantially different from that on clear days. Under haze conditions, most of the fine sulfates and nitrates coexisted with the ammonium in the forms of $(NH_4)_2SO_4$ and NH_4NO_3, respectively. Abundant NH_4NO_3 in fine particles reduced the deliquescence relative humidity, which provided the possibility of secondary inorganic aerosol formation through the heterogeneous aqueous uptake of precursory gases at moderate humidity. The secondary formation of droplet-mode SNA through heterogeneous aqueous reactions strongly relied on the relative humidity. Once the ambient humidity rose, the concentration of droplet-mode SNA apparently increased. The ratio of the droplet-mode SNA to total SNA and the fraction of SNA in droplet-mode particles also increased logarithmically with increasing RH. When the RH was above 50%, more than half of the SNA was produced in the droplet mode, and the SNA comprised over 45% of the droplet-mode aerosols. Case studies of haze episodes indicate that the secondary sulfates in the droplet mode were associated with the heterogeneous oxidation of SO_2 under catalysis by iron and manganese, which was emitted from the local steel smelting and coal combustion. The secondary formation of droplet-mode nitrates was correlated with the temperature-dependent thermodynamic equilibrium of NH_4NO_3 with gaseous nitric acid and ammonia. Under moderately high temperatures during summer, increased temperature led to greater decomposition of ammonium nitrates and thus suppressed the fine nitrate formation. By contrast, the ambient temperature in winter was generally low in Jinan. The rising temperature resulted in enhanced nitric acid formation and ammonia emission, and thus elevated the fine nitrate content.

Acknowledgments

The authors thank Jin Wang for his help in organizing the field study. We also thank Chao Yuan, Zheng Xu, Youping Shou and Jing Wang for their help in the laboratory work. This research was funded by the Shandong Provincial Environmental Protection Department (Project 2006045) and the National Basic Research Program of China (973 Project No. 2005CB422203).

References

Berresheim, H., Jaeschke, W., 1986. Study of metal aerosol systems as a sink for atmospheric SO_2. Journal of Atmospheric Chemistry 4, 311–334.

Buzcu, B., Yue, Z.W., Fraser, M.P., Nopmongcol, U., Allen, D.T., 2006. Secondary particle formation and evidence of heterogeneous chemistry during a wood smoke episode in Texas. Journal of Geophysical Research 111, D10S13. http://dx.doi.org/10.1029/2005jd006143.

Cheng, S., Yang, L., Zhou, X., Wang, Z., Zhou, Y., Gao, X., Nie, W., Wang, X., Xu, P., Wang, W., 2011a. Evaluating $PM_{2.5}$ ionic components and source apportionment in Jinan, China from 2004 to 2008 using trajectory statistical methods. Journal of Environmental Monitoring 13, 1662–1671.

Cheng, S., Yang, L., Zhou, X., Xue, L., Gao, X., Zhou, Y., Wang, W., 2011b. Size-fractionated water-soluble ions, situ pH and water content in aerosol on hazy days and the influences on visibility impairment in Jinan, China. Atmospheric Environment 45, 4631–4640.

Clegg, S.L., Brimblecombe, P., Wexler, A.S., 1998. Thermodynamic model of the system $H^+-NH_4^+-SO_4^{2-}-NO_3^--H_2O$ at tropospheric temperatures. The Journal of Physical Chemistry A 102, 2137–2154.

Feng, Y., Penner, J.E., 2007. Global modeling of nitrate and ammonium: interaction of aerosols and tropospheric chemistry. Journal of Geophysical Research 112, D01304. http://dx.doi.org/10.1029/2005jd006404.

Guo, S., Hu, M., Wang, Z.B., Slanina, J., Zhao, Y.L., 2010. Size-resolved aerosol water-soluble ionic compositions in the summer of Beijing: implication of regional secondary formation. Atmospheric Chemistry and Physics 10, 947–959.

Haywood, J., Bush, M., Abel, S., Claxton, B., Coe, H., Crosier, J., Harrison, M., Macpherson, B., Naylor, M., Osborne, S., 2008. Prediction of visibility and aerosol within the operational Met Office Unified Model. II: validation of model performance using observational data. Quarterly Journal of the Royal Meteorological Society 134, 1817–1832.

Hu, M., Zhao, Y., He, L., Huang, X., Tang, X., Yao, X., Chan, C., 2005. Mass size distribution of Beijing particulate matters and its inorganic water-soluble ions in winter and summer. Environmental Science 26, 1–6 (in Chinese with abstract in English).

Huang, X.F., Yu, J., He, L.Y., Yuan, Z., 2006. Water-soluble organic carbon and oxalate in aerosols at a coastal urban site in China: size distribution characteristics, sources and formation mechanisms. Journal of Geophysical Research 111, D22212. http://dx.doi.org/10.1029/2006JD007408.

John, W., Wall, S.M., Ondo, J.L., Winklmayr, W., 1990. Modes in the size distributions of atmospheric inorganic aerosol. Atmospheric Environment 24, 2349–2359.

Kang, C.M., Han, J.S., Sunwoo, Y., 2002. Hydrogen peroxide concentrations in the ambient air of Seoul, Korea. Atmospheric Environment 36, 5509–5516.

Kleeman, M.J., Riddle, S.G., Jakober, C.A., 2008. Size distribution of particle-phase molecular markers during a severe winter pollution episode. Environmental Science & Technology 42, 6469–6475.

Li, W., Zhou, S., Wang, X., Xu, Z., Yuan, C., Yu, Y., Zhang, Q., Wang, W., 2011. Integrated evaluation of aerosols from regional brown hazes over northern China in winter: concentrations, sources, transformation, and mixing states. Journal of Geophysical Research 116, D09301. http://dx.doi.org/10.1029/2010jd015099.

Liu, S., Hu, M., Slanina, S., He, L.Y., Niu, Y.W., Bruegemann, E., Gnauk, T., Herrmann, H., 2008. Size distribution and source analysis of ionic compositions of aerosols in polluted periods at Xinken in Pearl River Delta (PRD) of China. Atmospheric Environment 42, 6284–6295.

Marple, V.A., Rubow, K.L., Behm, S.M., 1991. A Microorifice Uniform Deposit Impactor (MOUDI): description, calibration and use. Aerosol Science and Technology 14, 434–446.

Martin, L.R., Good, T.W., 1991. Catalyzed oxidation of sulfur dioxide in solution: the iron–manganese synergism. Atmospheric Environment 25, 2395–2399.

Mather, T., Allen, A., Oppenheimer, C., Pyle, D., McGonigle, A., 2003. Size-resolved characterisation of soluble ions in the particles in the tropospheric plume of Masaya volcano, Nicaragua: origins and plume processing. Journal of Atmospheric Chemistry 46, 207–237.

Nie, W., Wang, T., Gao, X.M., Pathak, R.K., Wang, X.F., Gao, R., Zhang, Q.Z., Wang, W.X., 2010. Comparison of three filter-based and a continuous technique for measuring atmospheric fine sulfate and nitrate. Atmospheric Environment 44, 4396–4403.

Pathak, R.K., Wu, W.S., Wang, T., 2009. Summertime $PM_{2.5}$ ionic species in four major cities of China: nitrate formation in an ammonia-deficient atmosphere. Atmospheric Chemistry and Physics 9, 1711–1722.

Pio, C., Cerqueira, M., Harrison, R.M., Nunes, T., Mirante, F., Alves, C., Oliveira, C., Sanchez de la Campa, A., Artíñano, B., Matos, M., 2011. OC/EC ratio observations in Europe: re-thinking the approach for apportionment between primary and secondary organic carbon. Atmospheric Environment 45, 6121–6132.

Salma, I., Balashazy, I., Winkler-Heil, R., Hofmann, W., Zaray, G., 2002. Effect of particle mass size distribution on the deposition of aerosols in the human respiratory system. Journal of Aerosol Science 33, 119–132.

Schryer, D.R., 1982. Heterogeneous Atmospheric Chemistry. American Geophysical Union.

Seinfeld, J.H., Pandis, S.N., 2006. Atmospheric Chemistry and Physics: From Air Pollution to Climate Change, second ed. John Willey & Sons, Inc., New York.

Shen, J., Liu, X., Zhang, Y., Fangmeier, A., Goulding, K., Zhang, F., 2011. Atmospheric ammonia and particulate ammonium from agricultural sources in the North China Plain. Atmospheric Environment 45, 5033–5041.

Stelson, A.W., Seinfeld, J.H., 2007. Relative humidity and temperature dependence of the ammonium nitrate dissociation constant. Atmospheric Environment 41 (Suppl.), 126–135.

Sun, Y., Zhuang, G., Tang, A., Wang, Y., An, Z., 2006. Chemical characteristics of $PM_{2.5}$ and PM_{10} in haze–fog episodes in Beijing. Environmental Science & Technology 40, 3148–3155.

Tan, J., Duan, J., Chen, D., Wang, X., Guo, S., Bi, X., Sheng, G., He, K., Fu, J., 2009. Chemical characteristics of haze during summer and winter in Guangzhou. Atmospheric Research 94, 238–245.

Tang, I.N., Munkelwitz, H.R., 1993. Composition and temperature dependence of the deliquescence properties of hygroscopic aerosols. Atmospheric Environment 27, 467–473.

Turšič, J., Berner, A., Veber, M., Bizjak, M., Podkrajšek, B., Grgić, I., 2003. Sulfate formation on synthetic deposits under haze conditions. Atmospheric Environment 37, 3509–3516.

Wang, Y., Zhuang, G., Sun, Y., An, Z., 2006. The variation of characteristics and formation mechanisms of aerosols in dust, haze, and clear days in Beijing. Atmospheric Environment 40, 6579–6591.

Wang, X., Zhang, Y., Chen, H., Yang, X., Chen, J., Geng, F., 2009. Particulate nitrate formation in a highly polluted urban area: a case study by single-particle mass spectrometry in Shanghai. Environmental Science & Technology 43, 3061–3066.

Wexler, A.S., Seinfeld, J.H., 1991. Second-generation inorganic aerosol model. Atmospheric Environment 25, 2731–2748.

Wu, D., 2006. More discussions on the differences between haze and fog in city. Meteorological Monthly 32, 9–15 (in Chinese with abstract in English).

Yang, L., Wang, D., Cheng, S., Wang, Z., Zhou, Y., Zhou, X., Wang, W., 2007. Influence of meteorological conditions and particulate matter on visual range impairment in Jinan, China. Science of the Total Environment 383, 164–173.

Yao, X.H., Zhang, L., 2011. Sulfate formation in atmospheric ultrafine particles at Canadian inland and coastal rural environments. Journal of Geophysical Research 116, D10202. http://dx.doi.org/10.1029/2010jd015315.

Ye, X., Ma, Z., Zhang, J., Du, H., Chen, J., Chen, H., Yang, X., Gao, W., Geng, F., 2011. Important role of ammonia on haze formation in Shanghai. Environmental Research Letters 6, 024019.

Atmospheric Environment 44 (2010) 2415–2426

Contents lists available at ScienceDirect

Atmospheric Environment

journal homepage: www.elsevier.com/locate/atmosenv

Understanding of regional air pollution over China using CMAQ, part I performance evaluation and seasonal variation

Xiao-Huan Liu [a,b], Yang Zhang [b,*], Shu-Hui Cheng [a,b], Jia Xing [c], Qiang Zhang [d], David G. Streets [d], Carey Jang [e], Wen-Xing Wang [a], Ji-Ming Hao [c]

[a] *Shandong University, Jinan, Shandong Province 250100, PR China*
[b] *North Carolina State University, Raleigh, NC 27695, USA*
[c] *Tsinghua University, Beijing 100084, PR China*
[d] *Argonne National Laboratory, Argonne, IL 60439, USA*
[e] *The U.S. Environmental Protection Agency, Research Triangle Park, NC 27711, USA*

ARTICLE INFO

Article history:
Received 10 December 2009
Received in revised form
26 March 2010
Accepted 29 March 2010

Keywords:
CMAQ
Model evaluation
Seasonality
China
Sensitivity to horizontal grid resolution

ABSTRACT

The U.S. EPA Models-3 Community Multiscale Air Quality (CMAQ) modeling system with the process analysis tool is applied to China to study the seasonal variations and formation mechanisms of major air pollutants. Simulations show distinct seasonal variations, with higher surface concentrations of sulfur dioxide (SO_2), nitrogen dioxide (NO_2), and particulate matter with aerodynamic diameter less than or equal to 10 μm (PM_{10}), column mass of carbon monoxide (CO) and NO_2, and aerosol optical depth (AOD) in winter and fall than other seasons, and higher 1-h O_3 and troposphere ozone residual (TOR) in spring and summer than other seasons. Higher concentrations of most species occur over the eastern China, where the air pollutant emissions are the highest in China. Compared with surface observations, the simulated SO_2, NO_2, and PM_{10} concentrations are underpredicted throughout the year with NMBs of up to −51.8%, −32.0%, and −54.2%, respectively. Such large discrepancies can be attributed to the uncertainties in emissions, simulated meteorology, and deviation of observations based on air pollution index. Max. 1-h O_3 concentrations in Jan. and Jul. at 36-km are overpredicted with NMBs of 12.0% and 19.3% and agree well in Apr. and Oct. Simulated column variables can capture the high concentrations over the eastern China and low values in the central and western China. Underpredictions occur over the northeastern China for column CO in Apr., TOR in Jul., and AODs in both Apr. and Jul.; and overpredictions occur over the eastern China for column CO in Oct., NO_2 in Jan. and Oct., and AODs in Jan. and Oct. The simulations at 12-km show a finer structure in simulated concentrations than that at 36-km over higher polluted areas, but do not always give better performance than 36-km. Surface concentrations are more sensitive to grid resolution than column variables except for column NO_2, with higher sensitivity over mountain and coastal areas than other regions.

1. Introduction

During the past 30-year, China experienced a rapid economic development and industrial expansion, with a significant increase in anthropogenic emissions, especially over the eastern China, one of the regions with the highest emissions in the world and an increased exposure to ozone (O_3) and particulate matter with aerodynamic diameter less than and equal to 10 μm (PM_{10}). Hao et al. (2007) reported that among 522 cities with air quality monitoring sites, only 22 cities (or 4.2%) met the Class I of the National Ambient Air Quality Standard of China that was set to protect natural environments, 293 (or 56.1%) met the Class II standard that was set to protect humans and sensitive plants in urban residential, commercial/residential, cultural, or rural areas, and the remaining cities are in non-attainment and severely polluted.

A number of modeling studies have been recently conducted at a regional scale over East Asia or China. Tie et al. (2006) reported that rates of O_3 production and O_3 concentrations during summer in the eastern China were significantly lower than those in the

* Corresponding author. Department of Marine, Earth, and Atmospheric Sciences, Campus Box 8208, NCSU, Raleigh, NC 27606, USA. Tel.: +1 919 515 9688; fax: +1 919 515 7802.
E-mail address: yang_zhang@ncsu.edu (Y. Zhang).

1352-2310/$ – see front matter © 2010 Elsevier Ltd. All rights reserved.
doi:10.1016/j.atmosenv.2010.03.035

eastern U.S., because of considerably lower volatile organic compounds (VOCs). Yamaji et al. (2006) found that O_3 chemistry is NO_x-limited during summer time over East Asia. Liu et al. (2007) found that emissions of O_3 precursors and O_3 formation are the highest over the central and eastern China. Song et al. (2008) reported high aerosol optical depth (AOD) over Sichuan Basin, Bohai Bay, and the Yangtze River Delta (YRD) areas and AODs were underpredicted in spring due to the occurrence of dust storms and biomass plumes but overpredicted in other seasons due to overpredicted PM_{10} caused by overestimated NH_3 emissions. Using the Community Multiscale Air Quality (CMAQ), Wang et al. (2009) found that Asian pollution enhanced background concentrations of O_3 and SO_4^{2-} in the western U.S. by ~2.5% and ~20%, respectively.

Several modeling studies have also been performed at an urban scale over the Beijing area after its selection to be the host city for the 2008 Olympic Games. An et al. (2007) reported the contributions of non-Beijing sources of 39% to $PM_{2.5}$ and 30% to PM_{10} over the Beijing metropolitan area during a heavy air pollution episode 3–7 Apr. 2005. Streets et al. (2007) predicted that more than 30% of $PM_{2.5}$ and O_3 may be transported to the Olympic Stadium site from the sources outside Beijing during summer. Chen et al. (2007) also found that air quality improvement in Beijing requires emission reductions of sources outside of Beijing. Xu et al. (2008) found that the O_3 chemistry in the Beijing area was VOC-limited, which changed to NO_x-limited in its downwind areas during typical summer conditions.

The above studies focused largely on gas-phase air pollutants at the surface during a specific air pollution episode (mostly prior to 2005). None of them provide a comprehensive examination of the seasonality and formation mechanism of both O_3 and PM species over China. In this study, CMAQ is applied at 36-km over East Asia and a nested 12-km over the eastern China for Jan., Apr., Jul., and Oct. 2008 to study the seasonal variation and formation mechanisms of major pollutants such as O_3 and PM_{10}. Part I describes the model setup, evaluation datasets and protocols, simulated seasonal variations of major air pollutants and the results from model evaluation using surface and satellite observations. The sensitivity of model predictions to horizontal grid resolutions is also examined. Part II presents results from process analysis (PA) and additional sensitivity studies, identifies the most influential processes that lead to the formation and accumulation of major pollutants in China. The regime of O_3 chemistry and PM formation is diagnosed based on PA, then verified through sensitivity simulations. The policy implications of such regime analyses are discussed.

2. Model setup and evaluation protocols

CMAQ version 4.7 released in Dec. 2008 (Byun and Schere, 2006) is applied over East Asia for the 4-month simulations in 2008. Fig. S-1 in the Supplementary data shows the nested domains, with the 36-km domain over East Asia, and the 12-km domain over the eastern China. The vertical resolution includes 14 logarithmic structure layers from the surface to the tropopause, with the first model layer height of 36-m above the ground level. The meteorological fields are generated by the Pennsylvania State University/National Center for Atmospheric Research Mesoscale Modeling System Generation 5 (MM5) version 3.7 with four-dimensional data assimilation. The MM5 hourly output files are processed with the Meteorology-Chemistry Interface Processor version 3.0. More details about the meteorology simulation and evaluation can be found in Cheng et al. (in preparation). Emissions are generated by extrapolating the 2006 activity data to the year 2008 using the method described by Q. Zhang et al. (2009) and by updating those reported by Streets et al. (2006) and Zhang et al. (2007a, b). Initial conditions (ICONs) and boundary conditions (BCONs) are generated using the results from a global chemistry model of GEOS–CHEM. The 4-month simulations are performed separately, each with a 1-week spin-up period to minimize the influence of ICONs.

Table 1 summarizes the variables and observational data included in the model evaluation. The species include the surface concentrations of sulfur dioxide (SO_2), nitrogen dioxide (NO_2), and PM_{10} using observations derived from the Air Pollution Index (API) (http://www.zhb.gov.cn/quality/air.php3) in 23–24, 6–22, and 86 major cities, respectively, in China and hourly O_3 mixing ratios from Mt. Tai located in the eastern China and 3 sites (Ryori, Tsukuba, and Yonagunijima) in Japan. API is piecewise linearly related to the observed concentrations of SO_2, NO_2, and PM_{10}. The derived concentrations from API have been used for model evaluation (Jiang et al., 2004; Wang et al., 2009), despite some uncertainties. Satellite data are also used to evaluate column mass abundance of several species (i.e., carbon monoxide (CO), NO_2, tropospheric ozone residual (TOR)), and AODs to complement sparse surface data. Following the approach of Y. Zhang et al. (2009), the simulated total column mass of species and AODs are calculated using the 3-D gridded hourly average mixing ratios of CO, NO_2, O_3, and concentrations of $PM_{2.5}$ species predicted by CMAQ and vertically-resolved temperature and pressure estimated from MM5. Model evaluation is performed in terms of spatial distribution, temporal variation, column abundances, and overall domain-wide statistical trends using an evaluation protocol of Y. Zhang et al. (2009).

Table 1
Variables evaluated and observational data used in this study.

Datasets	Data type	Species	Data frequency	Number of sites	Data sources and notes
API	PM	PM_{10}	Daily	86	http://datacenter.mep.gov.cn/TestRunQian/air_dairy.jsp
API	Gas	SO_2, NO_2	Daily	Jan. NO_2: 6 Apr., Jul., Oct. NO_2: 22 Jan. SO_2: 23 Apr., Jul., Oct. SO_2: 24	Beijing: www.bjee.org.cn/api/; Shanghai: www.sepb.gov.cn/hjzhiliang/main.jsp; Guangzhou: www.gzepb.gov.cn/comm/apidate.asp; Guiyang: www.ghb.gov.cn/Gazette.asp; Wuhan: www.whepb.gov.cn/publish/whhbj/2008-07/17/kq/1200807171552400040.html; Xi'an: www.xianemc.gov.cn/airsearch.asp?tablename=airrb&;year1=2008 16 cities in Shandong: http://www.sdein.gov.cn/ Other sites: http://datacenter.mep.gov.cn/TestRunQian/air_dairy.jsp
WDCGG	GAS	O_3	Hourly	3	http://gaw.kishou.go.jp/cgi-bin/wdcgg/catalogue.cgi
Mt. Tai	GAS	O_3	Hourly	1	Shandong University, China; note that no observations were available at Mt. Tai in most days in Jan. and all days in Jul. 2008
OMI		Column NO_2, TOR	Monthly average	N/A	http://disc.sci.gsfc.nasa.gov/data/datapool/OMI http://hyperion.gsfc.nasa.gov/Dataservices/cloudslice
MODIS		AOD	Monthly average	N/A	http://ladsweb.nascom.nasa.gov/data/search.html
MOPITT		Column CO	Monthly average	N/A	http://eosweb.larc.nasa.gov/PRODOCS/mopitt/table_mopitt.html

API – Air Pollution Index; WDCGG – the World Data Center for Greenhouse Gases; OMI – the Ozone Monitoring Instrument; MODIS – the Moderate Resolution Imaging Spectroradiometer; MOPITT – the Measurements of Pollution in the Troposphere; TOR – tropospheric ozone residual, AOD – aerosol optical depth.

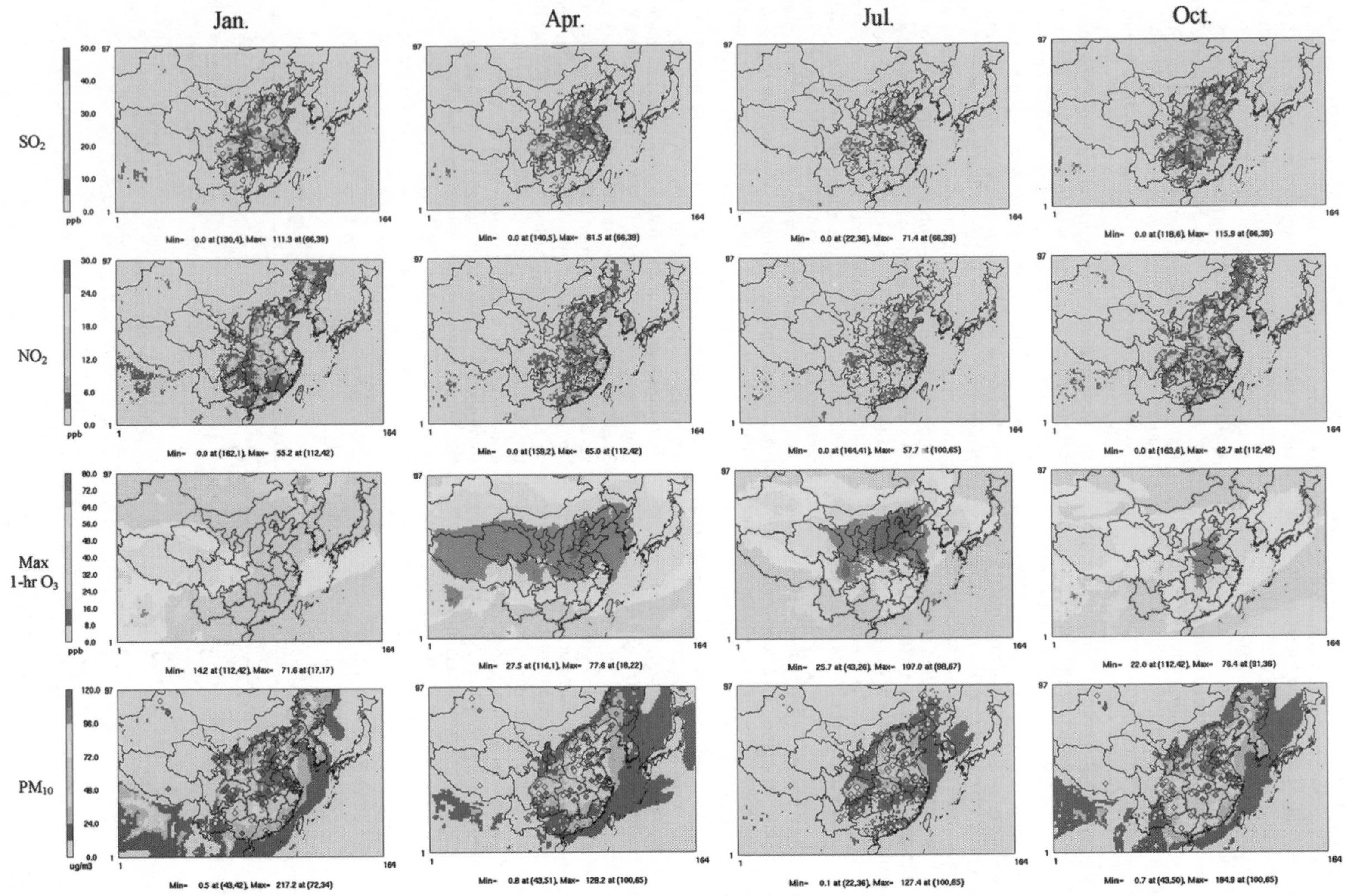

Fig. 1. Overlay of simulated and observed surface concentrations of 1-h max. O_3, PM_{10}, SO_2 and NO_2 in 2008 at a 36-km grid resolution. Diamond signs indicate the observations, including O_3 concentrations from Mt Tai, and 3 sites in Japan (note that no observations were available at Mt. Tai in Jan. and Jul.), PM_{10} concentrations are derived from the API data for 86 major cities in China, and the mixing ratios of SO_2 and NO_2 are derived from the API data for 6 cities in China: Beijing, Xi'an, Wuhan, Shanghai, Guangzhou, and Guiyang.

3. Model evaluation

3.1. Surface concentrations

3.1.1. Spatial distribution and performance statistics

Fig. 1 shows the observed vs. simulated spatial distributions of monthly-mean daily-average surface concentrations of SO_2 and NO_2, max. 1-h O_3, and daily-average PM_{10} at 36-km. The highest SO_2 concentration was simulated in Jan. in the eastern China, which is caused by the highest SO_2 emissions in winter resulted from an extensive coal combustion over the northern China, coupled with the low wind speed and dry weather conditions controlled by Siberian high in winter that are not conducive to dispersion and scavenging for pollutants. Low planetary boundary layer (PBL) height and temperature inversion in winter also prevent the dispersion of air pollutants. Emissions of SO_2 in other months are similar. The simulated SO_2 concentrations, however, in Oct. are higher than those in Apr. and Jul. due to several reasons. First, the weaker solar radiation in fall than in summer slows down the photochemical conversion from SO_2 to SO_4^{2-}. Second, the least amount of precipitation in fall among the four seasons as shown in Cheng et al. (in preparation) causes a slow wet scavenging and aqueous-phase oxidation rate of SO_2. In addition, higher wind speeds in Apr. than in Oct. would enhance the ventilation of SO_2. The lowest SO_2 levels are found in Jul., due to a strong solar radiation that favors its photochemical conversion and the heaviest precipitation among the four months that favors its aqueous-phase oxidation. Higher PBL height in Jul. also helps the dispersion of SO_2. In addition, the highest RHs and the highest NH_3 emissions in Jul. favor the formation of ammonium sulfate. These factors lead to the highest SO_4^{2-} concentration in Jul.

NO_2 concentrations show a similar seasonal variation trend to SO_2, with the maximum in Jan., followed by Oct., Apr., and Jul. The highest NO_2 concentrations in winter are resulted from the highest NO_2 emissions and the suppressed photochemical reaction rates and longer lifetime of NO_2. North China Plain, Sichuan Basin, and YRD are the areas with high concentrations of SO_2 and NO_2. The eastern China, in particular, the YRD area is the most-developed area with dense population, major industry (e.g., coal mining, and coal-fire power plants in surrounding provinces), and heavy traffics, which cause large emissions. Sichuan Basin is also highly-polluted due to large local emissions of air pollutants that are not dispersed efficiently under its special meteorological conditions (e.g., lack of sunny days, featured with cloudy, high RH, low wind speed and also temperature inversion) and complex terrain (the basin is surrounded by mountains that are 1000–3000 m above sea level).

Table 2 summarizes overall performance statistics for daily-average surface concentrations of SO_2 and NO_2, max. 1-h O_3, and daily-average PM_{10}. Compared with concentrations derived based on API in major cities, CMAQ underpredicts both SO_2 and NO_2 concentrations in all 4 months (with NMBs from −12.7% to −51.8% for SO_2 and from −6.5% to −32.0% for NO_2), due mainly to underestimated emissions. Q. Zhang et al. (2009) indicated that emissions in 2006 over China were highly uncertain at individual sites, which would affect the accuracy of the projected 2008 emission used here. Several other possible factors may include the inaccuracies in the simulated meteorology, uncertainties in the calculated dry deposition rates, as well as biases in the estimated observations derived from the API. For example, wind speeds are overpredicted by 40–86% (Cheng et al., in preparation), which may blow too much amount of pollutants out of the domain. For comparison, Wang et al. (2009) reported an NMB of 81.3% for surface NO_2 in China using CMAQ 4.4 at 120 km. Despite small NMBs here, NMEs and RMSEs are large for SO_2 and NO_2, indicating some compensation errors.

Table 2
Model performance of surface chemical concentrations at a 36-km horizontal grid resolution.

Mo.	SO_2						NO_2						Max. 1-h O_3						PM_{10} ($\mu g\ m^{-3}$)					
	Obs	Sim	Corr	RMSE	NMB	NME	Obs	Sim	Corr	RMSE	NMB	NME	Obs	Sim	Corr	RMSE	NMB	NME	Obs	Sim	Corr	RMSE	NMB	NME
Jan.	35.6	17.2	0.1	30.6	−51.8	69.4	26.8	25.0	0.6	15.5	−6.5	47.1	37.9	42.5	0.5	9.3	12.0	16.9	120.7	76.3	0.3	88.8	−36.8	54.9
Apr.	24.6	16.0	−0.02	26.4	−34.9	69.1	20.5	13.9	0.5	16.2	−32.0	61.9	61.3	62.3	0.6	14.9	1.6	17.9	104.0	50.4	0.2	73.4	−51.5	57.6
Jul.	15.3	11.3	0.01	16.3	−25.9	77.3	13.4	10.8	0.4	12.6	−19.7	66.6	43.0	51.1	0.5	19.5	19.3	36.6	84.9	38.9	0.3	57.1	−54.2	57.9
Oct.	23.7	20.7	0.06	26.2	−12.7	68.1	23.3	16.2	0.3	16.2	−30.3	54.0	47.0	47.5	0.7	10.7	1.1	17.1	99.4	69.5	0.4	59.3	−30.1	47.1

Obs – observation, ppb, Sim – simulation, ppb, Corr – correlation coefficient, RMSE – root mean square error, ppb, NMB – normalized mean bias, %, NME – normalized mean error, %.

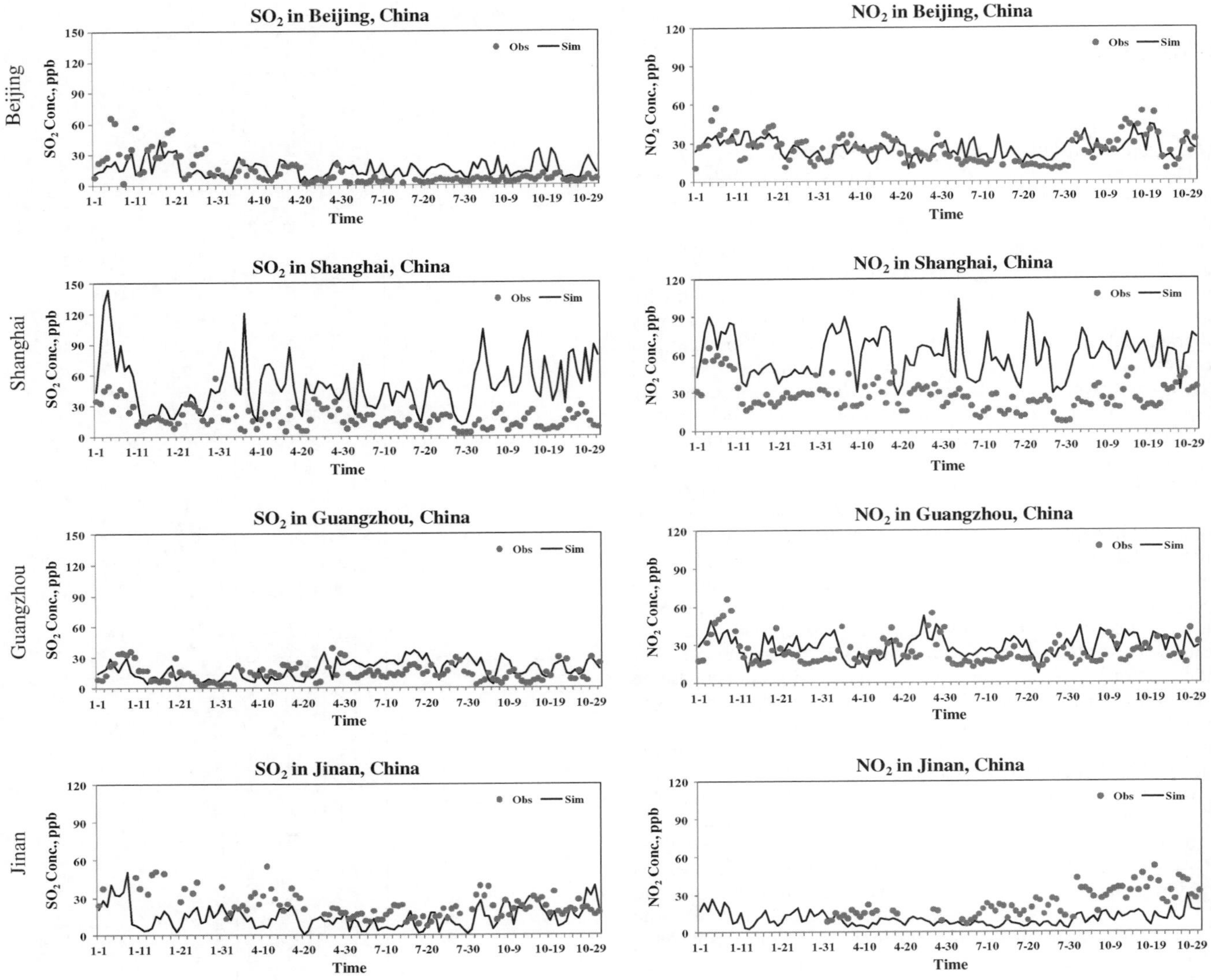

Fig. 2. Temporal variations of SO_2 and NO_2 at four sites (Beijing, Shanghai, Guangzhou, Jinan) in China.

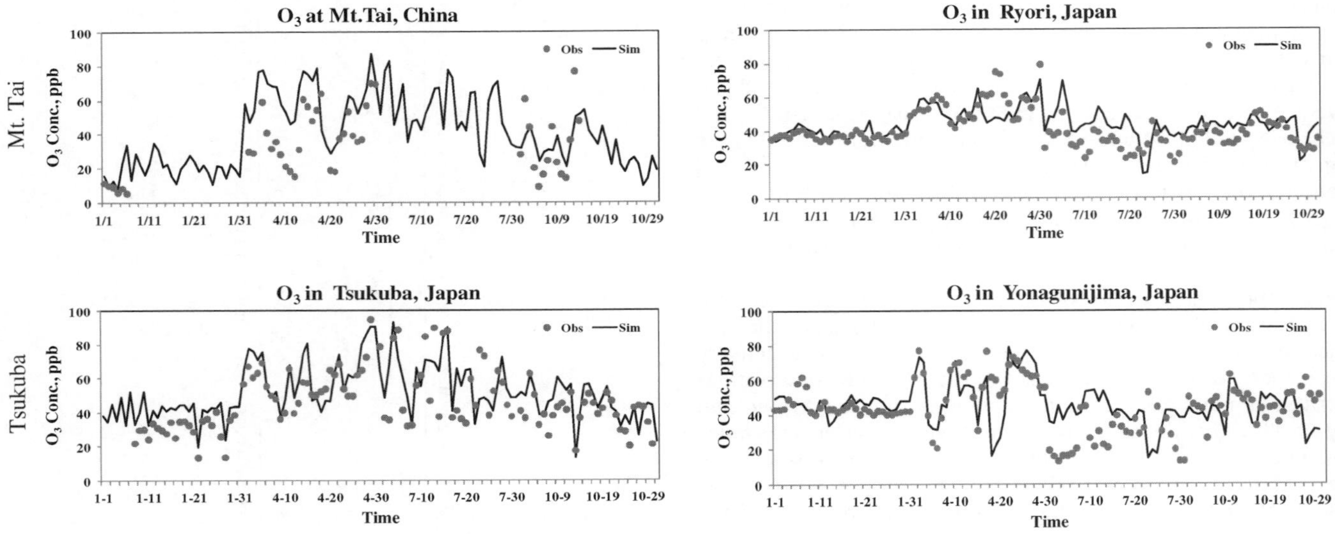

Fig. 3. Temporal variations of O_3 at four sites (Mt. Tai in China and Ryori, Tsukuba, and Yonagunijima in Japan).

The simulated distribution pattern for surface O_3 mixing ratios is different from that of SO_2, NO_2, and PM_{10}. The highest monthly-mean max. 1-h O_3 mixing ratios (up to 107 ppb) occur in Jul. over North China Plain (including Hebei, Shandong, Shanxi, and Henan province), and YRD, due to relatively higher VOC emissions and a strong photochemical production. O_3 mixing ratios over the southern China in Jul. are much lower than those of North China Plain, likely due to the monsoon intrusion of clean air mass from tropical Pacific (Yamaji et al., 2006; Lin et al., 2008). The lowest O_3 mixing ratios in Jul. are found over the Tibetan Plateau mainly due to lack of local sources for O_3 precursors. Compared with Jul., Apr. also has relatively high O_3 mixing ratios of >56 ppb over large areas. This is consistent with the findings of Monks (2000) that the spring surface O_3 maximum occurred widely across mid-latitudes in the Northern Hemisphere, due to relatively high O_3 precursor emissions and the accumulation of O_3 precursors during the winter and sufficiently strong sunlight for photochemical oxidations, high inflow O_3 from boundaries, as well as less precipitation and wet scavenging. Stratospheric–tropospheric exchange might be another factor to the O_3 spring maximum. Jan. has relatively lower O_3 mixing ratios (20–48 ppb) mainly because the production of O_3 is suppressed by the weak solar radiation and low temperatures (Yamaji et al., 2006). High O_3 of 56–64 ppb are found over the Tibetan Plateau in the western China in Jan., due likely in part to a weak or no titration of O_3 because of low NO mixing ratios (<50 ppt) in this region. Other studies reported several other factors such as the downward transport of O_3 from stratosphere (Wang and Li, 1998) or inflow of O_3 and its precursors from the boundary in winter (Yamaji et al., 2006). While CMAQ does not simulate the former process, the boundary conditions generated based on GEOS–CHEM may reflect the impact of inflow to some extent. Despite slightly higher O_3 mixing ratios in Apr. and Jul., the simulated spatial distribution patterns of surface O_3 in this study are similar to those in 2002 simulated by Yamaji et al. (2006). The simulated 1-h O_3 values agree well with limited observations at four sites (90–120 data pairs) with the best agreement in Oct. (an NMB of 1.1%) and the worst in Jul. (an NMB of 19.3%). The discrepancies between simulated and observed O_3 values are likely caused by the uncertainty of precursor emissions such as VOCs and NO_x (e.g., up to ±68% for emissions of VOCs (Q. Zhang et al., 2009)) and simulated meteorological fields.

The simulated PM_{10} also shows the highest concentrations in Jan., with >120 μg m^{-3} (up to 217 μg m^{-3}) over most North China Plain, Sichuan Basin, the east of Hubei province, north of Hunan province, and the Pearl River Delta (PRD) and YRD areas. Those over most regions in the southern China and Northeast Plain are less than 80 μg m^{-3} in winter. Compared with observations, PM_{10} concentrations are well predicted over North China Plain, except some coastal regions (e.g., Qingdao and Yantai in Shandong province) but significantly underpredicted over Northeast Plain, South China, and most areas in the central and western China, which leads to a domain-wide NMB of −36.8% in Jan. Simulated PM_{10} concentrations of >120 μg m^{-3} (up to 185 μg m^{-3}) are also found in Oct., but over a smaller area covering Beijing, Shanghai, and most areas of the Hebei, Henan, and Shandong provinces in Oct. than in Jan. Regional haze, which often occurred in fall and winter (e.g., Sept. 30–Oct. 6, 2008), partially contributes to the high observed PM_{10} concentrations in Oct. Compared with observations, the model underpredicts PM_{10} concentrations with an NMB of −30.1%. Although observed PM_{10} concentrations remained high in Apr., the simulated PM_{10} concentrations in Apr. are much lower than those in Jan. and Oct., with PM_{10} concentrations of 100–128 μg m^{-3} occurring at only a few sites in Hebei province. Simulated PM_{10} is significantly underpredicted with a domain-wide NMB of −51.5%. Lack of a dust emission module in CMAQ4.7 is one possible reason for this underprediction, although some coarse-mode PM species (i.e., SO_4^{2-}, NO_3^-, NH_4^+, Na^+, and Cl^-) are simulated in CMAQ v4.7 through gas/particle mass transfer. The lowest PM_{10} concentrations occur in Jul., due to heavy precipitation and other meteorology conditions such as stronger convective mixing, and higher PBL height than Jan. The simulated PM_{10} concentrations in almost all cities located outside the North China Plain are much lower than observations, resulting in a significant underprediction over the entire domain in Jul., with the largest NMB of −54.2% among all four months. For comparison, Wang et al. (2009) reported up to 85.6% underprediction for PM_{10} over China using CMAQ 4.4 at a grid resolution of 120 km, which is

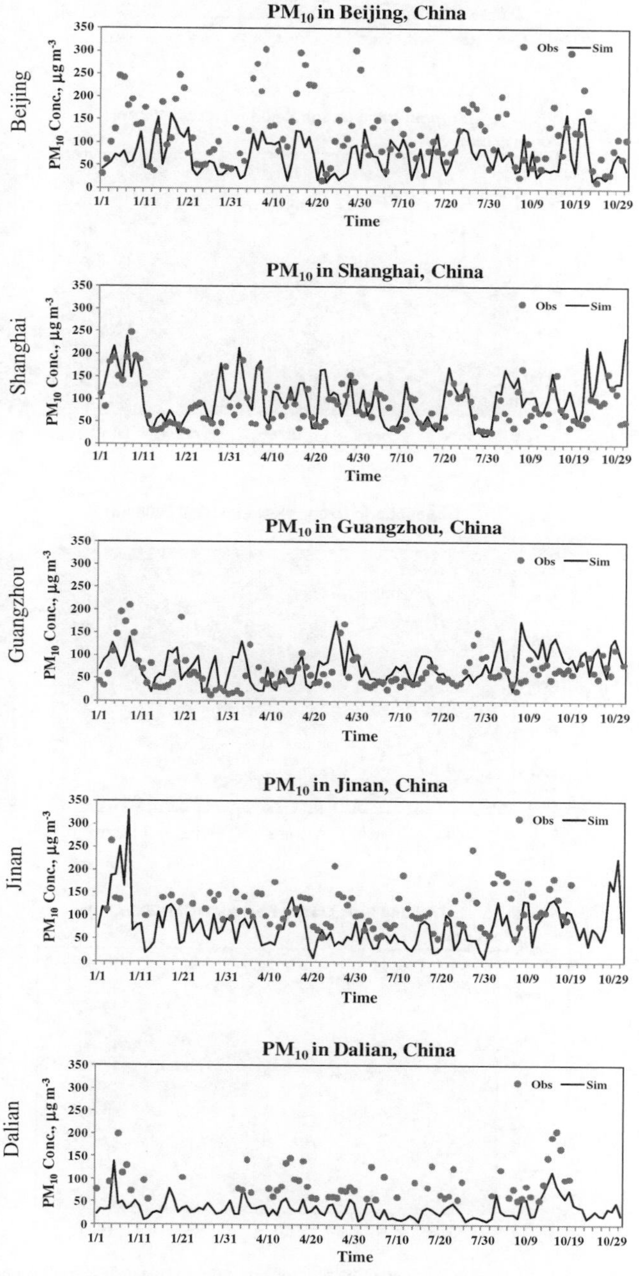

Fig. 4. Temporal variations of PM_{10} at five sites (Beijing, Shanghai, Guangzhou, Jinan, and Dalian) in China.

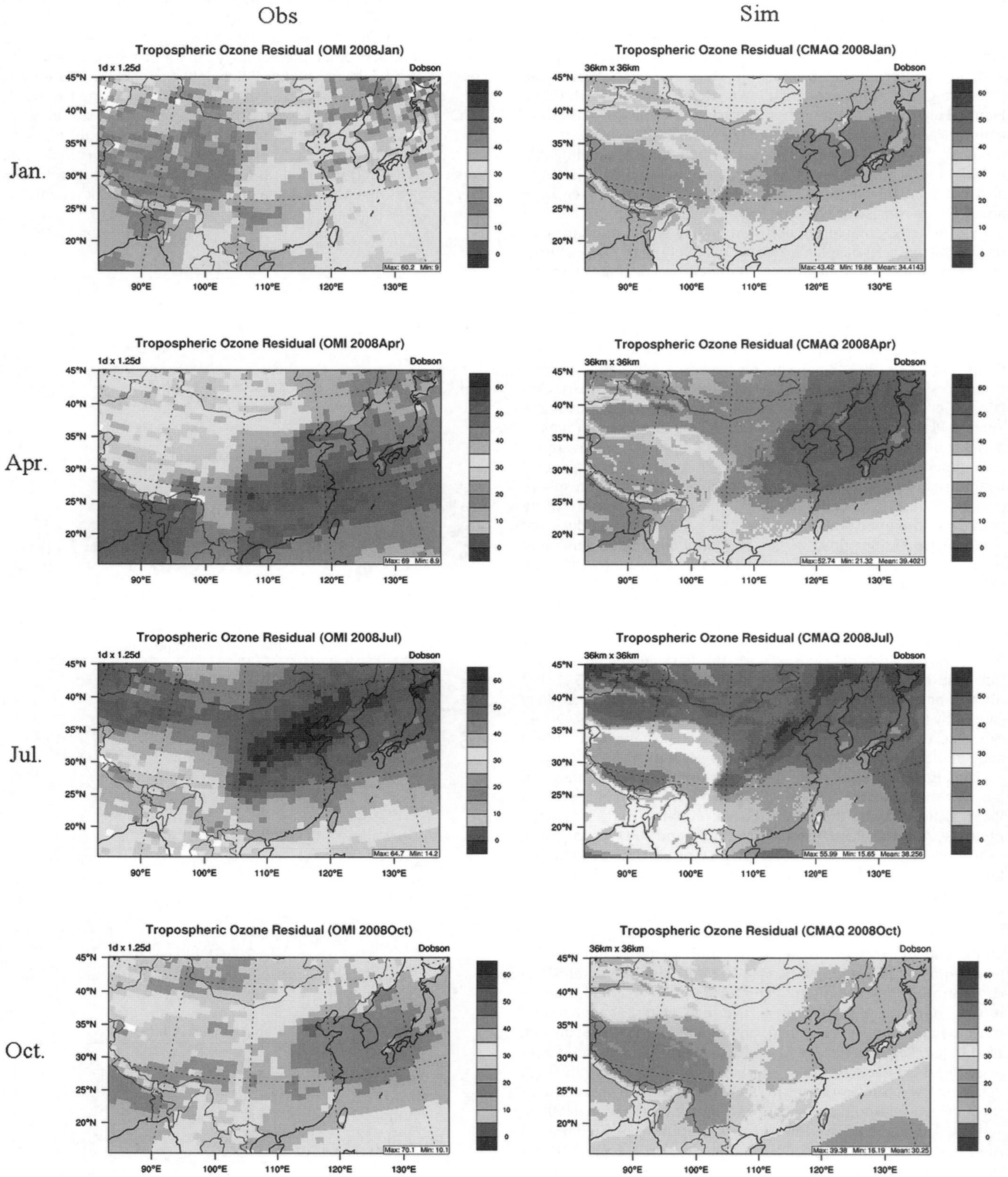

Fig. 5. Spatial distributions of monthly-mean TOR from observations (left) and simulations (right) in Jan., Apr., Jul. and Oct. 2008.

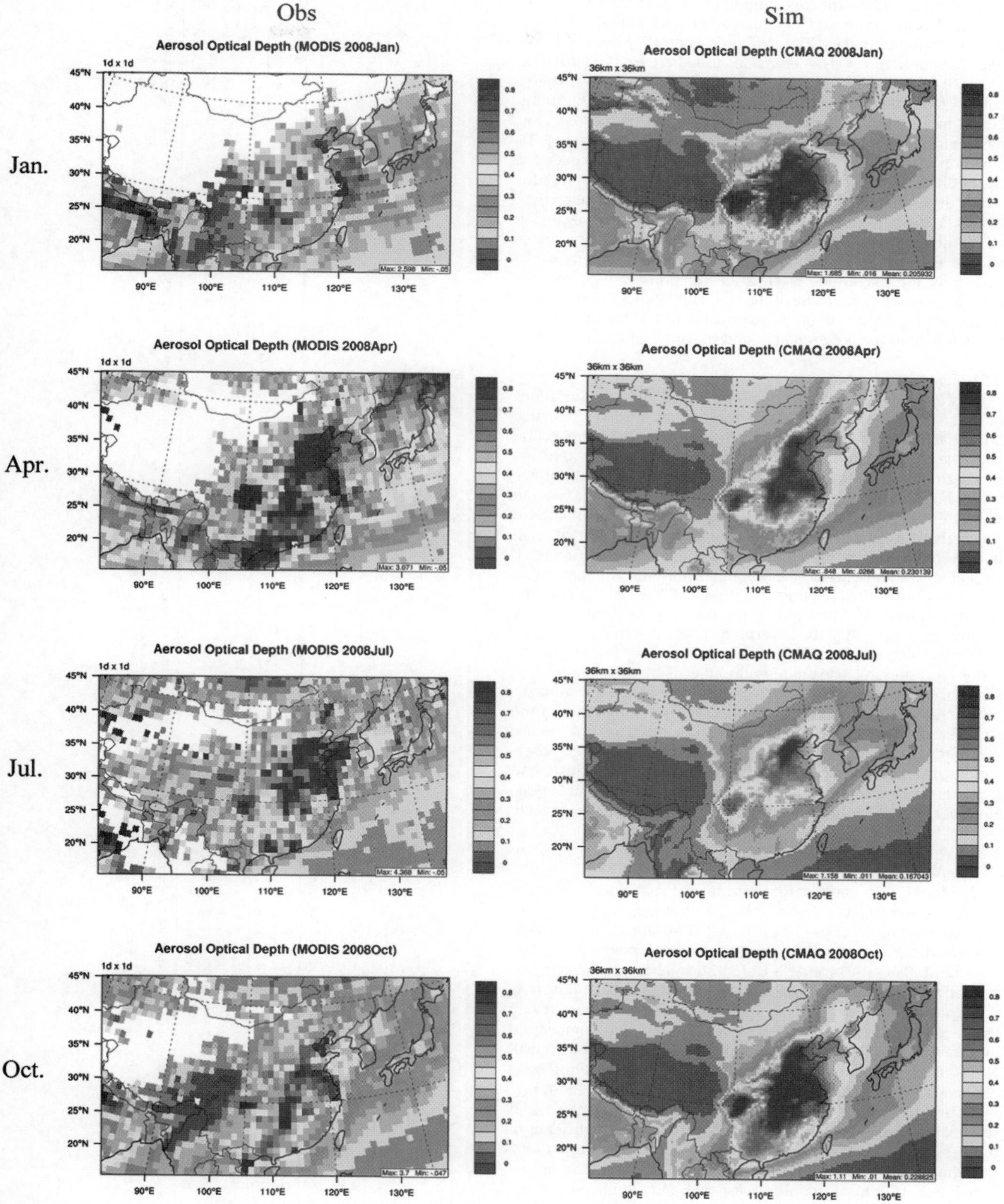

Fig. 6. Spatial distributions of monthly-mean AODs from observations (left) and simulations (right) in Jan., Apr., Jul. and Oct. 2008.

worse than this study. Factors contributing to the underpredictions may include underestimated emissions of primary PM species such as black carbon and organic carbon and emissions of the precursors for secondary PM such as SO_2 and NO_2, as well as the emissions of NH_3 (e.g., NH_3 emissions used here for the PRD areas are ~20% lower than the latest emission estimates by J.-Y. Zheng, South China University of technology, personal communications, 2009), in addition to inaccurate meteorological predictions (e.g., overpredicted wind speeds by about 40% in Jul.). The primary PM_{10} emission rates larger than 100 g s^{-1} are concentrated over the North China Plain, Sichuan Basin, and YRD and PRD in all months while those in the central and western China are smaller than 8.0 g s^{-1}. The high levels of PM_{10} occur over the source regions in all months with 40–95.4% primary PM_{10}, indicating a large influence of primary PM_{10} over these regions. However, the seasonal variations and magnitude of emissions of primary PM_{10} are not always consistent with those of the simulated PM_{10} concentrations because of the formation of the secondary PM and the influence of meteorology in the accumulation and transport of PM_{10}.

3.1.2. Temporal variation

Figs. 2–4 show the observed and simulated temporal variations of daily-average SO_2 and NO_2 at four urban sites (i.e., Beijing, Shanghai, Guangzhou, Jinan in China), max. 1-h O_3 at three rural sites (i.e., Ryori, Tsukuba, and Yonagunijima in Japan) and one mountain site (i.e., Mt. Tai in China), and daily-average PM_{10} at five urban sites (i.e., Beijing, Shanghai, Guangzhou, Jinan, and Dalian in China). CMAQ reproduces SO_2 concentrations well for all months at Guangzhou, in Jan. and Apr. at Beijing, and in Jul. and Oct. at Jinan but overpredicts those in all months at Shanghai and in Jul. and Oct. at Beijing and underpredicts those in Jan. and Apr. at Jinan and all months at other sites (Figures not shown). CMAQ reproduces the mixing ratios of NO_2 in terms of magnitude and variation trends at Beijing and Guangzhou, but overpredicts at Shanghai and underpredicts at Jinan on most days. Temporal variations of O_3 mixing ratios at all sites in Japan are reproduced well on most days, except slight overpredictions in late Apr. and Jul. at Tsukuba and Yonagunijima. At Mt. Tai, CMAQ well captures the O_3 mixing ratios of the first several days in Jan., but overpredicts those in Apr., either overpredicts or underpredicts those in Oct. Compared with observations, the PM_{10} concentrations simulated at Beijing, Jinan, and Dalian are underpredicted in all four months, particularly at Beijing. However, good performance is found at Shanghai and Guangzhou.

3.2. Column variables

Figs. S-2, S-3, 5 and 6 compare simulated spatial distribution of monthly-mean column abundance of CO and NO_2, TOR, and AOD, respectively, with observations from satellite data. Compared with satellite data, column CO is significantly underpredicted in Apr. with an NMB of −29.3% and NME of 31.1%, and best predicted in Oct. with an NMB of −1.0% and NME of 14.5% (see Table 3). In Jan. and Oct., underpredictions occur over the western and northern China as well as the oceanic area off the east coast of China and compensate overpredictions in the eastern and southern China, resulting in small NMBs. These biases may be caused by the uncertainties in the CO emissions and boundary conditions used (Y. Zhang et al., 2009).

CMAQ reproduces well the spatial distribution pattern and seasonality of column NO_2 with high correlation coefficients of 0.7–0.9. Higher column NO_2 is found over the eastern China including Beijing, Hebei, Henan, Shandong, YRD and PRD than the remaining areas in all months. Observed and simulated column NO_2 concentrations are maximum in Jan., followed by Oct., Apr., and Jul. The high winter values are likely resulted from a combined effect of decreased loss of NO_2 via its reaction with OH, reduced

Table 3
Model performance of column variables at a 36-km horizontal grid resolution.

Mo.	CO						NO_2						TOR						AOD					
	Obs	Sim	Corr	RMSE	NMB	NME	Obs	Sim	Corr	RMSE	NMB	NME	Obs	Sim	Corr	RMSE	NMB	NME	Obs	Sim	Corr	RMSE	NMB	NME
Jan.	19.6	19.1	0.6	5.8	−3.0	20.9	23.4	32.8	0.9	27.0	40.4	58.1	30.7	34.4	0.6	6.4	12.1	16.1	0.27	0.26	0.6	0.2	−4.0	45.7
Apr.	23.9	16.9	0.8	8.0	−29.3	31.1	20.1	21.4	0.8	16.4	6.3	41.5	40.6	39.4	0.3	8.7	−3.0	19.7	0.45	0.26	0.6	0.3	−43.0	44.9
Jul.	16.8	15.4	0.5	5.9	−8.4	29.7	17.4	15.4	0.7	15.4	−11.5	46.3	42.1	38.3	0.7	7.2	−9.1	14.2	0.32	0.17	0.6	0.2	−45.8	51.4
Oct.	18.4	18.2	0.8	3.5	−1.0	14.5	17.8	24.1	0.8	22.9	35.4	54.6	33.8	30.3	0.7	5.7	−10.6	15.0	0.25	0.24	0.7	0.1	−1.7	41.9

1. Obs – observation, Sim – simulation, Corr – correlation coefficient, RMSE – root mean square error, NMB – normalized mean bias, %, NME – normalized mean error, %, TOR – tropospheric ozone residual, AOD – aerosol optical depth.
2. The unit for Obs, Sim, and RMSE are in 1×10^{17} molecules cm^{-2} for CO, 1×10^{15} molecules cm^{-2} for NO_2, and Dobson for TOR.

Table 4
Model performance at horizontal grid resolutions of 36- and 12-km over the eastern China.

Surface chemical concentrations at 12- and 36-km																									
Mo.		SO_2						NO_2						1-h O_3						PM_{10}					
		Obs	Sim	Corr	RMSE	NMB	NME	Obs	Sim	Corr	RMSE	NMB	NME	Obs	Sim	Corr	RMSE	NMB	NME	Obs	Sim	Corr	RMSE	NMB	NME
Jan.	12-km	34.8	27.8	0.09	40.1	−20.2	76.6	29.3	36.8	0.5	28.5	25.5	70.2	41.1	45.9	0.3	11.2	11.6	18.1	119.9	103.6	0.3	78.7	−13.6	47.6
	36-km	34.8	15.9	0.10	30.9	−54.4	70.7	29.3	28.3	0.5	16.6	−3.4	47.7	41.1	44.8	0.3	10.8	9.1	17.0	119.9	84.7	0.4	79.8	−29.4	48.4
Apr.	12-km	24.7	27.7	−0.07	40.2	12.2	94.1	21.1	22.0	0.3	22.9	4.5	70.9	62.7	64.1	0.6	17.3	2.4	20.9	103.9	72.3	0.2	63.4	−30.4	47.7
	36-km	24.7	16.3	−0.03	26.9	−33.8	69.8	21.1	14.6	0.4	16.7	−31.0	62.1	62.7	64.6	0.6	18.3	3.2	21.5	103.9	58.7	0.2	65.8	−43.6	51.3
Jul.	12-km	15.3	21.2	−0.03	28.8	38.7	115.6	13.7	18.8	0.3	19.5	37.5	83.2	30.8	41.0	−0.4	22.9	33.0	58.8	86.1	60.0	0.3	50.8	−30.3	47.8
	36-km	15.3	11.4	0.01	16.6	−25.6	78.7	13.7	11.1	0.4	12.9	−18.7	67.2	30.8	40.0	−0.5	24.9	30.0	62.7	86.1	44.7	0.3	54.3	−48.1	52.8
Oct.	12-km	24.2	37.5	−0.06	53.1	54.6	114.3	23.9	25.6	0.3	21.9	6.8	54.3	54.3	48.2	0.7	15.4	−11.2	20.4	103.4	101.3	0.3	62.6	−2.1	42.9
	36-km	24.2	20.9	0.05	26.7	−13.5	68.6	23.9	16.6	0.3	16.6	−30.5	54.4	54.3	48.9	0.8	13.2	−9.9	17.5	103.4	79.6	0.3	56.5	−23.1	42.3
Column variables at 12- and 36-km																									
Mo.		CO						NO_2						TOR						AOD					
		Obs	Sim	Corr	RMSE	NMB	NME	Obs	Sim	Corr	RMSE	NMB	NME	Obs	Sim	Corr	RMSE	NMB	NME	Obs	Sim	Corr	RMSE	NMB	NME
Jan.	12-km	22.9	24.2	0.2	7.4	5.2	22.8	69.1	97.8	0.8	61.3	41.6	62.2	33.2	37.3	0.1	6.1	12.6	15.1	0.38	0.44	0.4	0.3	14.3	49.1
	36-km	22.9	24.6	0.3	7.5	7.5	22.7	68.7	93.3	0.8	50.7	35.9	52.9	33.2	37.2	0.1	6.0	12.2	14.9	0.38	0.46	0.5	0.3	20.6	51.6
Apr.	12-km	25.9	20.3	0.3	7.4	−21.2	26.9	44.6	55.2	0.7	37.5	23.8	54.8	45.0	42.6	−0.1	8.9	−5.5	17.8	0.61	0.43	0.5	0.3	−29.7	33.1
	36-km	25.9	20.6	0.3	7.3	−20.2	26.4	44.1	49.4	0.8	30.3	12.0	44.6	44.9	42.8	−0.1	8.9	−4.8	18.0	0.61	0.43	0.5	0.3	−28.9	32.3
Jul.	12-km	20.7	19.2	0.5	4.2	−7.2	16.2	37.1	38.2	0.6	33.4	2.9	51.4	47.9	40.3	0.9	8.7	−15.9	16.2	0.49	0.33	0.7	0.3	−32.5	38.8
	36-km	20.7	19.6	0.5	4.1	−5.2	15.2	36.4	34.8	0.6	28.3	−4.4	44.9	47.9	41.0	0.9	8.1	−14.5	15.0	0.49	0.33	0.7	0.3	−33.4	38.8
Oct.	12-km	21.5	22.7	0.7	3.7	5.3	12.7	41.7	68.0	0.7	52.2	63.1	76.4	37.8	33.2	0.7	5.3	−12.0	12.4	0.39	0.49	0.8	0.2	25.5	35.7
	36-km	21.5	22.9	0.7	4.0	6.55	13.8	40.9	60.6	0.8	43.9	48.1	62.5	37.7	33.5	0.7	5.1	−11.4	11.9	0.39	0.50	0.8	0.2	29.6	39.1

1. Obs – observation, ppb, Sim – simulation, ppb, Corr – correlation coefficient, RMSE – root mean square error, ppb, NMB – normalized mean bias, %, NME – normalized mean error, %, TOR – tropospheric ozone residual, AOD – aerosol optical depth.
2. The unit for Obs, Sim, and RMSE are in 1×10^{17} molecules cm^{-2} for CO, 1×10^{15} molecules cm^{-2} for NO_2, and Dobson for TOR.

vertical transport from the PBL to higher altitudes, and slightly increased emissions (Wang et al., 2009). The lower column NO_2 in Apr. and Jul. may be related to stronger photochemical reactions that convert it into other reactive nitrogen species during late spring and summer. Comparison with satellite data indicates an underprediction with an NMB of −11.5% in Jul. but overpredictions with NMBs of 6.3–40.4% in other months. For comparison, Han et al. (2009) reported >50% underprediction over North China and as much as 46% overprediction over South Korea for their simulated column NO_2 using CMAQ. Uncertainties in emission inventories of NO_x and related species such as isoprene may contribute the model biases.

As shown in Fig. 5, the observed TOR is the highest in the eastern China due to enhanced pollution and the lowest in the Tibetan Plateau that is the highest landmass in the world, with an average height of >4-km above sea level. The low observed TOR in the Tibetan Plateau can be attributed to two main factors including the thinness of the air column that can lead to ~2.5% (~10 Dobson Unit (DU)) reduction in total column O_3 and the large-scale uplift and descent of isentropic surfaces during the monsoon anticyclone circulation that shifted the vertical profiles of O_3 (Tian et al., 2008). Simulated and observed TOR in Jan., Apr., and Oct. are comparable, but all with overpredictions in the northeastern China and underpredictions over the southern China and Tibetan Plateau. The spatial distribution of observed TOR in Jul. is reproduced well but underpredictions occur over Hebei, Shandong, and Shanxi where the highest observations in Jul. occurred and over Tibetan Plateau. Domain-wide TOR is overpredicted in Jan. with an NMB of 12.1% and NME of 16.1%, and slightly underpredicted in other months, with NMBs of −10.6% to −3.0% and NMEs of 14.2–19.7%. The biases of TOR are strongly related to the uncertainties in the upper-layer boundary conditions of O_3 generated from GEOS–CHEM. Boundary O_3 mixing ratios from layers 10 to 14 used here are 41.0, 46.4, 58.6, and 36.5 ppb in Jan., Apr., Jul., and Oct., respectively. The uncertainty associated with the satellite retrieval algorithms may also influence the accuracy of observed TORs.

Compared with satellite AODs (Fig. 6), the model captures the high AODs over the North China Plain, YRD and PRD areas in the eastern China and Sichuan Basin in all four months, although it significantly overpredicts in Jan. and Oct., and underpredicts in Apr. While observed AODs are the highest in Apr., followed by Jul., Oct., and Jan., the simulated AODs are the highest in Jan., followed by Oct., Apr., and Jul. Domain-wide AODs are slightly underpredicted in Jan. and Oct., respectively. In Apr., simulated AODs in the north of YRD are reasonably good, underpredictions occur over the southern China, Japan, and Pacific ocean with a domain-wide NMB of −43.0% and NME of 44.9%. Underpredictions also occur throughout the whole domain in Jul. with an NMB of −45.8% and NME of 51.4%. The discrepancies between simulated and observed AODs might be caused by several factors, such as the underpredictions of PM concentrations as shown in Table 2, uncertainties in the boundary conditions used, and calculated AODs based on CMAQ PM simulations using an empirical equation (Wang et al., 2009; Y. Zhang et al., 2009).

3.3. Sensitivity of model predictions to horizontal grid resolution

Different grid resolution could influence modeling simulation performance due to a more detailed representation of emissions, land use, meteorological and chemical processes at finer grid resolutions. Some studies have shown that atmospheric modeling at finer resolutions may help improve the model performance (Lin et al., 2008), while some show that chemistry did not show linear relationship with horizontal grid resolution and the model did not always perform well at finer grid resolutions (e.g., Jang, 1995; Queen and Zhang, 2008; Wu et al., 2008). In addition, horizontal grid resolution can affect O_3 responses to changes in precursor emissions, with increased nonlinearity of O_3 production with respect to NO_x emissions at a finer grid scale (Cohan et al., 2006).

The spatial distribution patterns are almost identical between 12- and 36-km grid resolutions for all surface and column variables, although there are some differences over the areas with relatively higher pollutant concentrations, indicating that the model simulation at 12-km captures more detailed information over urban areas due to more detailed terrain and emissions (see Fig. S-4 and analysis in the Supplementary data). Table 4 shows the performance statistic over the eastern China at 12- and 36-km. For surface concentrations of NO_2 and PM_{10}, CMAQ performs better at 12-km than those at 36-km in all four months because of their higher concentrations at 12-km, with differences in NMBs > 10%. SO_2 concentrations at 12-km are closer to observations compared with those at 36-km in Jan. and Apr., but deviate more from observations in Jul. and Oct. at 36-km. Model performance for 1-h O_3 varies slightly at 12- and 36-km, with <3% difference NMBs. The model performance in terms of reproducing column variables at 12-km is slightly better than that at 36-km for column CO in Jan. and Oct., column NO_2 in Jul., and AODs in Jan., Jul., and Oct. but slightly worse in other months for those variables and for TOR for all months. The differences of model performance for domain-wide column variables at two grid resolutions are overall quite small, with less than 8% difference in NMBs except column NO_2 in Apr. and Oct.

4. Conclusions

CMAQ v4.7 is applied to simulate such pollution and its seasonality over East Asia in Jan., Apr., Jul., and Oct. 2008 and the results are evaluated with available surface and satellite data. Both predicted surface species (i.e., SO_2, NO_2, max. 1-h O_3, and PM_{10}) and column variables (e.g., column CO and NO_2, TOR, and AOD) show strong seasonality and spatial variations. The simulated concentrations of SO_2, NO_2, and PM_{10} in Jan. are the highest throughout the year, followed by Oct., Apr., and Jul., whereas the simulated mixing ratios of O_3 are the highest in Jul., followed by Apr., Oct., and Jan. While the model captures well the observed seasonal variations of PM_{10} concentrations, the model's capability in reproducing seasonal variations of surface concentrations of SO_2, NO_2, and O_3 cannot be accurately evaluated because of limited publically-available observational data. The simulated column CO, NO_2, and AODs are the highest in Jan., followed by Oct., Apr., and Jul., whereas the simulated TORs are the highest in Jul., followed by Apr., Jan., and Oct. Compared with satellite observations, the model captures well the seasonality of column NO_2 and AODs, but fails to reproduce those of column CO and TORs. High surface max. 1-h O_3 and TOR are found in Jul. and Apr. due to strong photochemical reactions for all layers and high boundary concentrations for upper layers, while other surface or column variables tend to be higher in Jan. and Oct. Most pollutants show very high concentrations over the eastern China surrounding Beijing, YRD, and PRD due to large emissions from dense population, rapid economic development, heavy traffic, and high energy consumption as well as stagnant meteorological conditions and complex terrain.

For surface species, SO_2, NO_2, and PM_{10} are all significantly underpredicted in all four months, with NMBs up to −51.8%, −32.0%, and −54.2%, respectively. These large discrepancies might be caused by several factors such as uncertainties in the emissions and simulated meteorology. The mixing ratios of max. 1-h O_3 in Jan. and Jul. are overpredicted with NMBs about 12.0% and 19.3%, respectively. Most simulated column variables could generally capture the spatial distribution patterns with high values over the eastern China and relatively lower values over the central and western China. Large underpredictions, however, are found for column CO in Apr., TOR in Jul., and AODs in both Apr. and Jul. over the northeastern China, while overprediction occur in column CO in Oct., column NO_2 in Jan. and Oct., AODs in Jan. and Oct. over the eastern China. The discrepancy of column variables may be related to the boundary conditions which affect their vertical profiles.

The spatial distribution patterns of surface concentrations and column variables at 12-km are overall similar to those at 36-km, but providing more detailed information over higher polluted areas, although it does not always lead to better model performance. The surface concentrations are more sensitive to grid resolution with relatively larger discrepancies (>5%) occurring over the complex terrain such as the mountain and coastal areas, mainly because different surface characteristics may result in different local circulation and meteorological conditions that in turn affect transport and chemistry of pollutants.

The model evaluation in this study indicates an overall acceptable performance especially for surface O_3 and column variables. The major reasons for the model biases include uncertainties in emissions and upper-layer boundary conditions (e.g., TORs) and model deficiencies (i.e., neglecting some coarse-mode species such as mineral dust). Other possible reasons include uncertainties simulated meteorology (such as the limitation of the use of a coarse grid resolution in capturing fine scale meteorological phenomena (Wang et al., 2009)) and uncertainties in the observations from API and satellite data.

Acknowledgements

The authors thank Xin-Yu Wen and Yao-Sheng Chen at North Carolina State University for post-processing and analysis of satellite data; Peng-Ju Xu at Shandong University for providing the O_3 monitoring data at Mt Tai. The O_3 data at 3 sites in Japan were provided by World Data Center for Greenhouse Gases. This work was funded by the U.S. NSF Career Award No. Atm-0348819, the U.S. EPA, and Shandong University in China. The meteorological simulations were funded by China Scholarship Council at Shandong University in China and the U.S. NSF Career Award No. Atm-0348819 at NCSU.

Appendix. Supplementary data

Supplementary data associated with this article can be found, in the online version, at doi:10.1016/j.atmosenv.2010.03.035.

References

An, X., Zhu, T., Wang, Z., Li, C., Wang, Y., 2007. A modeling analysis of a heavy air pollution episode occurred in Beijing. Atmospheric Chemistry and Physics 7, 3101–3114.

Byun, D.W., Schere, K.L., 2006. Review of the governing equations, computational algorithms, and other components of the Models-3 Community Multiscale Air Quality (CMAQ) Modeling System. Applied Mechanics Reviews 59, 51–77.

Chen, D., Cheng, S., Liu, L., Chen, T., Guo, X., 2007. An integrated MM5–CMAQ modeling approach for assessing trans-boundary PM_{10} contribution to the host city of 2008 Olympic summer games – Beijing, China. Atmospheric Environment 41, 1237–1250.

Cheng, S.-H., Zhang, Y., Chen, Y.-S., Wang, W.-X., Gilliam, R.C., Pleim, J. Application of MM5 in China: Model evaluation, seasonal variations, and sensitivity to horizontal grid resolutions. Atmospheric Environment, in preparation.

Cohan, D.S., Hu, Y., Russell, A.G., 2006. Dependence of ozone sensitivity analysis on grid resolution. Atmospheric Environment 40, 126–135.

Han, K.M., Song, C.H., Ahn, H.J., Park, R.S., Woo, J.H., Lee, C.K., Richter, A., Burrows, J.P., Kim, J.Y., Hong, J.H., 2009. Investigation of NO_x emissions and NO_x-related chemistry in East Asia using CMAQ-predicted and GOME-derived NO_2 columns. Atmospheric Chemistry and Physics 9, 1017–1036.

Hao, J.-M., He, K.-B., Duan, L., Li, J.-H., Wang, L.-T., 2007. Air pollution and its control in China. Frontiers of Environmental Science and Engineering in China 1, 129–142.

Jang, C.J., 1995. Sensitivity of ozone to model grid resolution – II, detailed process analysis for ozone chemistry. Atmospheric Environment 29, 3101–3114.

Jiang, D.-H., Zhang, Y., Hu, X., Zeng, Y., Tan, J.-G., Shao, D.-M., 2004. Progress in developing an ANN model for air pollution index forecasting. Atmospheric Environment 28, 7055–7064.

Lin, M., Holloway, T., Oki, T., Streets, D.G., Richter, A., 2008. Mechanisms controlling surface ozone over East Asia: a multiscale study coupling regional and global chemical transport models. Atmospheric Chemistry and Physics Discussions 8, 20239–20281.

Liu, L., Sundet, J.K., Liu, Y., Berntsen, T.K., Isaksen, I.S., 2007. A study of tropospheric ozone over China with a 3-D global CTM model. Terrestrial, Atmospheric and Oceanic Sciences 18, 515–545.

Monks, P.S., 2000. A review of the observations and origins of the spring ozone maximum. Atmospheric Environment 34, 3545–3561.

Queen, A., Zhang, Y., 2008. Examining the sensitivity of MM5–CMAQ predictions to explicit microphysics schemes and horizontal grid resolutions, part III—the impact of horizontal grid resolution. Atmospheric Environment 42, 3869–3881.

Song, C.H., Park, M.E., Lee, K.H., Ahn, H.J., Lee, Y., Kim, J.Y., Han, K.M., Kim, J., Ghim, Y.S., Kim, Y.J., 2008. An investigation into seasonal and regional aerosol characteristics in East Asia using model-predicted and remotely-sensed aerosol properties. Atmospheric Chemistry and Physics 8, 6627–6654.

Streets, D.G., Fu, J.S., Jang, C.J., Hao, J.-M., He, K.-B., Tang, X.-Y., Zhang, Y.-H., Wang, Z.-F., Li, Z.-P., Zhang, Q., Wang, L.-T., Wang, B.-Y., Yu, C., 2007. Air quality during the 2008 Beijing Olympic Games. Atmospheric Environment 41, 480–492.

Streets, D.G., Zhang, Q., Wang, L.-T., He, K.-B., Hao, J.-M., Wu, Y., Tang, Y.-H., Carmichael, G.R., 2006. Revisiting China's CO emissions after the Transport and Chemical Evolution over the Pacific (TRACE-P) mission: synthesis of inventories, atmospheric modeling, and observations. Journal of Geophysical Research 111 (D14306). doi:10.1029/2006JD007118.

Tian, W.-S., Chipperfield, M., Huang, Q., 2008. Effects of the Tibetan Plateau on total column ozone distribution. Tellus 60B, 622–635.

Tie, X.-X., Brasseur, G.P., Zhao, C.-S., Granier, C., Massie, S., Qin, Y., Wang, P.-C., Wang, G.-L., Yang, P.-C., Richter, A., 2006. Chemical characterization of air pollution in Eastern China and the Eastern United States. Atmospheric Environment 40, 2607–2625.

Wang, X.-H., Li, X.-S., 1998. A numerical study of the variations and distribution of tropospheric ozone and its precursors over China. Acta Meteorologica Sinica 56, 333–348.

Wang, K., Zhang, Y., Jang, C.J., Phillips, S., Wang, B.-Y., 2009. Modeling study of intercontinental air pollution transport over the trans-Pacific region in 2001 using the Community Multiscale Air Quality (CMAQ) modeling system. Journal of Geophysical Research 114 (D04307). doi:10.1029/2008JD010807.

Wu, S.-Y., Krishnan, S., Zhang, Y., Aneja, V., 2008. Modeling atmospheric transport and fate of ammonia in North Carolina, part I. Evaluation of meteorological and chemical predictions. Atmospheric Environment 42, 3419–3436.

Xu, J., Zhang, Y.-H., Fu, J.S., Zheng, S.-Q., Wang, W., 2008. Process analysis of typical summertime ozone episodes over the Beijing area. Science of the Total Environment 399, 147–157.

Yamaji, K., Ohara, T., Uno, I., Tanimoto, H., Kurokawa, J., Akimoto, H., 2006. Analysis of the seasonal variation of ozone in the boundary layer in East Asia using the Community Multi-scale Air Quality model: what controls surface ozone levels over Japan? Atmospheric Environment 40, 1856–1868.

Zhang, Q., Streets, D.G., He, K.-B., Klimont, Z., 2007a. Major components of China's anthropogenic primary particulate emissions. Environmental Research Letters 2 (045027). doi:10.1088/1748-9326/2/4/045027.

Zhang, Q., Streets, D.G., He, K.-B., Wang, Y.-X., Richter, A., Burrows, J.P., Uno, I., Jang, C.J., Chen, D., Yao, Z.L., Lei, Y., 2007b. NO_x emission trends for China, 1995–2004: the view from the ground and the view from space. Journal of Geophysical Research 112 (D22306). doi:10.1029/2007JD008684.

Zhang, Q., Streets, D.G., Carmichael, G.R., He, K.-B., Huo, H., Kannari, A., Klimont, Z., Park, I.S., Reddy, S., Fu, J.S., Chen, D., Duan, L., Lei, Y., Wang, L.-T., Yao, Z.-L., 2009. Asian emissions in 2006 for the NASA INTEX-B mission. Atmospheric Chemistry and Physics Discussions 9, 4081–4139.

Zhang, Y., Vijayaraghavan, K., Wen, X.-Y., Snell, H.E., Jacobson, M.Z., 2009. Probing into regional O_3 and PM pollution in the U.S., part I. A 1-year CMAQ simulation and evaluation using surface and satellite data. Journal of Geophysical Research 114, D22304. doi:10.1029/2009JD011898.

Atmospheric Environment 44 (2010) 3719–3727

Contents lists available at ScienceDirect

Atmospheric Environment

journal homepage: www.elsevier.com/locate/atmosenv

Understanding of regional air pollution over China using CMAQ, part II. Process analysis and sensitivity of ozone and particulate matter to precursor emissions

Xiao-Huan Liu [a,b], Yang Zhang [a,*], Jia Xing [c], Qiang Zhang [d], Kai Wang [b], David G. Streets [d], Carey Jang [e], Wen-Xing Wang [a], Ji-Ming Hao [c]

[a] Shandong University, Jinan, Shandong Province 250100, P.R China
[b] North Carolina State University, Raleigh, NC 27695, USA
[c] Tsinghua University, Beijing 100084, P.R China
[d] Argonne National Laboratory, Argonne, IL 60439, USA
[e] The U.S. Environmental Protection Agency, Research Triangle Park, NC 27711, USA

ARTICLE INFO

Article history:
Received 10 December 2009
Received in revised form
26 March 2010
Accepted 29 March 2010

Keywords:
CMAQ
Process analysis
Indicators for O_3 and $PM_{2.5}$ chemistry
China

ABSTRACT

Following model evaluation in part I, this part II paper focuses on the process analysis and chemical regime analysis for the formation of ozone (O_3) and particulate matter with aerodynamic diameter less than or equal to 10 μm (PM_{10}) in China. The process analysis results show that horizontal transport is the main contributor to the accumulation of O_3 in Jan., Apr., and Oct., and gas-phase chemistry and vertical transport contribute to the production and accumulation of O_3 in Jul. Removal pathways of O_3 include vertical and horizontal transport, gas-phase chemistry, and cloud processes, depending on locations and seasons. PM_{10} is mainly produced by primary emissions and aerosol processes and removed by horizontal transport. Cloud processes could either decrease or increase PM_{10} concentrations, depending on locations and seasons. Among all indicators examined, the ratio of $P_{HNO_3}/P_{H_2O_2}$ provides the most robust indicator for O_3 chemistry, indicating a VOC-limited O_3 chemistry over most of the eastern China in Jan., NO_x-limited in Jul., and either VOC- or NO_x-limited in Apr. and Oct. O_3 chemistry is NO_x-limited in most central and western China and VOC-limited in major cities throughout the year. The adjusted gas ratio, AdjGR, indicates that PM formation in the eastern China is most sensitive to the emissions of SO_2 and may be more sensitive to emission reductions in NO_x than in NH_3. These results are fairly consistent with the responses of O_3 and $PM_{2.5}$ to the reductions of their precursor emissions predicted from sensitivity simulations. A 50% reduction of NO_x or AVOC emissions leads to a reduction of O_3 over the eastern China. Unlike the reduction of emissions of SO_2, NO_x, and NH_3 that leads to a decrease in PM_{10}, a 50% reduction of AVOC emissions increases PM_{10} levels. Such results indicate the complexity of O_3 and PM chemistry and a need for an integrated, region-specific emission control strategy with seasonal variations to effectively control both O_3 and $PM_{2.5}$ pollution in China.

1. Introduction

As a result of a fast economic development in China, increasingly high anthropogenic emissions of nitrogen oxides (NO_x), sulfur dioxide (SO_2), and volatile organic compounds (VOCs) lead to the multi-pollutant pollution with high concentrations of ozone (O_3) and particulate matter with aerodynamic diameters less than or equal to 10 μm (PM_{10}) (Zhang et al., 1998; Xu et al., 2006). Three-dimensional (3-D) air quality models provide a fundamental tool to simulate the linkages among meteorology, emissions, and air pollution. Understanding of such linkages and the formation mechanism of major pollutants such as O_3 and PM_{10} is critical to air quality management and climate change mitigation due to their important chemical and climatic impacts. Such linkages and mechanisms are often complex, involving various chemical and physical processes and multiphase reactions; therefore they rely on detailed process analyses using advanced tools such as the process analysis (PA) tool embedded in 3-D air quality models. PA in the U.S. EPA Community Multiscale Air Quality (CMAQ) is a tool that calculates

* Corresponding author at: Department of Marine, Earth, and Atmospheric Sciences, Campus Box 8208, NCSU, Raleigh, NC 27695, USA. Tel.: +1 919 515 9688; fax: +1 919 515 7802.
E-mail address: yang_zhang@ncsu.edu (Y. Zhang).

1352-2310/$ – see front matter © 2010 Elsevier Ltd. All rights reserved.
doi:10.1016/j.atmosenv.2010.03.036

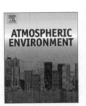

integrated rate and mass changes of a reaction and a process, thereby providing valuable information to the development of the effective emission control strategies. Process analysis has been widely applied to study the fate and formation of gaseous and PM pollutants (e.g., Jang et al., 1995; Jiang et al., 2003; Hogrefe et al., 2005; Kwok et al., 2005; Zhang et al., 2005, 2009a; Kimura et al., 2008; Gonçalves et al., 2008; Tonse et al., 2008; Xu et al., 2008; Yu et al., 2009; Liu et al., in review).

A number of photochemical indicators have been developed to indicate the sensitivity of O_3 to changes in its precursors' emissions to assess the effectiveness of VOCs or NO_x emission controls in reducing O_3 (e.g., Sillman, 1995; Tonnesen and Dennis, 2000a,b; Zhang et al., 2009a). One of the indicators, i.e., the ratio of the production of hydrogen peroxide (H_2O_2) and nitrate acid (HNO_3), can be determined through the PA tool. In this Part II paper, the PA tool embedded in CMAQ is used to identify the most influential processes and chemical reactions that lead to the formation and accumulation of surface O_3, PM_{10}, and components of PM_{10} such as SO_4^{2-}, NO_3^-, and SOA in China in Jan., Apr., Jul., and Oct. 2008. The PA products (e.g., odd oxygen production, OH chain length) provide the characteristics of spatial and seasonal variations of the total oxidation capacity over China. The ratios of $P_{H_2O_2}/P_{HNO_3}$ combined with several other photochemical indicators are used to indicate the NO_x vs. VOC-limited O_3 chemistry in China, which is further verified through sensitivity simulations. Several indicators for PM chemistry and additional sensitivity simulations with different emission reduction scenarios are used to examine the sensitivity of PM formation to changes in its precursor emissions. Such information provides useful perspectives for the development of local and regional emission control strategies and/or assessing their effectiveness over space and time in China.

2. Methodology for process analysis

PA embedded in CMAQ includes the Integrated Process Rates (IPR) that can identify dominant physical processes for O_3 and PM_{10} and the Integrated Reaction Rates (IRR) that can determine the most influential reactions for their precursors in the gas phase and the chemical regime for O_3 chemistry. Hourly IPRs for 33 species and IRRs for 187 gas-phase reactions in the 2005 Carbon Bond Mechanism (CB05) are calculated in the 4-month simulations at 36-km. The hourly IPRs are analyzed for major pollutants in the planetary boundary layer (PBL) (~0–2.9 km, corresponding to layers 1–10) to examine the relative importance of major atmospheric processes such as the emissions of primary species, horizontal transport, vertical transport, gas-phase chemistry, dry deposition, cloud processes, and aerosol processes. Horizontal transport is the sum of horizontal advection and diffusion, and vertical transport is the sum of vertical advection and diffusion. Aerosol processes represent the net effect of aerosol thermodynamics, new particle formation, condensation of sulfuric acid and organic carbon on preexisting particles, and coagulation within and between Aitken and accumulation modes of PM. Cloud processes represent the net effect of cloud attenuation of photolytic rates, aqueous-phase chemistry, below- and in-cloud mixing with chemical species, cloud scavenging, and wet deposition. The IPR results for O_3, PM_{10}, SO_4^{2-}, NO_3^-, and SOA are analyzed at 8 sites including 5 urban sites (Beijing, Shanghai, Guangzhou, Chengdu, and Jinan), 1 mountain site (Mt. Tai located in the North China Plain), 1 rural site (Xiaoping), and 1 background site (Waliguan). The main characteristics of these sites are described in the Supplementary information. The IRRs of 187 reactions are grouped into 34 products according to the reactions for radical initiation, propagation, production, and termination (see Table 1 in Zhang et al., 2009b). The grouping is based on the method used for CBM-IV in CMAQ (Byun and Ching, 1999) but modified for CB05 in this work. In this paper, the chemical production of total odd oxygen (Total_OxProd) (where $O_x = O_3$ + nitrogen dioxide (NO_2) + 2 × nitrogen trioxide (NO_3) + oxygen atom (O) + excited-state oxygen atom (O^1D) + peroxyacyl nitrate (PAN) + 3 × dinitrogen pentoxide (N_2O_5) + nitric acid (HNO_3) + pernitric acid (HNO_4) + unknown organic nitrate) that influences tropospheric oxidation capacity is examined. The chain length of hydroxyl radical (OH) (OH_CL) is the average number of times a newly-created OH radical will be recreated through radical chain propagation before it can be removed from the cycle (Seinfeld and Pandis, 2006). OH_CL provides a measure of an overall oxidation efficiency of the atmosphere (Zhang et al., 2009a). The ratio of production rates of hydrogen peroxide (H_2O_2) and nitric acid (HNO_3) ($P_{HNO_3}/P_{H_2O_2}$) from the IRR results (where P_{HNO_3} is the sum of HNO_3 production via reactions $OH + NO_2$, NO_3 + HC, and N_2O_5 hydrolysis) is used to decide the VOC- or NO_x-limited nature of O_3 chemistry at a given site during different seasons and whether the model is correctly predicting the responses of O_3 to VOC and/or NO_x emission controls. In addition to $P_{HNO_3}/P_{H_2O_2}$ from IRR output, several additional indicator species are calculated to determine O_3 chemistry. These include NO_y, the ratios of H_2O_2/HNO_3, $H_2O_2/(O_3 + HNO_3)$, O_3/NO_x, O_3/NO_y, $HCHO/NO_2$ and $HCHO/NO_y$ (Lu and Chang, 1998; Sillman, 1995; Sillman et al., 1997; Sillman and He, 2002; Zhang et al., 2009a,b). Three indicators: the degree of sulfate neutralization (DSN), gas ratio (GR), and adjusted gas ratio (AdjGR) are used to determine the sensitivity of PM formation. Their definitions and theoretical basis for both O_3 and PM chemistry indicators are provided in Table 2 in Zhang et al. (2009b) and references therein.

3. Results from process analysis

3.1. Analyses of Integrated Process Rates (IPR) in the PBL

Figs. 1 and 2 show the daily-mean hourly contributions of individual processes averaged in the PBL to the concentrations of O_3 and PM_{10}, respectively, at 3 sites: Beijing, Guangzhou, and Mt. Tai. Horizontal transport plays a dominant role in the accumulation of O_3 in Jan., Apr., and Oct., and gas-phase chemistry and vertical transport contribute to the production and accumulation of O_3 in Jul. at all sites. O_3 at Beijing is vented out of the PBL mainly through vertical transport and destroyed via NO titration in Jan., Apr., and Oct. and vented out via horizontal transport in Jul. O_3 at Guangzhou is mainly destroyed via NO titration in Jan., Apr., and Oct. and vented out through both horizontal and vertical transport in Jul. At Mt. Tai, the main processes contributing to O_3 loss may include vertical transport and gas-phase chemistry in Jan. and Apr. and horizontal transport and cloud processes in Jul. and Oct.

Local primary emissions or emissions from upwind areas are the dominant contributor to PM_{10} concentrations at all urban and mountain sites in all months (Fig. 2). A higher contribution of primary emissions at Guangzhou indicates that PM_{10} is composed mainly of primary PM at Guangzhou, which is different from that at the other two sites where both primary and secondary PM are important components. PM_{10} can be formed via aerosol processes such as homogeneous nucleation and condensation at all sites in all months except during some days in Jan. and Apr. at Guangzhou where NaCl is neutralized by large amounts of HNO_3 to release Cl^- from the particulate phase to the gas phase, causing a net loss of PM mass due to this gas/aerosol re-partitioning process. While cloud processes can increase PM_{10} formation due to the aqueous-phase oxidation of SO_2 during most days in Guangzhou in Jan. and Apr., they can also lead to a decrease in PM_{10} formation on some days at Guangzhou and other sites due to a dominance of cloud scavenging. Horizontal transport provides the main sink for PM_{10} at all sites in all months. Vertical transport may either serve as a sink or a source for PM_{10} accumulation.

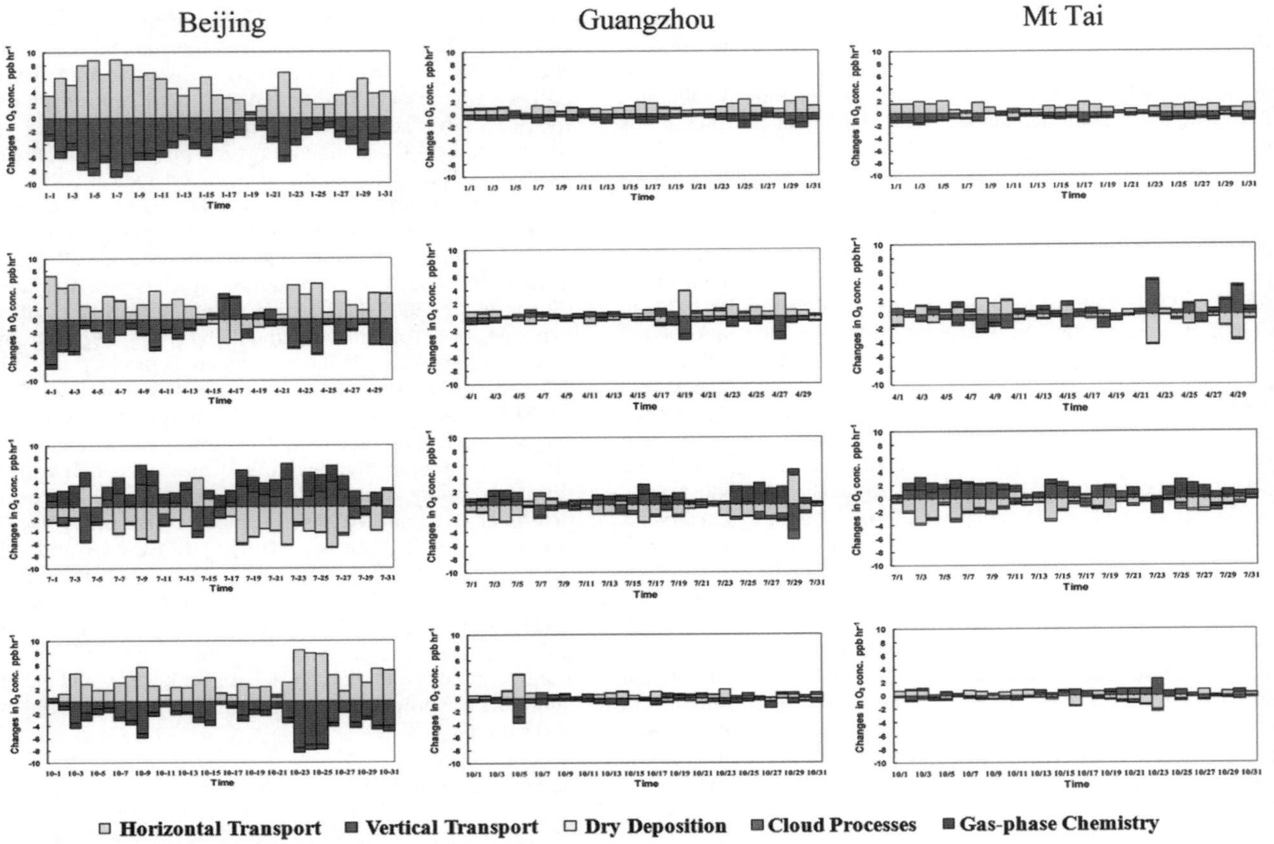

Fig. 1. Daily-mean hourly contributions of individual processes to the mixing ratios of O_3 in the PBL (0–2.9 km) at 3 sites (Beijing, Guangzhou, and Mt. Tai) in China in 2008.

Secondary aerosols are an important contributor to regional haze in China. Figures S-1–S-3 show daily-mean contributions of individual chemical and physical processes to SO_4^{2-}, NO_3^-, and SOA, respectively, at the 3 sites. The main processes contributing to the production/accumulation of SO_4^{2-} include cloud processes and emissions at Guangzhou in all months, cloud and aerosol processes at Mt. Tai in all months and at Beijing in Apr. and Jul., and emissions and horizontal transport at Beijing in Jan. and Oct. Horizontal transport is the dominant processes contributing to the depletion of SO_4^{2-} at Guangzhou and Mt. Tai in all months and at Beijing in Apr. and Jul. Vertical transport dominates the sink of SO_4^{2-} at Beijing on most of days in Jan. and Oct. NO_3^- and SOA concentrations at Beijing, Guangzhou, and Mt. Tai are enhanced primarily by aerosol processes in all months. Their loss is mainly caused by horizontal transport at these sites.

Figure S-4 contrasts the monthly-mean contributions of each process to O_3 and PM_{10} concentrations in different layers in the PBL and one layer above PBL (layer 10, ~2900 m) at one urban site (i.e., Beijing) and on rural site (i.e., Xiaoping) during summer. At Beijing, horizontal transport and vertical transport are the main sources of O_3 accumulation, and dry deposition and gas-phase chemistry are the main sinks of O_3 below layer 3 (surface to ~150 m). O_3 at layers ≥3 (~150–2000 m) mainly comes from gas-phase chemistry production, and horizontal transport contributes negatively from layer 4 to the top of PBL. The contribution of vertical transport to O_3 mixing ratios at different layers may either be negative or positive. The cloud processes contribute slightly to the O_3 increase below 650 m via convective mixing process that brings high O_3 aloft to lower atmosphere and to the O_3 loss near the top of PBL via cloud attenuation of photolytic rates and scavenging and wet deposition processes. As compared with Beijing, horizontal transport contributes to the accumulation of O_3 and vertical transport contributes to the loss of O_3 at higher altitudes (layers 1–7) and gas-phase chemistry contributes to O_3 production in all layers except for layer 10 at Xiaoping. At Beijing, PM_{10} comes from local emissions mainly below 300 m, horizontal transport helps transport particles from heavily-polluted areas to downwind areas, particularly between 150 and 300 m. Particles below ~74 m are also significantly uplifted to higher layers (100–700 m) via vertical transport, which is the opposite to its contribution to surface and near-surface O_3. PM_{10} is mainly produced via aerosol processes in the upper layers (>~450 m) where low temperatures favor the formation of its secondary components. Significant loss of PM_{10} due to aerosol processes occur at heights below 450 m but the net layer-weighted contribution of aerosol processes in the PBL is positive (see Fig. 2), indicating a net production of PM_{10} due to aerosol processes. Dry deposition is the main sink of PM_{10} at surface. Cloud processes contribute slightly to PM_{10} removal under ~1500 m. As compared with Beijing, horizontal transport contributes to the accumulation of PM_{10} and vertical transport contributes to its loss for layers 1–8, PM processes contribute to PM_{10} production in layers 1–2, 5, and 10, and cloud processes contribute to PM_{10} loss in all layers at Xiaoping. These results illustrate roles of various processes at different heights at different sites.

3.2. Integrated Reaction Rates (IRRs) and additional indicator analyses

3.2.1. IRR results

Fig. 3 shows the spatial distributions of monthly-mean total O_x chemical production rates, OH chain length, and the amount of OH reacted with anthropogenic and biogenic VOCs (AVOCs and BVOCs)

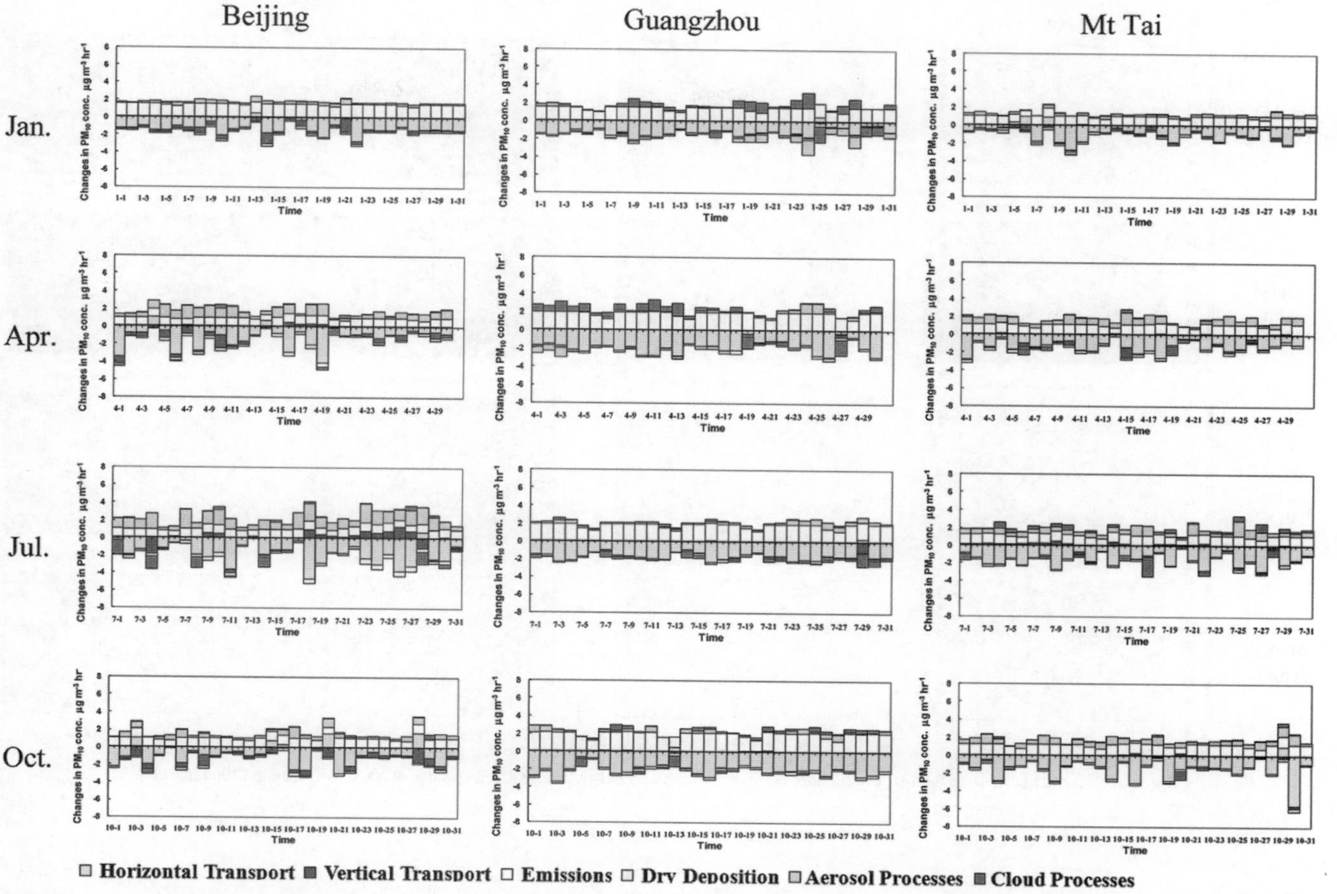

Fig. 2. Daily-mean hourly contributions of individual processes to the mass concentrations of PM_{10} in the PBL (0–2.9 km) at 3 sites (Beijing, Guangzhou, and Mt. Tai) in China in 2008.

over East Asia in the four months. Significant O_x production occurs over the eastern China throughout the year, indicating a stronger oxidation capacity over relatively-developed areas as compared with the central and western China. The O_x production is stronger over South China than North China due to a stronger solar radiation. The strongest oxidation capacity occurs during summer and the weakest during winter. By contrast, OH chain length shows the largest values in Jan. and the lowest in Jul., indicating a faster removal of OH from the photochemical reaction cycles during summer due to stronger oxidation rates of VOCs by OH radicals. Higher reaction rates of OH with AVOCs and BVOCs occur in summer. AVOCs are significant in the North China Plain and PRD areas where AVOC emissions are high due to dense population and rapid economic growth and BVOC emissions are relatively low due to fewer vegetations and a dry weather. BVOCs reacted with OH radicals are typically significant in the southeastern China where the vegetation coverage is dense and BVOC emissions are high. Figure S-5 compares these IRR products at 8 sites. Total O_x production is the highest in Jul. at all sites. The oxidation capacities in Apr. and Oct. are quite similar at all sites. The O_x production at urban sites, e.g., Beijing and Shanghai, are much higher than rural sites, especially in Jul., indicating that the precursor concentrations are another key factor that influences O_3 formation, in addition to solar radiation. The OH chain length shows the largest values in Jan. and the lowest in Jul. at all sites except at Guangzhou where the maximum OH chain length occurs in Jul. and the minimum occurs in Apr., with a stronger seasonal variation at rural and background sites. The rates of OH reacted with AVOCs (OH_AVOC) are significantly higher at urban sites than those at rural sites due to higher AVOC emissions in urban areas. OH amounts reacted with BVOCs (OH_BVOC) at the rural site (e.g., Xiaoping) are equal to or slightly higher than those in urban areas (e.g., Beijing, Shanghai), due to higher BVOC emissions in the rural areas. OH_ AVOCs are significantly higher than OH_BVOCs in urban areas but slightly higher or even lower than OH_BVOCs at rural sites in Jul., consistent with the findings of Zhang et al. (2009a).

3.2.2. Analysis of chemical indicators and sensitivity of model predictions to emissions

Fig. 4 shows the spatial distributions of monthly-mean values of $P_{H_2O_2}/P_{HNO_3}$ from the IRR output during the afternoon time (from 1 pm to 6 pm, local time) in the four months. The values of $P_{H_2O_2}/P_{HNO_3}$ less than 0.2 indicate a VOC-limited chemistry and higher values indicate a NO_x-limited chemistry (Tonnesen and Dennis, 2000b). In Jan., the ratios of $P_{H_2O_2}/P_{HNO_3}$ are below 0.2 over the northeastern China, North China plain, the YRD and PRD areas, and several big cities through China, indicating a VOC-limited O_3 atmosphere in the developed area of China where NO_x emissions from traffic and industry activities are high and vegetations are sparse, whereas other areas are in a regime with NO_x-limited O_3 chemistry due to the large amount of trees that produce high levels of BVOCs or less NO_x emissions into the atmosphere. In Jul., most of these VOC-limited areas change to NO_x-limited, although some urban areas (e.g., Beijing, Tianjin, Shanghai) remain to be VOC-limited, due to significantly higher NO_x emissions caused by power

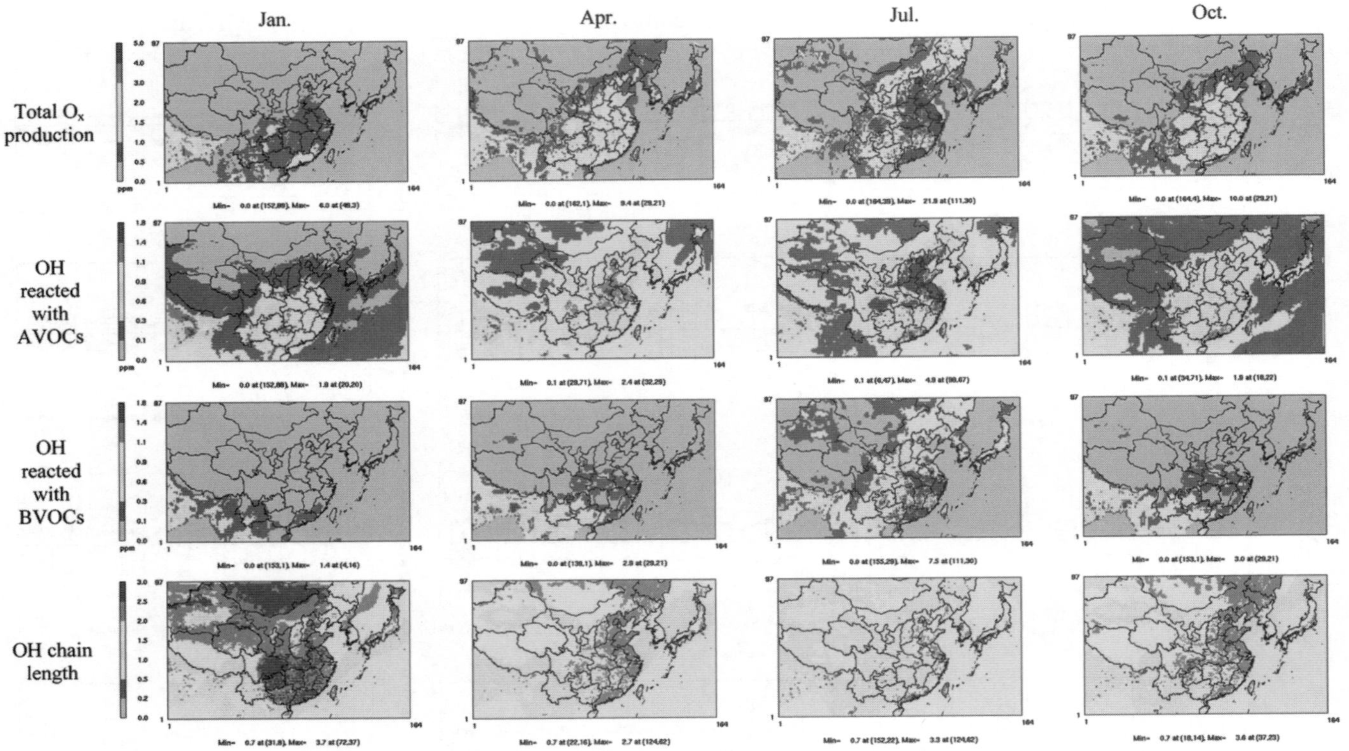

Fig. 3. Spatial distributions of total O_x production, OH chain length and OH reacted with AVOCs and BVOCs at surface in China in 2008.

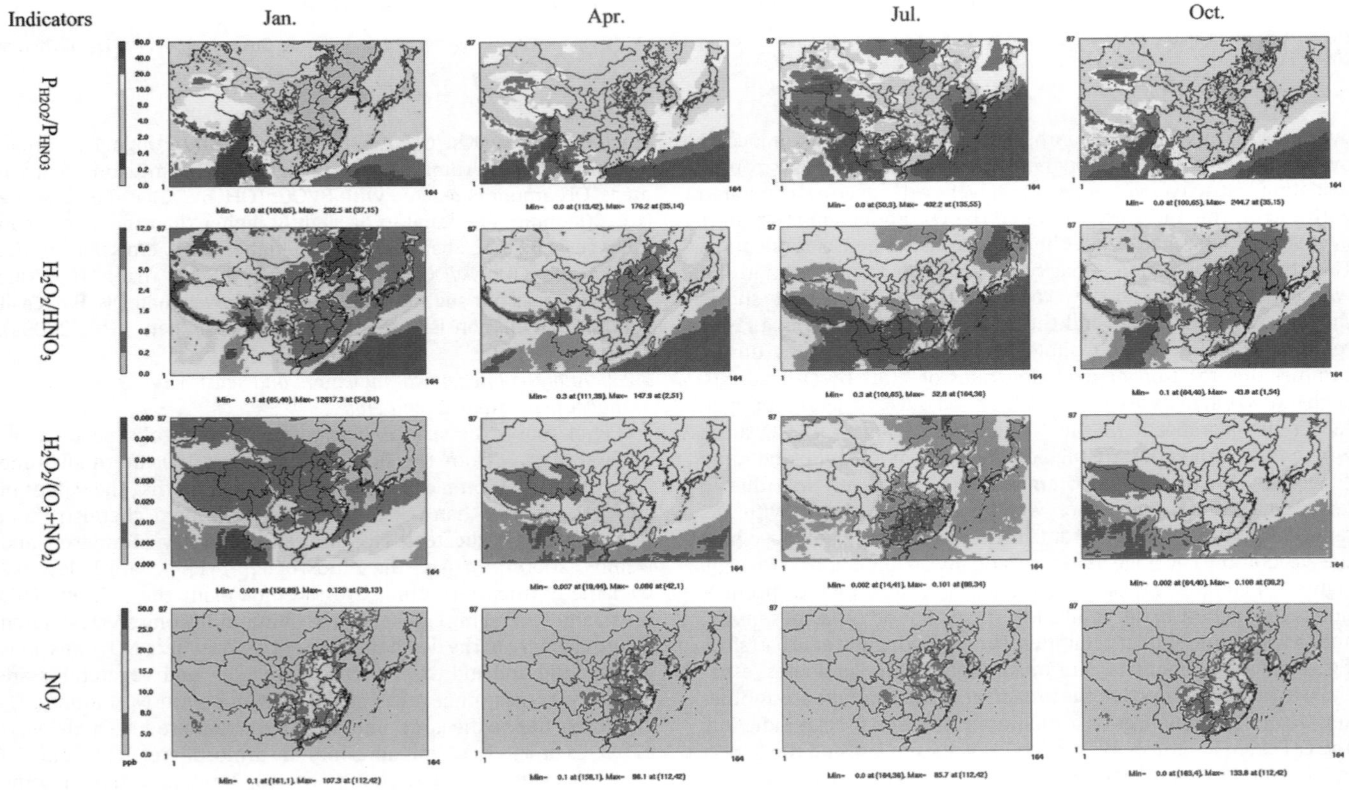

Fig. 4. Spatial distribution of monthly-mean ratios of $P_{H_2O_2}/P_{HNO_3}$, H_2O_2/HNO_3, $H_2O_2/(O_3 + NO_2)$, NO_y, O_3/NO_x, O_3/NO_y, $HCHO/NO_2$, and $HCHO/NO_y$ during afternoon (1–6 pm, local time) at surface in China in 2008.

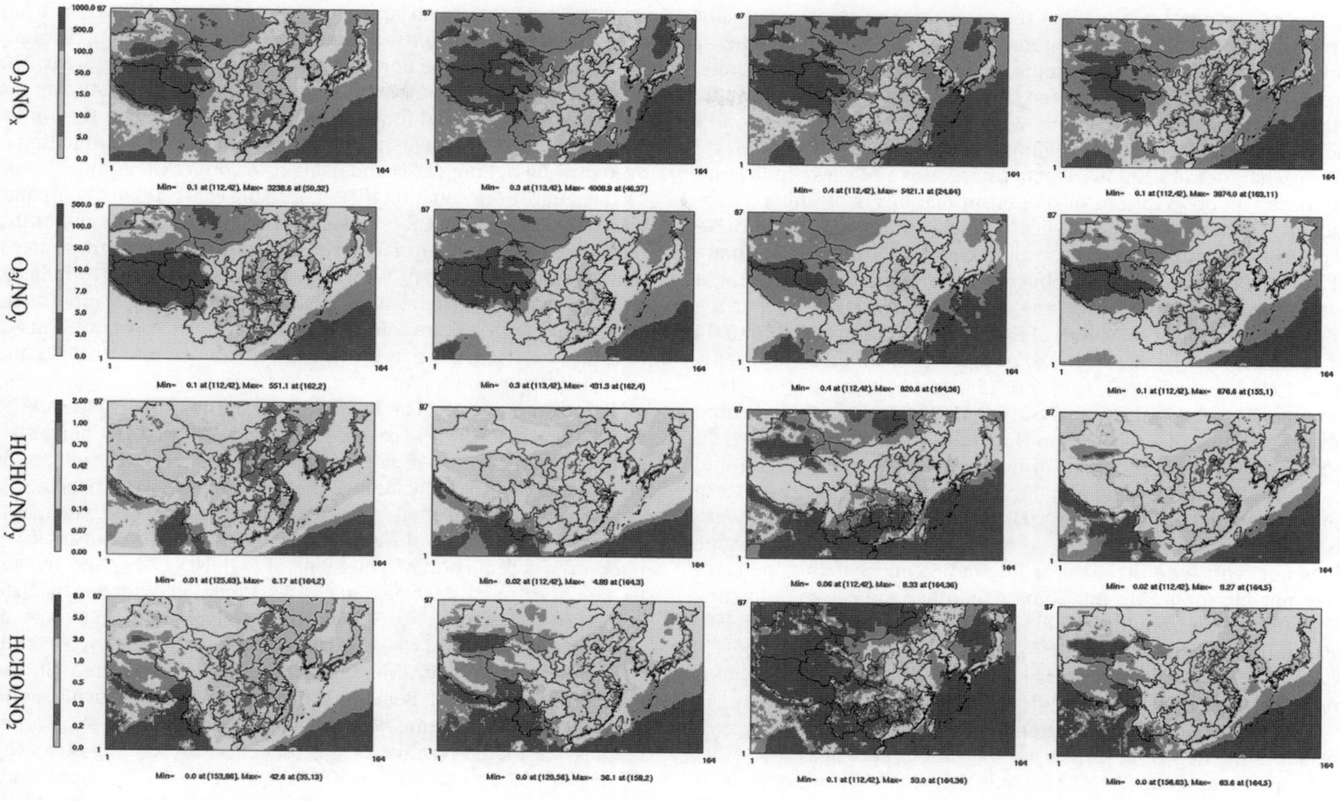

Fig. 4. (*continued*).

plants and vehicles over those areas. The values of $P_{H_2O_2}/P_{HNO_3}$ over the eastern portion of the domain in Apr. and Oct. are mostly higher than 0.2 except for the large metropolitan areas in the northeastern China, South Korea, and Japan.

Since $P_{H_2O_2}/P_{HNO_3}$ was developed based on model simulations over North America, two sensitivity simulations are conducted over China in Jul. 2008 to verify the results based on $P_{H_2O_2}/P_{HNO_3}$, one with a 50% reduction in domain-wide NO_x emissions, and one with a 50% reduction in AVOCs emissions. The absolute and percentage differences in simulated O_3 mixing ratios are shown in Fig. S-6. Reduction of emissions of AVOCs leads to an O_3 decrease over the central and eastern China, with 5–13% decrease over the North China Plain. NO_x reduction leads to a larger O_3 decrease over much larger areas, with the largest reduction of 16.3 ppb and 27.7% domain-wide. The larger response of O_3 to reduction of NO_x emissions than the reduction of AVOCs emissions indicates a NO_x-limited O_3 chemistry in most areas in July, consistent with the results using $P_{H_2O_2}/P_{HNO_3}$. O_3 increase is found over several large cities (i.e., cities located in the North China Plain, YRD, PRD, Taiwan, South Korea, and Japan), indicating that O_3 chemistry in these areas is VOC-limited, which is also consistent with the results based on the ratios of $P_{H_2O_2}/P_{HNO_3}$.

While $P_{H_2O_2}/P_{HNO_3}$ is a robust indicator for O_3 chemistry, other indicators such as the values of H_2O_2/HNO_3, $H_2O_2/(O_3 + NO_2)$, NO_y, O_3/NO_x, O_3/NO_y, $HCHO/NO_y$, and $HCHO/NO_2$ in the afternoon can also serve as effective indicators for O_3 chemistry (Tonnesen and Dennis, 2000b; Zhang et al., 2009a,b). The transition values are 0.2 for H_2O_2/HNO_3 (Lu and Chang, 1998; Sillman, 1995; Sillman et al., 1997; Tonnesen and Dennis, 2000b), 0.02 for $H_2O_2/(O_3 + NO_2)$ (Tonnesen and Dennis, 2000b), 20 for NO_y (Lu and Chang, 1998; Sillman, 1995), 15 for O_3/NO_x (Tonnesen and Dennis, 2000a,b), 7 for O_3/NO_y (Sillman et al., 1997), 0.28 for $HCHO/NO_y$ (Lu and Chang, 1998; Sillman, 1995), and 1 for $HCHO/NO_2$ (Tonnesen and Dennis, 2000b). Values less than these transition values indicate a VOC-limited chemistry and higher values indicate a NO_x-limited chemistry for all the above indicators except NO_y (for NO_y, a value greater than 20 means VOC-sensitive condition, otherwise NO_x chemistry).

As shown in Fig. 4, the values of $H_2O_2/(O_3 + NO_2)$, O_3/NO_x, O_3/NO_y, $HCHO/NO_y$ and $HCHO/NO_2$ over the eastern China in Jan. are all smaller than their receptive transition values, indicating a VOC-limited O_3 chemistry, which is consistent with the results based on the ratios of $P_{H_2O_2}/P_{HNO_3}$. Based on the transition values proposed by the original developers, however, some of these indicators, i.e., $H_2O_2/(O_3 + NO_2)$, $HCHO/NO_y$, and $HCHO/NO_2$ indicate larger areas under the VOC-limited condition as compared with those indicated by $P_{H_2O_2}/P_{HNO_3}$. The values of NO_y show that the NO_y concentrations larger than 20 ppb occur only in the areas surrounding Beijing, Tianjin, and Shanghai, indicating VOC-limited areas that are much smaller than those indicated by $P_{H_2O_2}/P_{HNO_3}$. The VOC-limited areas with an adjusted threshold value of NO_y of 5 ppb by Zhang et al. (2009a) are much more consistent with those indicated by $P_{H_2O_2}/P_{HNO_3}$. The ratios of H_2O_2/HNO_3 in Jan. are higher than 0.2 (indicating a NO_x-limited chemistry) over almost the entire domain, which is also inconsistent with $P_{H_2O_2}/P_{HNO_3}$. Five indicators (i.e., NO_y, O_3/NO_x, O_3/NO_y, $HCHO/NO_2$, and $HCHO/NO_y$) show results consistent with $P_{H_2O_2}/P_{HNO_3}$ in Jul., a NO_x-limited chemistry over nearly the entire China but a VOC-limited chemistry over several large cities. The ratios of H_2O_2/HNO_3 indicate a NO_x-limited O_3 chemistry over China including several big cities (e.g., Beijing) where a VOC-limited regime is denoted by $P_{H_2O_2}/P_{HNO_3}$ and the 5 indicators above. An adjusted value of 2.4 for H_2O_2/HNO_3 proposed by Zhang et al. (2009a) will, however, indicate a VOC-limited chemistry over those metropolitan areas. $H_2O_2/(O_3 + NO_2)$ indicates that O_3 chemistry is NO_x-limited over large cities and a VOC-limited over the Qinghai-Tibet Plateau in Jul., both are inconsistent with the results based on $P_{H_2O_2}/P_{HNO_3}$.

These results show that the transition values originally proposed by developers may not always be valid in China, due to several factors. For example, some of the original transition values were developed based on limited field measurements (Zhang et al., 2009b) or model simulations with different mechanisms and horizontal grid resolutions (Sillman and He, 2002; Wang et al., 2005; Zhang et al., 2009b). Furthermore, none of these indicators are based on observations or model simulations over China. These factors warrant a need to adjust the transition values of these indicators to better fit the atmospheric conditions in China. Zhang et al. (2009b) examined the robustness of those indicators through a full-year simulation over the North America and suggested adjusting the transition values of $H_2O_2/(O_3+NO_2)$ from 0.02 to 0.04 in summer, H_2O_2/HNO_3 from 0.2 to 2.4, NO_y from 20 to 5, O_3/NO_x from 15 to 60, and O_3/NO_y from 7 to 15 in Jan. and Aug. Based on the simulated spatial distribution of H_2O_2/HNO_3 in Jan., the adjusted transition value of 2.4 based on North American conditions would not give VOC-limited condition over both the eastern China and the North China Plain and NO_x-limited chemistry over the western China; however adjusting H_2O_2/HNO_3 from 0.2 to 1.6 would work in Jul. For indicator $H_2O_2/(O_3+NO_2)$, adjusting transition value does not work in either Jan. or Jul., indicating that this indicator may not be applicable for China. For other indicators, adjusting transition value from 20 to 5 for NO_y, from 0.28 to 0.14 for $HCHO/NO_y$, and from 1 to 0.3 in Jan., Apr., and Oct. and 0.5 in Jul. for $HCHO/NO_2$ will allow these indicators to predict an O_3 sensitivity that is more consistent with that predicted by $P_{H_2O_2}/P_{HNO_3}$.

Fig. 5 shows spatial distributions of three PM chemistry indicators including the degree of sulfate neutralization (DSN), gas ratio (GR), and adjusted gas ratio (AdjGR) for sensitivity of $PM_{2.5}$ formation to its precursors. Full sulfate neutralization with DSN ≥ 2 occurs largely in some areas in the eastern China in all months with the smallest areas in July, implying NH_4NO_3 formation over those areas. As indicated in Zhang et al. (2009b), however, a transition value of 1.5 for DSN is more appropriate than a value of 2 to indicate the formation of NH_4NO_3 because of a strong thermodynamic affinity between NH_4^+ and NO_3^- in winter and the lack of sufficient SO_4^{2-} to neutralize excess NH_3, in some areas in summer. Values of GR >1, 0–1, and <0 indicate NH_3-rich, neutral, and poor conditions, respectively. NH_3-rich condition occurs in much larger areas in Apr., Jul., and Oct. than in Jan. due to high NH_3 emissions, whereas NH_3-poor condition occurs only over a small area in the northwestern China and oceanic areas. The full neutralization assumed in GR deviation may underestimate the amount of free NH_3 and thus NH_4NO_3, particularly under the winter condition when such an assumption may not hold. This limitation is overcome by defining DSN and using it to correct GR as an adjusted GR (i.e., AdjGR, see equations (7)–(9), Zhang et al., 2009b). Compared with GR, all areas with GR <1 now have AdjGR ≥ 1 in all four months, reflecting a greater potential for NH_4NO_3 formation over those areas, which is more consistent with the adjusted DSN value to indicate NH_4NO_3 formation. High values of GR and AdjGR in the southern and eastern China in Jul. are due to very small molar concentrations of total nitrate (TNO_3, which is the total molar concentration of NO_3^- and HNO_3) in the denominator.

Based on Ansari and Pandis (1998) and the definition of the AdjGR, for areas with AdjGR >1 (i.e., most eastern and southern China), NH_3 is rich and sulfate is poor; PM formation is sensitive to the emissions of its major precursors including SO_2, NO_x, and NH_3, in particular, SO_2. In those areas, NO_3^- is more sensitive to changes in TNO_3 than in NH_3 because of abundance of free NH_3. Fig. S-7 shows the sensitivity of PM_{10} to 50% reduction of emissions of SO_2, NO_x, NH_3, and AVOCs. PM_{10} concentrations decrease due to SO_2 emission reduction by up to −27.8% (or −11.5 μg m^{-3}), NO_x emission reduction by up to −24.9% (or −17.5 μg m^{-3}), and NH_3 emission reduction by up to −16.8% (or −10.8 μg m^{-3}). While PM_{10} concentrations in some areas in the northern China (e.g., Beijing, Tianjin, Shangdong Province) are equally sensitive to emission reductions for SO_2 and NO_x, other regions are more sensitive to the reduction of SO_2 emissions (by −27.8% to −9% vs. −6%–0%). A 50% reduction of NO_x emissions leads to a larger percentage decrease in PM_{10} than a 50% reduction of NH_3 emissions (by −25.0% to −9% vs. −16.8% to −6%). These results are fairly consistent with the results indicated by AdjGR. AVOCs emission reduction in Jul. over the entire domain results in higher PM_{10} concentrations over developed areas up to 3.5 μg m^{-3} and 8.8%, due to an increase in the concentrations of the secondary PM species such as SO_4^{2-} and NO_3^- when more OH radicals become available for the oxidation of SO_2 and NO_x. Although the treatments of SOA formation are somewhat incomplete and also uncertain in CMAQ, the simulated

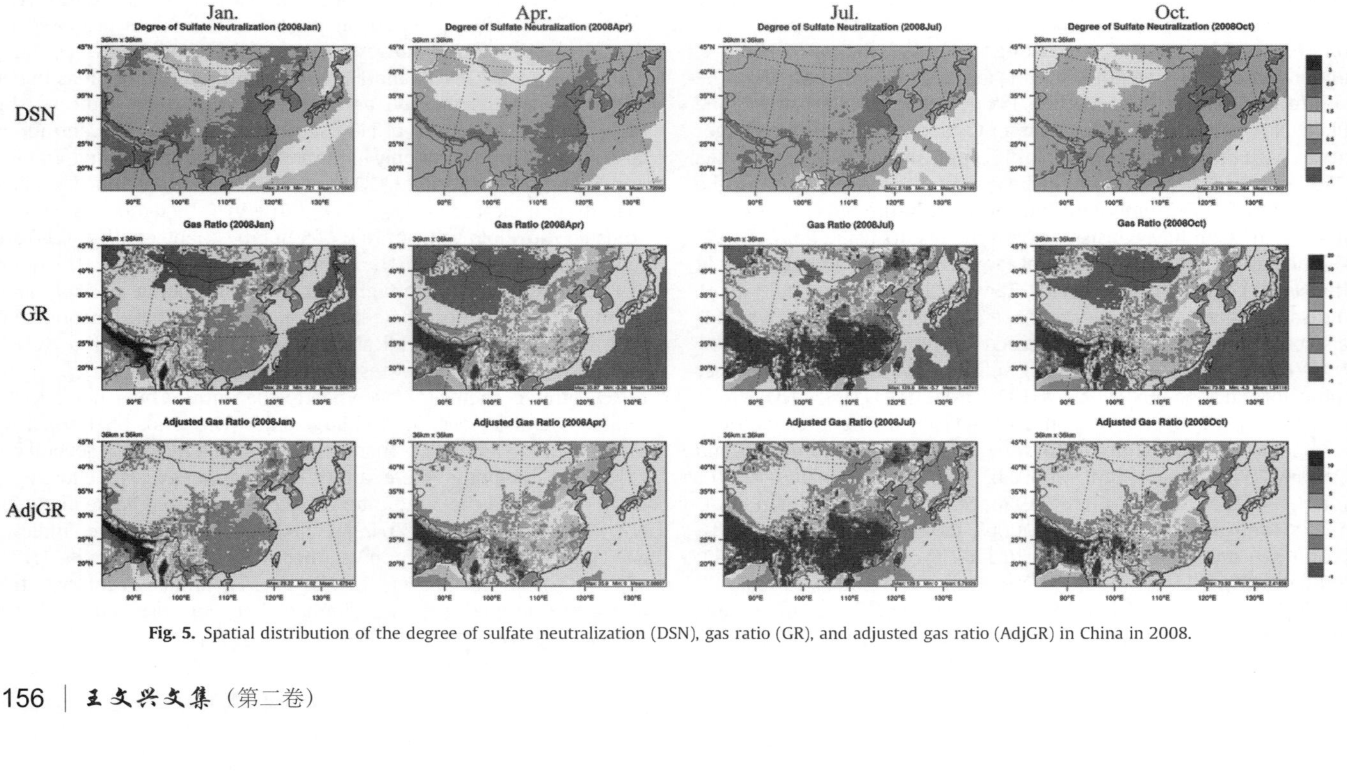

Fig. 5. Spatial distribution of the degree of sulfate neutralization (DSN), gas ratio (GR), and adjusted gas ratio (AdjGR) in China in 2008.

impact of AVOCs on O_3 and PM_{10} in this study are consistent with findings of Meng et al. (1997) and Pai et al. (2000) who reported that a 50% reduction in AVOC emissions may decrease O_3 by 31–34% but increase PM formation by 1–19% through increasing particulate nitrate formation in California, U.S. The sensitivity simulation results indicate that controlling SO_2 emission will be the most effective strategy to reduce PM_{10} pollution in most areas in July and additional control of emissions of NO_x and NH_3 can further reduce PM formation in the northern China, whereas controlling SOA precursors may lead to an enhanced PM formation, thus dis-benefiting the integrated control of ambient O_3 and PM_{10}.

4. Conclusions

The IPR and IRR embedded in CMAQ are applied to quantify the contributions of individual atmospheric processes to the formation and distributions of major pollutants and their seasonal variations in China in Jan., Apr., Jul., and Oct. 2008. The indicators for O_3 and PM chemistry are examined to understand their formation mechanisms via chemical transformations and provide a theoretical basis for the development of the integrated emission control strategies for their dominant precursors. The IPR analysis suggests that horizontal transport is a main process for the accumulation of O_3 in Jan., Apr., and Oct., and gas-phase chemistry and vertical transport contribute to the production and accumulation of O_3 in Jul. at all sites. O_3 can be removed through vertical and horizontal transport, gas-phase chemistry, and cloud processes, depending on locations and seasons. Primary emissions in local or upwind areas and aerosol processes are the main sources of PM_{10} and horizontal transport removes PM_{10} at all sites. Cloud processes could help decrease or increase PM_{10} concentrations, depending on locations and seasons. Cloud and aerosol processes are the dominant processes contributing to the formation of SO_4^{2-} and NO_3^-. Aerosol processes and in some cases vertical and horizontal transport contribute to the SOA production. The IRR results show that the strongest and weakest oxidation capacities occur in summer and winter, respectively, and a stronger O_x production is found over the eastern China than that over the central and western China. OH reacted with VOCs is the highest during summer and the lowest in winter, and OH mainly reacted with AVOCs in urban areas whereas such rates are typically slightly higher or even lower than those of OH with BVOCs at rural sites.

The ratios of $P_{H_2O_2}/P_{HNO_3}$ indicate a NO_x-limited O_3 chemistry over almost entire China during summer except several metropolitan areas, which changes to VOC-limited chemistry over the eastern China, as well as some major cities in most provinces under the cold weather in winter. Some provinces such as Shandong, Henan, and Jiangsu in the North Chain plain experience a transition from VOC-limited in Jan. to NO_x-limited conditions in Apr. and from NO_x-limited in Jul. back to VOC-limited conditions in Oct. Several megacities such as Beijing, Shanghai, Tianjin, and cities in the YRD and PRD areas are always under the VOC-limited conditions throughout the entire year due to large amounts of traffic and industrial emissions of NO_x. These results are consistent with results from two sensitivity simulations: one with a 50% reduction of NO_x emissions and the other with a 50% reduction of AVOC emissions in Jul., indicating the robustness of using the ratio of $P_{H_2O_2}/P_{HNO_3}$ to indicate O_3 chemistry regimes in China. The transition values originally proposed by the developers may not directly be applicable to China. They would need to be adjusted from 0.2 to 1.6 for H_2O_2/HNO_3 in winter, from 20 to 5 for NO_y, from 0.28 to 0.14 for $HCHO/NO_y$, and from 1 to 0.3 in Jan., Apr., and Oct. and from 1 to 0.5 in Jul. for $HCHO/NO_2$ to bring O_3 chemistry regimes more inline with results based on $P_{H_2O_2}/P_{HNO_3}$. Compared with GR, the adjGR provides a more robust indicator for PM chemistry regime. The AdjGR values of greater than 1 are found in most areas in the eastern China. This implies that PM formation is most sensitive to changes in the emissions of SO_2 and is more sensitive to the emission reductions in NO_x than in NH_3. Such a sensitivity to PM precursors is consistent with results from four sensitivity simulations, each with a 50% reduction of emissions of four PM_{10} precursors including SO_2, NO_x, NH_3, and AVOCs in Jul. The sensitivity simulations indicate that reducing 50% of SO_2, NO_x, and NH_3 emissions leads to up to −27.8%, −25.0%, and −16.8% reduction of PM_{10} concentrations. Reducing 50% of AVOC emissions, however, leads to an increase by up to 8.8% in PM_{10} concentrations, due to increased formation of secondary inorganic aerosols.

As shown in the Part I paper, the model simulations at 36-km overpredict O_3 concentrations but underpredict those of SO_2, NO_x, and PM_{10}. These model biases and errors may affect process analysis results to some extent. For example, since the underpredictions in PM_{10} concentrations are mainly caused by the underestimation in emissions, the process contributions to PM_{10} formation rates from primary PM emissions and emissions of PM gaseous precursors such as SO_2 and NO_x may be underestimated. Nevertheless, the results from this study provide useful insights into the governing processes that control the fate and transport of key pollutants in China. They indicate that different emission control strategies for air quality improvement (separate or integrated NO_x or VOC emission control) should be taken during different seasons and over different regions in the future to effectively control ambient O_3 and PM_{10} air pollution. In addition to emission controls, meteorological variables and their changes will also affect air quality improvement. Model simulations accounting for both emission control and climate change will be needed to develop climate-friendly emission control strategies for air quality improvement in the future in China.

Acknowledgements

The authors thank Ping Liu at North Carolina State University, U. S. for her help in setting up process analysis based on the CB05 mechanism in CMAQ. This work was funded by the U. S. NSF Career Award, No. Atm-0348819 and Shandong University in China. The meteorological simulations were funded by China Scholarship Council at Shandong University in China and the U.S. NSF Career Award No. Atm-0348819 at NCSU. The emissions used for model simulations were funded by the U.S. EPA at ANL.

Appendix. Supplementary information

Supplementary information associated with this paper can be found, in the online version, at doi:10.1016/j.atmosenv.2010.03.036

References

Ansari, A.S., Pandis, S.N., 1998. Response of inorganic PM to precursor concentrations. Environmental Science and Technology 32, 2706–2714.

Byun, D.W., Ching, J.K.S., 1999. Science Algorithms of the EPA Models-3 Community Multiscale Air Quality (CMAQ) Modeling System, EPA/600/R-99/030. Office of Research and Development, U.S. Environmental Protection Agency, Washington, D.C.

Gonçalves, M., Jimenez-Guerrero, P., Baldasano, J.M., 2008. Contribution of atmospheric processes affecting the dynamics of air pollution in south-western Europe during a typical summertime photochemical episode. Atmospheric Chemistry and Physics Discussions 8, 18457–18497.

Hogrefe, C., Lynn, B., Rosenzweig, C., Goldberg, R., Civerolo, K., Ku, J.-Y., Rosenthal, J., Knowlton, K., Kinney, P.L., 2005. Utilizing CMAQ process analysis to understand the impacts of climate change on ozone and particulate matter. The 4th Annual CMAS Models-3 User's Conference, Sept. 26–28, 2005, Chapel Hill, NC.

Jang, J.C., Jeffries, H.E., Tonnesen, S., 1995. Sensitivity of ozone to model grid resolution-II. Detailed process analysis for ozone chemistry. Atmospheric Environment 29, 3101–3114.

Jiang, G., Lamb, B., Westberg, H., 2003. Using back trajectories and process analysis to investigate photochemical ozone production in the Puget Sound region. Atmospheric Environment 37, 1489–1502.

Kimura, Y., Mcdonald-Buller, E., Vizuete, W., Allen, D.T., 2008. Application of a Lagrangian process analysis tool to characterize ozone formation in Southeast Texas. Atmospheric Environment 42, 5743–5759.

Kwok, R., Fung, J.C.H., Huang, J.-P., Lau, A.K.H., Lo, J., Wang, Z., Qin, Y., 2005. Comparison of CMAQ and SAQM using process analysis for Hong Kong ozone episodes. The 4th Annual CMAS Models-3 User's Conference, Sept 26–28, Chapel Hill, NC.

Liu, P., Zhang, Y., Yu, S.C., Schere, K.L., Use of a process analysis tool for diagnostic study on fine particulate matter predictions in the U.S. Part II: process analyses and sensitivity simulations, Atmospheric Environment, in review.

Lu, C.-H., Chang, J.S., 1998. On the indicator-based approach to assess ozone sensitivities and emissions features. Journal of Geophysical Research 103, 3453–3462.

Meng, Z., Dabdub, D., Seinfeld, J.H., 1997. Chemical coupling between atmospheric ozone and particulate matter. Science 277, 116–119.

Pai, P., Vijayaraghavan, K., Seigneur, C., 2000. Particulate matter modeling in the Los Angeles Basin using SAQM-AERO. Journal of the Air and Waste Management Association 50, 32–42.

Seinfeld, J.H., Pandis, S.N., 2006. Atmospheric Chemistry and Physics: From Air Pollution to Climate Change. John Wiley and Sons, Inc., 1203 pp

Sillman, S., He, D.-Y., 2002. Some theoretical results concerning O_3-NO_x-VOC chemistry and NO_x-VOC indicators. Journal of Geophysical Research 107 (D22). doi:10.1029/2001JD001123.

Sillman, S., He, D.-Y., Cardelino, C., Imhoff, R.E., 1997. The use of photochemical indicators to evaluate ozone-NO_x-hydrocarbon sensitivity: case studies from Atlanta, New York, and Los Angeles. Journal of the Air & Waste Management Association 47, 642–652.

Sillman, S., 1995. The use of NO_y, H_2O_2, and HNO_3 as indicators for ozone-NO_x-hydrocarbon sensitivity in urban locations. Journal of Geophysical Research 100 (D7), 4175–4188.

Tonnesen, G.S., Dennis, R.L., 2000a. Analysis of radical propagation efficiency to assess ozone sensitivity to hydrocarbons and NO_x 1. Local indicators of instantaneous odd oxygen production sensitivity. Journal of Geophysical Research 105 (D7), 9213–9225.

Tonnesen, G.S., Dennis, R.L., 2000b. Analysis of radical propagation efficiency to assess ozone sensitivity to hydrocarbons and NO_x 2. Long-lived species as indicators of ozone concentration sensitivity. Journal of Geophysical Research 105 (D7), 9227–9241.

Tonse, S.R., Brown, N.J., Harley, R.A., Jin, L., 2008. A process-analysis based study of the ozone weekend effect. Atmospheric Environment 42, 7728–7736.

Wang, X., Carmichael, G., Chen, D., Tang, Y., Wang, T., 2005. Impacts of different emission sources on air quality during March 2001 in the Pearl River Delta (PRD) region. Atmospheric Environment 39, 5227–5241.

Xu, J., Zhang, Y.-H., Wei, W., 2006. Numerical study for the impacts of heterogeneous reactions on ozone formation in Beijing urban area. Advances in Atmospheric Sciences 23, 605–614.

Xu, J., Zhang, Y.-H., Fu, J.S., Zheng, S.-Q., Wang, W., 2008. Process analysis of typical summertime ozone episodes over the Beijing area. Science of the Total Environment 399, 147–157.

Yu, S., Mathur, R., Kang, D., Schere, K., Tong, D., 2009. A study of the ozone formation by ensemble back trajectory-process analysis using the Eta-CMAQ forecast model over the northeastern U.S. during the 2004 ICARPT period. Atmospheric Environment 43, 355–363.

Zhang, Y.-H., Shao, K.-S., Tang, X.-Y., Li, J.-L., 1998. The study of urban photochemical smog pollution in China. Acta Scientiarum Naturalium Universitatis Pekinensis 34 (2–1), 392–400 (in Chinese).

Zhang, Y., Vijayaraghavan, K., Seigneur, C., 2005. Evaluation of three probing techniques in a three-dimensional air quality model. Journal of Geophysical Research 110 (D02305), doi:10.1029/2004JD005248.

Zhang, Y., Vijayaraghavan, K., Wen, X.-Y., Snell, H.E., Jacobson, M.Z., 2009a. Probing into regional O_3 and PM pollution in the U.S., part I. A 1-year CMAQ simulation and evaluation using surface and Satellite Data. Journal of Geophysical Research 114 (D22304), doi:10.1029/2009JD011898.

Zhang, Y., Wen, X.-Y., Wang, K., Vijayaraghavan, K., Jacobson, M.Z., 2009b. Probing into regional O3 and PM pollution in the U.S., Part II. An examination of formation mechanisms through a process analysis technique and sensitivity study. Journal of Geophysical Research 114 (D22305), doi:10.1029/2009JD011900.

Environ Monit Assess (2011) 179:443–456
DOI 10.1007/s10661-010-1747-2

Aerosol size distributions in urban Jinan: Seasonal characteristics and variations between weekdays and weekends in a heavily polluted atmosphere

Pengju Xu · Wenxing Wang · Lingxiao Yang · Qingzhu Zhang · Rui Gao · Xinfeng Wang · Wei Nie · Xiaomei Gao

Received: 24 February 2010 / Accepted: 4 October 2010 / Published online: 20 October 2010

Abstract Aerosol size distributions, trace gas, and $PM_{2.5}$ concentrations have been measured in urban Jinan, China, over 6 months in 2007 and 2008, covering spring, summer, fall, and winter time periods. Number concentrations of particles (10–2,500 nm) were 16,200, 13,900, 11,200, and 21,600 cm^{-3} in spring, summer, fall, and winter, respectively. Compared with other urban studies, Jinan has higher number concentrations of accumulation-mode particles (100–500 nm) and particles (10–2,500 nm), but lower concentrations of ultrafine particles (10–100 nm). The number, surface and volume concentrations, and size distributions of particles showed obvious seasonal variation and are also influenced by traffic emissions. Through correlation analysis, traffic emissions are proposed to be a more important contributor to Atkien-mode and accumulation-mode particles than coal firing. Around midday, the presence of nanoparticles and new particle formation is limited to pre-existing particles from traffic emissions and the mass transport of particles from suburban and rural areas. Compared with other studies in urban areas of Europe and the USA, the variation of particle number concentration and related gas concentration in Jinan between weekdays and weekends is smaller and the reasons has been deduced.

Keywords Aerosol · Particle number concentration · Seasonal variation · Diurnal variation · Weekday · Weekend

P. Xu · W. Wang (✉) · L. Yang · Q. Zhang · R. Gao · X. Wang · W. Nie · X. Gao
Environment Research Institute, Shandong University, Jinan 250100, China
e-mail: xupengju@gmail.com

Introduction

Atmospheric aerosols can modify the radiation budget of a geographic area directly by scattering and absorbing solar radiation or by reflecting it back into space (IPCC 2001; Engler et al. 2007). Aerosols can also have an indirect effect on radiation budget by influencing the properties and occurrence of clouds (Stott et al. 2000; Menon et al. 2002; Bellouin et al. 2005; Chung et al. 2005). Aerosol particles can be carriers and catalysts of atmospheric chemical reactions and are toxic to health, especially ultrafine particles ($D < 100$ nm) due to their large surface area-to-volume ratio and ability to penetrate deeper into the respiratory tract (Oberdörster 2000; Nemmar et al. 2003). Aerosol particles can also degrade atmospheric visibility which can influence human activities (Dockery and Pope 1994).

Not only is particle concentration of atmospheric aerosols important but many researchers have indicated that the size distribution of particles is an important parameter. Therefore, there has been considerable interest in the size distribution of atmospheric particles and mechanism of formation and growth of particles (Holmes 2007). These parameters and phenomena have been studied in many different atmospheric environments such as coastal, polar, remote boreal forest, and continental rural and urban areas (Wu et al. 2007). Kulmala et al. (2004a) and Holmes (2007) have reviewed many measurements taken around the world and pointed out that different characteristics of size distributions and mechanisms for new particle formation were observed in different areas.

Within urban areas, particle concentrations and size distributions are strongly influenced by traffic emissions (Hussein et al. 2004) and two mechanisms for new particle formation exist (Holmes 2007). The first mechanism involves the formation of particles either inside or within the first meter of exiting the exhaust pipe (Morawska et al. 1998). The second process appears to be associated with nucleation, probably arising from SO_2 oxidation, and subsequent condensational growth of particles (Holmes 2007). In recent years, many studies have been done on urban aerosol number size distributions, including long-term measurements. However, most of these projects were conducted in developed countries, and only a few studies have been carried out in the heavily polluted megacities of newly industrialized countries. The characteristics of particle size distributions in these megacities are usually different from those in cities in developed countries, because of larger populations, fast industrialization, and rapid urbanization (Yue et al. 2009).

Shandong Province, P. R. China, with an area of 150,000 km^2, is located in one of the most heavily air polluted areas in the world from gases (http://www.temis.nl/airpollution/no2.html; http://sacs.aeronomie.be/archive/month.php) and aerosols (http://ladsweb.nascom.nasa.gov/browse_images/l3_browser.html) arising from coal and vehicle usage without extensive emission controls and is the largest emitter of SO_2 in China (http://www.stats.gov.cn/tjsj/ndsj/2009/indexeh.htm). SO_2 emission in Shandong Province totaled more than 1.692 million tons in 2008 (http://www.stats.gov.cn/tjsj/ndsj/2009/indexeh.htm). The city of Jinan, in the center of the province, is the capital of Shandong Province, and has an urban population of about 3.6 million and a total population of more than six million in 2009 (http://www.jinan.gov.cn/col/col12/index.html). Like many cities in the northern area of China, Jinan is suffering serious air pollution from aerosol particles and SO_2 (Yang et al. 2007; http://www.stats.gov.cn/tjsj/ndsj/2009/indexch.htm). Yang et al. 2007 showed that annual mean mass concentration of $PM_{2.5}$ of urban Jinan was 138 $\mu g{\cdot}m^{-3}$, about 14 times of WHO Air Quality Guidelines (10 $\mu g{\cdot}m^{-3}$; WHO 2005). For a better understanding of aerosol size distributions of urban Jinan and characteristics of aerosol size distributions under serious air pollution, continuous measurements of particle number concentrations in the range of 10–2,500 nm are reported for all four seasons in urban Jinan, under heavily polluted conditions. Additionally, mass concentration of $PM_{2.5}$, SO_2, NO_x, and O_3 were measured simultaneously. The concentrations and size distributions of number, surface, and volume of particles were investigated. The seasonal, diurnal variations of particle number concentrations, and other related parameters are discussed. Additionally, variations between weekdays and weekends and correlations of several main parameters have been analyzed.

Methodology

Site description

The sampling site was on the roof of public teaching building in Hongjialou Campus of Shandong University located in the north-eastern part of urban Jinan (Fig. 1, 36°40′ N, 116°57′ E), about 20 m above ground level. There are no obvious industrial pollutant sources near the sampling site because all the districts nearby are residential or commercial. The sampling site is affected by two major roads with heavy traffic: east of the site is East Second Ring Road with a distance of about 240 m and south of the site is Huayuan Road with a distance of about 300 m.

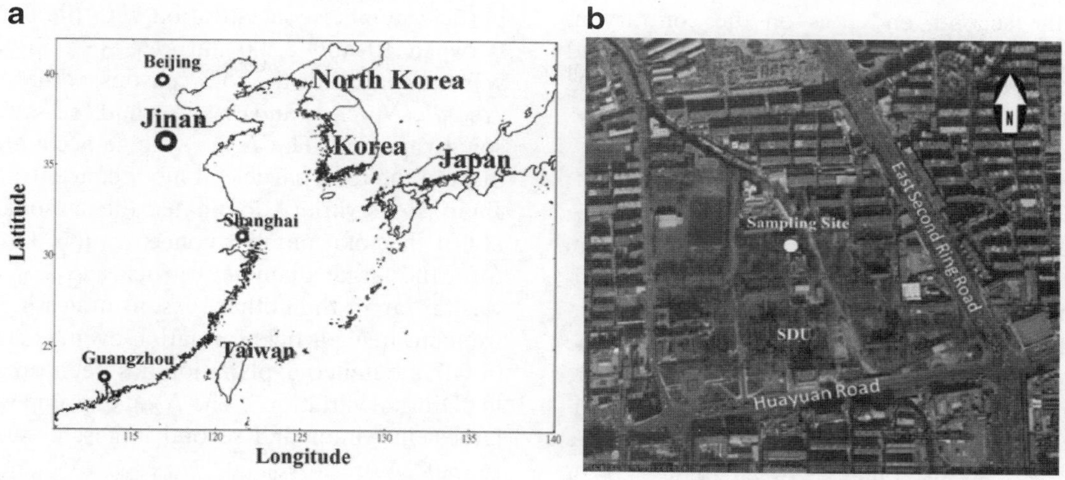

Fig. 1 Location of the atmospheric measurement site. **a** Location of Jinan. **b** Sampling site (*white dot* in the image from Google Earth™)

Instrumentation

A wide-range Particle Spectrometer™ (WPS-model 1000XP, MSP Corporation, USA) was used to measure size distribution data in the field study of this work. The instrument is a high-resolution aerosol spectrometer which combines the principles of differential mobility analysis (DMA), condensation particle counting (CPC), and laser light scattering (LPS). The DMA in the WPS™ has a cylindrical geometry with an annular space for the laminar aerosol and sheath air flows. These critical dimensions were optimized to obtain size classification of particles between 10 and 500 nm in up to 96 channels with a maximum voltage less than 10,000 V, using a sheath flow rate of 3 L·min^{-1}. The CPC portion of the instrument is of the thermal diffusion type, with a saturator maintained at 35°C. The LPS is a single-particle, wide-angle optical sensor used for measuring particle size from 0.35 to 10 μm using 24 additional channels. Particles are drawn into the aerosol inlet at a flow rate of 0.70 L·min^{-1} and focused with a 3 L·min^{-1} flow of sheath air towards the center of a laser beam generated by a laser diode. Before and after the observation campaign, the DMA was calibrated with NIST SRM 1691 and SRM 1963 PSL spheres (0.269 and 0.1007 μm mean diameter) to verify proper DMA transfer function and accurate particle sizing. The LPS was calibrated with four NIST traceable sizes of PSL (0.701, 1.36, 1.6, and 4.0 μm mean diameter). Each sample takes 8 min 16 s to be analyzed. Over the course of all measurements, the WPS experienced no malfunctions.

The mass concentration of $PM_{2.5}$ was sampled using Reference Ambient Air Samplers (RAAS) $PM_{2.5}$ samplers (TEI RAAS 2.5–400), and using Teflon filter weighing using a Sartorius ME5-F balance (0.001 mg). Ozone was monitored using a UV photometry analyzer (TEI model 49C), SO_2 by a pulsed UV fluorescence analyzer (TEI model 43C), and NO_x by a MoO/chemiluminescence analyzer (TEI Model 42i-TL). The TEI model 49C was fully calibrated using a TEI Model 49C Primary Standard UV photometric O_3 calibrator every month. The TEI Model 43C and TEI Model 42i-TL were fully calibrated each week using a TEI Model 111 zero air supply and a TEI Model 146C calibrator with mixed standard gas consisting of NO, SO_2, and CO. The TEI Model 42i-TL, 43C, and 49C were simultaneously zero checked every 3 to 4 days over the course of the measurements.

Meteorological data were obtained from an automatic meteorological station, located 10 m northeast of the sampling site. In spring, southwest wind is most frequent and northeast wind takes the second, and it is on the contrary in winter. In both summer and fall, northeast wind dominates and southwest wind takes the second but is far less frequent the northeast wind.

Sampling duration and data treatment

As the larger instrumental error originates from too small concentration values of particles with diameters larger than 2.5 μm and the particle number concentrations are dominated by ultrafine particles (Seinfeld and Pandis 1998), only particle data not larger than 2.5 μm was analyzed. Particles in the diameter from 10 nm to 2.5 μm were divided into six sub-ranges—10–20 nm (N_{10-20} or N_{nuc}, nuclei mode), 20–50 and 50–100 nm (N_{20-50}, N_{50-100}, Atkien mode), 100–200, 200–500, and 500–2,500 nm ($N_{100-200}$, $N_{200-500}$, $N_{500-2,500}$, accumulation mode).

In order to analyze seasonal differences, the available dataset was subdivided according to four seasons: spring (03/26/2008–05/03/2008), summer (05/31/2008–06/26/2008), fall (10/02/2008–10/26/2008), and winter (12/03/2007–01/07/2008, 02/02/2008–03/25/2008). Considering that 03/25/2008 was the end of the heating season (residential steam heat production ceases) in Jinan, we regarded the whole heating season as winter. After data screening, there are 7,308, 3,377, 3,763, and 12,014 WPS samples for spring, summer, fall, and winter, respectively.

Results and discussion

Number concentration and seasonal variety

Various descriptive statistics regarding the measured hour average number concentrations of the four seasonal periods are included in Table 1. The particle number concentration with the diameter between 10 and 2,500 nm ($N_{10-2,500}$) in winter was the largest of the four seasons, while N_{20-50}, N_{50-100}, $N_{100-200}$, and $N_{200-500}$ had the same seasonal variation. The N_{20-200} range accounted for most of the total particle number concentration in Jinan, so its variety dominated the seasonal variety of the total number concentration. Differing from the larger diameter particles, N_{10-20} in fall was far larger than other seasons indicating more frequent new particle formation events occurring in fall; a detailed explanation has been presented in "Diurnal variation". The $N_{200-500}$ value was the largest in winter and second largest in summer. In fall, N_{20-50}, N_{50-100}, $N_{100-200}$, $N_{200-500}$, and $N_{500-2,500}$ were the lowest and are attributed to good atmospheric dispersion.

Similar with Jinan (I have analyzed the particle sources of Jinan in "Number, surface, and volume concentration size distribution"), in Pittsburgh, a USA city, nucleation and vehicle emissions were the most important sources of particles (Stanier et al. 2004a, b; Zhang et al. 2004). Compared with the study on urban Pittsburgh (Table 2; Stanier et al. 2004a), urban Jinan possesses a lower number concentration of particles smaller than 20 nm (nucleation mode) and 50 nm but has a higher number concentration of accumulation mode particles and those between 50 and 100 nm. The mass concentrations of $PM_{2.5}$ are generally dominated by particles of accumulation mode, and higher number concentrations of accumulation mode indicated higher mass concentrations of $PM_{2.5}$ in urban Jinan. During the experimental period in Pittsburgh, the mass concentrations of $PM_{2.5}$ were 20 $\mu g{\cdot}m^{-3}$ in summer and 12 $\mu g{\cdot}m^{-3}$ in winter, compared to 173.2 and 159.6 $\mu g{\cdot}m^{-3}$ in summer and winter for Jinan, respectively. Otherwise, particle surface concentrations are dominated by the accumulation mode as well, so higher particle

Table 1 Descriptive statistics of the measured particle number concentrations ($\#{\cdot}cm^{-3}$) of the four seasons (diameter unit: nm)

Season	10–20	20–50	50–100	100–200	200–500	500–2,500	10–2,500
Spring	500	6,500	4,600	3,300	1,200	25	16,200
Summer	900	5,500	3,300	2,900	1,200	77	13,900
Fall	2,200	4,800	2,100	1,500	600	22	11,200
Winter	900	7,700	6,600	4,700	1,700	29	21,600

Table 2 Comparison of descriptive statistics of sub-ranges for Jinan and Pittsburgh, annual (Unit: #·cm^{-3})

Particle size range (nm)	10–20	20–50	50–100	100–200	200–500	500–1,000	1,000–2,500
Pittsburgh, USA	4,100	6,500	3,600	1,710	460	18	0.59
Jinan, China	1,100	6,100	4,100	3,100	1,200	40	6.7

number concentration of accumulation mode in urban Jinan indicates higher particle surface concentration. New particle formation and the existence of newly formed nanoparticles are strongly limited by surface concentrations of pre-existing aerosol particles (Kulmala et al. 2001, 2004a, b) So that may be an important reason for why lower nucleation mode particle number concentrations were observed in Jinan, under a high SO_2 concentration (annual mean 26 ppb).

For Tables 2 and 3, data were acquired using different types of instruments.

Compared with other studies in urban areas (Table 3), the value of N_{10-100} in Jinan was observed to be the least of six cities listed, while the value of $N_{100-500}$ was the highest, as was the mass concentrations of $PM_{2.5}$.

Number, surface, and volume concentration size distribution

Mean particle number, surface, and volume size distributions of the four seasons in Jinan are shown in Fig. 2; the surface and mass concentrations were calculated with the assumption that the particles are spherical. As can be seen in Fig. 2, there is obvious seasonal variation and the number concentrations are mainly determined by Atkien mode particles. The number distribution in winter shows a single peak around 49 nm dominated by Atkien mode particles; the number distribution in spring is similar to winter but has a slightly lower peak maximum (41 nm). A bimodal log-normal (Atkien mode and accumulation mode) distribution in summer and fall are also shown in Fig. 2a, and in fall, higher value for nucleation mode is observed. The differences of number size distribution in various seasons are determined by factors such as meteorology, the sources and sinks of particles, etc., so further information from particle size distributions cannot be determined. However, the fact that new particle formation events occur more frequently in fall than in other seasons in Jinan can be deduced.

Surface and volume (mass) size distributions (Fig. 2b, c) are mainly determined by the accumulation mode particles. Measurements in the fall not only showed the highest nucleation mode particle number concentration but also had the lowest surface concentration. Single peaks with inconspicuous secondary peaks were observed in spring, fall, and winter. The main peaks of surface and volume (mass) size distributions are located around 300 and 400 nm, respectively. In contrast, there is a conspicuous peak located near 1,095 nm for both surface and volume (mass) size distributions in summer. Venkataraman and Rao (2001) have observed biomass burning can produce a

Table 3 Comparison of descriptive statistics of sub size ranges for Jinan and other urban studies

Location		Number conc. (#·cm^{-3})		$PM_{2.5}$ (μg·m^{-3})	Source
		10–100 nm	100–500 nm		
Alkmaar, Netherlands	Winter	18,300	2,120	27.0	Ruuskanen et al. (2001)
Erfurt, Germany	Winter	17,700	2,270	41.9	Ruuskanen et al. (2001)
Helsinki, Finland	Winter	16,200	973	9.42	Ruuskanen et al. (2001)
Pittsburgh, USA	Annual	14,300	2,170	16	Stanier et al. (2004a)
Atlanta, USA	Annual	21,400		19.3	Woo et al. (2001)
Jinan, China	Spring	11,700	4,500	102.7	This work
Jinan, China	Summer	9,700	4,100	159.2	This work
Jinan, China	Fall	9,100	2,100	99.3	This work
Jinan, China	Winter	15,200	6,400	173.2	This work
Jinan, China	Annual	11,400	4,300	133.6	This work

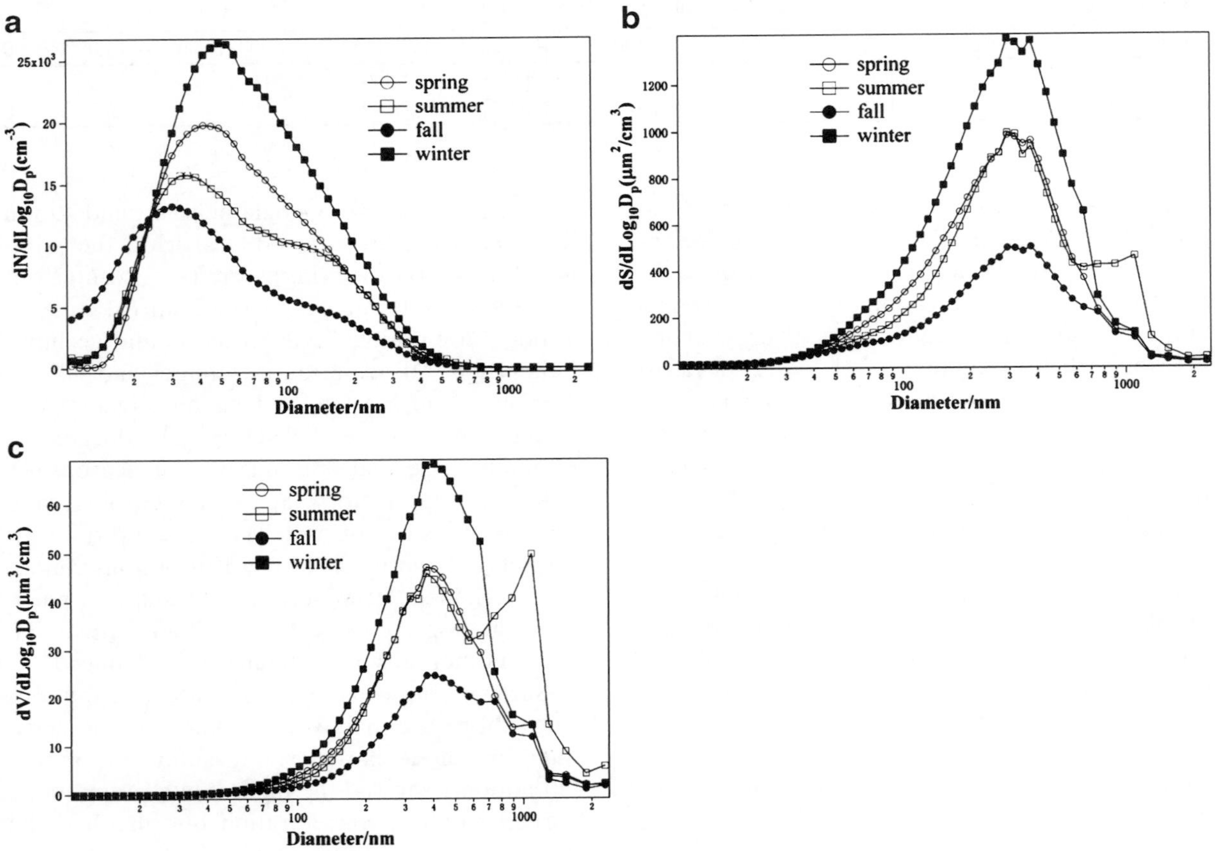

Fig. 2 Mean number (**a**), surface (**b**), and volume (mass) (**c**) size distributions in four seasons, Jinan

particle mass size distribution with a peak near 1 μm, and our summer measurement was operated during June, a frequent biomass burning period in the northern China (Fig. 3a), so the conspicuous peak located near 1,095 nm for both surface and volume (mass) size distributions in summer may have originated from agricultural biomass burning.

Diurnal variation

Figure 3 illustrates the diurnal variations of particle size distribution between 10 and 500 nm with diameter and geometric mean diameter (GMD) of particles (10–2,500 nm) on the *y*-axis, time of the day on the *x*-axis, and particle concentration ($dN/dLog_{10}D_p$) by shading contour for the various seasons. The simultaneous diurnal variations of the concentrations of NO_x, SO_2, and O_3, as well as aerosol surface area are shown in Fig. 4.

As seen in Fig. 4, in spring and winter, a high concentration of N_{20-200} was observed during day and night, indicating the accumulation of particles from traffic emissions under a poor atmospheric dispersion capacity and a high concentration of background particles in the Jinan urban area. In summer and fall, that the N_{20-200} was lower indicates a good dispersion capacity, especially in fall. Considering the diurnal of NO_x concentration, for all seasons (cf. Fig. 5), the contour area plots can be divided into three regions of high particle concentration: morning rush hour, evening rush hour and daytime around noon, with particles mainly arising from the accumulation mode. Compared with morning rush hour, the high particle number concentrations in evening rush hour lasted for longer time due to the poor dispersion capacity

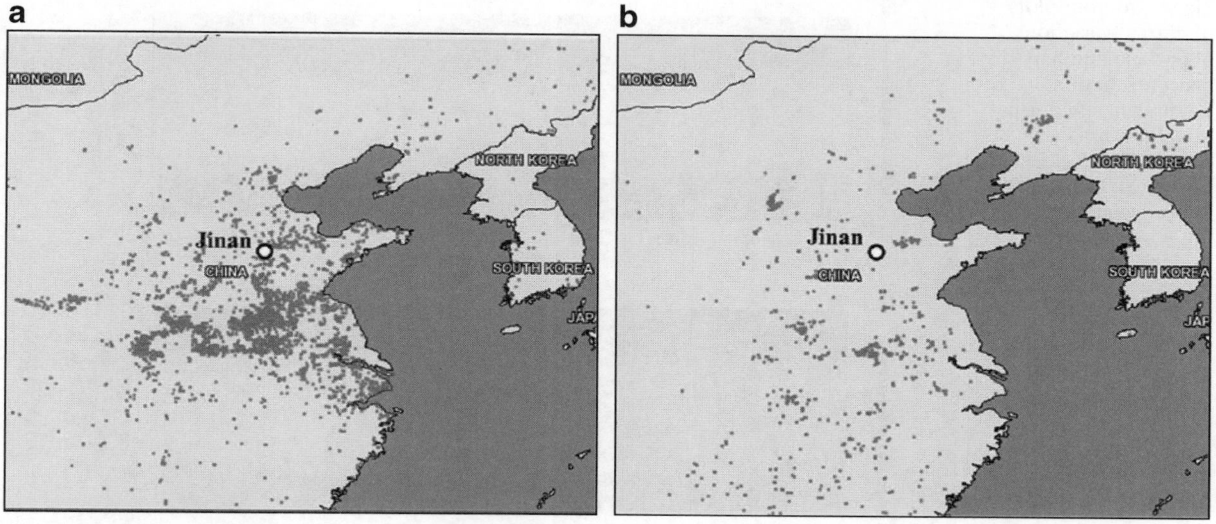

Fig. 3 MODIS images of fire spots, **a** summer (05/31/2008–06/26/2008), **b** fall (10/02/2008–10/26/2008), (Aqua & Terra, http://firefly.geog.umd.edu/firemap/)

at night. In winter, with a temperature inversion layer, lower wind speed, and a lower altitude of the atmospheric mixing layer, it is obvious that particle number concentrations are influenced by traffic emissions in rush hour.

An event of new particle formation was identified if there was a spontaneous burst of particle number concentrations in the nucleation mode (10–20 nm, the smallest channel of the WPS system) followed by their subsequent growth at the rate of a few nanometers per hour over a time span of hours. This event must also be, without obvious local emissions affecting the number concentrations of nucleation-mode particles (Gao et al. 2008). In this work, this criterion was used to identify the overall diurnal frequency of new particle formation in various seasons. In Fig. 5, although peaks of SO_2 concentration (indicating high concentrations of precursors for new particle formation and growth) and O_3 (indicating strong photochemical oxidation to a certain extent) occurred simultaneously in spring, summer, and fall, no obvious new particle formation was observed in spring, summer, and winter. Figure 4c shows that more frequent new particle formation events occurred in fall. New particle formation and the existence of newly formed nanoparticles are strongly limited by surface concentrations of pre-existing aerosol particles (Kulmala et al. 2001, 2004a, b). As can be seen in Fig. 5, compared with other seasons, fall has the lowest particle surface concentration at noon, so it may be concluded that high number concentrations of particles in spring, summer and winter limit new particle formation in urban Jinan.

The diurnal variations of surface concentration were highly accordant with that of NO_x concentration (Fig. 5), especially in the morning and evening rush hours. In the morning and evening, atmospheric dispersion capacity and the influence of air mass transport from the suburbs and rural areas are weaker, so it is reasonable that the surface concentration is dominated by traffic emissions. While both NO_x and surface concentrations are the lowest around noon, it cannot be confirmed that surface concentrations are also dominated by traffic emissions like the morni ng and evening, because both NO_x and surface concentrations may be affected strongly by atmospheric dispersion around midday. The correlation coefficients (R) between N_{nuc}, concentrations of NO_x and SO_2, and surface concentration around midday are shown in Table 4. Compared to the complete data set (Table 6 below), $R_{\text{surface vs NOx}}$ (the correlation coefficients between concentrations of particle surface area and NO_x) around midday

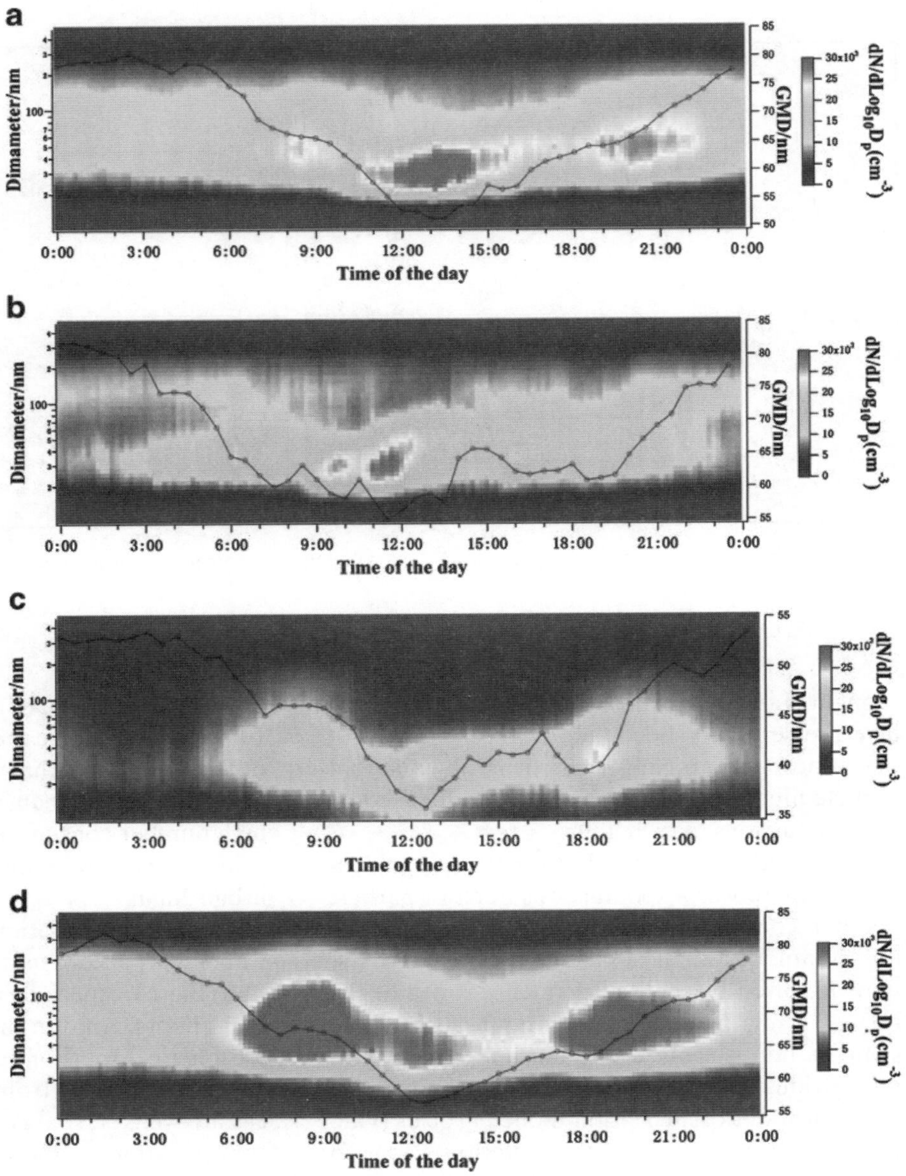

Fig. 4 Contour plots of diurnal variations of particle number concentration in 10–500 nm and diurnal variations of GMD (*black line*, 10–2,500 nm), spring (**a**), summer (**b**), fall (**c**), winter (**d**)

are lower and $R_{\mathrm{surface\ vs\ SO_2}}$ around midday are higher. This suggests that surface concentrations are affected more strongly by air mass transport from suburbs and rural areas, and more weakly affected by traffic emissions. Because most new particle formation events occurred around noon, it can be concluded that surface concentrations of pre-existing particles from traffic emissions and transporting from suburban and rural areas limit new particle formation in urban Jinan.

These data suggest that in the Jinan urban area, traffic emissions are important sources of particle number and surface concentrations. However, particle surface concentrations also decrease the latent concentration of ultrafine particles derived from new particle formation under the high SO_2 concentration levels in Jinan. Around midday, N_{20-60} was higher than during the morning and evening rush hours, although there is a smaller traffic flow and stronger dispersion capacity

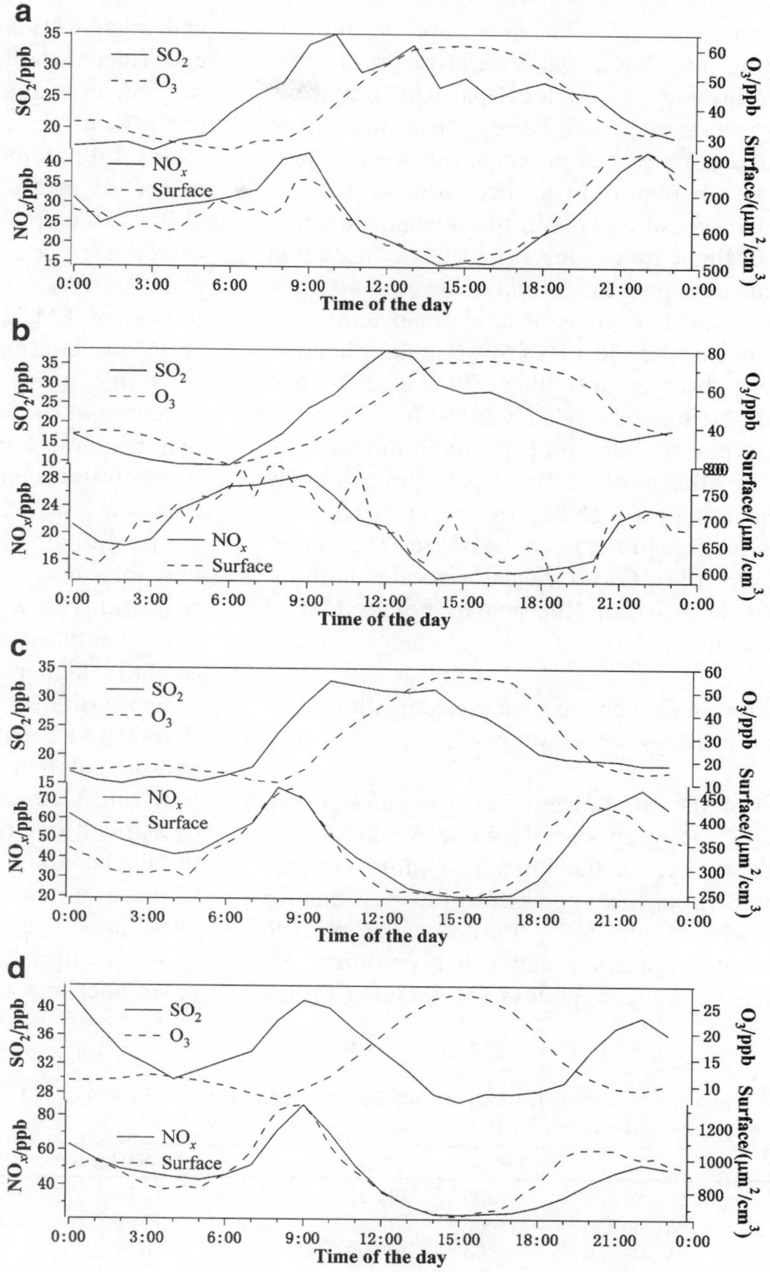

Fig. 5 Diurnal variations of SO_2, O_3, NO_x, and surface concentrations of $PM_{2.5}$, spring (**a**), summer (**b**), fall (**c**), winter (**d**)

Table 4 Correlation coefficients (R) between hour average of N_{nuc}, surface concentration of $PM_{2.5}$, and NO_x and SO_2 concentrations

	Spring			Summer			Fall			Winter		
	Surface	NO_x	SO_2	Surface	NO_x	SO_2	Surface	NO_x	SO_2	Surface	NO_x	SO_2
N_{nuc}	−0.28	−0.02	0.09	−0.34	−0.01	−0.05	−0.39	−0.18	0.07	−0.24	−0.02	−0.06
Surface	1.00	**0.53**	**0.56**	1.00	**0.56**	0.48	1.00	**0.59**	0.33	1.00	0.44	0.40

Spring 10:00–17:00; summer 9:00–16:00; fall 9:00–16:00; winter 10:00–16:00. $|R| \geq 0.5$ in bold

during midday. Also, N_{10-20} is lower at midday, indicating that N_{20-60} particles arose from other sources instead of new local particle formation. This suggests that atmospheric transport from rural and suburban areas of newly formed particles are an important particle source at noon. Although measurements in the upwind direction of urban Jinan are lacking, it may be deduced that frequent new particle formation events occur in rural or suburban areas around Jinan under the high concentration levels of SO_2; similar phenomena were observed in Beijing (Shi et al. 2007) and Pittsburgh (Stanier et al. 2004b). After transport to the urban of Jinan and thorough mixing with urban air, the newly formed particles gradually become larger through heterogeneous reaction or coagulate with pre-existing particles. The subsequent growth of GMD indicates good conditions for particle growth in the heavily polluted air of urban Jinan.

Variations in particle number concentration during weekdays versus weekends

Studies on the differences in particle number concentration between weekdays and weekends can provide insight on the impact of motor vehicle emissions on particle concentrations in urban air (Morawska et al. 2002). Variations of particle number concentration and other pollutant parameters between weekdays (Monday to Friday and other workdays, except holidays) and weekends (include holidays except workdays) were also studied, in order to investigate the influence of traffic flow rates.

Several parameters divided into weekdays and weekends in the four seasons are presented in Table 5. Compared to weekdays, the spring and winter averages of $N_{nuc}(N_{10-20})$, N_{Ait} (N_{20-100}), and N_{acc} ($N_{100-2,500}$), as well as surface concentration of $PM_{2.5}$, mass concentration of $PM_{2.5}$, and NO_x concentrations were lower on weekends. In contrast, summer and fall, variations between weekdays and weekends, were different than winter and spring. Generally, in urban areas, concentrations of NO_x are dominated by traffic emissions, so NO_x concentrations on weekends are potentially lower than on weekdays as traffic flow rates decrease. However, in Jinan, NO_x concentrations on weekends were higher than weekdays in summer and fall; SO_2 concentrations are similarly higher. Biomass burning can increase concentrations of NO_x, SO_2, and other pollutants (Crutzen and Andreae 1990; Scholes et al. 1996), and the summer and fall measurements taken in June and October, respectively, were influenced by agricultural biomass burning (Fig. 3a, b). It should be noted that agricultural production in China is not influenced by holidays and weekends; thus, it is expected that the sampling site was influenced by agricultural biomass burning in summer and fall, especially in the fall. For all

Table 5 Hour average particle number, surface, mass concentration of $PM_{2.5}$, SO_2, O_3, and NO_x concentrations in Jinan during weekdays and weekends

Seasons		N_{nuc}	N_{Ait}	N_{acc}	Surface	$PM_{2.5}$	SO_2	O_3	NO_x
Spring	Average	500	11,100	4,500	644	102.7	26.1	39.9	29.8
	Weekdays	600	11,700	4,600	653	100.1	28.1	39.8	32.1
	Weekends	500	9,900	4,200	625	110.0	22.2	40.3	25.8
Summer	Average	900	8,800	4,200	695	173.2	21	46.3	21.2
	Weekdays	1,000	8,600	4,000	655	167.1	20.0	39.2	20.7
	Weekends	800	9,200	4,800	778	186.4	23.0	58.6	22.0
Fall	Average	2,200	6,900	2,100	338	99.3	23.9	34	50.8
	Weekdays	2,200	7,300	2,300	360	95.7	19.0	36.4	48.4
	Weekends	2,200	6,100	1,900	296	104.7	30.2	30.9	53.8
Winter	Average	900	14,300	6,500	922	159.6	33.9	14.9	38.1
	Weekdays	900	14,600	6,500	924	170.7	30.8	14.3	37.5
	Weekends	800	14,100	6,300	859	125.7	33.9	14.7	36.5

N_{nuc}, N_{Ait}, and N_{acc}: Number concentrations of N_{10-20}, N_{20-100}, and $N_{100-2,500}$, unit:#·cm^{-3}. Surface concentration of $PM_{2.5}$, unit: μm^2·cm^{-3}. Mass concentration of $PM_{2.5}$, unit: μg·m^{-3}. SO_2, O_3, and NO_x, unit: ppb seasons, there was no obvious difference between N_{nuc} on weekdays and weekends. In addition, traffic emissions have been considered an important source of N_{nuc} and N_{Ait} (Morawska et al. 1998; Harris and Maricq 2001; Kittelson et al. 2004; Burtscher 2005), so it is an important reason for lower N_{nuc} and N_{Ait} on weekends, except N_{Ait} in summer.

The variation between weekdays and weekends of several parameters included in Table 5 are smaller than measurements in urban areas in developed countries (Pekkanen et al. 1997; Ruuskanen et al. 2001; Woo et al. 2001; Morawska et al. 2002; Wehner et al. 2002, Stanier et al. 2004a, b; Laakso et al. 2003; Voigtlander et al. 2006; Costabile et al. 2009). The same phenomenon was also observed in the Beijing urban area (Shi et al. 2007). Two reasons for this characteristic feature of the particles in Jinan may be attributed to social practices and terrain, respectively. First, in China, traffic, agricultural, and industrial activities do not decrease significantly on weekends compared to weekdays. Second, Jinan is surrounded by mountains on the east, south, and west sides, so the terrain favors the accumulation of pollutants, delays atmospheric mass transfer, and attenuates the variation between weekdays and weekends.

The correlation coefficient matrix subdivided into weekdays and weekends for several parameters for the four seasons is shown in Table 6. As mentioned above, surface concentrations are dominated by N_{nuc}, so N_{acc} and surface concentration have almost the same relationship with

Table 6 Correlation coefficients between particles and gases in spring, summer, fall, and winter, hour average, subdivided into weekdays and weekends

Seasons		Weekdays						Weekends					
		N_{nuc}	N_{Ait}	N_{acc}	Surface	NO_x	SO_2	N_{nuc}	N_{Ait}	N_{acc}	Surface	NO_x	SO_2
Spring	N_{nuc}	1.00						1.00					
	N_{Ait}	0.33	1.00					0.42	1.00				
	N_{acc}	−0.16	**0.56**	1.00				−0.16	**0.54**	1.00			
	Surface	−0.17	**0.60**	**0.97**	1.00			−0.16	**0.52**	**0.97**	1.00		
	NO_x	−0.10	0.48	**0.71**	**0.70**	1.00		−0.08	**0.51**	**0.76**	**0.74**	1.00	
	SO_2	0.09	0.44	**0.56**	**0.52**	0.36	1.00	0.31	0.49	**0.53**	0.46	0.39	1.00
	O_3	0.21	−0.24	**−0.55**	−0.47	**−0.72**	−0.20	0.24	−0.31	**−0.59**	**−0.53**	**−0.81**	−0.22
Summer	N_{nuc}	1.00						1.00					
	N_{Ait}	**0.53**	1.00					0.49	1.00				
	N_{acc}	−0.26	0.23	1.00				−0.34	0.25	1.00			
	Surface	−0.23	0.20	**0.91**	1.00			−0.32	0.28	**0.96**	1.00		
	NO_x	−0.10	0.30	**0.63**	**0.58**	1.00		−0.04	0.38	**0.60**	**0.63**	1.00	
	SO_2	0.07	0.27	0.22	0.21	0.30	1.00	0.27	0.39	0.23	0.25	0.10	1.00
	O_3	0.17	−0.01	−0.44	−0.37	**−0.65**	0.16	0.09	−0.18	−0.28	−0.31	**−0.78**	0.24
Fall	N_{nuc}	1.00						1.00					
	N_{Ait}	0.46	1.00					0.31	1.00				
	N_{acc}	−0.21	0.47	1.00				−0.14	**0.73**	1.00			
	Surface	−0.23	0.42	**0.98**	1.00			−0.14	**0.69**	**0.97**	1.00		
	NO_x	−0.09	0.49	**0.67**	**0.60**	1.00		−0.19	0.39	**0.70**	**0.71**	1.00	
	SO_2	0.21	0.18	0.15	0.09	0.14	1.00	0.22	0.23	0.37	0.40	0.09	1.00
	O_3	−0.09	−0.47	−0.33	−0.24	**−0.59**	0.08	0.36	−0.08	−0.24	−0.18	−0.48	0.22
Winter	N_{nuc}	1.00						1.00					
	N_{Ait}	0.25	1.00					0.16	1.00				
	N_{acc}	−0.17	**0.62**	1.00				−0.11	**0.58**	1.00			
	Surface	−0.21	**0.57**	**0.96**	1.00			−0.08	**0.57**	**0.98**	1.00		
	NO_x	0.02	0.38	0.49	0.39	1.00		−0.04	0.32	**0.53**	0.47	1.00	
	SO_2	−0.02	0.31	0.43	0.35	**0.78**	1.00	0.13	0.19	**0.53**	0.44	**0.69**	1.00
	O_3	0.37	−0.26	−0.40	−0.41	−0.17	−0.15	0.37	−0.27	−0.49	−0.48	−0.20	−0.22

$|R| \geq 0.5$ in bold

the other parameters. For all seasons, surface concentration and N_{Ait} have apparent correlations with NO_x, and have secondary correlations with SO_2, indicating they are more influenced by traffic emissions than by coal firing. Particle surface concentrations have negative correlations with N_{nuc}. On weekends, N_{nuc} and N_{Ait} have more apparent positive correlation with SO_2 than on weekdays, indicating that more ultrafine particles may be related with coal firing. There are no obvious differences for coefficients between weekdays and weekends.

Summary and conclusion

Particle number concentration and size distribution were measured in Jinan urban site in spring, summer, fall and winter, during 2007 and 2008, along with parameters including concentration measurements of SO_2, NO_x, and O_3, and the mass concentration of $PM_{2.5}$. Number concentrations of particles (10–2,500 nm) were 16,200, 13,900, 11,200, and 21,600 cm^{-3} in spring, summer, fall, and winter, respectively. Compared with other urban studies, Jinan has higher particle number concentrations of accumulation-mode particles (100–500 nm), and lower concentrations of ultrafine particles (10–100 nm), because the high concentration of pre-existing particles limits new particle formation and existence of the existence of nanoparticles. All the particle number, surface, and volume (mass) concentrations and size distributions showed seasonal variations in urban Jinan.

Diurnal variations of particle number concentrations (10–500 nm) with GMD (10–2,500 nm) and concentrations of NO_x, SO_2, and O_3 for four seasons have been studied and showed obvious seasonal variations. Infrequent new particle formation events were observed in spring, summer and winter, but not fall. The diurnal variation of particle surface concentration was highly correlated with NO_x in all seasons, indicating that the variation is predominantly determined by traffic emissions, especially during the morning and evening rush hours, and influenced by greater air mass transport from suburban and rural areas at midday. In Jinan, traffic emissions are main source of particle number concentrations, and air mass transport from suburban and rural areas turn more important at midday. The subsequent growth of GMD in the afternoon indicates the suitable conditions for particle growth in the heavily polluted air of urban Jinan.

Compared with other studies in urban areas, the variations of particle number and related gas concentrations, between weekdays and weekends are smaller in Jinan. Two reasons for this characteristic have been deduced from society social practices and terrain, respectively. Through correlation analysis, particle number concentrations are predominantly influenced by traffic emissions and coal firing for all seasons and these sources have different levels of impact depending on day of the week.

Acknowledgements This research was funded by Shandong Provincial Environmental Protection Department (2006045) and the National Basic Research Program (973 Program) of China (2005CB422203), assisted by Jinan Environmental Monitoring Center. We also thank Dr. J. David Van Horn (Visiting Professor from University of Missouri-Kansas City) for editorial assistance.

References

Bellouin, N., Boucher, O., Haywood, J., & Reddy, M. S. (2005). Global estimate of aerosol direct radiative forcing from satellite measurements. *Nature, 438*, 1138–1141.

Burtscher, H. (2005). Physical characterization of particulate emissions from diesel engines: A review. *Journal of Aerosol Science, 36*(7), 896–932.

Chung, C. E., Ramanathan, V., Kim, D., & Podgorny, I. A. (2005). Global anthropogenic aerosol forcing derived from satellite and ground-based observations. *Journal of Geophysical Research-Atmospheres, 110*, D24207. doi:10.1029/2005JD006356.

Costabile, F., Birmili, W., Klose, S., Tuch, T., Wehner, B., Wiedensohler, A., et al. (2009). Spatio-temporal variability and principal components of the particle number size distribution in an urban atmosphere. *Atmospheric Chemistry and Physics, 9*(9), 3163–3195.

Crutzen, P. J., & Andreae, M. O. (1990). Biomass burning in the tropics: Impact on atmospheric chemistry and biogeochemical cycles. *Science, 250*(4988), 1669–1678.

Dockery, D. W., & Pope, C. A. (1994). Acute respiratory effects of particulate air pollution. *Annual Review of Public Health, 15*, 107–132.

Engler, C., Lihavainen, H., Komppula, M., Kerminen, V. M., Kulmala, M., & Viisanen, Y. (2007). Continuous measurements of aerosol properties at the Baltic

Sea. *Tellus Series B-Chemical and Physical Meteorology, 59*(4), 728–741.
Gao, J., Wang, T., Zhou, X., Wu, W., & Wang, W. (2008). Measurement of aerosol number size distributions in the Yangtze River delta in China: Formation and growth of particles under polluted conditions. *Atmospheric Environment, 43*(4), 829–836.
Harris, S. J., & Maricq, M. M. (2001). Signature size distributions for diesel and gasoline engine exhaust particulate matter. *Journal of Aerosol Science, 32*(6), 749–764.
Holmes, N. S. (2007). A review of particle formation events and growth in the atmosphere in the various environments and discussion of mechanistic implications. *Atmospheric Environment, 41*(10), 2183–2201.
Hussein, T., Puustinen, A., Aalto, P. P., Makela, J. M., Hameri, K., & Kulmala, M. (2004). Urban aerosol number size distributions. *Atmospheric Chemistry and Physics, 4*, 391–411.
IPCC (Intergovernmental Panel on Climate Change) (2001). Climate change 2001: The scientific basis. In J. T. Houghton, Y. Ding, D. J. Griggs, M. Noguer, P. J. van der Linden, X. Dai, et al. (Eds.), *Contribution of working group I to the third assessment report of the intergovern-mental panel on climate change*. Cambridge: Cambridge University Press.
Kittelson, D. B., Watts, W. F., & Johnson, J. P. (2004). Nanoparticle emissions on Minnesota highways. *Atmospheric Environment, 38*(1), 9–19.
Kulmala, M., Maso, M., & Makela, J. M. (2001). On the formation, growth and composition of nucleation mode particles. *Tellus Series B-Chemical and Physical Meteorology, 53*(4), 479–490.
Kulmala, M., Petäjä, T., Mönkkönen, P., Koponen, I. K., Dal Maso, M., Aalto, P. P., et al. (2004a). On the growth of nucleation mode particles: Source rates of condensable vapor in polluted and clean environments. *Atmospheric Chemistry and Physics, 4(5)*, 409–416.
Kulmala, M., Vehkamäki, H., Petäjä, T., Dal Maso, M., Lauri, A., Kerminen, V. M., et al. (2004b). Formation and growth rates of ultrafine atmospheric particles: A review of observations. *Journal of Aerosol Science, 35*(2), 143–176.
Laakso, L., Hussein, T., Aarnio, P., Komppula, M., Hiltunen, V., Viisanen, Y., et al. (2003). Diurnal and annual characteristics of particle mass and number concentrations in urban, rural and Arctic environments in Finland. *Atmospheric Environment, 37*(19), 2629–2641.
Menon, S., Del Genio, A. D., Koch, D., & Tselioudis, G. (2002). GCM simulations of the aerosol indirect effect: Sensitivity to cloud parameterization and aerosol burden. *Journal of the Atmospheric Sciences, 59*(3), 692–713.
Morawska, L., Bofinger, N. D., Kocis, L., & Nwankwoala, A. (1998). Submicrometer and supermicrometer particles from diesel vehicle emissions. *Environmental Science & Technology, 32*(14), 2033–2042.
Morawska, L., Jayaratne, E. R., Mengersen, K., Jamriska, M., & Thomas, S. (2002). Differences in airborne particle and gaseous concentrations in urban air between weekdays and weekends. *Atmospheric Environment, 36*(27), 4375–4383.
Nemmar, A., Hoet, P. H., Dinsdale, D., Vermylen, J., Hoylaerts, M. F., & Nemery, B. (2003). Diesel exhaust particles in lung acutely enhance experimental peripheral thrombosis. *Circulation, 107*(8), 1202–1208.
Oberdörster, G. (2000). Toxicology of ultrafine particles: In vivo studies. *Philosophical Transactions of the Royal Society of London Series A—Mathematical Physical and Engineering Sciences, 358*(1775), 2719–2740.
Pekkanen, J., Timonen, K. L., Ruuskanen, J., Reponenc, A., Mirme, A., et al. (1997). Effects of ultrafine and fine particles in urban air on peak expiratory flow among children with asthmatic symptoms. *Environmental Research, 74*(1), 24–33.
Ruuskanen, J., Tuch, T., Ten Brink, H., Peters, A., Khlystov, A., Mirme, A., et al. (2001). Concentrations of ultrafine, fine and $PM_{2.5}$ particles in three European cities. *Atmospheric Environment, 35*(21), 3729–3738.
Scholes, R. J., Ward, D. E., & Justice, C. O. (1996). Emissions of trace gases and aerosol particles due to vegetation burning in southern hemisphere Africa. *Journal of Geophysical Research-Atmospheres, 101*(D19), 23677–23682.
Seinfeld, J. H., & Pandis, S. N. (1998). Atmospheric chemistry and physics: From air pollution to climate change. New York: Wiley.
Shi, Z. B., He, K. B., Yu, X. C., Yao, Z. L., Yang, F. M., Ma, Y. L., et al. (2007). Diurnal variation of number concentration and size distribution of ultrafine particles in the urban atmosphere of Beijing in winter. *Journal of Environmental Sciences-China, 19*(8), 933–938.
Stanier, C. O., Khlystov, A. Y., & Pandis, S. N. (2004a). Ambient aerosol size distributions and number concentrations measured during the Pittsburgh Air Quality Study (PAQS). *Atmospheric Environment, 38*(20), 3275–3284.
Stanier, C. O., Khlystov, A. Y., & Pandis, S. N. (2004b). Nucleation events during the Pittsburgh Air Quality Study: Description and relation to key meteorological, gas phase, and aerosol parameters. *Aerosol Science and Technology, 38*(12), 253–264.
Stott, P. A., Tett, S. F. B., Jones, G. S., Allen, M. R., Mitchell, J. F. B., & Jenkins, G. J. (2000). External control of 20th century temperature by natural and anthropogenic forcings. *Science, 290*, 2133–2137.
Venkataraman, C., & Rao, G. U. M. (2001). Emission factors of carbon monoxide and size-resolved aerosols from biofuel combustion. *Environmental Science & Technology, 35*(10), 2100–2107.
Voigtlander, J., Tuch, T., Birmili, W., & Wiedensohler, A. (2006). Correlation between traffic density and particle size distribution in a street canyon and the dependence on wind direction. *Atmospheric Chemistry and Physics, 6*, 4275–4286.

Wehner, B., Birmili, W., Gnauk, T., & Wiedensohler, A. (2002). Particle number size distributions in a street canyon and their transformation into the urban-air background: Measurements and a simple model study. *Atmospheric Environment, 36*(13), 2215–2223.

WHO (2005). *World health organization air quality guidelines*. Global Update, E87950.

Woo, K. S., Chen, D. R., Pui, D. Y. H., & McMurry, P. H. (2001). Measurement of Atlanta aerosol size distributions: Observations of ultrafine particle events. *Aerosol Science and Technology, 34*(1), 75–87.

Wu, Z. J., Hu, M., Liu, S., Wehner, B., Bauer, S., Andreas, M. B., et al. (2007). New particle formation in Beijing, China: Statistical analysis of a 1-year data set. *Journal of Geophysical Research-Atmospheres, 112*(D9), D09209.

Yang, L. X., Wang, D. C., Cheng, S. H., Wang, Z., Zhou, Y., Zhou, X. H., et al. (2007). Influence of meteorological conditions and particulate matter on visual range impairment in Jinan, China. *Science of the Total Environment, 383*, 164–173.

Yue, D. L., Hu, M., Wu, Z. J., Wang, Z. B., Guo, S., Wehner, B., et al. (2009). Characteristics of aerosol size distributions and new particle formation in the summer in Beijing. *Journal of Geophysical Research-Atmospheres, 114*, D00G12, doi:10.1029/2008JD010894.

Zhang, Q., Stanier, C. O., Canagaratna, M. R., Pandis, S. N., & Jimenez, J. L. (2004). Insights into the chemistry of new particle formation and growth events in Pittsburgh based on aerosol mass spectrometry. *Environmental Science & Technology, 38*(18), 4797–4809.

Science of the Total Environment 466–467 (2014) 357–368

Contents lists available at ScienceDirect

Science of the Total Environment

journal homepage: www.elsevier.com/locate/scitotenv

Airborne particulate polycyclic aromatic hydrocarbon (PAH) pollution in a background site in the North China Plain: Concentration, size distribution, toxicity and sources

Yanhong Zhu [a,b], Lingxiao Yang [a,b,*], Qi Yuan [a], Chao Yan [a], Can Dong [a], Chuanping Meng [a], Xiao Sui [a], Lan Yao [a], Fei Yang [a], Yaling Lu [a], Wenxing Wang [a,c]

[a] *Environment Research Institute, Shandong University, Jinan 250100, China*
[b] *School of Environmental Science and Engineering, Shandong University, Jinan 250100, China*
[c] *Chinese Research Academy of Environmental Sciences, Beijing 100012, China*

HIGHLIGHTS

- Size-resolved PAH characterizations were studied at a background site in the North China plain.
- Toxicity analysis indicated that the carcinogenic potency of particulate PAHs exists primarily in the <1.8 μm size ranges.
- The sources of PAHs were investigated by diagnostic ratios and PCA.
- Backward trajectory analysis showed that local anthropogenic emission dominated PAH levels.

ARTICLE INFO

Article history:
Received 26 March 2013
Received in revised form 5 July 2013
Accepted 9 July 2013
Available online 7 August 2013

Editor: Lidia Morawska

Keywords:
PAHs
Size distribution
Toxicity
Sources
Long-range transport
Background site

ABSTRACT

The size-fractionated characteristics of particulate polycyclic aromatic hydrocarbons (PAHs) were studied from January 2011 to October 2011 using a Micro-orifice Uniform Deposit Impactor (MOUDI) at the Yellow River Delta National Nature Reserve (YRDNNR), a background site located in the North China Plain. The average annual concentration of total PAHs in the YRDNNR (18.95 ± 16.51 ng/m^3) was lower than that in the urban areas of China; however, it was much higher than that in other rural or remote sites in developed countries. The dominant PAHs, which were found in each season, were fluorene (5.93%–26.80%), phenanthrene (8.17%–26.52%), fluoranthene (15.23%–27.12%) and pyrene (9.23%–16.31%). A bimodal distribution was found for 3-ring PAHs with peaks at approximately 1.0–1.8 μm and 3.2–5.6 μm; however, 4–6 ring PAHs followed a nearly unimodal distribution, with the highest peak in the 1.0–1.8 μm range. The mass median diameter (MMD) values for the total PAHs averaged 1.404, 1.467, 1.218 and 0.931 μm in spring, summer, autumn and winter, respectively. The toxicity analysis indicated that the carcinogenic potency of particulate PAHs existed primarily in the <1.8 μm size range. Diagnostic ratios and PCA analysis indicated that the PAHs in aerosol particles were mainly derived from coal combustion. In addition, back-trajectory calculations demonstrated that atmospheric PAHs were produced primarily by local anthropogenic sources.

1. Introduction

As the world's largest emitter of PAHs, China is suffering from severe PAH contamination from various sources (Zhang and Tao, 2008). Chinese PAH emissions accounted for approximately 22% of the total global PAH emissions in 2004 (Zhang and Tao, 2009). Moreover, it was estimated that 5.8% of the pollution level in China's land area, where 30% of the population lives, exceeded the national ambient benzo[a]pyrene (BaP) standard of 10 ng/m^3 (Xu et al., 2013) and that 1.6% of the lung cancer morbidity in China is caused by inhalation exposure to ambient air PAHs (Zhang and Tao, 2009).

The elucidation of the PAH size distributions is essential to estimate their inputs into ecosystems and the human respiratory system, to trace their origins and to understand their ageing process (Duan et al., 2007). Therefore, it is important to study the size distributions of the PAHs in atmospheric particles. In recent years, a number of studies on particle size distribution have been conducted in China (Bi et al., 2005; Wu et al., 2005; Duan et al., 2007; Hien et al., 2007). These studies have shown distinct differences in the PAH pollution level in urban and

* Corresponding author at: Environment Research Institute, Shandong University, Jinan 250100, China. Tel./fax: +86 531 88366072.
E-mail address: yanglingxiao@sdu.edu.cn (L. Yang).

0048-9697/$ – see front matter © 2013 Elsevier B.V. All rights reserved.
http://dx.doi.org/10.1016/j.scitotenv.2013.07.030

suburban areas, with high levels typically observed in the urban areas. The carcinogenic 5- and 6-ring particulate PAHs are often associated with particle sizes having a Dp < 2.0 μm (Kaupp and McLachlan, 2000; Kameda et al., 2005; Ravindra et al., 2008). The sources of PAHs were linked to residential biofuel, coal combustion and vehicular emissions. The previous studies on the size distribution of PAHs in China were conducted in several rapidly developed and highly polluted regions (i.e., Beijing, Tianjin and the Pearl River Delta); such studies were rarely conducted in other regions, such as background sites, which are suitable for monitoring the impact of long-range transported air pollutants and investigating the sources of PAHs.

The North China Plain was found to have a high PAH emission density in both urban and rural areas (Zhang et al., 2007), as previously validated by a pilot study of atmospheric PAHs (Liu et al., 2007). For a more comprehensive understanding of PAHs and to further promote the regulations pertaining to PAH contamination, seasonal atmospheric PAH data are necessary for background areas in this region. The Yellow River Delta National Nature Reserve was established in 1992 for the protection of a newly developed wetland ecosystem. The YRDNNR has a total area of 153,000 ha. This area provides a unique habitat for many species (birds, fish, amphibians, etc.) and is rarely influenced by human activities. However, the YRDNNR is located in the Bohai Sea region, including the Liaoning province, the Hebei province, the Shandong province, Beijing city and Tianjin city, which is an area that suffers from serious particulate pollution. The particle concentration in many cities in this region is more than ten times higher than that in similar cities in developed countries (Cheng et al., 2011). Therefore, the location of the YRDNNR also makes it an ideal site to investigate the transport of PAHs and other pollutants from a heavily polluted area to a background site. While previous studies investigated the origin and transboundary movement of inorganic constituents of aerosols, little attention has been given to organic constituents, specifically PAHs. The objectives of this study are the following: (1) to analyse the seasonal variations of PAHs in the Yellow River Delta National Nature Reserve, including the concentrations, compositional pattern and size distributions; (2) to assess the health risk of PAHs in the atmosphere in this coastal city; (3) to investigate the sources of PAHs using diagnostic ratios and principal component analysis (PCA); and (4) to estimate the effects of long-range transport by back-trajectory analysis.

2. Experimental procedure

2.1. The sampling site

Aerosol samples were collected on the rooftop of a four-story office building (20 m above ground level) at the Yi Qian Er Management Station (37°35′–38°12′N, 118°33′–119°20′E) in the Yellow River Delta National Nature Reserve (YRDNNR), which is located in the estuary of the Yellow River in Dongying, Shandong Province (Fig. 1). During the sampling periods, the relative humidity in the YRDNNR covered a range of 23%–80%, and the temperatures varied from −5.48 to 30.42 °C.

2.2. Sample collection

In this study, a micro-orifice uniform deposit impactor (MOUDI Model 110, MSP Corporation, USA), operating at a flow rate of 30 l/min, was used to collect the samples. The MOUDI is a 9-stage cascade impactor. The available particle cut-size diameters of MOUDI are 0.18, 0.32, 0.56, 1.0, 1.8, 3.2, 5.6, 10 and 18 μm. Before sampling, the quartz fibre filters were wrapped in aluminium foil and baked overnight in a muffle furnace at 600 °C. Before testing, the MOUDI was disassembled and cleaned with deionised water and n-hexane. To properly assemble and operate the instrument, the O-ring was coated with a small amount of Vaseline (Yang et al., 2005).

The sampling was conducted for 23.5 h periods, beginning at approximately 8:30 am on day one and ending at 8:00 am on day two. The sampling was conducted intermittently for seven days in January and February (winter) and in April (spring), and for five days in July (summer) and in October (autumn) in 2011. Two of the samples collections failed in summer and autumn due to mechanical problems with the sampler.

2.3. PAH analysis

Before and after sampling, all of the substrates were held at a constant temperature (20 ± 1 °C) and relative humidity (50 ± 2%) in an air-conditioned room for 24 h and then weighed on a microbalance (Model BP211D, Sartorius, Germany), which was accurate to 0.001 mg.

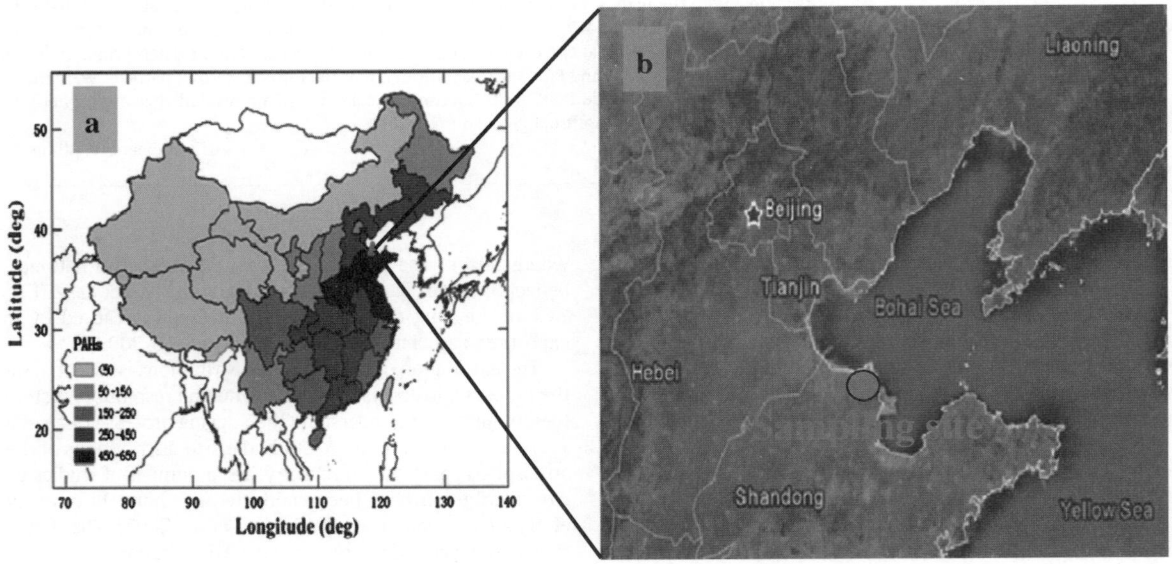

Fig. 1. (a) Location of the YRDNNR (marked with a red triangle) and the PAH emission density (g/km^2) in individual provinces and municipalities in 2004 (Zhang et al., 2011), (b) the location of our sampling site (marked with a red circle).

After weighing, the samples were extracted using accelerated solvent extraction (ASE 300, Dionex), which offers a high-speed extraction of PAHs in 15 min. A solvent combination of 4:1 dichloromethane/n-hexane was used for the extraction (Duan et al., 2007; Hong et al., 2007). The extracts were concentrated to approximately 2 ml by rotary evaporation with a thermostatic bath at $T = 30 \pm 1$ °C. The samples were dried under nitrogen stream, and dissolved in 1 ml of dichloromethane. An internal standard, hexamethylbenzene, was added to the samples immediately before analysis.

The samples were analysed with a gas chromatograph (GC) (Agilent 6890N) equipped with a mass selective detector (Agilent 5973 inert MSD) and a computer workstation. A 30 m × 0.25 mm × 0.5 μm DB 5 capillary column was used for the separation of the PAHs. The temperature programme used for GC analysis was as follows: the initial temperature of 60 °C was maintained for 1 min, and then increased to 150 °C at a rate of 20 °C/min; the temperature continued increasing to 270 °C at a rate of 7 °C/min and was eventually increased to 290 °C at 20 °C/min and held for 15 min, with a solvent delay time of 5 min (Ma et al., 2010). The injections were splitless, and the sample volume was 1 μl. High-purity helium was used as the carrier gas at a constant flow rate of 1 ml/min. The identification of the individual PAHs was based on the retention times of target ion peaks (within ±0.05 min of the retention of the calibration standard). The data acquisition and processing were controlled using HP Chemstation software.

The PAHs monitored in this study are abbreviated as follows: acenaphthylene (Acy, 3-ring), acenaphthene (Ace, 3-ring), fluorene (Fl, 3-ring), phenanthrene (Phe, 3-ring), anthracene (Ant, 3-ring), fluoranthene (Flu, 4-ring), pyrene (Pyr, 4-ring), benz[a]anthracene (BaA, 4-ring), chrysene (Chr, 4-ring), benzo[b]fluoranthene (BbF, 5-ring), benzo[k]fluoranthene (BkF, 5-ring), benzo[a]pyrene (BaP, 5-ring), indeno[1,2,3-cd]pyrene (IcdP, 6-ring), dibenz[a, h]anthracene (DBA, 5-ring) and benzo[ghi]perylene (BghiP, 6-ring).

2.4. Quality control

Field blanks, which accompanied the samples to and from the sampling sites, were used to determine the background contamination level. Five surrogate PAHs, consisting of naphthalene-D8, acenaphthene-D10, phenanthrene-D10, chrysene-D10 and perylene-D12 standards, were added to all of the samples to monitor the procedural performance and the matrix effects. None of the PAHs that were monitored in this study were detected in the blanks. The measured extraction recoveries were approximately 70% to 125% for the 16 PAHs. Because the recovery of naphthalene-D8 was low, the naphthalene result was not included in this study. Method detection limits were determined as the concentrations of the analytes in a sample that gave rise to peaks with a minimal signal to noise ratio (S/N) of 3. The LOD values for the PAHs ranged from 0.0050 ng/m^3 (Ace) to 0.1387 ng/m^3 (BghiP).

2.5. Back-trajectory calculation and cluster analysis

To identify the areas in the YRDNNR that are potential sources of PAHs, we calculated three-dimensional (height, longitude and latitude) five-day air mass trajectories every two hours during the study period. The HYSPLIT version 4.9 model (Draxler and Rolph, 2003) was used to acquire a database of back trajectories by calculating data using the Global Data Assimilation System (GDAS) from the NOAA Air Resources Laboratory's web server. According to the hierarchical cluster approach, these trajectories were classified into four different groups based on Ward's clustering method with a squared Euclidean distance. The computation was performed using the IGOR statistical software package.

3. Results and discussion

3.1. Overview of the measurement data

3.1.1. Seasonal variations of the particulate PAHs

The average concentrations of the individual PAHs in the particle phases (the sum of all stages) in different seasons in the YRDNNR are listed in Table 1. The average annual concentration of particulate PAHs in the YRDNNR was 18.95 ± 16.51 ng/m^3, which ranged from 4.77 to 66.17 ng/m^3. Compared recent results with those obtained from other cities in China, the concentration reported here (average values: 18.95 ± 16.51 ng/m^3) was much lower than that in the Beijing-Tianjin region from 2007 to 2008 (average value: 114 ng/m^3) (Wang et al., 2011) and in urban Guangzhou in 2007 (average value: 50.55 ng/m^3) (Duan et al., 2007). However, compared with other cities outside of China, such as Baltimore (0.33–6.52 ng/m^3) and New Brunswick (0.38–11.6 ng/m^3) in the USA (Gigliotti et al., 2000; Dachs et al., 2002), the particulate PAH concentrations at the YRDNNR were higher. Additionally, the particulate PAH concentrations at the YRDNNR were higher than that in other remote sites, such as Lake Superior in the USA (0.12 ng/m^3) (Baker and Eisenreich, 1990). Also, the particulate PAH concentrations at the YRDNNR were 6 times higher than that at Gosan (3.17 ± 3.31 ng/m^3) (Lee et al., 2006), a background site in Korea. Thus, although the YRDNNR had low emissions of air pollutants in the Bohai Sea region in China, the particulate PAH pollution in the atmosphere was higher than that in other rural and remote areas in developed countries.

The average concentration of $\sum$15PAHs in winter was the highest (34.17 ± 2.19 ng/m^3), then was those in spring (19.09 ± 1.00 ng/m^3) and autumn (9.71 ± 0.68 ng/m^3), and those in summer was lowest (7.43 ± 0.68 ng/m^3). The variation was caused by different meteorological conditions. Compared with the correlations between the PAH concentration with wind speed ($r = -0.49$) and relative humidity ($r = -0.42$), the ambient temperature exerted a higher negative correlation with the concentration of particulate PAHs ($r = -0.62$), indicating that the temperature was the major factor that influenced the concentrations of particulate PAHs. In summer, high temperatures and strong UV radiation increased the transition of PAHs from the particulate phase to the vapour phase and increased the photochemical decomposition of the PAHs. Conversely, in winter, the lower temperatures slowed the rate of photolysis and radical degradation; these factors combined

Table 1
The mean concentrations of particulate PAHs in the four seasons at the YRDNNR (ng/m^3).

	Spring	Summer	Autumn	Winter
Acy	0.26 ± 0.15	0.08 ± 0.04	0.10 ± 0.02	0.19 ± 0.05
Ace	0.29 ± 0.17	0.24 ± 0.19	0.28 ± 0.05	0.37 ± 0.18
Fl	2.16 ± 1.88	2.00 ± 2.29	1.17 ± 0.19	2.03 ± 1.12
Phe	3.44 ± 5.12	0.69 ± 1.03	2.57 ± 0.53	2.79 ± 1.75
Ant	1.12 ± 0.84	0.14 ± 0.16	0.06 ± 0.03	0.54 ± 0.66
Flu	3.20 ± 1.44	2.02 ± 1.77	1.48 ± 0.63	6.36 ± 3.93
Pyr	1.82 ± 0.97	1.21 ± 0.94	0.90 ± 0.47	4.40 ± 2.76
BaA	0.36 ± 0.11	0.13 ± 0.07	0.40 ± 0.22	1.87 ± 1.26
Chr	0.39 ± 0.14	0.07 ± 0.04	0.27 ± 0.23	1.70 ± 1.11
BbF	0.82 ± 1.13	0.05 ± 0.06	0.16 ± 0.24	0.04 ± 0.03
BkF	1.12 ± 0.51	0.20 ± 0.25	1.01 ± 0.88	7.23 ± 4.82
BaP	1.25 ± 1.10	0.29 ± 0.13	0.41 ± 0.27	2.15 ± 1.42
DBA	0.77 ± 1.10	0.10 ± 0.09	0.04 ± 0.04	0.09 ± 0.10
IcdP	1.31 ± 1.13	0.10 ± 0.08	0.45 ± 0.39	2.37 ± 1.41
BghiP	0.76 ± 0.45	0.12 ± 0.09	0.42 ± 0.32	2.06 ± 1.12
$\sum$PAHs	19.09 ± 1.00	7.43 ± 0.68	9.71 ± 0.68	34.17 ± 2.19
3 ring %	38.16	42.26	43.07	17.31
4 ring %	39.25	37.18	31.28	41.92
5–6 ring %	31.59	11.56	25.65	40.77
COMPAHs/$\sum$15PAHs	61.84	57.74	56.93	82.69
CANPAHs/$\sum$15PAHs	31.59	11.56	25.65	40.77

COMPAHs: combustion derived PAH concentrations.
CANPAHs: carcinogenic PAHs.

with an increase in domestic heating and consumption of fuels could result in higher levels of particulate PAHs (Hu et al., 2012) (Fig. 2).

3.1.2. PAH compositional pattern

The seasonal variation in the compositional patterns of PAHs was evaluated by analysing the relative proportions of individual PAH compounds. Although certain differences existed in the distribution patterns at different sampling periods, the main components of the PAHs in each season were Fl (5.93%–26.80%), Phe (8.17%–26.52%), Flu (15.23%–27.12%), and Pyr (9.23%–16.31%) (Fig. 3). These four dominant components were 3–4 ring PAHs that account for 45.58%–79.56% of the total 15 PAHs. This result was consistent with the previous report in urban and suburban areas (Omar et al., 2002; Zhou et al., 2005; Valavanidis et al., 2006), which related the PAH sources to coal combustion, vehicular emissions, and oil burning emissions.

A relatively high proportion of BkF was observed in autumn and winter. For example, in autumn and winter, BkF accounted for 10.36% and 21.16% of the total PAHs, whereas in spring and summer, BkF accounted for 5.85% and 2.73% of the total PAHs. BkF was believed to be less harmful than the other PAHs and was related to the increasing use of clean energy, such as natural gas (Wang et al., 2008), which is an important fuel in Dongying. In addition, in spring and autumn, the 3-ring PAHs were the dominant compounds, accounting for 38.16% and 43.07% of the total PAHs. In contrast, in summer and winter, the highest contributors were the 4-ring PAHs, accounting for 46.18% and 41.92% of the total PAHs. This result may be related to variations in the emission sources and/or the meteorological conditions in different seasons.

Flu, Pyr, Chr, BbF, BkF, BaA, BaP, IcdP and BghiP have been classified as combustion-derived PAHs (COMPAHs) by some authors (Bourotte et al., 2005; Del Rosario Sienra et al., 2005). In winter, the main combustion-derived PAH (COMPAH) concentrations accounted for 82.69% of the total PAHs, followed by spring (61.84%), summer (57.74%) and autumn (56.93%). In the four seasons, the combustion-derived PAHs represented more than half of the total PAHs, and in winter, they represented approximately 4/5 of the total PAHs. The concentration ratio of COMPAHs/ΣPAHs in the YRDNNR was higher than that for the urban aerosols in Greece (Gogou et al., 1996), an effect that is most likely due to the different fuel structure used.

Carcinogenic PAHs (CANPAHs), including BbF, BkF, BaA, BaP and IcdP (Bourotte et al., 2005; Del Rosario Sienra et al., 2005; Hong et al., 2007), were used to evaluate the carcinogenic potential of the PAHs to the exposed people. The results are shown in Table 1. In spring, summer, autumn and winter, the concentration ratios of CANPAHs/$\sum$ PAHs were 31.59%, 11.56%, 25.65%, 40.77%, respectively. This result was similar to the seasonal trend shown in the PAH concentrations, which indicated that the source of the CANPAHs was the same as the source of the PAHs in each season.

3.2. Size distributions of the PAHs

The size distributions of the PAHs are described using a histogram of the relative mass of dC/dlogDp versus the Dp on the logarithmic scale, which is a useful method for comparing the contributions of coarse and fine particles to the PAH concentrations (Allen et al., 1996; Oh et al., 2002). As shown in Fig. 4, the 3-ring PAHs exhibited a bimodal distribution with the major peaks in the ranges of 1.0–1.8 μm (accumulation mode) and 3.2–5.6 μm (coarse mode). The 4–6 ring PAHs showed a nearly unimodal distribution with the highest peak in the range of 1.0–1.8 μm (accumulation mode). In addition, the coarse fraction decreased for the 4-ring PAHs and nearly disappeared for the 5- and 6-ring PAHs (Duan et al., 2007). Fig. 4 shows that the higher ring number PAHs were adsorbed on the finer particles, whereas the more volatile PAHs were associated with the larger particles.

One explanation for these results is that the higher ring number PAHs, which have much lower subcooled liquid vapour pressures and Henry constants are expected to require much longer time to partition to coarse particles than the lower ring number PAHs and, thus, tend to remain in the fine particles (Allen et al., 1996). Moreover, the more unstable PAHs volatilised more rapidly from the fine particles than the non-volatile PAHs and sorbed onto the coarse particles. Another explanation for the results is that the chemical affinities between the PAHs and different size particles, along with different emission sources and different PAH reactivities on photooxidation resulted in the differences in the PAH distribution with respect to particle size (Allen et al., 1996; Venkataraman et al., 1999).

The mass median diameter (MMD), which is the diameter that divides the total mass by two (Yang et al., 2005), was calculated using an equation provided by Kavouras and Stephanou (2002). As shown in Table 2, the MMD values of the total PAHs were 1.404, 1.467, 1.218 and 0.931 μm for spring, summer, autumn and winter, respectively. In autumn and winter, the lower MMD values of the total PAHs reflected condensation mechanisms because the gas-particle condensation always enriches the aerosol fraction in the accumulation size range (Sicre et al., 1987; Aceves and Grimalt, 1993; Bi et al., 2005). Moreover, the differences in the MMD values of the individual PAH suggested that higher molecular weight PAHs (those containing 4, 5 and 6 benzene rings) were mainly distributed in the finer particles. In addition, the MMD values of certain PAHs (BaA, BaP, DBA and BghiP) were smaller than the total PAHs in each season, suggesting that these PAHs may arise from local sources in addition to the long-term transport in each season, specifically in winter (Bi et al., 2005).

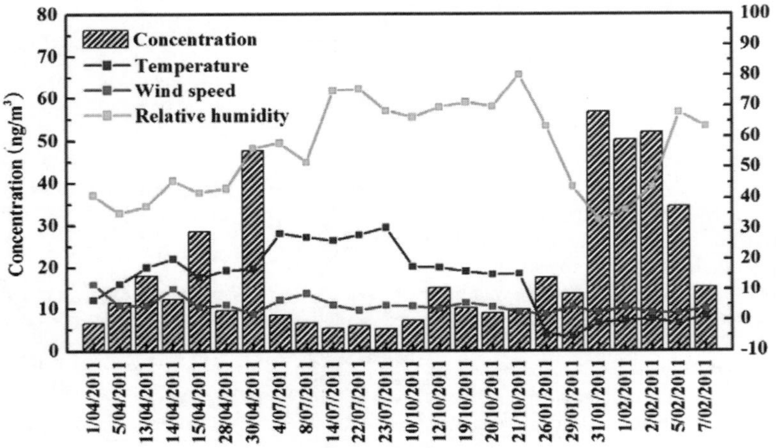

Fig. 2. The sampling time series of PAH concentration (ng/m^3), temperature (°C), wind speed (m/s) and relative humidity (%).

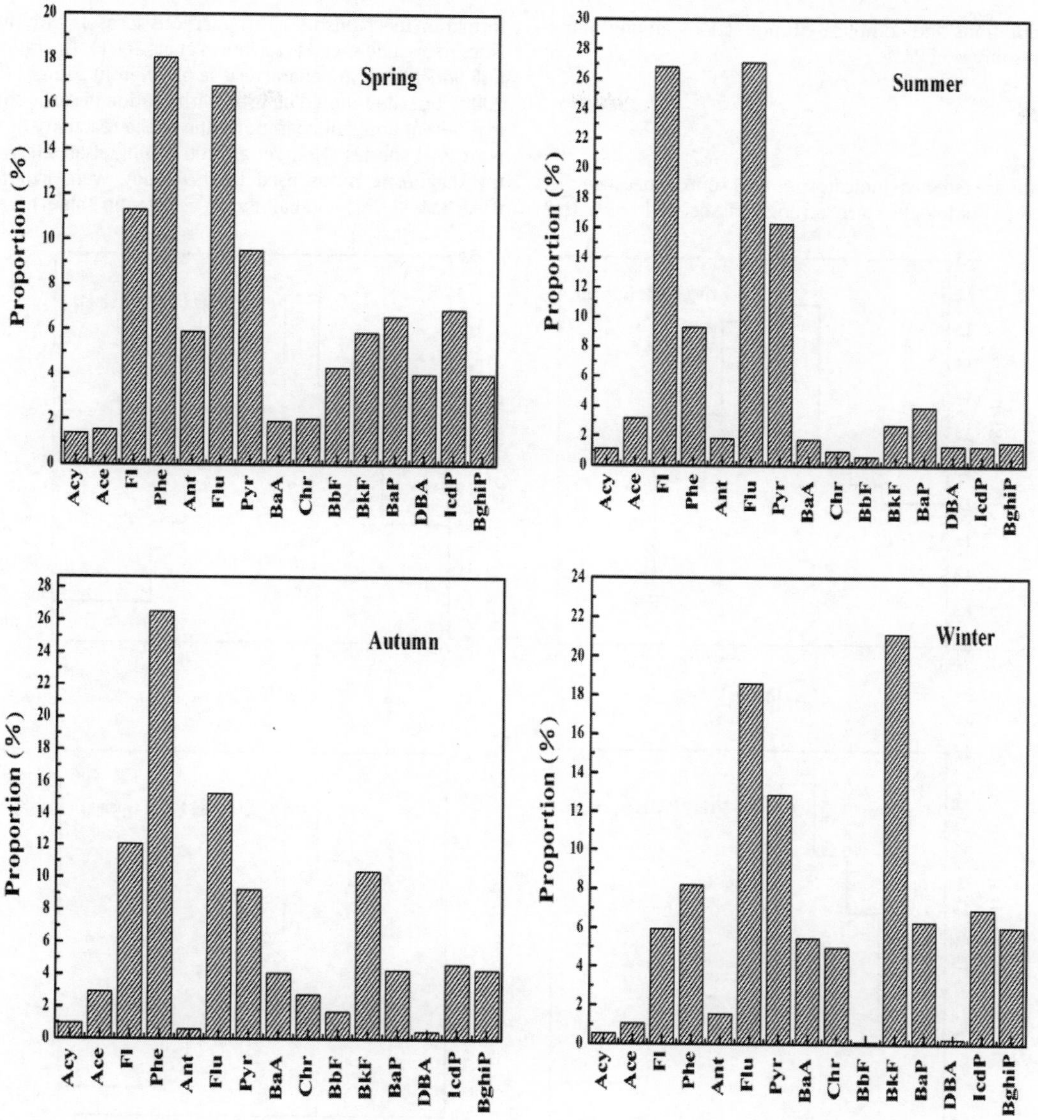

Fig. 3. Relative proportions of the individual particulate PAHs at the YRDNNR in the four seasons.

3.3. PAH toxicity evaluation

The development and establishment of toxic equivalency factors (TEFs) for PAHs could aid in the precise characterisation of the carcinogenic properties of PAH mixtures. The cancer potency of each PAH was assessed on the basis of its benzo[a]pyrene equivalent concentration (BaPeq). In this research, we calculated the toxic benzo[a]pyrene equivalent in different seasons (spring, summer, autumn and winter) based on the measured concentrations of the individual PAH compounds. The list of TEFs that was completed by Nisbet and LaGoy (1992) was adopted because it reflects the current knowledge on the toxic potency of each individual PAH specimen relative to its BaP concentration (Petry et al., 1996). Using the following equations, the toxic equivalency (TEQ) is calculated using the following formula (results in Table 3):

$$BaPeq_i = PAH_i * TEF_i; \quad \text{and} \quad TEQ = \sum(PAH_i * TEF_i)$$

where PAH_i is the concentration of the PAH congener i, TEF_i is the toxic equivalent factor for the PAH congener i, and TEQ is the toxic equivalent of the reference compound. Table 3 shows that the TEQ values of the sampling period varied from 0.45 to 3.44 ng/m^3, with a mean value of 1.84 ng/m^3, which was lower than the national standard of 10 ng/m^3 and significantly higher than the WHO standard (1 ng/m^3, Ventafridda et al., 1987). The higher TEQ value was found in winter. This value was also more than seven times higher than that in summer. The frequent rains that occurred in summer removed a large portion of the suspended particles from the air, and the particulate PAHs were the main contributors to the TEQ value. The concentrations of BaA, BkF, IcdP and DBA, which, like BaP, were considered to be carcinogenic, were relatively high, and their amounts were 31.8%–43.2% in the four seasons, suggesting a relatively higher human health risk in the YRDNNR when compared with 11.0%–45.4% in Xiamen, China (Hong et al., 2007).

The size distribution of the average BaPeq percentage in the TEQ value for all four seasons is shown in Fig. 5. As shown, 45.81%, 54.43%, 66.50% and 79.26% are attributable to particulates smaller than 1.8 μm for spring, summer autumn and winter, respectively. Fine particles tend to be emitted from the source and retained in the atmosphere for a long time. The PAHs that are associated with submicron particles will pose a stronger toxicity and do greater harm to human health. The slightly lower percentage in spring and summer may be due to meteorological conditions and complicated interactions among many different emission sources of PAHs.

3.4. Sources of PAHs

3.4.1. Diagnostic ratios

Diagnostic ratios are effective indicators of PAH sources because the distributions of the homologues are strongly associated with the formation mechanisms of carbonaceous aerosols with similar characteristics to organic species (Kavouras et al., 2001). The ratios of the specific individual PAHs are characteristic of different sources (Zhang and Tao, 2008), but they should be used with caution because some are variable in different ambient conditions due to the reactivity of some of the vapour PAH species (Hong et al., 2007; Ding et al., 2007). The results of the diagnostic ratios used in this study were IcdP/(BghiP + IcdP), BaA/(BaA + Chr) and BaP/BghiP, as listed in Table 4.

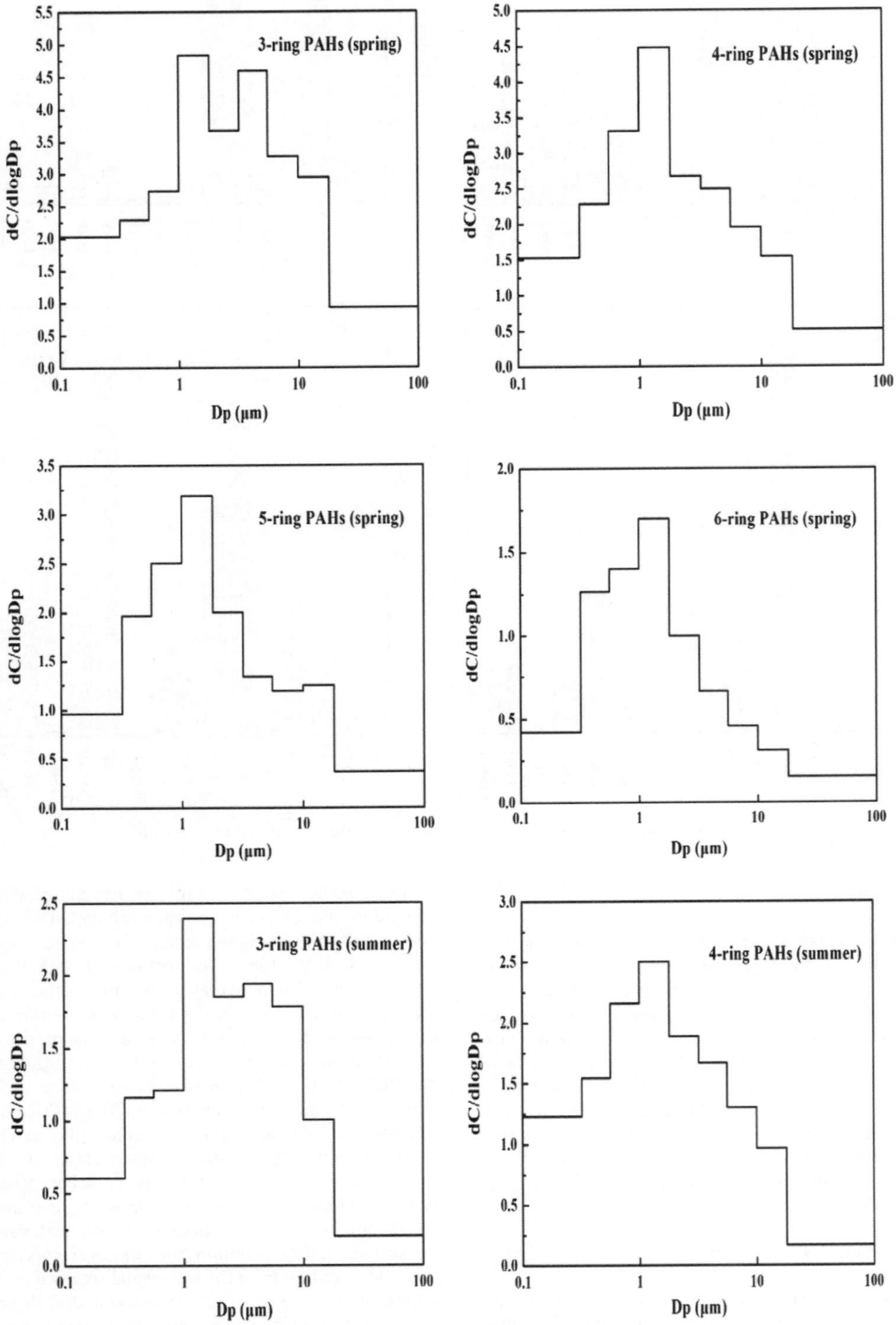

Fig. 4. Comparison of the distributions of the 3-, 4-, 5- and 6-ring PAHs in spring, summer, autumn and winter.

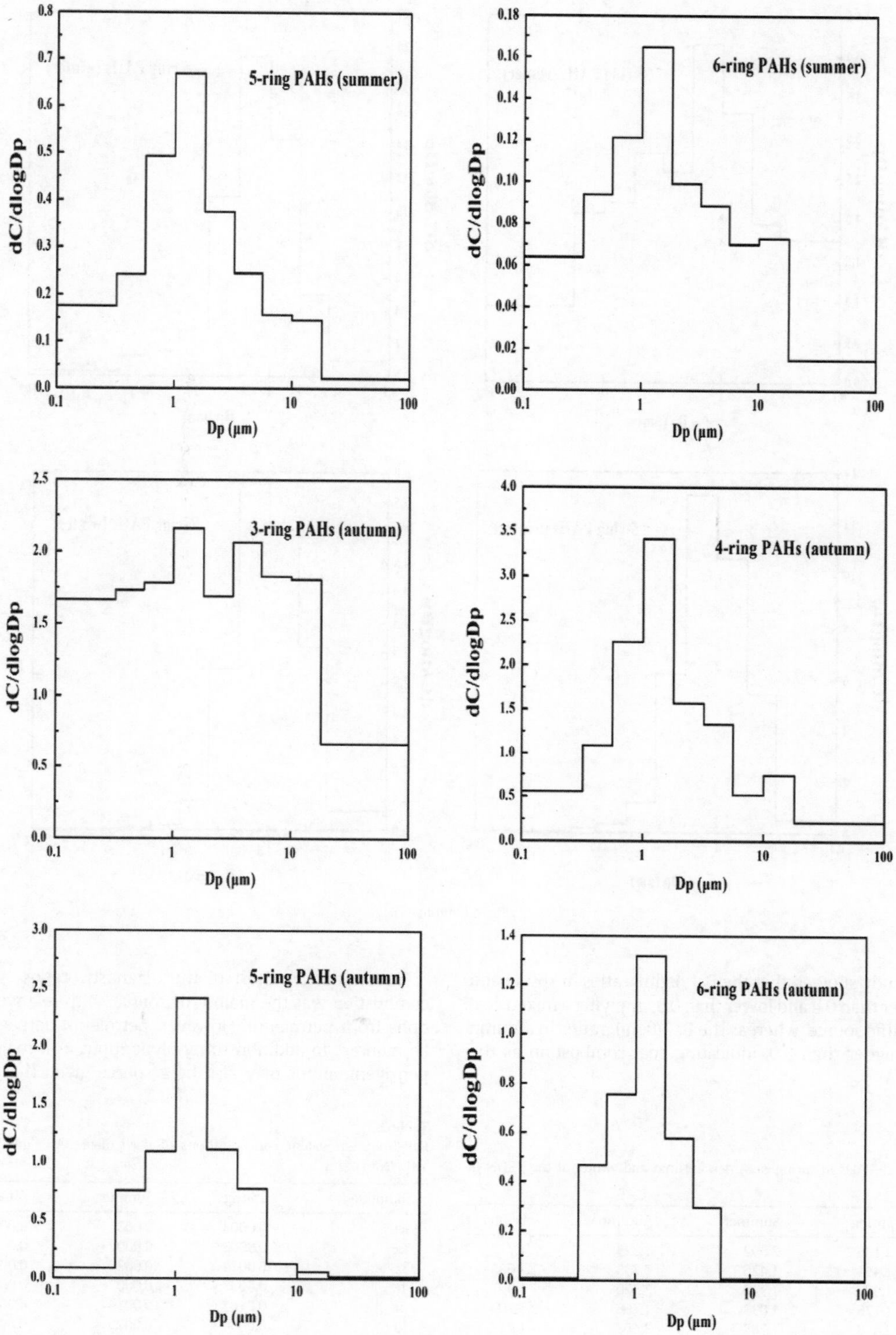

Fig. 4 (*continued*).

The ratio of IcdP/(BghiP + IcdP) higher than 0.5 signals coal/biomass combustion and lower than 0.5 indicates fuel combustion (Mandalakis et al., 2002; Hu et al., 2012). In Table 4, the IcdP/(BghiP + IcdP) ratios were all above 0.5 in spring, autumn and winter, indicating a strong contribution from coal and biomass combustion. In summer, the IcdP/(BghiP + IcdP) ratios were lower than 0.5, which implied that fuel combustion was the main source.

A BaA/(BaA + Chr) ratio that is higher than 0.35 signals pyrolytic sources, ratios lower than 0.2 indicate petrogenic sources, and ratios between 0.2 and 0.35 could indicate petroleum and combustion sources (Simcik et al., 1999). The values for all seasons varied from 0.48 to 0.60 (i.e., higher than 0.35), indicating that the PAHs could be of pyrolytic origin.

The value of the BaP/BghiP ratio ranges from 0.3 to 0.4 for traffic emissions and 0.9 to 6.6 for coal combustion (Park et al., 2002; Liu

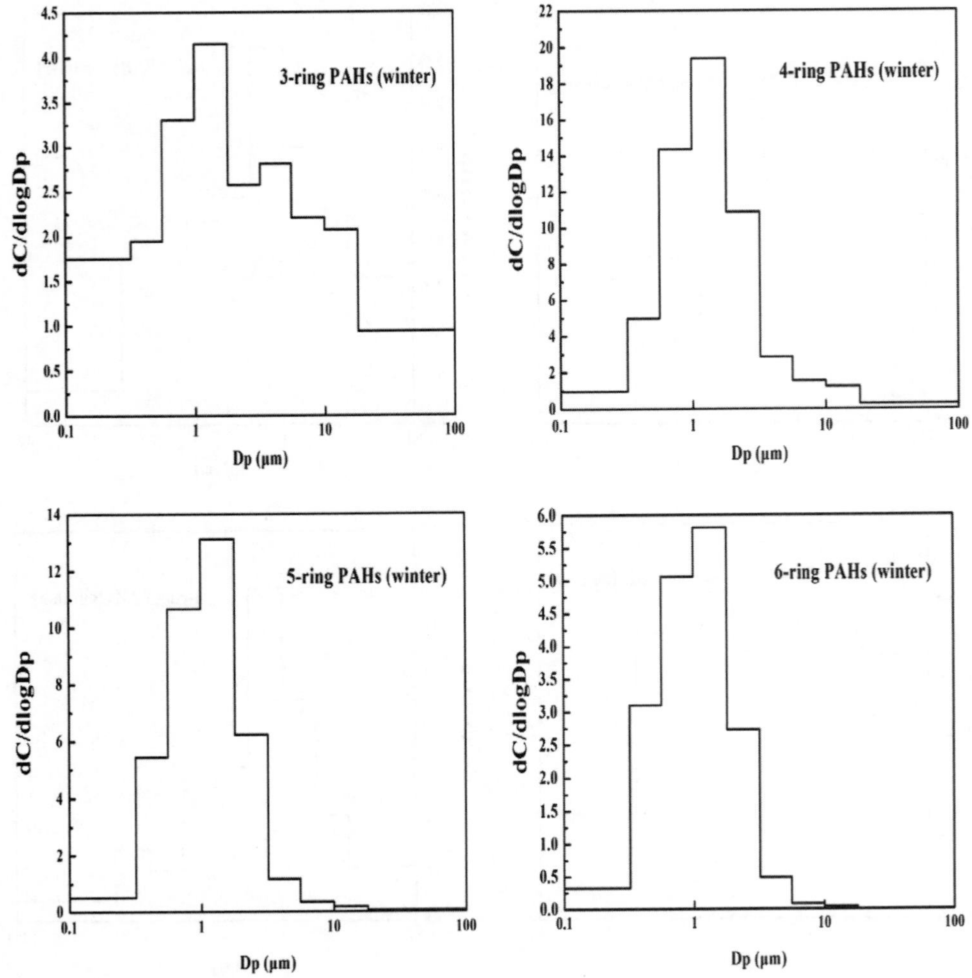

Fig. 4 (*continued*).

et al., 2007). This study showed that the BaP/BghiP ratios in spring and summer were higher than 0.4 and lower than 0.9, implying a mixed coal combustion and traffic source, whereas the BaP/BghiP ratios in autumn and winter were higher than 0.9, indicating coal combustion as the source.

In conclusion, most of the diagnostic ratios suggested that coal combustion was the main PAH source, with relatively lower contributions from petroleum. However, petroleum-derived pollution cannot be ignored. In addition to pyrolytic input as a major source of PAHs, petrogenic input may also be a source given that Shengli Oilfield is

Table 2
Mean MMD of particulate PAHs in spring, summer, autumn and winter at the YRDNNR (μm).

Compound	Spring	Summer	Autumn	Winter
Acy	2.136	2.092	1.823	1.717
Ace	1.431	1.426	2.755	2.685
Fl	1.977	1.720	2.320	1.967
Phe	1.176	1.033	1.766	1.709
Ant	2.058	1.787	3.408	1.413
Flu	1.477	1.897	1.946	1.025
Pyr	1.217	1.357	1.222	0.969
BaA	1.192	1.297	1.207	0.881
Chr	0.981	1.086	1.536	0.959
BbF	1.342	1.849	2.515	2.041
BkF	0.982	1.287	1.610	0.834
BaP	1.000	1.194	0.859	0.825
IcdP	1.013	1.000	1.163	0.950
DBA	0.900	0.549	0.874	0.797
BghiP	1.037	0.336	0.910	0.751
Total PAHs	1.404	1.467	1.218	0.931

Table 3
BaP toxic equivalent concentrations of particulate PAHs in different seasons at the YRDNNR (ng/m^3).

Compound[a]	Spring	Summer	Autumn	Winter
Acy	0.0003	0.0001	0.0001	0.0002
Ace	0.0003	0.0002	0.0003	0.0004
Fl	0.0021	0.0020	0.0011	0.0020
Phe	0.0033	0.0007	0.0025	0.0028
Ant	0.0107	0.0014	0.0007	0.0054
Flu	0.0030	0.0020	0.0014	0.0064
Pyr	0.0017	0.0012	0.0009	0.0044
BaA	0.0328	0.0131	0.0387	0.1866
Chr	0.0035	0.0007	0.0026	0.0170
BbF	0.0710	0.0047	0.0160	0.0036
BkF	0.1060	0.0203	0.1005	0.7230
BaP	1.0841	0.2938	0.4050	2.1477
IcdP	0.6515	0.1005	0.0430	0.0889
DBA	0.1120	0.0098	0.0448	0.2366
BghiP	0.0070	0.0012	0.0041	0.0206
TEQ	2.0894	0.4517	0.6616	3.4456

[a] Full names of compounds are indicated in the text.

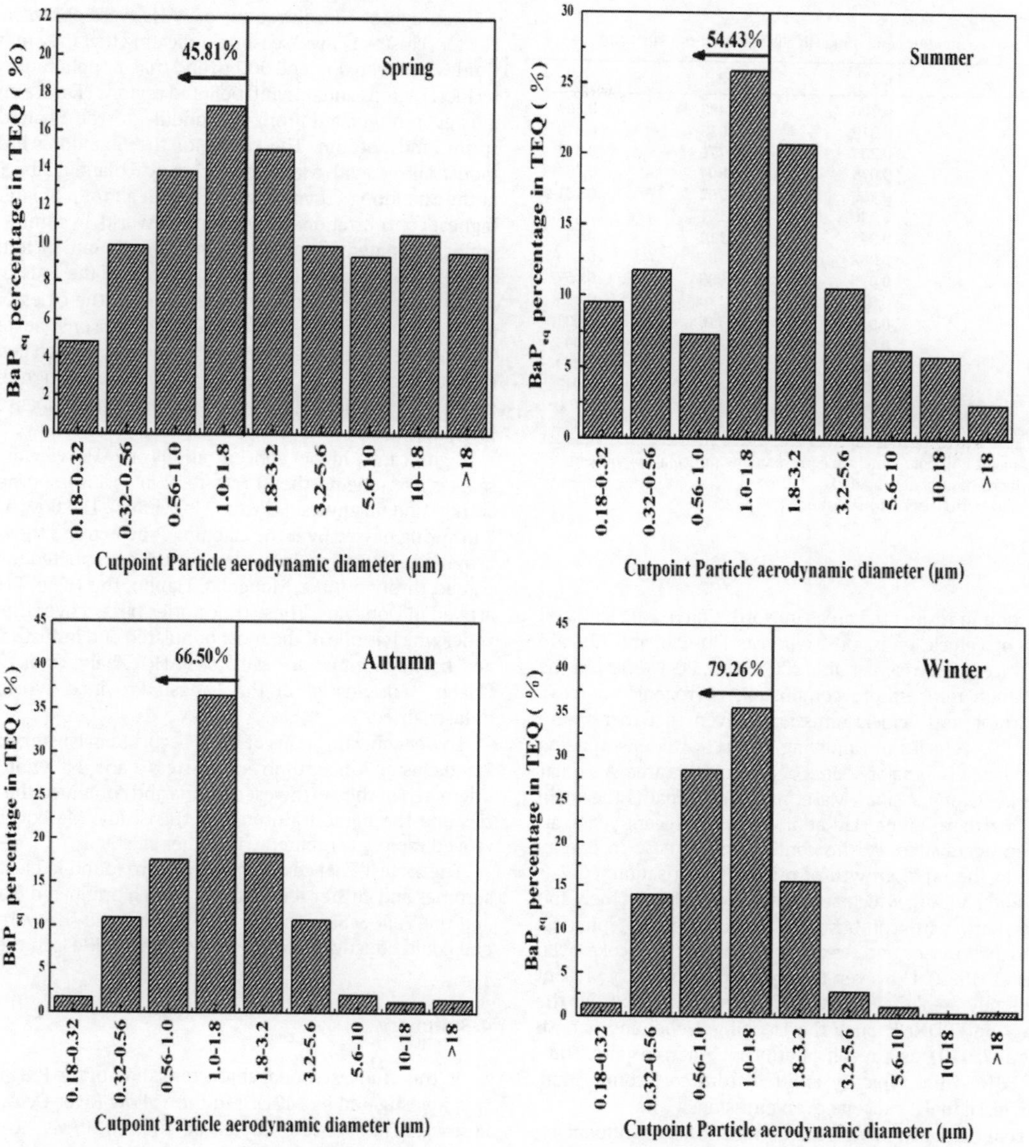

Fig. 5. Average BaPeq percentage for the particulate PAHs on each stage of MOUDI in the TEQ value for the four seasons.

near the YRDNNR, and oil exploration and refineries could contribute PAHs.

3.4.2. Source apportionment by PCA

In this study, PCA was applied to identify the sources of PAHs for the samples. The principal components (PCs) were extracted, accounting for more than 90% of the total variance. Table 5 shows that three PCs were extracted, and PC 1, PC 2 and PC 3 explained 58.90%, 25.56% and 8.37% of the total variance, respectively (92.83% in total).

Table 5 shows that PC 1 is associated with primarily high weight molecular (HMW) PAHs, such as Flu, Pyr, BaA, Chr, BkF, BaP, IcdP and BghiP. According to Liu et al. (2009) this PAH profile was commonly derived

Table 4
Diagnostic ratios of particulate PAHs in different seasons at the YRDNNR.

	Spring	Summer	Autumn	Winter	Reference source emissions
IcdP/(IcdP + BghiP)	0.59	0.46	0.51	0.53	<0.5 engine fuel combustion >0.5 coal/biomass combustion
BaA/(BaA + Chr)	0.48	0.59	0.60	0.52	<0.2 petrogenic 0.20–0.35 petroleum and combustion >0.35 pyrolytic sources
BaP/BghiP	0.64	0.48	0.99	1.04	0.3–0.4 traffic 0.9–6.6 coal combustion

Table 5
PCA analysis result for particulate PAH compounds in all seasons at the YRDNNR.

PAHs	PC 1	PC 2	PC 3
Acy	0.226	0.485	**0.569**
Ace	0.010	0.015	**0.819**
Fl	0.205	**0.776**	0.433
Phe	0.073	**0.804**	0.168
Ant	0.305	**0.702**	0.401
Flu	**0.930**	0.151	0.262
Pyr	**0.943**	0.110	0.234
BaA	**0.982**	0.102	0.045
Chr	**0.976**	0.096	0.057
BbF	0.077	0.170	0.085
BkF	**0.988**	0.116	0.001
BaP	**0.885**	0.428	0.043
DBA	0.006	0.361	0.063
IcdP	**0.896**	0.411	0.017
BghiP	**0.963**	0.118	0.100
Explained variance (%)	58.90	25.56	8.37

The bold data implied major PAH compounds in corresponding principal component.
Extraction method: principal component analysis.
Rotation method: varimax with Kaiser normalization.

from coal combustion in residential areas in North China. BaP, IcdP and BghiP are tracers of vehicle emissions, whereas Flu, Pyr and Chr are coal combustion tracers (Harrison et al., 1996). Thus, PC 1 reflected the characteristics of high temperature combustion (pyrogenic sources), including coal burning and vehicle emissions. Given that there were more than 5.78×10^5 vehicles in Dongying in 2011, it is unsurprising that automobile exhaust is a major source of PAHs in this area. A similar result was found in Beijing, China (Ma et al., 2010). Coal is the main energy source for northern China (Liu et al., 2009). In recent years, air pollution has been aggravated by the continuous increase in energy consumption and in the rapid growth of private motorisation. PC 2 is dominated by Fl, Phe and Ant, which are components of fossil fuels and are associated, in part, with combustion (Kavouras and Stephanou, 2002). PC 3 is enriched in Acy and Ace, which are tracers of coke oven sources (Simcik et al., 1999). However, there is no coke industry present in or around the study area. Coke oven emissions may arise from the areas surrounding the YRDNNR, such the Liaoning, Hebei and Jiangsu Provinces (Ma et al., 2010), as a result of atmospheric transport. Thus, in the YRDNNR, PAHs were typically emitted from coal combustion, vehicle emissions, fossil fuels and coke oven emissions.

In summary, coal combustion was the dominant factor influencing the concentration of PAHs, followed by vehicle emissions, fossil fuels and coke oven sources.

3.5. *Impact of air mass transport on the PAH levels*

To better understand the transport patterns of the sampled air masses and the impacts on aerosol pollution, we calculated three-dimensional, five-day back trajectories every 2 h, 50 m above ground level, and classified the trajectories into four different groups based on transport directions and speeds in both the horizontal and vertical scales (Xue et al., 2011; Gao et al., 2012). Fig. 6 shows the mean transport pathway with an occurrence percentage for each trajectory cluster.

From Fig. 6, it can be seen that the shortest/local transport pattern (cluster 1) was the most frequent, except in summer, accounting for 53% of the total trajectories in spring, 47% in autumn and 61% in winter. However, the air masses that originated from the Bohai Sea and the Yellow Sea were dominant in summer and accounted for 61% of the total trajectories. The second most frequent trajectory cluster was the northeast airflow in spring and autumn (cluster 2, 20% and cluster 2, 24%), whereas a short-range transport from the northwest was observed in winter (cluster 2, 17%).

The highest concentrations of PAHs were observed in the shortest cluster (cluster 1) in all seasons, indicating that they moved slowly at altitudes between 0 m and 500 m and that air pollutants were influenced by local regions and upwind polluted regions (Karaca and Camci, 2010). Cluster 1 originated from the middle of the Shandong Province in spring and autumn. The centre of the Shandong Province is highly industrialised, with a large petrochemical plant in Zibo. Jinan, the capital of the Shandong Province, was listed in a group of large cities with the highest concentrations of PAHs in the world. In summer, cluster 1 generally originated from the southeast of China, including Jiangsu and Shanghai. The increasing emission levels of the PAH pollutants can be attributed to the rapid industrialisation of the cities on the east coast of China. Compared with the other three seasons, cluster 1 was primarily from the northwest of Dongying in winter, which included Hebei and Beijing. The Hebei Province is a region with a high emission level of pollutants, and Beijing possesses the highest number of motor vehicles in China.

Significantly higher concentrations of PAHs were associated with air masses from the northeast (cluster 2 in spring), southeast (cluster 3 in spring) and northwest (cluster 2 in winter). The flow patterns of cluster 2 in spring passed over the Liaoning Province and the Bohai Gulf before arriving in Dongying. In winter, cluster 2 originated in Outer Mongolia, passed through Inner Mongolia, Tianjin, the Hebei Province and then arrived in Dongying. These trajectories passed over the Bohai economic zone, which is one of the most populated and industrial zones in China, and has high emission levels of PAHs (Zhang et al., 2009). In spring, cluster 3 originated in the Jiangsu Province, which is also highly industrialised.

Lower concentrations of PAHs occurred in cluster 4 in spring, cluster 2 and cluster 3 in autumn and clusters 3 and 4 in winter. These clusters originated in the northwest of China and contained the longest trajectories and the highest altitudes in this study. Moreover, the air parcels moved rapidly compared with other clusters.

The lowest PAH concentrations were found in clusters 2, 3 and 4 in summer and cluster 4 in autumn, which originated from the Bohai Sea and the Yellow Sea. The air masses from the sea were relatively clean and could effectively diffuse and dilute the PAHs in the urban area.

4. Summary

In this study, seasonal characterisation of the PAHs size distribution was investigated by MOUDI in the Yellow River Delta National Nature Reserve. The concentrations of the Σ15PAHs were 19.09 ng/m^3, 7.43 ng/m^3, 9.71 ng/m^3 and 34.17 ng/m^3 for spring, summer, autumn and winter, respectively, with the highest level in winter. The dominant PAH components in each season were Fl, Phe, Flu, and Pyr. A bimodal size distribution was found for the 3-ring PAHs with peaks in the fine size range (1.0–1.8 μm) and the coarse size range (3.2–5.6 μm) in each season. The 4–6 ring PAHs exhibited a nearly unimodal distribution, with the highest peak in the range of 1.0–1.8 μm. The MMD values for the total PAHs averaged 1.404, 1.467, 1.218 and 0.931 μm in spring, summer, autumn and winter, respectively, which verified that the fine particulate pollution in this area may pose adverse health risks. The average TEQ values of the PAHs in spring, summer, autumn and winter in the YRDNNR were 2.09, 0.45, 0.66 and 3.45 ng/m^3, respectively, with a higher value in winter. The size distribution of the average BaPeq percentage in the TEQ value for the four seasons (45.81%, 54.43%, 66.50% and 79.26% of TEQ) revealed that the largest BaPeq percentages were in the <1.8 μm size fraction, signifying an increased health risk. Diagnostic ratios and PCA analysis were used to locate the possible sources of PAHs, indicating coal combustion as the major source, with minor contributions from vehicular emissions and coke oven emissions. Based on the transboundary movement of air masses, we concluded that the atmospheric PAHs were primarily produced by local anthropogenic sources.

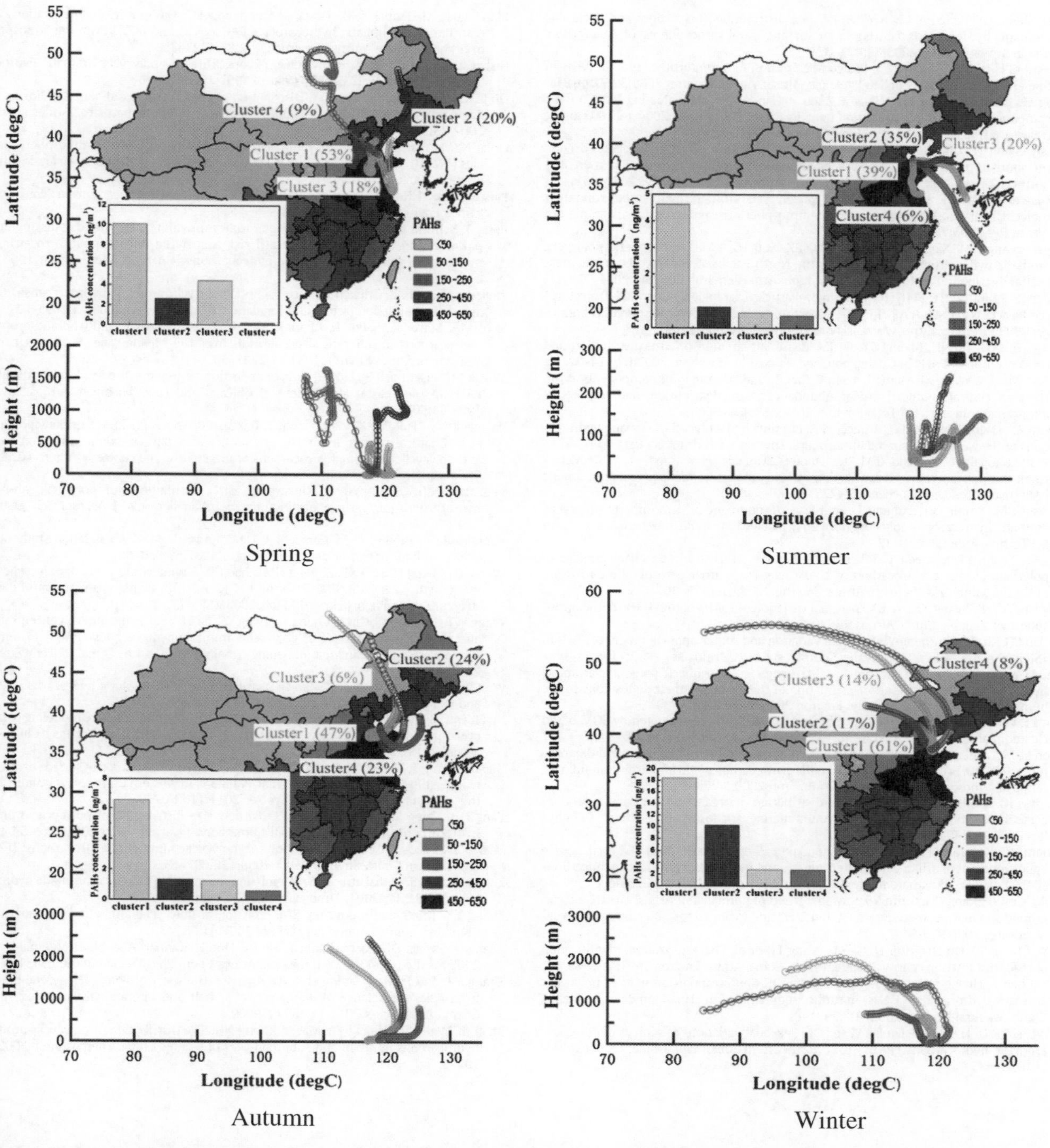

Fig. 6. Mean five-day backward trajectory clusters plotted on the Chinese PAH emission density (g/km^2) in individual provinces and municipalities in 2004 (Zhang et al., 2011). The open circles along the trajectories indicate 2-hour intervals.

Acknowledgements

This work was supported by the National Basic Research Program (973 Program) of China (2005CB422203), Key Project of Shandong Provincial Environmental Agency (2006045), Promotive Research Fund for Young and Middle-aged Scientists of Shandong Province (BS2010HZ010), Independent Innovation Foundation of Shandong University (2009TS024) and Special Research for Public-Beneficial Environment Protection (201009001-1).

References

Aceves M, Grimalt JO. Seasonally dependent size distributions of aliphatic and polynuclear hydrocarbons in urban aerosols from densely populated areas. Environ Sci Technol 1993;27:2896–908.

Allen JO, Dookeran NM, Smith KA, Sarofim AF, Taghizadeh K, Lafleur AL. Measurement of polycyclic aromatic hydrocarbons associated with size-segregated atmospheric aerosols in Massachusetts. Environ Sci Technol 1996;30:1023–31.

Baker JE, Eisenreich SJ. Concentrations and fluxes of polycyclic aromatic hydrocarbons and polychlorinated biphenyls across the air–water interface of Lake Superior. Environ Sci Technol 1990;24:342–52.

Bi XH, Sheng GY, Peng PA, Chen YJ, Fu JM. Size distribution of n-alkanes and polycyclic aromatic hydrocarbons (PAHs) in urban and rural atmospheres of Guangzhou, China. Atmos Environ 2005;39:477–87.

Bourotte C, Forti MC, Taniguchi S, Bícego MC, Lotufo PA. A wintertime study of PAHs in fine and coarse aerosols in São Paulo city, Brazil. Atmos Environ 2005;39:3799–811.

Cheng SH, Yang LX, Zhou XH, Wang Z, Zhou Y, Gao XM, et al. Evaluating PM2. 5 ionic components and source apportionment in Jinan, China from 2004 to 2008 using trajectory statistical methods. J Environ Monit 2011;13:1662–71.

Dachs J, Glenn TR, Gigliotti CL, Brunciak P, Totten LA, Nelson ED, et al. Processes driving the short-term variability of polycyclic aromatic hydrocarbons in the Baltimore and northern Chesapeake Bay atmosphere, USA. Atmos Environ 2002;36:2281–95.

Del Rosario Sienra M, Rosazza NG, Préndez M. Polycyclic aromatic hydrocarbons and their molecular diagnostic ratios in urban atmospheric respirable particulate matter. Atmos Res 2005;75:267–81.

Ding X, Wang XM, Xie ZQ, Xiang CH, Mai BX, Sun LG, et al. Atmospheric polycyclic aromatic hydrocarbons observed over the North Pacific Ocean and the Arctic area: spatial distribution and source identification. Atmos Environ 2007;41:2061–72.

Draxler R, Rolph G. HYSPLIT (HYbrid Single-Particle Lagrangian Integrated Trajectory) model access via NOAA ARL READY. Silver Spring, Md: NOAA Air Resources Laboratory; 2003 (website, http://www.arl.noaa.gov/ready/hysplit4.html).

Duan JC, Bi XH, Tan JH, Sheng GY, Fu JM. Seasonal variation on size distribution and concentration of PAHs in Guangzhou city, China. Chemosphere 2007;67:614–22.

Gao XM, Xue LK, Wang XF, Wang T, Yuan C, Gao R, et al. Aerosol ionic components at Mt. Heng in central southern China: Abundances, size distribution, and impacts of long-range transport. Sci Total Environ 2012;433:498–506.

Gigliotti CL, Dachs J, Nelson ED, Brunciak PA, Eisenreich SJ. Polycyclic aromatic hydrocarbons in the New Jersey coastal atmosphere. Environ Sci Technol 2000;34:3547–54.

Gogou A, Stratigakis N, Kanakidou M, Stephanou EG. Organic aerosols in Eastern Mediterranean: components source reconciliation by using molecular markers and atmospheric back trajectories. Org Geochem 1996;25:79–96.

Harrison RM, Smith DJT, Luhana L. Source apportionment of atmospheric polycyclic aromatic hydrocarbons collected from an urban location in Birmingham, UK. Environ Sci Technol 1996;30:825–32.

Hien TT, Thanh LT, Kameda T, Takenaka N, Bandow H. Distribution characteristics of polycyclic aromatic hydrocarbons with particle size in urban aerosols at the roadside in Ho Chi Minh City, Vietnam. Atmos Environ 2007;41:1575–86.

Hong HS, Yin HL, Wang XH, Ye CX. Seasonal variation of PM10-bound PAHs in the atmosphere of Xiamen, China. Atmos Res 2007;85:429–41.

Hu J, Liu CQ, Zhang GP, Zhang YL. Seasonal variation and source apportionment of PAHs in TSP in the atmosphere of Guiyang, Southwest China. Atmos Res 2012;118:271–9.

Kameda Y, Shirai J, Komai T, Nakanishi J, Masunaga S. Atmospheric polycyclic aromatic hydrocarbons: size distribution, estimation of their risk and their depositions to the human respiratory tract. Sci Total Environ 2005;340:71–80.

Karaca F, Camci F. Distant source contributions to PM10 profile evaluated by SOM based cluster analysis of air mass trajectory sets. Atmos Environ 2010;44:892–9.

Kaupp H, McLachlan MS. Distribution of polychlorinated dibenzo-P-dioxins and dibenzofurans (PCDD/Fs) and polycyclic aromatic hydrocarbons (PAHs) within the full size range of atmospheric particles. Atmos Environ 2000;34:73–83.

Kavouras IG, Stephanou EG. Particle size distribution of organic primary and secondary aerosol constituents in urban, background marine, and forest atmosphere. J Geophys Res 2002;107:4069.

Kavouras IG, Koutrakis P, Tsapakis M, Lagoudaki E, Stephanou EG, Baer DV, et al. Source apportionment of urban particulate aliphatic and polynuclear aromatic hydrocarbons (PAHs) using multivariate methods. Environ Sci Technol 2001;35:2288–94.

Lee JY, Kim YP, Kang CH, Ghim YS. Seasonal trend of particulate PAHs at Gosan, a background site in Korea between 2001 and 2002 and major factors affecting their levels. Atmos Res 2006;82:680–7.

Liu M, Cheng SB, Ou DN, Hou LJ, Gao L, Wang LL, et al. Characterization, identification of road dust PAHs in central Shanghai areas, China. Atmos Environ 2007;41:8785–95.

Liu Y, Chen L, Huang QH, Li WY, Tang YJ, Zhao JF. Source apportionment of polycyclic aromatic hydrocarbons (PAHs) in surface sediments of the Huangpu River, Shanghai, China. Sci Total Environ 2009;407:2931–8.

Ma WL, Li YF, Qi H, Sun DZ, Liu LY, Wang DG. Seasonal variations of sources of polycyclic aromatic hydrocarbons (PAHs) to a northeastern urban city, China. Chemosphere 2010;79:441.

Mandalakis M, Tsapakis M, Tsoga A, Stephanou EG. Gas–particle concentrations and distribution of aliphatic hydrocarbons, PAHs, PCBs and PCDD/Fs in the atmosphere of Athens (Greece). Atmos Environ 2002;36:4023–35.

Nisbet IC, LaGoy PK. Toxic equivalency factors (TEFs) for polycyclic aromatic hydrocarbons (PAHs). Regul Toxicol Pharmacol 1992;16:290–300.

Oh JE, Chang YS, Kim EJ, Lee DW. Distribution of polychlorinated dibenzo-p-dioxins and dibenzofurans (PCDD/Fs) in different sizes of airborne particles. Atmos Environ 2002;36:5109–17.

Omar NYM, Abas M, Ketuly KA, Tahir NM. Concentrations of PAHs in atmospheric particles (PM-10) and roadside soil particles collected in Kuala Lumpur, Malaysia. Atmos Environ 2002;36:247–54.

Park SS, Kim YJ, Kang CH. Atmospheric polycyclic aromatic hydrocarbons in Seoul, Korea. Atmos Environ 2002;36:2917–24.

Petry T, Schmid P, Schlatter C. The use of toxic equivalency factors in assessing occupational and environmental health risk associated with exposure to airborne mixtures of polycyclic aromatic hydrocarbons (PAHs). Chemosphere 1996;32: 639–48.

Ravindra K, Sokhi R, Grieken RV. Atmospheric polycyclic aromatic hydrocarbons: source attribution, emission factors and regulation. Atmos Environ 2008;42:2895–921.

Sicre MA, Marty JC, Saliot A, Aparicio X, Grimalt J, Albaiges J. Aliphatic and aromatic hydrocarbons in different sized aerosols over the Mediterranean sea: occurrence and origin. Atmos Environ 1987;21:2247–59.

Simcik MF, Eisenreich SJ, Lioy PJ. Source apportionment and source/sink relationships of PAHs in the coastal atmosphere of Chicago and Lake Michigan. Atmos Environ 1999;33:5071–9.

Valavanidis A, Fiotakis K, Vlahogianni T, Bakeas EB, Triantafillaki S, Paraskevopoulou V, et al. Characterization of atmospheric particulates, particle-bound transition metals and polycyclic aromatic hydrocarbons of urban air in the centre of Athens (Greece). Chemosphere 2006;65:760–8.

Venkataraman C, Thomas S, Kulkarni P. Size distributions of polycyclic aromatic hydrocarbons-gas/particle partitioning to urban aerosols. J Aerosol Sci 1999;30: 759–70.

Ventafridda V, Tamburini M, Caraceni A, De Conno F, Naldi F. A validation study of the WHO method for cancer pain relief. Cancer 1987;59:850–6.

Wang XF, Cheng HX, Xu XB, Zhuang GM, Zhao CD. A wintertime study of polycyclic aromatic hydrocarbons in PM (2.5) and PM (2.5–10) in Beijing: assessment of energy structure conversion. J Hazard Mater 2008;157:47.

Wang WT, Simonich S, Giri B, Chang Y, Zhang YG, Jia YL, et al. Atmospheric concentrations and air–soil gas exchange of polycyclic aromatic hydrocarbons (PAHs) in remote, rural village and urban areas of Beijing–Tianjin region, North China. Sci Total Environ 2011;409:2942–50.

Wu SP, Tao S, Zhang ZH, Lan T, Zuo Q. Distribution of particle-phase hydrocarbons, PAHs and OCPs in Tianjin, China. Atmos Environ 2005;39:7420–32.

Xu FL, Qin N, Zhu Y, He W, Kong XZ, Barbour MT, et al. Multimedia fate modeling of polycyclic aromatic hydrocarbons (PAHs) in Lake Small Baiyangdian, Northern China. Ecol Model 2013;252:246–57.

Xue LK, Wang T, Zhang JM, Zhang XC, Poon CN, Ding AJ, et al. Source of surface ozone and reactive nitrogen speciation at Mount Waliguan in western China: New insights from the 2006 summer study. J Geophys Res 2011;116:D07306.

Yang HH, Chien SM, Chao MR, Lin CC. Particle size distribution of polycyclic aromatic hydrocarbons in motorcycle exhaust emissions. J Hazard Mater 2005;125:154–9.

Zhang YX, Tao S. Seasonal variation of polycyclic aromatic hydrocarbons (PAHs) emissions in China. Environ Pollut 2008;156:657–63.

Zhang YX, Tao S. Global atmospheric emission inventory of polycyclic aromatic hydrocarbons (PAHs) for 2004. Atmos Environ 2009;43:812–9.

Zhang YX, Tao S, Cao J, Coveney RM. Emission of polycyclic aromatic hydrocarbons in China by county. Environ Sci Technol 2007;41:683–7.

Zhang Q, Streets DG, Carmichael GR, He KB, Huo H, Kannari A, et al. Asian emissions in 2006 for the NASA INTEX-B mission. Atmos Chem Phys Discuss 2009;9:5131–53.

Zhang YX, Tao S, Ma J, Simonich S. Transpacific transport of Benzo [a] pyrene emitted from Asia: importance of warm conveyor belt and interannual variations. Atmos Chem Phys Discuss 2011;11:18979–9009.

Zhou JB, Wang TG, Huang YB, Mao T, Zhong NN. Size distribution of polycyclic aromatic hydrocarbons in urban and suburban sites of Beijing, China. Chemosphere 2005;61: 792–9.

Atmospheric Environment 44 (2010) 278–282

Contents lists available at ScienceDirect

Atmospheric Environment

journal homepage: www.elsevier.com/locate/atmosenv

Short communication

Aircraft measurements of the vertical distribution of sulfur dioxide and aerosol scattering coefficient in China

Likun Xue [a,b], Aijun Ding [b,1], Jian Gao [a,b,2], Tao Wang [a,b,c,*], Wenxing Wang [a], Xuezhong Wang [c], Hengchi Lei [d], Dezhen Jin [e], Yanbin Qi [e]

[a] Environment Research Institute, Shandong University, Ji'nan, Shandong, PR China
[b] Department of Civil and Structural Engineering, The Hong Kong Polytechnic University, Hong Kong, PR China
[c] Chinese Research Academy of Environmental Sciences, Beijing, PR China
[d] Institute of Atmospheric Physics, Chinese Academy of Sciences, Beijing, PR China
[e] Weather Modification Office, Jilin Provincial Meteorological Bureau, Changchun, PR China

ARTICLE INFO

Article history:
Received 20 June 2009
Received in revised form
13 October 2009
Accepted 15 October 2009

Keywords:
Aircraft observation
Vertical profile
SO_2
PBL
OMI retrieval

ABSTRACT

Information on the vertical distribution of air pollution is important for understanding its sources and processes and validating satellite retrievals and chemical transport models. This paper reports the results of the measurements of sulfur dioxide (SO_2) and aerosol scattering coefficient (B_{sp}) obtained from several aircraft campaigns during summer and autumn 2007 in the north-eastern (NE), north-western (NW), and central-eastern (CE) regions of China. Their vertical profiles over the three regions with contrasting emission characteristics and climates are compared. Very high concentrations/values of SO_2 and B_{sp} (with a value of up to 51 ppbv and 950 Mm^{-1}, respectively) were recorded in the lower planetary boundary layer in CE China, indicating high SO_2 emissions in the region. The SO_2 column concentrations determined from the in-situ measurements were compared with Ozone Monitoring Instrument (OMI) SO_2 retrievals. The results show that the OMI data could distinguish the varying levels of SO_2 pollution in the study regions, but appeared to have underestimated the SO_2 column in the highly polluted region of CE China.

1. Introduction

Sulfur dioxide (SO_2) is an important trace gas in the atmosphere. At high concentrations it has a direct negative effect on human health (Ware et al., 1981); through the formation of acid rain, it damages forests and acidifies aquatic systems (Likens et al., 1972). Sulfate particles formed from the oxidation of SO_2 reduce visibility and change the climate directly by scattering sunlight and indirectly by affecting cloud optical properties (Penner et al., 1998; IPCC, 2007). SO_2 emissions in North America and Europe have decreased significantly since the 1980s (Smith et al., 2001; Vestreng et al., 2007), whereas in China, they have increased by 200–300% over the past three decades because of the dramatic increase in the consumption of coal due to rapid industrialization (Ohara et al., 2007; Li and Oberheitmann, 2009). Many cities in northern China suffer from high levels of SO_2 and sulfate particles, while southern China experiences persistent acidic rainfall (e.g., Wang and Wang, 1996; Hao and Wang, 2005; Chan and Yao, 2008).

Of the great number of studies of SO_2, sulfate, and optical properties (see Chan and Yao, 2008), only a few provide information on the vertical profiles in the troposphere (Wang et al., 2005; Dickerson et al., 2007). In view of the importance of such data in understanding vertical exchange/transport processes and evaluating satellite retrieval algorithms and chemical transport models, we carried out several aircraft measurement campaigns in three regions of China, including the northeast (NE), which is China's major agricultural base, the northwest (NW), which is sparsely populated and has an arid and semi-arid climate, and the central-eastern (CE) region, which is the largest flat area and thus the most populated and heavily industrialized region of China. This paper presents the overall characteristics of the vertical profiles of SO_2 and aerosol scattering coefficients (B_{sp}) and compares, in a preliminary fashion, in-situ measured and satellite-derived SO_2 columns over the three regions. A detailed analysis of a case of photochemical pollution observed on a flight in NE China has been presented in Ding et al. (2009).

* Corresponding author at: Department of Civil and Structural Engineering, The Hong Kong Polytechnic University, Hong Kong, PR China. Tel.: +852 2766 6059; fax: +852 2334 6389.
E-mail address: cetwang@polyu.edu.hk (T. Wang).
[1] Now at School of Atmospheric Sciences, Nanjing University, Nanjing, PR China.
[2] Now at Chinese Research Academy of Environmental Sciences, Beijing, PR Chiina.

1352-2310/$ – see front matter © 2009 Elsevier Ltd. All rights reserved.
doi:10.1016/j.atmosenv.2009.10.026

2. Description of the experiment

2.1. Study region

Three campaigns were conducted from June to October in 2007 over a large portion of China (see Fig. 1). In NE China, 16 flights were conducted from June 20 to July 13, over Changchun, the capital of Jilin province. The aircraft took off from an airport (125.2° E, 43.9° N, ~245 m ASL) located in the northwest of the city. A majority of fights were carried out around noon, with a mean spiral time of 11:15 LT (±4 h). An examination of global reanalysis data (figures not shown) suggests that during the study period, NE China was influenced by a low pressure system and the planetary boundary layer (PBL) air mostly came from the ocean in the south and from CE China in the southwest. In NW China, seven flights were carried out in August 2007 in Gansu province, which is a region with complex terrain and a dry climate with much fewer anthropogenic sources, except for an isolated large source of Lanzhou city. Our aircraft flied off Zhongchuan airport (103.6° E, 36.5° N, ~1860 m ASL), about 50 km northwest of Lanzhou (mean time 11:40 LT ±2.5 h). During August, the lower tropospheric air generally came from central China from the east. During September and October 2007, four flights were conducted from Ji'nan Yaoqiang airport (117.2° E, 36.9° N, ~20 m ASL), about 30 km northeast of Ji'nan city in Shandong province, with mean spiral time at 13:45 LT (±3 h); in addition, three flights were made over Wuxi (120.3° E, 31.6° N, ~10 m ASL) in the neighboring Jiangsu province (mean spiral time 13:00 LT ±3 h). Shandong is the largest emitter of SO_2 among China's 31 provinces and municipalities. During the measurement period, the region was under the strong influence of a sub-tropical high with weak north winds in the PBL, which created conditions favorable to the accumulation of air pollution.

2.2. Instrumentation

A twin-engine turboprop Yun-12 aircraft, similar to a Twin Otter, served as the sampling platform in these studies. The aircraft has a typical speed of 240 km h^{-1} and a maximum flight range of about 1340 km, with a ceiling altitude of 7000 m. The sampling inlet was mounted at the bottom of the aircraft, inside which an aft-facing inlet connected with the gas analyzers while an isokinetic forward-facing inlet fed the aerosol instruments. SO_2 was measured using a pulsed UV fluorescence analyzer (TEI, Model 43C TL) with a detection limit of 0.20 ppbv for 10-s integration and a precision of about 0.10 ppbv. Aerosol scattering coefficients were measured with an integrating nephelometer (EcoTech, M9003), which detects light scattering coefficients (B_{sp}) at 520 nm with a detection limit of <0.3 Mm^{-1} for 1-min integration and a light scattering angle of 10–170°. Calibrations were performed regularly on the ground and also on board during several flights. A data logger (Environmental System Corporation, Model 8816) was used to collect 5-s averaged data of SO_2, which represents a horizontal resolution of ~350 m or a vertical scale of ~20 m. In addition, air temperature, pressure, relative humidity (RH), liquid water content (LWC; measured using the on board forward scattering spectrometer probe FSSP-100, PMS Inc.), and global positioning system (GPS) data were also recorded during the flights. Temperature and pressure were available for only Changchun and Ji'nan.

In the present work, we focus on the vertical profiles of SO_2 and B_{sp} recorded in the ascent and descent stages over Changchun, Lanzhou, Ji'nan, and Wuxi. Note that B_{sp} was not measured over Wuxi. Each spiral typically lasted 10–30 min. All of the measurements used here were confined within a domain of 1° × 1° centered at the airports and within an altitude range from the surface to about 3 km (2 km for Ji'nan).

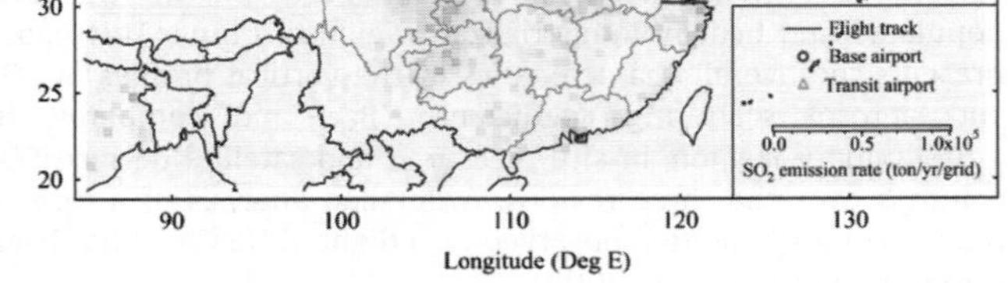

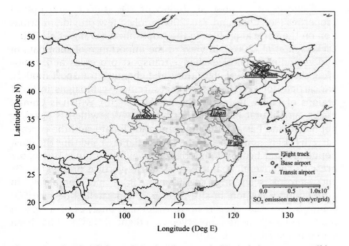

Fig. 1. Map showing flight tracks and anthropogenic SO_2 emission rates over China. Emission data were obtained from http://www.cgrer.uiowa.edu/EMISSION_DATA_new/index_16.html.

3. Results

3.1. Vertical profiles and intercorrelations

Fig. 2a shows the mean vertical profiles of SO_2 over Changchun, Lanzhou, Ji'nan, and Wuxi, with each point representing the mean concentration in a 300-m bin. Because all of the flights were over suburban/rural areas, the profiles are thought to be representative of the four regions. The figure shows that all of the profiles have a broader peak near the surface, which decreases with altitude. This is typical and can be explained by the interplay between surface emissions, turbulence mixing, and the dry deposition of SO_2. Moderate levels of SO_2 (4–6 ppbv) were observed over NE and NW China (i.e., Changchun and Lanzhou), but an extremely high SO_2 concentration existed in the PBL of CE China (i.e., Wuxi and Ji'nan), especially Ji'nan, where a mean value of up to 51 ppbv was observed in the lowest part of the atmosphere. Fig. 2a also shows that the mixing ratios of SO_2 were invariant with height in the free troposphere (altitude $>$ 1.5 km). The free tropospheric concentrations (0.5–3.0 ppbv), especially in eastern China, were much higher than those observed in continental Europe and the United States (Preunkert et al., 2007; Hains et al., 2008). The elevated SO_2 concentration in the free troposphere may be due to PBL/free troposphere exchange processes, such as warm conveyor belts and convection in warm seasons (Wild and Akimoto, 2001; Ding et al., 2009).

B_{sp} levels in NW and NE China (see Fig. 2b) were slightly larger than those observed in North America and western Europe (Guibert et al., 2005; Hains et al., 2008), while that of Ji'nan was 5–10 times higher, indicating the presence of high concentrations of particulate matter in the PBL of the North China Plain. Integrating B_{sp} in the lower troposphere gave an aerosol optical thickness (assuming a constant SSA of 0.9) of 0.42, 0.25, and 1.24 for Changchun, Lanzhou, and Ji'nan, respectively (see also Table 1). B_{sp} shows almost a linear decrease with altitude, in contrast to the more gradual decrease of SO_2 in the PBL and sharper decline in the free troposphere. This difference is likely due to the larger dry deposition velocity of SO_2 compared to that of fine aerosol (Xu and Carmichael, 1998). Although Changchun and Lanzhou had comparable SO_2 concentrations, the former had a higher B_{sp} concentration, which can be explained by the large aerosol loading and/or higher relative humidity in NE China.

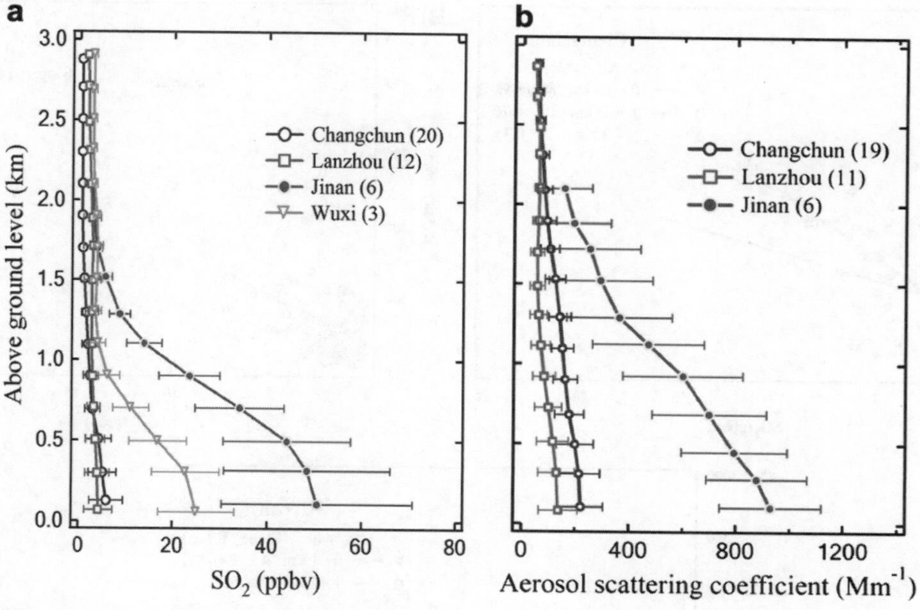

Fig. 2. Mean vertical profiles of (a) SO_2 and (b) B_{sp} over Changchun, Lanzhou, Ji'nan, and Wuxi. Solid bars represent half of the standard deviation for data points in 300-m bins. The number of profiles for each location is given in parentheses.

Fig. 2 shows the averaged vertical profiles. Individual case could be quite different from the mean profiles due to the variation in dynamic and transport processes. For example, we observed a layer of highly polluted air in the free troposphere over Jilin on 27 June, which was attributed to the transport associated with the warm conveyer belts (Ding et al., 2009). At Lanzhou, we also observed elevated SO_2 and B_{sp} values above the PBL, which was possibly due to the topographical lifting in that region.

Oxidation of SO_2 leads to the formation of sulfate aerosol, which is one of the principal contributors to particle scattering (Seinfeld and Pandis, 2006). It is of interest therefore to examine the relationship between SO_2 and B_{sp}. The scattering ability of sulfate aerosol depends not only on its mass but also RH because of its ability to absorb water vapor. Here we use the *f(RH)* functions from Malm and Day (2001) to calculate aerosol scattering coefficients ($B_{sp,dry}$) at dry conditions (RH $<$ 20%). The *f(RH)* measured from Great Smoky (cf. Table 3 in Malm and Day, 2001) was applied to the data from Changchun and Ji'nan, whereas that of Great Canyon (cf. Table 4) was chosen for Lanzhou because of the similar dry climate and a small fraction of inorganic ions in $PM_{2.5}$ at Lanzhou (Pathak et al., 2009) and at Great Canyon. Fig. 3a presents the scatter plots of $B_{sp,dry}$ vs. SO_2 in three altitude bands over Changchun. It shows that the $B_{sp,dry}/SO_2$ ratio increased with altitude (from 8 to 30), indicating a more aged air mass at higher altitudes. Fig. 3b and c shows the results for Lanzhou and Ji'nan, respectively (Note the enlarged scales in Fig. 3c). Those of Lanzhou are quite scattered, but those of Ji'nan show an altitude-dependence similar to that of Changchun. The much greater aerosol scattering (and by inference, poor visibility) in CE China reflects the high anthropogenic emissions of SO_2 and other pollutants in the region.

3.2. Comparison with OMI SO_2 data

One of the unique uses of aircraft vertical profiles is to evaluate/improve satellite retrievals. In this section, we compare the aircraft measurements with OMI (Ozone Monitoring Instrument) SO_2 products from PBL SO_2 column retrievals (column amount SO_2_PBL), which are processed with a band residual difference (BRD) algorithm (Krotkov et al., 2006). Considering the data sensitivity and possible uncertainty related to cloud and other factors, for the target locations we selected only the OMI data within the 1° × 1° grid, following the criteria of cloud fraction $<$ 0.2, solar zenith angle $<$ 60° and near-nadir viewing angles (cross-track positions 20–40) (Dr. N.A. Krotkov, personal communication). We integrated the aircraft profiles to estimate the SO_2 columns over the spiral regions. Temperature and pressure corrections were made with our measurements or with the default profiles of the U.S. Standard Atmosphere for the missing data at Lanzhou. We estimate that the bias from the use of the standard atmosphere was small ($<$0.1 DU for a variation of 100 hPa and 10 K). The mean SO_2

Table 1
Comparison of mean SO_2 columns derived from aircraft in-situ profiles and OMI satellite retrievals.

Location	Period	Aircraft			OMI Satellite		
		Number of profiles	SO_2 Column[a] DU	AOT[b]	Number of valid days[c]	SO_2 column DU	Slant column ozone[d] DU
Changchun (124.7 E – 125.7 E, 43.5 N – 44.5 N)	Jun 20 – Jul 20	(20)	0.58 (0.36)	0.42	(8)	0.52 (0.93)	724 (31)
Lanzhou (103.5 E – 104.5 E, 36 N – 37 N)	Jul 31 – Aug 31	(12)	0.64 (0.48)	0.25	(6)	0.98 (0.55)	602 (14)
Ji'nan (117 E – 118 E, 36 N – 37 N)	October	(6)	4.39 (2.04)	1.24	(6)	2.88 (2.04)	738 (70)

[a] Extrapolated from the surface to 3 km (0–2 km for Ji'nan).
[b] Estimated from the aircraft mean altitude profile of Bsp assuming a constant SSA of 0.9.
[c] The operational OMI SO_2 data with optimal viewing conditions (radioactive cloud fraction $<$ 0.2, solar zenith angle $<$ 60°, cross-track position 20–40).
[d] SCO is to account for the AMF errors from both observational geometry and total ozone (Ω). SCO = Ω • (sec(SZA) + sec(VZA)) [http://so2.umbc.edu/omi/].

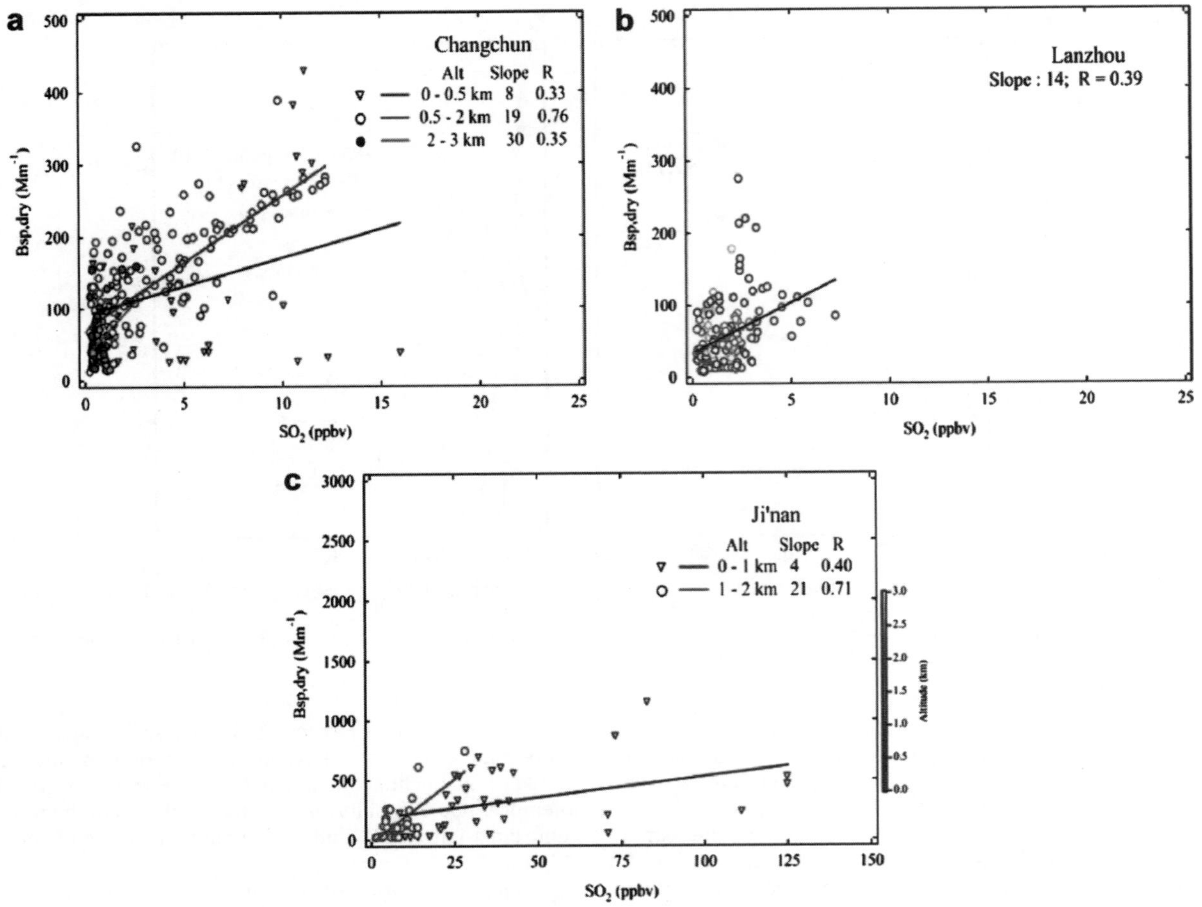

Fig. 3. Correlations between $B_{sp,dry}$ and SO_2 for three altitude bands over (a) Changchun and (b) Lanzhou and two altitude bands over (c) Ji'nan. All of the symbols are color coded with the flight altitude. Note that scales of (c) are enlarged by an equal proportion in order to show the high concentrations of SO_2 and B_{sp} in Ji'nan.

columns from OMI and aircraft are shown in Table 1 for Changchun, Lanzhou, and Ji'nan. The results suggest that OMI could distinguish different levels of anthropogenic SO_2 in the three regions; however, in the highly polluted North China Plain where Ji'nan is located, the OMI-retrieved SO_2 column was much lower than the aircraft result (2.88 Dobson Units [DU] compared with 4.39 DU, respectively).

Surface albedo, cloud cover, the SO_2 empirical profile, slant column ozone (SCO), aerosol loading, and optical properties (i.e., single scattering albedo) can affect SO_2 retrieval (Carn et al., 2007; Krotkov et al., 2008). To investigate the possible reasons for the underestimation of the satellite retrievals in the Ji'nan area, we examine one flight at 13:40 LT on October 25, which was close to the time of the satellite overpass (~13:18 LT). In this case, very high concentrations of SO_2 (>80 ppbv at 300 m and ~10 ppbv at 1600 m) were recorded during the ascent stage (figures not given). The aircraft-measured SO_2 column was 7 DU, integrating SO_2 the surface to 2 km (the highest altitude on that flight). The mean OMI SO_2 column was 5.35 DU obtained from a domain of 20 km × 20 km over the region of spiral (only one IFOV data with the solar zenith angle of about 53° and the cross-track position of about 20). An examination of the satellite cloud map suggests that there was little cloud cover on that day (cloud fraction: ~0.18). The aircraft-measured SO_2 profile was quite similar to the default profile in the OMI retrieval algorithm ($OMSO_2$ README File, http://so2.umbc.edu/omi/). Thus, the uncertainties from these two factors would be very small. Regarding the SCO of 799 DU, we estimate that it could have underestimated the SO_2 column by no more than 4% (Krotkov et al., 2008). It has been suggested that ultraviolet (UV) absorption by mineral dust and black carbon can significantly reduce the air mass factor (AMF) in the algorithms (Colarco et al., 2002; Torres et al., 2007; Krotkov et al., 2008). Although unpublished data obtained by researchers of Shandong University in Ji'nan urban area indicate absorbing nature of aerosols (with single scattering albedo of ~0.80), no data are available outside the city where the aircraft measurement took place. The OMI UV Aerosol Index (AI) on Oct 25 was 1.07. It does not provide definitive information on absorbing or scattering nature of aerosols in the planetary boundary layer, as previous research mainly used the AI to show the abundance of absorbing aerosol at high altitudes (i.e., in dust and biomass burning plumes), and the sensitivity of AI decreases at lower altitudes (Ahn et al., 2008). It is possible that the smaller value of OMI SO_2 column may be due to absorbing aerosol in the Ji'nan region, but data on the optical properties of aerosol are needed to confirm this contention.

4. Summary

The vertical profiles of SO_2 and aerosol scattering coefficients were measured in 2007 on board an aircraft in the north-eastern, north-western, and central-eastern parts of China. These regions have different air pollution levels and climates. The results revealed very high concentrations of SO_2 and B_{sp} in the plains of central-eastern

China, especially in Shandong, where an averaged PBL SO_2 of up to 51 ppbv was observed. A better correlation between $B_{sp,dry}$ and SO_2 and larger $B_{sp,dry}/SO_2$ ratios were found at higher altitudes. Comparison of the aircraft-measured and OMI-derived SO_2 columns suggests that OMI SO_2 retrieval could distinguish different pollution levels of anthropogenic SO_2 in the three study regions, but may have underestimated the SO_2 column in the highly polluted Ji'nan area. It is possible that UV-absorbing aerosols (black carbon) were present over the Ji'nan region which caused the underestimation. More aircraft studies in central-eastern China are needed to verify the results of this work, which are based on a limited number of flights.

Acknowledgements

The authors would like to thank Jilin Provincial Meteorological Bureau, Gansu Provincial Meteorological Bureau, and Shandong Provincial Meteorological Bureau for their cooperation and support during the aircraft campaigns. We thank Steven Poon and Xiaoqing Zhang for mounting the aircraft instruments. We are grateful to the aircraft crew members, especially Colonel Kuilin Li, for their help and piloting of the aircraft. We thank Professor Conglai Shen for developing the sample inlets, and Jiefang Yang for his operation on board. We also thank NASA for providing the OMI retrievals and Drs. N.A. Krotkov and X.L. Wei for their help in satellite data analysis. This work was funded by the National Basic Research Program (973 Program) of China (2005CB422203) and the Hong Kong Polytechnic University (1-BB94). We thank the two anonymous reviewers for their suggestions which have helped improve the original manuscript.

References

Ahn, C., Torres, O., Bhartia, P.K., 2008. Comparison of ozone monitoring instrument UV aerosol products with Aqua/Moderate resolution imaging spectroradiometer and multiangle imaging spectroradiometer observations in 2006. J. Geophys. Res. 113, D16S27. doi:10.1029/2007JD008832.

Carn, S.A., Krueger, A.J., Krotkov, N.A., Yang, K., Levelt, P.F., 2007. Sulfur dioxide emissions from Peruvian copper smelters detected by the Ozone Monitoring Instrument. Geophys. Res. Lett. 34, L09801. doi:10.1029/2006GL029020.

Chan, C.K., Yao, X., 2008. Air pollution in mega cities in China. Atmos. Environ. 42, 1–42.

Colarco, P.R., Toon, O.B., Torres, O., Rasch, P.J., 2002. Determining the UV imaginary index of refraction of Saharan dust particles from TOMS data using a three-dimensional model of dust transport. J. Geophys. Res. 107 (D16), 4289.

Dickerson, R.R., et al., 2007. Aircraft observations of dust and pollutants over northeast China: insight into the meteorological mechanisms of transport. J. Geophys. Res. 112, D24S90. doi:10.1029/2007JD008999.

Ding, A.J., et al., 2009. Transport of north China air pollution by mid-latitude cyclones: case study of aircraft measurements in summer 2007. J. Geophys. Res. 114, D08304. doi:10.1029/2008JD011023.

Guibert, S., Matthias, V., Schulz, M., Bosenberg, J., Eixmann, R., Mattis, I., Pappalardo, G., Perrone, M.R., Spinelli, N., Vaughan, G., 2005. The vertical distribution of aerosol over Europe: synthesis of one year of EARLINET aerosol lidar measurements and aerosol transport modeling with LMDzT-INCA. Atmos. Environ. 39, 2933–2943.

Hains, J.C., Taubman, B.F., Thompson, A.M., Stehr, J.W., Marufu, L.T., Doddridge, B.G., Dickerson, R.R., 2008. Origins of chemical pollution derived from mid-Atlantic aircraft profiles using a clustering technique. Atmos. Environ. 42, 1727–1741.

Hao, J.M., Wang, L.T., 2005. Improving urban air quality in China: Beijing case study. J. Air Waste Manag. Assoc. 55, 1298–1305.

IPCC, 2007. Climate change. In: Soloman, S., Qin, D., Manning, M., Chen, Z., Marquis, M., Avery, K.B., Tignor, M., Miller, H.L. (Eds.), The Physical Scientific Basis. Contribution of Working Group I to the Fourth Assessment Report of the Intergovernmental Panel on Climate Change. Cambridge University Press, Cambridge, United Kingdom, and New York NY, USA, 2007.

Krotkov, N.A., et al., 2008. Validation of SO_2 retrievals from the ozone Monitoring instrument (OMI) over NE China. J. Geophys. Res. 113, D16S40. doi:10.1029/2007JD008818.

Krotkov, N.A., Carn, S.A., Krueger, A.J., Bhartia, P.K., Yang, K., 2006. Band residual difference algorithm for retrieval of SO_2 from the Aura Ozone Monitoring Instrument (OMI). IEEE Trans. Geosci. Remote Sens. 44 (5), 1259–1266.

Li, Y., Oberheitmann, A., 2009. Challenges of rapid economic growth in China: reconciling sustainable energy use, environmental stewardship and social development. Energy Policy 37, 1412–1422.

Likens, G.E., Bormann, F.H., Johnson, N.M., 1972. Acid rain. Environment 14 (2), 33–40.

Malm, W.C., Day, D.E., 2001. Estimates of aerosol species scattering characteristics as a function of relative humidity. Atmos. Environ. 35, 2845–2860.

Ohara, T., Akimoto, H., Kurokawa, J., Horii, N., Yamaji, K., Yan, X., Hayasaka, T., 2007. An Asian emission inventory of anthropogenic emission sources for the period 1980-2020. Atmos. Chem. Phys. 7, 4419–4444.

Pathak, R.K., Wu, W.S., Wang, T., 2009. Summertime $PM_{2.5}$ ionic species in four major cities of China: nitrate formation in an ammonia-deficient atmosphere. Atmos. Chem. Phys. 9, 1711–1722.

Penner, J.E., Chuang, C.C., Grant, K., 1998. Climate forcing by carbonaceous and sulfate aerosols. Clim. Dyn. 14 (12), 839–851.

Preunkert, S., Legrand, M., Jourdain, B., Dombrowski-Etchevers, I., 2007. Acidic gases (HCOOH, CH_3COOH, HNO_3, HCl, and SO_2) and related aerosol species at a high mountain Alpine site (4360 m elevation) in Europe. J. Geophys. Res. 112, D23S12. doi:10.1029/2006JD008225.

Seinfeld, J.H., Pandis, S.N., 2006. Atmospheric Chemistry and Physics: From Air Pollution to Climate Change. Wiley, New York, NY.

Smith, S.J., Pitcher, H., Wigley, T.M.L., 2001. Global and regional anthropogenic sulfur dioxide emissions. Glob. Planetary Change 29, 99–119.

Torres, O., et al., 2007. Aerosols and surface UV products from Ozone Monitoring Instrument observations: an overview. J. Geophys. Res. 112, D24S47. doi:10.1029/2007JD008809.

Vestreng, V., Myhre, G., Fagerli, H., Reis, S., Tarrason, L., 2007. Twenty-five years of continuous sulphur dioxide emission reduction in Europe. Atmos. Chem. Phys. 7, 3663–3681.

Wang, Wei, Liu, Hongjie, Yue, Xin, Li, Hong, Chen, Jianhua, Tang, Dagang, 2005. Study on size distributions of airborne particles by aircraft observation in spring over eastern coastal areas of China. Adv. Atmos. Sci. 22, 328–336.

Wang, W.X., Wang, T., 1996. On acid rain formation in China. Atmos. Environ. 30, 4095–4099.

Ware, J.H., Tribodeau, L.A., Speizer, F.E., Colome, S., Ferris, B.G., 1981. Assessment of the health effects of atmospheric sulfur oxides and particulate matter: evidence from observational studies. Environ. Health Perspect. 42, 255–276.

Wild, O., Akimoto, H., 2001. Intercontinental transport of ozone and its precursors in a three-dimensional global CTM. J. Geophys. Res. 106 (D21), 27729–27744.

Xu, Y.W., Carmichael, G.R., 1998. Modeling the dry deposition velocity of sulfur dioxide and sulfate in Asia. J. Appl. Meteoro. 37, 1084–1099.

Atmospheric Environment 43 (2009) 829–836

Contents lists available at ScienceDirect

Atmospheric Environment

journal homepage: www.elsevier.com/locate/atmosenv

Measurement of aerosol number size distributions in the Yangtze River delta in China: Formation and growth of particles under polluted conditions

Jian Gao [a,b], Tao Wang [b,*], Xuehua Zhou [a], Waishing Wu [b], Wenxing Wang [a]

[a] Environment Research Institute, Shandong University, Ji'nan, Shandong 250100, China

[b] Department of Civil and Structural Engineering, The Hong Kong Polytechnic University, Hong Kong, China

ARTICLE INFO

Article history:
Received 24 March 2008
Received in revised form
22 October 2008
Accepted 22 October 2008

Keywords:
New particle formation
Sulfur dioxide
Photochemical activity
Shanghai

ABSTRACT

Particle size distribution is important for understanding the sources and effects of atmospheric aerosols. In this paper we present particle number size distributions (10 nm–10 μm) measured at a suburban site in the fast developing Yangtze River Delta (YRD) region (near Shanghai) in summer 2005. The average number concentrations of ultrafine (10–100 nm) particles were 2–3 times higher than those reported in the urban areas of North America and Europe. The number fraction of the ultrafine particles to total particle count was also 20–30% higher. The sharp increases in ultrafine particle number concentrations were frequently observed in late morning, and the particle bursts on 5 of the 12 nucleation event days can be attributed to the homogeneous nucleation leading to new particle formation. The new particle formation events were characterized with a larger number of nucleation-mode particles, larger particle surface area, and larger condensational sink than usually reported in the literature. These suggest an intense production of sulfuric acid from photo-oxidation of sulfur dioxide in the YRD. Overall, the growth rate of newly formed particles was moderate (6.4 ± 1.6 nm h^{-1}), which was comparable to that reported in the literature.

1. Introduction

The size distribution of atmospheric aerosols, together with chemical composition, provides important information on the sources and processes of aerosols and for assessing their health and climatic effects. Epidemiological studies have shown adverse health effects of atmospheric aerosols including respiratory irritation, changes in pulmonary function and associations between aerosol mass concentrations and mortality (Sament et al., 2000; Lippmann et al., 2000). Some studies have suggested that smaller particles can cause larger health effects due to their ability to penetrate deep into the lungs (Wichmann and Peters, 2000). The size distribution of particles has also been used to probe the formation and growth mechanisms of nanometer-size particles in the atmosphere, which is one of the foci of current aerosol research efforts (Kulmala et al., 2004; Holmes, 2006).

The formation and growth of particles have been observed in a wide range of environments (Kulmala et al., 2004; Yu et al., 2008), including forests (Kulmala et al., 1998), coastal areas (O'Dowd et al., 2002; Liu et al., 2008), and rural/remote environments (Weber et al., 1997; Birmili and Wiedensohler, 2000). The formation of new particles through homogeneous nucleation has been observed to occur more frequently in clean atmospheres. In urban atmospheres, where the surface area of preexisting particles is generally high due to prevalent primary emissions, the formation of new particles is less favored compared to the condensation of precursors on the pre-excising particle surface areas (Alam et al., 2003). Nevertheless, a number of studies have reported new particle formation in urban environments (Harrison et al., 2000; McMurry et al., 2000; Woo et al., 2001; Shi et al., 2001; Dunn et al., 2004; Stanier et al., 2004a,b; Wehner et al., 2004; Wu et al., 2007). These results suggest that new particle formation is far more facile than previously assumed and homogeneous nucleation in the polluted atmospheres is also conceivable. Nonetheless, more observational studies in urban/industrial areas are needed to characterize the formation and growth of particles under different loading and mix of particle precursors and sinks.

With a population of over 100 million, the Yangtze River Delta (YRD) is home to China's largest city, Shanghai (population: 15 million), and is a major industrial and commercial hub in China. Previous studies have indicated high concentrations of ozone and particulate matter in the YRD region (Wang et al., 2001, 2004; Yao et al., 2002; Chan and Yao, 2008). However, there is limited information on size distribution and the source/formation mechanism of aerosols. In the summer of 2005, number size distributions, $PM_{2.5}$ mass and trace gases concentrations were measured at

* Corresponding author. Tel.: +852 27666059; fax: +852 23346389.
E-mail address: cetwang@polyu.edu.hk (T. Wang).

1352-2310/$ – see front matter © 2008 Elsevier Ltd. All rights reserved.
doi:10.1016/j.atmosenv.2008.10.046

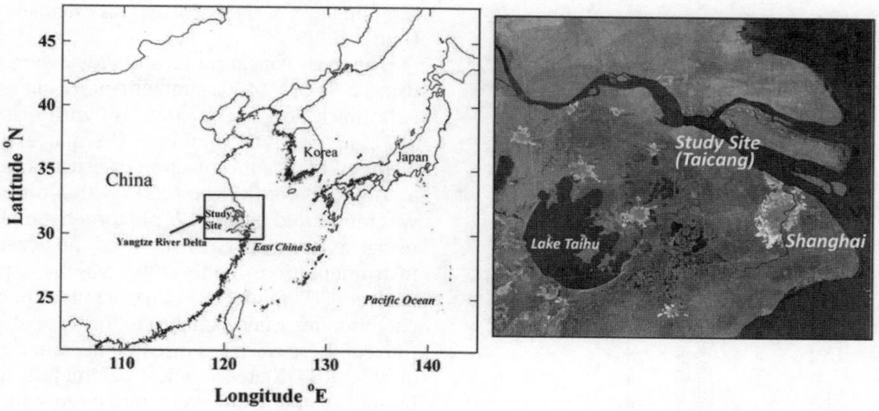

Fig. 1. The location of the observation site in Taicang near Shanghai. The Shaded areas in orange color are cities in the region.

a suburban site in the YRD, which was part of a large field campaign aimed at studying the impact of Chinese mega-cities on regional air quality. In this paper, we analyze the size distribution data in conjunction with other aerosol and trace gas data, focusing on the formation of new particles via homogeneous nucleation and their subsequent growth.

2. Methodology

2.1. Site information

The measurements were conducted on the rooftop of the Taicang Meteorological Bureau building, about 20 m above ground level. The site was 2 km northeast of downtown Taicang City (population: 180,000). As shown in Fig. 1, Taicang is 45 km northwest of Shanghai and 50 km southwest of Suzhou which has an urban population of 2 million.

A unique characteristic of the study area was the presence of a large number of coal-burning point sources. A large power plant with a capacity of 2170 MW was situated 18 km northeast. Further into Shanghai's industrial Boshan district, some 40 km east of the site, were several power plants with capacities over 1200 MW. Closer to the site, a coal-burning power plant with a capacity of 45 MW was 1.5 km east, several brick kilns were located in the north, and a 6 MW garbage-burning power plant was 5 km to the west. The high density of coal-burning sources in the region can be reflected by the concentrations of SO_2 measured at the site (see Section 3.3.1). Apart from the point sources, vehicular traffic in downtown Taicang may also have affected the measurements made at the study site. There was a local road with light traffic 20 m west of the observation site; another more trafficked road was about 100 m north of the site. The State Road (No. 204) was 2 km southwest, the Su-Kun-Tai Freeway was 5 km to the north, and the Riverside Freeway passed about 5 km to the east.

2.2. Instruments

A Wide-range Particle Spectrometer (WPS™, MSP Corporation model 1000XP) was used to measure size distributions in the range of 0.01–10 μm. It is a recently introduced commercial aerosol instrument with the capability of measuring aerosols over a wide diameter range. The instrument is an aerosol spectrometer that combines the principles of differential mobility analysis (DMA), condensation particle counting (CPC) and laser light scattering (LPS).

The DMA in the WPS™ has a cylindrical geometry with an annular space for the laminar aerosol and sheath air flows. These critical dimensions were optimized to obtain size classification of particles between 10 and 500 nm with a maximum voltage smaller than 10,000 V and a sheath flow rate of 3 L min^{-1}. The CPC is of the thermal diffusion type, with a saturator maintained at 35 °C. A feedback flow control system maintains a constant CPC flow rate at 0.30 L min^{-1}. It has a dual reservoir design to eliminate contamination of the working fluid with condensed sampling-air humidity. The LPS is a single-particle, wide-angle optical sensor used for measuring particle size from 0.35 to 10 μm. Before and after the field campaigns, the DMA was calibrated with NIST SRM 1691 and SRM 1963 PSL spheres (0.269 μm and 0.1007 μm mean diameter) to verify proper DMA transfer function and accurate particle sizing traceable to NIST. The LPS was calibrated with four NIST traceable

Table 1
Descriptive statistics of the measured number concentrations in different size ranges and of mass concentration of $PM_{2.5}$ and black carbon.

Size bins	Mean	Median	Max	Min	SD
10–20 (nm)	15,521	11,468	81,780	1241	12,362
20–30 (nm)	4812	3608	28,934	444	4039
30–50 (nm)	4422	3489	31,719	360	3903
50–100 (nm)	3756	2973	16,545	249	3018
100–200 (nm)	1368	997	6852	67	1221
200–500 (nm)	307	241	1791	2	306
0.5–1 (μm)	69	54	321	7	46
10–500 (nm)	29,990	24,226	132,421	1218	20,468
10–1000 (nm)	30,059	24,295	132,464	1245	20,462
1–2.5 (μm)	7	5	65	1	5
$PM_{2.5}$ (μg m^{-3})	67.3	56.0	519.5	7.5	45.3
BC (μg m^{-3})	5.4	4.5	23.8	0.9	3.5

Table 2
The comparison with the number concentration results in other continental sites.

Location	Number conc. (cm^{-3})		Source
	10–100 nm	100–500 nm	
Alkmaar, Netherlands	18,300	2120	Ruuskanen et al. (2001)
Erfurt, Germany	17,700	2270	Wichmann and Peters (2000)
Helsinki, Finland	16,200	973	Ruuskanen et al. (2001)
Pittsburgh, Urban, USA	14300	2170	Stanier et al. (2004a)
Pittsburgh, Rural, USA	6500	1900	Stanier et al. (2004a)
Atlanta, USA (25 months)	21,400		Woo et al. (2001)
Shanghai, China	28,511	1676	This work

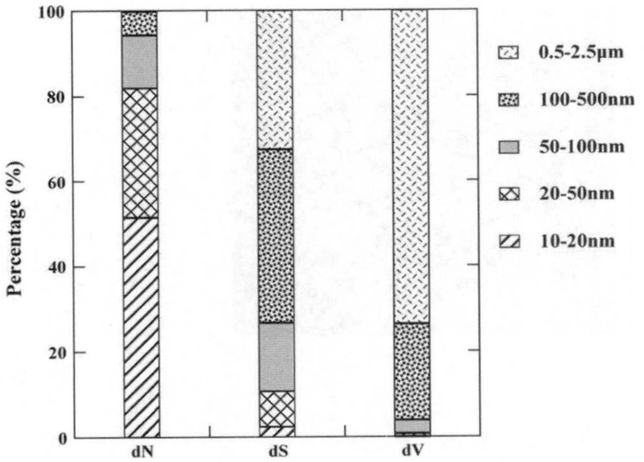

Fig. 2. The average fractions of the particle number, surface and volume concentrations in different size ranges.

sizes of PSL (0.701 μm, 1.36 μm, 1.6 μm, and 4.0 μm mean diameter). The DMA and the CPC can measure the aerosol size distribution in the 10 –500 nm range in up to 96 channels. The LPS covers the 350 –10,000 nm range in 24 additional channels. In the present study we chose the sample mode with 48 channels in DMA and 24 in LPS. Thus it takes about 8 min for one complete scanning of the entire size range with 5 s of scanning period for each channel. The sampling loss of the particles was calculated based on the theory of Hinds (1982).

The mass concentrations of $PM_{2.5}$ were continuously measured using a TEOM 1400a ambient particulate monitor (Rupprecht & Patashnick Co., Inc., Albany, NY) with instrument temperature at 50 °C in order to minimize thermal expansion of the tapered element. Mass concentration of black carbon was measured using an Athelometer (Magee AE21) with a time resolution of 5 min. O_3 was monitored using a UV photometry analyzer (TEI model 49), CO using a nondispersive infrared analyzer (Advanced Pollution Instrumentation, model 300), SO_2 by a pulsed UV fluorescence analyzer (TEI model 43S), and NO_y by a modified commercial MoO/chemiluminescence analyzer (TEI Model 42S). Detailed descriptions of these trace gas instruments are provided elsewhere (Wang et al., 2001). Meteorological parameters such as air temperature, humidity, and wind were measured continuously. The total UV value was measured by the EPPLAB Model TUVR paired with a model 450 Signal Conditioning Amplifier.

2.3. General weather conditions

The field study took place from 5 May to 2 June 2005. Late spring and early summer is a transition period for the winter and summer monsoons in East China. The Continental High and Pacific High interactively affected the region giving rise to relatively sunny and warm weather (mean air temperature: 21 °C) during the study period. Among 30 days of study, there were only three rainy days (6, 14 and 27 May) and three cloudy days (9, 23 May and 1 June).

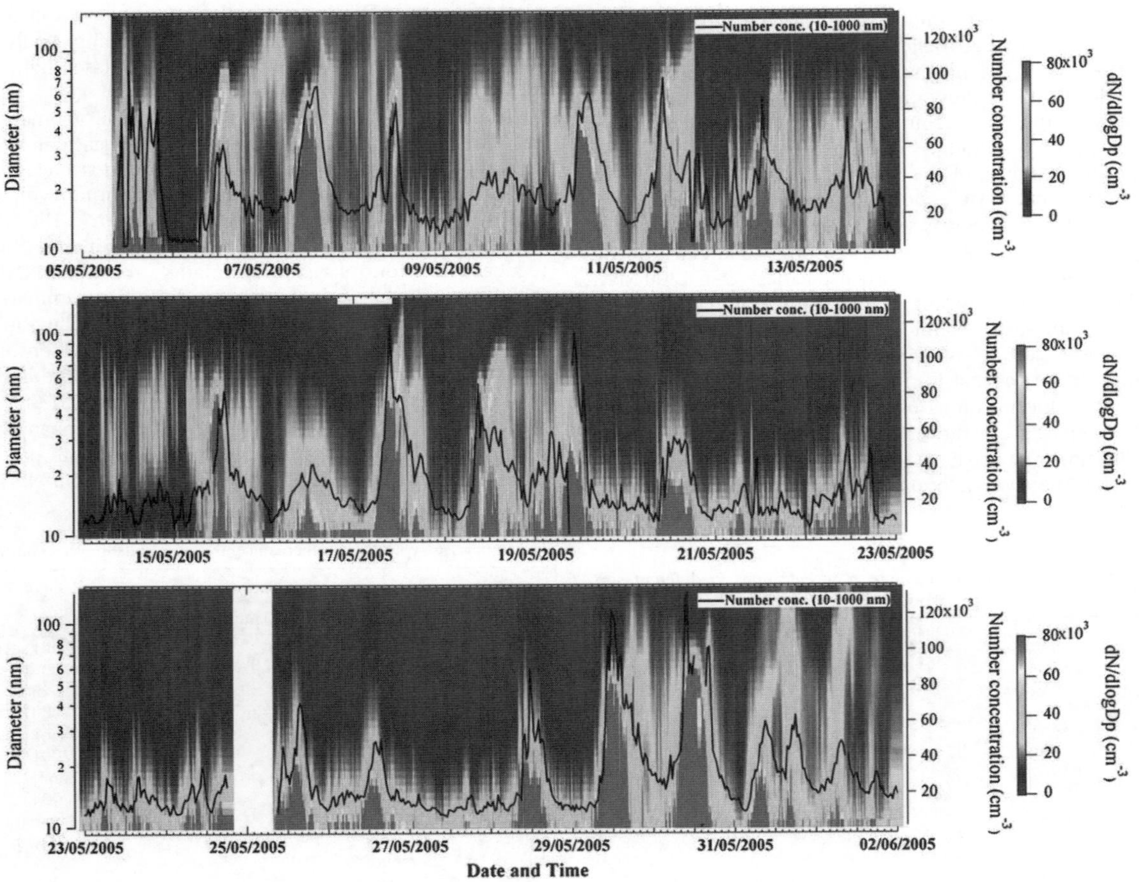

Fig. 3. Time series of number size distribution in 10–100 nm and number concentration of sub-micro particles.

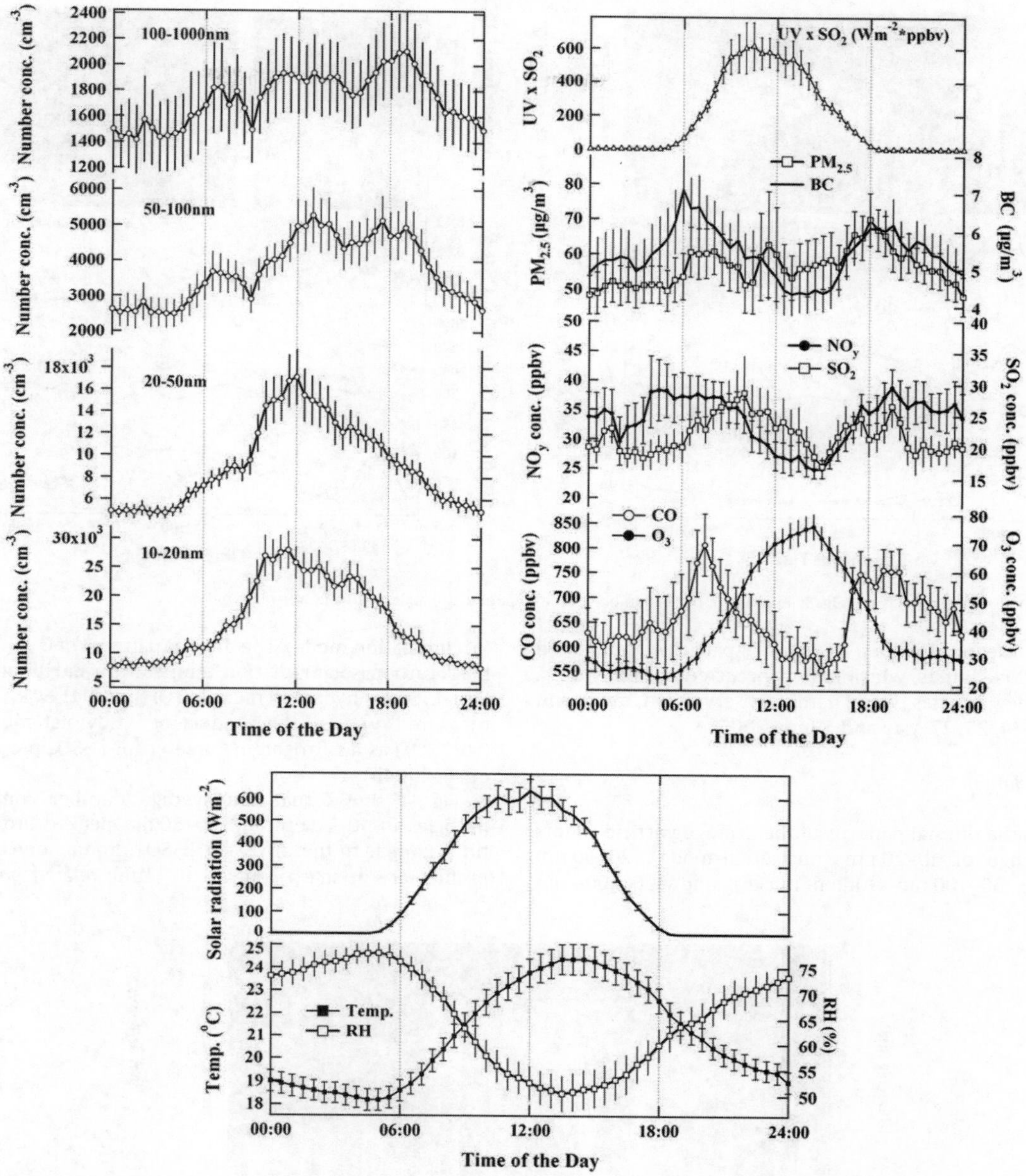

Fig. 4. The average diurnal variation of the number concentration in 10–20 nm, 20–50 nm, 50–100 nm and 100–1000 nm with relevant pollutants and meteorological parameters. Vertical bars are standard errors.

Surface wind observations show that the prevailing winds during this period were from the east and southeast. Back trajectory analysis indicates that over 60% of air masses passed across areas around Shanghai. Under such conditions, emissions from Shanghai and the surrounding areas were expected to be brought to the site. The air masses occasionally passed over cities like Suzhou when the winds switched to westerlies due to the influence of meso-scale circulation. Northerly winds were also observed on several occasions bringing in relatively well mixed regional air masses from central-eastern China.

3. Results and discussion

3.1. Number size distributions

Table 1 gives statistics of particle number in different size ranges. In the size range of 10–500 nm the average number concentration was 29,990 cm^{-3}, which is higher than the typical values (5000–25,000 cm^{-3}) reported for the urban/suburban sites of North America and Europe (Table 2). Also, high number concentrations (>15,000 cm^{-3}) were observed in the size range of 10–20 nm, suggesting abundant nucleation-mode particles in the YRD region. Fig. 2 shows the average relative contribution of various size fractions to the total particle counts, surface area and volume concentrations between 10 nm and 2.5 μm. On average 94% of the total particle count was found in the size range of 10 – 100 nm. This is much larger than that reported elsewhere, e.g., 72% in eastern Germany (Tuch et al., 1997) and 61% in Atlanta (Woo et al., 2001). Fig. 3 shows the time series of number size distributions in 10–100 nm and number concentration in submicron particles (10–1000 nm). The particle counts showed a large temporal variability with spikes higher than 110,000 cm^{-3} on several occasions (17, 19, 29 and 30 May 2005) exceeding the average number concentrations by a factor of four. The sharp

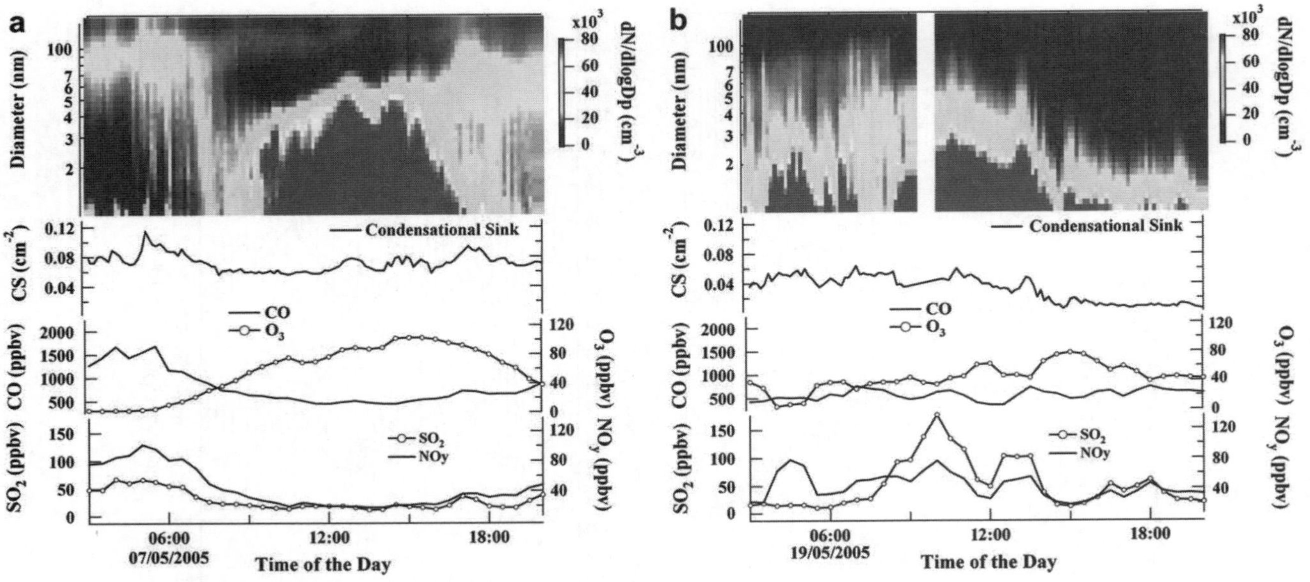

Fig. 5. Example of nucleation burst in (a) typical urban air and (b) a power-plant plume.

increases in nucleation-mode (10–20 nm) particles dominated most of the particle bursts, which mainly occurred on sunny days. Nucleation-mode particles were hardly observed on rainy and cloudy days (9, 14, 23, 27 May and 1 June 2005).

3.2. Diurnal variation

Fig. 4 shows the diurnal patterns of the average particle counts in the size range of 10–20 nm (nucleation-mode), 20–50 nm (Aitken nuclei), 50–100 nm (Aitken nuclei), and 100–1000 nm (accumulation mode). The diurnal patterns of O_3, SO_2, CO, NO_y, and black carbon, solar radiation, temperature and RH are also shown in Fig. 4. Since hydroxyl radicals (OH) and H_2SO_4 vapor were not measured, we use the product of ultraviolet radiation and SO_2 (UV^*SO_2) as a surrogate parameter for H_2SO_4 production (Stanier et al., 2004b).

Fig. 4 shows that the average number concentrations for particles in 10–20 nm and 20–50 nm peaked around noon local time, similar to the proxy for H_2SO_4 production (UV^*SO_2). These profiles are in sharp contrast to traffic related pollutant such as

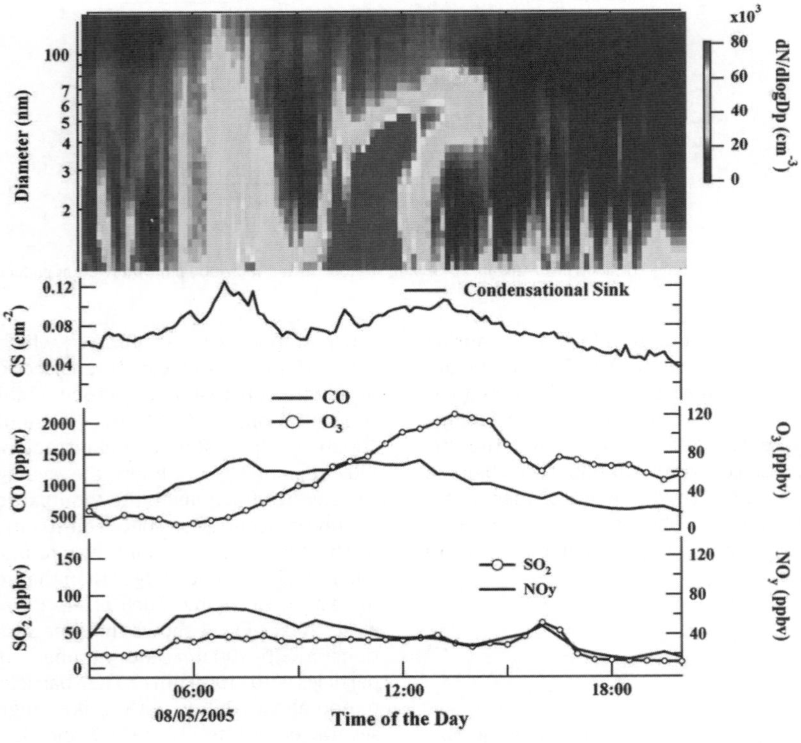

Fig. 6. Particle formation and growth on 8 May 2005.

CO, NO_y, and black carbon which exhibited typical morning and late afternoon peaks but an early afternoon minimum. These results suggest frequent formation of nucleation-mode particles in the study area. We noticed a time delay of about half an hour in reaching the respective noon peak for particles in 10–20 nm, 20–50 nm and 50–100 nm, which may be due to the time required for new particles to grow to the detectable size (10 nm) and for further growth of particles to larger sizes. Weber and McMurry (1996) showed a delay time of about 2 h in a remote environment.

Particles in the large size ranges (50–100 nm, 100–1000 nm) showed a somewhat different diurnal pattern than those for 10–20 nm and 20–50 nm. While their concentrations also indicated elevated levels at noontime, two additional peaks were also obvious: one in the morning (around 0630LT) and the other in the late afternoon (around 1800LT). The morning and late afternoon peaks correlated well with the traffic pollutants (see Fig. 4), suggesting vehicle exhausts were an important source of particles in the size range of 50–100 nm and larger. A similar feature has been reported in previous studies. The smaller particles (30–60 nm) have been suggested to come from light vehicles (Alam et al., 2003) and larger particles (~130 nm) from diesel vehicles (Morawska et al., 1998; Ristovski et al., 1998).

3.3. *Particle bursts and new particle formation events*

3.3.1. *Burst of nucleation-mode particles*

The previous section has shown initial evidence that the increases of the nucleation-mode particles were due to nucleation processes rather than to local emissions. Examination of the data reveals that in some cases there were only burst of nucleation particles without subsequent growth into larger particles, while in other cases both particle increases and growth occurred. During the 30-day observation, there were 7 days showing 'nucleation burst only'. They mainly occurred in polluted air masses characteristic of urban (vehicular) or power-plant plumes. Fig. 5(a) shows a case for 7 May. Very high concentrations of SO_2 (~60 ppbv), NO_y (~120 ppbv) and CO (~1500 ppbv) were observed in the early morning, and the sharp increase in particle number in 10–50 nm range occurred in the afternoon when the levels of primary pollutants reduced significantly and O_3 reached the daily maximum. Fig. 5(b) shows another case for 19 May indicating the site was impacted by a power-plant plume. In this case, the SO_2 concentration reached 177 ppbv at ~10 am, and an increase in nucleation-mode particles appeared about 30 min after that. The lack of significant growth in particle size in both cases is presumably due to a large amount of pre-existing aerosol in the polluted air masses. The events with both particle burst and growth are examined in the following section.

3.3.2. *New particle formation events*

In this section we focus on the five events of new particle formation observed during this study. We adopt the criterions for new particle formation which has been used in previous researches (Dal Maso et al., 2005; Kulmala et al., 2004; Wu et al., 2007). For each day, the contour plots of particle size distribution, meteorological variables, trace gases, and particle properties were examined. An event of new particle formation was identified if (1) there was a spontaneous burst of particle number in the nucleation-mode (10–20 nm, the smallest channel of the WPS system) followed by their subsequent growth at the rate of a few nanometers per hour over a time span of hours, and (2) the nucleation-mode particle concentrations uncorrelated with traffic or power plant related gases such as CO and SO_2. The nucleation with growth meeting these two criterions was likely to occur on a regional scale as the air was not dominated by a specific plume passing over the site. Backward trajectories indicate that air masses in these cases generally came from the north bringing relatively well mixed regional air. Fig. 6 shows an example on 8 May. During the study period there were 5 days with clear indications of 'regional' new particle formation. This frequency is comparable to that for urban areas in the US (Shi et al., 2007; Stanier et al., 2004b), but lower than that for urban Beijing (Wu et al., 2007).

Table 3 summarizes the concentrations of relevant chemical species, values of the meteorological parameters, net rate of increase of particle number concentration (dN/dt) and the aerosol condensational sink (CS, in units of cm^{-2}), and surface area of PM_{10}. (dN/dt) is a measure of the intensity of the new particle formation, and CS is a measure of the available surface area for condensation (Pirjola et al., 1999). Here surface area of PM_{10} was also calculated on the basis of the number distribution assuming spherical particles. The larger the CS and surface area are, the smaller the probability of homogeneous nucleation.

The particle formation events occurred when the ozone concentrations were in the range of 60–100 ppbv, and the BC and $PM_{2.5}$ mass concentrations were generally lower than the average values (5.4 $\mu g\,m^{-3}$ and 67.3 $\mu g\,m^{-3}$), except for a severe pollution event (BC: 9.0 $\mu g\,m^{-3}$ and $PM_{2.5}$: 117.7 $\mu g\,m^{-3}$). The average dN/dt in 10–15 nm diameter range was 7918 $cm^{-3}\,h^{-1}$, which is at the high end of dN/dt for 3–10 nm observed in Pittsburgh (Stanier et al., 2004b). Since our instrument could not measure the more numerous particles that are smaller than 10 nm, dN/dt in the size range of 3–10 nm at our site would be much larger than that in Pittsburgh.

Table 3 shows that particle surface area was in the range of 277–1026 cm^{-3} for the five events of new particle formation. Studies in some urban areas suggest that homogeneous nucleation was not favored under such high surface areas (Woo, 2003; Stanier et al., 2004b). The condensational sink at our study site was also higher than values reported by Hamed et al. (2007) and Stanier et al. (2004b). The formation of new particles in the presence of relatively high condensational sinks at our site suggests a very strong nucleation due to the presence of high concentrations of SO_2. Fig. 7 shows the scatter plot of CS versus UV^*SO_2 at our site and in Pittsburgh (Stanier et al., 2004b). It can be seen that most of the data points for our study fall below $CS:UV^*SO_2 = 1:1$, confirming our study area is a sulfur-rich environment providing abundant precursor to new particle formation despite the presence of large aerosol surface areas.

Table 3
Summary of chemical and meteorological conditions and growth rates during the new particle formation events.

Date	dN(10–20 nm) (cm^{-3})	Surface area ($\mu m^2\,cm^{-3}$)	O_3 (ppb)	CO (ppb)	SO_2 (ppb)	NO_y (ppb)	BC ($\mu g\,m^{-3}$)	$PM_{2.5}$ ($\mu g\,m^{-3}$)	Temp (°C)	RH (%)	Solar ($W\,m^{-2}$)	Grow Rate ($nm\,h^{-1}$)	UV^*SO_2 ($W\,m^{-2}ppb$)	CS (cm^{-2})	dN/dt ($cm^{-3}\,h^{-1}$)
5–6	18,530	277.1	33.0	396.3	26.7	22.6	nan	nan	15.5	65.6	474.4	7.3	523.5	0.024	4265
5–8	22,730	1026.7	71.4	1290.0	38.7	44.9	9.0	117.7	24.7	51.1	623.3	3.6	972.7	0.082	9000
5–11	31,105	686.1	55.7	541.3	17.9	26.2	4.5	48.7	22.2	59.3	691.3	6.4	473.2	0.054	8882
5–17	37,326	455.1	35.9	468.4	16.9	19.7	3.5	36.3	26.2	75.3	418.4	7.4	334.9	0.031	8450
5–18	27,163	321.4	52.3	485.7	27.5	19.1	2.2	20.5	18.3	62.4	483.7	7.2	464.2	0.030	8995

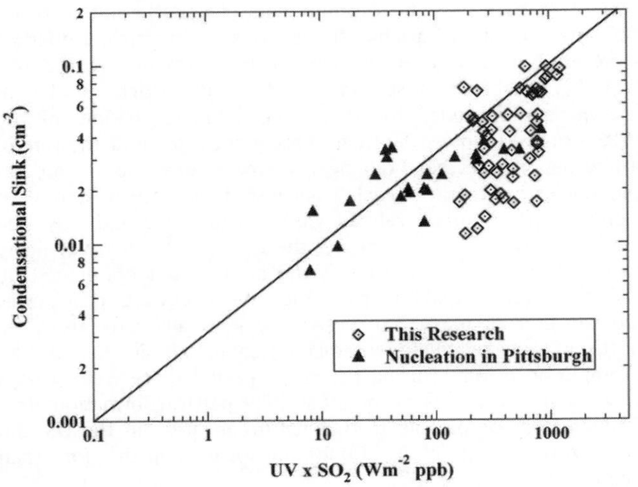

Fig. 7. Scatter plot of UV*SO$_2$ (a proxy for sulphuric acid production) versus condensational sink during five nucleation events at our study site and in Pittsburgh (Stanier et al., 2004b).

The growth rates, which were estimated as the growth in particle mean diameter as a function of time (nm h^{-1}), were 3.6–7.4 nm h^{-1} with an average of 6.4 nm h^{-1}. These values are within the range of reported values of 2–10 nm h^{-1} in other urban areas (Holmes, 2006) including Beijing (Wu et al., 2007) and the Pearl River Delta of China (Liu et al., 2008). Fig. 8 illustrates the number size distributions of a typical homogeneous nucleation event on 8 May 2005. The particle growth rate depends on the temperature and concentration of available condensable vapor (Kulmala et al., 2004), and it has been suggested that H_2SO_4 condensation typically accounts for only 10–30% of the observed growth (Weber et al., 1997; Boy et al., 2005) compared to the VOCs which account for more than 70% of the material for the particle growth. The relative importance of H_2SO_4 and VOCs in particle growth in our study is unclear, the former may have a large contribution given the abundance of SO_2, but more studies will be needed.

4. Summary

Both number concentration and fraction of ultrafine (10–100 nm) particles in total particle counts at this site in the Yangtze River delta were higher than those reported in the urban/suburban areas in North America and Europe. On average, the mean concentrations of particles in the 10–50 nm range reached maximum at noontime, different from that of traffic related pollutants. The formation of nucleation-mode particle and subsequent growth were distinctively observed on five days. The new particle formation events were characterized by intense homogeneous nucleation, a greater amount of sulfuric acid product, and moderate growth. The mean growth rate in these 5 events was 6.4 ± 1.6 nm h^{-1}, which was comparable to the values reported in previous investigations. Our study has provided new data on particle number and size and shed light on new particle formation under polluted conditions.

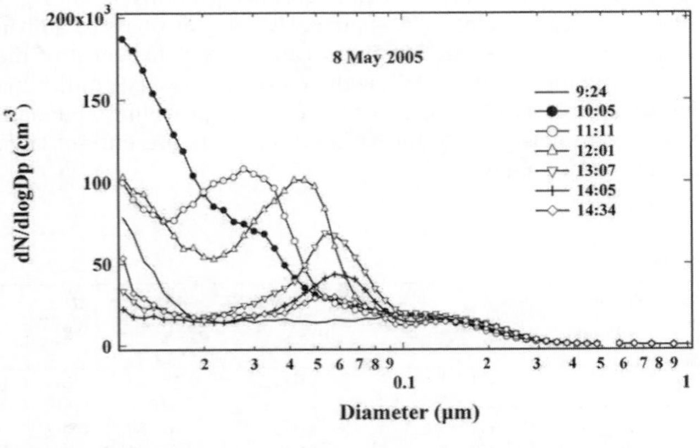

Fig. 8. Number size distribution of particles as a function of time during a new particle formation event on 8 May 2005.

Acknowledgement

We thank Mr. C.N. Poon for his contribution to the work in the field study, Y.H. Kwok for his assistance in processing the trace-gas data, Dr. Ravi Pathak for helpful comments, and Dr. Aijun Ding for help in weather analysis. We also thank the two anonymous referees for their comments which have helped improve the original manuscript. This research was funded by the Research Grants Council of Hong Kong (PolyU5144/04E), National Basic Research Program of China (2005CB422203) and Niche Area Development Scheme of the Hong Kong Polytechnic University (1-BB94).

References

Alam, A., Shi, J.P., Harrison, R.M., 2003. Observations of new particle formation in urban air. Journal of Geophysical Research 108 (D3), 4093. doi:10.1029/2001JD001417.

Birmili, W., Wiedensohler, A., 2000. New particle formation in the continental boundary layer: meteorological and gas phase parameter influence. Geophysical Research Letters 27 (20), 3325–3328.

Boy, M., Kulmala, M., Ruuskanen, T.M., Pihlatie, M., Reissell, A., Aalto, P.P., Keronen, P., Dal Maso, M., Hellen, H., Hakola, H., Jansson, R., Hanke, M., Arnold, F., 2005. Sulphuric acid closure and contribution to nucleation mode particle growth. Atmospheric Chemistry and Physics 5, 863–878.

Chan, C.K., Yao, X.H., 2008. Air pollution in mega cities in China. Atmospheric Environment 42, 1–42.

Dal Maso, M., Kulmala, M., Riipinen, I., Wagner, R., Hussein, T., Aalto, P.P., Lehtinen, K.E.J., 2005. Formation and growth of fresh atmospheric aerosols: eight years of aerosol size distribution data from SMEAR II, Hyytiälä, Finland. Boreal Environment Research 10, 323–336.

Dunn, M.J., Jimenez, J.L., Baumgardner, D., Castro, T., McMurry, P.H., Smith, J.N., 2004. Measurements of Mexico City nanoparticle size distributions: observations of new particle formation and growth. Geophysical Research Letters 31, L10102. doi:10.1029/2004GL019483.

Hamed, A., Joutsensaari, J., Mikkonen, S., Sogacheva, L., Dal Maso, M., Kulmala, M., Cavalli, F., Fuzzi, S., Facchini, M.C., Decesari, S., Mircea, M., Lehtinen, K.E.J., Laaksonen, A., 2007. Nucleation and growth of new particles in Po Valley, Italy. Atmospheric Chemistry and Physics 7, 355–376.

Harrison, R.M., Grenfell, J.L., Savage, N., Allen, A., Clemitshaw, K.C., Penkett, S., Hewitt, C.N., Davison, B., 2000. Observations of new particle production in the atmosphere of a moderately polluted site in eastern England. Journal of Geophysical Research D 105, 17819–17832.

Hinds, W.C., 1982. Aerosol Technology, Properties, Behavior, and Measurement of Airborne Particles. John Wiley & Sons, Inc., pp. 21–217.

Holmes, N.S., 2006. A review of particle formation events and growth in the atmosphere in the various environments and discussion of mechanistic implications. Atmospheric Environment 41, 2183–2201.

Kulmala, M., Toivonen, A., Mäkelä, J.M., Laaksonen, A., 1998. Analysis of the growth of nucleation mode particles in boreal forest. Tellus 50B, 449–462.

Kulmala, M., Vehkamaki, H., Petaja, T., Dal Maso, M., Lauri, A., Kerminen, V.M., Birmili, W., McMurry, P.H., 2004. Formation and growth rates of ultrafine atmospheric particles: a review of observations. Journal of Aerosol Science 35, 143–175.

Lippmann, M., Ito, K., Nffadas, A., Burnett, R.T., 2000. Association of Particulate Matter Components with Daily Mortality and Morbidity in Urban Populations. Research Report 95. Health Effects Institute, Cambridge, MA.

Liu, S., Hu, M., Wehner, B., Wiedensohler, A., Cheng, Y.F., 2008. Aerosol number size distribution and new particle formation at a rural/coastal site in Pearl River Delta (PRD) of China. Atmospheric Environment 42 (25), 6275–6283.

McMurry, P.H., Woo, K.S., Weber, R., Chen, D.R., Pui, D.Y., 2000. Size distributions of 3–10 nm atmospheric particles: implications for nucleation mechanisms. Philosophical Transactions of the Royal Society A 358, 2625–2642.

Morawska, L., Bofinger, N., Kocis, L., Nwankwoala, A., 1998. Submicrometer and supermicrometer particles from diesel vehicle emissions. Environmental Science & Technology 32 (14), 2033–2042.
O'Dowd, C.D., Hämeri, K., Mäkelä, J., Väkevä, M., Aalto, P., de Leeuw, G., Kunz, G.J., Becker, E., Hansson, H.C., Allen, A.G., Harrison, R.M., Berresheim, H., Kleefeld, C., Geever, M., Jennings, S.G., Kulmala, M., 2002. Coastal new particle formation: environmental conditions and aerosol physicochemical characteristics during nucleation bursts. Journal of Geophysical Research 107 (D19), 8107. doi:10.1029/2001JD000206.
Pirjola, L., Kulmala, M., Wilck, M., Bischoff, A., Stratmann, F., Otto, E., 1999. Effects of aerosol dynamics on the formation of sulphuric acid aerosols and cloud condensation nuclei. Journal of Aerosol Science 30, 1079–1094.
Ristovski, Z., Morawska, L., Bofinger, N.D., Hitchins, J., 1998. Submicrometer and supermicrometer particulate emission from spark ignition vehicles. Environmental Science & Technology 32, 3845–3852.
Ruuskanen, J., Tuch, Th., Ten Brink, H., Peters, A., Khystov, A., Mirme, A., Kos, G.P.A., Brunekreef, B., Wichmann, H.E., Buzorius, G., Vallius, M., Kreyling, W.G., Pekkanen, J., 2001. Concentrations of ultrafine, fine and $PM_{2.5}$ particles in three European cities. Atmospheric Environment 35, 3729–3738.
Sament, J.M., Dominici, F., Curriero, F.C., Coursac, I., Zeger, S.L., 2000. Fine particulate air pollution and mortality in 20 U.S. cities, 1987–1994. The New England Journal of Medicine 343, 1742–1749.
Shi, J.P., Evans, D.E., Khan, A.A., Harrison, R.M., 2001. Sources and concentration of nanoparticles (10 nm in diameter) in the urban atmosphere. Atmospheric Environment 35, 1193–1202.
Shi, Q., Sakurai, H., McMurry, P.H., 2007. Characteristics of regional nucleation events in urban East St. Louis. Atmospheric Environment 41, 4119–4127.
Stanier, C.O., Khlystov, A.Y., Pandis, S.N., 2004a. Ambient aerosol size distributions and number concentrations measured during the Pittsburgh Air Quality Study (PAQS). Atmospheric Environment 38, 3275–3284.
Stanier, C.O., Khlystov, A.Y., Pandis, S.N., 2004b. Nucleation events during the Pittsburgh Air Quality Study: description and relation to key meteorological, gas phase, and aerosol parameters. Aerosol Science and Technology 38 (S1), 253–264.
Tuch, T., Brand, P., Wichmann, H.E., Heyder, J., 1997. Variation of particle number and mass concentration in various size ranges of ambient aerosols in eastern Germany. Atmospheric Environment 31 (24), 4193–4197.
Wang, T., Cheung, V.T.F., Li, Y.S., Anson, M., 2001. Ozone and related gaseous pollutants in the boundary layer of eastern China: overview of the recent measurements at a rural site. Geophysical Research Letters 28, 2373–2376.
Wang, T., Wong, C.H., Cheung, T.F., Blake, D.R., Arimoto, R., Baumann, K., Tang, J., Ding, G.A., Yu, X.M., Li, Y.S., Streets, D.G., Simpson, I.J., 2004. Relationships of trace gases and aerosols and the emission characteristics at Lin'an, a rural site in eastern China, during spring 2001. Journal of Geophysical Research vol. 109, D19S05. doi:10.1029/2003JD004119.
Weber, R.J., McMurry, P.H., 1996. Fine particle size distributions at the Mauna Loa observatory, Hawaii. Journal of Geophysical Research – Atmospheres 101 (D9), 14767–14775.
Weber, R.J., Marti, J.J., McMurry, P.H., Eisele, F.L., Tanner, D.J., Jefferson, A., 1997. Measurements of new particle formation and ultrafine particle growth rates at a clean continental site. Journal of Geophysical Research D 102, 4375–4385.
Wehner, B., Wiedensohler, A., Tuch, T.M., Wu, Z.J., Hu, M., Slanina, J., Kiang, C.S., 2004. Variability of the aerosol number size distribution in Beijing, China: new particle formation, dust storms, and high continental background. Geophysical Research Letters 31, L22108. doi:10.1029/2004GL021596.
Wichmann, H.E., Peters, A., 2000. Epidemiological evidence of the effects of ultrafine particle exposure. Philosophical Transactions of the Royal Society A 358, 2751–2769.
Wu, Z., Hu, M., Liu, S., Wehner, B., Bauer, S., Ma ßling, A., Wiedensohler, A., Petaja, T., Dal Maso, M., Kulmala, M., 2007. New particle formation in Beijing, China: statistical analysis of a 1-year data set. Journal of Geophysical Research 112, D09209. doi:10.1029/2006JD007406.
Woo, K.S., Chen, D.R., Pui, D.Y.H., McMurry, P.H., 2001. Measurements of Atlanta aerosol size distributions: observations of ultrafine particle events. Aerosol Science and Technology 34, 75–87.
Woo, K.S., 2003. Measurement of Atmospheric Aerosols: size Distributions of Nanoparticles, Estimation of Size Distribution Moments and Control of Relative Humidity. PhD thesis. University of Minnesota.
Yao, X., Chan, C.K., Fang, M., Cadle, S., Chan, T., Mulawa, P., He, K., Ye, B., 2002. The water-soluble ionic composition of $PM_{2.5}$ in Shanghai and Beijing, China. Atmospheric Environment 36, 4223–4234.
Yu, F.Q., Wang, Z.F., Luo, G., Turco, R., 2008. Ion-mediated nucleation as an important global source of tropospheric aerosols. Atmospheric Chemistry and Physics 8, 2537–2554.

Atmospheric Environment 45 (2011) 4631–4640

Contents lists available at ScienceDirect

Atmospheric Environment

journal homepage: www.elsevier.com/locate/atmosenv

Size-fractionated water-soluble ions, situ pH and water content in aerosol on hazy days and the influences on visibility impairment in Jinan, China

Shu-hui Cheng[a], Ling-xiao Yang[a,b,*], Xue-hua Zhou[a], Li-kun Xue[a], Xiao-mei Gao[a], Yang Zhou[a], Wen-xing Wang[a,c]

[a] *Environment Research Institute, Shandong University, Jinan 250100, China*
[b] *School of Environmental Science and Engineering, Shandong University, Jinan 250100, China*
[c] *Chinese Research Academy of Environmental Sciences, Beiyuan, Beijing 100012, China*

ARTICLE INFO

Article history:
Received 7 April 2011
Accepted 21 May 2011

Keywords:
Size distribution
Water-soluble ions
AIM
Situ pH
Water content
Visibility impairment

ABSTRACT

To study the size-fractionated characteristics of aerosol chemical compounds including major water-soluble inorganic and organic ions, situ pH and water content as well as their influences on visibility, field sample collections using a Micro Orifice Uniform Deposit Impactor (MOUDI) combined with simulations by Aerosol Inorganic Model (AIM) were conducted on hazy and clear days in Jinan, China from April, 2006 to January, 2007. Average concentrations of TSP, PM_{10} and $PM_{1.8}$ on hazy days were found to be 1.49–5.13, 1.54–5.48 and 1.30–5.48 times those on clear days during the sampling periods, indicating that particulate pollution was very serious on hazy days. Size distributions of mass, SO_4^{2-}, NO_3^-, formate and acetate were all bimodal with fine mode predominant on hazy days, demonstrating that fine particles and secondary pollutants were more easily formed on hazy days. The average total aerosol concentration of H^+ ($[H^+]_{total}$), of which free H^+ concentration ($[H^+]_{ins}$) inside aerosol accounted for 30%, was 2.36–4.21 times that in other cities in China and US. The $[H^+]_{ins}$ on hazy days was 2.43–13.11 times that on clear days and the estimated situ pH was mainly influenced by RH and mole ratios of $[NH_4^+]/[SO_4^{2-}]$. Size distributions of situ pH were unimodal peaked at 0.56 μm on clear days in spring and autumn as well as on hazy days in autumn, while a trend of increasing was shown on hazy days in spring and summer. The normalized water content (NWC) was higher on hazy days in autumn and winter because of the easier uptake of water by aerosol. It was found that when NWC < 2, if $[NH_4^+]/[SO_4^{2-}] < 2.0$, aerosol chemical components were more sensitive to water content, while if $[NH_4^+]/[SO_4^{2-}] > 2.0$, relative humidity (RH) was more important. Size distributions of water content showed a trend of sharp increasing on hazy days in spring and autumn. Correlation and regression analysis indicated that visibility was significantly influenced by the concentrations of SO_4^{2-} and water content in the range of 1.0–1.8 μm.

1. Introduction

Haze has raised more and more serious attentions from environmental researchers since it can significantly diminish visibility and has an adverse effect upon human health. There are four major haze regions in China: Huang-Huai-Hai Plain, Yangtze River Delta, Pearl River Delta and Sichuan Basin. The Huang-Huai-Hai haze region, including Beijing, Tianjin, Hebei Province, Shandong Province and Liaoning Province, is one of the most developed areas in China. Accounting for only 5.47% of the total geographical area of China, it contributes 26.1% of the GDP, but also more than 16.8% of the total emissions of sulfate dioxide (SO_2) and total suspend particle (TSP) in China in 2007 (National Bureau of Statistics of China, 2008). Serious air pollution and high frequency of haze in this region has been reported (Sun et al., 2006; Yang et al., 2007; Chang et al., 2009). The emissions of haze precursors such as SO_2, NO_x and TSP in Shandong Province ranked first in China, surpassing the total emissions from Japan and Korea (Zhang et al., 2009). As a result, serious haze pollution and increasing annual number of haze days every year has been reported in Jinan, the capital of Shandong Province (Wang et al., 2009).

The formation of haze is related with atmospheric secondary pollutants reactions and transformations, which depend on the size distributions of aerosol chemical compounds in the atmosphere (Ludwig and Klemm, 1990; Anlauf et al., 2006). The existing researches on visibility impairment were mainly focused on chemical components in $PM_{2.5}$ and concluded that sulfate was the major

* Corresponding author. Environment Research Institute, Shandong University, Jinan 250100, China. Tel./fax: +86 531 88366072.
E-mail address: yanglingxiao@sdu.edu.cn (L.-x. Yang).

1352-2310/$ – see front matter © 2011 Elsevier Ltd. All rights reserved.
doi:10.1016/j.atmosenv.2011.05.057

contributor to the visibility impairment (Tsai and Cheng, 1999; Cheng and Tsai, 2000; Yang et al., 2007; Tan et al., 2009). Rare studies have been done on the influences of aerosol acidity and water content on visibility impairment in Jinan. However, the size-distributed chemical compounds of particle matter (PM) can affect the portioning of water-soluble, semi-volatile ions between gas, liquid and particles as well as the aerosol acidity in some extent (Fridlind and Jacobson, 2000). The acidity of aerosol is important for the acid formation and deposition, heterogeneous atmospheric processes. It also causes adverse health effects (Kerminen et al., 2001). Studies on size-fractioned aerosol acidity showed that the submicron particles were more acidic than or at least as acidic as larger particles because most sulfate and nitrate concentrated in submicron particles, while neutralization processes between crustal elements and acidic species happened in super-micron particles (Fridlind and Jacobson, 2000; Pszenny et al., 2004). The ability of aerosol to absorb water is significantly important to determine the extinction cross-section and aerosol mass concentrations. Hence, water content can be a major factor on visibility impairment. For example, the scattering cross-section for ammonium sulfate particles at 90% RH is five or more times higher than that of the dry particles (Malm and Day, 2001). Studies showed that aerosol water content played an important role in visibility impairment in Guangzhou urban area (Jung et al., 2009). Therefore, knowledge of the size-fractionated water-soluble ions, situ pH and water content in PM was of great importance to assess their impacts on visibility impairment.

It is difficult to measure the aerosol acidity due to the lower concentrations of water content. Aerosol Inorganic Model (AIM) (Clegg et al., 1998a; b) with gas-aerosol equilibrium disabled was a suitable method to estimate the situ pH and water content on submicron particles (Yao et al., 2006), which was applied to simulate situ pH and water content in this study. Based on the observations and simulations, the work presented in this study used the step-wise multiple linear regression to evaluate the influence of the size-distributed chemical components, aerosol acidity and water content, as well as meteorological parameters on visibility impairment in Jinan, where the concentrations of sulfate, nitrate and ammonium in $PM_{2.5}$ were almost the highest in the world (Cheng et al., 2011).

2. Experiment

2.1. Sampling and chemical analysis

To investigate the secondary chemical transformations of aerosol on hazy days and the influence of major chemical components on visibility impairment, size-fractionated particle matters in April, August, October and December of 2006, as well as January of 2007 were collected using an eight-stage Micro-Orifice Uniform Deposit Impactor (MOUDI) sampler (30 L min^{-1}, Model 100, MSP, USA), 50% cut off diameters of 18.0 (inlet), 10.0, 5.6, 3.2, 1.8, 1.0, 0.56, 0.32 and 0.18 μm. The sampling site was located on the roof (20 m above ground level) of a five-story building at the Jinan Environment Protection Monitoring Station (Fig. 1). The site was a typical urban site influenced by residential, traffic and commercial emissions in Jinan. The method on the extraction and analysis of water-soluble ions including SO_4^{2-}, NO_3^-, NH_4^+, Na^+, Cl^-, K^+, Ca^{2+}, Mg^{2+} and organic acids was reported by Yang et al. (2007).

Meanwhile, trace gases including NO_x and SO_2 were continuously measured by the Jinan Environment Protection Monitoring Station and meteorological parameters including temperature, RH, visibility, wind speed and wind direction were also monitored by the Meteorological Bureau of Shandong Province during the sampling period. According to <Ground Meteorological Factors Monitoring Standard in China>, the visibility above 10 km was defined as hazy days and otherwise clear days during sampling period.

2.2. AIM model

Aerosol acidity and water content are two important parameters which can influence gas-aerosol portioning, aerosol diameter, mass concentration and optical property. However the two parameters were difficult to be observed due to the lower concentrations.

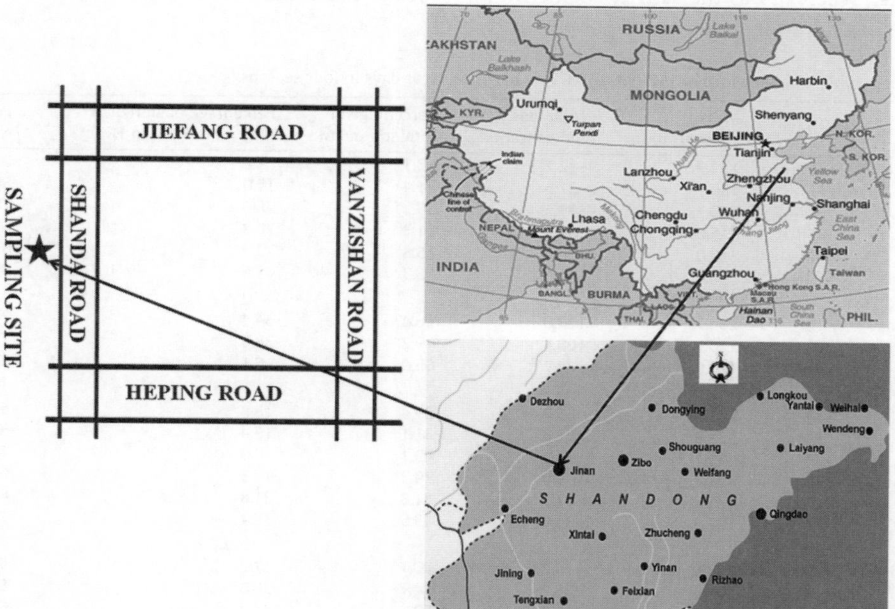

Fig. 1. Location of the sampling site in Jinan, Shandong Province, China.

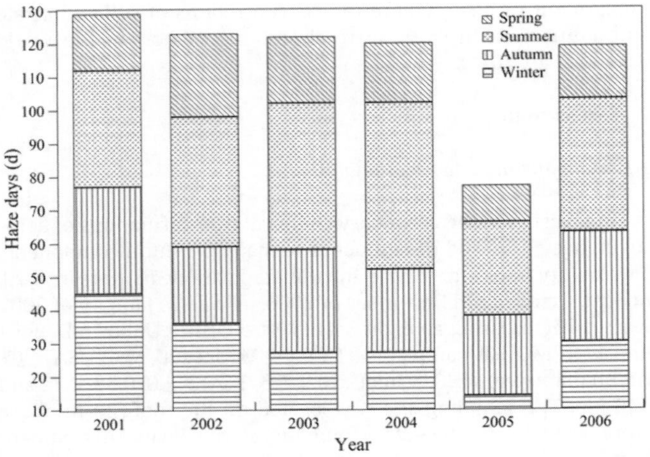

Fig. 2. Seasonal and inter-annual variations of haze days during 2001–2006.

The thermodynamic model such as Aerosol Inorganic Model (AIM) (Clegg et al., 1998a, b, http://www.aim.env.uea.ac.uk/aim/aim.htm) with gas-aerosol equilibrium disabled was a state-of-the-art model to accurately predict situ pH and water content (Yao et al., 2006). Since coarse particles contained large fraction of crustal elements and sulfate were almost completely neutralized, AIM-II and AIM-III were not applied to estimate situ pH and water content in the coarse aerosol (Yao et al., 2007). In this study, we just focused on the situ pH and water content inside particles with diameter ranging from 0.18 to 1.8 μm.

SO_4^{2-}, NO_3^- and NH_4^+ were the dominant water-soluble ions, the input H^+ was calculated as: $[H^+]measure = 2[SO_4^{2-}]measure + [NO_3^-]measure - [NH_4^+]measure$ (Pathak et al., 2004). Here, measured 24-h average temperature, RH, H^+, SO_4^{2-}, NO_3^- and NH_4^+ were used as the input of AIM model to estimate situ pH and water content in different sizes of fine particle. The situ pH was calculated as $pH = -Log(\gamma \times [H^+]_{ins}/\text{aerosol solution volume})$ inside the aerosol droplet, where γ was the activity coefficient of $[H^+]_{ins}$ which meant the free aerosol acidity in aqueous system (Yao et al., 2006).

3. Results and discussion

3.1. Seasonal variations of visibility

The visibility in Jinan showed a gradually declining trend during 1961–2005. A constant deterioration of visual range during 1961–1999 was ascribed to the dramatic economical development and the aggravation of air pollution, though the trend was slightly slowed in 2000–2005 due to the "Blue Sky" campaign initiated in 2000 in Jinan (Yang et al., 2007). Seasonal and inter-annual variations of the number of hazy days in 2001–2006 are shown in Fig. 2. The number of haze days actually decreased in 2001–2005 but again increased in 2006, possibly due to the end of the "Blue Sky" project and the sharp increase of real estate development in Jinan in 2006. The general decrease trends of haze days in 2000–2006 were also found in Shenyang and Shanghai. However, increasing trend was found in Chengdu, Guangzhou and Xi'an and almost constant appeared in Beijing (Chang et al., 2009).

The seasonal variations of the number of haze days were also depicted in Fig. 2 and it showed that the average number of hazy days in 2001–2006 in spring, summer, autumn and winter were 17.83, 39.33, 28.00 and 29.83 days respectively.

3.2. Particle mass and major chemical concentrations

Since there was no cut size of 2.5 μm in MOUDI, the cut size of 1.8 μm was used to divide PM_{10} into coarse and fine particles. Table 1 showed that on hazy days, the average concentrations of TSP and PM_{10} in all seasons were 1.05–1.66 and 1.47–2.76 times the second grade of National Ambient Air Quality Standards (NAAQS) in China (300 μg m^{-3} for TSP, 150 μg m^{-3} for PM_{10}), except for TSP in summer and $PM_{1.8}$ were 2.53–5.51 times the daily US NAAQS (35 μg m^{-3}) for $PM_{2.5}$. On clear days, the concentrations of TSP and PM_{10} could meet the NAAQS in China, except for PM_{10} in autumn, while the concentrations of $PM_{1.8}$ still exceeded US NAAQS for

Table 1
Average mass and major water-soluble inorganic ions of TSP, PM_{10}, $PM_{1.8}$ on hazy and clear days in four seasons.

			Mass concentration	Total ions concentration	Percentage of total ion in PM	Percentage of sum of SO_4^{2-}, NO_3^- and NH_4^+ in PM	Percentage of sum of SO_4^{2-}, NO_3^- and NH_4^+ in total ions
April	clear	TSP	179.7	32.1	18.1	9.6	53.7
		PM_{10}	130.7	27.5	21.4	12.0	56.9
		$PM_{1.8}$	40.4	14.5	38.3	26.6	71.5
	hazy	TSP	315.6	98.9	31.3	20.0	63.9
		PM_{10}	242.0	86.6	35.8	23.9	66.7
		$PM_{1.8}$	88.4	48.9	55.3	41.8	75.6
August	hazy	TSP	265.2	123.0	46.4	34.4	74.2
		PM_{10}	220.2	108.4	49.2	38.8	78.8
		$PM_{1.8}$	115.0	74.7	65.0	56.1	86.4
October	clear	TSP	223.8	65.7	29.4	15.8	53.6
		PM_{10}	178.4	55.6	31.1	17.4	55.7
		$PM_{1.8}$	59.7	29.3	49.1	34.8	70.9
	hazy	TSP	334.3	98.0	29.3	19.3	66.0
		PM_{10}	275.5	87.5	31.8	21.8	68.8
		$PM_{1.8}$	127.9	55.6	43.5	34.4	79.1
December & January	clear	TSP	96.8	29.2	30.1	17.7	58.9
		PM_{10}	75.5	26.2	34.7	20.6	59.4
		$PM_{1.8}$	35.2	18.4	52.3	34.7	66.3
	hazy	TSP	497.0	133.3	26.8	18.6	69.3
		PM_{10}	413.5	125.0	30.2	21.5	71.1
		$PM_{1.8}$	192.8	89.3	46.3	36.1	78.0

$PM_{2.5}$. The results indicated that particulate matter pollution on hazy days in Jinan was much more serious than that on clear days. The highest particulate mass concentrations appeared in winter (Nov. 15–Mar. 15) due to heavy consumption of fossil fuels like coal for heating purpose and poor meteorological conditions for atmospheric dispersion, while the lowest particulate mass concentrations appeared in summer possibly because of the scavenging effects of precipitation. The high correlation coefficients ($R^2 = 0.94$ and 0.93) between $PM_{1.8}$, $PM_{1.8-10}$ and PM_{10} in Jinan implied that they had the similar emission sources and were mainly influenced by the local emissions.

Table 1 depicted that the average percentages of total water-soluble ions in $PM_{1.8}$ were higher than those in PM_{10} in all four seasons, which demonstrated that water-soluble ions were likely to accumulate in the fine particles. The total amount of NO_3^-, SO_4^{2-} and NH_4^+ accounted for 53.6–74.2%, 55.7–78.8% and 66.3–86.4% of the total water-soluble ions in TSP, PM_{10} and $PM_{1.8}$ while the percentage increased from clear to hazy days, which demonstrated that NO_3^-, SO_4^{2-} and NH_4^+ were the major water-soluble ions in particulate matter and more secondary water-soluble ions could be formed and retained in the atmosphere on hazy days.

The average concentrations of formic, acetic and oxalic acids were 261.44, 314.20 and 1503.53 ng m^{-3} in TSP, 210.64, 256.25 and 1128.68 ng m^{-3} in PM_{10}, while 127.29, 129.42 and 646.37 ng m^{-3} in $PM_{1.8}$. The concentrations of formic acid in $PM_{1.8}$ and PM_{10} were comparable with those reported in Beijing, while the concentrations of acetic and oxalic acids in $PM_{1.8}$ were 0.66 and 0.83 times higher, respectively. The concentrations of those two acids in PM_{10} were more than two times higher than what reported in Beijing (Wang et al., 2007). The concentrations of oxalic acid in $PM_{1.8}$ in Jinan was at the same level with that in Nanjing (Wang et al., 2002) but much higher than that in New York (Khwaja, 1995) and Tokyo (Kawamura and Ikushima, 1993). Oxalic acid, predominantly organic acids in Jinan, had good correlations (R) with SO_4^{2-} (0.64), NO_3^- (0.80), NH_4^+ (0.77), K^+ (0.72), formic (0.70) and acetic (0.73) acids, implying that the emission sources of oxalate were complex and might be related with emissions from coal burning, traffic and biomass burning as well as secondary pollutants transformations (Wang et al., 2007). The higher concentrations of oxalic acid in winter could be related to the sharp increase of coal burning for heating. The formic acid was mainly from vehicular emissions while the acetic acid might be from photochemical reactions in the daytime (Souza et al., 1999).

3.3. Size distributions of mass and major chemical

3.3.1. Size distributions of mass concentrations

The size distributions of aerosol chemical components on hazy and clear days are shown in Fig. 3 and Fig. 4. The average mass size distributions on hazy days and clear days in four seasons were all bimodal with a coarse mode (3.2–5.6 μm) and a fine mode (0.56–1.0 μm), except that on clear days in spring it exhibited unimode with a peak at 3.2–5.6 μm. The fine mode and coarse mode on hazy days were comparable in summer, autumn and winter however on clear days the coarse mode were dominant, which implied that fine particles favored to accumulate on hazy days. In spring the coarse mode was dominant due to the sand storm, accounting for 19.44% of the total mass. The peak of fine mode on hazy days in winter was much higher than that in other seasons, demonstrating that the fine particle pollution in winter was more serious. The peak of fine mode on clear days in autumn was higher than that in other seasons, which could be due to the burning of large quantities of agricultural wastes in the suburban of Jinan. Mass size distributions similar to that in winter in Jinan were reported in Nanjing (Wang et al., 2003). Unimodal mass size distributions similar to those on clear days in Jinan were observed in Taiwan (Chen et al., 2004) where it was stated that the coarse particles were dominant during the Asia dust storm.

3.3.2. Size distributions of SO_4^{2-}, NO_3^- and NH_4^+

Average size distributions of SO_4^{2-} and NO_3^- both showed bimodal characteristics with fine mode at 0.18–1.0 μm and coarse mode at 3.2–5.6 μm on hazy and clear days in four seasons (Figs. 3 and 4), with fine mode dominant except for NO_3^- on hazy days in spring and summer. The size distribution of chloride was similar to NO_3^- on hazy days, which demonstrated that in autumn and winter, the reactions of nitric acid with sodium chloride or aerosol crustal elements on hazy days mainly existed in the fine mode but in spring and summer in the coarse mode. The correlation coefficients (R) between annual mean concentrations of SO_4^{2-} and NO_3^- were 0.83 and higher R existed in autumn (0.88) and winter (0.99), implying similar formation mechanisms for SO_4^{2-} and NO_3^- in both seasons. To find out the relative importance of mobile vs. stationary sources of sulfur and nitrogen in the atmosphere, the mass ratio of NO_3^-/SO_4^{2-} can be used as an indicator (Arimoto et al., 1996). The average mass ratio of NO_3^-/SO_4^{2-} on hazy and clear days were 0.43 and 0.42 respectively, which showed that the stationary source emissions were much more important than mobile source emissions in Jinan. Higher NO_3^-/SO_4^{2-} ratio appeared on hazy days in particles with diameter range of 1.8–10, 3.2–10, 3.2–10 and 1.0–5.6 μm in spring, summer, autumn and winter, respectively, while on clear days it appeared in the range of 0.56–5.6, 1.8–5.6 and 0.32–1.8 μm in spring, autumn and winter, respectively, which demonstrated that mobile source emissions made more contributions to the formation of coarse aerosol on hazy days but to fine aerosol on clear days. Average size distributions of NH_4^+ showed unimodal characterize on hazy and clear days in four seasons, with a peak at 0.32–1.0 μm.

The size-fractionated SO_4^{2-}, NO_3^- and NH_4^+ showed that secondary water-soluble ions mainly existed in fine mode and the seasonal variations were not apparent. The Sulfur Oxidation Ratio (SOR) and Nitrogen Oxidation Ratio (NOR) can be calculated as $SOR = \frac{S_{SO_4^{2-}}}{S_{SO_4^{2-}} + S_{SO_2}}$ and $NOR = \frac{N_{NO_3^-}}{N_{NO_3^-} + N_{NO_x}}$ and used to estimate the oxidation extent of SO_2 and NO_x. Table 2 depicted that the values of NOR in 0.18–18 μm were all lower than SOR in summer and autumn, however the opposite was observed in winter and spring except for the fine particles in spring. High temperature and RH were favorable for the liquid oxidation of SO_2 and the dissociation of NH_4NO_3, which caused higher SOR and lower NOR in summer. However, low temperature in winter inhibited the oxidation of SO_2 and the dissociation of NH_4NO_3, which might lead to the decrease of SOR and increase of NOR.

3.3.3. Size distributions of other inorganic ions and organic acids

The size distributions of Na^+ and Cl^- both showed unimodal on hazy days in four seasons while on clear days bimodal appeared in spring and autumn and unimodal in winter. The size distributions of Cl^- indicated that coal combustion for heating in winter led to the major peak of Cl^- in fine mode, while in summer it was influenced by long-range transport from the Yellow Sea. K^+, Ca^{2+} and Mg^{2+} all showed unimodal characteristics on hazy and clear days in four seasons. Most K^+ was from biomass burning and accumulated in fine mode, while Ca^{2+} and Mg^{2+} mainly came from soil source and concentrated in coarse mode.

Size distributions of formate and acetate were both bimodal (fine mode peaked at 0.56–1.8 μm and coarse mode peaked at 3.2–5.6 μm) on hazy days, while on clear days the fine mode diminished greatly for formate and disappeared for acetate. Oxalate showed bimodal (fine mode peaked at 0.32–1.8 μm, coarse mode

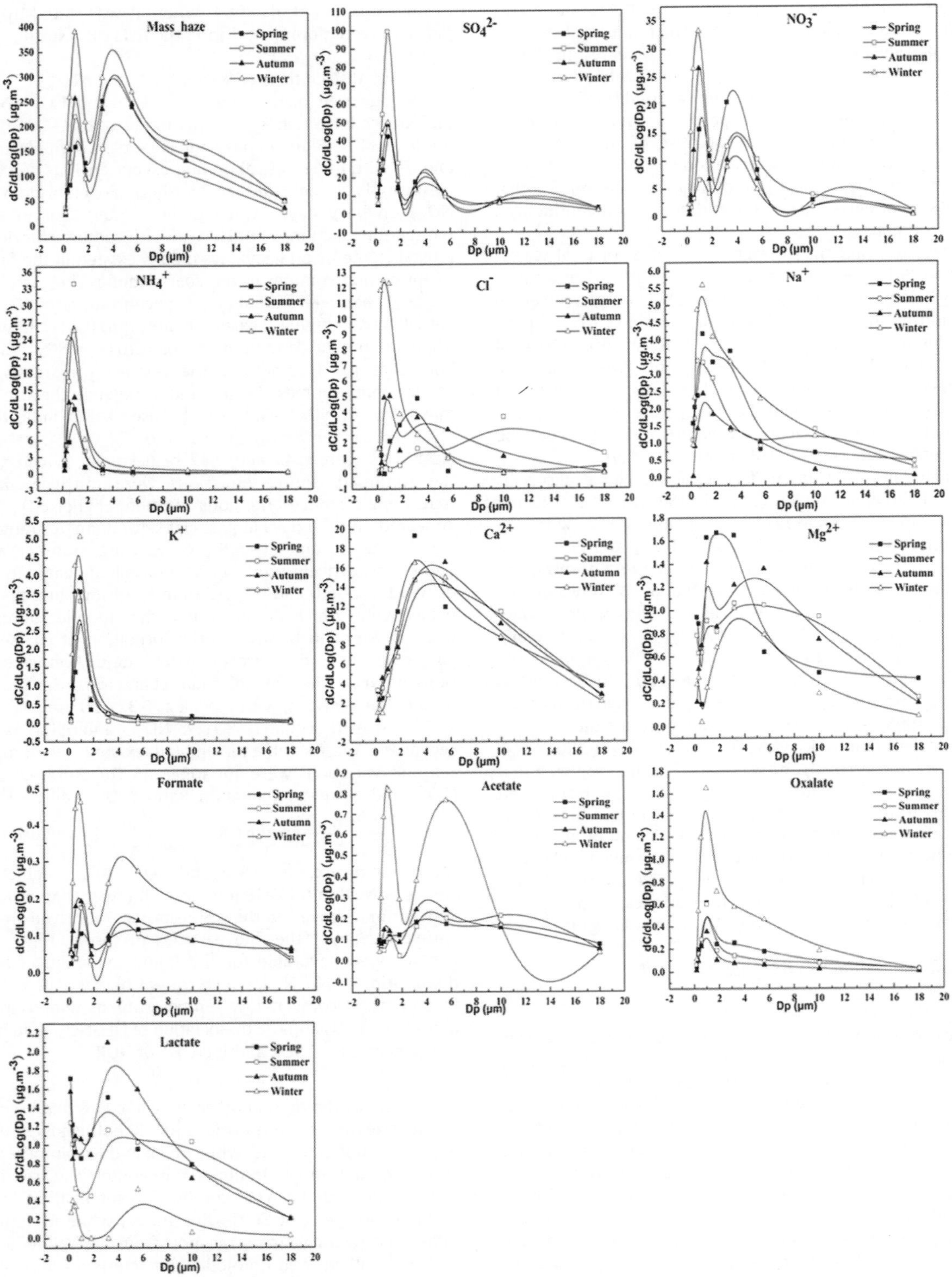

Fig. 3. The average size distributions of aerosol chemical compounds on hazy days in four seasons.

peaked at 3.2–5.6 μm) on clear days while changed to unimodal peaked at 0.56–1.0 μm on hazy days. The results showed that the secondary oxidation reaction of primary pollutants emitted from coal/biomass burning, soil and motor vehicles sources to form formate, acetate and oxalate (Wang et al., 2007) mainly existed in fine modal and could be enhanced on hazy days. Size distributions of lactate were unimodal with a peak at 1.8–3.2 μm, which demonstrated that lactate was mainly influence by dust.

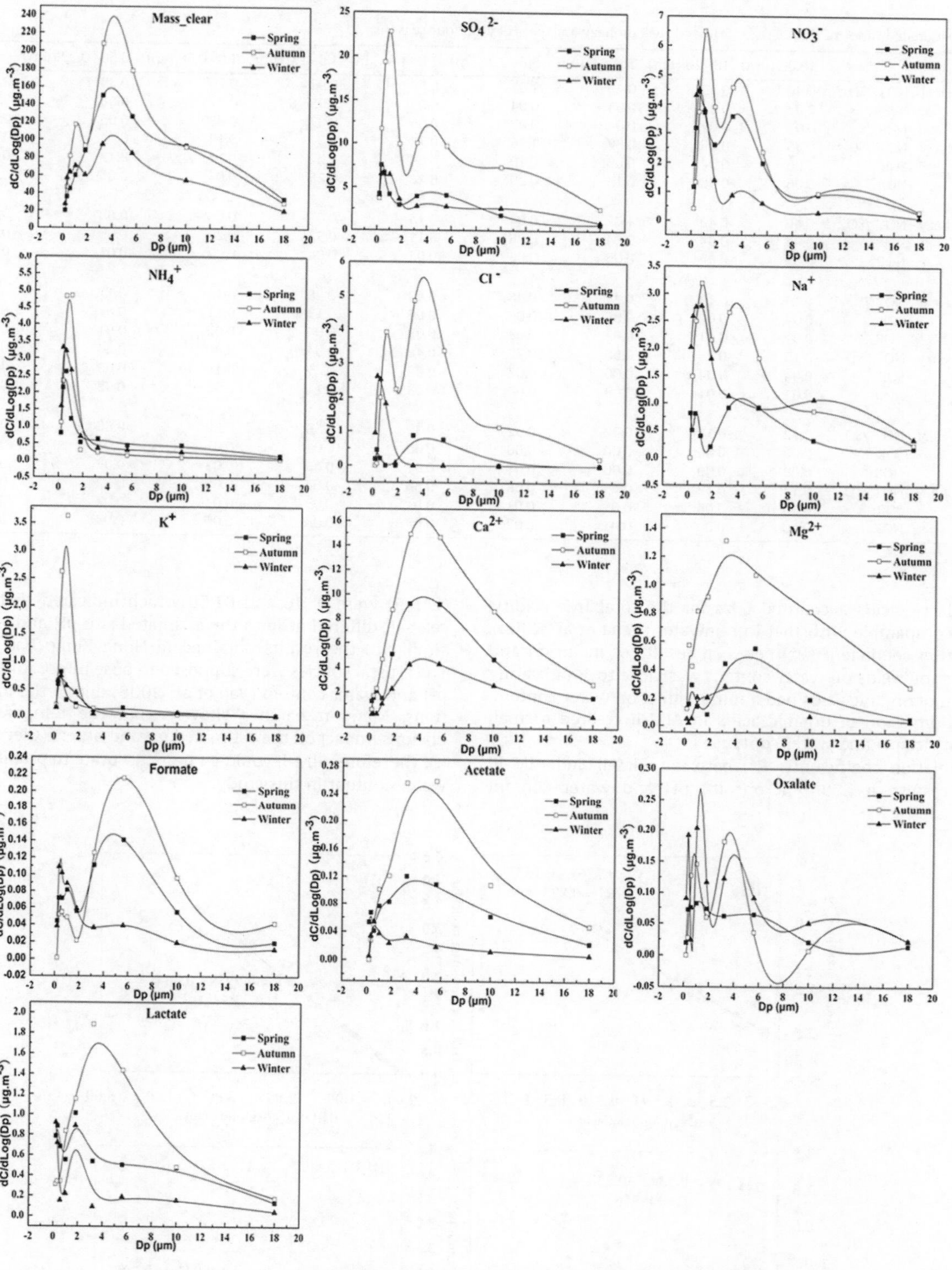

Fig. 4. The average size distributions of aerosol chemical compounds on clear days in four seasons.

3.4. AIM model results

3.4.1. Model evaluations

The influence of organic acids on the acidity of rainwater was estimated by Pena et al. (2002) who found that oxalic acid contributed 7.1% to the total free acidity and carboxylic acids accounted for 27.5%. Thus it was necessary to evaluate the influences of oxalic and lactic acids on the predictions of situ pH and water content. Fig. 5 depicts the impacts of inclusion or exclusion organic acids on situ pH and water content. The average situ pH including organic acids (−1.18 unit) was slightly lower than that without organic acids (−1.11 unit). In the estimated aerosol acidity,

Table 2
Average size-fractionated ratios of NO_3^-/SO_4^{2-}, SOR and NOR on hazy and clear days in four seasons.

			18.00 (μm)	18–10 (μm)	10–5.6 (μm)	5.6–3.2 (μm)	3.2–1.8 (μm)	1.8–1.0 (μm)	1.0–0.56 (μm)	0.56–0.32 (μm)	0.32–0.18 (μm)
April	clear	NO_3^-/SO_4^{2-}	0.40	0.55	0.49	1.02	0.88	0.93	0.70	0.38	0.30
		SOR	0.02	0.02	0.05	0.04	0.03	0.05	0.07	0.08	0.05
		NOR	0.02	0.02	0.04	0.07	0.05	0.07	0.08	0.05	0.02
	hazy	NO_3^-/SO_4^{2-}	0.35	0.46	0.87	1.16	0.74	0.37	0.14	0.14	0.08
		SOR	0.06	0.04	0.06	0.10	0.09	0.22	0.14	0.12	0.07
		NOR	0.06	0.06	0.15	0.29	0.18	0.25	0.07	0.06	0.02
August	hazy	NO_3^-/SO_4^{2-}	0.40	0.49	1.07	0.84	0.15	0.07	0.06	0.08	0.07
		SOR	0.04	0.04	0.04	0.06	0.11	0.32	0.20	0.09	0.04
		NOR	0.02	0.02	0.06	0.06	0.02	0.04	0.02	0.01	0.00
October	clear	NO_3^-/SO_4^{2-}	0.13	0.13	0.24	0.46	0.40	0.28	0.19	0.11	0.11
		SOR	0.04	0.04	0.06	0.06	0.06	0.13	0.11	0.07	0.02
		NOR	0.01	0.01	0.02	0.03	0.03	0.04	0.02	0.01	0.00
	hazy	NO_3^-/SO_4^{2-}	0.15	0.25	0.58	0.72	0.48	0.54	0.40	0.24	0.10
		SOR	0.04	0.04	0.06	0.07	0.07	0.21	0.14	0.08	0.03
		NOR	0.01	0.01	0.03	0.04	0.03	0.11	0.05	0.02	0.00
December &January	clear	NO_3^-/SO_4^{2-}	0.12	0.11	0.22	0.32	0.39	0.59	0.69	0.66	0.47
		SOR	0.00	0.00	0.00	0.00	0.00	0.01	0.01	0.01	0.01
		NOR	0.00	0.00	0.00	0.01	0.01	0.03	0.03	0.03	0.01
	hazy	NO_3^-/SO_4^{2-}	0.13	0.32	0.45	0.68	0.65	0.67	0.60	0.52	0.36
		SOR	0.00	0.00	0.01	0.01	0.01	0.03	0.03	0.02	0.00
		NOR	0.00	0.00	0.01	0.02	0.03	0.08	0.06	0.03	0.01

oxalic and lactic acids accounted 6.3% for the total free acidity, which was comparable with that in rainwater (Pena et al., 2002). The correlation coefficient (R^2) between results of inclusion and omission organic acids on water content was close to 1, indicating the inclusion of organic acids had a minor effect on water content. Overall, the omission of organic acids in AIM-II created a small deviation on situ pH and water content.

The correlation coefficients (R^2) between AIM-II and AIM-III simulation results at 298.15 K on situ pH and water content (Fig. 5) were 0.916 and 0.956 which indicated that exclusion of sea-salt differed little to the estimated situ pH and water content, similar to the results obtained in Hong Kong (Yao et al., 2006). The crustal species were supposed to pose minor influences on situ pH and water content (Yao et al., 2006) due to the low concentrations. Many researches showed that the temperature had an obvious impact on the estimated pH and others (Clegg et al., 1998a; b). Therefore, AIM-II could be more accurate to predict situ pH and water content in this study.

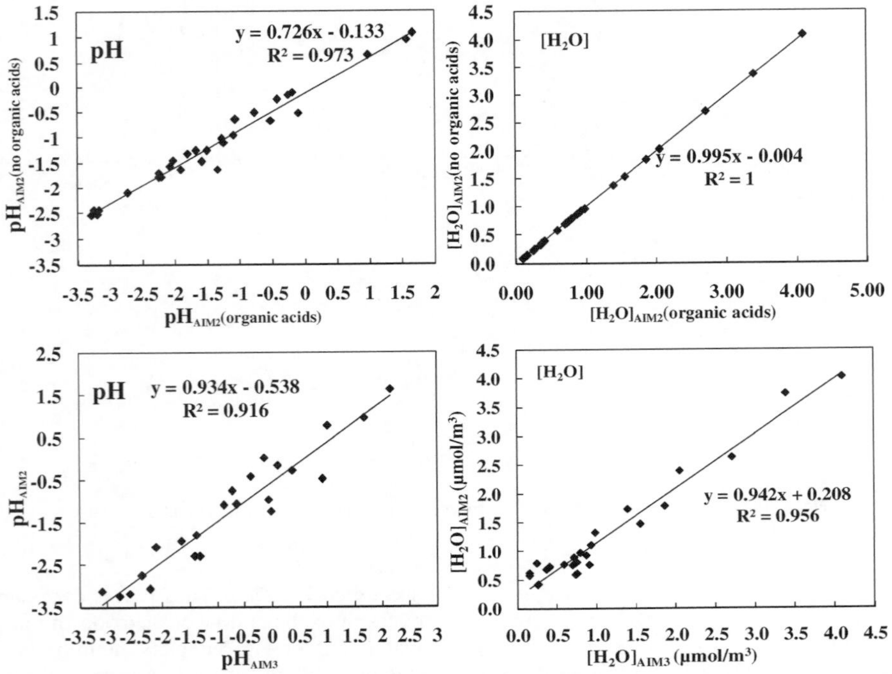

Fig. 5. Line correlations between organic acids and no organic acids of AIM2, AIM2 and AIM3 on situ pH and water content ([H_2O]). Missing values are due to [H^+] < 0, AIM model can't be used.

3.4.2. Situ pH and water content in $PM_{1.8}$

The slope and correlation coefficient (*R*) between the mole concentrations of $[NH_4^+]$ and $2[SO_4^{2-}]+[NO_3^-]$ were 1.03 and 0.92, indicating that SO_4^{2-} and NO_3^- were almost completely neutralized by NH_4^+ and most samples were ammonium-rich. The average $[H^+]_{total}$ in $PM_{1.8}$ in Jinan was 176.6 nmol m^{-3}, about 1.36–3.21 times higher than that in the ammonium-enrich samples of Lanzhou (65 nmol m^{-3}), Guangzhou (70 nmol m^{-3}), Hong Kong (75 nmol m^{-3}) and North US (45 nmol m^{-3}) in $PM_{2.5}$ (Speizer, 1989; Pathak et al., 2003, 2009). It was also found that $[H^+]_{total}$ on hazy days was about two times that on clear days. Model results revealed that most SO_4^{2-} in the aqueous system originally existed as bisulfate (HSO_4^-) in the atmospheric particles which was the predominant contributor to $[H^+]_{ins}$. The average $[H^+]_{ins}$ was 53.0 nmol m^{-3} which accounted for 30.0% of $[H^+]_{total}$. The $[H^+]_{ins}$ appeared higher in summer and autumn than in spring and winter, indicating that temperature and RH might influence its formation. The average ratio of $[H^+]_{ins}$ on hazy days to that on clear days was 4.67, 2.43 and 13.11 in spring, autumn and winter respectively, which might be due to the higher SOR and NOR on hazy days, demonstrating that the $[H^+]_{ins}$ was more easily to be formed on hazy days. The deviation of mole concentrations of $[HSO_4^-]$ from hazy to clear days was small, indicating that $[HSO_4^-]$ wasn't the major factor that caused the difference of situ pH on hazy and clear days. The average estimated situ pH (−1.11 unit) in 0.18–1.8 μm in Jinan was lower than that in Beijing (−0.52), Shanghai (−0.77), Lanzhou (−0.38), Guangzhou (0.61) and Hong Kong (0.25) in $PM_{2.5}$ (Pathak et al., 2004, 2009), indicating that the aerosol acidity was very strong in Jinan.

The double effect of water content on situ pH was due to the dissociation of HSO_4^- and the molarity of acidic solution in aqueous system. The average water content in $PM_{1.8}$ in Jinan was 3.11 μmol m^{-3} and water content had a good correlation with RH ($R = 0.83$), which indicated that RH mainly affected the water content inside particles. The mole ratios of $[NH_4^+]/[SO_4^{2-}]$ were often used as an indicator to evaluate the neutralization of acidic sulfate in the atmosphere and a relative indicator to describe the acidity of aerosols, therefore it might affect the situ pH and water content (Pathak et al., 2004). The total average mole ratio of $[NH_4^+]/[SO_4^{2-}]$ in Jinan was 1.63, indicating that most samples were ammonium rich. The average ratio of $[NH_4^+]/[SO_4^{2-}]$ in Jinan was similar with that in continental samples in Hong Kong and higher than that in marine samples (Pathak et al., 2004). The normalized water content (NWC) (the mole ratio of $[H_2O]_{AIM2}/[SO_4^{2-}]$), which showed the capacity of atmospheric aerosols to absorb water, was used to evaluate the influence of major water-soluble ions and RH on water content in the same study (Pathak et al., 2004). The values of NWC in Jinan on hazy days were higher than those on clear days in autumn and winter, indicating that hazy days were favorable for the uptake of water, while lower values of NWC appeared on hazy days in spring. The mole ratios of $[NH_4^+]/[SO_4^{2-}]$ were larger than 2.24 and NWC was zero in most samples in winter, which demonstrated that the aerosol was completely dry, so lower situ pH and water content were estimated due to the lower RH and the solid salt formations of $(NH_4)_3H(SO_4)_2$, NH_4NO_3 and $(NH_4)_2SO_4$. The results also showed that there was a clear transition point at NWC = 2 due to the solid formations associated with the mole ratios of $[NH_4^+]/[SO_4^{2-}]$: the mixtures of solid and liquid existed when NWC < 2. In this condition, if $[NH_4^+]/[SO_4^{2-}] < 2.0$, samples weren't completely neutralized and major water-soluble ions made a relatively important contribution to water content. If $[NH_4^+]/[SO_4^{2-}] > 2$, samples were completely neutralized and RH was a predominant factor to cause the lower water content. No solid existed at NWC > 2, hence it was difficult to evaluate which factor was more important.

3.4.3. Size distributions of situ pH and water content in aerosol

The seasonal variations of size-fractioned situ pH and water content are depicted in Fig. 6. The size distribution of situ pH showed unimodal characteristics in the fine mode peaked at 0.56 μm on clear

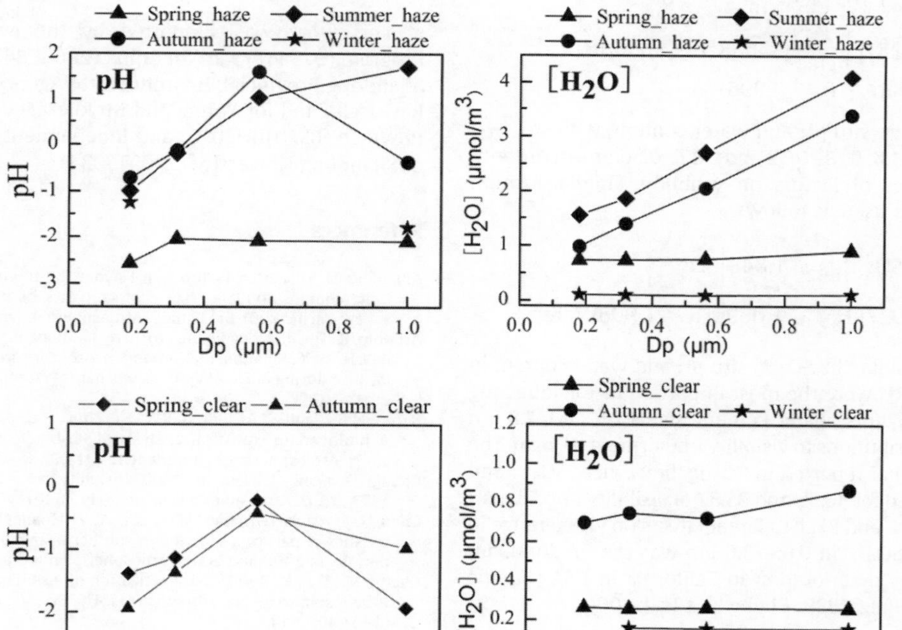

Fig. 6. Average size distributions of situ pH and water content ($[H_2O]$) on hazy and clear days in four seasons. Missing values are due to $[H^+] < 0$, AIM model can't be used.

days in spring and autumn. On haze days, it showed large seasonal variations. Similar characteristic was shown in autumn. Steady increasing trend was observed in summer and slight increased from 0.18 to 0.32 μm and then kept constant in spring. $[NH_4^+]/[SO_4^{2-}]$ had major peaks at 0.56 and 0.32 μm in autumn and winter and it kept increasing in spring and summer, similar to the size-fractionated situ pH, which confirmed that the size distributions of situ pH were related to the molar ratios of $[NH_4^+]/[SO_4^{2-}]$. Water content increased sharply with the diameter in summer and autumn, while remained almost constant in spring and winter. The lower values of NWC appeared in the diameter range of 0.32–1 μm and higher values in 0.18–0.32 μm and 1–1.8 μm indicated that the aerosol in 0.32–1 μm had a lower capacity to absorb water.

3.5. Relationship among chemical components, weather conditions and visibility

In order to further investigate the impact of SO_4^{2-}, NO_3^-, NH_4^+, situ pH and water content in 0.18–1.8 μm as well as meteorological variables on visibility, correlations analysis and regression models were used in this study.

Visibility had higher correlation coefficients with concentrations of SO_4^{2-} in 0.32–1.8 μm (−0.737 ~ −0.889), NH_4^+ in 0.56–1.8 μm (−0.834 ~ −0.934), and NO_3^- in 0.18–0.32 μm (−0.601), as well as situ pH in 1–1.8 μm (−0.834), water content in 0.18–1.8 μm (−0.939–0.963) and RH (−0.83) at a 95% confidence level. The significantly higher correlation coefficients of visibility with water content in 0.18–1.8 μm indicated that water content played an important role in reducing visibility.

All the aerosol major chemical compounds and meteorological parameters which had significant impacts on visibility were input to a stepwise multiple linear regression models to simulate the visibility. In this study, we developed a regression equation for urban visibility in Jinan as follows:

$$
\begin{aligned}
\text{Visibility} = & -5.119 - 3.103\left[SO_4^{2-}\right]_{0.32} + 1.706\left[SO_4^{2-}\right]_1 \\
& + 14.279\left[NO_3^-\right]_{0.18} + 0.363\text{RH} - 5.639\text{pH}_1 \\
& - 9.621\text{Water}_1 (R = 0.995)
\end{aligned}
$$

Here, pH and water were situ pH and water content, while $*_1$, $*_{0.18}$ and $*_{0.32}$ meant * in 1–1.8, 0.18–0.32 and 0.32–0.56 μm. In order to evaluate the influences of factors on visibility, standardization regression formula was used as follows:

$$
\begin{aligned}
\text{Visibility} = & -0.655\left[SO_4^{2-}\right]_{0.32} + 1.848\left[SO_4^{2-}\right]_1 \\
& + 0.297\left[NO_3^-\right]_{0.18} + 0.927\text{RH} - 1.069\text{pH}_1 - 1.908\text{Water}_1
\end{aligned}
$$

According to the results, the SO_4^{2-}, situ pH and water content in 1.0–1.8 μm as well as RH were the most important factors affecting visibility. Among the factors, water content and SO_4^{2-} 1.0–1.8 μm made comparable contributions to visibility impairment in Jinan. The result was similar to what reported in Guangzhou where SO_4^{2-} and water content accounted for 36.5% and 25.8% of visibility impairment (Jung et al., 2009). SO_4^{2-} and RH had larger effects on visual range in central Taiwan, while SO_4^{2-} in 0.65–3.6 μm was the predominant factors for the visibility deterioration in California in USA (Barone et al., 1978; Tsai and Cheng, 1999; Cheng and Tsai, 2000).

4. Summary

The concentrations of TSP, PM_{10} and $PM_{1.8}$ on hazy days were much higher than those on clear days. They were also 1.05–1.66 and 1.47–2.76 times China NAAQS for TSP and PM_{10}, 2.53–5.51 times US NAAQS for $PM_{2.5}$. Fine particles accounted for a higher portion of aerosol on hazy days than those on clear days, and secondary aerosol were favored to be formed on hazy days. Size distributions of mass, SO_4^{2-}, NO_3^-, formate and acetate were all bimodal while fine mode was dominant on hazy days. From clear days to hazy days, size distributions of K^+, NH_4^+, Mg^{2+}, Ca^{2+}, oxalate and lactate changed little, which were all unimodal.

AIM model was used to estimate situ pH and water content in particle with diameter ranged from 0.18 to 1.8 μm. Omission organic acids and sea-salt in AIM-II exerted a minor influence on estimation of situ pH and water content while temperature made a relatively important contribution to the model. Hence, AIM-II without organic acids and sea-salt was more suitable in this study. The estimated situ pH was mainly related to RH and mole ratios of $[NH_4^+]/[SO_4^{2-}]$, while $[HSO_4^-]$ differed little on hazy and clear days. NWC was mainly influenced by major water-soluble ions and RH. A clear transition point at NWC− = 2 was observed in Jinan. Seasonal variations of size-fractioned situ pH were apparent. The distribution of situ pH was unimodal with a peak at 0.56 μm appeared on clear days in spring and autumn and on hazy days in autumn, while increasing trend could be observed on hazy days in spring and summer. Size distribution of water content on hazy days showed sharp increasing trends in summer and autumn, while little variation in spring and winter as well as clear days in spring, autumn and winter.

Correlation analysis between visibility and aerosol chemical components together with meteorological parameters including wind speed, RH and temperature showed that situ pH in 1–1.8 μm, RH, as well as concentrations of SO_4^{2-} in 0.32–1.8 μm, NH_4^+ in 0.56–1.8 μm, NO_3^- in 0.18–0.32 μm and water content in 0.18–1.8 μm had higher correlations with visibility in Jinan. Stepwise multiple linear regression models indicated that SO_4^{2-} and water content with the diameter in 1–1.8 μm were the most important contributors to reduce visibility.

Acknowledgment

This work was supported by the National Basic Research Program (973 Program) of China (2005CB422203), a key project of Shandong Provincial Environmental Agency (2006045), Promotive Research Fund for Young and Middle-aged Scientists of Shandong Province (BS2010HZ010) and Independent Innovation Foundation of Shandong University (2009TS024).

References

Anlauf, K., Li, S., Leaitch, R., Brook, J., Hayden, K., Toom-Sauntry, D., Wiebe, A., 2006. Ionic composition and size characteristics of particles in the Lower Fraser valley: Pacific 2001 field study. Atmospheric Environment 40 (15), 2662–2675.

Arimoto, R., Duce, R., Savoie, D., Prospero, J., Talbot, R., Cullen, J., Tomza, U., Lewis, N., Ray, B., 1996. Relationships among aerosol constituents from Asia and the North Pacific during PEM-West A. Journal of Geophysical Research 101 (D1), 2011–2023.

Barone, J.B., Cahill, T.A., Eldred, R.A., Flocchini, R.G., Shadoan, D.J., Dietz, T.M., 1978. A multivariate statistical analysis of visibility degradation at four California cities. Atmospheric Environment 12 (11), 2213–2221 (1967).

Chang, D., Song, Y., Liu, B., 2009. Visibility trends in six megacities in China 1973–2007. Atmospheric Research 94 (2), 161–167.

Chen, S.-J., Hsieh, L.-T., Kao, M.-J., Lin, W.-Y., Huang, K.-L., Lin, C.-C., 2004. Characteristics of particles sampled in southern Taiwan during the Asian dust storm periods in 2000 and 2001. Atmospheric Environment 38 (35), 5925–5934.

Cheng, M., Tsai, Y., 2000. Characterization of visibility and atmospheric aerosols in urban, suburban, and remote areas. The Science of the Total Environment 263 (1–3), 101–114.

Cheng, S., Yang, L., Zhou, X., Wang, Z., Zhou, Y., Gao, X., Nie, W., Wang, X., Xu, P., Wang, W., 2011. Evaluating $PM_{2.5}$ ionic components and source apportionment in Jinan, China from 2004 to 2008 using trajectory statistical methods. Journal of Environmental Monitoring 13 (6), 1662–1671.

Clegg, S.L., Brimblecombe, P., Wexler, A.S., 1998a. Thermodynamic model of the system $H^+–NH_4^+–SO_4^{2-}–NO_3^-–H_2O$ at tropospheric temperatures. Journal of Physical Chemistry A 102 (12), 2137–2154.

Clegg, S.L., Brimblecombe, P., Wexler, A.S., 1998b. Thermodynamic model of the system $H^+-NH_4^+-Na^+-SO_4^{2-}-NO_3^--Cl^--H_2O$ at 298.15 K. Journal of Physical Chemistry A 102 (12), 2155–2171.
Fridlind, A., Jacobson, M., 2000. A study of gas-aerosol equilibrium and aerosol pH in the remote marine boundary layer during the First Aerosol Characterization Experiment(ACE 1). Journal of Geophysical Research 105 (17), 325. -C317.
Jung, J., Lee, H., Kim, Y.J., Liu, X., Zhang, Y., Gu, J., Fan, S., 2009. Aerosol chemistry and the effect of aerosol water content on visibility impairment and radiative forcing in Guangzhou during the 2006 Pearl River Delta campaign. Journal of Environmental Management 90 (11), 3231–3244.
Kawamura, K., Ikushima, K., 1993. Seasonal changes in the distribution of dicarboxylic acids in the urban atmosphere. Environmental Science & Technology 27 (10), 2227–2235.
Kerminen, V.M., Hillamo, R., Teinil, K., Pakkanen, T., Allegrini, I., Sparapani, R., 2001. Ion balances of size-resolved tropospheric aerosol samples: implications for the acidity and atmospheric processing of aerosols. Atmospheric Environment 35 (31), 5255–5265.
Khwaja, H.A., 1995. Atmospheric concentrations of carboxylic acids and related compounds at a semiurban site. Atmospheric Environment 29 (1), 127–139.
Ludwig, J., Klemm, O., 1990. Acidity of size-fractionated aerosol particles. Water, Air, & Soil Pollution 49 (1), 35–50.
Malm, W.C., Day, D.E., 2001. Estimates of aerosol species scattering characteristics as a function of relative humidity. Atmospheric Environment 35 (16), 2845–2860.
National Bureau of Statistics of China, 2008. China Statistical Yearbook. China Statistics Press, Beijing.
Pathak, R.K., Yao, X., Lau, A.K.H., Chan, C.K., 2003. Acidity and concentrations of ionic species of $PM_{2.5}$ in Hong Kong. Atmospheric Environment 37 (8), 1113–1124.
Pathak, R.K., Louie, P.K.K., Chan, C.K., 2004. Characteristics of aerosol acidity in Hong Kong. Atmospheric Environment 38 (19), 2965–2974.
Pathak, R.K., Wu, W.S., Wang, T., 2009. Summertime $PM_{2.5}$ ionic species in four major cities of China: nitrate formation in an ammonia-deficient atmosphere. Atmospheric Chemistry and Physics 9 (5), 1711–1722.
Pena, R.M., Garcia, S., Herrero, C., Losada, M., Vazquez, A., Lucas, T., 2002. Organic acids and aldehydes in rainwater in a northwest region of Spain. Atmospheric Environment 36 (34), 5277–5288.
Pszenny, A.A.P., Moldanová, J., Keene, W.C., Sander, R., Maben, J.R., Martinez, M., Crutzen, P.J., Perner, D., Prinn, R.G., 2004. Halogen cycling and aerosol pH in the Hawaiian marine boundary layer. Atmospheric Chemistry and Physics 4 (1), 147–168.
Souza, S.R., Vasconcellos, P.C., Carvalho, L.R.F., 1999. Low molecular weight carboxylic acids in an urban atmosphere: winter measurements in São Paulo City, Brazil. Atmospheric Environment 33 (16), 2563–2574.
Speizer, F.E., 1989. Studies of acid aerosols in six cities and in a new multi-city investigation: design issues. Environmental Health Perspectives 79, 61.
Sun, Y., Zhuang, G., Tang, A., Wang, Y., An, Z., 2006. Chemical characteristics of $PM_{2.5}$ and PM_{10} in haze–fog episodes in Beijing. Environmental Science and Technology 40 (10), 3148–3155.
Tan, J.-H., Duan, J.-C., Chen, D.-H., Wang, X.-H., Guo, S.-J., Bi, X.-H., Sheng, G.-Y., He, K.-B., Fu, J.-M., 2009. Chemical characteristics of haze during summer and winter in Guangzhou. Atmospheric Research 94 (2), 238–245.
Tsai, Y., Cheng, M., 1999. Visibility and aerosol chemical compositions near the coastal area in Central Taiwan. The Science of the Total Environment 231 (1), 37–51.
Wang, G., Niu, S., Liu, C., Wang, L., 2002. Identification of dicarboxylic acids and aldehydes of PM_{10} and $PM_{2.5}$ aerosols in Nanjing, China. Atmospheric Environment 36 (12), 1941–1950.
Wang, G., Wang, H., Yu, Y., Gao, S., Feng, J., Wang, L., 2003. Chemical characterization of water-soluble components of PM_{10} and $PM_{2.5}$ atmospheric aerosols in five locations of Nanjing, China. Atmospheric Environment 37 (21), 2893–2902.
Wang, Y., Zhuang, G., Chen, S., An, Z., Zheng, A., 2007. Characteristics and sources of formic, acetic and oxalic acids in $PM_{2.5}$ and PM_{10} aerosols in Beijing, China. Atmospheric Research 84 (2), 169–181.
Wang, Y., Sheng, C., Yang, X., Gao, H., Zhang, H., 2009. Spatial-temporal variations of hazes in Shandong Province and its relationship with climate elements. Advances in Climatic Change Research 5 (1), 24–28.
Yang, L.-x., Wang, D.-c., Cheng, S.-h., Wang, Z., Zhou, Y., Zhou, X.-h., Wang, W.-x., 2007. Influence of meteorological conditions and particulate matter on visual range impairment in Jinan, China. Science of the Total Environment 383 (1–3), 164–173.
Yao, X., Yan Ling, T., Fang, M., Chan, C.K., 2006. Comparison of thermodynamic predictions for in situ pH in $PM_{2.5}$. Atmospheric Environment 40 (16), 2835–2844.
Yao, X., Ling, T.Y., Fang, M., Chan, C.K., 2007. Size dependence of in situ pH in submicron atmospheric particles in Hong Kong. Atmospheric Environment 41 (2), 382–393.
Zhang, Q., Streets, D., Carmichael, G., He, K., Huo, H., Kannari, A., Klimont, Z., Park, I., Reddy, S., Fu, J., 2009. Asian emissions in 2006 for the NASA INTEX-B mission. Atmospheric Chemistry and Physics 9 (14), 5131–5153.

中国大气降水化学研究进展

王文兴* 许鹏举
(山东大学环境研究院 济南 250100)

摘 要 中国的酸雨区是继欧洲和北美之后的世界三大酸雨区之一。本文综述了近十年来有关中国降水化学的研究结果,主要内容和结论有:中国南方和西南地区已经成为世界上降水酸性最强的地区;中国降水化学组成仍属硫酸型,但正在向硫酸硝酸混合型转变;SO_4^{2-} 和 NO_3^- 以及 NH_4^+ 和 Ca^{2+} 分别是中国降水中主要阴阳离子,并且浓度远高于欧洲和北美;山上降水和高空云雾水的高酸性和高离子浓度说明这些地区高空已严重污染;讨论了中国各地区降水化学特征和酸雨形成的原因;最后指出了有待研究的科学问题。

关键词 降水化学 中国 酸雨 排放
中图分类号: X517 **文献标识码**: A **文章编号**: 1005-281X(2009)02/3-0266-16

Research Progress in Precipitation Chemistry in China

*Wang Wenxing** *Xu Pengju*
(Environment Research Institute, Shandong University, Jinan 250100, China)

Abstract The acid rain area in China is one of the three main acid rain areas in the world, next to Europe and North America. Most acid deposition investigations have been concerned with precipitation chemistry in China. The results for precipitation chemistry of China obtained by Chinese and foreign researchers are overviewed in this paper. Major conclusions are: High H^+ concentrations of precipitation have occurred in South and South Western China which have become areas with the highest acidity of precipitation in the world; The pollution of precipitation in China is still sulfur type with a trend to sulfuric-nitrous mixed type; Concentrations of dominant anions (SO_4^{2-} and NO_3^-) and cations (NH_4^+ and Ca^{2+}) in precipitation are much higher than that in Europe and North America; The high acidity and ionic concentration in precipitation at alpine sites and cloud—water in eastern and southern China indicates that upper air has been polluted seriously; Characteristics of precipitation chemistry in different areas of China and the origins of acid rain formation have been discussed; Finally, future research is suggested.

Key words precipitation chemistry; China; acid rain; emissions

Contents

1 Introduction
2 Geographical distributions and trends of precipitation acidity in china
2.1 Precipitation acidity
2.2 Trends of precipitation acidity
3 Ground precipitation chemistry
3.1 Precipitation acidity and water-soluble ion concentrations
3.2 Organic components in precipitation
3.3 Heavy metals in precipitation
3.4 Ground fogs
4 Chemistry of high altitude precipitation and cloud water
4.1 Precipitation chemistry at mountains

收稿: 2009 年 1 月
*通讯联系人 e-mail: wxwang99@hotmail.com

4.2 Chemical characteristics of precipitation in regional representative sites
4.3 High-altitude water in clouds and fogs
5 Formation mechanisms of acid rain
5.1 Emission trends of acidic precipitation related pollutants in China
5.2 Mechanisms of chemical transformations
5.3 Interactions between precipitation and aerosol and other pollutants
6 Future research on precipitation chemistry in China

1 引言

自然大气是多种气体与悬浮颗粒的混合物，其气体成分主要包括约 78.1%的 N_2、20.9%的 O_2、少量的 Ar、Ne、Kr、Xe 等几种惰性气体和水气、CO_2、O_3、CH_4、CO、NH_3、SO_2 及NO_x 等[1]。工业革命前，人类活动还不至于对大气组分造成明显的影响，工业革命后，随着人口和经济的快速增长，以化石燃料燃烧为主的人类活动向大气排放了大量气态和固态污染物，改变了大气组分的构成，从而产生了一系列的环境效应，其中人类活动对降水化学成分的影响是人们关注的热点之一。人类活动对降水化学成分的影响主要表现在离子浓度的升高，特别是降水酸性的增加，形成酸性降水。如果降水 pH< 5.6，就是通常所说的"酸雨"，包括酸性雨、雪、冰雹、露、雾、霜等多种形式。由于酸性降水对人类及生态系统产生了一系列重大影响，大气降水化学研究一直是大气化学的重要组成部分。

大气降水化学研究可以追溯到 1840s，英国化学家 Robert Angus Smith 在英格兰调查了酸沉降现象，并在 1872 年出版的 Air and Rain: the Beginnings of a Chemical Climatology 一书中叙述了世界工业发展先驱城市曼彻斯特郊区降水中含有高浓度 SO_4^{2-}，形成了"acid rain"，这就是"酸雨"这个词的由来[2]。在此后几十年间降水酸化问题并未引起人们的重视，直到 1950s，欧洲和北美先后由于长期酸性降水带来的酸性物质沉积，大面积内陆水域鱼类灭绝和森林衰亡，降水酸化问题才引起科学工作者们的重视[3]。1970s 欧洲和美国先后建立了大气酸沉降监测网，连续观测大气干、湿沉降（EMEP：www.emep.int，NADP：http://nadp.sws.uiuc.edu）。1990s 东亚也组建了东亚酸沉降监测网（EANET：www.eanet.cc）。我国大规模酸雨监测和研究始于 1970s 末，当时发现长江以南部分地区出现了酸雨[4]。为了查明中国酸雨污染的状况，国家环保部门于 1982 年建立了全国酸雨监测网[5]，中国气象局也于 1989 年建立了气象部门的全国酸雨监测网[6]，两大酸雨监测网为中国降水化学研究积累了大量数据，对我国酸雨控制和研究起了重要作用。

广义的大气降水化学研究领域包括酸雨前体物排放、污染物质传输、气相和多相化学过程、降水离子组成及其时空分布、降水化学对生态环境影响、降水污染控制对策等。由于降水化学组成特别是酸度对生态环境影响最为重要，降水酸度及化学组成研究是降水化学研究的重点。本文将主要综述中国近年来在降水酸化形势、化学组成特征、时空分布及酸化机理等领域的研究进展及存在的科学问题。

2 中国降水酸度的地理分布及变化趋势

2.1 降水酸度研究

中国气象局酸雨监测网 2006 年降水酸度的地理分布绘于图 1，由图可见，我国酸雨区目前主要分布于东北地区东南部、华北大部、西南和华南沿海地区及新疆北部地区，大体呈东北—西南走向。在欧、美、亚世界三大酸雨区中，中国的强酸雨区（pH< 4.5）面积最大，中国长江以南地区是全球强酸雨中心。

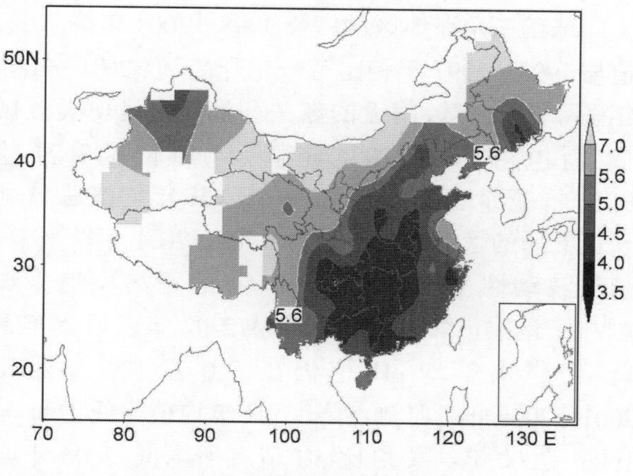

图 1　中国降水 pH 年均值等值线，2006 *

Fig. 1　Isopleths of annual mean pH of precipitation in China, 2006

与过去几年相比，2006 年全国区域酸雨特点如下：(1)南方酸雨区范围无明显变化，北方酸雨区继续扩展。北方 1994 年后出现的几个小块酸雨区呈现连成大片的趋势。(2)强酸雨区范围为 1994 年以

* 中国气象局酸雨监测网数据，由汤洁提供。

来最大，但酸度有所减弱。2006 年，我国强酸雨区范围为 1994 年以来最大，并呈现明显的向北扩展的趋势，强酸雨区降水酸度整体上有所减弱。但局部地区降水酸度出现增强的情况。总体来看，与过去几年相比，2006 年酸雨形势有所恶化，地理分布也产生一些变化，但也可能是气象因素所致。

由于我国降水监测点较少，又多受城市及工业区的影响，并且采取超低密度制图，所以图 1 给出的我国降水酸度分布只能反映我国降水酸度大体的形势，不能准确地反映局地降水酸度的细节。为了取得更准确的降水酸度的地理分布，必须按网格增加降水观测点。

2.2　降水酸度变化趋势研究

2000 年以来京津地区降水 pH 值呈较快下降的趋势，大气颗粒物质量浓度的持续下降可能是重要原因之一，机动车尾气排放量的急剧增加也可能是降水酸性增强的另一原因[7—9]。天津市区 1992—2004 年的降水 pH 值变化趋势显示，自 1992—2004 年，pH 值有逐年升高的趋势[10]，表明京津地区各区域趋势不尽相同。河南郑州和商丘至 2005 年降水 pH 值持续下降[11,12]。吉林长春 2005 年后降水酸化有加重的趋势，而同为吉林中部的四平降水酸度有所减小，两个城市的数据表明吉林中部基本上仍是非酸雨区[13]。

珠江三角洲地区的广州 1998—2004 年降水 pH 值与 1992—1997 年相比有一定升高，但 2001—2004 年呈降低的趋势，附近的韶关和汕头也呈现类似规律，而电白 2000—2004 年降水 pH 值呈升高趋势[14,15]，深圳地区 1986—2006 年 pH 年均值总体呈一直下降的趋势[16,17]。上海 1983—2004 年以来 pH 值呈波动性变化，1997—2004 年酸雨污染趋势加剧[18]。南宁市降水酸化程度从 2002 年以后逐年下降，但降水年均 pH 值仍在 5.0 左右[19]。四川 2001—2006 年各监测点降水 pH 年均值均低于 5，与 1996—2000 年结果相比 pH 值有所降低，2006 年四川重酸雨区比 2005 年增加了 4 个城市[20]。1997—2006 年，重庆各观测点酸雨变化趋势不尽相同，但整体上 pH 值缓慢下降[21]。湖南长沙 1996—2005 年 pH 平均值为 4.21，是全国降水酸度最强的地区之一，2000 年后也呈降水酸度升高的趋势[22]。

地处西北碱性大气环境的新疆乌鲁木齐市降水 pH 值自 2000 年后有下降趋势，2006 年 pH 年均值为 5.29[23]。新疆伊犁 2000 年后降水 pH 年均值明显下降，至 2004 年已经低至 4.45[24]。章典等[25]的研究表明，青藏高原降雨常呈弱碱性，1987—1988 年平均 pH 值为 8.36，而 1997—1999 年为 7.5。在格尔木和西宁的研究也表明，1990s 之前降水 pH 值在 8.0 附近，之后 pH 值虽有明显降低，但仍以碱性为主[26]。以上研究表明青藏高原存在着大面积碱性降水。

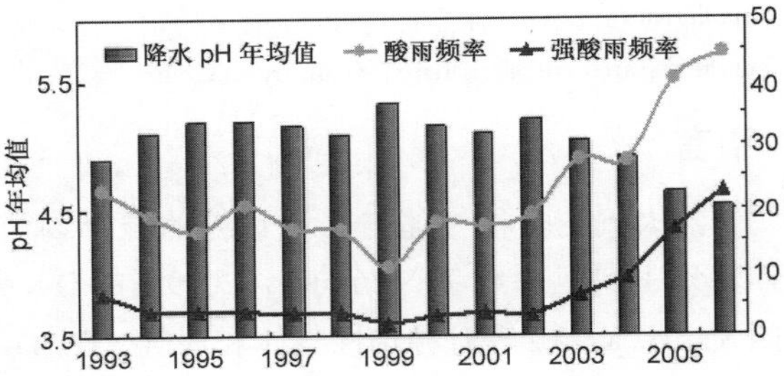

图 2　1993—2006 年中国降水 pH 年均值、酸雨频率年际变化 *

Fig. 2　Trends of annual mean pH of precipitation, acid rain frequency and severe acid rain frequency in China (1993—2006)

中国气象局酸雨监测网数据表明，十几年来我国酸雨发展状况大致经历两个阶段：1980s 到 1990s 中期为我国酸雨急剧发展期，1990s 中后期到 21 世纪初总的趋势是进入相对稳定期[6]。但 2002 年后，无论是降水酸度、酸雨频率还是强酸雨频率，均呈上升趋势，至 2006 年酸雨形势最为严重(见图 2)。

总之，虽然近年来我国各地区降水 pH 值变化趋势不尽相同，但总体上我国降水酸度趋势如图 2 所示以酸度升高为主。上述降水酸度趋势分析结果可以代表全国总体降水历史发展趋势，但不能反映局地变化。今后，全国降水和局地降水酸度的变化方向将决定于 SO_2 和 NO_x 酸性气体以及颗粒物和 NH_3 的排放量的相对削减和增长。

3　地面降水化学研究

地面降水化学组成数据是酸沉降研究和控制的基础，历来受到世界各国政府和研究者的重视。中国环境保护部酸雨监测网的站点、中国气象局监测网的少数站点和 EANET 在中国的站点进行了长期的降水化学组成分析，此外，还有很多研究者做了大量研究。

3.1　降水酸度及水溶性离子浓度研究

*　中国气象局酸雨监测网数据，由汤洁提供。

中国北方地区普遍存在大气污染严重、降水中 SO_4^{2-}、NO_3^- 浓度高的情况，但由于 Ca^{2+}、NH_4^+ 等碱性离子浓度比较高，所以降水酸度并不高。2004 年北京市区降水 pH 年均值为 5.52，SO_4^{2-} 和 NO_3^- 浓度均值分别高达 521μeq·L^{-1} 和 82.8μeq·L^{-1}，明显高于其它许多城市。SO_4^{2-}、NO_3^- 和 NH_4^+ 浓度在近十年中总体升高，SO_4^{2-}/NO_3^- 当量比有缓慢降低的趋势。由于北京地区总悬浮颗粒物(TSP)的有效控制，钙沉降量近十年来明显下降，这也是北京降水 pH 值下降的原因之一[8,27]。而北京北部上甸子本底站 2004 年降水 pH 年均值为 4.41，低于当年北京市区所测结果（5.52），上甸子降水 SO_4^{2-}（161.5 μeq·L^{-1}）、NO_3^-（72.2 μeq·L^{-1}）浓度也低于北京市区[9]。河南降水仍为典型的硫酸型，NH_4^+ 和 Ca^{2+} 浓度之和占阳离子总量的 65%[28]。王赞红等[29] 观测到河北石家庄市雪水 pH 平均值为 6.46，阴离子以 SO_4^{2-} 为主，NO_3^- 比较少，阳离子以 NH_4^+、Ca^{2+} 为主，降雪中不溶物形态表明颗粒主要来自燃料燃烧和裸地起尘。

作为干旱半干旱地区，中国西北及青藏高原地区空气中存在大量碱性浮尘，降水无人为污染状况下应该呈碱性。Xu 等[30] 研究了新疆乌鲁木齐 2001—2005 年降水 pH 值和化学成分，期间 pH 平均值为 6.86，降水中 SO_4^{2-} 浓度高达 414 μeq·L^{-1}，而 NO_3^- 浓度仅为 24μeq·L^{-1}，属于典型的硫酸型降水。章典等[25] 对青藏高原多个站点降水做了长期研究，1987—1988 年降水 pH 值为 8.36，1997—1999 年降到 7.5，但仍然呈弱碱性，属于典型的碱性降水。青藏高原作为北半球空气污染最轻的地区之一，虽然降水中 SO_4^{2-}、NO_3^- 浓度仍然很低，但 1980s 以后随人类活动增加，SO_4^{2-}、NO_3^- 浓度升高趋势明显，pH 值明显降低。拉萨东北 220km 处那木错地区 2005 年 8 月至 2006 年 8 月降水平均 pH 值为 6.38。该研究点降水更多地受附近那木错盐湖及干盐湖的影响，由于周边牧民习惯用牲畜粪便作燃料及牲畜粪便散发 NH_3 等影响，降水中 NO_3^- 与 NH_4^+ 浓度高于拉萨，这可能也是拉萨地区降水中 NO_3^- 当量浓度高于 SO_4^{2-}、出现我国少有的硝酸型降水的原因之一[31]。青海格尔木和西宁两地 1990—2000 年间降水与 1988 年相比，大部分离子浓度明显升高，特别是 NO_3^-、SO_4^{2-} 和 NH_4^+ 等与人为污染密切相关的离子[26]。甘肃嘉峪关市 2002 年降水 pH 值范围为 8.35—9.24，2003 年降水 pH 值范围为 7.7—8.72，2004 年降水 pH 值范围为 8.03—8.54，均呈碱性[32]。

2005 年，庄国顺等[33] 在上海所做的研究发现降水 pH 年均值只有 4.49，酸度比 1997 年增加了 15 倍，最低 pH 值只有 2.95，SO_4^{2-}、NO_3^- 浓度分别为 199.59 μeq·L^{-1}、49.80μeq·L^{-1}，对降水化学成分的进一步分析说明上海已经是世界上大气污染比较严重的城市之一。1990—2005 年间上海降水 SO_4^{2-}/NO_3^- 当量比逐年下降，仍属于混合型污染，上海降水中污染物主要来源于本地和中长尺度传输[18]。Tu 等[34] 研究了南京 1992—2003 年降水化学变化趋势，pH 值自 2000 年来下降趋势明显，2003 年低于 5，1992—2003 年间 SO_4^{2-} 和 NO_3^- 浓度分别为 241.78μeq·L^{-1}、39.55μeq·L^{-1}，南京降水中 SO_4^{2-} 浓度呈下降趋势，而 NO_3^- 浓度呈上升趋势，2003 年 SO_4^{2-}/NO_3^- 当量比降至 6 左右，可能是工业粉尘减排和城市裸露土壤减少导致降水中 Ca^{2+} 浓度逐年降低。2002—2003 年太湖地区 3 个研究点中有 2 个 pH 年均值低于 5，3 个研究点 SO_4^{2-} 和 NO_3^- 平均浓度分别为 162.36μeq·L^{-1}、41.90μeq·L^{-1}，并且 NO_x 影响有逐年增加的趋势[35]。在 Zhang 等[36] 2004 年的研究中，浙江金华降水 pH 年均值只有 4.54，SO_4^{2-} 和 NO_3^- 浓度分别为 116.91μeq·L^{-1} 和 36.98μeq·L^{-1}，并认为人为源分别占 SO_4^{2-} 和 NO_3^- 的 98.8% 和 99.9%。2002 年浙江舟山群岛酸雨频率为 88%，2003 年为 86%，SO_4^{2-} 和 NO_3^- 浓度分别为 178.3μeq·L^{-1} 和 28.20μeq·L^{-1}，SO_4^{2-}/NO_3^- 当量比约为 6，高于浙江内地，Cl^- 浓度也比较高，海盐影响明显[37]。Zhang 等[38] 于 2000—2002 年研究了黄海千里岩岛和东海嵊泗群岛降水的 pH 值和部分离子成分，两地降水 pH 年均值分别为 4.81 和 4.50，并具有明显的海洋源特征，湿沉降比干沉降能带入更多 NH_4^+、NO_3^-、NO_2^-、PO_4^{3-} 和 SiO_2^- 等营养物质。

2003—2004 年，刘君峰等[39] 在广州的研究表明降雨 pH 年均值为 4.37，酸雨频率高达 85%，期间 SO_4^{2-}/NO_3^- 当量比已经由 1985—1990 年间的 8.62 降低为 2003—2004 年间的 3.04，同时，降水中 SO_4^{2-} 和 Ca^{2+} 浓度也呈下降趋势，而 NO_3^- 和 NH_4^+ 浓度呈上升趋势。结合深圳的研究成果，说明珠三角地区的大气环境表现出由燃煤型污染向燃油型污染转变的趋势，同时进一步分析认为严重的降水酸化现象是局地污染与中、长距离输送叠加的结果[14,16,17,40]。赵卫红[41] 研究了福建 23 个城市 2004 年的降水数据，全省降水 pH 年均值＜5.6 的城市比例呈逐年上升趋势，酸雨区主要分布在闽东南沿海，酸雨类型主要为硫酸型。Wai 等[42] 研究了香港降雨中离子浓度与雨滴粒径的关系及机理，发现不同场次降雨的不同水溶性组分分布趋势相似。

四川各监测点 2001—2005 年降水中，SO_4^{2-} 是主要阴离子，Ca^{2+} 和 NH_4^+ 是主要阳离子，四川降水仍然是硫酸型[20]，而和我国其它大部分大城市类似，成都市区降水有从硫酸型迅速向混合型转变的趋势[43]。贵州 2002 年降水 pH 年均值为 5.38，个别城市低至 4.31，主要是硫酸型降水污染，且冬季酸雨出现频率比较高[44]。段浩等[45] 研究了 2003—2004 年四川南充市市区、近郊和远郊的降水，距离市区越近，酸雨频率、降雨酸度、SO_4^{2-} 浓度、SO_4^{2-}/NO_3^- 当量比越大，Ca^{2+}/SO_4^{2-} 当量比越小。湖南长沙 2005 年降水 pH 值为 4.02，酸雨频率为 98.0%，1996—2005 年间 SO_4^{2-}/NO_3^- 当量比逐年降低，2005 年为 3.93[22]。

由以上诸多研究者所得结果可以看出，我国北方地区特别是环渤海诸省市，虽然大气 SO_2、NO_x 浓度高，降水中 SO_4^{2-}、NO_3^- 浓度也很高，但因被 Ca^{2+} 和 NH_4^+ 中和而降水酸性不强。相反的是长江以南广大地区和四川盆地降水中 SO_4^{2-}、NO_3^- 虽系中等浓度，但降水酸性最强，形成了我国的主要酸雨区。青藏高原和西北广大地区酸性气体释放强度很小，气候干旱，大气碱性颗粒物浓度高，所以降水呈碱性或接近中性。我国城市及其近郊多受高架排放源影响，降水一般较乡村略酸。王文兴等[46,47] 对于中国降水化学出现的这些独特现象，综合考虑酸性气体排放、大气降水量、土壤酸碱性等因素进行了系统的研究，阐述了中国不同区域酸雨的形成因素。

3.2　降水有机组分研究

世界各地的降水中都存在有机酸，虽然通常认为降水酸度主要来自硫酸和硝酸等强酸，但是多年来的实测结果表明，有机弱酸也对降水酸度有贡献，有时在某些僻远地区，有机酸可能成为降水的主要致酸成分[48]。除了有机酸之外，降水中通常还含有其它有机物，对于这些有机物的研究也有助于降水酸度形成机制及其它大气污染相关机理的研究。

胡敏等[49] 分析了 2003 年 5—11 月在北京市区采集的降水样品，其中，甲酸（4.62 μeq·L^{-1}）、乙酸（4.60μeq·L^{-1}）和草酸（1.17μeq·L^{-1}）是主要的水溶性有机酸，占阴离子总量的 2%，夏季有机酸浓度比较高。胡敏等[50,51] 还研究了深圳 2004—2005 年降水中的有机酸，降水总有机碳（total organic carbon，TOC）平均浓度为 640.41μeq·L^{-1}，其中检测出 7 种低分子量有机酸，对降水总自由酸度贡献为 5.0%，高于北京（＜1%）。2004 年和 2005 年深圳降水中甲酸、乙酸和乙二酸三者浓度之和分别为 11.79μeq·L^{-1}和 11.83μeq·L^{-1}，分别占当年阴离子总量的 5% 和 8%，高于北京降水中有机酸占阴离子总量的比例，说明有机酸对深圳降水的影响更大，这是由于深圳地处亚热带林木繁茂地区，植被有机物排放量大，同时太阳辐射强烈、光化学过程活跃。2003—2004 年间广州降水中水溶性有机物（DOC）为 3.9mg·L^{-1}，约占总化学成分的 30.6%，在雨水 pH 值低于 5 时，雨水中的 DOC 值与 pH 值有明显的负相关趋势，这表明雨水中的有机物对低 pH 值雨水的酸度有较大贡献[40]。贵阳 2005 年秋冬季降水中有机酸占总阴离子的 1.4%，对降水自由酸度平均贡献为 23.2%，甲酸、乙酸和草酸为主要有机酸，质量分数分别为（0.2—4.5）×10^{-6}、（0.6—5.3）×10^{-6}和（0.1—4.9）×10^{-6}，比同纬度的美国洛杉矶高，可能有至少 3/4 的甲酸和 1/2 的乙酸来源于人类活动[52]。总的来看，与北方地区相比，我国南方植被排放有机物更多，光化学反应也更强，所以降水中有机酸浓度高一些，云南丽江玉龙雪山降水中有机酸甚至是第一位的 H^+ 提供者[53]。

Zhang 等[54] 研究了以华北平原为主的 15 个站点的溶解性有机氮（DON）沉降，其中降水中 DON 的平均浓度为 111μmol·L^{-1}，占 DON 全沉降的 68%，比世界其它地方的研究结果高得多。Mu 等[55] 研究了北京冬季降雪中的羰基硫（COS），COS 浓度为 16.8—145.2ng·L^{-1}，处于超饱和状态，并远高于 1987 年 Belviso 等在法国所做的研究。湿沉降是空气中多环芳烃（PAHs）的重要去除方式，空气中一半以上的 PAHs 是通过降水去除的。王静等[56] 分析了 2004 年三月浙江杭州的一场降雨，雨水中 PAHs 浓度为2 157ng·L^{-1}，Phen 的相对含量最高，其次为 Fl、Py 和 Flur，说明湿沉降主要去除 3 环化合物。全氟辛烷磺酸（PFOS）和全氟辛酸（PFOA）是很多全氟有机化合物在自然界和生物体内的最终降解产物，具有难降解性、生物累积性和多种毒性，目前在全球环境中已经被广泛检出。Liu 等[57] 研究了沈阳市降雪中的 PFOS 和 PFOA 的浓度，平均值分别为 2.0ng·L^{-1}和3.6ng·L^{-1}，郊区浓度高于市区，市中心区域内的 PFOS 和 PFOA 可能具有共同的来源。

降水中有机酸主要来自林木释放和人为源排放的有机物的大气光化学反应，我国南方林木茂密，光照强，所以通常南方降水中有机酸浓度高于北方。至于降水中如 POPs 等其它有机物，目前我国对其研究不多。鉴于有机物在降水化学中的重要作用，降水中有机物的监测分析及其液相化学转化尚待深入研究。

3.3　降水中重金属研究

降水中重金属元素的研究对深入研究大气圈中重金属元素的循环、酸沉降一重金属复合污染的化学作用机制以及维护该地区的大气环境质量具有重要意义，但由于降水中重金属含量低，对分析手段要求高，以前受到的重视程度不够，所以国内研究还比较少。

降水对大气中汞具有重要的清除作用。Tang 等[58]采集了南京城郊 2006 年 12 月份的雾水样品，其中汞浓度平均值为 5.471$\mu g \cdot L^{-1}$，在雾的稳定期汞浓度比较高，与其它污染物的相关性分析表明，Hg 浓度与 CO 的相关性最高，高达 0.939，与总烃和 PM_{10} 也呈正相关，表明这些污染物可能来自煤的燃烧。贵州是中国汞污染最严重的地区之一，Guo 等[59]于 2006 年在贵州乌江盆地的研究表明，5 个监测点降水中总汞年平均浓度和年沉降通量分别为 49.5$\mu g \cdot L^{-1}$ 和 34.7$\mu g \cdot m^{-2}$，甲基汞的年平均浓度和沉降通量分别为 0.25$ng \cdot L^{-1}$ 和 0.18$\mu g \cdot m^{-2}$。1996 年，贵阳雨水中总汞平均浓度为 33$ng \cdot L^{-1}$，年沉降通量为 39$\mu g \cdot m^{-2}$[60]。

王艳等[61]于 2006 年测定了泰山山上和山下降水中的重金属元素。在山上和山下降水中浓度分别为 Zn 92.94(70.41)$\mu g \cdot L^{-1}$、Fe 29.36(10.75)$\mu g \cdot L^{-1}$、Al 18.87(17.92)$\mu g \cdot L^{-1}$ 以及 Mn 10.11(11.81)$\mu g \cdot L^{-1}$，As、Pb 等相对含量较低，各种重金属浓度明显高于国内外其他站点，重金属污染较重，山上受区域影响及山下受局地污染源影响明显。Liu 等[62]研究了 2000—2002 年黄海千里岩岛和东海嵊泗群岛 Cu、Pb、Cd 和 Zn 的干湿沉降，两地均呈现降水中浓度与降雨量成反比的季节变化；Cu、Cd 和 Zn 的湿沉降量高于干沉降量，而 Pb 的干沉降量高于湿沉降量，两地比较，黄海千里岩岛重金属沉降量高于东海嵊泗群岛。龚香宜等[63]于 2004 年 7 月至 2005 年 7 月对东海福建沿海兴化湾的研究也证明了 Cu、Cd 和 Zn 湿沉降量高于干沉降量，而 Pb 干沉降量高于湿沉降量。兴化湾 Cu、Pb、Cd 和 Zn 的年湿沉降量分别为 14.69$mg \cdot m^{-2}$、1.96$mg \cdot m^{-2}$、1.01$mg \cdot m^{-2}$ 和 100.27$mg \cdot m^{-2}$，远高于 Liu 等[62]在黄海和东海所做的研究。

我国现有的两大降水监测网均未将重金属列为必测项目，只有个别研究者做过少量重金属研究工作。大气 Hg 是环境中 Hg 的重要来源，国外非常重视，例如美国就设立了全国大气 Hg 监测网，进行 Hg 的干湿沉降监测，我国应将 Hg 列入大气干湿沉降的必测项目。

3.4　地面雾水研究

雾与云相比，雾是在较低的气团中形成的，所以更能够反映局地排放对近地面的影响，污染的雾对人体健康及地表环境也有直接影响。

迪丽努尔·塔力甫等[64]于 2003—2005 年在距乌鲁木齐市 58km 的南山地区进行的研究表明，南山地区雾水总离子浓度在春季较高，SO_4^{2-} 浓度明显高于东京乘鞍山和美国洛杉矶，较高的 SO_4^{2-}（1 434 $\mu eq \cdot L^{-1}$）、NO_3^-（122$\mu eq \cdot L^{-1}$）浓度等说明人为污染严重。因为乌鲁木齐空气中碱性颗粒物较多，雾水 pH 值(6.35)高于东京乘鞍山和美国洛杉矶。闫琰等[65]对四川成都市区冬季一场雾的分析发现，雾水样品 pH 值在 6—7 之间，SO_4^{2-} 浓度（608$\mu eq \cdot L^{-1}$）最高，并且除 NO_3^-（24$\mu eq \cdot L^{-1}$）外雾水中各种离子浓度均高于雨水。罗清泉等[66]在 1991 年对重庆雾水的研究中，主城区、近郊区、远郊区雾水 pH 值分别为 4.27、4.94 和 7.22，2003 年分别为 4.59、5.28 和 6.63，SO_4^{2-} 是重庆雾水中主要的阴离子，Ca^{2+} 和 NH_4^+ 是主城区和远郊区主要的阳离子，Ca^{2+} 是近郊区主要阳离子，雾水离子浓度主城区 > 近郊区 > 远郊区；与 1991 年相比，重庆雾水中离子浓度呈下降趋势。Liu 等[67]于 2001 年 11 月至 2002 年 10 月研究了云南西双版纳热带雨林中的雾水和雨水，两者平均 pH 值分别为 6.78 和 6.13，同时期两者 pH 值相关性明显，雾水中 SO_4^{2-} 和 NO_3^- 浓度分别为 27.2$\mu eq \cdot L^{-1}$ 和 30.7$\mu eq \cdot L^{-1}$，各种离子浓度明显高于雨水中浓度。朱彬等[68]的研究表明西双版纳地区城区、郊区雾水 pH 值分别为 6.32、8.34，雾水中 F^- 浓度远高于我国其他地区，总离子浓度城区比郊区高很多，并与雾滴粒径呈反相关变化趋势。李一等[69]于 2006 年 12 月在南京城区分时段采集的一场雾水 pH 值范围为 4.24—7.27，平均为 5.63，电导率为 857.84$\mu S \cdot cm^{-1}$，SO_4^{2-}/NO_3^- 当量比为 4.95，与同时期雨水相比雾水中污染物浓度和 pH 值更高；同时期汤莉莉等[70]测得南京郊区雾水 pH 值为 5.71，电导率（810$\mu S \cdot cm^{-1}$）是全年雨水平均值的 4.8 倍，大气中污染物浓度变化与雾的生消基本同步。

地面雾水的离子浓度一般远高于雨水，城区高于郊区、乡村，雾水pH值受当地自然和社会环境的影响规律与地面降水基本一致。我国对地面雾水的研究很少，分析项目也多仅限于常规离子分析。

由以上研究可见，几年来研究者们在中国大气降水化学研究方面取得了很多成果。但由于中国地域辽阔，气候地理环境各异，降水化学组成及成因也相差很大，即使在同一地区，城市与郊区差别也很大。而目前我国降水化学监测和研究多在城市及其周边地区，区域代表性较差，在分析项目方面多侧重于酸度，对其它化学成分进行长期监测和研究较少，揭示降水化学组成特征方面仍有不足。所以，要全面了解我国不同地区不同季节的降水化学特征仍需要做大量工作。

4　高空降水及云雾水化学研究

4.1　高山站降水化学研究

高山降水化学研究能在一定程度上反映当地大气污染的背景值和污染物的长程传输，也具有比较好的区域代表性。为了查明我国高山地区酸沉降水平、来源、形成机制和区域性降水化学特征以及区域性酸性物质的传输，许多研究者在我国有代表性的山上进行了降水化学研究（见表1及图3）。

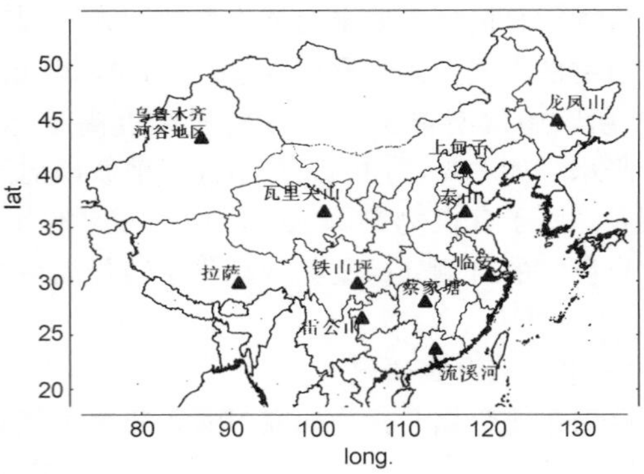

图3　部分区域代表性观测站分布

Fig. 3　Geographic distribution of regional representive sites

泰山是我国东部地区的制高点，泰山降水化学观测对研究该地区对流层内污染物的传输与转化以及东北亚地区酸性物质的输送具有重要的科学意义。王艳等[71—73]于2004年7月—2007年7月在泰山进行了长期的降水化学研究。在此期间泰山降水pH平均值为4.61，离子总浓度平均为428.99μeq·L^{-1}，降水污染比较严重，SO_4^{2-}、NH_4^+和Ca^{2+}是泰山降水中主要离子。SO_4^{2-}的平均浓度远高于NO_3^-，同时NH_4^+浓度高于Ca^{2+}，说明该区域NH_4^+污染源较强，大气及颗粒物中的硫氧化物、氨和铵盐对降水酸度起着主要作用。

表1　部分区域代表性观测站降水pH值及离子浓度（μeq·L^{-1}）

Table 1　pH and ion concentrations in precipitation at some regional representive sites (μeq·L^{-1})

site	duration	altitude/m	pH	SO_4^{2-}	NO_3^-	Cl^-	F^-	NH_4^+	Ca^{2+}	K^+	Na^+	Mg^{2+}	SO_4^{2-}/NO_3^-	$\Sigma+/\Sigma-$	total ions conc.	ref
泰山	2004.7—2007.7	1 530	4 61	146.1	36 33	17.18	5 99	98 78	69.17	8 44	23 98	9 09	4 0	0 9	439 6	71—73
雷公山	2003	1 630	4 44	75	26	4	1	33	25	4	3	5	2 9	0 6	212 3	76
铁山坪	2003	1 320	4 1	184	35	11	5	76	58	8	3	9	5 3	0.6	468 4	76
瓦里关山	1997	3 810	6 38	24	8 3	6 1	—	45 5	34	3 8	8 7	12 1	2 9	2 5	142 9	75
乌鲁木齐河谷	2003	3 551	6 99	52 97	9 63	16 47	—	25 17	174 2	4	19 02	18 14	5 5	3 2	319 7	74
丽江玉龙雪山	1993	3 240	4 89	9 14	3 19	4 65	—	4 71	6 14	0 69	0 70	2 14	2 9	0 9	44 3	53
拉萨	1998—2000	3 658	7 5	5 2	6 9	9 7	0 4	14 3	197 4	5 14	11 2	10 9	0 8	17 5	261 2	25
龙凤山	1991—1997	331	5 18	46 4	22 6	25 6	16 8	50	41	3 85	6 09	9 88	2 1	1 3	228 8	138
上甸子	1989—1997	287	5 33	90	88 6	38 6	34 2	115 6	144 6	12 1	9 57	22 2	1 0	1 5	560 1	138
临安	1985—1997	131	4 33	107	25	26 5	21 6	63 9	20	4 62	6 52	3 3	4 3	0 6	325 2	138
流溪河	2003	500	4 67	80	11	22	6	22	54	8	30	14	7 3	0 8	268 4	76
蔡家塘	2003	450	4 33	155	60	11	7	112	60	10	7	10	2 6	0 8	478 8	76
六冲关	2003	450	4 89	255	18	7	8	51	155	18	8	38	14 2	0 8	570 9	76

1. This table only shows concentrations of major ions, not includes other ion concentrations detected in the researches. 2. — means not detected.

2003年4月至2004年2月，Zhao等[74]在新疆乌鲁木齐西南方向的乌鲁木齐河谷地区进行了高山与地面降水的对比研究。地面点和高山点pH平均值分别为7.27、6.99，同时期乌鲁木齐市区pH值为6.86。Ca^{2+}和SO_4^{2-}是两个站点的主导离子，高山站点降水中Ca^{2+}、SO_4^{2-}浓度分别为174.19μeq·L^{-1}、52.97μeq·L^{-1}，地面站点分别为257.27μeq·L^{-1}、63.86μeq·L^{-1}。与高山站点相比，地面站点受人为污染影响更为明显。瓦里关本底站是由我国政府和全球环境基金(GEF)共同投资在青海省境内瓦里关山顶建设的内陆高原型大气本底基准观象台。汤洁等[75]研究了 1997 年瓦里关站的降水化学(见表 1),年平均 pH 值和电导率分别为 6.4 和 14.8μS·cm^{-1},降水中主要离子依次为NH_4^+、Ca^{2+}、SO_4^{2-}、Mg^{2+}等,与丽江地区相比各离子浓度稍高。SO_4^{2-}是该地降水中的主要致酸物质,SO_4^{2-}/NO_3^-当量比约为 3∶1。NH_4^+和Ca^{2+}浓度升高时降水的 pH 值也升高,说明在我国北方地区碱性气溶胶对降水的中和作用不容忽视。瓦里关和乌鲁木齐河谷地区降水化学特征明显地反映了我国西北及青藏高原降水中Ca^{2+}浓度较高的特点。

中挪合作"中国酸沉降观测与研究"项目(IMPACTS)于 2001—2003 年在中国南方和西南酸雨控制区内的重庆铁山坪、贵州六冲关和雷公山、湖南蔡家塘、广东流溪河进行了研究,其中铁山坪和雷公山是高山点,其余测点也具有比较好的区域代表性。铁山坪和雷公山降水明显酸化,SO_4^{2-}和NO_3^-浓度均比较高,特别是距离重庆市区较近的雷公山降水中SO_4^{2-}浓度年均值高达 184μeq·L^{-1}。这 5 个站点的NH_4^+浓度约是NO_3^-浓度的两倍,SO_2、SO_4^{2-}、NO_3^-、NH_4^+、Ca^{2+}远高于北美和欧洲类似站点。该研究证明了中国大气污染物浓度非常高,并且大气污染物长程传输对乡村背景站点的影响非常明显[76]。

四川盆地是中国酸雨污染最严重的地区之一。柳泽文孝等[77,78]从 1998 年开始在位于四川省中西部的峨眉山不同高程设置雨水采集点进行的研究表明,峨眉山冬季降水中SO_4^{2-}和NO_3^-浓度较高,从山顶到2 600m附近高程范围内降水($SO_4^{2-}+NO_3^-$)浓度变化不大,2 600m以下随着高度的降低而增加。缙云山自然保护区污染物浓度常常成为重庆市酸雨对照分析的本底。魏虹等[79]于 1998—1999 年对该地区降水进行了分析研究,结果表明,缙云山降水 pH 平均值为 5.23,电导率平均为 33.90 μS·cm^{-1},酸化程度低于重庆市区和南方其它酸雨区域。SO_4^{2-}占阴离子总量的 84.61%,SO_4^{2-}(212.85 μeq·L^{-1})和NO_3^-(14.02μeq·L^{-1})浓度略低于重庆市区,属于典型的硫酸型降水;NH_4^+(142.47μeq·L^{-1})和Ca^{2+}(80.78μeq·L^{-1})占阳离子总量的 77.36%,极大地中和了阴离子对雨水的酸化作用。

于涛[80]在位于广州市区内的白云山所做的研究证明了广州降水污染表现出由硫酸型向硝酸型转化的趋势。1994—2004 年间,白云山降水 pH 年均值整体稳定,平均值为 4.45,酸雨频率为 87.6%。白云山降水中SO_4^{2-}(2004 年为 245 μeq·L^{-1})浓度近年来呈下降趋势,而NO_3^-(2004 年为 94.6 μeq·L^{-1})呈升高趋势,SO_4^{2-}/NO_3^-当量比已经由 1994 年的 13 下降为 2004 年的 2.6。黄健等[81]于 1999 年 1—10 月分别在白云山海拔 50m、210m、270m 处采集了降水样品,发现随高度的增加降水 pH 值减小,酸雨和强酸雨频率增加,SO_4^{2-}和Ca^{2+}在降水中的比例下降,而NO_3^-和NH_4^+比例明显增加,对山上降水酸度的影响相对增大。吴兑等[82]分析了 1999 年 1 月和 2001 年 2—3 月粤北南岭大瑶山区海拔 815 m 处的降雨,发现酸雨频率达 88%,平均 pH 值为 4.73,电导率为 22.15μS·cm^{-1},雨水中离子浓度低于雾水,SO_4^{2-}(120 μeq·L^{-1})、Ca^{2+}(92 μeq·L^{-1})分别是雨水中浓度最高的阴阳离子,SO_4^{2-}/NO_3^-当量比约为 15。黄海洪等[83]对广西大明山海拔1 500m处降水的研究表明,大明山 1991—2001 年降水 pH 值为 5.13,低于广西平均值,酸雨频率为 64.7%,与附近地面点降水进行对比,地面点酸度略高。离子组成中,高山点与地面点降水中SO_4^{2-}(高山 132 μeq·L^{-1},地面 165 μeq·L^{-1})和Ca^{2+}(高山 193.2 μeq·L^{-1},地面 93.8 μeq·L^{-1})分别是首要的阴阳离子,地面点SO_4^{2-}的比例明显大于高山站,高山点和地面点SO_4^{2-}/NO_3^-当量比分别约为6 和 8。

1987—1993 年中美双方合作在云南丽江玉龙雪山进行了降水观测,1993 年结果见表 1,玉龙雪山降水中主要无机离子浓度均比较低,说明受人为污染影响较小,但自从 1987 年来也呈现降水 pH 值下降的趋势。由于玉龙雪山地处低纬度地区,周围植被茂密,有机酸是最重要的H^+提供者[53]。

由于山上观测研究费用高、工作条件差,我国对高山降水研究与地面降水研究相比工作甚少,长期系统的工作更少,缺乏系统观测数据,对当地生态环境保护特别是针叶林保护非常不利。

4.2　区域代表性站点降水化学特征研究

本文选取了部分近年来中国各地区有区域代表性站点的降水研究结果来说明全国降水化学背景的现状和区域特征,站点地理分布见图 3,降水化学分析结果见表 1。除中国西北地区外,其它背景站点降水 pH 值都小于 5.6,特别是中国南方降水酸化明显,而北方的泰山降水酸性也比较强,甚至超过泰山周围其它平原地区,说明这些站点降水受区域工业和人口密度影响较大,SO_2高排放强度使部分山上降水严重酸化。泰山和铁山坪 SO_4^{2-} 浓度高于我国东部和南部地区地面降水的一般水平，相当于欧洲和北美的 4—5 倍。各站点降水总离子浓度相差较大，也能较好地反映站点所在区域的大气污染状况；污染严重地区 SO_4^{2-} 和 NO_3^- 浓度都比较高，例如距离大城市较近的铁山坪(重庆)和上甸子(北京)。北方站点 $(Ca^{2+}+NH_4^+)/(SO_4^{2-}+NO_3^-)$ 当量比值普遍大于南方站点，特别是西北地区站点 Ca^{2+} 浓度更高，所以即使在 $(SO_4^{2-}+NO_3^-)$ 比较高的情况下，因为高浓度 Ca^{2+}、NH_4^+ 的中和作用，北方站点 pH 值也不会太低。东部距离海洋更近的站点 Cl^- 浓度高于内陆地区大部分站点，说明受海洋影响明显。

为了比较各地区降水中主要碱性离子 Ca^{2+} 和 NH_4^+ 对 SO_4^{2-} 和 NO_3^- 的中和作用，现引入 pA_i 概念[84]。与 pH 定义类似，pA_i 定义为降水中非海盐 SO_4^{2-} 和 NO_3^- 当量浓度和的以 10 为底的负对数，pA_i 与 pH 关系图可以较好地表征降水中 $(nss\text{-}SO_4^{2-}+NO_3^-)$、$(Ca^{2+}+NH_4^+)$ 和 pH 值三者之间的关系。EANET 位于中国的 3 个乡村监测站点 2001—2004 年降水 pAi 值与 pH 值关系见图 4[84]。西安渭水源和重庆缙云山降水 pA_i 都很低，说明两地降水中 $(nss\text{-}SO_4^{2-}+NO_3^-)$ 都比较高，但由于渭水源降水中 Ca^{2+} 等碱性离子的大量存在中和了降水中的酸性物质，所以渭水源 pH 值高于缙云山点。厦门小坪降水中 pAi 大于渭水源与缙云山，说明当地降水中 $(nss\text{-}SO_4^{2-}+NO_3^-)$ 比较低，而小坪诸多以夏季降水样品为多的数据点位于右下方，说明厦门降水酸度除受硫酸和硝酸影响外，受光化学反应生成的有机酸影响也比较大。

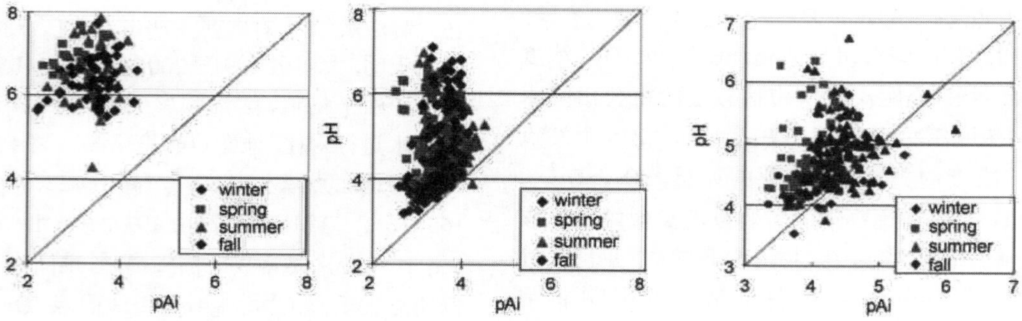

图 4 EANET 中国区三个站点 2001—2004 年降水 pH 值与 pA_i 值关系[84]

Fig. 4 pH versus pAi at selected EANET monitoring sites in China(2001—2004)[84]

4.3 高空云雾水研究

降水被污染的过程通常可分为云内雨除过程和云下冲刷过程，前者包括酸性气体、大气颗粒物等被云水吸收的过程，后者包括云下大气颗粒物和气体被水清除的过程。由于酸性降水是上述两个过程的产物，所以研究各个过程对酸性降水形成的影响都很重要。同时，针对云雾水化学的研究也有助于研究降水污染的长程传输。

航测是研究高空云雾水的重要方法之一，但由于航测对人员、技术和财力等要求比较高，近年来国内针对降水酸化而作的航测研究比较少。雷恒池等[85]根据 1985—1993 年在我国 10 个省市飞机观测的云水化学资料，分析了不同天气系统对我国云、雨水化学特征的影响。我国云水酸化程度从 1985 年以后逐年加强。研究期间，我国北方地区云水 pH 平均值 > 6.0，南方地区云水酸性较强，大部分地区 pH 平均值小于 5.0。北方地区与西南地区云水中 SO_4^{2-} 与 NH_4^+ 分别是主导的阴阳离子；而华中华南地区 NH_4^+ 是主要的阳离子，SO_4^{2-}/NO_3^- 当量比仅为约 1—3，不及华北高。我国西南地区和华中、华南地区云水酸化特征及酸雨形成过程有明显的差别。云水酸度空间分布方面，西南地区云下部比上部酸度高，受局地源的影响明显。华中、华南地区云水酸度随高度分布无明显差别。我国云雨酸化特征，既有区域性特征，又有局地性特征，不能仅用一种酸雨模型来研究我国的酸雨问题。任阵海等[86]于 1993 年 5 月在赣、湘地区航测采集了 71 份云水样品，其中 80.3% 的样品 pH 值 < 5.6，44% 的样品 pH 值 < 4.5，pH 值在 3.5—4.0 的样品数占 23%，pH 值在 5.0—6.5 的样品数占 25%，pH 值和化学组成具有明显的地域特征。

高山站点采集的云雾水也能在一定程度上反映低层云的化学特征，从而有助于了解当地降水化学的真实情况。林长城等[87]于 2003 年在福建九仙山进行了为期一年的云雾水和雨水监测，发现九仙山云雾水和雨水严重酸化，pH 年均值分别为 3.94 和 4.49，而 1989 年为 6.19 和 6.04。2003 年九仙山降水酸度远高于除厦门外的其它 13 个福建城市。九仙山云雾水和雨水受海洋因素影响明显，阴离子以 Cl^-、阳离子以 NH_4^+ 为主，云雾水的致酸离子与碱性离子比值明显大于雨水。吴兑等[88]在海拔 800 余米的南岭大瑶山云岩雾区对浓雾进行的研究发现酸雾频率达 51%，在雾水中浓度最高的阴离子是 SO_4^{2-}，其次是 NO_3^-；阳离子中 Ca^{2+}、NH_4^+ 浓度最高。雾水中离子浓度均远高于雨水中，但因为雨水中 NH_4^+ 和 Ca^{2+} 浓度低，雨水酸性更强，雾水中的 Ca^{2+}、SO_4^{2-}、Mg^{2+} 有明显的富集现象。

5　酸雨形成机理研究

5.1　全国酸性降水相关污染物排放历史趋势

SO_2 和 NO_x 是形成酸性降水最重要的致酸前体物，NH_3、烟尘及工业粉尘是中和降水酸性的重要物质。由图 5 可见，2002 年后 SO_2 和 NO_x 的排放量逐年增加，这意味着降水酸度会逐渐升高，但另一方面，气态 NH_3 的排放也一直呈上升态势，NH_3 对降水的中和作用也逐年增加，这一因素部分抵消了由 SO_2 和 NO_x 排放量增加引起的降水酸性升高，但 NH_3 排放量的增长速度低于 SO_2 和 NO_x，而富含碱性物质的烟尘和工业粉尘的减排也将大幅度促使降水酸度升高。大气酸性和碱性物质相对排放量的变化将决定降水酸度及其化学组成变化，这就需要了解当前各地降水中和状况和今后相关污染物排放的发展趋势，才能判断未来中国酸雨发展趋势，而恰恰这两个问题现在仍情况不明。

由图 5 可见，1950—1996 年，SO_2、NO_x 和 NH_3 年排放量均呈现总体增长的趋势。1996—2002 年，SO_2 和 NO_x 年排放量呈现明显的下降趋势，这主要是因为各减排措施的实施和产业结构调整。由于中国经济的快速发展，对化石燃料的需求持续快速增长，2002 年后 SO_2 和 NO_x 年排放量又开始快速增加。与 SO_2 和 NO_x 变化趋势相一致的是全国 pH 年均值和酸雨频率的变化规律（见图 1），这也表明 SO_2 和 NO_x 排放量与中国降水酸化的进程是一致的。自 1998 年以来，中国烟尘和工业粉尘年排放量一直呈现下降趋势，这与中国近年来的减排措施有关，烟尘和工业粉尘的减排也对中国近年来降水更加酸化有一定影响。

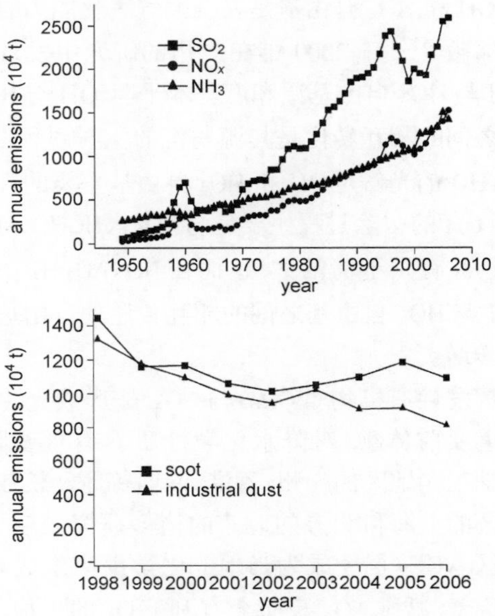

图 5　中国人为源 SO_2、NO_2、NH_3、烟尘和工业粉尘年排放量变化趋势

Fig. 5　Trends of annual emissions of anthropogenic SO_2, NO_2, NH_3, soot and industrial dust

数据来源：(1) 王文兴等[89,90]；(2) 中国统计年鉴；(3) 中国环境统计年报

5.2　化学转化机理研究

作为酸性降水形成的最重要前提物，SO_2、NO_x 以及其它与降水化学有关的污染物排入大气后，在大气中本来存在的气体和颗粒物的共同作用下，经过一系列复杂的物理化学过程进入降水中，并在水滴中继续进行化学反应，从而对降水的酸度和化学组成产生影响。上述过程包括均相反应、多相反应、云内转化和云下洗脱等，还有诸多过程和机理有待研究。

大气氧化性是降水化学转化研究的重要问题之一，对大气氧化性的定量描述具有重要意义。程艳丽等[91]推导出还原性污染物在大气中总的准一级氧化去除速率常数（$K_{por,T}$），用以定量表征大气氧化性，为对大气氧化性有关问题的深入探讨提供了量化参数，并以珠江三角洲为例，利用 $K_{por,T}$ 对该区域大气氧化性进行了数值模拟研究。结果表明，在珠江三角洲地区，大气氧化性具有明显的日变化规律和空间分布特征，$K_{por,T}$ 与 OH 自由基浓度具有一定的时空相似性，白天 OH 自由基的氧化作用占主导地位，SO_2 的非均相氧化反应在夜间起重要作用。

大气 OH 和 HO_2 自由基是表征大气氧化能力的关键物。邵敏等[92] 于 2000 年在广州的研究中发现大气中 OH 与 HO_2 自由基源和汇基本平衡，OH 与 HO_2 自由基之间的相互转化是其源与汇的主导过程；HNO_2 和 HCHO 的光解是 OH 和 HO_2 自由基重要的直接来源，而 OH 的去除过程主要是与 VOCs、HCHO、NO_2 和 CO 反应，有别于清洁大气，而在 NO、NO_2 存在条件下 OH 与 HO_2 自由基之间的相互传递使自由基浓度进一步增大[93]。

挥发性有机物(VOCs)与 NO_x 是形成光化学烟雾的重要前体物，对降水化学性质具有重要影响。针对珠三角地区的广州、香港、深圳等地及长江三角洲地区的上海和北方的北京的诸多研究表明，在上述地区，VOCs 已经成为当地光化学反应生成 O_3 的主导因素，减排 NO_x 反而会有利于 O_3 的生成，控制地面 O_3 浓度的重点应该转移到人为源 VOCs 的排放上面来[94—99]。Wang 等[100] 认为，在某些大城市及其周边区域，VOCs 已经成为 O_3 生成的主导因素，所以自然源 VOCs 对这些地区 O_3 的生成有一定贡献。在 NO_x 浓度比较高的华南平原地区，天然源 NO_x 的排放会抑制 O_3 的生成；而在 NO_x 为 O_3 生成主导因素的黄河中下游地区，天然源 NO_x 有利于 O_3 的生成。杜林等[101] 认为同时削减 VOCs 和 NO_x 排放才能更好地减少光化学污染的发生。此外，Bian 等[102] 在对天津市区 PM_{10} 对 O_3 生成影响的研究中发现 O_3 与 PM_{10} 呈非线性关系：当 PM_{10} 比较高时，O_3 与 PM_{10} 成比较敏感的负相关关系；而当 PM_{10} 比较低时，与 O_3 相关性降低。安俊琳等[103] 研究了城市大气中太阳紫外辐射与空气污染的相互关系，结果表明，太阳紫外辐射受云、低层大气气溶胶和污染气体特别是 NO_x 与 O_3 的影响比较大。

HONO 对 HO_x 的生成具有重要影响，但 HONO 的具体生成过程仍有诸多需要研究之处，很多国内学者对 HONO 的生成进行了研究。Hao 等[104] 的研究表明，当 40%< RH< 70%时，HONO/NO_2 最大值与 RH 成正相关；当 RH> 70%时，HONO/NO_2 最大值与 RH 成负相关。Su 等[105] 研究了广州新垦日间 HONO 与 O_3 对 OH 生成的影响，发现 HONO 导致的 OH 生成速率比 O_3 导致的 OH 生成速率高 3 倍。NO_2 与地面和颗粒物表面的异相反应可能也是 HONO 的源，张远航等[97] 和 Hao 等[104] 在相关研究中也提到了这一点。Su 等[106] 进一步的研究表明地面与气溶胶颗粒物表面的复杂反应对 HONO 的生成具有较大的贡献，且 HONO 生成率与 BC 具有良好相关性。Wang 等[107] 通过气态 HCl 与 $AgNO_2$ 反应制取了 HONO，并研究了 HONO 生成和消失的过程。房豪杰等[108] 研究了云滴、雾滴和雨滴中 HONO 与 CS_2 反应的机制，发现在 355nm 光催化下能产生两种以前从未报道过的瞬态中间产物：CS_2OH-HONO 和 CS_2NO^+。Quan 等[109] 通过模式演算了北京地区 NH_3 对 SO_2 转化和 S 沉降的影响，认为 NH_3 的存在可以使 SO_4^{2-} 浓度增大、SO_2 浓度降低，并能扩大 S 元素长距离传输的范围，从而使污染源地区的 S 沉降减少 10%—30%。

大气中的各种颗粒物为大气中复杂的多相反应提供了巨大的表面积，颗粒物表面不同的化学组分也对不同反应具有不同的催化作用，所以，大气颗粒物表面的多相反应也很受研究者的关注。含硫酸盐颗粒物对于降水中 SO_4^{2-} 浓度具有重要影响，然而 SO_2 通过颗粒物表面的多相反应生成硫酸盐的机制还有待深入研究。陈建民等[110, 111] 和 Li 等[112, 113] 针对 SO_2 在多种气溶胶颗粒物常见组分的表面参与多相反应的机制及影响因素进行了研究，认为多相反应可能是大气中硫酸盐形成的重要途径。颜敏等[114] 的研究表明，多相反应对空气中 NO_2、N_2O_5 和 NO_3 的消除即 NONO、硝酸盐、硫酸盐生成的影响与颗粒物质量浓度呈正相关，与颗粒物粒径呈负相关。COS 是大气中含量丰富的含硫化合物之一，对降水化学具有重要影响。COS 与颗粒物表面的多相反应是 COS 大气化学过程的重要去除途径，He 等[115—118] 和 Chen 等[119, 120] 针对羰基硫化物 COS 在颗粒物表面的多相反应进行了大量研究：COS 在颗粒物表面通过催化氧化，经由 $HSCO_2$ 和 HSO_3 类生成 SO_4^{2-} 和 CO_2，不同组分表面反应速率不同，比表面积、表面自由基和酸碱性对反应速率也有影响，大气中的矿物质颗粒物是 COS 的重要汇。

Liu 等[121] 分析了于 2000 年春季和夏季在北京采集的颗粒物粒子，认为颗粒物构成中 Ca-K-S 和其它含硫物质的形成与相对湿度和云量密切相关，而硫酸盐组分的长条结晶状形态可能是由于硫酸盐组分来源于诸如云中过程的液相氧化反应。Ouyang 等[122] 研究了云滴、雾滴和雨滴中 SO_4 自由基与 Cl^-、NO_3^-、HSO_3^- 和 HCO_3^- 等离子的反应，认为 Cl^- 是 SO_4 自由基最重要的去除离子。Yu 等[123] 收集了从北京至香港的多个颗粒物样品，发现草酸盐与硫酸盐含量具有很好的相关性，两者可能具有类似的生成途径。Du 等[124] 针对二乙基硫（DES）在大气光化学反应中的角色进行了研究，认为由于 DES 在大气中浓度很低，与 O_3 反应速率比较小，所以在全球硫循环中作用并不重要。

尽管国内外研究者对我国大气化学反应做了不少工作，但针对我国高浓度大气污染物及其特殊地理环境产生的颗粒物之间产生的相互作用，例如 SO_2 与海盐粒子、沙尘粒子之间的相互作用很少有人研究。此外，我国的化学模式也多是引用国外的，基本上没有国内开发的化学模式。另外，与 SO_2 和 NO_x 大气转化相关的基元反应，我国也应深入研究，例如，美国宾州大学 Lester 实验与理论计算发现 HOOO 自由基及氧化亚硝酸（$HONO_2$）的存在，可能对大气中 SO_2 和 NO_2 的转化亦即对酸性物质的传输产生重大影响[125, 126]，因此对降水化学有关的基元化学反应还应深入研究。

5.3　降水与气溶胶等污染物相互作用研究

气溶胶粒子可能在一定程度上促进降水的酸化，也可能缓冲和抑制降水的酸性，这取决于气溶胶的来源和化学组成。因此，分析和研究大气气溶胶的酸碱性和酸化缓冲能力不仅能在一定程度上阐述大气气溶胶的污染性质，还能评价它对降水的作用。降水过程也能清除气溶胶等大气污染物，从而改变大气污染状况。今后必须将大气气溶胶的研究列为降水化学研究的重要组成部分。

曾凡刚等[127] 于 2000 年 1 月分别在辽宁凤凰山、山东青岛田横岛和浙江舟山群岛研究了 PM_{10} 样品的酸度和酸化缓冲能力，结果表明，3 个观测点 PM_{10} 都不具有对酸雨的缓冲能力，而受人为污染影响比较大的地点 PM_{10} 酸性更强，与 1980s 结果相比，PM_{10} 酸化缓冲能力下降。王玮等[128] 在 2002 年春利用航测研究了广东珠海、辽宁大连近海地区的气溶胶颗粒物对酸雨的缓冲能力，这些地区的总悬浮颗粒物（TSP）总体来看已经呈酸性，对降水不具缓冲能力，北方地区 TSP 酸度低于南方地区。颗粒物粒径越小酸性越强，高空虽然颗粒物粒子数少，但酸性更强。毕晓辉等[129] 研究了天津城区 TSP 的酸雨缓冲能力，认为大气颗粒物浓度和降水量是影响该地区降水 pH 值的关键因素，天津颗粒物缓冲能力主要来源于开放源。模拟酸雨淋溶条件下 PM_{10} 中 Cu、Pb、Zn、Cd 均有不同程度的释放，Cd 和 Zn 的释放率明显高于 Cu 和 Pb，随模拟酸雨 pH 值降低，重金属的释放强度显著提高[130]。刘俊峰等[131] 研究了大气中影响降水 H^+ 浓度的主要因素同 H^+ 浓度的参数关系，并给出了各影响因素对 H^+ 浓度的单一以及综合参数化公式，为用三维气质模式研究酸雨问题创造了良好的条件。

赵海波等[132] 对降雨去除气溶胶的过程进行了模拟研究：当雨滴尺度谱和气溶胶尺度谱均满足对数正态分布时，降雨对中等尺度气溶胶的湿去除效果不理想，但可以有效地湿去除大尺度气溶胶和小尺度气溶胶；对于任何尺度的气溶胶，雨强的增加将有利于其被去除。胡敏等[133] 的研究发现降雨后颗粒物的质量浓度谱分布由降雨前的双模态分布变成单模态分布。王河涌等[134] 分析了一次降雨过程中大气中部分污染物的变化情况，降雨初期颗粒物对降水 pH 值有较大影响，而中后期 SO_2 对降水 pH 值起主导作用，SO_2、NO_2 和 PM_{10} 均在降雨初期急剧下降。

王勇等[135] 的研究表明 RH＞90％的雾对各种大气污染物都有明显的清除作用，RH＜80％的雾对各种大气污染物的污染程度都有明显的加重作用；而 80％＜RH＜90％的雾对 PM_{10} 污染有明显加重作用，对 SO_2 有一定的去除作用，对 NO_2 无明显影响。时宗波等[136] 研究了雾对大气颗粒物理化学特性的影响，发现雾期间出现了大量由液相反应生成的长条状 Ca-K-S 颗粒以及主要由硫酸盐组成的似圆状颗粒，雾期间大气中 0.2μm 以上颗粒物的数浓度比晴天低污染期间高 5—8 倍，并在 0.4μm 附近出现了新模态，雾中活跃的大气非均相反应极大地改变了颗粒物的形貌、化学组成、粒度分布等特性。李卫军等[137] 研究了北京冬季雾天和正常天气单个矿物颗粒物物理和化学特征，雾天矿物颗粒物数量-粒度分布峰值出现在 0.1μm—0.3μm 和 1μm—2.5μm，化学成分也与正常天气有较大差异，雾天中颗粒物表面 SO_2 向硫酸盐转化率比较高。

许多作者的工作都说明气溶胶对大气化学反应和降水的化学组成产生重大影响，影响的程度和性质决定于气溶胶粒子的化学组成和粒径分布。中国地域辽阔，气候、土壤各异，工业类型和分布不同，人口疏密不等，诸多因素综合起来决定了气溶胶的物理化学性质，由此看来，要阐明气溶胶对我国降水化学的影响，现有的工作远远不够，特别是大气颗粒物是我国首要控制的污染物，今后必将对降水化学组成产生重大影响，因此需要进行深入系统的研究。

6　今后降水化学研究方向

综上所述，我国研究者近年来在降水化学领域做了大量的工作，取得了一些重要研究成果，对促进我国酸雨控制起到了重要作用。但由于环境的变迁和科学的进步，我国降水化学亟待研究尚存及新出现的科学问题，为此提出以下建议。

（1）科学布设降水化学监测网点，提高数据的区域代表性

我国现有酸沉降监测网的监测点基本上都在城区和城郊，位于农村的监测点较少。由于我国城区和农村空气污染物浓度和降水化学组成相差很大，现有降水化学和大气化学数据基本上只能代表城区和郊区而不能代表广大农村，因而缺乏区域代表性。因此应根据各地区自然环境、工业交通和人口分布等状况科学布设监测网点以便得到具有区域代表性的降水化学参数。

（2）跟踪研究大气降水化学变化，揭示酸沉降控制效果

我国降水中主要酸性物质是硫酸和硝酸，现阶段硫酸起主导作用。现在国家正在采取严厉措施控制 SO_2 排放，这就需要了解 SO_2 排放量的削减与降水中 SO_4^{2-} 浓度降低的反馈量。另一方面目前我国 NO_x 排放量仍在增加，势必需要加强 NO_x 的排放控制。因此也需要研究降水中 NO_3^- 浓度变化以及 SO_4^{2-}/NO_3^- 当量比变化趋势，以观测降水酸化类型的转变。

（3）大气颗粒物和 NH_3 排放量的削减将促进降水酸度的升高

大气颗粒物是我国最重要、最普遍存在的空气污染物，也是首要控制的空气污染物。由于大气颗粒物对我国降水酸化起重要的中和作用，颗粒物的减排势必促进降水酸度的升高。另一方面大气中 NH_3 也对我国降水酸化具有重要中和作用，而 NH_3 的减排势在必行，这样又增加一个降水酸度升高的重要因素。所以，为了控制酸雨危害，必须研究大气颗粒物和 NH_3 减排带来的复杂连锁效应。

（4）深入研究大气化学基元反应，改进降水化学模式

经过国内外多年研究，已获得与降水化学有关的大气均相及多相化学反应的许多重要参数，开发了多种化学模式，但由于我国不同地区大气污染物浓度和化学组成的重大差异，尚需要研究我国不同大气环境条件下的有关化学过程。此外，随着科学的进步和国家新的需求，需要继续研究大气酸化的问题。

（5）增加降水化学分析参数，以满足大气环境保护需求

为了提供降水区域代表性化学参数，在科学建网布点的基础上除分析传统参数外，某些站点应分析包括重金属和 POPs 等在内的参数。

（6）严格执行降水化学质量控制和质量保证规范，确保数据质量

我国已颁布降水化学监测分析规范，其中包括监测点布设、采样、现场测量、样品送输、保存和分析、数据检验等全过程要求，确保全程中不得有任何步骤出差错，以保证数据质量。

参 考 文 献

[1]　唐孝炎(Tang X Y)，张远航(Zhang Y H)，邵敏(Shao M)．大气环境化学(Atmospheric Environmental Chemistry)．北京：高等教育出版社(Beijing：High Education Press)，2006．23
[2]　Smith A R. Air and Rain：the Beginnings of a Chemical Climatology. London：Longmans，Green，and Co.，1872．17—24
[3]　Ottar B. Air，Water，&Soil Pollution，1976，105—117
[4]　唐孝炎(Tang X Y)，张远航(Zhang Y H)，邵敏(Shao M)．大气环境化学(Atmospheric Environmental Chemistry)．北京：高等教育出版社(Beijing：High Education Press)，2006．367
[5]　王文兴(Wang W X)．中国环境科学(China Environmental Science)，1994，14(5)：323—329
[6]　丁国安(Ding G A)，徐晓斌(Xu X B)，王淑凤(Wang S F)等．应用气象学报(Journal of Applied Meteorological Science)，2004，15(Suppl.)：85—94
[7]　汤洁(Tang J)，徐晓斌(Xu X B)，巴金(Ba J)等．中国科学院研究生院学报(Journal of the Graduate School of the Chinese Academy of Sciences)，2007，24(5)：667—673
[8]　金蕾(Jin L)，徐谦(Xu Q)，林安国(Lin A G)等．环境科学学报(Acta Scientiae Circumstantiae)，2006，26(7)：1195—1202
[9]　徐敬(Xu J)，张小玲(Zhang X L)，徐晓斌(Xu X B)等．环境科学学报(Acta Scientiae Circumstantiae)，2008，28(5)：1001—1006
[10]　徐梅(Xu M)，郑勇(Zheng Y)，易笑园(Yi X Y)．气象科技(Meteorological Science and Technology)，2007，35(6)：792—796
[11]　秦福生(Qin F S)，马体顺(Ma T S)，张志红(Zhang Z H)等．河南气象(Henan Meteorology)，2006，(4)：48—49
[12]　孙民(Sun M)，王海翔(Wang H X)，孙杰(Sun J)等．安徽农业科学(Journal of Anhui Agri. Sci.)，2007，35(36)：11974—11976
[13]　晏晓英(Yan X Y)，王雅君(Wang Y J)，冯喜媛(Feng X Y)．吉林气象(Jilin Meteorology)，2007，(2)：44—46
[14]　陈伯通(Chen B T)，罗建中(Luo J Z)，冯爱坤(Feng A K)．环境污染与防治(Environmental Pollution &Control)，2006，28(2)：

112—115

[15] 秦鹏(Qin P)，杜尧东(Du R D)，刘锦銮(Liu J L)等. 热带气象学报(Journal of Tropical Meteorology)，2006，22(3)：297—300

[16] 王裕东(Wang Y D)，罗华铭(Luo H M). 城市环境与城市生态(Urban Environment & Urban Ecology)，2007，20(4)：38—41

[17] Huang Y L，Wang Y L，Zhang L P. Atmospheric Environment，2008，42：3740—3750

[18] 沙晨燕(Sha C Y)，何文珊(He W S)，童春富(Tong C F)等. 环境科学研究(Research of Environmental Sciences)，2007，20(5)：31—34

[19] 郑凤琴(Zheng F Q)，孙崇智(Sun C Z)，谢宏斌(Xie H B). 热带气象学报(Journal of Tropical Meteorology)，2007，23(6)：664—668

[20] 马丽雅(Ma L Y)，王斌(Wang B)，杨俊国(Yang J G). 环境科学与管理(Environmental Science and Management)，2008，133(14)：26—29

[21] 巴金(Ba J)，汤洁(Tang J)，王淑凤(Wang S F)等. 气象(Meteorology Monthly)，2008，34(9)：81—88

[22] 文涛(Wen T)，袁河清(Yuan H Q)，李萍(Li P). 华南师范大学学报(自然科学版)(Journal of South China Normal University (Natural Science Edition))，2008，(3)：89—94

[23] 刘新春(Liu X C)，何清(He Q)，艾力·买买提明(Aili M)等. 沙漠与绿洲气象(Desert and Oasis Meteorology)，2007，1(5)：10—14

[24] 詹红霞(Zhan H X)，杨金玲(Yang J L)，邱辉(Qiu H). 新疆气象(Bimonthly of Xinjiang Meteorology)，2006，29(2)：20—22

[25] 章典(Zhang D)，师长兴(Shi C X)，假拉(Jia L). 干旱区研究(Arid Zone Research)，2005，22(4)：471—475

[26] 章典(Zhang D)，师长兴(Shi C X)，假拉(Jia L). 环境科学学报(Acta Scientiae Circumstantiae)，2004，24(3)：555—557

[27] Tang A H，Zhuang G S，Wang，Y，et al. Atmospheric Environment，2005，39：3397—3406

[28] 丁卫东(Ding W D)，王建英(Wang J Y)，张兰真(Zhang L Z)，赵颖(Zhao Y)等. 中国环境监测(Environmental Monitoring in China)，2004，20(5)：11—14

[29] 王赞红(Wang Z H)，李纪标(Li J B). 河北师范大学学报(自然科学版)(Journal of Hebei Normal University (Natural Science Edition))，2006，30(2)：240—244

[30] Xu M，Lu A H，Xu F，et al. Atmospheric Environment，2008，42：1042—1048

[31] Li C L，Kang S C，Zhang Q G，et al. Atmospheric Research，2007，85：351—360

[32] 陈伟(Chen W). 中国环境监测(Environmental Monitoring in China)，2006，22(2)：72—74

[33] Huang K，Zhuang G S，Xu C，et al. Atmospheric Research，2008，89：149—160

[34] Tu J，Wang H S，Zhang Z F，et al. Atmospheric Research 2005，73：283—298

[35] 宋玉芝(Song Y Z)，秦伯强(Qin B Q)，杨龙元(Yang L Y)等. 南京气象学院学报(Journal of Nanjing Institute of Meteorology)，2005，28(5)：593—600

[36] Zhang M Y，Wang S J，Wu F C，et al. Atmospheric Research，2007，84：311—322

[37] 林雨霏(Lin Y F)，刘素美(Liu S M)，纪雷(Ji L)等. 环境科学(Environmental Science)，2005，26(5)：49—54

[38] Zhang G S，Zhang J，Liu S M. J. Atmos. Chem.，2007，57：41—57

[39] 刘君峰(Liu J F)，宋之光(Song Z G)，许涛(Xu T)等. 环境科学(Environmental Science)，2006，27(10)：1998—2002

[40] Huang D Y，Xu Y G，Peng P A，et al. Environmental Pollution，2009，157：35—41

[41] 赵卫红(Zhao W H). 环境科学与技术(Environmental Science & Technology)，2006，29(9)：41—44

[42] Wai K M，Tam C W F，Tanner P A. Atmospheric Environment，2005，39：7872—7879

[43] 梅自良(Mei Z L)，刘仲秋(Liu Z Q)，刘丽(Liu L)等. 四川环境(Sichuan Environment)，2005，24(3)：52—55

[44] 高一(Gao Y)，卫滇萍(Wei D P). 地球与环境(Earth and Environment)，2005，33(1)：59—62

[45] 段浩(Duan H)，苏智先(Su Z X)，李成柱(Li C Z)等. 中国环境监测(Environmental Monitoring in China)，2008，24(3)：79—83

[46] Wang W X，Wang T. Water，Air，and Soil Pollution，1995，85：2295—2300

[47] Wang W X，Wang T. Atmospheric Environment，1996，30(23)：4091—4093

[48] Galloway J N，Likens G E，Keene W C，et al. Journal of Geophysical Research-Atmospheres，1982，87：8771—8786

[49] 胡敏(Hu M)，张静(Zhang J)，吴志军(Wu Z J). 中国科学 B 辑：化学(Science in China Series B-Chemistry)，2005，35(2)：169—176

[50] 牛彧文(Niu Y W)，何凌燕(He L Y)，胡敏(Hu M). 环境科学(Environmental Science)，2008，29(4)：1014—1019

[51] 刘辰(Liu C)，何凌燕(He L Y)，牛彧文(Niu Y W)等. 环境科学研究(Research of Environmental Sciences)，2007，20(5)：20—25

[52] 徐刚(Xu G)，李心清(Li X Q)，黄荣生(Huang R S)等. 地球与环境(Earth and Environment)，2007，35(1)：46—50

[53] 刘嘉麟(Liu J L)，Keeney W C，霍义强(Huo Y Q)等. 中美科技合作全球内陆降水背景值研究(General report：China and United States cooperative study on precipitation background value in the global interior region). 北京：中国环境科学出版社(Beijing：China Environmental Science Press)，1995

[54] Zhang Y，Zheng L X，Liu X J，et al. Atmospheric Environment，2008，42：1035—1041

[55] Mu Y J，Wang M Z，Wu H，et al. Journal of Geophysical Research，2004，109：D13301

[56] 王静(Wang J)，朱利中(Zhu L Z). 中国环境科学(China Environmental Science)，2005，25(4)：471—474

[57] 刘薇(Liu W)，金一和(Jin Y H)，全燮(Quan X)等. 环境科学(Environmental Science)，2007，28(9)：2068—2073

[58] Tang L I，Niu S J，Fan S X，et al. Proceedings of the SPIE—The International Society for Optical Engineering，2008，6973：B9730

[59] Guo Y N，Feng X B，Li Z G，et al. Atmospheric Environment，

2008, 42(30): 7096—7103
[60] Feng X B, Sommar J, Lindqvist O, et al. Water, Air, and Soil Pollution, 2002, 139: 311—324
[61] 王艳(Wang Y), 刘晓环(Liu X H), 金玲仁(Jin L R)等. 环境科学(Environmental Science), 2007, 28(11): 2562—2568
[62] Liu C L, Zhang G S, Ren H B, et al. Chinese Journal of Oceanology and Limnology, 2005, 23(2): 230—237
[63] 龚香宜(Gong X Y), 祁士华(Qi S H), 吕春玲(Lü C L)等. 环境科学研究(Research of Environmental Sciences), 2006, 19(6): 31—34
[64] 迪丽努尔·塔力甫(Dilnur T), 阿不力克木·阿布力孜(Ablikim A). 干旱环境监测(Arid Environmental Monitoring), 2007, 21(2): 83—86
[65] 闫琰(Yan Y), 叶芝祥(Ye Z X), 闫军(Yan J)等. 四川环境(Sichuan Environment), 2008, 127(13): 16—40
[66] 罗清泉(Luo Q Q), 鲜学福(Xian X F). 西南农业大学学报(自然科学版)(Journal of Southwest Agricultural University (Natural Science)), 2005, 27(3): 393—396
[67] Liu W J, Zhang Y P, Li H M, et al. Water and Soil Pollution, 2005, 167(1/4): 295—309
[68] 朱彬(Zhu B), 李子华(Li Z H), 黄建平(Huang J P)等. 环境科学学报(Acta Scientiae Circumstantiae), 2000, 20(3): 316—321
[69] 李一(Li Y), 张国正(Zhang G Z), 濮梅娟(Pu M J)等. 中国环境科学(China Environmental Science), 2008, 28(5): 395—400
[70] 汤莉莉(Tang L L), 牛生杰(Niu S J), 樊曙先(Fan S X)等. 环境化学, 2008, 27(1): 105—109
[71] 王艳(Wang Y), 葛福玲(Ge F L), 刘晓环(Liu X H)等. 环境科学学报(Acta Scientiae Circumstantiae), 2006, 26(7): 1187—1194
[72] 王艳(Wang Y), 葛福玲(Ge F L), 刘晓环(Liu X H)等. 中国环境科学(China Environmental Science), 2006, 26(4): 422—426
[73] Wang Y, Wai K M, Gao J, et al. Atmospheric Environment, 2008, 42: 2959—2970
[74] Zhao Z P, Tian L D, Fischer E, et al. Atmospheric Environment, 2008, 42(39): 8934—8942
[75] 汤洁(Tang J), 薛虎圣(Xue H S), 于晓岚(Yu X L)等. 环境科学学报(Acta Scientiae Circumstantiae), 2000, 20(4): 420—425
[76] Aas W, Shao M, Lei J, et al. Atmospheric Environment, 2007, 41: 1706—1716
[77] 柳泽文孝(Yanagisawa F), 贾疏源(Jia S Y), 益田晴惠(Masuda H)等. 成都理工大学学报(自然科学版)(Journal of Chengdu University of Technology (Science & Technology Edition)), 2003, 30(1): 96—98
[78] 柳泽文孝(Yanagisawa F), 贾疏源(Jia S Y), 赤田尚史(Akata N F)等. 四川环境(Sichuan Environment), 2002, 21(1): 37—40
[79] 魏虹(Wei H), 王建力(Wang J L), 李建龙(Li J L). 农业环境科学学报(J. Agro-Environment Science), 2005, 24(2): 344—348
[80] 于涛(Yu T). 云南环境科学(Yunnan Environmental Science), 2006, 25(Suppl.): 134—135
[81] 黄健(Huang J), 李福娇(Li F J), 江奕光(Jiang Y G)等. 热带气象学报(Journal of Tropical Meteorology), 2003, 19(Suppl.): 126—135
[82] 吴兑(Wu D), 邓雪娇(Deng X J), 范绍佳(Fan S J)等. 中山大学学报(自然科学版)(Acta Scientiarum Naturalium Universitatis Sunyatseni), 2005, 144(16): 105—109
[83] 黄海洪(Huang H H), 董蕙青(Dong H Q), 陈竑(Chen H). 福建林学院学报(Journal of Fujian College of Forestry), 2003, 23(4): 331—334
[84] EANET. Periodic Report on the State of Acid Deposition in East Asia Part I: Regional Assessment, 2006, 97—98
[85] 雷恒池(Lei H C), 吴玉霞(Wu Y X), 肖辉(Xiao H)等. 高原气象(Plateau Meterology), 2001, 20(2): 127—131
[86] 任阵海等. 我国酸沉降及其生态环境影响研究. "八五"科技攻关项目(85-912-01)总结报告, 1996. 6
[87] 林长城(Lin C C), 赵卫红(Zhao W H), 蔡义勇(Cai Y Y)等. 福建农林大学学报(自然科学版)(Journal of Fujian Agriculture and Forestry University (Natural Science Edition)), 2007, 36(6): 622—626
[88] 吴兑(Wu D), 邓雪娇(Deng X J), 叶燕翔(Ye Y X)等. 气象学报, 2004, 62(4): 476—485
[89] 王文兴(Wang W X), 王纬(Wang W), 张婉华(Zhang W H)等. 中国环境科学(China Environmental Science), 1996, 16(3): 161—167
[90] 王文兴(Wang W X), 卢筱凤(Lu X F), 庞燕波(Pang Y B)等. 环境科学学报(Acta Scientiae Circumstantiae), 1997, 17(1): 2—7
[91] 程艳丽(Cheng Y L), 王雪松(Wang X S), 刘兆荣(Liu Z R)等. 中国科学 B 辑: 化学(Science in China Series B— Chemistry), 2008, 38(10): 938—946
[92] 邵敏(Shao M), 任信荣(Ren X R), 王会祥(Wang H X)等. 科学通报(Chinese Science Bulletin), 2004, 49(17): 1716—1721
[93] 任信荣(Ren X R), 王会祥(Wang H X), 邵可声(Shao K S)等. 环境科学(Environmental Science), 2004, 25(4): 28—31
[94] 程艳丽(Cheng Y L), 白郁华(Bai Y H). 环境科学学报(Acta Scientiae Circumstantiae), 2008, 28(4): 791—798
[95] 高东峰(Gao D F), 张远航(Zhang Y H), 曹永强(Cao Y Q). 环境科学研究(Research of Environmental Sciences), 2007, 20(1): 47—51
[96] Shao M, Zhang Y H, Zeng L M, et al. Journal of Environmental Management 2009, 90(1): 512—518
[97] Zhang Y H, Su H, Zhong L J, et al. Atmospheric Environment, 2008, 42: 6203—6218
[98] 徐峻(Xu J), 张远航(Zhang Y H). 环境科学学报(Acta Scientiae Circumstantiae), 2006, 26(6): 973—980
[99] 石玉珍(Shi Y Z), 王庚辰(Wang G C), 徐永福(Xu Y F). 气候与环境研究(Climatic and Environmental Research), 2008, 113(11): 84—92
[100] Wang Q G, Han Z W, Wang T J, et al. Science of the Total Environment, 2008, 395: 41—49
[101] 杜林(Du L), 徐永福(Xu Y F), 葛茂发(Ge M F)等. 环境科学(Environmental Science), 2007, 28(3): 482—488
[102] Bian H, Han S Q, Tie X X, et al. Atmospheric Environment, 2007, 41: 4672—4681

[103] 安俊琳(An J L)，王跃思(Wang Y S)，李昕(Li X)等. 环境科学(Environmental Science)，2008，29(4)：1053—1058
[104] Hao N，Zhou B，Chen D，et al. Journal of Environmental Science，2006，19(5)：910—915
[105] Su H，Cheng Y F，Shao M，et al. Journal of Geophysical Research-Atmospheres，2008，113(D14)：D14312
[106] Su H，Cheng Y F，Cheng P，et al. Atmospheric Environment，2008，42：6219—6232
[107] Wang W G，Ge M F，Yao L，et al. Chinese Science Bulletin，2007，52(22)：3056—3060
[108] 房豪杰(Fang H J)，侯惠奇(Hou H Q). 复旦学报(自然科学版)(Journal of Fudan University (Natural Science))，2004，43(6)：1098—1101
[109] Quan J N，Zhang X S. Atmospheric Research，2008，88：78—88
[110] Fu H B，Wang X，Wu H B，Chen J M，et al. Journal of Physical Chemistry-C，2007，111(16)：6077—6085
[111] Zhang X Y，Zhuang G S，Chen J M，et al. Journal of Physical Chemistry-B，2006，110(25)：12588—12596
[112] Li L，Chen Z M，Zhang Y H，et al. Journal of Geophysical Research-Atmospheres，2007，112(D18)：D18301
[113] Li L，Chen Z M，Zhang Y H，et al. Atmospheric Chemistry and Physics，2006，6：2453—2464
[114] 颜敏(Yan M)，王雪松(Wang X S)，刘兆荣(Liu Z R)等. 中国环境科学(China Environmental Science)，2008，28(9)：823—827
[115] Liu Y C，He H，Mu Y J. Atmospheric Environment，2008，42(5)：960—969
[116] Liu Y C，Liu J F，He H，et al. Chinese Science Bulletin，2007，52(15)：2063—2071
[117] Liu J F，Yu Y B，Mu Y J，et al. Journal of Physical Chemistry-B，2006，110(7)：3225—3230
[118] He H，Liu J F，Mu Y J，et al. Environmental Science & Technology，2005，39(24)：9637—9642
[119] Chen H H，Kong L D，Chen J M，et al. Environmental Science & Technology，2007，41(18)：6484—6490
[120] Wu H B，Wang X，Chen J M，et al. Chinese Science Bulletin，2004，49(12)：1231—1235
[121] Liu X D，Zhu J，van Espen P，et al. Atmospheric Environment，2005，39(36)：6909—6918
[122] Ouyang B，Fang H J，Zhu C Z，et al. Journal of Environmental Sciences—China，2005，17(5)：786—788
[123] Yu J Z，Huang X F，Xu J H，et al. Environmental Science & Technology，2005，39(1)：128—133
[124] Du L，Xu Y F，Ge M F，et al. Atmospheric Environment，2007，41：7434—7439
[125] Derro E L，Murray C，Sechler T D，et al. J. Phys. Chem.，2007，111：11592—11601
[126] Li E X J，Konen I M，Lester M I，et al. J. Phys. Chem，2006，110：5607—5612
[127] 曾凡刚(Zeng F G)，王玮(Wang W)，杨忠芳(Yang Z F)等. 中国环境监测(Environmental Monitoring in China)，2001，117(14)：13—17
[128] Wang W，Liu H J，Yue X，et al. Journal of Geophysical Research—Atmospheres，2006，111(D18)：D18207
[129] 毕晓辉(Bi X H)，冯银厂(Feng Y C)，朱坦(Zhu T)等. 中国环境科学(China Environmental Science)，2007，27(5)：579—583
[130] 冯茜丹(Feng Q D)，党志(Dang Z)，王焕香(Wang H X)等. 环境科学(Environmental Science)，2006，27(12)：2386—2391
[131] 刘俊峰(Liu J F)，李金龙(Li J L)，白郁华(Bai Y H). 环境科学研究(Research of Environmental Sciences)，2006，28(3)：15—18
[132] 赵海波(Zhao H B)，郑楚光(Zheng C G). 环境科学学报(Acta Scientiae Circumstantiae)，2005，25(12)：1590—1596
[133] 胡敏(Hu M)，刘尚(Liu S)，吴志军(Wu Z J)等. 环境科学(Environmental Science)，2006，27(11)：2293—2298
[134] 王河涌(Wang H Y)，张文平(Zhang W P)，胡欣(Hu X)等. 中国环境监测(Environmental Monitoring in China)，2004，20(5)：44—46
[135] 王勇(Wang Y)，胡晏玲(Hu Y L). 新疆环境保护(Environmental Protection of Xinjiang)，2006，28(3)：15—18
[136] 时宗波(Shi Z B)，贺克斌(He K B)，陈雁菊(Chen Y J)等. 环境科学(Environmental Science)，2008，29(3)：551—556
[137] 李卫军(Li W J)，邵龙义(Shao L Y)，时宗波(Shi Z B)等. 环境科学(Environmental Science)，2008，29(1)：253—258
[138] 杨东贞(Yang D Z)，周怀刚(Zhou H G)，张忠华(Zhang Z H). 应用气象学报(Journal of Applied Meterological Science)，2002，13(4)：430—439

Atmos. Chem. Phys., 10, 7603–7615, 2010
www.atmos-chem-phys.net/10/7603/2010/
doi:10.5194/acp-10-7603-2010

Atmospheric Chemistry and Physics

Air quality during the 2008 Beijing Olympics: secondary pollutants and regional impact

T. Wang[1,2,3], W. Nie[1,2], J. Gao[3], L. K. Xue[1,2], X. M. Gao[1,2], X. F. Wang[1,2], J. Qiu[1], C. N. Poon[1], S. Meinardi[4], D. Blake[4], S. L. Wang[3], A. J. Ding[1], F. H. Chai[3], Q. Z. Zhang[2], and W. X. Wang[2,3]

[1]Department of Civil and Structural Engineering, The Hong Kong Polytechnic University, Hong Kong, China
[2]Environment Research Institute, Shandong University, Jinan, China
[3]Chinese Research Academy of Environmental Sciences, Beijing, China
[4]Department of Chemistry, University of California at Irvine, Irvine, USA

Received: 9 April 2010 – Published in Atmos. Chem. Phys. Discuss.: 12 May 2010
Revised: 22 July 2010 – Accepted: 2 August 2010 – Published: 16 August 2010

Abstract. This paper presents the first results of the measurements of trace gases and aerosols at three surface sites in and outside Beijing before and during the 2008 Olympics. The official air pollution index near the Olympic Stadium and the data from our nearby site revealed an obvious association between air quality and meteorology and different responses of secondary and primary pollutants to the control measures. Ambient concentrations of vehicle-related nitrogen oxides (NO_x) and volatile organic compounds (VOCs) at an urban site dropped by 25% and 20–45% in the first two weeks after full control was put in place, but the levels of ozone, sulfate and nitrate in $PM_{2.5}$ increased by 16%, 64%, 37%, respectively, compared to the period prior to the full control; wind data and back trajectories indicated the contribution of regional pollution from the North China Plain. Air quality (for both primary and secondary pollutants) improved significantly during the Games, which were also associated with the changes in weather conditions (prolonged rainfall, decreased temperature, and more frequent air masses from clean regions). A comparison of the ozone data at three sites on eight ozone-pollution days, when the air masses were from the southeast-south-southwest sector, showed that regional pollution sources contributed >34–88% to the peak ozone concentrations at the urban site in Beijing. Regional sources also contributed significantly to the CO concentrations in urban Beijing. Ozone production efficiencies at two sites were low (~3 ppbv/ppbv), indicating that ozone formation was being controlled by VOCs. Compared with data collected in 2005 at a downwind site, the concentrations of ozone, sulfur dioxide (SO_2), total sulfur (SO_2+$PM_{2.5}$ sulfate), carbon monoxide (CO), reactive aromatics (toluene and xylenes) sharply decreased (by 8–64%) in 2008, but no significant changes were observed for the concentrations of $PM_{2.5}$, fine sulfate, total odd reactive nitrogen (NO_y), and longer lived alkanes and benzene. We suggest that these results indicate the success of the government's efforts in reducing emissions of SO_2, CO, and VOCs in Beijing, but increased regional emissions during 2005–2008. More stringent control of regional emissions will be needed for significant reductions of ozone and fine particulate pollution in Beijing.

Correspondence to: T. Wang
(cetwang@polyu.edu.hk)

1 Introduction

The air quality in Beijing has been of great concern to both the Chinese government and researchers, especially after the city won the bid to host the 29th Summer Olympic Games. To significantly improve the city's air quality during the Games (8–24 August 2008), in addition to the long-term control measures (UNEP, 2009), the Chinese government took drastic actions to reduce the emissions of air pollutants from industry, road traffic, and construction sites (UNEP, 2009; Wang et al., 2009a, 2010b). From 1 July, some 300 000 heavily polluting vehicles (the so called yellow-label vehicles) were banned from driving in the Beijing Municipality, which covers an area of 16 808 km^2, and starting from 20 July, half of the city's 3.5 million vehicles were taken off the roads through the alternative day-driving scheme. In addition, all construction activities were halted, power plants were asked to use cleaner fuels, and some polluting factories were ordered to reduce their activity. Additional control was implemented after the start of the Games in order to further

Published by Copernicus Publications on behalf of the European Geosciences Union.

reduce the emissions from vehicles and petrol-filling stations (Wang et al., 2009a). In addition to the strict controls on air pollution sources in Beijing, neighboring provinces also reduced their industrial output. A preliminary assessment suggests that these emission-reduction measures reduced the emissions of SO_2, NO_x, CO, VOCs, and PM_{10} by 14%, 38%, 47%, 30%, and 20% in the Beijing area, respectively (UNEP, 2009). Much larger reductions of SO_2 (41%), NO_x (47%), VOCs (57%), and PM_{10} (55%) are suggested in a more recent study (Wang et al., 2010b).

The large reductions in pollution emissions in the summer of 2008 in Beijing represents a human-perturbation experiment of unprecedented scale, and provides a rare opportunity to study the impact of pollution emissions on the air quality and atmospheric chemistry of Beijing and the surrounding regions. From an air-quality management point of view, it is of critical importance to know how the anticipated large reduction in emissions improved the city's air quality. During the summer of 2008, we measured trace gases and aerosols at three sites in and around Beijing before, during, and after the Games. Here, we report the first results from the analysis of this dataset, which provide new insights into the role of meteorology, the response of secondary pollutants to the pollution control, and the contribution of regional pollution to the air quality in Beijing.

A number of papers have been published on the results of surface and satellite measurements during the Beijing Olympics, all indicating sharp decreases in the concentrations of the measured pollutants in Beijing during the period of the Olympics. On-road measurements reported significant (12–70%) decreases in the ambient concentrations of CO, NO_x, SO_2, black carbon (BC), benzene, toluene, ethylbenzene, and xylenes (BTEX), and PM_1 during the Olympics (Wang et al., 2009a). Atmospheric measurements at other urban sites showed a decrease in the concentration of 35–43% for fine and coarse particulate matter (Wang et al., 2009b), 74% for BC (Wang et al., 2009c), 47–64% for BTEX (Liu et al., 2009), and 35% for total non-methane hydrocarbons (Wang et al., 2010a). These results were based on a comparison of the data obtained during the Olympics with those from non-Olympic periods (before and/or after the Olympics and Para-Olympics). The concentrations of O_3, CO, SO_2, and NO_y in plumes from urban Beijing transported to a rural site deceased by 21–61% in August 2008 compared to the same month in 2007 (Wang et al., 2009d). Analyses of satellite data from GOME-2, OMI, and MODIS, by comparing the results obtained during August 2008 with those in the same period in previous years, have shown a decrease of 43–59% in nitrogen dioxide (NO_2) column over Beijing (Mijling et al., 2009; Witte et al., 2009), 13% in boundary-layer SO_2, 12% in CO at the 700 hPa-level over a large region encompassing Beijing and its southern neighboring provinces (Witte et al., 2009), and 11% in aerosol optical thickness over Beijing (Cermak and Kutti, 2009).

With the aid of a chemical transport model or a statistical model, some studies have attempted to examine the relative role of meteorology and emission reduction in the improvement of the air qualities during the Olympics. Wang et al. (2009c) attributed 55% of the ozone decrease at a rural site during the Olympics from the same period in the previous year to the change in meteorology during the two years. Wang et al. (2009a) suggested a more dominant role of meteorological effects than the emission reductions in the variation in their observed particulate matter at an urban site. Cermak and Kutti (2009) also suggested a more important role of the meteorology in explaining the decrease in aerosol optical thickness. Mijling et al. (2009) attributed the 60% reduction in tropospheric NO_2 to the emission control.

Most of these published studies so far have focused on primary pollutants, and there are few results on the levels and variation of secondary pollutants and on the extent of regional contribution during the drastic emission control in Beijing. In addition, little attention has been given to the paradoxical response of secondary pollutants during the first two weeks after the full traffic control. The present study attempts to examine these important topics. We first show the relationship between weather and the general air quality in Beijing as indicated by the official air pollution index and our own measurements; we then estimate the regional contribution to ozone pollution on eight days when Beijing was influenced by air masses from the North China Plain in the south; we also examine the ozone production efficiencies, and lastly we compare the data collected at a downwind site in 2005 and 2008 to gain insight into the changes in the composition of urban and regional plumes, and discuss the changes in Beijing and regional emissions during the past several years.

2 Methodology

2.1 Measurement sites

Field studies were conducted at three sites in and outside the Beijing urban area that lie roughly on a south-north axis. The three sites are shown in Fig. 1 and are described in the following.

Xicicun (XCC) is situated near the border between Beijing and the Hebei province (39°28′ N, 116°7′ E), and is 53 km southwest of the center of Beijing (Tiananmen). When the winds come from the south or southwest, this site is upwind of the Beijing urban area. The site is located in farmland with few nearby sources of pollution. The ozone and CO data from this site are reported in this paper. The measurements were conducted between 20 July and 25 August.

The Chinese Research Academy of Environmental Sciences (CRAES) is located 4 km north of the 5th ring road, 15 km from the city center, and 5.8 km from the National Olympic Stadium (the "Bird's Nest"). This site is

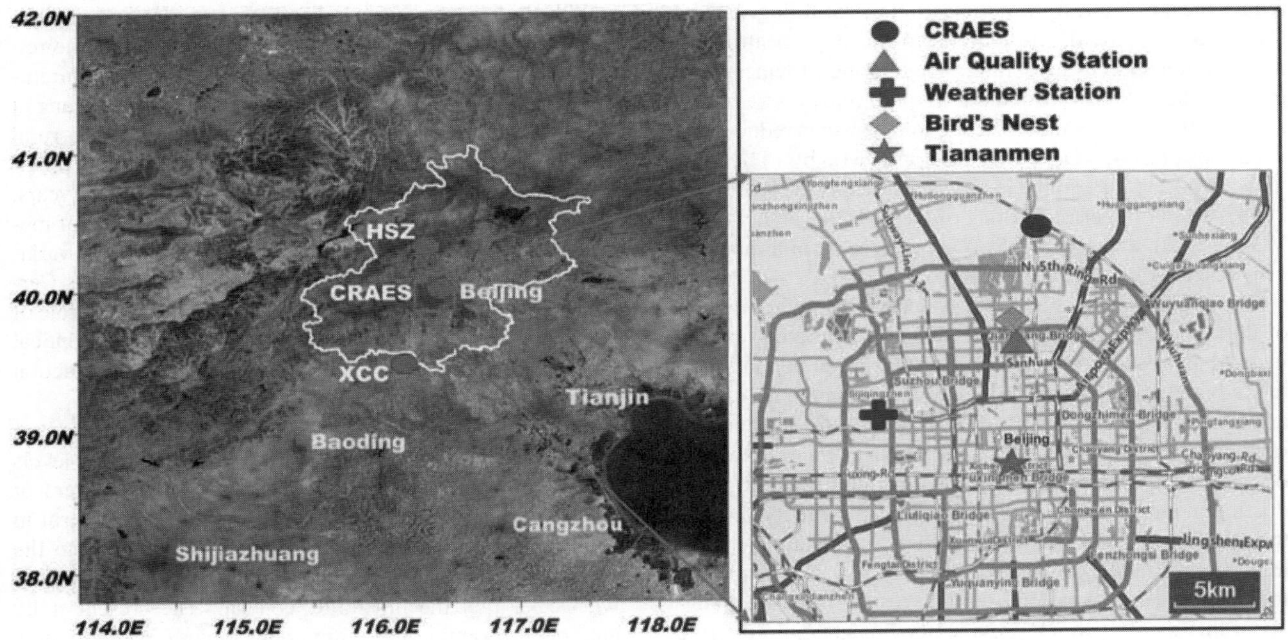

Fig. 1. Map showing the three measurement sites (XCC, CRAES, HSZ), the Beijing Municipality (the area within the white line), and the surrounding regions. Also shown are the location of the "Bird's Nest" National Olympic Stadium, a Beijing Municipal Environmental Protection Bureau's air quality station and a Beijing Municipal Meteorological Bureau's weather station whose data are used in this study.

immediately downwind of the maximum emissions from urban Beijing, and is the most heavily instrumented of the three sites. The site is located on the rooftop of a three-floor building in the Academy. Data on ozone, CO, NO_x, NO_y, VOCs from canisters, real-time $PM_{2.5}$ sulfate and nitrate taken between 10 July and 25 August are analyzed in this paper.

Heishanzhai (HSZ) is a rural mountainous area (40°22′ N, 116°18′ E, 280 m above sea level), approximately 50 km north of the center of Beijing. This site was used in our previous study in the summer of 2005, when high concentrations of ozone and secondary aerosol were observed (Wang et al., 2006; Pathak et al., 2009). In 2008, a different building was used for the measurements due to renovation work in the previous facility. This paper compares the O_3, CO, SO_2, NO_y, 24-h $PM_{2.5}$ mass, sulfate, nitrate, and NMHCs data during 10 July-August 25 2008 with the corresponding data from July of 2005.

2.2 Instrumentation

A brief description of the methods used to measure the gases and aerosols is given in the following. The reader is referred to relevant previous publications for further details. The limits of detections of the techniques are all sufficient to accurately measure the relatively high concentrations of gases and aerosols at the study sites.

Trace gases: O_3 was measured with a UV photometric analyzer (TEI model 49i), CO with a non-dispersive infrared analyzer (API model 300EU or API model 300E), and SO_2 with a pulsed UV fluorescence analyzer (TEI model 43C). Nitric oxide (NO) and NO_y were measured with a commercial chemiluminescence analyzer fitted with an externally placed molybdenum oxide (MoO) catalytic converter (Wang et al., 2001). NO_y is defined as the sum of NO, NO_2, HONO, HO_2NO_2, NO_3, PAN, HNO_3, N_2O_5, aerosol nitrate, and other organic nitrates etc, including nitrate in $PM_{2.5}$. A photolytic converter (Blue Light converter, Meteorologie Consult Gubh) coupled to a commercial NO analyzer was used to measure NO_2. The methods used to calibrate these instruments were the same as those reported by Wang et al. (2001). The NO_2 conversion efficiencies were determined by the gas-phase titration method, and an average efficiency of 35% was obtained.

Methane, NMHC, and halocarbon concentrations were determined by collecting whole-air samples in evacuated 2L electro-polished stainless steel canisters each equipped with a bellows valve. Between one and seven samples were collected each day, with more samples being collected on episode days. The sampling duration was 2 min. The canisters were shipped to the University of California at Irvine for chemical analysis using gas chromatography with flame ionization detection, electron capture detection, and mass spectrometer detection (Colman et al., 2001).

Aerosols: At the HSZ site, 24-h $PM_{2.5}$ samples were collected using a Thermo Andersen Chemical Speciation

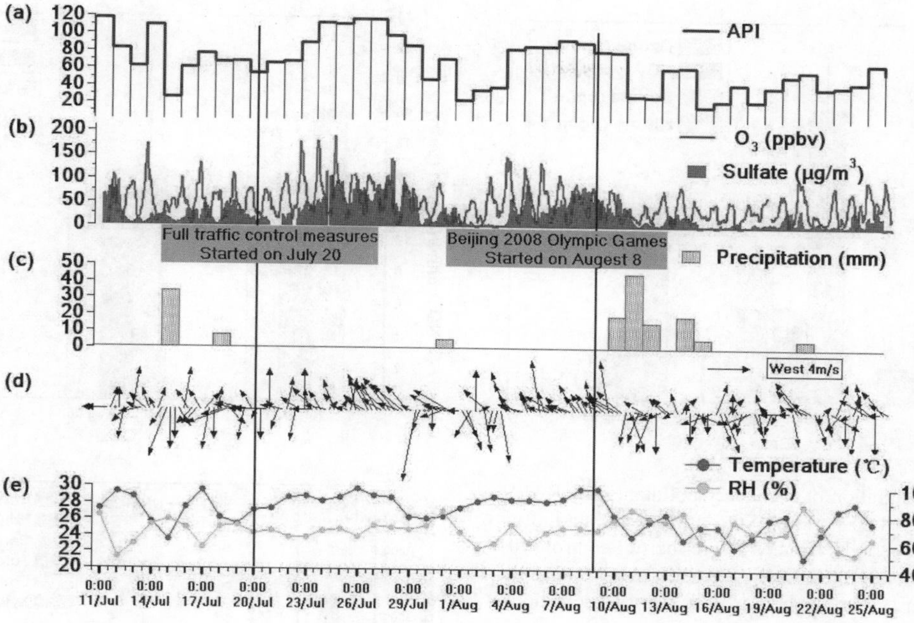

Fig. 2. Time series of **(a)** Air Pollution Index at a Beijing Municipal Environmental Protection Bureau's air-quality monitoring station at the Chaoyang National Olympics Sports Center, **(b)** hourly concentration of ozone and $PM_{2.5}$ sulfate measured at CRAES, **(c)** daily precipitation amount, **(d)** four-hourly wind vectors, **(e)** daily mean temperature and relative humidity. The meteorological data were obtained from a Beijing Municipal Meteorological Bureau's weather station (see Fig. 1 for its location).

Monitor (RAAS2.5-400, Thermo Electron Corporation) with Teflon filters (Teflo™, 2 μm pore size and 47 mm diameter, Pall Inc.) at a flow rate of 16.7 LPM (Wu and Wang, 2007). The $PM_{2.5}$ mass was determined using the standard gravimetric method, and the water soluble ions SO_4^{2-}, NO_3^-, F^-, Cl^-, NO_2^-, NH_4^+, K^+, Na^+, Mg^{2+}, and Ca^{2+} were analyzed using a Dionex ion chromatography 90 (Wu and Wang, 2007). At CRAES, real-time $PM_{2.5}$ ions were measured using an ambient ion monitor (URG 9000B, URG Corporation) (Wu and Wang, 2007). Another instrument (same model) was used in 2008, however, the negative artifact reported in the previous study was not observed.

2.3 Air pollution index, meteorological data, and back trajectories

In order to show the air quality at the main Olympic complex and its relationship with the secondary pollutants measured at our nearby CRAES, we obtained official Air Pollution Index (API) data (http://www.bjepb.gov.cn) at the Beijing Municipal Environmental Protection Bureau (BJEPB)'s air-quality monitoring station at Chaoyang Olympics Sports Center, located about 2 km south and southeast of the "Bird's Nest" (Fig. 1). The API is calculated based on the highest index of 24-h average concentrations of PM_{10}, SO_2, and NO_2 from noon of the present day to noon of the pervious day. An API of 0–50, 51–100, and 101–200 is classified as "excellent", "good", and "slightly polluted" condition, respectively (UNEP, 2009).

To help interpret the chemical data, we used surface meteorological data on precipitation, temperature, relative humidity, and wind speed and direction obtained from the Beijing Municipal Meteorological Bureau (BJMB) weather station located to the west of the city center (Fig. 1). The wind data were collected four times a day (02:00, 08:00, 14:00, and 20:00, local time), and the other data were daily averages. These data were obtained from Global Telecommunication Systems. In addition to surface winds, 48-h backward trajectories were calculated to identify the origin and transport pathway of large-scale air masses. The trajectories were calculated for four times a day (02:00, 08:00, 14:00, and 20:00, local time) using the NOAA ARL HYSPLIT model with GDAS (Global Data Assimilation System) data (http://ready.arl.noaa.gov/HYSPLIT.php), with the endpoint at the CRAES, and at an altitude of 100 m above ground level.

3 Results and discussion

3.1 Overall air quality and relation to weather conditions

Figure 2 shows the daily API, the hourly concentrations of ozone and sulfate at CRAES, and several meteorological

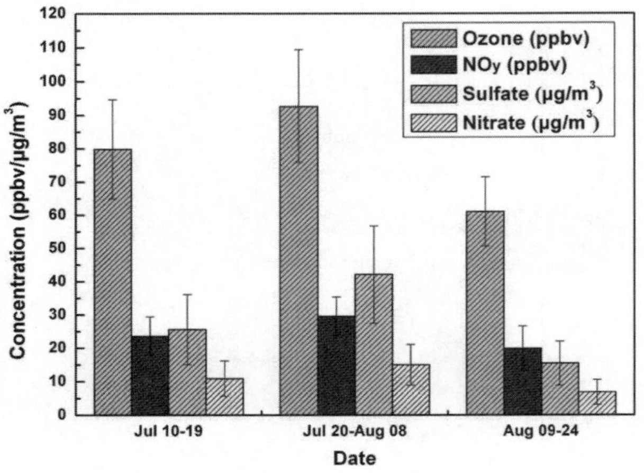

Fig. 3. Average concentration of secondary pollutants at CRAES: ozone (for the 11:00–18:00 period), NO_y, $PM_{2.5}$ sulfate and nitrate, during the three periods, namely before the full control (10–19 July), after the full control (20 July–8 August), and during the Olympics (9–24 August). Vertical bars are half standard deviations.

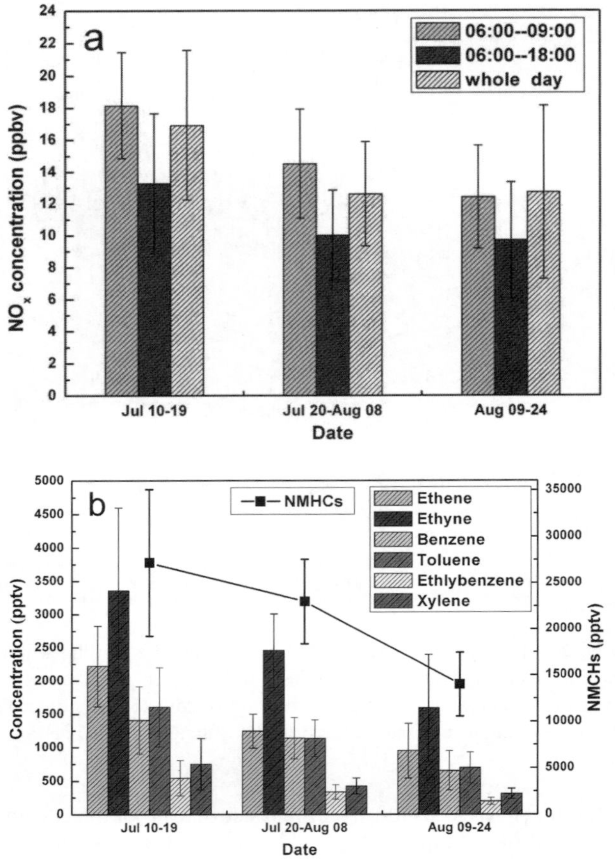

Fig. 4. Mean concentration of vehicle-related primary pollutants during the three periods at CRAES **(a)** NO_x during rush hours, daytime and whole day and **(b)** individual and total C_2-C_8 NMHCs. Vertical bars are half standard deviations. The number of VOC samples is 6, 14, 14 for period 1, 2, and 3, respectively.

parameters from BJMB from July 11 to August 25. During the observation period, PM_{10} was the dominant pollutant of the three reported pollutants (PM_{10}, SO_2, and NO_2) at the BJEPB site. Thus the officially reported air quality represented levels of coarse particulate matter. An API of 0–50, 51–100, and 101–200 corresponds to 0–50, 52–150, 152–350 μg/m^3 of PM_{10}, respectively (UNEP, 2009). However, because the computation of API does not include ozone, the API does not adequately reflect the situation of photochemical pollution. Thus the combined API and the secondary pollutants at our site shown in Fig. 2 better illustrate the variations in air quality (for both primary and secondary pollutants) in the Olympics complex and the adjacent areas.

The observation period can be divided into three parts: (1) before the full-scale control (11–19 July), (2) after the full-scale control but before the Olympics (20 July to 8 August), and (3) during the Olympics (9–24 August). Moderately high API (60–120) (and ozone and sulfate concentrations) was recorded in the first period. After the full traffic control came into effect, two multi-day pollution episodes occurred: one between 23 and 29 July, and one started three day before the Olympic openings and lasted for five days (4–9 August). The highest readings of API, ozone, and sulfate occurred during this period: nine days had a maximum 1-h ozone exceeding China's ambient air quality standard of 100 ppbv, with the highest value of 190 ppbv being recorded on 24 July; very high concentrations of sulfate (hourly values of 80–140 μg/m^3) were also observed. Good air quality was recorded on most days during the Games, as indicated by the lowest API values and the concentrations of secondary pollutants (see Fig. 2).

Figure 2 reveals that while the API captured the day-to-day variation of fine sulfate, which can be explained by the fact that sulfate is a part of PM_{10}, it did not adequately reflect the concentrations and variation of ozone: the three highest ozone days on 22–24 July were not indicated by the API, illustrating the deficiency of the current API in representing photochemical pollution.

Figure 3 gives the average concentrations and half of the standard deviations in the three periods for ozone (11:00–18:00, local time), sulfate, nitrate, and NO_y at CRAES. They represent secondary gases and aerosols. In aged air masses NO_y contains a large fraction of oxidation products of NO_x, such as in period 2 during which NO_x was only ~40% of NO_y (see Figs. 3 and 4).

The results shown in Figs. 2 and 3 are striking. The control measures were expected to reduce significantly emissions from vehicles, power generation, and other activities (e.g, Wang et al., 2010b). Indeed, NO_x measured at CRAES

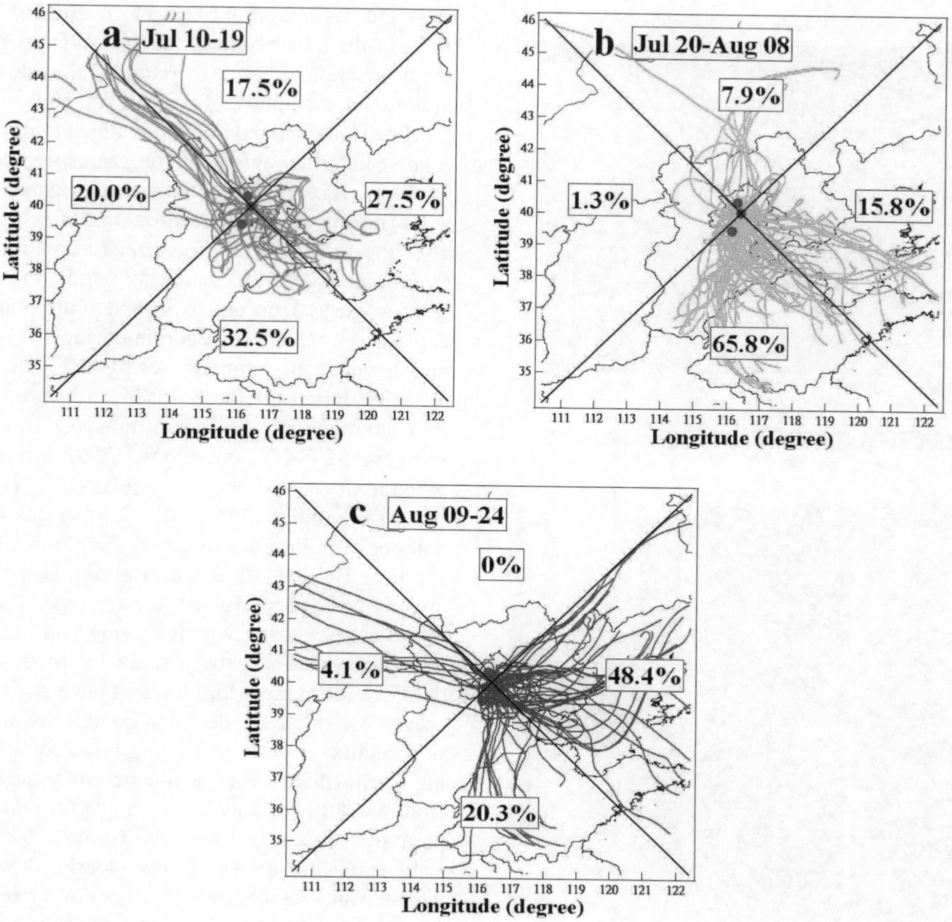

Fig. 5. Forty-eight hour backward trajectories during **(a)** 10–19 July **(b)** 20 July–8 August and **(c)** 9–24 August. Numbers are the percentage contributions from each of the four sectors. The percentage of non-defined (i.e. looping) trajectories is 2.5%, 9.2% and 27.2% for period 1, 2 and 3, respectively. The red points are the locations of the XCC, CRAES, and HSZ sites.

decreased by 20% and 25% in the morning (06:00–09:00, local time) and for the whole day, respectively (Fig. 4). Two-tailed *t* tests show that these differences are statistically significant at 99% confidence level (i.e., $P<0.01$). The canister samples collected in the afternoon also showed 20–45% decreases for ethene, ethyne, benzene, toluene, ethylbenzene and xylenes which are typical compounds from vehicular emissions. However, the levels of statistical significance for the changes in these VOCs are lower ($P>0.05$, that is, below 95% confidence level) than the continuously measured constituents due in part to the fewer samples of VOCs. The drop in NO_x and the apparent decreases in the VOCs at CRAES are consistent with other on-road and ambient measurements of NO_x and VOCs (Wang et al., 2009a, 2010a), indicating the effectiveness of the control measures on reducing vehicle emissions. The decreasing levels of toluene, ethylbenzene, and xylene were also due to the control of the usages of paints and solvents and of evaporation from petrol stations.

In contrast to the decreasing levels of NO_x and the VOCs, the average concentration of ozone, NO_y, $PM_{2.5}$ sulfate and nitrate at CRAES increased by 16% ($P<0.05$), 25% ($P<0.01$), 64% ($P<0.01$), and 37% ($P<0.01$), respectively after the full control (see Fig. 3). This result reveals that emission control implemented after July 20 was not sufficient to eliminate the occurrences of high concentrations of ozone and particulate under adverse meteorological conditions.

Was the observed increase in ozone concentrations after the full control a result of decreased titration by NO_x? According to Fig. 4, NO_x concentrations at CRAES decreased by 3–4 ppbv in period 2, which was much smaller than the increase of 13 ppbv in daytime O_3 concentration (see Fig. 3), indicating that the reduced titration by local NO was not the dominant cause of the O_3 increase in period 2. As to ozone production, both ambient data (Fig. 4 of the present study) and emission inventory (Wang et al., 2010b) indicate a larger reduction in the emission of reactive VOCs than in NO_x after

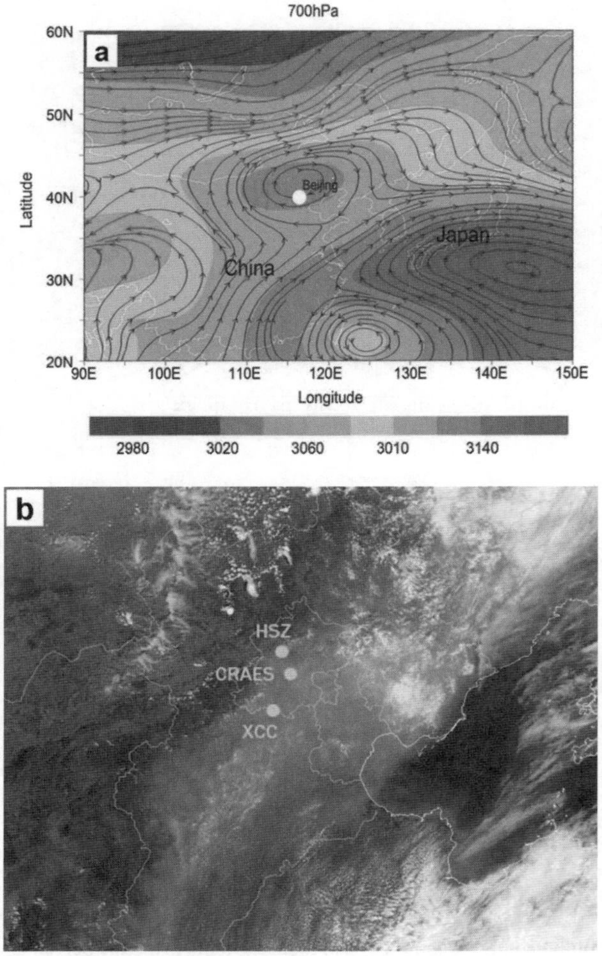

Fig. 6. **(a)** Mean wind and geopotential height for a multi-day pollution episode on 22–28 July 2008 **(b)** MODIS true-color imagery on 24 July.

20 July. The control measures are expected to reduce the production of ozone from Beijing, regardless whether the ozone formation is controlled by VOCs, NO_X, or both. Therefore, the observed increasing (both mean and peak) ozone concentrations must be mainly due to other reason(s). A 64% increase in the mean sulfate concentration in period 2 was also in sharp contrast of an anticipated 40% decrease in SO_2 emission from Beijing (c.f. Fig. 1a in Wang et al., 2010b). We show later in this section that regional pollution played a vital role in the air quality during 20 July–8 August.

The concurrent measurements of ozone and sulfate (Fig. 2) reveal another interesting phenomenon: the highest ozone concentrations preceded those of sulfate, indicating that somewhat different meteorological conditions promoted the formation of ozone and particulate sulfate. The high levels of ozone (and by inference other oxidants) may have accelerated the oxidation of SO_2 to form sulfate. The hazy conditions in the later phase of the episode may have suppressed the production of ozone by reducing the rate of photolysis reactions and via uptake of radicals and NO_X on moist aerosols.

Meteorological data shown in Fig. 2 (ambient temperature, relative humidity, wind direction and speed, and daily rainfall readings) reveal obvious impact of weather on the air quality. The increasing concentrations of secondary pollutants after the full control was associated with a lack of rainfall and the prevalence of southerly winds; the good air quality from August 10 through to the end of the Games can be partly explained by the persistent rain during 10–16 August, which also lowered the temperatures by a few degrees. Back trajectories during the three periods are shown in Fig. 5. They indicate much more frequent transport of regional pollution from the SE-S-SW directions (66%) during period 2 compared to the period before the full control (33%) and the period of Olympics (20%). It is known that there are a large number of pollution sources in the North China Plain to the south of Beijing. It is worth noting that the two episodes on July 23-29 and August 4-9 are typical summertime pollution cases, during which Beijing and the surrounding regions are influenced by a weak high-pressure system over the Hebei Province (Fig. 6a) that gave rise to hot and humid "sauna" weather. Under such conditions, widespread pollution occurs not only in Beijing but also over a large part of the North China Plain, as illustrated by the true-color image from MODIS for July 24 (Fig. 6b) (http://rapidfire.sci.gsfc.nasa.gov/subsets/?subset=FAS_China4.2008206.aqua.1km).

In sum, the above results clearly show the effects of weather and regional sources on the air quality during the study period. Additional modeling studies are needed to resolve the complex interplay between the changing emissions and meteorological conditions and to quantify local and regional contributions.

3.2 Regional contribution to ozone pollution in Beijing

It has been recognized that neighboring Tianjin, Hebei, Shanxi, and Shandong can have an important impact on Beijing's air quality (An et al., 2007; Chen et al., 2007; Streets et al., 2007; Jia et al., 2008; Wang et al., 2008; Lin et al., 2009). Model calculation by Streets et al. (2007) suggested that 35–60% of the simulated ozone at the Olympic Stadium site under July 2001 polluted conditions could be attributed to sources outside Beijing. Lin et al. (2009) observed elevated concentrations of O_3, CO, SO_2, and NO_X during summer at a polluted rural site (Gucheng), which is 110 km southwest of Beijing, and found that 80% of the northward trajectories from the site passed Beijing in August. Guo et al. (2010) compared particulate data at a site south of Beijing and another site inside Beijing in the summer of 2006 and showed that almost 90% fine sulfates were from regional contributions.

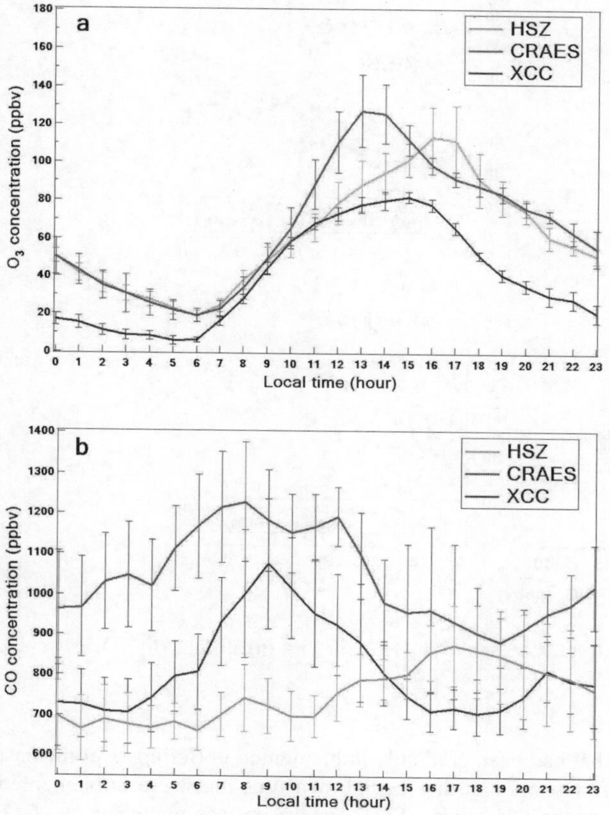

Fig. 7. Average diurnal variation of **(a)** ozone and **(b)** CO at three sites for eight ozone-pollution days when southerly winds prevailed. Vertical bars are half standard deviations.

The concurrent observations of ozone at three sites in the present study allow a direct examination of the regional contribution to ozone pollution in Beijing and the evaluation of Beijing plumes to downwind areas. To examine the regional contribution from the south, we selected days with ozone pollution (defined as 1-h ozone $\geq$100 ppbv at either of the three sites) that met the following criteria: (1) a day on which all four backward trajectories were from the SE-S-SW sector and (2) surface winds were also from that sector throughout the day, that is, no reversal in wind direction. Eight days were identified, including 23–28 July and 4–6 August. (There were no ozone data in the afternoon at CRAES on 25 July, and thus this day was not included.)

Figure 7a shows the average diurnal variation in ozone for the eight days at the three sites. By taking the difference between the maximum 1-h ozone concentration at the urban site and the ozone value at the same time at the upwind site, we can estimate the contribution from regional sources in the south in each case. The result shows the regional sources contributed 34–88% to the peak ozone at the urban site for the eight cases, with an average contribution of 62%. It should be pointed out that these estimates are likely to be the lower limits of the regional contribution as additional ozone was produced as the air was transported from the upwind site and the urban site. Figure 7b shows the mean diurnal plot for CO for the eight cases. The mean CO concentration at the upwind site was 69–91% of that at the urban site during morning and early afternoon, indicating strong regional contribution of CO (and possibly other ozone precursors). Emissions from cities, townships, villages, and agriculture fields in the North China Plain are believed to the sources.

Figure 7 shows that the ozone (and CO) peak at the downwind site lagged behind that at the urban site in time, which indicates that regional and Beijing plumes were transported to the mountain site in the afternoon (Wang et al., 2006). The lower ozone peak values at the downwind site (except in two cases) indicate that (1) ozone formation had reached maximum strength before reaching the site and/or (2) the ozone-rich plumes had been subjected to dilution during transport.

3.3 Ozone production efficiencies in urban and downwind areas

Ozone is formed by the oxidation of VOCs in the presence of NO_x and sunlight. Ozone is produced when NO_x is oxidized to various forms of NO_y, such as nitric acid, aerosol nitrate, and PAN. The ozone production efficiency (OPE), defined as the number of O_3 molecules produced for each NO_x molecule oxidized, can be expressed by the observed O_3 versus NO_z (NO_z=NO_y-NO_x) (Trainer et al., 1993). In polluted environments, O_3+NO_2 versus NO_z is often used to consider the effect of NO titration. Figure 8 shows the scatter plots at CRAES and HSZ for the hourly data collected in the afternoon (12:00–17:00, local time), which is when the maximum photochemistry occurs. (NO_z was not measured at the southern site.) Expanding the period to include the data from 10:00 in the morning showed very similar regression slopes with a smaller correlation coefficient (figure not shown).

Several interesting features were observed. First, the upper concentrations of ozone and NO_z are comparable at the two sites despite the more remote location of HSZ, suggesting that both sites experienced serious photochemical pollution. Second, a non-linear ozone-NO_z relationship was observed, with the slope ([O_3+NO_2]/[NO_z]) at NO_z <10 ppbv being larger (6.5±0.54 and 7.7±0.78 ppbv/ppbv) than that at $NO_z \geq 10$ ppbv (2.7±0.49 and 4.0±0.80 ppbv/ppbv). This can be explained by regional/diluted urban air being NO_x-limited and the VOC-limited conditions of the polluted air masses. (Note that the slopes were determined with the reduced major axis (RMA) method to take into account measurement uncertainties in both the x and y variables, Hirsch and Gilroy, 1984.) We also separately examined the scatter plots for the periods before and after the full traffic control, but did not find a significant difference in the respective slopes.

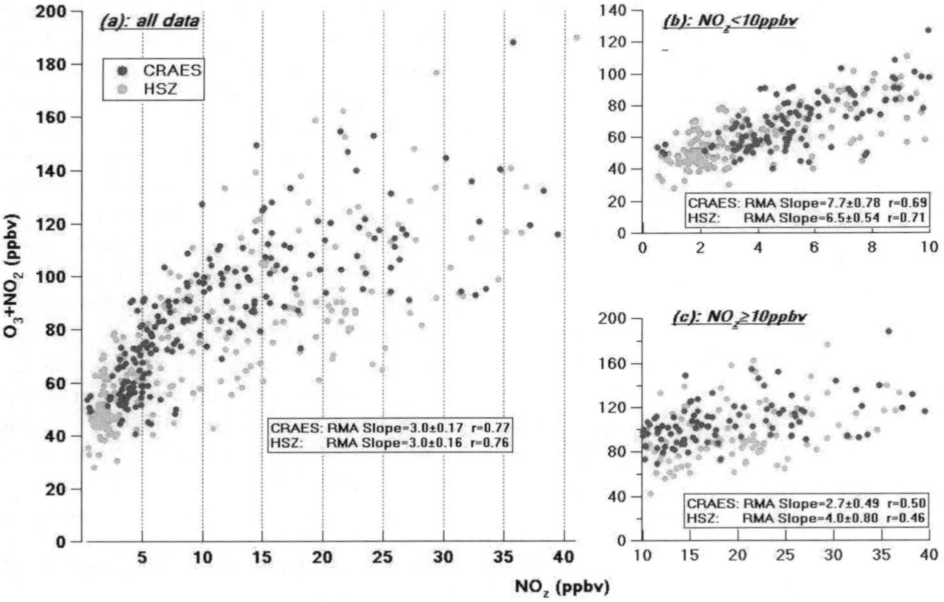

Fig. 8. Scatter plots of O_3+NO_2 versus NO_z in the afternoon (12:00–17:00) at CRAES (red) and HSZ (green): **(a)** all data, **(b)** NO_z<10 ppbv, and **(c)** $NO_z \geq 10$ ppbv.

Chou et al. (2009) observed an average OPE (defined as the regression slope of (([O_3]+[NO_2]+[NO_z]) versus [NO_z]) of 6.1 on the Peking University campus in August 2006. The OPE in the present study (using their definition of OPE and for the period of 09:00-13:00 local time) were 3.0±0.20 and 2.6±0.17 at the urban and the mountainous site, respectively, which are much smaller than their value. The data at the HSZ site in 2005 and 2008, however, showed a smaller difference in afternoon O_3-NO_y slope (3.7±0.28 vs. 2.8±0.15) (figure not shown). The low values of OPEs in our study indicated a strong VOCs-limited regime in the summer of 2008. The similar OPE at the urban and rural site is also interesting. Previous studies have often shown that as urban plumes advect to rural areas, the loss of NO_y components such as nitric acid and aerosol nitrate due to wet and dry depositions can lead to increased O_3/NO_z ratios (Nunnermacker et al., 1998; Wood et al., 2009). The similar values at the two sites in our study can be partially explained by the input of additional ozone precursors between the urban and the mountainous sites. The overall OPE in the present study is comparable to the observed values in Nashville (4.7) (Zaveri et al., 2003), but smaller than that (6–11) observed in Houston (Daum et al., 2004) and in Mexico City (6.2) (Wood et al., 2009).

3.4 Chemical changes in Beijing and regional plumes

In this section, we compare a set of chemical species that were measured in 2008 and 2005 at the HSZ site to see whether there were significant changes in the chemical mix of polluted air masses resulting from the control measures that had been gradually implemented in Beijing over the past few years and the special measures that were implemented for the Olympics. The comparison was made for O_3, CO, NO_y, SO_2, CO, 24-h $PM_{2.5}$ sulfate and nitrate, 24-h $PM_{2.5}$ mass, and C_2-C_8 NMHCs, and only for polluted air masses from the SE-S-SW directions, as determined by the back trajectories. The plume data were collected on 7–8 and 13–19 July in 2005 and on 8–10 and 22–28 July, 4–8 and 24–25 August in 2008. The plume data in 2008 were dominated by the two multi-day episodes that occurred after the full control. Table 1 shows the mean values, standard deviations, and number of samples for the aforementioned species, together with the statistics for temperature, relative humidity, wind speed, and solar radiation measured at the site for the period of plume impact. The results of a *t*-test of the levels of significance in the difference between the means in the two years are also given. The average values of CO and NO_x on 8–10 July in 2008 (before the full control) were a 10–15% larger compared with the remaining data in 2008 (after the full control), reflecting the effect of the traffic control. But because only two canister samples were collected on July 8-10, it was difficult to compare them with the data collected after the full control. Thus the plume data in 2008 were not further divided in Table 1.

Table 1 shows a similar mean temperature and sunlight intensity in the two periods (the mean temperature: ∼25°, and mean solar radiation: ∼130 W/m^2), but the 2008 period was more humid (RH: 87% versus 75%). Although the mean wind speed was larger in 2008, both were below 1 m/s,

Table 1. Comparison of chemical composition and meteorological parameters in plumes measured at a downwind site in north Beijing in 2005 and 2008.

Species/meteorological parameters	2005		2008		P (t-test)
	Mean±stdev	n	mean±stdev	n	
O_3 (ppbv)	59.4±30.5	212	54.5±28.4	446	<0.05
SO_2 (ppbv)	7.5±8.1	211	2.7±2.3	444	<0.01
CO (ppbv)	978±414	211	711±247	444	<0.01
NO_y (ppbv)	21.1±8.7	212	20.6±8.9	442	0.47
sulfate (μg/m^3)	49.8±28.0	8	48.0±22.7	19	0.87
nitrate (μg/m^3)	20.1±9.0	8	14.2±9.0	17	0.15
$PM_{2.5}$ (μg/m^3)	130.3±51.6	8	128.5±53.3	19	0.93
Ethane (pptv)	3378±1450	14	3236±1137	55	0.74
Ethyne (pptv)	2773±1261	14	2588±1064	55	0.62
Propane (pptv)	2060±1213	14	2017±925	55	0.90
Ethene (pptv)	1736±1462	14	1378±786	55	0.39
n-Butane (pptv)	1045±701	14	1185±522	55	0.49
Benzene (pptv)	1077±514	14	1119±409	55	0.78
i-Butane (pptv)	808±526	14	1101±436	55	0.07
i-Pentane (pptv)	1059±692	14	1022±422	55	0.85
Toluene (pptv)	1577±1184	14	776±443	54	<0.05
Ethylbenzene (pptv)	271±206	14	196±110	53	0.21
m+p-Xylene (pptv)	475±418	14	182±227	55	<0.05
o-Xylene (pptv)	173±137	14	52±68	55	<0.01
1,2,4-Trimethylbenzene(pptv)	151±87	14	18±48	55	<0.01
T (°C)	25.2±2.9	215	25.4±2.7	446	0.43
RH (%)	74.9±13.5	215	86.6±11.6	446	<0.01
WS (m/s)	0.68±0.45	215	0.34±0.26	446	<0.01
Solar radiation (w/m^2)	134.1±200.8	215	130.7±188.9	446	0.80

probably due to the effect of mountainous terrain. The surface wind speeds at HSZ may not be representative of the flow in the planetary boundary layer in the area. Thus in addressing air-mass transport, we used back trajectories and the wind in urban Beijing which showed higher speeds (2–4 m/s) on episode days (see Fig. 2).

For the chemical compounds, the average ozone concentration in the plumes decreased by 8.2% in 2008 compared with 2005 (59.4 ppbv versus 54.5 ppbv). The maximum 1-h value also decreased to 186 ppbv in 2008 from 199 ppbv in 2005, and the 90th percentile was 115 ppbv compared to 120 ppbv in 2005. The mean SO_2 value declined by 64% in 2008 (7.5 ppbv compared with 2.7 ppbv), and the CO levels decreased by 27%. In contrast, no statistically significant difference was observed for NO_y, $PM_{2.5}$ mass, and fine sulfate in 2008 compared with 2005. Of the NMHCs, reactive aromatics such as toluene, xylenes, and 1,2,4-trimethylbenzene decreased by 51%, 64%, and 88%, respectively, but the levels of C_2–C_5 alkanes, ethyne, and benzene did not change significantly. The decreasing concentrations of ozone, CO, and SO_2, in polluted air masses in our study are in agreement with the results of comparing 2007 and 2008 data at a different rural site that is about 42 km northeast of our site (Wang et al., 2009d). However, the similar NO_y values in our study in 2008 and 2005 compared with a 21% decrease in 2007–2008 may indicate an increase in NO_x emissions between 2005 and 2007. The possible changes in emission during 2005–2008 will be discussed later in this section. The above results indicate that the concentrations of some pollutants at this downwind site have not been reduced, in spite of concerted efforts in Beijing in the last three years before the Olympics and the drastic measures before the Olympics.

Ambient concentrations of the gases and aerosols in the two periods were influenced by the meteorological conditions as well as source emissions. As shown in Table 1, the plumes observed in 2008 were more humid which could lead to faster conversions of SO_2 and removal of aerosols and NO_y. This could partially explain the sharp (64%) reduction in the SO_2 level in 2008, but not for the decreases in CO and VOCs as they are not affected by wet removal and in-cloud processes. The comparable $PM_{2.5}$, NO_y, and long lived VOC levels in the two periods imply no reductions in their total emissions in the upwind regions.

We next examine the possible emission changes in the past three years (July 2005–June 2008) and during the special control period (July–August 2008) in both Beijing and the surrounding regions. Atmospheric concentrations of SO_2, CO, NO_x, most of VOCs, and PM_{10} in urban areas of Beijing have decreased in the recent years, indicating reduced emissions of these pollutants in Beijing (UNEP, 2009; Shao et al., 2009). During the special control period, the emissions in Beijing of all the above pollutants have decreased (UNEP, 2009; Wang et al., 2010b). Thus we believe the emissions of SO_2, CO, NO_x, PM, most of VOCs in Beijing decreased between the two data-collection periods in 2005 and 2008.

The emission trends in the North China Plain are less clear. National SO_2 emission in China reached the peak in 2006 and decreased by 8.9% during 2005-2008 (NBSC, 2006-2009). The emission of NO_x, on the other hand, is expected to continue the recent upward trend (Zhang et al., 2009) due to the slow implementation of advanced NO_x control in coal-fired power plants and to a relatively small decrease in NO_x emissions for new vehicles (Zhang et al., 2009; Zhao et al., 2008). The CO emissions are thought to have stabilized due to a sharp decrease in emissions from new cars (Zhang et al., 2009), while the VOC emissions increased by 29% during 2001-2006. During the special control period, the emissions of these pollutants are expected to have decreased, although no information on the amount of the reduction is available.

With the consideration of the emission situation, we interpret the observed concentration changes in 2005 and 2008 as follows. The decreases in the concentrations of SO_2 (and the total sulfur), CO, reactive aromatics in 2008 are mainly due to the long-term and special control measures implemented in Beijing: SO_2 from control of emissions in coal-fired power plants, CO from reduced emissions from vehicles and possibly also from open fires in the North China Plain, toluene, xylenes, and 1,2,4-trimethylbenzene from the control of emissions from vehicles and petrol stations and from solvent and paint use in the run up to and during the Olympics. On the other hand, the insignificant changes in aerosol ($PM_{2.5}$ mass and sulfate), NO_y, and longer lived NMHCs during the two periods imply increased emissions from sources outside Beijing and lacking stringent NO_x control in coal-fired power plants.

The increased regional emission during 2005–2008 inferred from the above analysis and the regional contribution to the very high concentrations of secondary air pollutants after the drastic control measures in the summer of 2008 suggest that more stringent control of the regional emissions will be needed in order to significantly improve the air quality (especially ozone and secondary aerosols) in Beijing and the surrounding regions.

4 Summary and conclusions

Atmospheric measurements from this work clearly demonstrate the strong impact of regional sources and meteorology on the variations of secondary pollutants (ozone, fine sulfate and nitrate) in the summer of 2008 in Beijing. The pollution reductions measures for the Beijing Olympics in July and August were successful in reducing atmospheric concentrations of primary pollutants such as NO_x and VOCs. However, high levels of ozone (with hourly values up to 190 ppbv) and secondary aerosols (with hourly sulfate up to 140 $\mu g/m^3$) still occurred at an urban site after the full control took effect, which was strongly associated with the transport of chemically processed air masses from the North China Plain. Regional sources were shown to have significant contribution to the concentrations of ozone and CO (approximately 62% and 77%, on average, respectively) at the urban site under southerly winds. Much improved air quality on most of the days during the Olympics also had apparent relationship with weather changes (persistent rainfall, lower temperature, and easterly air flow). Further modeling studies are needed to quantify the relative role of the emission reduction and weather changes and the contribution of local versus regional sources to the air quality changes.

By comparing with the data collected in July 2005 at the plume-impacted downwind site, we found similar concentrations of $PM_{2.5}$ mass, $PM_{2.5}$ sulfate, total reactive nitrogen and several long lived VOCs. We interpret this as evidence of growing regional emissions during the past several years. On the other hand, the sharp reductions in total CO, SO_2 (and total sulfur), and reactive aromatics suggest the success of the government's efforts in reducing emissions in Beijing by long-term and special measures. The increasing emissions from regional sources could make the pollution control effects in Beijing less effective in mitigating ozone and fine aerosol problems, which are regional in nature. Thus, more stringent controls of regional sources are needed to further improve the air quality in Beijing and the surrounding regions.

Acknowledgements. We thank Jing Wang, Rui Gao, Ravi Pathak, Youping Shou, Linlin Wang, Chao Yuan, Pengju Xu, Zheng Xu, Yangchun Yu, Waishing Wu, Xuehua Zhou, Joe Cheung, and Xuezhong Wang for their contributions to the field work. We thank NOAA Air Resources Laboratory for the provision of the HYSPLIT model. This study was funded by the Research Grants Council of the Hong Kong Special Administrative Region (Project No. PolyU 5294/07E), the National Basic Research Program of China (973 Project No. 2005CB422203), and the Hong Kong Polytechnic University (Project No. 1-BB94).

Edited by: D. Parrish

References

An, X., Zhu, T., Wang, Z., Li, C., and Wang, Y.: A modeling analysis of a heavy air pollution episode occurred in Beijing, Atmos. Chem. Phys., 7(12), 3103–3114, doi:10.5194/acp-7-3103-2007, 2007.

Cermak, J. and Knutti, R.: Beijing Olympics as an aerosol field experiment, Geophys. Res. Lett., 36, L10806, doi:10.1029/2009GL038572, 2009.

Chen, D. S., Cheng, S. Y., Liu, L., Chen, T., and Guo, X. R.: An integrated MM5-CMAQ modeling approach for assessing transboundary PM_{10} contribution to the host city of 2008 Olympic summer games – Beijing, China, Atmos. Environ., 41 (6), 1237–1250, 2007.

Chou, C. C. K., Tsai, C. Y., Shiu, C. J., Liu, S. C., and Zhu, T.: Measurement of NO_y during Campaign of Air Quality Research in Beijing 2006 (CAREBeijing-2006): Implications for the ozone production efficiency of NO_x, J. Geophys. Res.-Atmos., 114, D00G01, doi:10.1029/2008JD010446, 2009.

Colman, J. J., Swanson, A. L., Meinardi, S., Sive, B. C., Blake, D. R., and Rowland, F. S.: Description of the analysis of a wide range of volatile organic compounds in whole air samples collected during PEM-Tropics A and B, Anal. Chem., 73 (15), 3723–3731, 2001.

Daum, P. H., Kleinman, L. I., Springston, S. R., Nunnermacker, L. J., Lee, Y. N., Weinstein-Lloyd, J., Zheng, J., and Berkowitz, C. M.: Origin and properties of plumes of high ozone observed during the Texas 2000 Air Quality Study (TexAQS 2000), J. Geophys. Res.-Atmos., 109, D17306, doi:10.1029/2003JD004311, 2004.

Guo, S., Hu, M., Wang, Z. B., and Zhao, Y. L.: Size-resolved aerosol water-soluble ionic compositions in the summer of Beijing: implication of regional secondary formation, Atmos. Chem. Phys., 10, 947–959, doi:10.5194/acp-10-947-2010, 2010.

Hirsch, R. M. and Gilroy, E. J.: Methods of fitting a straight line to data: examples in water resources, Water Res. Bull., 20(5), 705–711, 1984.

Jia, Y. T., Rahn, K. A., He, K. B., Wen, T. X., and Wang, Y. S.: A novel technique for quantifying the regional component of urban aerosol solely from its sawtooth cycles, Geophys. Res. Lett., 113, D21309, doi:10.1029/2008JD010389, 2008.

Lin, W. L., Xu, X. B., and Zhang, X. C.: Characteristics of gaseous pollutants at Gucheng, a rural site southwest of Beijing, J. Geophys. Res.-Atmos., 114, D00G14, doi:10.1029/2008JD01 0339, 2009.

Liu, J. F., Mu, Y. J., Zhang, Y. J., Zhang, Z. M., Wang, X. K., Liu, Y. J., and Sun, Z. Q.: Atmospheric levels of BTEX compounds during the 2008 Olympic Games in the urban area of Beijing, Sci. Total Environ., 408, 109–116, 2009.

Mijling, B., van der A, R. J., Boersma, K. F., Van Roozendael, M., De Smedt, I., and Kelder, H. M.: Reductions of NO_2 detected from space during the 2008 Beijing Olympic Games, Geophys. Res. Lett., 36, L13801, doi:10.1029/2009GL038943, 2009.

NBSC, National Bureau of Statistics of China: China Statistical Yearbook, 2006–2009, China Statistics Press: Beijing, 2006–2009.

Nunnermacker, L. J., Imre, D., Daum, P. H., Kleinman, L., Lee, Y. N., Lee, J. H., Springston, S. R., Newman, L., Weinstein-Lloyd, J., Luke, W. T., Banta, R., Alvarez, R., Senff, C., Sillman, S., Holdren, M., Keigley, G. W., and Zhou, X.: Charaterization of the Nashville urban plume on July 3 and July 18, 1995, J. Geophys. Res., 103(D21), 28129–28148, 1998.

Pathak, R. K., Wu, W. S, and Wang, T.: Summertime $PM_{2.5}$ ionic species in four major cities of China: nitrate formation in an ammonia-deficient atmosphere, Atmos. Chem. Phys., 9, 1711–1722, doi:10.5194/acp-9-1711-2009, 2009.

Shao, M.,Wang, B., Lu, S. H., Liu, S. C., and Chang, C. C.: Trends in summertime non-methane hydrocarbons in Beijing City, 2004–2009, IGACtivity News Letter, 42, 18–25, 2009.

Streets, D. G., Fu, J. S., Jang, C. J., Hao, J. M., He, K. B., Tang, X.Y., Zhang, Y. H., Wang, Z. F., Li, Z. P., Zhang, Q., Wang, L. T., Wang, B. Y., and Yu, C.: Air quality during the 2008 Beijing Olympic Games, Atmos. Environ., 41(3), 480-492, 2007.

Trainer, M., Parrish, D. D., Buhr, M. P., Norton, R. B., Fehsenfeld, F. C., Anlauf, K. G., Bottenheim, J. W., Tang, Y. Z., Wiebe, H. A., Roberts, J. M., Tanner, R. L., Newman, L., Bowersox, V. C., Meagher, J. F., Olszyna, K. J., Rodgers, M. O., Wang, T., Berresheim, H., Demerjian, K. L., and Roychowdhury, U. K.: Correlation of ozone with NO_y in photochemically aged air, J. Geophys. Res.-Atmos., 98(D2), 2917–2915, 1993.

UNEP, United Nations Environmental Programme: Independent Environmental Assessment Beijing 2008 Plympic Games, Nairobi, Kenya, 2009, online available at: http://www.unep.org/pdf/BEIJING_REPORT_COMPLETE.pdf, last access: March 2010.

Wang, B., Shao, M., Lu, S. H., Yuan, B., Zhao, Y., Wang, M., Zhang, S. Q. and Wu. D.: Variation of ambient non-methane hydrocarbons in Beijing city in summer 2008, Atmos. Chem. Phys., 10, 5911–5923, doi:10.5194/acp-10-5911-2010, 2010a.

Wang, L. T., Hao, J. M., He, K. B., Wang, S. X., Li, J. H., Zhang, Q., Streets, D. G., Fu, J. S., Jang, C. J., Takekawa, H., and Chatani, S.: A modeling study of coarse particulate matter pollution in Beijing: Regional source contributions and control implications for the 2008 Summer Olympics, J. Air & Waste Manage. Assoc., 58 (8), 1057-1069, 2008.

Wang, M., Zhu, T., Zheng, J., Zhang, R. Y., Zhang, S. Q., Xie, X. X., Han, Y. Q., and Li, Y.: Use of a mobile laboratory to evaluate changes in on-road air pollutants during the Beijing 2008 Summer Olympics, Atmos. Chem. Phys., 9, 8247–8263, doi:10.5194/acp-9-8247-2009, 2009a.

Wang, S. X., Zhao, M., Xing, J., Wu, Y., Zhou, Y., Lei, Y., He, K. B., Fu, L. X., Hao, J. M.: Quantifying the Air Pollutants Emission Reduction during the 2008 Olympic Games in Beijing, Environ. Sci. Technol., 44(7), 2490–2496, 2010b.

Wang, T., Cheung, V. T. F., Anson, M., and Li, Y. S.: Ozone and related gaseous pollutants in the boundary layer of eastern China: Overview of the recent measurements at a rural site, Geophys. Res. Lett., 28(12), 2373–2376, 2001.

Wang, T, Ding, A. J., Gao, J., and Wu, W. S.: Strong ozone production in urban plumes from Beijing, China, Geophys. Res. Lett., 33, L21806, doi:10.1029/2006GL027689, 2006.

Wang, W. T., Primbs, T., Tao, S., and Simonich, S. L. M.: Atmospheric Particulate Matter Pollution during the 2008 Beijing Olympics, Environ. Sci. Technol., 43(14), 5314–5320, 2009b.

Wang, X., Westerdahl, D., Chen, L.C., Wu, Y., Hao, J. M., Pan, X. C., Guo, X. B., and Zhang, K. M.: Evaluating the air quality impacts of the 2008 Beijing Olympic Games: On-road emission factors and black carbon profiles, Atmos. Environ., 43(30), 4535–4543, 2009c.

Wang, Y., Hao, J., McElroy, M. B., Munger, J. W., Ma, H., Chen, D., and Nielsen, C. P.: Ozone air quality during the 2008 Beijing Olympics-effectiveness of emission restrictions, Atmos. Chem. Phys., 9, 5237–5251, doi:10.5194/acp-9-5237-2009, 2009d.

Witte, J. C., Schoeberl, M. R., Douglass, A. R., Gleason, J. F., Krotkov, N. A., Gille, J. C., Pickering, K. E., and Livesey, N.: Satellite observations of changes in air quality during the 2008 Beijing Olympics and Paralympics, Geophys. Res. Lett., 36, L17803, doi:10.1029/2009GL039236, 2009.

Wood, E. C., Herndon, S. C., Onasch, T. B., Kroll, J. H., Canagaratna, M. R., Kolb, C. E., Worsnop, D. R., Neuman, J. A., Seila, R., Zavala, M., and Knighton, W. B.: A case study of ozone production, nitrogen oxides, and the radical budget in Mexico City, Atmos. Chem. Phys., 9, 2499–2517, doi:10.5194/acp-9-2499-2009, 2009.

Wu, W. S. and Wang, T.: On the performance of a semi-continuous $PM_{2.5}$ sulphate and nitrate instrument under high loadings of particulate and sulphur dioxide, Atmos. Environ., 41(26), 5442–5451, 2007.

Zaveri, R. A., Berkowitz, C. M., Kleinman, L. I., Springston, S. R., Doskey, P. V., Lonneman, W. A., and Spicer, C. W.: Ozone production efficiency and NOx depletion in an urban plume: Interpretation of field observations and implications for evaluating O_3-NO_x-VOC sensitivity, J. Geophys. Res.-Atmos., 108(D19), 4436, doi:10.1029/2002JD003144, 2003.

Zhang, Q., Streets, D. G., Carmichael, G. R., Huo, H., Kannari, A., Klimont, Z., Park, I. S., Reddy, S., Fu, J. S., Chen, D., Duan, L., Lei, Y., Wang, I. T., and Yao, Z. L.: Asian emissions in 2006 for the NASA INTEX-B mission, Atmos. Chem. Phys., 9 (14), 5131–5153, doi:10.5194/acp-9-5131-2009, 2009.

Zhao, Y., Wang, S. X., Duan, L., Lei, Y., Cao, P. F., and Hao, J. M.: Primary air pollutant emissions of coal-fired power plants in China: Current status and future prediction, Atmos. Environ., 42(36), 8442–8452, 2008.

Geosci. Model Dev., 8, 3151–3162, 2015
www.geosci-model-dev.net/8/3151/2015/
doi:10.5194/gmd-8-3151-2015

Geoscientific Model Development
Open Access

Development of a chlorine chemistry module for the Master Chemical Mechanism

L. K. Xue[1], S. M. Saunders[2], T. Wang[3,1], R. Gao[1], X. F. Wang[1], Q. Z. Zhang[1], and W. X. Wang[1]

[1]Environment Research Institute, Shandong University, Ji'nan, Shandong, China
[2]School of Chemistry and Biochemistry, University of Western Australia, WA, Australia
[3]Department of Civil and Environmental Engineering, Hong Kong Polytechnic University, Hong Kong, China

Correspondence to: L. K. Xue (xuelikun@sdu.edu.cn)

Received: 27 May 2015 – Published in Geosci. Model Dev. Discuss.: 23 June 2015
Revised: 20 September 2015 – Accepted: 21 September 2015 – Published: 7 October 2015

Abstract. The chlorine atom (Cl•) has a high potential to perturb atmospheric photochemistry by oxidizing volatile organic compounds (VOCs), but the exact role it plays in the polluted troposphere remains unclear. The Master Chemical Mechanism (MCM) is a near-explicit mechanism that has been widely applied in the atmospheric chemistry research. While it addresses comprehensively the chemistry initiated by the OH, O_3 and NO_3 radicals, its representation of the Cl• chemistry is incomplete as it only considers the reactions for alkanes. In this paper, we develop a more comprehensive Cl• chemistry module that can be directly incorporated within the MCM framework. A suite of 205 chemical reactions describes the Cl•-initiated degradation of alkenes, aromatics, alkynes, aldehydes, ketones, alcohols, and some organic acids and nitrates, along with the inorganic chemistry involving Cl• and its precursors. To demonstrate the potential influence of the new chemistry module, it was incorporated into a MCM box model to evaluate the impacts of nitryl chloride ($ClNO_2$), a product of nocturnal halogen activation by nitrogen oxides (NO_X), on the following day's atmospheric photochemistry. With constraints of recent observations collected at a coastal site in Hong Kong, southern China, the modeling analyses suggest that the Cl• produced from $ClNO_2$ photolysis may substantially enhance the atmospheric oxidative capacity, VOC oxidation and O_3 formation, particularly in the early morning period. The results demonstrate the critical need for photochemical models to include more detailed chlorine chemistry in order to better understand the atmospheric photochemistry in polluted environments subject to intense emissions of NO_X, VOCs and chlorine-containing constituents.

1 Introduction

The chlorine atom (Cl•) acts as a major oxidant that can "jump-start" the photochemistry of the atmosphere (Finlayson-Pitts, 1993). It oxidizes various volatile organic compounds (VOCs) in a similar fashion to the hydroxyl radical (OH) but with reaction rates up to 2 orders of magnitude faster, and hence facilitates faster removal of VOCs and formation of ozone (O_3) and other oxidants (Atkinson et al., 1999; Chang et al., 2002; Tanaka et al., 2003b). In the troposphere, Cl• originates from a number of potential sources and the most recognized ones include the reaction of hydrochloric acid (HCl) with OH and photolysis of molecular chlorine (Cl_2), hypochlorous acid (HOCl), nitryl chloride ($ClNO_2$) and $ClONO_2$ (Riedel et al., 2014). These so-called Cl• precursors are either emitted from anthropogenic activities or formed through chemical activation of stable Cl-containing compounds (Tanaka et al., 2000). An example of the latter that has been recently demonstrated is the hydrolysis of dinitrogen pentoxide (N_2O_5) on Cl-containing aerosols producing $ClNO_2$ (Thornton et al., 2010). Despite ultra-trace ambient abundance, Cl• may play important roles in atmospheric chemistry in a variety of environments (e.g., polar, coastal and inland regions) wherever anthropogenic or natural chlorine sources exist.

The role that Cl• plays in VOC oxidation and O_3 formation remains a large uncertainty of tropospheric chemistry. Chemical mechanisms form the core of atmospheric models that are usually used to simulate the formation of air pollution and formulate science-based control strategies (Luecken et al., 2008; Stockwell et al., 2012). The degree by which the Cl• chemistry is accounted for in current major chemical mecha-

Published by Copernicus Publications on behalf of the European Geosciences Union.

nisms is inhomogeneous. To our knowledge, the organic and inorganic chemistry involving Cl• has not been represented in detail by most mechanisms. Tanaka et al. (2003a) developed a chlorine chemistry module containing 13 reactions for the carbon bond IV (or CB04) mechanism, and Sarwar et al. (2012) extended it for use in CB05 (including 25 reactions) to assess the impact of N_2O_5 hydrolysis on O_3 formation. Basic chemical modules describing the $ClNO_2$ formation and Cl• oxidation of VOCs have been incorporated in the SAPRC07 and its updates (Carter, 2010).

The Master Chemical Mechanism (MCM) is one of the most widely deployed chemical mechanisms, which near-explicitly describes the degradation of 143 primarily emitted VOCs (Jenkin et al., 2003, 2015; Saunders et al., 2003). In contrast to the comprehensive chemistry initiated by OH, O_3 and NO_3, the representation of the Cl• chemistry in the MCM remains incomplete. It considers only the reactions of Cl• with alkanes for which the oxidation by Cl• may play a dominant role (see Table S1 in the Supplement), while the reactions between Cl• and other VOC species are not represented. Recently, Riedel et al. (2014) added to the MCM Cl• reactions for 13 major reactive VOCs (i.e., ethene, propene, benzene, toluene, *o*-xylene, styrene, formaldehyde, methanol, ethanol, isopropanol, ethanal, propanal, acetone) with the aim to evaluate the impacts of $ClNO_2$ on the atmospheric photochemistry in the Los Angeles Basin. Nonetheless, this update only considers a small set of VOCs. There is still a need to further develop a more comprehensive Cl• chemical mechanism that can be applied to a wider range of tropospheric conditions.

A major goal of the present work is to develop a Cl• chemical mechanism that can be directly adopted in the MCM framework. Following the construction approach of MCM (Jenkin et al., 2003; Saunders et al., 2003), the existing chemical kinetics literature data were surveyed (mostly from the IUPAC database; http://iupac.pole-ether.fr/index.html) for the reactions of Cl• with various VOCs and compiled into a chemical module. This module contains 205 reactions and describes the Cl•-initiated degradation of all the MCM primary alkenes, alkynes, aromatics, aldehydes, ketones, alcohols, selected organic acids and nitrates as well as the inorganic chemistry of Cl• and its precursors. This new chemistry module introduces 22 additional chlorinated products that can be simulated with negligible increased cost of running time for the MCM models. It was then incorporated into a MCM-based chemical box model, with constraints of observations from a coastal site in southern China, to evaluate the impacts of $ClNO_2$ on the atmospheric photochemistry. With the observed maximum nighttime $ClNO_2$ (i.e., 1997 pptv), the modeling results suggest that the Cl• produced by photolysis of $ClNO_2$ plays a significant role in the next day's VOCs oxidation, O_3 formation and atmospheric oxidative capacity, especially during the early morning period.

2 Mechanism development

For the developed Cl• mechanism module to be readily incorporated within the MCM, the same approach, protocols and stoichiometry of the MCM are strictly followed (Jenkin et al., 2003, 2015; Saunders et al., 2003). The chemical reactions that are compiled in the mechanism module are summarized in Table 1, Table S2 and Figs. 1–3. A detailed description of the construction procedures is given below.

2.1 Inorganic reactions

Five types of inorganic reactions that are of potential significance to the production and fate of Cl• are considered. They include photolysis reactions of Cl• precursors, reactions of Cl• with inorganic species, reactions of OH with Cl-containing species recycling Cl• or its precursors, reactions of ClO, and heterogeneous reactions that have been recently found to be involved in chlorine activation (e.g., Thornton et al., 2010; Sarwar et al., 2012; Riedel et al., 2014). A total of 24 inorganic reactions are compiled, as outlined in Table 1. The rate coefficients and product yields are mostly taken from the latest IUPAC database.

2.2 Organic reactions

2.2.1 Aldehydes

The Cl• reactions are considered for formaldehyde, acetaldehyde, propanal, butanal, isobutyl aldehyde, pentanal, benzaldehyde, glyoxal, methylglyoxal and methacrolein (MACR) (see Table S2). Based on the available literature, the reactions are assumed to proceed primarily via H atom abstraction by Cl• to form an HCl molecule and a RO_2 radical, both of which are already present in the MCM framework. For formaldehyde, acetaldehyde and propanal, the chemical kinetic data including rate coefficients and product yields are adopted from the latest IUPAC database. For glyoxal and methylglyoxal, the rate constants are assumed to be same to those for formaldehyde and acetaldehyde according to the SAPRC mechanism (http://www.engr.ucr.edu/~carter/SAPRC/saprc07.pdf). For the other aldehydes for which kinetic data are unavailable, the approach of the MCM protocol is followed, to adapt the known experimental data to give reasonable estimates for the unknown kinetics (Saunders et al., 2003). The basic assumption is that they react with Cl• similar to OH with a rate constant of $k_{X+\mathrm{OH}}$ times a generic $k_{\mathrm{Cl}}/k_{\mathrm{OH}}$ ratio. The generic $k_{\mathrm{Cl}}/k_{\mathrm{OH}}$ ratio is estimated based on the known measured rate constants for acetaldehyde and propanal (note that formaldehyde is excluded here considering that the reaction rate of a C_1 species usually stands out from the remainder of a series). In this case, the generic $k_{\mathrm{Cl}}/k_{\mathrm{OH}}$ ratio of 6.08 (average) is adopted. Sensitivity tests were conducted by using the lower and upper limits of the $k_{\mathrm{Cl}}/k_{\mathrm{OH}}$ ratio, and suggested that the differences among the modeling results were negligible under typical polluted ur-

Table 1. Summary of inorganic reactions added to the MCM to represent Cl• chemistry.

Category	Reaction	k (cm^3 molecules^{-1} s^{-1} or s^{-1}) or J (s^{-1})	Remarks
Photolysis reactions	$Cl_2 \rightarrow Cl + Cl$	J_{Cl_2}	–
	$ClNO_2 \rightarrow NO_2 + Cl$	J_{ClNO_2}	–
	$ClONO_2 \rightarrow NO_3 + Cl$	$0.83 \times J_{ClONO_2}$	a
	$ClONO_2 \rightarrow NO_2 + ClO$	$0.17 \times J_{ClONO_2}$	a
	$HOCl \rightarrow OH + Cl$	J_{HOCl}	–
Cl + X	$Cl + O_3 \rightarrow ClO + O_2$	$2.8 \times 10^{-11} \times \exp(-250/T)$	b
	$Cl + HO_2 \rightarrow HCl + O_2$	3.5×10^{-11}	b
	$Cl + HO_2 \rightarrow ClO + OH$	$7.5 \times 10^{-11} \times \exp(-620/T)$	b
	$Cl + H_2O_2 \rightarrow HCl + HO_2$	$1.1 \times 10^{-11} \times \exp(-980/T)$	b
	$Cl + NO_3 \rightarrow NO_2 + ClO$	2.4×10^{-11}	b
	$Cl + ClONO_2 \rightarrow Cl_2 + NO_3$	$6.2 \times 10^{-12} \times \exp(145/T)$	b
			b
OH+ X	$OH + HCl \rightarrow Cl + H_2O$	$1.7 \times 10^{-12} \times \exp(-230/T)$	b
	$OH + Cl_2 \rightarrow HOCl + Cl$	$3.6 \times 10^{-12} \times \exp(-1200/T)$	b
	$OH + HOCl \rightarrow ClO + H_2O$	5.0×10^{-13}	b
	$OH + ClO \rightarrow HO_2 + Cl$	1.8×10^{-11}	b
	$OH + ClO \rightarrow HCl + O_2$	1.2×10^{-12}	b
			b
ClO + X	$ClO + NO_2 \rightarrow ClONO_2$	7.0×10^{-11}	b
	$ClO + HO_2 \rightarrow HOCl + O_2$	$2.2 \times 10^{-12} \times \exp(340/T)$	b
	$ClO + NO \rightarrow Cl + NO_2$	$6.2 \times 10^{-12} \times \exp(295/T)$	b
Hetero. reactions	$N_2O_5 \rightarrow NA + NA$	$0.25 \times C_{N_2O_5} \times \gamma_{N_2O_5} \times S_{AERO} \times (1 - \varphi_{CLNO_2})$	c
	$N_2O_5 \rightarrow NA + ClNO_2$	$0.25 \times C_{N_2O_5} \times \gamma_{N_2O_5} \times S_{AERO} \times \varphi_{CLNO_2}$	c
	$NO_3 \rightarrow$ products	$0.25 \times C_{NO_3} \times \gamma_{NO_3} \times S_{AERO}$	c
	$ClONO_2 \rightarrow Cl_2 + HNO_3$	$0.25 \times C_{ClONO_2} \times \gamma_{ClONO_2} S_{AERO}$	c
	$HOCl \rightarrow Cl_2$	$0.25 \times C_{HOCl} \times \gamma_{HOCl} \times S_{AERO}$	c

[a] The branching ratio is determined based on the Tropospheric Ultraviolet Visible (TUV) Radiation model calculations (http://cprm.acd.ucar.edu/Models/TUV/Interactive_TUV/). [b] The kinetic data are taken from the IUPAC database (http://iupac.pole-ether.fr/index.html). [c] C_X is the molecular speed of X; γ_X is the uptake coefficient of X on aerosols; S_{AERO} is the aerosol surface area concentration; φ_{CLNO_2} is the product yield of $ClNO_2$ from the heterogeneous reactions of N_2O_5. NA is nitrate aerosol.

ban conditions (see Sect. 2.3 and Table 2 for the details of sensitivity tests).

The reactions of Cl• with MACR are represented as follows in the new module. The mechanism is assumed to be the same to that for OH. According to the MCM v3.2, the OH oxidation of MACR proceeds via two routes: H abstraction from the aldehyde group (45 %) and OH addition to the C = C double bond (55 %). Hence, 45 % of MACR is oxidized by Cl• as a common aldehyde by using the aforementioned generic k_{Cl}/k_{OH} ratio, whilst the remainder can be treated as an alkene compound (being summed into a lumped species "OLEFIN"; see Sect. 2.2.7).

2.2.2 Ketones

The reactions of Cl• with ten primary ketones in the MCM are compiled following the same approach to that for aldehydes (Table S2). The reactions proceed via H atom abstraction by Cl• to form HCl and RO_2, which already exist in the MCM. Kinetics data including rate coefficient and product yield are adopted from the latest IUPAC database for acetone and methyl ethyl ketone (MEK). For other ketones, it is assumed that they react with Cl• similarly to OH with rate constants of k_{X+OH} times a generic k_{Cl}/k_{OH} ratio. The generic ratio of 23.9 was derived by averaging the k_{Cl}/k_{OH} values for acetone and MEK, for which experimental data are available. Sensitivity tests by adopting lower and upper limits of k_{Cl}/k_{OH} suggest that the difference in modeling results is minor under typical polluted urban conditions (see Sect. 2.3 and Table 2).

The OH oxidation of methyl vinyl ketone (MVK) takes place mainly by addition of OH to the C = C double bond. The same mechanism was adopted for the Cl• reaction. Specifically, MVK can be treated as an alkene and is lumped into the "OLEFIN" species (see Sect. 2.2.7).

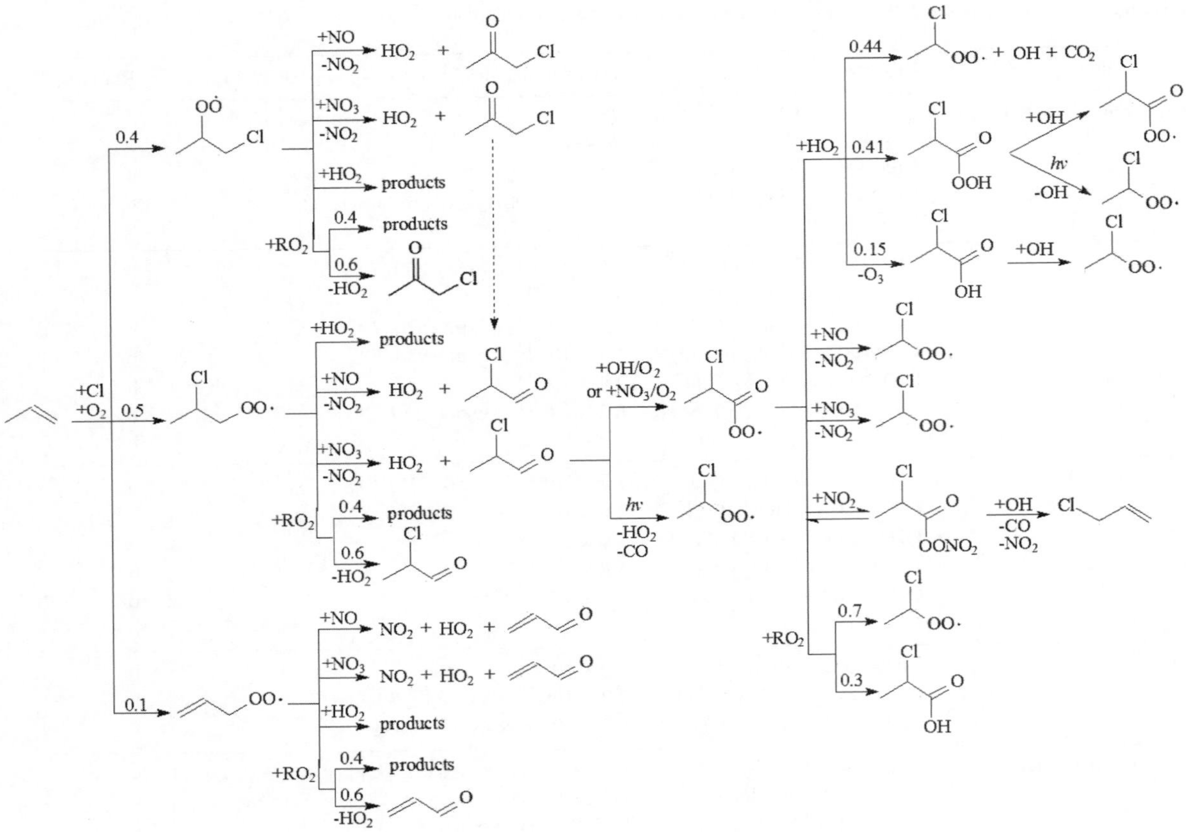

Figure 1. The oxidation mechanism of propene by Cl•. Note that (1) the species in red already exist in the MCM; (2) degradation of the species in blue is not further considered for simplicity; (3) the preceding plus and minus indicate the reactants and products respectively; (4) for simplicity, the degradation of CH_3COCH_2Cl was approximated to be the same as that of $CH_3CH(Cl)CHO$.

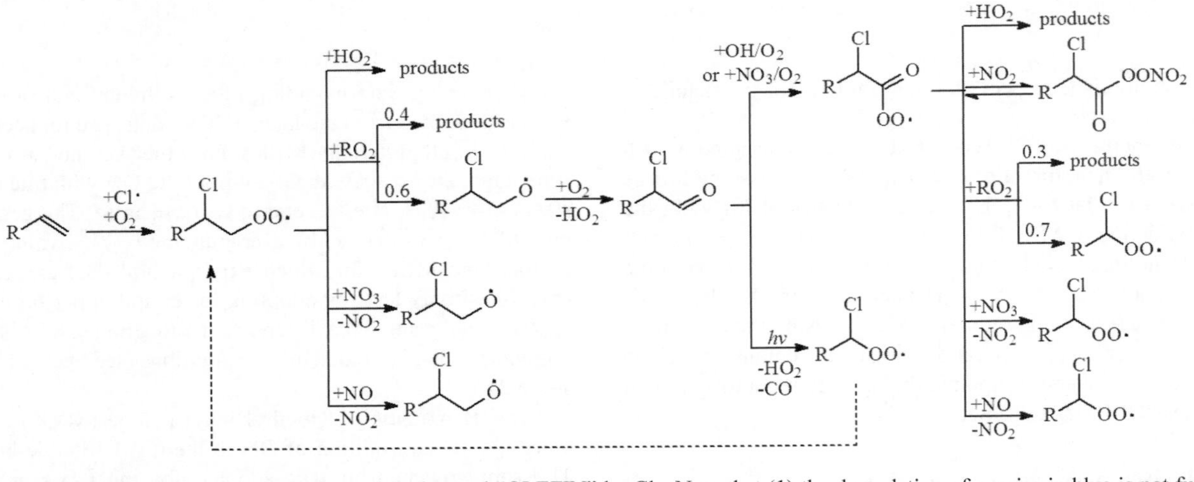

Figure 2. Simplified oxidation mechanism of the lumped "OLEFIN" by Cl•. Note that (1) the degradation of species in blue is not further considered for simplicity; (2) the preceding plus and minus indicate the reactants and products respectively; (3) for simplicity, the degradation of $RCH(Cl)O_2$ was approximated to be the same as that of $RCH(Cl)CH_2O_2$.

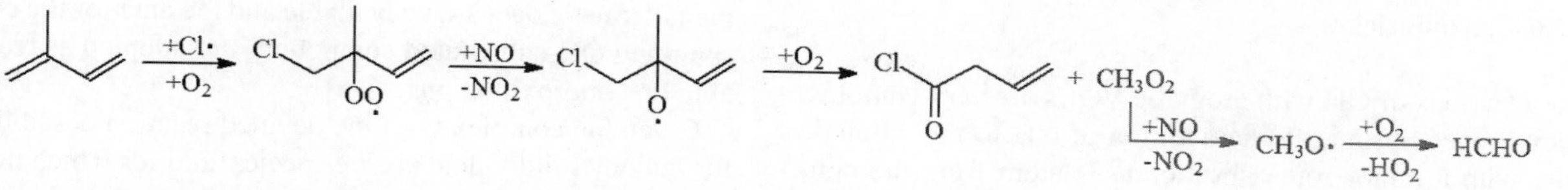

Figure 3. Simplified oxidation mechanism of isoprene by Cl•. Note that (1) the species in red already exist in the MCM; (2) degradation of the species in blue is not further considered for simplicity; (3) the preceding plus and minus indicate the reactants and products respectively.

Table 2. Summary of sensitivity test results[a].

Scenario	Difference against the base model[b]	[Cl] = [OH] / 50 Difference in daytime P(O_x)	[Cl] = [OH] / 50 Difference in daytime OC_{CL}	[Cl] = [OH] / 200 Difference in daytime P(O_x)	[Cl] = [OH] / 200 Difference in daytime OC_{CL}
S1	With base MCM, without the developed Cl• chemistry module	12.8 %	186 %	4.5 %	166 %
S2	Turn off the "OLEFIN" chemistry (including Cl• reactions of C_2H_4 and C_3H_6)	5.5 %	7.7 %	1.7 %	7.4 %
S3	Generic k_{Cl}/k_{OH} = 0 for aromatics (turn off)	2.2 %	3.1 %	< 1 %	5.6 %
	Generic k_{Cl}/k_{OH} = 220 for aromatics (upper limit)	< 1 %	< 1 %	< 1 %	1.2 %
	Generic k_{Cl}/k_{OH} = 150 for aromatics (lower limit)	< 1 %	< 1 %	< 1 %	1.0 %
S4	Generic k_{Cl}/k_{OH} = 0 for aldehydes (turn off)	< 1 %	1.6 %	< 1 %	< 1 %
	Generic k_{Cl}/k_{OH} = 8.5 for aldehydes (upper limit)	< 1 %	< 1 %	< 1 %	< 1 %
	Generic k_{Cl}/k_{OH} = 5.4 for aldehydes (lower limit)	< 1 %	< 1 %	< 1 %	< 1 %
S5	Generic k_{Cl}/k_{OH} = 0 for ketones (turn off)	< 1 %	3.0 %	< 1 %	2.0 %
	Generic k_{Cl}/k_{OH} = 36 for ketones (upper limit)	< 1 %	< 1 %	< 1 %	< 1 %
	Generic k_{Cl}/k_{OH} = 12 for ketones (lower limit)	< 1 %	1.6 %	< 1 %	< 1 %
S6	Generic k_{Cl}/k_{OH} = 0 for alcohols (turn off)	< 1 %	< 1 %	< 1 %	< 1 %
S7	Turn off the chemistry of Cl• + C_5H_8	< 1 %	< 1 %	< 1 %	< 1 %

[a] The daytime P(O_x) and OC_{CL} are the daytime average (09:00–18:00 local time) net production rate of O_x and oxidative capacity of the Cl• atom. The differences are with respect to the base model results with the full chlorine mechanism. [b] The base model was with the newly developed chlorine mechanism in the present study. The sensitivity model runs were generally the same to the base version, but with difference as specified here.

2.2.3 Alcohols

Based on the above approach, presented in the chlorine mechanism are the reactions of Cl• with 18 primary alcohols in the MCM (Table S2). The reactions are assumed to occur via abstraction of H atom by Cl• to form HCl. Depending on the position of the abstracted H atom, these reactions also yield either a RO_2 or a HO_2 together with a carbonyl compound. The rate coefficients and product yields are taken from the latest IUPAC database for methanol, ethanol, *n*-propanol, *i*-propanol and *n*-butanol. For cresol, the mechanism used in the SAPRC is adopted here (http://www.engr.ucr.edu/~carter/SAPRC/saprc07.pdf). For the other alcohols where no kinetics data are available, we assume that they react with Cl• similarly to OH with rate constants of k_{X+OH} times a generic k_{Cl}/k_{OH} ratio. The generic ratio of 17.1 was estimated by averaging the k_{Cl}/k_{OH} values for ethanol, *n*-propanol, *i*-propanol and *n*-butanol. Sensitivity studies show that the impact on modeling results of the treatment on other alcohols should be minor under typical polluted urban conditions (see also Sect. 2.3 and Table 2).

2.2.4 Organic acids and peroxides

Three major organic acids, i.e., formic acid, acetic acid and propanoic acid, are considered in the new chlorine mechanism. The reactions also proceed through H atom abstraction by Cl• to produce HCl and RO_2 (or HO_2), which already exist in the MCM. The kinetic data are taken from the latest IUPAC database for formic acid and acetic acid. For propanoic acid, we assume that it reacts with Cl• in the same way to OH but with a rate constant of k_{X+OH} times the k_{Cl}/k_{OH} ratio for acetic acid. In addition, the reaction of Cl• with methyl hydroperoxide (CH_3OOH) is also compiled according to the IUPAC data (see Table S2).

2.2.5 Organic nitrates

The Cl• reactions are only considered for five C_1–C_4 alkyl nitrates (see Table S2). These species are selected because of their relatively higher atmospheric abundance and the availability of experimental kinetics data. The reactions of Cl• with alkyl nitrates are assumed to proceed by H atom abstraction to form HCl, NO_2 and carbonyl compounds, all of which are present in the MCM. The rate coefficients are adopted from the IUPAC database.

2.2.6 Aromatics

The reactions of OH with aromatic VOCs are very complex. They proceed primarily via addition of OH to the aromatic ring, with a minor route abstracting H atom from the non-ring alkyl substitute (Jenkin et al., 2003). On the contrary, the addition of Cl• to the aromatic ring is very slow, with the rate being approximately two orders of magnitude slower than addition of OH (Tanaka et al., 2003a). Hence, the reactions of Cl• with aromatic VOCs are assumed to proceed via H atom abstraction by Cl• from non-ring alkyl substitutes, forming HCl and a RO_2 radical that already exists in the MCM. The reaction of Cl• with benzene is rather slow and thus is not considered in this module (Riedel et al., 2014). Styrene is treated as an alkene compound and is lumped into the "OLEFIN" species (see Sect. 2.2.7). The Cl• reactions for the other primary aromatic VOCs are presented in the new mechanism module as follows. For toluene and *o*-xylene, the experimental data of Shi and Bernhard (1997) are adopted. For other species, it is assumed that they react with Cl• similarly to OH (the H abstraction pathway only) with rate constants of $k_{X+\mathrm{OH}}$ times a generic $k_{\mathrm{Cl}}/k_{\mathrm{OH}}$ ratio. The generic ratio of 185 was derived by averaging the $k_{\mathrm{Cl}}/k_{\mathrm{OH}}$values for toluene and *o*-xylene (note that the $k_{X+\mathrm{OH}}$ only refer to the rate of the H atom abstraction route). Sensitivity studies by adopting lower and upper limits of $k_{\mathrm{Cl}}/k_{\mathrm{OH}}$ indicate that the difference in modeling results is minor under typical polluted urban conditions (see Sect. 2.3 and Table 2).

2.2.7 Alkenes

The reactions of Cl• with alkenes proceed primarily by addition of Cl• to the double bond, forming a Cl-substituted RO_2 radical that is generally new to the MCM. Further degradation of the reaction intermediates need to be considered if applicable. Reaction of Cl• with ethene leads to formation of $CH_2(Cl)CH_2O_2$ that is already present in the MCM. The rate coefficient of this reaction is adopted from the latest IUPAC database (Table S2).

The mechanism of the Cl•-initiated oxidation of propene is depicted in Fig. 1, with a general and brief description on this complex scheme as follows. First, propene reacts with Cl• via three routes each of which produces a new RO_2 radical. These RO_2 radicals then react individually with NO, NO_3, HO_2 and other RO_2 radicals to form products including carbonyls, HO_2 and compounds already in the MCM (several less-reactive products are not considered for further reactions). The new carbonyls further react with OH and NO_3 to produce new acyl peroxy radicals. Finally, acyl peroxy radicals react individually with NO, NO_2, NO_3, HO_2 and RO_2 to yield products that already exist in the mechanism (a minor product is not considered for further reaction). The rate coefficients and product yields are taken from Riedel et al. (2014) for the initiation reaction of propene with Cl•. For reactions of the chlorine-substituted intermediates, experimental kinetic data are unavailable and the data for the corresponding OH-substituted compounds are adopted as reasonable first approximations.

Given the complexity of the detailed reaction possibilities for the other individual alkene species, and for which no experimental data are available, a lumped method is applied, that is, a new model species "OLEFIN" is defined as the sum of all primary alkenes except for ethene and propene (note that styrene, MVK and 55 % of MACR are also included). The degradation of "OLEFIN" initiated by Cl• is assumed to be largely similar to that of propene, as illustrated in Fig. 2. Briefly, addition of Cl• to the double bond yields a Cl-substituted RO_2 radical, which then reacts with NO, NO_3, HO_2 and other RO_2 to form a Cl-substituted carbonyl and/or peroxide. The new carbonyl further reacts with OH and NO_3 to form an acyl peroxy radical, which finally reacts with NO, NO_2, NO_3, HO_2 and RO_2 to produce compounds already existing in the mechanism. Some less-reactive species (e.g., peroxides) are not considered for further reactions for simplicity. The reaction rates of Cl• + OLEFIN are assumed to be 11-fold faster than that of OH + OLEFIN, according to the $k_{\mathrm{Cl}}/k_{\mathrm{OH}}$ ratios for ethene, propene and 1-butene (Tanaka et al., 2003a). The rate constant of OH + OLEFIN reaction ($k_{\mathrm{OH+OLEFIN}}$) can be calculated by averaging the rate constants of individual species with consideration of their abundances. Sensitivity model test suggests that the impact of such a lumped approach on the modeling results should be of minor significance under typical urban conditions (see Sect. 2.3).

The degradation of isoprene is very complex and remains not fully elucidated (Jenkin et al., 2015). Given the inherent uncertainties in the mechanism, here a very simplified scheme from the CB04 mechanism is adopted, to represent the potential enhancement in ozone production by the Cl• oxidation. The reaction rate of Cl• with isoprene is assumed to be 4.75 times faster than that of OH (Tanaka et al., 2003a). The Cl•-initiated degradation mechanism of isoprene is illustrated in Fig. 3.

2.2.8 Alkyne

Only acetylene is considered for the Cl• reactions, with adoption of the mechanism used in the SAPRC07 (http://www.engr.ucr.edu/~carter/SAPRC/saprc07.pdf). Specifically, the reactions proceed through addition of Cl• to the triple bond, followed by further reactions of the intermediates with O_2 to form CO, HO_2 radical and a chlorinated aldehyde. The aldehyde is assumed to be relatively unreactive and is not further represented.

2.3 Sensitivity tests of the chlorine mechanism

A number of tests have been conducted with an observation-based MCM model to assess the sensitivity of model outputs to the reactions and estimated rate coefficients as defined above, where experimental kinetics data were unavailable. A high-pollution episode observed on 7 May 2005 at a downwind site of Shanghai, the largest city of China, was analyzed. This case was selected because very high VOC concentrations with significant contributions of both alkenes and aromatics were observed, and hence representative of typical polluted urban conditions. This episode has been analyzed in our previous study where details of the data are given (Xue et al., 2014a), and the concentrations of individual non-methane hydrocarbons are documented in Fig. S1 in the Supplement. The model was based on the MCM version 3.2, with incorporation of the newly developed Cl• mechanism, and was constrained by the observed diurnal data of O_3, CO, NO, C_1-C_{10} hydrocarbons and meteorological parameters (Xue et al., 2014a). The concentrations of Cl• were prescribed as a function of OH, which was simulated by the model. Two scenarios, i.e., a high Cl• case ([OH] / [Cl•] = 50) and a normal Cl• case ([OH] / [Cl•] = 200), were considered to represent a wide range of ambient conditions. The net O_3 production rate and oxidative capacity of Cl• (defined as the sum of the oxidation rates of VOCs by Cl•) were computed at a time resolution of 10 min within the model. Where experimental data are unavailable and generic rate coefficients were adopted, sensitivity tests were conducted by using both lower and upper limits of the generic ratio and/or switching off the target reactions. The differences in net O_3 production rate and oxidative capacity of Cl• between the sensitivity runs and base runs were examined.

All of the sensitivity test results are summarized in Table 2. Overall, the results indicate that the impacts of both generic rate estimation and the lumping approach for olefins on the modeling results are minor for the typical urban environments simulated (note that the difference may not be negligible for all polluted conditions, e.g., with high abundances of those heavy VOCs for which k_{Cl} was estimated using k_{Cl}/k_{OH} ratios.). In addition, another set of sensitivity tests were conducted to evaluate the performance of the new Cl• mechanism. Including this full mechanism in the model resulted in significant enhancement in oxidative capacity of Cl• as well as moderate enhancement in ozone production rates, in comparison with the model run with the base MCM (including reactions of Cl• with alkanes alone; Table 2). It is clear that photochemical models need to represent more detailed chlorine chemistry when applied to polluted conditions with abundant reactive VOCs.

3 Mechanism application: the role of nitryl chloride in daytime photochemistry

Following the Cl• mechanism sensitivity tests, the module was incorporated into another MCM box model to specifically evaluate the impact of $ClNO_2$ on atmospheric oxidative capacity and O_3 formation, an area of major uncertainty in current tropospheric chemistry research. Recent studies have confirmed the presence of high concentrations of $ClNO_2$, a product of nitrogen oxide induced halogen activation, in both coastal and inland regions, and have suggested its potential significance in enhancing O_3 formation (Osthoff et al., 2008; Phillips et al., 2012; Riedel et al., 2012; Thornton et al., 2010). Here we analyzed a high-pollution case observed at Hok Tsui, a coastal site of Hong Kong, in the summer of 2012, which provided the first ambient $ClNO_2$ observations in China. Elevated $ClNO_2$ concentrations of up to 1997 pptv (1 min data) were detected in a plume that originated from urban Hong Kong and the Pearl River Delta region during the night of 23–24 August. These measurement results have been reported in our previous work (Tham et al., 2014). In the present study, we focus on the consequence of $ClNO_2$ photolysis on the next day's photochemistry through a detailed modeling study. A full description of the observations and the target case is given in the Supplement.

The Observation-Based Model for investigating the Atmospheric Oxidative Capacity and Photochemistry (OBM-AOCP), which has been applied in many previous studies (Xue et al., 2013, 2014a, b, c), was here updated to include the Cl• chemistry module and used for the analyses. In addition to the comprehensive chemistry addressed by the MCM v3.2, dry deposition and dilution with evolution of the planetary boundary layer were also considered (Xue et al., 2014a; wet deposition was not considered here). The model was initialized by the measured nighttime concentrations of a full list of chemical species and meteorological parameters when the maximum $ClNO_2$ value was observed (see Table S3 for the initial model conditions; note that the heterogeneous production of $ClNO_2$ was turned off here as the model is initialized by the measured $ClNO_2$ data), and was then run for a 24 h period to simulate the chemical evolution of the prescribed plume. The model was run 5 times consecutively to stabilize the unmeasured species (e.g., radicals and reaction intermediates), and the daytime output of the last run was subject to further analysis. Two scenarios with and without $ClNO_2$ were conducted to examine the impact of $ClNO_2$ chemistry.

With 1997 pptv of initial $ClNO_2$, the model predicted an early morning (~08:30 local time) peak of Cl• of 8.2×10^4 molecule cm^{-3} that then decreased with time of the day (see Fig. 4a). Such level of Cl• accounts for up to 2.0 % of the abundance of OH, the predominant daytime oxidant. Considering the much faster reaction rates of Cl• with VOCs than OH, Cl• should be an important oxidant in the early morning (see below for a detailed quantification). The addition of $ClNO_2$ enhanced significantly the in situ O_3 production within the plume, with increases of 33.7 and 10.3 % in the early morning and daytime average (08:00–18:00) net O_3 production rates respectively (Fig. 4b). Despite the weakened role of Cl• oxidation after the morning period, the model simulated peak O_3 increased by 5 ppbv (or 6.8 %) compared to the non-$ClNO_2$ case (Fig. 4c). Such enhancement is comparable to that derived from a chemical transport modeling

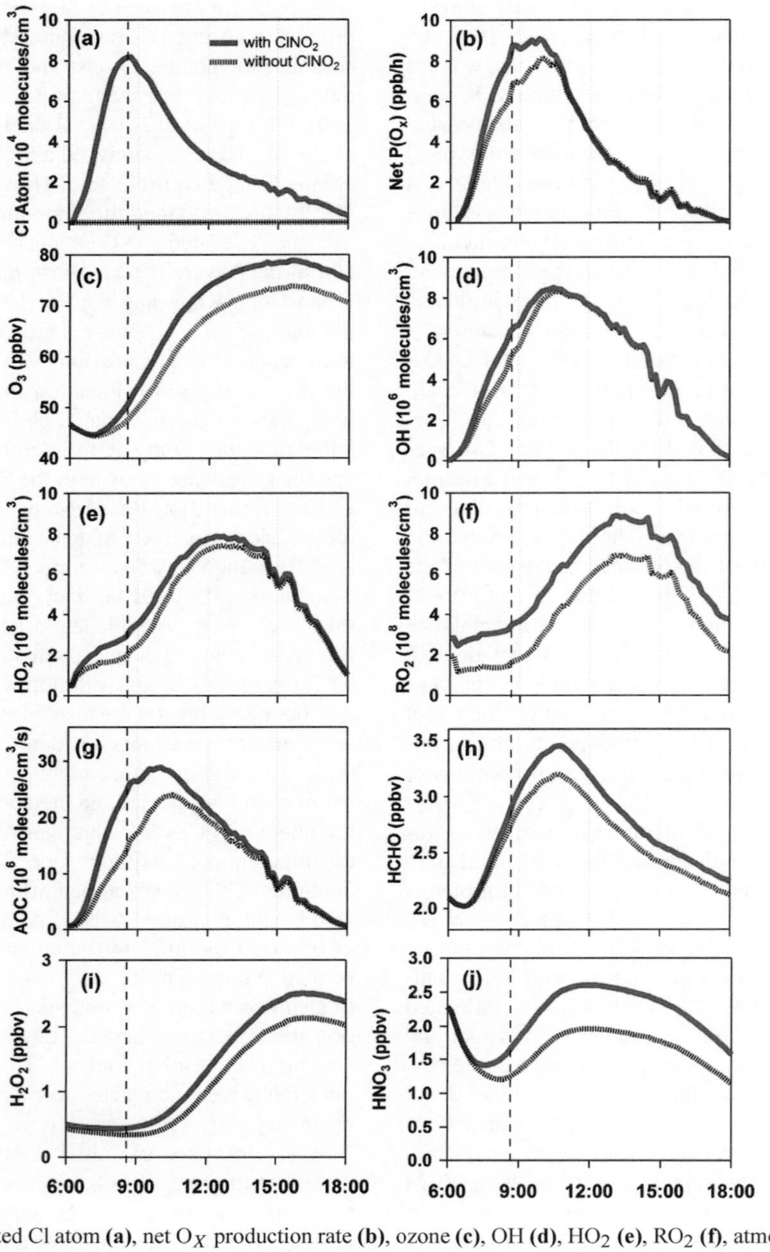

Figure 4. The model-simulated Cl atom **(a)**, net O_X production rate **(b)**, ozone **(c)**, OH **(d)**, HO_2 **(e)**, RO_2 **(f)**, atmospheric oxidative capacity (AOC; **g**), HCHO **(h)**, H_2O_2 **(i)** and HNO_3 **(j)** with and without initial concentration of $ClNO_2$, for the polluted plume observed at Hok Tsui, Hong Kong, on 24 August 2012. The vertical dashed lines indicate the "early morning case" when the peak of Cl atom occurs and the $ClNO_2$ impacts are evaluated in parallel with the "daytime average case".

study in North America (up to 6.6 ppbv or 10 % in summer; Sarwar et al., 2012). Evidently, the nighttime formation of $ClNO_2$ may pose a significant positive feedback to the next day's ozone formation in southern China.

The reactions of Cl• with VOCs produce RO_2 radicals, which are then recycled to HO_2 and OH. Figure 4d–f show the significant impacts of the addition of $ClNO_2$ on the model simulated OH, HO_2 and RO_2 radicals. In the early morning when the Cl• chemistry was the most active, the Cl• arising from $ClNO_2$ photolysis enhanced the concentrations of RO_2, HO_2 and OH by up to 120, 52.7 and 34.9 %, respectively. With photochemical processing, the Cl• became gradually exhausted while the other radical precursors (e.g., O_3 and OVOCs) accumulated, leading to decreasing contribu-

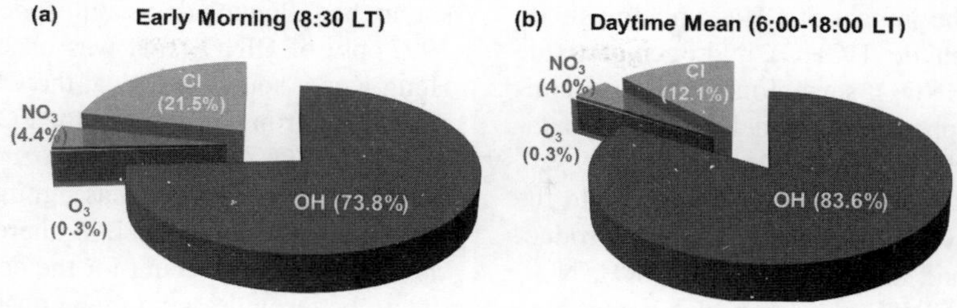

Figure 5. Breakdown of atmospheric oxidative capacity by individual oxidants in the early morning **(a)** and throughout the daytime **(b)** for the plume observed at Hok Tsui, Hong Kong, on 24 August 2012.

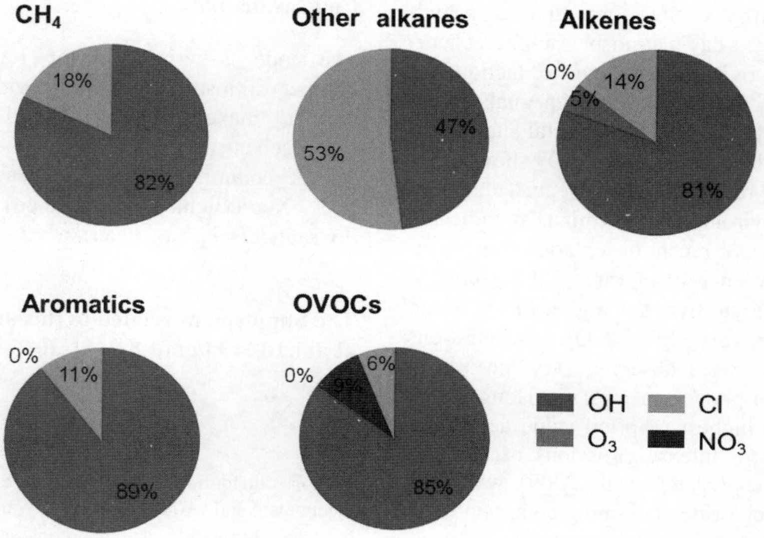

Figure 6. Oxidation of VOCs by individual oxidants in the early morning (08:30 local time) in the plume observed at Hok Tsui, Hong Kong, on 24 August 2012.

tions of Cl• to the radical production. In terms of the daytime average, nevertheless, the enhancements in the modeled RO_2 (45.1 %), HO_2 (12.2 %) and OH (6.6 %) resulting from the VOC oxidation by Cl• are still significant or considerable. These results suggest the important impact of Cl• on the RO_X radicals and hence atmospheric oxidative capacity.

We further quantified the contributions of Cl• to the atmospheric oxidative capacity (AOC). AOC is defined in the present study as the sum of the oxidation rates of CO and individual VOCs by all major oxidants, i.e., OH, O_3, NO_3 and Cl•, commonly used as a proxy of the intensity of oxidation capability of the atmosphere (Elshorbany et al., 2009), and was computed by the model. With inclusion of $ClNO_2$ (thus Cl•) in the model, the AOC was substantially strengthened with increases of 90.3 % in the early morning and of 25.4 % for the daytime average (Fig. 4g). Such large enhancements are attributable to the direct oxidation capacity of Cl• as well as the indirect effect of increasing RO_X radicals. Figure 5 depicts the breakdown of AOC by the individual oxidants. The analysis shows Cl• to be the second most important oxidant not only in the early morning (∼ 21.5 %) but also throughout the daytime (∼ 12.1 %). OH is clearly the predominant player in photochemical oxidation, while O_3 and NO_3 play a relatively minor role in this polluted plume observed in Hong Kong.

The role of Cl• in VOC oxidation was also assessed. The contributions of major oxidants to the degradation of individual VOC groups during the early morning period are summarized in Fig. 6. As expected, OH dominates the VOC oxidation contributing to 82, 47, 81, 89 and 85 % of oxidation of methane, other alkanes, alkenes, aromatics and OVOCs. Cl• is another important oxidant and in particular the principal one for alkanes. It oxidizes 18 % of methane and 53 % of other alkanes, and also accounts for 14, 11 and 6 % of the alkenes, aromatics and OVOCs oxidation. The significant role of Cl• in oxidizing alkanes agrees well with the previous studies (e.g., Young et al., 2014). Clearly, photolysis of $ClNO_2$ can facilitate the oxidation of VOCs (especially alkanes) in the early morning.

Figure 4h–j illustrate the impact of $ClNO_2$ photolysis on the formation of formaldehyde (HCHO), hydrogen peroxide (H_2O_2) and nitric acid (HNO_3), several major photochemical products in the atmosphere. The model-simulated maximum values of HCHO, H_2O_2 and HNO_3 increased by 8.0, 17.3 and 13.4 % with inclusion of $ClNO_2$ compared to the non-$ClNO_2$ case, respectively. By enhancing radical production (i.e., RO_X and Cl) and hence oxidation of SO_2, NO_X and VOCs, photolysis of $ClNO_2$ also has a high potential to promote formation of secondary aerosols, such as sulfate, nitrate and secondary organic aerosol (SOA), but has not been quantified here.

In summary, the nocturnal $ClNO_2$ formation has a high potential to perturb the next day's atmospheric photochemistry by promoting VOC oxidation, radical production and cycling, and O_3 formation. Although the present analyses are only based on a high-pollution case at a coastal site (sensitivity tests with a lower level of initial $ClNO_2$ show lower impacts, see Fig. S2), our results should be representative of other polluted coastal environments of China, as indicated by our follow-on studies. Our recent observations have confirmed the ubiquitous presence of elevated $ClNO_2$ both at other sites in Hong Kong and over the region of the North China Plain. In particular, very high $ClNO_2$ concentrations (1 min value of 4.7 ppbv) were observed very recently at a mountain site (~ 957 m above sea level) in Hong Kong, which appears to be the highest reported value across the world (Wang et al., 2015). Intense emissions of nitrogen oxides, VOCs and particles (Zhang et al., 2009), which interact with the abundant chlorine-containing compounds released from both anthropogenic and natural sources (e.g., sea spray), are conductive to the heterogeneous formation of $ClNO_2$ and in turn Cl• photochemistry in the coastal environments of China. A recent chemical transport modeling study also suggests the importance of $ClNO_2$ chemistry to ozone formation and the atmospheric oxidative capacity in the Northern Hemisphere, especially over China and western Europe (Sarwar et al., 2014). Detailed mechanisms describing the Cl• chemistry are crucial for current models to more accurately represent and lead to better understanding of atmospheric photochemistry and formation of air quality problems.

4 Summary and conclusions

A chemical mechanism has been developed for use in the MCM to address the chemistry of chlorine atom, a potential important oxidant in the atmosphere. It includes 205 reactions and describes the Cl•-initiated degradation of alkenes, alkynes, aromatics, aldehydes, ketones, alcohols, organic acids and organic nitrates, in combination with the inorganic chemistry of Cl• and its precursors. Application of this mechanism in a MCM box model suggests the important role of the nocturnal formation of $ClNO_2$, a major Cl• precursor, in the following day's atmospheric photochemistry. With 1997 pptv of $ClNO_2$ that was observed at a coastal site in Hong Kong, southern China, the Cl• produced from $ClNO_2$ photolysis strongly enhances the RO_X radical production and recycling, VOC oxidation, and ozone formation in the early morning period, and even has significant or moderate impacts throughout the daytime. It is therefore critical that photochemical models account for the detailed chemistry of chlorine to better understand the atmospheric oxidative capacity and ozone formation in polluted environments with abundant chlorine-containing compounds and VOCs.

Code availability

The code is written in the FACSIMILE language. A number of instructions are provided as comments in the code to make it easily adapted in the Master Chemical Mechanism framework. The code is freely available for the community and can be accessed by request from L. K. Xue (xuelikun@sdu.edu.cn) or S. M. Saunders (sandra.saunders@uwa.edu.au).

The Supplement related to this article is available online at doi:10.5194/gmd-8-3151-2015-supplement.

Author contributions. L. K. Xue and S. M. Saunders developed the mechanism and wrote the paper. T. Wang designed and provided the field data for model test and demonstration. R. Gao and Q. Z. Zhang provided Figs. 1–3. X. F. Wang and X. Wang gave helpful suggestions and polished the writing.

Acknowledgements. The authors would like to thank the University of Leeds for providing the Master Chemical Mechanism, the IUPAC for providing the chemical kinetics data, and the NCAR for providing the TUV model. We are grateful to Yee Jun Tham, Qiaozhi Zha and Steven Poon for their efforts collecting the field data which are used for the present modeling study, and to Prof. Aijun Ding for the helpful discussions. We also thank the two anonymous referees for their helpful comments to improve the original manuscript. This work was funded by the National Natural Science Foundation of China (Project No.: 41275123), Shandong University (11460075617010), and the Research Grant Council of Hong Kong (PolyU5125/12P).

Edited by: J. Williams

References

Atkinson, R., Baulch, D. L., Cox, R. A., Hampson, R. F., Kerr, J. A., Rossi, M. J., and Troe, J.: Evaluated kinetic and photochemical data for atmospheric chemistry, organic species, Supplement VII, J. Phys. Chem. Ref. Data, 28, 191–393, 1999.

Carter, W. P. L.: Development of the SAPRC-07 chemical mechanism, Atmos. Environ., 44, 5324–5335, 2010.

Chang, S. Y., McDonald-Buller, E., Kimura, Y., Yarwood, G., Neece, J., Russell, M., Tanaka, P., and Allen, D.: Sensitivity of urban ozone formation to chlorine emission estimates, Atmos. Environ., 36, 4991–5003, 2002.

Elshorbany, Y. F., Kurtenbach, R., Wiesen, P., Lissi, E., Rubio, M., Villena, G., Gramsch, E., Rickard, A. R., Pilling, M. J., and Kleffmann, J.: Oxidation capacity of the city air of Santiago, Chile, Atmos. Chem. Phys., 9, 2257–2273, doi:10.5194/acp-9-2257-2009, 2009.

Finlayson-Pitts, B. J.: Chlorine Atoms as a Potential Tropospheric Oxidant in the Marine Boundary-Layer, Res. Chem. Intermediat., 19, 235–249, 1993.

Jenkin, M. E., Saunders, S. M., Wagner, V., and Pilling, M. J.: Protocol for the development of the Master Chemical Mechanism, MCM v3 (Part B): tropospheric degradation of aromatic volatile organic compounds, Atmos. Chem. Phys., 3, 181–193, doi:10.5194/acp-3-181-2003, 2003.

Jenkin, M. E., Young, J. C., and Rickard, A. R.: The MCM v3.3 degradation scheme for isoprene, Atmos. Chem. Phys. Discuss., 15, 9709–9766, doi:10.5194/acpd-15-9709-2015, 2015.

Luecken, D. J., Phillips, S., Sarwar, G., and Jang, C.: Effects of using the CB05 vs. SAPRC99 vs. CB4 chemical mechanism on model predictions: Ozone and gas-phase photochemical precursor concentrations, Atmos. Environ., 42, 5805–5820, 2008.

Osthoff, H. D., Roberts, J. M., Ravishankara, A. R., Williams, E. J., Lerner, B. M., Sommariva, R., Bates, T. S., Coffman, D., Quinn, P. K., Dibb, J. E., Stark, H., Burkholder, J. B., Talukdar, R. K., Meagher, J., Fehsenfeld, F. C., and Brown, S. S.: High levels of nitryl chloride in the polluted subtropical marine boundary layer, Nat. Geosci., 1, 324–328, 2008.

Phillips, G. J., Tang, M. J., Thieser, J., Brickwedde, B., Schuster, G., Bohn, B., Lelieveld, J., and Crowley, J. N.: Significant concentrations of nitryl chloride observed in rural continental Europe associated with the influence of sea salt chloride and anthropogenic emissions, Geophys. Res. Lett., 39, L10811, doi:10.1029/2012GL051912, 2012.

Riedel, T. P., Bertram, T. H., Crisp, T. A., Williams, E. J., Lerner, B. M., Vlasenko, A., Li, S. M., Gilman, J., de Gouw, J., Bon, D. M., Wagner, N. L., Brown, S. S., and Thornton, J. A.: Nitryl Chloride and Molecular Chlorine in the Coastal Marine Boundary Layer, Environ. Sci. Technol., 46, 10463–10470, 2012.

Riedel, T. P., Wolfe, G. M., Danas, K. T., Gilman, J. B., Kuster, W. C., Bon, D. M., Vlasenko, A., Li, S.-M., Williams, E. J., Lerner, B. M., Veres, P. R., Roberts, J. M., Holloway, J. S., Lefer, B., Brown, S. S., and Thornton, J. A.: An MCM modeling study of nitryl chloride ($ClNO_2$) impacts on oxidation, ozone production and nitrogen oxide partitioning in polluted continental outflow, Atmos. Chem. Phys., 14, 3789–3800, doi:10.5194/acp-14-3789-2014, 2014.

Sarwar, G., Simon, H., Bhave, P., and Yarwood, G.: Examining the impact of heterogeneous nitryl chloride production on air quality across the United States, Atmos. Chem. Phys., 12, 6455–6473, doi:10.5194/acp-12-6455-2012, 2012.

Sarwar, G., Simon, H., Xing, J., and Mathur, R.: Importance of tropospheric $ClNO_2$ chemistry across the Northern Hemisphere, Geophys., Res. Lett., 41, 4050–4058, doi:10.1002/2014GL059962, 2014.

Saunders, S. M., Jenkin, M. E., Derwent, R. G., and Pilling, M. J.: Protocol for the development of the Master Chemical Mechanism, MCM v3 (Part A): tropospheric degradation of non-aromatic volatile organic compounds, Atmos. Chem. Phys., 3, 161–180, doi:10.5194/acp-3-161-2003, 2003.

Shi, J. C. and Bernhard, M. J.: Kinetic studies of Cl-atom reactions with selected aromatic compounds using the photochemical reactor-FTIR spectroscopy technique, Int. J. Chem. Kinet., 29, 349–358, 1997.

Stockwell, W. R., Lawson, C. V., Saunders, E., and Goliff, W. S.: A Review of Tropospheric Atmospheric Chemistry and Gas-Phase Chemical Mechanisms for Air Quality Modeling, Atmosphere, 3, 1–32, 2012.

Tanaka, P. L., Oldfield, S., Neece, J. D., Mullins, C. B., and Allen, D. T.: Anthropogenic sources of chlorine and ozone formation in urban atmospheres, Environ. Sci. Technol., 34, 4470–4473, 2000.

Tanaka, P. L., Allen, D. T., McDonald-Buller, E. C., Chang, S. H., Kimura, Y., Mullins, C. B., Yarwood, G., and Neece, J. D.: Development of a chlorine mechanism for use in the carbon bond IV chemistry model, J. Geophys. Res.-Atmos., 108, 4145, doi:10.1029/2002JD002432, 2003a.

Tanaka, P. L., Riemer, D. D., Chang, S. H., Yarwood, G., McDonald-Buller, E. C., Apel, E. C., Orlando, J. J., Silva, P. J., Jimenez, J. L., Canagaratna, M. R., Neece, J. D., Mullins, C. B., and Allen, D. T.: Direct evidence for chlorine-enhanced urban ozone formation in Houston, Texas, Atmos. Environ., 37, 1393–1400, 2003b.

Tham, Y. J., Yan, C., Xue, L. K., Zha, Q. Z., Wang, X. F., and Wang, T.: Presence of high nitryl chloride in Asian coastal environment and its impact on atmospheric photochemistry, Chinese Sci. Bull., 59, 356–359, 2014.

Thornton, J. A., Kercher, J. P., Riedel, T. P., Wagner, N. L., Cozic, J., Holloway, J. S., Dube, W. P., Wolfe, G. M., Quinn, P. K., Middlebrook, A. M., Alexander, B., and Brown, S. S.: A large atomic chlorine source inferred from mid-continental reactive nitrogen chemistry, Nature, 464, 271–274, 2010.

Wang, T., Tham, Y. J., Xue, L. K., Li, Q. Y., Zha, Q. Z., Wang, Z., Poon, S. C. N., Dube, W. P., Brown, S. S., Louie, P. K. K., Luk, C. W. Y., Blake, D. R., Tsui, W.: Observations of nitryl chloride and modeling its source and effect on ozone in the planetary boundary layer of southern China, in preparation, 2015.

Xue, L. K., Wang, T., Guo, H., Blake, D. R., Tang, J., Zhang, X. C., Saunders, S. M., and Wang, W. X.: Sources and photochemistry of volatile organic compounds in the remote atmosphere of western China: results from the Mt. Waliguan Observatory, Atmos. Chem. Phys., 13, 8551–8567, doi:10.5194/acp-13-8551-2013, 2013.

Xue, L. K., Wang, T., Gao, J., Ding, A. J., Zhou, X. H., Blake, D. R., Wang, X. F., Saunders, S. M., Fan, S. J., Zuo, H. C., Zhang, Q. Z., and Wang, W. X.: Ground-level ozone in four Chinese cities: precursors, regional transport and heterogeneous processes, Atmos. Chem. Phys., 14, 13175–13188, doi:10.5194/acp-14-13175-2014, 2014a.

Xue, L. K., Wang, T., Louie, P. K. K., Luk, C. W. Y., Blake, D. R., and Xu, Z.: Increasing External Effects Negate Local Efforts to Control Ozone Air Pollution: A Case Study of Hong Kong and Implications for Other Chinese Cities, Environ. Sci. Technol., 48, 10769–10775, 2014b.

Xue, L. K., Wang, T., Wang, X. F., Blake, D. R., Gao, J., Nie, W., Gao, R., Gao, X. M., Xu, Z., Ding, A. J., Huang, Y., Lee, S. C., Chen, Y. Z., Wang, S. L., Chai, F. H., Zhang, Q. Z., and Wang, W. X.: On the use of an explicit chemical mechanism to dissect peroxy acetyl nitrate formation, Environ. Pollut., 195, 39–47, 2014c.

Young, C. J., Washenfelder, R. A., Edwards, P. M., Parrish, D. D., Gilman, J. B., Kuster, W. C., Mielke, L. H., Osthoff, H. D., Tsai, C., Pikelnaya, O., Stutz, J., Veres, P. R., Roberts, J. M., Griffith, S., Dusanter, S., Stevens, P. S., Flynn, J., Grossberg, N., Lefer, B., Holloway, J. S., Peischl, J., Ryerson, T. B., Atlas, E. L., Blake, D. R., and Brown, S. S.: Chlorine as a primary radical: evaluation of methods to understand its role in initiation of oxidative cycles, Atmos. Chem. Phys., 14, 3427–3440, doi:10.5194/acp-14-3427-2014, 2014.

Zhang, Q., Streets, D. G., Carmichael, G. R., He, K. B., Huo, H., Kannari, A., Klimont, Z., Park, I. S., Reddy, S., Fu, J. S., Chen, D., Duan, L., Lei, Y., Wang, L. T., and Yao, Z. L.: Asian emissions in 2006 for the NASA INTEX-B mission, Atmos. Chem. Phys., 9, 5131–5153, doi:10.5194/acp-9-5131-2009, 2009.

Atmos. Chem. Phys., 16, 9891–9903, 2016
www.atmos-chem-phys.net/16/9891/2016/
doi:10.5194/acp-16-9891-2016

Atmospheric Chemistry and Physics
Open Access
EGU

Oxidative capacity and radical chemistry in the polluted atmosphere of Hong Kong and Pearl River Delta region: analysis of a severe photochemical smog episode

Likun Xue[1], Rongrong Gu[1], Tao Wang[2,1], Xinfeng Wang[1], Sandra Saunders[3], Donald Blake[4], Peter K. K. Louie[5], Connie W. Y. Luk[5], Isobel Simpson[4], Zheng Xu[1], Zhe Wang[2], Yuan Gao[2], Shuncheng Lee[2], Abdelwahid Mellouki[1], and Wenxing Wang[1]

[1]Environment Research Institute, Shandong University, Ji'nan, Shandong, China
[2]Department of Civil and Environmental Engineering, Hong Kong Polytechnic University, Hong Kong, China
[3]School of Chemistry and Biochemistry, University of Western Australia, WA, Australia
[4]Department of Chemistry, University of California at Irvine, Irvine, CA, USA
[5]Environmental Protection Department, the Government of Hong Kong Special Administrative Region, Hong Kong, China

Correspondence to: Likun Xue (xuelikun@sdu.edu.cn)

Received: 8 February 2016 – Published in Atmos. Chem. Phys. Discuss.: 15 February 2016
Revised: 11 June 2016 – Accepted: 17 July 2016 – Published: 8 August 2016

Abstract. We analyze a photochemical smog episode to understand the oxidative capacity and radical chemistry of the polluted atmosphere in Hong Kong and the Pearl River Delta (PRD) region. A photochemical box model based on the Master Chemical Mechanism (MCM v3.2) is constrained by an intensive set of field observations to elucidate the budgets of RO_x ($RO_x = OH+HO_2+RO_2$) and NO_3 radicals. Highly abundant radical precursors (i.e. O_3, HONO and carbonyls), nitrogen oxides (NO_x) and volatile organic compounds (VOCs) facilitate strong production and efficient recycling of RO_x radicals. The OH reactivity is dominated by oxygenated VOCs (OVOCs), followed by aromatics, alkenes and alkanes. Photolysis of OVOCs (except for formaldehyde) is the dominant primary source of RO_x with average daytime contributions of 34–47 %. HONO photolysis is the largest contributor to OH and the second-most significant source (19–22 %) of RO_x. Other considerable RO_x sources include O_3 photolysis (11–20 %), formaldehyde photolysis (10–16 %), and ozonolysis reactions of unsaturated VOCs (3.9–6.2 %). In one case when solar irradiation was attenuated, possibly by the high aerosol loadings, NO_3 became an important oxidant and the NO_3-initiated VOC oxidation presented another significant RO_x source (6.2 %) even during daytime. This study suggests the possible impacts of daytime NO_3 chemistry in the polluted atmospheres under conditions with the co-existence of abundant O_3, NO_2, VOCs and aerosols, and also provides new insights into the radical chemistry that essentially drives the formation of photochemical smog in the high-NO_x environment of Hong Kong and the PRD region.

1 Introduction

The hydroxyl radical (OH) and hydro/organic peroxy radicals (HO_2 and RO_2), collectively known as RO_x, play a central role in atmospheric chemistry and air pollution (Stone et al., 2012). They dominate the oxidative capacity of atmosphere, and hence govern the removal of primary contaminants and formation of secondary pollutants such as ozone (O_3) and secondary organic aerosol (Hofzumahaus et al., 2009). In the troposphere, they arise from photolysis of closed-shell molecules such as O_3, nitrous acid (HONO), formaldehyde (HCHO) and other carbonyls, as well as ozonolysis reactions of unsaturated volatile organic compounds (VOCs) (Dusanter et al., 2009; Lu et al., 2012; Volkamer et al., 2010). In the presence of nitrogen oxides (NO_x) and VOCs, the RO_x radicals can undergo efficient recycling (e.g. $OH \rightarrow RO_2 \rightarrow RO \rightarrow HO_2 \rightarrow OH$) and produce O_3 and oxygenated VOCs (OVOCs) (Sheehy et al., 2010).

Published by Copernicus Publications on behalf of the European Geosciences Union.

The radical recycling is terminated by their cross reactions with NO_x (under high-NO_x conditions) and RO_x themselves (under low-NO_x conditions), which results in the formation of nitric acid, organic nitrates and peroxides (Liu et al., 2012; Wood et al., 2009). Given the essential significance and complex processes involved, radical chemistry presents one of the core areas in the atmospheric chemistry research.

Understanding the sources and chemistry of RO_x has long been a focus of air quality studies over the past decades. It has been shown that although air pollution problems are visually quite similar, the radical chemistry, and in particular the relative importance of primary radical sources, is inhomogeneous in different metropolitan areas. For example, the dominant radical sources are O_3 photolysis in the South Coast Air Basin in California (2010 scenario) and Nashville, US (Martinez et al., 2003; Volkamer et al., 2010); HONO photolysis in New York City, US (Ren et al., 2003), Paris, France (Michoud et al., 2012) and Santiago, Chile (Elshorbany et al., 2009); HCHO photolysis in Milan, Italy (Alicke et al., 2002); and OVOC photolysis in Mexico City, Mexico (Volkamer et al., 2010), Beijing, China (Liu et al., 2012), Birmingham (summer scenario; Emmerson et al., 2005) and London in England (Emmerson et al., 2007) (note that HONO was not in situ measured but simulated by a box model in Emmerson et al. (2005, 2007), and hence the contributions of HONO photolysis might be underestimated). Therefore, identification of the principal radical sources is a fundamental step towards understanding the formation of air pollution and formulating science-based control strategies.

The nitrate radical (NO_3) is another important oxidant in the polluted atmosphere (Geyer et al., 2001). The NO_3-initiated degradation of VOCs presents an important source of RO_2, gaseous organic nitrates and nitrogen-containing aerosols (Rollins et al., 2012; Saunders et al., 2003). NO_3 has been recognized as a major player in nocturnal chemistry, but is usually neglected for the daytime chemistry given its fast photolysis in sunlight (Volkamer et al., 2010). Under certain conditions, e.g. with abundant O_3 and NO_2 (hence strong NO_3 production) and weak solar radiation (thus weak photolysis), however, NO_3 may also play a role in the daytime chemistry. Indeed, Geyer et al. (2003) observed by differential optical absorption spectroscopy (DOAS) ~5 pptv of NO_3 3 h before sunset in Houston, and indicated considerable contribution (10 %) of NO_3 chemistry to the daytime O_x loss. More studies are required to confirm the possible operation of NO_3 chemistry during daytime and to evaluate its impacts on the atmospheric oxidative capacity (AOC) and formation of O_3 and secondary aerosols.

Hong Kong and the adjacent Pearl River Delta (PRD) is the most industrialized region of southern China, and is suffering from serious photochemical air pollution (e.g. Ling et al., 2014; Zheng et al., 2010). A number of studies have been conducted in the last decade, most of which focused on either O_3-precursor relationships (Zhang et al., 2007; Zhang et al., 2008) or local vs. regional contributions (Wang et al., 2009; Li et al., 2012; Xue et al., 2014b), but few have attempted to understand the atmospheric oxidizing capacity and radical chemistry (Lu et al., 2014). Recent studies have observed the highest ever-reported concentrations of OH and HO_2 at a rural site in the northern PRD, which cannot be reproduced by the classic knowledge of atmospheric chemistry (Hofzumahaus et al., 2009). This indicates the strong oxidative capacity of atmosphere in this region as well as a deficiency in understanding the chemistry underlying the pollution.

As part of the Hong Kong Supersite programme aimed at elucidating the causes of regional smog and haze pollution, an intensive field campaign was conducted at a regional receptor site in summer 2011. A comprehensive set of measurements was taken, which facilitated the construction of a detailed observation-constrained box model to study the atmospheric photochemistry. In the present work, we analyze a severe photochemical episode occurring during 25–31 August 2011 to gain an understanding of atmospheric oxidative capacity and radical chemistry. We first provide an observational overview of the episode, and then evaluate the chemical budgets of both RO_x and NO_3 radicals. This study provides some new insights regarding: (1) the potential impact of NO_3 on the daytime photochemistry in polluted atmospheres and (2) the primary radical sources of RO_x in the high-NO_x environment of Hong Kong and the PRD region.

2 Methods

2.1 Experimental

The measurements were conducted at the Tung Chung air quality monitoring station (TC; 113.93° E, 22.30° N). It is located about 3 km south of the Hong Kong International airport, and is in a residential area of a new town in western Hong Kong (see Fig. S1 in the Supplement). This station is characterized as a polluted receptor site as it receives urban plumes from Hong Kong under easterly winds and regional air masses from the PRD region when northerly winds prevail, and is the location where the maximum O_3 levels are usually recorded in Hong Kong (Xue et al., 2014b). Details of this station and analyses of HONO and aerosol data have been described in our previous publications (Xu et al., 2015; Xue et al., 2014b; Zhou et al., 2014).

A 1-month campaign was carried out from 6 August to 7 September 2011, which covered two distinct types of meteorological conditions and air quality (see Fig. 1). For the majority of the campaign, Hong Kong was influenced by clean marine air masses and featured by good air quality (typical summer conditions as a result of the Asian monsoon). In contrast from 25–31 August, a heavy multi-day photochemical smog event hit Hong Kong with northerly winds prevailing during the daytime and elevated concentrations of various air pollutants were observed. In the present study, this episode was subject to a detailed modeling analysis to understand the atmospheric oxidative capacity and RO_x chemistry, made possible with the most comprehensive suite of measurements taken for the first time in Hong Kong.

A full suite of trace gases and meteorological parameters were simultaneously measured during this episode (as summarized in Supplement Table S1). Here a brief description is given of the measurements used in the present study. Major air quality target pollutants were routinely monitored with commercial analyzers: O_3 with a UV photometric analyzer (TEI model 49i); CO with a non-dispersive infrared equipment (API model 300EU); NO and NO_2 with a chemiluminescence analyzer (TEI model 42i) equipped with a selective blue light converter (Xu et al., 2013). NO_y was measured by another chemiluminescence instrument (TEI model 42cy) with an external molybdenum oxide (MoO) catalytic converter (Xue et al., 2011). HONO was measured in real-time by a long path absorption photometer (QUMA model LOPAP-03) (Xu et al., 2015). Nitryl chloride ($ClNO_2$) was detected using a custom-built chemical ionization mass spectrometer (CIMS; THS Instruments Inc., Atlanta) (Tham et al., 2014). Peroxyacetyl nitrate (PAN) was measured by the same CIMS instrument with a heated inlet, and the potential interference caused by high NO was corrected based on laboratory tests (Slusher et al., 2004; Wang et al., 2014). Hydrogen peroxide (H_2O_2) and organic peroxides were measured by an enzyme-catalyzed fluorescence instrument (Aerolaser AL-2021) (Guo et al., 2014). Particle number and size distributions in the range of 5 nm to 10 µm, which were used to calculate the aerosol surface density, were measured with a wide-range particle spectrometer (WPS; MSP model 1000XP) (Gao et al., 2009).

C_2–C_{10} non-methane hydrocarbons were measured at a time interval of 30 min by a commercial analyzer that combines gas chromatography (GC) with photoionization detection (PID) and flame-ionization detection (FID) (Syntech Spectras, model GC955 Series 600/800 POCP). The detection limits for the measured VOCs ranged from 0.001 to 0.19 ppbv. In addition, 24 h whole air canister samples were collected on selected days (e.g. 25 and 29 August) for the detection of C_1-C_{10} hydrocarbons by using GC with FID, electron capture detection (ECD) and mass spectrometry detection (MSD). The analyses were carried out at the laboratory of the University of California at Irvine, and the detection limit was 3 pptv for all measured species (Simpson et al., 2010; Xue et al., 2013). As evaluated in our previous study, both sets of hydrocarbon measurements agree very well apart from the alkenes. Here the real-time data tended to systematically overestimate the canister measurements (Xue et al., 2014b). Considering the generally lower detection limit of the canister observations, the high-resolution real-time data were corrected in the present study according to the canister data. C_1-C_8 carbonyls were measured by collecting air samples on DNPH-coated sorbent cartridges followed by high pressure liquid chromatography analysis (Xue et al., 2014c). For the carbonyls, a 24 h integrated sample was collected on 25 August, and eight 3 h samples were taken throughout the day on 31 August. The measured hydrocarbon and carbonyl species are listed in Table 1.

Meteorological parameters were monitored by a series of commercial sensors, including a probe for ambient temperature and relative humidity (Young RH/T probe) and an ultrasonic sensor for wind speed and direction (Gill WindSonic). Photolysis frequency of NO_2 (J_{NO_2}) was measured with a filter radiometer (Meteorologie Consult gmbh). All of the above techniques have been validated and applied in many previous studies, with detailed descriptions of the measurement principles, quality assurance and control procedures provided elsewhere (Guo et al., 2014; Xu et al., 2015; Xue et al., 2011, 2014a, b and c). See also Table S1 for a summary of the measurement techniques/instruments and time resolutions.

2.2 The OBM-AOCP model

The zero-dimensional chemical box model OBM-AOCP (Observation-Based Model for investigating the Atmospheric Oxidative Capacity and Photochemistry) has been utilized in many previous studies to evaluate O_3 production (Xue et al., 2013, 2014a, b), PAN formation (Xue et al., 2014c), and oxidative capacity (Xue et al., 2015). Briefly, the model is built on the Master Chemical Mechanism (MCM; v3.2), a nearly explicit gas phase mechanism describing the degradation of 143 primary VOCs (Jenkin et al., 2003; Saunders et al., 2003), and is updated to include both a heterogeneous chemistry scheme (including heterogeneous processes of NO_2, NO_3, N_2O_5, HO_2, and $ClONO_2$; Xue et al., 2014a) and a chlorine chemistry module that describes the reactions of Cl radical with various VOC compounds (Xue et al., 2015; note that the basic MCM only considers the reactions of Cl radical with alkanes). In addition to the chemistry, dry deposition and dilution mixing within the boundary layer are also included in the model (Xue et al., 2014a). The mixing layer height affecting the deposition rate and dilution mixing was assumed to vary from 300 m at night to 1500 m in the afternoon. Sensitivity model runs with different maximum mixing heights (1000 and 2000 m) indicated that its impacts on the modeling results (e.g. simulated HO_x concentrations and OH production rate) were negligible. A detailed description of the model set up is provided in the Supplement.

The model is capable of simulating the concentrations of highly reactive species (e.g. radicals) and quantitatively evaluating several key aspects of atmospheric photochemistry such as oxidant formation (e.g. O_3 and PAN), VOC oxidation and radical budgets. In our model, the rates of over 15600 reactions out of the full MCM (v3.2) are individually and instantaneously computed and grouped into a relatively small number of major routes. The calculation of ozone and PAN production rates have been described elsewhere (Xue et al., 2014a, c). Here the emphasis is placed on the computation of AOC and RO_x budget. AOC is cal-

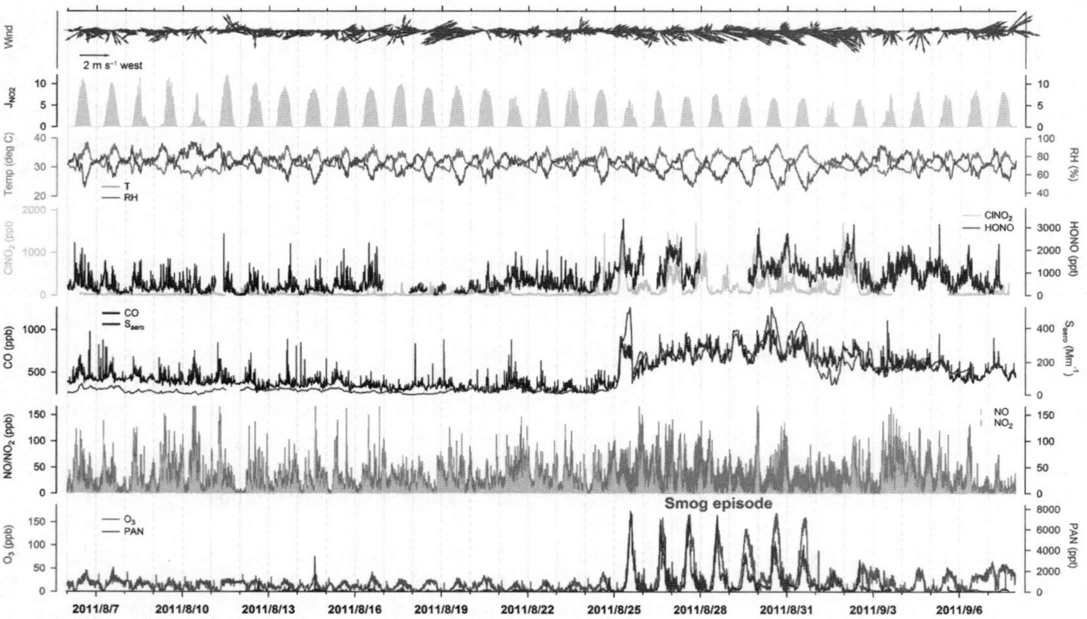

Figure 1. Time series of air pollutants and meteorological parameters observed at Tung Chung from 6 August to 7 September 2011. S_{aero} stands for the scattering coefficient of $PM_{2.5}$. The data gaps were mainly due to the calibration and maintenance of the instruments.

Table 1. 24 h average concentrations of hydrocarbons and oxygenated VOCs measured at Tung Chung on 25 and 31 August 2011*.

Species	25 Aug	31 Aug	Species	25 Aug	31 Aug
Methane	2.264	2.275	benzene	1008	569
Ethane	1192	525	toluene	9465	3557
Propane	2717	1589	ethylbenzene	1718	700
n-Butane	3751	1361	*o*-xylene	979	328
i-Butane	2614	929	*m*-xylene	2082	935
n-Pentane	1175	561	*p*-xylene	813	239
i-Pentane	1569	817	propylbenzene	63	24
n-Hexane	1161	1039	*i*-propylbenzene	54	20
n-Heptane	519	297	*o*-ethyltoluene	140	82
n-Octane	150	410	*m*-ethyltoluene	338	167
n-Nonane	133	–	*p*-ethyltoluene	143	113
2-Methylpentane	1123	–	1,2,3-trimethylbenzene	204	46
3-Methylpentane	842	–	1,2,4-trimethylbenzene	515	338
Ethene	1861	681	1,3,5-trimethylbenzene	124	49
Propene	537	482	formaldehyde	9890	8968
1-butene	196	136	acetaldehyde	4250	3990
i-Butene	224	282	propanal	940	670
trans-2-Butene	68	36	acetone	590	10670
cis-2-butene	54	23	butanal	640	269
1,3-Butadiene	72	57	pentanal	1420	1596
1-Pentene	50	16	hexanal	200	506
Isoprene	779	65	benzaldehyde	890	660
α-Pinene	92	48	methyl ethyl ketone	260	1027
β-Pinene	36	21	acrolein	30	BDL
Ethyne	2903	265	crotonaldehyde	30	510

* The units are pptv except for methane which is in ppmv. "–" indicates no data available, and "BDL" indicates below detection limit.

culated as the sum of oxidation rates of CO and VOCs by the principal oxidants, namely OH, O_3, NO_3 and Cl (Xue et al., 2015). The partitioning of the AOC among individual oxidants or VOC groups can be also assessed. The chemical budgets of OH, HO_2, and RO_2 are quantified by grouping a huge number of relevant reactions into dozens of major production, cycling and loss routes. The principal radical sources in the polluted atmosphere generally include photolysis of O_3, HONO, H_2O_2 and OVOCs as well as reactions of O_3+VOCs, NO_3+VOCs and Cl+VOCs. The radical sinks mainly include the RO_x-NO_x and RO_x-RO_x cross reactions. Besides, a number of other minor reaction pathways were also computed to facilitate a thorough investigation of the RO_x chemistry (see Figs. 5 and 7).

The measurement data of O_3, NO, NO_2, HONO, $ClNO_2$, H_2O_2, PAN, CO, C_1-C_{10} HCs, C_1-C_8 carbonyls, aerosol surface area and radius, temperature, RH and J_{NO_2} were averaged or interpolated to a time resolution of 10 min for the model constraints. For carbonyls, the diurnal profiles measured on 31 August 2011, throughout which eight 3 h samples collected were adopted and scaled to the 24 h average data observed on 25 August (see Fig. S4 for the measured profiles of selected carbonyls). An initial concentration of 0.5 ppm of H_2 was assumed in the model. Photolysis frequencies were calculated as a function of solar zenith angle within the model (Saunders et al., 2003) and further scaled with the measured J_{NO_2} values. The model starts from 00:00 local time (LT) and runs for a 24 h period. Prior to the formal calculation, the model was run for 5 days with constraints of the campaign-average data to reach steady states for the unconstrained compounds (e.g. radicals). The final outputs were extracted and subject to further analyses.

3 Results and discussion

3.1 Observational overview

The measured concentrations of major pollutants and meteorological parameters at TC are depicted in Fig. 1. During 25–31 August 2011, Hong Kong was hit by a prolonged photochemical smog episode, with concentrations of various air pollutants exceeding the ambient air quality standard (the zoomed-in figure of this episode is given in the supplement). Peak O_3 mixing ratios of over 150 ppbv were observed almost every day within the 1-week period, except for 29 August when the peak was 135 ppbv. As another indicator of photochemical smog, the concentrations of PAN were also very high with the peak values exceeding 4 ppbv every day (except for 3.7 ppbv on 29 August). The maximum hourly values of O_3 and PAN were recorded at 162 and 6.95 ppbv, respectively. Extremely high levels of NO_x (peak of ~ 150 ppbv), CO (peak of ~ 1000 ppbv) and particulate matter (as indicative of > 500 Mm^{-1} of aerosol scattering coefficient) were also determined. Overall, inspection of observational data reveals the markedly poor air quality and serious photochemical pollution over the region during the episode.

Table 1 lists the 24 h average concentrations of hydrocarbons and carbonyls measured on 25 and 31 August 2011. It is clearly seen that the VOC levels, in particular for reactive aromatics and aldehydes, were also very high during the episode. On 25 August, for instance, the 24 h average values of toluene, summed xylenes, formaldehyde and acetaldehyde were as high as 9.47, 3.87, 9.89, and 4.25 ppbv, which were 3–30 folders higher than those measured during the non-episode period of the campaign (figures not shown). HONO and $ClNO_2$, two precursors of OH and Cl radicals, were also measured. Elevated HONO (up to 2–3 ppbv) and moderate $ClNO_2$ (up to 0.5–1 ppbv) were usually found at night, and what is more interesting is that the daytime HONO levels were also significant (over 1 ppbv in general; see Fig. 1). Such daytime HONO levels cannot be explained by the known gas-phase source and indicates the existence of other unknown source(s) (Xu et al., 2015), yet exploring the unknown HONO sources is beyond the scope of the present study. High abundances of O_3, HONO and carbonyls would definitely lead to strong production of RO_x radicals, and the abundant VOCs would facilitate efficient radical propagation (e.g. OH→RO_2). Therefore, strong atmospheric oxidative capacity and intensive in situ photochemistry can be expected from the above analyses.

The dynamic cause of this episode was a distant tropical cyclone that introduced warm stagnant weather and facilitated accumulation of air pollutants in Hong Kong and the PRD region. The weather condition featured high temperatures (30–35 °C) and relatively low RH (40–80 %; see Fig. S2). During the daytime, the prevailing surface winds were consistently from the northwest with relatively low wind speeds (~ 2 $m\,s^{-1}$), suggesting the transport of processed air masses from the upwind PRD region to the site. This was further confirmed by the 48 h backward trajectories calculated by the HYSPLIT model (Draxler and Rolph, 2016), which indicated that for most days the air masses had spent a large portion of time over the PRD region prior to arriving at TC (Fig. S3).

There was an exception on 25 August when the air flow was switching from southerly maritime air to northerly PRD regional air masses (see Fig. S3). This case is believed to be more influenced by the local air in Hong Kong, because (1) northerly winds during the daytime were somewhat weak compared to the other cases (see Fig. S2); (2) the backward trajectories also indicated less impact from the PRD region (Fig. S3); and (3) the CO / NO_y ratio on that day was significantly lower than those on the following days (Fig. S5), which is consistent with the previous finding that the PRD air masses have higher CO / NO_y ratios than those from Hong Kong (Wang et al., 2003). The evolution of the CO / NO_y ratio clearly indicates the transition from local (25 August) to regional air masses (27–31 August) throughout the 1-week

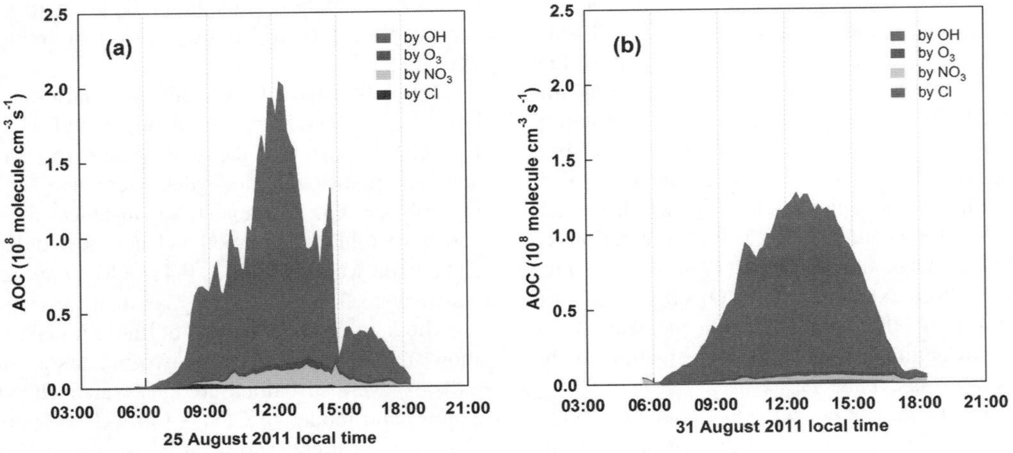

Figure 2. Daytime atmospheric oxidative capacity (AOC) and contributions of major oxidants at Tung Chung on **(a)** 25 August and **(b)** 31 August 2011.

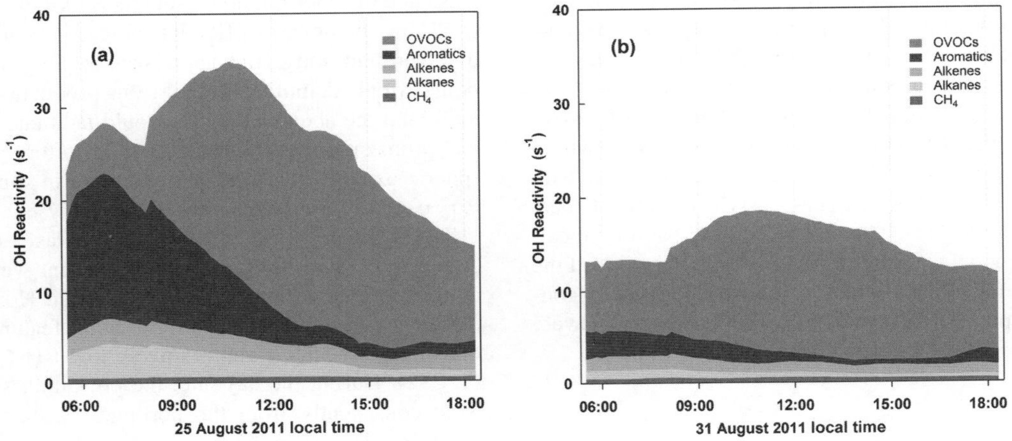

Figure 3. Partitioning of the daytime OH reactivity by oxidation of major VOC groups at Tung Chung on **(a)** 25 August and **(b)** 31 August 2011.

episode (Fig. S5). In the following discussion, detailed modeling analyses are conducted for the 25 and 31 August cases, which are representative of local Hong Kong and regional PRD pollution, respectively.

3.2 Atmospheric oxidative capacity

The strong oxidative capacity of the atmosphere during the pollution episodes was confirmed by quantifying the loss rates of CO and VOCs via reactions with OH, O_3, NO_3 and Cl, as shown in Fig. 2. The calculated AOC was up to 2.04×10^8 and 1.27×10^8 molecules $cm^{-3}\,s^{-1}$, with daytime averages (06:00–18:00 LT) of 7.26×10^7 and 6.30×10^7 molecules $cm^{-3}\,s^{-1}$, on 25 and 31 August, respectively. As such, the total number of CO and VOC molecules depleted throughout the daytime was 3.14×10^{12} and 2.72×10^{12} per cm^{-3} of air for both cases. Such levels of AOC at TC are much higher than those determined from a rural site in Germany (Geyer et al., 2001), but a bit lower than that assessed from a polluted area in Santiago, Chile (Elshorbany et al., 2009).

OH was, as expected, the predominant oxidant accounting for 89 and 93 % of the AOC on 25 and 31 August, respectively. NO_3 was the second important oxidant with contributions of 7 and 3 % for both cases. In particular, NO_3 contributed to 43 % of the AOC at 15:00 LT on 25 August under a weak solar radiation condition. The major fuels for NO_3 oxidation were OVOCs (i.e. 77–90 %) and alkenes (10–23 %). In comparison, O_3 and Cl (produced from $ClNO_2$ photolysis) had minor contributions due to the relatively lower abundances of alkenes and Cl radicals (i.e. the modelled peak value of Cl was $\sim 1 \times 10^4$ atoms cm^{-3}). Overall, the OH-dominated AOC at TC is in line with the previous studies at other urban locales (Elshorbany et al., 2009; Bannan et al., 2015), and the present analysis suggests that the NO_3 radical may play an important role in the daytime oxidation under certain conditions (see a detailed evaluation in Sect. 3.4).

We further assessed the loss rates of major VOC groups due to OH oxidation, from which the partitioning of OH reactivity among different VOCs can be elucidated. The results are presented in Fig. 3. OVOCs clearly dominate the OH reactivity with daytime average contributions of 60 and 75 % and with maximums in the afternoon of over 80 % for both cases. Aromatics are the second largest contributor comprising on average 22 and 10 % of the daytime OH reactivity. For the Hong Kong local case on 25 August, especially, aromatics made up the majority (i.e. 40–60 %) of the OH reactivity in the early morning period when there were much fresher air masses. In comparison, alkenes and alkanes only accounted for a small fraction (8–10 %) of the OH reactivity at TC. These results are in fair agreement with the previous studies of Lou et al. (2010) and Whalley et al. (2016), which indicated the dominance of secondary OVOCs in the observed OH reactivity in the PRD region and central London.

As shown above, the partitioning of principal oxidants and OH reactivity is quite similar for both cases. In comparison with the regional case on 31 August 2011, nonetheless, the Hong Kong local case (i.e. 25 August 2011) showed higher AOC levels and more contribution from aromatic VOCs to the OH reactivity. Such difference should be due to the fresher air masses and hence more reactive VOC species during the local case. In the following section, a detailed budget analysis of the radical initiation, recycling and termination processes is presented.

3.3 RO_x budget analysis

3.3.1 The Hong Kong local case

Figure 4 presents the primary daytime sources of OH, HO_2 and RO_2 at TC on 25 August 2011, and the detailed daytime RO_x budget is schematically illustrated in Fig. 5. HONO photolysis is not only the predominant source of OH in the early morning but also a major source throughout the daytime. Photolysis of O_3 becomes an important OH source at midday, the strength of which is comparable to that of HONO photolysis. In terms of the daytime average (06:00–18:00 LT), HONO photolysis is the dominant OH source with an average OH production rate of 1.5 ppbv h^{-1}, followed by O_3 photolysis (0.9 ppbv h^{-1}). In addition, ozonolysis reactions of unsaturated VOCs are another considerable OH source with a mean production rate of 0.2 ppbv h^{-1}, whilst other sources (e.g. photolysis of H_2O_2, HNO_3 and OVOCs) are generally negligible.

For HO_2, the most important source is the photolysis of OVOCs (including not only the measured carbonyls but also the oxidation products generated within the model), with a daytime average production rate of 2.7 ppbv h^{-1}. Specifically, photolysis of formaldehyde produces HO_2 at a rate of 0.8 ppbv h^{-1}, while the remaining majority (1.9 ppbv h^{-1}) is from the photolysis of the other OVOCs. Such source strength of OVOC photolysis was comparable to those determined in the metropolitan areas of Beijing (Liu et al., 2012) and Mexico City (Volkamer et al., 2010). In addition, another source that needs to be considered is reactions of O_3 with unsaturated VOCs, which produce HO_2 at 0.1 ppbv h^{-1} on average during the daytime.

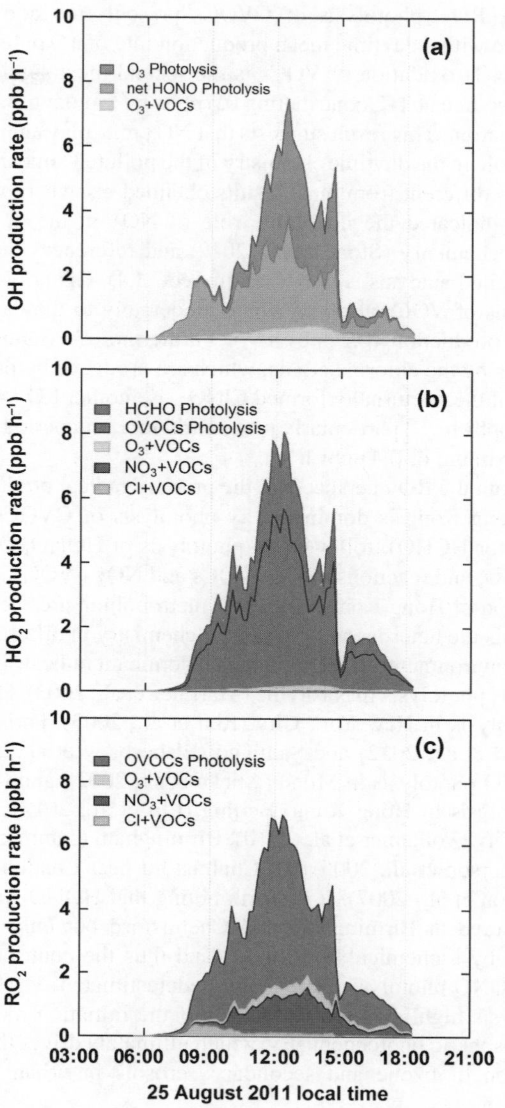

Figure 4. Primary daytime sources of **(a)** OH, **(b)** HO_2 and **(c)** RO_2 radicals at Tung Chung on 25 August 2011. The term "net HONO photolysis" represents the contribution of net HONO (i.e. subtracting the formation rate of HONO from NO+OH = HONO).

For RO_2, photolysis of OVOCs presents the dominant source with a daytime mean production rate of 1.9 ppbv h^{-1}. The NO_3 oxidation of VOCs is the second-most significant RO_2 source at TC, contributing 0.5 ppbv h^{-1} of daytime RO_2 production. This result suggests that NO_3 may play an important role in the daytime chemistry of the polluted atmosphere, and is different from most results obtained elsewhere which have indicated the negligible role of NO_3 in the daytime photochemistry (Stone et al., 2012; and references therein; a detailed analysis is presented in Sect. 3.4). Ozonolysis reactions of VOCs also contribute moderately to the daytime RO_2 production (0.2 ppbv h^{-1}). Furthermore, oxidation of VOCs by the chlorine atoms, which are produced by photolysis of the nocturnally formed $ClNO_2$, is another RO_2 source (0.1 ppbv h^{-1}), particularly in the early morning period (with a maximum of 0.4 ppbv h^{-1}).

From the RO_x perspective, the primary radical production in Hong Kong is dominated by photolysis of OVOCs (except for HCHO), followed by photolysis of HONO, O_3 and HCHO, and reactions of O_3+VOCs and NO_3+VOCs. Comparison of Hong Kong with other metropolitan areas clearly reveals the heterogeneity in radical chemistry in different urban environments. For example, the dominant radical sources are O_3 photolysis in Nashville (Martinez et al., 2003), HONO photolysis in New York City (Ren et al., 2003), Paris (Michoud et al., 2012) and Santiago (Elshorbany et al., 2009), HCHO photolysis in Milan (Alicke et al., 2002), and OVOC photolysis in Hong Kong, Beijing (Liu et al., 2012), Mexico City (Volkamer et al., 2010), Birmingham (summer case; Emmerson et al., 2005) and Chelmsford near London (Emmerson et al., 2007). It is worth noting that HONO was not measured at Birmingham and Chelmsford but only simulated by a chemical box model, and thus the contributions of HONO photolysis were likely underestimated. The above analysis highlights the variability of the initiation mode of atmospheric photochemistry, which ultimately drives the formation of ozone and secondary aerosols in urban atmospheres.

Efficient recycling of radicals can be also illustrated in Fig. 5. Oxidation of CO and VOCs by OH produces HO_2 and RO_2 with daytime average rates of 3.3 and 8.0 ppbv h^{-1}, respectively. Reactions of RO_2+NO and HO_2+NO in turn result in strong production of RO (9.0 ppbv h^{-1}) and OH (12.5 ppbv h^{-1}), with O_3 formed as a by-product. It is evident that these recycling processes dominate the total production of OH, HO_2 and RO_2 radicals. It is common that the radical propagation is efficient and amplifies the effect of the newly produced radicals in the polluted atmospheres with the co-existence of abundant NO_x and VOCs (Elshorbany et al., 2009; Liu et al., 2012). As to the termination processes, the RO_x radical sink is clearly dominated by their reactions with NO_x. Specifically, reactions of OH+NO_2 and RO_2+NO_2, forming HNO_3 and organic nitrates, contributed approximately 2.8 and 2.5 ppbv h^{-1} of the radical loss on daytime average at TC. This is in line with the understanding

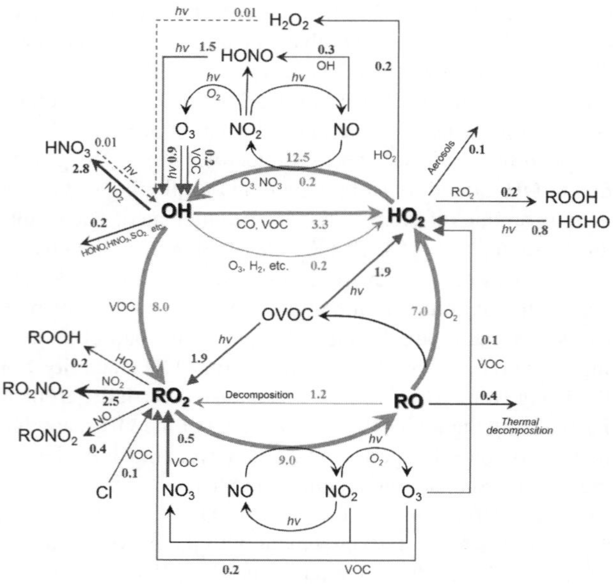

Figure 5. Daytime average RO_x budget at Tung Chung on 25 August 2011. The unit is ppb h^{-1}. The red, blue and green lines indicate the production, destruction and recycling pathways of radicals, respectively.

that reactions with NO_x usually dominate the radical sink in high-NO_x environments.

3.3.2 The PRD regional case

The detailed radical budget for the regional case on 31 August 2011 is illustrated in Figs. 6 and 7. Overall, the chemical budget of RO_x radicals was essentially the same as that of the Hong Kong local case. Specifically, the most significant primary source is photolysis of OVOCs except for HCHO, which produces both HO_2 and RO_2 equally at a daytime average rate of 1.3 ppbv h^{-1}. The other important radical sources include photolysis of HONO (1.7 ppbv h^{-1} as OH), O_3 (1.5 ppbv h^{-1} as OH) and HCHO (1.2 ppbv h^{-1} as HO_2), ozonolysis reactions of unsaturated VOCs (0.3 ppbv h^{-1} for the sum of RO_x), and reactions of NO_3+VOCs (0.2 ppbv h^{-1} as RO_2) and Cl+VOCs (0.1 ppbv h^{-1} as RO_2). For the termination processes, reactions of OH+NO_2 and RO_2+NO_2 present the major radical loss pathways, with daytime average rates of 3.5 and 1.5 ppbv h^{-1}, respectively.

Despite the abovementioned general similarity, two aspects are noteworthy about the difference between the two cases. First, the primary radical source strength was significantly higher on 25 August than 31 August, suggesting the stronger oxidation capacity of the atmosphere during the local case. Second, the source strengths of photolysis of HONO, O_3 and HCHO were higher on 31 August than 25 August, whilst the sources of OVOCs photolysis, O_3+VOCs and NO_3+VOCs showed an opposite picture. Such differ-

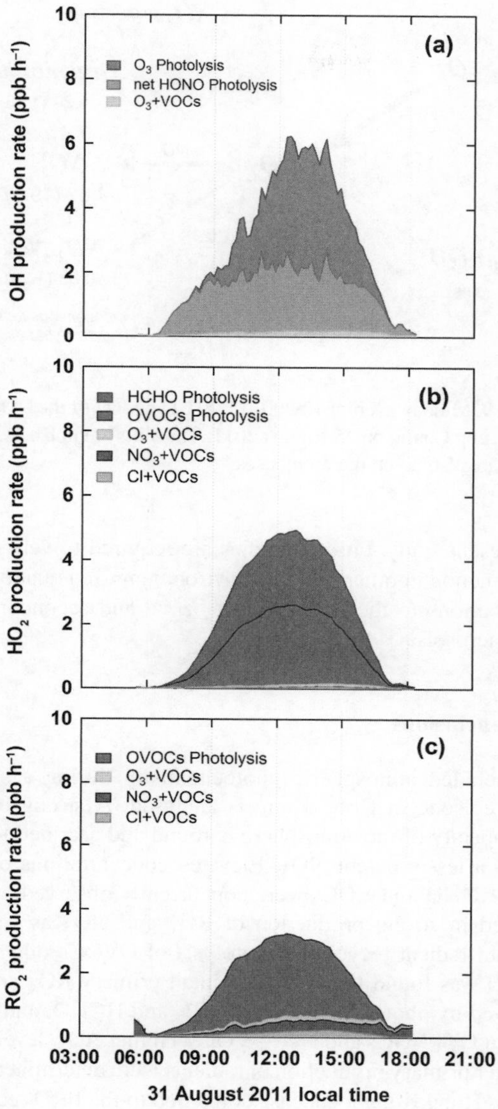

Figure 6. The same as Fig. 4 but for the case of 31 August 2011.

ence in the partitioning of radical sources between both cases should be ascribed to the higher VOC levels (with more fresh emissions) and weaker solar radiation (possibly attenuated by the high aerosol loading; see Sect. 3.4) on 25 August.

3.4 Evidence of daytime NO_3 chemistry

The NO_3 radical can initiate the oxidation of VOCs and lead to formation of RO_2 and nitrogen-containing organic aerosols (Rollins et al., 2012; Saunders et al., 2003). These processes are usually considered to mainly occur at night and be negligible during the daytime due to the fast photolysis of NO_3. In the present study, we observed an interesting case that provided evidence of the operation of daytime NO_3-

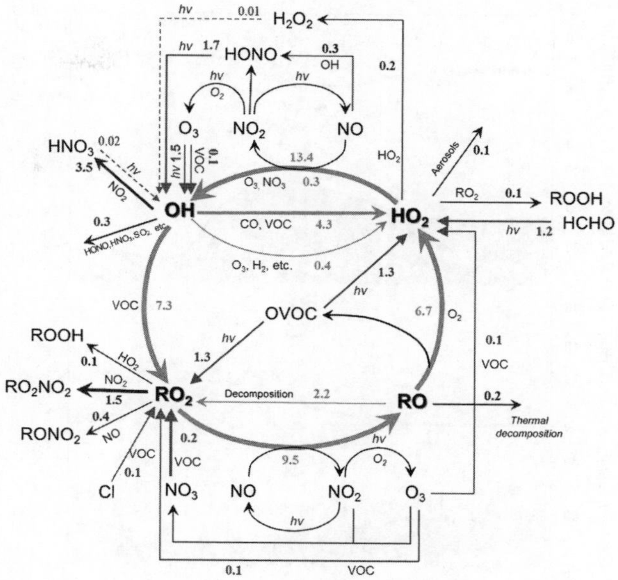

Figure 7. The same as Fig. 5 but for the case of 31 August 2011.

initiated chemistry. The detailed measurement data of chemical and meteorological parameters in this case (i.e. 25 August 2011) are depicted in Fig. 8. During this episode, the air was characterized by high concentrations of O_3 (up to 170 ppbv), NO_2 ($\sim$25 ppbv as the afternoon average) and VOCs (see Table 1). Meanwhile, the solar irradiation arriving at the surface was weaker than other days, as evidenced by the relatively lower values of J_{NO_2} (with a peak of $6.0\times10^{-3}\,s^{-1}$) compared to clear days with $\sim10\times10^{-3}\,s^{-1}$ (see Fig. 1). The ambient relative humidity (RH) in the afternoon was in the range of 60–70 %, implying that there was little cloud on the site, whilst the aerosol scattering coefficient was very high (up to 525 Mm^{-1}; compared to $28\pm12\,Mm^{-1}$ on clear days). Hence, the attenuated solar radiation is possibly attributed to the abundant aerosol loadings. Under such conditions, the model produced an afternoon peak of NO_3 of $\sim$7 pptv at 13:30–15:00 LT (except for the maximum of 11.3 pptv at 14:50 LT that was coincident with an extremely low solar radiation condition).

To further understand the causes and impacts of the daytime NO_3 chemistry, a detailed budget analysis was conducted with the OBM-AOCP model. The midday average (09:00–15:00 LT) production and destruction rates of NO_3 from the individual reaction pathways are documented in Fig. 9. The co-existence of high concentrations of O_3 and NO_2 resulted in a very strong NO_3 production with an average strength of 11.0 ppb h^{-1}. Given its high reactivity, NO_3 once formed, can be readily photolysed as well as react with NO and VOCs. For this case, about 80 % (i.e. 8.8 ppb h^{-1}) of NO_3 reacted with NO to convert back to NO_2. Due to the weak solar radiation, photolysis only accounted for

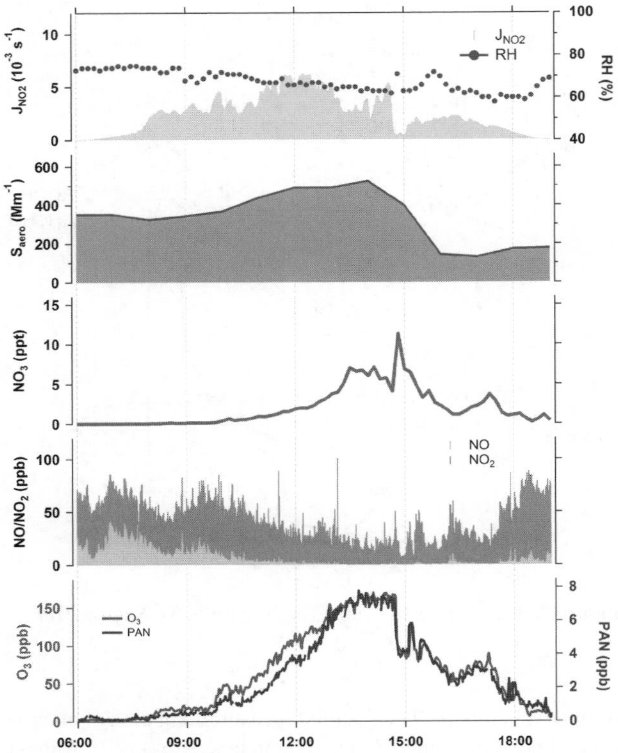

Figure 8. Time series of chemical and meteorological parameters observed at Tung Chung on 25 August 2011.

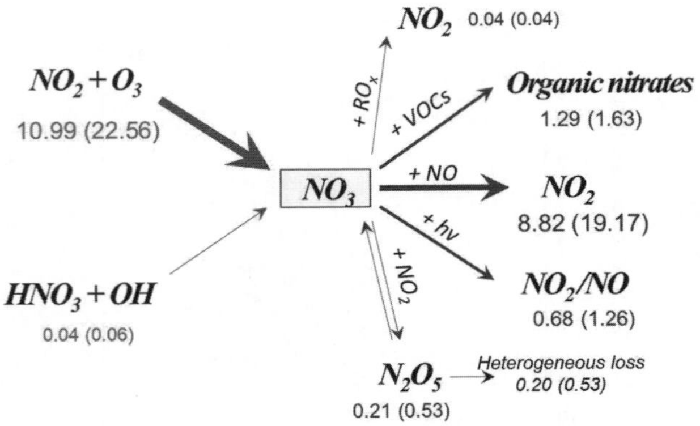

Figure 9. Midday average (09:00–15:00 LT) budget of the NO_3 radical at Tung Chung on 25 August 2011. The units are ppb h^{-1}. Peak values are also given in parentheses.

6.2 % (or 0.7 ppb h^{-1}) of the NO_3 loss. In comparison, reactions of NO_3 with VOCs contributed 11.7 % (or 1.3 ppb h^{-1}) to the total loss at midday. During this episode, therefore, NO_3 appeared to be the second-most important oxidant (see Sect. 3.2) and the reactions of NO_3 with VOCs presented a considerable RO_2 source during the daytime (Sect. 3.3). In addition, the NO_3-initiated degradation of VOCs could also lead to formation of secondary organic nitrate aerosols, but was not simulated in the present study.

The above analysis indicates the possible importance of NO_3-initiated oxidation in the daytime atmospheric photochemistry under specific conditions. This analysis is solely derived from an observation-based modeling study of a unique pollution case in Hong Kong. Nevertheless, we hypothesize that it may also take place in other polluted urban atmospheres, especially in the large cities of China. It is known that eastern China now suffers from widespread and severe photochemical smog during the summer, which features elevated concentrations of O_3, NO_x, VOCs, and fine particulate matter (Xue et al., 2014a). The intense air pollution usually induces "smoldering" weather with poor visibility and hence attenuated solar irradiation (Ding et al., 2013). All these unfavourable conditions would facilitate the operation of daytime NO_3 chemistry as found in Hong Kong in the present study. Further studies are required to verify this phenomenon in other polluted environments and quantify its contributions to the formation of ozone and secondary organic aerosols.

4 Conclusions

The detailed atmospheric photochemistry during a severe smog episode in Hong Kong is analysed. A strong oxidative capacity of the atmosphere is found and ascribed to OH and to a lesser extent NO_3. Elevated concentrations of O_3, NO_2, HONO and VOCs were concurrently observed, which resulted in strong production of RO_x and NO_3 as well as efficient radical recycling. Photolysis of OVOCs other than HCHO was found to be the dominant primary RO_x source, followed by photolysis of HONO, O_3 and HCHO, and reactions of O_3+VOCs and NO_3+VOCs. Higher AOC levels and stronger primary production of radicals were determined during the Hong Kong local case compared to the PRD regional case. Although the primary radical sources were essentially the same, photolysis of OVOCs (except for HCHO) and reactions of O_3+VOCs and NO_3+VOCs were stronger for the Hong Kong local case, which was ascribed to the higher VOC levels. In comparison, the source strengths of photolysis of HONO, O_3 and HCHO were higher during the regional case.

On 25 August 2011, a unique case when heavy air pollution attenuated the solar irradiation reaching the surface in Hong Kong, NO_3 was identified as an important oxidant in the daytime chemistry. VOC oxidation by NO_3 represented the second largest source of RO_2, with a daytime average production rate of 0.5 ppbv h^{-1}. The NO_3-initiated degradation of VOCs would enhance the formation of O_3 and nitrogen-containing organic aerosols. This study indicates the potential operation of the daytime NO_3 chemistry in polluted urban atmospheres characterized by the co-existence of abundant O_3, NO_2, VOCs and particles. Further studies, especially direct observations of the NO_3 radical, are required to verify this interesting phenomenon in other environments and to evaluate its contribution to the O_3 and secondary organic aerosol formation.

5 Data availability

The underlying research data can be accessed upon contact with the corresponding author (L. K. Xue; xuelikun@sdu.edu.cn).

The Supplement related to this article is available online at doi:10.5194/acp-16-9891-2016-supplement.

Acknowledgements. The authors appreciate Steven Poon, Yee Jun Tham, Shengzhen Zhou, Wei Nie and Jia Guo for their contributions to the field study; the University of Leeds for providing the Master Chemical Mechanism; and the NOAA Air Resources Laboratory for providing the web-based HYSPLIT model. We thank the two anonymous referees for their helpful comments to improve the quality of our original manuscript. The field observations were funded by the Environment and Conservation Fund of Hong Kong (project no.: 7/2009), and the data analyses were supported by the National Natural Science Foundation of China (project no.: 41505111) and Qilu Youth Talent Programme of Shandong University.

Disclaimer. The opinions expressed in this paper are those of the authors and do not necessarily reflect the views or policies of the Government of the Hong Kong Special Administrative Region, nor does mention of trade names or commercial products constitute an endorsement or recommendation of their use.

Edited by: D. Heard
Reviewed by: two anonymous referees

References

Alicke, B., Platt, U., and Stutz, J.: Impact of nitrous acid photolysis on the total hydroxyl radical budget during the Limitation of Oxidant Production/Pianura Padana Produzione di Ozono study in Milan, J. Geophys. Res.-Atmos., 107, 8196, doi:10.1029/2000JD000075, 2002.

Bannan, T. J., Booth, A. M., Bacak, A, Muller, J. B. A., Leather, K. E., Breton, M. L., Jones, B., Young, D., Coe, H., Allan, J., Visser, S., Slowik, J. G., Furger, M., Prévôt, A. S. H., Lee, J., Dunmore, R. E., Hopkins, J. R., Hamilton, J. F., Lewis, A. C., Whalley, L. K., Sharp, T., Stone, D., Heard, D. E., Fleming, Z. L., Leigh, R., Shallcross, D. E., and Percival, C. J.: The first UK measurements of nitryl chloride using a chemical ionization mass spectrometer in central London in the summer of 2012, and an investigation of the role of Cl atom oxidation, J. Geophys. Res.-Atmos., 120, 5638–5657, 2015.

Ding, A. J., Fu, C. B., Yang, X. Q., Sun, J. N., Petäjä, T., Kerminen, V.-M., Wang, T., Xie, Y., Herrmann, E., Zheng, L. F., Nie, W., Liu, Q., Wei, X. L., and Kulmala, M.: Intense atmospheric pollution modifies weather: a case of mixed biomass burning with fossil fuel combustion pollution in eastern China, Atmos. Chem. Phys., 13, 10545–10554, doi:10.5194/acp-13-10545-2013, 2013.

Draxler, R. R. and Rolph, G. D.: HYSPLIT (HYbrid Single-Particle Lagrangian Integrated Trajectory) Model access via NOAA ARL READY Website, http://ready.arl.noaa.gov/HYSPLIT.php (last access: 26 May 2016), NOAA Air Resources Laboratory, Silver Spring, MD, 2016.

Dusanter, S., Vimal, D., Stevens, P. S., Volkamer, R., Molina, L. T., Baker, A., Meinardi, S., Blake, D., Sheehy, P., Merten, A., Zhang, R., Zheng, J., Fortner, E. C., Junkermann, W., Dubey, M., Rahn, T., Eichinger, B., Lewandowski, P., Prueger, J., and Holder, H.: Measurements of OH and HO_2 concentrations during the MCMA-2006 field campaign – Part 2: Model comparison and radical budget, Atmos. Chem. Phys., 9, 6655–6675, doi:10.5194/acp-9-6655-2009, 2009.

Elshorbany, Y. F., Kurtenbach, R., Wiesen, P., Lissi, E., Rubio, M., Villena, G., Gramsch, E., Rickard, A. R., Pilling, M. J., and Kleffmann, J.: Oxidation capacity of the city air of Santiago, Chile, Atmos. Chem. Phys., 9, 2257–2273, doi:10.5194/acp-9-2257-2009, 2009.

Emmerson, K. M., Carslaw, N., and Pilling, M. J.: Urban atmospheric chemistry during the PUMA campaign 2: Radical budgets for OH, HO_2 and RO_2, J. Atmos. Chem., 52, 165–183, 2005.

Emmerson, K. M., Carslaw, N., Carslaw, D. C., Lee, J. D., McFiggans, G., Bloss, W. J., Gravestock, T., Heard, D. E., Hopkins, J., Ingham, T., Pilling, M. J., Smith, S. C., Jacob, M., and Monks, P. S.: Free radical modelling studies during the UK TORCH Campaign in Summer 2003, Atmos. Chem. Phys., 7, 167–181, doi:10.5194/acp-7-167-2007, 2007.

Gao, J., Wang, T., Zhou, X. H., Wu, W. S., and Wang, W. X.: Measurement of aerosol number size distributions in the Yangtze River delta in China: Formation and growth of particles under polluted conditions, Atmos. Environ., 43, 829–836, 2009.

Geyer, A., Alicke, B., Konrad, S., Schmitz, T., Stutz, J., and Platt, U.: Chemistry and oxidation capacity of the nitrate radical in the continental boundary layer near Berlin, J. Geophys. Res.-Atmos., 106, 8013–8025, 2001.

Geyer, A., Alicke, B., Ackermann, R., Martinez, M., Harder, H., Brune, W., di Carlo, P., Williams, E., Jobson, T., Hall, S., Shetter, R., and Stutz, J.: Direct observations of daytime NO_3: Implications for urban boundary layer chemistry, J. Geophys. Res.-Atmos., 108, 4368, doi:10.1029/2002JD002967, 2003.

Guo, J., Tilgner, A., Yeung, C., Wang, Z., Louie, P. K. K., Luk, C. W. Y., Xu, Z., Yuan, C., Gao, Y., Poon, S., Herrmann, H., Lee, S., Lam, K. S., and Wang, T.: Atmospheric Peroxides in a Polluted Subtropical Environment: Seasonal Variation, Sources and Sinks, and Importance of Heterogeneous Processes, Environ. Sci. Technol., 48, 1443–1450, 2014.

Hofzumahaus, A., Rohrer, F., Lu, K. D., Bohn, B., Brauers, T., Chang, C. C., Fuchs, H., Holland, F., Kita, K., Kondo, Y., Li, X., Lou, S. R., Shao, M., Zeng, L. M., Wahner, A., and Zhang, Y. H.: Amplified Trace Gas Removal in the Troposphere, Science, 324, 1702–1704, 2009.

Jenkin, M. E., Saunders, S. M., Wagner, V., and Pilling, M. J.: Protocol for the development of the Master Chemical Mechanism, MCM v3 (Part B): tropospheric degradation of aromatic volatile organic compounds, Atmos. Chem. Phys., 3, 181–193, doi:10.5194/acp-3-181-2003, 2003.

Li, Y., Lau, A. K. H., Fung, J. C. H., Zheng, J. Y., Zhong, L. J., and Louie, P. K. K.: Ozone source apportionment (OSAT) to differentiate local regional and super-regional source contributions in the Pearl River Delta region, China, J. Geophys. Res.-Atmos., 117, D15305, doi:10.1029/2011JD017340, 2012.

Ling, Z. H., Guo, H., Lam, S. H. M., Saunders, S. M., and Wang, T.: Atmospheric photochemical reactivity and ozone production at two sites in Hong Kong: Application of a Master Chemical Mechanism-photochemical box model, J. Geophys. Res.-Atmos., 119, 10567–10582, doi:10.1002/2014JD021794, 2014.

Liu, Z., Wang, Y., Gu, D., Zhao, C., Huey, L. G., Stickel, R., Liao, J., Shao, M., Zhu, T., Zeng, L., Amoroso, A., Costabile, F., Chang, C.-C., and Liu, S.-C.: Summertime photochemistry during CAREBeijing-2007: RO_x budgets and O_3 formation, Atmos. Chem. Phys., 12, 7737–7752, doi:10.5194/acp-12-7737-2012, 2012.

Lou, S., Holland, F., Rohrer, F., Lu, K., Bohn, B., Brauers, T., Chang, C. C., Fuchs, H., Häseler, R., Kita, K., Kondo, Y., Li, X., Shao, M., Zeng, L., Wahner, A., Zhang, Y., Wang, W., and Hofzumahaus, A.: Atmospheric OH reactivities in the Pearl River Delta – China in summer 2006: measurement and model results, Atmos. Chem. Phys., 10, 11243–11260, doi:10.5194/acp-10-11243-2010, 2010.

Lu, K. D., Rohrer, F., Holland, F., Fuchs, H., Bohn, B., Brauers, T., Chang, C. C., Häseler, R., Hu, M., Kita, K., Kondo, Y., Li, X., Lou, S. R., Nehr, S., Shao, M., Zeng, L. M., Wahner, A., Zhang, Y. H., and Hofzumahaus, A.: Observation and modelling of OH and HO_2 concentrations in the Pearl River Delta 2006: a missing OH source in a VOC rich atmosphere, Atmos. Chem. Phys., 12, 1541–1569, doi:10.5194/acp-12-1541-2012, 2012.

Lu, K. D., Rohrer, F., Holland, F., Fuchs, H., Brauers, T., Oebel, A., Dlugi, R., Hu, M., Li, X., Lou, S. R., Shao, M., Zhu, T., Wahner, A., Zhang, Y. H., and Hofzumahaus, A.: Nighttime observation and chemistry of HO_x in the Pearl River Delta and Beijing in summer 2006, Atmos. Chem. Phys., 14, 4979–4999, doi:10.5194/acp-14-4979-2014, 2014.

Martinez, M., Harder, H., Kovacs, T. A., Simpas, J. B., Bassis, J., Lesher, R., Brune, W. H., Frost, G. J., Williams, E. J., Stroud, C. A., Jobson, B. T., Roberts, J. M., Hall, S. R., Shetter, R. E., Wert, B., Fried, A., Alicke, B., Stutz, J., Young, V. L., White, A. B., and Zamora, R. J.: OH and HO_2 concentrations, sources, and loss rates during the Southern Oxidants Study in Nashville, Tennessee, summer 1999, J. Geophys. Res.-Atmos., 108, 4617, doi:10.1029/2003JD003551, 2003.

Michoud, V., Kukui, A., Camredon, M., Colomb, A., Borbon, A., Miet, K., Aumont, B., Beekmann, M., Durand-Jolibois, R., Perrier, S., Zapf, P., Siour, G., Ait-Helal, W., Locoge, N., Sauvage, S., Afif, C., Gros, V., Furger, M., Ancellet, G., and Doussin, J. F.: Radical budget analysis in a suburban European site during the MEGAPOLI summer field campaign, Atmos. Chem. Phys., 12, 11951–11974, doi:10.5194/acp-12-11951-2012, 2012.

Ren, X. R., Harder, H., Martinez, M., Lesher, R. L., Oliger, A., Simpas, J. B., Brune, W. H., Schwab, J. J., Demerjian, K. L., He, Y., Zhou, X. L., and Gao, H. G.: OH and HO_2 chemistry in the urban atmosphere of New York City, Atmos. Environ., 37, 3639–3651, 2003.

Rollins, A. W., Browne, E. C., Min, K. E., Pusede, S. E., Wooldridge, P. J., Gentner, D. R., Goldstein, A. H., Liu, S., Day, D. A., Russell, L. M., and Cohen, R. C.: Evidence for NO_x Control over Nighttime SOA Formation, Science, 337, 1210–1212, 2012.

Saunders, S. M., Jenkin, M. E., Derwent, R. G., and Pilling, M. J.: Protocol for the development of the Master Chemical Mechanism, MCM v3 (Part A): tropospheric degradation of non-aromatic volatile organic compounds, Atmos. Chem. Phys., 3, 161–180, doi:10.5194/acp-3-161-2003, 2003.

Sheehy, P. M., Volkamer, R., Molina, L. T., and Molina, M. J.: Oxidative capacity of the Mexico City atmosphere – Part 2: A RO_x radical cycling perspective, Atmos. Chem. Phys., 10, 6993–7008, doi:10.5194/acp-10-6993-2010, 2010.

Simpson, I. J., Blake, N. J., Barletta, B., Diskin, G. S., Fuelberg, H. E., Gorham, K., Huey, L. G., Meinardi, S., Rowland, F. S., Vay, S. A., Weinheimer, A. J., Yang, M., and Blake, D. R.: Characterization of trace gases measured over Alberta oil sands mining operations: 76 speciated C_2-C_10 volatile organic compounds (VOCs), CO_2, CH_4, CO, NO, NO_2, NO_y, O_3 and SO_2, Atmos. Chem. Phys., 10, 11931–11954, doi:10.5194/acp-10-11931-2010, 2010.

Slusher, D. L., Huey, L. G., Tanner, D. J., Flocke, F. M., and Roberts, J. M.: A thermal dissociation-chemical ionization mass spectrometry (TD-CIMS) technique for the simultaneous measurement of peroxyacyl nitrates and dinitrogen pentoxide, J. Geophys. Res.-Atmos., 109, D19315, doi:10.1029/2004JD004670, 2004.

Stone, D., Whalley, L. K., and Heard, D. E.: Tropospheric OH and HO_2 radicals: field measurements and model comparisons, Chem. Soc. Rev., 41, 6348–6404, 2012.

Tham, Y. J., Yan, C., Xue, L. K., Zha, Q. Z., Wang, X. F., and Wang, T.: Presence of high nitryl chloride in Asian coastal environment and its impact on atmospheric photochemistry, Chinese Sci. Bull., 59, 356–359, 2014.

Volkamer, R., Sheehy, P., Molina, L. T., and Molina, M. J.: Oxidative capacity of the Mexico City atmosphere – Part 1: A radical source perspective, Atmos. Chem. Phys., 10, 6969–6991, doi:10.5194/acp-10-6969-2010, 2010.

Wang, T., Poon, C. N., Kwok, Y. H., and Li, Y. S.: Characterizing the temporal variability and emission patterns of pollution plumes in the Pearl River Delta of China, Atmos. Environ., 37, 3539–3550, 2003.

Wang, T., Wei, X. L., Ding, A. J., Poon, C. N., Lam, K. S., Li, Y. S., Chan, L. Y., and Anson, M.: Increasing surface ozone concentrations in the background atmosphere of Southern China, 1994–2007, Atmos. Chem. Phys., 9, 6217–6227, doi:10.5194/acp-9-6217-2009, 2009.

Wang, X., Wang, T., Yan, C., Tham, Y. J., Xue, L., Xu, Z., and Zha, Q.: Large daytime signals of N_2O_5 and NO_3 inferred at 62 amu in a TD-CIMS: chemical interference or a real atmospheric phenomenon?, Atmos. Meas. Tech., 7, 1–12, doi:10.5194/amt-7-1-2014, 2014.

Whalley, L. K., Stone, D., Bandy, B., Dunmore, R., Hamilton, J. F., Hopkins, J., Lee, J. D., Lewis, A. C., and Heard, D. E.: Atmospheric OH reactivity in central London: observations, model predictions and estimates of in situ ozone production, Atmos. Chem. Phys., 16, 2109–2122, doi:10.5194/acp-16-2109-2016, 2016.

Wood, E. C., Herndon, S. C., Onasch, T. B., Kroll, J. H., Canagaratna, M. R., Kolb, C. E., Worsnop, D. R., Neuman, J. A., Seila, R., Zavala, M., and Knighton, W. B.: A case study of ozone production, nitrogen oxides, and the radical budget in Mexico City, Atmos. Chem. Phys., 9, 2499–2516, doi:10.5194/acp-9-2499-2009, 2009.

Xu, Z., Wang, T., Xue, L. K., Louie, P. K. K., Luk, C. W. Y., Gao, J., Wang, S. L., Chai, F. H., and Wang, W. X.: Evaluating the uncertainties of thermal catalytic conversion in measuring atmospheric nitrogen dioxide at four differently polluted sites in China, Atmos. Environ., 76, 221–226, 2013.

Xu, Z., Wang, T., Wu, J. Q., Xue, L. K., Chan, J., Zha, Q. Z., Zhou, S. Z., Louie, P. K. K., and Luk, C. W. Y.: Nitrous acid (HONO) in a polluted subtropical atmosphere: Seasonal variability, direct vehicle emissions and heterogeneous production at ground surface, Atmos. Environ., 106, 100–109, 2015.

Xue, L. K., Wang, T., Zhang, J. M., Zhang, X. C., Deliger, Poon, C. N., Ding, A. J., Zhou, X. H., Wu, W. S., Tang, J., Zhang, Q. Z., and Wang, W. X.: Source of surface ozone and reactive nitrogen speciation at Mount Waliguan in western China: New insights from the 2006 summer study, J. Geophys. Res.-Atmos., 116, D07306, doi:10.1029/2010JD014735, 2011.

Xue, L. K., Wang, T., Guo, H., Blake, D. R., Tang, J., Zhang, X. C., Saunders, S. M., and Wang, W. X.: Sources and photochemistry of volatile organic compounds in the remote atmosphere of western China: results from the Mt. Waliguan Observatory, Atmos. Chem. Phys., 13, 8551–8567, doi:10.5194/acp-13-8551-2013, 2013.

Xue, L. K., Wang, T., Gao, J., Ding, A. J., Zhou, X. H., Blake, D. R., Wang, X. F., Saunders, S. M., Fan, S. J., Zuo, H. C., Zhang, Q. Z., and Wang, W. X.: Ground-level ozone in four Chinese cities: precursors, regional transport and heterogeneous processes, Atmos. Chem. Phys., 14, 13175–13188, doi:10.5194/acp-14-13175-2014, 2014a.

Xue, L. K., Wang, T., Louie, P. K. K., Luk, C. W. Y., Blake, D. R., and Xu, Z.: Increasing External Effects Negate Local Efforts to Control Ozone Air Pollution: A Case Study of Hong Kong and Implications for Other Chinese Cities, Environ. Sci. Technol., 48, 10769–10775, 2014b.

Xue, L. K., Wang, T., Wang, X. F., Blake, D. R., Gao, J., Nie, W., Gao, R., Gao, X. M., Xu, Z., Ding, A. J., Huang, Y., Lee, S. C., Chen, Y. Z., Wang, S. L., Chai, F. H., Zhang, Q. Z., and Wang, W. X.: On the use of an explicit chemical mechanism to dissect peroxy acetyl nitrate formation, Environ. Pollut., 195, 39–47, 2014c.

Xue, L. K., Saunders, S. M., Wang, T., Gao, R., Wang, X. F., Zhang, Q. Z., and Wang, W. X.: Development of a chlorine chemistry module for the Master Chemical Mechanism, Geosci. Model Dev., 8, 3151–3162, doi:10.5194/gmd-8-3151-2015, 2015.

Zhang, J., Wang, T., Chameides, W. L., Cardelino, C., Kwok, J., Blake, D. R., Ding, A., and So, K. L.: Ozone production and hydrocarbon reactivity in Hong Kong, Southern China, Atmos. Chem. Phys., 7, 557–573, doi:10.5194/acp-7-557-2007, 2007.

Zhang, Y. H., Su, H., Zhong, L. J., Cheng, Y. F., Zeng, L. M., Wang, X. S., Xiang, Y. R., Wang, J. L., Gao, D. F., Shao, M., Fan, S. J., and Liu, S. C.: Regional ozone pollution and observation-based approach for analyzing ozone-precursor relationship during the PRIDE-PRD2004 campaign, Atmos. Environ., 42, 6203–6218, 2008.

Zheng, J. Y., Zhong, L. J., Wang, T., Louie, P. K. K., and Li, Z. C.: Ground-level ozone in the Pearl River Delta region: Analysis of data from a recently established regional air quality monitoring network, Atmos. Environ., 44, 814–823, 2010.

Zhou, S. Z., Wang, T., Wang, Z., Li, W. J., Xu, Z., Wang, X. F., Yuan, C., Poon, C. N., Louie, P. K. K., Luk, C. W. Y., and Wang, W. X.: Photochemical evolution of organic aerosols observed in urban plumes from Hong Kong and the Pearl River Delta of China, Atmos. Environ., 88, 219–229, 2014.

Atmos. Chem. Phys., 16, 10637–10650, 2016
www.atmos-chem-phys.net/16/10637/2016/
doi:10.5194/acp-16-10637-2016

Atmospheric Chemistry and Physics
Open Access

Significant increase of summertime ozone at Mount Tai in Central Eastern China

Lei Sun[1], Likun Xue[1], Tao Wang[2,1], Jian Gao[3], Aijun Ding[4], Owen R. Cooper[5,6], Meiyun Lin[7,8], Pengju Xu[9], Zhe Wang[2], Xinfeng Wang[1], Liang Wen[1], Yanhong Zhu[1], Tianshu Chen[1], Lingxiao Yang[1,10], Yan Wang[10], Jianmin Chen[1,10], and Wenxing Wang[1]

[1]Environment Research Institute, Shandong University, Ji'nan, Shandong, China
[2]Department of Civil and Environmental Engineering, Hong Kong Polytechnic University, Hong Kong, China
[3]Chinese Research Academy of Environmental Sciences, Beijing, China
[4]Institute for Climate and Global Change Research and School of Atmospheric Sciences, Nanjing University, Nanjing, Jiangsu, China
[5]Cooperative Institute for Research in Environmental Sciences, University of Colorado, Boulder, Colorado, USA
[6]NOAA Earth System Research Laboratory, Boulder, Colorado, USA
[7]Atmospheric and Oceanic Sciences, Princeton University, Princeton, New Jersey, USA
[8]NOAA Geophysical Fluid Dynamics Laboratory, Princeton, New Jersey, USA
[9]School of Geography and Environment, Shandong Normal University, Ji'nan, Shandong, China
[10]School of Environmental Science and Engineering, Shandong University, Ji'nan, Shandong, China

Correspondence to: Likun Xue (xuelikun@sdu.edu.cn)

Received: 17 March 2016 – Published in Atmos. Chem. Phys. Discuss.: 24 March 2016
Revised: 11 July 2016 – Accepted: 11 August 2016 – Published: 26 August 2016

Abstract. Tropospheric ozone (O_3) is a trace gas playing important roles in atmospheric chemistry, air quality and climate change. In contrast to North America and Europe, long-term measurements of surface O_3 are very limited in China. We compile available O_3 observations at Mt. Tai – the highest mountain over the North China Plain – during 2003–2015 and analyze the decadal change of O_3 and its sources. A linear regression analysis shows that summertime O_3 measured at Mt. Tai has increased significantly by 1.7 ppbv yr^{-1} for June and 2.1 ppbv yr^{-1} for the July–August average. The observed increase is supported by a global chemistry-climate model hindcast (GFDL-AM3) with O_3 precursor emissions varying from year to year over 1980–2014. Analysis of satellite data indicates that the O_3 increase was mainly due to the increased emissions of O_3 precursors, in particular volatile organic compounds (VOCs). An important finding is that the emissions of nitrogen oxides (NO_x) have diminished since 2011, but the increase of VOCs appears to have enhanced the ozone production efficiency and contributed to the observed O_3 increase in central eastern China. We present evidence that controlling NO_x alone, in the absence of VOC controls, is not sufficient to reduce regional O_3 levels in North China in a short period.

1 Introduction

Ozone (O_3) in the troposphere is a trace gas of great importance for climate and air quality. It is the principal precursor of the hydroxyl radical (OH) which plays a central role in atmospheric chemistry (Seinfeld and Pandis, 2006), and the third most important greenhouse gas contributing to the warming of the Earth (IPCC, 2013). At ground level, high levels of O_3 have adverse effects on human health and ecosystem productivity (National Research Council, 1991; Monks et al., 2015). In the troposphere, the ambient O_3 burden is the product of the flux from the stratosphere (e.g., Stohl et al., 2003; Lin et al., 2015a), dry deposition, and net photochemical production involving the reactions of nitrogen oxides (NO_x) with carbon monoxide (CO) and volatile organic compounds (VOCs) in the presence of sunlight (Crutzen, 1973; Ma et al., 2002). Increases in anthro-

Published by Copernicus Publications on behalf of the European Geosciences Union.

pogenic emissions of ozone precursors have contributed to changes in the tropospheric O_3 abundances, both globally and regionally (The Royal Society, 2008; Cooper et al., 2014; Monks et al., 2015; Lin et al., 2015b). Decadal shifts in climate and circulation regimes can also contribute to changes in tropospheric O_3, as found in the 40-year record at Mauna Loa Observatory in Hawaii (Lin et al., 2014). Conversely, the changes in tropospheric O_3 may also pose significant feedbacks to the environment and climate (e.g., Shindell et al., 2012; Stevenson et al., 2013). Therefore, the long-term changes (or trends) of tropospheric O_3 has long been a topic of great interest in the atmospheric sciences.

Since the 1970s, long-term measurements of surface O_3 (and O_3 precursors) have been increasingly carried out worldwide, mostly in North America and Europe (e.g., Logan et al., 2012; Oltmans et al., 2013; Cooper et al., 2014). The existing knowledge of tropospheric O_3 trends has been recently reviewed (UNEP and WMO, 2011; Cooper et al., 2014; Monks et al., 2015). Overall, upward trends have been recorded around the world since the 1970s, but trends over the past two decades have varied regionally. In Europe, the surface O_3 in rural or remote areas, usually regarded as the regional baseline O_3, rose until the year 2000 but has since leveled off or decreased (Logan et al., 2012; Oltmans et al., 2013; Parrish et al., 2012). In the eastern US, summertime O_3 at most rural and urban stations has decreased over 1990–2010 (Lefohn et al., 2010; Cooper et al., 2012). In the western US, extreme ozone events have decreased in urban areas, particularly in southern California (Warneke et al., 2012), but springtime O_3 at remote mountain sites has shown large interannual variability due to stratospheric influence (Lin et al., 2015a), with little overall trends or small increases (Cooper et al., 2012; Fine et al., 2015). Analysis of available observations and chemistry-climate model hindcast indicates that springtime O_3 in the free troposphere over western North America has increased significantly by 0.4 $ppbv\,yr^{-1}$ over 1995–2014 (Lin et al., 2015b).

In comparison with North America and Europe, investigations of long-term O_3 trends are scarce in China, where rapid urbanization and industrialization has occurred over the past 3 decades. Significant increasing trends of surface O_3 have been derived from a handful of long-term monitoring stations over China, including Beijing (Tang et al., 2009; Zhang et al., 2014), Hong Kong (Xue et al., 2014), Taiwan (Lin et al., 2010) and Mount Waligian (Xu et al., 2016). Based on the MOZAIC commercial aircraft measurements, Ding et al. (2008) derived an O_3 increase of $\sim 2\,\%$ per year from 1995 to 2005 for the lower troposphere above Beijing by analyzing the difference between observations from 1995 to 2000 and 2000 to 2005. Wang et al. (2009) reported the first long-term continuous observations of Chinese surface O_3 at a regional background site in southern China (Hok Tsui), and indicated an average increase of 0.58 $ppbv\,yr^{-1}$ during 1994–2007. Xu et al. (2016) recently reported another continuous O_3 record (1994–2013) at Mt. Waliguan, a global atmospheric watch station in western China, and found significant positive trends with 0.15–0.27 $ppbv\,yr^{-1}$ for daytime O_3 and 0.13–0.29 $ppbv\,yr^{-1}$ for nighttime O_3. Despite the valuable information obtained from the abovementioned efforts, additional studies are required to improve our understanding of tropospheric O_3 trends across the rapidly developing China. In particular, long-term observations covering more than 10 years remain very limited over the highly polluted regions of central eastern China.

In recent years, China has implemented a series of stringent air quality control measures. Following the successful reductions in sulfur dioxide (SO_2) emissions since 2006 (Lu et al., 2010), China has recently launched a national programme to reduce NO_x emissions during its "Twelfth Five-Year Plan" (2011–2015) (China State Council, 2011). However, to our knowledge, fewer controls have been placed on VOC emissions in China. Rather, anthropogenic VOC emissions have continued to increase (Bo et al., 2008; Wang et al., 2014). Therefore, it is of great interest and critical importance to evaluate the effect of the current control policy (i.e., controlling NO_x with little action for VOCs) on regional O_3 and other secondary air pollution problems in China.

Mt. Tai (36.25° N, 117.10° E; 1534 m altitude) is the highest mountain in the center of the North China Plain (NCP; see Fig. 1) – a fast developing region facing severe air pollution. During daytime, the summit is well within the planetary boundary layer (PBL), and the site is therefore regionally representative of the region (Kanaya et al., 2013). Since 2003, several field measurement campaigns have been conducted at this site, with surface O_3 of major interest. In this paper, we analyze all available O_3 and its precursor observations at Mt. Tai to understand the summertime O_3 characteristics, including trends. To place the short observational record into a long-term context, we analyze multi-decadal hindcast simulations (1980–2014) conducted with the GFDL-AM3 chemistry-climate model (Lin et al., 2014, 2015a, b). In the following sections, we first present the seasonal and diurnal ozone variations followed by the climatological air mass transport pattern in summer. We then derive the O_3 trend (or systematic change) during 2003–2015 using linear regression. We finally elucidate the key factors affecting the O_3 trends by examining satellite and in situ observed trace gas data. Our analysis demonstrates a significant increase of summertime surface O_3 at this important regional site in northern China, and indicates the urgent need of VOC control in China to reduce regional O_3 pollution.

2 Observational data set

The observations analyzed in the present study are summarized in Table 1. It comprises several sets of field observations from different periods. The earliest O_3 measurements at Mt. Tai were made from July to November 2003 (Gao et al., 2005), and the longest observations lasted for 3 years

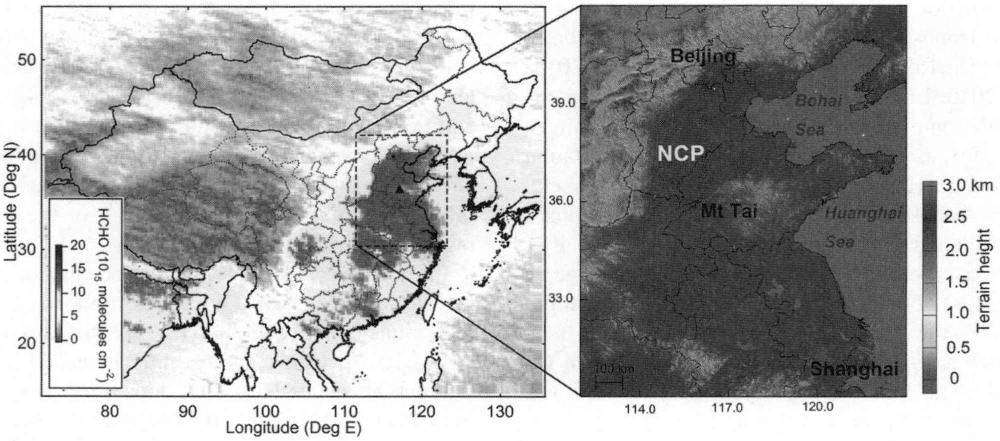

Figure 1. Geographical map showing the North China Plain and the location of Mt. Tai. The left map is color-coded by the HCHO column density in the summer period (JJA; 2003–2014) retrieved from SCIAMACHY and GOME-2(B).

Table 1. The observations at Mt. Tai analyzed in the present study.

Year	Month	Observed species	Data source
2003	July–November	O_3, CO	Gao et al. (2005) and our study
2004	June–August	O_3	Kanaya et al. (2013)[a]
2005	June–August	O_3	Kanaya et al. (2013)[a]
2006	June–December	O_3, NO_2^*, NO	Our study
2007	March–December	O_3, NO, NO_2^*, CO	Our study [b]
2008	January–December	O_3, NO, NO_2^*, CO	Our study
2009	January–June	O_3, NO, NO_2^*, CO	Our study
2014	June–August	O_3, NO_x^*, CO	Our study [c]
2015	June–August	O_3, NO, NO_2^*, CO	Our study [c]

[a] Monthly average data were taken from Kanaya et al. (2013). [b] Valid NO and NO_2^* data were only available in March–April and July–December, and CO was available in March–April in 2007. [c] Note that the intensive measurement periods were 6 June–3 July and 24 July–26 August in 2014 and 14 June–8 August in 2015.

from June 2006 to June 2009 (despite a data gap in January–February 2007 owing to instrument maintenance). In recent years, we conducted intensive measurements at Mt. Tai during June–August of 2014 and 2015. In addition, another set of 3-year measurements was carried out from March 2004 to May 2007, as an international joint effort between Japanese and Chinese scientists (Kanaya et al., 2013); monthly average O_3 data are taken from this work. As most of the measurements are available for the period of June–August each year, we focus on the summertime O_3 in this study.

Two study sites have been used for field observations at Mt. Tai. One was the Mt. Tai Meteorological Observatory (Site 1) at the summit with an altitude of 1534 m a.s.l., and the other was in a hotel (Site 2) that is ∼ 1 km to the northwest of Site 1 and slightly lower (1465 m a.s.l.; see Fig. S1 in the Supplement for the site locations). Such elevations position these sites either within the PBL in the afternoon in summer, or in the free troposphere during the night. Although Mt. Tai is a famous tourism spot, both sites are located in the less frequently visited zones. Hence the impact of local anthropogenic emission should be small, and the data collected are believed to be regionally representative. Details of these sites have been described elsewhere (Site 1: Gao et al., 2005; Kanaya et al., 2013; Site 2: Guo et al., 2012; Shen et al., 2012). For our data set, most of the measurements were taken at Site 1, and only the intensive campaign in 2014 took place at Site 2.

All measurements were implemented using standard techniques, which were detailed in the previous publications (e.g., Gao et al., 2005; Xue et al., 2011). Briefly, O_3 was measured using a commercial ultraviolet photometric instrument (*Thermo Environment Instruments (TEI), Model 49C*) with a detection limit of 2 ppbv and a precision of 2 ppbv. CO was monitored with a gas filter correlation, non-dispersive infrared analyzer (*Teledyne Advanced Pollution Instrumentation, Model 300E*), with automatic zeroing every 2 hours. This technique has a detection limit of 30 ppbv and a precision of 1 % for a level of 500 ppbv. NO and NO_2^* were measured by a chemiluminescence analyzer equipped with an internal MoO catalytic converter (*TEI, Model 42C*), with a detection limit of 0.4 ppbv and precision of 0.4 ppbv. Intercomparison with a highly selective photolytic NO_2 detection approach indicated that the NO_2^* measured with MoO conversion significantly overestimated NO_2 (i.e., up to 130 % in afternoon hours), and NO_2^* actually represented a major fraction (60–80 %) of NO_y at Mt. Tai (Xu et al., 2013). During the measurements, the O_3 analyzer was calibrated routinely (i.e., quarterly for the 3-year observations in 2006–2009, and before and after the other campaigns) by an ozone primary standard (*TEI, Model 49PS*). For CO and NO_x^*, zero and span calibrations were performed weekly during 2006–2009 and every 3 days during the intensive campaigns. Meteorological data including temperature, relative humidity (RH), and wind vectors were obtained from the Mt. Tai Meteorological Observatory, where Site 1 is located.

It is noteworthy that some portion of this long-term data set has been reported previously. Gao et al. (2005) analyzed the measurements of O_3 and CO from July to November 2003 and examined their diurnal variations and relations to backward trajectories. Kanaya et al. (2013) reported the observations in 2004–2007 and examined the processes influencing the seasonal variations and regional pollution episode. The major objective of the present study is to compile all of the available O_3-related observations at Mt. Tai and to establish the trend (or systematic change), if any, of ambient O_3 levels in the past decade at this unique site, regionally representative of the North China Plain.

3 Results and discussion

3.1 Seasonal and diurnal variations from more recent data

Figure 2 depicts the seasonal variation of surface O_3 at Mt. Tai derived from the more recent year-round observations from 2006 to 2009. Overall, O_3 shows higher levels in the warm season, i.e. April–October, compared to the cold season, i.e. November–March, with two peaks in June and October. The elevated O_3 levels in April and May should be affected by the stratosphere-troposphere exchange process which usually occurs at its maximum in the spring season (Yamaji et al., 2006). In addition to the high temperatures and intense solar radiation (especially in June), biomass burning is believed to be another factor shaping the O_3 maximums in June and October, both of which are major harvest seasons of wheat and corn in northern China. The significant impacts of biomass burning on air quality over the North China Plain during June have been evaluated by a number of studies (Lin et al., 2009; Yamaji et al., 2010; Suthawaree et al., 2010). It is also noticeable that the O_3 concentrations in July and August at Mt. Tai are substantially lower than those in June. This is attributed in part to the more humid weather and greater

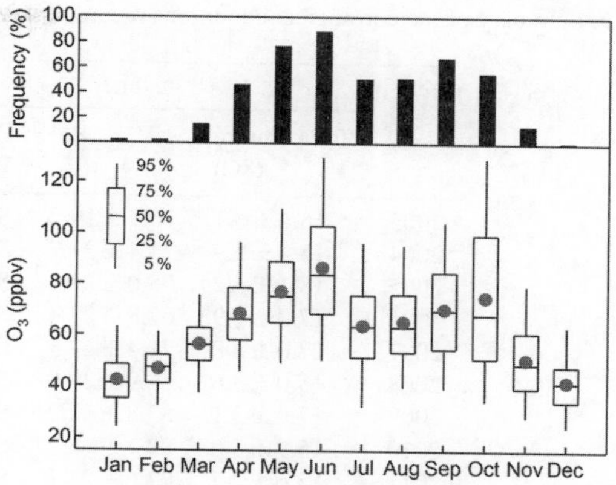

Figure 2. Seasonal variation of surface O_3 mixing ratios at Mt. Tai derived from the continuous observations from 2006 to 2009. Red dots indicate the monthly average O_3 concentrations. Shown in the upper panel is the frequency of the MDA8 O_3 exceeding the Chinese national ambient air quality standard, i.e. 75 ppbv (Class II).

precipitation in July and August in this region (see Table 2 for the RH condition). Inspection of meteorological conditions day by day also indicates a frequency of cloudy days (i.e., with RH $\geq$ 95 %) of $\sim$ 25 % in June and of $\sim$ 51 % during July–August. In the following analyses, therefore, we assess the ozone characteristics separately for June and July–August.

Shown in the upper panel of Fig. 2 is the frequency of the maximum daily 8-hour average O_3 mixing ratios (MDA8 O_3) exceeding the 75-ppbv National Ambient Air Quality Standard (Class II). Although located in a relatively remote mountain-top area, the observed O_3 pollution at Mt. Tai was rather serious in the warm season, with frequencies of the O_3-exceedence days of over 45 % throughout April–October. In particular, the occurrence of O_3-exceedence days was as high as 89 % in June. These results demonstrate the severe O_3 pollution situation across the North China Plain.

Figure 3 illustrates the well-defined diurnal variations of surface O_3 with a trough in the early morning and a broad peak lasting from afternoon to early evening, which is commonly observed at polluted rural sites. As the summit of Mt. Tai is well above the PBL at nighttime, the O_3 concentrations during the latter part of the night (e.g., 02:00–05:00 LT) are usually considered to reflect the regional baseline O_3 (defined hereafter as regional O_3 without impact of local photochemical formation). Comparing the diurnal profiles in June and July–August clearly reveals the significantly higher regional baseline O_3 in June (with a mean difference of $\sim$ 17 ppbv). On the other hand, the daytime O_3 build-up, defined as the increase in O_3 concentrations from the early-morning minimum to the late-afternoon maximum, may reflect the potential of regional O_3 formation. For the Mt. Tai case, the average daytime O_3 build-up was 22 ppbv in June and 15 ppbv in July–August, indicating the stronger photo-

Table 2. Summary of meteorological conditions recorded at Mt. Tai in June and July–August over 2003–2015*.

Year	June			July–August		
	Temperature (°C)	RH (%)	Prevailing WD	Temperature (°C)	RH (%)	Prevailing WD
2003	15.4 ± 3.1	70.9 ± 19.2	SW	17.3 ± 2.9	88.3 ± 16.3	SW
2004	15.4 ± 3.9	74.4 ± 21.2	SW	17.0 ± 2.5	86.8 ± 15.5	SSW
2005	18.1 ± 3.2	65.0 ± 22.0	SSW	17.4 ± 2.9	86.8 ± 16.3	SW
2006	17.2 ± 2.9	67.8 ± 21.4	SSW	18.3 ± 2.0	88.3 ± 16.3	SSW
2007	17.0 ± 2.8	71.4 ± 27.3	E	17.8 ± 2.3	85.2 ± 21.1	E
2008	15.0 ± 3.6	76.9 ± 20.0	S	17.0 ± 2.3	86.3 ± 14.6	S
2009	17.9 ± 3.1	57.8 ± 22.2	SSW	17.4 ± 2.9	78.7 ± 21.8	S
2010	15.9 ± 3.6	80.1 ± 18.7	SW	18.2 ± 2.7	91.0 ± 16.9	SW
2011	16.6 ± 2.7	72.6 ± 21.1	SSW	17.6 ± 2.5	87.6 ± 16.1	S
2012	16.8 ± 3.2	71.0 ± 23.8	SW	18.2 ± 2.4	89.5 ± 17.9	E
2013	16.1 ± 3.1	79.8 ± 19.6	SW	19.3 ± 2.4	88.2 ± 15.6	SW
2014	15.3 ± 2.9	79.9 ± 17.6	SW	16.5 ± 2.3	86.7 ± 14.6	SW
2015	15.8 ± 2.8	70.5 ± 21.0	SW	18.0 ± 2.1	86.1 ± 16.9	SW

* Average and standard deviations are provided.

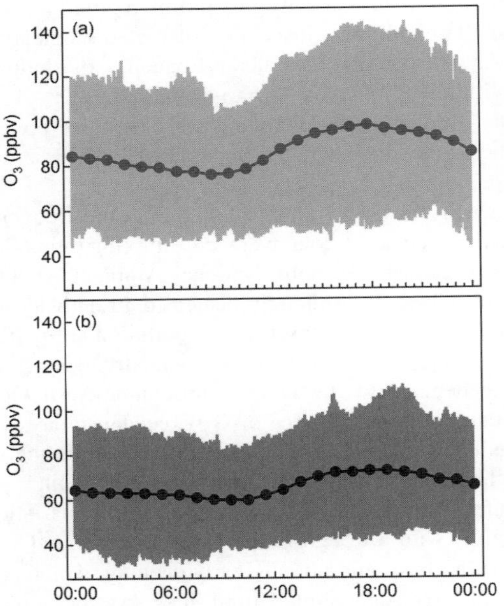

Figure 3. Average diurnal variations of surface O_3 at Mt. Tai in **(a)** June and **(b)** July-August derived from the continuous observations from 2006 to 2009. The shaded area indicates the 5th and 95th percentiles of the data.

chemical ozone production in June. Hence, the more intense photochemistry and higher regional baseline bring about the more serious O_3 pollution in June at Mt. Tai.

Another remarkable feature of surface O_3 at Mt. Tai is the relatively high nighttime levels, with average concentrations of 75–85 ppbv in June and 60–70 ppbv during July–August. This should be the composite result of the residual O_3 produced in the preceding afternoon in the boundary layer, less O_3 loss from NO titration, and long-range transport of processed regional plumes. The transport of regional plumes was evidenced by the coincident evening NO_2^* (including NO_2 and some higher oxidized nitrogen compounds) maximums and relatively low NO levels (indicative of the aged air mass), as shown in Fig. 4. Inspection of the time series day by day also reveals the frequent transport of photochemically aged air masses containing elevated concentrations of O_3 (over 100 ppbv), CO and NO_2^* to the study site during the late evening (figures not shown). Similarly, the MOZAIC aircraft measurements have also found ~60 ppbv on average of O_3 at around 1500 m a.s.l. over Beijing at 05:00–06:00 LT in summer (i.e., May–July; Ding et al., 2008), which is comparable to what we observed at Mt. Tai. These results imply the existence of the O_3-laden air in the nocturnal residual layer over the North China Plain region. Moreover, Ding et al. (2008) also showed in their Fig. 11 that the O_3 enhancement extended from the surface up to about 2 km, further evidence that during the daytime Mt. Tai should be within the boundary layer and is sampling at a vertical level where ozone enhancements are expected.

3.2 Impact of long-range transport

Long-range transport associated with synoptic weather and large-scale circulations is an important factor for the variation of O_3 in rural areas (Wang et al., 2009; Ding et al., 2013; Lin et al., 2014; Zhang et al., 2016). To elucidate the history of air masses sampled at Mt. Tai, we analyzed the summertime climatological air mass transport pattern during 2003–2015 with the aid of cluster analysis of back trajectories. The NCEP reanalysis data and GDAS archive data (http://ready.arl.noaa.gov/archives.php) were used to compute tra-

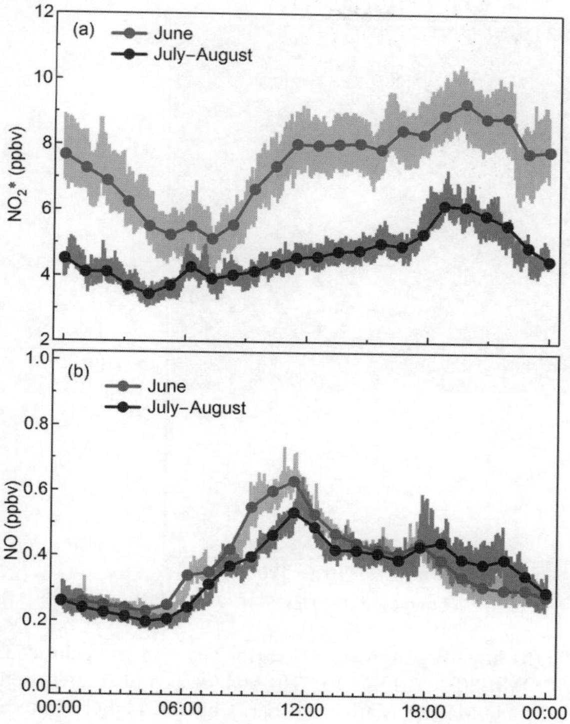

Figure 4. Average diurnal variations of **(a)** NO_2^* and **(b)** NO at Mt. Tai in June and July–August derived from the continuous observations from 2006 to 2009. The shaded area indicates the standard error of the mean.

jectories during 2003–2004 and 2005–2015. The detailed methodology has been documented by Wang et al. (2009) and Xue et al. (2011). Briefly, three-dimensional 72-hour back trajectories were computed four times a day (i.e., 02:00, 08:00, 14:00 and 20:00 LT) for June–August with the Hybrid Single-Particle Lagrangian Integrated Trajectory model (HYSPLIT, v4.9; Draxler et al., 2009), with an endpoint of 300 m above ground level exactly over Mt. Tai. All the trajectories were then categorized into a small number of major groups with the HYSPLIT built-in cluster analysis approach. Total spatial variance (TSV) and the variance between each trajectory component were calculated to determine the optimum number of clusters (Draxler et al., 2009).

A total of five air mass types were extracted for the summer period, with four identified for June and July–August respectively. These air mass types are named according to the regions they traversed, and are described as follows: "Marine and East China" (M&EC) – air masses from the southeast passing over the ocean and polluted central eastern China; "Northeast China" (NEC) – air masses from the north passing over Northeast China; "Central China" (CC) – air masses from the south moving slowly over central China; "Southeast China" (SEC) – air masses from the south moving fast from southeast China; "Mongolia and North China" (M&NC) – air masses from the northwest passing over Mongolia and central northern China.

The above identified major types of air masses are presented in Fig. 5. In June, M&EC was most frequent (57 %), followed by CC (26 %), NEC (9 %) and M&NC (8 %; only identified in June). During the July–August period, the most frequent air mass type was still M&EC (36 %), then CC (29 %) and NEC (29 %), with a minor fraction of SEC (6 %; only identified in July–August). Overall, the transport patterns in June and July–August are quite similar, and it is evident that southerly and easterly air flows (e.g., M&EC and CC) dominated the air mass transport to Mt. Tai in summer. Such patterns are believed to be driven by the summer Asian monsoon (Ding et al., 2008).

The chemical signatures of the different air masses were also inspected and summarized in Table 3. The air masses of M&EC, CC and NEC, which passed over several polluted regions of eastern China, contained higher abundances of O_3 (with averages of 89–94 ppbv in June and 64–77 ppbv in July–August), CO and NO_2^*. In comparison, the more aged air masses of M&NC and SEC showed relatively lower concentrations of O_3 (78 ± 21 ppbv for M&NC in June and 58 ± 17 ppbv for SEC in July–August) and its precursors (except for NO_2^* in the SEC air mass). In view of the higher frequency and higher O_3 levels of the M&EC, CC and NEC air masses, it could be concluded that the regions with the greatest influence on O_3 at Mt. Tai in summer are primarily located in the southern and eastern parts of central eastern China.

3.3 Observed ozone trend

Figure 6 presents the monthly average hourly O_3 and MDA8 O_3 mixing ratios in June and July–August whenever available from 2003 to 2015 at Mt. Tai. The least square linear regression analysis reveals the significant increase of surface O_3 at Mt. Tai since 2003. Monthly mean O_3 values based on hourly data increased at rates of 1.7 ± 1.0 ppbv yr^{-1} (±95 % confidence intervals) in June and 2.1 ± 0.9 ppbv yr^{-1} in July–August, and the increases were statistically significant ($p < 0.01$). For the monthly means based on MDA8 O_3, the fewer available data points (as we only have monthly average data during 2004–2005 from Kanaya et al., 2013) likely reduced the significance of the trend in June, with a positive by statistically insignificant increase (rate $= 1.4 \pm 1.9$ ppbv yr^{-1}; $p = 0.12$). However the site had a significant positive trend in July–August (rate $= 2.2 \pm 1.2$ ppbv yr^{-1}; $p < 0.01$). Therefore we conclude that summertime surface O_3 levels at Mt. Tai have increased over the period 2003–2015.

Given the fact that Mt. Tai is above the PBL at night when there is no photochemistry, the ambient O_3 levels before dawn (e.g., 02:00–05:00 LT) are representative of the regional baseline O_3. The diurnal variation in Fig. 3 shows a slight but steady decrease in O_3 concentrations overnight,

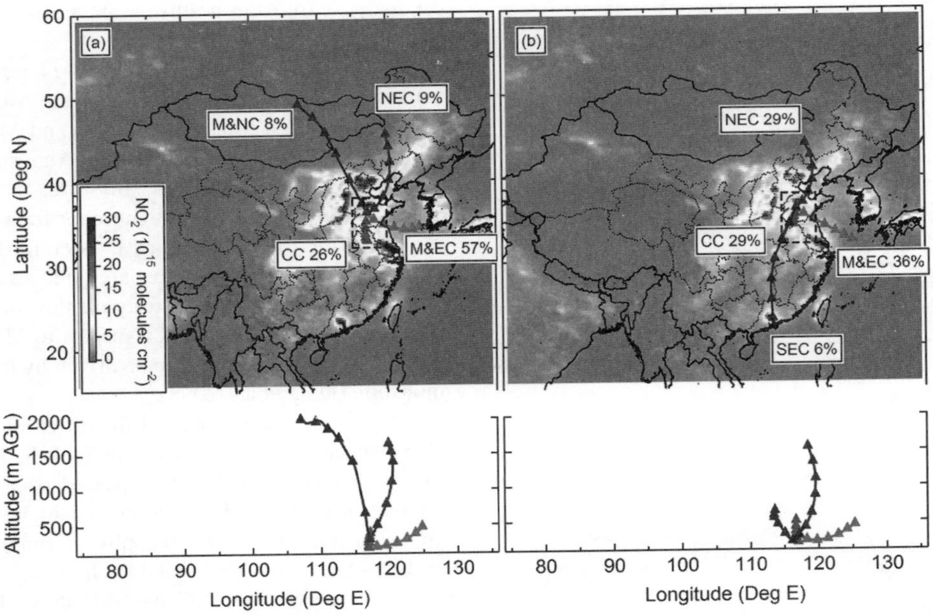

Figure 5. Climatological air mass transport pattern at Mt. Tai in **(a)** June and **(b)** July–August over 2003–2015. The maps are color-coded by the NO_2 column density retrieved from SCIAMACHY (2003–2011) and GOME-2(B) (2013–2015). The box (dashed line) refers to the domain for which the satellite retrievals were averaged. Five major air masses: (1) M&EC: Marine and East China, (2) NEC: Northeast China, (3) CC: Central China, (4) SEC: Southeast China, (5) M&NC: Mongolia and North China.

Table 3. Statistics of O_3, NO_2^* and CO in different air mass categories.[a].

June				July–August			
Air mass[b]	O_3	NO_2^*	CO	Air mass[b]	O_3	NO_2^*	CO
M&EC	92 ± 27	7.2 ± 5.3	500 ± 300	M&EC	64 ± 22	3.6 ± 2.9	370 ± 180
CC	89 ± 24	6.3 ± 5.0	550 ± 300	CC	77 ± 21	3.3 ± 2.5	440 ± 180
NEC	94 ± 25	7.0 ± 4.5	380 ± 180	NEC	73 ± 22	4.5 ± 4.2	380 ± 180
M&NC	78 ± 21	4.0 ± 3.3	280 ± 180	SEC	58 ± 17	4.1 ± 2.4	350 ± 80

[a] The unit is ppbv; average and standard deviations are provided. [b] Refer to Fig. 5 and Sect. 3.2 for the derivation and description of the air mass types.

which should arise from dry deposition. It was assumed that dry deposition was essentially the same every year and did not affect the derived trends. Figure 7 shows the monthly averaged late-night O_3 mixing ratios in June and July–August available from 2003 to 2015 at Mt. Tai. Again, positive trends were found. The rate of increase was quantified at 1.9 ± 1.8 ppbv yr^{-1} ($p = 0.04$, significant) in June and 1.1 ± 1.2 ppbv yr^{-1} ($p = 0.06$, insignificant) during July–August. The increase of regional baseline likely explains the observed O_3 rise at Mt. Tai in June and also accounts for the majority of the increase during July–August. These results indicate the significant increase of surface O_3 in summer on the regional scale across northern China.

Table 4 compares the surface and lower tropospheric ozone trends available in East Asia in recent decades. Two aspects are particularly noteworthy from this comparison. First, most studies have deduced significant positive trends demonstrating the broad increase of tropospheric O_3 over East Asia, especially in China. This pattern is distinct from that found in Europe and the eastern U.S., where O_3 levels have begun to decrease or level off since the 1990s or 2000s (e.g., Cooper et al., 2014; Lefohn et al., 2010; Oltmans et al., 2013; Parrish et al., 2012). The O_3 increase in East Asia is expected due to the rapid economic growth and increasing anthropogenic emissions of O_3 precursors in the past 3 decades (e.g., Ohara et al., 2007). Second, the magnitude of O_3 increase is quite heterogeneous in different regions and the fastest rise was found in the North China Plain. For example, the rates of O_3 increase were in the range of 0.54–0.58 ppbv yr^{-1} in Hong Kong (1994–2007, Wang et al., 2009; 2002–2013, Xue

Table 4. Summary of surface and lower tropospheric ozone trends recorded in East Asia.

Station	Site type	Period	Rate of change (ppbv yr^{-1})	Reference
Mt. Tai	rural	2003–2015 (summer)	1.7 ± 1.0 (June) 2.1 ± 0.9 (July–August)	This study
Beijing	rural (MOZAIC)	1995–2005	~1 (annual average) ~3 (summer afternoon)	Ding et al. (2008)
Hong Kong (Hok Tsui)	rural	1994–2007	0.58	Wang et al. (2009)
Hong Kong	urban & suburban	2002–2013 (autumn)	0.54 ± 0.49	Xue et al. (2014)
Lin'an	rural	1991–2006	2.7 % (summer daily maximum)	Xu et al. (2008)
Waliguan	remote	1994–2013 (summer)	0.15 ± 0.19	Xu et al. (2016)
Taiwan (Yangming)	rural	1994–2007	0.54 ± 0.21	Lin et al. (2010)
Mt. Happo, Japan	rural	1991–2011 (summer)	0.64 ± 0.40	Parrish et al. (2012)
Tokyo, Japan	urban & suburban	1990–2010	0.31 ± 0.02	Akimoto et al. (2015)
Nagoya, Japan	urban & suburban	1990–2010	0.22 ± 0.05	Akimoto et al. (2015)
Osaka/Kyoto, Japan	urban & suburban	1990–2010	0.37 ± 0.03	Akimoto et al. (2015)
Fukuoka, Japan	urban & suburban	1990–2010	0.37 ± 0.04	Akimoto et al. (2015)
South Korea	124 urban sites average	1999–2010	0.26	Seo et al. (2014)
South Korea	56 urban sites average	1990–2010	0.48 ± 0.07 (annual average) 0.55 ± 0.13 (summer)	Lee et al. (2013)

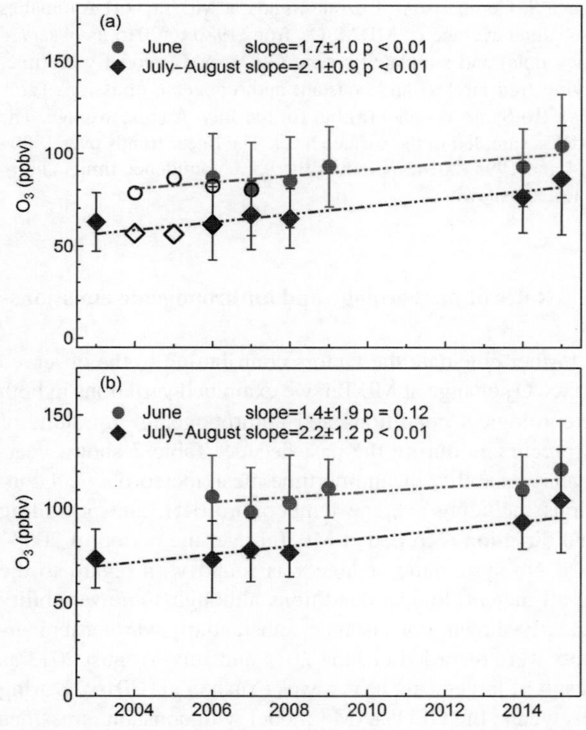

Figure 6. Monthly averaged **(a)** 1-hour and **(b)** MDA8 O_3 mixing ratios at Mt. Tai in June and July–August over 2003–2015. Error bars indicate the standard deviation of the mean. The black open circles and squares represent the data taken from Kanaya et al. (2013). The fitted lines are derived from the least square linear regression analysis with the slopes (±95 % confidence intervals) and p values annotated.

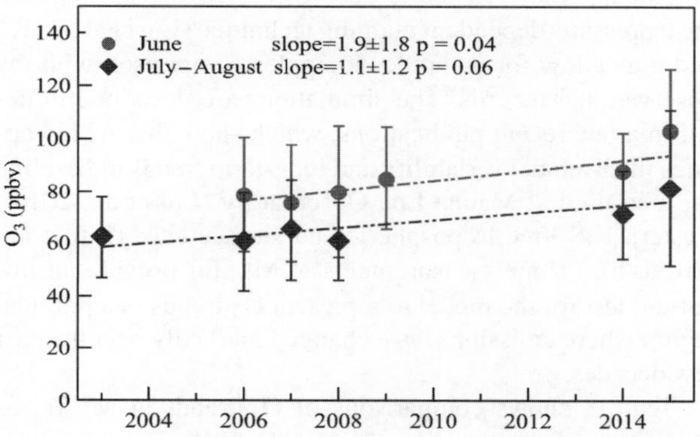

Figure 7. Monthly averaged nighttime O_3 mixing ratios (inferring the regional background O_3) at Mt. Tai in June and July–August over 2003–2015. Error bars indicate the standard deviation of the mean. The fitted lines are derived from the least square linear regression analysis with the slopes (±95 % confidence intervals) and p values annotated.

et al., 2014), 0.54 ppbv yr^{-1} in Taiwan (1994–2007, Lin et al., 2010), 0.26–0.55 ppbv yr^{-1} over South Korea (1990–2010, Lee et al., 2013; 1999–2010, Seo et al., 2014), 0.22–0.37 ppbv yr^{-1} in major Japanese metropolitan areas (1990–2010; Akimoto et al., 2015), 0.64 ppbv yr^{-1} at Mt. Happo, Japan (1991–2011, summer scenario; Parrish et al., 2014), and 0.15 ppbv yr^{-1} at Mt. Waliguan, a GAW station in western China (1994–2013, summer scenario; Xu et al., 2016). According to the MOZAIC aircraft observations, in comparison, Ding et al. (2008) have reported the PBL O_3 increases of ~1 ppbv yr^{-1} for the annual average and ~3 ppbv yr^{-1} for the summer afternoon peaks over the period of 1995–

2005. Zhang et al. (2014) analyzed their field measurements at an urban site in Beijing during 2005–2011 and quantified an increasing rate of 2.6 ppbv yr^{-1} for the daytime average O_3 in summer. Comparable rates of O_3 increase (1.7–2.1 ppbv yr^{-1}) were determined in the present study from the measurements of longer time coverage and at a more regionally representative mountain site, affirming the significant rise of surface O_3 levels over the North China Plain region. Furthermore, the magnitude of O_3 increase in this region is also among the highest records currently reported in the world (Cooper et al., 2012, 2014; Lin et al., 2014; Parrish et al., 2014).

3.4 Comparison with multi-decadal chemistry-climate simulations

We draw on a multi-decadal GFDL-AM3 chemical transport model simulation to provide context for the trends derived from the short observation record. In contrast to the free-running models used in Cooper et al. (2014) that generate their own metrology, the GFDL-AM3 simulations used in the present study are relaxed to the NCEP/NCAR reanalysis using a pressure-dependent nudging technique (Lin et al., 2012) and thus allow for an apples-to-apples comparison with the observational records. The simulations are described in detail in a few recent publications, which show that AM3 captures inter-annual variability and long-term trends of baseline O_3 measured at Mauna Loa Observatory (Lin et al., 2014), western U.S. free tropospheric and surface sites (Lin et al., 2015a, b). Ozone measurements at Mt. Tai provide an important test for the model to represent O_3 trends in a polluted region where emissions have changed markedly over the past few decades.

Figure 8 shows comparisons of O_3 trends at Mt Tai as observed and simulated by the GFDL-AM3 model for June and July-August, respectively. The model has a mean state ozone bias of 10–20 ppbv as in the other global models (Fiore et al., 2009). For illustrative purposes, both observations and model results in Fig. 8 are shown as anomalies. With anthropogenic emissions varying over time (based on Lamarque et al., 2010, with annual interpolation after 2000 to RCP 8.5), AM3 simulates significant MDA8 O_3 increases of 1.04 ± 0.45 ppbv yr^{-1} over 1995–2014 for June and 1.65 ± 0.41 ppbv yr^{-1} for July–August. A greater rate of O_3 increase is simulated for July–August with warmer temperatures compared to June, consistent with the trends derived from the shorter observational records. With constant emissions, AM3 gives no significant long-term ozone trends over the entire 1980–2014 period despite large inter-annual variability (see gray lines in Fig. 8). The model indicates that changes in regional emissions have raised surface ozone over the North China Plain by ~30 ppbv in June and ~45 ppbv in July–August over the past 35 years, with an accelerating trend in the most recent 20 years.

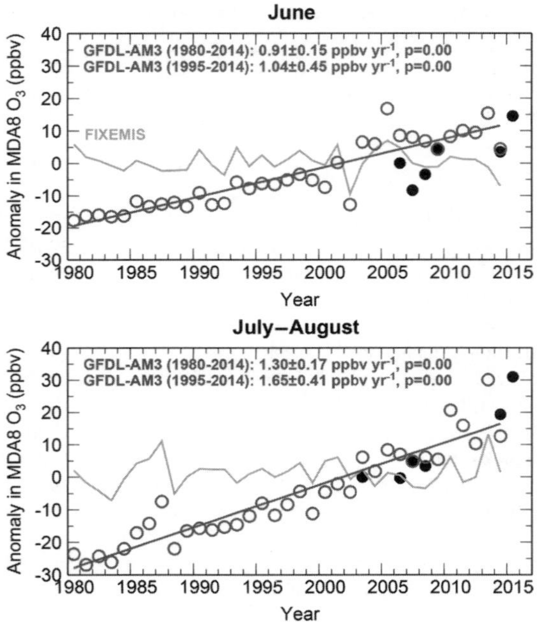

Figure 8. Comparison of ozone trends at Mt. Tai. **(a)** Anomalies in the June average of MDA8 O_3 from 1980 to 2015 as observed (black dots) and simulated by the GFDL-AM3 model with time-varying (red circles) and constant anthropogenic emissions (gray lines). **(b)** Same as panel **(a)** but for the July–August average. The model is sampled in the surface level. The linear trends over 1980–2014 and 1995–2014, including the 95 % confidence limits and p-values, are shown.

3.5 Roles of meteorology and anthropogenic emissions

To further elucidate the factors contributing to the observed surface O_3 change at Mt. Tai, we examined variations in both meteorological conditions and anthropogenic emissions of O_3 precursors during the past decade. Table 2 shows year-to-year variability in summertime mean meteorological conditions including average temperature, RH, and prevailing wind direction recorded at Mt. Tai over the period of 2003–2015. No systematic change was found with regard to the overall meteorological conditions although some variability is clearly shown. For instance, substantially warmer temperatures were recorded in June 2015 and July–August 2013 as a result of large-scale heat waves (Yuan et al., 2016). During these years, the GFDL-AM3 model with constant emissions simulates high-O_3 anomalies (up to ~10 ppbv) relative to the climatological mean (Fig. 8), indicating that heat waves can enhance summertime O_3 pollution in Central Eastern China. Unfortunately, there were no observations available in 2005 and 2013. Observations show greater O_3 levels during July–August of 2014–2015 compared to 2004–2009 (Fig. 8) but no significant change in temperature were found between the two time periods (Table 2), indicating the key role of regional emission changes in contributing to the observed ozone increase at Mt. Tai.

We also explored the air mass transport pattern deduced from cluster analysis of back trajectories (see Sect. 3.2) year by year over 2003–2015 (Table S1). Despite the large year-to-year variability, again, no systematic change in the air mass transport pattern was indicated during the target period. Although the impact of meteorology on tropospheric O_3 is very complex and might not be quantified by such a simple analysis, the significant increase of surface O_3 observed at Mt. Tai should not primarily arise from the change in meteorological conditions.

We analyzed the satellite retrievals of formaldehyde (HCHO) and NO_2 to track the variations in the abundances of O_3 precursors (i.e., NO_x and VOCs) during the study period. Considering that HCHO is a major oxidation product of a variety of VOC species and due to the availability of the satellite-retrieved products, HCHO was selected as an indicator of the VOC abundances. The satellite data were obtained from SCIAMACHY for 2003–2011 and GOME-2(B) from 2013 onwards, with the Level-2 products taken from the TEMIS archive (Tropospheric Emission Monitoring Internet Service; http://www.temis.nl/index.php). Considering the geographical representativeness of Mt. Tai, a larger domain (32–38° N, 115–120° E; see Fig. 5) was selected to process the regional mean satellite data. The monthly averaged HCHO and NO_2 column densities in June and July–August from 2003–2015 are documented in Fig. 9. Significant positive trends are seen for HCHO, with rates of $2.7 \pm 2.2\,\%$ ($p = 0.02$) for June and $2.2 \pm 1.4\,\%$ ($p < 0.01$) for July–August, indicative of the strong increase of VOCs in this region. This result agrees very well with the emission inventory estimates which showed significant increases of anthropogenic VOC emissions in China in the past decades (Bo et al., 2008; Wang et al., 2014), and is consistent with the lack of nationwide VOC controls. All of these results evidence the increase of atmospheric VOC abundances over the North China Plain.

What is more interesting is the two-phase variation of the NO_2 column, showing a significant increase first from 2003 to 2011 (June: $4.8 \pm 3.4\,\%$, $p = 0.01$; July–August: $7.7 \pm 3.6\,\%$, $p < 0.01$) and a decrease afterwards (see Fig. 9b). These satellite observations agree very well with the bottom-up emission inventory estimates, which clearly showed a break point occurring in 2011 in the anthropogenic NO_x emissions of China (see Fig. S2). China has just launched a national NO_x control programme during its "Twelfth Five-Year Plan" (i.e., 2011–2015) (China State Council, 2011). The strict control measures are very efficient and have resulted in an immediate reduction of NO_x emissions, as affirmed by both emission inventories and satellite retrievals. Furthermore, the reduced levels of NO_x in the most recent 5 years were also evidenced by our limited in situ NO_x^* measurements at Mt. Tai. As shown in Fig. 10, the ambient NO_2^* levels in 2014 and 2015 were indeed substantially lower than those measured in the previous years before 2010.

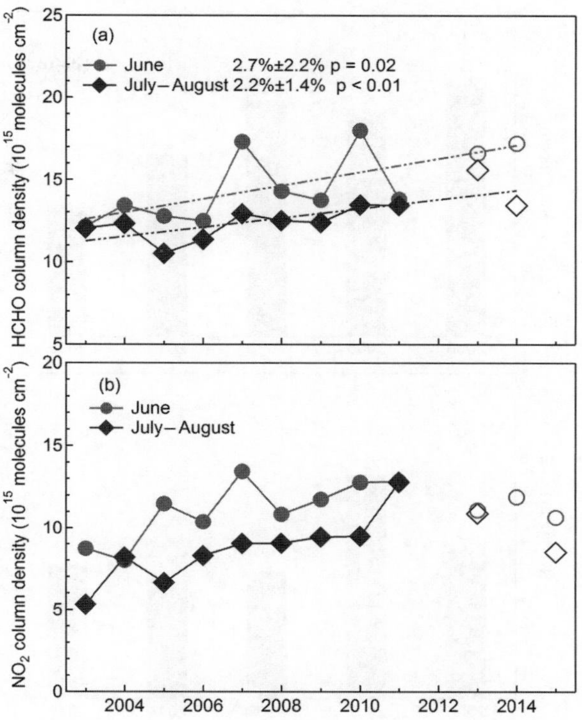

Figure 9. Monthly average column density of **(a)** formaldehyde and **(b)** NO_2 retrieved from SCIAMACHY (2003–2011; solid markers) and GOME-2(B) (2013–2015; open markers) for the target domain (32–38° N, 115–120° E). For formaldehyde, the fitted lines are derived from the least square linear regression analysis with the slopes (±95 % confidence intervals) and p values also shown.

From the above analyses, the O_3 increase between 2003 and 2011 is easy to understand in light of the consistent increase of both NO_x and VOCs. For the later period, i.e. after 2011, in comparison, opposite trends have taken place with NO_x decreasing but VOCs still increasing. The observed continuing O_3 rise suggests that the reduction of NO_x is not adequate to reduce the ambient O_3 levels, with a background of increasing VOCs. We then evaluated the ozone production efficiency (OPE) for the air masses sampled at Mt. Tai. OPE is usually derived from the regression slope of the scatter plots of O_3 versus NO_z (Trainer et al., 1993), and is a useful metric to infer how efficient O_3 is produced per oxidation of unit of NO_x (e.g., Wang et al., 2010; Xue et al., 2011). As NO_y (and thus NO_z) is not routinely measured in the present study, NO_2^* is used instead of NO_z to infer the OPE values. It should be reasonable considering that our measured NO_2^* significantly overestimated true NO_2 and actually contained a large fraction of NO_z, especially in the afternoon period when NO_z was at its maximum with NO_2 at the minimum (Xu et al., 2013). The scatter plots of O_3 ver-

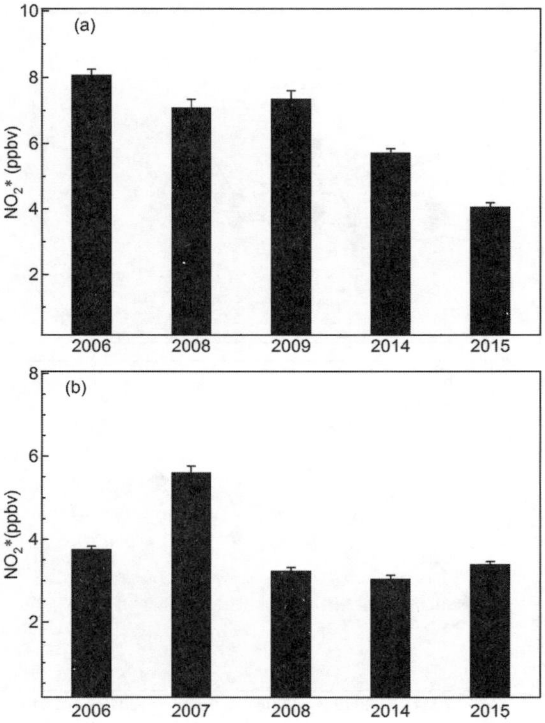

Figure 10. Monthly averaged NO_2^* concentrations measured at Mt. Tai in **(a)** June and **(b)** July–August during 2006–2015. Error bars indicate the standard error of the mean. Note that the data point in June 2014 is for NO_x^* instead of NO_2^*.

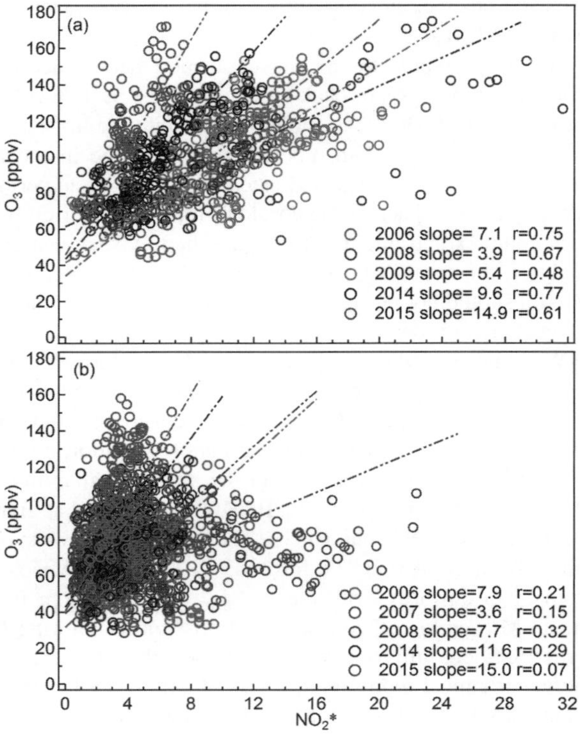

Figure 11. Scatter plots of O_3 versus NO_2^* at Mt. Tai in **(a)** June and **(b)** July–August during 2006–2015. Only the afternoon data (i.e., 12:00–18:00 local time) were used for this analysis. Note that the NO_2^* data in June 2014 stand for NO_x^* ($NO_x^* = NO + NO_2^*$), of which NO usually presents a minor fraction. The slopes are determined by the reduced major axis (RMA) method.

sus NO_2^* during the afternoon hours (i.e., 12:00–18:00 LT) available from 2006–2015 at Mt. Tai are presented in Fig. 11. The OPE values in 2014–2015 (i.e., 9.6–15.0) were significantly higher than those determined during 2006–2009 (i.e., 3.6–7.9). This demonstrates the greater ozone production efficiency in recent years with increasing VOCs. These results indicate that although NO_x in China has been reduced since 2011, little action on VOC control has led to increased emissions of VOCs, which could make O_3 formation more efficient per unit of NO_x. As a consequence, ambient O_3 levels have been rising in northern China. We conclude that control of VOCs is urgently needed, in addition to the ongoing strict NO_x control, to mitigate regional O_3 pollution in China.

4 Conclusions

We have compiled all field observations of O_3 and its precursors ever collected at Mt. Tai in the North China Plain and found a significant increase of summertime surface O_3 over 2003–2015 at this regionally representative site. The ozone increase derived from these valuable but sparse observations is supported by a chemistry-climate model simulation with emissions varying from 1980 to 2014. The marked increase of surface O_3 levels over central Eastern China is primarily attributable to rising anthropogenic emissions of O_3 precursors. We provide evidence that the changes in VOCs emissions contribute to the observed ozone increase since 2011 when NO_x has reduced efficiently as a result of a strict national emission control programme. This study provides some evidence that the current Chinese control programme, focusing on NO_x alone with little action for VOCs, is not sufficient for mitigating regional O_3 pollution, and calls on the implementation of VOC controls as soon as possible. Similar to the U.S., China phased in its air pollution control measures resulting in decreasing SO_2 emission after 2006 and NO_x emission after 2011. It is foreseen that the national VOC control will be launched very soon. Thus, follow-up long-term measurements are required to evaluate the response of ambient O_3 to the upcoming VOC control and to provide observational constraints to evaluate global and regional chemical transport models.

5 Data availability

The underlying research data can be accessed upon contact with the corresponding author (Likun Xue, xuelikun@sdu.edu.cn).

The Supplement related to this article is available online at doi:10.5194/acp-16-10637-2016-supplement.

Acknowledgements. The authors thank Steven Poon and Wei Nie for their contributions to the field study, and the staff of the Mt. Tai Meteorological Observatory for the logistics and help during the field measurements. We are also grateful to the NOAA Air Resources Laboratory for providing the HYSPLIT model and meteorological data, and the European Space Agency for the free distribution of SCIAMACHY and GOME-2(B) satellite data through the TEMIS website (http://www.temis.nl/index.php). This work was supported by the National Natural Science Foundation of China (project no.: 41275123) and the Qilu Youth Talent Programme of Shandong University.

Edited by: S. Brown
Reviewed by: two anonymous referees

References

Akimoto, H., Mori, Y., Sasaki, K., Nakanishi, H., Ohizumi, T., and Itano, Y.: Analysis of monitoring data of ground-level ozone in Japan for long-term trend during 1990–2010: Causes of temporal and spatial variation, Atmos. Environ., 102, 302–310, 2015.

Bo, Y., Cai, H., and Xie, S. D.: Spatial and temporal variation of historical anthropogenic NMVOCs emission inventories in China, Atmos. Chem. Phys., 8, 7297–7316, doi:10.5194/acp-8-7297-2008, 2008.

China State Council: Twelfth Five-Year Plan on National Economy and Social Development of the People's Republic of China, available at: http://www.gov.cn/2011lh/content_1825838.htm (last access: 20 December, 2015), 2011 (in Chinese).

Cooper, O. R., Gao, R.-S., Tarasick, D., Leblanc, T., and Sweeney, C.: Long-term ozone trends at rural ozone monitoring sites across the United States, 1990–2010, J. Geophys. Res.-Atmos., 117, D22307, doi:10.1029/2012jd018261, 2012.

Cooper, O. R., Parrish, D. D., Ziemke, J., Balashov, N. V., Cupeiro, M., Galbally, I. E., Gilge, S., Horowitz, L., Jensen, N. R., Naik, V., Oltmans, S. J., Schwab, J., Shindell, D. T., Thompson, A. M., Thouret, V., Wang, Y., and Zbinden, R. M.: Global distribution and trends of tropospheric ozone: An observation-based review, Elementa Sci. Anth., 2, 000029, doi:10.12952/journal.elementa.000029, 2014.

Crutzen, P.: A discussion ofthe chemistry of some minor constituents in the stratosphere and troposphere, Pure Appl. Geophys., 106–108, 1385–1399, 1973.

Ding, A. J., Wang, T., Thouret, V., Cammas, J.-P., and Nédélec, P.: Tropospheric ozone climatology over Beijing: analysis of aircraft data from the MOZAIC program, Atmos. Chem. Phys., 8, 1–13, doi:10.5194/acp-8-1-2008, 2008.

Ding, A. J., Fu, C. B., Yang, X. Q., Sun, J. N., Zheng, L. F., Xie, Y. N., Herrmann, E., Nie, W., Petäjä, T., Kerminen, V.-M., and Kulmala, M.: Ozone and fine particle in the western Yangtze River Delta: an overview of 1 yr data at the SORPES station, Atmos. Chem. Phys., 13, 5813–5830, doi:10.5194/acp-13-5813-2013, 2013.

Draxler, R. R., Stunder, B., Rolph, G., and Taylor, A.: HYSPLIT4 user's guide, available at: http://www.arl.noaa.gov/documents/reports/hysplit_user_guide.pdf (last access: 27 December 2015), 2009.

Fine, R., Miller, M. B., Burley, J., Jaffe, D. A., Pierce, R. B., Lin, M., and Gustin, M. S.: Variability and sources of surface ozone at rural sites in Nevada, USA: Results from two years of the Nevada Rural Ozone Initiative, Sci. Total Environ., 530–531, 471–482, 2015.

Fiore, A. M., Dentener, F. J., Wild, O., Cuvelier, C., Schultz, M. G., Hess, P., Textor, C., Schulz, M., Doherty, R. M., Horowitz, L. W., MacKenzie, I. A., Sanderson, M. G., Shindell, D. T., Stevenson, D. S., Szopa, S., van Dingenen, R., Zeng, G., Atherton, C. S., Bergmann, D. J., Bey, I., Carmichael, G. R., Collins, W. J., Duncan, B. N., Faluvegi, G., Folberth, G. A., Gauss, M., Gong, S., Hauglustaine, D., Holloway, T., Isaksen, I. S. A., Jacob, D. J., Jonson, J. E., Kaminski, J. W., Keating, T. J., Lupu, A., Marmer, E., Montanaro, V., Park, R. J., Pitari, G., Pringle, K. J., Pyle, J. A., Schroeder, S., Vivanco, M. G., Wind, P., Wojcik, G., Wu, S., and Zuber, A.: Multimodel estimates of intercontinental source receptor relationships for ozone pollution, J. Geophys. Res.-Atmos., 114, D04301, doi:10.1029/2008JD010816, 2009.

Gao, J., Wang, T., Ding, A. J., and Liu, C. B.: Observational study of ozone and carbon monoxide at the summit of mount Tai (1534 m a.s.l.) in central-eastern China, Atmos. Environ., 39, 4779–4791, 2005.

Guo, J., Wang, Y., Shen, X., Wang, Z., Lee, T., Wang, X., Li, P., Sun, M., Collett Jr., J. L., Wang, W., and Wang, T.: Characterization of cloud water chemistry at Mount Tai, China: Seasonal variation, anthropogenic impact, and cloud processing, Atmos. Environ., 60, 467–476, 2012.

IPCC: Climate Change 2013: The Assessment Reports of the Intergovernmental Panel on Climate Change, edited by: Stocker, T. F., Qin, D., Plattner, G.-K., Tignor, M. M. B., Allen, S. K., Boschung, J., Nauels, A., Xia, Y., Bex, V., and Midgley, P. M., Cambridge University Press, Cambridge, UK, 1552 pp., 2013.

Kanaya, Y., Akimoto, H., Wang, Z.-F., Pochanart, P., Kawamura, K., Liu, Y., Li, J., Komazaki, Y., Irie, H., Pan, X.-L., Taketani, F., Yamaji, K., Tanimoto, H., Inomata, S., Kato, S., Suthawaree, J., Okuzawa, K., Wang, G., Aggarwal, S. G., Fu, P. Q., Wang, T., Gao, J., Wang, Y., and Zhuang, G.: Overview of the Mount Tai Experiment (MTX2006) in central East China in June 2006: studies of significant regional air pollution, Atmos. Chem. Phys., 13, 8265–8283, doi:10.5194/acp-13-8265-2013, 2013.

Lamarque, J.-F., Bond, T. C., Eyring, V., Granier, C., Heil, A., Klimont, Z., Lee, D., Liousse, C., Mieville, A., Owen, B., Schultz, M. G., Shindell, D., Smith, S. J., Stehfest, E., Van Aardenne, J., Cooper, O. R., Kainuma, M., Mahowald, N., McConnell, J. R., Naik, V., Riahi, K., and van Vuuren, D. P.: Historical (1850–2000) gridded anthropogenic and biomass burning emissions of reactive gases and aerosols: methodology and application, Atmos. Chem. Phys., 10, 7017–7039, doi:10.5194/acp-10-7017-2010, 2010.

Lee, H.-J., Kim, S.-W., Brioude, J., Cooper, O. R., Frost, G. J., Kim, C.-H., Park, R. J., Trainer, M., and Woo, J.-H.: Transport of NO_x in East Asia identified by satellite and in situ measurements and Lagrangian particle dispersion model simulations, J. Geophys. Res.-Atmos., 119, 2574–2596, 2013.

Lefohn, A. S., Shadwick, D., and Oltmans, S. J.: Characterizing changes in surface ozone levels in metropolitan and rural areas in the United States for 1980–2008 and 1994–2008, Atmos. Environ., 44, 5199–5210, 2010.

Lin, M., Holloway, T., Oki, T., Streets, D. G., and Richter, A.: Multi-scale model analysis of boundary layer ozone over East Asia, Atmos. Chem. Phys., 9, 3277–3301, doi:10.5194/acp-9-3277-2009, 2009.

Lin, M., Fiore, A. M., Horowitz, L. W., Cooper, O. R., Naik, V., Holloway, J., Johnson, B. J., Middlebrook, A. M., Oltmans, S. J., Pollack, I. B., Ryerson, T. B., Warner, J. X., Wiedeninmyer, C., Wilson, J., and Wyman, B.: Transport of Asian ozone pollution into surface air over the western United States in spring, J. Geophys. Res.-Atmos., 117, D00V07, doi:10.1029/2011JD016961, 2012.

Lin, M., Horowitz, L. W., Oltmans, S. J., Fiore, A. M., and Fan, S.: Tropospheric ozone trends at Mauna Loa Observatory tied to decadal climate variability, Nat. Geosci., 7, 136–143, 2014.

Lin, M., Fiore, A. M., Horowitz, L. W., Langford, A. O., Oltmans, S. J., Tarasick, D., and Rieder, H. E.: Climate variability modulates western US ozone air quality in spring via deep stratospheric intrusions, Nat. Commun., 6, 7105, doi:10.1038/ncomms8105, 2015a.

Lin, M., Horowitz, L. W., Cooper, O. R., Tarasick, D., Conley, S., Iraci, L. T., Johnson, B., Leblanc, T., Petropavlovskikh, I., and Yates, E. L.: Revisiting the evidence of increasing springtime ozone mixing ratios in the free troposphere over western North America, Geophys. Res. Lett., 42, 8719–8728, 2015b.

Lin, Y.-K., Lin, T.-H., and Chang, S.-C.: The changes in different ozone metrics and their implications following precursor reductions over northern Taiwan from 1994 to 2007, Environ. Monit. Assess., 169, 143–157, 2010.

Logan, J. A., Staehelin, J., Megretskaia, I. A., Cammas, J.-P., Thouret, V., Claude, H., De Backer, H., Steinbacher, M., Scheel, H.-E., Stübi, R., Fröhlich, M., and Derwent, R.: Changes in ozone over Europe: Analysis of ozone measurements from sondes, regular aircraft (MOZAIC) and alpine surface sites, J. Geophys. Res.-Atmos., 117, D09301, doi:10.1029/2011jd016952, 2012.

Lu, Z., Streets, D. G., Zhang, Q., Wang, S., Carmichael, G. R., Cheng, Y. F., Wei, C., Chin, M., Diehl, T., and Tan, Q.: Sulfur dioxide emissions in China and sulfur trends in East Asia since 2000, Atmos. Chem. Phys., 10, 6311–6331, doi:10.5194/acp-10-6311-2010, 2010.

Ma, J., Zhou, X., and Hauglustaine, D.: Summertime tropospheric ozone over China simulated with a regional chemical transport model 2. Source contributions and budget, J. Geophys. Res.-Atmos., 107, 4612, doi:10.1029/2001jd001355, 2002.

Monks, P. S., Archibald, A. T., Colette, A., Cooper, O., Coyle, M., Derwent, R., Fowler, D., Granier, C., Law, K. S., Mills, G. E., Stevenson, D. S., Tarasova, O., Thouret, V., von Schneidemesser, E., Sommariva, R., Wild, O., and Williams, M. L.: Tropospheric ozone and its precursors from the urban to the global scale from air quality to short-lived climate forcer, Atmos. Chem. Phys., 15, 8889–8973, doi:10.5194/acp-15-8889-2015, 2015.

National Research Council: Rethinking the Ozone Problem in Urban and Regional Air Pollution, Natl. Acad. Press, Washington, D. C., USA, 1991.

Ohara, T., Akimoto, H., Kurokawa, J., Horii, N., Yamaji, K., Yan, X., and Hayasaka, T.: An Asian emission inventory of anthropogenic emission sources for the period 1980–2020, Atmos. Chem. Phys., 7, 4419–4444, doi:10.5194/acp-7-4419-2007, 2007.

Oltmans, S. J., Lefohn, A. S., Shadwick, D., Harris, J. M., Scheel, H. E., Galbally, I., Tarasick, D. W., Johnson, B. J., Brunke, E.-G., Claude, H., Zeng, G., Nichol, S., Schmidlin, F., Davies, J., Cuevas, E., Redondas, A., Naoe, H., Nakano, T., and Kawasato, T.: Recent tropospheric ozone changes – A pattern dominated by slow or no growth, Atmos. Environ., 67, 331–351, 2013.

Parrish, D. D., Law, K. S., Staehelin, J., Derwent, R., Cooper, O. R., Tanimoto, H., Volz-Thomas, A., Gilge, S., Scheel, H.-E., Steinbacher, M., and Chan, E.: Long-term changes in lower tropospheric baseline ozone concentrations at northern mid-latitudes, Atmos. Chem. Phys., 12, 11485–11504, doi:10.5194/acp-12-11485-2012, 2012.

Parrish, D. D., Lamarque, J. F., Naik, V., Horowitz, L., Shindell, D. T., Staehelin, J., Derwent, R., Cooper, O., Tanimoto, H., Volz-Thomas, A., Gilge, S., Scheel, H.-E., Steinbacher, M., and Chan, E.: Long-term changes in lower tropospheric baseline ozone concentrations: Comparing chemistry-climate models and observations at northern midlatitudes, J. Geophys. Res.-Atmos., 119, 5719–5736, doi:10.1002/2013jd021435, 2014.

The Royal Society: Ground-level Ozone in the 21st century: Future Trends, Impacts and Policy Implications, Royal Society policy document 15/08, RS1276, available at: https://royalsociety.org/~/media/Royal_Society_Content/policy/publications/2008/7925.pdf (last access: 21 December 2015), 2008.

Seinfeld, J. H. and Pandis, S. N.: Atmospheric Chemistry and Physics: From Air Pollution to Climate Change, Wiley, New York, USA, 2006.

Seo, J., Youn, D., Kim, J. Y., and Lee, H.: Extensive spatiotemporal analyses of surface ozone and related meteorological variables in South Korea for the period 1999–2010, Atmos. Chem. Phys., 14, 6395–6415, doi:10.5194/acp-14-6395-2014, 2014.

Shen, X., Lee, T., Guo, J., Wang, X., Li, P., Xu, P., Wang, Y., Ren, Y., Wang, W., Wang, T., Li, Y., Carn, S. A., and Collett Jr., J. L.: Aqueous phase sulfate production in clouds in eastern China, Atmos. Environ., 62, 502–511, 2012.

Shindell, D., Kuylenstierna, J. C., Vignati, E., van Dingenen, R., Amann, M., Klimont, Z., Anenberg, S. C., Muller, N., Janssens-Maenhout, G., Raes, F., Schwartz, J., Faluvegi, G., Pozzoli, L., Kupianinen, K., H-Isaksson, L., Emberson, L., Streets, D., Ramanathan, V., Hicks, K., Oanh, N. T., Milly, G., Williams, M., Demkine, V., and Fowler, D.: Simultaneously mitigating near-term climate change and improving human health and food security, Science, 335, 183–189, 2012.

Stevenson, D. S., Young, P. J., Naik, V., Lamarque, J.-F., Shindell, D. T., Voulgarakis, A., Skeie, R. B., Dalsoren, S. B., Myhre, G., Berntsen, T. K., Folberth, G. A., Rumbold, S. T., Collins, W. J., MacKenzie, I. A., Doherty, R. M., Zeng, G., van Noije, T. P. C.,

Strunk, A., Bergmann, D., Cameron-Smith, P., Plummer, D. A., Strode, S. A., Horowitz, L., Lee, Y. H., Szopa, S., Sudo, K., Nagashima, T., Josse, B., Cionni, I., Righi, M., Eyring, V., Conley, A., Bowman, K. W., Wild, O., and Archibald, A.: Tropospheric ozone changes, radiative forcing and attribution to emissions in the Atmospheric Chemistry and Climate Model Intercomparison Project (ACCMIP), Atmos. Chem. Phys., 13, 3063–3085, doi:10.5194/acp-13-3063-2013, 2013.

Stohl, A., Bonasoni, P., Cristofanelli, P., Collins, W. J., Feichter, J., Frank, A., Forster, C., Gerasopoulos, E., Gäggeler, H., James, P., Kentarchos, T., Kromp-Kalb, H., Kruger, B., Land, C., Meloen, J., Papayannis, A., Priller, A., Seibert, P., Sprenger, M., Roelofs, G. J., Scheel, H. E., Schnabel, C., Siegmund, P., Tobler, L., Trickl, T., Wernli, H., Wirth, V., Zanis, P., and Zerefos, C.: Stratosphere-troposphere exchange: A review, and what we have learned from STACCATO, J. Geophys. Res.-Atmos., 108, 8516, doi:10.1029/2002jd002490, 2003.

Suthawaree, J., Kato, S., Okuzawa, K., Kanaya, Y., Pochanart, P., Akimoto, H., Wang, Z., and Kajii, Y.: Measurements of volatile organic compounds in the middle of Central East China during Mount Tai Experiment 2006 (MTX2006): observation of regional background and impact of biomass burning, Atmos. Chem. Phys., 10, 1269–1285, doi:10.5194/acp-10-1269-2010, 2010.

Tang, G., Li, X., Wang, Y., Xin, J., and Ren, X.: Surface ozone trend details and interpretations in Beijing, 2001–2006, Atmos. Chem. Phys., 9, 8813–8823, doi:10.5194/acp-9-8813-2009, 2009.

Trainer, M., Parrish, D. D., Buhr, M. P., Norton, R. B., Fehsenfeld, F. C., Anlauf, K. G., Bottenheim, J. W., Tang, Y. Z., Wiebe, H. A., Roberts, J. M., Tanner, R. L., Newman, L., Bowersox, V. C., Meagher, J. F., Olszyna, K. J., Rodgers, M. O., Wang, T., Berresheim, H., Demerjian, K. L., and Roychowdhury, U. K.: Correlation of ozone with NO_y in photochemically aged air, J. Geophys. Res.-Atmos., 98, 2917–2925, doi:10.1029/92jd01910, 1993.

UNEP and WMO: Integrated Assessment of black carbon and tropospheric ozone: Summary for decision makers, United Nations Environment Programme, Nairobi, Kenya, 2011.

Wang, S. X., Zhao, B., Cai, S. Y., Klimont, Z., Nielsen, C. P., Morikawa, T., Woo, J. H., Kim, Y., Fu, X., Xu, J. Y., Hao, J. M., and He, K. B.: Emission trends and mitigation options for air pollutants in East Asia, Atmos. Chem. Phys., 14, 6571–6603, doi:10.5194/acp-14-6571-2014, 2014.

Wang, T., Wei, X. L., Ding, A. J., Poon, C. N., Lam, K. S., Li, Y. S., Chan, L. Y., and Anson, M.: Increasing surface ozone concentrations in the background atmosphere of Southern China, 1994–2007, Atmos. Chem. Phys., 9, 6217–6227, doi:10.5194/acp-9-6217-2009, 2009.

Wang, T., Nie, W., Gao, J., Xue, L. K., Gao, X. M., Wang, X. F., Qiu, J., Poon, C. N., Meinardi, S., Blake, D., Wang, S. L., Ding, A. J., Chai, F. H., Zhang, Q. Z., and Wang, W. X.: Air quality during the 2008 Beijing Olympics: secondary pollutants and regional impact, Atmos. Chem. Phys., 10, 7603–7615, doi:10.5194/acp-10-7603-2010, 2010.

Warneke, C., de Gouw, J., Holloway, J., Peischl, J., Ryerson, T., Atlas, E., Blake, D., Trainer, M., and Parrish, D.: Multiyear trends in volatile organic compounds in Los Angeles, California: five decades of decreasing emissions, J. Geophys. Res.-Atmos., 117, D00V17, doi:10.1029/2012JD017899, 2012.

Xu, W., Lin, W., Xu, X., Tang, J., Huang, J., Wu, H., and Zhang, X.: Long-term trends of surface ozone and its influencing factors at the Mt Waliguan GAW station, China – Part 1: Overall trends and characteristics, Atmos. Chem. Phys., 16, 6191–6205, doi:10.5194/acp-16-6191-2016, 2016.

Xu, X., Lin, W., Wang, T., Yan, P., Tang, J., Meng, Z., and Wang, Y.: Long-term trend of surface ozone at a regional background station in eastern China 1991–2006: enhanced variability, Atmos. Chem. Phys., 8, 2595–2607, doi:10.5194/acp-8-2595-2008, 2008.

Xu, Z., Wang, T., Xue, L. K., Louie, P. K. K., Luk, C. W. Y., Gao, J., Wang, S. L., Chai, F. H., and Wang, W. X.: Evaluating the uncertainties of thermal catalytic conversion in measuring atmospheric nitrogen dioxide at four differently polluted sites in China, Atmos. Environ., 76, 221–226, 2013.

Xue, L. K., Wang, T., Zhang, J. M., Zhang, X. C., Deligeer, Poon, C. N., Ding, A. J., Zhou, X. H., Wu, W. S., Tang, J., Zhang, Q. Z., and Wang, W. X.: Source of surface ozone and reactive nitrogen speciation at Mount Waliguan in western China: New insights from the 2006 summer study, J. Geophys. Res.-Atmos., 116 , D07306, doi:10.1029/2010jd014735, 2011.

Xue, L. K., Wang, T., Louie, P. K. K., Luk, C. W. Y., Blake, D. R., and Xu, Z.: Increasing external effects negate local efforts to control ozone air pollution: a case study of Hong Kong and implications for other Chinese cities, Environ. Sci. Tech., 48, 10769–10775, 2014.

Yamaji, K., Ohara, T., Uno, I., Tanimoto, H., Kurokawa, J., and Akimoto, H.: Analysis of the seasonal variation of ozone in the boundary layer in East Asia using the Community Multiscale Air Quality model: What controls surface ozone levels over Japan?, Atmos. Environ., 40, 1856–1868, 2006.

Yamaji, K., Li, J., Uno, I., Kanaya, Y., Irie, H., Takigawa, M., Komazaki, Y., Pochanart, P., Liu, Y., Tanimoto, H., Ohara, T., Yan, X., Wang, Z., and Akimoto, H.: Impact of open crop residual burning on air quality over Central Eastern China during the Mount Tai Experiment 2006 (MTX2006), Atmos. Chem. Phys., 10, 7353–7368, doi:10.5194/acp-10-7353-2010, 2010.

Yuan, W., Cai, W., Chen, Y., Liu, S., Dong, W., Zhang, H., Yu, G., Chen, Z., He, H., Guo, W., Liu, D., Liu S., Xiang, W., Xie, Z., Zhao, Z., and Zhou, G.: Severe summer heatwave and drought strongly reduced carbon uptake in Southern China, Sci. Rep., 6, 18813, doi:10.1038/srep18813, 2016.

Zhang, Q., Yuan, B., Shao, M., Wang, X., Lu, S., Lu, K., Wang, M., Chen, L., Chang, C.-C., and Liu, S. C.: Variations of ground-level O_3 and its precursors in Beijing in summertime between 2005 and 2011, Atmos. Chem. Phys., 14, 6089–6101, doi:10.5194/acp-14-6089-2014, 2014.

Zhang, Y., Ding, A., Mao, H., Nie, W., Zhou, D., Liu, L., Huang, X., and Fu, C.: Impact of synoptic weather patterns and inter-decadal climate variability on air quality in the North China Plain during 1980–2013, Atmos. Envion., 124, 119–128, 2016.

JOURNAL OF GEOPHYSICAL RESEARCH, VOL. 116, D07306, doi:10.1029/2010JD014735, 2011

Source of surface ozone and reactive nitrogen speciation at Mount Waliguan in western China: New insights from the 2006 summer study

L. K. Xue,[1,2] T. Wang,[1,2] J. M. Zhang,[2] X. C. Zhang,[3] Deliger,[4] C. N. Poon,[2] A. J. Ding,[5] X. H. Zhou,[1] W. S. Wu,[2] J. Tang,[3] Q. Z. Zhang,[1] and W. X. Wang[1]

Received 9 July 2010; revised 6 December 2010; accepted 20 January 2011; published 14 April 2011.

[1] Surface ozone (O_3), carbon monoxide (CO), and total and speciated reactive nitrogen compounds (NO_y, NO, NO_2, PAN, HNO_3, and particulate NO_3^-) were measured at Mount Waliguan (WLG; 36.28°N, 100.90°E, 3816 m above sea level (asl)) in the summer of 2006 to further understand the sources of ozone and reactive nitrogen and to investigate the partitioning of reactive nitrogen over the remote Qinghai-Tibetan Plateau. The mean mixing ratios of O_3, CO, NO_y, and daytime NO were 59 ppbv, 149 ppbv, 1.44 ppbv, and 71 pptv, respectively, which (except for NO_y) were higher than those measured from a previous campaign in summer 2003, which is consistent with more frequent transport of anthropogenic pollution from central and eastern China in the measurement period of 2006 (55%) than that of 2003 (25%). The abnormally high values of NO_y observed in 2003 were suspected to be due to the positive interference from ammonia (NH_3) to the particular catalytic converter used in that study. Varied diurnal patterns were observed for the various NO_y components. The ozone production efficiencies ($\Delta O_3/\Delta NO_z$), which were estimated from the slope of the O_3-NO_z scatterplot, were 7.7–11.3 for the polluted plumes from central and eastern China. The speciation of reactive nitrogen was investigated for the first time in the remote free troposphere in western China. PAN and particulate NO_3^- were the most abundant reactive nitrogen species at WLG, with average proportions of 32% and 31%, followed by NO_x (24%) and HNO_3 (20%). The relatively large contribution of particulate NO_3^- to NO_y was due to the presence of high concentrations of NH_3 and crustal particles, which favor the formation of particulate nitrate. An analysis of backward trajectories for the recent 10 years revealed that air masses from central and eastern China dominated the airflow at WLG in summer, suggesting strong impact of anthropogenic forcing on the surface ozone and other trace constituents on the Plateau.

Citation: Xue, L. K., et al. (2011), Source of surface ozone and reactive nitrogen speciation at Mount Waliguan in western China: New insights from the 2006 summer study, *J. Geophys. Res.*, *116*, D07306, doi:10.1029/2010JD014735.

1. Introduction

[2] Ozone is an important trace gas in the troposphere for its roles in determining the oxidative capacity of the atmosphere, affecting human and vegetation health, and influencing the radiation budget of the atmosphere [e.g., *National Research Council*, 1991; *Intergovernmental Panel on Climate Change*, 2007]. The budget of ozone in the troposphere is determined by the downward transport of stratospheric air [*Stohl et al.*, 2003], dry deposition on the earth's surface, and photochemistry in the troposphere involving volatile organic compounds (VOCs) and nitrogen oxides (NO_x) [*Crutzen*, 1973].

[3] Reactive oxidized nitrogen compounds play a central role in the chemistry of the troposphere. They are emitted primarily as NO, followed by oxidation to NO_2 and other forms of oxidized nitrogen. Total reactive nitrogen (NO_y) is defined as the sum of NO_x (NO_x = NO + NO_2) and its atmospheric oxidation products and reactive intermediates (collectively abbreviated NO_z). NO_x, peroxyacetyl nitrate (PAN), nitric acid (HNO_3), and aerosol nitrate (NO_3^-) are the most abundant NO_y species, although their relative abundance can vary significantly [*Zellweger et al.*, 2003,

[1]Environment Research Institute, Shandong University, Ji'nan, China.
[2]Department of Civil and Structural Engineering, Hong Kong Polytechnic University, Hong Kong, China.
[3]Centre for Atmosphere Watch and Services, Key Laboratory for Atmospheric Chemistry, Chinese Academy of Meteorological Sciences, Beijing, China.
[4]China GAW Baseline Observatory, Qinghai Meteorological Bureau, Xining, China.
[5]Institute for Climate and Global Change Research, Nanjing University, Nanjing, China.

0148-0227/11/2010JD014735

references therein]. NO_x is usually dominant near its sources, whereas PAN tends to be more abundant in regionally polluted air masses that have undergone more active organic photochemistry. HNO_3 is the dominant component of NO_y in more remote areas of the troposphere [e.g., *Atlas and Ridley*, 1996; *Val Martin et al.*, 2008]. Besides, owing to its temperature dependent thermal decomposition and insoluble nature, PAN is usually more dominant than HNO_3 at colder and/or humid environments (i.e., at high altitudes and latitudes).

[4] In the Northern Hemisphere, emissions of reactive nitrogen are dominated by the anthropogenic sources in urban and industrial regions. After emitted, nitrogen oxides are photochemically processed and can be exported out of the planetary boundary layer, thereby affecting the ozone budget in distant downwind regions [*Wang et al.*, 1998]. The export of NO_x away from source regions is facilitated by the export of PAN, which can be transported long distances at cold temperatures and decomposed to NO_x in warmer environments [*Singh and Salas*, 1983; *Singh et al.*, 1986]. Similarly, the export of HNO_3 followed by its photolysis to NO_x, may also be an important source of NO_x even in the lower troposphere [*Neuman et al.*, 2006]. Thus, the measurement of NO_y and its constituents in the remote troposphere provides important information for understanding the impact of anthropogenic forcing on the global ozone budget.

[5] The Qinghai-Tibetan Plateau is the highest landmass on Earth, with an average altitude of over 4,000 m asl. The atmosphere lying over the Plateau is probably the least affected by human activities in the Asian continent due to the sparse population and minimal industrial activity in western China. A number of studies have been conducted at WLG on the northeastern edge of the Plateau on reactive trace gases, aerosols, and greenhouse gases [e.g., *Tang et al.*, 1995; *Ma et al.*, 2003; *Zhou et al.*, 2004; *Wang et al.*, 2006; *Mu et al.*, 2007; *Kivekäs et al.*, 2009]. A broad summer maximum of surface ozone has been observed at WLG [*Tang et al.*, 1995], which is in contrast to the common spring maximum pattern in most remote areas of the Northern Hemisphere [*Monks*, 2000]. Although field observations provided strong evidence of frequent intrusions of stratospheric air in summer [*Tang et al.*, 1995; *Ma et al.*, 2005; *Wang et al.*, 2006], chemical transport modeling studies suggested that anthropogenic effects from central and eastern China made a greater contribution to the summer ozone maximum at WLG [*Zhu et al.*, 2004; *Li et al.*, 2009]. In our 2003 study [*Wang et al.*, 2006], ozone rich air was found to be associated with low levels of water vapor and low concentrations of CO, which is a tracer of anthropogenic emissions, and meteorological simulations suggested an upper tropospheric/lower stratospheric source [*Ding and Wang*, 2006]. The 2003 study also observed abnormally high concentrations of NO_y at WLG compared with other remote environments, which was attributed to enhanced microbial processes in the soil over the Plateau [*Wang et al.*, 2006].

[6] To better understand the sources of summertime surface ozone and reactive nitrogen at WLG, another intensive campaign was conducted in the summer of 2006. Concurrent measurements of O_3, CO, NO_y, and its individual constituents (NO, NO_2, PAN, HNO_3, and NO_3^-) were taken. Compared with the previous summer campaign in 2003, the transport of anthropogenic pollution from central and eastern China was frequently observed during the study. Here, we present the results of this study. We first compare the data from 2006 with those from 2003 and examine the ozone production efficiencies in polluted air masses. We then examine the partitioning of NO_y and compare our results with those from other similar remote locations in the world. It is worth pointing out that this is the first time that NO_y partitioning data from the remote free troposphere in western China has been reported. We also conducted an analysis of the recent 10 years' back trajectories in order to derive the climatology of air mass transport in summer at WLG and to compare the 2006 and 2003 results with this long term mean.

2. Experiment and Methodologies

2.1. Site Description

[7] The field campaign was carried out at the WLG Observatory (36.28°N, 100.90°E, 3816 m asl) from 22 July to 16 August 2006. The station is one of 24 baseline observatories of the World Meteorological Organization's (WMO) Global Atmospheric Watch (GAW) program, which was developed to monitor long term trends in gaseous and aerosol parameters in the global free troposphere. The Observatory is situated on an isolated mountain peak at the northeastern edge of the Qinghai-Tibetan Plateau, and is far away from the main industrial and populated regions of China (see Figure 1). The closest cities with more than 10^6 inhabitants are Xining and Lanzhou, which are located 90 km to the northeast and 260 km to the east, respectively. More specific details on the site have been described elsewhere [e.g., *Ma et al.*, 2003; *Wang et al.*, 2006].

2.2. Measurement Techniques

[8] The measurement instruments were housed in a temperature controlled laboratory on the second floor of the station. The sampling system, instrumentation, and calibration procedures were similar to those used in 2003 [*Wang et al.*, 2006; *J. Zhang et al.*, 2009]. Only a brief summary is presented here, with an emphasis on several newly added instruments for the measurement of nitrogen containing species.

[9] O_3, CO, NO, NO_2, and PAN were sampled at 4.1 m above the rooftop of the building through a PFA tube (inside diameter: 9.6 mm; length: 10.2 m; total flow rate = 9.0 L/min), which was connected to a PFA made manifold in the laboratory. NO_y, HNO_3, and NO_3^- were sampled via two inlet boxes that housed catalytic converters and were placed at the sample intake point (2.1 m above the rooftop). The sample line connecting the catalytic box and the instrument has a length of 8.4 m and an inside diameter of 3.2 mm, with a flow rate of 1.5 L/min.

[10] O_3 was detected using a UV photometric analyzer (Thermal Environmental Instruments (TEI) Model 49C) with a time resolution of 1 min, a detection limit of 2 ppbv and a precision of 2 ppbv. CO was measured with a gas filter correlation nondispersive infrared analyzer (Advanced Pollution Instrumentation (API) Model 300) with a detection limit of 30 ppbv for a 2 min average and a precision of ~1% for a level of 500 ppbv. During the study, the baseline was determined every 2 h by passing ambient air to the internal CO scrubber for 15 min [*Wang et al.*, 2006].

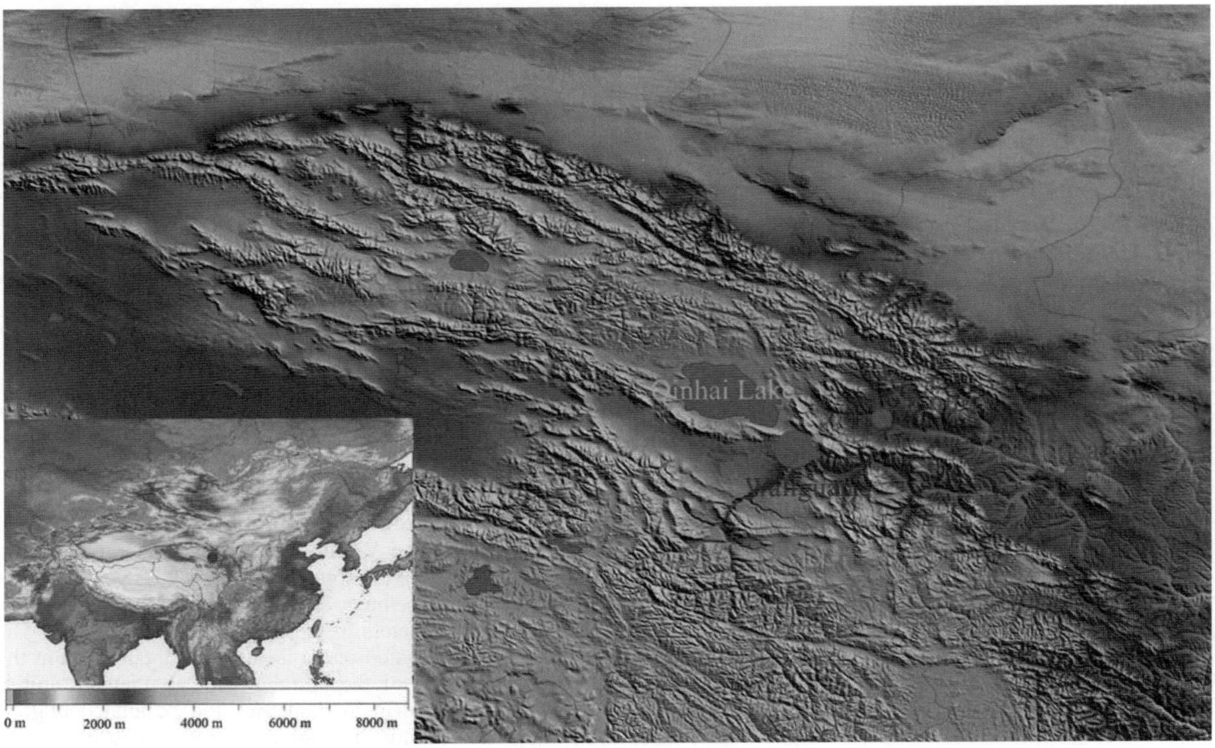

Figure 1. Topographical map (35°N–40°N, 95°E–105°E) showing the study site and surrounding regions.

[11] NO and NO_2 were measured using a chemiluminescence analyzer (Eco Physics, Model CLD 770 AL ppt) coupled with a photolytic converter (Eco Physics, Model PLC 760) by which NO_2 was converted to NO using a 300W, high-pressure, Xenon Arc lamp [*Wang et al.*, 2001]. The analyzer had a detection limit of 15 pptv for an integration time of 5 min, with a 2 σ precision of 10% and an uncertainty of 15%. The conversion efficiency of the photolytic converter was determined every 3 days by an NO_2 standard generated by the gas phase titration of NO with O_3, and ranged from 35% to 40% over the study period.

[12] NO_y was detected by another chemiluminescence analyzer (TEI, Model 42S) equipped with an externally placed molybdenum oxide (MoO) catalytic converter. NO_y was converted to NO on the surface of MoO at 350°C, and was then measured by the chemiluminescence detector [*Wang et al.*, 2006]. The detection limit of the analyzer was estimated to be 50 pptv for the time resolution of 1 min, with a 2 σ precision of 4% and an uncertainty of about 10%. The conversion efficiency of the MoO catalyst for NO_y species was checked daily by an n-propyl nitrate (NPN) standard that indicated the near complete (~100%) conversion of NPN throughout the campaign.

[13] HNO_3 and NO_3^- were measured by a third chemiluminescence analyzer (TEI, Model 42CY) in combination with two MoO converters. The measurement procedures were similar to that for NO_y, except that before entering the MoO converter, ambient air passed through a Teflon filter (which measures NO_y-NO_3^-) and another nylon filter to measure NO_y-HNO_3-NO_3^-. Laboratory and field studies have shown that gaseous HNO_3 can pass the upstream Teflon filter and would be efficiently collected by the downstream nylon filter [*Goldan et al.*, 1983; *Anlauf et al.*, 1986]. However, the efficiency with which HNO_3 penetrates the Teflon filter is highly dependent on the ambient RH and the particle loading on the filter. At WLG, the RH was generally low with daytime/nighttime mean values of 59% (±17%)/71% (±16%) during this study. Archived photos of weather conditions on the site reveal that there were few cloudy/foggy days during the campaign, with only several rainy events during which all the instruments were turned off. The Teflon and nylon filters were routinely changed in the morning every two days. Given the mean $PM_{2.5}$ concentration of 9.2μg/m^3 measured in the summer of 2003 [*Wang et al.*, 2006], assuming an increase of ~20% in 2006 (the same as CO) and a $PM_{2.5}$/TSP ratio of 0.6 [*Li et al.*, 2000], the particulate loading on the Teflon filter after two days was estimated as ~160 μg. Such particle loadings usually result in little retention (~10%) of HNO_3 [*Appel et al.*, 1981].The HNO_3 transmission efficiency is also expected to decrease with time after the filter replacement. Overall, we estimate the measurement uncertainties to be 20% for HNO_3 and 15% for NO_3^-. The detection limit of the detector was 50 pptv for 1 min data and 15 pptv for hourly averages.

[14] PAN was measured with a commercially available automatic PAN analyzer equipped with a calibration unit (Meteorologie Consult GmbH). The analytical method was based on gas chromatographic separation with sub-

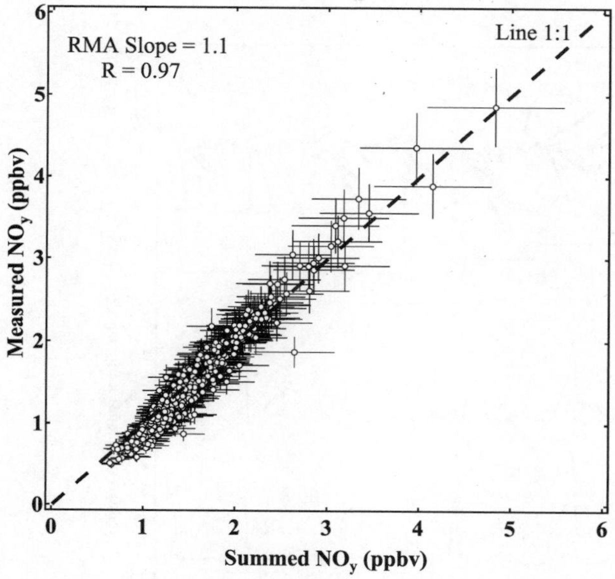

Figure 2. Comparison of the measured NO_y and the sum of individually measured constituents (NO_x, PAN, HNO_3, and NO_3^-). The error bars stand for the uncertainties within the specific measurements.

sequent electron capture detection (ECD). PC software (Adam32 v1.42) was used to control the sampling protocol and to collect the 10 min data. The analyzer had a detection limit of 50 pptv, with a 2 σ precision of 6% and an uncertainty of 15%. The instrument was calibrated weekly using the calibration unit. *J. Zhang et al.* [2009] provide further details of this instrument.

[15] The comparison of measured NO_y and the sum of its individual constituents can be used to check the accuracy of NO_y measurements. Generally, the summed NO_y is comparable to or reasonably smaller than the NO_y concentration [e.g., *Williams et al.*, 1997; *Zellweger et al.*, 2000]. In this study, the measured and summed NO_y values showed a good agreement (see Figure 2, r = 0.97), with an average $[NO_y]_{sum}/[NO_y]_{meas}$ ratio of 1.07. NO_x, PAN, HNO_3, and NO_3^- are believed to be the main reactive nitrogen compounds at WLG, accounting for ~100% of the total NO_y. The slightly higher value of the sum than the measured NO_y is due to the uncertainties of the different instruments used for the measurements.

[16] A data logger (Environmental Systems Corporation, Model 8816) was used to control the calibrations and to collect data (except for PAN), which were averaged over 1 min intervals. The data presented in this paper are hourly averaged values (except that 10 min data was used for the ozone production efficiency calculation). Various meteorological parameters are routinely monitored at the WLG Observatory [*Wang et al.*, 2006].

2.3. Calculation and Categorization of Backward Trajectories

[17] In the analysis of data collected in 2003, backward trajectories were calculated and further classified into several different groups to investigate the origins and transport pathways of air masses sampled at WLG [*Wang et al.*, 2006]. For the comparison with the 2003 study, we performed the same analysis for the 2006 study. Three-dimensional 10 day backward trajectories, terminated at 500 m above the ground level (AGL) of the peak of WLG, were calculated hourly using the Hybrid Single Particle Lagrangian Integrated Trajectory (HYSPLIT Model, version 4.8, 2010, from R. R. Draxler and G. D. Rolph, http://ready.arl.noaa.gov/HYSPLIT.php) with the GDAS meteorological data. We then applied cluster analysis to segregate the calculated trajectories into a number of groups using the hierarchical Ward's method with a squared Euclidean measure. The positions of the endpoints (i.e., longitude, latitude, pressure) at 3 h intervals along the 240 h trajectories were selected as the clustering variables. The cluster analysis was performed on a PC using the statistical software SPSS.

3. Results and Discussion

3.1. Data Overview

3.1.1. Air Mass Transport Regimes

[18] Time series of the trace gases and meteorological parameters for the study period is shown in Figure 3, together with the air mass categories deduced from cluster analysis of the backward trajectories. As expected, two main types of air masses were sampled during the study period. One was the air mass containing high levels of O_3 but low concentrations of CO and water (see Figure 3, e.g., 23–26 July), which is characteristic of air from the upper troposphere/lower stratosphere. The other was the air mass influenced by anthropogenic pollution, which was characterized by sharp rises in the levels of combustion tracer(s), specifically CO and/or NO_y (see Figure 3, e.g., 29 July and 4 and 7–11 August). Air masses of this type also contained relatively high amount of water but had generally lower O_3 mixing ratios.

[19] To further identify the origins and transport pattern of the sampled air masses, three dimensional 10 day backward trajectories were calculated and then classified into five groups using the cluster method. Figure 4 shows the mean trajectories of the five clusters plotted on the Asian anthropogenic CO emissions [*Q. Zhang et al.*, 2009]. Open circles on the trajectories represent 6 h intervals, with the marker size indicating the percentage of each cluster. The five air mass groups are described as follows. CA denotes air masses coming from the northwest, originating in Central Asia, passing over Kazakhstan and China's Xinjiang region and subsiding from the higher altitudes to the measurement site. SM denotes air masses coming from the north originating in western Siberia and Mongolia and passing in the lower troposphere over the Inner Mongolia and Gansu provinces of China. CSC denotes air masses coming from the southeast passing through the planetary boundary layer at lower speeds over southern and central China. CEC denotes air masses passing over eastern and central China at fast speeds in the lower troposphere. CWC denotes air masses originating in Xinjiang and passing over the Gansu and Qinghai provinces of China. It can be seen from the anthropogenic CO emissions that the trajectories coming from the west pass over a large remote area with very few anthropogenic emissions.

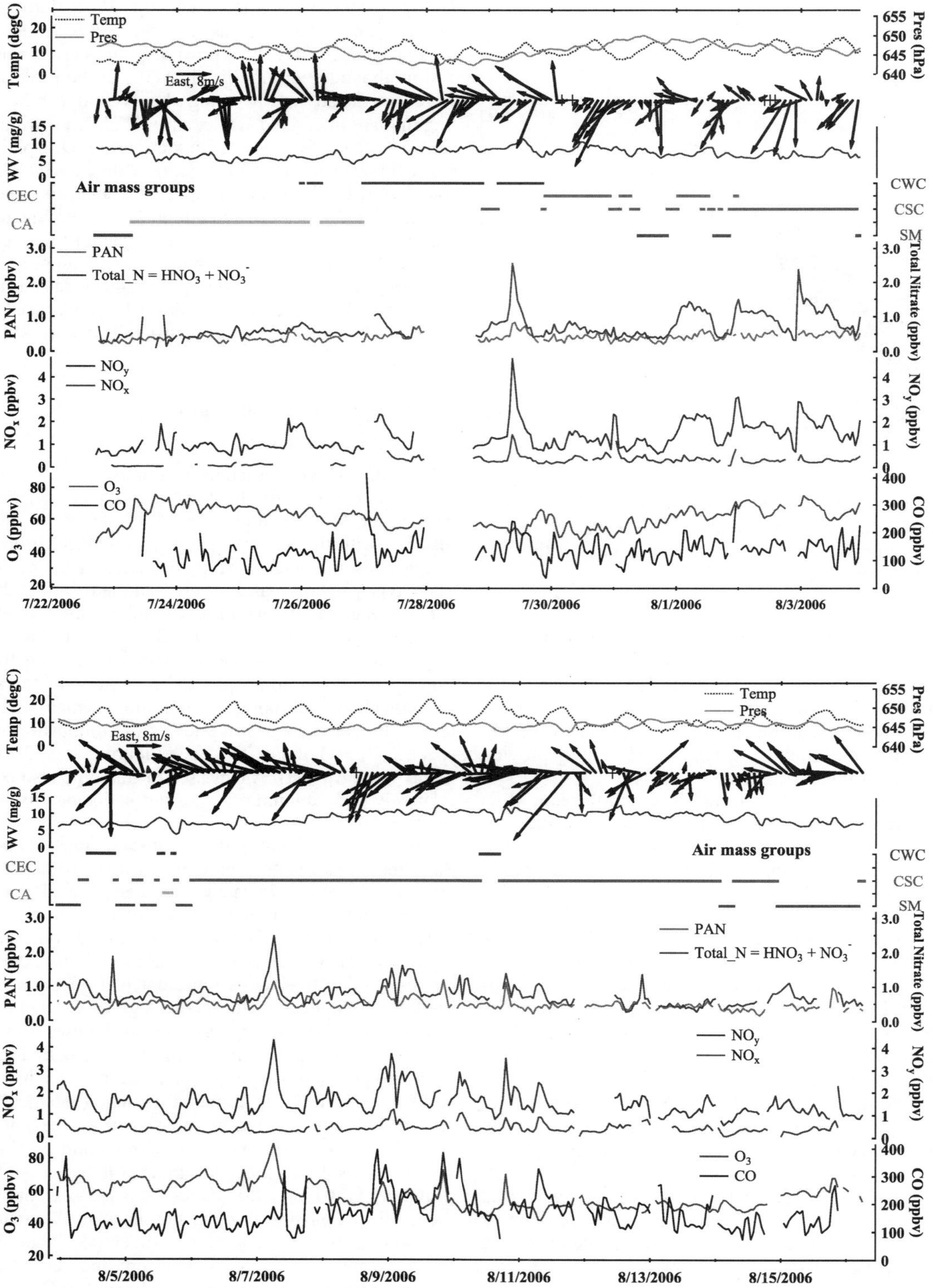

Figure 3. Time series of hourly data of trace gases and meteorological parameters measured at WLG between 22 July and 16 August 2006. The air mass groups are deduced from cluster analysis of the 10 day backward trajectories: SM, Siberia/Mongolia; CA, central Asia; CSC, central/southern China; CEC, central/eastern China; and CWC, central/western China.

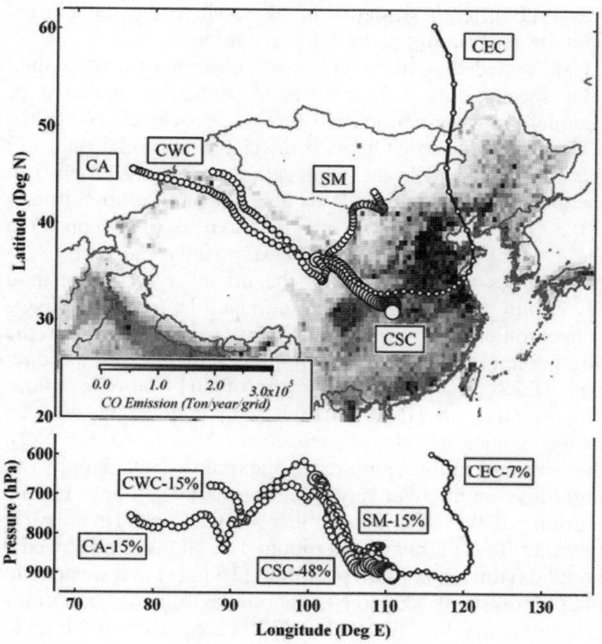

Figure 4. Mean 10 day backward trajectory clusters plotted on the Asian anthropogenic CO emissions (http://www.cgrer.uiowa.edu/EMISSION_DATA_new/data/intex b_emissions/). The open circles on the trajectories indicate 6 h intervals. The size of the markers indicates the percentage of each cluster.

[20] Among the five groups identified, CSC occurred the most, accounting for 48% of the total, followed by CA (15%), SM (15%), CWC (15%), and CEC (7%). The majority of air masses (e.g., CSC and CEC, 55%) had passed over the populated regions of China before arriving at the site, suggesting extensive transport of anthropogenic pollution to the Plateau during the study period. This is quite different from the transport pattern encountered in the summer study of 2003, in which most air masses came from the west (e.g., CA 37% and EU Trop 13%) and only a small fraction passed over central and southern China (e.g., SC 25%) [*Wang et al.*, 2006]. These results indicate significant year to year variability in the large scale dynamics.

[21] To learn about the typical air mass transport pattern at WLG in summer, we calculated 10 day backward trajectories every 3 h for the summer months (15 July to 17 Aug) in the recent 10 years (2000–2009). The cluster approach was then applied to the whole of trajectories to picture the climatological transport pathways of air masses. Three types of transport pathways were identified, representing three major directions from which air parcels came to the site. Table 1 gives percentages of three air mass types for the period of 2000–2009 and for the 2 years during which the field campaigns took place. Over the recent 10 years, the transport regime at WLG in summer was dominated by the air coming from the east (~49%) that may have passed over the populated regions of China, with 26% of air masses from the remote west and 25% from the north. Comparing against this 'long term' averaged pattern, the summer 2006 had more frequent transport from the east (63%) and less from the west (20%) whereas 2003 had less impact from the east (37%) but much more from the west (44%). The reduced transport from the east in 2003 may have been due to a weaker summer monsoon in 2003 [e.g., *Ogi et al.*, 2005; *Saigusa et al.*, 2010].

3.1.2. Concentrations in 2006

[22] Table 2 summarizes the statistics of trace gases for the whole data set and for subdivisions of the upslope (0800–1959 local time (LT)) and downslope (2000–0759 LT) observations according to changes in vertical winds (see Figure 5). The mean levels of trace gases (except for NO_y) were higher than those measured in the previous campaign in summer 2003. In the 2006 study, the average concentrations of O_3, CO, and NO (daytime data) were 59 ppbv, 149 ppbv, and 71 pptv, respectively, compared with the respective mean values of 54 ppbv, 125 ppbv, and 47 pptv in 2003 [*Wang et al.*, 2006]. The *t* tests suggest the differences in the levels of trace gases between the two periods to be statistically significant ($p < 0.01$). However, the values of NO_y (mean = 1.44 ppbv) were much lower than those in 2003 (mean = 3.60 ppbv), which is further discussed in section 3.3.1. A discussion of the relative abundance of the individual NO_y components is given in section 3.3.2.

[23] We also examined the chemical characteristics of the five air mass groups by sorting the hourly data on trace gases and water vapor according to the trajectory clusters. The statistical results are given in Table 3. The air masses from the western remote regions (CA) contained the lowest levels of CO (mean = 110 ppbv) and NO_y (mean = 1.04 ppbv), the least amount of water (mean = 5.9 mg/g), but the highest O_3 mixing ratios (mean = 66 ppbv). Indeed, the CA air mass was in good correspondence with the measurements potentially influenced by the upper tropospheric/lower stratospheric air as examined in Figure 3 (e.g., 23–26 July). In contrast, the other air masses that were possibly influenced by anthropogenic pollution had high concentrations of CO (mean = 136–166 ppbv) and NO_y (mean = 1.18–1.65 ppbv), a relatively higher amount of water (mean = 7.3–8.8 mg/g), and lower O_3 mixing ratios (mean = 57–60 ppbv).

3.1.3. Diurnal Variations

[24] At WLG, weak diurnal variations were observed for O_3, CO, and reactive nitrogen oxides. Figure 5 shows the average diurnal patterns of trace gases and surface winds

Table 1. Percentage of Three Major Types of Air Masses at WLG in the Summer Months of 2000–2009[a]

Period	East[b]	West[b]	North[b]
2000–2009	49%	26%	25%
2003	37%	44%	19%
2006	63%	20%	17%

[a]Ten day backward trajectories terminating at 500 m (AGL) above the peak of WLG were calculated every 3 h by the HYSPLIT model for the period of 15 July to 17 August in the year of 2000–2009. The same cluster approach (see section 2.3) was used to segregate the trajectories into specific clusters.

[b]Three major air masses: "East" contains air masses coming from the east and southeast that have passed over the populated regions of China (i.e., CSC and CEC in Figure 4); "West" contains air masses coming from the west, passing over the remote regions of central Asia and Xinjiang (i.e., CA); and "North" contains air masses coming from the north and mainly passing over Siberia and Mongolia regions (i.e., SM).

Table 2. Statistics on Hourly Data of Trace Gases Measured at WLG[a]

	All Data		Upslope[b]		Downslope[b]	
Species	Mean	N	Mean	N	Mean	N
O_3	59 (8)	558	58 (7)	280	61 (8)	278
CO	149 (57)	506	142 (50)	267	156 (64)	239
NO_y	1.44 (0.61)	540	1.37 (0.56)	273	1.51 (0.66)	267
NO	40 (47)	421	71 (45)	223	4 (14)	198
NO_2	0.32 (0.18)	478	0.28 (0.13)	257	0.36 (0.22)	221
PAN	0.44 (0.14)	524	0.44 (0.13)	265	0.45 (0.15)	259
HNO_3	0.26 (0.09)	511	0.26 (0.11)	264	0.26 (0.08)	247
NO_3^-	0.48 (0.32)	525	0.43 (0.32)	260	0.53 (0.31)	265
NH_3*[c]	7.9 (4.0)	535	8.0 (3.8)	269	7.8 (4.2)	266

[a]The mean (standard deviation) values are in ppbv, except for NO, which is in pptv. N is the number of hourly data points.

[b]Upslope equals 0800–1959 local time (Beijing time); downslope equals 2000–0759 local time.

[c]NH_3* equals NH_3 + NH_4^+, which was measured by the NO_y detector with MoO conversion at 450°C.

(both horizontal and vertical) in the 2006 study. Similar to other mountaintop sites, local anabatic (represented by positive speeds) and catabatic winds are clearly shown at WLG. For trace gases, the mean O_3 mixing ratios showed a trough around noon (1000–1400 LT) with a minimum of ~56 ppbv, and enhanced levels at nighttime with a maximum of ~62 ppbv. The average concentrations of CO and NO_y were lower (CO = 120 ppbv; NO_y = 1.0 ppbv) in the afternoon and higher (CO = 150–175 ppbv; NO_y = 1.6–1.7 ppbv) in the morning and at night. The diurnal patterns of CO and O_3 were similar to those measured in the 2003 summer study [*Wang et al.*, 2006], but the NO_y profile was quite different from that in 2003, which showed a broad daytime maximum pattern with much higher values.

[25] Varied diurnal patterns were observed for individual NO_y species due to their different properties in transport, chemical transformation, and removal processes. NO_3^- had a similar profile to NO_y, with lower levels (~0.25 ppbv) in the afternoon and higher concentrations (0.6–0.7 ppbv) in the morning and at night. PAN also had an afternoon trough (~0.37 ppbv), and showed an apparent peak (~0.56 ppbv) in the evening (~2000 LT) which was mainly caused by several plumes transported from the urban areas of Lanzhou [*J. Zhang et al.*, 2009]. In contrast, the average HNO_3 concentrations showed a maximum (~0.32 ppbv) in the afternoon (1400–1600 LT), implying the in situ production of HNO_3 with higher levels of OH and/or evaporation of NO_3^- to HNO_3 at high temperatures (as evinced by the coinciding sharp decrease in NO_3^-). As for NO_2, the mixing ratios exhibited a maximum (~0.45 ppbv) at nighttime, with a relatively smaller peak (~0.35 ppbv) in the morning (0900 LT) and lower levels (~0.25 ppbv) in the early morning (0600 LT) and afternoon. The NO levels showed a broad daytime maximum pattern (~115 pptv) that was due to the photolysis of NO_2 to NO. These daytime concentrations also imply that the NO levels at WLG are sufficiently high to sustain a net ozone photochemical production [*Duderstadt et al.*, 1998].

3.2. Ozone Production Efficiency in Polluted Air Masses

[26] It is of interest to investigate how efficient O_3 is formed during the oxidation of NO_x in the air masses with recent influence of anthropogenic emissions from the east.

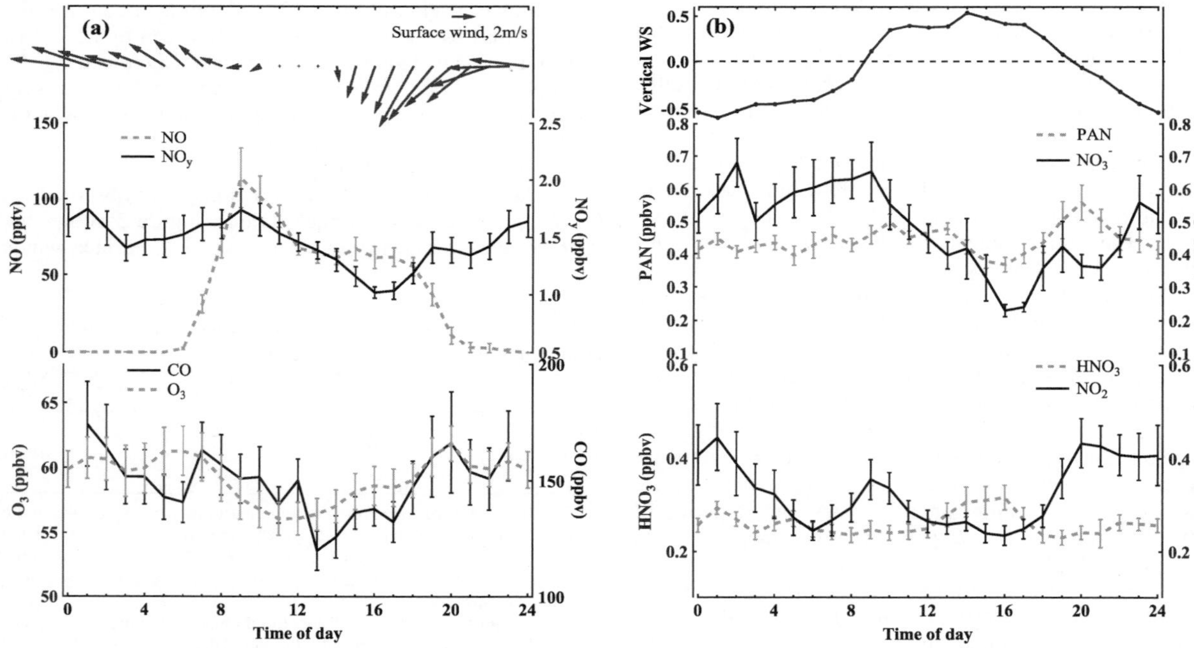

Figure 5. Average diurnal profiles of (a) NO, NO_y, and other trace gases and (b) reactive nitrogen compounds at WLG. Diurnal patterns of surface winds (both horizontal and vertical) are also given. The error bars indicate the standard errors of the measurement data.

Table 3. Classification of Trace Gases and Water Vapor in the Different Air Mass Groups[a]

Air Mass	O_3	CO	NO_y	NO_x	PAN	HNO_3	NO_3^-	NH_3*	WV
CA	66 (4)	110 (40)	1.04 (0.44)	0.10 (0.06)	0.36 (0.07)	0.20 (0.05)	0.40 (0.32)	4.4 (1.0)	5.9 (1.0)
SM	60 (6)	138 (51)	1.18 (0.45)	0.32 (0.17)	0.40 (0.14)	0.24 (0.08)	0.37 (0.19)	5.7 (1.7)	7.3 (0.9)
CSC	58 (8)	166 (60)	1.65 (0.59)	0.40 (0.19)	0.49 (0.15)	0.28 (0.10)	0.53 (0.32)	9.7 (4.5)	8.8 (1.8)
CEC	58 (6)	136 (48)	1.47 (0.57)	0.38 (0.14)	0.41 (0.09)	0.27 (0.06)	0.52 (0.29)	8.1 (2.2)	8.1 (1.2)
CWC	57 (5)	146 (50)	1.43 (0.72)	0.41 (0.22)	0.45 (0.13)	0.28 (0.12)	0.51 (0.40)	7.5 (2.8)	8.1 (1.5)

[a]Values are given as means (standard deviations) in ppbv, except for water vapor (WV), which is in mg/g.

Ozone production efficiency (OPE) is defined as the number of ozone molecules produced per molecule of NO_x oxidized, and is empirically inferred from the regression slope of the observed scatterplot of O_3 versus NO_z [*Trainer et al.*, 1993]. It's worth noting that the slope is also affected by the relative removal rates of O_3 and NO_z (especially the rapid deposition of HNO_3) and by the mixing of different air masses [*Zanis et al.*, 2007]. Thus, the slope of an O_3-NO_z scatterplot generally provides only the upper limit of the gross ozone production efficiency.

[27] To exclude the interference from mixing with highly processed free tropospheric air, we only selected several polluted plumes with recent anthropogenic influences using the measured atmospheric NO_y/CO ratios, which have been used to discriminate between disturbed and undisturbed free tropospheric conditions at Jungfraujoch [*Zellweger et al.*, 2003]. The NO_y/CO ratio generally decreases with the lifespan of an air parcel due to the longer life time of CO than NO_y. *Jaeglé et al.* [1998] reported an average NO_y/CO ratio of ~0.1 in air close to anthropogenic sources and of ~0.005 in the upper troposphere. In the present study, the polluted air masses were identified if the following criteria were fulfilled. (1) The air masses had passed over the populated central and eastern regions of China before arriving at the site (belonging to CSC and CEC air mass groups); (2) there was no shift in wind directions during the pollution events; (3) the NO_y/CO ratios were higher than 0.01; and (4) concurrent measurements of O_3, NO_y and NO_x were available. Five plumes with enhancements of ozone (and other pollutants) were finally extracted for the further analysis. They were observed during 3, 6, 7, 9, and 10 August (see Figures 3 and 6).

[28] Figure 6 shows the O_3-NO_z scatterplots for 10 min data for all the five polluted plumes. The slopes were in the range of 7.7–11.3 ppbv/ppbv, with a mean value of 9.9 (±1.7). Note that the slopes were determined with the reduced major axis (RMA) method to take into account measurement uncertainties in both the x and y variables [*Hirsch and Gilroy*, 1984]. These values are somewhat higher than those observed in urban and rural areas, and are comparable to the observations made at other remote continental sites. For example, *Wang et al.* [2010] observed OPE values of ~3 at both an urban and a downwind mountainous site of Beijing in summer 2008. *Chin et al.* [1994] reported an average OPE value of 5.5 in the U.S. boundary layer with values being more than 2 times higher in the west (9.1) compared to the east (4.2). *Wood et al.* [2009] derived an OPE value of 6.2 at a mountaintop site within the Mexico City Basin during a stagnant ozone episode with an extremely large ozone production rates. At Jungfraujoch, *Zanis et al.* [2007] calculated OPE values of 7–9 for polluted air masses in summer months, compared to the higher values of 15–20 for highly processed free tropospheric air.

3.3. Reactive Nitrogen Chemistry

3.3.1. Lower Concentrations of NO_y Than in 2003

[29] One objective of the present study was to reexamine the atmospheric levels and source(s) of reactive nitrogen at WLG. In the previous study in 2003, we found that the concentrations of NO_y at WLG were abnormally high (means = 3.60–3.83 ppbv) compared with those at other remote sites [*Wang et al.*, 2006]. Soil emissions enhanced by the presence of animal waste were proposed to be an important factor. During the 2006 study, the NO_y levels were much lower (mean = 1.44 ppbv) than those observed in 2003 despite the more frequent transport of polluted air masses. In this section, we discuss the possible reasons for the difference in NO_y levels between the two studies by examining both changes in emissions and measurement interference.

[30] Soil emissions of NO_x (as NO) are closely related to temperature, precipitation, and soil type [*Williams et al.*, 1988; *Yienger and Levy*, 1995]. The amount of NO released increases with the surface temperature, precipita-

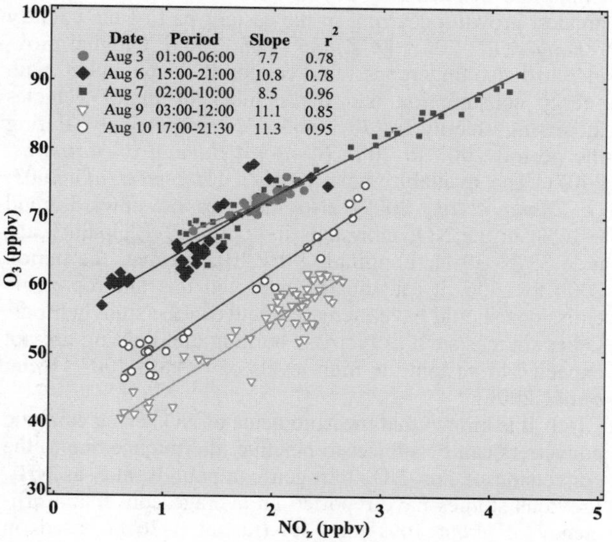

Figure 6. Scatterplots of ozone versus NO_z for the 10 min data for the identified five plumes with recent anthropogenic influences from central and eastern China. The regression lines were obtained with the reduced major axis method.

Table 4. Comparison of the NO_y Budget at WLG With Those Obtained From Other Remote, High-Altitude Sites in the Northern Hemisphere

Site	Location	Altitude (m asl)	Period	NO_y Partitioning				NO_y (pptv)	Reference[a]
				NO_x	PAN	HNO_3	NO_3^-		
WLG	36.28°N, 100.90°E	3816	22 Jul to 16 Aug 2006	0.24	0.32	0.20	0.31	1439	1
Jungfraujoch	46.55°N, 7.98°E	3580	18 Jul to 23 Aug 1997	0.22	0.36	0.07	0.17	986	2
Idaho Hill	39.5°N, 105.37°W	3070	24 Sep to 4 Oct 1993	0.66	0.19	0.09	0.06	1759	3
Mauna Loa	19.54°N, 155.58°W	3400	15 Jul to 15 Aug 1992	0.14	0.04	0.35	0.13	203	4
Pico Mountain	38.47°N, 28.40°W	2200	20 Jul 2002 to 25 Aug 2005 (summer)	0.12	0.21	0.71		365	5
Summit, Greenland	72.33°N, 38.75°W	3210	3–16 Jul 1999	0.33	0.55			50~350	6
Summit, Greenland	72.55°N, 38.4°W	3200	May-Jul 1995			0.01		850	7
Goose Bay, Labrador	53.3°N, 60.4°W	2000–4000	Jul-Aug 1990	0.15	0.77	0.29	0.04	330	8
Western Pacific Ocean	25°N–42°N	3000–7000	Sep-Oct 1991	0.18	0.20	0.25	0.13	376	9
TRACE P (costal)	20°N–40°N, 90°E–130° E	2000–7000	Feb-Apr 2001	0.06	0.24	0.69		475	10
INTEX A (North America)	25°N–55°N, 140°W–40°W	2000–4000	1 Jul to 15 Aug 2004	0.08	0.26	0.60	0.14	809	11

[a]References: 1, this study; 2, *Zellweger et al.* [2000]; 3, *Williams et al.* [1997]; 4, *Atlas and Ridley* [1996]; 5, *Val Martin et al.* [2008]; 6, *Ford et al.* [2002]; 7, *Munger et al.* [1999]; 8, *Singh et al.* [1994]; 9, *Singh et al.* [1996]; 10, *Talbot et al.* [2003]; and 11, *Singh et al.* [2007].

tion frequency, and nitrate and ammonium content of the soil. In the 2006 study, the average temperature (±standard deviation) at WLG was 10.9(±3.4)°C, which was even slightly higher than that of 9.3(±3.5)°C in the summer campaign of 2003. According to our on-site observations, there was no significant decrease in rainfall events between the study periods (five events in 2006 versus six events in 2003). On the Qinghai Tibetan Plateau, animal feces are a major source of the N substances in the soil. In 2006, there were about 23 million livestock in Qinghai province, which is also similar to the number in 2003 [*Qinghai Bureau of Statistics*, 2007]. It can thus be concluded that the difference in NO_y levels between the 2 years is unlikely to be due to changes in soil emissions.

[31] As to anthropogenic emissions, the NO_x emissions in China had increased by 55% from 2001 to 2006, with the greatest growth occurring in the eastern part of the country [*Zhang et al.*, 2007; *Q. Zhang et al.*, 2009]. Qinghai province had also undergone rapid economic development, with energy consumption and the number of motor vehicles increasing sharply by 70% and 43%, respectively, during the period 2003 to 2006 [*Qinghai Bureau of Statistics*, 2007]. The available emission data [*Streets et al.*, 2003; *Q. Zhang et al.*, 2009] also indicates an upward trend (~16%) in the NO_x emissions in "Qinghai" (domain: Latitude 33°N–39°N, Longitude 90°E–102°E) over the period 2000 to 2006. It can thus be expected that anthropogenic emissions should have increased between the study periods. Other sources such as biomass burning and lighting are not expected to sustain the high levels of NO_y in 2003 [*Wang et al.*, 2006].

[32] It is known that measurements of NO_y using catalytic converters can be subject to positive interference due to the conversion of non-NO_y nitrogen compounds such as NH_3. Previous studies have reported an average conversion efficiency of about 10% for NH_3 (range: 0–26%) based on testing a suite of MoO converters [*Williams et al.*, 1998; *Fitz et al.*, 2003]. The conversion efficiency of NH_3 has been found to be independent of temperature, but to rise as the converter ages (note that in both of our studies, all of the converters were newly installed at the beginning of the measurement period). To further determine the interference from NH_3 in the NO_y measurements, we tested four MoO converters of different ages in a recent field study by spiking them 49 ppbv of NH_3. Out of the four converters, three converted NH_3 at an efficiency of 0–14%, which is consistent with the previously reported results. However, a new converter showed an extremely high conversion efficiency of ~38% at 350°C but this decreased to ~11% at 325°C. This suggests that new converters may also be subject to strong interference by NH_3.

[33] At WLG, the concentrations of NH_3* (measured as NH_3+NH_4^+ by MoO conversion at 450°C) were relatively high, with an average value (±standard deviation) of 7.9 (±4.0) ppbv during the present study. A passive sampling study also showed a summertime NH_3 level of ~4 ppbv [*Meng et al.*, 2010]. If we assume a very high (30%) conversion efficiency in the catalytic converter used in the 2003 study and the same levels of NH_3* as in 2006, the positive interference could have been as large as 2.4 ppbv in 2003, which is sufficiently high to explain the discrepancy in NO_y levels between the two studies. We have no solid evidence to indicate that this actually occurred, there is, however, a possibility that the very high NO_y levels in 2003 were in large part due to NH_3 interference in the particular converter used in that study.

[34] Although the mean NO_y level in the 2006 study was much lower than those in the previous campaigns in 2003, it was still higher than the mean levels measured at other remote sites (see Table 4, and see also Table 2 of *Wang et al.* [2006]). For example, values of 1.15 ppbv were observed at Mount Cimone (Italy, 2165 m asl), 0.99 ppbv at Mount Jungfraujoch (Switzerland, 3580 m asl), 0.81 ppbv and 0.48 ppbv on INTEX A and TRACE P aircrafts. Much lower concentrations were measured at Mauna Loa (0.20 ppbv, Hawaii, 3400 m asl) and Pico Mountain (0.37 ppbv, Azores, 2200 m asl). However, it was lower than the mean concentrations measured at Idaho Hill (1.76 ppbv, U.S., 3070 m asl) and Niwot Ridge (2.0 ppbv, U.S., 3050 m asl), both of which are relatively close to urban sources.

Table 5. Average Proportions of Individual NO_y Species at WLG[a]

Data Set	N	NO_x/NO_y	PAN/NO_y	NO_3^-/NO_y	HNO_3/NO_y
All data	404	24% (10%)	32% (8%)	31% (9%)	20% (9%)
Upslope[b]	221	25% (9%)	34% (8%)	29% (9%)	20% (10%)
Downslope[b]	183	23% (11%)	30% (8%)	34% (8%)	19% (7%)
CA	33	12% (6%)	38% (7%)	38% (10%)	23% (7%)
SM	61	24% (10%)	34% (9%)	30% (7%)	20% (6%)
CSC	225	25% (9%)	31% (7%)	31% (10%)	18% (8%)
CEC	40	27% (10%)	30% (8%)	34% (8%)	21% (8%)
CWC	45	29% (7%)	35% (8%)	30% (8%)	23% (13%)

[a]Values are given as mean ratios (standard deviations). N is the number of available hourly data points.

[b]Upslope equals 0800–1959 local time (Beijing time); downslope equals 2000–0759 local time.

3.3.2. NO_y Speciation

[35] In this section, we investigate the partitioning of NO_y at WLG, and compare the results with measurements taken at other remote high-altitude sites in the Northern Hemisphere. Table 5 summarizes the proportions of individual NO_y species at WLG for the whole data set and for the five air mass groups. Table 4 compares the NO_y partitioning data from WLG with those obtained from other remote areas of the Northern Hemisphere. At WLG, PAN and NO_3^- were the most abundant reactive nitrogen compounds, with average proportions (±standard deviation) of 32% (±8%) and 31% (±9%), followed by NO_x (24% ± 10%) and HNO_3 (20% ± 9%). The sum of NO_x, PAN, HNO_3, and NO_3^- accounted for the whole of the measured total NO_y. Our previous measurements using sampling canisters in 2003 indicated that the mixing ratios of seven C_1-C_5 alkyl nitrates at WLG were very low (mean = 10.8 pptv) accounting for < 1% of the measured NO_y.

[36] It is evident from Table 5 that although reactive nitrogen is principally emitted as NO, it largely exists in its secondary reservoir forms at WLG. This is consistent with the fact that air parcels are well processed photochemically during long range transport to the WLG site. The low NO_x/NO_y ratios (24% ± 10%) at WLG are comparable to those measured at other remote continental sites, such as ~22% at Jungfraujoch [*Zellweger et al.*, 2000] and ~30% at Mount Cimone [*Fischer et al.*, 2003], whereas much lower values (6–14%) have been observed in the more remote marine free troposphere (e.g., Mauna Loa, Pico Mountain, and TRACE P aircraft; see Table 4). Of the various air masses sampled at WLG, the CA air mass coming from the sparsely inhabited western regions had a lower ratio of NO_x/NO_y (12% ± 6%) than the other four air mass categories (24–29%).

[37] Among the secondary NO_y species, inorganic nitrate (HNO_3 20% plus NO_3^- 31%) was more abundant at WLG. This result is in good agreement with other measurements taken at remote, midlatitude sites such as Mauna Loa, Pico Mountain, and on the TRACE-P and INTEX-A aircraft, where inorganic nitrate made up 48–74% of NO_y (see Table 4). In contrast, higher PAN/NO_y ratios (55–77%) have been widely observed at high altitudes and latitudes in the Northern Hemisphere (e.g., Summit, Greenland; Goose Bay, Labrador). The temperature dependent decomposition of PAN is believed to be an important factor governing this latitudinal distribution of NO_y partitioning.

[38] An interesting result regarding the NO_y budget at WLG is that inorganic nitrate largely existed in the particle phase, with an average particulate to total nitrate ratio, i.e., $NO_3^-\backslash(NO_3^- + HNO_3)$, of 0.62 (±0.13). It is distinctly different from the observations made at other remote sites (e.g., Mauna Loa, Pico Mountain, and the TRACE P and INTEX A missions; see Table 4), which found that the majority of inorganic nitrate was present in the gaseous phase. Instead, the pattern at WLG is quite similar to those usually observed at urban and pollution affected locations, where high levels of gaseous NH_3 and/or mineral particles are present [*Song and Carmichael*, 2001]. Previous measurements using filter based methods have also demonstrated the dominance of aerosol NO_3^- in the nitrate partitioning at WLG [*Xue*, 2002; *Ma et al.*, 2003]. *Ma et al.* [2003] reported a typical particulate to total nitrate ratio of ~0.9 for an autumn-winter campaign, with NO_3^- being present in both fine particles and coarse modes. They suggested that nitrate in fine particles may be produced by the reaction of gaseous HNO_3 with NH_3, whereas coarse nitrate may be generated by the condensation of HNO_3 on the surface of mineral particles. Livestock husbandry and agriculture are the main activities on the Plateau, both of which are major emission sources of NH_3. As indicated in section 3.3.1, the concentrations of NH_3 are indeed very high at WLG. Furthermore, particles at WLG have been shown to be predominantly from natural sources, such as soil and crust [*Gao and Anderson*, 2001]. The large abundances of NH_3 and crustal aerosols provide sufficient neutralizers for gaseous HNO_3, thus favoring the formation of particulate nitrate and explaining its greater presence in NO_y at WLG.

4. Summary and Conclusions

[39] In summer 2006, speciated reactive nitrogen compounds (NO, NO_2, PAN, HNO_3, and NO_3^-) were continuously measured in conjunction with total reactive nitrogen (NO_y), ozone, and carbon monoxide at WLG on the Qinghai-Tibetan Plateau. The data were analyzed to help further understand the origins of surface ozone and reactive nitrogen, and to investigate nitrogen partitioning in the remote free troposphere of western China.

[40] The mean concentrations of O_3, CO, and daytime NO were 59 ppbv, 149 ppbv, and 71 pptv, respectively, which were significantly higher than those measured during the previous study in summer 2003. More frequent anthropogenic impact from central and eastern China was observed at WLG in summer of 2006 (~55%) than that of 2003 (~25%). Varied diurnal patterns were observed for the different NO_y constituents. Ozone production efficiencies of 7.7–11.3 were calculated for the air masses recently influenced by the anthropogenic emissions in central and eastern China.

[41] Despite the more frequent transport of anthropogenic pollution, the NO_y levels in the 2006 study (mean = 1.44 ppbv) were substantially lower than those observed in 2003 (mean = 3.60 ppbv). Changes in emissions from soil and anthropogenic sources are unlikely to account for the large discrepancy between the two study periods. Based on the recent test results on NH_3 interference in MoO converters, we suspect that the abnormally high NO_y values in 2003 were due to NH_3 interference in the particular converter used.

[42] NO_y partitioning was investigated for the first time in the remote free troposphere in western China. PAN and particulate NO_3^- were the dominant NO_y species in summer at WLG, and comprised on average 32% and 31% of the measured NO_y, followed by NO_x (24%) and HNO_3 (20%). Compared with results obtained from other remote, high-altitude sites in the Northern Hemisphere, particulate nitrate made a greater contribution to NO_y, which can be attributed to the high abundance of NH_3 and mineral particles over the Plateau.

[43] Further analysis of backward trajectories for the recent 10 years indicated that WLG was frequently (~50% of air masses) influenced by the air from the east, suggesting an important role of anthropogenic emissions in central and eastern China in shaping the summertime surface ozone and other atmospheric trace constituents at WLG and over the Tibetan Plateau.

[44] **Acknowledgments.** We are grateful to the staff at the WLG Observatory for their help throughout the field campaign, to Hongchao Zuo for assistance with securing shipment and customs clearance for the instruments, and to NOAA Air Resources Laboratory (ARL) for providing the HYSPLIT model. The work was funded by the National Basic Research Program (973 Program) of China (2005CB422203) and the Hong Kong Polytechnic University (1-BB94).

References

Anlauf, K., H. A. Wiebe, and P. Fellin (1986), Characterization of several integrative sampling methods for nitric acid, sulphur dioxide and atmospheric particles, *J. Air Pollut. Control Assoc.*, *36*, 715–723.

Appel, B., Y. Tokiwa, and M. Haik (1981), Sampling of nitrates in ambient air, *Atmos. Environ.*, *15*, 283–289, doi:10.1016/0004-6981(81)90029-9.

Atlas, E., and B. A. Ridley (1996), The Mauna Loa Observatory Photochemistry Experiment: Introduction, *J. Geophys. Res.*, *101*(D9), 14,531–14,541, doi:10.1029/96JD01203.

Chin, M., D. J. Jacob, J. W. Munger, D. D. Parrish, and B. G. Doddridge (1994), Relationship of ozone and carbon monoxide over North America, *J. Geophys. Res.*, *99*(D7), 14,565–14,573, doi:10.1029/94JD00907.

Crutzen, P. (1973), A discussion of the chemistry of some minor constituents in the stratosphere and troposphere, *Pure Appl. Geophys.*, *106*, 1385–1399, doi:10.1007/BF00881092.

Ding, A., and T. Wang (2006), Influence of stratosphere to troposphere exchange on the seasonal cycle of surface ozone at Mount Waliguan in western China, *Geophys. Res. Lett.*, *33*, L03803, doi:10.1029/2005GL024760.

Duderstadt, K., et al. (1998), Photochemical production and loss rates of ozone at Sable Island, Nova Scotia during the North Atlantic Regional Experiment (NARE) 1993 summer intensive, *J. Geophys. Res.*, *103*(D11), 13,531–13,555, doi:10.1029/98JD00397.

Fischer, H., et al. (2003), Ozone production and trace gas correlations during the June 2000 MINATROC intensive measurement campaign at Mt. Cimone, *Atmos. Chem. Phys.*, *3*, 725–738, doi:10.5194/acp-3-725-2003.

Fitz, D., K. Bumiller, and A. Lashgari (2003), Measurement of NO_y during the SCOS97 NARSTO, *Atmos. Environ.*, *37*, Suppl. 2, 119–134, doi:10.1016/S1352-2310(03)00385-6.

Ford, K. M., B. M. Campbell, P. B. Shepson, S. B. Bertman, R. E. Honrath, M. Peterson, and J. E. Dibb (2002), Studies of peroxyacetyl nitrate (PAN) and its interaction with the snowpack at Summit, Greenland, *J. Geophys. Res.*, *107*(D10), 4102, doi:10.1029/2001JD000547.

Gao, Y., and J. R. Anderson (2001), Characteristics of Chinese aerosols determined by individual particle analysis, *J. Geophys. Res.*, *106*(D16), 18,037–18,045, doi:10.1029/2000JD900725.

Goldan, P., W. C. Kuster, D. L. Albritton, F. C. Fehsenfeld, P. S. Connell, R. B. Norton, and B. J. Huebert (1983), Calibration and tests of the filter collection method for measuring clean air, ambient levels of nitric acid, *Atmos. Environ.*, *17*, 1355–1364, doi:10.1016/0004-6981(83)90410-9.

Hirsch, R. M., and E. J. Gilroy (1984), Methods of fitting a straight line to data: Examples in water resources, *Water Resour. Bull.*, *20*(5), 705–711.

Intergovernmental Panel on Climate Change (2007), *Climate Change 2007: The Physical Science Basis. Contribution of Working Group I to the Fourth Assessment Report of the Intergovernmental Panel on Climate Change*, Cambridge Univ. Press, Cambridge, U. K.

Jaeglé, L., D. J. Jacob, Y. Wang, A. J. Weinheimer, B. A. Ridley, T. L. Campos, G. W. Sachse, and D. E. Hagen (1998), Sources and chemistry of NO_x in the upper troposphere over the United States, *Geophys. Res. Lett.*, *25*(10), 1705–1708, doi:10.1029/97GL03591.

Kivekäs, N., J. Sun, M. Zhan, V. M. Kerminen, A. Hyvärinen, M. Komppula, Y. Viisanen, N. Hong, Y. Zhang, M. Kulmala, X. C. Zhang, Deli Geer, and H. Lihavainen (2009), Long term particle size distribution measurements at Mount Waliguan, a high altitude site in inland China, *Atmos. Chem. Phys.*, *9*, 5461–5474, doi:10.5194/acp-9-5461-2009.

Li, J., Z. Wang, H. Akimoto, J. Tang, and I. Uno (2009), Modeling of the impacts of China's anthropogenic pollutants on the surface ozone summer maximum on the northern Tibetan Plateau, *Geophys. Res. Lett.*, *36*, L24802, doi:10.1029/2009GL041123.

Li, S., J. Tang, H. Xue, and D. Toom Sauntry (2000), Size distribution and estimated optical properties of carbonate, water soluble organic carbon, and sulfate in aerosols at a remote high altitude site in western China, *Geophys. Res. Lett.*, *27*(8), 1107–1110, doi:10.1029/1999GL010929.

Ma, J., J. Tang, S. M. Li, and M. Z. Jacobson (2003), Size distributions of ionic aerosols measured at Waliguan Observatory: Implication for nitrate gas to particle transfer processes in the free troposphere, *J. Geophys. Res.*, *108*(D17), 4541, doi:10.1029/2002JD003356.

Ma, J., X. Zheng, and X. Xu (2005), Comment on "Why does surface ozone peak in summertime at Waliguan?" by Bin Zhu et al., *Geophys. Res. Lett.*, *32*, L01805, doi:10.1029/2004GL021683.

Meng, Z., X. B. Xu, T. Wang, X. Y. Zhang, X. L. Yu, S. F. Wang, W. L. Lin, Y. Z. Chen, Y. A. Jiang, and X. Q. An (2010), Ambient sulfur dioxide, nitrogen dioxide, and ammonia at ten background and rural sites in China during 2007–2008, *Atmos. Environ.*, *44*, 2625–2631, doi:10.1016/j.atmosenv.2010.04.008.

Monks, P. S. (2000), A review of the observations and origins of the spring ozone maximum, *Atmos. Environ.*, *34*, 3545–3561, doi:10.1016/S1352 2310(00)00129 1.

Mu, Y., X. Pang, J. Quan, and X. Zhang (2007), Atmospheric carbonyl compounds in Chinese background area: A remote mountain of the Qinghai Tibetan Plateau, *J. Geophys. Res.*, *112*, D22302, doi:10.1029/2006JD008211.

Munger, J. W., D. J. Jacob, S. M. Fan, A. S. Colman, and J. E. Dibb (1999), Concentrations and snow atmosphere fluxes of reactive nitrogen at Summit, Greenland, *J. Geophys. Res.*, *104*(D11), 13,721–13,734, doi:10.1029/1999JD900192.

National Research Council (1991), *Rethinking the Ozone Problem in Urban and Regional Air Pollution*, Natl. Acad. Press, Washington, D. C.

Neuman, J. A., et al. (2006), Reactive nitrogen transport and photochemistry in urban plumes over the North Atlantic Ocean, *J. Geophys. Res.*, *111*, D23S54, doi:10.1029/2005JD007010.

Ogi, M., J. Yamazaki, and Y. Tachibana (2005), The summer northern annular mode and abnormal summer weather in 2003, *Geophys. Res. Lett.*, *32*, L04706, doi:10.1029/2004GL021528.

Qinghai Bureau of Statistics (2007), *Qinghai Statistical Yearbook 2007* (in Chinese), China Stat. Press, Beijing.

Saigusa, N., et al. (2010), Impact of meteorological anomalies in the 2003 summer on gross primary productivity in East Asia, *Biogeosciences*, *7*, 641–655, doi:10.5194/bg-7-641-2010.

Singh, H. B., and L. J. Salas (1983), Peroxyacetyl nitrate in the free troposphere, *Nature*, *302*, 326–328, doi:10.1038/302326a0.

Singh, H. B., L. J. Salas, and W. Viezee (1986), Global distribution of peroxyacetyl nitrate, *Nature*, *321*, 588–591, doi:10.1038/321588a0.

Singh, H., et al. (1994), Summertime distribution of PAN and other reactive nitrogen species in the northern high latitude atmosphere of eastern Canada, *J. Geophys. Res.*, *99*(D1), 1821–1835, doi:10.1029/93JD00946.

Singh, H. B., et al. (1996), Reactive nitrogen and ozone over the western Pacific: Distribution, partitioning, and sources, *J. Geophys. Res.*, *101*(D1), 1793–1808, doi:10.1029/95JD01029.

Singh, H. B., et al. (2007), Reactive nitrogen distribution and partitioning in the North American troposphere and lowermost stratosphere, *J. Geophys. Res.*, *112*, D12S04, doi:10.1029/2006JD007664.

Song, C., and G. R. Carmichael (2001), A three dimensional modeling investigation of the evolution processes of dust and sea salt particles in East Asia, *J. Geophys. Res.*, *106*(D16), 18,131–18,154, doi:10.1029/2000JD900352.

Stohl, A., et al. (2003), Stratosphere troposphere exchange: A review, and what we have learned from STACCATO, *J. Geophys. Res.*, *108*(D12), 8516, doi:10.1029/2002JD002490.

Streets, D. G., et al. (2003), An inventory of gaseous and primary aerosol emissions in Asia in the year 2000, *J. Geophys. Res.*, *108*(D21), 8809, doi:10.1029/2002JD003093.

Talbot, R., et al. (2003), Reactive nitrogen in Asian continental outflow over the western Pacific: Results from the NASA Transport and Chemical

Evolution over the Pacific (TRACE P) airborne mission, *J. Geophys. Res.*, *108*(D20), 8803, doi:10.1029/2002JD003129.
Tang, J., Y. P. Wen, X. B. Xu, X. D. Zheng, S. Guo, and Y. C. Zhao (1995), China Global Atmosphere Watch Baseline Observatory and its measurement program, in *CAMS Annual Report 1994–95*, pp. 56–65, China Meteorol. Press, Beijing.
Trainer, M., et al. (1993), Correlation of ozone with NO_y in photochemically aged air, *J. Geophys. Res.*, *98*(D2), 2917–2925, doi:10.1029/92JD01910.
Val Martin, M., R. E. Honrath, R. C. Owen, and Q. B. Li (2008), Seasonal variation of nitrogen oxides in the central North Atlantic lower free troposphere, *J. Geophys. Res.*, *113*, D17307, doi:10.1029/2007JD009688.
Wang, T., V. T. F. Cheung, K. S. Lam, G. L. Kok, and J. M. Harris (2001), The characteristics of ozone and related compounds in the boundary layer of the South China coast: Temporal and vertical variations during autumn season, *Atmos. Environ.*, *35*, 2735–2746, doi:10.1016/S1352-2310(00)00411-8.
Wang, T., H. L. A. Wong, J. Tang, A. Ding, W. S. Wu, and X. C. Zhang (2006), On the origin of surface ozone and reactive nitrogen observed at a remote mountain site in the northeastern Qinghai Tibetan Plateau, western China, *J. Geophys. Res.*, *111*, D08303, doi:10.1029/2005JD006527.
Wang, T., et al. (2010), Air quality during the 2008 Beijing Olympics: Secondary pollutants and regional impact, *Atmos. Chem. Phys.*, *10*, 7603–7615, doi:10.5194/acp-10-7603-2010.
Wang, Y., J. A. Logan, and D. J. Jacob (1998), Global simulation of tropospheric O_3 NO_x hydrocarbon chemistry: 2. Model evaluation and global ozone budget, *J. Geophys. Res.*, *103*(D9), 10,727–10,755, doi:10.1029/98JD00157.
Williams, E. J., D. D. Parrish, M. P. Buhr, F. C. Fehsenfeld, and R. Fall (1988), Measurement of soil NO_x emissions in central Pennsylvania, *J. Geophys. Res.*, *93*(D8), 9539–9546, doi:10.1029/JD093iD08p09539.
Williams, E. J., J. M. Roberts, K. Baumann, S. B. Bertman, S. Buhr, R. B. Norton, and F. C. Fehsenfeld (1997), Variations in NO_y composition at Idaho Hill, Colorado, *J. Geophys. Res.*, *102*(D5), 6297–6314, doi:10.1029/96JD03252.
Williams, E. J., et al. (1998), Intercomparison of ground based NO_y measurement techniques, *J. Geophys. Res.*, *103*(D17), 22,261–22,280, doi:10.1029/98JD00074.
Wood, E. C., et al. (2009), A case study of ozone production, nitrogen oxides, and the radical budget in Mexico City, *Atmos. Chem. Phys.*, *9*, 2499–2516, doi:10.5194/acp-9-2499-2009.
Xue, H. S. (2002), Observational study of chemical characteristics of aerosol and trace gases of HNO_3, SO_2, at Mt. Waliguan (in Chinese), Master's thesis, Beijing Univ., Beijing.
Yienger, J., and H. Levy II (1995), Empirical model of global soil biogenic NO_x emissions, *J. Geophys. Res.*, *100*(D6), 11,447–11,464, doi:10.1029/95JD00370.
Zanis, P., A. Ganser, C. Zellweger, S. Henne, M. Steinbacher, and J. Staehelin (2007), Seasonal variability of measured ozone production efficiencies in the lower free troposphere of central Europe, *Atmos. Chem. Phys.*, *7*, 223–236, doi:10.5194/acp-7-223-2007.
Zellweger, C., M. Ammann, B. Buchmann, P. Hofer, M. Lugauer, R. Rüttimann, N. Streit, E. Weingartner, and U. Baltensperger (2000), Summertime NO_y speciation at the Jungfraujoch, 3580 m above sea level, Switzerland, *J. Geophys. Res.*, *105*(D5), 6655–6667, doi:10.1029/1999JD901126.
Zellweger, C., J. Forrer, P. Hofer, S. Nyeki, B. Schwarzenbach, E. Weingartner, M. Ammann, and U. Baltensperger (2003), Partitioning of reactive nitrogen (NO_y) and dependence on meteorological conditions in the lower free troposphere, *Atmos. Chem. Phys.*, *3*, 779–796, doi:10.5194/acp-3-779-2003.
Zhang, J., et al. (2009), Continuous measurement of peroxyacetyl nitrate (PAN) in suburban and remote areas of western China, *Atmos. Environ.*, *43*, 228–237, doi:10.1016/j.atmosenv.2008.09.070.
Zhang, Q., et al. (2007), NO_x emission trends for China, 1995–2004: The view from the ground and the view from space, *J. Geophys. Res.*, *112*, D22306, doi:10.1029/2007JD008684.
Zhang, Q., et al. (2009), Asian emissions in 2006 for the NASA INTEX B mission, *Atmos. Chem. Phys.*, *9*, 5131–5153, doi:10.5194/acp-9-5131-2009.
Zhou, L., D. E. J. Worthy, P. M. Lang, M. K. Ernst, X. C. Zhang, Y. P. Wen, and J. L. Li (2004), Ten years of atmospheric methane observations at a high elevation site in western China, *Atmos. Environ.*, *38*, 7041–7054, doi:10.1016/j.atmosenv.2004.02.072.
Zhu, B., H. Akimoto, Z. Wang, K. Sudo, J. Tang, and I. Uno (2004), Why does surface ozone peak in summertime at Waliguan?, *Geophys. Res. Lett.*, *31*, L17104, doi:10.1029/2004GL020609.

Deliger, China GAW Baseline Observatory, Qinghai Meteorological Bureau, Xining, Qinghai 810001, China.
A. J. Ding, Institute for Climate and Global Change Research, Nanjing University, Nanjing, Jiangsu 210093, China.
C. N. Poon, T. Wang, W. S. Wu, and J. M. Zhang, Department of Civil and Structural Engineering, Hong Kong Polytechnic University, Hung Hom, Kowloon, Hong Kong, China. (cetwang@polyu.edu.hk)
J. Tang and X. C. Zhang, Centre for Atmosphere Watch and Services, Key Laboratory for Atmospheric Chemistry, Chinese Academy of Meteorological Sciences, Beijing 100081, China.
W. X. Wang, L. K. Xue, Q. Z. Zhang, and X. H. Zhou, Environment Research Institute, Shandong University, Ji'nan, Shandong 250100, China.

Environ. Sci. Technol. **2008,** *42,* 7301–7308

A Quantum Mechanical Study on the Formation of PCDD/Fs from 2-Chlorophenol as Precursor

QINGZHU ZHANG,* SHANQING LI, XIAOHUI QU, XIANGYAN SHI, AND WENXING WANG*

Environment Research Institute, Shandong University, Jinan 250100, P. R. China

Received June 11, 2008. Accepted August 01, 2008.. Revised manuscript received August 01, 2008

The most direct route to the formation of polychlorinated dibenzo-*p*-dioxins and dibenzofurans (PCDD/Fs) in combustion and thermal processes is the gas-phase reaction of chemical precursors such as chlorinated phenols. Detailed insight into the mechanism and kinetics properties is a prerequisite for understanding the formation of PCDD/Fs. In this paper, we carried out molecular orbital theory calculations for the homogeneous gas-phase formation of PCDD/Fs from 2-chlorophenol (2-CP). The profiles of the potential energy surface were constructed, and the possible formation pathways are discussed. The single-point energy calculation was carried out at the MPWB1K/6−311+G(3f,2p) level. Several energetically favorable formation pathways were revealed for the first time. The rate constants of crucial elementary steps were deduced over a wide temperature range of 600~1200 K using canonical variational transition-state theory (CVT) with small curvature tunneling contribution (SCT). The rate-temperature formulas were fitted. The ratio of PCDD to PCDF formed shows strong dependency on the reaction temperature and chlorophenoxy radicals (CPRs) concentration.

1. Introduction

Polychlorinated dibenzo-*P*-dioxins (PCDDs) and dibenzofurans (PCDFs) have been at the forefront of public and regulatory concern because of their known toxicity, persistence and bioaccumulation. Major sources of PCDD/Fs in the environment are the combustion of waste materials as well as many other high-temperature processes commonly used in industrial settings. Reports of PCDD/F emissions from municipal waste incinerators have caused much public alarm and led to severe difficulties in constructing both municipal and hazardous waste incinerators.

The most direct route to the formation of PCDD/Fs is the gas-phase reaction of chemical precursors (*1–3*). Chlorophenols (CPs) are the most direct precursors of dioxin and among the most abundant aromatic compounds found in incinerator gas emissions (*4–6*). The homogeneous gas-phase formation of PCDD/Fs from CPs was suggested to make a significant contribution to the observed PCDD/F yields in full-scale incinerators (*7, 8*). The understanding of the reaction mechanism is crucial for any attempt to prevent PCDD/F formation. To obtain more insight into the actual formation pathways and explain the results of experimental observations, we initiated a theoretical study, employing a high-accuracy quantum chemical method, for the formation of PCDD/Fs using chlorophenol as precursor. Monochlorophenol (2-chlorophenol) was selected as a model precursor.

Formation of PCDD/Fs from the 2-chlorophenol (2-CP) precursor has been extensively studied experimentally, and continues to receive considerable attention. Evans (*9*) studied the thermal degradation of 2-CP under pyrolytic conditions. Dibenzo-p-dioxin (DD), dibenzofuran (DF), and 1-chlorodibenzo-p-dioxin (1-MCDD) were observed between 575 and 900 °C, but 4,6-dichlorodibenzofuran (4,6-DCDF) was not detected. In other work (*10*), Evans studied the oxidative degradation of 2-CP. Observed products in order of yield are as follows: 4,6-DCDF > DD > 1-MCDD, 4-MCDF, DF. In 2002, Sidhu and Edwards (*11*) investigated the role of phenoxy radicals in PCDD/Fs formation by studying the oxidation of 2-CP at different precursor concentrations: 4, 33, and 684 ppm. Observation shows that the concentration of radicals present in the oxidation system has a significant effect on the PCDD/F product distribution and ultimately the PCDD/PCDF ratio. The dominant PCDD/F product was found to be 1-MCDD at low inlet concentration (4 ppm) of 2-CP, whereas the major product of oxidation is 4,6-DCDF at high inlet concentration (33 and 684 ppm) of precursor. Several possible formation mechanisms were proposed to explain the observed products. In a recent paper, Altarawneh (*12*) carried out quantum chemical investigation into the formation mechanism of PCDD/Fs. Two pathways for molecule−molecule interactions, three pathways for molecule-radical interactions, and nine pathways for radical−radical interactions were investigated for the formation of DD, 1-MCDD, 4,6-DCDF, and 4-MCDF from the different possible combinations of 2-CP and 2-chlorophenoxy radical (2-CPR). The study of Altarawneh (*12*) indicated that 1-MCDD should be the favored product in 2-CP pyrolysis, in agreement with experimental findings. However, because the proposed formation pathways of 4,6-DCDF requires crossing a large activation barrier (53.0 and 48.1 kcal/mol), Altarawneh could not give a reasonable explanation of the experimental observation of Evans (*10*) that 4,6-DCDF is the major product under oxidative conditions.

The kinetic models that account for the contribution of the gaseous route in the production of PCDD/Fs in combustion processes use the rate constants of the elementary reactions (*13*). However, due to the absence of direct experimental and theoretical values, the rate constants of many elementary steps were assigned to be the values reported in the literature for analogous reactions (*13, 14*). However, where there are uncertainties, the numerical values have been adjusted somewhat to bias the mechanism in favor of PCDD/Fs formation, i.e., worst case modeling (*13, 14*). In this study, the rate constants of crucial elementary reactions, especially the controlling step in the dioxin formation, have been deduced using canonical variational transition-state theory (CVT) (*15*) with small curvature tunneling (SCT) (*16*) contribution.

2. Computational Methods

High-accuracy molecular orbital calculations were carried out using the Gaussian 03 package (*17*) on an SGI Origin 2000 supercomputer. The geometrical parameters were optimized at the MPWB1K/6-31+G(d,p) level. The MPWB1K method (*18*) is an HDFT model with excellent performance for thermochemistry, thermochemical kinetics, hydrogen bonding, and weak interactions. The vibrational frequencies

* Address correspondence to aither author. E-mail: zqz@sdu.edu.cn (Q.Z.); wxwang@sdu.edu.cn (W.W.). Fax: 86-531-8836 9788 (Q.Z. and W.W.).

10.1021/es801599n CCC: $40.75
Published on Web 09/04/2008

were also calculated at the same level in order to determine the nature of the stationary points, the zero-point energy (ZPE), and the thermal contributions to the free energy of activation. Each transition state was verified to connect the designated reactants with products by performing an intrinsic reaction coordinate (IRC) analysis (*19*). For a more accurate evaluation of the energetic parameters, a more flexible basis set, 6-311+G(3df,2p), was employed to determine the energies of the various species. By means of the Polyrate 9.3 program (*20*), direct dynamics calculations were carried out using the canonical variational transition state theory (CVT) (*15*) with small curvature tunneling contribution (SCT) (*16*).

3. Results and Discussion

To clarify the reliability of the theoretical calculations, we optimized the geometries and calculated the vibrational frequencies of phenol and chlorobenzene. The results at the MPWB1K/6-31+G(d,p) level agree well with the available experimental values, and the maximum relative error is less than 0.9% for the geometrical parameters and less than 8.0% for the vibrational frequencies. From this result, it might be inferred that the same accuracy could be expected for the species involved in the formation of PCDD/Fs from 2-CP. Furthermore, previous study shows that MPWB1K is an excellent method for prediction of transition state geometries (*18*).

3.1. Formation of 2-CPRs. The homogeneous gas-phase formation of PCDD/Fs from 2-CP was proposed that involve chlorophenoxy radical−radical coupling, radical−molecule recombination of chlorophenoxy and chlorophenol. Previous research (*12, 21*) has shown that the radical−molecule pathway requires chlorine and hydroxyl displacement as first steps, and these steps are not energetically favored. So the radical−molecule pathway is not competitive with the radical−radical pathway. The dimerization of 2-chlorophenoxy radicals (2-CPRs) is the major pathway for the formation of PCDD/Fs. Thus, we first studied the formation of 2-CPRs. In municipal waste incinerators, 2-CPRs can be formed through loss of the phenoxyl-hydrogen via unimolecular, bimolecular, or possibly other low-energy pathways (including heterogeneous reactions). The unimolecular reaction includes the decomposition of 2-CP with the cleavage of the O−H bond. The bimolecular reactions include attack by H or Cl under pyrolytic conditions and by H, OH, O (^{3}P) or Cl under high-temperature oxidative conditions.

For the unimolecular decomposition of 2-CP, breaking the O−H bond is a highly endothermic process. There is no transition state in the reaction because the potential curve is attractive along the O−H distance. The bimolecular reactions of 2-CP with H, OH, O (^{3}P), and Cl proceed via a direct hydrogen abstraction mechanism. The potential barriers (ΔE) and the reaction enthalpies (ΔH, 0 K) were calculated at the MPWB1K/6−311+G(3df,2p) level. For the reaction with Cl, the energy of the transition state without ZPE correction is 0.32 kcal/mol higher than the total energy of the reactants. However, the energy of the transition state including ZPE is 2.32 kcal/mol lower than the total energy of the reactants.

$$\text{2-CP} \rightarrow \text{2-CPR} + \text{H} \qquad \Delta H = 85.91 \text{ kcal/mol} \qquad (1)$$

$$\text{2-CP} + \text{H} \rightarrow \text{2-CPR} + \text{H}_2 \qquad \Delta E = 13.80 \text{ kcal/mol} \quad \Delta H = -12.01 \text{ kcal/mol} \qquad (2)$$

$$\text{2-CP} + \text{OH} \rightarrow \text{2-CPR} + \text{H}_2\text{O} \qquad \Delta E = 3.20 \text{ kcal/mol} \quad \Delta H = -26.91 \text{ kcal/mol} \qquad (3)$$

$$\text{2-CP} + \text{O}(^3\text{P}) \rightarrow \text{2-CPR} + \text{OH} \qquad \Delta E = 7.51 \text{ kcal/mol} \quad \Delta H = -11.35 \text{ kcal/mol} \qquad (4)$$

$$\text{2-CP} + \text{Cl} \rightarrow \text{2-CPR} + \text{HCl} \qquad \Delta E = -2.32 \text{ kcal/mol} \quad \Delta H = -14.96 \text{ kcal/mol} \qquad (5)$$

3.2. Formation of PCDD/Fs from 2-CPRs. *3.2.1. Formation of PCDDs from 2-CPRs.* Three PCDD congeners, DD, 1-MCDD, and 1,6-DCDD, can be experimentally observed in the oxidation and pyrolysis of 2-CP (*9−11*). Four possible formation pathways, depicted in Figure 1a, are proposed for the formation of three dioxins in this study. The formation pathways involve four elementary processes: dimerization of 2-CPRs, Cl, or H abstraction, ring closure, and intra-annular elimination of Cl or H. In pathway 1 and pathway 3, the ring closure and elimination of Cl occur in a one-step reaction. All of the Cl or H abstraction processes are highly exothermic with low-energy barriers. The ring closure process involves a high barrier and is strongly endoergic, and it is the rate determining step for pathways 1 and 3. Unimolecular elimination of H is the kinetically controlling process for pathways 2 and 4 due to the high potential barrier. The thermodynamically favored routes to PCDDs formation are pathways 1 and 3. DD and 1-MCDD are the major PCDD products, in agreement with experimental findings (*9−12*).

The formation of DD and 1-MCDD were also studied by Altarawneh (*12*). For a comparison, the formation mechanism proposed by Altarawneh is presented in Figure 1b. Comparison of Figure 1a and b shows that the pathways proposed by Altarawneh (*12*) are similar to pathways 1 and 3. But two obvious differences were observed: (1) The coupling of the oxygen-centered radical mesomer with the ortho carbon−hydrogen centered radical mesomer requires crossing a potential barrier of 9.0 kcal/mol in the study of Altarawneh (*12*), whereas this process is barrierless in our study; (2) The potential barriers reported by Altarawneh are much lower than the values obtained by us for all of the Cl or H abstraction processes. For example, the potential barrier of Cl abstraction from IM1 (denoted as D8 in the study of Altarawneh (*12*)) by H atom is 0.7 kcal/mol in the study of Altarawneh (*12*), whereas the value we obtained is 6.69 kcal/mol. The discrepance may arise from the different calculation levels. The energies of Altarawneh (*12*) are calculated at the B3LYP level, while our energy calculations are carried out at the MPWB1K level. It is well-known that the B3LYP method systematically underestimates barrier heights (*18*). The kinetic calculations are most sensitive to the energies. In order to justify the performance of the MPWB1K method for the potential barriers, we have carried out additional potential barrier calculations employing higher level, G3MP2, for the Cl abstraction from IM1 by H atom. The G3MP2 calculation is rather expensive and extremely computationally demanding. It took more than 12 days to calculate the G3MP2 energy of TS1 on a personal computer with an Intel Core 2 Duo E6750 processor and 4GB main memory. The scratch file used up to 80 GB of disk space. The similar calculation using MPWB1K/6-311+G(3df,2p) requires less than 11 h. The two methods, MPWB1K/6-311+G(3df,2p) and G3MP2, produce consistent potential barriers within 0.9 kcal/mol.

In order to further compare with the formation mechanism proposed by Altarawneh (*12*), we checked the structures of reactants, transition states, intermediates, and products involved in the formation pathways of DD and 1-MCDD. Figure 2 shows the configurations of dimerization products (IM1 versus D8, IM2 versus D5), intermediates (IM3 versus preD3, IM5 versus preD5), and transition states (TS4 versus TS(preD3 → DD+Cl) located by Altarawneh (*12*) and us. The configurations of D8, preD3, D5, PreD5, and TS(preD3 → DD+Cl) are from the Cartesian coordinates offered by Altarawneh in the Supporting Information (*12*). It can be seen from Figure 2 that the configurations of IM1, IM2, IM3, and IM5 are obviously different from the configurations of D8, preD3, D5, and PreD5 (*12*). In IM1, IM2, IM3, and IM5, C(1) atom is at the cis-position of C(3) with respect to the C(2)−O bond, and C(2) atom is at the cis-position of C(4) atom with respect to the O−C(3) bond. This cis-configuration

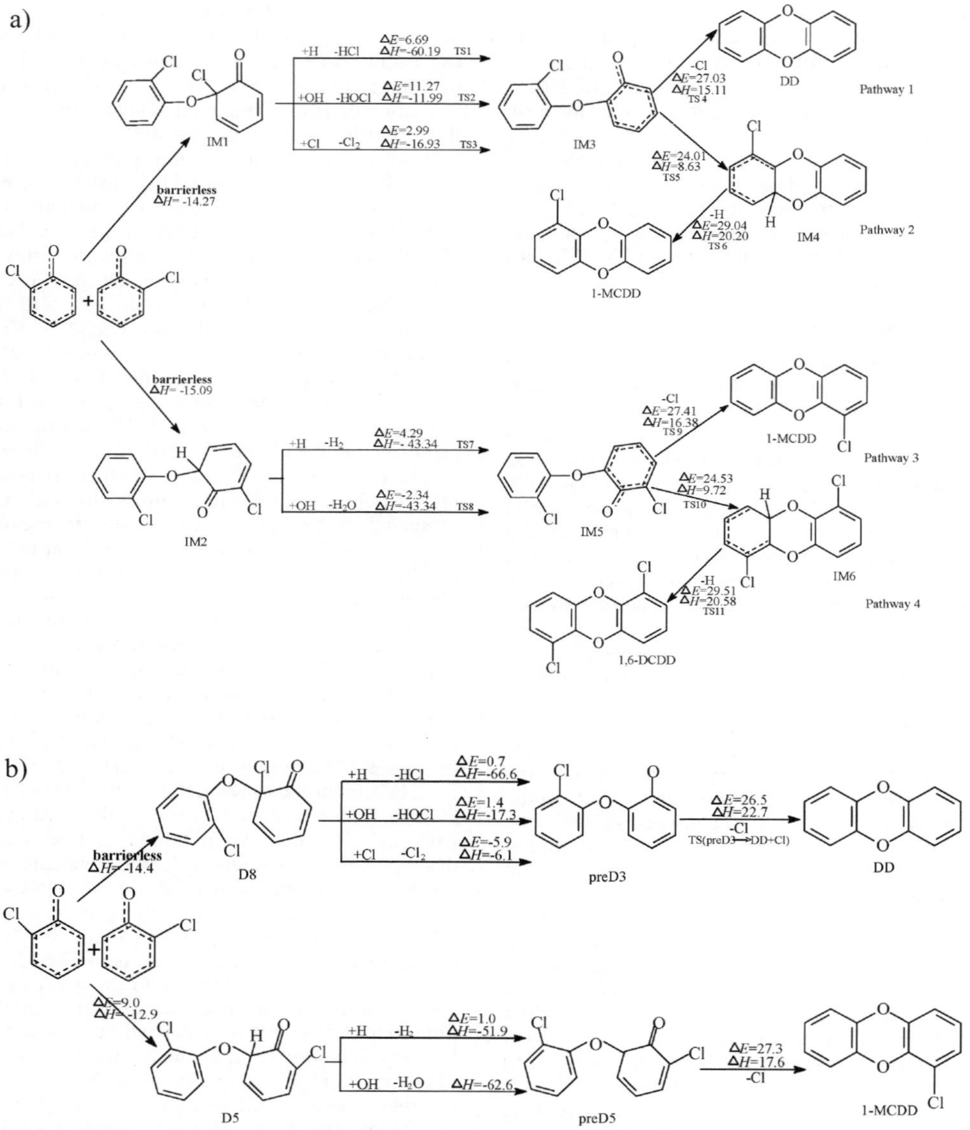

FIGURE 1. a. Formation routes of PCDDs from the 2-CP precursor. Δ*H* is calculated at 0 K. b. Formation routes of DD and 1-MCDD proposed by Altarawneh (*12*).

easily proceeds to the ring closure. However, in D8, preD3, D5, and PreD5, C(1) atom is at the trans-position of C(3), and C(2) atom is at the trans-position of C(4). The ring will be difficult to close from this trans-configuration due to the large steric hindrance. Figure 2 shows that the transition states (TS4 and TS(preD3 → DD+Cl)) of the ring closure process have the same configurations. The difference of the forming C(1)−O bond length (1.943 versus 1.935 Å) in the two transition states results from the different calculation levels. IRC calculation verified that the transition state (TS4 or TS(preD3 → DD+Cl)) connects to the intermediate IM3 not preD3. Therefore, the cis-formed IM1, IM2, IM3, and IM5 are reasonable intermediates for the formation of DD and 1-MCDD.

3.2.2. Formation of PCDFs from 2-CPRs. Experimental observation (*10*) shows that 4,6-DCDF is the major PCDF under oxidation of 2-CP. Two reaction routes were proposed by Altarawneh (*12*) to account for its formation. However, because the two formation routes require crossing large barriers (53.0 and 48.1 kcal/mol), Altarawneh could not give a reasonable explanation for the experimental observation of Evans (*10*) that 4,6-DCDF is the major product rather than DD and 1-MCDD under oxidative conditions. Therefore, there may be other possible 4,6-DCDF formation mechanisms. In this study, two new reaction pathways are proposed for the formation of 4,6-DCDF, as presented in Figure 3.

The first formation route involves five elementary steps: coupling of 2-CPRs, H abstraction by H atom or OH radical, tautomerization (H-shift), ring closure, and elimination of OH. The ortho−ortho coupling has a potential barrier of 8.27 kcal/mol. Both H abstraction processes have low barriers and are strongly exothermic. In the study of Altarawneh (*12*), the H-shift process has a significant barrier of 48.1 kcal/mol. In our study, H-shift could proceed via two different transition states: a five-membered ring transition state, denoted TS15, and a four-membered ring transition state, denoted TS16. The structures of TS15 and TS16 are shown in Figure 4. The energy of TS15 is 30.93 kcal/mol lower than that of TS16. Thus, the H-shift step via TS15 has a low potential barrier, only 18.45 kcal/mol. The ring closure process involves a high barrier of 29.17 kcal/mol and is strongly endoergic by 9.56

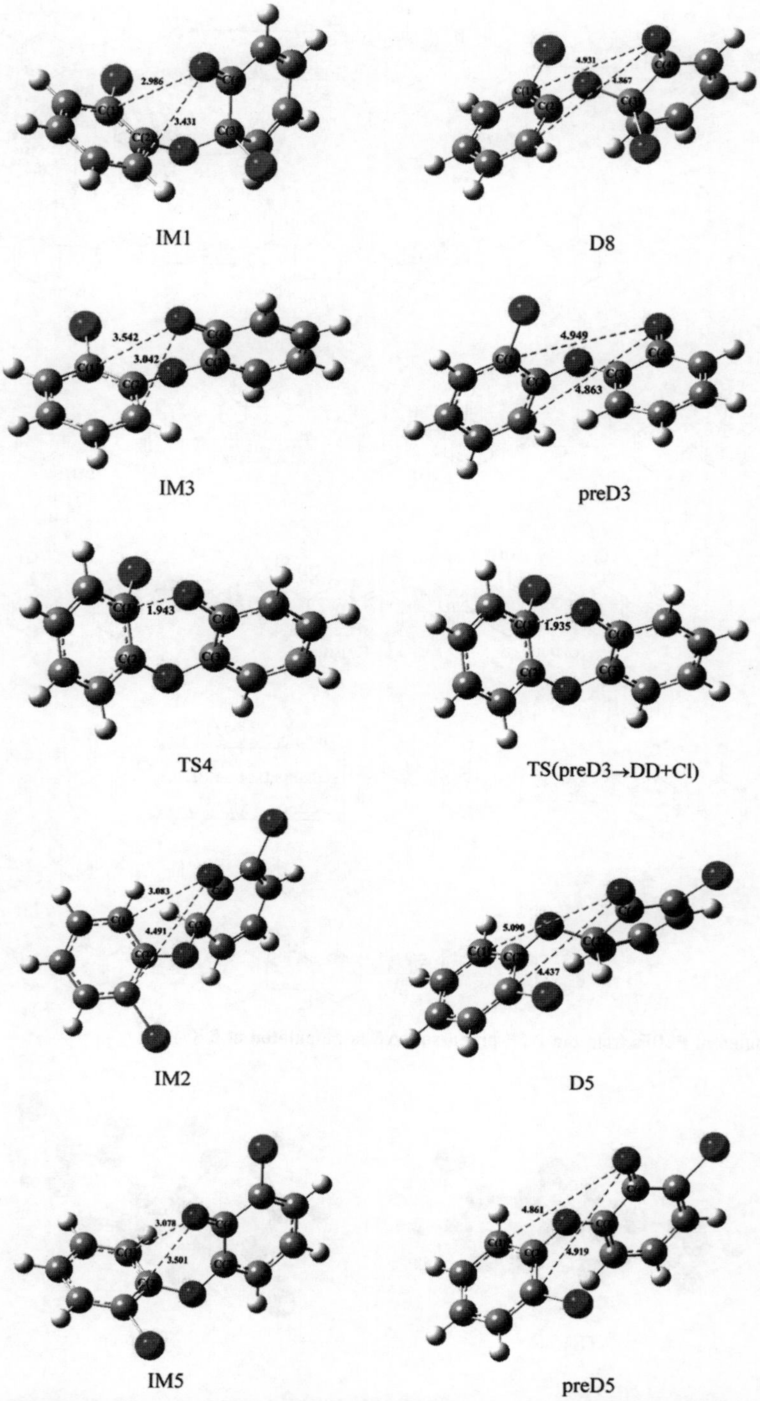

FIGURE 2. Configurations of dimerization products (IM1 versus D8, IM2 versus D5), intermediates (IM3 versus preD3, IM5 versus preD5), and transition states (TS4 versus TS(preD3 → DD+Cl)) located by Altarawneh and in this work.

kcal/mol, which is the rate controlling step for the formation of 4,6-DCDF in our study.

Similar to the first formation route of 4,6-DCDF, the second one also involves five elementary processes: dimerization of 2-CPRs, tautomerization (double H-transfer), H abstraction, ring closure, and OH desorption. In the study of Altarawneh (*12*), H transfer reactions from the keto–keto dimer IM7 to the enol–enol dimer IM11 involved two elementary steps. The first step is a single H atom transfer to the neighboring oxygen keto atom to form an enol–keto dimer. This process has a very high barrier of 53.0 kcal/mol. Transformation of the enol–keto dimer to the enol–enol dimer could proceed via two different energy barriers, 18.7 and 42.1 kcal/mol. In our study, the double hydrogen atom migration to the keto oxygen atoms can occur in a one-step reaction via the transition state TS19. The structure of TS19 is shown in Figure 4. The calculated vibrational frequencies contained only one imaginary component, 1546i cm^{-1}, confirming the first-order saddle point configuration. This concerted double hydrogen atom transfer process involves

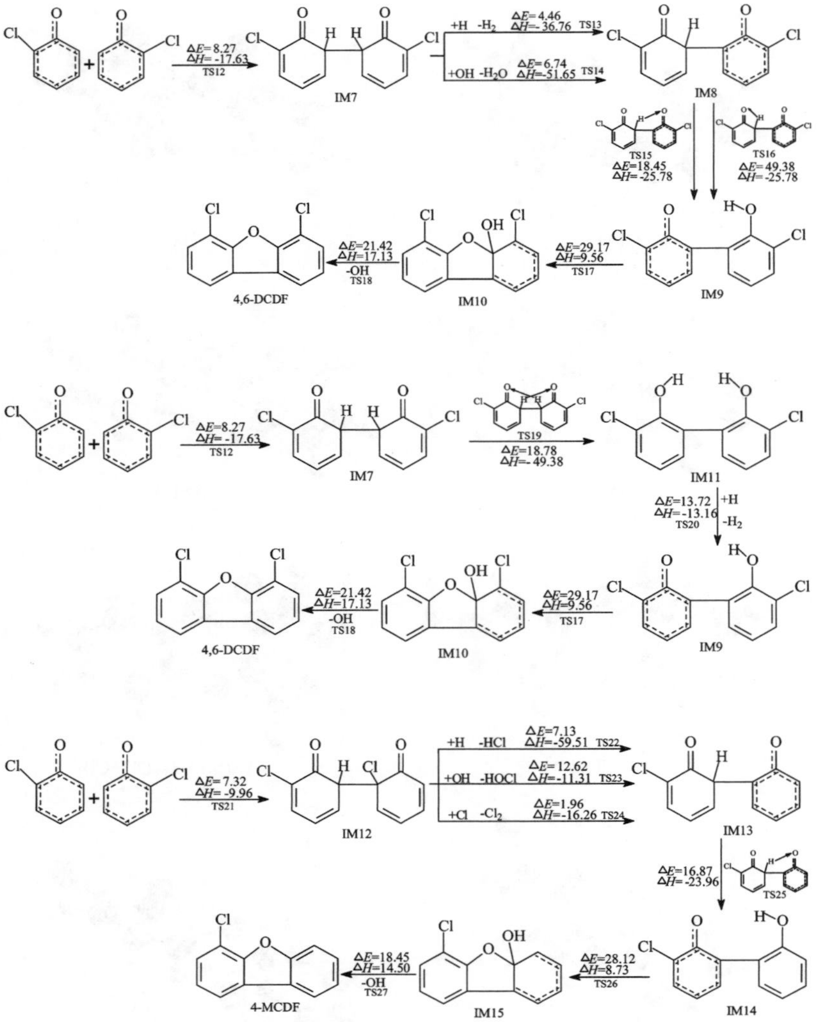

FIGURE 3. Formation routes of PCDFs from the 2-CP precursor. Δ*H* is calculated at 0 K.

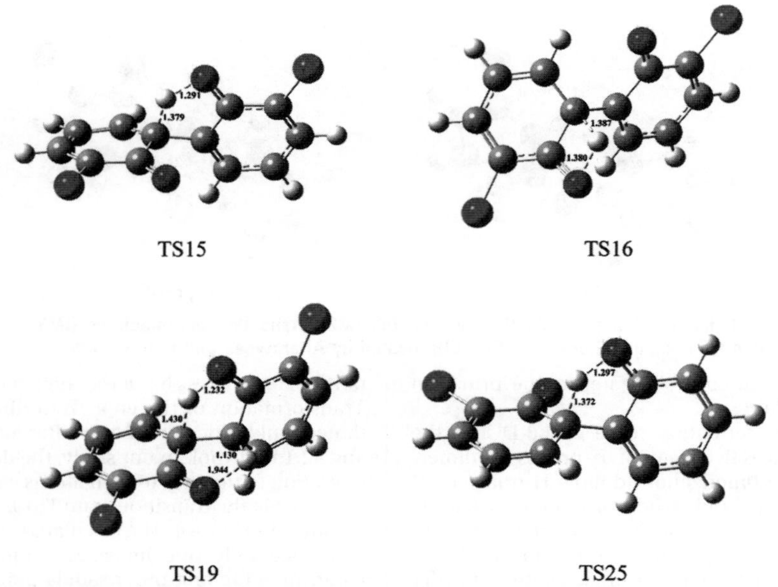

FIGURE 4. Configurations of the transition states of H-shift involved in the formation of PCDFs.

a barrier of 18.78 kcal/mol and is strongly exothermic by 49.38 kcal/mol. Altarawneh's study (*12*) shows that direct dehydration of the enol–enol dimer IM11 to form 4,6-DCDF is energetically unfavorable, with a very large barrier, 68.7 kcal/mol. In the combustion environment, active radicals will greatly facilitate H abstraction from the enol–enol dimer IM11. Then, 4,6-DCDF could be formed through ring closure and OH desorption reactions.

4-MCDF has been observed in the oxidation of 2-CP (*11*). A possible formation mechanism is proposed in this study. Similar to the formation of 4,6-DCDF, tautomerization from the keto–keto dimer to the enol–enol dimer requires crossing a high barrier of 47.7 kcal/mol in the study of Altarawneh (*12*), whereas the process has a low barrier of 16.87 kcal/mol in our study. Another possible formation route of 4-MCDF is the recombination of 2-chlorophenoxy radical with phenoxy radical.

The *o,o'*-dihydroxybiphenyl (DOHB) intermediate IM7 can be regarded as a prestructure for 4,6-DCDF, and IM12 is a prestructure of 4-MCDF. Figure 3 shows that the formation of IM7 is more exoergic than IM12 formation (by 7.67 kcal/mol at 0 K). Thus, IM7 is thermodynamically more stable than IM12. This can be explained by the steric effect. The coupling of two carbon (hydrogen)-centered radical mesomers forms IM7. IM12 is formed by the recombination of the carbon (hydrogen)-centered radical mesomer with the carbon (chlorine)-centered radical mesomer. Cl atom is larger than H atom, and has a larger steric effect. The sterically demanding formation of DOHB is inhibited by the voluminous chlorine atoms. If both ortho-positions of the phenol are substituted with chlorine, the formation of PCDFs is almost completely inhibited.

Comparison of Figure 1a and Figure 3 shows that the oxygen–carbon couplings appear to be barrierless, whereas the ortho–ortho couplings have a barrier of more than 7 kcal/mol. The preferred formation routes of DD and 1-MCDD involve three elementary steps. The formation pathways of 4,6-DCDF involve five elementary processes. Thus, the formation of PCDDs is preferred over the formation of PCDFs that is consistent with the experimental data: DD and 1-MCDD are the major dioxin products rather than 4,6-DCDF in the pyrolysis of 2-CP and are also major products of oxidation at low inlet concentration of precursor. However, at high inlet concentration of precursor, the dominant PCDD/F product is 4,6-DCDF under the oxidation conditions (*10, 11*). This can be explained by the concentration of 2-CPRs based on the study of Sidhu (*11*). At high inlet concentration of precursor and under the oxidation conditions, the carbon-centered mesomer of 2-CPR increases more than that of the oxygen-centered mesomer, as the carbon-centered mesomer is a more stable form of 2-CPR (*22*). Since PCDF formation involves only the more stable carbon-centered mesomer of 2-CPR, the increase in PCDF yields is more than the observed increase in PCDD.

3.3. Kinetics Calculations. Canonical variational transition state theory (CVT) (*15*) with small-curvature tunneling (SCT) (*16*) correction has been successfully performed for several analogous reactions (*23, 24*) and is an efficient method to calculate the rate constants. In this study, we used this method to calculate the rate constants of crucial elementary reactions over a wide temperature range of 600~1200 K, which covers the possible formation temperature of PCDD/Fs in municipal waste incinerators. Due to the absence of the available experimental values, it is difficult to make a direct comparison of the calculated CVT/SCT rate constants with the experimental values for all the elementary reactions. An alternative approach to clarifying the reliability of the kinetics calculation is to compare the CVT/SCT rate constants with the available literature rate constant values for structurally similar compounds.

TABLE 1. Arrhenius Formulas (Units Are s^{-1} and cm^3 $Molecule^{-1}$ s^{-1} for Unimolecular and Bimolecular Reactions, Respectively) for Elementary Reactions Involved in the Formation of PCDD/Fs from the 2-CP Precursor over the Temperature Range of 600~1200 K.

reactions	Arrhenius formulas
2-CP+H → 2-CPR+H_2	$k(T) = (1.10 \times 10^{-11}) \exp(-6437.64/T)$
2-CP+OH → 2-CPR+H_2O	$k(T) = (3.43 \times 10^{-12}) \exp(-2661.33/T)$
2-CP+O(^{3}P) → 2-CPR+OH	$k(T) = (4.75 \times 10^{-12}) \exp(-3513.60/T)$
2-CP+Cl → 2-CPR+HCl	$k(T) = (2.09 \times 10^{-11}) \exp(118.87/T)$
IM1+H → IM3+HCl	$k(T) = (3.12 \times 10^{-12}) \exp(-2891.38/T)$
IM1+OH → IM3+HOCl	$k(T) = (1.02 \times 10^{-11}) \exp(-6442.22/T)$
IM1+Cl → IM3+Cl_2	$k(T) = (1.04 \times 10^{-11}) \exp(-1861.28/T)$
IM3 → DD+Cl	$k(T) = (3.58 \times 10^{11}) \exp(-13706.87/T)$
IM3 → IM4	$k(T) = (4.26 \times 10^{11}) \exp(-11885.95/T)$
IM4 → 1-MCDD+H	$k(T) = (2.35 \times 10^{13}) \exp(-15078.04/T)$
IM2+H → IM5+H_2	$k(T) = (2.00 \times 10^{-20}) \exp(-2885.70/T)$
IM2+OH → IM5+H_2O	$k(T) = (6.81 \times 10^{-21}) \exp(-487.51/T)$
IM5 → 1-MCDD+Cl	$k(T) = (5.05 \times 10^{11}) \exp(-13763.30/T)$
IM5 → IM6	$k(T) = (3.74 \times 10^{9}) \exp(-6677.26/T)$
IM6 → 1,6-DCDD+H	$k(T) = (2.22 \times 10^{13}) \exp(-15257.27/T)$
2-CPR+2-CPR → IM7	$k(T) = (3.86 \times 10^{-15}) \exp(-5527.18/T)$
IM7+H → IM8+H_2	$k(T) = (1.28 \times 10^{-11}) \exp(-2222.77/T)$
IM7+OH → IM8+H_2O	$k(T) = (1.14 \times 10^{-12}) \exp(-4313.22/T)$
IM8 → IM9 via TS15	$k(T) = (1.39 \times 10^{9}) \exp(-2074.71/T)$
IM8 → IM9 via TS16	$k(T) = (6.28 \times 10^{3}) \exp(-5740.21/T)$
IM9 → IM10	$k(T) = (2.11 \times 10^{12}) \exp(-14722.61/T)$
IM10 → 4,6-DCDF	$k(T) = (4.11 \times 10^{13}) \exp(-11176.13/T)$
IM7 → IM11	$k(T) = (3.71 \times 10^{10}) \exp(-6831.91/T)$
IM11+H → IM9+H2	$k(T) = (1.25 \times 10^{-13}) \exp(-106.32/T)$
2-CPR+2-CPR → IM12	$k(T) = (1.56 \times 10^{-16}) \exp(-4904.96/T)$
IM12+H → IM13+HCl	$k(T) = (9.27 \times 10^{-12}) \exp(-4210.06/T)$
IM12+OH → IM13+HOCl	$k(T) = (7.35 \times 10^{-12}) \exp(-7395.10/T)$
IM12+Cl → IM13+Cl_2	$k(T) = (3.29 \times 10^{-11}) \exp(-2140.44/T)$
IM13 → IM14	$k(T) = (2.11 \times 10^{10}) \exp(-4657.53/T)$
IM14 → IM15	$k(T) = (2.04 \times 10^{12}) \exp(-14193.42/T)$
IM15 → 4-MCDF+OH	$k(T) = (2.34 \times 10^{13}) \exp(-9511.18/T)$

There are no previous data for the reactions of 2-CP+H → 2-CPR+H_2 and 2-CP+OH → 2-CPR+H_2O. So, the calculated rate constants were compared with the literature values (*25–27*) for phenol → C_6H_5O+H_2 and phenol+OH → C_6H_5O+H_2O, respectively. The CVT/SCT rate constants for

2-CP+H → 2-CPR+H_2 and 2-CP+OH → 2-CPR+H_2O are over 1 order of magnitude lower than the literature values (*25–27*) for phenol+H → C_6H_5O+H_2 and phenol+OH → C_6H_5O+H_2O, respectively. It is well-known that the chlorine substitution in the ortho position increases the strength of the O—H bond in chlorophenol and decreases its reactivity (*26*). The dissociation energies (0 K) of the O—H bonds in 2-CP and phenol are 85.91 and 83.95 kcal/mol. Thus, our CVT/SCT rate constants for 2-CP+H → 2-CPR+H_2 and 2-CP+OH → 2-CPR+H_2O are reasonable.

For the reactions with Cl atom, only one paper (*28*) is on record for the reactions of phenol, 2-CP, 3-CP, and 4-CP with Cl. At 296 K, the rate constant (*28*) for phenol+Cl → C_6H_5O+HCl is 1.93×10^{-10} cm^3 $molecule^{-1}$ s^{-1}. The values (*28*) for 3-CP+Cl → 3-CPR+HCl and 4-CP+Cl → 4-CPR+HCl are 1.56×10^{-10} and 2.37×10^{-10} cm^3 $molecule^{-1}$ s^{-1}. However, the value (*28*) for 2-CP+Cl → 2-CPR+HCl is 7.32×10^{-12} cm^3 $molecule^{-1}$ s^{-1}, which is much lower than those of the reactions of phenol, 3-CP and 4-CP with Cl. The CVT/SCT rate constant we calculated for 2-CP+Cl → 2-CPR+HCl is 4.29×10^{-11} cm^3 $molecule^{-1}$ s^{-1}, which is almost 6 times higher than the experimental value of 7.32×10^{-12} cm^3 $molecule^{-1}$ s^{-1}, but is lower than those of the reactions of phenol, 3-CP, and 4-CP with Cl. In order to check the reactivity and strength of the O—H bond in phenol and chlorophenols, we calculated the dissociation energy of the O—H bond. The values (0 K) are 83.95, 85.91, 85.01, and 82.76 kcal/mol for the O—H bond dissociation energy in phenol, 2-CP, 3-CP, and 4-CP, respectively. The O—H bond dissociation energy in 2-CP is only 0.9 kcal/mol higher than that in 3-CP. However, the experimental rate constant of 3-CP+Cl → 3-CPR+HCl is 21.31 times larger than that of 2-CP+Cl → 2-CPR+HCl (*28*). The dissociation energy of the O—H bond in 3-CP is 2.25 kcal/mol higher than that in 4-CP. The experimental rate constant of 4-CP+Cl → 4-CPR+HCl is only 1.52 times larger than that of the reaction of 3-CP+Cl → 3-CPR+HCl (*28*). It seems likely that the experiment underestimates the rate constant of the reaction of 2-CP+Cl → 2-CPR+HCl.

The unimolecular reactions involved in the formation of PCDD/Fs are the decomposition or the isomerization of the activated intermediates. There are no available experimental rate constants for the activated intermediates or structurally similar compounds, largely due to the short lifetime and the lack of efficient detection schemes for these activated species. Thus, it is difficult to make a comparison of the calculated CVT/SCT rate constants with the experimental values for the unimolecular decomposition or the isomerization of the activated intermediates. The CVT/SCT method has been successfully performed for many unimolecular reactions (*29, 30*). Fernández-Ramos (*30*) calculated the rate constants for the dissociation and elimination channels in the thermal decomposition of methyl nitrite by direct dynamics method. In particular, the CVT/SCT calculations predicted rate constants that are in excellent agreement with those determined experimentally, indicating that the CVT/SCT method is able to predict reliable rate constants for the unimolecular reactions (*30*). We hope our CVT/SCT calculations may provide a good estimate for the crucial elementary reactions involved in the formation of PCDD/Fs from 2-CP.

Knowledge of the temperature dependence would be useful for the kinetic models (*13, 14*). The pre-exponential factor, the activation energy, and the rate constants are important input parameters in the kinetics models that account for the contribution of the gaseous route in the production of PCDD/Fs in combustion processes. Thus, the calculated CVT/SCT rate constants are fitted over the temperature range of 600~1200 K and Arrhenius formulas are given in Table 1.

Acknowledgments

This work was supported by NSFC (National Natural Science Foundation of China, project no. 20507013, 20737001). We thank Professor Donald G. Truhlar for providing the POLYRATE 9.3 program. We also thank Dr. Pamela Holt for proofreading the manuscript.

Supporting Information Available

MPWB1K/6–31G(d,p) optimized geometries and calculated imaginary frequencies for the transition states involved in the formation of PCDD/Fs from 2-CP as precursor. The material is available free of charge via the Internet at http://pubs.acs.org.

Literature Cited

(1) Dickson, L. C.; Lenoir, D.; Hutzinger, O. Quantitative comparison of de novo and precursor formation of polychlorinated dibenzo-p-dioxins under simulated municipal solid waste incinerator postcombustion conditions. *Environ. Sci. Technol.* **1992**, *26*, 1822–1828.

(2) Addink, R.; Olie, K. Mechanisms of formation and destruction of polychlorinated dibenzo-p-dioxins and dibenzofurans in heterogeneous systems. *Environ. Sci. Technol.* **1995**, *29*, 1425–1435.

(3) Milligan, M. S.; Altwicker, E. R. Chlorophenol reactions on fly ash. 1. Adsorption/desorption equilibria and conversion to polychlorinated dibenzo-*p*-dioxins. *Environ. Sci. Technol.* **1996**, *30*, 225–229.

(4) Karasek, F. W.; Dickson, L. C. Model studies of polychlorinated dibenzo-p-dioxin formation during municipal refuse incineration. *Science* **1987**, *237*, 754–756.

(5) Shaub, W. M.; Tsang, W. Dioxin formation in incinerators. *Environ. Sci. Technol.* **1983**, *17*, 721–730.

(6) Altwicker, E. R. Relative rates of formation of polychlorinated dioxins and furans from precursor and de novo reactions. *Chemosphere* **1996**, *33*, 1897–1904.

(7) Hung, H.; Buekens, A. Comparison of dioxin formation levels in laboratory gas-phase flow reactors with those calculated using the Shaub-Tsang mechanism. *Chemosphere* **1999**, *38*, 1595–1602.

(8) BabushokV. TsangW. 7th International Congress on Combustion By-Products: Origins, Fate, and Health Effects, Research Triangle Park, North Carolina, 2001; p 36.

(9) Evans, C. S.; Dellinger, B. Mechanisms of dioxin formation from the high-temperature pyrolysis of 2-chlorophenol. *Environ. Sci. Technol.* **2003**, *37*, 1325–1330.

(10) Evans, C. S.; Dellinger, B. Mechanisms of dioxin formation from the high-temperature oxidation of 2-chlorophenol. *Environ. Sci. Technol.* **2005**, *39*, 122–127.

(11) Sidhu, S.; Edwards, P. Role of phenoxy radicals in PCDD/F formation. *Int. J. Chem. Kinet.* **2002**, *34*, 531–541.

(12) Altarawneh, M.; Dlugogorski, B. Z.; Kennedy, E. M.; Mackie, J. C. Quantum chemical investigation of formation of polychlorodibenzo-p-dioxins and dibenzofurans from oxidation and pyrolysis of 2-chlorophenol. *J. Phys. Chem. A* **2007**, *111*, 2563–2573.

(13) Shaub, W. M.; Tsang, W. Dioxin formation in incinerators. *Environ. Sci. Technol.* **1983**, *17*, 721–730.

(14) Khachatryan, L.; Asatryan, R.; Dellinger, B. Development of expanded and core kinetic models for the gas phase formation of dioxins from chlorinated phenols. *Chemosphere* **2003**, *52*, 695–708.

(15) Baldridge, M. S.; Gordor, R.; Steckler, R.; Truhlar, D. G. Ab initio reaction paths and direct dynamics calculations. *J. Phys. Chem.* **1989**, *93*, 5107–5119.

(16) Liu, Y.-P.; Lynch, G. C.; Truong, T. N.; Lu, D.-H.; Truhlar, D. G.; Garrett, B. C. Molecular modeling of the kinetic isotope effect for the [1,5]-sigmatropic rearrangement of cis-1,3-pentadiene. *J. Am. Chem. Soc.* **1993**, *115*, 2408–2415.

(17) Frisch, M. J.; Trucks, G. W.; Schlegel, H. B.; Scuseria, G. E.; Robb, M. A.; Cheeseman, J. R.; Montgomery, J. A., Jr.; Vreven, T.; Kudin, K. N.; Burant, J. C.; Millam, J. M.; Iyengar, S. S.; Tomasi, J.; Barone, V.; Mennucci, B.; Cossi, M.; Scalmani, G.; Rega, N.; Petersson, G. A.; Nakatsuji, H.; Hada, M.; Ehara, M.; Toyota, K.; Fukuda, R.; Hasegawa, J.; Ishida, M.; Nakajima, T.; Honda, Y.; Kitao, O.; Nakai, H.; Klene, M.; Li, X.; Knox, J. E.; Hratchian, H. P.; Cross, J. B.; Bakken, V.; Adamo, C.; Jaramillo, J.; Gomperts, R.; Stratmann, R. E.; Yazyev, O.; Austin, A. J.; Cammi, R.; Pomelli, C.; Ochterski, J. W.; Ayala, P. Y.; Morokuma, K.; Voth, G. A.;

Salvador, P.; Dannenberg, J. J.; Zakrzewski, V. G.; Dapprich, S.; Daniels, A. D.; Strain, M. C.; Farkas, O.; Malick, D. K.; Rabuck, A. D.; Raghavachari, K.; Foresman, J. B.; Ortiz, J. V.; Cui, Q.; Baboul, A. G.; Clifford, S.; Cioslowski, J.; Stefanov, B. B.; Liu, G.; Liashenko, A.; Piskorz, P.; Komaromi, I.; Martin, R. L.; Fox, D. J.; Keith, T.; Al-Laham, M. A.; Peng, C. Y.; Nanayakkara, A.; Challacombe, M.; Gill, P. M. W.; Johnson, B.; Chen, W.; Wong, M. W.; Gonzalez, C.; Pople, J. A. *Gaussian 03*, revision C.02; Gaussian, Inc.: Wallingford, CT, 2004.

(18) Zhao, Y.; Truhlar, D. G. Hybrid meta density functional theory methods for therochemistry, thermochemical kinetics, and noncovalent interactions: the MPW1B95 and MPWB1K models and comparative assessments for hydrogen bonding and van der waals interactions. *J. Phys. Chem. A* **2004**, *108*, 6908–6918.

(19) Fukui, K. The path of chemical reactions - the IRC approach. *Acc. Chem. Res.* **1981**, *14* (12), 363–368.

(20) Steckler, R.; Chuang, Y. Y.; Fast, P. L.; Corchade, J. C.; Coitino, E. L.; Hu, W. P.; Lynch, G. C.; Nguyen, K.; Jackells, C. F.; Gu, M. Z.; Rossi, I.; Clayton, S.; Melissas, V.; Garrett, B. C.; Isaacson, A. D.; Truhlar, D. G. *POLYRATE Version 9.3*; University of Minnesota: Minneapolis., 2002.

(21) Louw, R.; Ahonkhai, S. I. Radical/radical va radical/molecule reactions in the formation of PCdd/Fs from (chloro)phenols in incinerators. *Chemosphere* **2002**, *46*, 1273–1278.

(22) Mackie, J. C.; Doolan, K. R.; Nelson, P. F. Kinetics of the thermal decomposition of methoxybenzene (anisole). *J. Phys. Chem.* **1989**, *93*, 664–670.

(23) Zhang, Q.-Z.; Gu, Y.-S.; Wang, S.-K. Theoretical studies on the variational transitional state theory rate constants for the hydrogen abstraction reaction of O (^{3}P) with CH_3Cl and CH_2Cl_2. *J. Chem. Phys.* **2003**, *119*, 4339–4345.

(24) Zhang, Q.-Z.; Gu, Y.-S.; Wang, S.-K. Theoretical investigation on the mechanism and thermal rate constants for the reaction of atomic O (^{3}P) with CHF_2Cl. *J. Phys. Chem. A* **2003**, *107*, 3069–3075.

(25) Baulch, D. L.; Cobos, C. J.; Cox, R. A.; Esser, C.; Frank, P.; Just, Th.; Kerr, J. A.; Pilling, M. J.; Troe, J.; Walker, R. W.; Warnatz, J. Evaluated kinetic data for combustion modelling. *J. Phys. Chem. Ref. Data* **1992**, *21*, 411–429.

(26) Han, J.; Deming, R. L.; Tao, F. —M. Theoretical study of molecular structures and properties of the complete series of chlorophenols. *J. Phys. Chem. A* **2004**, *108*, 7736–7743.

(27) He, Y. Z.; Mallard, W. G.; Tsang, W. Kinetics of hydrogen and hydroxyl radical attack on phenol at high temperatures. *J. Phys. Chem.* **1988**, *92*, 2196–2201.

(28) Platz, J.; Nielsen, O. J.; Wallington, T. J.; Ball, J. C.; Hurley, M. D.; Straccia, A. M.; Schneider, W. F.; Sehested, J. Atmospheric chemistry of the phenoxy radical, C_6H_5O: UV spectrum and kinetics of its reaction with NO, NO_2, and O_2. *J. Phys. Chem. A* **1998**, *102*, 7964–7974.

(29) Luo, Q.; Li, Q. —S. Direct ab initio dynamics study of the unimocular reaction of CH_2FO. *J. Phys. Chem. A* **2004**, *108*, 5050–5056.

(30) Antonio, F.-R.; Emilio, M. —N.; Rios, M. A.; Jesus, R. —O.; Vazquez, S. A.; Estevez, C. M. Direct dynamics study of the dissociation and elimination channels in the thermal decomposition of methyl nitrite. *J. Am. Chem. Soc.* **1998**, *120*, 7594–7601.

ES801599N

Environ. Sci. Technol. **2010**, *44*, 6745–6751

Dioxin Formations from the Radical/Radical Cross-Condensation of Phenoxy Radicals with 2-Chlorophenoxy Radicals and 2,4,6-Trichlorophenoxy Radicals

FEI XU, WANNI YU, RUI GAO, QIN ZHOU, QINGZHU ZHANG,* AND WENXING WANG

Environment Research Institute, Shandong University, Jinan 250100, P. R. China

Received May 27, 2010. Revised manuscript received July 22, 2010. Accepted July 22, 2010.

It is important to understand the role of phenol in the dioxin formations because it is present in the high amount in municipal waste incinerators (MWIs). The formation mechanism of dioxins from the cross-condensation of PhRs with 2-CPRs and 2,4,6-TCPRs was investigated by using hybrid density functional theory (DFT) and compared with the dioxin formation mechanism from the self-condensation of single chlorophenol precursors. The geometrical parameters were optimized at the MPWB1K level with the 6-31+G(d,p) basis set without symmetry constraints. Single-point energy calculations were carried out at the MPWB1K/6-311+G(3df,2p) level of theory. The rate constants were deduced by using canonical variational transition-state (CVT) theory with small curvature tunneling (SCT) contribution over the temperature range of 600−1200 K. The Arrhenius formulas were reported for the first time. Results show that phenol is responsible for the formation of dioxin congeners. This work, together with results already published from our group, provides a comprehensive investigation of the homogeneous gas-phase formation of dioxins from (chloro)phenol precursors and should help to clarify the formation mechanism of dioxins in real waste combustion and to develop more effective control strategies.

1. Introduction

Of the several groups of chlorinated materials found in the environment, none has given rise to more public concern than dioxins—the set of polychlorinated dibenzo-*p*-dioxin (PCDDs) and polychlorinated dibenzofurans (PCDFs). Dioxins are considered as typical persistent organic pollutants (POPs) with the carcinogenic, teratogenic, and mutagenic effects (*1*). They are also suspected to be environment endocrine disruptors that disturb the balance of hormones and damage the metabolism, immunity, and reproduction of exposed organisms (*2−4*). PCDD/Fs have never been intentionally synthesized for commercial purposes. Studies on the dioxin formations are of interest because they can serve as a basis for minimizing dioxin emissions.

It is well-established that combustion processes, especially those of municipal solid waste, are the principal origin of dioxins (*5−7*). In municipal waste incinerators (MWIs), dioxin byproducts can be formed by two general formation pathways, precursor pathway and *de novo* synthesis. The former is $10^2\sim10^5$ times faster than the latter (*8, 9*). The relative yields of dioxins produced from precursors are 72−99 000 times higher than those formed by *de novo* synthesis (*8, 9*). The formation pathway via precursors accounts for the majority of dioxin emissions from combustion sources (*10, 11*). Among the variety of precursors, chlorophenols (CPs) are structurally similar to dioxins and relatively easy to form dioxins during thermal treatment. Additionally, they are among the most abundant aromatic compounds found in MWI flue gases. CPs have been demonstrated to be the predominant precursors of dioxins in MWIs and are implicated as key intermediates in *de novo* pathway (*12−14*). Much attention has been devoted to the dioxin formations from the CP precursors. However, phenol is typically much more abundant than CPs in municipal waste incinerators. For example, the concentration of phenol in the MWI flue gases is 30−100 times higher than the sum of total CPs (*15*). Numerous studies have shown that phenol has the greatest propensity for the formation of PCDFs, especially DF and less chlorinated PCDF congeners (*16, 17*). Steric and electronic effects associated with chlorine substitution suppress the PCDF formations. The cross-condensation of phenol with CPs with ortho-chlorine substituent is also responsible for the distribution of PCDD homologues. It is somewhat surprising that there is relatively little information available in the literature on the dioxin formations from the phenol precursor.

Precursor formation pathway can occur via homogeneous gas-phase reactions and heterogeneous metal-mediated reactions. The homogeneous gas-phase formation of dioxins was proposed that involve (chloro)phenoxy radical/radical condensation, radical/molecule recombination of (chloro)phenoxy and (chloro)phenol. It has been shown that the radical/molecule mechanism requires chlorine and hydroxyl displacement as first steps, and these steps are not energetically favored (*18, 19*). The radical/radical condensation is the dominant pathway in the homogeneous gas-phase formation of dioxins. The radical/radical condensation plays a significant role in the heterogeneous metal-mediated formation pathway as well. Transition-metal species promote the formation of surface-associated (chloro)phenoxy radicals that react to produce dioxins through radical/radical reactions (*20*). (Chloro)phenols readily form (chloro)phenoxy radicals, which are neutral ambient radicals capable of reacting at the phenolic oxygen atom as well as at ortho and para carbon sites, under the combustion conditions. Due to their significant resonance stabilization, (chloro)phenoxy radicals could build up considerable concentration in the combustion environment affording their condensation to dioxin congeners.

Despite the large volume of research data related to the dioxin formations, the specific formation mechanism of PCDD/Fs remains unclear. This is due in part to their extreme toxicity and the lack of efficient detection schemes for radical intermediate species. Quantum calculation is especially suitable for establishing the feasibility of a reaction pathway. In this paper, therefore, we present a rather comprehensive computational study on the dioxin formations from the cross-condensation of phenoxy radicals (PhRs) with 2-chlorophenoxy radicals (2-CPRs) and 2,4,6-trichlorophenoxy radicals (2,4,6-TCPRs). 2-chlorophenol (2-CP) and 2,4,6-trichlorophenol (2,4,6-TCP) are among the most abundant CP congeners found in MWIs (*16*). This work is a continuation

* Corresponding author fax: 86-531-8836 1990; e-mail: zqz@sdu.edu.cn.

10.1021/es101794v
Published on Web 08/09/2010

of our studies on the dioxin formations. In the recently published papers, we investigated the PCDD/F formations from the self-condensation of single chlorophenol precursors (*21–23*). A second motivation for this work is to evaluate the rate constants of the elementary reactions involved in the dioxin formations. The absence of the kinetic parameters, such as the pre-exponential factors, the activation energies, and the rate constants, of the elementary reactions is the most difficult challenge in further improving dioxin formation models.

2. Computational Methods

The present study is carried out in two stages. In the first stage, by means of the Gaussian 03 suite of programs (*24*), DFT calculations are performed on an SGI 2000 supercomputer. It is well-known that MPWB1K is an excellent method for prediction of transition state geometries and thermochemical kinetics, based on the modified Perdew and Wang exchange functional (MPW) and Becke's 1995 correlation functional (B95) (*25*). The geometrical parameters and harmonic vibrational frequencies of reactants, intermediates, transition states, and products were optimized at the MPWB1K level with a standard 6-31+G(d,p) basis set. Stationary points were characterized as minima or transition states by diagonalizing their Hessian matrices and confirming that there are zero or one negative eigenvalue, respectively. The minimum energy path (MEP) was obtained by the intrinsic reaction coordinate (IRC) theory to confirm that the transition state really connects to minima along the reaction path. At some point along the MEP, the matrices of force constants were computed in order to do the following calculations of the canonical variational rate constants. To yield more reliable reaction heats and barrier heights, a more flexible basis set, 6-311+G(3df,2p), was used for single-point energy calculations. All the relative energies quoted and discussed in this paper include zero-point energy (ZPE) corrections.

In the second stage, the electronic structure information is input in Polyrate-Version 9.3 to calculate canonical variational transition-state (CVT) theory rate constants and their temperature dependence (*26–28*). The CVT rate constant, $k^{CVT}(T)$, at a fixed temperature (T) that minimized the generalized transition-state theory rate constant, $k^{GT}(T, s)$, with respect to the dividing surface at s is expressed as

$$k^{CVT}(T) = \min_{s} k^{GT}(T,s) \tag{1}$$

The generalized transition-state theory rate constant $k^{GT}(T, s)$ for T and a dividing surface at s is

$$k^{GT}(T, s) = \frac{\sigma k_B T}{h}\frac{Q^{GT}(T, s)}{\Phi^R(T)}e^{-V_{MEP}(s)/k_B T} \tag{2}$$

where, σ is the symmetry factor accounting for the possibility of more than one symmetry-related reaction path, k_B is Boltzmann's constant, h is Planck's constant. $\Phi^R(T)$ is the reactant partition function per unit volume, excluding symmetry numbers for rotation, and $Q^{GT}(T, s)$ is the partition function of a generalized transition state at s with a local zero of energy at $V_{MEP}(s)$ and with all rotational symmetry numbers set to unity. The rotational partition functions were calculated classically, and the vibrational modes were treated as quantum-mechanically separable harmonic oscillators.

3. Results and Discussion

It is vital to clarify the reliability of the theoretical calculations, especially for a continuous work. The optimized geometries and the calculated vibrational frequencies of phenol, DD, and 1-MCDD at the MPWB1K/6-31+G(d,p) level are in good agreement with the available experimental values (*29–32*), and the relative error remains within 1.5% for the geometrical parameters and 8.0% for the vibrational frequencies except for the lowest frequency of DD, with the relative error up to 11.5%. In order to verify the reliability of the energies, we calculated the reaction enthalpies for the reactions of Ph+2-CP→DD+H_2+HCl and Ph+2-CP→1-MCDD+$2H_2$ at the MPWB1K/6-311+G(3df,2p)//MPWB1K/6-31+G(d,p) level. The calculated values of 17.79 and 35.18 kcal/mol at 298.15 K and 1.0 atm show good consistency with the corresponding experimental values of 18.12 and 33.87 kcal/mol obtained from the measured standard formation enthalpies ($\Delta H_{f,0}$) of phenol, 2-CP, DD, 1-MCDD, HCl (*33–36*), especially if the experimental uncertainties are taken into consideration.

Under typical incinerator conditions, (chloro)phenoxy radicals can be produced from (chloro)phenols through loss of the phenoxyl-hydrogen via unimolecular, bimolecular, or possibly other low-energy pathways (including heterogeneous reactions). The unimolecular reaction contains the decomposition of (chloro)phenols with the cleavage of the O—H bond. The bimolecular reactions involve the phenolic-hydrogen abstraction homogeneously in the gas phase from (chloro)phenols by the active radicals, H, OH, O (^{3}P), and Cl, which are abundant in the combustion environment. The heterogeneous reactions include the reactions catalyzed by transition-metal oxides and chlorides. The formation of (chloro)phenoxy radicals from (chloro)phenols has been investigated in detail in the literature (*37–39*).

3.1. Formation of PCDDs. *3.1.1. Formation of PCDDs from the Cross-Condensation of PhRs with 2-CPRs.* Five possible reaction pathways, displayed in Figure 1, are proposed for the formation of PCDDs from the cross-condensation of PhRs with 2-CPRs. It can be seen from Figure 1 that pathway 1, pathway 2, and pathway 5 are similar, they involve four elementary steps: oxygen—carbon coupling, Cl or H abstraction, ring closure, and intra-annular elimination of H. The intra-annular elimination of H is the rate determining step. Pathway 3 includes six elementary reactions: oxygen—carbon coupling, H abstraction, Smiles rearrangement (two elementary steps), ring closure, and intra-annular elimination of H (the rate determining step). Pathway 4 contains three elementary processes: oxygen—carbon coupling, H abstraction, ring closure and intra-annular elimination of Cl. The ring closure and intra-annular elimination of Cl are found to occur in one step and are the rate determining step. Apparently, pathway 4 covers relatively less elementary steps. Furthermore, the rate determining step involved in pathway 4 has a lower barrier height and is less endoergic than those involved in pathway 1, pathway 2, pathway 3, and pathway 5, respectively. So, pathway 4 is thermodynamically preferred route, resulting in the formation of DD, consistent with the experimental study (*40*).

The rate determining step to the PCDD formation is intra-annular elimination of Cl or H. Mechanisms described above and our previous study (*21*) show that the thermodynamically preferred routes proceed through intra-annular elimination of Cl. In general, there are two kinds of oxygen—carbon coupling modes, the coupling of the phenolic oxygen with the ortho carbon bonded to chlorine of a second (chloro)phenoxy radical (O/σ-CCl for short), and the coupling of the phenolic oxygen with the ortho carbon bonded to hydrogen of a second (chloro)phenoxy radical (O/σ-CH for short). Thus, two kinds of thermodynamically preferred routes can be identified for the PCDD formations. The first one contains the elementary steps of O/σ-CCl coupling, Cl abstraction, intra-annular elimination of Cl, or intra-annular elimination of Cl after a Smiles rearrangement by two chlorine losses. Here, the elementary step of intra-annular elimination of Cl involves the ring closure and intra-annular elimination of Cl because they are found to occur in one-step reaction.

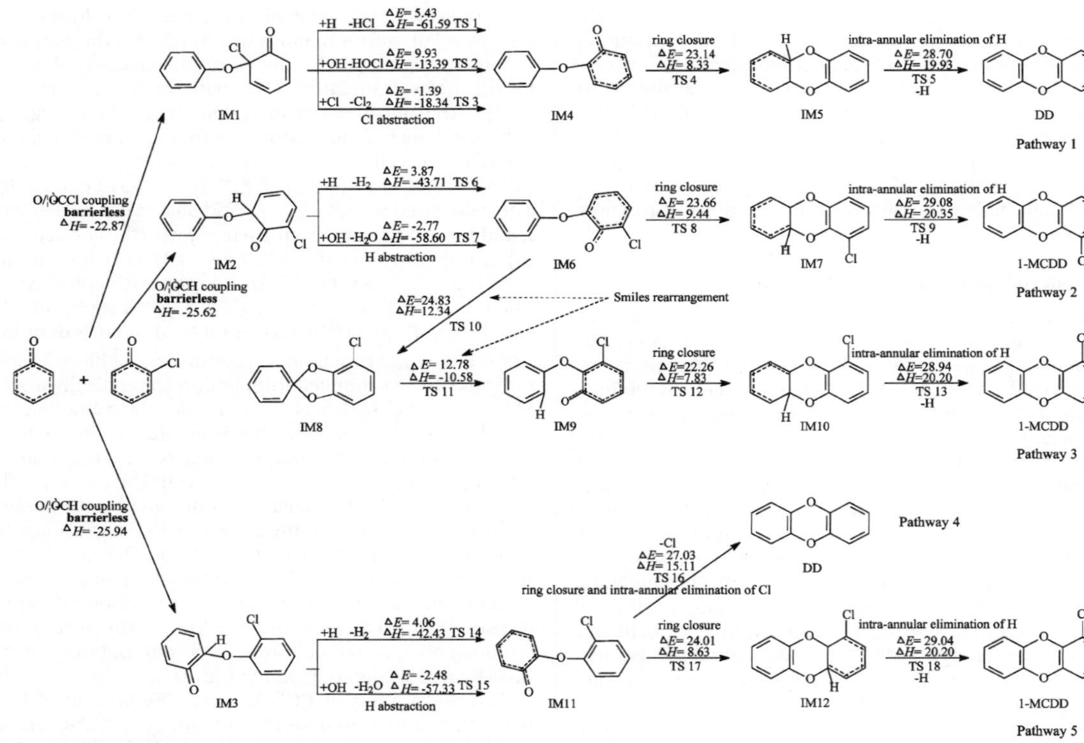

FIGURE 1. PCDD formation routes embedded with the potential barriers ΔE (in kcal/mol) and reaction heats ΔH (in kcal/mol) from the cross-condensation of PhRs with 2-CPRs. ΔH is calculated at 0 K.

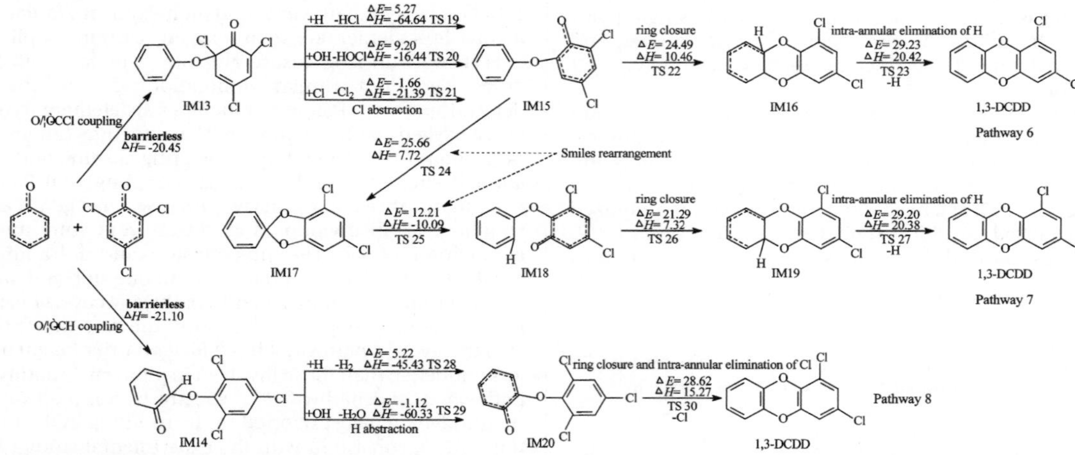

FIGURE 2. PCDD formation routes embedded with the potential barriers ΔE (in kcal/mol) and reaction heats ΔH (in kcal/mol) from the cross-condensation of PhRs with 2,4,6-TCPRs. ΔH is calculated at 0 K.

The second one includes the elementary processes of O/σ-CH coupling, H abstraction, intra-annular elimination of Cl or intra-annular elimination of Cl after a Smiles rearrangement by the loss of one chlorine atom. As shown in Figure 1, the O/σ-CH coupling is more exothermic compared to the O/σ-CCl coupling. In order to further justify the result, we have carried out additional study on the oxygen−carbon coupling from the self-condensation of 19 CPRs and the cross-condensation of PhR with other 18 CPRs. Results indicate that the O/σ-CH coupling is favored over the O/σ-CCl coupling with a exception of the oxygen−carbon coupling from the self-condensation of 2,5-dichlorophenoxy radicals (2,5-DCPRs). Furthermore, the elementary steps of H abstraction have a lower barrier compared to the corresponding Cl abstraction steps. For example, the barrier height for the H abstraction from IM2 by H atom is 3.87 kcal/mol, whereas the value is 5.43 kcal/mol for the Cl abstraction from IM1 by H atom. These results may provide an explanation for the experimental observation that one chlorine loss is favored over two chlorine losses in the PCDD formations (*40, 41*).

Among all the (chloro)phenol precursors, only the self-condensation of 2-CPRs and the cross-condensation of PhRs with 2-CPRs can produce DD, which is present in a high concentration in MWIs. Comparison of the mechanism displayed in Figure 1 with a previous study (*23*) shows that the formation of DD from the cross-condensation of PhRs with 2-CPRs by one chlorine loss is preferred over the DD formation from the self-condensation of 2-CPRs by two chlorine losses. In addition, phenol is much more abundant

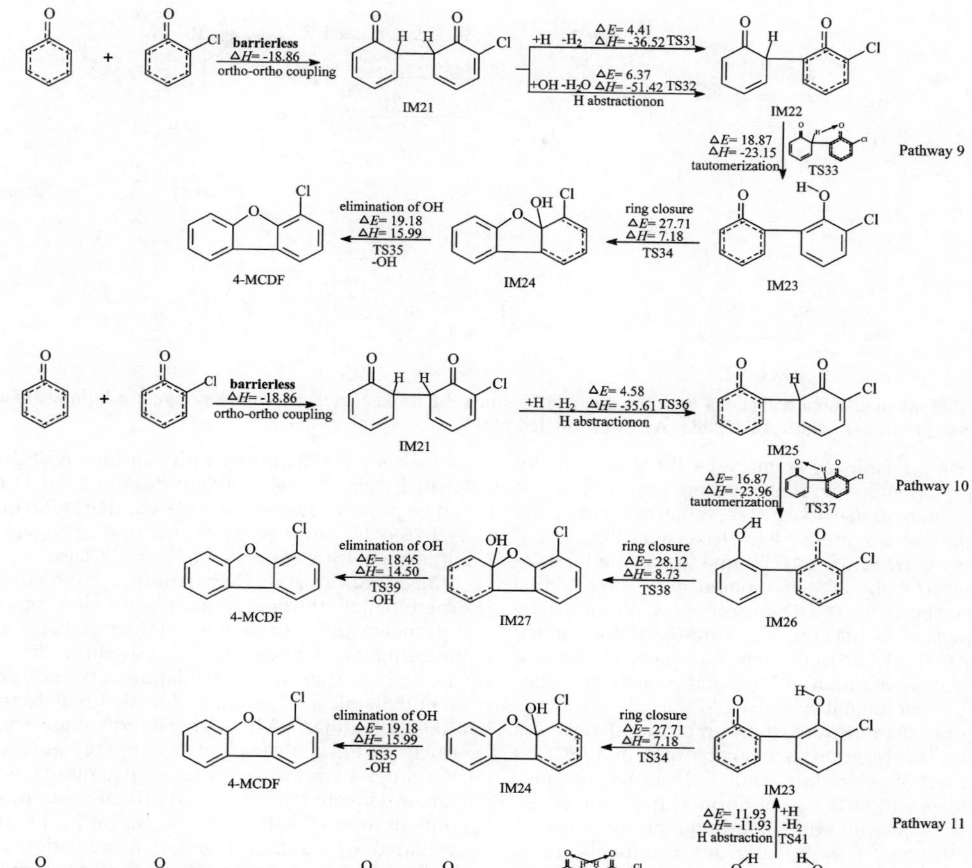

FIGURE 3. PCDF formation routes embedded with the potential barriers ΔE (in kcal/mol) and reaction heats ΔH (in kcal/mol) from the cross-condensation of PhRs with 2-CPRs. ΔH is calculated at 0 K.

than 2-CP in MWIs (*15*). Thus, phenol plays a crucial role in the formation of DD.

3.1.2. Formation of PCDDs from the Cross-Condensation of PhRs with 2,4,6-TCPRs. Similar to the cross-condensation of PhRs with 2-CPRs, three possible PCDD formation pathways are postulated from the cross-condensation of PhRs with 2,4,6-TCPRs. The formation schemes embedded with the potential barriers and reaction heats are depicted in Figure 2. Pathway 6 involves four elementary steps: O/σ-CCl coupling, Cl abstraction, ring closure, and intra-annular elimination of H (the rate determining step). Pathway 7 contains the elementary reactions of O/σ-CCl coupling, Cl abstraction, Smiles rearrangement (two elementary steps), ring closure, and intra-annular elimination of H (the rate

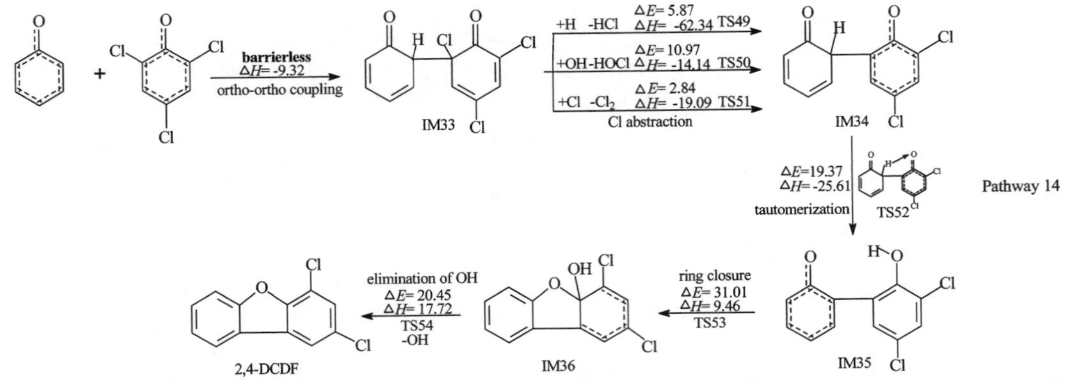

FIGURE 4. PCDF formation route embedded with the potential barriers ΔE (in kcal/mol) and reaction heats ΔH (in kcal/mol) from the cross-condensation of PhRs with 2,4,6- TCPRs. ΔH is calculated at 0 K.

determining step). Pathway 8 includes three elementary processes: O/σ-CH coupling, H abstraction, ring closure and intra-annular elimination of Cl (they are found to occur in one step and are the rate determining step). It is evident from Figure 2 that pathway 8 involving one chlorine loss is the thermodynamically most feasible PCDD formation route. This reaffirms the conclusion above that PCDDs are preferentially formed by O/σ-CH coupling, H abstraction, ring closure and intra-annular elimination of Cl. 1,3-DCDD is the only PCDD product obtained from the cross-condensation of PhRs and 2,4,6-TCPRs, supported by the experimental evidence (*40, 42*).

Previous study demonstrated that only chlorophenols with chlorine at the ortho position were capable of forming PCDDs (*21*). So, the self-dimerization of phenol can not produce PCDDs. However, PCDDs can be formed from the cross-condensation of phenol with chlorophenols with ortho-chlorine substituent. That is, only one ortho-chlorine is needed to produce PCDDs. Thus, phenol can contribute to PCDD isomer distributions in MWIs.

3.2. Formation of PCDFs. Two PCDF congeners, 4-MCDF and DF, can be formed from the cross-condensation of PhRs with 2-CPRs. Four possible formation pathways are illustrated in Figure 3 for 4-MCDF. Clearly, pathway 9 and pathway 10 are similar, and they cover five elementary steps: ortho−ortho coupling, H abstraction, tautomerization (H-shift), ring closure, and elimination of OH. The ring closure process requires crossing a large potential barrier and is strongly endoergic, and it is the rate determining step. Pathway 11 is analogous to pathway 12, which involves five elementary processes: ortho−ortho coupling, tautomerization (double H-transfer), H abstraction, ring closure (the rate determining step), and elimination of OH. One possible formation pathway, depicted in Figure 3, is proposed for DF. The intermediate IM21 can be regarded as a prestructure for 4-MCDF, and IM29 is a prestructure of DF. As shown in Figure 3, the formation of IM21 is more exothermic than the formation of IM29. Furthermore, the rate determining step involved in the formation of 4-MCDF has a lower barrier and is less endothermic compared to that involved in the formation of DF. Thus, the formation of 4-MCDF is preferred over the formation of DF. Comparison of the mechanism presented in Figure 3 with a previous study (*23*) tells us that the PCDF isomers formed from the cross-condensation of PhRs with 2-CPRs is favored over isomers formed from the self-condensation of 2-CPRs. The result supports the experimental result that phenol plays a significant role in the distributions of PCDF homologues (*16*).

Due to the symmetry of 2,4,6-TCPR, only one PCDF, 2,4-DCDF, can be produced from the cross-condensation of PhRs with 2,4,6-TCPRs. The formation route involves five elementary processes: ortho−ortho coupling, Cl abstraction, tautomerization (H-shift), ring closure (the rate determining step), and elimination of OH. Because both ortho-positions of phenol are substituted with chlorine, no PCDFs can be formed from the self-condensation of 2,4,6-TCPRs. PCDF can be produced from the cross-condensation of PhRs with 2,4,6-TCPRs, however. It means that only one ortho-site without chlorine is needed to form PCDFs.

3.3. Rate Constant Calculations. It is difficult to measure experimentally the rate constants of the elementary reactions, especially relative to the radical intermediates, involved in the formation of dioxins. In such a situation, direct dynamics calculations, that is, the calculation of the rate constants or other dynamical information directly from electronic structure calculations without the intermediate stage of constructing a full analytical potential energy surface, can be an alternative. In this work, the rate constants of the elementary reactions involved in the formation of dioxins from the cross-condensation of PhRs with 2-CPRs and 2,4,6-TCPRs were evaluated by canonical variation transition-state (CVT) theory, which is among the most promising current avenues of approach in theoretical chemical kinetics, over the temperature range of 600−1200 K. Quantum tunneling effect is calculated by means of the small curvature tunneling (SCT) approximation. The CVT/SCT method has been successfully performed for the elementary reactions involved in the formation of PCDD/Fs from the self-condensation of 2-CPRs, 2,4-DCPRs, and 2,4,6-TCPRs, respectively (*21−23*).

The reliability of the CVT/SCT method has been clarified in our published studies (*37, 38*). The CVT/SCT rate constants of $C_6H_5OH+H \rightarrow C_6H_5O+H_2$, $C_6H_5OH+OH \rightarrow C_6H_5O+H_2O$ are in good agreement with the corresponding experimental values, respectively (*37, 38*). In the kinetic models of the dioxin formations, the rate constants for the elementary step of

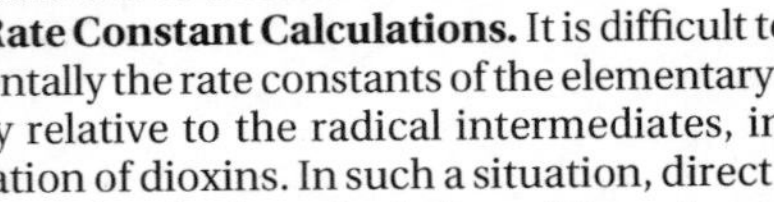

was assigned to be the values for the reaction of $CH_3Cl+H \rightarrow CH_3+HCl$ (*43, 44*). The CVT/SCT rate constants for

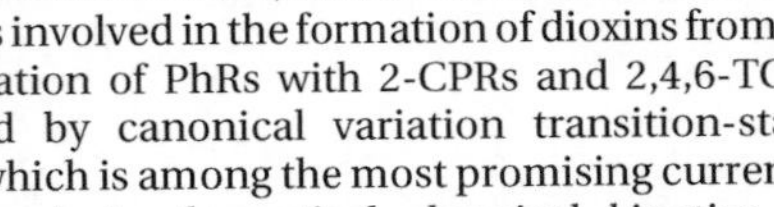

are reasonable compared to the experimental values for the reaction of $CH_3Cl+H \rightarrow CH_3+HCl$ (*21, 45*). For example, at 800 K, the CVT/SCT rate constant for

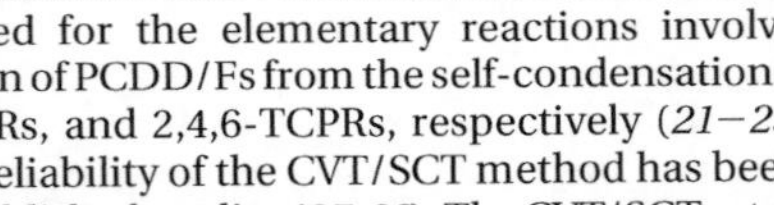

TABLE 1. Arrhenius Formulas (Units Are s^{-1} and cm^3 $Molecule^{-1}$ s^{-1} for Unimolecular and Bimolecular Reactions, Respectively) for Elementary Reactions Involved in the Thermodynamically Preferred Formation Pathway of PCDD/Fs from the Cross-Condensation of PhRs with 2-CPRs and 2,4,6-TCPRs over the Temperature Range of 600−1200 K

reactions	Arrhenius formulas
IM3+H → IM11+H_2	$k(T) = (1.39 \times 10^{-11})\exp(-2532.2/T)$
IM11 → DD+Cl	$k(T) = (3.58 \times 10^{11})\exp(-13706.9/T)$
IM14+H → IM20+H_2	$k(T) = (1.36 \times 10^{-11})\exp(-3364.0/T)$
IM20 → 1,3-DCDD+Cl	$k(T)=(7.92 \times 10^{11})\exp(-14283.4/T)$
IM21+H → IM22+H_2	$k(T) = (1.80 \times 10^{-11})\exp(-3053.7/T)$
IM21+OH → IM22+H_2O	$k(T) = (5.44 \times 10^{-13})\exp(-4840.4/T)$
IM22 → IM23	$k(T)=(3.08 \times 10^{13})\exp(-9948.6/T)$
IM23 → IM24	$k(T) = (1.50 \times 10^{12})\exp(-13859.2/T)$
IM24 → 4-MCDF+OH	$k(T) = (3.42 \times 10^{13})\exp(-10068.7/T)$
IM21+H → IM25+H_2	$k(T) = (1.03 \times 10^{-11})\exp(-2091.1/T)$
IM25 → IM26	$k(T) = (2.11 \times 10^{10})\exp(-4657.5/T)$
IM26 → IM27	$k(T) = (2.04 \times 10^{12})\exp(-14193.4/T)$
IM27 → 4-MCDF+OH	$k(T)=(2.34 \times 10^{13})\exp(-9511.2/T)$
IM21 → IM28	$k(T) = (1.48 \times 10^{12})\ \exp(-8692.8/T)$
IM28+H → IM23+H_2	$k(T)=(7.22 \times 10^{-13})\exp(-7878.6/T)$
IM28+H → IM26+H_2	$k(T) = (4.63 \times 10^{-13})\exp(-7932.1/T)$
IM29+H → IM30+HCl	$k(T) = (5.40 \times 10^{-11})\exp(-4101.1/T)$
IM29+OH → IM30+HOCl	$k(T)=(1.00 \times 10^{-12})\exp(-7198.3/T)$
IM29+Cl → IM30+Cl_2	$k(T)=(2.45 \times 10^{-11})\exp(-2672.3/T)$
IM30 → IM31	$k(T) = (1.89 \times 10^{12})\exp(-9098.9/T)$
IM31 → IM32	$k(T) = (5.18 \times 10^{12})\exp(-16190.1/T)$
IM32 → DF+OH	$k(T) = (2.77 \times 10^{13})\exp(-9374.2/T)$
IM33+H → IM34+HCl	$k(T) = (6.34 \times 10^{-11})\exp(-3883.6/T)$
IM33+OH → IM34+HOCl	$k(T) = (1.50 \times 10^{-12})\exp(-6595.0/T)$
IM33+Cl → IM34+Cl_2	$k(T) = (7.65 \times 10^{-11})\exp(-2910.0/T)$
IM34 → IM35	$k(T) = (2.89 \times 10^{12})\exp(-10093.1/T)$
IM35 → IM36	$k(T) = (5.04 \times 10^{12})\exp(-15753.8/T)$
IM36 → 2,4-DCDF+OH	$k(T) = (1.52 \times 10^{13})\exp(-10196.3/T)$

is 2.08×10^{-13} cm^3 $molecule^{-1}$ s^{-1}, whereas the experimental value for the reaction of $CH_3Cl+H \to CH_3+HCl$ is 1.77×10^{-13} cm^3 $molecule^{-1}$ s^{-1}. The potential barrier for the reaction of

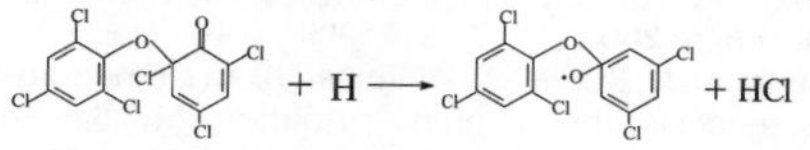

is lower than that of $CH_3Cl+H \to CH_3+HCl$.

To be used more effectively, the calculated CVT/SCT rate constants are fitted, and Arrhenius formulas are given in Table 1 for elementary reactions involved in the thermodynamically preferred formation pathway of PCDD/Fs from the cross-condensation of PhRs with 2-CPRs and 2,4,6-TCPRs. The pre-exponential factor, the activation energy, and the rate constants can be obtained from these Arrhenius formulas.

Acknowledgments

This work was supported by NSFC (National Natural Science Foundation of China, project No. 20737001, 20777047, 20977059), Shandong Province Outstanding Youth Natural Science Foundation (project No. JQ200804), the Research Fund for the Doctoral Program of Higher Education of China (project No. 200804220046) and Independent Innovation Foundation of Shandong University (IIFSDU, project No. 2009JC016). The authors thank Professor Donald G. Truhlar for providing the POLYRATE 9.3 program.

Supporting Information Available

The total energies (in a.u.), the zero-point energies (ZPE, in a.u.), and the imaginary frequencies (in cm^{-1}) for the transition states. The geometries in terms of Cartesian coordinate (in Angstrom) for the reactants, products, intermediates, and transition states. The reaction enthalpies ΔH_0, the changes of Gibbs free energies ΔG for the elementary reactions involved in the formation of dioxins from the cross-condensation of PhRs with 2-CPRs and 2,4,6-TCPRs at 298.15 K and 1.0 atm. This material is available free of charge via the Internet at http://pubs.acs.org.

Literature Cited

(1) Schecter, A. Dioxin and Health; Plenum Press: New York, 1994.

(2) Wang, S. L.; Chang, Y. C.; Chao, H. R.; Li, C. M.; Li, L. A.; Lin, L. Y.; Päpke, O. Body burdens of polychlorinated dibenzo-*p*-dioxins, dibenzofurans, and biphenyls and their relations to estrogen metabolism in pregnant women. *Environ. Health Perspect.* **2006**, *114* (5), 740–745.

(3) Chao, H. R.; Wang, S. L.; Lin, L. Y.; Lee, W. J.; Päpke, O. Placental transfer of polychlorinated dibenzo-*p*-dioxins, dibenzofurans, and biphenyls in Taiwanese mothers in relation to menstrual cycle characteristics. *Food Chem. Toxicol.* **2007**, *45* (2), 259–265.

(4) Viluksela, M.; Raasmaja, A.; Lebosfsky, M.; Stahl, B. U.; Rozman, K. K. Tissue-specific effects of 2,3,7,8-tetrachlorodibenzo-*p*-dioxin (TCDD) on the activity of 5′-deiodinases I and II in rats. *Toxicol. Lett.* **2004**, *147* (2), 133–142.

(5) Harris, J. C.; Anderson, P. C.; Goodwin, B. E.; Rechsteiner, C. E. Dioxin Emissions from Combustion Sources: A Review of the Current State of Knowledge. Final Report to ASME; ASME: New York, NY, 1980.

(6) Addink, R.; Altwicker, E. R. Formation of polychlorinated dibenzo-*p*-dioxins/dibenzofurans from soot of benzene and *o*-dichlorobenzene combustion. *Environ. Sci. Technol.* **2004**, *38* (19), 5196–5200.

(7) Yasuhara, A.; Katami, T.; Okuda, T.; Ohno, N.; Adriaens, P. Formation of dioxins during the combustion of newspapers in the presence of sodium chloride and poly(vinyl chloride). *Environ. Sci. Technol.* **2001**, *35* (7), 1373–1378.

(8) Addink, R.; Olie, K. Mechanisms of formation and destruction of polychlorinated dibenzo-*p*-dioxins and dizenzofurans in heterogeneous systems. *Environ. Sci. Technol.* **1995**, *29* (6), 1425–1435.

(9) Tuppurainen, K.; Halonen, I.; Ruokojärvi, P.; Tarhanen, J.; Ruuskanen, J. Formation of PCDDs and PCDFs in municipal waste incineration and its inhibition mechanisms: A review. *Chemosphere* **1998**, *36* (7), 1493–1511.

(10) Luijk, R.; Akkerman, D.; Slot, P.; Olie, K.; Kepteijn, F. Mechanism of formation of polychlorinated dibenzo-*p*-dioxins and dibenzofurans in the catalyzed combustion of carbon. *Environ. Sci. Technol.* **1994**, *28* (2), 312–321.

(11) Dickson, L. C.; Lenoir, D.; Hutzinger, O. Quantitative comparison of *de novo* and precursors formation of polychlorinated dibenzo-*p*-dioxins under simulated municipal solid waste incinerator post-combustion conditions. *Environ. Sci. Technol.* **1992**, *26* (9), 1822–1828.

(12) Karasek, F. W.; Dickson, L. C. Model studies of polychlorinated dibenzo-*p*-dioxin formation during municipal refuse incineration. *Science* **1987**, *237* (4816), 754–756.

(13) Shaub, W. M.; Tsang, W. Dioxin formation in incinerators. *Environ. Sci. Technol.* **1983**, *17* (12), 721–730.

(14) Altwicker, E. R. Relative rates of formation of polychlorinated dioxins and furans from precursor and *de novo* reactions. *Chemosphere* **1996**, *33* (10), 1897–1904.

(15) Zimmermann, R.; Blumenstock, M.; Heger, H. J.; Schramm, K. W.; Kettrup, A. Emission of nonchlorinated and chlorinated aromatics in the flue gas of incineration plants during and after transient disturbances of combustion conditions: delayed emission effects. *Environ. Sci. Technol.* **2001**, *35* (6), 1019–1030.

(16) Ryu, J. Y.; Mulholland, J. A.; Kim, D. H.; Takeuchi, M. Homologue and isomer patterns of polychlorinated dibenzo-*p*-dioxins and dibenzofurans from phenol precursors: Comparison with municipal waste incinerator data. *Environ. Sci. Technol.* **2005**, *39* (12), 4398–4406.

(17) Ryu, J. Y.; Mulholland, J. A.; Oh, J. E.; Nakahata, D. T.; Kim, D. H. Prediction of polychlorinated dibenzofuran congener distribution from gas-phase phenol condensation pathways. *Chemosphere* **2004**, *55* (11), 1447–1455.

(18) Altarawneh, M.; Dlugogorski, B. Z.; Kennedy, E. M.; Mackie, J. C. Quantum chemical investigation of formation of polychlorodibenzo-*p*-dioxins and dibenzofurans from oxidation and pyrolysis of 2-chlorophenol. *J. Phys. Chem. A* **2007**, *111* (13), 2563–2573.

(19) Louw, R.; Ahonkhai, S. I. Radical/radical vs radical/molecule reactions in the formation of PCDD/Fs from (chloro)phenols in incinerators. *Chemosphere* **2002**, *46* (9–10), 1273–1278.

(20) Alderman, S. L.; Farquar, G. R.; Poliakoff, E. D.; Dellinger, B. An infrared and X-ray spectroscopic study of the reactions of 2-chlorophenol, 1,2-dichlorobenzene, and chlorobenzene with model Cuo/Silica fly ash surfaces. *Environ. Sci. Technol.* **2005**, *39* (19), 7396–7401.
(21) Zhang, Q. Z.; Yu, W. N.; Zhang, R. X.; Zhou, Q.; Gao, R.; Wang, W. X. Quantum chemical and kinetic study on dioxin formation from the 2,4,6-TCP and 2,4-DCP precursors. *Environ. Sci. Technol.* **2010**, *44* (9), 3395–3403.
(22) Qu, X. H.; Wang, H.; Zhang, Q. Z.; Shi, X. Y.; Xu, F.; Wang, W. X. Mechanistic and kinetic studies on the homogeneous gas-phase formation of PCDD/Fs from 2,4,5-trichlorophenol. *Environ. Sci. Technol.* **2009**, *43* (11), 4068–4075.
(23) Zhang, Q. Z.; Li, S. Q.; Qu, X. H.; Wang, W. X. A quantum mechanical study on the formation of PCDD/Fs from 2-chlorophenol as precursor. *Environ. Sci. Technol.* **2008**, *42* (19), 7301–7308.
(24) Frisch, M. J.; Trucks, G. W.; Schlegel, H. B.; Scuseria, G. E.; Robb, M. A.; Cheeseman, J. R.; Montgomery, J. A., Jr.; Vreven, T.; Kudin, K. N.; Burant, J. C.; Millam, J. M.; Iyengar, S. S.; Tomasi, J.; Barone, V.; Mennucci, B.; Cossi, M.; Scalmani, G.; Rega, N.; Petersson, G. A.; Nakatsuji, H.; Hada, M.; Ehara, M.; Toyota, K.; Fukuda, R.; Hasegawa, J.; Ishida, M.; Nakajima, T.; Honda, Y.; Kitao, O.; Nakai, H.; Klene, M.; Li, X.; Knox, J. E.; Hratchian, H. P.; Cross, J. B.; Bakken, V.; Adamo, C.; Jaramillo, J.; Gomperts, R.; Stratmann, R. E.; Yazyev, O.; Austin, A. J.; Cammi, R.; Pomelli, C.; Ochterski, J. W.; Ayala, P. Y.; Morokuma, K.; Voth, G. A.; Salvador, P.; Dannenberg, J. J.; Zakrzewski, V. G.; Dapprich, S.; Daniels, A. D.; Strain, M. C.; Farkas, O.; Malick, D. K.; Rabuck, A. D.; Raghavachari, K.; Foresman, J. B.; Ortiz, J. V.; Cui, Q.; Baboul, A. G.; Clifford, S.; Cioslowski, J.; Stefanov, B. B.; Liu, G.; Liashenko, A.; Piskorz, P.; Komaromi, I.; Martin, R. L.; Fox, D. J.; Keith, T.; Al-Laham, M. A.; Peng, C. Y.; Nanayakkara, A.; Challacombe, M.; Gill, P. M. W.; Johnson, B.; Chen, W.; Wong, M. W.; Gonzalez, C.; Pople, J. A. *Gaussian 03*, revision C.02; Gaussian, Inc.: Wallingford, CT, 2004.
(25) Zhao, Y.; Truhlar, D. G. Hybrid meta density functional theory methods for thermochemistry, thermochemical kinetics, and noncovalent interactions: the MPW1B95 and MPWB1K models and comparative assessments for hydrogen bonding and van der waals interactions. *J. Phys. Chem. A* **2004**, *108* (33), 6908–6918.
(26) Baldridge, M. S.; Gordor, R.; Steckler, R.; Truhlar, D. G. Ab initio reaction paths and direct dynamics calculations. *J. Phys. Chem.* **1989**, *93* (13), 5107–5119.
(27) Gonzalez-Lafont, A.; Truong, T. N.; Truhlar, D. G. Interpolated variational transition-state theory: practical methods for estimating variational transition-state properties and tunneling contributions to chemical reaction rates from electronic structure calculations. *J. Chem. Phys.* **1991**, *95* (12), 8875–8894.
(28) Garrett, B. C.; Truhlar, D. G. Generalized transition state theory. Classical mechanical theory and applications to collinear reactions of hydrogen molecules. *J. Phys. Chem.* **1979**, *83* (8), 1052–1079.
(29) *Landolt-Bornstein: Group II: Atomic and Molecular Physics Vol. 7: Structure Data of Free Polyatomic Molecules*; Hellwege, K. H.; Hellwege, A. M. Ed.; Springer-Verlag: Berlin. 1976.
(30) Senma, M.; Taira, Z.; Taga, T.; Osaki, K. Dibenzo-*p*-dioxin, $C_{12}H_8O_2$. *Crystallogr. Struct. Comm.* **1973**, *2* (2), 311–314.
(31) Leon, L. A.; Notario, R.; Quijano, J.; Sanchez, C. Structures and enthalpies of formation in the gas phase of the most toxic polychlorinated dibenzo-*p*-dioxins: a DFT study. *J. Phys. Chem. A* **2002**, *106* (28), 6618–6627.
(32) Gastilovich, E. A.; Klimenko, V. G.; Korolkova, N. V.; Nurmukhametov, R. N. Spectroscopic data on nuclear configuration of dibenzo-*p*-dioxin in S0, S1, and T1 electronic states. *Chem. Phys.* **2002**, *282* (2), 265–275.
(33) Cox, J. D. The heats of combustion of phenol and the three cresols. *Pure Appl. Chem.* **1961**, *2* (1−2), 125–128.
(34) Burcat, A.; Ruscic, B. Ideal gas thermochemical database with updates from active thermochemical tables. ftp://ftp.technion.ac.il/pub/upported/aetdd/thermodynamics/BURCAT.THR.
(35) Lukyanova, V. A.; Kolesov, V. P.; Avramenko, N. V.; Vorobieva, V. P.; Golovkov, V. F. Standard enthalpy of formation of dibenzo-*p*-dioxin. *Zh. Fiz. Khim.* **1997**, *71* (3), 406–408.
(36) Kolesov, V. P.; Papina, T. S.; Lukyanova, V. A. The enthalpies of formation of some polychlorinated dibenzodioxins. In *Abstracts of the 14th IUPAC Conference on Chemical Thermodynamics. Osaka, Japan, Aug 25−30*1996; p 329.
(37) Zhang, Q. Z.; Qu, X. H.; Xu, F.; Shi, X. Y.; Wang, W. X. Mechanism and thermal rate constants for the complete series reactions of chlorophenols with H. *Environ. Sci. Technol.* **2009**, *43* (11), 4105–4112.
(38) Xu, F.; Wang, H.; Zhang, Q. Z.; Zhang, R. X.; Qu, X. H.; Wang, W. X. Kinetic properties for the complete series reactions of chlorophenols with OH radicals - relevance for dioxin formation. *Environ. Sci. Technol.* **2010**, *44* (4), 1399–1404.
(39) Sun, Q.; Altarawneh, M.; Dlugogorski, B. Z.; Kennedy, E. M.; Mackie, J. C. Catalytic effect of CuO and other transition metal oxides in formation of dioxins: theoretical investigation of reaction between 2,4,5-trichlorophenol and CuO. *Environ. Sci. Technol.* **2007**, *41* (16), 5708–5715.
(40) Ryu, J. - Y.; Mulholland, J. A.; Takeuchi, M.; Kim, D.-H.; Hatanaka, T. $CuCl_2$-catalyzed PCDD/F formation and congener patterns from phenols. *Chemosphere* **2005**, *61* (9), 1312–1326.
(41) Ryu, J.-Y.; Mulholland, J. A. Dioxin and furan formation on $CuCl_2$ from chlorinated phenols with one ortho chlorine. *Proc. Combust. Inst.* **2002**, *29* (2), 2455–2461.
(42) Ryu, J. - Y.; Mulholland, J. A.; Chu, B. Metal-mediated chlorinated dibenzo-*p*-dioxin and dibenzofuran formation from phenols. *Chemosphere* **2005**, *58* (7), 977–988.
(43) Khachatryan, L.; Burcat, A.; Dellinger, B. An elementary reaction-kinetic model for the gas-phase formation of 1,3,6,8- and 1,3,7,9-tetrachlorinated dibenzo-*p*-dioxins from 2, 4,6-trichlorophenol. *Combust. Flame* **2003**, *132* (3), 406–421.
(44) Khachatryan, L.; Asatryan, R.; Dellinger, B. Development of expanded and core kinetic models for the gas phase formation of dioxins from chlorinated phenols. *Chemosphere* **2003**, *52* (4), 695–708.
(45) Westenberg, A. A.; DeHaas, N. Rates of H + CH_3X reactions. *J. Chem. Phys.* **1975**, *62* (8), 3321–3325.

ES101794V

Chemosphere 92 (2013) 382–390

Contents lists available at SciVerse ScienceDirect

Chemosphere

journal homepage: www.elsevier.com/locate/chemosphere

Formation of bromophenoxy radicals from complete series reactions of bromophenols with H and OH radicals

Rui Gao, Fei Xu, Shanqing Li, Jingtian Hu, Qingzhu Zhang *, Wenxing Wang

Environment Research Institute, Shandong University, Jinan 250100, PR China

HIGHLIGHTS

- ► We studied the complete series reactions of 19 BPs congeners with H and OH radicals.
- ► We investigated the formation of the bromophenoxy radicals (BPRs) theoretically.
- ► The *ortho* bromine increases strength of O–H bond in BPs and decreases its reactivity.
- ► The phenoxyl-hydrogen abstraction from BPs by OH is more efficient than H radicals.

ARTICLE INFO

Article history:
Received 9 October 2012
Received in revised form 29 December 2012
Accepted 3 January 2013
Available online 10 February 2013

Keywords:
Bromophenols
Bromophenoxy radicals
H radicals
OH radicals
Reaction mechanism
Rate constants

ABSTRACT

The bromophenoxy radicals (BPRs) are key intermediate species involved in the formation of polybrominated dibenzo-*p*-dioxin/dibenzofurans (PBDD/Fs). In this work, the formation of BPRs from the complete series reactions of 19 bromophenol (BP) congeners with H and OH radicals were investigated theoretically by using the density functional theory (DFT) method and the direct dynamics method. The geometries and frequencies of the reactants, transition states, and products were calculated at the MPWB1K/6-31 + G(d,p) level, and the energetic parameters were further refined by the MPWB1K/6-311 + G(3df,2p) method. The rate constants were evaluated by the canonical variational transition-state (CVT) theory with the small curvature tunneling (SCT) contribution over a wide temperature range of 600–1200 K. The present study indicates that the reactivity of the O–H bonds in BPs as well as the formation potential of BPRs from BPs is strongly related to the bromine substitution pattern. The obtained results can be used for future estimates of PBDD/F emissions quantity based on the well estimated PCDD/F inventory.

1. Introduction

Due to similar structures and chemical properties, polybrominated dibenzo-*p*-dioxin/dibenzofurans (PBDD/Fs) show similar toxicities, biological properties and geochemical behavior to their chlorinated counterparts, polychlorinated dibenzo-*p*-dioxin/dibenzofurans (PCDD/Fs) (Mennear and Lee, 1994; Birnbaum et al., 2003; Olsman et al., 2007; Li et al., 2008; Samara et al., 2009, 2010). Recently, PBDD/Fs have attracted considerable attention because of the more and more extensive use of brominated flame retardants (BFRs) (Evans and Dellinger, 2006; Wang et al., 2008; Haglund, 2010; Hou et al., 2011). PBDD/Fs can be formed unintentionally in the manufacture of brominated flame retardants. In addition, products with brominated flame retardants will sooner or later end up as waste in metal recycling plants or municipal solid incineration. So, a large quantity of PBDD/Fs is emitted from electronic waste recycling and the pyrolysis or combustion of waste materials containing brominated flame retardants (Lai et al., 2007; Li et al., 2007; Wang and Chang-Chien, 2007; Wang et al., 2010a, 2010b; Ma et al., 2009). In China, at least 70 million telephones, 5 million TVs, 4 million refrigerators, and 6 million washing machines have been abandoned annually since 2003 (Li et al., 2007). Furthermore, each year, more than 1 million tons of electronic waste from US, Europe, and other areas is shipped to China (Li et al., 2007). Cheap and primordial methods, like roasting, pyrolysis and combustion, are often used to dismantle the electronic waste, which contains considerable amounts of brominated flame retardants, to recover valuable metals, plastics, and electronic devices (Li et al., 2007). These methods result in severe PBDD/F emissions (Li et al., 2007; Ma et al., 2009).

Excellent linear relationship between the PBDD/F and PCDD/F concentrations revealed their similar formation mechanism under the pyrolysis or combustion conditions (Wang et al., 2010a, 2010b). Chlorophenols (CPs) are structurally similar to PCDD/Fs and the most direct precursors of PCDD/Fs (Karasek and Dickson, 1987; Milligan and Altwicker, 1995). Similarly, bromophenols (BPs) are structurally similar to PBDD/Fs and have been

* Corresponding authors. Fax: +86 531 8836 1990.
E-mail address: zqz@sdu.edu.cn (Q. Zhang).

0045-6535/$ - see front matter © 2013 Published by Elsevier Ltd.
http://dx.doi.org/10.1016/j.chemosphere.2013.01.032

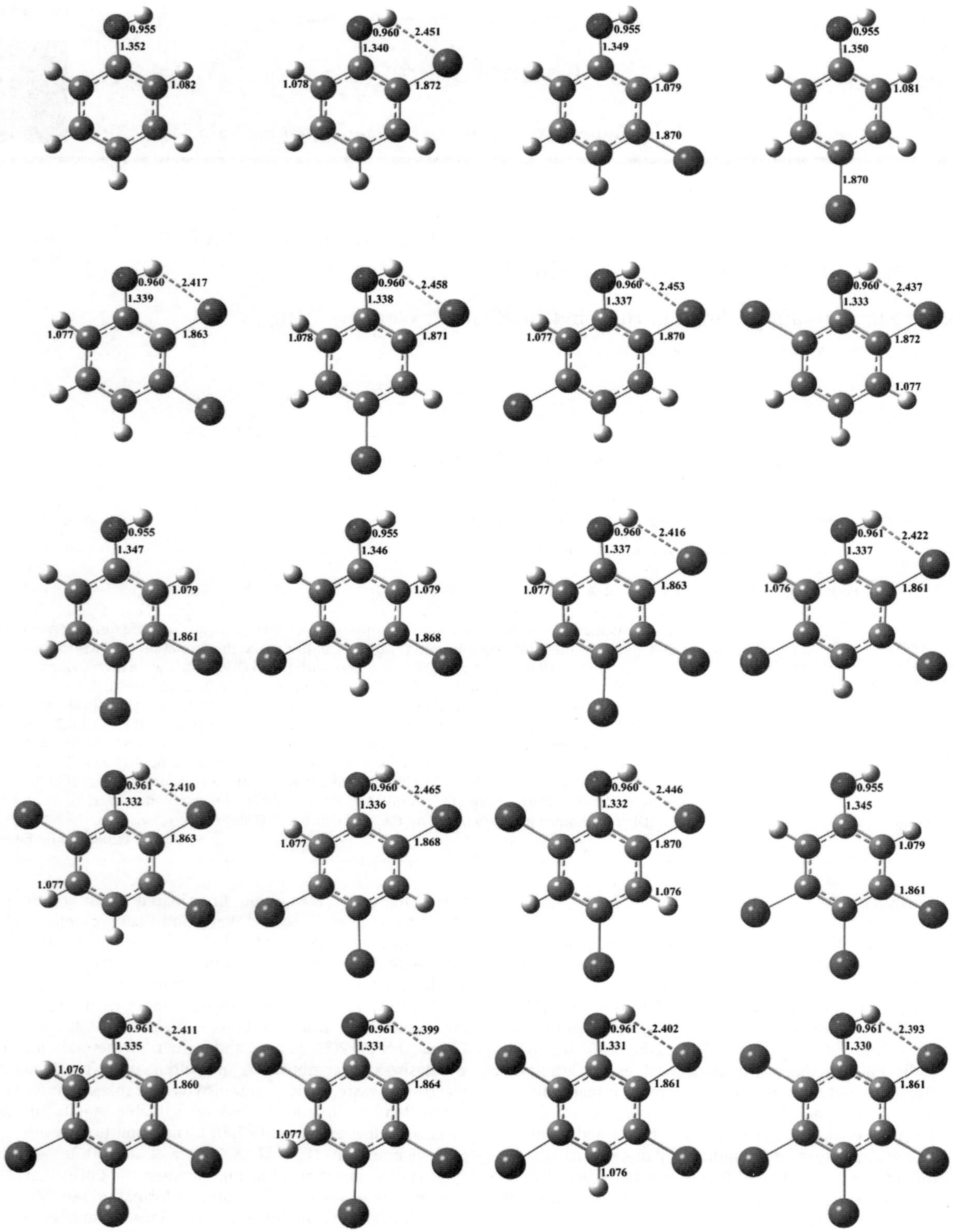

Fig. 1. MPWB1K/6-31 + G(d,p) optimized geometries for bromophenols and phenol. Distances are in angstroms. Gray sphere, C; White sphere, H; Red sphere, O; Blue sphere, Br. (For interpretation of the references to colour in this figure legend, the reader is referred to the web version of this article.)

demonstrated to be the predominant precursors or key intermediates in essentially all proposed pathways of the formation of PBDD/Fs (Evans and Dellinger, 2005a, 2005c, 2006). Utilized as flame retardants, BPs have been consistently used in large quantities in many products such as electronic products and household textiles, as well as additives or intermediates for the yield of other brominated flame retardants. The production volume of 2,4,6-TBP, the most widely manufactured BP, was estimated to be approximately

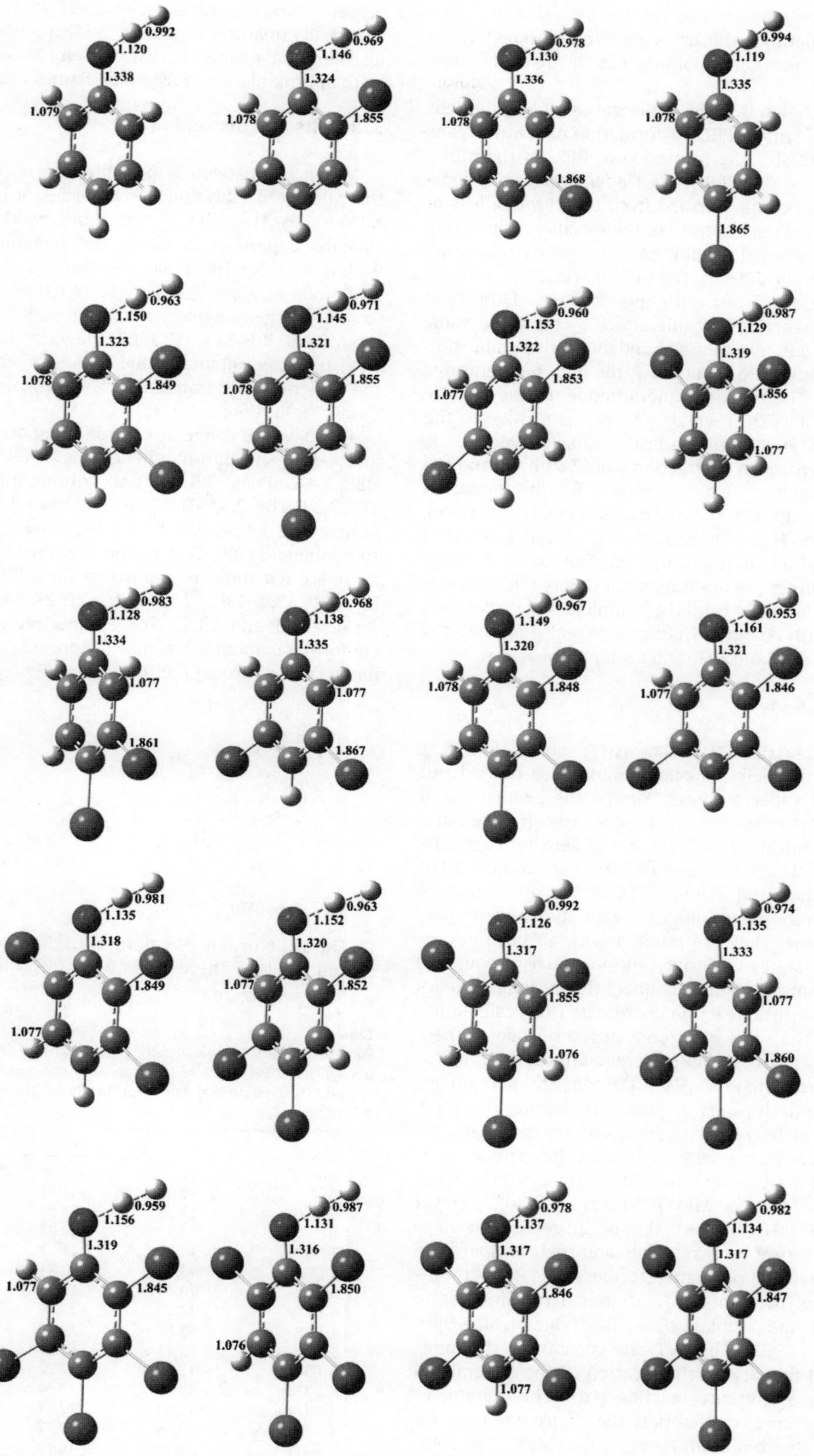

Fig. 2. MPWB1K/6-31 + G(d,p) optimized geometries for the transition states of the phenoxyl-hydrogen abstraction from BPs by H. Distances are in angstroms. Gray sphere, C; White sphere, H; Red sphere, O; Blue sphere, Br. (For interpretation of the references to colour in this figure legend, the reader is referred to the web version of this article.)

2500 tons year^{-1} in Japan and 9500 tons year^{-1} worldwide (IUCLID, 2003). In addition, BPs may be released into the environment as major degradation products of other brominated flame retardants (Sim et al., 2009). In contrast to the situation for PCDD/Fs, relatively few researches have focused on the formation of PBDD/Fs, in spite of the fact that more PBDD/Fs were found to be formed from BPs than PCDD/Fs from corresponding CPs (Evans and Dellinger, 2003, 2005b).

Recent works have shown that the dimerization of bromophenoxy radicals (BPRs) is the major PBDD/F formation pathway (Evans and Dellinger, 2003, 2005b). The formation of BPRs is the initial and key step in the formation of PBDD/Fs. Under pyrolysis or combustion conditions, BPRs can be formed from BPs through loss of the phenoxyl-hydrogen via unimolecular, bimolecular, or possibly other low-energy pathways (including heterogeneous reactions) (Evans and Dellinger, 2003, 2005b). The unimolecular reaction includes the decomposition of BPs with the cleavage of the O–H bond. The bimolecular reactions include attack by the active radicals H, OH, O (^{3}P), Cl and Br, which are abundant in the combustion environment. As yet, very little is known at the high temperatures about these reactions. The reaction kinetic model of the PCDD/F formation revealed that PCDD/F yields are most sensitive to the reactions of CPs with H and OH radicals (Khachatryan et al., 2003). The phenoxyl-hydrogen abstraction from CPs by H and OH radicals is the dominant propagation pathway for the formation of chlorophenoxy radicals (CPRs) (Khachatryan et al., 2003). So, the reactions of BPs with H and OH radicals are naturally expected to play the most central role in the formation of BPRs. Here, therefore, we performed a direct density functional theory (DFT) kinetic study on the formation of BPRs from the complete series reactions of 19 BP congeners with H and OH radicals. We also studied the reactions of phenol with H and OH radicals for comparison.

2. Computational methods

By means of the Gaussian 03 programs (Frisch et al., 2003), high-accuracy quantum chemical calculations were carried out on a SGI Origin 2000 supercomputer. Firstly, the geometries of the stationary points (reactants, complexes, transition states, and products) were fully optimized with the aid of density functional theory (DFT) (Yang et al., 2010; Qu et al., 2012) at the MPWB1K/6-31 + G(d,p) level (Zhao and Truhlar, 2004). For all stationary points, harmonic vibrational frequency calculations were performed to verify whether they are minima with all positive frequencies or transition states with only one imaginary frequency. To confirm that the transition states connect the designated reactants with products, intrinsic reaction coordinate (IRC) calculation was conducted using the same electronic structure theory. Then, the minimum energy paths (MEP) were computed in mass-weighted Cartesian coordinates. Also, the energy derivatives, including gradients and Hessians at geometries along the MEP, were obtained to calculate the curvature of the reaction path. Finally, to yield more accurate energetic information, single point energies for the stationary points and the extra energies along the MEP were calculated at the MPWB1K/6-311 + G(3df,2p) level on the basis of the MPWB1K/6-31 + G(d,p) optimized geometries.

To obtain the theoretical rate constants and their temperature dependence, the Polyrate 9.3 program (Steckler et al., 2002) is employed with the aid of the canonical variational transition-state (CVT) theory (Garrett and Truhlar, 1979; Baldridge et al., 1989; Gonzalez-Lafont et al., 1991). The CVT rate constant is the minimized value obtained by varying the location of the generalized transition state along a reference reaction path. This minimizes the error due to "recrossing" trajectories. The choice we made for the reference path is the minimum energy path (MEP). The rotational and translational partition functions were calculated classically. The vibrational modes were treated as quantum mechanical separable harmonic oscillators. In order to include quantum effect for motion along the reaction coordinate, the CVT rate constant was multiplied by a ground-state transmission coefficient. In the present work, the transmission coefficient was calculated by using the small curvature tunneling (SCT) method, based on the centrifugal-dominant small-curvature semi-classical adiabatic ground-state approximation (Fernandez-Ramos et al., 2007).

3. Results and discussion

When comparison is possible, the optimized geometries and the calculated vibrational frequencies of phenol and 2-BP at the MPWB1K/6-31 + G(d,p) level show good consistency with the available experimental values, and the largest discrepancy remains within 0.9% for the bond lengths and 8.0% for the frequencies (Callomon et al., 1976). For the reaction of phenol + OH → phenoxy + H_2O, the reaction enthalpy of −28.74 kcal mol^{-1} deduced at the MPWB1K/6-311 + G(3df,2p) level and at 298.15 K agrees well with the experimental value of −31.25 kcal mol^{-1}, derived from the experimental standard enthalpies of formation (Cox, 1961; Tsang, 1996; Chase, 1998).

BPs have 19 congeners, three monobromophenols (2-BP, 3-BP and 4-BP), six dibromophenols (2,3-DBP, 2,4-DBP, 2,5-DBP, 2,6-DBP, 3,4-DBP and 3,5-DBP), six tribromophenols (2,3,4-TBP, 2,3,5-TBP, 2,3,6-TBP, 2,4,5-TBP 2,4,6-TBP and 3,4,5-TBP), three tetrabromophenols (2,3,4,5-TeBP, 2,3,4,6-TeBP and 2,3,5,6-TeBP), and pentabromophenol (PBP). Due to the asymmetric bromine substitution, there are *syn* and *anti* conformers for 2-BP, 3-BP 2,3-DBP, 2,4-DBP, 2,5-DBP, 3,4-DBP, 2,3,4-TBP, 2,3,5-TBP, 2,3,6-TBP, 2,4,5-TBP, 2,3,4,5-TeBP and 2,3,4,6-TeBP, respectively. For a given BP, the *syn* conformer is about 3 kcal mol^{-1} more stable than the corresponding *anti* form. So, throughout this paper, BPs denote the *syn* conformers.

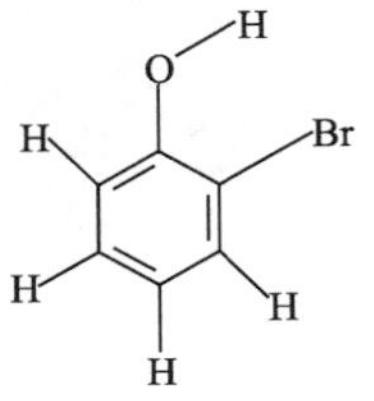

syn BP

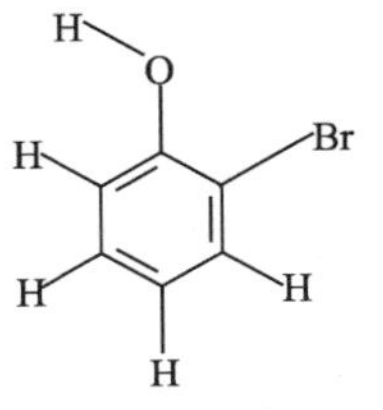

anti BP

The structures of BPs along with the structure of phenol are presented in Fig. 1. There exists weak intramolecular hydrogen bond-

Table 1
Potential barriers ΔE (in kcal mol^{-1}), reaction heats ΔH (in kcal mol^{-1}, 0 K), imaginary frequencies v (in cm^{-1}) of the transition states, and the O–H bond dissociation energies D_0(O–H) (in kcal mol^{-1}) for the phenoxyl-hydrogen abstraction from BPs by H.

	ΔE	ΔH	v	D_0(O–H)
Phenol	11.73	−13.98	2166i	83.95
2-BP	13.96	−11.81	2213i	86.12
3-BP	12.33	−12.92	2191i	85.01
4-BP	11.52	−14.83	2179i	83.10
2,3-DBP	14.25	−11.22	2210i	86.71
2,4-DBP	13.71	−12.81	2234i	85.11
2,5-DBP	14.44	−11.01	2211i	86.92
2,6-DBP	13.56	−12.99	2192i	84.94
3,4-DBP	12.02	−13.89	2200i	84.04
3,5-DBP	12.85	−11.90	2198i	86.03
2,3,4-TBP	13.88	−12.22	2232i	85.71
2,3,5-TBP	14.59	−10.44	2207i	87.49
2,3,6-TBP	13.89	−12.63	2205i	85.30
2,4,5-TBP	13.88	−12.07	2235i	85.86
2,4,6-TBP	13.17	−14.17	2199i	83.76
3,4,5-TBP	12.45	−12.82	2206i	85.11
2,3,4,5-TeBP	14.23	−10.56	2227i	87.37
2,3,4,6-TeBP	13.60	−13.57	2210i	84.35
2,3,5,6-TeBP	14.15	−12.00	2208i	85.93
PBP	13.84	−13.17	2212i	84.76

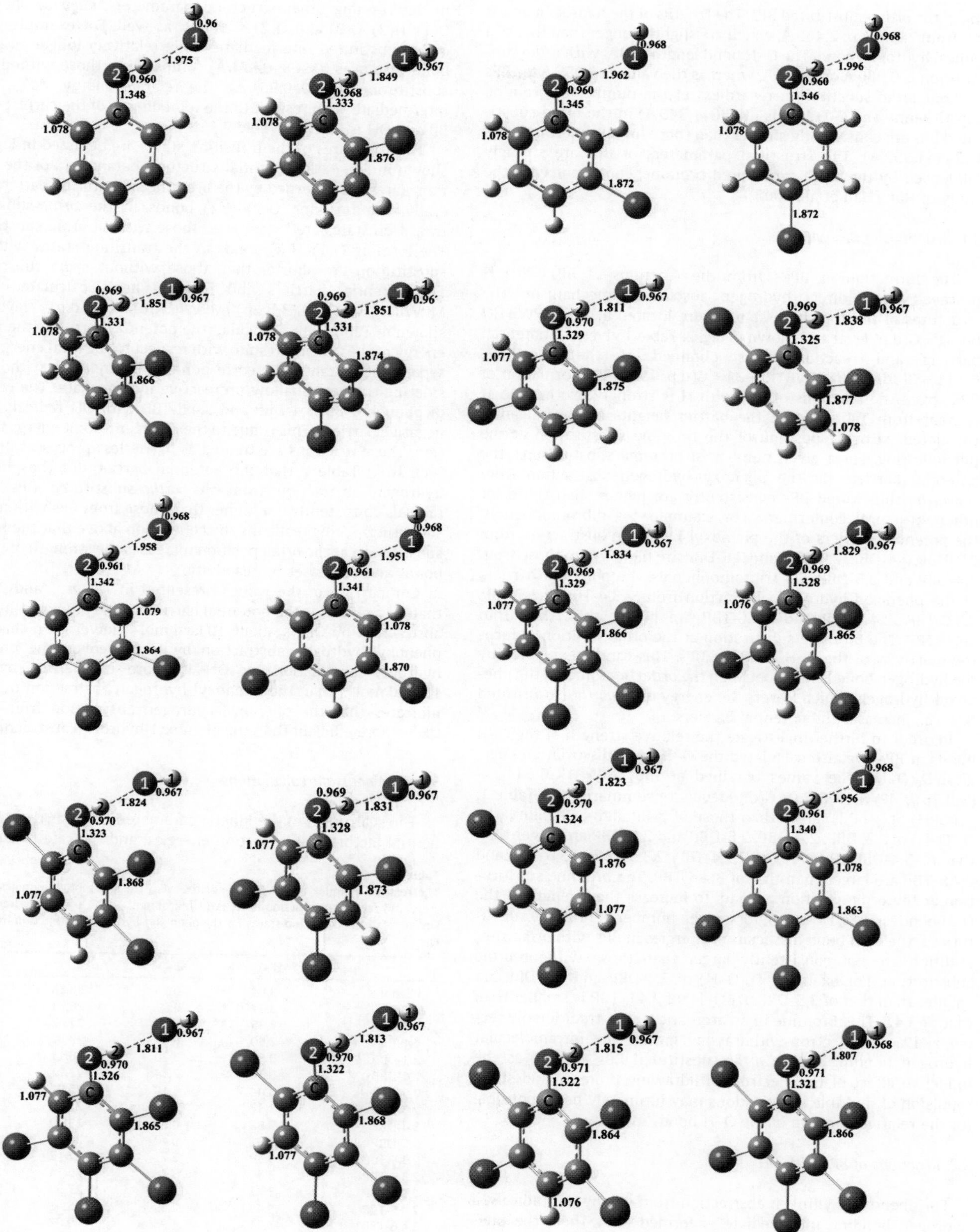

Fig. 3. MPWB1K/6-31 + G(d,p) optimized geometries for the prereactive intermediates of the phenoxyl-hydrogen abstraction from BPs by OH. Distances are in angstroms. Gray sphere, C; White sphere, H; Red sphere, O; Blue sphere, Br. (For interpretation of the references to colour in this figure legend, the reader is referred to the web version of this article.)

ing in the *ortho*-substituted BPs. The lengths of the hydrogen bonds are from 2.393 Å to 2.465 Å, which are slightly longer than that of a typical hydrogen bond. The O–H bond length in BPs with *ortho* substitution is 0.960 Å or 0.961 Å, whereas the value is 0.955 Å for BPs without *ortho* substitution regardless of the number of bromine substituents. The C–O bonds (1.330–1.340 Å) in the *ortho*-substituted BPs are consistently shorter than those for all *nonortho* forms (1.345–1.352 Å). The structural parameters of BPs are strongly influenced by the *ortho*-substituted bromine, consistent with the study of Han (Han et al., 2005).

3.1. Reactions of BPs with H

The formation of BPRs from the reactions of BPs with H proceeds via a direct hydrogen abstraction mechanism. The structures of the transition states were located at the MPWB1K/60-31 + G(d,p) level and shown in Fig. 2. Table 1 gives the potential barriers and reaction heats obtained at the MPWB1K/6-311 + G(3df,2p)//MPWB1K/6-31 + G(d,p) level. The formation of BPRs from the reactions of BPs with H is strongly exothermic. It is clear from Table 1 that the barrier heights are significantly correlated with the position of the bromine substitution at the phenolic ring. For a given number of bromine substitutions, the potential barriers for the phenoxyl-hydrogen abstraction from the *ortho*-substituted BPs consistently are higher than those for other structural conformers. For example, for dibromophenols, the potential barriers of the phenoxyl-hydrogen abstraction from 2,3-DBP, 2,4-DBP, 2,5-DBP and 2,6-DBP are higher than those from 3,4-DBP and 3,5-DBP. For tribromophenols, the potential barriers of the phenoxyl-hydrogen abstraction from 2,3,4-TBP, 2,3,5-TBP, 2,3,6-TBP, 2,4,5-TBP and 2,4,6-TBP are higher than that from 3,4,5-TBP. The bromine substitution at the *ortho* position reduces the reactivity of the O–H bonds in BPs. This can be explained by the hydrogen bond formed between the *ortho* Br atom and the phenoxyl-hydrogen, which lowers the energy of the *ortho*-substituted BPs, i.e., increases the potential barriers.

In order to further investigate the relative strength of the O–H bonds in BPs, we also calculated the O–H bond dissociation energies, D_0(O–H). The values obtained at the MPWB1K/6-311 + G (3df,2p)//MPWB1K/6-31 + G(d,p) level are summarized in Table 1. D_0(O–H) of 2-BP is larger than those of 3-BP and 4-BP. Similarly, D_0(O–H) of 2,3-DBP, 2,4-DBP, 2,5-DBP and 2,6-DBP are larger than that of 3,4-DBP. D_0(O–H) of 2,3,4-TBP, 2,3,5-TBP, 2,3,6-TBP and 2,4,5-TBP are larger than that of 3,4,5-TBP. The bromine substitution at the *ortho* position appears to increase the strength of the O–H bonds in BPs. However, for a given number of bromine substitutions, the O–H bond dissociation energies in BPs with *ortho* substitution are not consistently larger than those without *ortho* substitution. For example, D_0(O–H) of 2,4-DBP and 2,6-DBP are smaller than that of 3,5-DBP. D_0(O–H) of 2,4,6-TBP is smaller than that of 3,4,5-TBP. Bromine in an aromatic ring is traditionally recognized as an electron-withdrawing group. The intramolecular hydrogen bonding in the *ortho*-substituted BPs as well as the inductive effect of the electron-withdrawing bromine and steric repulsion of multiple substitutions may ultimately be responsible for the relative strength of the O–H bonds in BPs.

3.2. Reactions of BPs with OH

For phenoxyl-hydrogen abstraction from BPs by OH radicals, a hydrogen bonding intermediate is formed first. Then, the phenoxyl-hydrogen is abstracted via a transition state. The structures of the intermediates are displayed in Fig. 3. The hydrogen bonds in the intermediates with *ortho* substitution are systematically shorter (1.807–1.851 Å) than those without *ortho* substitution (1.951–1.996 Å). The *ortho* bromine substitution plays an important role in determining other structural parameters, such as the H(1)–O(1), H(2)–O(2) and O(2)–C bonds, as well. For example, all the *ortho*-substituted intermediates have relatively longer H(2)–O(2) bond distances (0.968–0.971 Å) compared to those without *ortho* substitution (0.960–0.961 Å). The relative energy, $\triangle E_{IM}$, of the intermediate with respect to the total energy of the corresponding BP and OH is listed in Table 2.

The structures of the transition states are depicted in Fig. 4. As shown in Fig. 4, the essential structural parameters of the transition states are governed by the bromine substitution pattern. Generally, the breaking O(2)–H(2) bonds in the *ortho*-substituted transition states are longer than those without *ortho* substitution. The forming O(1)–H(2) bonds in the transition states with *ortho* substitution are shorter than those without *ortho* substitution. The potential barriers and reaction heats calculated at the MPWB1K/6-311 + G(3df,2p)//MPWB1K/6-31 + G(d,p) level are shown in Table 2. In particular, the potential barrier is the relative energy of the transition state with respect to the total energy of the separated reactants (the corresponding BP and OH), without considering the very shallow prereactive intermediate. The reactions of phenol, 4-BP, 3,4-DBP and 3,4,5-TBP with OH radicals exhibit negative barriers. This is due to the ZPE (zero-point energy) correction. These reactions are treated as barrierless processes. It can be seen from Table 2 that the potential barriers for the phenoxyl-hydrogen abstraction from the *ortho*-substituted BPs by OH radicals consistently are higher than those from BPs without *ortho* substitution. This reaffirms the conclusion above that the bromine substitution at the *ortho* position increases the strength of the O–H bonds and decreases its reactivity.

Comparison of the values presented in Tables 1 and 2 shows that for a given BP, the potential barrier for the phenoxyl-hydrogen abstraction by OH is about 10 kcal mol^{-1} lower than that of the phenoxyl-hydrogen abstraction by H. Moreover, the phenoxyl-hydrogen abstraction by OH is more exothermic by about 15 kcal mol^{-1} than the phenoxyl-hydrogen abstraction by H. This indicates that the phenoxyl-hydrogen abstraction from BPs by OH is more efficient than the phenoxyl-hydrogen abstraction by H.

3.3. Rate constant calculations

The scarceness of the kinetic parameters, such as the pre-exponential factors, the activation energies, and the rate constants, of

Table 2
The relative energies of the intermediates ΔE_{IM} (in kcal mol^{-1}), potential barriers ΔE_{TS} (in kcal mol^{-1}), reaction heats ΔH (in kcal mol^{-1}, 0 K), imaginary frequencies v (in cm^{-1}) of the transition states for the phenoxyl-hydrogen abstraction from BPs by OH.

	ΔE_{IM}	ΔE_{TS}	ΔH	v
Phenol	−3.04	−0.24	−28.88	1306i
2-BP	−3.27	2.87	−26.71	1660i
3-BP	−3.35	0.16	−27.82	1601i
4-BP	−3.60	−0.82	−29.73	1350i
2,3-DBP	−3.37	3.20	−26.11	1775i
2,4-DBP	−3.36	2.54	−27.71	1646i
2,5-DBP	−3.04	3.18	−25.91	1807i
2,6-DBP	−3.32	2.40	−27.89	1643i
3,4-DBP	−3.58	−0.69	−28.79	1656i
3,5-DBP	−3.61	0.44	−26.80	1837i
2,3,4-TBP	−3.39	2.71	−27.12	1750i
2,3,5-TBP	−3.33	3.44	−25.34	1900i
2,3,6-TBP	−3.25	2.67	−27.53	1680i
2,4,5-TBP	−3.43	2.52	−26.96	1758i
2,4,6-TBP	−3.21	1.93	−29.07	1642i
3,4,5-TBP	−3.72	−0.34	−27.71	1761i
2,3,4,5-TeBP	−3.29	2.96	−25.46	1837i
2,3,4,6-TeBP	−3.21	1.96	−28.47	1744i
2,3,5,6-TeBP	−3.25	2.76	−26.90	1774i
PBP	−3.19	2.23	−28.07	1708i

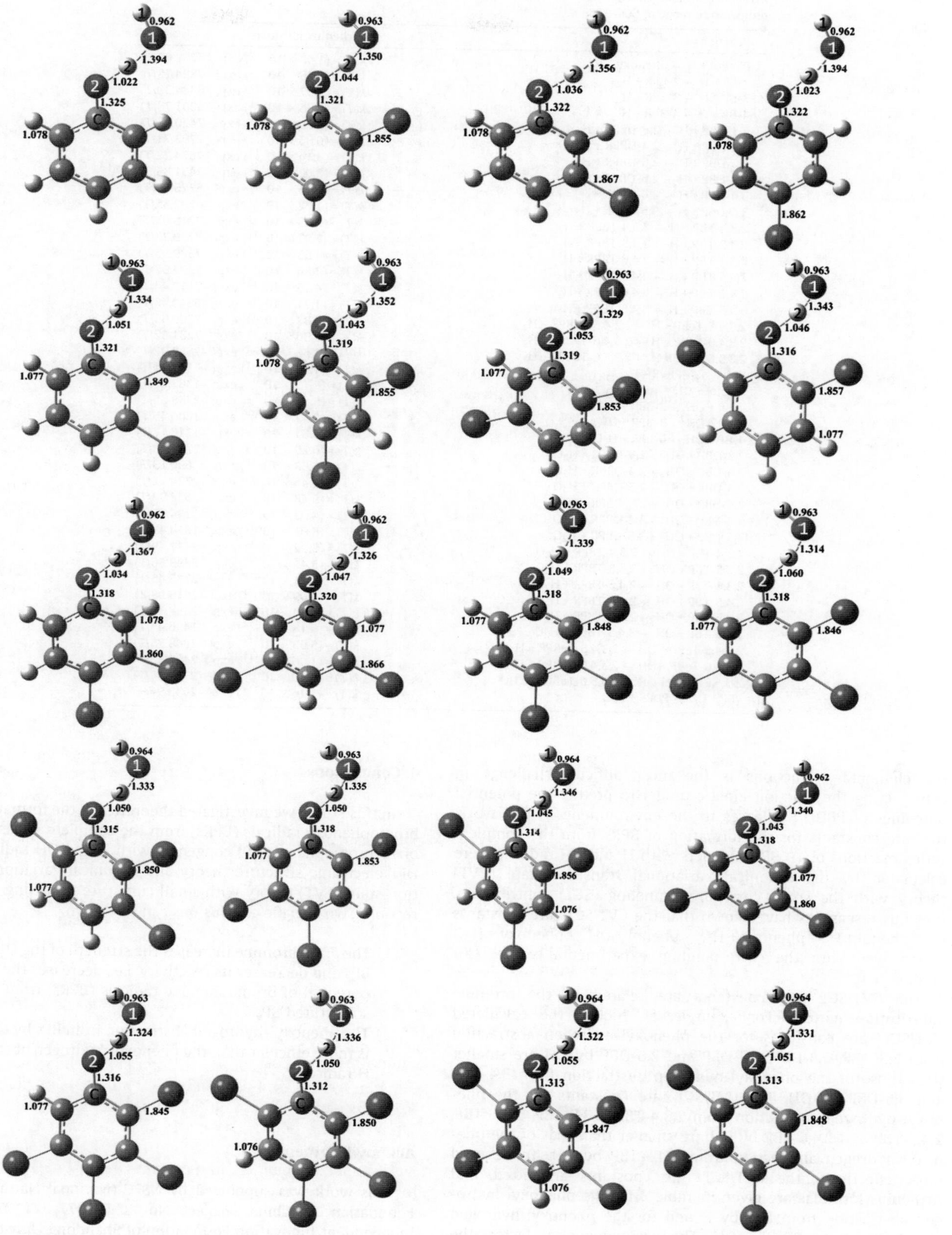

Fig. 4. MPWB1K/6-31 + G(d,p) optimized geometries for the transition states of the phenoxyl-hydrogen abstraction from BPs by OH. Distances are in angstroms. Gray sphere, C; White sphere, H; Red sphere, O; Blue sphere, Br. (For interpretation of the references to colour in this figure legend, the reader is referred to the web version of this article.)

Table 3
Arrhenius formulas (in $cm^3\ mol^{-1}\ s^{-1}$) for the phenoxyl-hydrogen abstraction from BPs by H and OH over the temperature range of 600–1200 K.

Reactions	Arrhenius formulas
Phenol + H → phenoxy + H_2	$k(T) = (1.98 \times 10^{-11})\ \exp(-4767.69/T)$
2-BP + H → 2-BPR + H_2	$k(T) = (3.38 \times 10^{-11})\ \exp(-7384.78/T)$
3-BP + H → 3-BPR + H_2	$k(T) = (2.37 \times 10^{-11})\ \exp(-6384.39/T)$
4-BP + H → 4-BPR + H_2	$k(T) = (2.33 \times 10^{-11})\ \exp(-6001.73/T)$
2,3-DBP + H → 2,3-DBPR + H_2	$k(T) = (5.20 \times 10^{-11})\ \exp(-7420.48/T)$
2,4-DBP + H → 2,4-DBPR + H_2	$k(T) = (6.83 \times 10^{-11})\ \exp(-7292.71/T)$
2,5-DBP + H → 2,5-DBPR + H_2	$k(T) = (3.14 \times 10^{-12})\ \exp(-7674.22/T)$
2,6-DBP + H → 2,6-DBPR + H_2	$k(T) = (8.02 \times 10^{-12})\ \exp(-7431.33/T)$
3,4-DBP + H → 3,4-DBPR + H_2	$k(T) = (2.15 \times 10^{-11})\ \exp(-6275.30/T)$
3,5-DBP + H → 3,5-DBPR + H_2	$k(T) = (1.72 \times 10^{-11})\ \exp(-6792.55/T)$
2,3,4-TBP + H → 2,3,4-TBPR + H_2	$k(T) = (7.43 \times 10^{-11})\ \exp(-7215.03/T)$
2,3,5-TBP + H → 2,3,5-TBPR + H_2	$k(T) = (7.97 \times 10^{-11})\ \exp(-7759.38/T)$
2,3,6-TBP + H → 2,4,6-TBPR + H_2	$k(T) = (3.55 \times 10^{-11})\ \exp(-7389.95/T)$
2,4,5-TBP + H → 2,4,5-TBPR + H_2	$k(T) = (3.56 \times 10^{-12})\ \exp(-7394.97/T)$
2,4,6-TBP + H → 2,4,6-TBPR + H_2	$k(T) = (4.15 \times 10^{-12})\ \exp(-7112.44/T)$
3,4,5-TBP + H → 3,4,5-TBPR + H_2	$k(T) = (1.21 \times 10^{-12})\ \exp(-6549.46/T)$
2,3,4,5-TeBP + H → 2,3,4,5-TeBPR + H_2	$k(T) = (5.82 \times 10^{-11})\ \exp(-7200.65/T)$
2,3,4,6-TeBP + H → 2,3,4,6-TeBPR + H_2	$k(T) = (3.30 \times 10^{-11})\ \exp(-7281.52/T)$
2,3,5,6-TeBP + H → 2,3,5,6-TeBPR + H_2	$k(T) = (3.56 \times 10^{-11})\ \exp(-7582.32/T)$
PBP + H → PBPR + H_2	$k(T) = (1.12 \times 10^{-11})\ \exp(-7179.71/T)$
Phenol + OH → phenoxy + H_2O	$k(T) = (8.13 \times 10^{-12})\ \exp(-1367.37/T)$
2-BP + OH → 2-BPR + H_2O	$k(T) = (5.34 \times 10^{-12})\ \exp(-2758.55/T)$
3-BP + OH → 3-BPR + H_2O	$k(T) = (5.47 \times 10^{-12})\ \exp(-1600.45/T)$
4-BP + OH → 4-BPR + H_2O	$k(T) = (5.61 \times 10^{-12})\ \exp(-1216.17/T)$
2,3-DBP + OH → 2,3-DBPR + H_2O	$k(T) = (6.20 \times 10^{-12})\ \exp(-2751.03/T)$
2,4-DBP + OH → 2,4-DBPR + H_2O	$k(T) = (6.68 \times 10^{-12})\ \exp(-2690.13/T)$
2,5-DBP + OH → 2,5-DBPR + H_2O	$k(T) = (4.45 \times 10^{-12})\ \exp(-2724.12/T)$
2,6-DBP + OH → 2,6-DBPR + H_2O	$k(T) = (9.48 \times 10^{-12})\ \exp(-2638.97/T)$
3,4-DBP + OH → 3,4-DBPR + H_2O	$k(T) = (4.92 \times 10^{-12})\ \exp(-1234.12/T)$
3,5-DBP + OH → 3,5-DBPR + H_2O	$k(T) = (6.91 \times 10^{-12})\ \exp(-1810.45/T)$
2,3,4-TBP + OH → 2,3,4-TBPR + H_2O	$k(T) = (5.82 \times 10^{-12})\ \exp(-2717.49/T)$
2,3,5-TBP + OH → 2,3,5-TBPR + H_2O	$k(T) = (5.41 \times 10^{-12})\ \exp(-2893.56/T)$
2,3,6-TBP + OH → 2,4,6-TBPR + H_2O	$k(T) = (1.01 \times 10^{-11})\ \exp(-2775.45/T)$
2,4,5-TBP + OH → 2,4,5-TBPR + H_2O	$k(T) = (6.09 \times 10^{-12})\ \exp(-2618.98/T)$
2,4,6-TBP + OH → 2,4,6-TBPR + H_2O	$k(T) = (8.93 \times 10^{-12})\ \exp(-2368.10/T)$
3,4,5-TBP + OH → 3,4,5-TBPR + H_2O	$k(T) = (6.18 \times 10^{-12})\ \exp(-1496.45/T)$
2,3,4,5-TeBP + OH → 2,3,4,5-TeBPR + H_2O	$k(T) = (5.53 \times 10^{-12})\ \exp(-2848.75/T)$
2,3,4,6-TeBP + OH → 2,3,4,6-TeBPR + H_2O	$k(T) = (1.06 \times 10^{-11})\ \exp(-2437.39/T)$
2,3,5,6-TeBP + OH → 2,3,5,6-TeBPR + H_2O	$k(T) = (9.80 \times 10^{-12})\ \exp(-2770.54/T)$
PBP + OH → PBPR + H_2O	$k(T) = (1.16 \times 10^{-11})\ \exp(-2643.94/T)$

the elementary reactions is the most difficult challenge in constructing the reaction kinetic model to predict the potential outcomes of PBDD/F releases to the environment. In this work, the rate constants for the formation of BPRs from the complete series reactions of 19 BP congeners with H and OH radicals were calculated by the canonical variational transition-state (CVT) theory with the small curvature tunneling (SCT) contribution. Previous researches have shown that the CVT/SCT rate constants of phenol + H → phenoxy + H_2, phenol + OH → phenoxy + H_2O match well with the corresponding experimental values (Xu et al., 2010).

The CVT/SCT rate constants are related to the bromine substitution pattern. For example, at 1000 K, the calculated CVT/SCT rate constants for the phenoxyl-hydrogen abstraction from 2,3-DBP, 2,4-DBP, 2,5-DBP and 2,6-DBP by OH are smaller than those of the phenoxyl-hydrogen abstraction from 3,4-DBP and 3,5-DBP by OH. The CVT/SCT rate constants for the phenoxyl-hydrogen abstraction from 2,3,4-TBP, 2,3,5-TBP, 2,3,6-TBP, 2,4,5-TBP and 2,4,6-TBP by OH are smaller than that of the phenoxyl-hydrogen abstraction from 3,4,5-TBP by OH. To be used more effectively, the CVT/SCT rate constants are fitted, and Arrhenius formulas are given in Table 3 for the phenoxyl-hydrogen abstraction from BPs by H and for the phenoxyl-hydrogen abstraction from BPs by OH. The pre-exponential factor, the activation energy, and the rate constants can be obtained from these Arrhenius formulas.

4. Conclusions

In this study, we investigated theoretically the formation of the bromophenoxy radicals (BPRs) from the complete series reactions of 19 bromophenol (BP) congeners with H and OH radicals using DFT electronic structure theory and canonical variational transition-state (CVT) theory with small curvature tunneling (SCT) correction. Two specific conclusions can be drawn:

(1) The *ortho* bromine increases the strength of the O–H bond in BPs and decreases its reactivity, i.e., decreases the formation potential of bromophenoxy radicals (BPRs) from the *ortho*-substituted BPs.
(2) The phenoxyl-hydrogen abstraction from BPs by OH radicals is more efficient than the phenoxyl-hydrogen abstraction by H radicals.

Acknowledgments

This work was supported by NSFC (National Natural Science Foundation of China, Project No. 21177077, 21177076) and Independent Innovation Foundation of Shandong University (IIFS-DU, Project No. 2012JC030). The authors thank Professor Donald G. Truhlar for providing the POLYRATE 9.3 program.

Appendix A. Supplementary material

Supplementary data associated with this article can be found, in the online version, at http://dx.doi.org/10.1016/j.chemosphere.2013.01.032.

References

Baldridge, M.S., Gordor, R., Steckler, R., Truhlar, D.G., 1989. Ab initio reaction paths and direct dynamics calculations. J. Phys. Chem. 93, 5107–5119.

Birnbaum, L.S., Staskal, D.F., Diliberto, J.J., 2003. Health effects of polybrominated dibenzo-p-dioxins (PBDDs) and dibenzofurans (PBDFs). Environ. Int. 29, 855–860.

Callomon, J.H., Hellwege, K.H., Hellwege, A.M., 1976. Landolt-Börnstein: Group II: Atomic and Molecular Physics. Structure Data of Free Polyatomic Molecules, vol. 7. Springer-Verlag, Berlin.

Chase, M.W., 1998. NIST JANAF Thermochemical Tables. J. Phys. and Chem. Ref. Data Monogr., fourth ed. American Institute of Physics, College Park, MD.

Cox, J.D., 1961. The heats of combustion of phenol and the three cresols. Pure Appl. Chem. 2, 125–128.

Evans, C.S., Dellinger, B., 2003. Mechanisms of dioxin formation from the high-temperature pyrolysis of 2-bromophenol. Environ. Sci. Technol. 37, 5574–5580.

Evans, C.S., Dellinger, B., 2005a. Formation of bromochlorodibenzo-p-dioxins and furans from the high-temperature pyrolysis of a 2-chlorophenol/2-bromophenol mixture. Environ. Sci. Technol. 39, 7940–7948.

Evans, C.S., Dellinger, B., 2005b. Mechanisms of dioxin formation from the high-temperature oxidation of 2-bromophenol. Environ. Sci. Technol. 39, 2128–2134.

Evans, C.S., Dellinger, B., 2005c. Surface-mediated formation of polybrominated dibenzo-p-dioxins and dibenzofurans from the high-temperature pyrolysis of 2-bromophenol on a CuO/silica surface. Environ. Sci. Technol. 39, 4857–4863.

Evans, C.S., Dellinger, B., 2006. Formation of bromochlorodibenzo-p-dioxins and dibenzofurans from the high-temperature oxidation of a mixture of 2-chlorophenol and 2-bromophenol. Environ. Sci. Technol. 40, 3036–3042.

Fernandez-Ramos, A., Ellingson, B.A., Garret, B.C., Truhlar, D.G., 2007. Variational transition state theory with multidimensional tunneling. In: Lipkowitz, K.B., Cundari, T.R. (Eds.), Reviews in Computational Chemistry. Wiley-VCH, Hoboken NJ, pp. 125–232.

Frisch, M.J., Trucks, G.W., Schlegel, H.B., Scuseria, G.E., Robb, M.A., Cheeseman, J.R., Zakrzewski, V.G., Montgomery, J.A., Stratmann, R.E., Burant, J.C., Dapprich, S., Millam, J.M., Daniels, A.D., Kudin, K.N., Strain, M.C., Farkas, O., Tomasi, J., Barone, V., Cossi, M., Cammi, R., Mennucci, B., Pomelli, C., Adamo, C., Clifford, S., Ochterski, J., Petersson, G.A., Ayala, P.Y., Cui, Q., Morokuma, K., Malick, D.K., Rabuck, A.D., Raghavachari, K., Foresman, J.B., Cioslowski, J., Ortiz, J.V., Baboul, A.G., Stefanov, B.B., Liu, G., Liashenko, A., Piskorz, P., Komaromi, I., Gomperts, R., Martin, R.L., Fox, D.J., Keith, T., Al-Laham, M.A., Peng, C.Y., Nanayakkara, A., Challacombe, M., Gill, P.M.W., Johnson, B., Chen, W., Wong, M.W., Andres, J.L., Gonzalez, C., Head-Gordon, M., Replogle, E.S., Pople, J.A., 2003. Gaussian 03, Revision A.1. Gaussian Inc., Pittsburgh, PA.

Garrett, B.C., Truhlar, D.G., 1979. Generalized transition state theory. Classical mechanical theory and applications to collinear reactions of hydrogen molecules. J. Phys. Chem. 83, 1052–1079.

Gonzalez-Lafont, A., Truong, T.N., Truhlar, D.G., 1991. Interpolated variational transition-state theory: practical methods for estimating variational transition-state properties and tunneling contributions to chemical reaction rates from electronic structure calculations. J. Chem. Phys. 95, 8875–8894.

Haglund, P., 2010. On the identity and formation routes of environmentally abundant tri- and tetrabromodibenzo-p-dioxins. Chemosphere 78, 724–730.

Han, J., Lee, H.J., Tao, F.M., 2005. Molecular structures and properties of the complete series of bromophenols: density functional theory calculations. J. Phys. Chem. A 109, 5186–5192.

Hou, H.F., Chen, B.Y., Zhang, X.S., Wang, Z.Y., 2011. Comparative studies on some properties of polyfluorinated, polychlorinated and polybrominated dibenzo-p-dioxins. Acta. Chimica. Sinica. 69, 617–626.

IUCLID, 2003. IUCLID Data Set: 2,4,6-Tribromophenol, 2,4,6-Tribromophenol and Other Simple Brominated Phenols. International Uniform Chemical Information Database, European Chemicals Bureau, Ispra.

Karasek, F.W., Dickson, L.C., 1987. Model studies of polychlorinated dibenzo-p-dioxin formation during municipal refuse incineration. Science 237, 754–756.

Khachatryan, L., Burcat, A., Dellinger, B., 2003. An elementary reaction-kinetic model for the gas-phase formation of 1, 3, 6, 8- ans 1, 3, 7, 9-tetrachlorinated dibenzo-p-dioxins from 2, 4, 6-trichlorophenol. Combust. Flame. 132, 406–421.

Lai, Y.C., Lee, W.J., Li, H.W., 2007. Inhibition of polybrominated dibenzo-p-dioxin and dibenzofuran formation from the pyrolysis of printed circuit boards. Environ. Sci. Technol. 41, 957–962.

Li, H.R., Yu, L.P., Sheng, G.P., Fu, J.M., Peng, P.A., 2007. Severe PCDD/F and PBDD/F pollution in air around an electronic waste dismantling area in China. Environ. Sci. Technol. 41, 5641–5646.

Li, H.R., Feng, J.L., Sheng, G.Y., Lü, S.L., Fu, J.M., Peng, P.A., Man, R., 2008. The PCDD/F and PBDD/F pollution in the ambient atmosphere of Shanghai, China. Chemosphere 70, 576–583.

Ma, J., Addink, R., Yun, S., Cheng, J.P., Wang, W.H., Kannan, K., 2009. Polybrominated dibenzo-p-dioxins/dibenzofurans and polybrominated diphenyl ethers in soil, vegetation, workshop-floor dust, and electronic shredder residue facility and in soils from a chemical industrial complex in eastern China. Environ. Sci. Technol. 43, 7350–7356.

Mennear, J.H., Lee, C.C., 1994. Polybrominated dibenzo-p-dioxins and dibenzofurans: literature review and health assessment. Environ. Health. Perspect. 102, 265–274.

Milligan, M.S., Altwicker, E.R., 1995. Chlorophenol reactions on fly ash. 1. Adsorption/desorption equilibria and conversion to polychlorinated dibenzo-p-dioxins. Environ. Sci. Technol. 30, 225–229.

Olsman, H., Engwall, M., Kammann, U., Klempt, M., Otte, J., van Bavel, B., Hollert, H., 2007. Relative differences in aryl hydrocarbon receptor-mediated response for 18 polybrominated and mixed halogenated dibenzo-p-dioxins and -furans in cell lines from different species. Environ. Toxicol. Chem. 26, 2448–2454.

Qu, R.J., Liu, H.X., Zhang, Q., Flamm, A., Yang, X., Wang, Z.Y., 2012. The effect of hydroxyl groups on the stability and thermodynamic properties of polyhydroxylated xanthones as calculated by density functional theory. Thermochim. Acta 527, 99–111.

Samara, F., Gullett, B.K., Harrison, R.O., Chu, A., Clark, G.C., 2009. Determination of relative assay response factors for toxic chlorinated and brominated dioxins/furans using an enzyme immunoassay (EIA) and a chemically-activated luciferase gene expression cell bioassay (CALUX). Environ. Int. 35, 588–593.

Samara, F., Wyrzykowska, B., Tabor, D., Touati, D., Gullett, B.K., 2010. Toxicity comparison of chlorinated and brominated dibenzo-p-dioxins and dibenzofurans in industrial source samples by HRGC/HRMS and enzyme immunoassay. Environ. Int. 36, 247–253.

Sim, W.J., Lee, S.H., Lee, I.S., Choi, S.D., Oh, J.E., 2009. Distribution and formation of chlophenols and bromophenols in marine and riverine environments. Chemosphere 77, 552–558.

Steckler, R., Chuang, Y.Y., Fast, P.L., Corchade, J.C., Coitino, E.L., Hu, W.P., Lynch, G.C., Nguyen, K., Jackells, C.F., Gu, M.Z., Rossi, I., Clayton, S., Melissas, V., Garrett, B.C., Isaacson, A.D., Truhlar, D.G., 2002. POLYRATE Version 9.3. University of Minnesota, Minneapolis.

Tsang, W., 1996. Heats of formation of organic free radicals by kinetic methods. In: Energetics of Organic Free Radicals, first ed. Blackie Academic and professional, London, pp. 22–58.

Wang, L.C., Chang-Chien, G.P., 2007. Characterizing the emissions of polybrominated dibenzo-p-dioxins and dibenzofurans from municipal and industrial waste incinerators. Environ. Sci. Technol. 41, 1159–1165.

Wang, L.C., Tsai, C.H., Chang-Chien, G.P., Hung, C.H., 2008. Characterization of polybrominated dibenzo-p-dioxins and dibenzofurans in different atmospheric environments. Environ. Sci. Technol. 42, 75–80.

Wang, L.C., His, H.C., Wang, Y.F., Lin, S.L., Chang-Chien, G.P., 2010a. Distribution of polybrominated diphenyl ethers (PBDEs) and polybrominated dibenzo-p-dioxins and dibenzofurans (PBDD/Fs) in municipal solid waste incinerators. Environ. Pollut. 158, 1595–1602.

Wang, L.C., Wang, Y.F., His, H.C., Chang-Chien, G.P., 2010b. Characterizing the emissions of polybrominated diphenyl ethers (PBDEs) and polybrominated dibenzo-p-dioxins and dibenzofurans (PBDD/Fs) from metallurgical processes. Environ. Sci. Technol. 44, 1240–1246.

Xu, F., Wang, H., Zhang, Q.Z., Zhang, R.X., Qu, X.H., Wang, W.X., 2010. Kinetic properties for the complete series reactions of chlorophenols with OH radicals – relevance for dioxin formation. Environ. Sci. Technol. 44, 1399–1404.

Yang, X., Liu, H., Hou, H.F., Flamm, A., Zhang, X.S., Wang, Z.Y., 2010. Studies of thermodynamic properties and relative stability of a series of polyfluorinated dibenzo-p-dioxins by density functional theory. J. Hazard. Mater. 181, 969–974.

Zhao, Y., Truhlar, D.G., 2004. Hybrid meta density functional theory methods for thermochemistry, thermochemical kinetics, and noncovalent interactions: the MPW1B95 and MPWB1K models and comparative assessments for hydrogen bonding and van der waals interactions. J. Phys. Chem. A 108, 6908–6918.

Environ. Sci. Technol. **2010**, *44*, 1399–1404

Kinetic Properties for the Complete Series Reactions of Chlorophenols with OH Radicals—Relevance for Dioxin Formation

FEI XU,† HUI WANG,‡
QINGZHU ZHANG,*,† RUIXUE ZHANG,†
XIAOHUI QU,† AND WENXING WANG†

Environment Research Institute, Shandong University, Jinan 250100, P. R. China, and Department of Environmental Science and Engineering, Tsinghua University, Beijing 100084, P. R. China

Received October 20, 2009. Revised manuscript received December 5, 2009. Accepted January 5, 2010.

The chlorophenoxy radical (CPR) is a key intermediate species in the formation of polychlorinated dibenzo-*p*-dioxins (PCDDs) and dibenzofurans (PCDFs). In municipal waste incinerators, the reactions of chlorophenols with OH radicals play the most central role in the formation of chlorophenoxy radicals. In this paper, molecular orbital theory calculations have been performed to investigate the formation of chlorophenoxy radicals from the complete series reactions of 19 chlorophenol congeners with OH radicals. The single-point energy calculation was carried out at the MPWB1K/6-311+G(3df,2p) level on the basis of the MPWB1K/6-31+G(d,p) optimized geometries. The kinetic modeling of the PCDD/PCDF (PCDD/F for short) formation demands the knowledge of the rate parameters for the formation of chlorophenoxy radicals from chlorophenols. So, the kinetic properties of the reactions of chlorophenols with OH radicals were deduced over a wide temperature range of 600−1200 K using canonical variational transition-state theory (CVT) with small curvature tunneling contribution (SCT). This study shows that the chlorine substitution at the ortho position in chlorophenol not only has a significant effect on the structures of chlorophenols, prereactive intermediates, the transition states, and chlorophenoxy radicals, but also plays a decisive role in determining the rate parameters.

1. Introduction

Polychlorinated dibenzo-*p*-dioxins (PCDDs) and dibenzofurans (PCDFs) are among the most toxic known environmental pollutants (*1*). Despite international efforts to control and regulate persistent halogenated organic pollutants, total annual global deposition of PCDD/PCDFs (PCDD/Fs for short) from the atmosphere is 13 000 kg/yr (*2*). Among various PCDD/F emission sources, municipal waste incinerations (MWIs) have been recognized as the most significant sources of dioxins release to the environment. In UK, U.S., and Japan, municipal waste incinerations are responsible for 30−56, 38, and 87%, respectively, of the total PCDD/F emissions (*3*−*5*). The total amount of PCDD/Fs emitted from municipal waste incinerations to the atmosphere in China was estimated to be 19.64 g TEQ year^{-1} in 2006 (*6*). Chlorophenols (CPs) are structurally similar to PCDD/Fs and among the most abundant aromatic compounds found in MWI exhaust gases (*7, 8*). CPs have been demonstrated to be the predominant precursors of PCDD/Fs in municipal waste incinerations. It is now known that the potential contributions of the gas-phase pathways to the PCDD/F formation were underestimated by the reaction kinetic model proposed by Shaub and Tsang (*9*−*11*). The homogeneous gas-phase formation of PCDD/Fs from chlorophenol precursors was suggested to make a significant contribution to the observed PCDD/F yields in full-scale incinerators (*10*−*13*).

Recent works have shown that the dimerization of chlorophenoxy radicals (CPRs) is the major pathway in the gas-phase formation of PCDD/Fs from chlorophenol precursors (*10, 11, 14*−*16*). The formation of chlorophenoxy radicals is the initial and key step in the formation of PCDD/Fs. In municipal waste incinerators, chlorophenoxy radicals can be formed through loss of the phenoxyl-hydrogen via unimolecular, bimolecular, or possibly other low-energy pathways (including heterogeneous reactions). The unimolecular reaction includes the decomposition of chlorophenols with the cleavage of the O−H bond. The bimolecular reactions include attack by H, OH, O (^{3}P), Cl, and O_2. The reaction kinetic model of the PCDD/F formation indicated that among various formation pathways of chlorophenoxy radicals from chlorophenols, PCDD/F yields are most sensitive to the reaction of phenoxyl-hydrogen abstraction from chlorophenol by OH radicals (*11*). The reactions of chlorophenols with OH radicals are the dominant propagation pathways for the formation of chlorophenoxy radicals (*11*). The reaction kinetic models that account for the contribution of the gaseous route in the production of PCDD/Fs in combustion processes use the rate constants of the elementary reactions. However, only the theoretical rate constants for the reactions of 2-CP, 2,4,5-TCP with OH radicals are on record (*17*−*19*). There are no existing experimental or theoretical data on the rate constants in the literature for the reactions of other 17 chlorophenol congeners with OH radicals. This is in spite of the fact that their role in the formation of PCDD/Fs is widely discussed (*11*). Thus, the rate constants for the formation of chlorophenoxy radicals from the reactions of chlorophenols with OH radicals were assigned to be the values reported in the literature for the formation of phenoxy radicals from the reaction of phenol with OH radicals in the reaction kinetic model of the PCDD/F formation (*9, 11, 20, 21*). However, where there are uncertainties, the numerical values have been adjusted somewhat to bias the mechanism in favor of the PCDD/F formation, i.e., worst case modeling (*9, 20*).

In a recent contribution from this laboratory, we investigated the formation of chlorophenoxy radicals from the reactions of chlorophenols with atomic H (*22*). As part of our ongoing work in the field, this paper presents mechanistic and kinetic studies on the formations of chlorophenoxy radicals from the reactions of chlorophenols with OH radicals. To compare with the formation of chlorophenoxy radicals, we also studied the formation of phenoxy radicals from the reaction of phenol with OH. The rate constants were calculated using canonical variational transition-state theory (CVT) (*23*−*25*) with small curvature tunneling contribution (SCT) (*26*) over the temperature range of 600−1200 K, which covers the possible formation temperature of PCDD/Fs in municipal waste incinerators. The effect of the chlorine substitution pattern on the structures of the prereactive

* Corresponding author e-mail: zqz@sdu.edu.cn; fax: 86-531-8836 1990.
† Shandong University.
‡ Tsinghua University.

10.1021/es9031776
Published on Web 01/21/2010

intermediates, transition states, and chlorophenoxy radicals is discussed.

2. Computational Methods

The quantum chemical computations were performed on an SGI 2000 supercomputer using the Gaussian 03 package (*27*) for the complete series reactions of 19 chlorophenol congeners with OH radicals. As a reasonable compromise between accuracy and computational time, the geometries of chlorophenols, intermediates, transition states, and chlorophenoxy radicals were fully optimized by employing the MPWB1K method (*28*) with a standard 6-31+G(d,p) basis set. Harmonic vibrational frequency calculations were made at the same level to determine the nature of the stationary points, the zero-point energy (ZPE), and the thermal contributions to the free energy of activation. Our recently published study on the reactions of chlorophenols with atomic H indicated that MPWB1K/6-31+G(d,p) is an excellent method for prediction of the geometrical parameters and the vibrational frequencies of phenol and chlorophenols. Each transition state was verified to connect the designated reactants with products by performing an intrinsic reaction coordinate (IRC) analysis (*29*). The minimum energy path (MEP) was constructed starting from the transition state geometry and going downhill to both the asymptotic reactant and product channel with a gradient stepsize of 0.02 amu$^{1/2}$ bohr. The force constant matrices of the stationary points and selected nonstationary points near the transition state along the MEP were also calculated to do the following kinetic calculation. For a more accurate evaluation of the energetic parameters, a more flexible basis set, 6-311+G(3df,2p), was employed to determine the energies of the various species.

The rate constants were calculated by means of canonical variational transition-state (CVT) theory. The CVT theory is based on the idea of varying the dividing surface along a reference path to minimize the rate constant (*23–25*). The CVT rate constant for temperature T is given by:

$$k^{\mathrm{CVT}}(T) = \min_{s} k^{\mathrm{GT}}(T, s) \qquad (1)$$

where

$$k^{\mathrm{GT}}(T, s) = \frac{\sigma k_{\mathrm{B}} T}{h} \frac{Q^{\mathrm{GT}}(T, s)}{\Phi^{\mathrm{R}}(T)} e^{-V_{MEP}(s)/k_{\mathrm{B}}T} \qquad (2)$$

where $k^{\mathrm{GT}}(T, s)$ is the generalized transition state theory rate constant at the dividing surface s, σ is the symmetry factor accounting for the possibility of more than one symmetry-related reaction path, k_{B} is Boltzmann's constant, h is Planck's constant, $\Phi^{\mathrm{R}}(T)$ is the reactant partition function per unit volume, excluding symmetry numbers for rotation, and $Q^{\mathrm{GT}}(T, s)$ is the partition function of a generalized transition state at s with a local zero of energy at $V_{\mathrm{MEP}}(s)$ and with all rotational symmetry numbers set to unity. All the kinetic calculations have been carried out using the Polyrate 9.3 program (*30*). To include quantum tunneling effects for motion along the reaction coordinate, CVT rate constants were multiplied by a transmission coefficient. In particular, we employed the small curvature tunneling (SCT) method (*26*), based on the centrifugal-dominant small-curvature semiclassical adiabatic ground-state approximation, to calculate the transmission coefficient. The rotational partition functions were calculated classically, and the vibrational modes were treated as quantum-mechanically separable harmonic oscillators.

3. Results and Discussion

3.1. Reaction Mechanism. Due to the different substitution pattern of phenol, chlorophenols have 19 congeners. They include three monochlorophenols (2-CP, 3-CP, and 4-CP), six dichlorophenols (2,3-DCP, 2,4-DCP, 2,5-DCP, 2,6-DCP, 3,4-DCP, and 3,5-DCP), six trichlorophenols (2,3,4-TCP, 2,3,5-TCP, 2,3,6-TCP, 2,4,5-TCP 2,4,6-TCP, and 3,4,5-TCP), three tetrachlorophenols (2,3,4,5-TeCP, 2,3,4,6-TeCP, and 2,3,5,6-TeCP), and pentachlorophenol (PCP). As introduced in our recently published study (*22*), there are two geometric conformers resulting from the two main orientations of the hydroxyl-hydrogen due to the asymmetric chlorine substitution for 12 chlorophenol congeners (2-CP, 3-CP, 2,3-DCP, 2,4-DCP, 2,5-DCP, 3,4-DCP, 2,3,4-TCP, 2,3,5-TCP, 2,3,6-TCP, 2,4,5-TCP, 2,3,4,5-TeCP, 2,3,4,6-TeCP). The conformer with the hydroxyl-hydrogen facing the closest neighboring Cl is labeled as the syn conformer and otherwise the anti conformer. For a given chlorophenol, the syn conformer is about 3 kcal/mol more stable than the corresponding anti form. So throughout this paper, chlorophenols denote the syn conformers. The structures of chlorophenols optimized at the MPWB1K/6-31+G(d,p) level were presented in our recent study (*22*). The effect of the chlorine substitution pattern on the structures of chlorophenols was discussed (*22*).

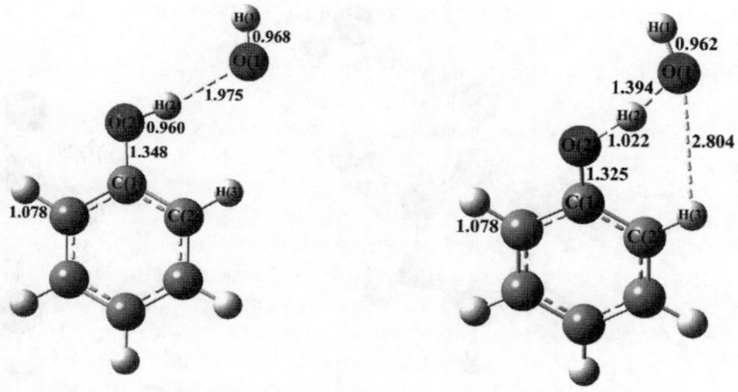

FIGURE 1. MPWB1K/6-31+G(d,p) optimized geometries for the prereactive intermediate, transition state, and phenoxy radical involved in the reaction of phenol with OH radicals. Distances are in angstroms. Gray = C, white = H, red = O.

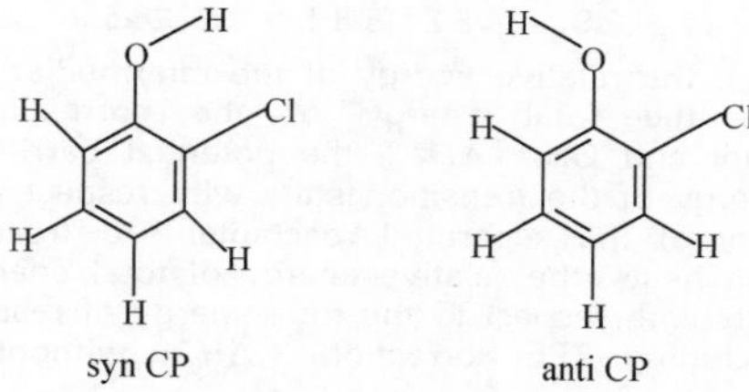

For phenoxyl-hydrogen abstraction from chlorophenols by OH radicals, a hydrogen bonding intermediate is formed first. Then, the phenoxyl-hydrogen is abstracted via a transition state. The structures of the intermediates for the reactions of phenol and 2-CP with OH are presented in Figures 1 and 2, respectively. The structures of the intermediates for the reactions of other chlorophenols with OH radicals are shown in the Supporting Information. The hydrogen bonds in the intermediates with ortho chlorine substitution are systematically shorter (1.907–1.926 Å) than those without ortho substitution (1.953–1.975 Å). This may be due to the different conformations of the intermediates. In the ortho-substituted intermediates, H(1) atom is at the trans-position of O(2) with respect to the O(1)–H(2) bond. In contrast, H(1) atom is at the cis-position of O(2) in the intermediates without ortho substitution. For the same reason, the ortho substitution also has an effect on other

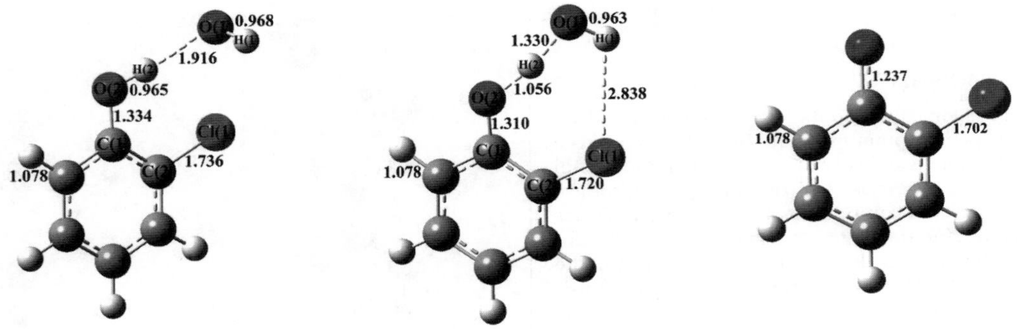

Prereactive intermediate Transition state 2-CPR

FIGURE 2. MPWB1K/6-31+G(d,p) optimized geometries of the prereactive intermediate, transition state, and 2-CPR involved in the reaction of 2-CP with OH radicals. Distances are in angstroms. Gray = C, white = H, red = O, green = Cl.

TABLE 1. Relative Energies of the Intermediates ΔH_0^o(int.) (in kcal/mol, Including ZPE Correction), Potential Barriers $\Delta H_0^{\neq,o}$ (in kcal/mol), Reaction Heats ΔH_0^o(reaction) (in kcal/mol, 0 K, Including ZPE Correction), and Imaginary Frequencies ν (in cm^{-1}) of the Transition States

	ΔH_0^o(int.)[a]	$\Delta H_0^{\neq,o\,b}$	ΔH_0^o (reaction)[c]	ν
phenol	−3.0	-0.2^d (0.7^e)	−28.9	1306i
2-CP	−2.6	3.2^d (4.9^e)	−26.9	2014i
3-CP	−3.3	0.2^d (1.4^e)	−27.8	1575i
4-CP	−3.4	-0.8^d (0.3^e)	−30.1	1317i
2,3-DCP	−2.4	3.6^d (5.5^e)	−26.1	2115i
2,4-DCP	−2.6	2.8^d (4.6^e)	−28.2	2060i
2,5-DCP	−2.6	3.5^d (5.5^e)	−26.1	2241i
2,6-DCP	−2.7	2.8^d (5.0^e)	−28.2	2146i
3,4-DCP	−3.6	-0.4^d (1.1^e)	−29.1	1595i
3,5-DCP	−3.8	0.6^d (2.1^e)	−26.7	1825i
2,3,4-TCP	−2.5	3.3^d (5.0^e)	−27.5	2109i
2,3,5-TCP	−2.9	3.9^d (5.9^e)	−25.3	2277i
2,3,6-TCP	−2.6	3.4^d (5.4^e)	−27.7	2230i
2,4,5-TCP	−2.9	3.2^d (5.0^e)	−27.5	2209i
2,4,6-TCP	−2.7	2.7^d (4.5^e)	−29.5	2187i
3,4,5-TCP	−3.7	-0.0^d (1.5^e)	−28.0	1780i
2,3,4,5-TeCP	−2.8	3.5^d (5.5^e)	−26.7	2259i
2,3,4,6-TeCP	−2.8	3.2^d (5.0^e)	−28.9	2183i
2,3,5,6-TeCP	−2.8	3.7^d (5.7^e)	−27.2	2329i
PCP	−2.9	3.2^d (5.3^e)	−28.5	2317i

[a] ΔH_0^o(int.), the relative energy of the intermediate with respect to the total energy of the corresponding chlorophenol and OH. [b] $\Delta H_0^{\neq,o}$, the potential barrier, the relative energy of the transition state with respect to the total energy of the separated reactants. [c] ΔH_0^o(reaction), the reaction heats, the relative energy of total energy of the products with respect to the total energy of reactants. [d] $\Delta H_0^{\neq,o}$ including ZPE correction. [e] $\Delta H_0^{\neq,o}$ without ZPE correction.

structural parameters, such as the H(1)−O(1), H(2)−O(2), and O(2)−C(1) bonds. For example, all the ortho-substituted intermediates have relative longer H(2)−O(2) bond distances (0.965−0.966 Å) compared to those without ortho substitution (0.960−0.961 Å).

The relative energy (including ZPE correction), ΔH_0^o(int.), of the intermediate with respect to the total energy of the corresponding chlorophenol and OH is summarized in Table 1. The intermediate involved in the reaction of 2-CP with OH was also studied by Altarawneh (*17*). The relative energy of −2.6 kcal/mol obtained in our study at the MPWB1K/6-311+G(3df,2p)//MPWB1K/6-31+G(d,p) level is in excellent agreement with the value of −2.5 kcal/mol (the relative energy of PR1 in the paper of Altarawneh) calculated by Altarawneh at the BB1K/6-311+G(3df,2p)//BB1K/6-31G(d) level (*17*). As shown in Table 1, the relative energies of the intermediates are correlated with the chlorine substitution pattern. ΔH_0^o(int.) are −2.9 to −2.4 kcal/mol for the ortho-substituted intermediates, whereas the values are −3.8 to −3.0 kcal/mol for those without ortho substitution.

Figures 1 and 2 present the essential structural parameters calculated at the MPWB1K/6-31+G(d,p) level of theory for the transition states corresponding to the reactions of phenol and 2-CP with OH radicals, respectively. The structures of the transition states for the reactions of other chlorophenols with OH radicals are shown in the Supporting Information. Clearly, the conformations of the transition states with ortho chlorine substitution are different from the conformations without ortho substitution. In the ortho transition states, H(1) atom is at the trans-position of O(2) with respect to the O(1)−H(2) bond. This trans-conformation results in a weak intramolecular hydrogen bonding between H(1) and Cl(1). The lengths of the hydrogen bonds are from 2.781 to 2.891 Å, which is slightly longer than that of a typical hydrogen bond. Contrarily, H(1) atom is at the cis-position of O(2) in the nonortho transition states. The cis-structure also leads to a weak intramolecular hydrogen bonding between O(1) and H(3). The lengths of the hydrogen bonds are from 2.720 to 2.804 Å. The hydrogen bond can lower the energy of the transition state, i.e., lower the reaction potential barrier.

The increase in length of the O(2)−H(2) bond being broken and the elongation of the O(1)−H(2) bond being formed with respect to its equilibrium value in the reactants and the products are the most important aspect of the geometric structure of the transition state. In the ortho-substituted transition states, the O(2)−H(2) bonds being broken are 7.0%−9.6% longer than the corresponding equilibrium value in chlorophenols, and the O(1)−H(2) bonds being formed are elongated by 39.4%−46.9%. In the transition states without ortho substitution, the O(2)−H(2) bonds being broken are stretched by 10.2%−11.8%, and the O(1)−H(2) bonds being formed are 36.2%−39.6% longer than the corresponding equilibrium value in H_2O molecule. All the transition states under study are reactant-like and appear earlier on the reaction path, as can be anticipated for exothermic reactions (*31*). Additionally, the ortho-substituted transition states are more reactant-like than those without ortho substitution.

All of the transition states have one and only one imaginary frequency. The values of the imaginary frequencies are shown in Table 1. It is clear from Table 1 that the values of the imaginary frequencies are strongly influenced by the chlorine substitution pattern. The imaginary frequencies are 2014i−2329i cm^{-1} for the ortho-substituted transition states, whereas the values are 1306i−1825i cm^{-1} for those without ortho substitution. The large value of the imaginary frequency will narrow the width of the potential barrier. It is known that a narrow barrier is favored in the quantum tunneling effect.

TABLE 2. Parameters *C* and *D* Involved in Arrhenius Formula of $k = C\exp(-D/T)$ (in cm^3 $molecule^{-1}$ s^{-1}) for the Formation of Chlorophenoxy Radicals (CPRs) from the Reactions of Chlorophenols with OH Radicals over the Temperature Range of 600−1200 K

reaction	*C*	*D*
Phenol + OH → C_6H_5O + H_2O	8.13×10^{-12}	1367.37
2-CP + OH → 2-CPR + H_2O	3.43×10^{-12}	2661.33
3-CP + OH → 3-CPR + H_2O	3.12×10^{-12}	1512.16
4-CP + OH → 4-CPR + H_2O	3.60×10^{-12}	1014.82
2,3-DCP + OH → 2,3-DCPR + H_2O	2.48×10^{-13}	2030.49
2,4-DCP + OH → 2,4-DCPR + H_2O	3.43×10^{-12}	2948.92
2,5-DCP + OH → 2,5-DCPR + H_2O	9.93×10^{-13}	3472.64
2,6-DCP + OH → 2,6-DCPR + H_2O	9.54×10^{-13}	2168.31
3,4-DCP + OH → 3,4-DCPR + H_2O	2.32×10^{-12}	1214.28
3,5-DCP + OH → 3,5-DCPR + H_2O	3.36×10^{-12}	1254.00
2,3,4-TCP + OH → 2,3,4-TCPR + H_2O	9.39×10^{-13}	3123.53
2,3,5-TCP + OH → 2,3,5-TCPR + H_2O	7.48×10^{-13}	3705.89
2,3,6-TCP + OH → 2,4,6-TCPR + H_2O	4.27×10^{-13}	2149.90
2,4,5-TCP + OH → 2,4,5-TCPR + H_2O	1.51×10^{-12}	2099.61
2,4,6-TCP + OH → 2,4,6-TCPR + H_2O	9.58×10^{-13}	2752.80
3,4,5-TCP + OH → 3,4,5-TCPR + H_2O	5.20×10^{-12}	1225.42
2,3,4,5-TeCP + OH → 2,3,4,5-TeCPR + H_2O	5.86×10^{-13}	3464.44
2,3,4,6-TeCP + OH → 2,3,4,6-TeCPR + H_2O	8.92×10^{-13}	3354.83
2,3,5,6-TeCP + OH → 2,3,5,6-TeCPR + H_2O	5.88×10^{-13}	3571.13
PCP + OH → PCPR + H_2O	5.71×10^{-13}	3347.88

The potential barriers, $\Delta H_0^{\neq,o}$, and the reaction heats, ΔH_0^o(reaction), calculated at the MPWB1K/6-311+G(3df,2p)//MPWB1K/6-31+G(d,p) level are listed in Table 1. In particular, the potential barrier is the relative energy of the transition state with respect to the total energy of the separated reactants (the corresponding chlorophenol and OH), without considering the very shallow prereactive intermediate. The reactions of phenol, 4-CP, 3,4-DCP, and 3,4,5-TCP with OH radicals exhibit negative barriers. This is due to the ZPE (zero-point energy) correction. For example, for phenol + OH → phenoxy + H_2O, the energy of the transition state without the ZPE correction is higher by 0.7 kcal/mol than the total energy of the separated reactants (phenol and OH). However, the ZPE of the transition state is 0.9 kcal/mol lower than the total ZPE of the separated reactants. So, the potential barrier without the ZPE correction is positive. In contrast, the barrier including the ZPE correction has negative sign for the reactions of phenol, 4-CP, 3,4-DCP, and 3,4,5-TCP with OH radicals.

The potential barrier and reaction heat for 2-CP + OH → 2-CPR + H_2O were also calculated by Altarawneh at the BB1K/6-311+G(3df,2p)//BB1K/6-31G(d) level (*17*). The values of 3.0 and −27.1 kcal/mol reported by Altarawneh (*17*) are in excellent agreement with the values of 3.2 and −26.9 kcal/mol obtained from this work. It can be seen from Table 1 that the potential barriers for the phenoxyl-hydrogen abstraction from the ortho-substituted chlorophenols consistently are higher than those from chlorophenols without ortho substitution. As discussed in our recently published study (*22*), the chlorine substitution at the ortho position in chlorophenol increases the strength of the O−H bonds and decreases its reactivity. It is interesting to compare the phenoxyl-hydrogen abstraction from chlorophenol by H and OH radicals. For a given chlorophenol, the potential barrier for the phenoxyl-hydrogen abstraction by OH is about 10 kcal/mol lower than that of H abstraction by H (*22*). In addition, the phenoxyl-hydrogen abstraction by OH is more exothermic by about 15 kcal/mol than the phenoxyl-hydrogen abstraction by H (*22*). This indicates that the phenoxyl-hydrogen abstraction from chlorophenol by OH can occur more readily than the phenoxyl-hydrogen abstraction by H.

The structures of chlorophenoxy radicals are displayed in the Supporting Information. Chlorophenoxy radical is delocalized, which is a hybrid of one oxygen-centered and three carbon-centered radicals (two ortho and one para carbon sites). The C−O bonds in chlorophenoxy radicals are longer than the C−O double bond and shorter than the C−O single bond. Additionally, the C−O bond lengths also vary with the positions and the number of chlorine substitutions. The chlorophenoxy radicals without ortho substitution have C−O bond lengths of 1.242−1.247 Å. The C−O bonds in the ortho-substituted chlorophenoxy radicals are consistently shorter than 1.24 Å, and decrease with the increase in the number of the ortho chorine substitutions. The C−O bonds are 1.233−1.237 Å and 1.225−1.229 Å in the chlorophenoxy radicals with one and two ortho chlorine substitutions, respectively. This may be attributed to the fact that the induction of chlorine at the ortho position in the phenolic ring is the most effective (*32*).

3.2. Kinetic Calculations. The rate constant calculations were carried out using canonical variational transition state theory (CVT) with small-curvature tunneling (SCT) contribution, which has been successfully performed to evaluate the rate constants for the formation of chlorophenoxy radicals from the reactions of chlorophenols with atomic H (*22*) over the temperature range of 600−1200 K. In particular, direct inspection of the transition state low-frequency mode indicates that the mode of the lowest frequency is a hindered internal rotation instead of a small-amplitude vibration. The mode was removed from the vibration partition function for the transition state and the corresponding hindered rotor partition function $Q_{HR}(T)$, calculated by the method devised by Truhlar (*33*), was included in the expression of the rate constant.

An early experiment studied the reaction of phenol + OH → phenoxy + H_2O (*34*). A relative rate constant of 9.96 × 10^{-12} cm^3 $molecule^{-1}$ s^{-1} was reported over the temperature range of 1000−1150 K on the basis of the rate constant for the reaction of CO + OH → CO_2 + H (*34*). The deviation between our CVT/SCT rate constant and the experimental value remains within a factor of less than 5. Due to the great difference in reactivity between carbon monoxide and phenol, this experimental rate constant for the reaction of phenol + OH → phenoxy + H_2O has large error limits (*34*). Taking into account the significant uncertainty for the experimental value of 9.96 × 10^{-12} cm^3 $molecule^{-1}$ s^{-1}, we think the present CVT/SCT calculations predict reasonable rate constants. The kinetic parameters for the reaction of 2-CP + OH → 2-CPR + H_2O were also studied by Altarawneh using the CVT/SCT method (*17*). The CVT/SCT rate constants calculated by Altarawneh are in excellent agreement with our CVT/SCT values. For example, at 1200 K, the CVT/SCT rate constant calculated by Altarawneh is 2.51 × 10^{-13} cm^3 $molecule^{-1}$ s^{-1} (*17*), and agrees well with the value of 3.73 × 10^{-13} cm^3 $molecule^{-1}$ s^{-1} from this study. Due to the absence of the available experimental values, it is difficult to make a direct comparison of the calculated CVT/SCT rate constants with the experimental data for the reactions of other 18 chlorophenols with OH radicals. We hope that our CVT/SCT calculations may provide a good estimate for the formations of chlorophenoxy radicals from the reactions of chlorophenols with OH radicals.

The CVT/SCT rate constants are strongly affected by the chlorine substitution pattern. Generally, the CVT/SCT rate constants for the phenoxyl-hydrogen abstraction from the ortho-substituted chlorophenols are smaller than those from chlorophenols without ortho subsitution for a given number of chlorine substitutions at a given temperature. For example, at 1000 K, the calculated CVT/SCT rate constants are 3.26 × 10^{-14}, 1.80 × 10^{-13}, 3.08 × 10^{-14}, and 1.09 × 10^{-13} cm^3 $molecule^{-1}$ s^{-1} for the phenoxyl-hydrogen abstraction from 2,3-DCP, 2,4-DCP, 2,5-DCP, and 2,6-DCP, whereas the values are 6.89 × 10^{-13} and 9.59 × 10^{-13} cm^3 $molecule^{-1}$ s^{-1} for the phenoxyl-hydrogen abstraction from 3,4-DCP and 3,5-DCP, respectively. The similar observation also can be found in monochlorophenols as well as in trichlorophenols.

For a given chlorophenol, the CVT/SCT rate constants for the phenoxyl-hydrogen abstraction by OH are consistently larger than those of the phenoxyl-hydrogen abstraction by H over the whole studied temperature range. For example, at 600 K, the CVT/SCT rate constant of the phenoxyl-hydrogen abstraction from 2,3,4-TCP by OH is 5.15 × 10^{-15} cm^3 molecule s^{-1}, whereas the value is 7.45 × $10^{-17}$$cm^3$ molecule s^{-1} for the phenoxyl-hydrogen abstraction by H. This is consistent with previous experimental observation: OH is the most reactive center among various oxidative radicals existing in municipal waste incinerations (*11*).

The calculated CVT/SCT rate constants for the phenoxyl-hydrogen abstraction from chlorophenols by OH radicals are expressed in the Arrhenius form of $k = C\exp(-D/T)$ (in cm^3 $molecule^{-1}$ s^{-1}). The parameters C and D are given in Table 2. The pre-exponential factor, the activation energy, and the rate constants can be obtained from these Arrhenius formulas.

Acknowledgments

This work was supported by NSFC (National Natural Science Foundation of China, project 20737001, 20777047, 20977059),

Shandong Province Outstanding Youth Natural Science Foundation (project JQ200804), and the Research Fund for the Doctoral Program of Higher Education of China (project 200804220046). We thank Professor Donald G. Truhlar for providing the POLYRATE 9.3 program.

Supporting Information Available

MPWB1K/6-31+G(d,p) optimized structures of the prereactive intermediates, transition states, and chlorophenoxy radicals. This material is available free of charge via the Internet at http://pubs.acs.org.

Literature Cited

(1) Hays, S. M.; Aylward, L. L. Dioxin risks in perspective: past, present, and future. *Regul. Toxicol. Pharmacol.* **2003**, *37* (2), 202–217.

(2) Baker, J. I.; Hites, R. Global mass balance for polychlorinated dibenzo-p-dioxins and dibenzofurans. *Environ. Sci. Technol.* **1996**, *30* (6), 1797–1804.

(3) Alcock, R. E.; Gemmill, R.; Jones, K. C. Improvements to the UK PCDD/F and PCB atmospheric emission inventory following an emission measurement programme. *Chemosphere* **1999**, *38* (4), 759–770.

(4) U.S. EPA. *Database of sources of environmental releases of dioxin-like compounds in the United States*; EPA/600/C-01/012; Washington, DC, 2001.

(5) Japan Ministry of the Environment. *Environmental monitoring report on the persistent organic pollutants (POPs) in Japan*, 2002; pp 40–55.

(6) Ni, Y. W.; Zhang, H. J.; Zhang, X. P.; Zhang, Q.; Chen, J. P. Emissions of PCDD/Fs from municipal solid waste incinerators in China. *Chemosphere* **2009**, *75* (9), 1153–1158.

(7) Karasek, F. W.; Dickson, L. C. Model studies of polychlorinated dibenzo-p-dioxin formation during municipal refuse incineration. *Science* **1987**, *237* (4816), 754–756.

(8) Altwicker, E. R. Relative rates of formation of polychlorinated dioxins and furans from precursor and de novo reactions. *Chemosphere* **1996**, *33* (10), 1897–1904.

(9) Shaub, W. M.; Tsang, W. Dioxin formation in incinerators. *Environ. Sci. Technol.* **1983**, *17* (12), 721–730.

(10) Babushok, V. I.; Tsang, W. Gas-phase mechanism for dioxin formation. *Chemosphere* **2003**, *51* (10), 1023–1029.

(11) Khachatryan, L.; Burcat, A.; Dellinger, B. An elementary reaction-kinetic model for the gas-phase formation of 1,3,6,8- and 1,3,7,9-tetrachlorinated dibenzo-p-dioxins from 2,4,6-trichlorophenol. *Combust. Flame* **2003**, *132* (3), 406–421.

(12) Hung, H.; Buekens, A. Comparison of dioxin formation levels in laboratory gas-phase flow reactors with those calculated using the Shaub-Tsang mechanism. *Chemosphere* **1999**, *38* (7), 1595–1602.

(13) Babushok, V.; Tsang, W. *7th International Congress on Combustion By-Products: Origins, Fate, and Health Effects*; Research Triangle Park, NC, 2001, p 36.

(14) Altarawneh, M.; Dlugogorski, B. Z.; Kennedy, E. M.; Mackie, J. C. Quantum chemical investigation of formation of polychlorodibenzo-p-dioxins and dibenzofurans from oxidation and pyrolysis of 2-chlorophenol. *J. Phys. Chem. A* **2007**, *111* (13), 2563–2573.

(15) Sidhu, S.; Edwards, P. Role of phenoxy radicals in PCDD/F formation. *Int. J. Chem. Kinet.* **2002**, *34* (9), 531–541.

(16) Born, J. G. P.; Louw, R.; Mulder, P. Formation of dibenzodioxins and dibenzofurans in homogenous gas-phase reactions of phenols. *Chemosphere* **1989**, *19* (1–6), 401–406.

(17) Altarawneh, M.; Dlugogorski, B. Z.; Kennedy, E. M.; Mackie, J. C. Quantum chemical and kinetic study of formation of 2-chlorophenoxy radical from 2-chlorophenol: Unimolecular decomposition and bimolecular reactions with H, OH, Cl, and O_2. *J. Phys. Chem. A* **2008**, *112* (16), 3680–3692.

(18) Zhang, Q. Z.; Li, S. Q.; Qu, X. H.; Shi, X. Y.; Wang, W. X. A quantum mechanical study on the formation of PCDD/Fs from 2-chlorophenol as precursor. *Environ. Sci. Technol.* **2008**, *42* (19), 7301–7308.

(19) Qu, X. H.; Wang, H.; Zhang, Q. Z.; Shi, X. Y.; Xu, F.; Wang, W. X. Mechanistic and kinetic studies on the homogeneous gas-phase formation of PCDD/Fs from 2,4,5-trichlorophenol. *Environ. Sci. Technol.* **2009**, *43* (11), 4068–4075.

(20) Khachatryan, L.; Asatryan, R.; Dellinger, B. Development of expanded and core kinetic models for the gas phase formation of dioxins from chlorinated phenols. *Chemosphere* **2003**, *52* (14), 695–708.

(21) Khachatryan, L.; Asatryan, R.; Dellinger, B. An elementary reaction kinetic model of the gas-phase formation of polychlorinated dibenzofurans from chlorinated phenols. *J. Phys. Chem. A* **2004**, *108* (44), 9567–9572.

(22) Zhang, Q. Z.; Qu, X. H.; Xu, F.; Shi, X. Y.; Wang, W. X. Mechanism and thermal rate constants for the complete series reactions of chlorophenols with H. *Environ. Sci. Technol.* **2009**, *43* (11), 4105–4112.

(23) Baldridge, M. S.; Gordon, R.; Steckler, R.; Truhlar, D. G. Ab initio reaction paths and direct dynamics calculations. *J. Phys. Chem.* **1989**, *93* (13), 5107–5119.

(24) Gonzalez-Lafont, A.; Truong, T. N.; Truhlar, D. G. Interpolated variational transition-state theory: Practical methods for estimating variational transition-state properties and tunneling contributions to chemical reaction rates from electronic structure calculations. *J. Chem. Phys.* **1991**, *95* (12), 8875–8894.

(25) Garrett, B. C.; Truhlar, D. G. Generalized transition state theory. Classical mechanical theory and applications to collinear reactions of hydrogen molecules. *J. Phys. Chem.* **1979**, *83* (8), 1052–1079.

(26) Liu, Y. P.; Lynch, G. C.; Truong, T. N.; Lu, D. H.; Truhlar, D. G.; Garrett, B. C. Molecular modeling of the kinetic isotope effect for the [1,5]-sigmatropic rearrangement of cis-1,3-pentadiene. *J. Am. Chem. Soc.* **1993**, *115* (6), 2408–2415.

(27) Frisch, M. J.; Trucks, G. W.; Schlegel, H. B.; Gill, P. W. M.; Johnson, B. G.; Robb, M. A.; Cheeseman, J. R.; Keith, T. A.; Petersson, G. A.; Montgomery, J. A.; et al. *GAUSSIAN 03*; Pittsburgh, PA, 2003.

(28) Zhao, Y.; Truhlar, D. G. Hybrid meta density functional theory methods for therochemistry, thermochemical kinetics, and noncovalent interactions: the MPW1B95 and MPWB1K models and comparative assessments for hydrogen bonding and van der Waals interactions. *J. Phys. Chem. A* **2004**, *108* (33), 6908–6918.

(29) Fukui, K. The path of chemical reactions - the IRC approach. *Acc. Chem. Res.* **1981**, *14* (12), 363–368.

(30) Steckler, R.; Chuang, Y. Y.; Fast, P. L.; Corchado, J. C.; Coitino, E. L.; Hu, W. P.; Lynch, G. C.; Nguyen, K.; Jackels, C. F.; Gu, M. Z.; et al. *POLYRATE Version 9.3*; University of Minnesota, Minneapolis, 2002.

(31) Hammond, G. S. A correlation of reaction rates. *J. Am. Chem. Soc.* **1955**, *77* (2), 334–338.

(32) Han, J.; Deming, R. L.; Tao, F. M. Theoretical study of molecular structures and properties of the complete series of chlorophenols. *J. Phys. Chem. A* **2004**, *108* (38), 7736–7743.

(33) Truhlar, D. G. A simple approximation for the vibrational partition function of a hindered internal rotation. *J. Comput. Chem.* **1991**, *12* (2), 266–270.

(34) He, Y. Z.; Mallard, W. G.; Tsang, W. Kinetics of hydrogen and hydroxyl radical attack on phenol at high temperatures. *J. Phys. Chem.* **1988**, *92* (8), 2196–2201.

ES9031776

Environ. Sci. Technol. **2011**, *45*, 643–650

Mechanism and Direct Kinetic Study of the Polychlorinated Dibenzo-*p*-dioxin and Dibenzofuran Formations from the Radical/Radical Cross-Condensation of 2,4-Dichlorophenoxy with 2-Chlorophenoxy and 2,4,6-Trichlorophenoxy

FEI XU, WANNI YU, QIN ZHOU, RUI GAO, XIAOYAN SUN, QINGZHU ZHANG,* AND WENXING WANG

Environment Research Institute, Shandong University, Jinan 250100, People's Republic of China

Received August 5, 2010. Revised manuscript received November 27, 2010. Accepted November 29, 2010.

A direct density functional theory (DFT) kinetic calculation is carried out for the homogeneous gas-phase formation of polychlorinated dibenzo-*p*-dioxins and dibenzofurans (PCDD/Fs) from the cross-condensation of 2,4-dichlorophenoxy radical (2,4-DCPR) with 2-chlorophenoxy radical (2-CPR) and 2,4,6-trichlorophenoxy radical (2,4,6-TCPR). The possible formation mechanism is investigated and compared with the PCDD/F formation mechanism from the self-condensation of 2,4-DCPR, 2-CPR, and 2,4,6-TCPR. The rate constants and their temperature dependence of the crucial elementary reactions are computed by the canonical variational transition-state theory with the small curvature tunneling contribution over the temperature range of 600–1200 K. This study shows that the multichlorine substitutions suppress the PCDD/F formations. Because of a lack of experimental kinetic data, the present theoretical results are expected to be useful and reasonable to estimate the kinetic properties, such as the pre-exponential factors, the activation energies, and the rate constants, of the elementary reactions involved in the formation of PCDD/Fs.

1. Introduction

Incineration as a municipal solid waste management strategy has a number of advantages, namely, reduction of volume and weight and reuse of the energy in the waste. Major disadvantages of incineration processes are, however, the emissions of toxic chlorinated aromatic compounds, such as dioxins—the class of polychlorinated dibenzo-*p*-dioxin (PCDDs) and polychlorinated dibenzofurans (PCDFs). Dioxins are notorious for their biochemical and toxic effects (*1–3*). PCDD/PCDF (PCDD/F for short) emissions from municipal waste incinerators (MWIs) have raised serious concerns globally and led to severe difficulties in constructing both municipal and hazardous waste incinerators (*4, 5*). A mechanistic understanding of the formation of PCDD/Fs is of practical value as well as fundamental interest.

PCDD/F formation rates from precursors that are similar in structure, such as chlorophenols (CPs), have been found to be significantly faster than rates from particulate carbon, or de novo synthesis, under typical incinerator conditions (*6–8*). Also, it is well-known that CPs are among the most abundant aromatic compounds found in MWI flue gases (*8, 9*). Precursor pathways from CPs are the most important for the formation of PCDD/F congeners. Considerable studies have been conducted on the PCDD/F formations from CPs under various experimental conditions (*10–16*). However, most of the research focused on the self-condensation of single CP precursors. The cross-condensation of different CP pairs is responsible for the distribution of PCDD/F homologues as well. For example, the most abundant P_5CDD isomers, 1,2,4,6,8-P_5CDD, 1,2,4,7,9-P_5CDD, 1,2,3,6,8-P_5CDD, and 1,2,3,7,9-P_5CDD, are produced from the cross-condensation of 2,4,6-trichlorophenol (2,4,6-TCP) with 2,3,4,6-tetrachlorophenol (2,3,4,6-TeCP) (*17*). In addition, the study of Ryu showed that PCDF isomers formed from CPs with different numbers of Cl atoms are favored over isomers formed from CPs with similar numbers of Cl atoms, due to steric effects associated with a parallel plane approach geometry of reacting phenoxy radicals (*18*). However, little attention has been given to the formation of PCDD/Fs from the cross-condensation of different CP pairs.

Chlorophenoxy radicals (CPRs) have been identified as key intermediates in essentially all proposed pathways of the formation of PCDD/Fs. The radical/radical condensation of CPRs plays a crucial role in the dioxin formations, especially in the homogeneous gas-phase formation of PCDD/Fs (*19, 20*). The formation of CPRs from CPs, which is the initial step in the formation of PCDD/Fs, has been investigated in detail in the literature (*21–23*). Owing to their significant resonance stabilization, a considerable concentration of CPRs could build up in the combustion environment to enable condensation to occur.

It is difficult to obtain experimental results related to the formation pathway of PCDD/Fs due to a lack of efficient detection schemes for radical intermediate species. Quantum chemical calculation is a widely adopted tool to find the favorite reaction pathways and reaction sites. Furthermore, several research studies pointed out that the scarceness of rate parameters for the elementary reactions involved in the formation of PCDD/Fs is the most difficult challenge in further improving PCDD/F formation models and suggested that additional attention be paid to estimating these parameters (*18, 24–27*). In this paper, therefore, we focus our interest on a quantum mechanical and direct kinetic study of the homogeneous gas-phase formation of PCDD/Fs from the radical/radical cross-condensation of 2,4-dichlorophenoxy radical (2,4-DCPR) with 2-chlorophenoxy radical (2-CPR) and 2,4,6-trichlorophenoxy radical (2,4,6-TCPR). This work complements and expands our previously published studies on the PCDD/F formations from the self-condensation of 2,4-DCPR, 2-CPR, and 2,4,6-TCPR (*28, 29*).

2. Computational Methods

All geometries are fully optimized at the MPWB1K/6-31+G(d,p) level of theory (*30*) using the Gaussian 03 suite of programs (*31*). For all stationary points, frequency calculations are performed to verify whether they are minima with all positive frequencies or transition states with only one imaginary frequency. The intrinsic reaction coordinate (IRC) calculations are carried out at the MPWB1K/6-31+G(d,p)

* Corresponding author e-mail: zqz@sdu.edu.cn; fax: 86-531-8836 1990; phone: 86-531-8836 9788.

10.1021/es102660j
Published on Web 12/13/2010

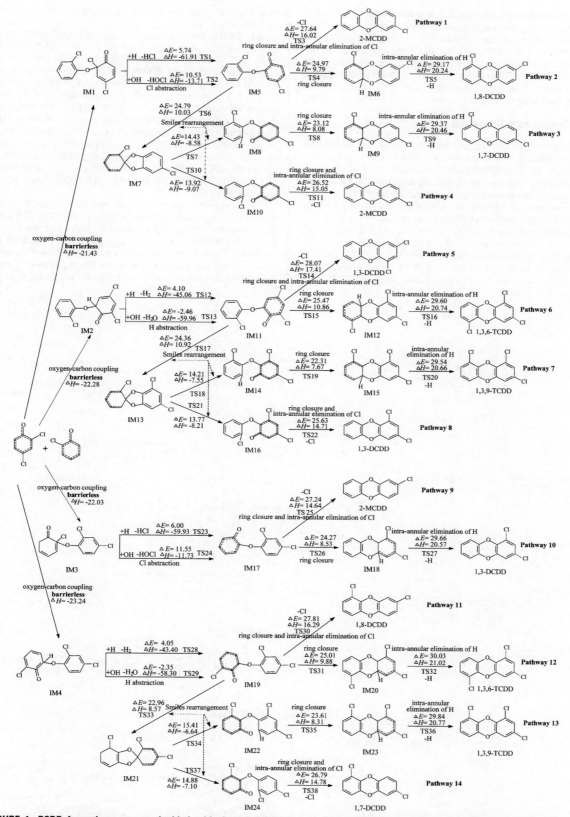

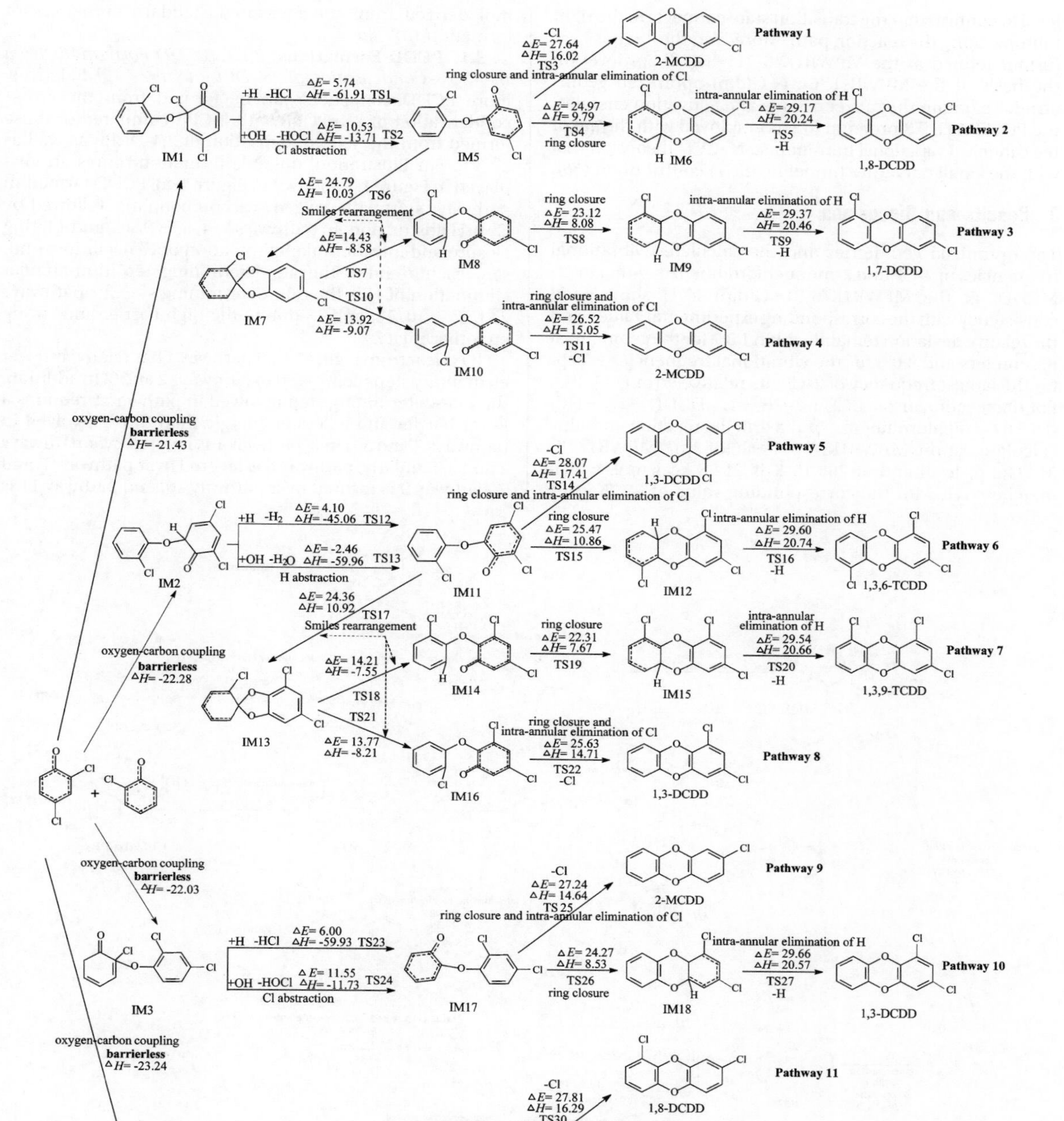

FIGURE 1. PCDD formation routes embedded with the potential barriers ΔE (kcal/mol) and reaction heats ΔH (kcal/mol) from the cross-condensation of 2,4-DCPR with 2-CPR. ΔH is calculated at 0 K.

level to confirm that the transition state connects to the right minima along the reaction path. Moreover, the energies are further refined at the MPWB1K/6-311+G(3df,2p) level on the basis of the MPWB1K/6-31+G(d,p)-optimized geometries. To obtain the rate constants and activation energies, the POLYRATE 9.3 program (*32*) is employed with the aid of the canonical variational transition-state (CVT) theory (*33–35*) with the small curvature tunneling (SCT) contribution (*36*).

3. Results and Discussion

The optimized geometries and the calculated vibrational frequencies of 2-CP and 2-monochlorodibenzo-*p*-dioxin (2-MCDD) at the MPWB1K/6-31+G(d,p) level show good consistency with the corresponding experimental values, and the relative deviation remains within 1.0% for the geometrical parameters and 8.0% for the vibrational frequencies except for the largest frequency of 2-CP (its relative error is 11.4%). For the reaction of 2,4-DCP + 2-CP → 1,3-DCDD + H_2 + HCl (DCDD = dichlorodibenzo-*p*-dioxin), the reaction enthalpy calculated at the MPWB1K/6-311+G(3df,2p)//MPWB1K/6-31+G(d,p) level and at 298.15 K is 21.88 kcal/mol, which matches well with the corresponding value of 22.78 kcal/mol derived from the measured standard enthalpies of formation (*37–40*).

3.1. PCDD Formations. *3.1.1. PCDD Formations from the Cross-Condensation of 2,4-DCPR with 2-CPR.* Relatively more PCDD congeners can be formed from the cross-condensation of 2,4-DCPR with 2-CPR compared to those formed from the self-condensation of 2,4-DCPR as well as 2-CPR. For illustration, possible formation routes are displayed in Figure 1. As shown in Figure 1, all PCDD formation pathways start with oxygen–carbon coupling, followed by Cl or H abstraction. In pathways 1, 4, 5, 8, 9, 11, and 14, ring closure and intra-annular elimination of Cl occur in a one-step reaction and are the rate-determining step. Intra-annular elimination of H is the rate-determining step for pathways 2, 3, 6, 7, 10, 12, and 13 due to the high barrier and strong endothermicity.

It is clear from Figure 1 that pathway 1 has relatively fewer elementary steps compared to pathways 2 and 3. In addition, the rate-determining step involved in pathway 1 requires a lower barrier and is less endoergic than those involved in pathways 2 and 3. Thus, pathway 1 is favored over pathways 2 and 3. Similarly, pathway 5 is favored over pathways 6 and 7, pathway 9 is favored over pathway 10, and pathway 11 is

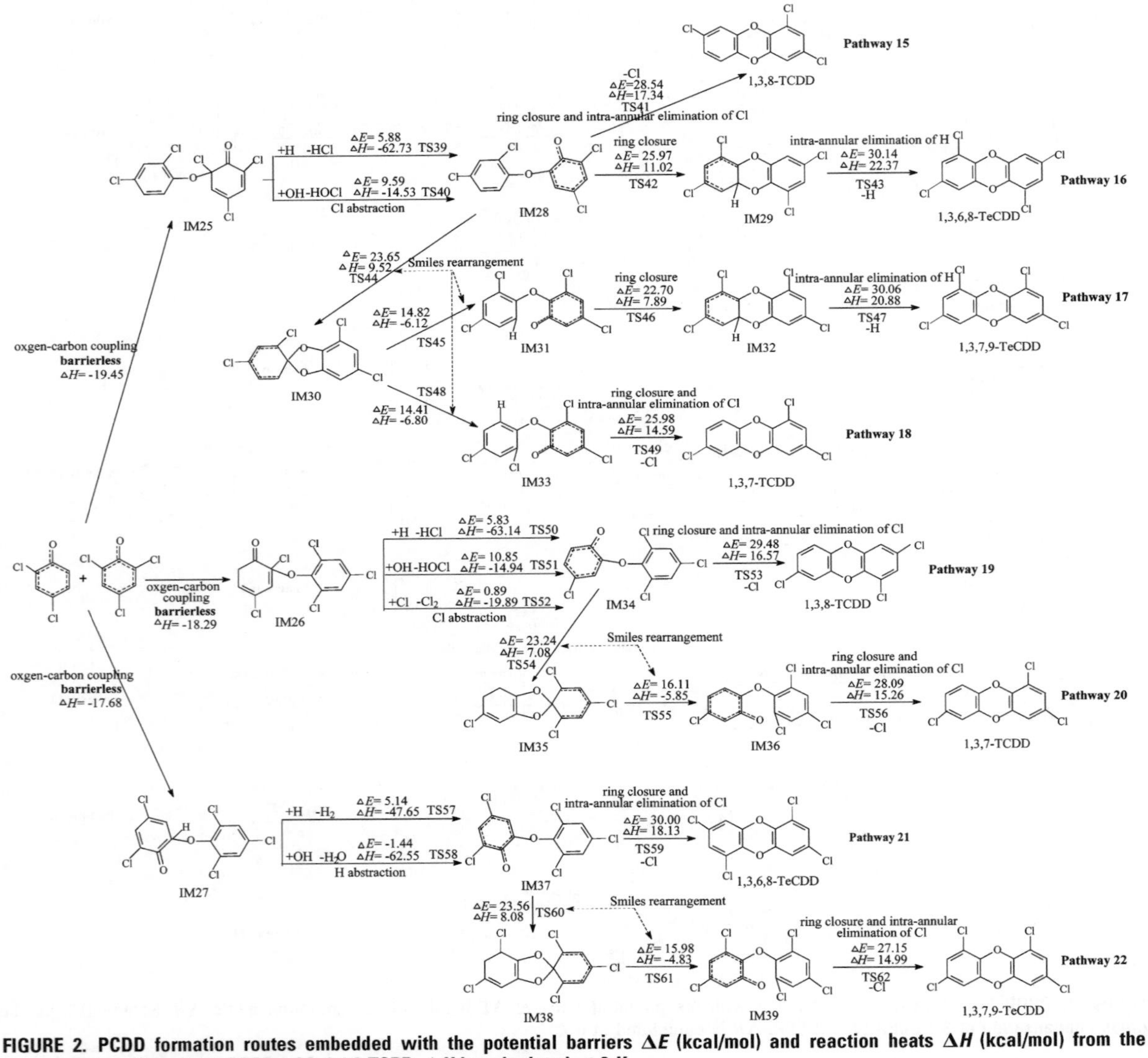

FIGURE 2. PCDD formation routes embedded with the potential barriers Δ*E* (kcal/mol) and reaction heats Δ*H* (kcal/mol) from the cross-condensation of 2,4-DCPR with 2,4,6-TCPR. Δ*H* is calculated at 0 K.

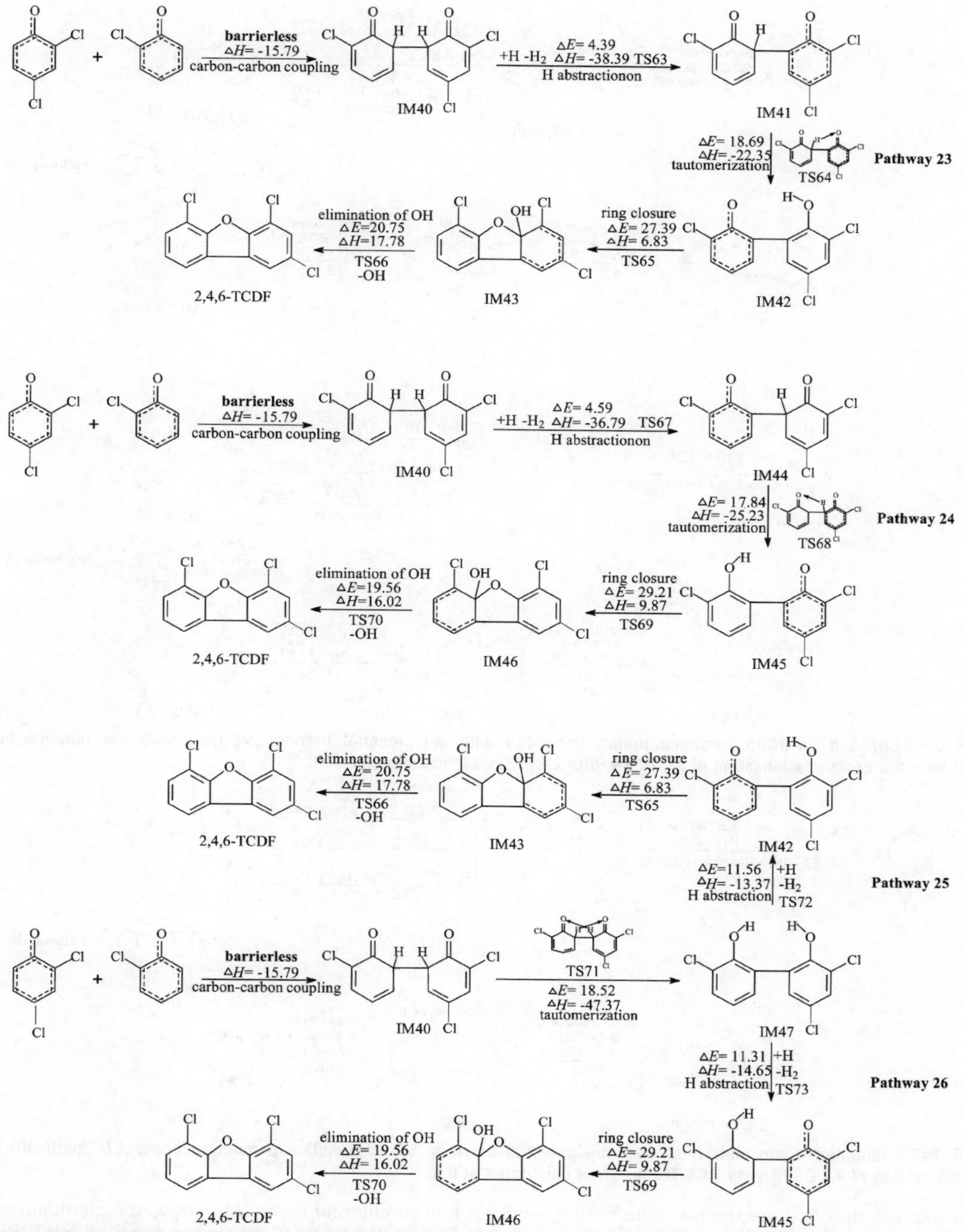

FIGURE 3. 2,4,6-TCDF formation routes embedded with the potential barriers ΔE (kcal/mol) and reaction heats ΔH (kcal/mol) from the cross-condensation of 2,4-DCPR with 2-CPR. ΔH is calculated at 0 K.

favored over pathways 12 and 13. Pathway 8 includes two more elementary steps than pathway 5. However, the rate-determining step involved in pathway 5 has a higher barrier and is more endothermic compared to that involved in pathway 8. Therefore, pathways 5 and 8 should be competitive. For the same reason, pathways 1 and 4 should be competitive and pathways 11 and 14 should be competitive. Therefore, the thermodynamically favored PCDD formation pathways are pathways 1, 4, 5, 8, 9, 11, and 14. This reaffirms the previous conclusion that the thermodynamically favored PCDD formation pathways occur through intra-annular elimination of Cl (*28*). The resulting 2-MCDD, 1,3-DCDD, 1,7-DCDD, and 1,8-DCDD are the dominant PCDD products from the cross-condensation of 2,4-DCPR with 2-CPR.

3.1.2. PCDD Formations from the Cross-Condensation of 2,4-DCPR with 2,4,6-TCPR. Figure 2 depicts the PCDD formation mechanism from the cross-condensation of 2,4-DCPR with 2,4,6-TCPR. Similar to the cross-condensation of 2,4-DCPR with 2-CPR, all PCDD formation pathways start with oxygen−carbon coupling, followed by Cl or H abstraction. The oxygen−carbon coupling is a barrierless and strongly exothermic process. All of the Cl or H abstraction steps are highly exothermic. In pathways 15, 18, 19, 20, 21, and 22, ring closure and intra-annular elimination of Cl occur

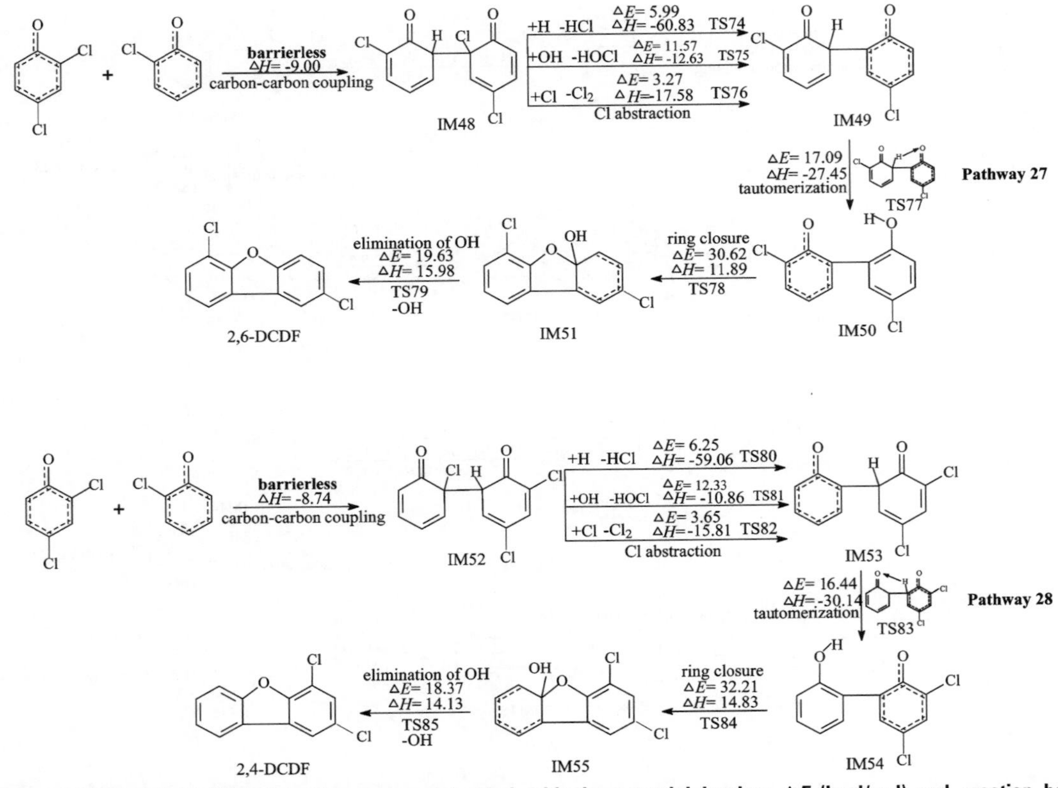

FIGURE 4. 2,6-DCDF and 2,4-DCDF formation routes embedded with the potential barriers ΔE (kcal/mol) and reaction heats ΔH (kcal/mol) from the cross-condensation of 2,4-DCPR with 2-CPR. ΔH is calculated at 0 K.

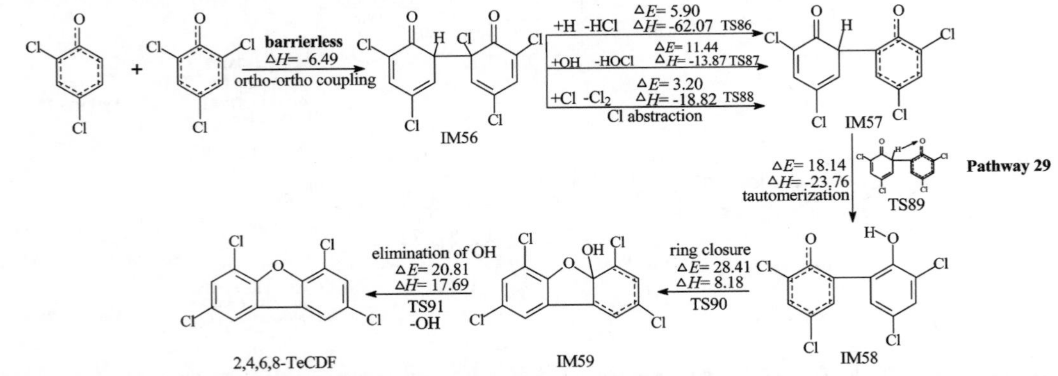

FIGURE 5. PCDF formation route embedded with the potential barriers ΔE (kcal/mol) and reaction heats ΔH (kcal/mol) from the cross-condensation of 2,4-DCPR with 2,4,6-TCPR. ΔH is calculated at 0 K.

in a one-step reaction and are the rate-determining step. Intra-annular elimination of H is the rate-determining step for pathways 16 and 17. As seen from Figure 2, PCDDs are preferentially formed from pathways 15, 18, 19, 20, 21, and 22, resulting in the formation of 1,3,7-TCDD, 1,3,8-TCDD, 1,3,6,8-TeCDD and 1,3,7,9-TeCDD (TCDD = trichlorodibenzo-*p*-dioxin, and TeCDD = tetrachlorodibenzo-*p*-dioxin), consistent with the study of Ryu (*41*).

In general, 1,3,6,8-TeCDD and 1,3,7,9-TeCDD, the most abundant PCDD congeners found in municipal waste incinerators, are considered to be formed from the self-condensation of 2,4,6-TCP. Comparison of the mechanism displayed in Figure 2 with a previous study (*28*) shows that the formation of 1,3,6,8-TeCDD and 1,3,7,9-TeCDD from the cross-condensation of 2,4-DCPR with 2,4,6-TCPR by one chlorine loss is favored over the formation of 1,3,6,8-TeCDD and 1,3,7,9-TeCDD from the self-condensation of 2,4,6-TCPR by two chlorine losses. Moreover, the abundances of 2,4-DCP and 2,4,6-TCP are almost equal in MWI flue gases. Therefore, the cross-condensation of 2,4-DCPR with 2,4,6-TCPR should make a significant contribution to the distribution of 1,3,6,8-TeCDD and 1,3,7,9-TeCDD in MWIs.

Comparison of the reaction pathways presented in Figures 1 and 2 with previous research (*28, 29*) clearly shows that the substitution pattern of chlorophenols has a significant influence on the PCDD formation mechanism, especially on the oxygen−carbon coupling of CPR. The exothermicities of the oxygen−carbon coupling for 2-CPR + 2-CPR, 2-CPR + 2,4-DCPR, 2,4-DCPR + 2,4-DCPR, 2,4-DCPR + 2,4,6-TCPR, and 2,4,6-TCPR + 2,4,6-TCPR are 23.04−24.50, 21.43−23.04, 20.26−21.14, 17.68−19.45, and 17.06 kcal/mol. The exothermicity of the oxygen−carbon coupling decreases with increasing number of chlorine substitutions. The ranking of the PCDD formation potential is as follows: 2-CPR + 2-CPR

TABLE 1. Arrhenius Formulas[a] for the Elementary Reactions Involved in the Thermodynamically Preferred Formation Pathways of PCDDs from the Cross-Condensation of 2,4-DCPR with 2-CPR and 2,4,6-TCPR over the Temperature Range of 600−1200 K

reaction	Arrhenius formula
IM1 + H → IM5 + HCl	$k(T) = (3.98 \times 10^{-11}) \exp(-3424.8/T)$
IM1 + OH → IM5 + HOCl	$k(T) = (2.02 \times 10^{-11}) \exp(-6931.5/T)$
IM5 → 2-MCDD + Cl	$k(T) = (3.38 \times 10^{11}) \exp(-13833.2/T)$
IM5 → IM7	$k(T) = (1.37 \times 10^{12}) \exp(-12584.0/T)$
IM7 → IM10	$k(T) = (9.23 \times 10^{12}) \exp(-6968.1/T)$
IM10 → 2-MCDD + Cl	$k(T) = (7.67 \times 10^{11}) \exp(-12836.3/T)$
IM2 + H → IM11 + H_2	$k(T) = (1.06 \times 10^{-11}) \exp(-2663.5/T)$
IM11 → 1,3-DCDD + Cl	$k(T) = (2.55 \times 10^{11}) \exp(-13936.9/T)$
IM11 → IM13	$k(T) = (5.94 \times 10^{11}) \exp(-11789.7/T)$
IM13 → IM16	$k(T) = (4.01 \times 10^{13}) \exp(-8721.6/T)$
IM16 → 1,3-DCDD + Cl	$k(T) = (5.18 \times 10^{11}) \exp(-12620.1/T)$
IM3 + H → IM17 + HCl	$k(T) = (1.89 \times 10^{-11}) \exp(-3359.7/T)$
IM3 + OH → IM17 + HOCl	$k(T) = (1.09 \times 10^{-11}) \exp(-7442.4/T)$
IM17 → 2-MCDD + Cl	$k(T) = (2.54 \times 10^{11}) \exp(-13578.8/T)$
IM4 + H → IM19 + H_2	$k(T) = (1.57 \times 10^{-11}) \exp(-2198.1/T)$
IM19 → 1,8-DCDD + Cl	$k(T) = (2.52 \times 10^{11}) \exp(-13832.6/T)$
IM19 → IM21	$k(T) = (3.60 \times 10^{11}) \exp(-11373.4/T)$
IM21 → IM24	$k(T) = (1.04 \times 10^{13}) \exp(-7680.4/T)$
IM24 → 1,7-DCDD + Cl	$k(T) = (9.48 \times 10^{11}) \exp(-12786.6/T)$
IM25 + H → IM28 + HCl	$k(T) = (5.51 \times 10^{-11}) \exp(-3723.3/T)$
IM25 + OH → IM28 + HOCl	$k(T) = (6.43 \times 10^{-12}) \exp(-6453.2/T)$
IM28 → 1,3,8-TCDD + Cl	$k(T) = (5.28 \times 10^{11}) \exp(-14332.4/T)$
IM28 → IM30	$k(T) = (1.26 \times 10^{12}) \exp(-11828.3/T)$
IM30 → IM33	$k(T) = (3.25 \times 10^{11}) \exp(-4806.5/T)$
IM33 → 1,3,7-TCDD + Cl	$k(T) = (5.02 \times 10^{11}) \exp(-13802.3/T)$
IM26 + H → IM34 + HCl	$k(T) = (3.84 \times 10^{-11}) \exp(-3976.2/T)$
IM26 + OH → IM34 + HOCl	$k(T) = (1.18 \times 10^{-11}) \exp(-6457.9/T)$
IM26 + Cl → IM34 + Cl_2	$k(T) = (5.02 \times 10^{-11}) \exp(-1630.7/T)$
IM34 → 1,3,8-TCDD + Cl	$k(T) = (1.55 \times 10^{12}) \exp(-14918.6/T)$
IM34 → IM35	$k(T) = (1.83 \times 10^{12}) \exp(-11452.0/T)$
IM35 → IM36	$k(T) = (1.87 \times 10^{13}) \exp(-8305.6/T)$
IM36 → 1,3,7-TCDD + Cl	$k(T) = (3.50 \times 10^{11}) \exp(-13975.9/T)$
IM27 + H → IM37 + H_2	$k(T) = (5.71 \times 10^{-11}) \exp(-3486.3/T)$
IM37 → 1,3,6,8-TeCDD + Cl	$k(T) = (5.52 \times 10^{10}) \exp(-12386.7/T)$
IM37 → IM38	$k(T) = (3.07 \times 10^{12}) \exp(-9365.3/T)$
IM38 → IM39	$k(T) = (1.70 \times 10^{13}) \exp(-8569.5/T)$
IM39 → 1,3,7,9-TeCDD + Cl	$k(T) = (5.25 \times 10^{11}) \exp(-13581.6/T)$

[a] Units are s^{-1} and cm^3 $molecule^{-1}$ s^{-1} for unimolecular and bimolecular reactions.

TABLE 2. Arrhenius Formulas[a] for the Elementary Reactions Involved in the Formation of PCDFs from the Cross-Condensation of 2,4-DCPR with 2-CPR and 2,4,6-TCPR over the Temperature Range of 600−1200 K

reaction	Arrhenius formula
IM40 + H → IM41 + H_2	$k(T) = (2.62 \times 10^{-11}) \exp(-2999.8/T)$
IM41 → IM42	$k(T) = (3.32 \times 10^{12}) \exp(-9755.8/T)$
IM42 → IM43	$k(T) = (1.77 \times 10^{12}) \exp(-13710.2/T)$
IM43 → 2,4,6-TCDF + OH	$k(T) = (3.21 \times 10^{13}) \exp(-10884.1/T)$
IM40 + H → IM44 + H_2	$k(T) = (2.65 \times 10^{-11}) \exp(-3183.1/T)$
IM44 → IM45	$k(T) = (2.24 \times 10^{12}) \exp(-9338.0/T)$
IM45 → IM46	$k(T) = (1.79 \times 10^{12}) \exp(-14580.8/T)$
IM46 → 2,4,6-TCDF + OH	$k(T) = (2.78 \times 10^{13}) \exp(-10216.7/T)$
IM40 → IM47	$k(T) = (1.34 \times 10^{12}) \exp(-5447.9/T)$
IM47 + H → IM42 + H_2	$k(T) = (5.38 \times 10^{-13}) \exp(-7565.9/T)$
IM47 + H → IM45 + H_2	$k(T) = (3.84 \times 10^{-13}) \exp(-7111.2/T)$
IM48 + H → IM49 + HCl	$k(T) = (4.11 \times 10^{-11}) \exp(-3960.5/T)$
IM48 + OH → IM49 + HOCl	$k(T) = (1.16 \times 10^{-12}) \exp(-7015.5/T)$
IM48 + Cl → IM49 + Cl_2	$k(T) = (2.85 \times 10^{-11}) \exp(-3066.9/T)$
IM49 → IM50	$k(T) = (3.65 \times 10^{12}) \exp(-8966.3/T)$
IM50 → IM51	$k(T) = (5.01 \times 10^{12}) \exp(-15529.8/T)$
IM51 → 2,6-DCDF + OH	$k(T) = (2.83 \times 10^{13}) \exp(-10275.9/T)$
IM52 + H → IM53 + HCl	$k(T) = (3.28 \times 10^{-11}) \exp(-4080.9/T)$
IM52 + OH → IM53 + HOCl	$k(T) = (2.01 \times 10^{-12}) \exp(-7495.6/T)$
IM52 + Cl → IM53 + Cl_2	$k(T) = (8.74 \times 10^{-11}) \exp(-3228.8/T)$
IM53 → IM54	$k(T) = (1.32 \times 10^{12}) \exp(-8654.6/T)$
IM54 → IM55	$k(T) = (7.60 \times 10^{12}) \exp(-16507.9/T)$
IM55 → 2,4-DCDF + OH	$k(T) = (3.49 \times 10^{13}) \exp(-9758.7/T)$
IM56 + H → IM57 + HCl	$k(T) = (4.83 \times 10^{-11}) \exp(-3817.4/T)$
IM56 + OH → IM57 + HOCl	$k(T) = (1.34 \times 10^{-12}) \exp(-6855.4/T)$
IM56 + Cl → IM57 + Cl_2	$k(T) = (3.63 \times 10^{-11}) \exp(-2799.4/T)$
IM59 → 2,4,6,8-TeCDF + OH	$k(T) = (3.17 \times 10^{13}) \exp(-10982.1/T)$

[a] Units are s^{-1} and cm^3 $molecule^{-1}$ s^{-1} for unimolecular and bimolecular reactions.

> 2-CPR + 2,4-DCPR > 2,4-DCPR + 2,4-DCPR > 2,4-DCPR + 2,4,6-TCPR > 2,4,6-TCPR + 2,4,6-TCPR. This means that the PCDD formations are favored from less chlorinated phenols. The result is supported by the experimental evidence that the PCDD yields and homologue fractions decrease with increasing number of chlorine substituents (*17, 41*). Although *o*-chlorine is needed for the formation of PCDDs (*28*), multichlorine substitutions suppress the PCDD formations.

3.2. Formation of PCDFs. Three PCDF congeners, 2,4,6-TCDF, 2,6-DCDF, and 2,4-DCDF (TCDF = trichlorodibenzofuran, DCDF = dichlorodibenzofuran), can be formed from the cross-condensation of 2,4-DCPR with 2-CPR. Four possible formation pathways are depicted in Figure 3 to explain the formation of 2,4,6-TCDF. Pathways 23 and 24 are similar, and they involve five elementary steps: carbon−carbon coupling, H abstraction, tautomerization (H shift), ring closure, and elimination of OH. The ring closure process has a large barrier and is strongly endoergic, and it is the rate determining-step. Pathway 25 is similar to pathway 26, which also involves five elementary processes: carbon−carbon coupling, tautomerization (double H transfer), H abstraction, ring closure (the rate-determining step), and elimination of OH. Pathways 27 and 28 in Figure 4 illustrate 2,6-DCDF and 2,4-DCDF being formed from the cross-condensation of 2,4-DCPR with 2-CPR. The intermediate IM40 can be regarded as a prestructure for 2,4,6-TCDF. IM48 is a prestructure of 2,6-DCDF, and IM52 is a prestructure of 2,4-DCDF. As seen from Figure 3, the formation of IM40 is more exothermic than the formations of IM48 and IM52. Furthermore, the rate-determining step involved in the formation of 2,4,6-TCDF has a lower barrier and is less endothermic compared to those involved in the formation of 2,6-DCDF and 2,4-DCDF. Thus, the formation of 2,4,6-TCDF is preferred over the formation of 2,6-DCDF and 2,4-DCDF. Due to the symmetry of 2,4,6-TCPR, only one possible PCDF formation pathway, displayed in Figure 5, is proposed for the cross-condensation of 2,4-DCPR with 2,4,6-TCPR.

Comparison with the previous studies (*28, 29*) tells us that the PCDF formation mechanism is controlled largely by the substitution pattern of chlorophenols. Steric and electronic effects associated with chlorine substitution suppress the dimerization of CPR. The ranking of the PCDF formation potential is as follows: PhR + 2-CPR > 2-CPR + 2,4-DCPR > 2,4-DCPR + 2,4-DCPR > 2,4-DCPR + 2,4,6-TCPR. This is consistent with the experimental observation that the PCDF formations are favored from less chlorinated phenols (*18, 41*).

3.3. Rate Constant Calculations. On the basis of the MPWB1K/6-311+G(3df,2p)//MPWB1K/6-31+G(d,p) energies, canonical variational transition-state rate calculations augmented by the small curvature tunneling corrections (CVT/SCT) are carried out in the temperature range of 600−1200 K. Our recently published studies show that the CVT/SCT rate constants of $C_6H_5OH + H \rightarrow C_6H_5O + H_2$, $C_6H_5OH + OH \rightarrow C_6H_5O + H_2O$ are in good agreement with the corresponding experimental values (*21, 22*). In particular, direct inspection of the transition-state low-frequency mode indicates that the modes of the lowest frequency involved in TS12, TS21, TS24, TS28, TS40, TS51, TS53, TS57, TS63, TS68, TS72, TS73, TS77, and TS83 are hindered internal rotation instead. These modes are removed from the vibration partition function for the transition state, and the corresponding hindered rotor partition function, $Q_{HR}(T)$, calculated by the method devised by Truhlar (*42*), is included in the expression of the rate constant.

To be used more effectively, the CVT/SCT rate constants are fitted, and Arrhenius formulas are given in Table 1 for the elementary reactions involved in the thermodynamically preferred formation pathways of PCDDs and in Table 2 for the elementary reactions involved in the formation of PCDFs from the cross-condensation of 2,4-DCPR with 2-CPR and 2,4,6-TCPR.

Acknowledgments

This work was supported by the National Natural Science Foundation of China (Project 20737001) and Shandong Province Outstanding Youth Natural Science Foundation (Project JQ200804). We thank Professor Donald G. Truhlar for providing the POLYRATE 9.3 program.

Supporting Information Available

Total energies (au), zero-point energies (au), and imaginary frequencies (cm^{-1}) for the transition states and geometries in terms of Cartesian coordinates (Å) for the reactants, products, intermediates, and transition states. This material is available free of charge via the Internet at http://pubs.acs.org.

Literature Cited

(1) Schecter, A. *Dioxin and Health*; Plenum Press: New York, 1994.

(2) Wang, S. L.; Chang, Y. C.; Chao, H. R.; Li, C. M.; Li, L. A.; Lin, L. Y.; Päpke, O. Body burdens of polychlorinated dibenzo-*p*-dioxins, dibenzofurans, and biphenyls and their relations to estrogen metabolism in pregnant women. *Environ. Health Perspect.* **2006**, *114* (5), 740–745.

(3) Choudhry, G. G., Keith, L. H., Rappe, C., Eds. *Chlorinated Dioxins and Dibenzofurans in the Total Environment*; Butterworth: Boston, 1983.

(4) Wu, Y. L.; Lin, L. F.; Hsieh, L. T.; Wang, L. C.; Chang-Chien, G. P. Atmospheric dry deposition of polychlorinated dibenzo-*p*-dioxins and dibenzofurans in the vicinity of municipal solid waste incinerators. *J. Hazard. Mater.* **2009**, *162* (9), 521–529.

(5) Vogg, H.; Metzger, M.; Stieglitz, L. Recent findings on the formation and decomposition of PCDD/PCDF in municipal solid waste incinerators. *Waste Manage. Res.* **1987**, *5* (3), 285–294.

(6) Addink, R.; Olie, K. Mechanisms of formation and destruction of polychlorinated dibenzo-*p*-dioxins and dizenzofurans in heterogeneous systems. *Environ. Sci. Technol.* **1995**, *29* (6), 1425–1435.

(7) Tuppurainen, K.; Halonen, I.; Ruokojärvi, P.; Tarhanen, J.; Ruuskanen, J. Formation of PCDDs and PCDFs in municipal waste incineration and its inhibition mechanisms: A review. *Chemosphere* **1998**, *36* (7), 1493–1511.

(8) Altwicker, E. R. Relative rates of formation of polychlorinated dioxins and furans from precursor and *de novo* reactions. *Chemosphere* **1996**, *33* (10), 1897–1904.

(9) Karasek, F. W.; Dickson, L. C. Model studies of polychlorinated dibenzo-*p*-dioxin formation during municipal refuse incineration. *Science* **1987**, *237* (4816), 754–756.

(10) Evans, C. S.; Dellinger, B. Mechanisms of dioxin formation from the high-temperature pyrolysis of 2-chlorophenol. *Environ. Sci. Technol.* **2003**, *37* (7), 1325–1330.

(11) Evans, C. S.; Dellinger, B. Mechanisms of dioxin formation from the high-temperature oxidation of 2-chlorophenol. *Environ. Sci. Technol.* **2005**, *39* (1), 122–127.

(12) Sidhu, S.; Edwards, P. Role of phenoxy radicals in PCDD/F formation. *Int. J. Chem. Kinet.* **2002**, *34* (9), 531–541.

(13) Altarawneh, M.; Dlugogorski, B. Z.; Kennedy, E. M.; Mackie, J. C. Quantum chemical investigation of formation of polychlorodibenzo-*p*-dioxins and dibenzofurans from oxidation and pyrolysis of 2-chlorophenol. *J. Phys. Chem. A* **2007**, *111* (13), 2563–2573.

(14) Sidhu, S. S.; Maqsud, L.; Dellinger, B.; Mascolo, G. The homogeneous, gas-phase formation of chlorinated and brominated dibenzo-*p*-dioxin from 2,4,6-trichloro- and 2,4,6-tribromophenols. *Combust. Flame* **1995**, *100* (1−2), 11–20.

(15) Mulholland, J. A.; Akki, U.; Yang, Y.; Ryu, J. Y. Temperature dependence of DCDD/F isomer distributions from chlorophenol precursors. *Chemosphere* **2001**, *42* (5−7), 719–727.

(16) Lomnicki, S.; Dellinger, B. A detailed mechanism of the surfaced-mediated formation of PCDD/F from the oxidation of 2-chlorophenol on a CuO/silica surface. *J. Phys. Chem. A* **2003**, *107* (22), 4387–4395.

(17) Ryu, J. Y.; Mulholland, J. A.; Kim, D. H.; Takeuchi, M. Homologue and isomer patterns of polychlorinated dibenzo-*p*-dioxins and dibenzofurans from phenol precursors: Comparison with

municipal waste incinerator data. *Environ. Sci. Technol.* **2005**, *39* (12), 4398–4406.
(18) Ryu, J. Y.; Mulholland, J. A.; Oh, J. E.; Nakahata, D. T.; Kim, D. H. Prediction of polychlorinated dibenzofuran congener distribution from gas-phase phenol condensation pathways. *Chemosphere* **2004**, *55* (11), 1447–1455.
(19) Louw, R.; Ahonkhai, S. I. Radical/radical vs radical/molecule reactions in the formation of PCDD/Fs from (chloro)phenols in incinerators. *Chemosphere* **2002**, *46* (9−10), 1273–1278.
(20) Altarawneh, M.; Dlugogorski, B. Z.; Kennedy, E. M.; Mackie, J. C. Quantum chemical investigation of formation of polychlorodibenzo-*p*-dioxins and dibenzofurans from oxidation and pyrolysis of 2-chlorophenol. *J. Phys. Chem. A* **2007**, *111* (13), 2563–2573.
(21) Zhang, Q. Z.; Qu, X. H.; Xu, F.; Shi, X. Y.; Wang, W. X. Mechanism and thermal rate constants for the complete series reactions of chlorophenols with H. *Environ. Sci. Technol.* **2009**, *43* (11), 4105–4112.
(22) Xu, F.; Wang, H.; Zhang, Q. Z.; Zhang, R. X.; Qu, X. H.; Wang, W. X. Kinetic properties for the complete series reactions of chlorophenols with OH radicals—Relevance for dioxin formation. *Environ. Sci. Technol.* **2010**, *44* (4), 1399–1404.
(23) Sun, Q.; Altarawneh, M.; Dlugogorski, B. Z.; Kennedy, E. M.; Mackie, J. C. Catalytic effect of CuO and other transition metal oxides in formation of dioxins: Theoretical investigation of reaction between 2,4,5-trichlorophenol and CuO. *Environ. Sci. Technol.* **2007**, *41* (16), 5708–5715.
(24) Khachatryan, L.; Asatryan, R.; Dellinger, B. Development of expanded and core kinetic models for the gas phase formation of dioxins from chlorinated phenols. *Chemosphere* **2003**, *52* (4), 695–708.
(25) Khachatryan, L.; Asatryan, R.; Dellinger, B. An elementary reaction kinetic model of the gas-phase formation of polychlorinated dibenzofurans from chlorinated phenols. *J. Phys. Chem. A* **2004**, *108* (44), 9567–9572.
(26) Khachatryan, L.; Burcat, A.; Dellinger, B. An elementary reaction-kinetic model for the gas-phase formation of 1,3,6,8- and 1,3,7,9-tetrachlorinated dibenzo-*p*-dioxins from 2,4,6-trichlorophenol. *Combust. Flame* **2003**, *132* (3), 406–421.
(27) Shaub, W. M.; Tsang, W. Dioxin formation in incinerators. *Environ. Sci. Technol.* **1983**, *17* (12), 721–730.
(28) Zhang, Q. Z.; Yu, W. N.; Zhang, R. X.; Zhou, Q.; Gao, R.; Wang, W. X. Quantum chemical and kinetic study on dioxin formation from the 2,4,6-TCP and 2,4-DCP precursors. *Environ. Sci. Technol.* **2010**, *44* (9), 3395–3403.
(29) Zhang, Q. Z.; Li, S. Q.; Qu, X. H.; Wang, W. X. A quantum mechanical study on the formation of PCDD/Fs from 2-chlorophenol as precursor. *Environ. Sci. Technol.* **2008**, *42* (19), 7301–7308.
(30) Zhao, Y.; Truhlar, D. G. Hybrid meta density functional theory methods for therochemistry, thermochemical kinetics, and noncovalent interactions: the MPW1B95 and MPWB1K models and comparative assessments for hydrogen bonding and van der waals interactions. *J. Phys. Chem. A* **2004**, *108* (33), 6908–6918.
(31) Frisch, M. J.; Trucks, G. W.; Schlegel, H. B.; Scuseria, G. E.; Robb, M. A.; Cheeseman, J. R.; Montgomery, J. A., Jr.; Vreven, T.; Kudin, K. N.; Burant, J. C.; Millam, J. M.; Iyengar, S. S.; Tomasi, J.; Barone, V.; Mennucci, B.; Cossi, M.; Scalmani, G.; Rega, N.; Petersson, G. A.; Nakatsuji, H.; Hada, M.; Ehara, M.; Toyota, K.; Fukuda, R.; Hasegawa, J.; Ishida, M.; Nakajima, T.; Honda, Y.; Kitao, O.; Nakai, H.; Klene, M.; Li, X.; Knox, J. E.; Hratchian, H. P.; Cross, J. B.; Bakken, V.; Adamo, C.; Jaramillo, J.; Gomperts, R.; Stratmann, R. E.; Yazyev, O.; Austin, A. J.; Cammi, R.; Pomelli, C.; Ochterski, J. W.; Ayala, P. Y.; Morokuma, K.; Voth, G. A.; Salvador, P.; Dannenberg, J. J.; Zakrzewski, V. G.; Dapprich, S.; Daniels, A. D.; Strain, M. C.; Farkas, O.; Malick, D. K.; Rabuck, A. D.; Raghavachari, K.; Foresman, J. B.; Ortiz, J. V.; Cui, Q.; Baboul, A. G.; Clifford, S.; Cioslowski, J.; Stefanov, B. B.; Liu, G.; Liashenko, A.; Piskorz, P.; Komaromi, I.; Martin, R. L.; Fox, D. J.; Keith, T.; Al-Laham, M. A.; Peng, C. Y.; Nanayakkara, A.; Challacombe, M.; Gill, P. M. W.; Johnson, B.; Chen, W.; Wong, M. W.; Gonzalez, C.; Pople, J. A. *Gaussian 03*; Gaussian: Pittsburgh, PA, 2003.
(32) Steckler, R.; Chuang, Y. Y.; Fast, P. L.; Corchade, J. C.; Coitino, E. L.; Hu, W. P.; Lynch, G. C.; Nguyen, K.; Jackells, C. F.; Gu, M. Z.; Rossi, I.; Clayton, S.; Melissas, V.; Garrett, B. C.; Isaacson, A. D.; Truhlar, D. G. *POLYRATE*, version 9.3; University of Minnesota: Minneapolis, 2002.
(33) Baldridge, M. S.; Gordor, R.; Steckler, R.; Truhlar, D. G. Ab initio reaction paths and direct dynamics calculations. *J. Phys. Chem.* **1989**, *93* (13), 5107–5119.
(34) Gonzalez-Lafont, A.; Truong, T. N.; Truhlar, D. G. Interpolated variational transition-state theory: Practical methods for estimating variational transition-state properties and tunneling contributions to chemical reaction rates from electronic structure calculations. *J. Chem. Phys.* **1991**, *95* (12), 8875–8894.
(35) Garrett, B. C.; Truhlar, D. G. Generalized transition state theory. Classical mechanical theory and applications to collinear reactions of hydrogen molecules. *J. Phys. Chem.* **1979**, *83* (8), 1052–1079.
(36) Fernandez-Ramos, A.; Ellingson, B. A.; Garret, B. C.; Truhlar, D. G. Variational transition state theory with multidimensional tunneling. In *Reviews in Computational Chemistry*; Lipkowitz, K. B., Cundari, T. R., Eds.; Wiley-VCH: Hoboken, NJ, 2007.
(37) Burcat, A.; Ruscic, B. Ideal Gas Thermochemical Database with Updates from Active Thermochemical Tables. ftp://ftp.technion.ac.il/pub/upported/aetdd/thermodynamics/BURCAT-.THR.
(38) Ribeiro da Silva, M. A. V.; Ferrao, M. L. C. C. H.; Fang, J. Standard enthalpies of combustion of the six dichlorophenols by rotating-bomb calorimetry. *J. Chem. Thermodyn.* **1994**, *26* (8), 839–846.
(39) Papina, T. S.; Kolesov, V. P.; Vorobieva, V. P.; Golovkov, V. F. The standard molar enthalpy of formation of 2-chlorodibenzo-*p*-dioxin. *J. Chem. Thermodyn.* **1996**, *28* (3), 307–311.
(40) Wang, L.; Heard, D. E.; Pilling, M. J.; Seakins, P. W. A Gaussian-3X prediction on the enthalpies of formation of chlorinated phenols and dibenzo-*p*-dioxins. *J. Phys. Chem. A* **2008**, *112* (8), 1832–1840.
(41) Ryu, J. Y.; Mulholland, J. A.; Takeuchi, M.; Kim, D. H.; Hatanaka, T. $CuCl_2$-catalyzed PCDD/F formation and congener patterns from phenols. *Chemosphere* **2005**, *61* (9), 1312–1326.
(42) Truhlar, D. G. A simple approximation for the vibrational partition function of a hindered internal rotation. *J. Comput. Chem.* **1991**, *12* (2), 266–270.

ES102660J

Environ. Sci. Technol. **2009**, *43*, 4068–4075

Mechanistic and Kinetic Studies on the Homogeneous Gas-Phase Formation of PCDD/Fs from 2,4,5-Trichlorophenol

XIAOHUI QU,[†] HUI WANG,[‡] QINGZHU ZHANG,*[,†] XIANGYAN SHI,[†] FEI XU,[†] AND WENXING WANG[†]

Environment Research Institute, Shandong University, Jinan 250100, P. R. China, and Department of Environmental Science and Engineering, Tsinghua University, Beijing 100084, P. R. China

Received October 8, 2008. Revised manuscript received March 12, 2009. Accepted April 20, 2009.

An understanding of the reaction mechanism of polychlorinated dibenzo-*p*-dioxins and dibenzofurans (PCDD/Fs) formation is crucial for any attempt to prevent PCDD/Fs formation. Among the polychlorophenols, 2,4,5-trichlorophenol (2,4,5-TCP) has the minimum number of Cl atoms needed to form 2,3,7,8-tetrachlorinated dibenzo-*p*-dioxin (2,3,7,8-TeCDD), which is the most toxic among all 210 PCDD/F isomers. Experiments on the formation of PCDD/Fs from the 2,4,5-TCP precursor have been hindered by the strong toxicity of 2,3,7,8-TeCDD. In this work, we carried out molecular orbital theory calculations for the homogeneous gas-phase formation of PCDD/Fs from the 2,4,5-TCP precursor. Several energetically favorable formation pathways were revealed for the first time. The rate constants of crucial elementary steps were deduced over a wide temperature range of 600~1200 K, using canonical variational transition state theory with small curvature tunneling contribution. The rate temperature formulas were fitted. This study shows that the formation of polychlorinated dibenzo-*p*-dioxins (PCDDs) from the 2,4,5-TCP precursor is preferred over the formation of polychlorinated dibenzofurans (PCDFs). The chlorine substitution pattern has a significant effect on the dimerization of chlorophenoxy radicals.

1. Introduction

Polychlorinated dibenzo-*p*-dioxins (PCDDs) and dibenzofurans (PCDFs) are notorious for their acute and chronic toxicity and their carcinogenic, teratogenic, and mutagenic effects (*1*). PCDD/Fs were never intentionally synthesized for commercial purposes but are formed as byproducts from the synthesis of chlorinated aromatic compounds during bleach processes in the pulp and paper industries and combustion (natural and anthropogenic) processes and even as microbial byproducts in activated sludge basins fed with chlorophenol-containing waste (*2–4*). In particular, a large quantity of PCDD/Fs is emitted from the combustion of waste materials in municipal incinerators (*5–7*). Reports of PCDD/F emissions from municipal waste incinerators have caused much public alarm and have led to severe difficulties in constructing municipal and hazardous waste incinerators.

Despite many studies, the origin of PCDD/Fs emitted from waste incinerators continues to be a cause of considerable controversy (*8–10*). It is now known that the potential contributions of the gas-phase pathways to PCDD/Fs formation were underestimated by the reaction kinetic model proposed by Shaub and Tsang (*10–12*). The homogeneous gas-phase formation of PCDD/Fs from chlorophenol precursors was suggested to make a significant contribution to the observed PCDD/F yields in full-scale incinerators (*13–15*). Detailed insight into the mechanism and kinetic properties is a prerequisite for understanding the formation of PCDD/Fs. In a previous contribution from this laboratory (*16*), we have investigated the formation of PCDD/Fs from the monochlorophenol precursor, 2-chlorophenol (2-CP). As part of our ongoing work in the field, this paper presents mechanistic and kinetic studies on the homogeneous gas-phase formation of PCDD/Fs from the more highly chlorinated phenols, 2,4,5-trichlorophenol (2,4,5-TCP), and compares the formation mechanism of PCDD/Fs from the 2,4,5-TCP precursor to the formation mechanism of PCDD/Fs from the 2-CP precursor.

There are four reasons for initiating this work. First, it was suspected that an increasing degree of chlorine substitution of the precursor chlorophenols not only had an effect on the reactivity of chlorophenols, but also had an effect on the resulting PCDD/PCDF ratio (*17*). Second, trichlorophenols (TCPs) are manufactured for commercial use as biocides, disinfectants, wood preservatives, impregnating agents, paint components, cooling agents, and chemicals for processing of paper pulp and leather. The concentrations of 2,4,5-TCP were reported to be 0.1 $\mu g/Nm^3$ and 11.5 ng/g at the furnace outlets and in fly ash samples of municipal waste incinerators (*18*). The concentration of 2,4,5-TCP in the urine of the incineration workers was found to be significantly higher than the corresponding values in people who had no known occupational contact to the substances that are thought to be produced in garbage incineration (1.2 versus 0.8 μg/l) (*19*). Third, among the polychlorophenols, 2,4,5-TCP has the minimum number of Cl atoms to form 2,3,7,8-TeCDD, which is the most toxic among all of the 210 PCDD/F isomers (*20*). Experiments on the formation of PCDD/Fs from 2,4,5-TCP are dangerous because of the high toxicity of 2,3,7,8-TeCDD. In such a situation, quantum chemical calculation provides a safe alternative to studying these highly toxic compounds. Only one ab initio study is on record for the formation mechanism of PCDDs from 2,4,5-TCP (*21*). However, the formation of PCDDs via the dimerization of chlorophenoxy radicals (CPRs), which was recently suggested to be the major pathway in the homogeneous gas-phase formation of PCDD/Fs from chlorophenol precursors (*10–13*), was not investigated in this ab initio study (*21*). Furthermore, the formation of PCDFs from the 2,4,5-TCP precursor has not been studied (*21*). Fourth, the kinetic models that account for the contribution of the gaseous route in the production of PCDD/Fs in combustion processes use the rate constants of the elementary reactions (*11*). Becasue of the absence of direct experimental and theoretical values, the rate constants of many elementary steps were assigned to be the values reported in the literature for analogous reactions (*11, 12*). However, where there are uncertainties, the numerical values have been adjusted somewhat to bias the mechanism in favor of PCDD/Fs formation, i.e., worst case modeling (*11, 12*).

* Corresponding author fax: 86-531-8836 1990; e-mail: zqz@sdu.edu.cn.
[†] Shandong University.
[‡] Tsinghua University.

10.1021/es802835e CCC: $40.75
Published on Web 05/04/2009

2. Computational Methods

By means of the Gaussian 03 programs (*22*), high-accuracy molecular orbital calculations were carried out for the homogeneous gas-phase formation of PCDD/Fs from the 2,4,5-TCP precursor. The choice of computational levels and basis sets requires a compromise between accuracy and computational time. The geometrical parameters of reactants, transition states, intermediates, and products were optimized at the MPWB1K/6-31+G(d,p) level. The MPWB1K method is a hybrid density functional theory (HDFT) model with excellent performance for thermochemistry, thermochemical kinetics, hydrogen bonding, and weak interactions (*23*). The vibrational frequencies were also calculated at the same level in order to determine the nature of the stationary points, zero-point energy (ZPE), and thermal contributions to the free energy of activation. The minimum energy paths (MEPs) were obtained in mass-weighted Cartesian coordinates. The force constant matrices of the stationary points and selected nonstationary points near the transition state along the MEPs were also calculated in order to do the following kinetic calculations. The single-point energy calculations were carried out at the MPWB1K/6-311+G(3df,2p) level. The kinetic calculations are most sensitive to the energies. The reliability of the MPWB1K/6-311+G(3df,2p) level for the potential barriers was clarified in our recent study (*16*) on the formation of PCDD/Fs from the 2-CP precursor. The profiles of the potential energy surface were constructed at the MPWB1K/6-311+G(3df,2p)//MPWB1K/6-31+G(d,p) level, including ZPE correction.

By means of the Polyrate 9.3 program (*24*), direct dynamics calculations were carried out using the canonical variational transition state theory (CVT). Canonical variational transition state theory (*25–27*) is based on the idea of varying the dividing surface along a reference path to minimize the rate constant. The CVT rate constant for temperature *T* is given by

$$k^{CVT}(T) = \min_{s} k^{GT}(T,s) \qquad (1)$$

where

$$k^{GT}(T,s) = \frac{\sigma k_B T}{h} \frac{Q^{GT}(T,s)}{\Phi^R(T)} e^{-V_{MEP}(s)/k_B T} \qquad (2)$$

where, $k^{GT}(T, s)$ is the generalized transition state theory rate constant at the dividing surface s, σ is the symmetry factor accounting for the possibility of more than one symmetry-related reaction path, k_B is Boltzmann's constant, h is Planck's constant, $\Phi^R(T)$ is the reactant partition function per unit volume (excluding symmetry numbers for rotation), and $Q^{GT}(T, s)$ is the partition function of a generalized transition state at s with a local zero of energy at $V_{MEP}(s)$ and with all rotational symmetry numbers set to unity. The level of the tunneling calculation is the small curvature tunneling (SCT) (*28*) method, based on the centrifugal dominant small curvature semiclassical adiabatic ground state (CD-SCSAG) approximation. The rotational partition functions were calculated classically, and the vibrational modes were treated as quantum-mechanically separable harmonic oscillators. The error of the kinetic calculation may be mainly from the small curvature tunneling (SCT) method, especially for the heavy–light–heavy (HLH) mass combination reactions. The large curvature tunneling approximation may be especially desirable to model this kind of reaction in detail. Methods for large curvature cases have been developed (*29*), but they require more information about the potential energy surface than was determined in the present study. So, the SCT approximation becomes an alternative to the large curvature approximation. The new CD-SCSAG approximation for small curvature tunneling is an improvement over the original SCSAG method in that it accounts for the effect of mode–mode coupling on the extent of corner cutting through each vibrational degree of freedom, orthogonal to the reaction path (*30*). Furthermore, previous studies (*31, 32*) show that the tunneling correction plays an important role, mainly in a low-temperature range for the calculation of the rate constants. Fortunately, the rate constants were calculated over a high-temperature range (600~1200 K) in this work. The error from the tunneling calculation does not significantly influence the reliability of our results.

3. Results and Discussion

The first step in this study was to identify the level of theoretical approximation that was not only able to produce accurate results but was also computationally feasible and economical for currently available hardware and software. The geometric parameters and the vibrational frequencies of phenol, chlorobenzene, and 2,3,7,8-TeCDD were calculated at the MPWB1K/6-31+G(d,p) level. The results agree well with the available experimental values (*33–36*), and the maximum relative error is less than 1.4% for the geometrical parameters and less than 8.0% for the vibrational frequencies. From this results, we infer that the same accuracy can be expected for the species involved in the formation of PCDD/Fs from the 2,4,5-TCP precursor. Furthermore, a previous study shows that MPWB1K is an excellent method for prediction of transition state geometries (*23*).

For 2,4,5-TCP, there are two geometric conformers resulting from the two main orientations of the hydroxyl-hydrogen due to the asymmetric chlorine substitutions. The conformer with the hydroxyl-hydrogen facing the closest neighboring Cl is labeled the *syn* conformer, and the other is the *anti* conformer. The *syn* conformer is more stable by 3.24 kcal/mol than the corresponding *anti* form. So throughout this paper, 2,4,5-TCP denotes the *syn* conformer.

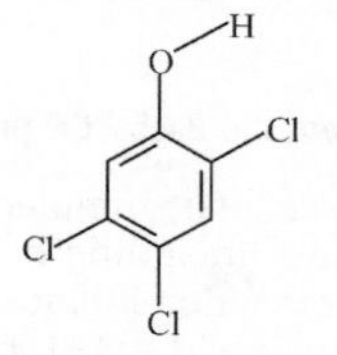

syn 2,4,5-TCP

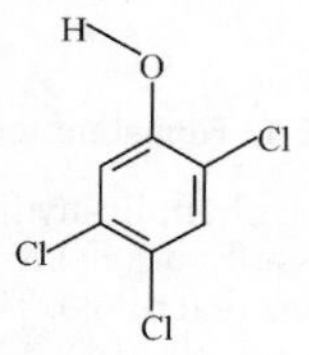

anti 2,4,5-TCP

3.1. Formation of 2,4,5-Trichlorophenoxy Radicals.

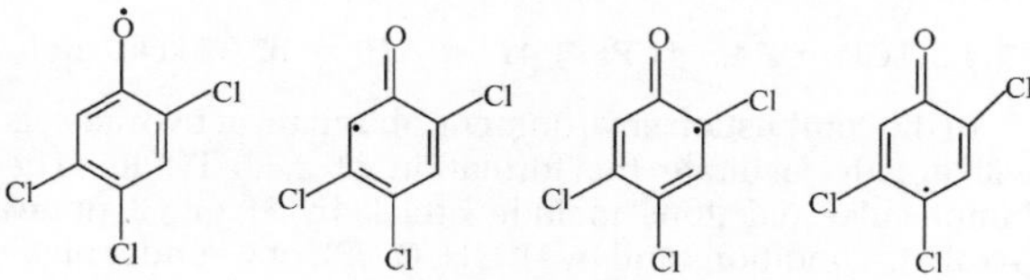

The chlorophenoxy radical (CPR), in which the electrons are delocalized, is a hybrid of one oxygen-centered and three carbon-centered radicals (2 *ortho* and 1 *para* carbon sites). The resonance structures provide about 16 kcal/mol of resonance stabilization energy for phenoxy radicals (*37*). The resonance stabilization energy of CPR should be approximately identical to the value of phenoxy radical. The high resonance stabilization energy means that CPRs could build up sufficient concentrations to enable self-condensation to occur. Recent works (*13, 38*) have shown that the dimerization of CPRs is the major pathway in the homogeneous gas-phase formation of PCDD/Fs. Thus, we first studied the formation of 2,4,5-trichlorophenoxy radicals (2,4,5-TCPRs). In municipal waste incinerators, 2,4,5-TCPRs can be formed through

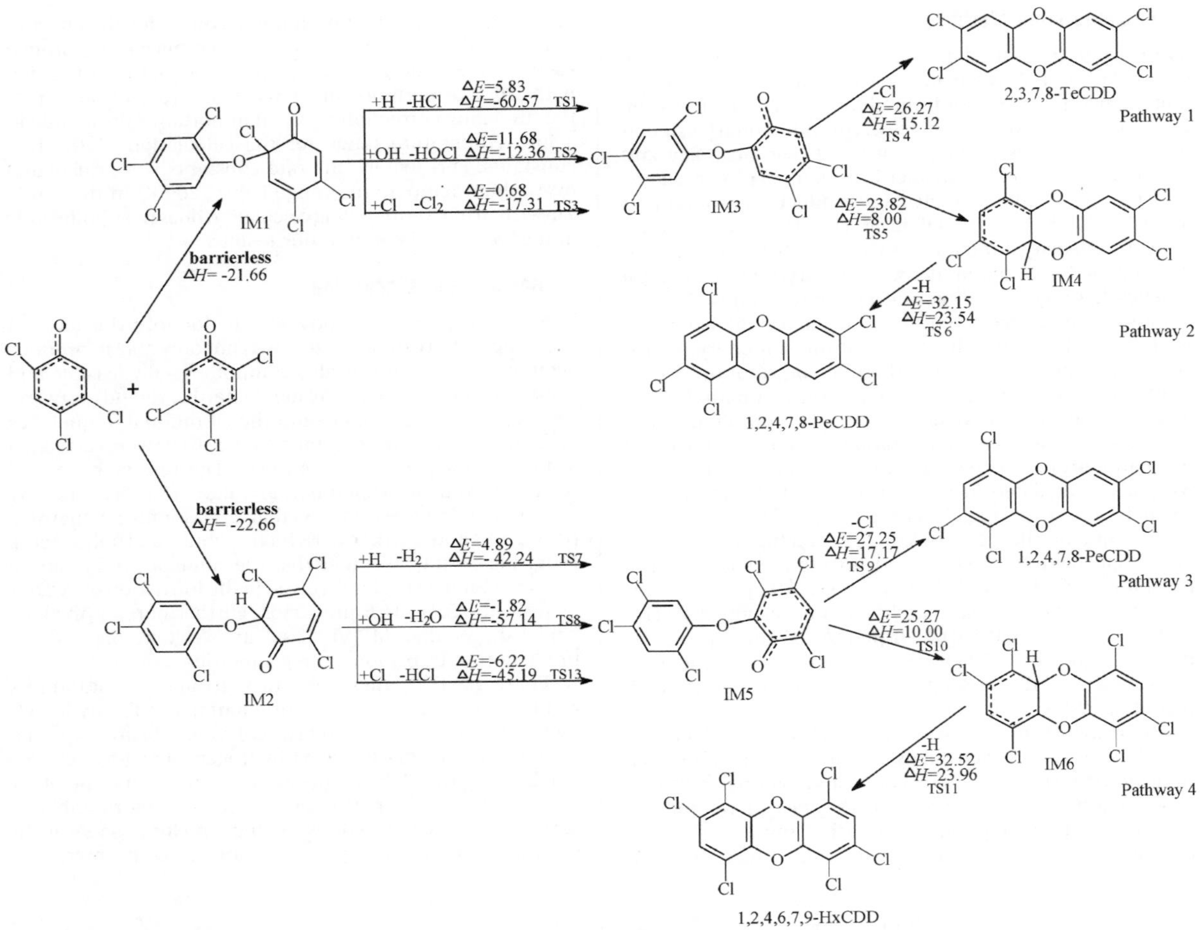

FIGURE 1. Formation routes of PCDDs from the 2,4,5-TCP precursor. Δ*H* is calculated at 0 K.

loss>of phenolic-hydrogen via unimolecular, bimolecular, or possibly other low-energy pathways (including heterogeneous reactions). The unimolecular reaction includes the decomposition of 2,4,5-TCP with cleavage of the O−H bond. The profile of the potential energy surface was scanned by varying the O−H bond length. We found no energy exceeding the O−H bond dissociation threshold along the reaction coordinate. This shows that there is no transition state in the decomposition process.

$$\text{2,4,5-TCP} \rightarrow \text{2,4,5-TCPR} + \text{H} \qquad \Delta H = 85.37\ \text{kcal/mol}$$

In the combustion environment, abundant active radicals will greatly facilitate the formation of 2,4,5-TCPRs. The bimolecular reactions include attack by H or Cl under pyrolytic conditions and by H, OH, O (^{3}P), or Cl under high-temperature oxidative conditions. Studies show that the formation of 2,4,5-TCPRs from the bimolecular reactions of 2,4,5-TCP with H, OH, O (^{3}P), and Cl proceeds via a direct phenolic hydrogen abstraction mechanism. The potential barriers (Δ*E*) and the reaction enthalpies (Δ*H*, 0 K) were calculated at the MPWB1K/6-311+G(3df,2p)//MPWB1K/6-31+G(d,p) level. The reaction enthalpies at 600 and 1000 K are presented in the Supporting Information. Particularly, for the phenolic hydrogen abstraction from 2,4,5-TCP by Cl, the energy of the transition state without ZPE (zero-point energy) correction is 0.94 kcal/mol higher than the total energy of the original reactants (2,4,5-TCP and Cl). However, the energy of the transition state, including ZPE, is 1.93 kcal/mol lower than the total energy of the original reactants. This is because the zero-point energy of the transition state for the phenolic hydrogen abstraction from 2,4,5-TCP by Cl is 2.87 kcal/mol lower than that of 2,4,5-TCP.

$$\text{2,4,5-TCP} + \text{H} \rightarrow \text{2,4,5-TCPR} + H_2$$
$$\Delta E = 13.96\ \text{kcal/mol} \quad \Delta H = -12.56\ \text{kcal/mol}$$
$$\text{2,4,5-TCP} + \text{OH} \rightarrow \text{2,4,5-TCPR} + H_2O$$
$$\Delta E = 3.23\ \text{kcal/mol} \quad \Delta H = -27.46\ \text{kcal/mol}$$
$$\text{2,4,5-TCP} + O(^3P) \rightarrow \text{2,4,5-TCPR} + \text{OH}$$
$$\Delta E = 7.67\ \text{kcal/mol} \quad \Delta H = -11.89\ \text{kcal/mol}$$
$$\text{2,4,5-TCP} + \text{Cl} \rightarrow \text{2,4,5-TCPR} + \text{HCl}$$
$$\Delta E = -1.93\ \text{kcal/mol} \quad \Delta H = -15.51\ \text{kcal/mol}$$

The potential barriers of the phenolic hydrogen abstraction from 2,4,5-TCP are a little higher than those from 2-CP. For example, the potential barrier of 2,4,5-TCP + H → 2,4,5-TCPR + H_2 is 13.93 kcal/mol at the MPWB1K/6-311+G(3df,2p)//MPWB1K/6-31+G(d,p) level, and the value for 2-CP + H → 2-CPR + H_2 is 13.80 kcal/mol (*16*). The phenolic hydrogen abstraction from 2,4,5-TCP is a little more exothermic than the phenolic hydrogen abstraction from 2-CP. For example, the reaction of 2,4,5-CP + OH → 2-CPR + H_2O is exothermic by 27.46 kcal/mol (0 K), whereas the reaction of 2-CP + OH → 2-CPR + H_2O is exothermic by 26.91 kcal/mol (0 K) at the MPWB1K/6-311+G(3df,2p)//MPWB1K/6-31+G(d,p) level (*16*). The degree of chlorination has an effect on the potential barrier and reaction enthalpies for the phenolic hydrogen abstraction from chlorophenols.

3.2. Formation of PCDD/Fs from 2,4,5-Trichlorophenoxy Radicals. *3.2.1. Formation of PCDDs from 2,4,5-*

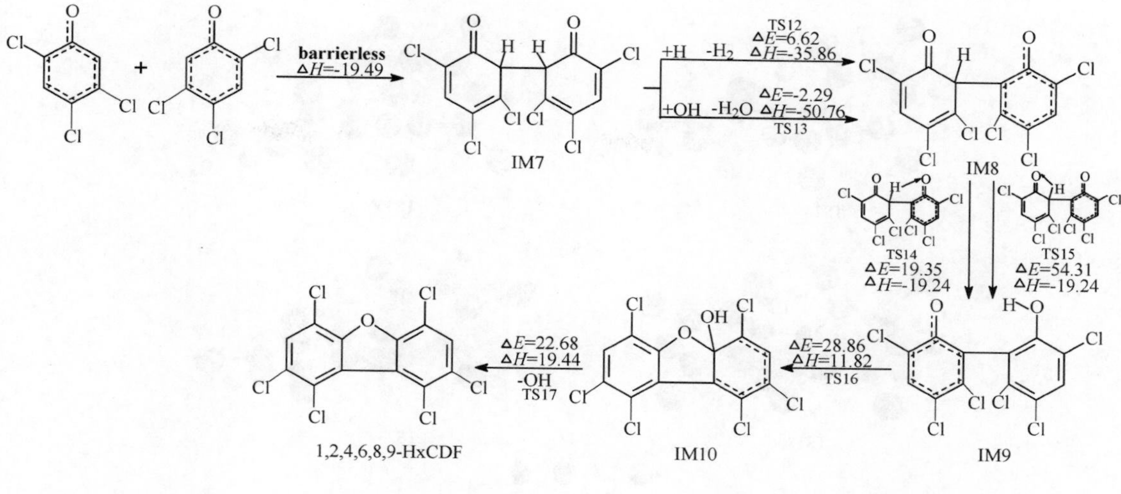

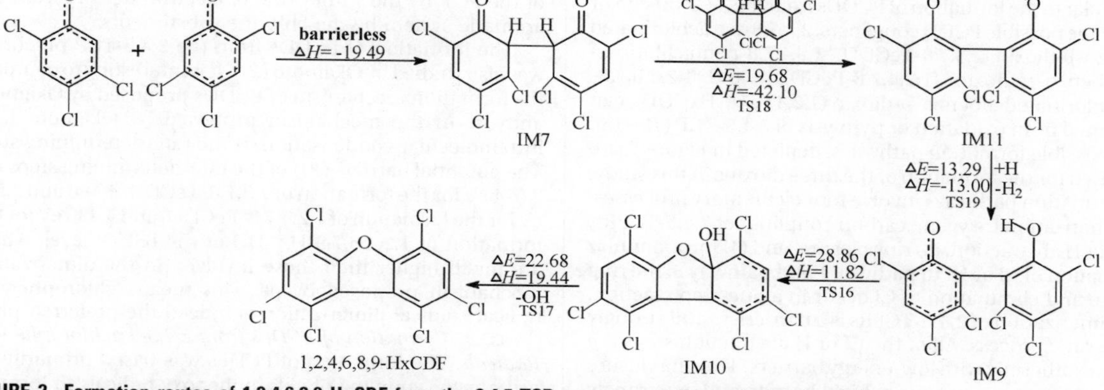

FIGURE 2. Formation routes of 1,2,4,6,8,9-HxCDF from the 2,4,5-TCP precursor. Δ*H* is calculated at 0 K.

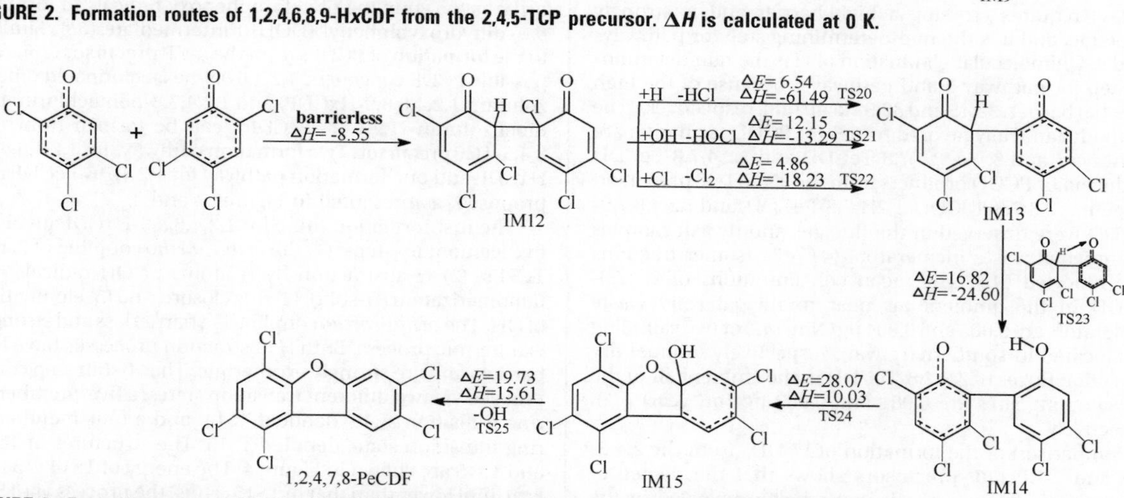

FIGURE 3. Formation route of 1,2,4,7,8-PeCDF from the 2,4,5-TCP precursor. Δ*H* is calculated at 0 K.

Trichlorophenoxy Radicals. Previous research (*39, 40*) has shown that the formation of PCDDs can be attributed to the formation of *o*-phenoxy-phenol (POP) intermediates. There are two possible oxygen–carbon coupling modes to form POPs from the self-condensation of 2,4,5-TCPRs. Thus, two POP intermediates, denoted IM1 and IM2, were identified in this study. In order to evaluate the nature of the entrance channel for the two coupling reactions, we examined the potential along the reaction coordinate, especially to determine whether there is a well-defined transition state or the coupling proceeds via a loose transition state without a barrier. The profile of the potential energy surface was scanned against the newly formed C–O bond length. We find no energy exceeding the C–O bond dissociation threshold along the reaction coordinate. This shows that both coupling reactions appear to be barrierless.

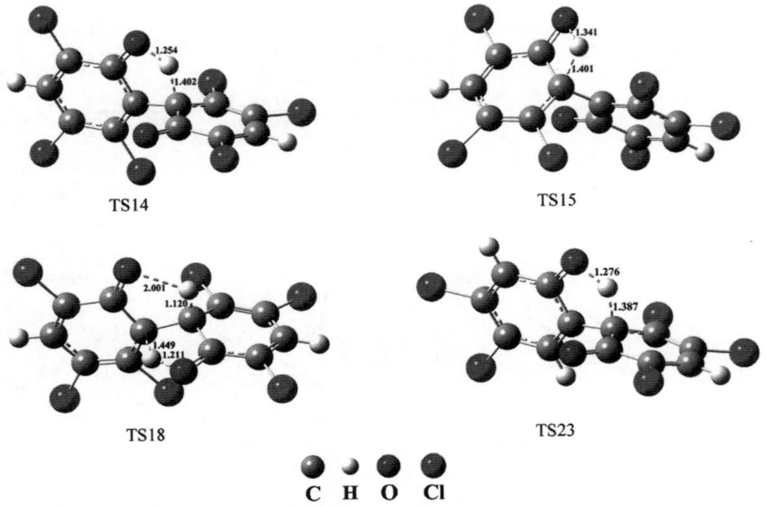

FIGURE 4. Configurations of the transition states of H-shift involved in the formation of PCDFs.

Similar to the formation of PCDDs from the 2-CP precursor (*16*), three possible PCDD congeners, 2,3,7,8-tetrachlorinated dibenzo-*p*-dioxin (2,3,7,8-TeCDD), 1,2,4,7,8-pentachlorinated dibenzo-*p*-dioxin (1,2,4,7,8-PeCDD), and 1,2,4,6,7,9-hexachlorinated dibenzo-*p*-dioxin (1,2,4,6,7,9-HxCDD), can be formed from oxidation or pyrolysis of 2,4,5-TCP (*13–15*). Four possible formation pathways, depicted in Figure 1, are proposed for the formation of the three dioxins in this study. The formation pathways involve four elementary processes: (1) dimerization (oxygen–carbon coupling) of 2,4,5-TCPRs, (2) Cl or H abstraction, (3) ring closure, and (4) intra-annular elimination of Cl or H. In pathway 1 and pathway 3, the ring closure and elimination of Cl occur in a one-step reaction. The dimerization of 2,4,5-TCPRs is a barrierless and strongly exothermic process. All of the Cl or H abstraction steps are highly exothermic with low-energy barriers. The ring closure process requires crossing a high barrier and is strongly endoergic, and it is the rate-determining step for pathways 1 and 3. Unimolecular elimination of H is the rate-determining step for pathway 2 and pathway 4 because of the high potential barrier, 32.15 and 32.52 kcal/mol, respectively. The thermodynamically favored routes for PCDD formation are pathways 1 and 3. So, 2,3,7,8-TeCDD and 1,2,4,7,8-PeCDD are the major PCDD products from the 2,4,5-TCP precursor. Congeners 2,3,7,8-TeCDD, 1,2,4,7,8-PeCDD, and 1,2,4,6,7,9-H*x*CDD were observed in the flue gas and fly ash samples of municipal waste incinerators (*41, 42*). Isomer fractions were reported (*41*). The mean concentrations of 2,3,7,8-TeCDD in the ambient air near municipal solid waste incinerators are 0.005 and 0.004 pg/Nm^3 (*42*) at two sampling sites located in southern Taiwan, respectively. Annual dry deposition fluxes of 2,3,7,8-TeCDD in the ambient air of the two sampling sites are 0.081 and 0.089 ng/(m^2 year) (*42*), respectively.

Comparison of the formation of PCDDs from the 2,4,5-TCP and 2-CP (*16*) precursors shows that the degree of chlorination affects each elementary step involved in the formation of PCDDs. In particular, the degree of chlorination has a significant effect on the dimerization of chlorophenoxy radicals (CPRs). The oxygen–carbon coupling of 2,4,5-TCPRs is more exothermic than the oxygen–carbon coupling of 2-CPRs. It appears that the trend toward POP formation increases with increasing degree of chlorination. (This appears to be the case on the basis of the limited number of chlorophenol isomers studied.) This is probably due to the fact that the nucleophilic attack of the phenolic oxygen atom at the second phenol is facilitated during the formation of the POP by the withdrawal of electron density from the aromatic system by the chlorine substituents.

The formation of PCDDs from the 2,4,5-TCP precursor was also studied by Okamoto (*21*). It is interesting to compare the formation mechanism of PCDDs proposed by Okamoto and us. In the mechanism proposed by Okamoto (*21*), intramolecular condensation, is the rate-determining step. The potential barriers (*21*) of the rate-determining steps are 2.574 eV for the formation of 2,3,7,8-TeCDD, 4.486 and 2.528 eV for the formation of 1,2,4,7,8-PeCDD, and 4.436 eV for the formation of 1,2,4,6,7,9-HxCDD at the B3LYP level, which are much higher than those involved in the dimerization mechanism proposed by us. This means chlorophenoxy radical–radical dimerization is indeed the preferred path.

3.2.2. Formation of PCDFs from 2,4,5-Trichlorophenoxy Radicals. The formation of PCDFs was based primarily on *ortho-ortho* coupling of chlorophenoxy radicals to form an *o,o'*-dihydroxybiphenyl (DOHB) intermediate (*38*). Similar to the formation of PCDFs from the 2-CP precursor (*16*), two possible PCDF congeners, 1,2,4,6,8,9-hexachlorinated dibenzofuran (1,2,4,6,8,9-HxCDF) and 1,2,4,7,8-pentachlorinated dibenzofuran (1,2,4,7,8-PeCDF), can be formed from the 2,4,5-TCP precursor. Two formation pathways for 1,2,4,6,8,9-H*x*CDF and one formation pathway for 1,2,4,7,8-PeCDF are proposed, as presented in Figures 2 and 3.

The first formation route for 1,2,4,6,8,9-H*x*CDF involves five elementary steps: (1) the *ortho–ortho* coupling of 2,4,5-TCPRs, (2) H abstraction by H atoms or OH radicals, (3) tautomerization (H-shift), (4) ring closure, and (5) elimination of OH. The *ortho–ortho* coupling is a barrierless and strongly exothermic process. Both H abstraction processes have low barriers and are strongly exothermic. The H-shift step could proceed via two different transition states: a five-membered ring transition state, denoted TS14, and a four-membered ring transition state, denoted TS15. The structures of TS14 and TS15 are shown in Figure 4. The energy of TS14 is 34.96 kcal/mol lower than that of TS15. Thus, the process via TS14 is energetically favorable for the H-shift step. The ring closure process requires crossing a large barrier, is strongly endoergic, and is the rate-determining step.

The second formation pathway of 1,2,4,6,8,9-H*x*CDF also involves five elementary processes: (1) dimerization of 2,4,5-TCPRs, (2) tautomerization (double H-transfer), (3) H abstraction, (4) ring closure, and (5) OH desorption. The double hydrogen atom migration to the key to oxygen atoms proceeding via the transition state TS18, whose geometrical structure is presented in Figure 4. The calculated vibrational

TABLE 1. CVT/SCT Rate Constants for the Phenolic Hydrogen Abstraction from 2,4,5-TCP along with the Available Experimental Rate Constants for the Phenolic Hydrogen Abstraction from Phenol (*44−46*)

reactions	*T* (K)	*k* (cm^3 molecule^{-1} s^{-1})	reference
2,4,5-TCP + H → 2,4,5-TCPR + H_2	1000	1.10×10^{-14}	this study
phenol + H → phenoxy + H_2	1000	3.72×10^{-13}	ref *44*
2,4,5-TCP + OH → 2,4,5-TCPR + H_2O	1000	1.85×10^{-13}	this study
phenol + OH → phenoxy + H_2O	1000	9.48×10^{-12}	ref (*45*)
2,4,5-TCP + Cl → 2,4,5-TCPR + HCl	296	8.95×10^{-12}	this study
phenol + Cl → phenoxy + HCl	296	1.93×10^{-10}	ref *46*

TABLE 2. Arrhenius Formulas[a] for Elementary Reactions Involved in the Formation of PCDD/Fs from the 2,4,5-TCP Precursor over the Temperature Range of 600∼1200 K

reactions	Arrhenius formulas
2,4,5-TCP + H → 2,4,5-TCPR + H_2	$k(T) = (5.97 \times 10^{-12}) \exp(-6298.20/T)$
2,4,5-TCP + OH → 2,4,5-TCPR + H_2O	$k(T) = (1.51 \times 10^{-12}) \exp(-2099.61/T)$
2,4,5-TCP + O(^{3}P) → 2,4,5-TCPR + OH	$k(T) = (2.96 \times 10^{-13}) \exp(-3386.30/T)$
2,4,5-TCP + Cl → 2,4,5-TCPR + HCl	$k(T) = (4.87 \times 10^{-11}) \exp(-710.73/T)$
IM1 + OH → IM3 + HOCl	$k(T) = (1.88 \times 10^{-11}) \exp(-1461.15/T)$
IM1 + Cl → IM3 + Cl_2	$k(T) = (8.90 \times 10^{-12}) \exp(-7185.73/T)$
IM3 → 2,3,7,8-TeCDD	$k(T) = (2.34 \times 10^{11}) \exp(-13069.04/T)$
IM3 → IM4	$k(T) = (2.36 \times 10^{9}) \exp(-7389.52/T)$
IM4 → 1,2,4,7,8-PeCDD	$k(T) = (2.90 \times 10^{13}) \exp(-16719.64/T)$
IM2 + H → IM5 + H_2	$k(T) = (4.08 \times 10^{-12}) \exp(-1552.54/T)$
IM2 + OH → IM5 + H_2O	$k(T) = (3.11 \times 10^{-12}) \exp(-226.80/T)$
IM5 → 1,2,4,7,8-PeCDD	$k(T) = (5.70 \times 10^{9}) \exp(-10065.22/T)$
IM5 → IM6	$k(T) = (3.77 \times 10^{11}) \exp(-12656.49/T)$
IM6 → 1,2,4,6,7,9-HxCDD	$k(T) = (2.13 \times 10^{13}) \exp(-16809.22/T)$
IM7 + H → IM8 + H_2	$k(T) = (1.40 \times 10^{-11}) \exp(-1622.86/T)$
IM7 + OH → IM8 + H_2O	$k(T) = (3.37 \times 10^{-12}) \exp(18.30/T)$
IM8 → IM9 via TS14	$k(T) = (3.37 \times 10^{10}) \exp(-586.27/T)$
IM9 → IM10	$k(T) = (4.18 \times 10^{10}) \exp(-6141.44/T)$
IM10 → 1,2,4,6,8,9-HxCDF	$k(T) = (5.50 \times 10^{13}) \exp(-11874.30/T)$
IM7 → IM11	$k(T) = (1.86 \times 10^{9}) \exp(-1419.06/T)$
IM11 + H → IM9 + H_2	$k(T) = (6.72 \times 10^{-11}) \exp(-7238.72/T)$
IM12 + H → IM13 + HCl	$k(T) = (1.13 \times 10^{-10}) \exp(-4022.09/T)$
IM12 + OH → IM13 + HOCl	$k(T) = (8.58 \times 10^{-13}) \exp(-7098.10/T)$
IM12 + Cl → IM13 + Cl_2	$k(T) = (2.29 \times 10^{-11}) \exp(-3448.01/T)$
IM13 → IM14	$k(T) = (5.79 \times 10^{9}) \exp(-2159.64/T)$
IM14 → IM15	$k(T) = (3.27 \times 10^{12}) \exp(-14307.63/T)$
IM15 → 1,2,4,7,8-PeCDF	$k(T) = (3.23 \times 10^{13}) \exp(-10349.92/T)$

[a] Units are s^{-1} and cm^3 molecule^{-1} s^{-1} for unimolecular and bimolecular reactions, respectively.

frequencies contained one and only one imaginary component, 1532i cm^{-1}, confirming the first-order saddle point configuration. Similarly, an energetically feasible formation pathway, depicted in Figure 3, is proposed for 1,2,4,7,8-PeCDF. Both 1,2,4,6,8,9-HxCDF and 1,2,4,7,8-PeCDF were observed in municipal waste incinerator flue gas and fly ash samples (*41*).

Comparison of the formation of PCDFs from the 2,4,5-TCP and 2-CP (*16*) precursors shows that degree of chlorination has a significant effect on the *ortho–ortho* coupling of CPRs. The *ortho–ortho* coupling of 2,4,5-TCPRs proceeds via a barrierless process, whereas the *ortho–ortho* coupling of 2-CPRs requires crossing a potential barrier of 8.27 kcal/mol (*16*).

Comparison of Figures 1 and 2 shows that the preferred formation routes of PCDDs from the 2,4,5-TCP precursor involve three elementary steps, whereas the formation pathways of PCDFs involve five elementary processes. The POP intermediates, IM1 and IM2, can be regarded as prestructures for PCDDs. DOHB intermediates, IM7 and IM12, are prestructures of PCDFs. This study shows that the formation of POPs is more exoergic than the formation of DOHBs. It means that POPs are thermodynamically more stable than DOHBs. Furthermore, the rate-determining step (ring closure process) involved in the formation of PCDFs requires crossing a larger barrier than that involved in the formation of PCDDs. Thus, the formation of PCDDs is preferred over the formation of PCDFs from the 2,4,5-TCP precursor. Further direct experimental observation would be anticipated to verify the conclusion. However, this conclusion may be supported indirectly by the results of laboratory experiments: PCDDs are formed at lower temperatures than PCDFs, and PCDDs are the major dioxin products rather than PCDFs in the pyrolysis of 2-CP (*39*).

3.3. Kinetic Calculations. Canonical variational transition state theory (CVT) (*25–27*), with small curvature tunneling (SCT) (*28*) correction, has been successfully performed for the elementary reactions involved in the formation of PCDD/Fs from the 2-CP precursor (*16*) and is an efficient method to calculate the rate constants (*43*). In this study, we used this method to calculate the rate constants of crucial elementary reactions, especially the rate-determining steps involved in the PCDD/F formations from the 2,4,5-TCP precursor over a wide temperature range of 600∼1200 K, which covers the possible formation temperature of PCDD/Fs in municipal waste incinerators.

Because of the absence of available experimental values, it is difficult to make a direct comparison of the calculated CVT/SCT rate constants with the experimental values for all of the elementary reactions. An alternative approach to clarifying the reliability of the kinetics calculation is to compare the CVT/SCT rate constants with the available literature rate constants for structurally similar compounds. There are no available experimental or theoretical rate constants for the phenolic hydrogen abstraction from 2,4,5-TCP. So, the calculated rate constants were compared with the literature values (*44–46*) for the phenolic hydrogen abstraction from phenol. As shown in Table 1, the CVT/SCT rate constants for the phenolic hydrogen abstraction from 2,4,5-TCP are over 1 order of magnitude lower than the literature values (*44–46*) for the corresponding phenolic hydrogen abstraction from phenol. For example, at 1000 K, our CVT/SCT value for 2,4,5-TCP + H → 2,4,5-TCPR + H_2 is 1.10×10^{-14} cm^3 $molecule^{-1}$ s^{-1}, whereas the rate constant for phenol + H → phenoxy + H_2 is 3.72×10^{-13} cm^3 $molecule^{-1}$ s^{-1} (*44*). A new study (*47*) from our group shows that the chlorine substitution in the *ortho* position increases the strength of the O–H bond in chlorophenol and decreases its reactivity. At the MPWB1K/6-311+G(3df,2p) level, the potential barrier for 2,4,5-TCP + H → 2,4,5-TCPR + H_2 is 13.96 kcal/mol, while the value for phenol + H → phenoxy + H_2 is 11.73 kcal/mol. The dissociation energies (0 K) of the O–H bonds in 2,4,5-TCP and phenol are 85.37 and 83.95 kcal/mol, respectively. Thus, our CVT/SCT rate constants for the phenolic hydrogen abstraction from 2,4,5-TCP are reasonable. In order to further check the validity of our computational scheme, we also calculated the rate constants for phenol + H → phenoxy + H_2, using canonical variational transition state theory (CVT) with small curvature tunneling (SCT) contribution. The CVT/SCT rate constants are in good agreement with the experimental values (*44*), with the maximum relative deviation less than 3 times that for phenol + H → phenoxy + H_2. For example, at 1000 K, the calculated CVT/SCT rate constant, 1.68×10^{-13} cm^3 $molecule^{-1}$ s^{-1}, perfectly matches the experimental value of 3.72×10^{-13} cm^3 $molecule^{-1}$ s^{-1} (*44*). From these good agreements, it is inferred that the same accuracy could be expected for the other crucial elementary reactions involved in the formation of PCDD/Fs from the 2,4,5-TCP precursor.

Regulatory decisions and risk analyses often rely on the use of mathematical models to predict the potential outcomes of contaminant releases to the environment. A better knowledge of the temperature dependence would be useful for the kinetic models that account for the contribution of the gaseous route in the production of PCDD/Fs in combustion processes (*11, 12*). The pre-exponential factor, activation energy, and rate constants are important input parameters in the kinetic models (*11, 12*). Thus, the calculated CVT/SCT rate constants are fitted over the temperature range of 600~1200 K, and Arrhenius formulas are given in Table 2.

Acknowledgments

This work was supported by NSFC (National Natural Science Foundation of China, project No. 20737001, 20777047), Shandong Province Outstanding Youth Natural Science Foundation (project No. JQ200804) and the Research Fund for the Doctoral Program of Higher Education of China (project No. 200804220046). The authors thank Professor Donald G. Truhlar for providing the POLYRATE 9.3 program. The authors also thank Dr. Pamela Holt for proofreading the manuscript.

Supporting Information Available

MPWB1K/6-31+G(d,p) optimized geometries and calculated frequencies for the transition states involved in the formation of PCDD/Fs from 2,4,5-TCP as precursor and reaction enthalpies (△*H*) of all of the elementary steps involved in the formation of PCDD/Fs from 2,4,5-TCP as precursor at 600 and 1000 K. This information is available free of charge via the Internet at http://pubs.acs.org.

Literature Cited

(1) Schecter, A. *Dioxin and Health*; Plenum Press: New York, 1994.

(2) Stefan, V.; Achim, Z.; Reinhard, N. Formation of polychlorinated dibenzo-*p*-dioxins and polychlorinated dibenzofurans during the photolysis of pentachlorophenol-containing water. *Environ. Sci. Technol.* **1994**, *28* (6), 1145–1149.

(3) Harris, J. C.; Anderson, P. C.; Goodwin, B. E.; Rechsteiner, C. E. *Dioxin Emissions from Combustion Sources: A Review of the Current State of Knowledge*; Final Report to American Society of Mechanical Engineers (ASME); ASME: New York, 1980.

(4) Addink, R.; Altwicker, E. R. Formation of polychlorinated dibenzo-*p*-dioxins/dibenzofurans from soot of benzene and *o*-dichlorobenzene combustion. *Environ. Sci. Technol.* **2004**, *38* (19), 5196–5200.

(5) Yasuhara, A.; Katami, T.; Okuda, T.; Ohno, N.; Adriaens, P. Formation of dioxins during the combustion of newspapers in the presence of sodium chloride and poly(vinyl chloride). *Environ. Sci. Technol.* **2001**, *35* (7), 1373–1378.

(6) Addink, R.; Govers, H. A. J.; Olie, K. Isomer distributions of polychlorinated dibenzo-*p*-dioxins/dibenzofurans formed during De Novo synthesis on incinerator fly ash. *Environ. Sci. Technol.* **1998**, *32* (13), 1888–1893.

(7) Chen, C. K.; Lin, C.; Lin, Y. C.; Lin, Y. C.; Wang, L. C.; Chang-Chien, G. P. Polychlorinated dibenzo-*p*-dioxins/dibenzofuran mass distribution in both start-up and normal condition in the whole municipal solid waste incinerator. *J. Hazard. Mater.* **2008**, *160* (1), 37–44.

(8) Fiedler, H. Thermal formation of PCDD/PCDF: A survey. *Environ. Eng. Sci.* **1998**, *15* (1), 49–58.

(9) Froese, K. L.; Hutzinger, O. Polychlorinated benzene, phenol, dibenzo-*p*-dioxin, and dibenzofuran in heterogeneous combustion reactions of acetylene. *Environ. Sci. Technol.* **1996**, *30* (3), 998–1008.

(10) Khachatryan, L.; Burcat, A.; Dellinger, B. An elementary reaction-kinetic model for the gas-phase formation of 1,3,6,8- and 1,3,7,9-tetrachlorinated dibenzo-*p*-dioxins from 2,4,6-trichlorophenol. *Combust. Flame* **2003**, *132* (3), 406–421.

(11) Shaub, W. M.; Tsang, W. Dioxin formation in incinerators. *Environ. Sci. Technol.* **1983**, *17* (12), 721–730.

(12) Khachatryan, L.; Asatryan, R.; Dellinger, B. Development of expanded and core kinetic models for the gas phase formation of dioxins from chlorinated phenols. *Chemosphere* **2003**, *52* (4), 695–708.

(13) Evans, C. S.; Dellinger, B. Mechanisms of dioxin formation from the high-temperature oxidation of 2-chlorophenol. *Environ. Sci. Technol.* **2005**, *39* (1), 122–127.

(14) Huang, H.; Buekens, A. Comparison of dioxin formation levels in laboratory gas-phase flow reactors with those calculated using the Shaub–Tsang mechanism. *Chemosphere* **1999**, *38* (7), 1595–1602.

(15) Babushok, V.; Tsang, W. *Origins, Fate, and Health Effects*; 7th International Congress on Combustion By-Products; MBD, Inc.: Research Triangle Park, NC, 2001.

(16) Zhang, Q. Z.; Li, S. L.; Qu, X. H.; Wang, W. X. A quantum mechanical study on the formation of PCDD/Fs from 2-chlorophenol as precursor. *Environ. Sci. Technol.* **2008**, *42* (19), 7301–7308.

(17) Sidhu, S.; Edwards, P. Role of phenoxy radicals in PCDD/F formation. *Int. J. Chem. Kinet.* **2002**, *34* (9), 531–541.

(18) Weber, R.; Hagenmaier, H. PCDD/PCDF formation in fluidized bed incineration. *Chemosphere* **1999**, *38* (11), 2643–2654.

(19) Angerer, J.; Heinzow, B.; Reimann, D. O.; Knorz, W.; Lehnert, G. Internal exposure to organic substances in a municipal waste incinerator. *Int. Arch. Occup. Environ. Health* **1992**, *64* (4), 265–273.

(20) Kimbrough, R. D. Toxicity of chlorinated hydrocarbons and related compounds: Review including chlorinated dibenzodioxins and chlorinated dibenzofurans. *Arch. Environ. Health* **1972**, *25* (2), 125–131.

(21) Okamoto, Y.; Tomonari, M. Formation pathways from 2,4,5-trichlorophenol to polychlorinated dibenzo-*p*-dioxins (PCDDs): An ab initio study. *J. Phys. Chem. A* **1999**, *103* (38), 7686–7691.

(22) Frisch, M. J.; Trucks, G. W.; Schlegel, H. B.; Gill, P. W. M.; Johnson, B. G.; Robb, M. A.; Cheeseman, J. R.; Keith, T. A.; Petersson, G. A.; Montgomery, J. A.; Raghavachari, K.; Allaham, M. A.; Zakrzewski, V. G.; Ortiz, J. V.; Foresman, J. B.; Cioslowski, J.; Stefanov, B. B.; Nanayakkara, A.; Challacombe, M.; Peng, C. Y.; Ayala, P. Y.; Chen, W.; Wong, M. W.; Andres, J. L.; Replogle, E. S.; Gomperts, R.; Martin, R. L.; Fox, D. J.; Binkley, J. S.; Defrees, D. J.; Baker, J.; Stewart, J. P.; Head-Gordon, M.; Gonzales, C.; Pople, J. A. *Gaussian 03*; Gaussian, Inc.: Wallingford, CT, 2003.

(23) Zhao, Y.; Truhlar, D. G. Hybrid meta density functional theory methods for thermochemistry, thermochemical kinetics, and

noncovalent interactions: The MPW1B95 and MPWB1K models and comparative assessments for hydrogen bonding and van der Waals interactions. *J. Phys. Chem. A* **2004**, *108* (33), 6908–6918.

(24) Steckler, R.; Chuang, Y. Y.; Fast, P. L.; Corchade, J. C.; Coitino, E. L.; Hu, W. P.; Lynch, G. C.; Nguyen, K.; Jackells, C. F.; Gu, M. Z.; Rossi, I.; Clayton, S.; Melissas, V.; Garrett, B. C.; Isaacson, A. D.; Truhlar, D. G. *POLYRATE,* version 9.3; University of Minnesota: Minneapolis, MN, 2002.

(25) Baldridge, M. S.; Gordor, R.; Steckler, R.; Truhlar, D. G. Ab initio reaction paths and direct dynamics calculations. *J. Phys. Chem.* **1989**, *93* (13), 5107–5119.

(26) Gonzalez-Lafont, A.; Truong, T. N.; Truhlar, D. G. Interpolated variational transition state theory: Practical methods for estimating variational transition state properties and tunneling contributions to chemical reaction rates from electronic structure calculations. *J. Chem. Phys.* **1991**, *95* (12), 8875–8894.

(27) Garrett, B. C.; Truhlar, D. G. Generalized transition state theory. Classical mechanical theory and applications to collinear reactions of hydrogen molecules. *J. Phys. Chem.* **1979**, *83* (8), 1052–1079.

(28) Fernandez-Ramos, A.; Ellingson, B. A.; Garret, B. C.; Truhlar, D. G. Variational Transition State Theory with Multidimensional Tunneling. In *Reviews in Computational Chemistry*; Lipkowitz, K. B., Cundari, T. R. Eds; Wiley-VCH: Hoboken, NJ, 2007.

(29) Garrett, B. C.; Truhlar, D. G.; Wagner, A. F.; Dunning, T. H., Jr. Variational transition state theory and tunneling for a heavy-light-heavy reaction using an ab initio potential energy surface. ^{37}Cl+H(D) ^{35}Cl→H(D) ^{37}Cl+^{35}Cl. *J. Chem. Phys.* **1983**, *78* (7), 4400–4413.

(30) Skodje, R. T.; Truhlar, D. G.; Garrett, B. C. Vibrationally adiabatic models for reactive tunneling. *J. Chem. Phys.* **1982**, *77* (12), 5955–5976.

(31) Zhang, Q. Z.; Wang, S. K.; Gu, Y. S. A theoretical investigation on the mechanism and kinetics for the reaction of atomic O (^{3}P) with CH_3CHCl_2. *J. Chem. Phys.* **2003**, *119* (21), 11172–11179.

(32) Zhang, Q. Z.; Wang, S. K.; Zhou, J. H.; Gu, Y. S. Ab initio and kinetic calculation for the abstraction reaction of atomic O (^{3}P) with SiH_4. *J. Phys. Chem. A* **2002**, *106* (1), 115–121.

(33) Hellwege, K. H.; Hellwege A. M., Eds. Structure Data of Free Polyatomic Molecules. *Landolt-Bornstein*; Group II, Atomic and Molecular Physics; Springer-Verlag: Berlin, Germany, 1976; Volume *7*.

(34) Roussy, G.; Michel, F. Spectres de rotation de la molécule de chlorobenzéne: I. Variétés isotopiques monosubstituées et structure r_s. *J. Mol. Struct.* **1976**, *30* (2), 399–407.

(35) Boer, F. P.; Neuman, M. A.; Van Remoortere, F. P.; North, P. P.; Rinn, H. W. *Adv. Chem. Ser.* **1973**, *120*, 1.

(36) Grainger, J.; Reddy, V. V.; Patterson, D. G., Jr. Molecular geometry/toxicity correlations for laterally tetrachlorinated dibenzo-*p*-dioxins by Fourier transform infrared spectroscopy. *Chemospere* **1989**, *18* (1–6), 981.

(37) Mallard, W. G., Ed. National Institute of Standards and Technology (NIST) Chemistry Web Book: NIST Standard Reference; National Institute of Standards and Technology: Gaithersburg, MD, Database 69, 1998.

(38) Altarawneh, M.; Dlugogorski, B. Z.; Kennedy, E. M.; Mackie, J. C. Quantum chemical investigation of formation of polychlorodibenzo-*p*-dioxins and dibenzofurans from oxidation and pyrolysis of 2-chlorophenol. *J. Phys. Chem. A* **2007**, *111* (13), 2563–2573.

(39) Evans, C. S.; Dellinger, B. Mechanisms of dioxin formation from the high-temperature pyrolysis of 2-chlorophenol. *Environ. Sci. Technol.* **2003**, *37* (7), 1325–1330.

(40) Mulholland, J. A.; Akki, U.; Yang, Y.; Ryu, J. Y. Temperature dependence of DCDD/F isomer distributions from chlorophenol precursors. *Chemosphere* **2001**, *42* (5–7), 719–727.

(41) Ryu, J. Y.; Mulholland, J. A.; Kim, D. H.; Takeuchi, M. Homologue and isomer patterns of polychlorinated dibenzo-*p*-dioxins and dibenzofurans from phenol precursors: comparison with municipal waste incinerator data. *Environ. Sci. Technol.* **2005**, *39* (12), 4398–4406.

(42) Wu, Y. L.; Lin, L. F.; Hsieh, L. T.; Wang, L. C.; Chang-Chien, G. P. Atmospheric dry deposition of polychlorinated dibenzo-*p*-dioxins and dibenzofurans in the vicinity of municipal solid waste incinerators. *J. Hazard. Mater.* **2009**, *162* (1), 521–529.

(43) Melissas, V. S.; Truhlar, D. G. Interpolated variational transition state theory and tunneling calculations of the rate constant of the reaction OH + CH_4 at 223−2400 K. *J. Chem. Phys.* **1993**, *99* (2), 1013–1027.

(44) Baulch, D. L.; Cobos, C. J.; Cox, R. A.; Esser, C.; Frank, P.; Just, Th.; Kerr, J. A.; Pilling, M. J.; Troe, J.; Walker, R. W.; Warnatz, J. Evaluated kinetic data for combustion modeling. *J. Phys. Chem. Ref. Data* **1992**, *21* (3), 411–429.

(45) He, Y. Z.; Mallard, W. G.; Tsang, W. Kinetics of hydrogen and hydroxyl radical attack on phenol at high temperatures. *J. Phys. Chem.* **1988**, *92* (8), 2196–2201.

(46) Platz, J.; Nielsen, O. J.; Wallington, T. J.; Ball, J. C.; Hurley, M. D.; Straccia, A. M.; Schneider, W. F.; Sehested, J. Atmospheric chemistry of the phenoxy radical, C_6H_5O: UV spectrum and kinetics of its reaction with NO, NO_2, and O_2. *J. Phys. Chem. A* **1998**, *102* (41), 7964–7974.

(47) Zhang, Q. Z.; Qu, X. H.; Xu F.; Shi X. Y.; Wang, W. X. Mechanism and thermal rate constants for the complete series reactions of chlorophenols with H. *Environ. Sci. Technol.* **2009**, in press.

ES802835E

Environmental Science & Technology

ARTICLE

pubs.acs.org/est

Mechanism and Direct Kinetics Study on the Homogeneous Gas-Phase Formation of PBDD/Fs from 2-BP, 2,4-DBP, and 2,4,6-TBP as Precursors

Wanni Yu, Jingtian Hu, Fei Xu, Xiaoyan Sun, Rui Gao, Qingzhu Zhang,* and Wenxing Wang

Environment Research Institute, Shandong University, Jinan 250100, P. R. China

Supporting Information

ABSTRACT: This study investigated the homogeneous gas-phase formation of polybrominated dibenzo-*p*-dioxin/dibenzofurans (PBDD/Fs) from 2-BP, 2,4-DBP, and 2,4,6-TBP as precursors. First, density functional theory (DFT) calculations were carried out for the formation mechanism. The geometries and frequencies of the stationary points were calculated at the MPWB1K/6-31+G(d,p) level, and the energetic parameters were further refined by the MPWB1K/6-311+G(3df,2p) method. Then, the formation mechanism of PBDD/Fs was compared and contrasted with the PCDD/F formation mechanism from 2-CP, 2,4-DCP, and 2,4,6-TCP as precursors. Finally, the rate constants of the crucial elementary reactions were evaluated by the canonical variational transition-state (CVT) theory with the small curvature tunneling (SCT) correction over a wide temperature range of 600−1200 K. Present results indicate that only BPs with bromine at the ortho position are capable of forming PBDDs. The study, together with works already published from our group, clearly shows an increased propensity for the dioxin formations from BPs over the analogous CPs. Multibromine substitutions suppress the PBDD/F formations.

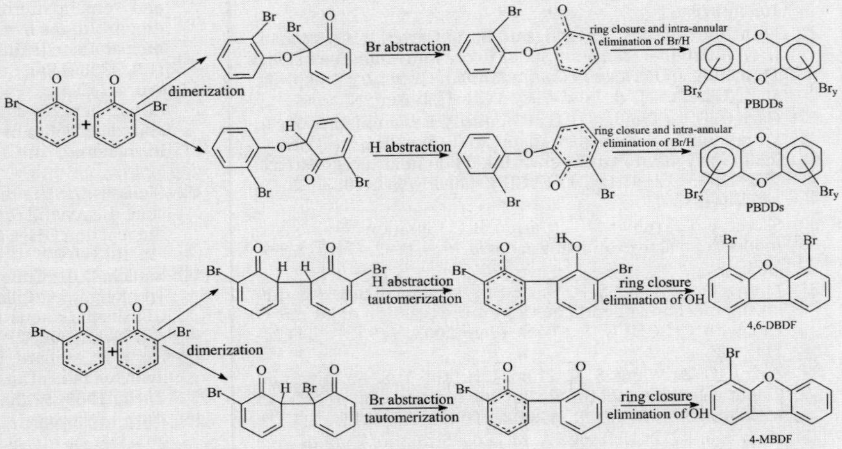

■ INTRODUCTION

Polybrominated dibenzo-*p*-dioxin/dibenzofurans (PBDD/Fs) are polychlorinated dibenzo-*p*-dioxin/dibenzofurans (PCDD/Fs) analogues in which all of the chlorine atoms are substituted by bromine atoms. Therefore, they have similar physicochemical properties, toxicity, and geochemical behavior in the environment.[1−6] PBDD/Fs are known to occur unnaturally and to be produced unintentionally as unwanted byproduct. Recently, the level of environmental concern regarding PBDD/Fs has been raised[7−9] because of the rapid increase in the use of brominated flame retardants (BFRs). PBDD/Fs can be formed in the process of manufacturing BFRs. In particular, a large quantity of PBDD/Fs is emitted from electronic waste recycling and the pyrolysis or combustion of waste materials containing BFRs.[10−15]

The high correlation between the PBDD/F and PCDD/F concentrations revealed their similar formation mechanism in the pyrolysis or combustion system.[12,13] Chlorophenols (CPs) are key intermediates in essentially all proposed pathways of the formation of PCDD/Fs.[16,17] Similarly, bromophenols (BPs) have been demonstrated to be the precursors of PBDD/Fs.[8,18−20] BPs are extensively used as flame retardants, intermediates for the yield of other flame retardants, and pesticides for wood preservation. In addition, BPs may be released into the environment as major degradation products of other BFRs.[21] Under pyrolysis or combustion conditions, PBDD/F can be formed from BPs by two general reactions: homogeneous gas-phase reactions and heterogeneous metal-mediated reactions. It is generally believed that gas-phase mechanism accounts for about 30% of the total PBDD/F emissions, and surface-mediated mechanism is responsible for the remainder.[18]

Compared to the situation for PCDD/Fs, there are very few studies on the formation of PBDD/Fs. This is largely due to shortage of commercial available PBDD/Fs' standards for quantification.[6,11,22,23] Here, therefore, we conducted a direct density functional theory (DFT) kinetics study on the

Received: October 22, 2010
Accepted: January 26, 2011
Revised: January 19, 2011
Published: February 10, 2011

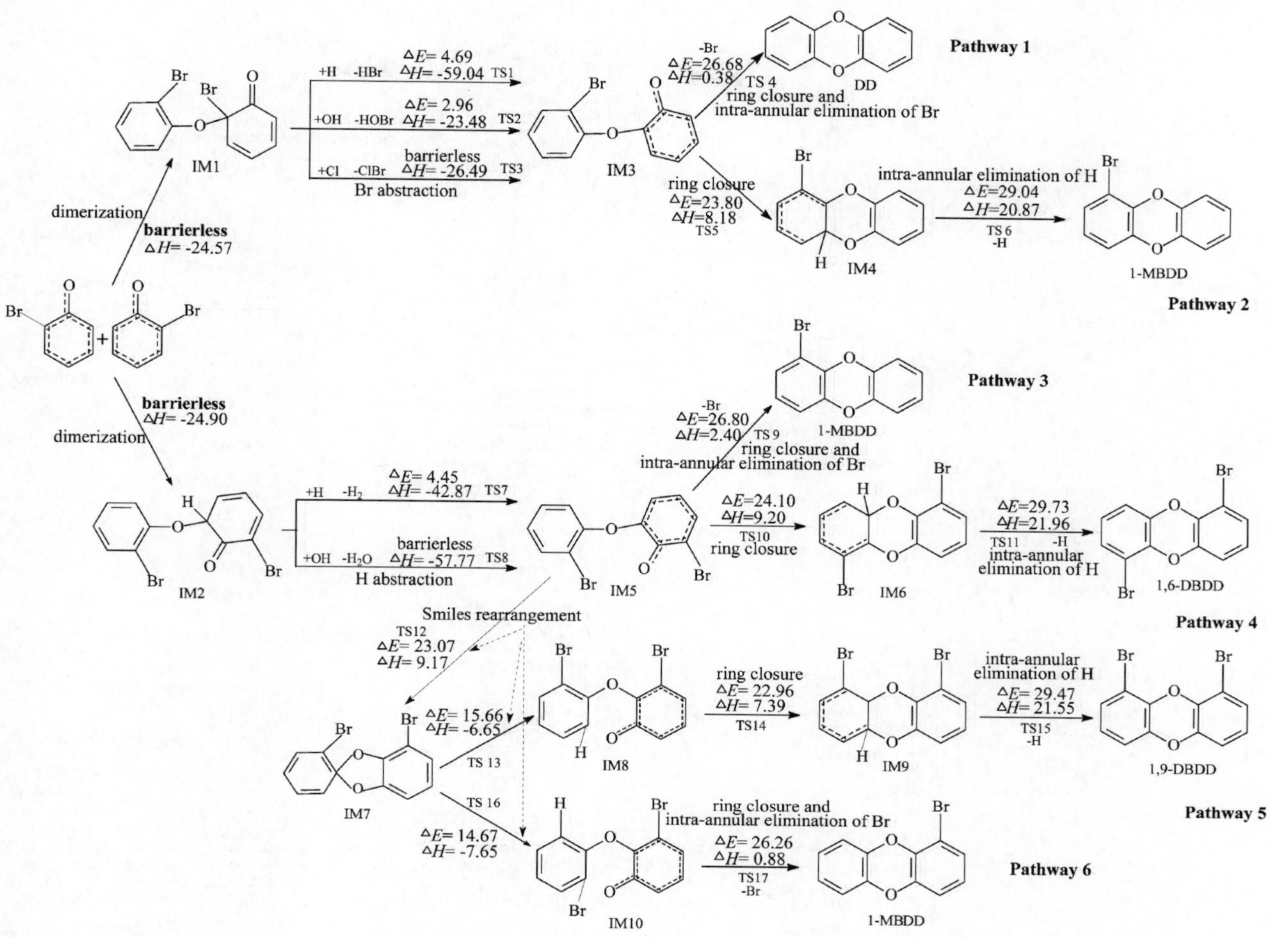

Figure 1. PBDD formation routes embedded with the potential barriers ΔE (in kcal/mol) and reaction heats ΔH (in kcal/mol) from 2-BP as precursor. ΔH is calculated at 0 K.

homogeneous gas-phase formation of PBDD/Fs from 2-BP, 2,4-DBP, and 2,4,6-TBP as precursors, which are the most widely manufactured BPs and the most abundant BPs found in waste incinerators. The primary objective of the study is to elucidate the homogeneous gas-phase formation mechanism of PBDD/Fs from BPs. A second objective is to deduce the rate constants of the key elementary reactions involved in the PBDD/F formations. Further interest is planed to focus on the homogeneous gas-phase formation of 2,3,7,8-TeBDD, which is the most toxic among all 210 PBDD/F isomers, as well as the heterogeneous metal-mediated formation of PBDD/Fs.

■ COMPUTATIONAL METHODS

The density functional theory (DFT) calculations were performed using the Gaussian 03 package.[24] All geometries were fully optimized at the MPWB1K/6-31+G(d,p) level.[25] The nature of various stationary points was determined by frequency calculations. To check whether the obtained transition states connect the right minima, the intrinsic reaction coordinate (IRC) calculations were carried out. To improve the reaction heat and potential barrier, the single point energies were calculated at the MPWB1K/6-311+G(3df,2p) level. By means of the Polyrate 9.3 program,[26] the rate constants were computed by the canonical variational transition-state (CVT) theory[27−29] with the small curvature tunneling (SCT) contribution.[30]

■ RESULTS AND DISCUSSION

The optimized geometries of 2-BP and DD at the MPWB1K/6-31+G(d,p) level are in reasonable accordance with the corresponding experimental values, and the largest discrepancy remains within 1.5% for the bond lengths.[31,32] For the reaction of 2-BP+2-BP→DD+H_2+Br_2, the reaction enthalpy of 29.22 kcal/mol calculated at the MPWB1K/6-311+G(3df,2p) level and at 298.15 K agrees well with the experimental value of 30.52 kcal/mol, which is derived from the experimental standard enthalpies of formation.[33−35]

Formation of 2-BPRs, 2,4-DBPRs, and 2,4,6-TBPRs. It has shown that the homogeneous gas-phase formation of PBDD/Fs is attributed to the dimerization of bromophenoxy radicals (BPRs).[36,37] Thus, the formation of BPRs is the initial step in the formation of PBDD/Fs. Under pyrolysis or combustion conditions, BPRs can be readily formed from BPs through

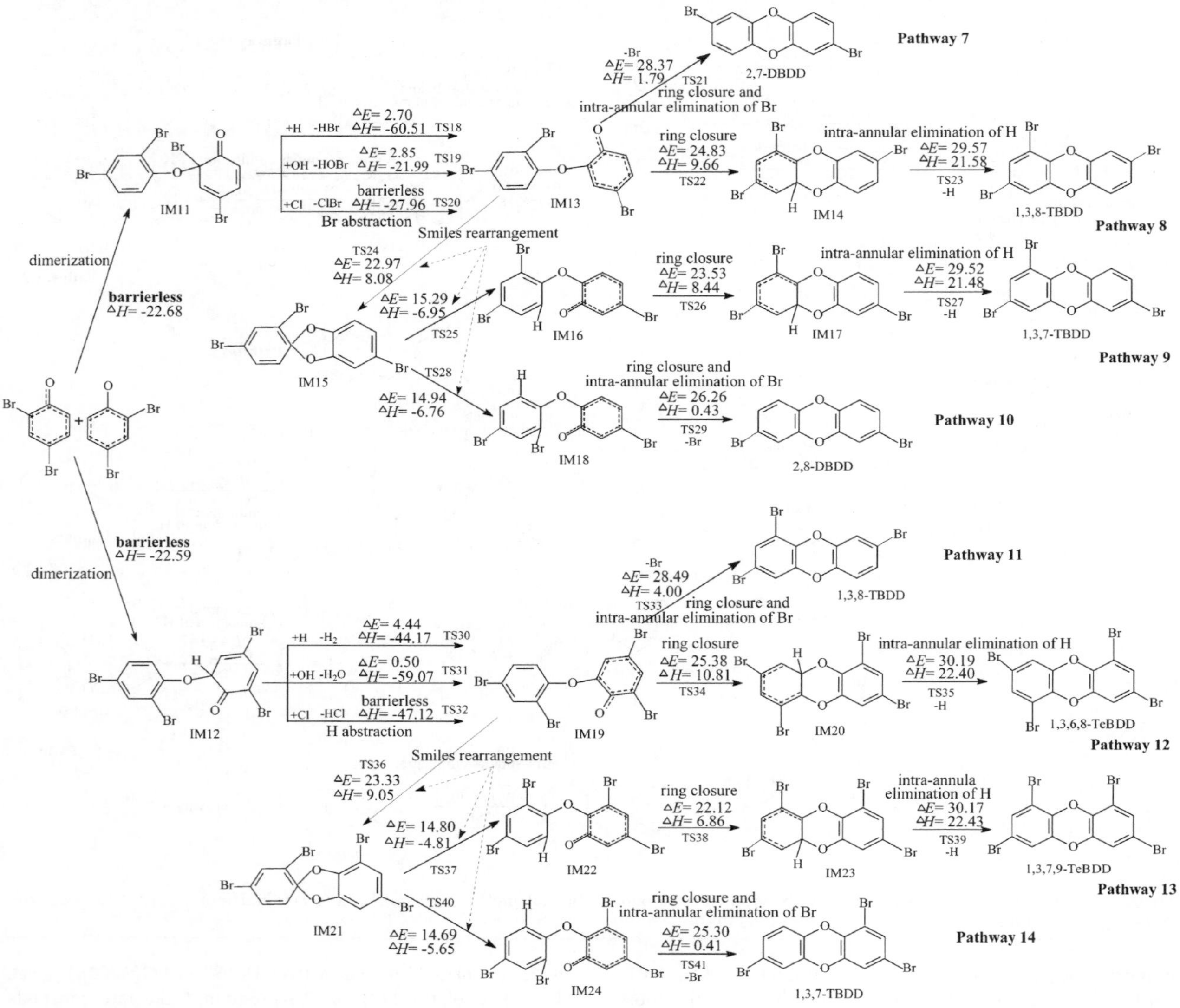

Figure 2. PBDD formation routes embedded with the potential barriers ΔE (in kcal/mol) and reaction heats ΔH (in kcal/mol) from 2,4-DBP as precursor. ΔH is calculated at 0 K.

loss of the phenoxyl-hydrogen by unimolecular, bimolecular, or possibly other low-energy pathways (including heterogeneous reactions).[36,37] The unimolecular reaction occurs via the decomposition of BPs with the cleavage of the O−H bond. The bimolecular reactions proceed through the phenoxyl-hydrogen abstraction from BPs by the active radicals, H, OH, O (^{3}P), Cl, and Br, which are abundant in the combustion environment.[36,37] The potential barriers (ΔE) and reaction heats (ΔH, 0 K) calculated at the MPWB1K/6-311+G(3df,2p) level are presented in the Supporting Information.

Formation of PBDDs. *Formation of PBDDs from the Dimerization of 2-BPRs.* Six possible PBDD formation pathways, illustrated in Figure 1, are proposed from 2-BP. As seen from Figure 1, all PBDD formation pathways start with oxygen−carbon coupling, followed by Br or H abstraction. The oxygen−carbon coupling is a barrierless and strongly exothermic process. All of the Br or H abstraction steps are highly exothermic with low-energy barriers. In pathways 1, 3, and 6, ring closure and intra-annular elimination of Br occur in a one-step reaction and are the rate determining step. There is a Smiles rearrangement after H abstraction in pathways 5 and 6, respectively. Intra-annular elimination of H is the rate determining step for pathways 2, 4, and 5 due to the high barrier and strong endothermicity.

Obviously, pathway 1 covers relatively less elementary steps compared to pathway 2. Furthermore, the rate determining step involved in pathway 1 has a lower barrier and is much less endoergic than that involved in pathway 2. Thus, pathway 1 is favored over pathway 2. Similarly, pathway 3 is favored over pathway 4. Pathway 6 is favored over pathway 5. Therefore, the thermodynamically favorable PBDD formation pathways are pathways 1, 3, and 6, leading to the formation of DD and 1-MBDD, consistent with the experimental observation.[36,37]

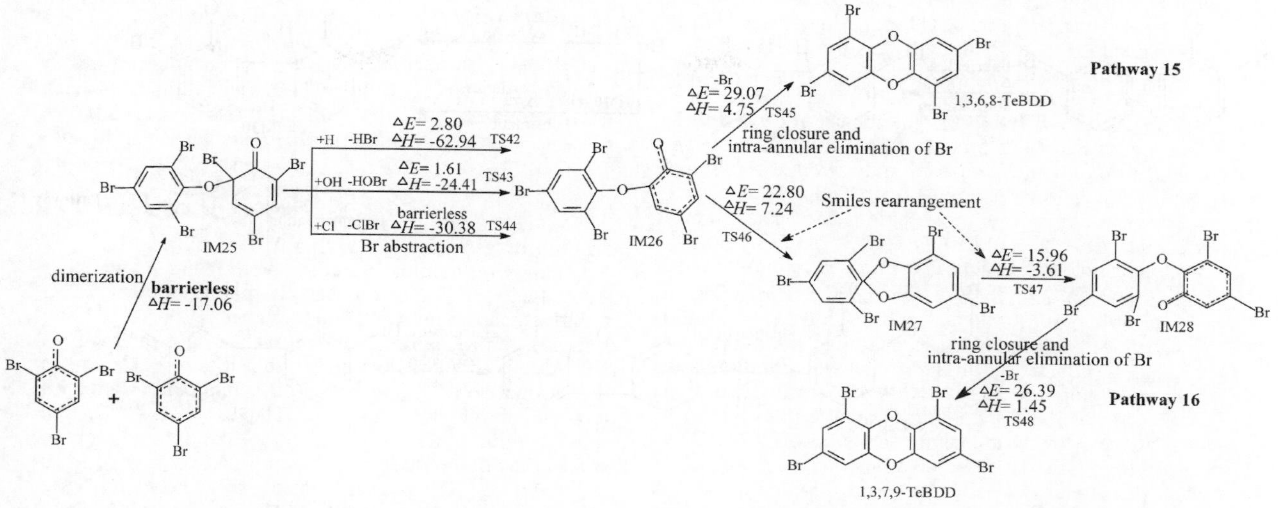

Figure 3. PBDD formation routes embedded with the potential barriers ΔE (in kcal/mol) and reaction heats ΔH (in kcal/mol) from 2,4,6-TBP as precursor. ΔH is calculated at 0 K.

Comparison of the PBDD formations from 2-BP with the previous study for 2-CP[38] clearly shows that intra-annular elimination of Br is much less endothermic than the analogous intra-annular elimination of Cl because of the weaker C−Br bond versus C−Cl bond, resulting in the formation of DD being more favorable from 2-BP than 2-CP. The result is supported by the experimental evidence that the maximum yield for the DD formation from 2-BP is 20 times higher than that from 2-CP.[36,37]

Formation of PBDDs from the Dimerization of 2,4-DBPRs and 2,4,6-TBPRs. Eight possible reaction pathways are postulated to explain the formation of PBDDs from 2,4-DBP. The formation schemes embedded with the potential barriers and reaction heats are depicted in Figure 2. Similar to the mechanism suggested for 2-BP, all PBDD formation pathways start with oxygen−carbon coupling, followed by Br or H abstraction. In pathways 7, 10, 11, and 14, ring closure and intra-annular elimination of Br occur in a one-step reaction and are the rate determining step. Intra-annular elimination of H is the rate determining step for pathways 8, 9, 12, and 13. It is clear from Figure 2 that PBDDs are preferentially formed from pathways 7, 10, 11, and 14. The resulting 2,7-DBDD, 2,8-DBDD, 1,3,7-TBDD, and 1,3,8-TBDD are the dominant PBDD products.

Due to the symmetry of 2,4,6-TBP, only two PBDD formation pathways, pathways 15 and 16, were identified and displayed in Figure 3. Pathway 16 contains two more elementary steps than pathway 15. Nevertheless, the rate determining step involved in pathway 15 requires a higher barrier and is more endothermic than that involved in pathway 16. So, pathways 15 and 16 should be competitive. The resulting 1,3,6,8-TeBDD and 1,3,7,9-TeBDD were experimentally detected in the pyrolysis of 2,4,6-TBP.[23,39] Comparison with the previous work[40] suggests that the formation of PBDDs from 2,4,6-TBP is relatively easier compared to the formation of the analogous PCDDs from 2,4,6-TCP because Br is a better leaving group than Cl.

Mechanisms presented in Figure 1, Figure 2, and Figure 3 show that the rate determining step to the PBDD formations is intra-annular elimination of Br or H. Intra-annular elimination of H has a higher barrier and is much more endothermic than intra-annular elimination of Br. The thermodynamically favorable PBDD formation pathways occur through intra-annular elimination of Br. The Br atom, which is eliminated, is the substituent at the ortho position in BPs. This implies that only BPs with bromine at the ortho position are capable of forming PBDDs. The result is similar to that observed for the PCDD formations from CPs.[41] Comparison of the PBDD formations from 2-BP, 2,4-DBP, and 2,4,6-TBP clearly indicates that the formation mechanism is controlled largely by the substitution pattern of BPs. The exothermicity of the oxygen−carbon coupling decreases with increasing number of bromine substitutions. Although one ortho-bromine is needed for the formation of PBDDs, multi-bromine substitutions suppress the dimerization of BPRs, i.e., suppress the formation of PBDDs.

Formation of PBDFs. 4,6-DBDF and 4-MBDF were experimentally observed in the high-temperature oxidation of 2-BP.[37] Two reaction routes, presented in Figure 4, are offered to interpret the formation of 4,6-DBDF. Pathway 17 involves five elementary processes: carbon−carbon coupling, H abstraction, tautomerization (H-shift), ring closure, and elimination of OH. The ring closure process has a large barrier and is strongly endoergic, and it is the rate determining step. Pathway 18 also involves five elementary processes: carbon−carbon coupling, tautomerization (double H-transfer), H abstraction, ring closure (the rate determining step), and elimination of OH. Pathway 19 in Figure 4 illustrates 4-MBDF being formed from 2-BP. The *o,o′*-dihydroxybiphenyl intermediate IM29 can be regarded as a prestructure of 4,6-DBDF, and IM34 is a prestructure of 4-MBDF. It can be seen from Figure 4 that the formation of IM29 is more exothermic than the formation of IM34. Moreover, the rate determining step involved in the formation of 4,6-DBDF has a lower barrier and is less endothermic than that involved in the formation of 4-MBDF. Thus, the formation of 4,6-DBDF is preferred over the formation of 4-MBDF.

Similar to the formation of PBDFs from 2-BP, two PBDF congeners, 2,4,6,8-TeBDF and 2,4,8-TBDF, can be formed

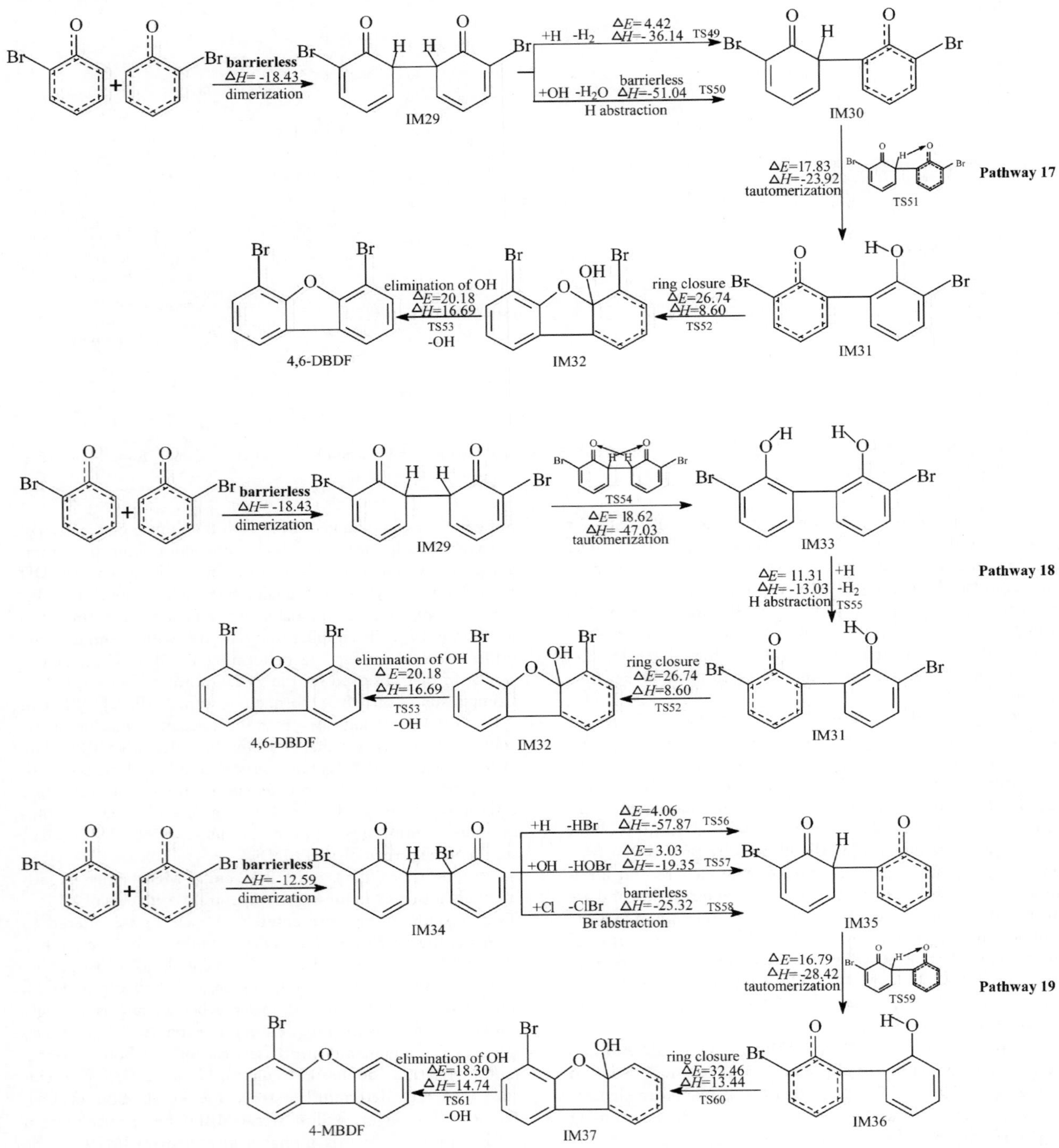

Figure 4. PBDF formation routes embedded with the potential barriers ΔE (in kcal/mol) and reaction heats ΔH (in kcal/mol) from 2-BP as precursor. ΔH is calculated at 0 K.

from 2,4-DBP. The formation mechanism is schemed in Figure 5. Evidently, the formation of 2,4,6,8-TeBDF is preferred over the formation of 2,4,8-TBDF. Comparison of the reaction pathways presented in Figure 4 and Figure 5 shows that the substitution pattern of BPs has a significant effect on the formation mechanism of PBDFs. The arbon−carbon coupling of 2-BPRs is more exothermic compared to the carbon−carbon coupling of 2,4-DBPRs. Because both ortho-positions are substituted with the voluminous bromine atoms, the carbon−carbon coupling of 2,4,6-TCPRs is sterically inhibited. So, no PBDFs can be formed from 2,4,6-TBP.

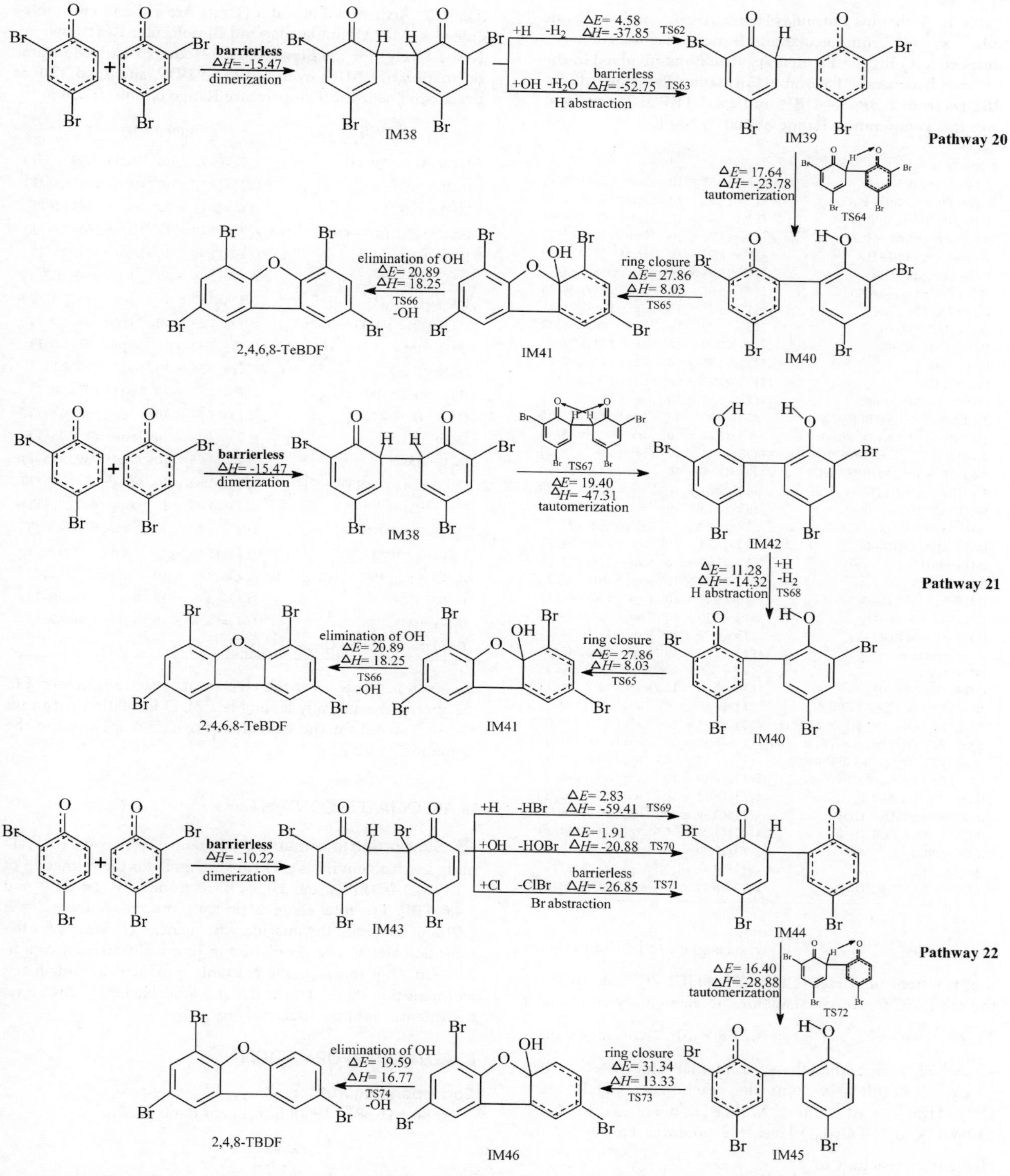

Figure 5. PBDF formation routes embedded with the potential barriers ΔE (in kcal/mol) and reaction heats ΔH (in kcal/mol) from 2,4-DBP as precursor. ΔH is calculated at 0 K.

Rate Constant Calculations. Previous researches have shown that the CVT/SCT rate constants of $C_6H_5OH+H \rightarrow C_6H_5O+H_2$, $C_6H_5OH+OH \rightarrow C_6H_5O+H_2O$ match well with the corresponding experimental values.[42,43] In the reaction kinetic model of the PCDD/F formations, the rate constant for the elementary reaction of

Table 1. Arrhenius Formulas (Units Are s^{-1} and cm^3 molecule^{-1} s^{-1} for Unimolecular and Bimolecular Reactions, Respectively) for the Elementary Reactions Involved in the Thermodynamically Favorable Formation Pathways of PBDDs from 2-BP, 2,4-DBP, and 2,4,6-TBP as Precursors over the Temperature Range of 600−1200 K

reactions	Arrhenius formulas
2-BP+H→2-BPR+H_2	$k(T)=(3.38\times10^{-11})\exp(-7384.78/T)$
2-BP+OH→2-BPR+H_2O	$k(T)=(5.34\times10^{-12})\exp(-2758.55/T)$
2-BP+O(^{3}P)→2-BPR+OH	$k(T)=(1.66\times10^{-11})\exp(-6404.68/T)$
2-BP+Cl→2-BPR+HCl	$k(T)=(1.71\times10^{-10})\exp(-1237.87/T)$
2-BP+Br→2-BPR+HBr	$k(T)=(1.93\times10^{-11})\exp(-3910.39/T)$
IM1+H→IM3+HBr	$k(T)=(5.85\times10^{-10})\exp(-3577.55/T)$
IM1+OH→IM3+HOBr	$k(T)=(2.07\times10^{-11})\exp(-3171.08/T)$
IM3→DD+Br	$k(T)=(5.67\times10^{11})\exp(-13390.79/T)$
IM2+H→IM5+H_2	$k(T)=(1.79\times10^{-11})\exp(-2692.33/T)$
IM5→1-MBDD+Br	$k(T)=(7.09\times10^{11})\exp(-13832.79/T)$
IM5→IM7	$k(T)=(2.55\times10^{12})\exp(-11951.40/T)$
IM7→IM10	$k(T)=(2.33\times10^{13})\exp(-7701.85/T)$
IM10→1-MBDD+Br	$k(T)=(7.40\times10^{11})\exp(-13268.05/T)$
2,4-DBP+H→2,4-DBPR+H_2	$k(T)=(6.83\times10^{-11})\exp(-7292.71/T)$
2,4-DBP+OH→2,4-DBPR+H_2O	$k(T)=(6.68\times10^{-12})\exp(-2690.13/T)$
2,4-DBP+O(^{3}P)→2,4-DBPR+OH	$k(T)=(4.71\times10^{-11})\exp(-5455.78/T)$
2,4-DBP+Cl→2,4-DBPR+HCl	$k(T)=(8.98\times10^{-11})\exp(-1231.28/T)$
2,4-DBP+Br→2,4-DBPR+HBr	$k(T)=(2.61\times10^{-10})\exp(-3518.55/T)$
IM11+H→IM13+HBr	$k(T)=(6.65\times10^{-11})\exp(-1967.34/T)$
IM11+OH→IM13+HOBr	$k(T)=(4.78\times10^{-12})\exp(-3419.21/T)$
IM13→2,7-DBDD+Br	$k(T)=(3.44\times10^{11})\exp(-14304.24/T)$
IM13→IM15	$k(T)=(1.41\times10^{12})\exp(-11816.21/T)$
IM15→IM18	$k(T)=(1.22\times10^{13})\exp(-7919.38/T)$
IM18→2,8-DBDD+Br	$k(T)=(4.41\times10^{11})\exp(-13097.41/T)$
IM12+H→IM19+H_2	$k(T)=(7.30\times10^{-13})\exp(-1141.98/T)$
IM19→1,3,8-TBDD+Br	$k(T)=(4.77\times10^{10})\exp(-14274.45/T)$
IM19→IM21	$k(T)=(1.90\times10^{12})\exp(-12032.47/T)$
IM21→IM24	$k(T)=(1.66\times10^{12})\exp(-7633.92/T)$
IM24→1,3,7-TBDD+Br	$k(T)=(2.40\times10^{11})\exp(-12767.14/T)$
2,4,6-TBP+H→2,4,6-TBPR+H_2	$k(T)=(4.15\times10^{-12})\exp(-7112.44/T)$
2,4,6-TBP+OH→2,4,6-TBPR+H_2O	$k(T)=(8.93\times10^{-12})\exp(-2368.10/T)$
2,4,6-TBP+O(^{3}P)→2,4,6-TBPR+OH	$k(T)=(6.30\times10^{-12})\exp(-5373.21/T)$
2,4,6-TBP+Cl→2,4,6-TBPR+HCl	$k(T)=(1.32\times10^{-10})\exp(-1032.16/T)$
2,4,6-TBP+Br→2,4,6-TBPR+HBr	$k(T)=(6.39\times10^{-10})\exp(-5381.05/T)$
IM25+H→IM26+HBr	$k(T)=(8.22\times10^{-11})\exp(-2370.52/T)$
IM25+OH→IM26+HOBr	$k(T)=(3.39\times10^{-12})\exp(-2564.09/T)$
IM26→1,3,6,8-TeBDD+Br	$k(T)=(1.06\times10^{12})\exp(-14774.34/T)$
IM26→IM27	$k(T)=(5.30\times10^{12})\exp(-11692.33/T)$
IM27→IM28	$k(T)=(1.95\times10^{13})\exp(-8337.48/T)$
IM28→1,3,7,9-TeBDD+Br	$k(T)=(6.91\times10^{11})\exp(-15447.33/T)$

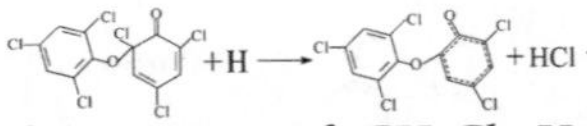 was assigned to be the value for the reaction of $CH_3Cl+H\rightarrow CH_3+HCl$.[44,45] At 1000 K, the CVT/SCT rate constant for the elementary reaction of 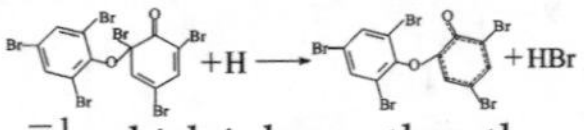 is 7.68×10^{-12} cm^3 molecule^{-1} s^{-1}, which is larger than the experimental value of 4.49×10^{-12} cm^3 molecule^{-1} s^{-1} for the reaction of $CH_3Br+H\rightarrow CH_3+HBr$.[46] At the MPWB1K/6-311+G(3df,2p)//MPWB1K/6-31+G(d,p) level, the potential barrier for the elementary reaction of 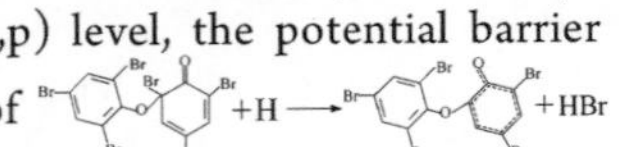 is 2.80 kcal/mol, whereas the value for the reaction of $CH_3Br+H\rightarrow CH_3+HBr$ is 6.89 kcal/mol. The CVT/SCT rate constant for the elementary reaction of 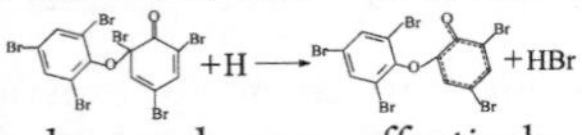 appears to be reasonable. To be used more effectively, the CVT/SCT rate constants are fitted, and Arrhenius formulas

Table 2. Arrhenius Formulas (Units Are s^{-1} and cm^3 molecule^{-1} s^{-1} for Unimolecular and Bimolecular Reactions, Respectively) for the Elementary Reactions Involved in the Formation of PBDFs from 2-BP, 2,4-BDP, and 2,4,6-TBP as Precursors over the Temperature Range of 600−1200 K

reactions	Arrhenius formulas
IM29+H→IM30+H_2	$k(T)=(9.75\times10^{-11})\exp(-1923.18/T)$
IM30→IM31	$k(T)=(3.77\times10^{12})\exp(-10251.93/T)$
IM31→IM32	$k(T)=(1.11\times10^{12})\exp(-11614.16/T)$
IM32→4,6-DBDF+OH	$k(T)=(3.10\times10^{13})\exp(-10683.59/T)$
IM29→IM33	$k(T)=(1.98\times10^{12})\exp(-9558.46/T)$
IM33+H→IM31+H_2	$k(T)=(2.44\times10^{-11})\exp(-5097.33/T)$
IM34+H→IM35+HBr	$k(T)=(2.17\times10^{-11})\exp(-2927.17/T)$
IM34+OH→IM35+HOBr	$k(T)=(1.63\times10^{-12})\exp(-2968.01/T)$
IM35→IM36	$k(T)=(1.39\times10^{12})\exp(-8079.76/T)$
IM36→IM37	$k(T)=(8.55\times10^{12})\exp(-16679.13/T)$
IM37→4-MBDF+OH	$k(T)=(2.77\times10^{13})\exp(-9572.70/T)$
IM38+H→IM39+H_2	$k(T)=(4.87\times10^{-11})\exp(-3260.61/T)$
IM39→IM40	$k(T)=(2.26\times10^{12})\exp(-8890.25/T)$
IM40→IM41	$k(T)=(1.75\times10^{12})\exp(-13935.26/T)$
IM41→2,4,6,8-TeBDF+OH	$k(T)=(1.89\times10^{13})\exp(-10713.53/T)$
IM38→IM42	$k(T)=(1.43\times10^{12})\exp(-9947.05/T)$
IM42+H→IM40+H_2	$k(T)=(2.17\times10^{-12})\exp(-5819.31/T)$
IM43+H→IM44+HBr	$k(T)=(3.65\times10^{-11})\exp(-1536.89/T)$
IM43+OH→IM44+HOBr	$k(T)=(2.61\times10^{-12})\exp(-2651.62/T)$
IM44→IM45	$k(T)=(2.04\times10^{12})\exp(-8320.15/T)$
IM45→IM46	$k(T)=(5.39\times10^{12})\exp(-16047.96/T)$
IM46→2,4,8-TBDF+OH	$k(T)=(2.87\times10^{13})\exp(-10379.37/T)$

are given in Table 1 for the elementary reactions involved in the thermodynamically favorable PBDD formation pathways and in Table 2 for the elementary reactions involved in the formation of PBDFs.

■ ASSOCIATED CONTENT

S Supporting Information. Potential barriers ΔE (in kcal/mol) and reaction heats ΔH (in kcal/mol) for the formation of 2-BPRs, 2,4-DBPRs, and 2,4,6-TBPRs from 2-BP, 2,4-DBP, and 2,4,6-TBP. The total energies (in a.u.), the zero-point energies (ZPE, in a.u.), and the imaginary frequencies (in cm^{-1}) for the transition states. The geometries in terms of Cartesian coordinate (in Angstrom) for the reactants, products, intermediates, and transition states. This material is available free of charge via the Internet at http://pubs.acs.org.

■ AUTHOR INFORMATION

Corresponding Author

*Fax: 86-531-8836 1990. E-mail: zqz@sdu.edu.cn.

■ ACKNOWLEDGMENT

This work was supported by NSFC (National Natural Science Foundation of China, project No. 20737001), Shandong Province Outstanding Youth Natural Science Foundation (project No. JQ200804), and Independent Innovation Foundation of Shandong University (IIFSDU, project No. 2009JC016).

The authors thank Professor Donald G. Truhlar for providing the POLYRATE 9.3 program.

■ REFERENCES

(1) Samara, F.; Gullett, B. K.; Harrison, R. O.; Chu, A.; Clark, G. C. Determination of relative assay response factors for toxic chlorinated and brominated dioxins/furans using an enzyme immunoassay (EIA) and a chemically-activated luciferase gene expression cell bioassay (CALUX). *Environ. Int.* **2009**, *35* (3), 588–593.

(2) Samara, F.; Wyrzykowska, B.; Tabor, D.; Touati, D.; Gullett, B. K. Toxicity comparison of chlorinated and brominated dibenzo-*p*-dioxins and dibenzofurans in industrial source samples by HRGC/HRMS and enzyme immunoassay. *Environ. Int.* **2010**, *36* (3), 247–253.

(3) Mennear, J. H.; Lee, C. C. Polybrominated dibenzo-*p*-dioxins and dibenzofurans: literature review and health assessment. *Environ. Health Perspect.* **1994**, *102* (1), 265–274.

(4) Birnbaum, L. S.; Staskal, D. F.; Diliberto, J. J. Health effects of polybrominated dibenzo-*p*-dioxins (PBDDs) and dibenzofurans (PBDFs). *Environ. Int.* **2003**, *29* (6), 855–860.

(5) Olsman, H.; Engwall, M.; Kammann, U.; Klempt, M.; Otte, J.; van Bavel, B.; Hollert, H. Relative differences in aryl hydrocarbon receptor-mediated response for 18 polybrominated and mixed halogenated dibenzo-*p*-dioxins and -furans in cell lines from different species. *Environ. Toxicol. Chem.* **2007**, *26* (11), 2448–2454.

(6) Li, H. R.; Feng, J. L.; Sheng, G. Y.; Lü, S. L.; Fu, J. M.; Peng, P. A.; Man, R. The PCDD/F and PBDD/F pollution in the ambient atmosphere of Shanghai, China. *Chemosphere* **2008**, *70* (4), 576–583.

(7) Wang, L. C.; Tsai, C. H.; Chang-Chien, G. P.; Hung, C. H. Characterization of polybrominated dibenzo-*p*-dioxins and dibenzofurans in different atmospheric environments. *Environ. Sci. Technol.* **2008**, *42* (1), 75–80.

(8) Evans, C. S.; Dellinger, B. Formation of bromochlorodibenzo-*p*-dioxins and dibenzofurans from the high-temperature oxidation of a mixture of 2-chlorophenol and 2-bromophenol. *Environ. Sci. Technol.* **2006**, *40* (9), 3036–3042.

(9) Haglund, P. On the identity and formation routes of environmentally abundant tri- and tetrabromodibenzo-*p*-dioxins. *Chemosphere* **2010**, *78* (6), 724–730.

(10) Wang, L. C.; Wang, Y. F.; His, H. C.; Chang-Chien, G. P. Characterizing the emissions of polybrominated diphenyl ethers (PBDEs) and polybrominated dibenzo-*p*-dioxins and dibenzofurans (PBDD/Fs) from metallurgical processes. *Environ. Sci. Technol.* **2010**, *44* (4), 1240–1246.

(11) Ma, J.; Addink, R.; Yun, S.; Cheng, J. P.; Wang, W. H.; Kannan, K. Polybrominated dibenzo-*p*-dioxins/dibenzofurans and polybrominated diphenyl ethers in soil, vegetation, workshop-floor dust, and electronic shredder residue facility and in soils from a chemical industrial complex in eastern China. *Environ. Sci. Technol.* **2009**, *43* (19), 7350–7356.

(12) Wang, L. C.; His, H. C.; Wang, Y. F.; Lin, S. L.; Chang-Chien, G. P. Distribution of polybrominated diphenyl ethers (PBDEs) and polybrominated dibenzo-*p*-dioxins and dibenzofurans (PBDD/Fs) in municipal solid waste incinerators. *Environ. Pollut.* **2010**, *158* (5), 1595–1602.

(13) Wang, L. C.; Chang-Chien, G. P. Characterizing the emissions of polybrominated dibenzo-*p*-dioxins and dibenzofurans from municipal and industrial waste incinerators. *Environ. Sci. Technol.* **2007**, *41* (4), 1159–1165.

(14) Lai, Y. C.; Lee, W. J.; Li, H. W. Inhibition of polybrominated dibenzo-*p*-dioxin and dibenzofuran formation from the pyrolysis of printed circuit boards. *Environ. Sci. Technol.* **2007**, *41* (3), 957–962.

(15) Li, H. R.; Yu, L. P.; Sheng, G. P.; Fu, J. M.; Peng, P. A. Severe PCDD/F and PBDD/F pollution in air around an electronic waste dismantling area in China. *Environ. Sci. Technol.* **2007**, *41* (16), 5641–5646.

(16) Karasek, F. W.; Dickson, L. C. Model studies of polychlorinated dibenzo-*p*-dioxin formation during municipal refuse incineration. *Science* **1987**, *237* (4816), 754–756.

(17) Milligan, M. S.; Altwicker, E. R. Chlorophenol reactions on fly ash. 1. adsorption/desorption equilibria and conversion to polychlorinated dibenzo-*p*-dioxins. *Environ. Sci. Technol.* **1995**, *30* (1), 225–229.

(18) Evans, C.; Dellinger, B. Surface-mediated formation of polybrominated dibenzo-*p*-dioxins and dibenzofurans from the high-temperature pyrolysis of 2-bromophenol on a CuO/silica surface. *Environ. Sci. Technol.* **2005**, *39* (13), 4857–4863.

(19) Evans, C.; Dellinger, B. Formation of bromochlorodibenzo-*p*-dioxins and furans from the high-temperature pyrolysis of a 2-chlorophenol/2-bromophenol mixture. *Environ. Sci. Technol.* **2005**, *39* (20), 7940–7948.

(20) IUCLID, 2,4,6-tribromophenol and other simple brominated phenols, Data Set for 2,4,6-tribromophenol. Ispra, European Chemicals Bureau, International Uniform Chemical Information Database, 2003.

(21) Sim, W. J.; Lee, S. H.; Lee, I. S.; Choi, S. D.; Oh, J. E. Distribution and formation of chlophenols and bromophenols in marine and riverine environments. *Chemosphere* **2009**, *77* (4), 552–558.

(22) Na, Y. C.; Kim, K. J.; Park, C. S.; Hong, J. K. Formation of tetrahalogenated dibenzo-*p*-dioxins (TXDDs) by pyrolysis of a mixture of 2,4,6-trichlorophenol and 2,4,6-tribromophenol. *J. Anal. Appl. Pyrolysis* **2007**, *80* (1), 254–261.

(23) Sidhu, S. S.; Maqsud, L.; Dellinger, B. The homogeneous, gas-phase formation of chlorinated and brominated dibenzo-*p*-dioxin from 2,4,6-trichloro- and 2,4,6-tribromophenols. *Combust. Flame* **1995**, *100* (1−2), 11–20.

(24) Frisch, M. J.; Trucks, G. W.; Schlegel, H. B.; Gill, P. W. M.; Johnson, B. G.; Robb, M. A.; Cheeseman, J. R.; Keith, T. A.; Petersson, G. A.; Montgomery, J. A.; et al. GAUSSIAN 03, Pittsburgh, PA, 2003.

(25) Zhao, Y.; Truhlar, D. G. Hybrid meta density functional theory methods for therochemistry, thermochemical kinetics, and noncovalent interactions: The MPW1B95 and MPWB1K models and comparative assessments for hydrogen bonding and van der waals interactions. *J. Phys. Chem. A* **2004**, *108* (33), 6908–6918.

(26) Steckler, R.; Chuang, Y. Y.; Fast, P. L.; Corchade, J. C.; Coitino, E. L.; Hu, W. P.; Lynch, G. C.; Nguyen, K.; Jackells, C. F.; Gu, M. Z.; Rossi, I.; Clayton, S.; Melissas, V.; Garrett, B. C.; Isaacson, A. D.; Truhlar, D. G. *POLYRATE Version 9.3*; University of Minnesota: Minneapolis, 2002.

(27) Baldridge, M. S.; Gordor, R.; Steckler, R.; Truhlar, D. G. Ab initio reaction paths and direct dynamics calculations. *J. Phys. Chem.* **1989**, *93* (13), 5107–5119.

(28) Gonzalez-Lafont, A.; Truong, T. N.; Truhlar, D. G. Interpolated variational transition-state theory: Practical methods for estimating variational transition-state properties and tunneling contributions to chemical reaction rates from electronic structure calculations. *J. Chem. Phys.* **1991**, *95* (12), 8875–8894.

(29) Garrett, B. C.; Truhlar, D. G. Generalized transition state theory. Classical mechanical theory and applications to collinear reactions of hydrogen molecules. *J. Phys. Chem.* **1979**, *83* (8), 1052–1079.

(30) Fernandez-Ramos, A.; Ellingson, B. A.; Garret, B. C.; Truhlar, D. G. Variational Transition State Theory with Multidimensional Tunneling. In *Reviews in Computational Chemistry*; Lipkowitz, K. B., Cundari, T. R., Eds; Wiley-VCH: Hoboken, NJ, 2007.

(31) Senma, M.; Taira, Z.; Taga, T.; Osaki, K. Dibenzo-*p*-dioxin, $C_{12}H_8O_2$. *Crystallogr. Struct. Commun.* **1973**, *2* (2), 311–314.

(32) Landolt-Bornstein: Group II: Atomic and Molecular Physics Vol. 7: Structure Data of Free Polyatomic Molecules. Hellwege, K. H., Hellwege, A. M., Ed.; Springer-Verlag: Berlin, 1976.

(33) Ribeiro da Silva, M. A. V.; Lobo Ferreira, A. I. M. C. Gas phase enthalpies of formation of monobromophenols. *J. Chem. Thermodyn.* **2009**, *41* (10), 1104–1110.

(34) Lukyanova, V. A.; Kolesov, V. P.; Avramenko, N. V.; Vorobieva, V. P.; Golovkov, V. F. Standard enthalpy of formation of dibenzo-*p*-dioxin. *Zh. Fiz. Khim.* **1997**, *71* (3), 406–408.

(35) Chase, M. W., Jr. NIST-JANAF Themochemical Tables, 4th ed.; *J. Phys. Chem. Ref. Data*, Monograph 9, **1998**; 1-1951.

(36) Evans, C. S.; Dellinger, B. Mechanisms of dioxin formation from the high-temperature pyrolysis of 2-bromophenol. *Environ. Sci. Technol.* **2003**, *37* (24), 5574–5580.

(37) Evans, C. S.; Dellinger, B. Mechanisms of dioxin formation from the high-temperature oxidation of 2-bromophenol. *Environ. Sci. Technol.* **2005**, *39* (7), 2128–2134.

(38) Zhang, Q. Z.; Li, S. Q.; Qu, X. H.; Wang, W. X. A quantum mechanical study on the formation of PCDD/Fs from 2-chlorophenol as precursor. *Environ. Sci. Technol.* **2008**, *42* (19), 7301–7308.

(39) Na, Y. C.; Hong, J. K.; Kim, K. J. Formation of polybrominated dibenzo-*p*-dioxins/furans (PBDDs/Fs) by the pyrolysis of 2,4-dibromophenol, 2,6-dibromophenol, and 2,4,6-tribromophenol. *Bull. Korean Chem. Soc.* **2007**, *28* (4), 547–552.

(40) Zhang, Q. Z.; Yu, W. N.; Zhang, R. X.; Zhou, Q.; Gao, R.; Wang, W. X. Quantum chemical and kinetic study on dioxin formation from the 2,4,6-TCP and 2,4-DCP precursors. *Environ. Sci. Technol.* **2010**, *44* (9), 3395–3403.

(41) Weber, R.; Hagenmaier, H. Mechanism of the formation of polychlorinated dibenzo-*p*-dioxins and dibenzofurans from chlorophenols in gas phase reactions. *Chemosphere* **1999**, *38* (3), 529–549.

(42) Zhang, Q. Z.; Qu, X. H.; Xu, F.; Shi, X. Y.; Wang, W. X. Mechanism and thermal rate constants for the complete series reactions of chlorophenols with H. *Environ. Sci. Technol.* **2009**, *43* (11), 4105–4112.

(43) Xu, F.; Wang, H.; Zhang, Q. Z.; Zhang, R. X.; Qu, X. H.; Wang, W. X. Kinetic properties for the complete series reactions of chlorophenols with OH radicals—relevance for dioxin formation. *Environ. Sci. Technol.* **2010**, *44* (4), 1399–1404.

(44) Khachatryan, L.; Asatryan, R.; Dellinger, B. Development of expanded and core kinetic models for the gas phase formation of dioxins from chlorinated phenols. *Chemosphere* **2003**, *52* (4), 695–708.

(45) Khachatryan, L.; Burcat, A.; Dellinger, B. An elementary reaction-kinetic model for the gas-phase formation of 1,3,6,8- and 1,3,7,9-tetrachlorinated dibenzo-*p*-dioxins from 2,4,6-trichlorophenol. *Combust. Flame* **2003**, *132* (3), 406–421.

(46) Baulch, D. L.; Duxbury, J.; Grant, S. J.; Montague, D. C. Evaluated kinetic data for high temperature reactions. Volume 4 Homogeneous gas phase reactions of halogen- and cyanide- containing species. *J. Phys. Chem. Ref. Data* **1981**, *10* (Supplement 1), 1–721.

Environ. Sci. Technol. **2009,** *43,* 4105–4112

Mechanism and Thermal Rate Constants for the Complete Series Reactions of Chlorophenols with H

QINGZHU ZHANG,*,† XIAOHUI QU,† HUI WANG,‡ FEI XU,† XIANGYAN SHI,† AND WENXING WANG*,†

Environment Research Institute, Shandong University, Jinan 250100, P. R. China, and Department of Environmental Science and Engineering, Tsinghua University, Beijing 100084, P. R. China

Received January 19, 2009. Revised manuscript received March 23, 2009. Accepted April 8, 2009.

Reactions of chlorophenols with atomic H are important initial steps for the formation of polychlorinated dibenzo-*p*-dioxins and dibenzofurans (PCDD/Fs) in incinerators. Detailed insight into the mechanism and kinetic properties of crucial elementary steps is a prerequisite for understanding the formation of PCDD/Fs. In this paper, the complete series reactions of 19 chlorophenol congeners with atomic H have been studied theoretically using the density functional theory (DFT) method and the direct dynamics method. The profiles of the potential energy surface were constructed at the MPWB1K/6-311+G(3df,2p)//MPWB1K/6-31+G(d,p) level. Modeling of the PCDD/Fs formation requires kinetic information about the elemental reactions. The rate constants were deduced over a wide temperature range of 600~1200 K using canonical variational transition-state theory (CVT) with small curvature tunneling contribution (SCT). The rate-temperature formulas were fitted for the first time. This study shows that the substitution pattern of the phenol has a significant effect on the strength and reactivity of the O—H bonds in chlorophenols. Intramolecular hydrogen bonding plays a decisive role in determining the reactivity of the O—H bonds for *ortho*-substituted phenols.

1. Introduction

Polychlorinated dibenzo-*p*-dioxins (PCDDs) and dibenzofurans (PCDFs) are notorious for their acute and chronic toxicity and their carcinogenic, teratogenic, and mutagenic effects (*1*). Major sources of PCDD/Fs in the environment are the combustion of waste materials as well as many other high-temperature processes commonly used in industrial settings (*2–4*). The most direct route to the formation of PCDD/Fs is the gas-phase reaction of chemical precursors (*5–7*). Chlorophenols (CPs) are the most direct precursors of dioxin and among the most abundant aromatic compounds found in incinerator gas emissions (*8–10*). The homogeneous gas-phase formation of PCDD/Fs from chlorophenol precursors was suggested to make a significant contribution to the observed PCDD/F yields in full-scale incinerators (*11, 12*). Chlorophenols are almost ubiquitous in the environment due to their agricultural and industrial uses as pesticides, disinfectants, wood preservatives, personal care formulations, and many other products. Five chlorophenols are listed by the U.S. Environmental Protection Agency as priority pollutants, including PCP and 2,4,6-TCP, which are present in the environment in significant quantities.

The gas-phase formation of PCDD/Fs from chlorophenol precursors was proposed that involve chlorophenoxy radical–radical coupling, radical-molecule recombination of chlorophenoxy and chlorophenol. The previous researches (*13, 14*) have shown that the radical-molecule pathway requires chlorine and hydroxyl displacement as first steps, and these steps are not energetically favored. So, the radical-molecule pathway is not competitive with the radical–radical pathway. The dimerization of chlorophenoxy radicals (CPRs) is the major pathway for the formation of PCDD/Fs (*13, 14*). Thus, the formation of chlorophenoxy radicals is a crucial elementary step involved in the formation of PCDD/Fs. In municipal waste incinerators, chlorophenoxy radicals can be formed through loss of the phenoxyl–hydrogen via unimolecular, bimolecular, or possibly other low-energy pathways (including heterogeneous reactions). The unimolecular reaction includes the decomposition of chlorophenols with the cleavage of the O—H bond. The bimolecular reactions include attack by H or Cl under pyrolytic conditions and attack by H, OH, O (^{3}P), or Cl under high-temperature oxidative conditions. However, very little work has been done at the high temperatures relevant to these reactions. The kinetic models that account for the contribution of the gaseous route in the production of PCDD/Fs in combustion processes use the rate constants of the elementary reactions (*9, 15*). Due to the absence of direct experimental and theoretical values, the rate constants of the reactions of chlorophenols with H, OH, O (^{3}P), or Cl were assigned to be the values reported in the literature for the reactions of phenol with H, OH, O (^{3}P), or Cl (*9, 15*). However, where there are uncertainties, the numerical values have been adjusted somewhat to bias the mechanism in favor of PCDD/Fs formation, i.e., worst case modeling (*9, 15*).

The reactions with atomic H, the simplest free-radical species, are of particular interest since these reactions are desirable not only to provide an uncomplicated probe of chemical reactivity but also to throw light on the formation mechanism of the chlorophenoxy radicals (CPRs). Here, we present the first systematic study on the complete series reactions of 19 chlorophenol congeners with atomic H. At first, we examined the reaction mechanism at high-accuracy of the density functional theory (DFT). In a second step, the rate constants were calculated using canonical variational transition-state theory (CVT) (*16–18*) with small curvature tunneling contribution (SCT) (*19*) over a wide temperature range of 600~1200 K. The calculated values were compared with the available experimental results. The rate-temperature formulas were fitted. Third, the effect of the substitution pattern of the phenol on the strength and reactivity of the O—H bonds in chlorophenols is discussed.

2. Computational Methods

By means of the Gaussian 03 programs (*20*), high-accuracy quantum chemical calculations were carried out on an SGI Origin 2000 supercomputer. The geometrical parameters of reactants, transition states, and products were optimized at the MPWB1K level with a standard 6-31+G(d,p) basis set. The MPWB1K method (*21*) is a hybrid DFT model with excellent performance for thermochemistry, thermochemical kinetics, hydrogen bonding, and weak interactions. The vibrational frequencies were also calculated at the same level

* Address corresponding to either author. E-mail: zqz@sdu.edu.cn (Q. Z.); wxwang@sdu.edu.cn (W. W.). Fax: 86-531-8836 1990.
† Shandong University.
‡ Tsinghua University.

10.1021/es9001778 CCC: $40.75
Published on Web 04/30/2009

in order to determine the nature of the stationary points, the zero-point energy (ZPE), and the thermal contributions to the free energy of activation. Each transition state was verified to connect the designated reactants with products by performing an intrinsic reaction coordinate (IRC) analysis (*22*). The minimum energy paths (MEP) were obtained in mass-weighted Cartesian coordinates. The single-point energy calculations were carried out at the MPWB1K/6-311+G(3df,2p) level. The kinetic calculations are most sensitive to the energies. The reliability of the MPWB1K/6-311+G(3df,2p) level for the potential barriers was clarified in our recent study (*23*) on the formation of PCDD/Fs from the 2-CP precursor. The profiles of the potential energy surface were constructed at the MPWB1K/6-311+G(3df,2p)//MPWB1K/6-31+G(d,p) level including ZPE correction. By means of the Polyrate 9.3 program (*24*), direct dynamics calculations were carried out using the canonical variational transition state theory (CVT) (*16*–*18*). The level of tunneling calculation is the small curvature tunneling (SCT) (*19*) method.

3. Results and Discussion

3.1. Reaction Mechanism. Due to the different substitution pattern of phenol, chlorophenols have 19 congeners. They include three monochlorophenols (2-CP, 3-CP and 4-CP), six dichlorophenols (2,3-DCP, 2,4-DCP, 2,5-DCP, 2,6-DCP, 3,4-DCP, and 3,5-DCP), six trichlorophenols (2,3,4-TCP, 2,3,5-TCP, 2,3,6-TCP, 2,4,5-TCP 2,4,6-TCP, and 3,4,5-TCP), three tetrachlorophenols (2,3,4,5-TeCP, 2,3,4,6-TeCP, and 2,3,5,6-TeCP), and pentachlorophenol (PCP). There are two geometric conformers resulting from the two main orientations of the hydroxyl–hydrogen due to the asymmetric chlorine substitution for 12 chlorophenol congeners (2-CP, 3-CP 2,3-DCP, 2,4-DCP, 2,5-DCP, 3,4-DCP, 2,3,4-TCP, 2,3,5-TCP, 2,3,6-TCP, 2,4,5-TCP, 2,3,4,5-TeCP, 2,3,4,6-TeCP). The conformer with the hydroxyl–-hydrogen facing the closest neighboring Cl is labeled as the *syn* conformer and otherwise the *anti* conformer. There exists weak intramolecular hydrogen bonding in the *syn* conformers. The lengths of the hydrogen bonds are from 2.342 to 2.408 Å, which are slightly longer than that of a typical hydrogen bond. No such intramolecular hydrogen bonding forms in the *anti* conformers except those with chlorine substitutions at both *ortho*-positions. The energies of the *syn* conformers are about 3 kcal/mol lower than that of the corresponding *anti* forms, suggesting a stabilization effect because of intramolecular hydrogen bonding. So throughout this paper, chlorophenols denote the *syn* conformers. The structures of chlorophenols along with the structure of phenol are presented in Figure 1.

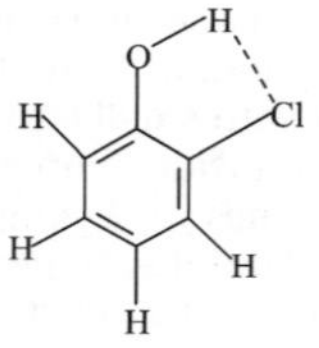

syn CP

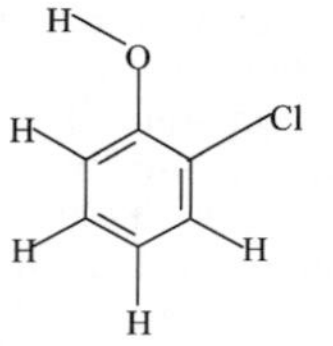

anti CP

Figure 1 shows that all the O–H bonds in *ortho*-substituted phenols are longer that those without *ortho* substitution. The O–H bond length in chlorophenols with *ortho* substitution is 0.959 or 0.960 Å at the MPWB1K/6-31+G(d,p) level, whereas the value is 0.955 Å for chlorophenols without *ortho* substitution. The O–H bond lengths are significantly correlated with the position of the chlorine substitutions, but not with the number of chlorine substitutions. The structures of 19 chlorophenol congeners and phenol were also studied by Han using density functional theory and ab initio molecular calculations (*25*). The O–H bond lengths calculated by Han at the B3LYP/6-311++G(d,p) and MP2/6-311++G(d,p) levels are longer than the corresponding values in our study obtained at the MPWB1K/6-31+G(d,p) level. The O–H bonds in chlorophenols are 0.963–0.968 Å at the B3LYP/6-311++G(d,p) level and 0.962–0.967 Å at the MP2/6-311+G(d,p) level (*25*), whereas the values are 0.955–0.960 Å at the MPWB1K/6-31+G(d,p) level. For phenol, the O–H bond length of 0.955 Å obtained in our study at the MPWB1K/6-31++G(d,p) level is in better agreement with the experimental value of 0.956 Å (*26*) than the value of 0.963 Å calculated by Han at the B3LYP/6-311++G(d,p) and MP2/6-311++G(d,p) levels (*25*). However, the general conclusions from our study that the O–H bond lengths in chlorophenols are strongly influenced by the intramolecular hydrogen bonding and the O–H bonds in *ortho*-substituted phenols are longer that those without *ortho* substitution are in accordance with the conclusions obtained from Han's study (*25*).

The formation of chlorophenoxy radicals from the bimolecular reaction of chlorophenols with atomic H proceeds via a direct hydrogen abstraction mechanism. In order to compare with the formation of chlorophenoxy radicals, we also studied the formation of phenoxy radicals from the reaction of phenol with H. At the MPWB1K/6-31+G(d,p) level, the transition states were located. The structures of the transition states and chlorophenoxy radicals are shown in Figure 2 and Figure 3, respectively. All of the transition states have one and only one imaginary frequency. The values of the imaginary frequencies are large, which implies that the quantum tunneling contribution may be significant and may play an important role in the calculation of the rate constants.

The calculated potential barriers and reaction enthalpies at the MPWB1K/6-311+G(3df,2p)//MPWB1K/6-31+G(d,p) level are listed in Table 1. The formation of chlorophenoxy radicals from the reactions of chlorophenols with H is strongly exothermic. The kinetic calculations are most sensitive to the energies, especially to the potential barriers. In order to justify the performance of the MPWB1K method for the potential barriers, we have also carried out additional potential barrier calculations employing the BB1K and B3LYP methods. The values obtained at the BB1K and B3LYP levels are presented in Table 1. The potential barriers calculated at the B3LYP/6-311+G(3df,2p)//B3LYP/6-31+G(d,p) level are 9–11 kcal/mol lower than the corresponding values obtained at the MPWB1K/6-311+G(3df,2p)//MPWB1K/6-31+G(d,p) and BB1K/6-311+G(3df,2p)//BB1K/6-31+G(d,p) levels. It is well-known that the B3LYP method usually underestimates barrier heights (*21*). Table 1 shows that the two methods, MPWB1K/6-311+G(3df,2p)//MPWB1K/6-31+G(d,p) and BB1K/6-311+G(3df,2p)//BB1K/6-31+G(d,p), produce consistent potential barriers within 0.6 kcal/mol. This study is a subsequent work from our group who had published the two similar works on the formation of PCDD/Fs from the 2-CP and 2,4,5-TCP precursors at the MPWB1K/6-311+G(3df,2p)//MPWB1K/6-31+G(d,p) level (*23, 27*). So, as part of our ongoing work in the field, we chose the energies calculated at the MPWB1K/6-311+G(3df,2p)//MPWB1K/6-31+G(d,p) level for the following kinetic calculation. The potential barriers for the phenoxyl–hydrogen abstraction from phenol, 2-CP, 3-CP, 4-CP, 2,5-CP, 2,3,5-TCP and TCP by H were also calculated by Altarawneh at the BB1K/6-311+G(3df,2p)//BB1K/6-31G(d) (*28*). The difference in the potential barriers between the MPWB1K/6-311+G(3df,2p)//MPWB1K/6-31+G(d,p) and BB1K/6-311+G(3df,2p) //BB1K/6-31G(d) approaches is no more than 1.61 kcal/mol.

Table 1 shows that the potential barriers of the phenoxyl–hydrogen abstraction from chlorophenols except for 4-CP are higher than that from phenol. This means that the phenoxyl–hydrogen abstraction from chlorophenols except for 4-CP is more difficult than from phenol. Since the

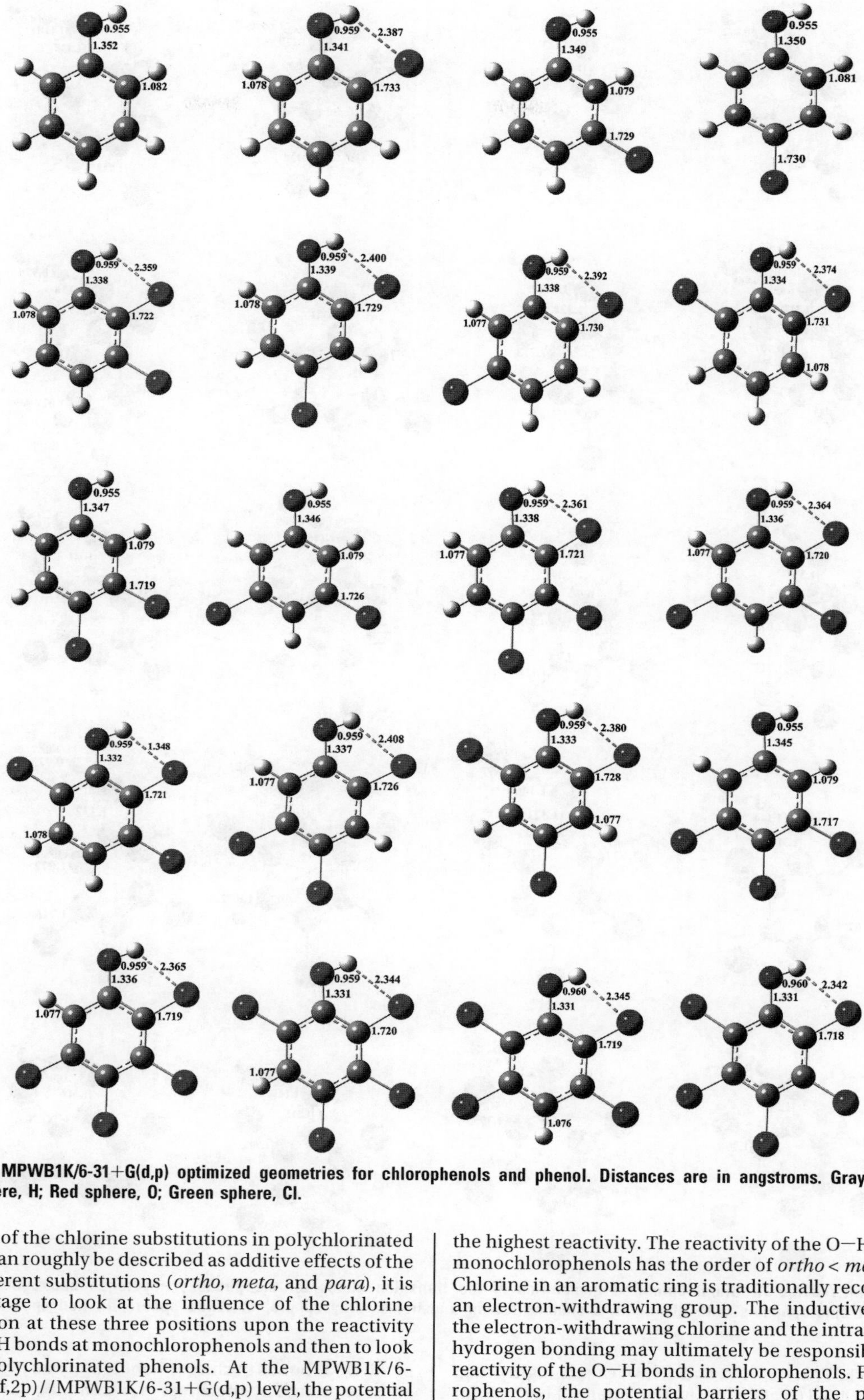

FIGURE 1. MPWB1K/6-31+G(d,p) optimized geometries for chlorophenols and phenol. Distances are in angstroms. Gray sphere, C; White sphere, H; Red sphere, O; Green sphere, Cl.

influence of the chlorine substitutions in polychlorinated phenols can roughly be described as additive effects of the three different substitutions (*ortho*, *meta*, and *para*), it is of advantage to look at the influence of the chlorine substitution at these three positions upon the reactivity of the O—H bonds at monochlorophenols and then to look at the polychlorinated phenols. At the MPWB1K/6-311+G(3df,2p)//MPWB1K/6-31+G(d,p) level, the potential barrier of the phenoxyl—hydrogen abstraction from 2-CP is 1.29 and 2.39 kcal/mol higher than those from 3-CP and 4-CP, respectively. This means that the O—H bond in 2-CP has the lowest reactivity, and the O—H bond in 4-CP has the highest reactivity. The reactivity of the O—H bonds in monochlorophenols has the order of *ortho* < *meta* < *para*. Chlorine in an aromatic ring is traditionally recognized as an electron-withdrawing group. The inductive effect of the electron-withdrawing chlorine and the intramolecular hydrogen bonding may ultimately be responsible for the reactivity of the O—H bonds in chlorophenols. For dichlorophenols, the potential barriers of the phenoxyl—hydrogen abstraction from 2,3-DCP, 2,4-DCP, 2,5-DCP, and 2,6-DCP are higher than those from 3,4-DCP and 3,5-DCP. For trichlorophenols, the potential barriers of the phenoxyl—hydrogen abstraction from 2,3,4-TCP, 2,3,5-

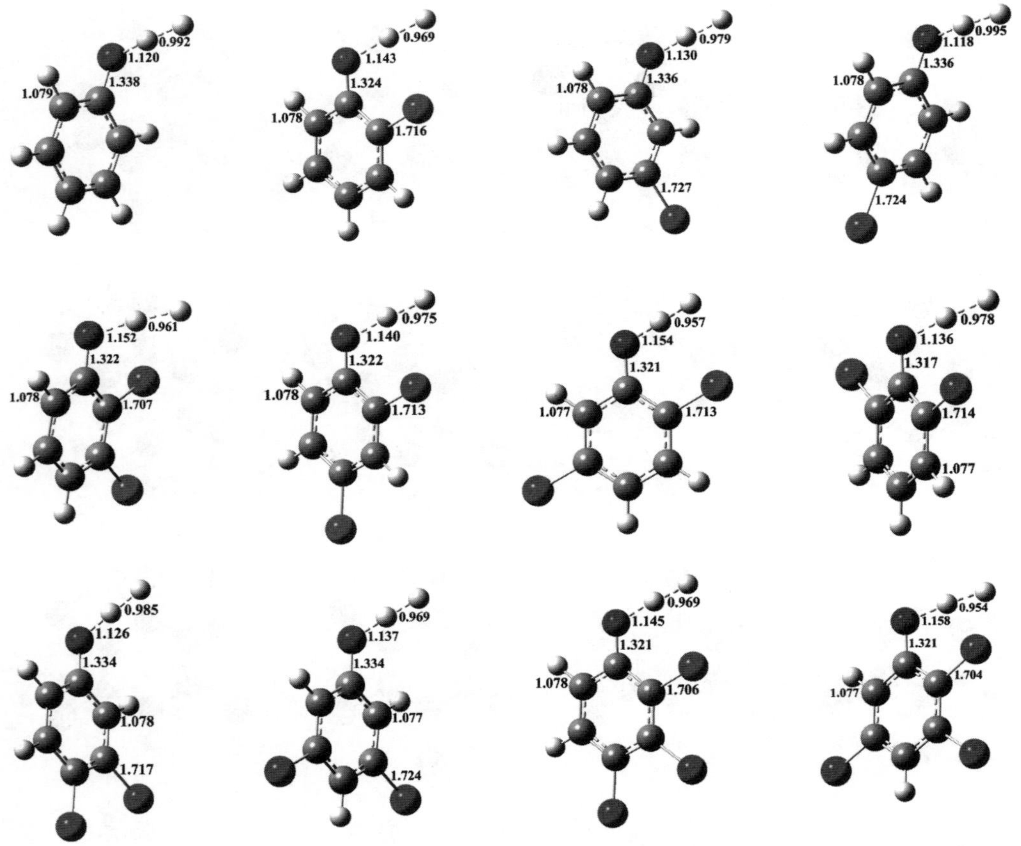

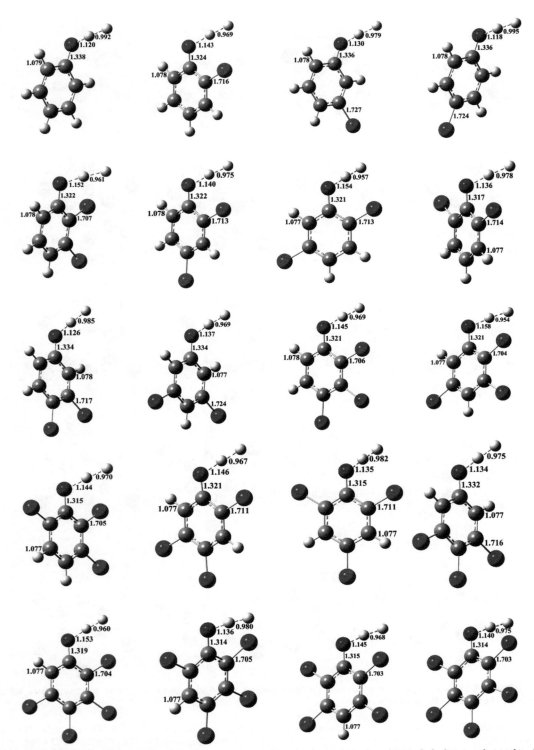

FIGURE 2. MPWB1K/6-31+G(d,p) optimized geometries of the transition states for the phenoxyl−hydrogen abstraction from the reactions of chlorophenols with atomic H. Distances are in angstroms. Gray sphere, C; White sphere, H; Red sphere, O; Green sphere, Cl.

TCP, 2,3,6-TCP, 2,4,5-TCP, and 2,4,6-TCP are higher than that from 3,4,5-TCP. Obviously, the potential barriers for the phenoxyl−hydrogen abstraction from chlorophenols with intramolecular hydrogen bonding consistently are higher than from other structural conformers for a given number of chlorine substitutions. The relative reactivity of the O−H bonds in chlorophenols appears to be assisted and dominated by the effect of intramolecular hydrogen bonding. It appears to reduce the reactivity of O−H bonds. The relatively strong reactivity of the O−H bond in 4-CP can be explained by the inductive effect of the electron-withdrawing chlorine. The induction of chlorine at the

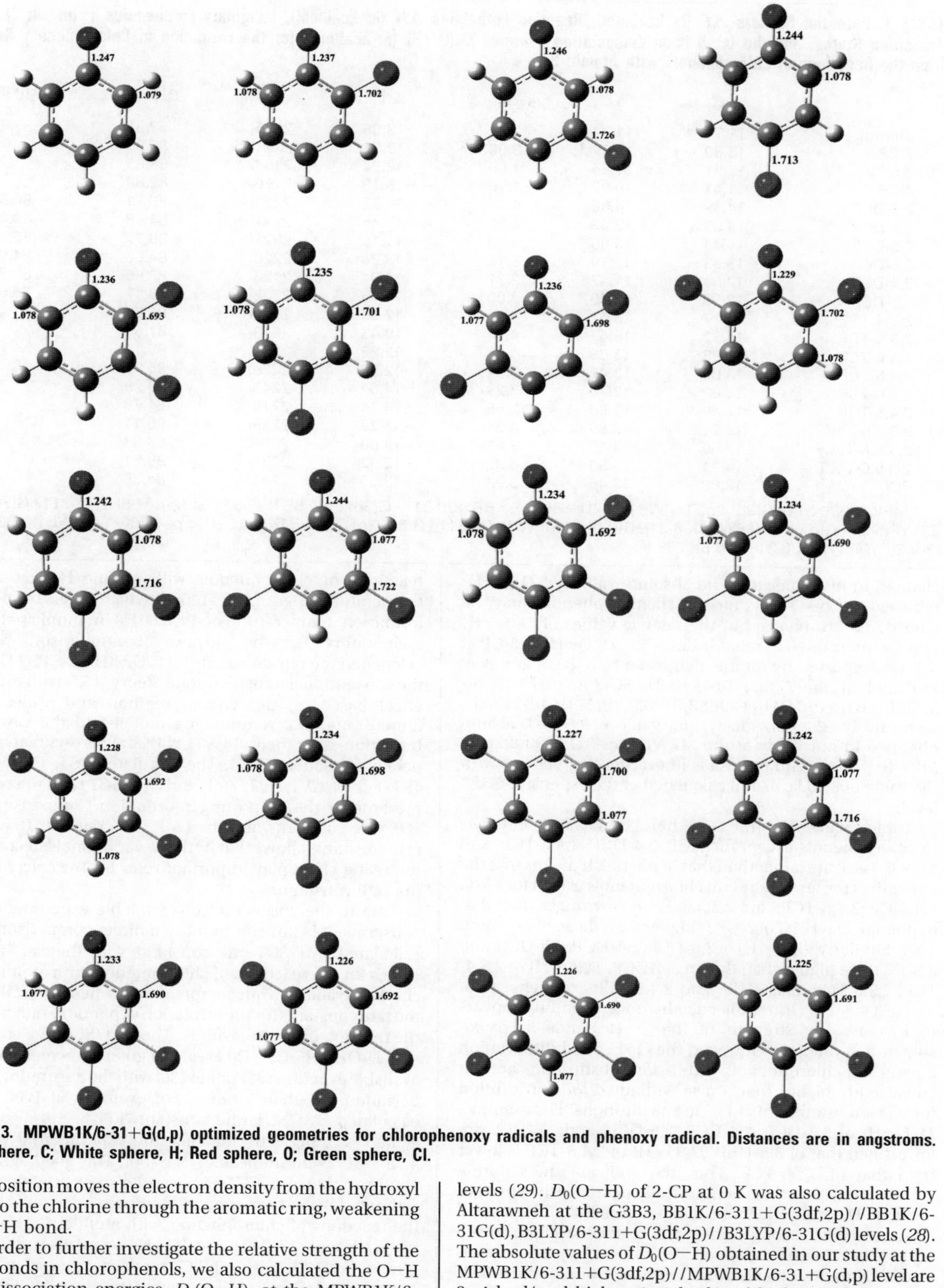

FIGURE 3. MPWB1K/6-31+G(d,p) optimized geometries for chlorophenoxy radicals and phenoxy radical. Distances are in angstroms. Gray sphere, C; White sphere, H; Red sphere, O; Green sphere, Cl.

para-position moves the electron density from the hydroxyl group to the chlorine through the aromatic ring, weakening the O—H bond.

In order to further investigate the relative strength of the O—H bonds in chlorophenols, we also calculated the O—H bond dissociation energies, D_0(O—H), at the MPWB1K/6-311+G(3df,2p)//MPWB1K/6-31+G(d,p) level. The values (19 chlorophenols and phenol at 0 K, six dichlorophenols and phenol at 298.15 K) are presented in Table 1 and compared with the data in the literatures. D_0(O—H) of six dichlorophenols at 298.15 K was calculated by Gomes at the UB3LYP/DZVP, ROB3LYP/DZVP and ROB3LYP/6-311++G(2df,2p) levels (*29*). D_0(O—H) of 2-CP at 0 K was also calculated by Altarawneh at the G3B3, BB1K/6-311+G(3df,2p)//BB1K/6-31G(d), B3LYP/6-311+G(3df,2p)//B3LYP/6-31G(d) levels (*28*). The absolute values of D_0(O—H) obtained in our study at the MPWB1K/6-311+G(3df,2p)//MPWB1K/6-31+G(d,p) level are 3−4 kcal/mol higher, 1−2 kcal/mol higher and 3−4 kcal/mol lower than the corresponding values obtained by Gomes at the UB3LYP/DZVP, ROB3LYP/DZVP, and ROB3LYP/6-311++G(2df,2p) levels, respectively (*29*). But, the relative values of D_0(O—H), D_0(O—H)(dichlorophenol)-D_0(O—H)-(phenol), obtained in our study are in good agreement with the corresponding relative values obtained by Gomes. Gomes

TABLE 1. Potential Barriers ΔE (in kcal/mol), Reaction Enthalpies ΔH (in kcal/mol), Imaginary Frequencies ν (in cm^{-1}) of the Transition States, and the O—H bond dissociation energies D_0(O—H) (in kcal/mol) for the Formation of Chlorophenoxy Radicals from the Reactions of Chlorophenols with Atomic H

	ΔE^a	ΔE^b	ΔE^c	ΔH^a	ν^d	D_0(O—H)[e]	D_0(O—H)[e]
phenol	11.73	11.24	2.11	−13.98	2166i	83.95	83.79
2-CP	13.80	13.37	3.90	−12.01	2221i	85.91	
3-CP	12.51	11.94	1.00	−12.94	2192i	84.99	
4-CP	11.41	10.92	1.75	−15.13	2176i	82.80	
2,3-DCP	14.49	14.06	4.37	−11.22	2225i	86.71	86.61
2,4-DCP	13.58	13.09	3.53	−13.35	2227i	84.58	84.47
2,5-DCP	14.44	13.89	4.30	−11.16	2221i	86.77	86.68
2,6-DCP	13.33	13.19	3.71	−13.26	2226i	84.66	84.56
3,4-DCP	12.01	11.47	2.21	−14.04	2198i	83.89	83.71
3,5-DCP	12.99	12.46	3.00	−11.76	2202i	86.17	86.02
2,3,4-TCP	14.17	13.57	4.02	−12.56	2230i	85.37	
2,3,5-TCP	14.92	14.47	4.74	−10.43	2213i	87.50	
2,3,6-TCP	13.88	13.65	4.09	−12.82	2238i	85.11	
2,4,5-TCP	13.96	13.47	3.81	−12.56	2228i	85.37	
2,4,6-TCP	12.81	12.70	3.29	−14.64	2243i	83.29	
3,4,5-TCP	12.40	11.96	2.55	−13.14	2210i	84.79	
2,3,4,5-TeCP	14.21	13.89	4.21	−11.79	2235i	86.14	
2,3,4,6-TeCP	13.60	13.19	3.81	−14.00	2238i	83.93	
2,3,5,6-TeCP	14.11	13.96	4.35	−12.32	2238i	85.61	
PCP	13.68	13.46	3.93	−13.55	2244i	84.38	

[a] MPWB1K/6-311+G(3df,2p)//MPWB1K/6-31+G(d,p). [b] BB1K/6-311+G(3df,2p)// BB1K/6-31+G(d,p). [c] B3LYP/6-311+G(3df,2p)//B3LYP/6-31+G(d,p). [d] MPWB1K/6-31+G(d,p). [e] MPWB1K/6-311+G(3df,2p)//MPWB1K/6-31+G(d,p), 0 K. [f] MPWB1K/6-311+G-(3df,2p)//MPWB1K/6-31+G(d,p), 298 K.

claimed in his paper that the absolute values of D_0(O—H) reported in his study for six dichlorophenols may be considered erroneous, but the relative values of D_0(O—H) may be interpreted as reliable data (*29*). D_0(O—H) of 2-CP at 0 K was reported by Altarawnch to be 86.8, 83.5, and 80.9 kcal/mol at the G3B3, BB1K/6-311+G(3df,2p)//BB1K/6-31G(d), B3LYP/6-311+G(3df,2p)//B3LYP/6-31G(d) levels, respectively (*28*). Obviously, the value of 85.91 kcal/mol obtained in our study at the MPWB1K/6-311+G(3df,2p)//MPWB1K/6-31+G(d,p) level is in excellent agreement with the value of 86.8 kcal/mol calculated at the expensive G3B3 level.

Table 1 shows that the O—H bond dissociation energies in chlorophenols except for 4-CP, 3,4-DCP, 2,4,6-TCP, and 2,3,4,6-TeCP are larger than that in phenol. It means that the strength of the O—H bonds in chlorophenols except for 4-CP, 3,4-DCP, 2,4,6-TCP, and 2,3,4,6-TeCP is stronger than that in phenol. D_0(O—H) of 2-CP is higher than those of 3-CP and 4-CP. Similarly, D_0(O—H) of 2,3-DCP, 2,4-DCP, 2,5-DCP, and 2,6-DCP are higher than that of 3,4-DCP. D_0(O—H) of 2,3,4-TCP, 2,3,5-TCP, 2,3,6-TCP, and 2,4,5-TCP are higher than that of 3,4,5-TCP. Intramolecular hydrogen bonding appears to increase the strength of the O—H bonds in *ortho*-substituted phenols. However, the O—H bond dissociation energies in chlorophenols with *ortho* substitutions are not consistently higher than those without *ortho* substitution for a given number of chlorine substitutions. For example, D_0(O—H) of 2,3-DCP, 2,4-DCP, 2,5-DCP, and 2,6-DCP are lower than that of 3,5-DCP. D_0(O—H) of 2,4,6-TCP is lower than that of 3,4,5-TCP. This may indicate the chlorine substitutions at *ortho* position would decrease the strength of the O—H bonds, and the resulting steric effect and inductive effect may also strongly affect the relative strength of the O—H bonds in chlorophenols.

3.2. Kinetics Calculations. Canonical variational transition state theory (CVT) (*16–18*) with small-curvature tunneling (SCT) (*19*) contribution has been successfully performed for the elementary reactions involved in the formation of PCDD/Fs from the 2-CP and 2,4,5-TCP precursors (*23, 27*) and is an efficient method to calculate the rate constants. In this study, we used this method to calculate the rate constants for the formation of chlorophenoxy radicals from the reactions of chlorophenols with atomic H over a wide temperature range of 600~1200 K, which covers the possible formation temperature of PCDD/Fs in municipal waste incinerators. For the purpose of comparison, we also calculated the rate constants for 2-CP+H→2-CPR+H_2 using the conventional transition state theory (TST) with tunneling effect based on the Wigner method and Miller's one-dimensional Eckart tunneling model and the variational transition state theory (CVT) with the zero-curvature tunneling correction (ZCT). The calculated TST, TST/Wigner, TST/Eckart, CVT, CVT/ZCT, and CVT/SCT rate constants are presented in the Supporting Information. Comparison of the TST rate constants and the CVT, TST/Wigner, TST/Eckart rate constants shows that both the variational effect and the tunneling effect play important roles for the calculation of the rate constants.

Due to the absence of the available experimental rate constants, it is difficult to make a direct comparison of the calculated CVT/SCT rate constants with the experimental values for the reactions of chlorophenols with atomic H. To check the validity of our computational scheme, we calculated the rate constants for the formation of phenoxy radicals from the reaction of phenol with H. The CVT/SCT rate constants of C_6H_5OH+H→C_6H_5O+H_2 are in good agreement with the available experimental values (*30*) with the maximum relative deviation less than 3 times. For example, at 1000 K, The calculated CVT/SCT rate constant, 1.68×10^{-13} cm^3 molecule^{-1} s^{-1}, perfectly matches the experimental value of 3.72×10^{-13} cm^3 molecule^{-1} s^{-1} (*30*). This good agreement confirms that our CVT/SCT results may provide a good estimate for the formation of chlorophenoxy radicals from the reactions of chlorophenols with atomic H. The kinetic properties of 2-CP+H→2-CPR+H_2 were also studied by Altarawneh (*28*). The CVT/SCT rate constants calculated by Altarawneh are about 5 times higher than our CVT/SCT ones. For example, at 1000 K, the CVT/SCT rate constant calculated by Altarawneh is 7.66×10^{-14} cm^3 molecule^{-1} s^{-1}, whereas the CVT/SCT value in our study is 1.76×10^{-14} cm^3 molecule^{-1} s^{-1}. This may be due to that the potential barrier calculated by Altarawneh at the BB1K/6-311+G(3df,2p)//BB1K/6-31+G(d,p) level is 0.7 kcal/mol lower than that obtained in

TABLE 2. Arrhenius Formulas (in cm^3 $molecule^{-1}$ s^{-1}) for the Formation of Chlorophenoxy Radicals (CPRs) from the Reactions of Chlorophenols with Atomic H over the Temperature Range of 600~1200 K

reactions	Arrhenius formulas
$C_6H_5OH+H \rightarrow C_6H_5O+H_2$	$k(T) = (1.98 \times 10^{-11}) \exp(-4767.69/T)$
$2\text{-CP+H} \rightarrow 2\text{-CPR+H}_2$	$k(T) = (1.10 \times 10^{-11}) \exp(-6437.64/T)$
$3\text{-CP+H} \rightarrow 3\text{-CPR+H}_2$	$k(T) = (2.36 \times 10^{-12}) \exp(-6579.12/T)$
$4\text{-CP+H} \rightarrow 4\text{-CPR+H}_2$	$k(T) = (4.61 \times 10^{-12}) \exp(-5602.30/T)$
$2,3\text{-DCP+H} \rightarrow 2,3\text{-DCPR+H}_2$	$k(T) = (3.82 \times 10^{-12}) \exp(-7768.93/T)$
$2,4\text{-DCP+H} \rightarrow 2,4\text{-DCPR+H}_2$	$k(T) = (5.01 \times 10^{-11}) \exp(-7238.73/T)$
$2,5\text{-DCP+H} \rightarrow 2,5\text{-DCPR+H}_2$	$k(T) = (2.14 \times 10^{-12}) \exp(-7005.45/T)$
$2,6\text{-DCP+H} \rightarrow 2,6\text{-DCPR+H}_2$	$k(T) = (2.78 \times 10^{-12}) \exp(-6922.27/T)$
$3,4\text{-DCP+H} \rightarrow 3,4\text{-DCPR+H}_2$	$k(T) = (2.24 \times 10^{-12}) \exp(-5933.90/T)$
$3,5\text{-DCP+H} \rightarrow 3,5\text{-DCPR+H}_2$	$k(T) = (2.61 \times 10^{-12}) \exp(-6678.71/T)$
$2,3,4\text{-TCP+H} \rightarrow 2,3,4\text{-TCPR+H}_2$	$k(T) = (1.91 \times 10^{-11}) \exp(-7472.63/T)$
$2,3,5\text{-TCP+H} \rightarrow 2,3,5\text{-TCPR+H}_2$	$k(T) = (4.56 \times 10^{-11}) \exp(-7893.54/T)$
$2,3,6\text{-TCP+H} \rightarrow 2,4,6\text{-TCPR+H}_2$	$k(T) = (3.56 \times 10^{-12}) \exp(-6881.01/T)$
$2,4,5\text{-TCP+H} \rightarrow 2,4,5\text{-TCPR+H}_2$	$k(T) = (4.36 \times 10^{-12}) \exp(-6270.24/T)$
$2,4,6\text{-TCP+H} \rightarrow 2,4,6\text{-TCPR+H}_2$	$k(T) = (2.61 \times 10^{-12}) \exp(-6393.41/T)$
$3,4,5\text{-TCP+H} \rightarrow 3,4,5\text{-TCPR+H}_2$	$k(T) = (1.24 \times 10^{-12}) \exp(-4592.88/T)$
$2,3,4,5\text{-TeCP+H} \rightarrow 2,3,4,5\text{-TeCPR+H}_2$	$k(T) = (4.99 \times 10^{-12}) \exp(-7457.09/T)$
$2,3,4,6\text{-TeCP+H} \rightarrow 2,3,4,6\text{-TeCPR+H}_2$	$k(T) = (1.96 \times 10^{-12}) \exp(-7039.66/T)$
$2,3,5,6\text{-TeCP+H} \rightarrow 2,3,5,6\text{-TeCPR+H}_2$	$k(T) = (1.28 \times 10^{-12}) \exp(-6776.23/T)$
$\text{PCP+H} \rightarrow \text{PCPR+H}_2$	$k(T) = (3.18 \times 10^{-12}) \exp(-7008.45/T)$

our study at the MPWB1K/6-311+G(3df,2p)//MPWB1K/6-31+G(d,p) level.

For 19 chlorophenol congeners, the CVT/SCT rate constants vary systematically depending on the substitution pattern of phenol and temperature. Generally, the CVT/SCT rate constants for the phenoxyl−hydrogen abstraction from chlorophenols are lower than that from phenol at a given temperature. For example, at 1000 K, the calculated CVT/SCT rate constant for $2\text{-CP+H} \rightarrow 2\text{-CPR+H}_2$ is 1.76×10^{-14} cm^3 $molecule^{-1}$ s^{-1}, whereas the value is 1.68×10^{-13} cm^3 $molecule^{-1}$ s^{-1} for $C_6H_5OH+H \rightarrow C_6H_5O+H_2$. The substitution pattern of phenol strongly affects the rate constants at a given temperature. For example, at 600 K, the CVT/SCT rate constants are 7.45×10^{-17}, 8.82×10^{-17}, 3.72×10^{-17}, 1.26×10^{-16}, 6.15×10^{-17} cm^3 $molecule^{-1}$ s^{-1} for the phenoxyl−hydrogen abstraction from 2,3,4-TCP, 2,3,5-TCP, 2,3,6-TCP, 2,4,5-TCP, and 2,4,6-TCP, whereas the value is 5.87×10^{-16} cm^3 $molecule^{-1}$ s^{-1} for the phenoxyl−hydrogen abstraction from 3,4,5-TCP. However, the rate constants for the phenoxyl−hydrogen abstraction from chlorophenols with intramolecular hydrogen bonding are not consistently larger than those from other structural conformers for a given number of chlorine substitutions at a given temperature. For example, at 700 K, the rate constant for the phenoxyl−hydrogen abstraction from 2,4-DCP is about 3 times larger than that from 3,4-DCP.

Regulatory decisions and risk analyses often rely on the use of mathematical models to predict the potential outcomes of contaminant releases to the environment. A better knowledge of the temperature dependence would be useful for the kinetic models that account for the contribution of the gaseous route in the production of PCDD/Fs in combustion processes (*9*, *15*). So, the calculated CVT/SCT rate constants are fitted over the temperature range of 600~1200 K and Arrhenius formulas are given in Table 2. The pre-exponential factor, the activation energy, and the rate constants can be obtained from these Arrhenius formulas.

Acknowledgments

This work was supported by NSFC (National Natural Science Foundation of China, project No. 20737001, 20777047), Shandong Province Outstanding Youth Natural Science Foundation (project No. JQ200804) and the Research Fund for the Doctoral Program of Higher Education of China (project No. 200804220046). We thank Professor Donald G. Truhlar for providing the POLYRATE 9.3 program.

Supporting Information Available

TST, TST/Wigner, TST/Eckart, CVT, CVT/ZCT, and CVT/SCT rate constants as function of the reciprocal of the temperature (*T*) over the temperature range of 600~1200 K for $2\text{-CP+H} \rightarrow 2\text{-CPR+H}_2$. This material is available free of charge via the Internet at http://pubs.acs.org.

Literature Cited

(1) Schecter, A. *Dioxin and Health*; Plenum Press: New York, 1994.

(2) Yasuhara, A.; Katami, T.; Okuda, T.; Ohno, N.; Adriaens, P. Formation of dioxins during the combustion of newspapers in the presence of sodium chloride and poly(vinyl chloride). *Environ. Sci. Technol.* **2001**, *35* (7), 1373–1378.

(3) Addink, R.; Govers, H. A. J.; Olie, K. Isomer distributions of polychlorinated dibenzo-p-dioxins/dibenzofurans formed during De Novo synthesis on incinerator fly ash. *Environ. Sci. Technol.* **1998**, *32* (13), 1888–1893.

(4) Olie, K.; Vermeulen, P. L.; Hutzinger, O. Chlorodibenzop-dioxins and chlorodibenzofurans are trace components of fly ash and flue gas of some municipal incinerators in the Netherlands. *Chemosphere* **1977**, *6* (8), 455–459.

(5) Dickson, L. C.; Lenoir, D.; Hutzinger, O. Quantitative comparison of de novo and precursor formation of polychlorinated dibenzo-p-dioxins under simulated municipal solid waste incinerator postcombustion conditions. *Environ. Sci. Technol.* **1992**, *26* (9), 1822–1828.

(6) Addink, R.; Olie, K. Mechanisms of formation and destruction of polychlorinated dibenzo-*p*-dioxins and dibenzofurans in heterogeneous systems. *Environ. Sci. Technol.* **1995**, *29* (6), 1425–1435.

(7) Milligan, M. S.; Altwicker, E. R. Chlorophenol reactions on fly ash. 1. Adsorption/desorption equilibria and conversion to polychlorinated dibenzo-*p*-dioxins. *Environ. Sci. Technol.* **1996**, *30* (1), 225–229.

(8) Karasek, F. W.; Dickson, L. C. Model studies of polychlorinated dibenzo-p-dioxin formation during municipal refuse incineration. *Science* **1987**, *237* (4828), 754–756.

(9) Shaub, W. M.; Tsang, W. Dioxin formation in incinerators. *Environ. Sci. Technol.* **1983**, *17* (12), 721–730.

(10) Altwicker, E. R. Relative rates of formation of polychlorinated dioxins and furans from precursor and de novo reactions. *Chemosphere* **1996**, *33* (10), 1897–1904.

(11) Hung, H.; Buekens, A. Comparison of dioxin formation levels in laboratory gas-phase flow reactors with those calculated using the Shaub-Tsang mechanism. *Chemosphere* **1999**, *38* (7), 1595–1602.

(12) Babushok, V.; Tsang, W. *7th International Congress on Combustion By-Products: Origins, Fate, and Health Effects*; Research Triangle Park, NC, 2001; p 36.

(13) Altarawneh, M.; Dlugogorski, B. Z.; Kennedy, E. M.; Mackie, J. C. Quantum chemical investigation of formation of polychlorodibenzo-*p*-dioxins and dibenzofurans from oxidation and pyrolysis of 2-chlorophenol. *J. Phys. Chem. A* **2007**, *111* (13), 2563–2573.

(14) Evans, C. S.; Dellinger, B. Mechanisms of dioxin formation from the high-temperature oxidation of 2-chlorophenol. *Environ. Sci. Technol.* **2005**, *39* (1), 122–127.

(15) Khachatryan, L.; Asatryan, R.; Dellinger, B. Development of expanded and core kinetic models for the gas phase formation

of dioxins from chlorinated phenols. *Chemosphere* **2003**, *52* (14), 695–708.
(16) Baldridge, M. S.; Gordor, R.; Steckler, R.; Truhlar, D. G. Ab initio reaction paths and direct dynamics calculations. *J. Phys. Chem.* **1989**, *93* (13), 5107–5119.
(17) Gonzalez-Lafont, A.; Truong, T. N.; Truhlar, D. G. Interpolated variational transition-state theory: Practical methods for estimating variational transition-state properties and tunneling contributions to chemical reaction rates from electronic structure calculations. *J. Chem. Phys.* **1991**, *95* (12), 8875–8894.
(18) Garrett, B. C.; Truhlar, D. G. Generalized transition state theory. Classical mechanical theory and applications to collinear reactions of hydrogen molecules. *J. Phys. Chem.* **1979**, *83* (8), 1052–1079.
(19) Liu, Y,-P.; Lynch, G. C.; Truong, T. N.; Lu, D.-H.; Truhlar, D. G.; Garrett, B. C. Molecular modeling of the kinetic isotope effect for the [1,5]-sigmatropic rearrangement of cis-1,3-pentadiene. *J. Am. Chem. Soc.* **1993**, *115* (6), 2408–2415.
(20) Frisch, M. J.; Trucks, G. W.; Schlegel, H. B.; Gill, P. W. M.; Johnson, B. G.; Robb, M. A.; Cheeseman, J. R.; Keith, T. A.; Petersson, G. A.; Montgomery, J. A.; Raghavachari, K.; Allaham, M. A.; Zakrzewski, V. G.; Ortiz, J. V.; Foresman, J. B.; Cioslowski, J.; Stefanov, B. B.; Nanayakkara, A.; Challacombe, M.; Peng, C. Y.; Ayala, P. Y.; Chen, W.; Wong, M. W.; Andres, J. L.; Replogle, E. S.; Gomperts, R.; Martin, R. L.; Fox, D. J.; Binkley, J. S.; Defrees, D. J.; Baker, J.; Stewart, J. P.; Head-Gordon, M.; Gonzales, C.; Pople, J. A. Gaussian 03, Pittsburgh, PA, 2003.
(21) Zhao, Y.; Truhlar, D. G. Hybrid meta density functional theory methods for therochemistry, thermochemical kinetics, and noncovalent interactions: the MPW1B95 and MPWB1K models and comparative assessments for hydrogen bonding and van derwaals interactions. *J. Phys. Chem. A* **2004**, *108* (33), 6908–6918.
(22) Fukui, K. The path of chemical reactions - the IRC approach. *Acc. Chem. Res.* **1981**, *14* (12), 363–368.
(23) Zhang, Q. Z.; Li, S. Q.; Qu, X. H.; Wang, W. X. A quantum mechanical study on the formation of PCDD/Fs from 2-chlorophenol as precursor. *Environ. Sci. Technol.* **2008**, *42* (19), 7301–7308.
(24) Steckler, R.; Chuang, Y. Y.; Fast, P. L.; Corchade, J. C. Coitino, E. L.; Hu, W. P.; Lynch, G. C.; Nguyen, K.; Jackells, C. F.; Gu, M. Z.; Rossi, I.; Clayton, S.; Melissas, V. Garrett, B. C.; Isaacson, A. D.; Truhlar, D. G. *POLYRATE Version 9.3*: University of Minnesota: Minneapolis 2002.
(25) Han, J.; Deming, R. L.; Tao, F. M. Theoretical study of molecular structures and properties of the complete series of chlorophenols. *J. Phys. Chem. A* **2004**, *108* (38), 7736–7743.
(26) *Landolt-Bornstein: Group II: Atomic and Molecular Physics Volume 7: Structure Data of Free Polyatomic Molecules*; Hellwege, K. H., Hellwege, A. M., Eds.; Springer-Verlag: Berlin, 1976.
(27) Qu, X. H.; Wang, H.; Zhang, Q. Z.; Shi, X. Y.; Xu, F.; Wang, W. X. Mechanistic and kinetic studies on the homogeneous gas-phase formation of PCDD/Fs from 2,4,5-trichlorophenol *Environ. Sci. Technol.*, accepted.
(28) Altarawneh, M.; Dlugogorski, B. Z.; Kennedy, E. M.; Mackie, J. C. Quantum chemical and kinetic study of formation of 2-chlorophenoxy radical from 2-chlorophenol: unimolecular decomposition and bimolecular reactions with H, OH, Cl, and O_2. *J. Phys. Chem. A* **2008**, *112* (16), 3680–3692.
(29) Gomes, J. R. B.; Ribeiro da Silva, M. A. V. Gas-phase thermodynamic properties of dichlorophenols determined from density functional theory calculations. *J. Phys. Chem. A* **2003**, *107* (6), 869–874.
(30) He, Y. Z.; Mallard, W. G.; Tsang, W. Kinetics of hydrogen and hydroxyl radical attack on phenol at high temperatures. *J. Phys. Chem.* **1988**, *92* (8), 2196–2201.

ES9001778

Journal of Hazardous Materials 295 (2015) 104–111

Contents lists available at ScienceDirect

Journal of Hazardous Materials

journal homepage: www.elsevier.com/locate/jhazmat

PBCDD/F formation from radical/radical cross-condensation of 2-Chlorophenoxy with 2-Bromophenoxy, 2,4-Dichlorophenoxy with 2,4-Dibromophenoxy, and 2,4,6-Trichlorophenoxy with 2,4,6-Tribromophenoxy

Xiangli Shi[a], Wanni Yu[a,b], Fei Xu[a], Qingzhu Zhang[a,*], Jingtian Hu[a], Wenxing Wang[a]

[a] *Environment Research Institute, Shandong University, Jinan 250100, PR China*
[b] *College of Resources and Environment, Linyi University, Linyi 276000, PR China*

HIGHLIGHTS

- We studied the formation of PBCDD/Fs from the reaction of three CPRs with BPRs.
- The substitution pattern of halogenated phenols determines those of PBCDD/Fs.
- The substitution of halogenated phenols influence the coupling of phenoxy radicals.
- The rate constants of the crucial elementary steps were evaluated.

ARTICLE INFO

Article history:
Received 9 September 2014
Received in revised form 25 March 2015
Accepted 2 April 2015
Available online 3 April 2015

Keywords:
2-Chlorophenoxy radicals
2-Bromophenoxy radicals
2,4-Dichlorophenoxy radicals
2,4-Dibromophenoxy radicals
2,4,6-Trichlorophenoxy radicals
2,4,6-Tribromophenoxy radicals

ABSTRACT

Quantum chemical calculations were carried out to investigate the homogeneous gas-phase formation of mixed polybrominated/chlorinated dibenzo-*p*-dioxins/benzofurans (PBCDD/Fs) from the cross-condensation of 2-chlorophenoxy radical (2-CPR) with 2-bromophenoxy radical (2-BPR), 2,4-dichlorophenoxy radical (2,4-DCPR) with 2,4-dibromophenoxy radical (2,4-DBPR), and 2,4,6-trichlorophenoxy radical (2,4,6-TCPR) with 2,4,6-tribromophenoxy radical (2,4,6-TBPR). The geometrical parameters and vibrational frequencies were calculated at the MPWB1K/6-31+G(d,p) level, and single-point energy calculations were performed at the MPWB1K/6-311+G(3df,2p) level of theory. The rate constants of the crucial elementary reactions were evaluated by the canonical variational transition-state (CVT) theory with the small curvature tunneling (SCT) correction over a wide temperature range of 600–1200 K. Studies show that the substitution pattern of halogenated phenols not only determines the substitution pattern of the resulting PBCDD/Fs, but also has a significant influence on the formation mechanism of PBCDD/Fs, especially on the coupling of the halogenated phenoxy radicals.

1. Introduction

Polybrominated dibenzo-*p*-dioxin/dibenzofurans (PBDD/Fs) are polychlorinated dibenzo-*p*-dioxin/dibenzofuran (PCDD/F) analogues in which all of the chlorine atoms are substituted by the bromine atoms. Mixed polybrominated/chlorinated dibenzo-*p*-dioxins/benzofurans (PBCDD/Fs) are also PCDD/F analogues in which some of the chlorine atoms are substituted by the bromine atoms. These three groups of compounds are all toxicologically significant environmental contaminants, and classed as unintentionally produced persistent organic pollutants (UP-POPs) [1,2]. PBCDD/Fs may be more toxic [3,4]. Combustions of waste materials as well as many other high-temperature processes used in industrial settings are considered as their dominant sources [1,5–7]. Currently, electrical and electronic products commonly contain both chlorinated materials and brominated materials, including the chlorinated plastic polyvinyl chloride (PVC) and a wide range of brominated flame retardants (BFRs). So, the growth of the electronic waste recycling industry in developing countries has drawn the world's attention as a new source of PBCDD/Fs. A study found that mixed PBCDD/Fs presented at far higher concentrations than their chlorinated analogues in soil from e-waste facility

* Corresponding author. Tel.: +86 186 6079 9598; fax: +86 531 8836 1990.
E-mail address: zqz@sdu.edu.cn (Q. Zhang).

http://dx.doi.org/10.1016/j.jhazmat.2015.04.007
0304-3894/

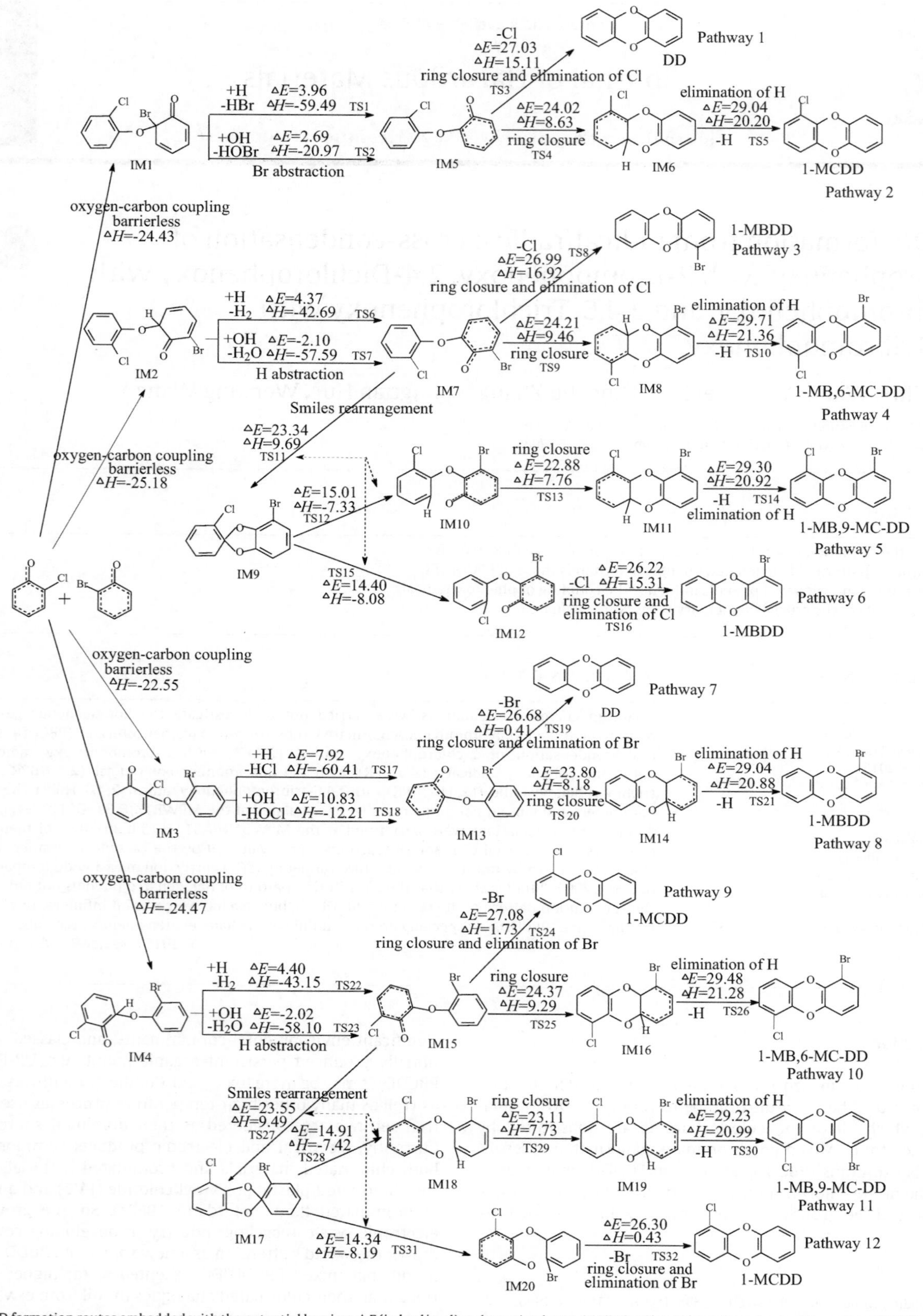

Fig. 1. PBCDD formation routes embedded with the potential barriers ΔE (in kcal/mol) and reaction heats ΔH (in kcal/mol) from the cross-condensation of 2-CPR with 2-BPR. ΔH is calculated at 0 K.

[8]. PBCDD/Fs have also been detected in fly ash and flue gas from municipal waste incinerators and industrial waste incinerators [9–12]. Additional attention and more research should be paid to the formation of PBCDD/Fs. PBCDD/Fs are believed to share a similar formation mechanism with PCDD/Fs as well as PBDD/Fs [10]. Dioxins can be formed by two general formation pathways, precursor pathway and de novo synthesis. The formation pathway via precursors accounts for the majority of dioxin emissions from combustion sources. Chlorophenols (CPs) are generally considered as the predominant precursors of PCDD/Fs and implicated as key intermediates in the de novode novo synthesis [13–17]. And bromophenols (BPs) are structurally similar to PBDD/Fs and have been demonstrated to be the predominant precursors of PBDD/Fs [18,19]. So, PBCDD/Fs can be formed from the combustion or high-temperature pyrolysis of the CP/BP mixture [21]. However, relatively little is known about the formation of PBCDD/Fs, probably due to the difficulties in the separation of a number of dioxin isomers using chromatographic techniques and the lack of commercially available authentic standards [6,21]. In addition, it is also difficult to separate the different pathways from the cross-condensation of CP/BP and the self-condensation of CPs as well as BPs by experimentally technologies [18–21,25].

It is well-known that the formation pathway via precursors such as halogenated phenols, catechol, halogenated biphenyl ethers and biphenyls accounts for the majority of dioxin emissions from combustion sources [26–28]. Halogenated phenoxy radicals

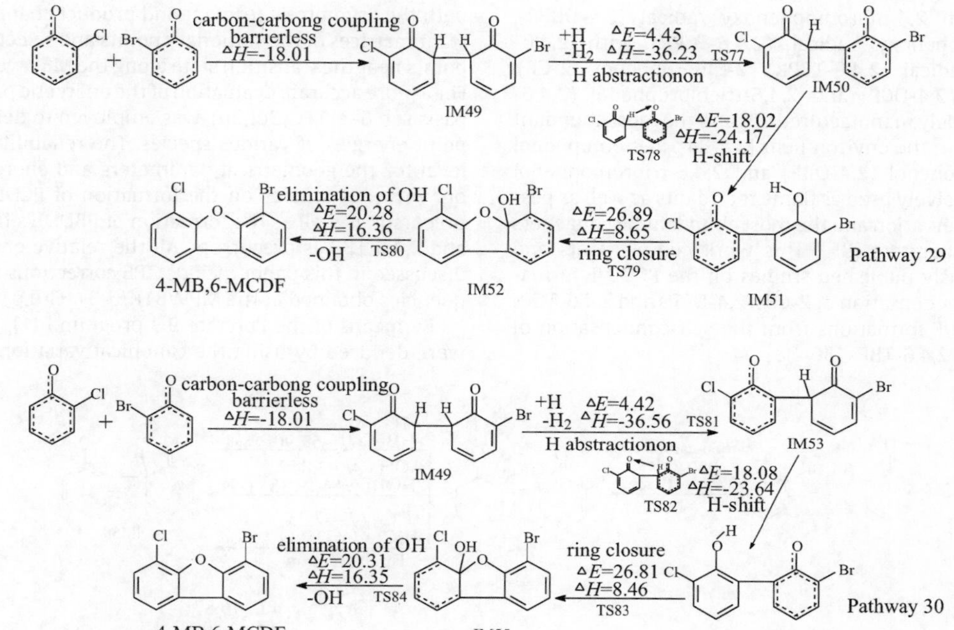

Fig. 2. 4-MB,6-MCDF formation routes embedded with the potential barriers ΔE (in kcal/mol) and reaction heats ΔH (in kcal/mol) from the cross-condensation of 2-CPR with 2-BPR. ΔH is calculated at 0 K.

have been identified as key intermediates in the formation of halogenated dioxins/furans [23,24,19,20,25]. The radical/radical condensation of halogenated phenoxy radicals plays a crucial role in the dioxin/furan formations, especially in the homogeneous gas-phase formation of dioxins/furans [23,24,19,20,25]. Under combustion or pyrolysis conditions, halogenated phenoxy radicals can be formed from CPs or BPs through loss of the phenoxyl-hydrogen via unimolecular, bimolecular, or possibly other low-energy pathways (including heterogenous reactions). In the previously published papers, we investigated the formation of chlorophenoxy radicals (CPRs) and bromophenoxy radicals (BPRs) from CPs and BPs [29–31]. In the present work, we initiated a rather comprehensive theoretical study on the PBCDD/F formations from the cross-condensation of 2-chlorophenoxy radical (2-CPR) with 2-bromophenoxy radical (2-BPR), 2,4-dichlorophenoxy radical (2,4-DCPR) with 2,4-dibromophenoxy radical (2,4-DBPR), and 2,4,6-trichlorophenoxy radical (2,4,6-TCPR) with 2,4,6-tribromophenoxy radical (2,4,6-TBPR). 2-Chlorophenol (2-CP), 2,4-dichlorophenol (2,4-DCP) and 2,4,6-trichlorophenol (2,4,6-TCP) are the most widely manufactured CPs and the most abundant CP congeners found in the environment [32–34]. 2-Bromophenol (2-BP), 2,4-dibromophenol (2,4-DBP) and 2,4,6-tribromophenol (2,4,6-TBP) are extensively used as flame retardants as well as pesticides for wood preservation and the most abundant BP congeners found in waste incinerators [35]. This work complements and expands our previously published studies on the PCDD/F formations from the self-condensation of 2-CPs, 2,4-DCPs and 2,4,6-TCPs as well as the PBDD/F formations from the self-condensation of 2-BPs, 2,4-DBPs, and 2,4,6-TBPs [36–38].

2. Computational methods

High-accuracy molecular orbital calculations were carried out with the aid of the Gaussian 09 package [39] on a supercomputer. The geometries of the reactants, intermediates, transition states and products were optimized by using density functional theory (DFT) at the MPWB1K/6-31+G(d,p) level [40]. The vibrational frequencies were also calculated at the same level in order to determine the nature of the stationary points, the zero-point energy (ZPE), and the thermal contributions to the free energy of activation. Besides, the intrinsic reaction coordinate (IRC) analysis was performed to confirm that each transition state connects to the right minima along the reaction path. At the MPWB1K/6-31+G(d,p) level, the minimum energy paths (MEPs) were constructed, starting from the, respectively, transition state geometries and going downhill to both the asymptotic reactant and product channel. The force constant matrices of the stationary points and selected non-stationary points near the transition state along the MEP were also calculated. For a more accurate evaluation of the energetic parameters, a larger basis set, 6-311+G(3df,2p), was employed to determine the single point energies of various species. The reliability of the MPWB1K level for the geometrical parameters and energies was tested in our previous studies on the formation of PCDD/Fs from 2-CP as precursor as well as the formation of PBDD/Fs from 2-BP, 2,4-DBP and 2,4,6-TBP as precursors. All the relative energies quoted and discussed in this paper include ZPE corrections with unscaled frequencies obtained at the MPWB1K/6-31+G(d,p) level.

By means of the Polyrate 9.3 program [41], the rate constants were deduced by using the canonical variational transition-state

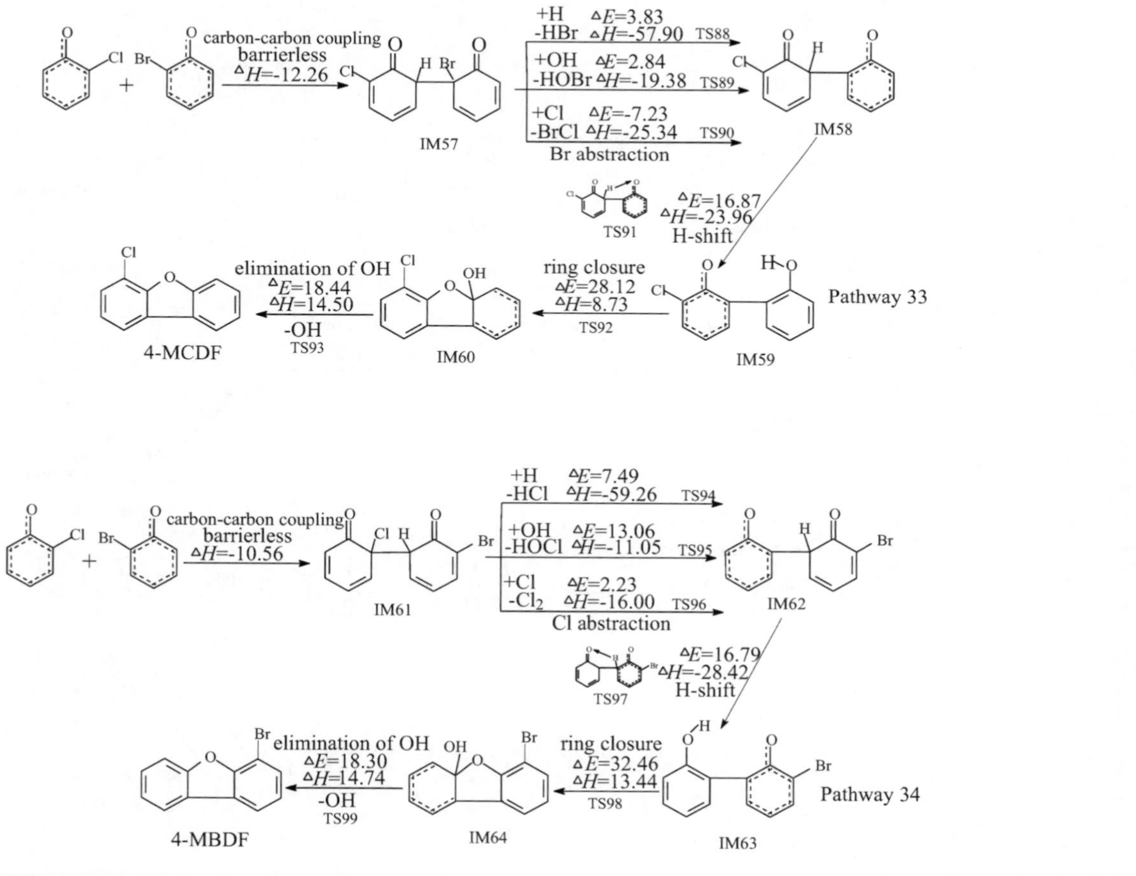

Fig. 3. 4-MCDF and 4-MBDF formation routes embedded with the potential barriers ΔE (in kcal/mol) and reaction heats ΔH (in kcal/mol) from the cross-condensation of 2-CPR with 2-BPR. ΔH is calculated at 0 K.

(CVT) theory [42–44]. The CVT rate constant for temperature T is given by:

$$k^{\mathrm{CVT}}(T) = \min_{s} k^{\mathrm{GT}}(T, s) \tag{1}$$

where

$$k^{\mathrm{GT}}(T, s) = \frac{\sigma k_{\mathrm{B}} T}{h} \times \frac{Q^{\mathrm{GT}}(T, s)}{\Phi^{\mathrm{R}}(T)} e^{-V_{\mathrm{MEP}}(s)/k_{\mathrm{B}} T} \tag{2}$$

where $k^{\mathrm{GT}}(T,s)$ is the generalized transition state theory rate constant at the dividing surface s, σ is the symmetry factor accounting for the possibility of more than one symmetry-related reaction path, k_{B} is Boltzmann's constant, h is Planck's constant, $\Phi^{\mathrm{R}}(T)$ is the reactant partition function per unit volume, excluding symmetry numbers for rotation, and $Q^{\mathrm{GT}}(T, s)$ is the partition function of a generalized transition state at s with a local zero of energy at $V_{\mathrm{MEP}}(s)$ and with all rotational symmetry numbers set to unity. The rotational partition functions were calculated classically, and the vibrational modes were treated as quantum-mechanically separable harmonic oscillators. In order to include the quantum effect for motion along the reaction coordinate, the CVT rate constant was multiplied by a ground-state transmission coefficient, which was calculated by using the small curvature tunneling (SCT) method based on the centrifugal-dominant small-curvature semi-classical adiabatic ground-state approximation [45].

3. Results and discussion

3.1. The formation of PBCDDs

The unpaired electron in the halogenated phenoxy radicals is delocalized, and radical character appears at the phenolic oxygen, the *para* carbon, and the *ortho* carbon [16]. There are 15 mesomers that could result from the combination of 2-CPR and 2-BPR on the phenolic oxygen, on H, Cl, and Br-substituted *ortho* carbons and on H-bearing *para* carbons [16]. However, only the combinations of the phenolic oxygen with the H, Cl, and Br-substituted *ortho* carbon produce the keto-ether structures, which can be regarded as predioxin structures [22].

As shown in Fig. 1, relatively more dioxin formation pathways can be identified from the cross-condensation of 2-CPR with 2-BPR compared to the self-condensation of 2-CPRs as well as 2-BPRs [36,38]. All pathways start with oxygen–carbon coupling, followed by Br (H or Cl) abstraction, ring closure, and intra-annular elimination of Cl (H or Br). It is noted that there is a Smiles rearrangement after H abstraction in pathways 5, 6, 11, and 12. The oxygen–carbon coupling is a barrierless and strongly exothermic process. Br (H or Cl) abstraction steps are highly exothermic with low-energy barriers. In pathways 1, 3, 6, 7, 9 and 12, ring closure and intra-annular elimination of Cl or Br occur in a one-step reaction and are the rate determining step due to the high barrier and strong endothermicity. For pathways 2, 4, 5, 8, 10 and 11, intra-annular elimination of H is the rate determining step.

Comparison of the reaction pathways presented in Fig. 1 shows that pathway 1 covers relatively less elementary steps than pathway 2. Furthermore, the rate determining step involved in pathway 1 has a lower barrier and is much less endoergic than that involved in pathway 2. Thus, pathway 1 is favored over pathway 2. For the same reason, pathway 3 is favored over pathway 4. Pathway 6 is favored over pathway 5. It means that the pathways via intra-annular elimination of Cl are favored over those via intra-annular elimination of H because of the weaker C—Cl bond versus the C—H bond. Similarly, pathway 7 is favored over pathway 8, pathway 9 is favored over pathway 10, and pathway 12 is favored over pathway 11 due to the weaker C—Br bond versus the C—H bond. Pathway 9 is favored over pathway 3, and pathway 12 is favored over pathway 6 because intra-annular elimination of Br is much easier than intra-annular elimination of Cl. Comparison of pathways 1 with 7 shows that Br abstraction has a lower barrier than Cl abstraction, however, intra-annular elimination of Cl is much more endothermic than intra-annular elimination of Br due to the stronger C—Cl bond versus the C—Br bond. Hence, pathways 1 and 7 should be competitive. Previous study showed that only CPs with chlorine at the *ortho* position are capable of forming PCDDs because the energetically feasible PCDD formation pathways proceed through intra-annular elimination of Cl [37]. From the present result, it might be inferred that the formation of PCDDs from the cross-condensation of CP with 2-BP is relatively easier compared to the self-condensation of CPs because Br is a better leaving group than Cl. The resulting dioxin products are DD, 1-MCDD and 1-MBDD, which were observed experimentally from the high-temperature pyrolysis of a 2-CP/2-BP mixture [8–20]. It can be inferred from this work, together with works already published from our group, that the formation potential of 1-MCDD increases from the cross-condensation of 2-CPR and 2-CBR than that from the self-condensation of 2-CPRs, whereas the formation potential of 1-MBDD decreases from the cross-condensation of 2-CPR and 2-CBR compared to the self-condensation of 2-BPRs.

Relatively more dioxin congeners can be formed from the cross-condensation of 2,4-DCPR with 2,4-DBPR compared to the

Table 1
Arrhenius formulas (units are s^{-1} and cm^3 $molecule^{-1}$ s^{-1} for unimolecular and bimolecular reactions) for the elementary reactions involved in the thermodynamically preferred formation pathways of PBCDDs from the cross-condensation of 2-CPR with 2-BPR, 2,4-DCPR with 2,4-DBPR, and 2,4,6-TCPR with 2,4,6-TBPR over the temperature range of 600–1200 K.

Reactions	Arrhenius formulas
IM1+H → IM5+HBr	$k(T)=(4.94\times10^{-11})\exp(-1660.90/T)$
IM1+OH → IM5+HOBr	$k(T)=(1.46\times10^{-11})\exp(-2912.02/T)$
IM5 → DD+Cl	$k(T)=(4.27\times10^{11})\exp(-15753.69/T)$
IM2+H → IM7+H_2	$k(T)=(6.18\times10^{-12})\exp(-2407.18/T)$
IM9 → IM12	$k(T)=(1.40\times10^{13})\exp(-7445.69/T)$
IM12 → 1-MBDD+Cl	$k(T)=(7.67\times10^{11})\exp(-13234.84/T)$
IM3+H → IM13+HCl	$k(T)=(8.78\times10^{-11})\exp(-4749.31/T)$
IM3+OH → IM13+HOCl	$k(T)=(1.34\times10^{-11})\exp(-6466.20/T)$
IM13 → DD+Br	$k(T)=(7.29\times10^{11})\exp(-13737.33/T)$
IM4+H → IM15+H_2	$k(T)=(1.79\times10^{-11})\exp(-2526.65/T)$
IM15 → 1-MCDD+Br	$k(T)=(5.95\times10^{11})\exp(-13766.71/T)$
IM15 → IM17	$k(T)=(2.26\times10^{12})\exp(-12192.77/T)$
IM17 → IM20	$k(T)=(1.54\times10^{13})\exp(-7428.07/T)$
IM20 → 1-MCDD+Br	$k(T)=(1.39\times10^{12})\exp(-13261.72/T)$
IM21+H → IM25+HBr	$k(T)=(8.90\times10^{-11})\exp(-1340.43/T)$
IM21+OH → IM25+HOBr	$k(T)=(3.08\times10^{-11})\exp(-3339.01/T)$
IM25 → 2-MB,8-MCDD+Cl	$k(T)=(4.62\times10^{11})\exp(-13444.71/T)$
IM22+H → IM27+H_2	$k(T)=(8.95\times10^{-11})\exp(-2359.42/T)$
IM27 → 1,3-DB,8-MCDD	$k(T)=(5.95\times10^{11})\exp(-14265.60/T)$
IM27 → IM29	$k(T)=(2.58\times10^{11})\exp(-11142.77/T)$
IM29 → IM32	$k(T)=(1.28\times10^{12})\exp(-7976.69/T)$
IM32 → 1,3-DB,7-MCDD+Cl	$k(T)=(1.53\times10^{12})\exp(-13976.74/T)$
IM23+H → IM33+HCl	$k(T)=(8.21\times10^{-11})\exp(-3357.22/T)$
IM23+OH → IM33+HOCl	$k(T)=(3.34\times10^{-12})\exp(-5281.35/T)$
IM33 → 2-MB,7-MCDD+Br	$k(T)=(4.96\times10^{11})\exp(-14200.10/T)$
IM24+H → IM35+H_2	$k(T)=(2.29\times10^{-11})\exp(-2239.08/T)$
IM35 → 2-MB,7,9-DCDD+Br	$k(T)=(5.18\times10^{11})\exp(-14709.26/T)$
IM35 → IM37	$k(T)=(1.03\times10^{11})\exp(-10240.90/T)$
IM37 → IM40	$k(T)=(8.43\times10^{13})\exp(-10465.03/T)$
M40 → 2-MB,6,8-DCDD+Br	$k(T)=(6.15\times10^{11})\exp(-13526.27/T)$
IM41+H → IM43+HBr	$k(T)=(5.16\times10^{-11})\exp(-1432.44/T)$
IM41+OH → IM43+HOBr	$k(T)=(3.12\times10^{-11})\exp(-2443.48/T)$
IM43 → 1,3-DB,6,8-DCDD+Cl	$k(T)=(1.11\times10^{12})\exp(-15026.38/T)$
IM43 → IM44	$k(T)=(1.89\times10^{12})\exp(-11551.44/T)$
IM44 → IM45	$k(T)=(2.49\times10^{13})\exp(-6885.77/T)$
IM45 → 1,3-DB,7,9-DCDD+Cl	$k(T)=(2.11\times10^{11})\exp(-12783.69/T)$
IM42+H → IM46+HCl	$k(T)=(8.12\times10^{-11})\exp(-3224.32/T)$
IM42+OH → IM46+HOCl	$k(T)=(2.90\times10^{-12})\exp(-6745.67/T)$
IM46 → 1,3-DB,6,8-DCDD+Br	$k(T)=(7.37\times10^{11})\exp(-15233.12/T)$
IM46 → IM47	$k(T)=(1.66\times10^{12})\exp(-11484.64/T)$
IM47 → IM48	$k(T)=(3.89\times10^{12})\exp(-10644.87/T)$

Table 2
Arrhenius formulas (units are s^{-1} and cm^3 $molecule^{-1}$ s^{-1} for unimolecular and bimolecular reactions) for the elementary reactions involved in the formation of PBCDFs from the cross-condensation of 2-CPR with 2-BPR, and 2,4-DCPR with 2,4-DBPR over the temperature range of 600–1200 K.

Reactions	Arrhenius formulas
IM49 + H → IM50 + H_2	$k(T) = (6.80\times10^{-11})\exp(-3276.90/T)$
IM50 → IM51	$k(T) = (2.87\times10^{12})\exp(-9022.85/T)$
IM51 → IM52	$k(T) = (2.56\times10^{12})\exp(-13660.58/T)$
IM52 → 4-MB,6-MCDF + OH	$k(T) = (2.53\times10^{13})\exp(-10579.84/T)$
IM49 + H → IM53 + H_2	$k(T) = (1.74\times10^{-11})\exp(-2468.73/T)$
IM53 → IM54	$k(T) = (2.99\times10^{12})\exp(-9073.80/T)$
IM54 → IM55	$k(T) = (2.29\times10^{12})\exp(-13522.27/T)$
IM55 → 4-MB,6-MCDF + OH	$k(T) = (2.18\times10^{13})\exp(-10339.82/T)$
IM49 → IM56	$k(T) = (1.62\times10^{12})\exp(-9548.38/T)$
IM56 + H → IM51 + H_2	$k(T) = (4.23\times10^{-11})\exp(-6542.43/T)$
IM57 + H → IM58 + HBr	$k(T) = = (5.07\times10^{-11})\exp(-2854.64/T)$
IM57 + OH → IM58 + HOBr	$k(T) = (2.16\times10^{-12})\exp(-2564.58/T)$
IM58 → IM59	$k(T) = (8.93\times10^{11})\exp(-7584.36/T)$
IM59 → IM60	$k(T) = (1.97\times10^{12})\exp(-14245.60/T)$
IM60 → 4-MCDF + OH	$k(T) = (2.40\times10^{13})\exp(-9914.24/T)$
IM61 + H → IM62 + HCl	$k(T) = (8.89\times10^{-11})\exp(-4524.11/T)$
IM61 + OH → IM62 + HOCl	$k(T) = (3.63\times10^{-12})\exp(-7941.21/T)$
IM63 → IM64	$k(T) = (2.98\times10^{12})\exp(-13301.56/T)$
IM64 → 4-MBDF + OH	$k(T) = (1.04\times10^{13})\exp(-6446.86/T)$
IM65 + H → IM66 + H_2	$k(T) = (1.75\times10^{-11})\exp(-3185.59/T)$
IM66 → IM67	$k(T) = (2.96\times10^{12})\exp(-9410.88/T)$
IM67 → IM68	$k(T) = (2.20\times10^{12})\exp(-14479.95/T)$
IM68 → 2,4-DB,6,8-DCDF + OH	$k(T) = (2.47\times10^{13})\exp(-11114.05/T)$
IM65 + H → IM69 + H_2	$k(T) = (4.82\times10^{-11})\exp(-2593.53/T)$
IM69 → IM70	$k(T) = (8.62\times10^{11})\exp(-8856.47/T)$
IM70 → IM71	$k(T) = (1.56\times10^{12})\exp(-14114.26/T)$
IM71 → 2,4-DB,6,8-DCDF + OH	$k(T) = (2.90\times10^{13})\exp(-11192.86/T)$
IM65 → IM72	$k(T) = (6.63\times10^{11})\exp(-10050.85/T)$
IM72 + H → IM67 + H_2	$k(T) = (9.20\times10^{-11})\exp(-5697.86/T)$
IM72 + H → IM70 + H_2	$k(T) = (6.58\times10^{-11})\exp(-6893.70/T)$
IM73 + H → IM74 + HBr	$k(T) = (8.36\times10^{-11})\exp(-1423.82/T)$
IM73 + OH → IM74 + HOBr	$k(T) = (2.24\times10^{-12})\exp(-2335.99/T)$
IM74 → IM75	$k(T) = (2.11\times10^{11})\exp(-7048.13/T)$
IM75 → IM76	$k(T) = (2.27\times10^{12})\exp(-13827.93/T)$
IM76 → 2-MB,6,8-DCDF + OH	$k(T) = (2.31\times10^{13})\exp(-10564.78/T)$
IM77 + H → IM78 + HCl	$k(T) = (4.17\times10^{-11})\exp(-3346.24/T)$
IM77 + OH → IM78 + HOCl	$k(T) = (3.94\times10^{-13})\exp(-7233.47/T)$
IM78 → IM79	$k(T) = (4.03\times10^{12})\exp(-8698.96/T)$
IM80 → 2,4-DB,8-MCDF + OH	$k(T) = (6.80\times10^{2})\exp(-9790.31/T)$

self-condensation of 2,4-DCPRs as well as 2,4-DBPRs [37,38]. The formation schemes embedded with the potential barriers, and reaction heats are depicted in Fig. S1 of Supporting information. Similar to the cross-condensation of 2-CPR with 2-BPR, all PBCDD formation pathways start with oxygen–carbon coupling, followed by Br (H or Cl) abstraction, ring closure, and intra-annular elimination of Cl (H or Br). In pathways 13, 15, 18, 19, 21 and 24, ring closure and intra-annular elimination of Cl or Br occur in a one-step reaction and are the rate determining step. For pathways 14, 16, 17, 20, 22 and 23, intra-annular elimination of H is the rate determining step due to the high barrier and strong endothermicity. Comparison of the formation pathways depicted in Fig. 1S of Supporting information shows that pathways 13, 15 and 18 are favored over pathways 14, 16, 17, 20, 22, and 23, whereas pathways 21 and 24 are favored over pathways 15 and 18, and pathways 13 and 21 should be competitive. 2-MB,8-MCDD, 1,3-DB,8-MCDD, 1,3-DB,7-MCDD, 2-MB,7-MCDD, 2-MB,7,9-DCDD and 2-MB,6,8-DCDD are possible PBCDD products from the cross-condensation of 2,4-DCPR and 2,4-DBPR. And it can be inferred from the present study that the formations of 2-MB,7,9-DCDD and 2-MB,6,8-DCDD are preferred over the formations of 1,3-DB,8-MCDD and 1,3-DB,7-MCDD.

Similar to the formation of PBCDDs from the cross-condensation of 2,4-DCPR with 2,4-DBPR, two PBCDD congeners, 1,3-DB,6,8-DCDD and 1,3-DB,7,9-DCDD, can be formed from the cross-condensation of 2,4,6-TCPR with 2,4,6-TBPR. The formation mechanism is schemed in Fig. S2 of Supporting information. The barrier of Br abstraction is lower compared to Cl abstraction. However, intra-annular elimination of Cl is much more endothermic than intra-annular elimination of Br. Taken one with another, the four reaction pathways presented in Fig. S2 of Supporting information should be possible to occur for the cross-condensation of 2,4,6-TCPR with 2,4,6-TBPR. 1,3,6,8- and 1,3,7,9-halogen substituted dioxin isomers were detected experimentally from the high-temperature pyrolysis of a 2,4,6-TCP/2,4,6-TBP mixture [21]. Comparison of the PBCDD formations from the cross-condensation of 2-CPR with 2-BPR, 2,4-DCPR with 2,4-DBPR, and 2,4,6-TCPR with 2,4,6-TBPR indicates that substitution pattern of halogenated phenols not only determines the substitution pattern of the resulting PBCDDs, but also has a significant influence on the formation mechanism of PBCDDs, especially on the oxygen–carbon coupling of the halogenated phenoxy radicals. The ranking of the exothermicity for the oxygen–carbon coupling is as follows: 2-CPR + 2-BPR > 2,4-DCPR + 2,4-DBPR > 2,4,6-TCPR + 2,4,6-TBPR.

3.2. The formation of PBCDFs

Three furan congeners, 4-MB,6-MCDF, 4-MCDF and 4-MBDF, can be yielded from the cross-condensation of 2-CPR with 2-BPR. Four reaction pathways, pathways 29, 30, 31 and 32, are offered in Fig. 2 to interpret the formation of 4-MB,6-MCDF. Pathway 33 is proposed for 4-MCDF, and pathway 34 is proposed for 4-MBDF, illustrated in Fig. 3. Figs. 2 and 3 show that pathways 29, 30, 33 and 34 involve five elementary processes: carbon–carbon coupling, H (Br or Cl) abstraction, H-shift, ring closure, and elimination of OH. The ring closure process has a large barrier, is strongly endoergic, and is the rate determining step. Pathways 31 and 32 involves five elementary processes as well: carbon–carbon coupling, double H-shift, H abstraction, ring closure (the rate determining step), and elimination of OH. Comparison of the formation mechanism described in Figs. 2 and 3 shows that the carbon–carbon coupling involved in pathways 29, 30, 31 and 32 is more exothermic compared to pathways 33 and 34. Furthermore, the ring closure process involved in pathways 29, 30, 31 and 32 has a lower barrier than that involved in pathways 33 and 34. Hence, pathways 29, 30, 31 and 32 are the thermodynamically preferred PBCDF formation pathways compared to pathways 33 and 34. And the formation of 4-MB,6-MCDF dominates over the formations of 4-MCDF and 4-MBDF. Similarly, pathway 33 is the thermodynamically preferred compared to pathway 34, and the formation of 4-MCDF is favored over the formation of 4-MBDF.

Similar to the formation of furans from the cross-condensation of 2-CPR with 2-BPR, three PBCDF congeners, 2,4-DB,6,8-DCDF, 2-MB,6,8-DCDF and 2,4-DB,8-MCDF, can be formed from the cross-condensation of 2,4-DCPR with 2,4-DBPR. The formation routes embedded with the potential barriers and reaction heats are schemed in Figs. S3 and S4 of Supporting information. Obviously, pathways 35, 36, 37 and 38 are the more energetically favorable compared to pathways 39 and 40. So, the formation of 2,4-DB,6,8-DCDF is preferred over the formations of 2-MB,6,8-DCDF and 2,4-DB,8-MCDF from the cross-condensation of 2,4-DCPR with 2,4-DBPR. The research above shows that the formation of PBCDFs is based primarily on the *ortho–ortho* coupling of halogenated phenoxy radicals. If both *ortho*-positions of phenol are substituted with voluminous chlorine or bromine, the *ortho–ortho* coupling of halogenated phenoxy radicals is sterically inhibited. So, no PBCDFs can be yielded from the cross-condensation of 2,4,6-TCPR with 2,4,6-TBPR.

3.3. Rate constant calculations

Regulatory decision and risk analysis often rely on the use of mathematical models to predict the potential outcomes of contaminant releases to the environment. The mathematical model to predict the dioxin/furan emission places a high demand on accurate kinetic parameters of the elementary reactions involved in the dioxin/furan formations. So, in this work, the rate constants of key elementary reactions involved in the formation of PBCDD/Fs were calculated by using canonical variational transition-state (CVT) theory with small curvature tunneling (SCT) contribution over a wide temperature range of 600–1200 K on the basis of the MPWB1K/6-311+G(3df,2p)//MPWB1K/6-31+G(d,p) energies. Actually, the CVT/SCT method has been successfully used to deal with the elementary reactions involved in the formation of PCDD/Fs from 2-CP, 2,4-DCP and 2,4,6-TCP as precursors as well as the formation of PBDD/Fs from 2-BP, 2,4-DBP and 2,4,6-TBP as precursors [36–38]. For the purpose of comparison, the conventional transition state theory (TST) rate constants and the canonical variational transition state (CVT) rate constants without the tunneling correction were calculated. The CVT rate constants are smaller than the TST ones, especially for the abstraction reactions. It enables us to conclude that the variational effect should play an important role in the calculation of rate constants. To be used more effectively, the CVT/SCT rate constants are fitted, and Arrhenius formulas are given in Table 1 for the elementary reactions involved in the thermodynamically favorable PBCDD formation pathways and in Table 2 for the elementary reactions involved in the formation of PBCDFs. The pre-exponential factors and the activation energies can be obtained from these Arrhenius formulas.

4. Conclusions

This study investigated theoretically the formation mechanism of PBCDD/Fs from the cross-condensation of 2-CPR with 2-BPR, 2,4-DCPR with 2,4-DBPR, and 2,4,6-TCPR with 2,4,6-TBPR. The rate constants of the key elementary reactions involved in the formation of PBCDD/Fs were calculated by using the CVT/SCT method, and the rate-temperature formulas were obtained. Several specific conclusions can be drawn:

(1) The formation potential of 1-MCDD increases from the cross-condensation of 2-CPR and 2-CBR than that from the self-condensation of 2-CPRs, whereas the formation potential of 1-MBDD decreases from the cross-condensation of 2-CPR and 2-CBR compared to the self-condensation of 2-BPRs.
(2) The substitution pattern of halogenated phenols not only determines the substitution pattern of the resulting PBCDD/Fs, but also has a significant influence on the formation mechanism of PBCDD/Fs, especially on the coupling of the halogenated phenoxy radicals.

Supporting information available

PBCDD/F formation routes from the cross-condensation of 2,4-DCPR with 2,4-DBPR, and 2,4,6-TCPR with 2,4,6-TBPR. The imaginary frequencies (in cm^{-1}), the zero-point energies (ZPE, in a.u.) and the total energies (in a.u.) for the transition states involved in the formation of PBCDD/Fs from the cross-condensation of 2-CPR with 2-BPR, 2,4-DCPR with 2,4-DBPR, and 2,4,6-TCPR with 2,4,6-TBPR. The material is available free of charge via the Internet at http://www.elsevier.com/.

Acknowledgments

The work was financially supported by NSFC (National Natural Science Foundation of China, project No. 21337001), National High Technology Research and Development Program 863 Project (No. 2012AA06A301) and Independent Innovation Foundation of Shandong University (IIFSDU, project No. 2012JC030). The authors thank Professor Donald G. Truhlar for providing the POLYRATE 9.3 program.

Appendix A. Supplementary data

Supplementary data associated with this article can be found, in the online version, at http://dx.doi.org/10.1016/j.jhazmat.2015.04.007.

References

[1] B. Du, M.H. Zheng, Y.R. Huang, A.M. Liu, H.H. Tian, L.L. Li, N. Li, T. Ba, Y.W. Li, S.P. Dong, W.B. Liu, G.J. Su, Mixed polybrominated/chlorinated dibenzo-*p*-dioxins and dibenzofurans in stack gas emissions from industrial thermal processes, Environ. Sci. Technol. 44 (15) (2010) 5818–5823.
[2] UNEP Chemicals, Standardized Toolkit for Identification and Quantification of Dioxin and Furan Releases, Inter-organization Programme for the Sound Management of Chemicals (IOMC), Geneva, Switzerland (2005).
[3] F. Samara, B.K. Gullett, R.O. Harrison, G.C. Clark, Determination of relative assay response factors for toxic chlorinated and brominated dioxins/furans using an enzyme immunoassay (EIA) and a chemically-activated luciferase gene expression cell bioassay (CALUX), Environ. Int. 35 (3) (2009) 588–593.
[4] H. Olsman, M. Engwall, U. Kammann, M. Klempt, J. Otte, B. van Bavel, H. Hollert, Relative differences in aryl hydrocarbon receptor-mediated response for 18 polybrominated and mix halogenated dibenzo-*p*-dioxins and -furans in cell lines from four different species, Environ. Toxicol. Chem. 26 (11) (2007) 2448–2454.
[5] M. Zennegg, X. Yu, M. Wong, R. Weber, Fingerprints of chlorinated: brominated and mixed halogenated dioxins at two e-waste recycling sites in guiyu/China, Organohalogen Compd. 71 (2009) 2263–2267.
[6] S. Gunilla, M. Stellan, Formation of PBCDD and PBCDF during flue gas cooling, Environ. Sci. Technol. 38 (3) (2004) 825–830.
[7] K. Kawamoto, Potential formation of PCDD/Fs and related bromine-substituted compounds from heating processes for ashes, J. Hazard. Mater. 168 (2-3) (2009) 641–648.
[8] X. Yu, Z. Markus, E. Magnus, R. Anna, L. Maria, M.H. Wong, W. Rolands, E-waste recycling heavily contaminates a Chinese city with chlorinated: brominated and mixed halogenated dioxins, Organohalogen Compd. 70 (2008) 813–816.
[9] Z.W. Tang, Q.F. Huang, Y.F. Yang, PCDD/Fs in fly ash from waste incineration in China: a need for effective risk management, Environ. Sci. Technol. 47 (11) (2013) 5520–5521.
[10] R. Weber, B. Kuch, Relevance of BFRs and thermal conditions on the formation pathways of brominated and brominated-chlorinated dibenzodioxins and dibenzofurans, Environ. Int. 29 (6) (2003) 699–710.
[11] L.Q. Huang, H.Y. Tong, J.R. Donnelly, Characterization of dibromopolychlorodibenzo-para-dioxins and dibromopolychlorodibenzofurans in municipal waste incinerator fly-ash using gas-chromatography mass-spectrometry, Anal. Chem. 64 (9) (1992) 1034–1040.
[12] W. Chatkittikunwong, C.S. Creaser, Bromo-dibenzo-*p*-dioxins, bromochloro-dibenzo-*p*-dioxins and chloro-dibenzo-*p*-dioxins and dibenzofurans in incinerator fly-ash, Chemosphere 29 (3) (1994) 559–566.
[13] F.W. Karasek, L.C. Dickson, Model studies of polychlorinated dibenzo-*p*-dioxin formation during municipal refuse incineration, Science 237 (4816) (1987) 754–756.
[14] M.S. Milligan, E.R. Altwicker, Chlorophenol reactions on fly ash. 1. Adsorption/desorption equilibria and conversion to polychlorinated dibenzo-*p*-dioxins, Environ. Sci. Technol. 30 (1) (1995) 225–229.
[15] M. Altarawneh, B.Z. Dlugogorski, Mechanisms of pollutant formation in fires, Proc. of the Seventh International Seminar on Fire and Explosion Hazards (ISFEH7) (2013), http://dx.doi.org/10.3850/978-981-08-7724-8_0x-0x, ISBN: 978-981-08-7724-8.
[16] A. Saeed, M. Altarawneh, B.Z. Dlugogorski, Reactions of 2-chlorophenol and 2-bromophenol: mechanisms of formation of mixed halogenated dioxins and furans (PXDD/Fs), Organohalogen Compd. 76 (2014) 345–348.
[17] S. Nganai, B. Dellinger, S. Lomnicki, PCDD/PCDF ratio in the precursor formation model over CuO surface, Environ. Sci. Technol. 48 (23) (2014) 13864–13870.
[18] C.S. Evans, B. Dellinger, Formation of bromochlorodibenzo-*p*-dioxins and dibenzofurans from the high-temperature oxidation of a mixture of 2-chlorophenol and 2-bromophenol, Environ. Sci. Technol. 40 (9) (2006) 3036–3042.
[19] C.S. Evans, B. Dellinger, Surface-mediated formation of polybrominated dibenzo-*p*-dioxins and dibenzofurans from the high-temperature pyrolysis of 2-bromophenol on a CuO/silica surface, Environ. Sci. Technol. 39 (13) (2005) 4857–4863.
[20] C.S. Evans, B. Dellinger, Formation of bromochlorodibenzo-*p*-dioxins and furans from the high-temperature pyrolysis of a 2-chlorophenol/2-bromophenol mixture, Environ. Sci. Technol. 39 (20) (2005) 7940–7948.

[21] Y.C. Na, K.J. Kim, C.S. Park, J. Hong, Formation of tetrahalogenated dibenzo-*p*-dioxins (TXDDs) by pyrolysis of a mixture of 2,4,6-trichlorophenol and 2,4,6-tribromophenol, J. Anal. Appl. Pyrolysis 80 (1) (2007) 254–261.
[22] M. Altarawneh, B.Z. Dlugogorski, E.M. Kennedy, J.C. Mackie, Quantum chemical investigation of formation of polychlorodibenzo-*p*-dioxins and dibenzofurans from oxidation and pyrolysis of 2-chlorophenol, J. Phys. Chem. A 111 (13) (2007) 2563–2573.
[23] V.I. Babushok, W. Tsang, Gas-phase mechanism for dioxin formation, Chemosphere 51 (10) (2003) 1023–1029.
[24] C.S. Evans, B. Dellinger, Mechanisms of dioxin formation from the high-temperature pyrolysis of 2-bromophenol, Environ. Sci.Technol. 37 (24) (2003) 5574–5580.
[25] C.S. Evans, B. Dellinger, Mechanisms of dioxin formation from the high-temperature oxidation of 2-bromophenol, Environ. Sci. Technol. 39 (7) (2005) 2128–2134.
[26] M. Altarawneh, B.Z. Dlugogorsk, E.M. Kennedy, J.C. Mackie, Theoretical study of unimolecular decomposition of catechol, J. Phys. Chem. A 114 (2) (2010) 1060–1067.
[27] M. Altarawneh, B.Z. Dlugogorsk, Mechanisms of transformation of polychlorinated diphenyl ethers into polychlorinated dibenzo-p-dioxins and dibenzofurans, Chemosphere 114 (2014) (2014) 129–135.
[28] M. Altarawneh, B.Z. Dlugogorsk, Formation of polybrominated dibenzofurans from polybrominated biphenyls, Chemosphere 119 (2015) (2014) 1048–1053.
[29] Q.Z. Zhang, X.H. Qu, F. Xu, X.Y. Shi, W.X. Wang, Mechanism and thermal rate constants for the complete series reactions of chlorophenols with H, Environ. Sci. Technol. 43 (11) (2009) 4105–4112.
[30] F. Xu, H. Wang, Q.Z. Zhang, R.X. Zhang, X.H. Qu, W.X. Wang, Kinetic properties for the complete series reactions of chlorophenols with OH radicals-relevance for dioxin formation, Environ. Sci. Technol. 44 (4) (2010) 1399–1404.
[31] R. Gao, F. Xu, S.Q. Li, J.T. Hu, Q.Z. Zhang, W.X. Wang, Formation of bromophenoxy radicals from complete series reactions of bromophenols with H and OH radicals, Chemosphere 92 (4) (2013) 382–390.
[32] R. Addink, P.A.J.P. Cnubben, K. Olie, Formation of polychlorinated dibenzo-*p*-dioxins/dibenzofurnans on fly ash from precursors and carbon model compounds, Carbon 33 (10) (1995) 1463–1471.
[33] P.W. Cains, L. Mccausland, A.R. Fernandes, P. Dyke, Polychlorinated dibenzo-*p*-dioxins and dibenzofurans formation in incineration: effects of fly ash and carbon source, Environ. Sci. Technol. 31 (3) (1997) 776–785.
[34] R. Weber, H. Hagenmaier, PCDD/PCDF formation in fluidized bed incineration, Chemosphere 38 (11) (1999) 2643–2654.
[35] IUCLID, 2,4,6-Ttribromophenol and Other Simple Brominated Phenols, Data Set for 2,4,6-tribromophenol, Ispra, European Chemicals Bureau, International Uniform Chemical Information Database, 2003.
[36] Q.Z. Zhang, S.Q. Li, X.H. Qu, W.X. Wang, A quantum mechanical study on the formation of PCDD/Fs from 2-chlorophenol as precursor, Environ. Sci. Technol. 42 (19) (2008) 7301–7308.
[37] Q.Z. Zhang, W.N. Yu, R.X. Zhang, Q. Zhou, R. Gao, W.X. Wang, Quantum chemical and kinetic study on dioxin formation from the 2,4,6-TCP and 2,4-DCP precursors, Environ. Sci. Technol. 44 (9) (2010) 3395–3403.
[38] W.N. Yu, J.T. Hu, F. Xu, X.Y. Sun, R. Gao, Q.Z. Zhang, W.X. Wang, Mechanism and direct kinetics study on the homogeneous gas-phase formation of PBDD/Fs from 2-BP, 2,4-DBP, and 2,4,6-TBP as precursors, Environ. Sci. Technol. 45 (5) (2011) 1917–1925.
[39] M. Frisch, G. Trucks, H.B. Schlegel, G. Scuseria, M. Robb, J. Cheeseman, G. Scalmani, V. Barone, B. Mennucci, G. Petersson, Gaussian 09, Revision A. 02, 270, Gaussian. Inc., Wallingford CT, 2009, pp. 271.
[40] Y. Zhao, D.G. Truhlar, Hybrid meta density functional theory methods for therochemistry, thermochemical kinetics, and noncovalent interactions: the MPW1B95 and MPWB1K models and comparative assessments for hydrogen bonding and van der waals interactions, J. Phys. Chem A 108 (33) (2004) 6908–6918.
[41] R. Steckler, Y.Y. Chuang, P.L. Fast, J.C. Corchade, E.L. Coitino, W.P. Hu, G.C. Lynch, K. Nguyen, C.F. Jackells, M.Z. Gu, I. Rossi, S. Clayton, V. Melissas, B.C. Garrett, A.D. Isaacson, D.G. Truhlar, POLYRATE Version 9.3, University of Minnesota, Minneapolis, 2002.
[42] M.S. Baldridge, R. Gordor, R. Steckler, D.G. Truhlar, Ab initio reaction paths and direct dynamics calculations, J. Phys. Chem. 93 (13) (1989) 5107–5119.
[43] A. Gonzalez-Lafont, T.N. Truong, D.G. Truhlar, Interpolated variational transition-state theory: practical methods for estimating variational transition-state properties and tunneling contributions to chemical reaction rates from electronic structure calculations, J. Chem. Phys. 95 (12) (1991) 8875–8894.
[44] B.C. Garrett, D.G. Truhlar, Generalized transition state theory. Classical mechanical theory and applications to collinear reactions of hydrogen molecules, J. Phys. Chem. 83 (8) (1979) 1052–1079.
[45] A. Fernandez-Ramos, B.A. Ellingson, B.C. Garret, D.G. Truhlar, Variational transition state theory with multidimensional tunneling, in: K.B. Lipkowitz, T.R. Cundari (Eds.), Reviews in Computational Chemistry, Wiley-VCH, Hoboken, NJ, 2007.

Atmospheric Environment 63 (2012) 14–21

Contents lists available at SciVerse ScienceDirect

Atmospheric Environment

journal homepage: www.elsevier.com/locate/atmosenv

Mechanism and kinetic studies for OH radical-initiated atmospheric oxidation of methyl propionate

Xiaoyan Sun [a], Yueming Hu [b], Fei Xu [a], Qingzhu Zhang [a,*], Wenxing Wang [a]

[a] Environment Research Institute, Shandong University, PR China
[b] School of Chemistry and Chemical Engineering, Shandong University, Jinan 250100, PR China

HIGHLIGHTS

► Mechanism and kinetics for OH-initiated reaction of methyl propionate were studied.
► Detailed reaction mechanism was proposed.
► The overall rate constant have been obtained.
► The atmospheric lifetime determined by OH radicals is about 15.5 days.

ARTICLE INFO

Article history:
Received 5 July 2012
Received in revised form
18 August 2012
Accepted 21 August 2012

Keywords:
Methyl propionate
OH radicals
Atmospheric oxidation
Reaction mechanism
Kinetic parameters

ABSTRACT

DFT molecular orbital theory calculations were carried out to investigate OH radical-initiated atmospheric oxidation of methyl propionate. Geometry optimizations of the reactants as well as the intermediates, transition states and products were performed at the B3LYP/6-31G(d,p) level. As the electron correlation and basis set effect, the single-point energies were computed by using various levels of theory, including second-order Møller–Plesset perturbation theory (MP2) and the coupled-cluster theory with single and double excitations including perturbative corrections for the triple excitations (CCSD(T)). The detailed oxidation mechanism is presented and discussed. The results indicate that the formation of 3-oxo-methyl propionate ($HC(O)CH_2C(O)OCH_3$) is thermodynamically feasible and the isomerization of alkoxy radical IM17 ($CH_3CH(O)C(O)OCH_3$) can occur readily under the general atmospheric conditions. Canonical variational transition-state (CVT) theory with small curvature tunneling (SCT) contribution was used to predict the rate constants. The overall rate constants were determined, $k(T)(CH_3CH_2COOCH_3 + OH) = (1.35 \times 10^{-12})\exp(-174.19/T)$ cm^3 $molecule^{-1}$ s^{-1}, over the possible atmospheric temperature range of 180–370 K.

1. Introduction

Methyl propionate ($CH_3CH_2COOCH_3$) is used as industrial solvent and synthesis reagent during the manufacture of cellulose derivative, fragrances and flavoring agents (Budavari and O'Neil, 1996; Lewis, 1993; Cavalli et al., 2000). Just in China, 30 thousand tons of methyl propionate was produced in 2010 (China survey, 2011). It can also be detected as a volatile component from some fruits (Snyder, 1992; Bartley and Schwede, 1989; Dirinck et al., 1984) and identified from various waste exudates (Wilkins and Larsen, 1996). Methyl propionate can be released to the atmosphere and exists mainly in the gas phase because of the high vapor pressure (84 mmHg at 298 K) (Bidleman, 1988; Daubert and Danner, 1989). The increasing use of methyl propionate as the replacement for traditional solvents leads to increasing concentration in the atmosphere where it can undergo transport and chemical transformation (Cavalli et al., 2000). To assess the atmospheric behavior of pollutants, it is critical to know their atmospheric reactions.

The atmospheric removal or transformation of gaseous methyl propionate involves wet and dry deposition, photolysis, and oxidation reactions with OH, NO_3 and O_3. Reaction with Cl atoms may also be important in certain locations during certain times of the year (Finlayson-Pitts, 1993; McGillen et al., 2006; Eladio et al., 2003). The wet and dry deposition of gaseous methyl propionate is of relatively minor importance as a removal pathway because of its low Henry's law coefficients (6.1 M atm^{-1}) (Hine and Mookerjee, 1975). Photolysis of esters appears to be insignificant in the lower atmosphere (Calvert and Pitts, 1966). As a sort of saturated ester,

* Corresponding author. Fax: +86 531 8836 1990.
E-mail address: zqz@sdu.edu.cn (Q. Zhang).

1352-2310/$ – see front matter © 2012 Elsevier Ltd. All rights reserved.
http://dx.doi.org/10.1016/j.atmosenv.2012.08.045

methyl propionate is therefore expected to react with ozone hardly. As discussed in the former references (Atkinson, 1991; McGillen et al., 2006; Smith and Plane, 1995), except for typical species include olefins that are C_4 and higher, terpenes, many aromatics, and sulfur-containing alkanes, NO_3 radicals initiated abstraction is a negligible tropospheric loss process compared to OH radicals. Among the various oxidants, OH radicals play the most essential role in determining the oxidation power of the atmosphere. The reaction of methyl propionate with OH radicals is considered to be a dominant removal process. The atmospheric oxidation of methyl propionate may significantly contribute to the formation of second contamination belong to the components of the photochemical smog in urban areas.

Current available database concerning the atmospheric chemistry of methyl propionate is limited. Calve performed the experimental study on OH-initiated reactions of a series of methyl esters using the pulsed laser photolysis-laser induced fluorescence (PLP-RF) technique over the temperature range of 253–372 K (Calve et al., 1997). Arrhenius expression for the reaction of methyl propionate with OH was derived as follows: $k = (1.45 \pm 0.42) \times 10^{-12}$ $\exp - (148 \pm 86)/T$ cm^3 $molecule^{-1}$ s^{-1}, and the rate constant was given as $(0.83 \pm 0.09) \times 10^{-12}$ cm^3 $molecule^{-1}$ s^{-1} at 298 K. Absolute rate coefficients were measured for the reactions of a series of esters with Cl at the room temperature and over the pressure range of 15–60 Torr, using the PLP-RF technique (Alberto et al., 1998). The rate constant for methyl propionate with Cl was determined as $(1.98 \pm 0.26) \times 10^{-11}$ cm^3 $molecule^{-1}$ s^{-1}. Cavalla studied the atmospheric oxidation of methyl propionate by OH and Cl radicals in the presence of NO_x in 740 Torr of air at 296 ± 2 K using smog chamber FTIR techniques (Cavalli et al., 2000). Relative rate techniques were used to measure the rate constants, $k(OH + CH_3CH_2C(O)OCH_3) = (9.29 \pm 1.13) \times 10^{-13}$ cm^3 $molecule^{-1}$ s^{-1}, $k(Cl + CH_3CH_2C(O)OCH_3) = (1.51 \pm 0.22) \times 10^{-11}$ cm^3 $molecule^{-1}$ s^{-1}. Andersen investigated the products for the reaction of methyl propionate with OH radicals (Andersen et al., 2011).

However, there is a notable shortage of direct experimental data associated with the reaction mechanism, largely due to lack of efficient detection schemes for radical intermediate species. Quantum calculation is especially suitable for establishing whether a reaction pathway is feasible or not. In this paper, we have carried out a theoretical study of the application of quantum calculations for the OH-initiated atmospheric oxidation of methyl propionate in order to find out favorable reaction pathways and sites. Elucidation of the reaction mechanism is very challenging due to the inherent complexity. The potential energy surface is useful to explain the experimentally observed degradation products, thermochemical properties, and rate coefficients.

2. Computational method

The electronic structure calculations were performed with the Gaussian 03 software package (Frisch et al., 2003). Geometries of the reactants, intermediates, transition states, and products were optimized with the aid of Becke's three-parameter hybrid method employing the LYP correction function (B3LYP) in conjunction with the split valence polarized basis set 6-31G(d,p). The harmonic vibrational frequencies were also calculated at the same level to determine the nature of the stationary points, the zero-point energy (ZPE), and the thermal contribution to the free energy of activation. The intrinsic reaction coordinate (IRC) analysis was carried out to confirm that each transition state connects to the right minima along the reaction path (Fukui, 1981). The DFT geometries were then used in the single-point energy calculation at the frozen-core second-order Møller–Plesset perturbation theory (MP2) and the coupled-cluster theory with single and double excitations including perturbative corrections for the triple excitations (CCSD(T)) with various basis sets. The single-point energies were further corrected. The procedure involved determination of a correction factor associated with the basis set effect at the MP2 level and subsequent correction to the energy calculated at a higher level of electron correlation with a moderate size basis set. A correction factor, CF, was determined from the energy difference between the MP2/6-31G(d) and MP2/6-311++G(d,p) levels. The values of calculated energies at the CCSD(T)/6-31G(d) level were then corrected by the MP2 level correction factors, corresponding to the CCSD(T)/6-31G(d)+CF level of theory (Lei and Zhang, 2001; McGivern et al., 2000).

Rate constants were estimated by using the canonical variational transition-state (CVT) theory (Baldridge et al., 1989; Gonzalez-Lafont et al., 1991; Garrett and Truhlar, 1979) with the small curvature tunneling (SCT) (Fernandez-Ramos et al., 2007) method. The CVT rate constant for temperature T is given by:

$$k^{CVT}(T) = \min_{s} k^{GT}(T, s) \tag{1}$$

where

$$k^{GT}(T, s) = \frac{\sigma k_B T}{h} \frac{Q^{GT}(T, s)}{\Phi^R(T)} e^{-V_{MEP}(s)/k_B T} \tag{2}$$

where $k^{GT}(T, s)$ is the generalized transition state theory rate constant at the dividing surface s, σ is the symmetry factor accounting for the possibility of more than one symmetry-related reaction path, k_B is Boltzmann's constant, h is Planck's constant, $\Phi^R(T)$ is the reactant partition function per unit volume, excluding symmetry numbers for rotation, and $Q^{GT}(T, s)$ is the partition function of a generalized transition state at s with a local zero of energy at $V_{MEP}(s)$ and with all rotational symmetry numbers set to unity. To include quantum tunneling effects for motion along the reaction coordinate, CVT rate constants were multiplied by a transmission coefficient. In particular, we employed the small curvature tunneling (SCT) method, based on the centrifugal-dominant small-curvature semi-classical adiabatic ground-state approximation, to calculate the transmission coefficient. The rotational partition functions were calculated classically, and the vibrational modes were treated as quantum-mechanically separable harmonic oscillators. The rate constant calculations were carried out with POLYRATE 9.3 program (Steckler et al., 2002).

3. Results and discussion

The reliability of the theoretical calculations was confirmed firstly. The geometries and vibrational frequencies of $CH_3COOCH_2CH_3$ and CH_3CHO were calculated at the B3LYP/6-31G(d,p) level. The results show good consistency with the corresponding experimental values, and the relative deviation remains within 1.0% for the geometrical parameters and 4.0% for the vibrational frequencies (Kuchitsu, 1998; Hollenstien and Gunthard, 1971). For the reaction of $C_4H_{10} + OH \rightarrow C_4H_9 + H_2O$, the reaction enthalpy of -16.15 kcal mol^{-1} obtained at the CCSD(T)/6-31G(d)+CF level and 0 K agrees well with the experimental value of -18.55 kcal mol^{-1} (IUPAC, 2006).

As shown in Fig. 1, there are four geometric conformers resulting from the internal rotation of the O–C_4 and C_1–C_2 bonds for methyl propionate. Structure A is 8.21, 1.28 and 9.40 kcal mol^{-1} more stable than structures B, C and D. So throughout this paper, methyl propionate denotes structure A.

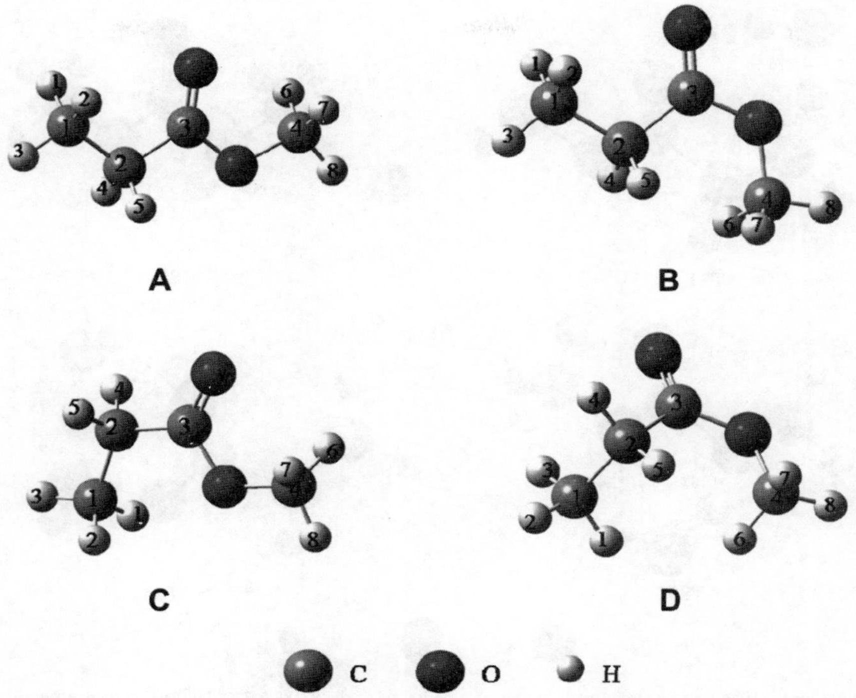

Fig. 1. The four geometric conformers of methyl propionate.

3.1. Reaction mechanism

3.1.1. The reaction with OH

The reaction of methyl propionate with OH proceeds via a direct hydrogen abstraction mechanism. For convenience, the four carbon atoms in methyl propionate are labeled as C_1, C_2, C_3 and C_4. As shown in Fig. 1, structure A of methyl propionate has C_s symmetry. H_1 and H_2 are equivalent. H_4 and H_5 are equivalent. H_6 and H_7 are equivalent. So, there are five kinds of H atoms in the methyl propionate molecule. Therefore, five primary processes were identified: H abstractions from the C_1–H_1, C_1–H_3, C_2–H_4, C_4–H_6 and C_4–H_8 bonds. Since the O–C_4 bond is rotatable in methyl propionate, the H abstraction from the C_4–H_6 and C_4–H_8 bonds occurs via the same transition state. Thus, only four transition states, TS1, TS2, TS3 and TS4, were located. They were identified with one and only one negative eigenvalue of the Hessian matrix and, therefore, one imaginary frequency.

The geometrical structures of the four transition states are shown in Fig. 2. The reaction schemes embedded with the potential barriers and reaction heats are depicted in Fig. 3. The potential barrier for the H abstraction from the C_1–H_3 is about 3 kcal mol^{-1} higher than those of the H abstractions from the C_1–H_2, C_2–H_4 and C_4–H_6 bonds. As shown in Fig. 2, there exists an intramolecular hydrogen bond in the transition states, TS1, TS3 and TS4, respectively. The hydrogen bond can lower the energy of the transition state, i.e., lower the reaction potential barrier. This indicates that the H abstractions from the C_1–H_2, C_2–H_4 and C_4–H_6 bonds can occur more readily than the H abstraction from the C_1–H_3 bond. All of the H abstraction processes are strongly exothermic.

3.1.2. Secondary reactions

The study above shows that the H abstractions from methyl propionate by OH radicals are energetically favorable and can occur readily under the general atmospheric conditions. The products of the H abstractions, IM1, IM2 and IM3, are important radical intermediates produced in the degradation process of methyl propionate initiated by OH radicals. In the atmosphere, they are in the "ocean" of reactive O_2 molecules, which represent 21% of the atmosphere. The conventional view is that these open-shell activated radical intermediates could react further with O_2/NO as their removal from the troposphere. Published work on the products of atmospheric oxidation of methyl propionate in smog chambers supports this point (Cavalli et al., 2000; Andersen et al., 2011).

3.1.2.1. Atmospheric reaction pathway of IM1. The reaction of IM1 with O_2 produces an organic peroxy radical, IM4. To evaluate the nature of the entrance channel for the formation of IM4, we examined the potential along the reaction coordinate, especially to determine whether there is a well-defined transition state or if the process proceeds via a loose transition state without a barrier. The profile of the potential energy surface was scanned by varying the newly formed C_1–O bond length. We found no energy exceeding the C_1–O bond dissociation threshold along the reaction coordinate. This shows that the reaction of IM1 with O_2 proceeds via a barrierless association. The process is strongly exothermic by 63.55 kcal mol^{-1}.

In the troposphere, IM4 will react immediately with ubiquitous NO. The process is exoergic, leading to a vibrationally excited intermediate (denoted IM5), which promptly reacts via unimolecular decomposition. The reaction scheme is described in Fig. 4. Unimolecular decomposition of IM5 results in the formation of NO_2 and an alkoxy radical IM6. A transition state, TS5, was identified as associated with the decomposition. Calculations indicate that the process has an apparent potential barrier of 10.48 kcal mol^{-1} and is endothermic by 11.19 kcal mol^{-1}.

IM6 is open-shell activated intermediate and can further react with NO_2 and O_2 or via the unimolecular decomposition. As seen from Fig. 4, NO_2 not only can undergo addition to the alkoxy radical IM6 without a potential barrier, but also can abstract H atom from IM6. The abstraction has low potential barrier and is strongly

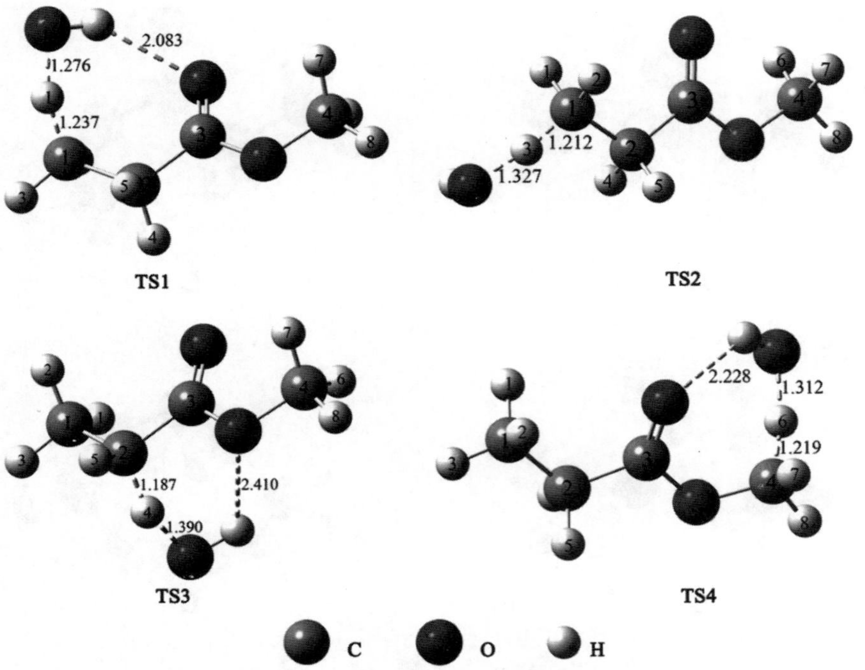

Fig. 2. Transition states for the H abstractions from methyl propionate optimized at the B3LYP/6-31G(d,p) level of theory. Distances are in angstrom.

exothermic. Theoretically, the reaction of IM6 with NO_2 can easily occur in the atmosphere. However, the atmospheric concentration of NO_2 is much lower than that of O_2. The reaction of IM6 with NO_2 is not expected to play an important role for the removal of IM6 from the atmosphere. The reaction of IM6 with O_2 will lead to the formation of 3-oxo-methyl propionate. In the previous studies (Cavalli et al., 2000; Andersen et al., 2011), it is uncertain whether the process can occur because of the absence of a reference spectrum for 3-oxo-methyl propionate. Fig. 4 shows that the reaction of IM6 with O_2 is strongly exothermic by 30.89 kcal mol^{-1} without

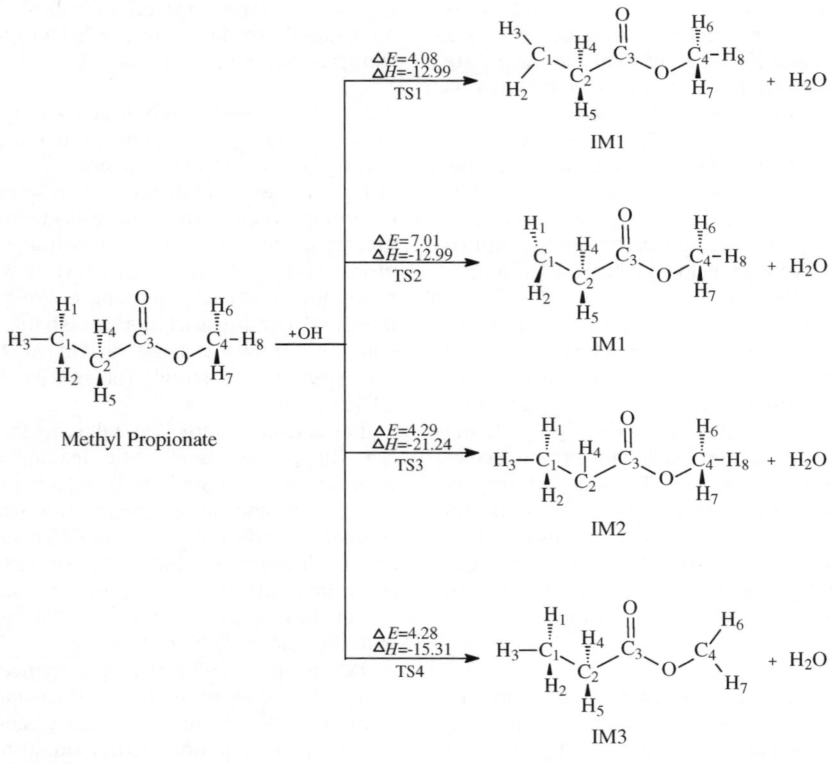

Fig. 3. H abstraction schemes embedded with the potential barriers ΔE (in kcal mol^{-1}) and reaction heats ΔH (in kcal mol^{-1}, 0 K).

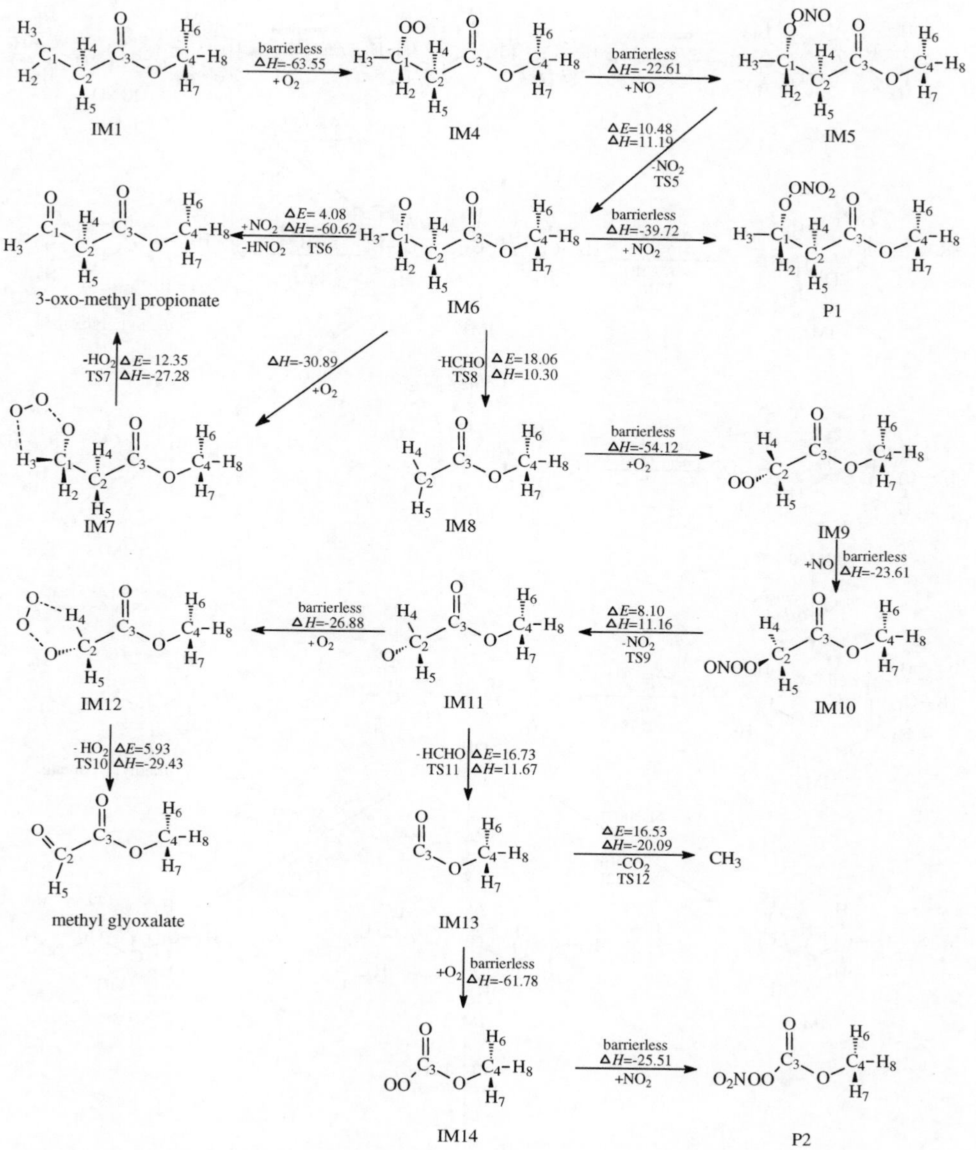

Fig. 4. Secondary reaction schemes embedded with the potential barriers ΔE (in kcal mol^{-1}) and reaction heats ΔH (in kcal mol^{-1}, 0 K) from IM1.

a barrier to form IM7. 3-oxo-methyl propionate can be produced from IM7 via the transition state TS7, with the energy barrier of 12.35 kcal mol^{-1}. The process is strongly exothermic. The formation of 3-oxo-methyl propionate is energetically feasible under the general atmospheric conditions.

Unimolecular decomposition of IM6 leads to the formation of formaldehyde and IM8. In the troposphere, IM8 can further react with O_2/NO to form IM11. Then, methyl glyoxalate and HO_2 are produced from the reaction of IM11 with O_2. The formation of methyl glyoxalate is supported by the previous studies in the experiment (Cavalli et al., 2000; Andersen et al., 2011). The subsequent reaction of IM11 can occur via unimolecular decomposition to form IM13 and HCHO. IM13 is unstable and can further react via the reaction with O_2/NO_2 or the unimolecular decomposition. The decomposition of IM13 results in the formation of CO_2 and a methyl radical. Fig. 4 shows that the reaction of IM13 with O_2/NO_2 is favored over the decomposition of IM13. The calculated result is supported by the experimental observation that the CO_2 formation was found to be insignificant (Cavalli et al., 2000).

3.1.2.2. Atmospheric reaction pathway of IM2. Possible reaction routes from IM2 are displayed in Fig. 5. IM17 is formed from the reaction of IM2 with O_2/NO. As shown in Fig. 5, four reaction channels were proposed from IM17. The first one is the unimolecular decomposition via the C_1–C_2 bond cleavage to produce methyl glyoxylate and methyl radical. The barrier height is calculated to be 8.90 kcal mol^{-1}. The process is endothermic by 5.26 kcal mol^{-1}. The second channel is the decomposition of IM17 via the C_2–C_3 bond cleavage to produce IM13 and CH_3CHO. Comparison of the two decomposition channels shows that the second decomposition is favored over the first one, consistent with the result in the previous study by Cavalli et al. (2000). The third channel is the reaction of IM17 with O_2, followed by the loss of HO_2, to yield methyl pyruvate. The fourth channel is the isomerization of IM17 via 1,5-H shift. Due to lack of the infrared spectrum of the product, it is uncertain whether the channel can occur under the general atmospheric conditions (Cavalli et al., 2000; Andersen et al., 2011). 1,5-H shift isomerization is found to proceed via a six-membered transition state, TS17, with an energy barrier of

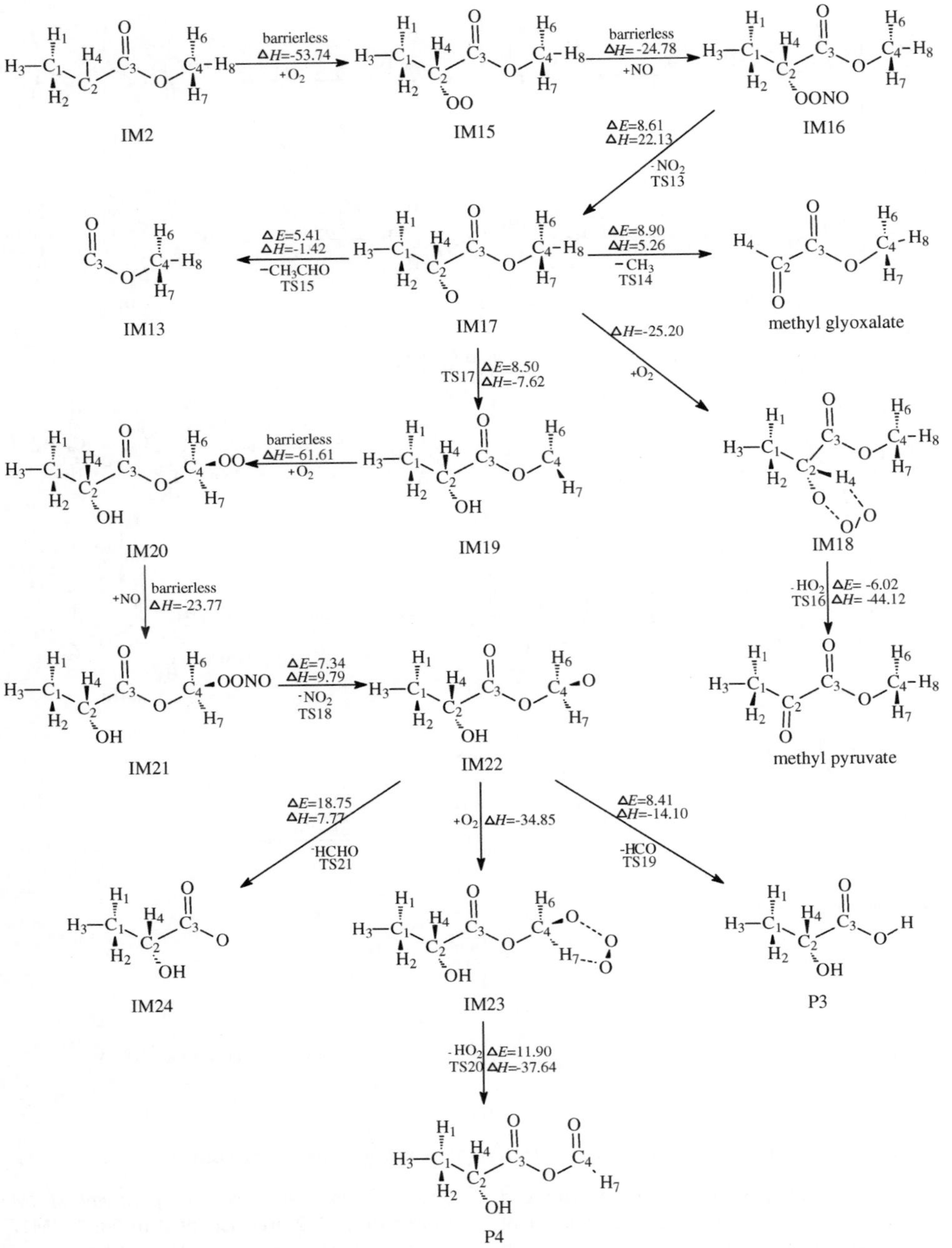

Fig. 5. Secondary reaction schemes embedded with the potential barriers ΔE (in kcal mol^{-1}) and reaction heats ΔH (in kcal mol^{-1}, 0 K) from IM2.

8.50 kcal mol^{-1} to produce IM19. The reaction is exothermic by 7.62 kcal mol^{-1}. So, this study shows that the isomerization of IM17 is energetically feasible.

IM19 can further react with O_2/NO to form IM22. Three reaction channels were identified from IM22. The first channel occurs via intramolecular H migration and simultaneously C_4—O bond cleavage to yield 2-hydroxypropanoic (P3) and HCO. The barrier height is 8.41 kcal mol^{-1}. The process is exothermic by 14.10 kcal mol^{-1}. The second channel is the reaction with O_2, followed by the loss of HO_2, to produce P4. The third channel is the decomposition of IM22 via the loss of HCHO. Calculations show that the third decomposition channel is energetically unfavorable due to the high barrier and strong endothermicity.

3.1.2.3. Atmospheric reaction pathway of IM3. Similar to IM1 and IM2, IM3 can react with O_2/NO to from IM26. Then, the energy-rich IM26 decomposes to yield IM27 via the loss of NO_2. The reaction schemes embedded with the potential barriers and reaction heats are depicted in Fig. 6. Three reaction channels were determined from IM27. The first channel is the reaction with O_2, followed by the loss of HO_2, to yield propionic formic anhydride. The second channel proceeds via an α ester-rearrangement and simultaneously C_4—O bond decomposition to form propionic acid and formyl radical HCO. The potential barrier is 16.53 kcal mol^{-1}. The third channel is the decomposition of IM27 via the loss of HCHO. It can be seen from Fig. 6 that the channel is energetically unfavorable due to the high barrier and strong endothermicity, consistent with the

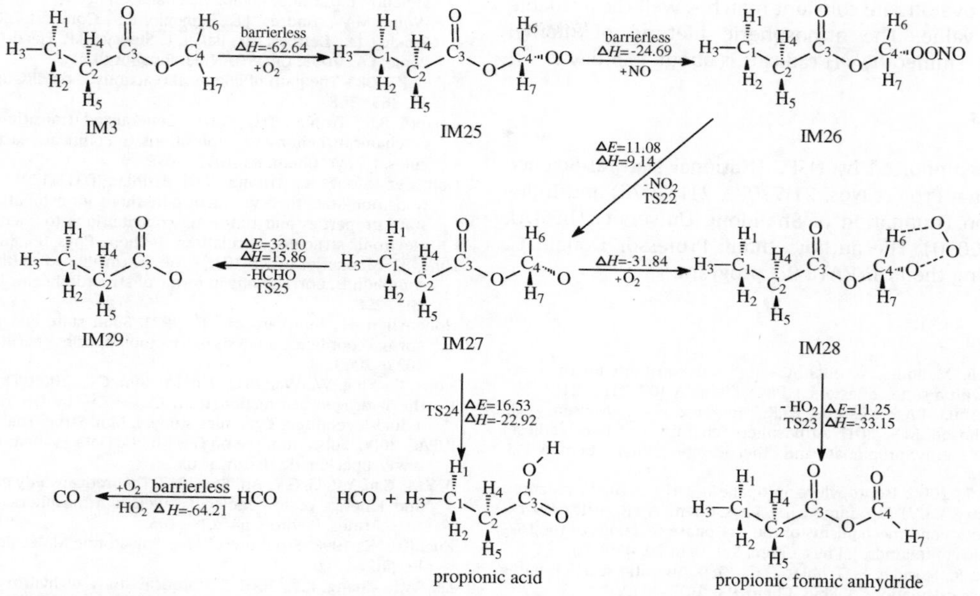

Fig. 6. Secondary reaction schemes embedded with the potential barriers ΔE (in kcal mol^{-1}) and reaction heats ΔH (in kcal mol^{-1}, 0 K) from IM3.

experimental result (Cavalli et al., 2000; Andersen et al., 2011) that the decomposition of the alkoxy radical IM27 via C—O bond scission can not occur.

3.2. Rate constant calculations

On the basis of the CCSD(T)/6-31G(d) + CF//B3LYP/6-31G(d,p) energies, the individual and overall rate constants for the H abstractions from methyl propionate were calculated with the aid of canonical variational transition-state rate calculations augmented by the small curvature tunneling corrections (CVT/SCT) over the temperature range from 180 to 370 K. Actually, the CVT/SCT method has been successfully used to deal with many bimolecular reactions (Hong et al., 2009; Cao et al., 2011; Ji et al., 2012; Xu et al., 2010; Annia and Armando, 2006). The individual rate constants for the H abstractions from the C_1-H_1, C_1-H_3, C_2-H_4 and C_4-H_6 bonds are noted as k_1, k_2, k_3, and k_4, respectively. The overall rate constants for the reaction of methyl propionate with OH radicals are noted as k, $k = 2k_1 + k_2 + 2k_3 + 3k_4$.

The calculated CVT/SCT overall rate constants are fitted over the temperature range of 180–370 K, and Arrhenius formula is given in units of cm^3 molecule^{-1} s^{-1}: $k(T)(CH_3CH_2COOCH_3 + OH) = (1.35 \times 10^{-12})\exp(-174.19/T)$. At 298 K, the calculated CVT/SCT overall rate constant for the reactions of methyl propionate with OH radicals is 7.49×10^{-13} cm^3 molecule^{-1} s^{-1} and agrees with the available experimental values excellently (Calve et al., 1997; Cavalli et al., 2000; Andersen et al., 2011). From the good agreement, it might be inferred that the CVT/SCT individual rate constants are reasonable. Arrhenius formulas for the H abstractions from the C_1-H_1, C_1-H_3, C_2-H_4 and C_4-H_6 bonds over the temperature range of 180–370 K are as follows (in cm^3 molecule^{-1} s^{-1}):

$$k_1(T) = \left(2.19 \times 10^{-13}\right)\exp(-153.38/T)$$

$$k_2(T) = \left(3.52 \times 10^{-14}\right)\exp(-363.16/T)$$

$$k_3(T) = \left(3.34 \times 10^{-13}\right)\exp(-177.67/T)$$

$$k_4(T) = \left(0.79 \times 10^{-13}\right)\exp(-222.02/T)$$

Due to having the highest potential barrier, the rate constant of the H abstraction from the C_1-H_3 bond is so small that it can be negligible.

According to the rate constant of the reaction of methyl propionate with OH radicals and a typical OH concentration (C_{OH}) of 9.4×10^5 molecule cm^{-3} (Prinn et al., 2001), using the expression:

$$\tau_{OH} = \frac{1}{k_{(OH+CH_3CH_2COOCH_3)} \times c_{OH}}$$

The atmospheric lifetime of methyl propionate determined by OH radicals is about 15.5 days, consistent with the experimental values (14 and 16 days) (Calve et al., 1997; Cavalli et al., 2000).

4. Conclusions

A theoretical study is presented on the reaction mechanism of OH radical-initiated atmospheric oxidation of methyl propionate. The rate constants were calculated by using the CVT/SCT method. Several specific conclusions can be drawn from this study:

(1) The reaction of methyl propionate with OH radicals is the prototype of simple metathesis reaction in which a hydrogen atom is transferred via an apparent barrier in the reaction coordinate. The H abstractions from the C_1-H_1, C_2-H_4 and C_4-H_6 are the dominant reaction pathways. The H abstraction from the C_1-H_3 bond can be negligible because of the high barrier.
(2) The products of the H abstractions, IM1, IM2 and IM3, are open-shell activated radical intermediates and could react further with O_2/NO as their removal from the troposphere. Several secondary pollutants can be formed. The formation of 3-oxo-methyl propionate ($HC(O)CH_2C(O)OCH_3$) is proved to be feasible and isomerization of alkoxy radical IM17 ($CH_3CH(O)C(O)OCH_3$) can occur readily under the general atmospheric conditions.

(3) The calculated overall rate constant matches well the available experimental value. The atmospheric lifetime of methyl propionate determined by OH radicals is about 15.5 days.

Acknowledgments

This work was supported by NSFC (National Natural Science Foundation of China, Project Nos. 21177077, 21177076) and Independent Innovation Foundation of Shandong University (IIFSDU, project No. 2012JC030). The authors thank Professor Donald G. Truhlar for providing the POLYRATE 9.3 program.

References

Alberto, N., Georges, L.B., Mellouki, A., 1998. Absolute rate constants for the reactions of Cl atoms with a series of esters. J. Phys. Chem. A 102, 3112–3117.

Andersen, V.F., Berhanu, T.A., Nilsson, E.J.K., Jørgensen, S., Nielsen, O.J., Wallington, T.J., Johnson, M.S., 2011. Atmospheric chemistry of two biodiesel model compounds: methyl propionate and ethyl acetate. J. Phys. Chem. A 115, 8906–8919.

Annia, G., Armando, C.T., 2006. Isopropylcyclopropane + OH gas phase reaction: a quantum chemistry + CVT/SCT approach. J. Phys. Chem. A 110, 1917–1924.

Atkinson, R., 1991. Kinetics and mechanisms of the gas-phase reactions of the NO_3 radical with organic compounds. J. Phys. Chem. Ref. Data 20, 459–507.

Baldridge, M.S., Gordor, R., Steckler, R., Truhlar, D.G., 1989. Ab initio reaction paths and direct dynamics calculations. J. Phys. Chem. 93, 5107–5119.

Bartley, J.P., Schwede, A.M., 1989. Production of volatile compounds in ripening kiwi fruit (*Actinidia chinensis*). J. Agric. Food Chem. 37, 1023–1025.

Bidleman, T.F., 1988. Atmospheric processes. Environ. Sci. Technol. 22, 361–367.

Budavari, S., O'Neil, M.J., 1996. An Encyclopedia of Chemicals, Drugs, and Biologicals. Merck and Co, New Jersey.

Calve, S.L., Bras, G.L., Mellouki, A., 1997. Kinetic studies of OH reactions with a series of methyl esters. J. Phys. Chem. A 101, 9137–9141.

Calvert, J.G., Pitts, J.N., 1966. Photochemistry. Wiley, New York.

Cao, H.J., He, M.X., Han, D.D., Sun, Y.H., Xie, J., 2011. Theoretical study on the mechanism and kinetics of the reaction of 2, 2′, 4, 4′-tetrabrominated diphenyl ether (BDE-47) with OH radicals. Atmos. Environ. 45, 1525–1531.

Cavalli, F., Barnes, I., Becker, K.H., 2000. Atmospheric oxidation mechanism of methyl propionate. J. Phys. Chem. A 104, 11310–11317.

Daubert, T.E., Danner, R.P., 1989. Physical and Thermodynamic Properties of Pure Chemicals: Data Compilation. Hemisphere Pub Corp, NewYork.

Dirinck, P., et al., 1984. Analysis of Volatiles. Walter Degruyter & Co., Berlin.

Eladio, M., Knipping, Donald, D., 2003. Impact of chlorine emissions from sea-salt aerosol on coastal urban ozone. Environ. Sci. Technol. 37, 184–275.

Fernandez-Ramos, A., Ellingson, B.A., Garret, B.C., Truhlar, D.G., 2007. Variational transition state theory with multidimensional tunneling. In: Lipkowitz, K.B., Cundari, T.R. (Eds.), Reviews in Computational Chemistry. Wiley-VCH, Hoboken, NJ.

Finlayson-Pitts, B.J., 1993. Indications of photochemical histories of pacific air masses from measurements of atmospheric trace species at point arena, California. J. Geophys. Res. 98, 14991–14993.

Frisch, M.J., Trucks, G.W., Schlegel, H.B., Gill, P.W.M., Johnson, B.G., Robb, M.A., Cheeseman, J.R., Keith, T.A., Petersson, G.A., Montgomery, J.A., Raghavachari, K., Allaham, M.A., Zakrzewski, V.G., Ortiz, J.V., Foresman, J.B., Cioslowski, J., Stefanov, B.B., Nanayakkara, A., Challacombe, M., Peng, C.Y., Ayala, P.Y., Chen, W., Wong, M.W., Andres, J.L., Replogle, E.S., Gomperts, R., Martin, R.L., Fox, D.J., Binkley, J.S., Defrees, D.J., Baker, J., Stewart, J.P., Head-Gordon, M., Gonzales, C., Pople, J.A., 2003. GAUSSIAN 03. Pittsburgh, PA.

Fukui, K., 1981. The path of chemical reactions – the IRC approach. Acc. Chem. Res. 14, 363–368.

Garrett, B.C., Truhlar, D.G., 1979. Generalized transition state theory. Classical mechanical theory and applications to collinear reactions of hydrogen molecules. J. Phys. Chem. 83, 1052–1079.

Gonzalez-Lafont, A., Truong, T.N., Truhlar, D.G., 1991. Interpolated variational transition-state theory: practical methods for estimating variational transition-state properties and tunneling contributions to chemical reaction rates from electronic structure calculations. J. Chem. Phys. 95, 8875–8894.

Hine, J., Mookerjee, P.K., 1975. The intrinsic hydrophilic character of organic compounds. Correlations in terms of structural contributions. J. Org. Chem. 40, 292–298.

Hollenstien, H., Gunthard, Hs. H., 1971. Solid state and gas infrared spectra and normal coordinate analysis of 5 isotopic species of acetaldehyde. Spec. Acta 27A, 2027–2060.

Hong, G., Ying, W., Wan, S.Q., Liu, J.Y., Sun, C.C., 2009. Theoretical investigation of the hydrogen abstraction from $CF_3CH_2CF_3$ by OH radicals, F, and Cl atoms: a dual-level direct dynamics study. J. Mol. Struc. Theochem. 913, 107–116.

IUPAC, 2006. Subcommittee on Gas Kinetic Data Evaluation. Available from: http://www.iupackinetic.ch.cam.ac.uk.

Ji, Y.M., Gao, Y.P., Li, G.Y., An, T.C., 2012. Theoretical study of the reaction mechanism and kinetics of low-molecular-weight atmospheric aldehydes (C_1–C_4) with NO_2. Atmos. Environ. 54, 288–295.

Kuchitsu, K., 1998. Structure of Free Polyatomic Molecules – Basic Data. Springer, Berlin.

Lei, W.F., Zhang, R.Y., 2001. Theoretical study of hydroxyisoprene alkoxy radicals and their decomposition pathways. J. Phys. Chem. A 105, 3808–3815.

Lewis Sr., R.J., 1993. Hawley's Condensed Chemical Dictionary. Van Nostrand Rheinhold Co., New York. Chemical Watch Fact Sheet: Methyl Propionate. Available from: www.chinadiaoyan.com.

McGillen, M.R., Percival, C.J., Duran, T.R., Reyna, G.S., Shallcross, D.E., 2006. Can topological indices be used to predict gas-phase rate coefficients of importance to tropospheric chemistry? Free radical abstraction reactions of alkanes. Atmos. Environ. 40, 2488–2500.

McGivern, W.S., Suh, I., Clinkenbeard, A.D., Zhang, R.Y., Simon, W.N., 2000. Experimental and computational study of the OH-Isoprene reaction: isomeric branching and low-pressure behavior. J. Phys. Chem. A 104, 6609–6616.

Prinn, R.G., Huang, J., Weiss, R.F., Cunnold, D.M., Fraser, P.J., Simmonds, P.G., McCulloch, A., Harth, C., Salameh, P., Doherty, S.O., Wang, R.H.J., Proter, L., Miller, B.R., 2001. Evidence for substantial variations of atmospheric hydroxyl radicals in the past two decades. Science 292, 1882–1887.

Smith, N., Plane, J.M.C., 1995. Nighttime radical chemistry in the San Joaquin valley. Atmos. Environ. 29, 2887–2897.

Snyder, R., 1992. Ethel Browning's Toxicity and Metabolism of Industrial Solvents. In: Alcohols and Esters, vol. 3. Elsevier, New York.

Steckler, R., Chuang, Y.Y., Fast, P.L., Corchade, J.C., Coitino, E.L., Hu, W.P., Lynch, G.C., Nguyen, K., Jackells, C.F., Gu, M.Z., Rossi, I., Clayton, S., Melissas, V., Garrett, B.C., Isaacson, A.D., Truhlar, D.G., 2002. POLYRATE Version 9.3. University of Minnesota, Minneapolis.

Wilkins, K., Larsen, K., 1996. Volatile organic compounds from garden waste. Chemosphere 32, 2049–2055.

Xu, F., Wang, H., Zhang, Q.Z., Zhang, R.X., Qu, X.H., Wang, W.X., 2010. Kinetic properties for the complete series reactions of chlorophenols with OH radicals-relevance for dioxin formation. Environ. Sci. Technol. 44, 1399–1404.

Science of the Total Environment 490 (2014) 639–646

Contents lists available at ScienceDirect

Science of the Total Environment

journal homepage: www.elsevier.com/locate/scitotenv

Mechanistic and kinetic studies on the OH-initiated atmospheric oxidation of fluoranthene

Juan Dang, Xiangli Shi, Qingzhu Zhang *, Jingtian Hu, Jianmin Chen, Wenxing Wang

Environment Research Institute, Shandong University, Jinan 250100, PR China

HIGHLIGHTS

- We studied a comprehensive mechanism of OH-initiated oxidation of fluoranthene.
- We reported the formation pathways of fluoranthone, fluoranthenequinone and epoxide.
- The rate constants of the crucial elementary steps were evaluated.

ARTICLE INFO

Article history:
Received 2 April 2014
Accepted 30 April 2014
Available online 2 June 2014

Editor: Pavlos Kassomenos

Keywords:
Fluoranthene
OH radicals
Oxidation mechanism
Oxidation products
Rate constants

ABSTRACT

The atmospheric oxidation of polycyclic aromatic hydrocarbons (PAHs) can generate toxic derivatives which contribute to the carcinogenic potential of particulate organic matter. In this work, the mechanism of the OH-initiated atmospheric oxidation of fluoranthene (Flu) was investigated by using high-accuracy molecular orbital calculations. All of the possible oxidation pathways were discussed, and the theoretical results were compared with the available experimental observation. The rate constants of the crucial elementary reactions were evaluated by the Rice–Ramsperger–Kassel–Marcus (RRKM) theory. The main oxidation products are a range of ring-retaining and ring-opening chemicals containing fluoranthols, fluoranthones, fluoranthenequinones, nitro-fluoranthenes, dialdehydes and epoxides. The overall rate constant of the OH addition reaction is 1.72×10^{-11} cm^3 $molecule^{-1}$ s^{-1} at 298 K and 1 atm. The atmospheric lifetime of Flu determined by OH radicals is about 0.69 days. This work provides a comprehensive investigation of the OH-initiated oxidation of Flu and should help to clarify its atmospheric conversion.

1. Introduction

Polycyclic aromatic hydrocarbons (PAHs), an important fraction of semi-volatile organic compounds (SVOCs), are formed as byproducts of any incomplete combustion from traffic exhausts, industrial activities, domestic heating, forest fires and biomass burnings (Chen et al., 2013; Christie et al., 2012; Seinfeld and Pandis, 2006). The atmospheric emission of 16 priority PAHs in all Asian countries accounted for 53.5% of the total global emissions (504 Gg), with the highest emission from China (106 Gg) and India (67 Gg) during 2007 (Shen et al., 2013). PAHs are hydrophobic, stable, and sparingly soluble in water (Chaudhry, 1994). They can be metabolized to reactive electrophilic intermediates that can form DNA adducts, which may induce mutations and ultimately tumors (Ramírez et al., 2011). Due to the potential mutagenicity and carcinogenicity, their ubiquitous presence in air, water, soil and vegetation is of major concern (Boström et al., 2002; Xue and Warshawsky, 2004). The European Union has set a target value of 1 ng m^{-3} of benzo[a]pyrene used as a main indicator of carcinogenic PAHs (EU, 2005).

Fluoranthene (Flu), a member of non-alternant PAHs, is one of the most abundant PAHs in the environment (Monte et al., 2012; Wetzel et al., 1994). Because of its high concentration in ambient air and potential carcinogenicity, it is suggested as a complementary indicator to benzo[a]pyrene (Boström et al., 2002). During spring of 1994–2000, the measurement in Stockholm has shown ambient air concentrations of Flu ranging from 8 to 25 ng m^{-3} (Boström et al., 2002). Air samples collected from eight locations in the Laurentian Great Lakes region revealed that the maximum concentration of Flu is up to 9350 pg m^{-3} (Galarneau et al., 2006). Atmospheric monitoring in Eordea Basin, west Macedonia, Greece and Kurashiki City has also detected Flu and other PAHs, the gas phase concentration of Flu scattered over the range of 0.623–30.3 ng m^{-3} (Terzi and Samara, 2004). On account of its prevalent presence in air, it is critical to understand the fate of gaseous Flu. In general, the tropospheric removal of Flu involves wet and dry deposition, and oxidation reactions with OH, NO_3 and O_3. The

* Corresponding author. Fax: +86 531 8836 1990.
E-mail address: zqz@sdu.edu.cn (Q. Zhang).

http://dx.doi.org/10.1016/j.scitotenv.2014.04.134
0048-9697/

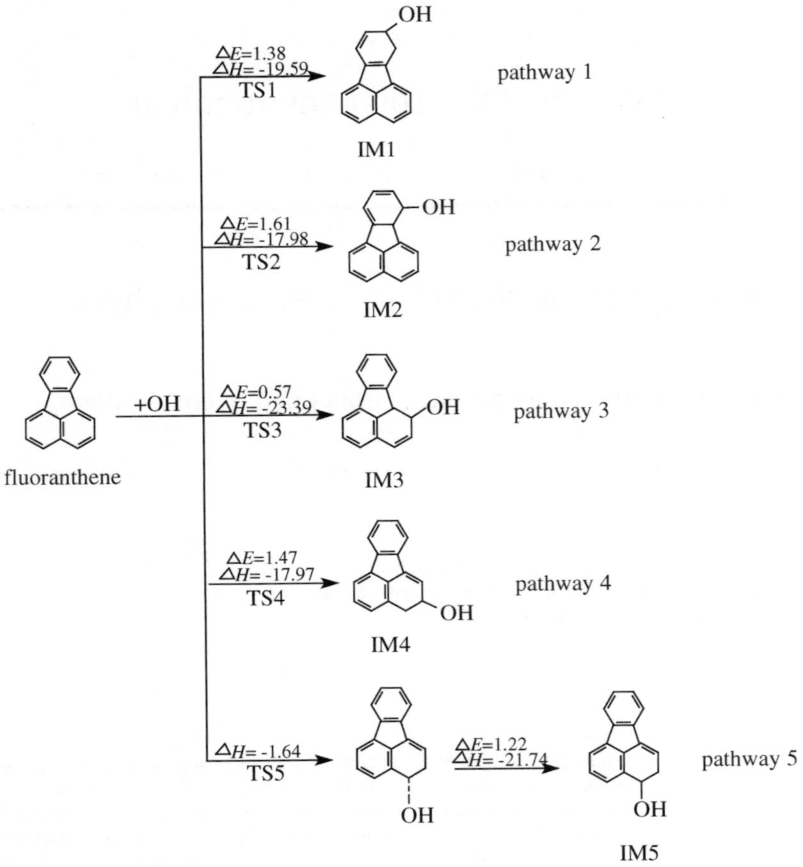

Fig. 1. The OH addition reaction scheme of Flu embedded with the potential barrier ΔE (in kcal/mol) and reaction heat ΔH (in kcal/mol). ΔH is calculated at 0 K.

wet and dry deposition of gaseous Flu is relatively insignificance as a removal pathway. Among the various oxidants, OH radicals play the most essential role in determining the oxidation power of the atmosphere (Keyte et al., 2013). The reaction with OH radicals is considered to be a dominant removal process for gaseous Flu. However, there exists a notable absence of direct experimental data concerning the reaction mechanism, largely due to the lack of efficient detection schemes for intermediate radical species. In this paper, using quantum mechanics and the RRKM theory, we carried out a theoretical study on the OH-initiated atmospheric oxidation reaction of Flu in the presence of O_2/NO and HO_2, which is helpful to clarify the atmospheric fate of Flu.

2. Computational method

The electronic structure calculations were performed with the Gaussian 03 software package (Frisch et al., 2003). Geometries of the reactants, intermediates, transition states, and products were optimized at the BB1K/6-31 + G(d,p) level, which has yielded satisfying results in the previous research (Qu et al., 2006). The harmonic frequency calculations were also performed at the same level in order to determine the nature of the stationary points, the zero-point energy (ZPE), and the thermal contribution to the free energy of activation. Besides, the intrinsic reaction coordinate (IRC) analysis was carried out to confirm that each transition state connects to the right minima along the reaction path. For a more accurate evaluation of the energetic parameters, a more flexible basis set 6-311 + G(3df,2p), was employed to determine the single point energies of various species.

By means of the MESMER program (Glowacki et al., 2012), the rate constants of the crucial elementary reactions were deduced by using Rice–Ramsperger–Kassel–Marcus (RRKM) theory (Robinson and Holbrook, 1972). The RRKM rate constant is given by:

$$k(E) = \frac{W(E)}{h\rho(E)} \qquad (1)$$

where, $W(E)$ is the rovibrational sum of states at the transition state, $\rho(E)$ is the density of states of reactants, and h is Planck's constant. Then, canonical rate constant $k(T)$ is determined by using the usual equation:

$$k(T) = \frac{1}{Q(T)} \int k(E)\rho(E) \exp(-\beta E) dE \qquad (2)$$

where, Q(T) is the reactant partition function.

3. Results and discussion

Due to the lack of experimental information on the thermochemical parameters for the present reaction system, it is difficult to compare the calculated results with experimental data directly. To verify the reliability of the computational results, we optimized the geometries and calculated the vibrational frequencies of benzene, phenol and naphthalene. The results agree well with the available experimental values, and the maximum relative errors are less than 3.0% for geometrical parameters and less than 7.2% for vibrational frequencies

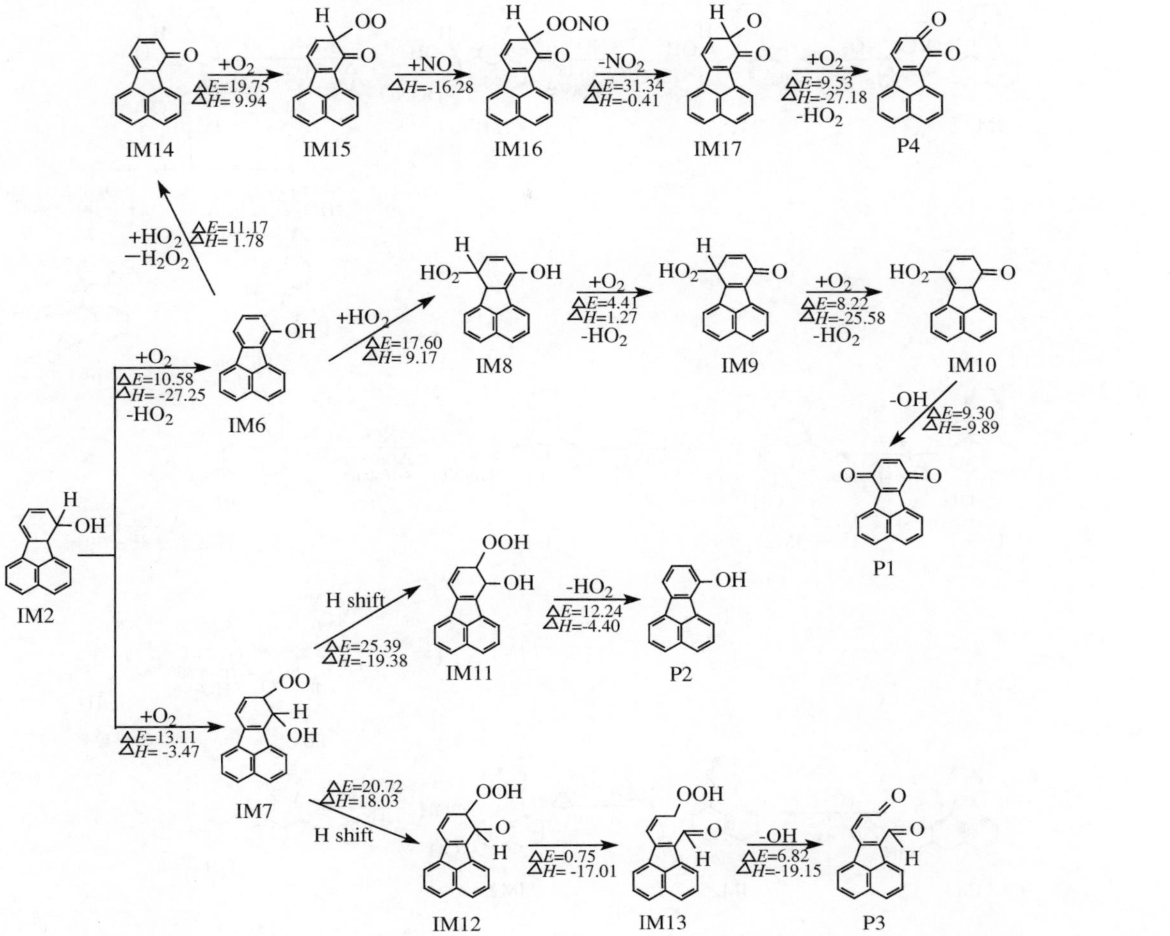

Fig. 2. Secondary reaction scheme from the OH-Flu adduct, IM2, embedded with the potential barrier ΔE (in kcal/mol) and reaction heat ΔH (in kcal/mol) in the presence of O_2 and NO. ΔH is calculated at 0 K.

(Herzberg, 1966; Shimanouchi, 1972; Hellwege and Hellwege, 1976; Martin et al., 1996).

3.1. OH additions

Previous experimental studies suggested that the addition of OH to C=C in PAHs plays a dominant role compared with H abstractions for anthracene, naphthalene and monocyclic aromatic compounds under the general atmospheric conditions (Ananthula et al., 2006; Lorenz and Zellner, 1983; Perry et al., 1977). Thus, only the addition reaction of Flu with OH was discussed in this paper. Analysis of the molecular structure of Flu shows that there are five different kinds of C atoms, leading to five primary processes of the OH addition. The OH addition reaction schemes embedded with the potential barriers ($\triangle E$) and reaction heats ($\triangle H$, 0 K) are presented in Fig. 1. The geometrical structures of the transition states are shown in Fig. 5.

All of the OH addition pathways are highly exothermic with low energy barriers, which indicate that they can occur readily under the general atmospheric conditions. As shown in Fig. 1, for pathway 5, a van der Waals complex is formed firstly, whereas other pathways proceed via direct OH addition mechanism. At the BB1K/6-311 + G(3df,2p) level, the potential barriers are 0.57 ~ 1.61 kcal/mol, and the reaction heats are from −23.39 to −17.97 kcal/mol. The resulting adducts, OH-Flu, from these OH addition reactions will further react with O_2/NO or HO_2 as their removal.

3.2. Secondary reactions

The discussion above shows that the OH addition to Flu is energetically favorable reaction pathways under the general atmospheric conditions. The five OH-Flu adducts (IM1, IM2, IM3, IM4 and IM5) are important intermediates. The conventional view is that these open-shell activated radical intermediates could be further oxidized by ubiquitous O_2/NO or HO_2. Several important secondary pollutants are produced from the secondary reactions with O_2/NO or HO_2. All of the possible oxidation pathways were discussed in this paper. Besides, the OH-Flu adducts can react with NO_2 to form OH–NO_2-Flu adducts via barrierless associations. The OH–NO_2-Flu adducts may subsequently undergo unimolecular decomposition to yield nitro-fluoranthenes through the direct loss of water. Our research shows that water plays an important role in the formation of nitro-fluoranthenes. This study has been published in Environ. Sci. Technol (Zhang et al., 2014).

3.2.1. O_2 abstraction channels of OH-Flu adducts

As shown in Figs. 2 and S1 (Supporting Information), the OH-Flu adducts are unstable energy-rich radicals which can readily react with molecular oxygen to form fluoranthols. The potential barriers of H abstraction from the C−H bonds are 12.49 ~ 9.61 kcal/mol at the BB1K/6-311 + G(3df,2p) level. The transition states lie at 5.48 ~ 13.77 kcal/mol below the sum energy of Flu + OH + O_2. The processes are strongly exothermic by 21.09 ~ 27.25 kcal/mol. The overall

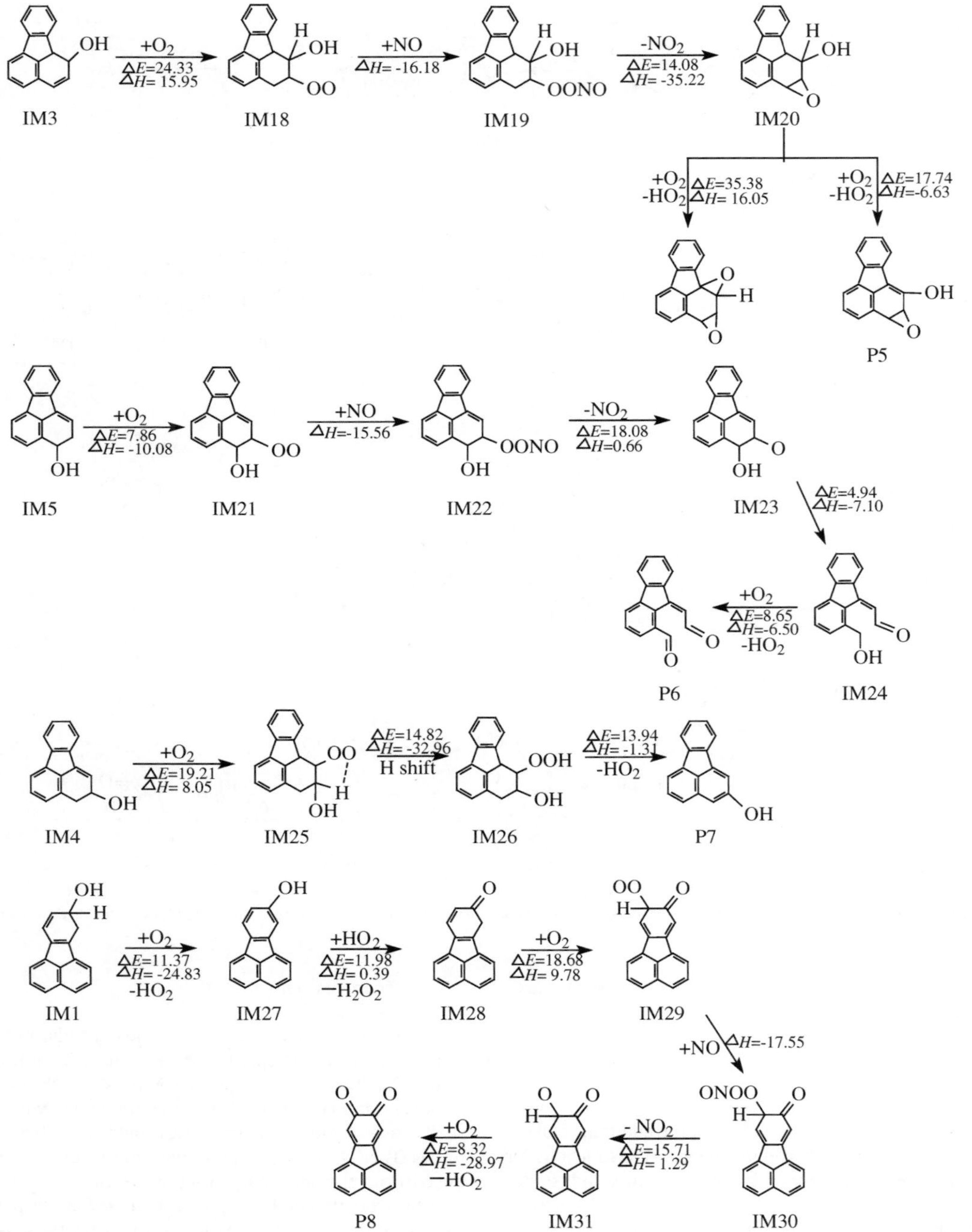

Fig. 3. Secondary reaction scheme from the OH-Flu adducts (IM1, IM3, IM4, IM5), embedded with the potential barrier ΔE (in kcal/mol) and reaction heat ΔH (in kcal/mol) in the presence of O_2 and NO.

reactions, Flu + OH + O_2 → fluoranthol + HO_2, are strongly exothermic by 44.42 ~ 45.63 kcal/mol. This shows that these H abstraction reactions can easily occur. For the intermediate fluoranthen-7-ol generated from the H abstraction reaction of IM2 and O_2, it can further react with HO_2, followed by twice H abstraction by O_2. The calculated profiles of the potential energy surface show that these steps are exoergic and low-barrier reactions. Subsequent reaction of the resulting radical IM10 is the unimolecular decomposition via the cleavage of O–OH bond to form fluoranthenequinone. Calculations indicate that the process has a low potential barrier of 9.30 kcal/mol and is exothermic by 9.89 kcal/mol. Similar products have been detected in the laboratory experiments of naphthalene and OH radical (Kautzman et al., 2009).

3.2.2. O_2 addition channels of OH-Flu adducts

3.2.2.1. The formation of dialdehyde and epoxide. The formation schemes of dialdehyde from OH-Flu and O_2 are displayed in Figs. 2 and S2 (Supporting Information). For IM2, O_2 add to the carbon from the anti-position of hydroxyl, followed by H atom shift via a six-membered ring transition state. The barrier height of H shift is

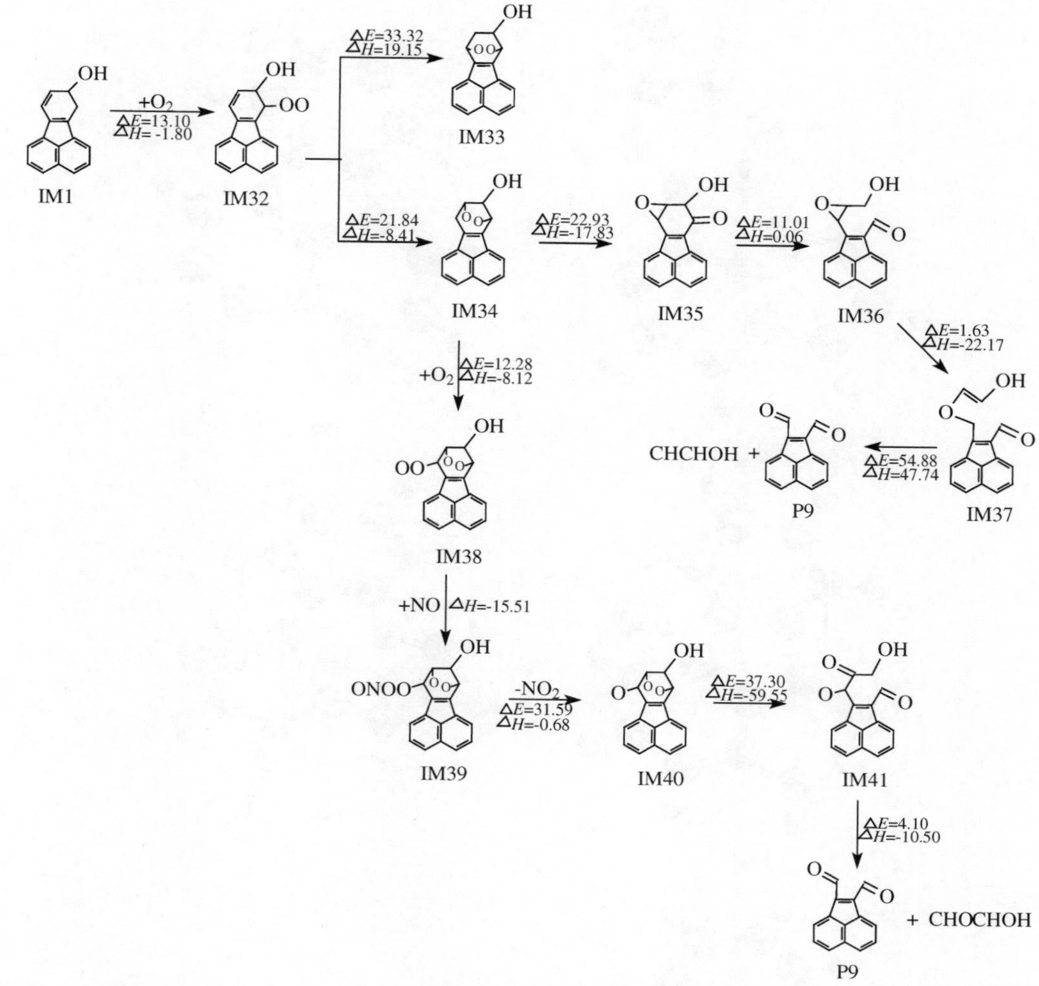

Fig. 4. The reaction scheme of the bicyclic peroxy radical embedded with the potential barrier ΔE (in kcal/mol) and reaction heat ΔH (in kcal/mol) in the presence of O_2 and NO. ΔH is calculated at 0 K.

calculated to be 21.48 kcal/mol and the process is endothermic by 19.00 kcal/mol. The resulting intermediate is unstable and easy to proceed with a ring-opening reaction. The process has a potential barrier of 0.78 kcal/mol and is strongly exothermic by 17.64 kcal/mol. Through the cleavage of O – OH bond, it ultimately results in OH and dialdehyde. Similar products have been observed in the reaction chamber experiments of gaseous phenanthrene and OH radical (Lee and Lane, 2010). Other OH-Flu adducts (IM1, IM3, IM4 and IM5) can also react with O_2 to form dialdehydes via similar mechanisms of IM1. In particular, O_2 can add to IM4 from two positions, which are on the left and right side of the carbon linked with the hydroxyl. As shown in Figs. 2 and S2, the process of H shift is the rate-determining step due to its high potential barrier.

Based on the study above, the other subsequent reaction of the OH-O_2-Flu adducts is expected to react with NO, and then the rupture of O – ONO bond occurs. The reaction of NO addition is barrierless and exothermic. For the first three pathways (Route1–3 in Fig. S3), there are two kinds of H abstractions by O_2 either from the hydroxyl or from the C – H bond of benzene ring. By comparison, the latter is easily to occur due to the lower potential barrier. It ultimately results in the formation of NO_2 and epoxide. Comparison of the reaction scheme presented in Figs. 3 and S3 of Supporting Information shows that the subsequent reactions of IM18 have lower barriers and release more heat, which indicate that P5 is the dominant product. For the last three reaction routes (Route4–6 in Fig. S3), these processes lead to the formation of NO_2 and dialdehydes because of the extra ring-opening reactions. By comparison of the potential barriers and reaction heats of these three reaction routes, the reaction of IM5 is favored over those of IM4. So it is concluded that P6 is the main product.

3.2.2.2. The formation of fluoranthol, fluoranthone and fluoranthenequinone. The formation of fluoranthols from the reaction of OH-Flu adducts and O_2 are depicted in Figs. 3 and S2. Similarly, O_2 can attack on the C atoms with an unpaired electron in OH-Flu via an addition mechanism, and then the H atom of the benzene ring transfer to the O – O bond via a six-membered ring transition state. Subsequently the C – OOH bond cleavage occurs, leading to the formation of HO_2 and fluoranthol. As shown in Fig. 3, the subsequent reaction of IM25 has the lowest barrier and is strongly exothermic, which means this pathway is thermodynamically most favorable. As seen from Figs. S4 and 3, fluoranthols can react with HO_2 to produce fluoranthone and H_2O_2. Subsequently it react with O_2/NO and O_2, yielding HO_2 and fluoranthenequinone. Calculations show that the reaction channel of IM27 can easily proceed due to the low barrier. Similar products have been detected in the experiments of phenanthrene and OH radical (Lee and Lane, 2010).

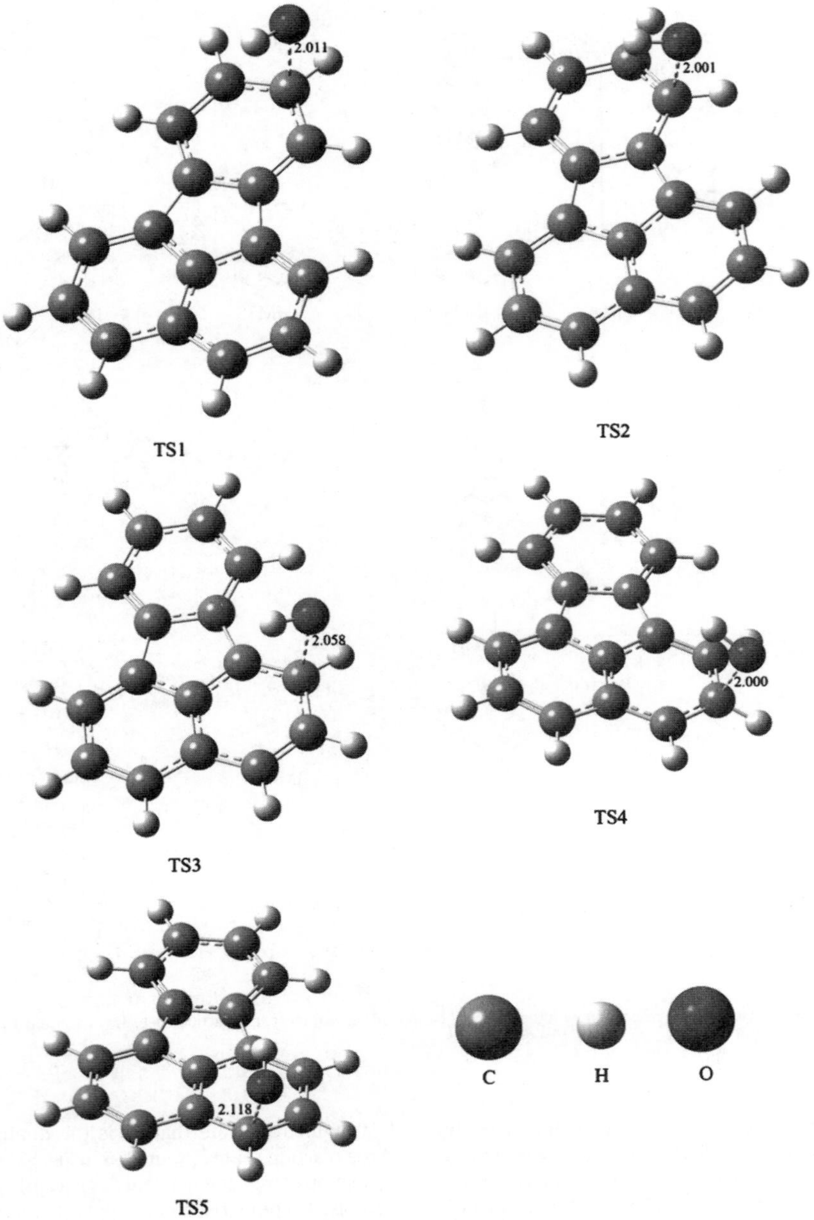

Fig. 5. Transition states for the OH addition reactions of Flu optimized at the BB1K/6-31 + G(d,p) level of theory. Distances are in angstrom.

3.2.2.3. The reaction of bicyclic peroxy radical. Firstly, OH-Flu adducts react with O_2 to produce different bicyclic peroxy radicals. As shown in Figs. 4 and S5, two O atoms and the connected C atoms can form a four-membered, a five-membered or a six-membered ring. By comparison of the potential barriers and reaction heats of these reactions, a five-membered ring is easier to form. Due to the structural similarity, just take IM1 as an example. The subsequent reaction of IM34 has two pathways. The first one involves four elementary steps: the O – O bond cleavage, the C – C bond cleavage, the C – C bond rupture and the C – O bond rupture. The calculated profiles of the potential energy surface show that the last step is less likely to occur because of the high potential barrier of 54.88 kcal/mol. The former three steps are energetically favorable and can proceed readily under the general atmospheric conditions. The second pathway includes five elementary reactions: the addition of O_2 to form organic peroxy radicals, the reaction of resulting peroxy radicals with NO, the cleavage of O – ONO bond to form alkoxy radicals, the synchronous cleavage of O – O bond and C – C bond and the rupture of C – C bond to form dialdehyde and CHOCHOH. Using gas chromatography/electron ionization-time-of-flight mass spectrometry (GC/EI-TOFMS), the atmospheric oxidation product of naphthalene, phthaldialdehyde was detected, which is structurally similar to the dialdehyde formed by the reactions of OH-Flu adducts and O_2/NO (Kautzman et al., 2009).

3.3. Rate constant calculations

The prediction of accurate rate constants of the elementary reactions involved in the degradation of PAHs is crucial for the transport and fate of PAHs in the atmosphere (Keyte et al., 2013). On the basis of the BB1K/6-311 + G(3df,2p)//BB1K/6-31 + G(d,p) energies, the rate constants of the elementary reactions involved in the OH-initiated oxidation degradation of Flu were evaluated by using Rice–Ramsperger–Kassel–Marcus

Table 1
The rate constants (cm^3 $molecule^{-1}$ s^{-1}) of the crucial elementary reactions involved in the OH-initiated oxidation degradation of Flu at 298 K and 1 atm.

Reactions	Rate constants
Flu + OH → OH-Flu	(k) 1.72×10^{-11}
Flu + OH → IM1	(k_1) 1.74×10^{-13}
Flu + OH → IM2	(k_2) 1.05×10^{-13}
Flu + OH → IM3	(k_3) 1.73×10^{-12}
Flu + OH → IM4	(k_4) 1.26×10^{-13}
Flu + OH → IM5	(k_5) 6.46×10^{-12}
IM2 + O_2 → IM7	2.91×10^{-24}
IM7 → IM11	1.95×10^{-10}
IM11 → P2	5.31×10^{3}
IM7 → IM12	3.32×10^{-6}
IM12 → IM13	1.10×10^{8}
IM13 → P3	1.48×10^{7}
IM18 → IM19	4.95×10^{-13}
IM19 → IM20	2.68×10^{3}
IM5 → IM21	1.89×10^{-20}
IM21 → IM22	5.39×10^{-13}
IM22 → IM23	1.77
IM4 → IM25	2.50×10^{-25}
IM25 → IM26	1.96×10^{-13}
IM26 → P7	1.75×10^{5}
IM1 → IM32	2.61×10^{-24}
IM32 → IM33	4.56×10^{-13}
IM32 → IM34	1.08×10^{-4}
IM34 → IM35	1.73×10^{-4}
IM35 → IM36	4.08×10^{4}
IM36 → IM37	1.10×10^{8}
IM37 → P9	1.05×10^{-29}
IM34 → IM38	4.83×10^{-5}
IM38 → IM39	9.05
IM39 → IM40	1.35×10^{-10}
IM40 → IM41	6.38×10^{-9}
IM41 → P9	3.24×10^{8}

(RRKM) theory (Robinson and Holbrook, 1972) at 298 K and 1 atm. The RRKM method has been used to deal with several reactions (Glowacki et al., 2012; Zhou et al., 2011). The calculated rate constants of some important elementary reactions are presented in Table 1. The individual rate constants for the OH addition to Flu are noted as k_1, k_2, k_3, k_4, and k_5, respectively. The overall rate constant of the OH addition reaction is noted as k, $k = (k_1 + k_2 + k_3 + k_4 + k_5) \times 2$. By the calculation, the overall rate constant k is 1.72×10^{-11} cm^3 $molecule^{-1}$ s^{-1} at 298 K and 1 atm. which is consistent with the experimental value of 1.1×10^{-11} cm^3 $molecule^{-1}$ s^{-1} at 298 K and 1 atm (Brubaker and Hites, 1998). From the good agreement with the experimental value, it can be inferred that the rate constants of other elementary reactions listed in Table 1 are reasonable. We expect that the RRKM values will assist in the construction of detailed kinetic model describing the transport and fate of Flu in the atmosphere.

According to the rate constant of the reaction of Flu with OH radicals and a typical OH concentration (C_{OH}) of 9.75×10^5 molecule cm^{-3} (Prinn et al., 1995), using the expression:

$$\tau_{OH} = \frac{1}{k_{(OH+Flu)} \times c_{OH}}.$$

The atmospheric lifetime of Flu determined by OH radicals is calculated about 0.69 days. Depending on the literature, Brubaker and Hites estimated that the atmospheric lifetime of Flu is 26 h in an experimental study of gas-phase OH reactions (Brubaker and Hites, 1998). The bias of atmospheric lifetime mainly results from the rate constant difference between the calculated value (1.72×10^{-11} cm^3 $molecule^{-1}$ s^{-1}) and the experimental value (1.1×10^{-11} cm^3 $molecule^{-1}$ s^{-1}).

4. Conclusions

A theoretical study is presented on the reaction mechanism of OH radical-initiated atmospheric oxidation of Flu. The rate constants were calculated by using the RRKM method. Two specific conclusions can be drawn from this study:

(1) The OH-initiated atmospheric oxidation of Flu generates a range of ring-retaining and ring-opening products containing fluoranthols, fluoranthones, fluoranthenequinones, nitro-fluoranthenes, dialdehydes and epoxides.
(2) The calculated overall rate constant matches well with the available experimental value. The atmospheric lifetime of Flu by OH radicals is about 0.69 days.

Acknowledgment

The work was financially supported by NSFC (National Natural Science Foundation of China, project Nos. 21337001, 21377073 and 21177076), Taishan Grand (No. ts20120522) and Independent Innovation Foundation of Shandong University (IIFSDU, project No. 2012JC030).

Appendix A. Supplementary data

H abstraction reaction scheme of OH-Flu adducts by O_2. O_2 addition reaction scheme of OH-Flu adducts. NO addition reaction scheme of OH-Flu-O_2 adducts. The formation reaction scheme of fluoranthenequinones. The reaction scheme of the bicyclic peroxy radical. The material is available free of charge via the Internet at http://www.elsevier.com/. Supplementary data associated with this article can be found, in the online version, at http://dx.doi.org/10.1016/j.scitotenv.2014.04.134.

References

Ananthula R, Yamada T, Taylor PH. Kinetics of OH radical reaction with anthracene and anthracene-d10. J Phys Chem A 2006;110:3559–66.
Boström CE, Gerde P, Hanberg A, Jernström B, Johansson C, Kyrklund T, et al. Cancer risk assessment, indicators, and guidelines for polycyclic aromatic hydrocarbons in the ambient air. Environ Health Perspect 2002;110:451–88.
Brubaker WW, Hites RA. OH reaction kinetics of polycyclic aromatic hydrocarbons and polychlorinated dibenzo-p-dioxins and dibenzofurans. J Phys Chem A 1998;102:915–21.
Chaudhry GR. Biological degradation and bioremediation of toxic chemicals. Portland, OR, USA: Dioscorides Press; 1994.
Chen F, Hu W, Zhong Q. Emissions of particle-phase polycyclic aromatic hydrocarbons (PAHs) in the Fu Gui-shan Tunnel of Nanjing, China. Atmos Res 2013;124:53–60.
Christie S, Raper D, Lee DS, Williams PI, Rye L, Blakey S, et al. Polycyclic aromatic hydrocarbon emissions from the combustion of alternative fuels in a gas turbine engine. Environ Sci Technol 2012;46(11):6393–400.
EU. Directive 2004/107/EC of the European Parliament and of the Council of 15 December 2004 relating to arsenic, cadmium, mercury, nickel and polycyclic aromatic hydrocarbons in ambient air. Off J L 2005;23. [of 26.01.2005].
Frisch MJ, Trucks GW, Schlegel HB, Scuseria GE, Robb MA, Cheeseman JR, et al. Gaussian 03, Revision A.1. Pittsburgh, PA: Gaussian Inc.; 2003.
Galarneau E, Bidleman TF, Blanchard P. Seasonality and interspecies differences in particle/gas partitioning of PAHs observed by the Integrated Atmospheric Deposition Network (IADN). Atmos Environ 2006;40:182–97.
Glowacki DR, Liang CH, Morley C, Pilling MJ, Robertson SH. MESMER: an open-source master equation solver for multi-energy well reactions. J Phys Chem A 2012;116:9545–60.
Hellwege KH, Hellwege AM. Landolt–Bornstein: group II: atomic and molecular physics. Structure Data of Free Polyatomic Molecules, vol. 7. Berlin: Springer-Verlag; 1976.
Herzberg G. Electronic spectra and electronic structure of polyatomic molecules. New York: Van Nostrand; 1966.
Kautzman KE, Surratt JD, Chan MN, Chan AWH, Hersey SP, Chhabra PS, et al. Chemical composition of gas- and aerosol-phase products from the photooxidation of naphthalene. J Phys Chem A 2009;114:913–34.
Keyte IJ, Harrison RM, Lammel G. Chemical reactivity and long-range transport potential of polycyclic aromatic hydrocarbons — a review. Chem Soc Rev 2013;42:9333–91.

Lee JY, Lane DA. Formation of oxidized products from the reaction of gaseous phenanthrene with the OH radical in a reaction chamber. Atmos Environ 2010;44:2469–77.
Lorenz K, Zellner R. Kinetics of the reactions of hydroxyl radicals with benzene, benzene-d6 and naphthalene. Ber Bunsen Ges 1983;87:629–36.
Martin JML, El-Yazal J, Francois J-P. Structure and vibrational spectrum of some polycyclic aromatic compounds studied by density functional theory. 1. naphthalene, azulene phenanthrene, and anthracene. J Phys Chem 1996;100:15358–67.
Monte MJS, Notario R, Pinto Sónia P, Ferreira Lobo, Ana IMC, da Silva Ribeiro, et al. Thermodynamic properties of fluoranthene: an experimental and computational study. J Chem Thermodyn 2012;49:159–64.
Perry RA, Atkinson R, Pitts JN. Kinetics and mechanism of the gas phase reaction of hydroxyl radicals with aromatic hydrocarbons over the temperature range 296–473 K. J Phys Chem 1977;81:296–304.
Prinn RG, Weiss RF, Miller BR, Huang J, Alyea FN, Cunnold DM, et al. Atmospheric trends and lifetime of CH_3CCl_3 and global OH concentrations. Science 1995;269:187–92.
Qu XH, Zhang QZ, Wang WX. Mechanism for OH-initiated photooxidation of naphthalene in the presence of O_2 and NOx: a DFT study. Chem Phys Lett 2006;429:77–85.
Ramírez N, Cuadras A, Rovira E, Marcé RM, Borrull F. Risk assessment related to atmospheric polycyclic aromatic hydrocarbons in gas and particle phases near industrial sites. Environ Health Perspect 2011;119:1110–6.
Robinson PJ, Holbrook KA. Unimolecular reactions. John Wiley & Sons; 1972.
Seinfeld JH, Pandis SN. Atmospheric chemistry and physics; Wiley-Interscience: New York; 2006 [ISBN 9471720186].
Shen HZ, Huang Y, Wang R, Zhu D, Li W, Shen GF, et al. Global atmospheric emissions of polycyclic aromatic hydrocarbons from 1960 to 2008 and future predictions. Environ Sci Technol 2013;47(12):6415–24.
Shimanouchi, T. Tables of Molecular Vibrational Frequencies, Consolidated Volume 1, 1972, NSRDS NBS-39.
Terzi E, Samara C. Gas-particle partitioning of polycyclic aromatic hydrocarbons in urban, adjacent coastal, and continental background sites of western Greece. Environ Sci Technol 2004;38:4973–8.
Wetzel A, Alexander T, Brandt S, Haas R, Werner D. Reduction by fluoranthene of copper and lead accumulation in *Triticum aestivum L.* Environ Contam Toxicol 1994;53:856–62.
Xue WL, Warshawsky D. Metabolic activation of polycyclic and heterocyclic aromatic hydrocarbons and DNA damage: a review. Toxicol Appl Pharmacol 2004;206(1):73–93.
Zhang QZ, Gao R, Xu F, Zhou Q, Jiang GB, Wang T, et al. Role of water molecule in the gas-phase formation process of nitrated polycyclic aromatic hydrocarbons in the atmosphere. 2014;48(9):5051–7.
Zhou J, Chen JW, Liang C-H, Xie Q, Wang Y-N, Zhang SY, et al. Quantum chemical investigation on the mechanism and kinetics of PBDE photooxidation by OH: a case study for BDE-15. Environ Sci Technol 2011;45:4839–45.

Chemosphere 156 (2016) 101–110

Contents lists available at ScienceDirect

Chemosphere

journal homepage: www.elsevier.com/locate/chemosphere

Mechanical and kinetic study on gas-phase formation of dinitro-naphthalene from 1- and 2-nitronaphthalene

Zixiao Huang, Qingzhu Zhang*, Wenxing Wang

Environment Research Institute, Shandong University, Jinan 250100, PR China

HIGHLIGHTS

- Formation of dinitro-naphthalene via both OH and NO_3 were investigated firstly.
- Water molecule plays an important role during the formation of dinitro-naphthalene.
- The overall rate constants were calculated for 1- and 2-nitronaphthalene at 298 K.
- The Arrhenius formulas were fitted for the OH addition.

ARTICLE INFO

Article history:
Received 7 December 2015
Received in revised form
24 April 2016
Accepted 25 April 2016
Available online 9 May 2016

Handling Editor: Ralf Ebinghaus

Keywords:
Dinitro-naphthalene
Formation
Gas phase
Reaction mechanism
Rate constants

ABSTRACT

Nitrated polycyclic aromatic hydrocarbons have received an increasing number of considerations because of their higher mutagens than parent PAHs. In this paper, the formation of dinitro-naphthalene was investigated mechanistically using 1- and 2-nitronaphthalene as precursors with the aid of high-accuracy quantum chemistry calculation. The geometrical parameters, as well as vibrational frequencies, were calculated at the BB1K/6-31+G(d,p) level. Water molecule plays an important role in the formation of dinitro-naphthalene. The rate constants were deduced by canonical variational transition-state theory with small curvature tunneling contribution over the temperature range of 273–333 K. Meanwhile, the Arrhenius formulas were fitted for the OH addition of both 1- and 2-nitronaphthalene. The calculated overall rate constants for 1-nitronaphthalene and 2-nitronaphthalene at 298 K and 1 atm are 7.43×10^{-13} and 7.48×10^{-13} cm^3 $molecule^{-1}$ s^{-1}, respectively. The rate constants of NO_3 addition to 1-nitronaphthalene and 2-nitronaphthalene by RRKM method at 298 K and 1 atm are 3.55×10^{-15} and 3.47×10^{-15} cm^3 $molecule^{-1}$ s^{-1}, respectively. This study provides a comprehensive investigation of the formation process of dinitro-naphthalenes, initiated by OH and NO_3 radicals and should facilitate to illuminate its atmospheric source. Oxygen may probably be competitive with the second NO_2 addition step when the concentration of NO_2 is at low level.

1. Introduction

Nitrated polycyclic aromatic hydrocarbons (nitro-PAHs), as derivatives of parent PAHs, have been recognized as the toxic air pollutants (Schauer et al., 2003; Perrini et al., 2005; Hien et al., 2007; Albinet et al., 2008; Kawanaka et al., 2008; Benbrahim-Tallaa et al., 2012). Though the concentration of nitro-PAHs is generally lower than that of parent PAHs at 1–2 orders of magnitude, these atmospheric pollutants have been at the forefront of public concern because of their known 100,000 times more mutagenic and 10 times more carcinogenic than PAHs (Durant et al., 1996). Nitro-PAHs have been demonstrated to account for over 50% of the total direct-acting mutagenicity and the total carcinogenicity activity of ambient air (Pitts et al., 1978; Salmeen et al., 1984; Finlayson-Pitts and Pitts, 1999; Lu et al., 1999). Among nitro-PAHs, 5 dinitro-PAHs (3,7- and 3,9-dinitro-fluoranthene, 1,6-, 1,8-dinitro-pyrene, and 3,6-dinitro-benzo[*a*]pyrene) are found extremely high mutagenic potency ($\geq$100,000 revertants $nmol^{-1}$) in the *Salmonella typhimurium* microsome test(Tokiwa et al., 1994). This result elaborates some kind of dinitro-PAHs more mutagenic to bacterial assays, compared with mononitro-PAHs. Furthermore, the mutagenicity of PAHs and their oxo- and nitro-derivatives has also been modelled by means of Quantitative Structure-Activity Relationships (QSAR) in human cells (Papa et al., 2008). However, in

* Corresponding author.
E-mail address: zqz@sdu.edu.cn (Q. Zhang).

http://dx.doi.org/10.1016/j.chemosphere.2016.04.108
0045-6535/© 2016 Published by Elsevier Ltd.

the aspect of carcinogenicity, nitro-PAHs and parent PAH are not totally similar depending on the route of administration (Organization, 2003). In addition, nitro-PAHs occur in a wide range of global environment (Finlayson-Pitts and Pitts, 1999), even have been detected in the Antarctic air particulate (Vincenti et al., 2001).

Nitro-PAHs can be formed in the atmosphere by both direct emission, primarily by diesel engines through electrophilic nitration process, and secondary formation of the atmospheric reaction of parent PAHs (Ringuet et al., 2012). Secondary formation has been revealed a significant contribution to various nitro-PAHs through the observation of specific isomers for nitro-PAHs in ambient air (Arey et al., 1986; Gibson et al., 1986; Atkinson and Arey, 1994; Bamford and Baker, 2003; Reisen and Arey, 2005). Nitro-PAHs can be generated via homogeneous reactions of PAHs with OH at day-time and NO_3 radicals at night-time in the presence of NO_2 (Nielsen, 1984; Ramdahl et al., 1986; Atkinson et al., 1987; Atkinson and Arey, 1994). In addition, nitro-PAHs can also be produced by the heterogeneous reactions of parent PAHs in particles with N_2O_5 or HNO_3 molecule (Nielsen, 1983; Kamens et al., 1990). The previous research indicates that the homogeneous reactions are generally faster contrasted with the heterogeneous reactions (Esteve et al., 2006). Thus, the formation process in gas phase has been recognized as a significant source of nitro-PAHs in the atmosphere. However, to our best knowledge, previous researches mostly focused on the atmospheric formation of mononitro-PAHs, and few studies on the atmospheric formation of dinitro-PAHs.

Due to the limitation of experiment condition, some short-lived reaction intermediates and transition states are difficult to be detected. Therefore, as a supplement to the experiment, quantum chemical calculation can be of considerable help in identifying the mechanism of specific reaction route. In this paper, we investigated the formation of dinitro-naphthalene arising from the OH-initiated and NO_3-initiated atmospheric reaction of 1- and 2-nitronaphthalene by quantum chemical calculation. Recent years, researches with density functional theory have yielded satisfying results in PAH derivatives (Bekbolet et al., 2009; Dang et al., 2015a,b; Onchoke et al., 2016). The atmospheric model places a high demand on accurate kinetic parameters (Feilberg et al., 1999). So, the rate constants of key elementary reactions initiated by OH radical were evaluated by using canonical variational transition-state (CVT) theory with small curvature tunneling (SCT) contribution on the basis of the electronic structure calculations. Because the reactions initiated by NO_3 radical are barrierless, the rate constants initiated by NO_3 radical were carried out by Rice-Ramsperger-Kassel-Marcus (RRKM) theory.

2. Computational methods

Utilizing Gaussian 09 program (Frisch et al., 2009), high-accuracy molecular orbital calculations were performed on a supercomputer. The geometries of the reactants, intermediates, transition states (Qu et al., 2009; Xu et al., 2010b, 2010c) and products were carried out at the BB1K level with the 6-31+G(d,p) (Zhao et al., 2004), which has an excellent results in the previous study (Zhang et al., 2014). The vibrational frequencies were calculated at the same level to verify whether the structures obtained are true minima or first-order saddle points. To confirm each transition state connecting to the right minima along the reaction path, the intrinsic reaction coordinate (IRC) calculations were further carried out at the 6-31+G(d,p) level. Single-point energy calculations were performed at a more flexible basis set, 6-311+G(3df,2p), to obtain a more accurate evaluation of the energetic parameters.

The rate constants initiated by OH radical were performed by the means of POLYRATR 9.3 program (Steckler et al., 2002). This program is based on the canonical variational transition-state (CVT) theory. The CVT rate constant for temperature T is given by:

$$k^{CVT}(T) = \min_s k^{GT}(T, s) \tag{1}$$

where

$$k^{GT}(T, s) = \frac{\sigma k_B T}{h} \frac{Q^{GT}(T, s)}{\Phi^R(T)} e^{-V_{MEP}(s)/k_B T} \tag{2}$$

where, $k^{GT}(T, s)$ is the generalized transition state theory rate constant at the dividing surface s, σ is the symmetry factor accounting for the possibility of more than one symmetry-related reaction path, k_B is Boltzmann's constant, h is Planck's constant, $\Phi^R(T)$ is the reactant partition function per unit volume (excluding symmetry numbers for rotation) and $Q^{GT}(T, s)$ is the partition function of a generalized transition state at s with a local zero of energy at $V_{MEP}(s)$ and with all rotational symmetry numbers set to unity. To include quantum tunneling effects for motion along the reaction coordinate, the CVT rate constants were multiplied by transmission coefficient. In the present work, the small curvature tunneling (SCT) method, based on the centrifugal-dominant small-curvature semi-classical adiabatic ground-state approximation, was employed to calculate the transmission coefficient.

The rate constants initiated by NO_3 radical were carried out with the aid of Rice-Ramsperger-Kassel- Marcus (RRKM) theory (Robinson and Holbrook, 1972). The RRKM rate constant is given by:

$$k(E) = \frac{W(E)}{h\rho(E)} \tag{3}$$

where $W(E)$ is the rovibrational sum of states at the transition state, $\rho(E)$ is density of states of reactants, and h is Planck's constant. Then, the canonical rate constant $k(T)$ is determined from the equation below:

$$k(T) = \frac{1}{Q(T)} \int k(E)\rho(E)\exp(-\beta E)dE \tag{4}$$

where $Q(T)$ is the reactant partition function.

3. Results and discussion

It is difficult to directly compare the theoretical results with experimental values because of the lack of the experimental data of present reaction. Therefore, we optimized geometries and calculated vibrational frequencies of nitro-benzene, 1-nitron-aphthalene and 1,5-dinitro-naphthalene at the BB1K/6-31+G(d,p) level, for the sake of verifying the reliability of our computational results. The relative deviation remains within 3% for bond length and 8% for the vibrational frequencies (Johnson, 2005; Arivazhagan et al., 2009; Govindarajan and Karabacak, 2012).

To describe the results conveniently, the carbon atoms in naphthalene are numbered as follows:

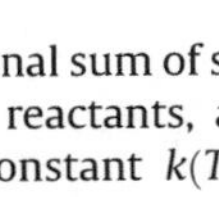

C–H and C=C bonds exist in 1- and 2-nitronaphthalene, which makes it possible for both H abstraction from C–H bonds and OH addition to C=C bonds. Obviously, both 1- and 2-nitronaphthalene have seven H abstraction positions. The potential barrier and reaction heat were calculated at BB1K/6-311+G(3df,2p) level. Results show that the potential barriers spread from 5.17 to 9.26 kcal mol^{-1}. Moreover, the reaction heats of H abstraction are from −5.29

to −1.15 kcal mol^{-1}, which illustrated all these H abstraction pathways are exothermic. Due to the higher potential barrier and less reaction heat, H abstraction is energetically less favorable. The detailed H abstraction reaction pathways of 1- and 2-nitronaphthalene by OH radical were displayed in Figs. S3 and S4 in Supplementary Material, respectively. Calculated rate constants of H abstraction reaction of 1-nitronaphthalene were displayed in Table S1, and that of 2-nitronaphthalene were also described in Table S2.

3.1. Formation mechanism

Previous experimental data indicate that there is no evidence for the reactions of 1- and 2-nitronaphthalene with the NO_3 radicals (Atkinson et al., 1989). We note here that this chamber experiment was carried out under dry condition. In the real atmospheric condition, however, a large amount of water molecule exists in ambient. Water is the third most abundant species in the atmosphere after N_2 and O_2 (DeMore et al., 1992). The concentration of water, corresponding to 50% of relative humidity at 298 K, can reach up to 7.64×10^{17} molecules cm^{-3}. Therefore, because of the existence of water, it has the possibility to generate dinitro-naphthalene from 1-nitro-naphthalene and 2-nitro-naphthalne with NO_3 radicals. Dinitro-PAHs can be formed from the reactions of mononitro-PAHs with OH, NO_3, and N_2O_5 (Atkinson et al., 1989). Nevertheless, the rate constant of mono-nitro-naphthalene with N_2O_5 in gas phase is extremely low at about 10^{-18} cm^3 $molecule^{-1}$ s^{-1} (Atkinson et al., 1989). Thus, we mainly concern about the formation mechanism of dinitro-naphthalene from the oxidation reactions of 1-nitro-naphthalene and 2-nitro-naphthalne initiated by OH and NO_3 radicals. The formation pathways of dinitro-naphthalene initiated by OH radicals were shown Fig. 1 and Fig. 2 embedded with the potential barrier and reaction heat. The reaction pathways initiated by NO_3 radical were displayed in Fig. 3 and Fig. 4. Due to the presence of nitro-group, the ten C atoms in 1-nitro-naphthalene and 2-nitro-naphthalne are all unequal, so that eight different OH or NO_3 addition positions were found. The OH or NO_3 addition to C9 and C10 were found to have high barriers and/or be strong endothermicity, so the OH or NO_3 addition to C9 and C10 are unfavorable. In order to compare with the formation of dinitro-naphthalene from mono-naphthalene, we also studied the formation of dinitro-naphthalene from naphthalene, as shown in Fig. S1.

3.2. Reaction with OH radicals

The OH addition to 1-nitro-naphthalene and 2-nitro-naphthalene are strong exothermic from 12.43 to 29.90 kcal mol^{-1}. The potential barriers of OH addition for both 1-nitironaphthalene and 2-nitronaphthalene are quite low from 0.79 to 5.23 kcal mol^{-1}, shown in Figs. 1 and 2. By comparing both reaction heats and barriers, it indicates that IM4, IM5, IM8 are predominant adducts for 1-nitronaphthalene, while IM23, IM26, IM27 are predominant adducts for 2-nitronaphthalene.

The following addition process of NO_2 to OH-nitronaphthalene adduct is proved to be barrierless and strong exothermic. Notice that only the OH—NO_2-nitronaphthalene adducts formed through the addition of NO_2 to the *para-* or *ortho*-positions of OH can react to form dinitro-naphthalene. Furthermore, the OH—NO_2-nitronaphthalene adducts formed through the NO_2 addition to the *meta*-positions of OH are difficult to react to form dinitro-naphthalene because of the strong endothermic or an obvious isomerization. The adducts IM14, IM15, IM16, IM21, IM22 initiated from 1-nitronaphthalene and IM31, IM35, IM36, IM37, IM38 initiated from 2-nitronaphthalene are easy to form because of the strong exothermicity without a barrier.

The NO_2 group is on the different side of OH group relative to the benzene ring, shown in Figs. 1 and 2 symbolled by [a]. *Ortho*-OH—NO_2-nitronaphthalene can lose a water molecule directly via a four-membered ring transition state. This unimolecular decomposition has extremely high barriers as shown in Figs. 1 and 2. For example, the direct unimolecular decomposition of IM22 has a barrier of 44.37 kcal mol^{-1}. This makes the reaction difficult to occur under the general atmospheric conditions. However, due to the abundance of the ambient water, the *ortho*-OH—NO_2-nitronaphthalene can loss water via a bimolecular reaction (OH—NO_2-nitronaphalene + H_2O) with the aid of the water molecule. In the bimolecular reaction, a six-membered ring transition state was formed. The water molecule plays the role as bridge in this configuration to accept the hydrogen atom from the aromatic ring and donate a different hydrogen atom to the phenolic group. For example, the barrier of the water loss from IM22 via the bimolecular reaction is 25.43 kcal mol^{-1} which is 18.94 kcal mol^{-1} lower than the direct loss water process via the unimolecular decomposition. As shown in Fig. 2, IM33 can lose water to generate 2,3-dinitro-naphthalene by both unimolecular decomposition and bimolecular process. The barrier of bimolecular process is 16.19 kcal mol^{-1}, which is 26.52 kcal mol^{-1} lower than the unimolecular decomposition, 42.71 kcal mol^{-1}. The results suggest that the water molecule plays a positive catalytic effect on the loss of water from the OH—NO_2-nitronaphalene adducts and promotes the formation of dinitro-naphthalenes.

Notice that the NO_2 group can be on the same side of the OH group relative to the benzene ring, which is shown in Figs. 1 and 2 symbolled by [s]. We also wonder whether the water loss process could occur for the OH—NO_2-nitronaphalene adducts with the NO_2 and OH group at the same side relative to the benzene ring. To answer this question, 1-nitronaphthalene was taken for an example by both unimolecular decomposition and bimolecular reaction (Fig. S2 of supporting information). However, the barriers of the water loss from the *ortho*-OH—NO_2-nitronaphalene adducts with the NO_2 and OH group at the same side are much higher than that of the water loss from the *ortho*-OH—NO_2-nitronaphalene adducts with the NO_2 and OH group at the different side. *Para*-OH—NO_2-nitronaphthalene with the NO_2 and OH group at the same side can lose water via OH—NO_2-nitronaphthalene + H_2O + H_2O → dinitro-naphthalene + $3H_2O$ with the introduction of two water molecules. Because of the short length between H atom and OH group in *para*-position of IM31, it still losses water by IM31 + H_2O → 1,3-dinitro-naphthalene + $2H_2O$. Owing to the low probability for the collision of these three molecules (OH—NO_2-nitronaphthalene + H_2O + H_2O), the termolecular mechanism is unreliable in this water loss reaction. Water dimer is ubiquitous in ambient, measured in a high concentration of 6×10^{14} molecules cm^{-3}. Therefore, the reaction of *para*-OH—NO_2-nitronaphthalene with water dimer, for instance, OH—NO_2-nitronaphthalene + $(H_2O)_2$ → dinitro-naphthalene + $3H_2O$ is the most possible mechanism. It suggests that the pathway with lower reaction barrier has more possibility to occur in ambient. Hence, based on the addition reactions of OH radical and NO_2 molecule, the favorable pathways are pathway 11, 23, 25 initiated from 1-nitronaphthalene and pathway 26, 38 initiated from 2-nitronaphthalene. As a result, the dinitrated products from 1-nitro-naphthalene or 2-nitro-naphthalene initiated by OH radicals are 1,2-, 1,3-, 1,5-, 1,6-, 1,7-dinitro-naphthalene.

It is of great important significance to take O_2 into account, due to the competitive relationship of O_2 and NO_2 for the further reaction of OH-nitronaphthalene adducts. To our best knowledge, only a few studies on the reactions of OH-nitronaphthalene adducts with NO_2 or O_2 have been reported. However, we have found

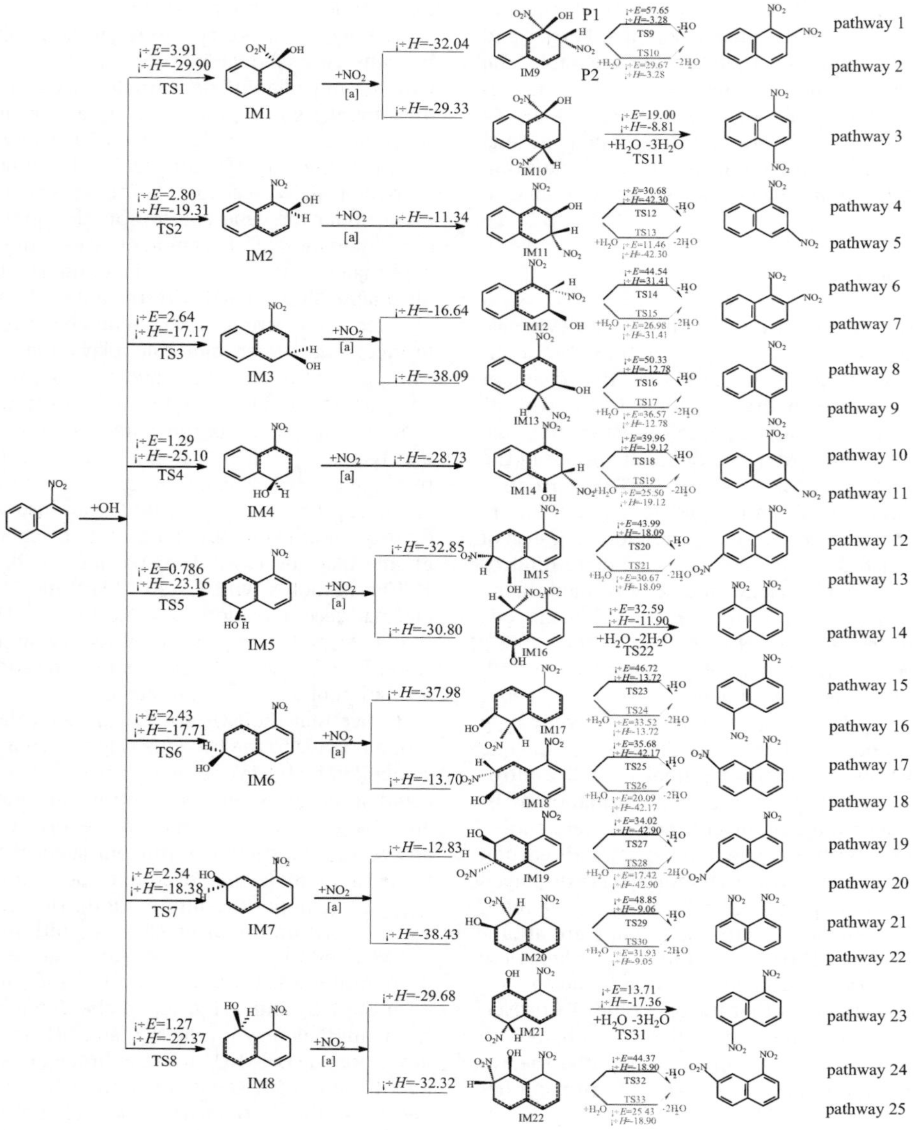

Fig. 1. Dinitro-naphthalenes formation routes embedded with the potential barrier ΔE (in kcal mol^{-1}) and reaction heat ΔH (in kcal mol^{-1}) from the OH-initiated oxidation of 1-nitronaphthalene in the presence of NO_2. ΔH is calculated at 0 K.

several experimental studies on OH-aromatic adducts. It has been proved that OH-PAH adducts can react more feasibly with NO_2 under the atmospheric condition. This is because the consistently observed nitro-PAH in gas phase, as well as detected isomers, shown it is consistent with their formation from atmospheric reactions of PAHs with OH radicals in gas phase (Arey et al., 1967, 1989; Reisen et al., 2003; Reisen and Arey, 2005). Furthermore, for OH-aromatic adducts, the reactions with O_2 and NO_2 are of equal importance in the atmosphere at certain NO_2 maxing ratios. For example, it is 3.3 ppmV for toluene, 0.06 ppmV for naphthalene and 0.6 ppmV for biphenyl (Nishino et al., 2008). It is said for monocyclic aromatic hydrocarbons adducts, the reaction with O_2 is dominant (Koch et al., 2007). However, it depends on the NO_2 mixing ratio for OH-PAH adducts. When the NO_2 mixing ratio is high, for example, in some polluted area, NO_2 dominates the reaction. The competitive reactions of O_2 and NO_2 with OH-aromatic adducts are also reported in computational studies (Zhang et al., 2012). At NO_2 concentration of 60 ppbv, 2-methylnaphthalene-1-OH radical prevails to react with NO_2 and the products, nitronaphthalenes, are of great concern because of their strong mutagenicity (Wu et al., 2015). Therefore, whether O_2 or NO_2 dominates the further reaction of the OH-nitronaphthalene adducts, it depends on the nature of the OH-nitronaphthalene adducts, especially the difference with OH-aromatic adducts. More important, it is also related to the NO_2 concentration. Therefore, we calculated the addition rate constants of OH-1-nitronaphthalene with O_2 molecule for estimate. The pathway and rate constants for the reaction of OH-1-nitronaphthalene with O_2 molecule were displayed in Fig. S5. Details are discussed in Chapter 3.4. More experimental study should be done to figure out whether NO_2 or O_2 is dominant when NO_2 concentration at both low and high level.

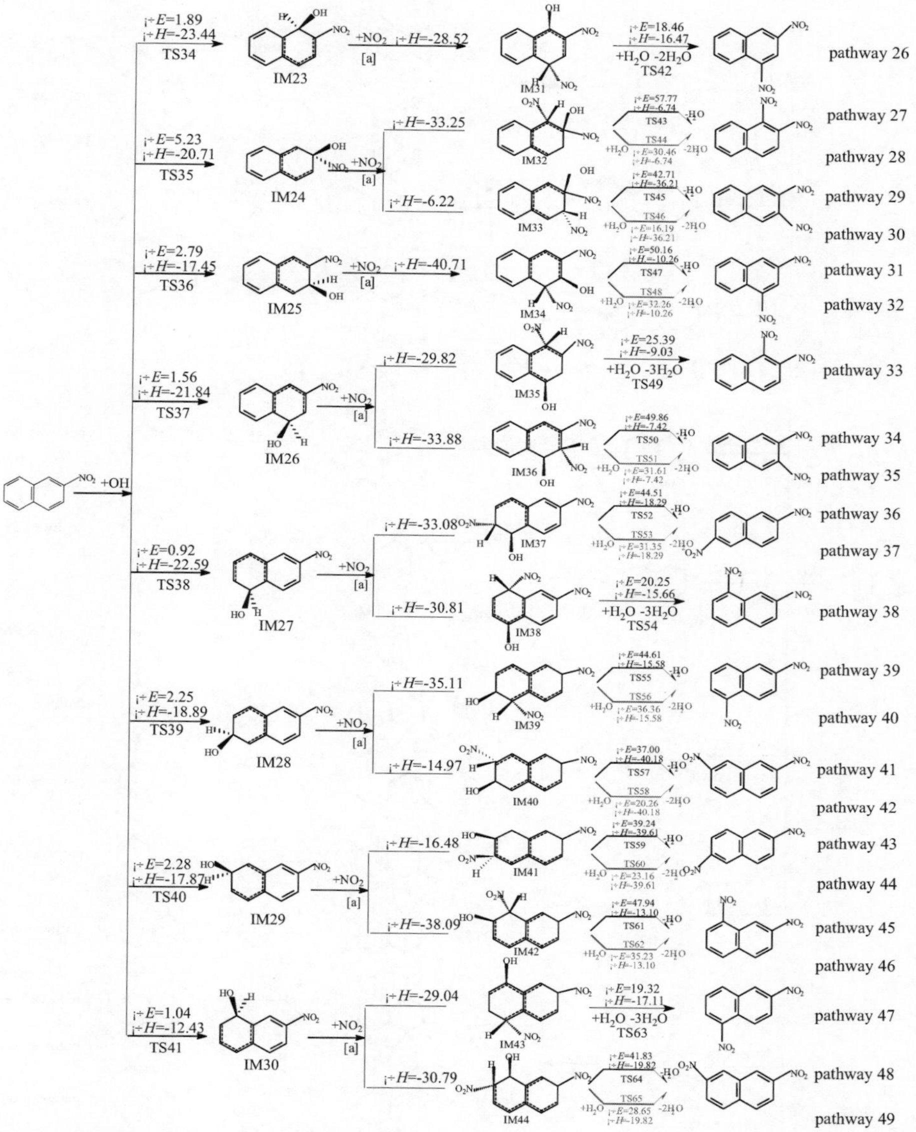

Fig. 2. Dinitro-naphthalenes formation routes embedded with the potential barrier ΔE (in kcal mol^{-1}) and reaction heat ΔH (in kcal mol^{-1}) from the OH-initiated oxidation of 2-nitronaphthalene in the presence of NO_2. ΔH is calculated at 0 K.

3.3. Reaction with NO_3 radicals

Calculations show that the addition of NO_3 to 1-naphthalene or 2- naphthalene is barrierless and strong exothermic (shown in Figs. 3 and 4). Moreover, it shows that IM45, IM48, IM49, IM52 from 1-nitronaphthlaene and IM67, IM70, IM71, IM74 from 2-nitronaphthalene are more favorable to be formed. Similar to the NO_2 addition to the OH-nitronaphthalene adducts, the NO_2 addition to the NO_3-nitronaphthalene adducts is barrierless and strong exothermic. The heats released from these reactions are extremely high, at the range of 21.50–34.21 kcal mol^{-1}. After the NO_2 addition, IM53, IM54, IM58, IM59, IM60, IM65, IM66 initiated from 1-nitronaphthalene and IM75, IM79, IM80, IM81, IM82, IM87, IM88 initiated from 2-nitronaphthalene are generated favorably. For further nitric acid loss process, *ortho*-intermediates perform unimolecular decomposition exclusively. Pathway 56, 57 initiated by 1-nitronaphthalene and pathway 75, 84 initiated by 2-nitronaphthalene are favorable pathways because the barriers of these reactions are lower. *Para*-intermediates can lose a nitric acid molecule via the unimolecular decomposition directly or the bimolecular reaction with the aid of a water molecule, NO_3–NO_2-nitronaphthalene + H_2O → dinitro-naphthalene + HNO_3 + H_2O. The calculation shows that the barrier of the latter reaction is lower than the former one. Similar to the water loss process initiated by OH radicals, the six-member ring transition state is formed in the process of loss nitric acid directly and the eight-member ring transition state is formed in the bimolecular reaction with the aid of a water molecule. The favorable pathways for bimolecular reactions are pathway 59, 65 initiated by 1-nitronaphthalene and pathway 68, 73, 77, 83 initiated by 2-nitronaphthalene, respectively. In addition, the water molecules act as a bridge to accept the hydrogen atom on aromatic ring and donate a different hydrogen atom to the NO_3 group. It is clear that only one water molecule is needed in the denitration of *para*-NO_3-NO_2-nitronaphthalene rather than dimer.

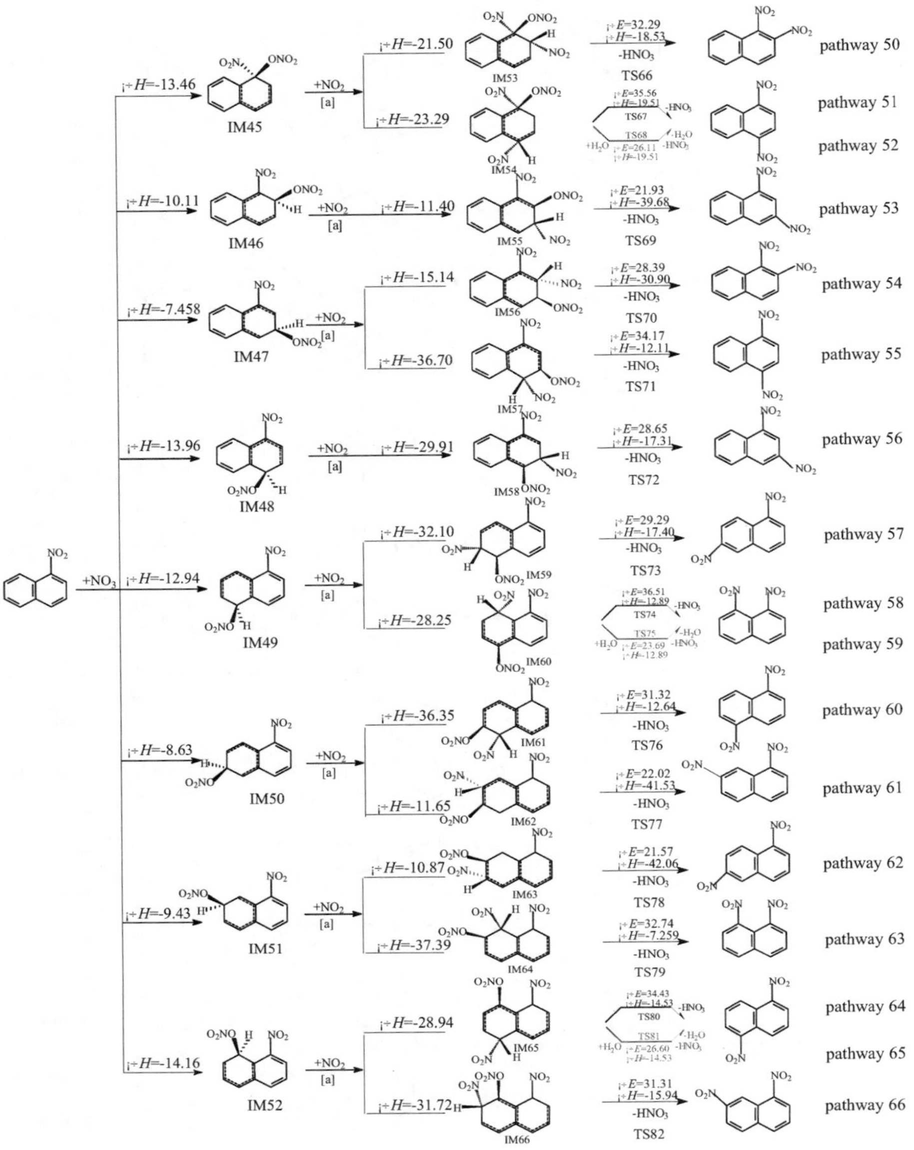

Fig. 3. Dinitro-naphthalenes formation routes embedded with the potential barrier ΔE (in kcal mol^{-1}) and reaction heat ΔH (in kcal mol^{-1}) from the NO_3-initiated oxidation of 1-nitronaphthalene in the presence of NO_2. ΔH is calculated at 0 K.

As a result, the dinitrated products of 1-nitronaphthalene or 2-nitronaphthalene initiated by NO_3 radicals are 1,2-, 1,3-, 1,5-, 1,6-, 1,7-, 1,9-, 2,6-, 2,7-dinitronaphthalene.

The competitive reaction of NO_3-nitronaphthalene with O_2 molecule is similar with OH initiated reaction, as discussed above. Pervious theoretical study has estimated nitroxycyclohexadienyl is to have priority in reacting with NO_2 rather than O_2, on the basis of some estimates (Ghigo et al., 2006).

3.4. Rate constant calculation

The concentrations of ambient 1-nitronaphthalene, 2-nitronaphthalene, OH and NO_3 have a significant influence on the formation rate and yield of dinitro-naphthalene. The reaction rate constants of 1-nitronaphthalene and 2-nitronaphthalene with OH and NO_3 radicals have been recorded in the previous study (Atkinson et al., 1989). The rate constants for OH radical are $(5.4 \pm 1.8) \times 10^{-12}$ cm^3 molecule^{-1} s^{-1} for 1-nitronaphthalene and $(5.6 \pm 0.9) \times 10^{-12}$ cm^3 molecule^{-1} s^{-1} for 2-nitronaphthalene. The rate constants for NO_3 radical are $\leq 7.2 \times 10^{-15}$ cm^3 molecule^{-1} s^{-1} for both 1- and 2-nitronaphthalene. They indicate that dinitro-naphthalenes are reaction products of 1-nitronaphthalene and 2-nitronaphthalene with N_2O_5 instead of NO_3 radicals. However, it is worth to noticing that this experiment is performed under dry purified air condition. Therefore, in the actual atmosphere, because the existence of water molecular in high concentration, it has the possibility to generate dinitro-naphthalene from 1-nitronaphthalene and 2-nitronaphthalne with NO_3 radicals. The rate constants obtained experimentally is the overall rate constants of the addition reaction, which means the individual rate constants are still absent. On the basis of the energies at the BB1K/6-311+G(3df,2p)//BB1K/6-31+G(d,p), we calculated the individual and overall rate constants for the addition reactions of 1-nitronaphthalene and 2-nitronaphthalene initiated by OH and

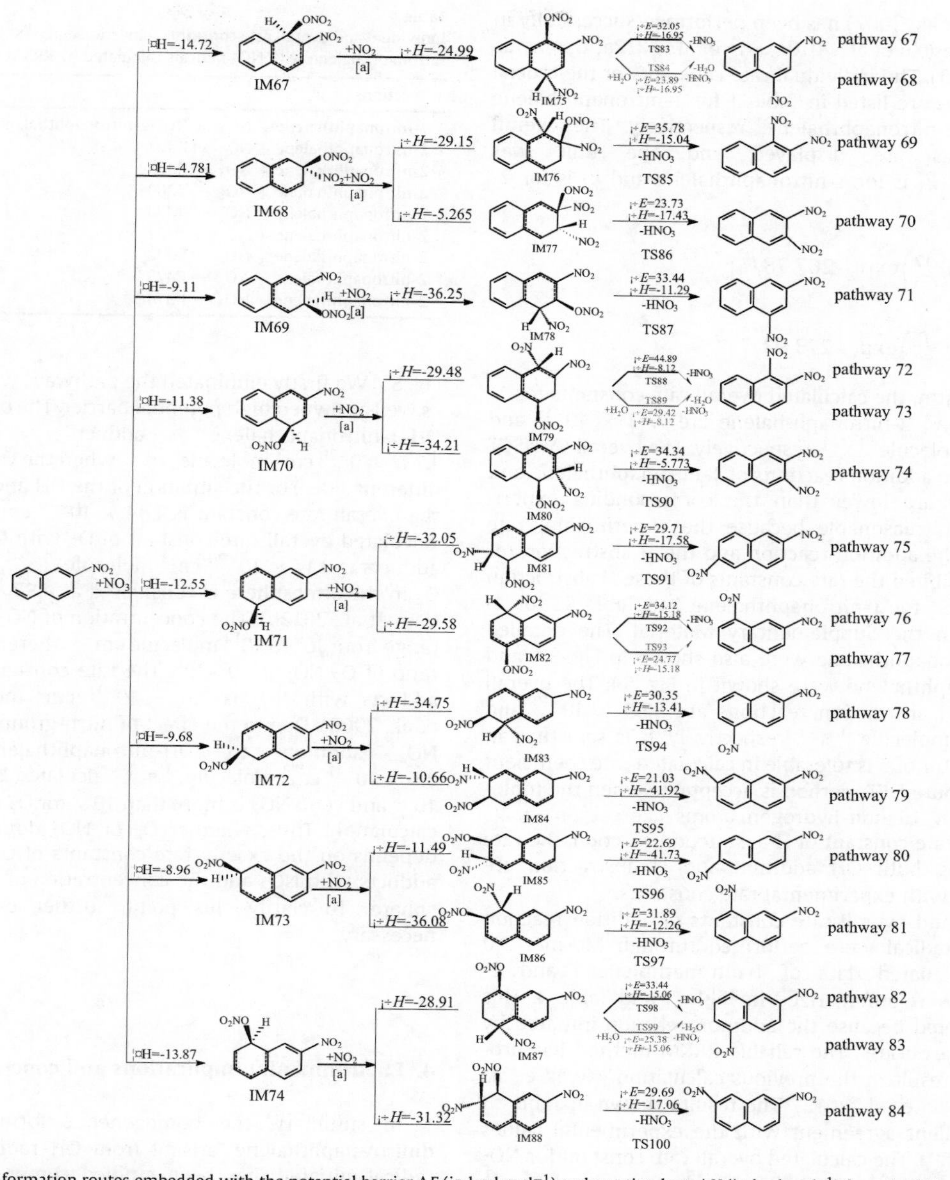

Fig. 4. Dinitro-naphthalenes formation routes embedded with the potential barrier ΔE (in kcal mol^{-1}) and reaction heat ΔH (in kcal mol^{-1}) from the NO_3-initiated oxidation of 2-nitronaphthalene in the presence of NO_2. ΔH is calculated at 0 K.

NO_3 radicals. The calculation initiated by OH radical was with the aid of the CVT/SCT method over the temperature range of 273–333 K. The reliability of the CVT/SCT method (Garrett and Truhlar, 1979; Baldridge et al., 1989; Gonzalez-Lafont et al., 1991;

Table 1
Individual and overall rate constants (cm^3 $molecule^{-1}$ s^{-1}) of addition reaction of 1-nitronaphthalene with OH radicals calculated by CVT/SCT method at 298 K and 1 atm.

Reactions	Rate constants
1-nitronaphthalene + OH → OH-1-nitronaphthalene	7.43×10^{-13}
1-nitronaphthalene + OH → IM1	4.54×10^{-14}
1-nitronaphthalene + OH → IM2	8.87×10^{-14}
1-nitronaphthalene + OH → IM3	8.86×10^{-14}
1-nitronaphthalene + OH → IM4	1.03×10^{-13}
1-nitronaphthalene + OH → IM5	1.22×10^{-13}
1-nitronaphthalene + OH → IM6	9.24×10^{-14}
1-nitronaphthalene + OH → IM7	9.47×10^{-14}
1-nitronaphthalene + OH → IM8	1.08×10^{-13}

Table 2
Individual and overall rate constants (cm^3 $molecule^{-1}$ s^{-1}) of addition reaction of 2-nitronaphthalene with OH radicals calculated by CVT/SCT method at 298 K and 1 atm.

Reactions	Rate constants
2-nitronaphthalene + OH → OH-2-nitronaphthalene	7.48×10^{-13}
2-nitronaphthalene + OH → IM23	1.03×10^{-13}
2-nitronaphthalene + OH → IM24	9.72×10^{-15}
2-nitronaphthalene + OH → IM25	8.61×10^{-14}
2-nitronaphthalene + OH → IM26	1.09×10^{-13}
2-nitronaphthalene + OH → IM27	1.19×10^{-13}
2-nitronaphthalene + OH → IM28	1.00×10^{-13}
2-nitronaphthalene + OH → IM29	9.87×10^{-14}
2-nitronaphthalene + OH → IM30	1.22×10^{-13}

Fernandez-Ramos et al., 2007) has been performed successfully in our published studies(Xu et al., 2010a; Sun et al., 2012a; Gao et al., 2013; Shi et al., 2015). The individual rate constants of the crucial elementary reactions are listed in Table 1 for 1-nitronaphthalene and Table 2 for 2-nitronaphthalene, respectively. The overall Arrhenius formulas are displayed and the unit was cm^3 $molecule^{-1}$ s^{-1}(k_1 is for 1-nitronaphthalene and k_2 is for 2-nitronaphthalene):

$$k_1(T) = \left(1.83 \times 10^{-12}\right)\exp(-267.78/T)$$

$$k_2(T) = \left(1.88 \times 10^{-12}\right)\exp(-273.93/T)$$

At 298 K and 1 atm, the calculated overall rate constants for 1-nitronaphthalene and 2-nitronaphthalene are 7.43×10^{-13} and 7.48×10^{-13} cm^3 $molecule^{-1}$ s^{-1}, respectively. The overall CVT/SCT rate constants of the addition reactions of 1-nitronaphthalene and 2-nitronaphthalene are lower than the corresponding experimental values. It is reasonable because the experimental rate constants include the addition reaction and the H abstraction reaction. We also exhibited the rate constants of these H abstraction reactions in Table S1 for 1-nitronaphthalene and Table S2 for 2-nitronaphthalene in the Supplementary Material. The detailed pathways of 1-nitronaphthalene were also shown in Fig. S3, and those for 2-nitronaphthalene were shown in Fig. S4. The overall rate constants of H abstraction reactions are 3.69×10^{-13} and 3.68×10^{-13} cm^3 $molecule^{-1}$ s^{-1}, respectively. It is said that an uncertainty of a factor of 5 is tolerable in calculated rate coefficient at 298 K. Besides, pure DFT method is acceptable when the molecule contains about 15 non-hydrogen atoms (Vereecken et al., 2015). Adding the rate constant of H abstraction reaction, the rate constants including both OH addition and H abstraction are tolerant compared with experimental rate constants.

The individual and overall rate constants of addition reaction initiated by NO_3 radical were performed through Mesmer 3.0 program. The calculated data of 1-nitronathphalene and 2-nitronaphthalene were summarized in Table 3 and Table 4. This software was adopted because the addition reaction initiated by NO_3 radical was barrierless. The reliable RRKM method has provided a satisfying result in the previous calculation (Kwok et al., 1994, 1997; Glowacki et al., 2012). The results shown in Tables 3 and 4 are in excellent agreement with the experimental values (Atkinson et al., 1989). The calculated overall rate constant for NO_3 radical of 1-nitronaphthalene is 3.55×10^{-15} cm^3 $molecule^{-1}$ s^{-1}, and that for NO_3 radical of 2-nitronaphthalene is 3.47×10^{-15} cm^3 $molecule^{-1}$ s^{-1}.

The further reaction of O_2 with OH-nitronaphthalen adducts is competitive with NO_2. It is critical to clarify whether NO_2 or O_2 dominants. The rate constants of addition reactions for OH-nitronaphthalene adducts with O_2 were carried out and shown in Fig. S5. We firstly eliminated the pathways which are endothermal as well as owned high potential barrier. The overall rate constant of OH-1-nitronaphthalene adducts with O_2 is 1.73×10^{-20} cm^3 $molecule^{-1}$ s^{-1} when the OH group and O_2 are in different side. For the situation of the OH and O_2 in the same side, the overall rate constant is 1.44×10^{-20} cm^3 $molecule^{-1}$ s^{-1}. The calculated overall rate constant of O_2 with OH-1-nitronaphthlane adducts is 3.17×10^{-20} cm^3 $molecule^{-1}$ s^{-1}. The concentration of O_2 in the atmosphere is considered as 4.92×10^{18} molecule cm^{-3} (Sun et al., 2012b). The concentration of NO_2 in the atmosphere is range from 10^9 to 10^{12} molecule cm^{-3}. Therefore, the concentration ratio of O_2/NO_2 is 10^6~10^9. The rate constant of OH-naphthalene adducts with NO_2 is 3.6×10^{-11} cm^3 $molecule^{-1}$ s^{-1}(Nishino et al., 2008). Due to the effect of nitro-group, the rate constant for NO_2 addition to OH-nitronaphthalene is less than 3.6×10^{-11} cm^3 $molecule^{-1}$ s^{-1}. Calculated kO_2/kNO_2 is more than 10^{-9} and vO_2/vNO_2 is more than 10^{-3} for O_2 addition, based on our calculation. Thus, whether O_2 or NO_2 dominate the reaction, it depends on the exactly rate constants of OH-1-nitronaphthalene adducts with NO_2 and the concentration of NO_2 in the real atmosphere. To clarify this point, further experimental study is necessary.

Table 3
Individual and overall rate constants (cm^3 $molecule^{-1}$ s^{-1}) of addition reaction of 1-nitronaphthalene with NO_3 radicals calculated by RRKM theory at 298 K and 1 atm.

Reactions	Rate constants
1-nitronaphthalene + NO_3 → NO_3-1-nitronaphthalene	3.55×10^{-15}
1-nitronaphthalene + NO_3 → IM45	5.57×10^{-16}
1-nitronaphthalene + NO_3 → IM46	4.39×10^{-16}
1-nitronaphthalene + NO_3 → IM47	1.55×10^{-16}
1-nitronaphthalene + NO_3 → IM48	5.97×10^{-16}
1-nitronaphthalene + NO_3 → IM49	5.28×10^{-16}
1-nitronaphthalene + NO_3 → IM50	2.73×10^{-16}
1-nitronaphthalene + NO_3 → IM51	3.95×10^{-16}
1-nitronaphthalene + NO_3 → IM52	6.02×10^{-16}

Table 4
Individual and overall rate constants (cm^3 $molecule^{-1}$ s^{-1}) of addition reaction of 2-nitronaphthalene with NO_3 radicals calculated by RRKM theory at 298 K and 1 atm.

Reactions	Rate constants
2-nitronaphthalene + NO_3 → NO_3-2-nitronaphthalene	3.47×10^{-15}
2-nitronaphthalene + NO_3 → IM67	5.95×10^{-16}
2-nitronaphthalene + NO_3 → IM68	9.91×10^{-18}
2-nitronaphthalene + NO_3 → IM69	4.00×10^{-16}
2-nitronaphthalene + NO_3 → IM70	5.56×10^{-16}
2-nitronaphthalene + NO_3 → IM71	5.83×10^{-16}
2-nitronaphthalene + NO_3 → IM72	4.06×10^{-16}
2-nitronaphthalene + NO_3 → IM73	3.23×10^{-16}
2-nitronaphthalene + NO_3 → IM74	5.92×10^{-16}

4. Environmental implications and conclusions

In summary, the homogeneous formation mechanism of dinitro-naphthalene, arising from OH radical-initiated and NO_3 radical-initiated, has been studied comprehensively. A possible formation mechanism of dinitro-naphthalenes from mono-naphthalenes is proposed in which water plays a catalytic role. the dinitrated products from 1-nitro-naphthalene or 2-nitro-naphthalene initiated by OH radicals are 1,2-, 1,3-, 1,5-, 1,6-, 1,7-dinitro-naphthalene. The dinitrated products of 1-nitronaphthalene or 2-nitronaphthalene initiated by NO_3 radicals are 1,2-, 1,3-, 1,5-, 1,6-, 1,7-, 1,9-, 2,6-, 2,7-dinitronaphthalene. Since the theoretical calculation is infinitude approach but have difficulty in reaching the reaction in actual environmental, more experimental studies need to be executed to prove these fully elucidated reaction pathways.

This study provides a comprehensive perspective on whether toxic mononitrated polycyclic aromatic hydrocarbons (PAH) can convert to (PAH) dinitrated derivatives. Besides, O_2 with OH- or NO_3-nitronaphthalene adducts may be competitive with NO_2 addition at low concentration of NO_2, which needs to be further studied. Though the concentrations of these nitro-PAHs are lower than their parent PAH in the atmosphere, these second formed pollutants should not be neglect due to their mutagenicity and carcinogenicity.

Acknowledgments

The work was financially supported by NSFC (National Natural Science Foundation of China, project Nos. 21337001, 21377073, 21477066) and the Research Fund for the Doctoral Program of Higher Education of China (project No. 20130131110058).

Appendix A. Supplementary data

Supplementary data related to this article can be found at http://dx.doi.org/10.1016/j.chemosphere.2016.04.108.

References

Albinet, A., Leoz-Garziandia, E., Budzinski, H., Villenave, E., Jaffrezo, J.L., 2008. Nitrated and oxygenated derivatives of polycyclic aromatic hydrocarbons in the ambient air of two French alpine valleys: Part 1: concentrations, sources and gas/particle partitioning. Atmos. Environ. 42, 43–54.

Arey, J., Zielinska, B., Atkinson, R., Winer, A.M., 1967. Polycyclic aromatic hydrocarbon and nitroarene concentrations in ambient air during a wintertime high-NO x episode in the Los Angeles basin. Atmos. Environ. (1967) 21, 1437–1444.

Arey, J., Zielinska, B., Atkinson, R., Winer, A.M., Ramdahl, T., Pitts, J.N., 1986. The formation of nitro-PAH from the gas-phase reactions of fluoranthene and pyrene with the OH radical in the presence of NO x. Atmos. Environ. (1967) 20, 2339–2345.

Arey, J., Atkinson, R., Zielinska, B., McElroy, P.A., 1989. Diurnal concentrations of volatile polycyclic aromatic hydrocarbons and nitroarenes during a photochemical air pollution episode in Glendora, California. Environ. Sci. Technol. 23, 321–327.

Arivazhagan, M., Krishnakumar, V., Xavier, R.J., Ilango, G., Balachandran, V., 2009. FTIR, FT-Raman, scaled quantum chemical studies of the structure and vibrational spectra of 1,5-dinitronaphthalene. Spectrochim. Acta Part A Mol. Biomol. Spectrosc. 72, 941–946.

Atkinson, R., Arey, J., 1994. Atmospheric chemistry of gas-phase polycyclic aromatic hydrocarbons: formation of atmospheric mutagens. Environ. Health Perspect. 102, 117–126.

Atkinson, R., Arey, J., Zielinska, B., Pitts, J.N., Winer, A.M., 1987. Evidence for the transformation of polycyclic organic matter in the atmosphere. Atmos. Environ. (1967) 21, 2261–2262.

Atkinson, R., Aschmann, S.M., Arey, J., Barbara, Z., Schuetzle, D., 1989. Gas-phase atmospheric chemistry of 1- and 2-nitronaphthalene and 1,4-naphthoquinone. Atmos. Environ. 23, 2679–2690.

Baldridge, K.K., Gordon, M.S., Steckler, R., Truhlar, D.G., 1989. Ab initio reaction paths and direct dynamics calculations. J. Phys. Chem. 93, 5107–5119.

Bamford, H.A., Baker, J.E., 2003. Nitro-polycyclic aromatic hydrocarbon concentrations and sources in urban and suburban atmospheres of the Mid-Atlantic region. Atmos. Environ. 37, 2077–2091.

Bekbolet, M., Çınar, Z., Kılıç, M., Uyguner, C.S., Minero, C., Pelizzetti, E., 2009. Photocatalytic oxidation of dinitronaphthalenes: theory and experiment. Chemosphere 75, 1008–1014.

Benbrahim-Tallaa, L., Baan, R.A., Grosse, Y., Lauby-Secretan, B., El Ghissassi, F., Bouvard, V., Guha, N., Loomis, D., Straif, K., 2012. Carcinogenicity of diesel-engine and gasoline-engine exhausts and some nitroarenes. Lancet Oncol. 13, 663–664.

Dang, J., Shi, X., Zhang, Q., Hu, J., Wang, W., 2015a. Insights into the mechanism and kinetics of the gas-phase atmospheric reaction of 9-chloroanthracene with NO 3 radical in the presence of NO x. RSC Adv. 5, 84066–84075.

Dang, J., Shi, X., Zhang, Q., Hu, J., Wang, W., 2015b. Mechanism and kinetic properties for the OH-initiated atmospheric oxidation degradation of 9, 10-Dichlorophenanthrene. Sci. Total Environ. 505, 787–794.

DeMore, W.B., Sander, S., Golden, D., Hampson, R., Kurylo, M., Howard, C., Ravishankara, A., Kolb, C., Molina, M., 1992. Chemical Kinetics and Photochemical Data for Use in Stratospheric Modeling.

Durant, J.L., Busby Jr., W.F., Lafleur, A.L., Penman, B.W., Crespi, C.L., 1996. Human cell mutagenicity of oxygenated, nitrated and unsubstituted polycyclic aromatic hydrocarbons associated with urban aerosols. Mutat. Res. Genet. Toxicol. 371, 123–157.

Esteve, W., Budzinski, H., Villenave, E., 2006. Relative rate constants for the heterogeneous reactions of NO2 and OH radicals with polycyclic aromatic hydrocarbons adsorbed on carbonaceous particles. Part 2: PAHs adsorbed on diesel particulate exhaust SRM 1650a. Atmos. Environ. 40, 201–211.

Feilberg, A., Kamens, R., Strommen, M., Nielsen, T., 1999. Modeling the formation, decay, and partitioning of semivolatile nitro-polycyclic aromatic hydrocarbons (nitronaphthalenes) in the atmosphere. Atmos. Environ. 33, 1231–1243.

Fernandez-Ramos, A., Ellingson, B.A., Garrett, B.C., Truhlar, D.G., 2007. Variational transition state theory with multidimensional tunneling. Rev. Comput. Chem. 23, 125.

Finlayson-Pitts, B.J., Pitts Jr., J.N., 1999. Chemistry of the Upper and Lower Atmosphere: Theory, Experiments, and Applications. Academic Press.

Frisch, M., Trucks, G., Schlegel, H.B., Scuseria, G., Robb, M., Cheeseman, J., Scalmani, G., Barone, V., Mennucci, B., Petersson, G., 2009. Gaussian 09, Revision A. 02. Gaussian. Inc., Wallingford, CT 200.

Gao, R., Xu, F., Li, S., Hu, J., Zhang, Q., Wang, W., 2013. Formation of bromophenoxy radicals from complete series reactions of bromophenols with H and OH radicals. Chemosphere 92, 382–390.

Garrett, B.C., Truhlar, D.G., 1979. Generalized transition state theory. Classical mechanical theory and applications to collinear reactions of hydrogen molecules. J. Phys. Chem. 83, 1052–1079.

Ghigo, G., Causà, M., Maranzana, A., Tonachini, G., 2006. Aromatic hydrocarbon nitration under tropospheric and combustion conditions. A theoretical mechanistic study. J. Phys. Chem. A 110, 13270–13282.

Gibson, T., Korsog, P., Wolff, G., 1986. Evidence for the transformation of polycyclic organic matter in the atmosphere. Atmos. Environ. (1967) 20, 1575–1578.

Glowacki, D.R., Liang, C.-H., Morley, C., Pilling, M.J., Robertson, S.H., 2012. MESMER: an open-source master equation solver for multi-energy well reactions. J. Phys. Chem. A 116, 9545–9560.

Gonzalez-Lafont, A., Truong, T.N., Truhlar, D.G., 1991. Interpolated variational transition-state theory: practical methods for estimating variational transition-state properties and tunneling contributions to chemical reaction rates from electronic structure calculations. J. Chem. Phys. 95, 8875–8894.

Govindarajan, M., Karabacak, M., 2012. FT-IR, FT-Raman and UV spectral investigation; computed frequency estimation analysis and electronic structure calculations on 1-nitronaphthalene. Spectrochim. Acta Part A Mol. Biomol. Spectrosc. 85, 251–260.

Hien, T.T., Thanh, L.T., Kameda, T., Takenaka, N., Bandow, H., 2007. Nitro-polycyclic aromatic hydrocarbons and polycyclic aromatic hydrocarbons in particulate matter in an urban area of a tropical region: Ho Chi Minh City, Vietnam. Atmos. Environ. 41, 7715–7725.

Johnson III, R.D., 2005. NIST Computational Chemistry Comparison and Benchmark Database. NIST Standard Reference Database Number 101, Release 15b. 2011. cccbdb. nist. gov.

Kamens, R.M., Guo, J., Guo, Z., McDow, S.R., 1990. Polynuclear aromatic hydrocarbon degradation by heterogeneous reactions with N 2 O 5 on atmospheric particles. Atmos. Environ. Part A. Gen. Top. 24, 1161–1173.

Kawanaka, Y., Matsumoto, E., Wang, N., Yun, S.-J., Sakamoto, K., 2008. Contribution of nitrated polycyclic aromatic hydrocarbons to the mutagenicity of ultrafine particles in the roadside atmosphere. Atmos. Environ. 42, 7423–7428.

Koch, R., Knispel, R., Elend, M., Siese, M., Zetzsch, C., 2007. Consecutive reactions of aromatic-OH adducts with NO, NO2 and O2: benzene, naphthalene, toluene, m- and p-xylene, hexamethylbenzene, phenol, m-cresol and aniline. Atmos. Chem. Phys. 7, 2057–2071.

Kwok, E.S., Harger, W.P., Arey, J., Atkinson, R., 1994. Reactions of gas-phase phenanthrene under simulated atmospheric conditions. Environ. Sci. Technol. 28, 521–527.

Kwok, E.S., Atkinson, R., Arey, J., 1997. Kinetics of the gas-phase reactions of indan, indene, fluorene, and 9, 10-dihydroanthracene with OH radicals, NO3 radicals, and O3. Int. J. Chem. Kinet. 29, 299–309.

Lu, Y.-M., Ding, X.-C., Ye, S.-H., Jin, X.-P., 1999. Mutagenicity of various organic fractions of diesel exhaust particles. Environ. Health Prev. Med. 4, 9–12.

Nielsen, T., 1983. Isolation of polycyclic aromatic hydrocarbons and nitro derivatives in complex mixtures by liquid chromatography. Anal. Chem. 55, 286–290.

Nielsen, T., 1984. Reactivity of polycyclic aromatic hydrocarbons towards nitrating species. Environ. Sci. Technol. 18, 157–163.

Nishino, N., Atkinson, R., Arey, J., 2008. Formation of nitro products from the gas-phase OH radical-initiated reactions of toluene, naphthalene, and biphenyl: effect of NO2 concentration. Environ. Sci. Technol. 42, 9203–9209.

Onchoke, K.K., Chaudhry, S.N., Ojeda, J.J., 2016. Vibrational and electronic spectra of 2-nitrobenzanthrone: an experimental and computational study. Spectrochim. Acta Part A Mol. Biomol. Spectrosc. 153, 402–414.

Organization, W.H, 2003. Environmental Health Criteria 229, Selected Nitro-and Nitro-oxy-polycyclic Aromatic Hydrocarbons. World Health Organization, Geneva, pp. 47–48.

Papa, E., Pilutti, P., Gramatica, P., 2008. Prediction of PAH mutagenicity in human cells by QSAR classification. SAR QSAR Environ. Res. 19, 115–127.

Perrini, G., Tomasello, M., Librando, V., Minniti, Z., 2005. Nitrated polycyclic aromatic hydrocarbons in the environment: formation, occurrences and analysis. Ann. Chim. 95, 567–577.

Pitts, J., Van Cauwenberghe, K., Grosjean, D., Schmid, J., Fitz, D., Belser, W., Knudson, G., Hynds, P., 1978. Atmospheric reactions of polycyclic aromatic hydrocarbons: facile formation of mutagenic nitro derivatives. Science 202, 515–519.

Qu, X., Wang, H., Zhang, Q., Shi, X., Xu, F., Wang, W., 2009. Mechanistic and kinetic studies on the homogeneous gas-phase formation of PCDD/Fs from 2,4,5-trichlorophenol. Environ. Sci. Technol. 43, 4068–4075.

Ramdahl, T., Zielinska, B., Arey, J., Atkinson, R., Winer, A.M., Pitts, J.N., 1986. Ubiquitous occurrence of 2-nitrofluoranthene and 2-nitropyrene in air. Nature 321, 425–427.

Reisen, F., Arey, J., 2005. Atmospheric reactions influence seasonal PAH and nitro-PAH concentrations in the Los Angeles Basin. Environ. Sci. Technol. 39, 64–73.

Reisen, F., Wheeler, S., Arey, J., 2003. Methyl-and dimethyl-/ethyl-nitronaphthalenes measured in ambient air in Southern California. Atmos. Environ. 37, 3653–3657.

Ringuet, J., Leoz-Garziandia, E., Budzinski, H., Villenave, E., Albinet, A., 2012. Particle size distribution of nitrated and oxygenated polycyclic aromatic hydrocarbons (NPAHs and OPAHs) on traffic and suburban sites of a European megacity: Paris (France). Atmos. Chem. Phys. 12, 8877–8887.

Robinson, P.J., Holbrook, K.A., 1972. Unimolecular Reactions. Wiley-Interscience, New York.

Salmeen, I.T., Pero, A.M., Zator, R., Schuetzle, D., Riley, T.L., 1984. Ames assay chromatograms and the identification of mutagens in diesel particle extracts. Environ. Sci. Technol. 18, 375–382.

Schauer, C., Niessner, R., Poschl, U., 2003. Polycyclic Aromatic Hydrocarbons (pah) and Nitro-pah in Urban,Rural,and Alpine Aerosols:Local Differences,Decadal and Seasonal Trends,and Sampling Artefacts.

Shi, X., Yu, W., Xu, F., Zhang, Q., Hu, J., Wang, W., 2015. PBCDD/F formation from radical/radical cross-condensation of 2-chlorophenoxy with 2-bromophenoxy, 2,4-dichlorophenoxy with 2,4-dibromophenoxy, and 2,4,6-trichlorophenoxy with 2,4,6-tribromophenoxy. J. Hazard. Mater. 295, 104–111.

Steckler, R., Chuang, Y., Fast, P., Corchade, J., Coitino, E., Hu, W., Lynch, G., Nguyen, K., Jackells, C., Gu, M., 2002. POLYRATE Version 9.3. University of Minnesota, Minneapolis.

Sun, X., Hu, Y., Xu, F., Zhang, Q., Wang, W., 2012a. Mechanism and kinetic studies for OH radical-initiated atmospheric oxidation of methyl propionate. Atmos. Environ. 63, 14–21.

Sun, X., Zhang, C., Zhao, Y., Bai, J., Zhang, Q., Wang, W., 2012b. Atmospheric chemical reactions of 2,3,7, 8-tetrachlorinated dibenzofuran initiated by an OH radical: mechanism and kinetics study. Environ. Sci. Technol. 46, 8148–8155.

Tokiwa, H., Sera, N., Nakashima, A., Nakashima, K., Nakanishi, Y., Shigematu, N., 1994. Mutagenic and carcinogenic significance and the possible induction of lung cancer by nitro aromatic hydrocarbons in particulate pollutants. Environ. Health Perspect. 102, 107–110.

Vereecken, L., Glowacki, D.R., Pilling, M.J., 2015. Theoretical chemical kinetics in tropospheric chemistry: methodologies and applications. Chem. Rev. 115, 4063–4114.

Vincenti, M., Maurino, V., Minero, C., Pelizzetti, E., 2001. Detection of nitro-substituted polycyclic aromatic hydrocarbons in the Antarctic airborne particulate. Int. J. Environ. Anal. Chem. 79, 257–272.

Wu, R., Li, Y., Pan, S., Wang, S., Wang, L., 2015. The atmospheric oxidation mechanism of 2-methylnaphthalene. Phys. Chem. Chem. Phys. 17, 23413–23422.

Xu, F., Wang, H., Zhang, Q., Zhang, R., Qu, X., Wang, W., 2010a. Kinetic properties for the complete series reactions of chlorophenols with OH radicals—relevance for dioxin formation. Environ. Sci. Technol. 44, 1399–1404.

Xu, F., Yu, W., Gao, R., Zhou, Q., Zhang, Q., Wang, W., 2010b. Dioxin formations from the radical/radical cross-condensation of phenoxy radicals with 2-chlorophenoxy radicals and 2,4,6-trichlorophenoxy radicals. Environ. Sci. Technol. 44, 6745–6751.

Xu, F., Yu, W., Zhou, Q., Gao, R., Sun, X., Zhang, Q., Wang, W., 2010c. Mechanism and direct kinetic study of the polychlorinated dibenzo-p-dioxin and dibenzofuran formations from the radical/radical cross-condensation of 2, 4-dichlorophenoxy with 2-chlorophenoxy and 2,4,6-trichlorophenoxy. Environ. Sci. Technol. 45, 643–650.

Zhang, Z., Lin, L., Wang, L., 2012. Atmospheric oxidation mechanism of naphthalene initiated by OH radical. A theoretical study. Phys. Chem. Chem. Phys. 14, 2645–2650.

Zhang, Q., Gao, R., Xu, F., Zhou, Q., Jiang, G., Wang, T., Chen, J., Hu, J., Jiang, W., Wang, W., 2014. Role of water molecule in the gas-phase formation process of nitrated polycyclic aromatic hydrocarbons in the atmosphere: a computational study. Environ. Sci. Technol. 48, 5051–5057.

Zhao, Y., Lynch, B.J., Truhlar, D.G., 2004. Development and assessment of a new hybrid density functional model for thermochemical kinetics. J. Phys. Chem. A 108, 2715–2719.

Atmospheric Environment 60 (2012) 460–466

Contents lists available at SciVerse ScienceDirect

Atmospheric Environment

journal homepage: www.elsevier.com/locate/atmosenv

Mechanism and kinetic study on the gas-phase reactions of OH radical with carbamate insecticide isoprocarb

Chenxi Zhang [a], Wenbo Yang [a], Jing Bai [a], Yuyang Zhao [a], Chen Gong [a], Xiaomin Sun [a,b,*], Qingzhu Zhang [a], Wenxing Wang [a]

[a] *Environment Research Institute, Shandong University, Jinan 250100, PR China*

[b] *State Key Laboratory of Solid Lubrication, Lanzhou Institute of Chemical Physics, Chinese Academy of Science, Lanzhou 730000, PR China*

HIGHLIGHTS

- ► OH radicals are more easily added to aromatic ring than to C=O bond.
- ► The rate constant of the MIPC with OH is about 5.1×10^{-12} cm^3 $molecule^{-1}$ s^{-1}.
- ► The OH addition reactions decreases with the temperature increasing.
- ► The H abstraction reactions increases with the temperature increasing.

ARTICLE INFO

Article history:
Received 28 May 2012
Received in revised form
5 July 2012
Accepted 7 July 2012

Keywords:
Isoprocarb
OH radical
Microscopic mechanism
Kinetic study

ABSTRACT

As one of the most important carbamate insecticides, isoprocarb [2-(1-methylethyl) phenyl methylcarbamate, MIPC] is widely used in agricultural and cotton spraying. The atmospheric chemical reaction mechanism and kinetics of MIPC with OH radical have been researched using the density functional theory in this paper. The study shows that OH radical is more easily added to the C atoms of aromatic ring than to carbon-oxygen double bond, while the H atom is abstracted more difficulty from −CONH− group and aromatic ring than from the $-CH_3-$ group and the −CH− group. At room temperature, the total rate constant of MIPC with OH radical is about $5.1 \times 10^{-1}2^{cm3}$ $molecule^{-1}$ s^{-1}. OH radical addition reaction and H atom abstraction reaction are both important for the OH-initiated reaction of MIPC. The energy-rich adducts (MIPC-OH) and the MIPC's radical isomers are open-shell activated radicals and can be further oxidized in the atmosphere.

1. Introduction

Since 1950s, insecticides have been used to increase yield and control pests in agriculture (Sadiki and Poissant, 2008). The residues of insecticides in environment have attracted great concern from the scientific community due to the possible negative effects on human health and the equilibrium of ecosystem (Burrows et al., 2002; Kawahara et al., 2005). The majority of insecticides were acetylcholinesterase inhibitors, 55% of them belonging to the group of organophosphates and 11% to carbamates, whereas the rest of them were pyrethroids, chlorinated hydrocarbons or other insecticides (Schulze et al., 2002). Once insecticides are used, they can move to the atmosphere as vapor by volatilization. In the atmosphere, insecticides may be transported far from their source regions to remote locations such as the Arctic or high elevation ecosystems (Hung et al., 2005; Hageman et al., 2006; Primbs et al., 2008). To assess the atmospheric behavior of these pollutants, it is critical to understand their atmospheric reactions. However, previous research of insecticides focused primarily on organophosphate insecticides including their concentrations, toxicity, transport and degradation mechanisms in the atmosphere (Kwong, 2002; Jaga and Dharmani, 2003; Zhang et al., 2007; Ozcan and Aydin, 2009; Zhou et al., 2009), the related research on carbamate insecticides was seldom reported.

Isoprocarb [2-(1-methylethyl) phenyl methylcarbamate, MIPC], one of the most important carbamates insecticides, is used worldwide in agricultural and cotton spraying since 1970 (Ni et al., 2008). Like most carbamate insecticides, residues of MIPC in the environment become important pollutants which may exert toxic effects on human beings and animals. In this study, MIPC was chosen to describe the degradation mechanism of carbamate

* Corresponding author. Environment Research Institute, Shandong University, Jinan 250100, PR China. Fax: +86 531 8836 1990.
E-mail addresses: sxmwch@sdu.edu.cn, sdzhangcx@163.com (X. Sun).

1352-2310/$ – see front matter © 2012 Elsevier Ltd. All rights reserved.
http://dx.doi.org/10.1016/j.atmosenv.2012.07.015

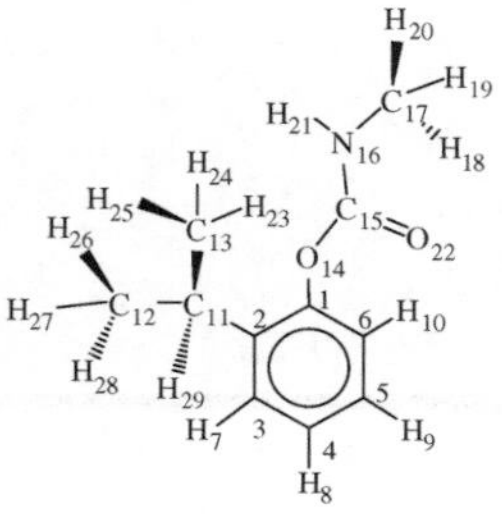

MIPC

Scheme 1. The labeled numbers in the structure of MIPC.

insecticides in atmosphere. With respect to O_3 and NO_3 radicals, the OH radical is the most significant oxidizing species in atmosphere. It is produced from photolysis of ozone by solar ultraviolet radiation, $O_3 + h\nu \rightarrow O_2 + O(^1D)$, followed by $O(^1D) + H_2O \rightarrow 2OH$. OH radical is also substantially recycled, for instance by the reaction $NO + HO_2 \rightarrow OH + NO_2$ (Lin et al., 2008). The OH radical reaction with MIPC is a dominant removal process in the atmosphere.

The experimental methods are difficult to detect and deal with the intermediate radicals due to the rather low vapor pressure, 1.0×10^{-6} mmHg. In addition, the experimental methods are time-consuming, costly, and equipment dependent. Theoretical study can provide accurate predictions for the reaction mechanism through calculating energies and can generate kinetic data for key elementary reaction steps (Xie et al., 2009; Yang et al., 2010; Shi et al., 2011; Zhou et al., 2011). In this work, we employed the density functional theory (DFT) calculation to predict the mechanism and kinetics about the OH-initiated atmospheric reaction of MIPC. On the basis of the quantum chemical information, the rate constants were calculated using canonical variational transition state theory (CVT) with small-curvature tunneling (SCT) over a temperature range of 200–400 K (Baldridge et al., 1989; Gonzalez-Lafont et al., 1991; Liu et al., 1993).

2. Computational methods

The geometrical parameters and vibrational frequencies of all stationary points on the reaction potential energy surface have been calculated using the MPWB1K method with a standard 6–31 + G(d,p) basis set (Zhao and Truhlar, 2004). The reliability of our computational approach adopted for organophosphorus insecticides has been addressed by Zhou and Bao et al., who tested the available experimental data with their computational results (Zhou et al., 2009; Bao et al., 2012). And the maximum relative errors are less than 2.0%. To yield more reliable reaction heat and barrier height, single-point calculations with the largest practical basis set of 6–311 + G(3df,2p) have been performed. Each transition state was verified to connect the designated reactants and products by performing an intrinsic reaction coordinate (IRC) analysis (Fukui, 1981). All the work was carried out using the Gaussian 03 programs (Frisch et al., 2003).

The CVT with SCT contributions is an effective method to calculate rate constants (Yu et al., 2011; Zhang et al., 2011). In this paper, this method was used to calculate the rate constants of elementary reactions over a suitable temperature range. Along the minimum energy path, about 40 points were selected. All the kinetic calculations have been carried out using the POLYRATE 9.7 program (Corchado et al., 2007).

3. Results and discussion

3.1. Initial reactions with OH radical

For convenience, the atom number of MIPC is labeled in Scheme 1. Obviously, the addition of OH radical to the aromatic ring and C=O bond are two kinds of possible reaction channels for the reaction of MIPC with OH radical. In addition, OH is a strongly nucleophilic radical. Thus, H atom abstraction from MIPC should be taken into consideration. The reaction pathways of OH radical addition and H atom abstraction are depicted in Fig. 1 and Fig. 2, in both of which the potential barriers (E^*) and the reaction heat (ΔH) are also marked. The optimized structures of the transition states and the intermediates involved in the reactions of MIPC with OH radical are shown in Fig. 3 and Fig. 4, respectively.

3.1.1. OH radical addition pathways

For the addition to aromatic ring, all the energies of transition states are lower than those of the reactants (MIPC + OH) except that of C_6 addition, which implies the existence of van der Waals (vdW) complex between MIPC and OH radical. The energy of vdW complex is lower than that of the separated reactants and below

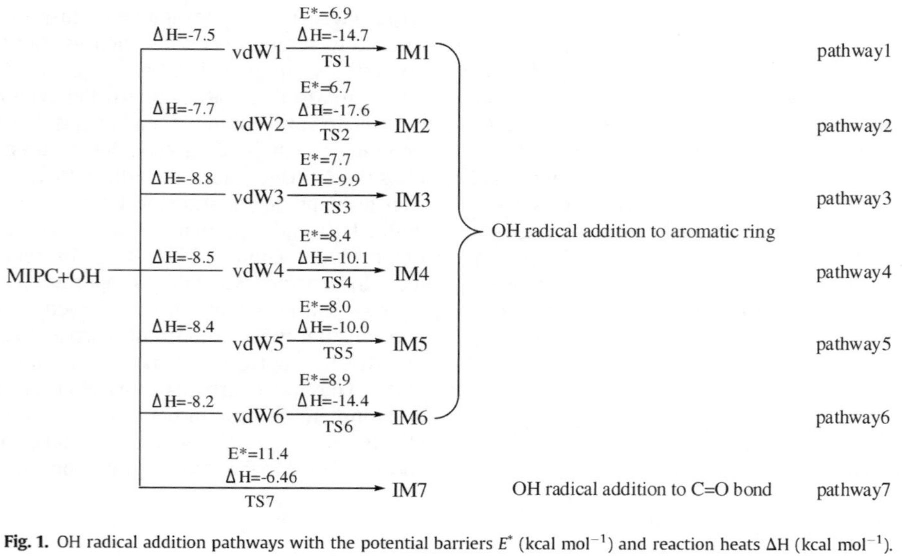

Fig. 1. OH radical addition pathways with the potential barriers E^* (kcal mol^{-1}) and reaction heats ΔH (kcal mol^{-1}).

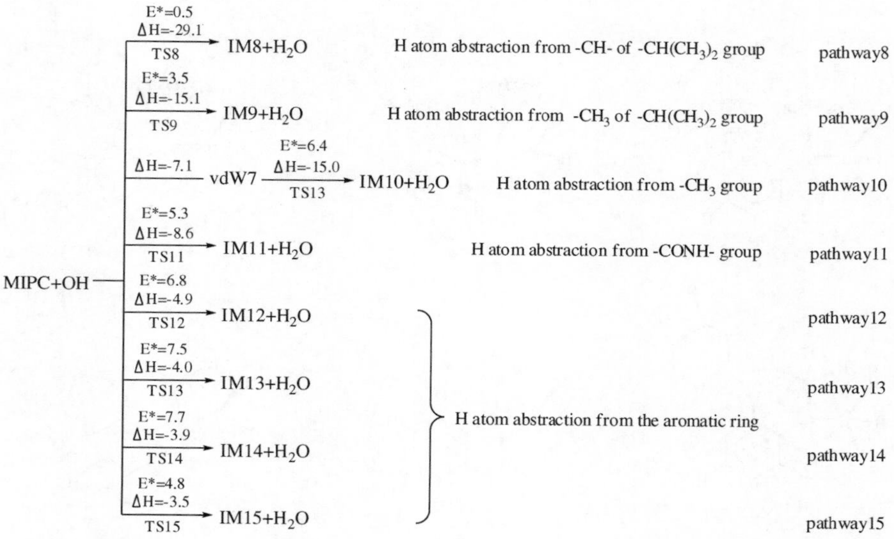

Fig. 2. H atom abstraction pathways with the potential barriers E^* (kcal mol^{-1}) and reaction heats ΔH (kcal mol^{-1}).

the transition structure. Six standard hydrogen-bonds are formed with the bond length ranging from 1.855 to 1.908 Å in these complexes (Supporting Information Figure S1). With the potential barriers of 6.9–8.4 kcal mol^{-1}, OH radicals are added to the aromatic ring. The length of the newly formed C–O bonds in the six transition-states ranges from 1.927 Å to 1.975 Å and is 0.505–0.584 Å longer than that in the corresponding MIPC-OH radical adducts. All these processes are strongly exothermic, releasing at least 9.9 kcal mol^{-1} of energy.

Besides, OH radical can be added to the C atom of C_{15}=O_{22} bond, generating IM7 via a potential barrier of 11.4 kcal mol^{-1}, which is higher than that of the addition to aromatic ring. This process is exothermic and gives out 6.5 kcal mol^{-1} of energy, less than that of the addition to the aromatic ring. By contrast, OH additions to the aromatic ring are energetically favorable reaction pathways.

The energy-rich adducts can undergo further reactions via unimolecular decomposition such as IM1, IM2 and IM7. The schematic energy profile of the potential energy surface (PES) is drawn in Fig. 5. IM1 will open up the C_1–O_{14} bond to generate the products: 2-isopropylphenol (denoted as P1) and $CH_3NHC(O)O$ radical (denoted as P2). This process crosses a potential barrier of 13.1 kcal mol^{-1} and is slightly endothermic, giving out 1.9 kcal mol^{-1} of energy. 2-Isopropylphenol, a kind of alkylphenols, is an important intermediate in the chemical industry of pharmaceuticals, dyes, and antioxidants. Its intravenous median lethal dose (LD50) for mice is about 100 mg kg^{-1} of body weight (TCI, 2006), while lethal dose of MIPC is 66 mg kg^{-1}. The toxicity is reduced sharply. The decomposition of 2-isopropylphenol can occur via dealkylation and rearrangement to produce phenol, propene, and 2-propylphenol over the temperature range of 613–713 K and water densities of 0–0.6 g cm^{-3} (Sato et al., 2002).

For the unimolecular decomposition of IM2, the C_2–C_{11} bond in the transition state TS17 is elongated by 0.721 Å. 2-Hydroxyphenyl methylcarbamate (denoted as P3) and isopropyl (denoted as P4) are produced via an apparent barrier of 20.3 kcal mol^{-1}, and this reaction is predicted to be endothermic, releasing 9.6 kcal mol^{-1} of energy. Isopropyl can be added to O_2 quickly and produce peroxyalkyl radical, which is recognized as a component and promoter of photochemical smog.

With a barrier of 9.9 kcal mol^{-1}, unimolecular decomposition of IM7 results in the formation of the following products: 2-isopropylphenoxy (denoted as P5) and methyl-carbamic acid (denoted as P6). The O_{14}–C_{15} bond fission is strongly exothermic, with 28.9 kcal mol^{-1} of energy released. The unimolecular decomposition of IM7 is most likely to occur under atmospheric conditions. Methyl-carbamic acid is the simplest member of alkylcarbamic acids (R–NHCOOH), which might play a role as a precursor in the formation of glycin (Kayi et al., 2011).

The adducts, denoted as MIPC-OH, are open-shell activated radicals and will be further oxidized in the atmosphere. The following pathways of MIPC-OH are of great significance, which will be investigated in the future.

3.1.2. H atom abstraction pathways

As seen in Fig. 2, eight H-abstraction sites exist in MIPC structure: two in the isopropyl group, one in the methyl group, one in the acylamino group, and four in the aromatic ring.

In the isopropyl group, i.e., the $-CH(CH_3)_2$ group, the OH radical can abstract an H atom from either –CH– group or the $-CH_3$ group. A transition state (TS8) was found in the abstraction of the H_{29} atom from –CH– group. This process has a small potential barrier of 0.5 kcal mol^{-1} and is strongly exothermic, giving out 29.1 kcal mol^{-1} of energy. For the H atom abstraction from the $-CH_3$ group, OH radical abstracts H_{28} atom to generate IM9 via a potential barrier of 3.5 kcal mol^{-1}. This reaction is strongly exothermic by 15.1 kcal mol^{-1}, which shows that the H abstraction from the –CH– group occurs more easily than the H abstraction from the $-CH_3-$ group does.

H atom abstraction from the methyl group, i.e. the $-CH_3$ group, proceeds via the formation of a pre-reaction vdW complex (vdW7). After 7.1 kcal mol^{-1} of energy released, two hydrogen bonds (H–O_{22} and H_{18}–O) are formed in vdW7, which have the bond length of 1.866 and 2.398 Å (Figure S1), respectively. With a barrier of 6.4 kcal mol^{-1}, H_{18} atom transfers from C_{17} to OH radical. The whole H atom abstraction reaction is strongly exothermic by 22.1 kcal mol^{-1}. Therefore, the abstraction from the $-CH_3$ group can occur easily in the atmosphere.

As for the H atom abstraction from the acylamino group, i.e., the –CONH– group, it crosses a barrier of 4.3 kcal mol^{-1} and is exothermic by 8.6 kcal mol^{-1}, which is shown in Fig. 2.

The OH radical can also abstract the H atom and then form the aromatic ring. The abstraction reaction of all the four H atoms: H_7,

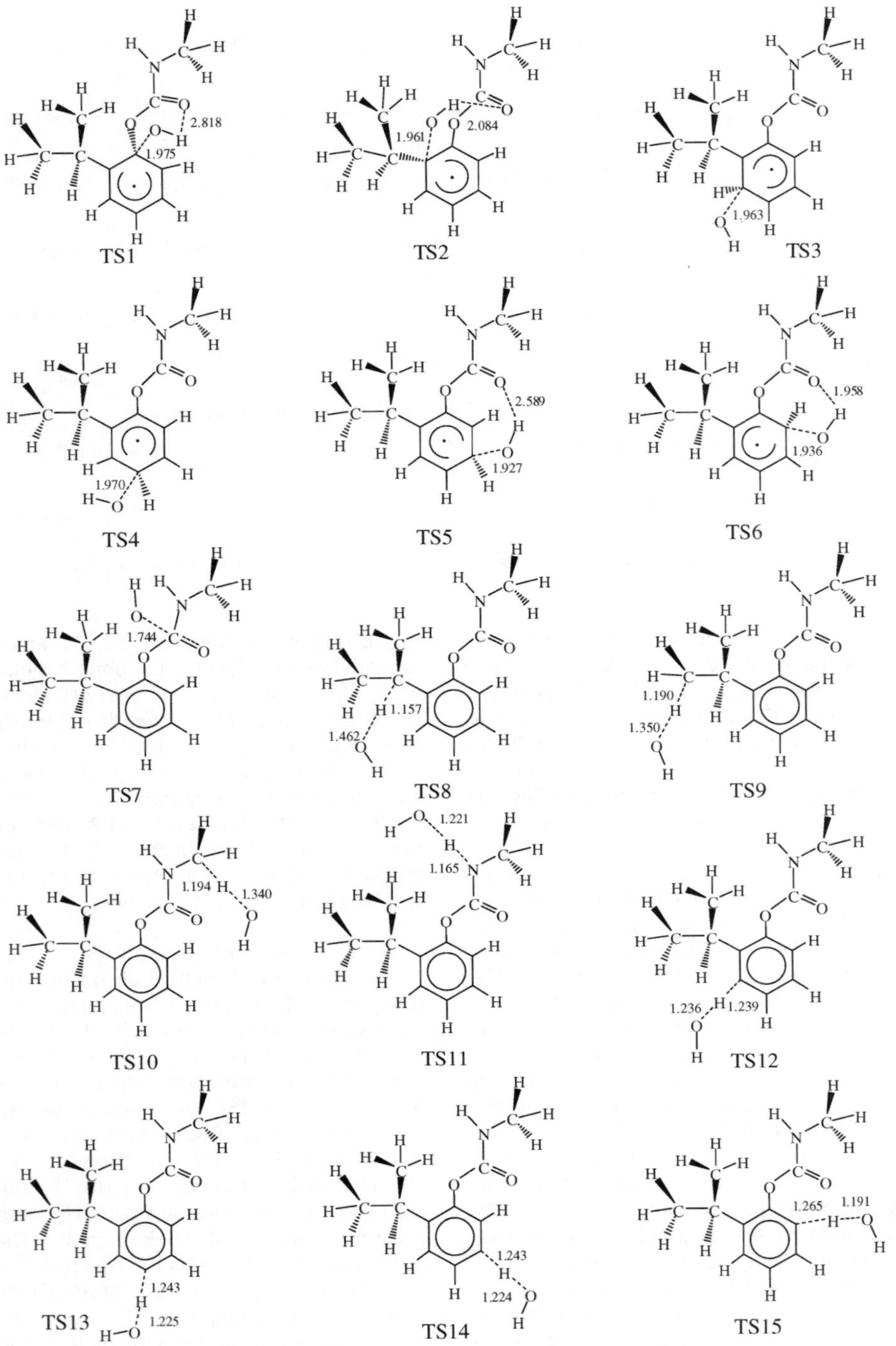

Fig. 3. The main geometry parameters of the transition states in the reaction of MIPC with OH radical at the MPWB1K/6–31 + G(d,p) level.

H_8, H_9 and H_{10}, are considered. The length of breaking C–H bonds in the four transition-states is elongated by 0.144–0.145 Å, longer than the equilibrium value of 1.078 Å. The E^* of these four reactions are similar (6.8–7.7 kal mol^{-1}), except the barrier of H_8-abstraction (4.8 kcal mol^{-1}), which can be attributed to the formation of an intramolecular hydrogen bond with the bond length of 2.127 Å.

Comparison of the eight H atom abstraction channels shows that the H atom is abstracted more easily from the $-CH(CH_3)_2$ group and the $-CH_3-$ group than from the –CONH– group and the aromatic ring. That is, H atom abstractions from the alkyl group can occur readily and are expected to play an important role in the H atom abstraction of MIPC in the atmosphere. The MIPC' radical isomers (IM7–IM15) are quite active and can be further oxidized in the atmosphere, which will be investigated in the future.

3.2. Kinetics

In this study, the kinetic calculations were carried out using the CVT/SCT method. A suitable range of 200–400 K is chosen to study the relationship between the temperature and rate constants. The

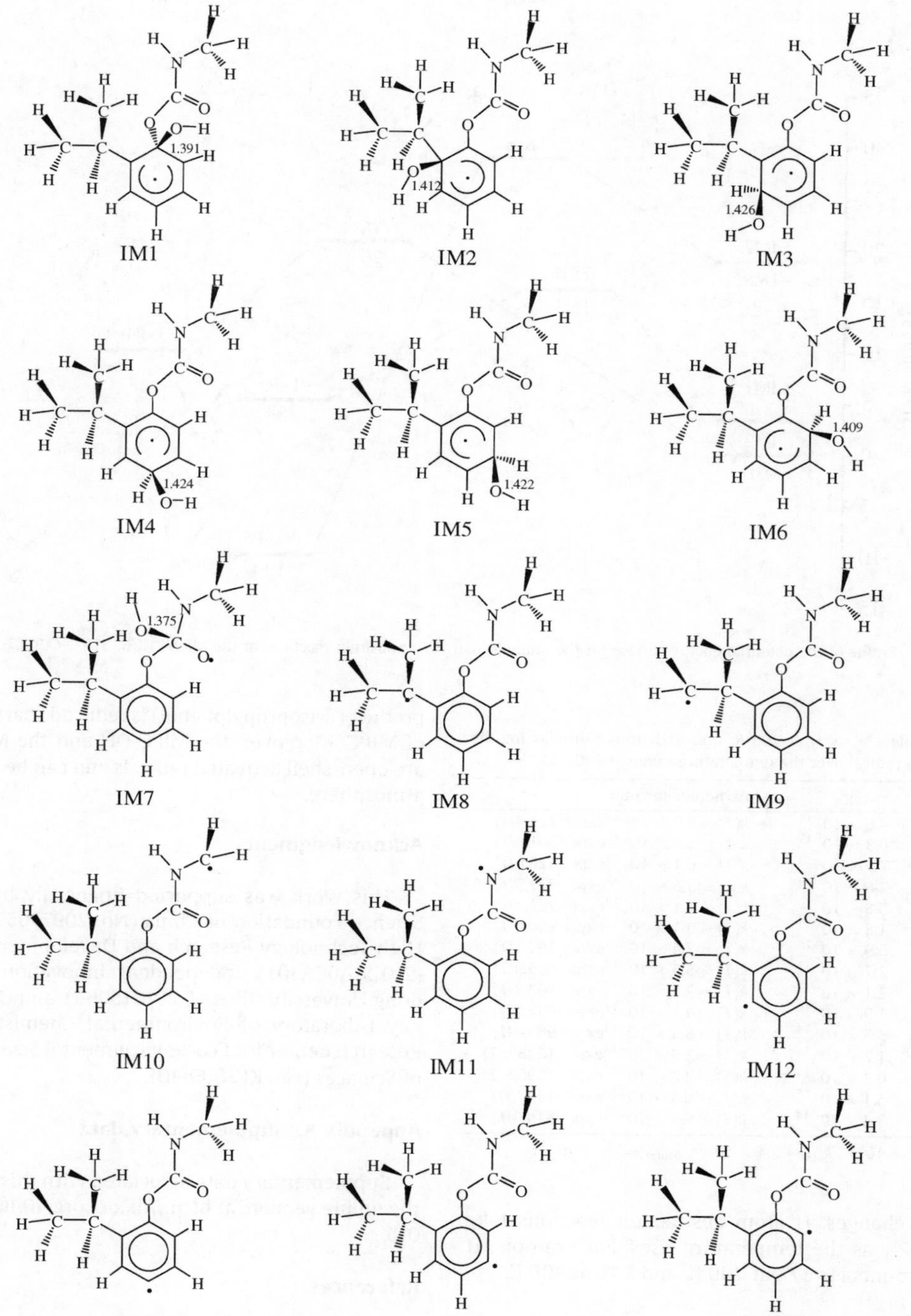

Fig. 4. The main geometry parameters of intermediates in the reaction of MIPC with OH radical at the MPWB1K/6–31 + G(d,p) level.

calculated CVT/SCT rate constants for the OH reaction with MIPC are fitted in the Arrhenius formula $k = A\exp(-E/T)$. The formulas are presented in Table 1, and the rate constants of 298 K are given as well. For the multichannel reaction of MIPC with OH radical, the rate constant of overall OH radical addition reaction is denoted as k_{add}. The rate constant for H atom abstraction is denoted as k_{abs}. The total rate constant for the MIPC + OH reaction is labeled as k_{total}, where $k_{total} = k_{add} + k_{abs}$. The branching ratios ($R$) for the whole reaction are k_{add}/k_{total} and k_{abs}/k_{total}. The variations of k_{add}, k_{abs}, k_{total}, k_{add}/k_{total} and k_{abs}/k_{total} at 200–400 K are shown in Table S1.

At 298 K, the k_{total} of OH with MIPC is 5.1×10^{-12} cm^3 $molecule^{-1}$ s^{-1}, which matches the result of 9.4×10^{-12} cm^3 $molecule^{-1}$ s^{-1} estimated by Estimation Programs Interface (EPI) Suite (SRC: AopWin, 2011). For the average OH radical concentration of 9.7×10^5 molecule cm^{-3} (Prinn et al., 1995), the atmospheric lifetime (τ) of MIPC can be approximated by $\tau = 1/(k_{OH}[OH])$, i.e., the lifetime of MIPC is determined to be 2.3 days. Both OH radical addition reactions and H atom abstraction are important for MIPC at room temperature. But the proportions of OH radical addition reaction and H atom abstraction reaction will vary

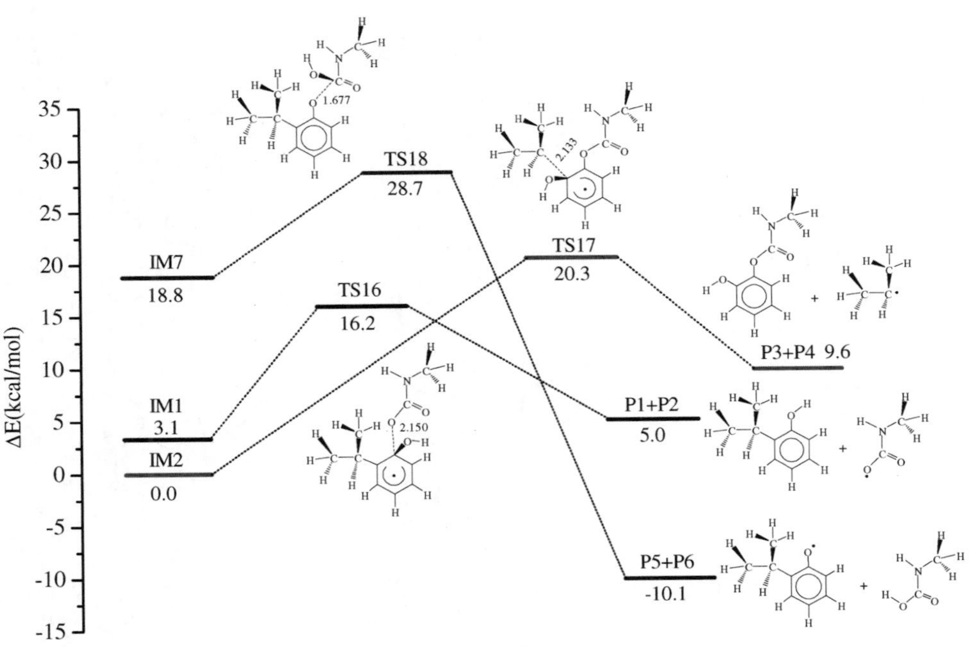

Fig. 5. Profile of the potential energy surface for the unimolecular decomposition reactions at the MPWB1K/6–311 + G(3df,2p) level.

Table 1
Rate constants k (cm^3 $molecule^{-1}$ s^{-1}) at 298 K, and Arrhenius formulas for the reactions of MIPC with OH radical over the temperature range of 200–400 K.

Reactions	$k_{298\ K}$	Arrhenius formulas
MIPC + OH → IM1	4.6×10^{-13}	$k(T) = 5.0 \times 10^{-12}\exp(-711.4/T)$
MIPC + OH → IM2	6.8×10^{-13}	$k(T) = 3.7 \times 10^{-12}\exp(-501.6/T)$
MIPC + OH → IM3	7.5×10^{-13}	$k(T) = 6.1 \times 10^{-12}\exp(-626.6/T)$
MIPC + OH → IM4	2.6×10^{-13}	$k(T) = 3.5 \times 10^{-12}\exp(-777.5/T)$
MIPC + OH → IM5	3.6×10^{-13}	$k(T) = 4.1 \times 10^{-12}\exp(-723.3/T)$
MIPC + OH → IM6	1.8×10^{-13}	$k(T) = 3.1 \times 10^{-12}\exp(-841.8/T)$
MIPC + OH → IM7	5.9×10^{-16}	$k(T) = 7.4 \times 10^{-15}\exp(-752.5/T)$
MIPC + OH → IM8	2.7×10^{-13}	$k(T) = 5.5 \times 10^{-12}\exp(-893.9/T)$
MIPC + OH → IM9	2.1×10^{-13}	$k(T) = 2.7 \times 10^{-12}\exp(-763.3/T)$
MIPC + OH → IM10	1.9×10^{-12}	$k(T) = 4.5 \times 10^{-11}\exp(-937.8/T)$
MIPC + OH → IM11	3.3×10^{-14}	$k(T) = 8.3 \times 10^{-13}\exp(-963.4/T)$
MIPC + OH → IM12	1.2×10^{-14}	$k(T) = 3.9 \times 10^{-13}\exp(-1028.8/T)$
MIPC + OH → IM13	6.3×10^{-15}	$k(T) = 2.5 \times 10^{-13}\exp(-1095.5/T)$
MIPC + OH → IM14	5.1×10^{-15}	$k(T) = 6.5 \times 10^{-13}\exp(-1447.7/T)$
MIPC + OH → IM15	3.4×10^{-14}	$k(T) = 5.3 \times 10^{-15}\exp(-814.9/T)$

Note: At 298 K, $k_{add} = 2.7 \times 10^{-12}$, $k_{abs} = 2.4 \times 10^{-12}$, $k_{total} = 5.1 \times 10^{-12}$.

as the temperature changes. H atom abstraction reactions take place evidently quickly as the temperature rises. For example, H atom abstraction accounts for 37% at 200 K, and 54% at 400 K.

4. Conclusions

From the perspective of thermodynamics and kinetics, OH radicals are more easily added to aromatic ring than to C=O bond, while H atom is abstracted more difficultly from –CONH– group and aromatic ring than from the $-CH(CH_3)_2$ group and the $-CH_3-$ group. At room temperature, the total rate constant of the MIPC with OH radical is about 5.1×10^{-12} cm^3 $molecule^{-1}$ s^{-1}. Both OH radical addition reactions and H atom abstraction reactions are important for the initial reaction of OH radicals with MIPC. As the temperature rises, OH radical addition reaction speed will decrease, while H atom abstraction reaction speed increases.

The MIPC-OH adducts, such as IM1, IM2 and IM7, can undergo reactions via unimolecular decomposition. The toxicity of the product (2-isopropylphenol) is reduced sharply with respect to that of MIPC. Moreover, the MIPC-OH and the MIPC's radical isomers, are open-shell activated radicals and can be further oxidized in the atmosphere.

Acknowledgments

This work was supported financially by the National Nature Science Foundation of China (No. 20977059, 20903062), National High Technology Research and Development Program 863 Project (2012AA06A301), Independent Innovation Foundation of Shandong University (IIFSDU, 2010TS064), and Open Project from State Key Laboratory of Environmental Chemistry and Ecotoxicology, Research Center for Eco-Environmental Sciences, Chinese Academy of Sciences (No. KF2009-10).

Appendix A. Supplementary data

Supplementary data associated with this article can be found, in the online version, at http://dx.doi.org/10.1016/j.atmosenv.2012.07.015.

References

Baldridge, M.S., Gordon, R., Steckler, R., Truhlar, D.G., 1989. Ab initio reaction paths and direct dynamics calculations. J. Phys. Chem. 93, 5107–5119.
Bao, Y., Yang, W.B., Zhang, C.X., Hu, J.T., Sun, X.M., 2012. Mechanism and kinetics study on the OH-initiated oxidation of organophosphorus pesticide trichlorfon in atmosphere. Sci. Total. Environ. 419, 144–150.
Burrows, H.D., Canle, L.M., Santaballa, J.A., Steenken, S., 2002. Reaction pathways and mechanisms of photodegradation of pesticides. J. Photochem. Photobiol. B: Biol. 67, 71–108.
Corchado, J.C., Chuang, Y.Y., Fast, P.L., Villa, J., Hu, W.P., Liu, Y.P., Lynch, G.C., Nguyen, K.A., Jackels, C.F., Melissas, V.S., Lynch, B.J., Rossi, I., Coitino, E.L., Ramos, A.F., Pu, J., Albu, T.V., Garrett, R.B.C., Truhlar, D.G., 2007. Polyrate Version 9.7.
Frisch, M.J., Trucks, G.W., Schlegel, H.B., Pople, J.A., et al., 2003. Gaussian 03, Revision B. 03. Gaussian, Inc, Pittsburgh PA.
Fukui, K., 1981. The path of chemical reactions – the IRC approach. Acc. Chem. Res. 14, 363–368.
Gonzalez-Lafont, A., Truong, T.N., Truhlar, D.G., 1991. Interpolated variational transition-state theory: practical methods for estimating variational transition-

state properties and tunneling contributions to chemical reaction rates from electronic structure calculations. J. Chem. Phys. 95, 8875–8894.
Hageman, K.J., Simonich, S.L., Campbell, D.H., Wilson, G.R., Landers, D.H., 2006. Atmospheric deposition of current-use and historic-use pesticides in snow at national parks in the Western United States. Environ. Sci. Technol. 40, 3174–3180.
Hung, H., Blanchard, P., Halsall, C.J., Bidleman, T.F., Stern, G.A., Fellin, P., Muir, D.C.G., Barrie, L.A., Jantunen, L.M., Helm, P.A., Ma, J., Konoplev, A., 2005. Temporal and spatial variabilities of atmospheric polychlorinated biphenyls (PCBs), organochlorine (OC) pesticides and polycyclic aromatic hydrocarbons (PAHs) in the Canadian Arctic: results from a decade of monitoring. Sci. Total Environ. 342, 119–144.
Jaga, K., Dharmani, C., 2003. Sources of exposure to and public health implications of organophosphate pesticides. Pan. Am. J. Public Health 14, 171–185.
Kawahara, J., Horikoshi, R., Yamaguchi, T., Kumagai, K., Yanagisawa, Y., 2005. Air pollution and young children's inhalation exposure to organophosphorus pesticide in an agricultural community in Japan. Environ. Int. 31, 1123–1132.
Kayi, H., Kaiser, R.I., Head, J.D., 2011. A theoretical investigation of the low energy conformers of the isomers glycine and methylcarbamic acid and their role in the interstellar medium. Phys. Chem. Chem. Phys. 13, 15774–15784.
Kwong, T.C., 2002. Organophosphate pesticides: biochemistry and clinical toxicology. Ther. Drug Monit. 24, 144–149.
Lin, W., Zhu, T., Song, Y., Zou, H., Tang, M.Y., Tang, X.Y., Hu, J.X., 2008. Photolysis of surface O_3 and production potential of OH radicals in the atmosphere over the Tibetan Plateau. J. Geophys. Res. 113, D02309.
Liu, Y.P., Lynch, G.C., Truong, T.N., Lu, D.H., Truhlar, D.G., Garrett, B.C., 1993. Molecular modeling of the kinetic isotope effect for the (1,5)-sigmatropic rearrangement of cis-1,3-pentadiene. J. Am. Chem. Soc. 115, 2408–2415.
Ni, Y.N., Liu, G.L., Kokot, S., 2008. Fluorescence spectrometric study on the interactions of isoprocarb and sodium 2-isopropylphenate with bovine serum albumin. Talanta 76, 513–521.
Ozcan, S., Aydin, M.E., 2009. Organochlorine pesticides in urban air: concentrations, sources, seasonal trends and correlation with meteorological parameters. Clean 37, 343–348.
Primbs, T., Wilson, G., Schmedding, D., Higginbotham, C., Simonich, S.M., 2008. Influence of Asian and Western United States agricultural areas and fires on the atmospheric transport of pesticides in the Western United States. Environ. Sci. Technol. 42, 6519–6525.
Prinn, R.G., Weiss, R.F., Miller, B.R., Huang, J., Alyea, F.N., Cunnold, D.M., Fraser, P.J., Hartley, D.E., Simmonds, P.G., 1995. Atmospheric trends and lifetime of CH_3CCl_3 and global OH concentrations. Science 269, 187–192.
Sadiki, M., Poissant, L., 2008. Atmospheric concentrations and gas-particle partitions of pesticides: comparisons between measured and gas-particle partitioning models from source and receptor sites. Atmos. Environ. 42, 8288–8299.
Sato, T., Sekiguchi, G., Saisu, M., Watanabe, M., Adschiri, T., Arai, K., 2002. Dealkylation and rearrangement kinetics of 2-isopropylphenol in supercritical water. Ind. Eng. Chem. Res. 41, 3124–3130.
Schulze, H., Scherbaum, E., Anastassiades, M., Vorlova, S., Schmid, R.D., Bachmann, T.T., 2002. Development, validation, and application of an acetylcholinesterase-biosensor test for the direct detection of insecticide residues in infant food. Biosens. Bioelectron. 17, 1095–1105.
Shi, J.Q., Liu, H.X., Sun, L., Hou, H.F., Xu, Y., Wang, Z.Y., 2011. Theoretical study on hydrophilicity and thermodynamic properties of polyfluorinated dibenzofurans. Chemosphere 84, 296–304.
SRC, Syracuse Research Corporation, 2011. AopWin Estimation Software, Ver. 4.10. North Syracuse, NY.
TCI America, 2006. Material Safety Data Sheet – 2-Isopropylphenol. Portland OR.
Xie, Q., Chen, J.W., Shao, J.P., Chen, C.E., Zhao, H.X., Hao, C., 2009. Important role of reaction field in photodegradation of deca-bromodiphenyl ether: theoretical and experimental investigations of solvent effects. Chemosphere 76, 1486–1490.
Yang, X., Liu, H., Hou, H.F., Flamm, A., Zhang, X.S., Wang, Z.Y., 2010. Studies of thermodynamic properties and relative stability of a series of polyfluorinated dibenzo-p-dioxins by density functional theory. J. Hazard. Mater. 181, 969–974.
Yu, W.N., Hu, J.T., Xu, F., Sun, X.Y., Gao, R., Zhang, Q.Z., Wang, W.X., 2011. Mechanism and direct kinetics study on the homogeneous gas-phase formation of PBDD/Fs from 2-BP, 2,4-DBP, and 2,4,6-TBP as precursors. Environ. Sci. Technol. 45, 1917–1925.
Zhang, C.X., Sun, T.L., Sun, X.M., 2011. Mechanism for OH-initiated degradation of 2,3,7,8-tetrachlorinated dibenzo-p-dioxins in the presence of O_2 and NO/H_2O. Environ. Sci. Technol. 45, 4756–4762.
Zhang, Q.Z., Qu, X.H., Wang, W.X., 2007. Mechanism of OH-initiated atmospheric photooxidation of dichlorvos: a quantum mechanical study. Environ. Sci. Technol. 41, 6109–6116.
Zhao, Y., Truhlar, D.G., 2004. Hybrid meta density functional theory methods for thermochemistry, thermochemical kinetics, and noncovalent interactions: the MPW1B95 and MPWB1K models and comparative assessments for hydrogen bonding and van der Waals interactions. J. Phys. Chem. A 108, 6908–6918.
Zhou, J., Chen, J.W., Liang, C.H., Xie, Q., Wang, Y.N., Zhang, S.Y., Qiao, X.L., Li, X.H., 2011. Quantum chemical investigation on the mechanism and kinetics of PBDE photooxidation by OH: a case study for BDE-15. Environ. Sci. Technol. 45, 4839–4845.
Zhou, Q., Shi, X.Y., Xu, F., Zhang, Q.Z., He, M.X., Wang, W.X., 2009. Mechanism of OH-initiated atmospheric photooxidation of the organophosphorus insecticide $(C_2H_5O)_3PS$. Atmos. Environ. 43, 4163–4170.

Atmospheric Environment 43 (2009) 4163–4170

Contents lists available at ScienceDirect

Atmospheric Environment

journal homepage: www.elsevier.com/locate/atmosenv

Mechanism of OH-initiated atmospheric photooxidation of the organophosphorus insecticide $(C_2H_5O)_3PS$

Qin Zhou, Xiangyan Shi, Fei Xu, Qingzhu Zhang*, Maoxia He, Wenxing Wang

Environment Research Institute, Shandong University, Jinan 250100, PR China

ARTICLE INFO

Article history:
Received 17 November 2008
Received in revised form
4 May 2009
Accepted 25 May 2009

Keywords:
TEPT
OH radicals
Atmospheric photooxidation
Reaction mechanism
Quantum chemical study

ABSTRACT

O,O,O-triethyl phosphorothioate ($(C_2H_5O)_3PS$, TEPT) is a widely used organophosphorus insecticide. TEPT may be released into the atmosphere where it can undergo transport and chemical transformations, which include reactions with OH radicals, NO_3 radicals and O_3. The mechanism of the atmospheric reactions of TEPT has not been fully understood due to the short-lifetime of its oxidized radical intermediates, and the extreme difficulty in detection of these species experimentally. In this work, we carried out molecular orbital theory calculations for the OH radical-initiated atmospheric photooxidation of TEPT. The profile of the potential energy surface was constructed, and the possible channels involved in the reaction are discussed. The theoretical study shows that OH addition to the P=S bond and H abstractions from the CH_3CH_2O moiety are energetically favorable reaction pathways. The dominant products TEP and SO_2 arise from the secondary reactions, the reactions of OH-TEPT adducts with O_2. The experimentally uncertain dominant product with molecular weight 170 is mostly due to $(C_2H_5O)_2P(S)OH$ and not $(C_2H_5O)_2P(O)SH$.

1. Introduction

Organophosphorus pesticides (OPs) were introduced to substitute for organochlorine pesticides and other chlorinated hydrocarbon pesticides because they have the ability to degrade faster and more easily in the environment (Chambers, 1992; Jaga and Dharmani, 2003). OPs have been used in agricultural and household pest control for more than 40 years (Karczmar, 1970). Due to their widespread use, serious environmental problems are emerging, and they pose an important risk to human health (Westlake et al., 1981; Mcinnes et al., 1996; Hai et al., 1997; Kawahara et al., 2005). OPs belong to the most toxic chemicals in the world (Levin and Rodnitzky, 1976; Donarski et al., 1989; Kwong, 2002). They are responsible for 80% of pesticide-related hospitalizations (Taylor, 2001) and sometimes cause death. In the modern age, nearly everyone is, or has been, exposed to OPs in their home or work environment (Barr et al., 2004).

Pesticides and their precursors may enter the atmosphere as drift during spraying crops or from deposited residue by volatilization and wind erosion where they can undergo transport and chemical transformations (Glotfelty et al., 1990; Van den Berg et al., 1999). Organophosphorus pesticides have been frequently observed in the atmosphere, especially in rain and air in urban and agricultural areas (Coupe et al., 2000; Yao et al., 2008). The frequency of detection in air samples is more than 50% in Mississippi (Coupe et al., 2000). Analysis of particle- and gas-phase fractions of air samples revealed that most OPs exist mainly in the gas phase under the atmospheric conditions (Yao et al., 2008). This has greatly increased the potential for human exposure to these highly toxic materials. To assess the atmospheric behavior of pollutants, it is critical to understand their atmospheric reactions. OH radicals play the most central role in determining the oxidation power of the atmosphere. The reactions initiated by OH radicals have been regarded as the dominant atmospheric loss process of many pesticides (Tuazon et al., 1986; Atkinson et al., 1988; Goodman et al., 1988). Despite their importance, the OH radical-initiated atmospheric reactions of organophosphorus pesticides have received relatively little attention.

In several studies, *O,O,O*-triethyl phosphorothioate ($(C_2H_5O)_3PS$, TEPT) was selected as a "model" alkyl organophosphorus pesticide (Verschoyle and Cabral, 1982; Zhang et al., 2002; Basheer et al., 2007). Two experimental studies are on record for the reaction of TEPT with OH radicals. In 2006, Aschmann and Atkinson (2006) measured the rate constants using relative rate methods at the normal atmospheric pressure over the temperature range of 296–348 K, and reported the Arrhenius expression. Since TEPT contains a P=S bond, it is highly reactive toward OH radicals, but reacts only slowly with NO_3 radicals and O_3. The calculated atmospheric lifetimes are 0.7–1.8 h, 23–45 days and >275 days, respectively. In 2007, Tuazon and his coworkers (Tuazon et al., 2007) used *in situ* atmospheric pressure ionization tandem mass

* Corresponding author. Fax: +86 531 8836 1990.
E-mail address: zqz@sdu.edu.cn (Q. Zhang).

1352-2310/$ – see front matter © 2009 Elsevier Ltd. All rights reserved.
doi:10.1016/j.atmosenv.2009.05.044

spectrometry (API-MS), gas chromatography (GC) and *in situ* Fourier transform infrared spectroscopy (FT-IR) to investigate the products formed from the OH radical-initiated reaction of TEPT. The dominant products observed are $(C_2H_5O)_3PO$ (TEP, 54–62% yield), SO_2 (67 ± 10% yield), CH_3CHO (22–40% yield) and one uncertain compound. The API-MS analyses indicate that the molecular weight of this uncertain product is 170, and it was attributed to $(C_2H_5O)_2P(O)SH$ or $(C_2H_5O)_2P(S)OH$. Possible reaction mechanisms were proposed to explain the observed products. However, there is a shortage of direct experimental data associated with the reaction mechanism, largely due to the lack of efficient detection schemes for radical intermediate species.

Quantum calculation is especially suitable for establishing the feasibility of a reaction pathway. In this paper, we have carried out a theoretical study on the OH radical-initiated atmospheric photooxidation reaction of TEPT in order to find favorable reaction pathways and sites. Possible secondary reaction pathways were also studied to find the formation mechanism of secondary pollutants from the OH radical-initiated atmospheric reaction of TEPT. Elucidation of the reaction mechanism is very challenging due to its inherent complexity. The potential energy surface is useful to explain the experimentally observed branching ratios, thermochemical properties, and rate coefficients.

2. Computational method

High-accuracy molecular orbital calculations were carried out for the OH radical-initiated atmospheric photooxidation of TEPT. The Gaussian 03 package (Frisch et al., 2003) was used on an SGI Origin 2000 supercomputer. The geometrical parameters of reactants, transition states, intermediates and products were optimized at the MPWB1K (Becke, 1996; Adamo and Barone, 1998; Zhao and Truhlar, 2004) level with a standard 6-31G(d,p) basis set. The MPWB1K method that is based on the modified Perdew and Wang 1991 exchange functional (Perdew, 1991) and Becke's 1995 meta correlation functional (Becke, 1996) is a hybrid density functional theory (HDFT) model with excellent performance for thermochemistry, thermochemical kinetics, hydrogen bonding and weak interactions. Furthermore, previous study shows that MPWB1K is an excellent method for prediction of transition state geometries (Zhao and Truhlar, 2004). The mean unsigned errors (MUEs) of transition state geometries optimized at the MPWB1K level, which are the averages of the absolute deviations of calculated values from database values, are 0.02 Å for the five reactions in the SPG15/01 database (Zhao and Truhlar, 2004). The vibrational frequencies were also calculated at the MPWB1K/6-31G(d,p) level in order to determine the nature of the stationary points, the zero-point energy (ZPE), and the thermal contributions to the free energy of activation. Each transition state was verified to connect the designated reactants with products by performing an intrinsic reaction coordinate (IRC) analysis (Fukui, 1981). For a more accurate evaluation of the energetic parameters, a more flexible basis set, 6-311 + G(3df,2p), was employed to determine the energies of the various species. The profile of the potential energy surface was constructed at the MPWB1K/6-311 + G(3df,2p)//MPWB1K/6-31G(d,p) level including ZPE correction.

3. Results and discussion

Due to the absence of experimental information on the thermochemical parameters for the present reaction system, it is difficult to make a direct comparison of the calculated results with experimental data. Thus, we optimized the geometries of $(CH_3O)_3PO$, PS and CH_3CHO. The calculated results at the MPWB1K/6-31G(d,p) level agree well with the available experimental values (Hollenstein and Gunthard, 1971; Huber and Herzberg, 1979; Van Wazer and Ewig, 1986) and the maximum relative error is less than 2.0%.

3.1. The reaction of TEPT with OH radicals

Addition of OH to the P=S bond is a possible pathway for the reaction of TEPT with OH radicals. Since OH is a strongly nucleophilic radical, H abstraction from the alkyl group in TEPT should be another possible reaction pathway. Altogether, four possible reaction pathways, R1–R4, were identified. The profile of the potential energy surface for the reaction of TEPT with OH radicals is shown in Fig. 1. The reaction scheme can be described as follows:

$$\mathrm{TEPT+OH \rightarrow IM-POH \rightarrow TS-POH \rightarrow (C_2H_5O)_2P(S)OH + C_2H_5O} \quad \text{R1 association–elimination}$$

$$\mathrm{\rightarrow IM-SOH \rightarrow TS-SOH \rightarrow P1 + C_2H_5} \quad \text{R2 association–elimination}$$

$$\mathrm{\rightarrow TS-abs1 \rightarrow IM-abs1 + H_2O} \quad \text{R3 H abstraction from} - CH_2 - \text{group}$$

$$\mathrm{\rightarrow TS-abs2 \rightarrow IM-abs2 + H_2O} \quad \text{R4 H abstraction from} - CH_3 \text{ group}$$

3.1.1. Association–elimination pathways

The association–elimination pathways are analyzed first. Two different adduct isomers, addition of the OH radical to the P atom (denoted IM-POH) and addition of OH to the S atom (denoted IM-

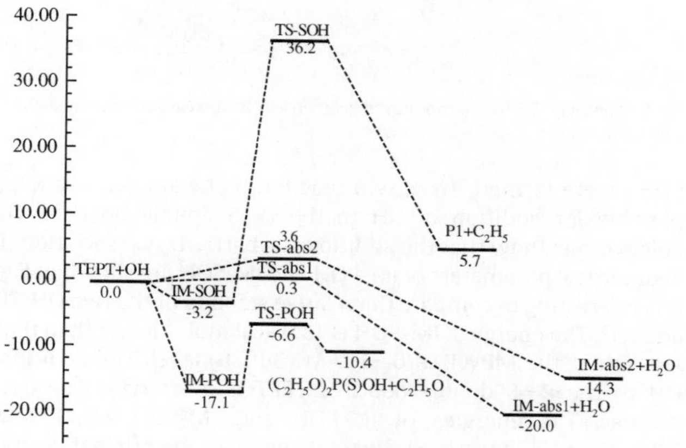

Fig. 1. Profile of the potential energy surface for the reaction of TEPT with OH radicals at the MPWB1K/6-311 + G(3df,2p) level.

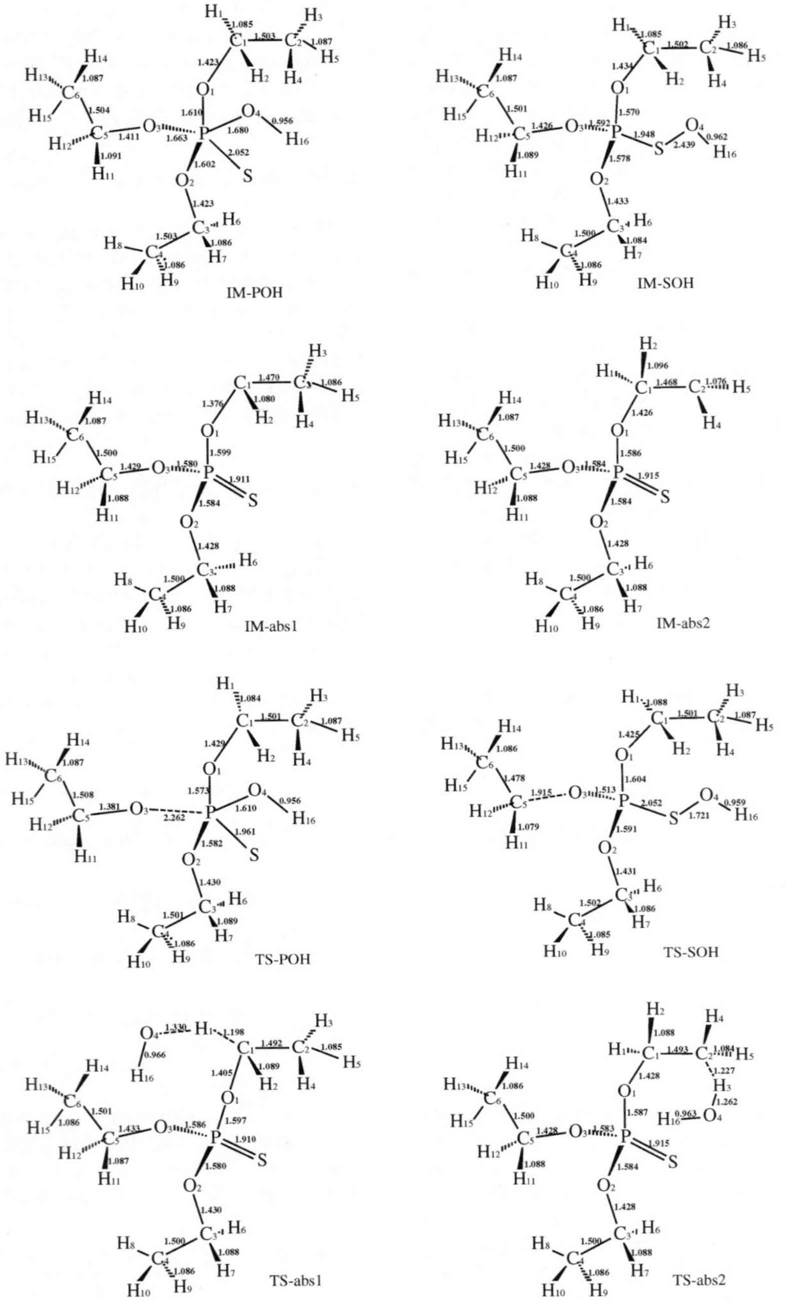

Fig. 2. MPWB1K/6-31G(d,p) optimized geometries for the reactant, intermediates, transition states and products involved in the reaction of TEPT with OH radicals. Distances are in angstroms.

SOH), were formed. Thus, two reaction pathways, R1 and R2, are possible for addition of OH to the P=S double bond in TEPT. Calculations show that the addition is a barrierless association. The geometrical parameters of IM-POH and IM-SOH are shown in Fig. 2. It is interesting to compare the relative stability of the two OH-TEPT adducts. The energy of IM-POH is 13.9 kcal mol^{-1} lower than that of IM-SOH at the MPWB1K/6-311 + G(3df,2p) level. The addition of OH to the P=S double bond of TEPT is a strongly exothermic process. The energies of IM-POH and IM-SOH are 17.1 and 3.2 kcal mol^{-1} lower than the total energy of the original reactants (TEPT and OH), respectively. The high reaction energies are retained as the internal energy of the adducts. The energy-rich adducts, IM-POH and IM-SOH, can react via unimolecular decomposition and/or with atmospheric O_2.

Unimolecular decomposition of IM-POH results in the products $(C_2H_5O)_2P(S)OH$ and C_2H_5O via an apparent barrier, 10.4 kcal mol^{-1}, at the MPWB1K/6-311 + G(3df,2p) level. The total energies of $(C_2H_5O)_2P(S)OH$ and C_2H_5O are 10.4 kcal mol^{-1} lower than the total energy of the original reactants (TEPT and OH). This process would occur readily under general atmospheric conditions. $(C_2H_5O)_2P(S)OH$ and C_2H_5O should be possible products for the reaction of TEPT with OH radicals. One dominant product with molecular weight 170 was experimentally observed in the chamber in which the OH radical-initiated oxidation reaction of TEPT was simulated under general atmospheric conditions. The molecular weight of $(C_2H_5O)_2P(S)OH$ is 170. C_2H_5O radicals will subsequently be oxidized by O_2 to yield CH_3CHO, which was identified and quantified (22–40% yield) in the reaction chamber (Tuazon et al., 2007). CH_3CHO can not only participate in HO_x production, but also produce PA-radicals and is thus an important precursor for PAN in the atmosphere.

The unimolecular decomposition of IM-SOH with cleavage of the C_5–O_3 bond results in the formation of P1 $[(C_2H_5O)_2P(O)SOH]$ and C_2H_5 radical via the transition state TS-SOH. Fig. 1 shows that this process has a high barrier of 39.4 kcal mol^{-1} and is endothermic by 8.9 kcal mol^{-1} at the MPWB1K/6-311 + G(3df,2p) level. Thus, this decomposition process is energetically unfavorable. In the troposphere, IM-SOH will mainly be removed by reaction with O_2.

The formation of $(C_2H_5O)_2P(O)SH$ (molecular weight 170) via the association–elimination mechanism was also investigated. The formation of $(C_2H_5O)_2P(O)SH$ from the decomposition of IM-POH involves H_{16}-migration from O_4 to S and cleavage of the P–O_3 bond. The potential barrier of H_{16}-migration is 28.1 kcal mol^{-1} at the MPWB1K/6-311 + G(3df,2p) level. Thus, the direct formation of $(C_2H_5O)_2P(O)SH$ from IM-POH is energetically unfavorable. Similarly, there is no energetically feasible formation pathway for $(C_2H_5O)_2P(O)SH$ from the decomposition of IM-SOH.

3.1.2. H-abstraction pathways

Two kinds of H atoms exist in TEPT structure. Therefore, two primary pathways, R3 and R4, were identified: H abstraction from the $-CH_2-$ group, and H abstraction from the $-CH_3$ group.

H abstraction from the $-CH_2-$ moiety proceeds via the transition state TS-abs1. The structures of TS-abs1 are shown in Fig. 2. The transition vector clearly shows the motion of H_1 between C_1 and O_4, with an imaginary frequency of 1119i cm^{-1}. This H-abstraction reaction has a low potential barrier of 0.3 kcal mol^{-1} and is strongly exothermic by 20.0 kcal mol^{-1} at the MPWB1K/6-311 + G(3df,2p) level.

H abstraction from the $-CH_3$ group requires crossing a barrier of 3.6 kcal mol^{-1} and is strongly exothermic by 14.3 kcal mol^{-1}. Comparison of the two H-abstraction channels shows that the H abstraction from the $-CH_2-$ group is easier than the H abstraction from the $-CH_3$ group. This indicates that the H atoms in the $-CH_2-$ group are more activated than the H atoms in the $-CH_3$ group. H abstractions from the alkyl group in TEPT can occur readily and are expected to play an important role for the OH radical-initiated degradation of TEPT in the atmosphere. The products, denoted IM-abs1 and IM-abs2, are open-shell activated radicals and will be further oxidized in the atmosphere.

3.2. Secondary reactions

The discussion above shows that OH addition to the P=S bond and H abstractions from the CH_3CH_2O moiety are energetically favorable reaction pathways for the reaction of TEPT with OH radicals. IM-POH, IM-SOH, IM-abs1 and IM-abs2 are important intermediates produced in the degradation process of TEPT initiated by OH radicals. The conventional view is that these radical intermediates could be oxidized further by ubiquitous O_2 or O_2/NO and removed from the troposphere. Published work (Aschmann and Atkinson, 2006; Tuazon et al., 2007) on the products of oxidation of TEPT in smog chambers via hydroxyl chemistry supports this point. Several important secondary pollutants are produced from the reactions of IM-POH, IM-SOH, IM-abs1 and IM-abs2 with O_2 or O_2/NO.

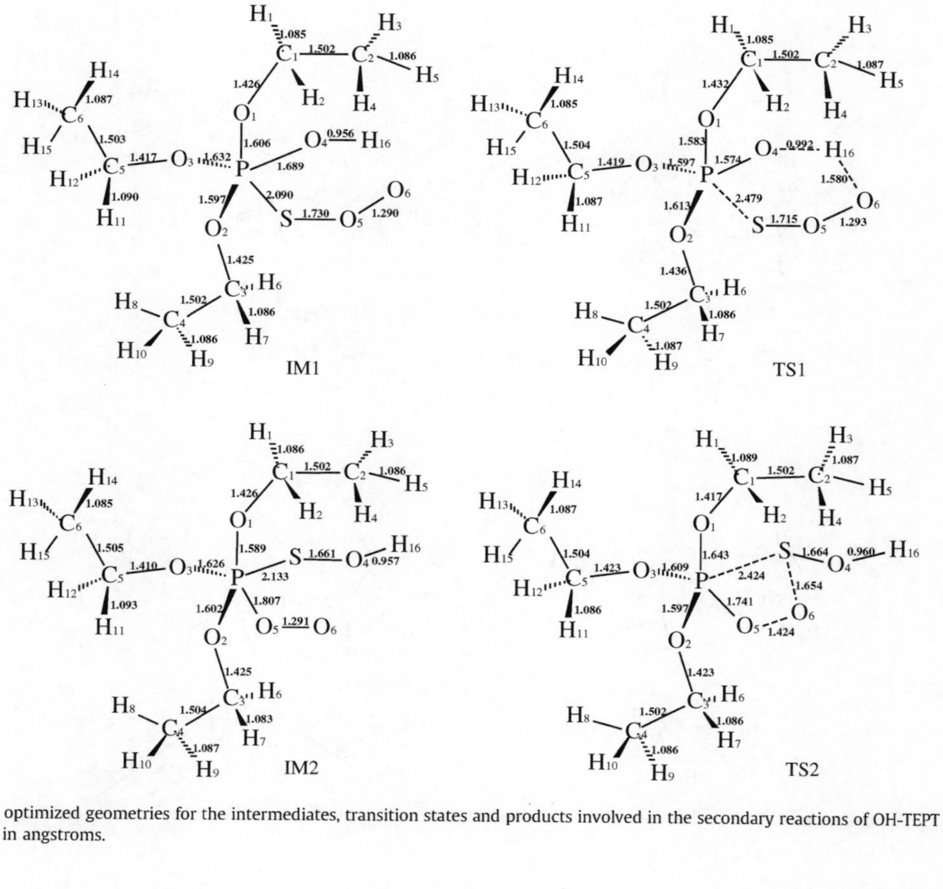

Fig. 3. MPWB1K/6-31G(d,p) optimized geometries for the intermediates, transition states and products involved in the secondary reactions of OH-TEPT adducts, IM-POH and IM-SOH, with O_2. Distances are in angstroms.

3.2.1. Atmospheric reactions of OH-TEPT adducts, IM-POH and IM-SOH

The calculated profile of the potential energy surface shows that the reaction of IM-POH with O_2 is a barrierless association. The structure of the OH–O_2-TEPT adduct, denoted IM1, is depicted in Fig. 3. The process is exothermic by 2.5 kcal mol^{-1}. IM1 further reacts in a direct decomposition to form $(C_2H_5O)_3PO$ (TEP) and SOOH. TS1 represents the transition state for this unimolecular decomposition. The potential barrier of this decomposition is 9.1 kcal mol^{-1}. The process is strongly exothermic by 24.8 kcal mol^{-1}. Similarly, the other OH-TEPT adduct, IM-SOH, can also be oxidized by O_2 in the atmosphere to produce TEP and HOSO. The reaction pathway scheme can be clarified as follows.

$$\mathrm{IM-SOH}+O_2 \rightarrow \mathrm{IM2} \qquad \Delta H = -8.9\ \mathrm{kcal\ mol^{-1}}$$

$$\mathrm{IM2} \rightarrow \mathrm{TS2} \rightarrow (C_2H_5O)_3PO+\mathrm{HOSO} \qquad \Delta E = 20.1\mathrm{kcal\ mol^{-1}} \qquad \Delta H = -107.1\ \mathrm{kcal\ mol^{-1}}$$

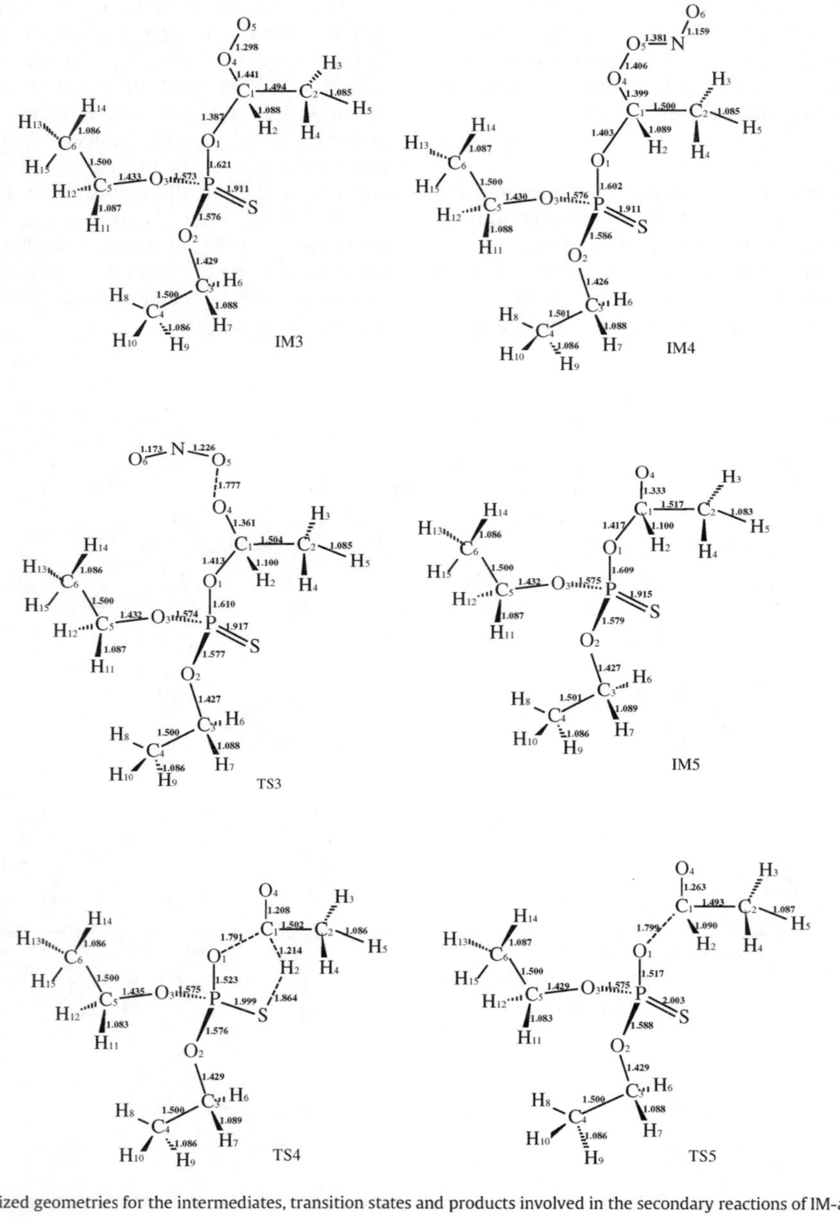

Fig. 4. MPWB1K/6-31G(d,p) optimized geometries for the intermediates, transition states and products involved in the secondary reactions of IM-abs1 with O_2/NO. Distances are in angstroms.

The study above shows that the formation of $(C_2H_5O)_3PO$ (TEP) would occur readily from the OH radical-initiated atmospheric reaction of TEPT. TEP was detected in the chamber with a high formation yield (54–62%) (Tuazon et al., 2007). TEP is a harmful substance with narcotic effects and certain neurotoxic properties. It has relatively high vapor pressure and is subject to long-range air transportation. And also it can further react with OH, O_3 and NO_3. A recent study (Aschmann et al., 2008) shows that new particle formation occurs from the reaction of TEP with OH radicals. Assuming that the aerosol had the same density as the organophosphorus reactant, the aerosol yield, defined as {(aerosol formed, corrected for wall loses)/(organophosphorus compound reacted)}, is estimate to be 6%.

SOOH and HOSO are activated radicals and will subsequently be oxidized by O_2 to yield SO_2, which was identified and quantified (67% yield) in the reaction chamber (Tuazon et al., 2007). TEP and SO_2 are formed from one reaction pathway, the oxidation of OH-TEPT adducts, which is in accordance with the experimental observation that the formation of TEP is accompanied by the formation of SO_2 (Tuazon et al., 2007). Interestingly, the yield of TEP (54–62%) from the OH radical-initiated atmospheric reaction of TEPT is less than that of SO_2 (67%) (Tuazon et al., 2007). This is due to the further reaction of TEP with OH, O_3 and NO_3.

3.2.2. Atmospheric reaction pathway of IM-abs1

H abstraction from the $-CH_2-$ group of TEPT is the energetically feasible reaction pathway for photochemical oxidation of TEPT initiated by OH radicals, leading to the products IM-abs1 and H_2O. IM-abs1 is an activated radical and can further react with the ubiquitous oxygen molecules in the atmosphere to form an organic peroxy radical, IM3, via a barrierless association. The process is

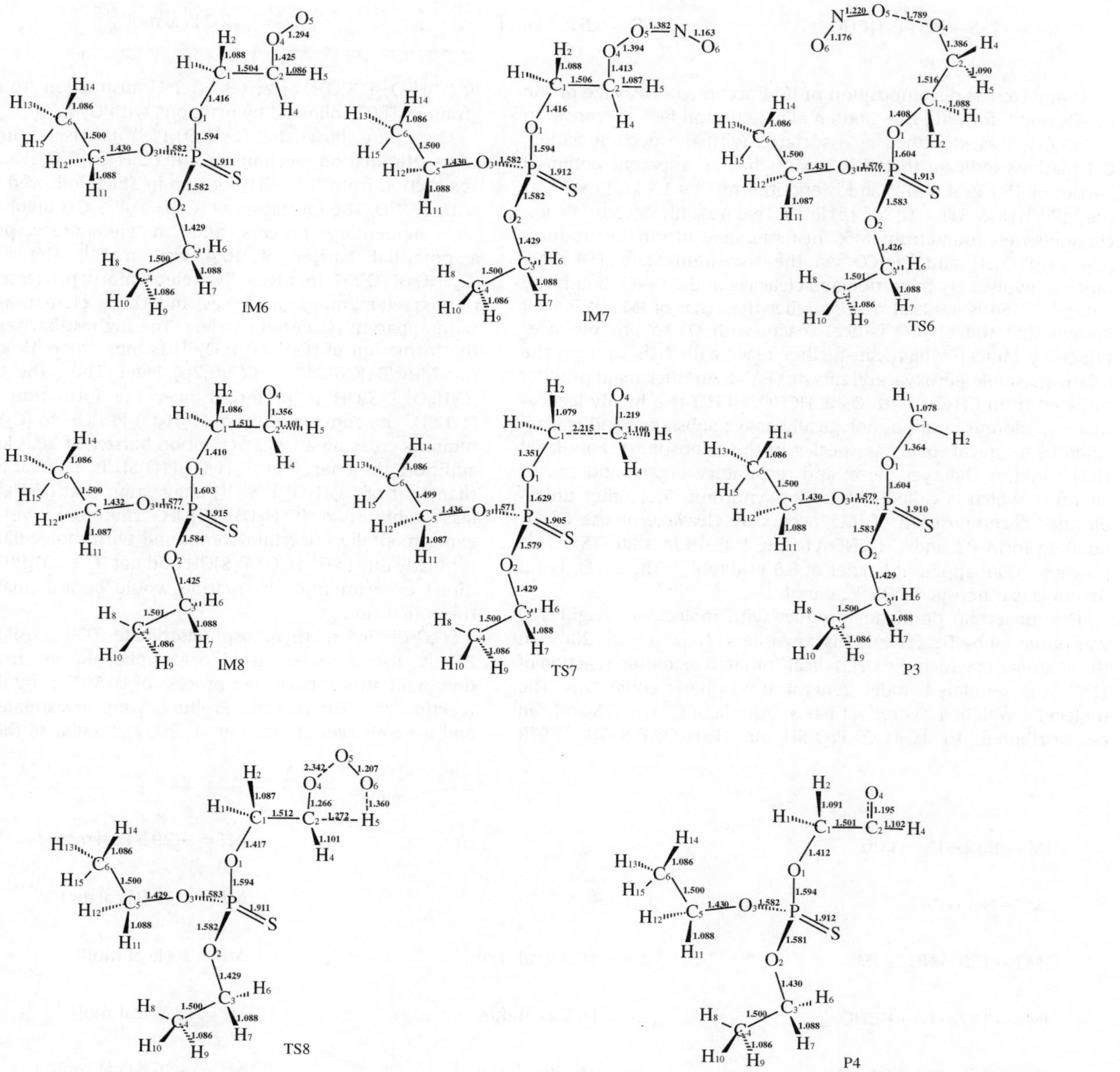

Fig. 5. MPWB1K/6-31G(d,p) optimized geometries for the intermediates, transition states and products involved in the secondary reactions of IM-abs2 with O_2/NO. Distances are in angstroms.

strongly exothermic by 31.5 kcal mol^{-1}. In the troposphere, IM3 will react immediately with ubiquitous NO. The entrance channel of the reaction is exoergic, leading to a vibrationally excited intermediate (denoted IM4), which promptly reacts via unimolecular decomposition. The equilibrium structure of IM4 is illustrated in Fig. 4. The reaction scheme can be described as follows:

$$\mathrm{IM-abs1} + O_2 \rightarrow \mathrm{IM3} \qquad \Delta H = -31.5\ \mathrm{kcal\ mol^{-1}}$$

$$\mathrm{IM3} + \mathrm{NO} \rightarrow \mathrm{IM4} \qquad \Delta H = -19.1\ \mathrm{kcal\ mol^{-1}}$$

$$\mathrm{IM4} \rightarrow \mathrm{TS3} \rightarrow \mathrm{IM5} + NO_2 \qquad \Delta E = 15.1\ \mathrm{kcal mol^{-1}} \qquad \Delta H = 1.3\ \mathrm{kcal\ mol^{-1}}$$

$$\mathrm{IM5} \rightarrow \mathrm{TS4} \rightarrow (C_2H_5O)_2P(O)SH + CH_3CO \qquad \Delta E = 4.9\ \mathrm{kcal\ mol^{-1}} \qquad \Delta H = -1.1\ \mathrm{kcal\ mol^{-1}}$$

$$\mathrm{IM5} \rightarrow \mathrm{TS5} \rightarrow \mathrm{P2} + CH_3CHO \qquad \Delta E = 6.5\ \mathrm{kcal\ mol^{-1}} \qquad \Delta H = -2.2\ \mathrm{kcal\ mol^{-1}}$$

Unimolecular decomposition of IM4 occurs via cleavage of the O_4–O_5 bond, forming NO_2 and an alkoxy radical IM5. A transition state, TS3, was identified as associated with the decomposition. Calculations indicate that this process has an apparent potential barrier of 15.1 kcal mol^{-1} and is endothermic by 1.3 kcal mol^{-1} at the MPWB1K/6-311 + G(3df,2p) level. Two possible decomposition channels were found from IM5. The first one results in the products $(C_2H_5O)_2P(O)SH$ and CH_3CO via the transition state TS4. This process involves H_2-migration and cleavage of the C_1–O_1 bond. The energy of TS4 is 4.9 kcal mol^{-1} higher than that of IM5. It is well known that the CH_3CO radical reacts with O_2 to produce acetylperoxy radical, which can further react with NO_2 to form the relatively stable peroxyacetyl nitrate (PAN). Another main product initiated from CH_3CO with O_2 is HCHO. HCHO is a highly lachrymatory, odorous, and physiologically active substance and is classified as a typical toxic gas species in the atmosphere. Low-level HCHO injures the eyes, nose, and respiratory organs and causes allergies, which is called sick house syndrome. The other unimolecular decomposition of IM5 occurs by cleavage of the C_1–O_1 bond to form P2 and CH_3CHO via the transition state TS5. This process has an apparent barrier of 6.5 kcal mol^{-1}. The C_1–O_1 bond fission is exothermic by 2.2 kcal mol^{-1}.

One uncertain dominant product with molecular weight 170 was observed by Tuazon and his coworkers (Tuazon et al., 2007) in the chamber in which the OH radical-initiated oxidation reaction of TEPT was simulated under general atmospheric conditions. The molecular weight 170 product has a formula of $C_4H_{11}O_3PS$ and can be attributed to $(C_2H_5O)_2P(O)SH$ or $(C_2H_5O)_2P(S)OH$. FT-IR spectrometry cannot conclusively determine the product due to interferences from other products, including diethyl phosphate (a product from the OH radical reaction with TEP). Tuazon et al. (2007) suggested that the formation of $(C_2H_5O)_2P(O)SH$ or $(C_2H_5O)_2P(S)OH$ arises from an association–elimination process: $OH + (C_2H_5O)_3PS \leftrightarrow (C_2H_5O)_3P(S)OH \rightarrow C_2H_5O + (C_2H_5O)_2P(O)SH$ or $(C_2H_5O)_2P(S)OH$, or arises after H abstraction from the $-CH_2-$ group in TEPT, followed by reactions with O_2/NO.

Our study shows that $(C_2H_5O)_2P(S)OH$ arises from the association–elimination mechanism, and $(C_2H_5O)_2P(O)SH$ arises after H abstraction from the $-CH_2-$ group in TEPT, followed by reactions with O_2/NO. The formation of $(C_2H_5O)_2P(S)OH$ involves a barrierless elementary process and an elementary process with a potential barrier of 10.4 kcal mol^{-1}. The formation of $(C_2H_5O)_2P(O)SH$ involves five elementary processes: two barrierless elementary processes and three elementary processes with apparent potential barriers. The highest barrier involved in the formation of $(C_2H_5O)_2P(O)SH$ is more than 15 kcal mol^{-1} at the MPWB1K/6-311 + G(3df,2p) level. Thus, the formation of $(C_2H_5O)_2P(S)OH$ is preferred over the formation of $(C_2H_5O)_2P(O)SH$. The conversion from $(C_2H_5O)_2P(S)OH$ to $(C_2H_5O)_2P(O)SH$ requires crossing a large activation barrier of 26.8 kcal mol^{-1}. In addition, the energy of $(C_2H_5O)_2P(O)SH$ is 3.3 kcal mol^{-1} higher than that of $(C_2H_5O)_2P(S)OH$ suggesting that $(C_2H_5O)_2P(O)SH$ is less stable than $(C_2H_5O)_2P(S)OH$. Therefore, we suggest the experimentally uncertain compound with molecular weight 170 is mostly due to $(C_2H_5O)_2P(S)OH$ and not $(C_2H_5O)_2P(O)SH$. Further direct experimental observation would be anticipated to verify the conclusion.

O,O-diethyl methylphosphonothioate ($(C_2H_5O)_2P(S)CH_3$, DEMPT) is also a widely used organophosphorus insecticide. The dominant atmospheric loss process of DEMPT is by the gas-phase reaction with OH radicals. Products were investigated by Tuazon and his coworker (Tuazon et al., 2007). Similar to the OH radical-

$$\mathrm{IM-abs2} + O_2 \rightarrow \mathrm{IM6} \qquad \Delta H = -29.3\ \mathrm{kcal\ mol^{-1}}$$

$$\mathrm{IM6} + \mathrm{NO} \rightarrow \mathrm{IM7} \qquad \Delta H = -17.9\ \mathrm{kcal\ mol^{-1}}$$

$$\mathrm{IM7} \rightarrow \mathrm{TS6} \rightarrow NO_2 + \mathrm{IM8} \qquad \Delta E = 18.4\ \mathrm{kcal\ mol^{-1}} \qquad \Delta H = 3.3\ \mathrm{kcal\ mol^{-1}}$$

$$\mathrm{IM8} \rightarrow \mathrm{TS7} \rightarrow \mathrm{P3} + \mathrm{HCHO} \qquad \Delta E = 16.3\ \mathrm{kcal\ mol^{-1}} \qquad \Delta H = 13.3\ \mathrm{kcal\ mol^{-1}}$$

$$\mathrm{IM8} + O_2 \rightarrow \mathrm{TS8} \rightarrow \mathrm{P4} + HO_2 \qquad \Delta E = 16.8\ \mathrm{kcal\ mol^{-1}} \qquad \Delta H = -26.8\ \mathrm{kcal\ mol^{-1}}$$

initiated reaction of TEPT, an uncertain dominant product with molecular weight 140 was observed by Tuazon and his coworkers (Tuazon et al., 2007) in the chamber in which the OH radical-initiated atmospheric reaction of DEMPT was simulated. The molecular weight 140 product has a formula of $C_3H_9O_2PS$ and can be attributed to $C_2H_5OP(O)(CH_3)SH$ or $C_2H_5OP(S)(CH_3)OH$. According to the mechanism of the OH radical-initiated oxidation reaction of TEPT, we suggest the experimentally uncertain compound with molecular weight 140 is mostly due to $C_2H_5OP(S)(CH_3)OH$ and not $C_2H_5OP(O)(CH_3)SH$.

3.2.3. Atmospheric reaction pathway of IM-abs2

The atmospheric reaction pathways of IM-abs2 with O_2/NO are similar to the reactions of IM-abs1. The equilibrium structures of intermediates, transition states and products involved in the atmospheric reaction pathways of IM-abs2 with O_2/NO are illustrated in Fig. 5. The results show that the removal reaction of IM-abs2 with O_2/NO is more favored than its unimolecular decomposition. The reaction scheme is shown as follows:

4. Conclusions

The atmospheric oxidation of TEPT was investigated theoretically by high-accuracy molecular orbital calculations. The study shows that OH addition to the P=S bond and H abstractions from the CH_3CH_2O moiety are energetically favorable reaction pathways. The dominant products TEP and SO_2 arise from the secondary reactions, the atmospheric reactions of OH-TEPT adducts with O_2. $(C_2H_5O)_2P(S)OH$ arises from the association–elimination mechanism, and $(C_2H_5O)_2P(O)SH$ arises after H abstraction from the $-CH_2-$ group in TEPT, followed by reactions with O_2/NO. The formation of $(C_2H_5O)_2P(S)OH$ is preferred over the formation of $(C_2H_5O)_2P(O)SH$. Therefore, we suggest the experimentally uncertain compound with molecular weight 170 is mostly due to $(C_2H_5O)_2P(S)OH$ and not $(C_2H_5O)_2P(O)SH$. Under the general atmospheric conditions, the OH radical-initiated atmospheric photooxidation of TEPT would occur readily. Unfortunately, several degradation products have some direct or indirect harmful effects on the environment and the human health.

Acknowledgements

This work was supported by NSFC (National Natural Science Foundation of China, project No. 20737001, 20777047), Shandong Province Outstanding Youth Natural Science Foundation (project No. JQ200804) and the Research Fund for the Doctoral Program of Higher Education of China (project No. 200804220046). The authors thank Dr. Pamela Holt for proofreading the manuscript.

References

Adamo, C., Barone, V., 1998. Exchange functionals with improved long-range behavior and adiabatic connection methods without adjustable parameters: the mPW and mPW1PW models. J. Chem. Phys. 108, 664–675.

Aschmann, S.M., Atkinson, R., 2006. Kinetic and product study of the gas-phase reactions of OH radicals, NO_3 radicals, and O_3 with $(C_2H_5O)_2P(S)CH_3$ and $(C_2H_5O)_3PS$. J. Phys. Chem. A 110, 13029–13035.

Aschmann, S.M., Long, W.D., Atkinson, R., 2008. Rate constants for the gas-phase reactions of OH radicals with dimethyl phosphonate over the temperature range of 278–351 K and for a series of other organophosphorus compounds at ~280 K. J. Phys. Chem. A 112, 4793–4799.

Atkinson, R., Aschmann, S.M., Goodman, M.A., Winer, A.M., 1988. Kinetics of the gas-phase reactions of the OH radical with $(C_2H_5O)_3PO$ and $(CH_3O)_2P(S)Cl$ at 296 ± 2 K. Int. J. Chem. Kinet. 20, 273–281.

Barr, D.B., Bravo, R., Weerasekera, G., Caltabiano, L.M., Whitehead Jr., R.D., Olsson, A.O., Caudill, S.P., Schober, S.E., Pirkle, J.L., Sampson, E.J., Jackson, R.J., Needham, L.L., 2004. Concentrations of dialkyl phosphate metabolites of organophosphorus pesticides in the U.S. population. Environ. Health Perspect. 112, 186–200.

Basheer, C., Alnedhary, A.A., Madhava Rao, B.S., Lee, H.K., 2007. Determination of organophosphorous pesticides in wastewater samples using binary-solvent liquid-phase microextraction and solid-phase microextraction: a comparative study. Anal. Chim. Acta 605, 147–152.

Becke, A.D., 1996. Density-functional thermochemistry. IV. A new dynamical correlation functional and implications for exact-exchange mixing. J. Chem. Phys. 104, 1040–1046.

Chambers, W.H., 1992. Organophosphorus compounds: an overview. In: Chambers, J.E., Levi, P.E. (Eds.), Organophosphates Chemistry, Fate, and Effects. Academic Press, San Diego, CA, pp. 3–17.

Coupe, R.H., Manning, M.A., Foreman, W.T., Goolsby, D.A., Majewski, M.S., 2000. Occurrence of pesticides in rain and air in urban and agricultural areas of Mississippi, April–September 1995. Sci. Total Environ. 248, 227–240.

Donarski, W.J., Dumas, D.P., Heitmeyer, D.P., Lewis, V.E., Raushel, F.M., 1989. Structure–activity relationships in the hydrolysis of substrates by the phosphotriesterase from Pseudomonas diminuta. Biochemistry 28, 4650–4655.

Frisch, M.J., Trucks, G.W., et al., 2003. GAUSSIAN 03, Pittsburgh, PA.

Fukui, K., 1981. The path of chemical reactions - the IRC approach. Acc. Chem. Res. 14, 363–368.

Glotfelty, D.E., Majewski, M.S., Seiber, J.N., 1990. Distribution of several organophosphorus insecticides and their oxygen analogues in a foggy atmosphere. Environ. Sci. Technol. 24, 353–357.

Goodman, M.A., Aschmann, S.M., Atkinson, R., Winer, A.M., 1988. Kinetics of the atmospherically important gas-phase reactions of a series of trimethyl phosphorothioates. Arch. Environ. Contam. Toxicol. 17, 281–288.

Hai, D.Q., Varga, S.I., Matkovics, B., 1997. Organophosphate effects on antioxidant system of carp (*Cyprinus carpio*) and catfish (*Ictalurus nebulosus*). Comp. Biochem. Physiol. 117C, 83–88.

Hollenstein, H., Gunthard, H.H., 1971. Solid state and gas infrared spectra and normal coordinate analysis of 5 isotopic species of acetaldehyde. Spectrochim. Acta A 27, 2027–2060.

Huber, K.P., Herzberg, G., 1979. Molecular Spectra and Molecular Structure. IV. Constants of Diatomic Molecules. Van Nostrand Reinhold Co.

Jaga, K., Dharmani, C., 2003. Sources of exposure to and public health implications of organophosphate pesticides. Pan Am. J. Public Health 14, 171–185.

Introduction. In: Karczmar, A.G. (Ed.), Anticholinesterase Agents, Sect 13 of International Encyclopaedia of Pharmacology and Therapeutics, vol. 1. Pergamon Press, Oxford, New York, Toronto, Sydney, Braunschweig, pp. 1–35.

Kawahara, J., Horikoshi, R., Yamaguchi, T., Kumagai, K., Yanagisawa, Y., 2005. Air pollution and young children's inhalation exposure to organophosphorus pesticide in an agricultural community in Japan. Environ. Int. 31, 1123–1132.

Kwong, T.C., 2002. Organophosphate pesticides: biochemistry and clinical toxicology. Ther. Drug Monit. 24, 144–149.

Levin, H.S., Rodnitzky, R.L., 1976. Behavioral effects of organophosphate pesticides in man. Clin. Toxicol. 9, 391–405.

Mcinnes, P.F., Andersen, D.E., Hoff, D.J., Hooper, M.J., Kinkel, L.L., 1996. Monitoring exposure of nestling songbirds to agricultural application of an organophosphorus insecticide using cholinesterase activity. Environ. Toxicol. Chem. 15, 544–552.

Perdew, J.P., 1991. In: Ziesche, P., Eschig, H. (Eds.), Electronic Structure of Solids 91. Akademie Verlag, Berlin, p. 11.

Taylor, P., 2001. In: Hardman, J.G., Limbird, L.E. (Eds.), Goodman and Gilman's the Pharmaceutical Basis of Therapeutics, tenth ed. McGraw-Hill, New York, p. 175.

Tuazon, E.C., Aschmann, S.M., Atkinson, R., 2007. Products of the gas-phase reactions of OH radicals with $(C_2H_5O)_2P(S)CH_3$ and $(C_2H_5O)_3PS$. J. Phys. Chem. A 111, 916–924.

Tuazon, E.C., Atkinson, R., Aschmann, S.M., Arey, J., Winer, A.M., Pitts Jr., J.N., 1986. Atmospheric loss processes of 1,2-dibromo-3-chloropropane and trimethyl phosphate. Environ. Sci. Technol. 20, 1043–1046.

Van den Berg, F., Kubiak, R., Benjey, W.G., Majewski, M.S., Yates, S.R., Reeves, G.L., Smelt, J.H., Van der Linden, A.M.A., 1999. Emission of pesticides into the air. Water Air Soil Pollut. 115, 195–218.

Van Wazer, J.R., Ewig, C.S., 1986. Ab initio structures of phosphorus acids and esters. 2. Methyl phosphinate, dimethyl phosphonate, and trimethyl phosphate. J. Am. Chem. Soc. 108, 4354–4360.

Verschoyle, R.D., Cabral, J.R.P., 1982. Investigation of the acute toxicity of some trimethyl and triethyl phosphorothioates with particular reference to those causing lung damage. Arch. Toxicol. 51, 221–231.

Westlake, G.E., Bunyan, P.J., Martin, A.D., Stanley, P.I., Steed, L.C., 1981. Organophosphate poisoning. Effects of selected organophosphate pesticides on plasma enzymes and brain esterases of Japanese quail (Coturnix coturnix japonica). J. Agric. Food Chem. 29, 772–778.

Yao, Y., Harner, T., Blanchard, P., Tuduri, L., Waite, D., Poissant, L., Murphy, C., Belzer, W., Aulagnier, F., Sverko, E., 2008. Pesticides in the atmosphere across Canadian agricultural regions. Environ. Sci. Technol. 42, 5931–5937.

Zhang, Z.L., Hong, H.S., Zhou, J.L., Yu, G., 2002. Occurrence and behaviour of organophosphorus insecticides in the River Wuchuan, southeast China. J. Environ. Monit. 4, 498–504.

Zhao, Y., Truhlar, D.G., 2004. Hybrid meta density functional theory methods for thermochemistry, thermochemical kinetics, and noncovalent interactions: the MPW1B95 and MPWB1K models and comparative assessments for hydrogen bonding and van der Waals interactions. J. Phys. Chem. A 108, 6908–6918.

Environmental Science & Technology

Article

pubs.acs.org/est

OH-Initiated Oxidation Mechanisms and Kinetics of 2,4,4′-Tribrominated Diphenyl Ether

Haijie Cao,[†] Maoxia He,*[,†] Dandan Han,[†] Jing Li,[†] Mingyue Li,[†] Wenxing Wang,[†] and Side Yao[‡]

[†]Environment Research Institute, Shandong University, Jinan 250100, P. R. China

[‡]Shanghai Institute of Applied Physics, Chinese Academy of Sciences, P. O. Box 800-204, Shanghai 201800, P. R. China

Supporting Information

ABSTRACT: 2,4,4′-Tribromodiphenyl ether (BDE-28) was selected as a typical congener of polybrominated diphenyl ethers (PBDEs) to examine its fate both in the atmosphere and in water solution. All the calculations were obtained at the ground state. The mechanism result shows that the oxidations between BDE-28 and OH radicals are highly feasible especially at the less-brominated phenyl ring. Hydroxylated dibrominated diphenyl ethers (OH-PBDEs) are formed through direct bromine-substitution reactions (P1~P3) or secondary reactions of OH-adducts (P4~P8). Polybrominated dibenzo-*p*-dioxins (PBDDs) resulting from *o*-OH-PBDEs are favored products compared with polybrominated dibenzofurans (PBDFs) generated by bromophenols and their radicals. The complete degradation of OH adducts in the presence of O_2/NO, which generates unsaturated ketones and aldehydes, is less feasible compared with the *H*-abstraction pathways by O_2. Aqueous solution reduces the feasibility between BDE-28 and the OH radical. The rate constant of BDE-28 and the OH radical is determined to be 1.79×10^{-12} cm^3 $molecule^{-1}$ s^{-1} with an atmospheric lifetime of 6.7 days.

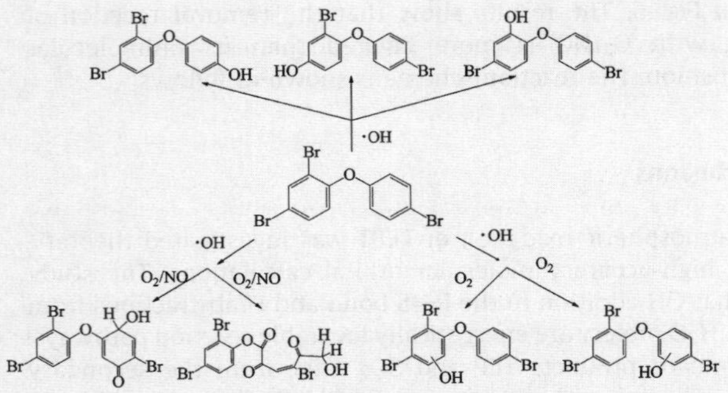

■ INTRODUCTION

Polybrominated diphenyl ethers (PBDEs) have been incorporated into a variety of products such as plastics, electronic devices, building materials, and furniture as fire retardants. They are added to products without covalent bonds and are easily released into the environment. PBDEs have been detected in many samples, such as air,[1] water, sediment,[2] marine animals,[3] and humans.[4−6] 2,4,4′-Tribromodiphenyl ether (BDE-28), 2,2′,4,4′-tetrabromodiphenyl ether (BDE-47), 2,2′,4,4′,5-pentabromodiphenyl ether (BDE-99), and 2,2′,4,4′,5,5′-hexabromodiphenyl ether (BDE-153) have been detected in most samples.[7−11] Today, PBDEs have become ubiquitous worldwide contaminants and have generated deep concern due to their confirmed ability to bioaccumulate[12,13] and be transported over long distances.[14,15]

PBDEs are toxic to the nervous system, immune system, endocrine system, liver, and thyroid.[16,17] A prior study on cats that were suffering from feline hyperthyroidisms, and generally exposed to high concentrations of PBDEs, indicates that PBDEs in dust are significantly correlated with serum thyroxin concentration.[18] PBDEs can also be metabolized into methoxylated PBDEs and hydroxylated PBDEs (OH-PBDEs) which are more toxic than the parent compounds.[19−21] Studies on the concentrations of PBDEs in the environment show an obvious increasing trend in many areas and the trends are expected to continue in the following years[22,23] which raises concern over this serious environmental problem and a need for effective removal methods.

Existing studies of the recalcitrant PBDEs have focused primarily on debromination degradation under ultraviolet light and pyrolysis. Pyrolysis of PBDEs leads to the formation of the notorious polybrominated dibenzo-*p*-dioxins (PBDDs) and polybrominated dibenzofurans (PBDFs) which have been detected in quantitative samples from municipal waste incinerators. The apparent formation of PBDFs was also observed in plastic samples.[24] OH-PBDEs are proposed to be another precursor for photochemical formation of PBDDs with high yields.[25,26]

The transformations of PBDEs and their derivatives have been observed using various catalytic treatments (electrolysis, hydrothermal treatment, catalytic hydrogenation, etc.).[27−29] Mono- and dibrominated diphenyl ethers are inclined to transform into OH-PBDEs with OH radicals.[30] Luo et al. combined reduction and oxidation of BDE-47 in aqueous solution and observed the final degradation products of BDE-47 as phenol and small organic acids.[31] DecaBDE treated with zerovalent iron can slowly generate lower brominated PBDEs.[32] Ultraviolet photolysis can highly promote the debromination process.[33] The authors also pointed out that the reaction with the OH radical predominates in the UV/H_2O_2/AP system. These experiments have provided evidence supporting the debromination and bimolecular

Received: January 8, 2013
Revised: June 17, 2013
Accepted: June 26, 2013
Published: June 26, 2013

oxidation of PBDEs. However, transformation processes under real world conditions are hard to simulate experimentally since the results depend upon many factors (standard samples, catalyzing methods, concentrations, etc). Hence, the precise mechanism for the transformation of PBDEs to PBDDs/PBDFs and the impact of bromine on the property of the PBDEs is vague, and their environmental fates are still unclear. Quantum methods can determine reaction properties precisely and serve as an accurate benchmark for photolysis of PBDEs. Existing theoretical studies about PBDEs have been successful and their results are consistent with the experimental data.[34,35] Moreover, the properties of some PBDEs were calculated for both ground and excited states,[36] and the results indicate that the structures and activities of PBDEs do vary with the pattern of bromine substitution.

To better characterize the atmospheric removal process of PBDEs, BDE-28, which has been detected in many samples including fish, water, and blood plasma,[37−39] was selected as the probe compound. BDE-28 has no anthropogenic source and is mostly generated through debromination of BDE-47 and other highly brominated PBDEs. BDE-28 is easily released into the atmosphere and is suggested to be active to react with oxidants. Briefly, the present combined quantum mechanics and molecular dynamics study has been pursued to gain insight into the feasibility of transformation from PBDEs to OH-PBDEs, PBDD/Fs, and other removal processes of PBDEs.

■ COMPUTATIONAL METHODS

The Gaussian 03 program[40] was used to perform all the calculations on the geometries, energies, frequencies of stationary points, and transition states (TS). The MPWB1K functional[41] was employed in this paper. This method has yielded satisfying results in previous research.[42−44] Geometry optimization and energy calculation were obtained under a basis set of 6-31+G(d,p). In order to obtain more accurate energies, a higher basis set of 6-311+G(3df,2p) using the same method was employed. For each transition state, the intrinsic reaction coordinate (IRC) has been calculated to identify the connections between reactants, transition states, and products. Aqueous calculations were performed using the polarizable continuum model (PCM),[45−47] which created the solute cavity via a set of overlapping spheres. In this paper, water was selected as a solvent, and its dielectric constant was set at 80.0. To examine the thermodynamic properties in water, both reaction heats and Gibbs free energies, including solvation energies in solution, were calculated. As such, the equilibrium constants for elementary reactions were available using the Gibbs free energies.

Master equation calculations were employed to study the kinetic properties using Rice−Ramsperger−Kassel−Marcus (RRKM) theory[48] with the recently developed MESMER program.[49] The Mesmer program has been successfully applied in previous research.[50,51] The microcanonical rate coefficients have been calculated using the RRKM expression:

$$k(E) = \frac{W(E)}{h\rho(E)} \tag{1}$$

where $W(E)$ is the rovibrational sum of states at the TS, $\rho(E)$ is density of states of reactants, and h is Planck's constant. Then, canonical rate coefficients $k(T)$ are determined using the usual equation:

$$k(T) = \frac{1}{Q(T)} \int k(E)\, \rho(E)\, \exp(-\beta E)\, \mathrm{d}E \tag{2}$$

where $Q(T)$ is the reactant partition function.

■ RESULTS AND DISCUSSION

Reaction Mechanism in the Gas Phase. First, the reliability of several DFT methods and MP2 are checked. For B3LYP, MPWB1K, BB1K, M05, and MP2, the reaction enthalpies of biphenyl + H_2 → phenylcyclohexane are −32.40, −48.16, −44.85, −47.10, and −35.24 kcal/mol at the basis of 6-311+G(3df,2p). Apparently, MPWB1K and M05 provide satisfying results compared with experimental values (−47.02 ± 1.2 kcal/mol).[52,53] However, the energy enthalpy of M05 shows an abnormal value (−100 kcal/mol) at the basis set of 6-31+G(d,p). In addition, previous research also proves that MPWB1K is one of the best DFT methods in calculation energies.[54] Therefore, we have selected the MPWB1K method.

Frontier Molecular Orbital Analysis. For BDE-28, the most stable structure (Figure S1) of the three typical conformations of PBDEs are chosen for investigation.[55] BDE-28 is anisomerous, and the dihedral formed by two phenyl rings is 74.06°. The ether bonds are 1.357 Å and 1.365 Å, longer than the corresponding ones of BDE-7 (1.355 Å and 1.360 Å) but shorter than that of BDE-47 (1.361 Å and 1.368 Å), indicating an enhanced stabilizing effect caused by bromine atoms. The highest occupied molecular orbital (HOMO) and the lowest unoccupied molecular orbital (LUMO) of the reactants are shown in Figure S3. For BDE-28, the LUMO and LUMO+1 mainly consist of the antibonding orbital of ring 1. The lone electron pair of the Br atom and π electrons of ring 2 contribute most to the HOMO. And the HOMO−1 locates at ring 1 (mainly the lone electron pair of Br atoms and π electrons of phenyl ring 1). From Figure S3, the energy difference between the HOMO of BDE-28 and the LUMO of the OH radical is 4.13 eV, indicating an electron transfer from the HOMO of BDE-28 to the LUMO of the OH radical. The OH radical is inclined to react with the C atoms of ring 2. The phenomenon shows that bromine atoms will decrease the activity of the phenyl ring.

OH-Initiated Reactions. The OH radical is chosen as the initial oxidant for its high oxidability and abundance in the air. Our previous studies suggest that *OH*-addition pathways play a dominant role compared to hydrogen abstraction pathways.[34,35,56] Thus, only OH addition pathways are discussed in this paper. The phenyl ring with two bromine atoms of BDE-28 is defined as ring 1, and the other phenyl ring is ring 2 for convenience. The reaction heats and the energy barriers of initial reactions are shown in Figure 1.

All the initial reactions are exothermic. Pathways 2, 4, and 10 are direct *Br*-substitution pathways followed by the formation of 6-hydroxyl-4,4′-dibromodiphenyl ether (6-OH-BDE-15, P1), 4-hydroxyl-2,4′-dibromodiphenyl ether (4-OH-BDE-8, P2), and 4′-hydroxyl-2,4-dibromodiphenyl ether (4′-OH-BDE-7, P3), respectively. The rest of the pathways will generate *OH* adducts. Obviously, OH radicals prefer to interact with nonbromine substituted carbon atoms, and *Br*-substitution pathways are not important under natural conditions because of their energy barriers of 5.12−7.15 kcal/mol compared to other hydroxyl addition pathways. Additionally, 2,4-dibromophenoxy (ring 1-O-, R1-O-) shows an *ortho-,para*-directing effect while 4-bromophenoxy (ring 2-O-, R2-O-) acts as a *meta*-directing group. Among these reactions, pathway 8 is the most favorable. Overall, ring 2 is more active than ring 1, which has supported our previous conclusions that bromine atoms will deactivate the same ring. Interestingly, an obvious difference is found between BDE-28 and its congener. For BDE-15, the OH additions to the *ortho*-carbon atoms are barrierless, but the full addition reactions

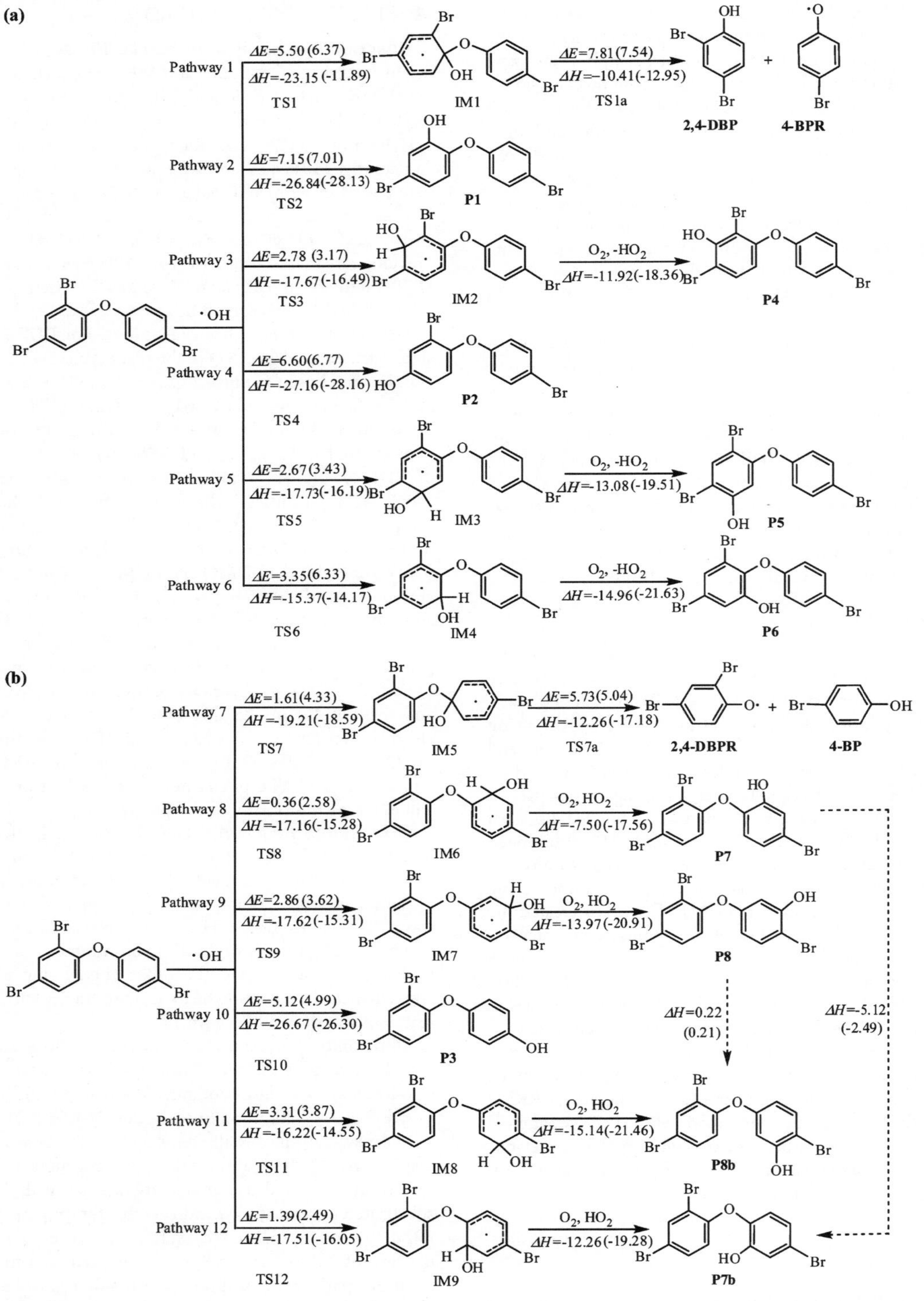

Figure 1. Reaction heats (ΔH) and energy barriers (ΔE) of the initial and subsequent elementary reactions. Values within parentheses are respective values considering the solvent effect. Units of ΔE and ΔH are kcal/mol.

have to overcome a high or moderate energy barrier.[34] This is attributed to the bromine atom in ring 2.

The OH adducts are unstable radical compounds. From Figure 1, IM1 decomposes to 2,4-dibromophenol (2,4-DBP) and the 4-bromophenol radical (4-BPR), while IM5 decomposes to the 2,4-dibromophenol radical (2,4-DBPR) and 4-bromophenol (4-BP). Intermediates IM2–IM9 will react with an atmospheric O_2 molecule as well as other aromatic *OH* adducts. The O_2 molecule has three spin states (two singlet states and one triplet state). The triplet state (the ground state of O_2) is the most

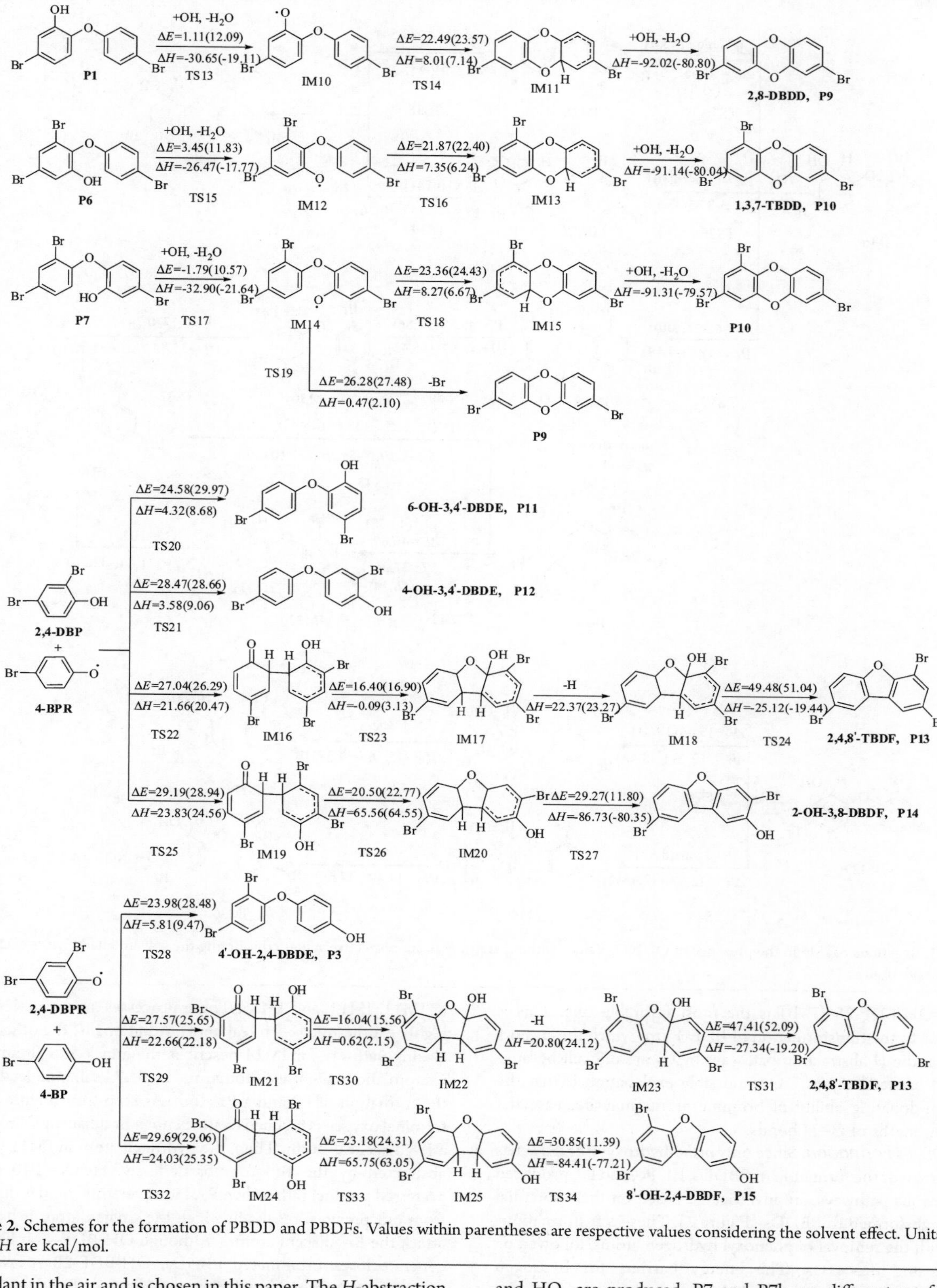

Figure 2. Schemes for the formation of PBDD and PBDFs. Values within parentheses are respective values considering the solvent effect. Units of ΔE and ΔH are kcal/mol.

abundant in the air and is chosen in this paper. The *H*-abstraction pathways by O_2 are barrierless and exothermic, making the *OH*-addition pathways the rate-determining steps of the formation of OH-BDE-28s. Then, stable products (3-OH-BDE-28 (P4), 5-OH-BDE-28 (P5), 6-OH-BDE-28 (P6), P7, P8, P7b, and P8b) and HO_2 are produced. P7 and P7b are different conformers of 2′-OH-BDE-28, while P8 and P8b are different conformers of 3′-OH-BDE-28. Both OH-DBDEs (P1−P3, resulting from direct bromine-substitution) and OH-BDE-28s (P4−P8b, resulting from the stepwise mechanism) are available products,

dx.doi.org/10.1021/es400088v | Environ. Sci. Technol. 2013, 47, 8238−8247

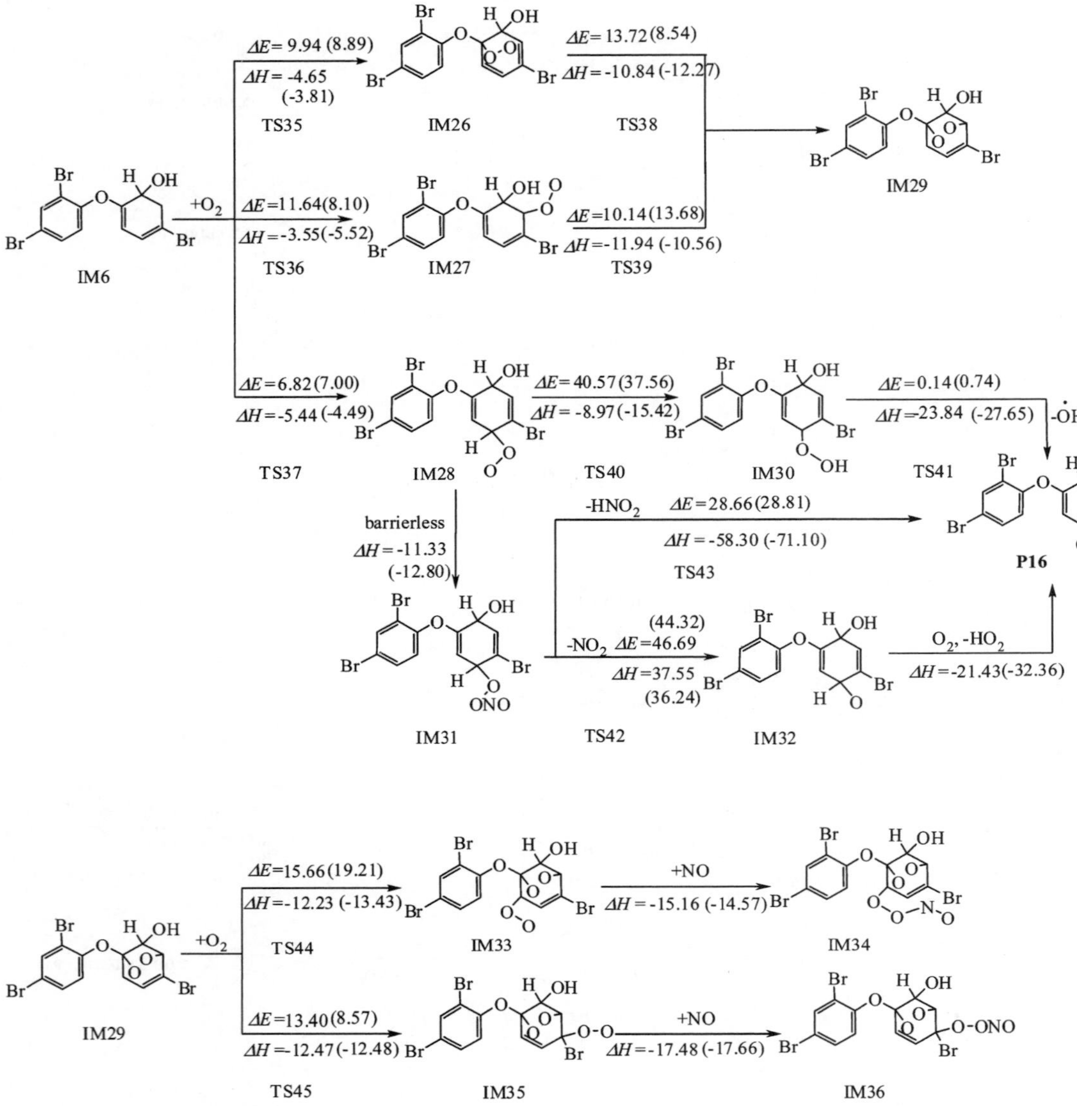

Figure 3. Reactions of IM6 in the presence of O_2/NO. Values within parentheses are respective values considering the solvent effect. Units of ΔE and ΔH are kcal/mol.

but 2′-OH-BDE-28 (P7b) is the most favorable one. This is converse to the transformation of BDE-15 for which the authors find that the *H*-abstraction pathways by O_2 are difficult because of high barrier heights.[34] One possible explanation is that the electron-donating ability of bromine atoms may decrease the bond strengths of C−H bonds.

PBDD/Fs Formation. Since only *ortho*-hydroxyl PBDEs act as precursors for the formation of PBDDs, P1, P6, and P7, produced via previous pathways, are investigated based on their potential to be transformed to PBDDs (Figure 2). These transformations start with the removal of phenoxyl hydrogen atoms, followed by ring-closing processes. The energy barriers for hydrogen abstraction reactions are definitely different. In particular, the abstraction of phenoxyl hydrogen of P7 is barrierless. The ring-closing reactions are endothermic and proceed with energy barriers of ∼23 kcal/mol for attacking the *H*-adjacent C atoms. Once produced, OH radicals will subsequently abstract H atoms in IM11, IM13, and IM15. These processes are barrierless and highly exothermic (∼91 kcal/mol). Additionally, the other ring-closing pathway for IM14 describes the attack at a *Br*-adjacent C atom and is slightly endothermic by 0.47 kcal/mol, leading to the formation of P9 and a Br atom. As seen from Figure 2, the terminal oxygen is inclined to attack not a *Br*-adjacent C atom but a *H*-adjacent C atom. Then, the hydrogen atom in IM11 will be abstracted by the Br atom barrierlessly. Hence, P10 is the preferred product rather than P9. However, this trend is different from our previous results in which the oxygen atom is likely to attack the *Br*-adjacent atom.[57] Although OH-BDE-47 is likely to generate lower-brominated PBDDs, OH-BDE-28 is favorable to transfer to tribrominated dibenzo-*p*-dioxins (TBDDs). This is consistent with high frequency detection of TBDDs but rare detection of DBDDs in experiments.

Recombination of the bromophenols and phenol radicals, which are produced in the early steps, can generate other

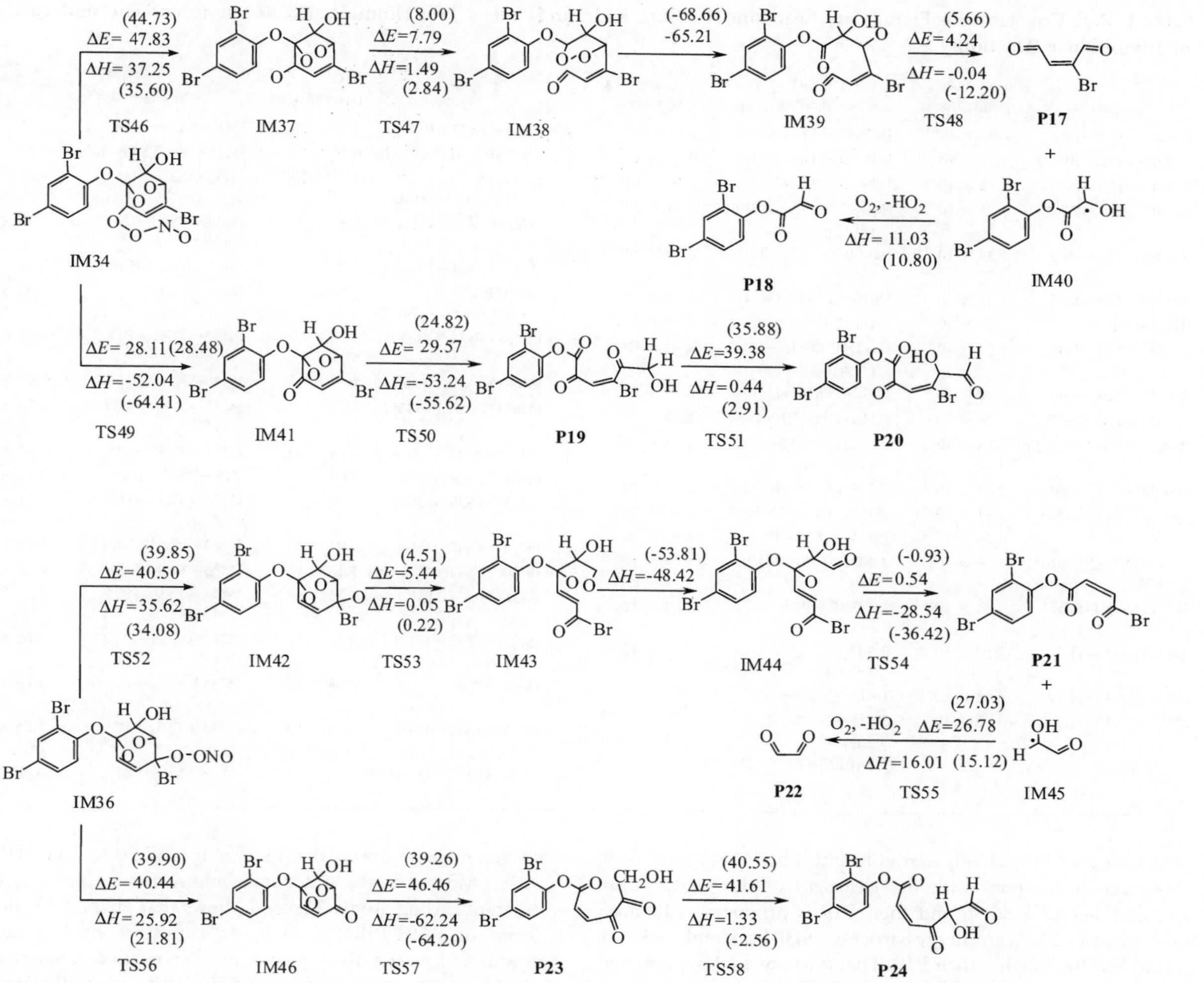

Figure 4. Further reactions of IM10. Values within parentheses are respective values considering the solvent effect. Units of ΔE and ΔH are kcal/mol.

OH-PBDEs or PBDFs. P11 and P12 are formed through the terminal oxygen atom attacking different Br atoms. Apparently, P11 is a more favorable product than P12. That is to say, the *ortho*-Br of 2,4-DBP is easier to release than the *para*-Br. The formation of PBDFs, initiated with the recombination of C—C in 2,4-DBP and 4-BPR, is highly endothermic, followed by ring-closing processes. The abstractions of H atoms in IM17 and IM22 are also barrierless but highly endothermic by ~22 kcal/mol. However, the loss of water molecules in IM18 and IM23 encountered high energy barriers such as 49.48 and 47.41 kcal/mol. The formation of P14 and P15 carries on with modest energy barriers. Hence, dibromodibenzofurans (DBDFs) are more favorable products than tribromodibenzofurans (TBDFs). Comparing the full pathways of the formation of PBDD and PBDFs, PBDDs are the preferred product according to their lower energy barriers.

Further Reaction of IM6 Oxidized by O_2/NO. Experiments show that many aromatic compounds can completely degrade to small molecules.[58,59] In this section, the authors try to ascertain the potential of complete decomposition of BDE-28 in the air. IM6, as the most favorable *OH* adduct, is selected for the above purpose.

Because of the delocalization of the Π bond, IM6 has three active sites (C(1′), C(3′), and C(5′)) to interact with the O_2 molecule (Figure 3). All the O_2 addition processes of IM6 have positive energy barriers and are exothermic. The terminal oxygen atoms of IM26 and IM27 prefer to attack the phenyl carbon atoms to form the five-member intermediate IM29. IM29 has two active sites (C(4′) and C(6′) atoms) and is supposed to form two peroxy compounds, followed by NO addition to the terminal oxygen atom. The process of O_2 attacking the C(5′) atom generates IM28. Both self-decomposition and bimolecular reaction of IM28 produce P16 finally. For IM31, the removal of HNO_2 is obviously easier than the removal of NO_2.

Decompositions of NO Adducts. Both IM34 and IM36 can decompose easily. The reaction heats and the energy barriers of elementary reactions are depicted in Figure 4. A detailed discussion of their decomposition is as follows.

IM34 can decompose by the removal of NO_2 and generate IM37 as well as general decomposition path of most -O-ONO- compounds. Then intramolecular decomposition of IM37 will proceed moderately via a series of step-by-step bond breaking reactions. During the full process of IM34 decomposition referred to above, the removal of NO_2 is the rate-determining step with an energy barrier of 47.83 kcal/mol, which is conceptually unreasonable. Thus, an alternative mechanism is pursued. Our results indicate that the loss of HNO_2 of IM34 only needs to

Table 1. Rate Constants for Elementary Reactions at 298.15 K, 1 atm (Units s^{-1} for Unimolecular Reactions and cm^3 $molecule^{-1}$ s^{-1} for Bimolecular Reactions)

reaction	rate constant	reaction	rate constant
R+OH→TS1→IM1	5.01×10^{-16}	IM7+O_2→P8+HO_2	2.63×10^{-15}
R+OH→TS2→P1+Br	3.38×10^{-18}	IM8+O_2→P8b+HO_2	4.41×10^{-15}
R+OH→TS3→IM2	5.12×10^{-14}	IM9+O_2→P7b+HO_2	2.10×10^{-15}
R+OH→TS4→P2+Br	1.71×10^{-17}	IM5→TS1→2,4-DBPR +4BP	2.44×10^{7}
R+OH→TS5→IM3	6.34×10^{-14}	P1+OH→TS13→IM10 +H_2O	2.16×10^{-14}
R+OH→TS6→IM4	6.31×10^{-15}	IM10→TS14→IM11	2.13×10^{-5}
Ring1+OH	1.21×10^{-13}	IM11+OH→P9+H_2O	6.44×10^{-12}
R+OH→TS7→IM5	1.52×10^{-13}	P6+OH→TS15→IM12 +H_2O	4.49×10^{-18}
R+OH→TS8→IM6	1.06×10^{-12}	IM12→TS16→IM13	2.20×10^{-5}
R+OH→TS9→IM7	4.76×10^{-14}	IM13+OH→P10+H_2O	6.22×10^{-12}
R+OH→TS10→P3+Br	2.16×10^{-15}	P7+OH→TS17→IM14 +H_2O	1.11×10^{-12}
R+OH→TS11→IM8	2.34×10^{-14}	IM14→TS18→IM15	9.51×10^{-6}
R+OH→TS12→IM9	3.81×10^{-13}	IM15+OH→P10+H_2O	6.19×10^{-12}
Ring2+OH	1.67×10^{-12}	IM14→TS19→P9	5.91×10^{-8}
IM1→TS1→2,4-DBP +4BPR	1.15×10^{6}	2,4-DBP+4BPR→P11 +Br	6.14×10^{-28}
IM2+O_2→P4+HO_2	1.44×10^{-15}	2,4-DBP+4BPR→P12 +Br	7.04×10^{-35}
IM3+O_2→P5+HO_2	2.50×10^{-15}	2,4-DBP +4BPR→TS22→IM16	1.39×10^{-21}
IM4+O_2→P6+HO_2	4.02×10^{-15}	IM16→TS23→IM17	0.91
IM6+O_2→P7+HO_2	1.02×10^{-16}	IM18→TS24→P13	1.79×10^{-24}
		2,4-DBP +4BPR→TS25→IM19	1.78×10^{-23}

reaction	rate constant	reaction	rate constant
IM20→TS27→P14+H_2	3.29×10^{-10}	IM33+NO→IM34	2.92×10^{-12}
2,4-DBPR+4BP→P3+Br	6.10×10^{-26}	IM28+O_2→TS45→IM35	4.00×10^{-19}
2,4-DBPR +4BP→TS29→IM21	2.32×10^{-21}	IM35+NO→IM36	4.69×10^{-12}
IM21→TS30→IM22	1.82	IM34→TS46→IM37 +NO_2	3.92×10^{-23}
IM23→TS31→P13+H_2O	4.12×10^{-23}	IM37→TS47→IM38	7.93×10^{6}
2,4-DBPR +4BP→TS32→IM24	1.37×10^{-23}	IM38→IM39	7.00×10^{-12}
IM25→TS34→P15+H_2	4.74×10^{-11}	IM39→TS48→P17 +IM40	1.73×10^{8}
IM6+O_2→TS35→IM26	1.19×10^{-22}	IM40+O_2→P18+HO_2	7.61×10^{-15}
IM6+O_2→TS36→IM27	7.94×10^{-25}	IM34→TS49→IM41 +HNO_2	1.13×10^{-8}
IM6+O_2→TS37→IM28	2.54×10^{-20}	IM41→TS50→P19	7.26×10^{-10}
IM26→TS38→IM29	8.04×10^{-15}	P19→TS51→P20	4.64×10^{-17}
IM27→TS39→IM29	1.78×10^{-9}	IM36→TS52→IM42 +NO_2	1.31×10^{-17}
IM28→TS40→IM30	1.02×10^{-20}	IM42→TS53→IM43	6.84×10^{7}
IM30→TS41→P16+OH	1.45×10^{11}	IM43→IM44	6.00×10^{-12}
IM28+NO→IM31	1.73×10^{-15}	IM44→TS54→P21 +IM45	1.15×10^{10}
IM31→TS42→IM32 +NO_2	5.22×10^{-23}	IM45+O_2→P22+HO_2	4.65×10^{-18}
IM32→P16	3.69×10^{-12}	IM36→TS56→IM46 +NO_2+Br	1.28×10^{-17}
IM31→TS43→P16 +HNO_2	3.38×10^{-9}	IM46→TS57→*P23*	6.65×10^{-24}
IM29+O_2→TS44→IM33	7.52×10^{-21}	*P23*→TS58→P24	3.08×10^{-19}

overcome a 28.11 kcal/mol barrier height. Then IM41 is formed. IM41 can decompose with the cleavages of the O−O bond and C(1′)−C(2′) bond, and then P19 is produced. P19 may isomerize to P20 via an energy barrier of 39.38 kcal/mol, and P20 is 0.44 kcal/mol higher than P19. That is to say, P19 is the major product. Similar to IM34, two decomposition pathways are considered for IM36. The energy barriers of the removal of NO_2 and NO_2 + Br are 40.50 and 40.44 kcal/mol, respectively. Although the following reactions are easy to go on, IM36 is difficult to decompose because of the high energy barriers of rate determining steps.

To conclude, the favorable products of the complete oxidation of BDE-28 are 2-bromo-5-(2,4-dibromophenoxy)-4-hydroxycyclohexa-2,5-dien-1-one (P16) and 2,4-dibromophenyl-(3E)-4-bromo-6-hydroxy-2,5-dioxohex-3-enoate (P19). These processes have to overcome high barrier heights in the ground state, so efficiency catalysts are required to improve the reaction's feasibility.

Solvent Effect for the Title Reaction. In this section, water is selected as the solvent, since in most experimental research, PBDEs have been explored in soil or solution. To compare with the results in the gas phase, the same processes are calculated in the aqueous solution. The reaction heats and the energy barriers in water are available in Figure 1−4 (values in brackets). The changes of reactions in Gibbs free energy (ΔG_{gas}, ΔG_{aq}), free energies of activation ($\Delta G^{\neq}_{gas}$, $\Delta G^{\neq}_{aq}$), and the equilibrium rate constants (K_{gas}, K_{aq}) for elementary reactions are summarized in Table S1 in the Supporting Information.

Water solution obviously increases the energy barriers for most initial steps but stabilizes the products on the whole (Figure 1). The priority of ring 2 is reduced, and the dissolved oxygen in water makes the transfer of BDE-28 to OH-BDE-28s easier. Moreover, the water molecule shows various effects on different carbon atoms. Figure 2 shows that phenoxyl hydrogen atoms of *o*-OH-PBDEs are hard to abstract via OH radicals in water. However, the aqueous solution exhibits a minor effect on other processes for the formation of PBDDs and PBDFs. The energy barrier of the first oxygen molecule addition is decreased, and the O_2 addition to the *ipso*-C atom faces the highest energy barrier. This effect is more obvious in the participation of the second O_2 molecule.

The aqueous study exhibits basically similar trends with the gas phase analysis except for the above differences. Briefly, degradation of BDE-28 in the wastewater treatment and other places is feasible by oxidant catalysts, which generate OH radicals. However, the pollution control progress must be cautious, out of the consideration of the newly generated OH-PBDEs.

Kinetic Study. All the rate coefficients are calculated at 298.15 K and 1 atm. For the bimolecular reactions, the rate constants are divided by the concentrations of corresponding excess reactants. For example, the rate constants of the initial reactions have been divided by the atmospheric concentration of OH radicals (9.75×10^5 molecules cm^{-3}).[60] Microcanonical rate constants for the barrierless reactions were obtained using the inverse laplace transform method. The rate constants for elementary reactions are presented in Table 1.

The total rate constant of BDE-28 and the OH radical is 1.79×10^{-12} cm^3 $molecule^{-1}$ s^{-1}, of which the rate constant of ring 2 and the OH radical accounts for 93.2%. A comparison between our results and experimental values is unavailable because of a lack of experimental data. Alternatively, we have chosen the rate constants of congeners for indirect comparison. Raff and Hites have determined the rate constant of BDE-7 with OH radicals ($3.88_{-0.71}^{+0.87}$ cm^3 $molecule^{-1}$ s^{-1}, 298.15 K).[30] Previous theoretical results show that the rate constants of BDE-7, BDE-15, and BDE-47 with OH radicals are 3.76×10^{-12} cm^3 $molecule^{-1}$ s^{-1}, 7.02×10^{-12} cm^3 $molecule^{-1}$ s^{-1}, and 8.29×10^{-13} cm^3 $molecule^{-1}$ s^{-1} at 298.15 K in the atmosphere, respectively.[34,35,56] Considering the effect generated by the bromine substituted degree, we think our results are reasonable. Apparently, the rate constants of BDE-28 and the OH radical are smaller than those of BDE-7 and BDE-15, but larger than that of BDE-47. This provides evidence for the negative effect generated by bromine atoms. The association of BDE-28 and OH radicals has shown a similar phenomenon to BDE-7; that is, the phenyl ring with less bromine has higher reactivity.

The atmospheric lifetime τ has been calculated through the formula $\tau = 1/k_a[OH]$. k_a is the total rate constant of the initial reactions. $[OH]$ is the atmospheric concentration of the OH radical (9.7×10^5 molecules cm^{-3}). Thus, the atmospheric lifetime of BDE-28 is 6.7 days.

The decompositions of IM1 and IM5 proceed with high efficiency compared to the *H*-abstraction processes by the O_2 molecule. However, we believe the *H*-abstraction processes are preferred because of the abundance of O_2 molecules in the air. The rate constants for the formation of PBDDs are larger than those of PBDFs. Thus, the major product is PBDDs not PBDFs at room temperature, which is consistent with experimental observations.[25,26] Decomposition of IM6 shows that O_2 prefers to interact with C(5′), leading to the formation of IM28. The kinetic conclusion is consistent with the mechanism.

Environmental Relevance. In summary, this study proposes a diverse transformation of BDE-28, determining the major products. The ground-state oxidation mechanism and kinetics indicate that BDE-28 reacts feasibly with the OH radical, especially in the lower-brominated phenyl ring. OH-DBDEs (P1−P3) are formed through a direct bromine-substitution process, and OH-BDE-28s (P4−P8b) are generated via the stepwise mechanism. Among these OH-PBDEs, OH-BDE-28s are preferred thermodynamically. The subsequent O_2 additions to the phenyl ring of the OH adduct are difficult, compared to the *H*-abstraction pathways in a ground state, which is quite different from the actions of BDE-15 in prior work.[34] PBDEs are easily transferred to OH-PBDEs, and the subsequent reaction of OH-PBDEs will continue moderately to generate notorious PBDD/Fs. The kinetic results show that BDE-28 has a shorter lifetime than BDE-47 but longer than BDE-7 and BDE-15, providing strong evidence for the deactivation effect of bromine atoms. At room temperature and atmospheric pressure, PBDDs are more favorable than PBDFs both thermodynamically and kinetically. Overall, oxidation of BDE-28 exhibits some differences with its congeners. We expect these results to serve as supplemental data and contribute to the risk assessment of PBDEs.

■ ASSOCIATED CONTENT

Ⓢ Supporting Information

Additional information, including the structure, frequencies of stationary points and transition states, and Gibbs free energy of elementary reactions as noted in the text, is available free of charge via the Internet at http://pubs.acs.org.

■ AUTHOR INFORMATION

Corresponding Author

*Fax: 86-531-8836 1990. E-mail: hemaox@sdu.edu.cn.

Notes

The authors declare no competing financial interest.

■ ACKNOWLEDGMENTS

This work was supported financially by the National Natural Science Foundation of China (21077067, 21073220, and 21177076). We thank Dr. Struan H. Robertson for providing the Mesmer program.

■ REFERENCES

(1) Birgul, A.; Katsoyiannis, A.; Gioia, R.; Crosse, J.; Earnshaw, M.; Ratola, N.; Jones, K. C.; Sweetman, A. J. Atmospheric polybrominated diphenyl ethers (PBDEs) in the United Kingdom. *Environ. Pollut.* **2012**, *169*, 105−111, DOI: 10.1016/j.envpol.2012.05.005.

(2) Moon, H. B.; Choi, M.; Yu, J.; Jung, R. H.; Choi, H. G. Contamination and potential sources of polybrominated diphenyl ethers (PBDEs) in water and sediment from the artificial Lake Shihwa, Korea. *Chemosphere* **2012**, *88* (7), 837−43, DOI: 10.1016/j.chemosphere.2012.03.091.

(3) Hoenicke, R.; Oros, D. R.; Oram, J. J.; Taberski, K. M. Adapting an ambient monitoring program to the challenge of managing emerging pollutants in the San Francisco Estuary. *Environ. Res.* **2007**, *105* (1), 132−144, DOI: 10.1016/j.envres.2007.01.005.

(4) Hites, R. A. Electron impact and electron capture negative ionization mass spectra of polybrominated diphenyl ethers and methoxylated polybrominated diphenyl ethers. *Environ. Sci. Technol.* **2008**, *42* (7), 2243−2252, DOI: 10.1021/es072064g.

(5) Muir, D. C. G.; Backus, S.; Derocher, A. E.; Dietz, R.; Evans, T. J.; Gabrielsen, G. W.; Nagy, J.; Norstrom, R. J.; Sonne, C.; Stirling, I.; Taylor, M. K.; Letcher, R. J. Brominated flame retardants in polar bears (Ursus maritimus) from Alaska, the Canadian Arctic, East Greenland, and Svalbard. *Environ. Sci. Technol.* **2006**, *40* (2), 449−455, DOI: 10.1021/Es051707u.

(6) Petreas, M.; Nelson, D.; Brown, F. R.; Goldberg, D.; Hurley, S.; Reynolds, P. High concentrations of polybrominated diphenylethers (PBDEs) in breast adipose tissue of California women. *Environ. Int.* **2011**, *37* (1), 190−197, DOI: 10.1016/j.envint.2010.09.001.

(7) Bohlin, P.; Jones, K. C.; Tovalin, H.; Strandberg, B. Observations on persistent organic pollutants in indoor and outdoor air using passive polyurethane foam samplers. *Atmos. Environ.* **2008**, *42* (31), 7234−7241, DOI: 10.1016/j.atmosenv.2008.07.012.

(8) Butt, C. M.; Diamond, M. L.; Truong, J.; Ikonomou, M. G.; Ter Schure, A. F. H. Spatial distribution of polybrominated diphenyl ethers in southern Ontario as measured in indoor and outdoor window organic films. *Environ. Sci. Technol.* **2004**, *38* (3), 724−731, DOI: 10.1021/Es034670r.

(9) Dodder, N. G.; Strandberg, B.; Hites, R. A. Concentrations and spatial variations of polybrominated diphenyl ethers and several organochlorine compounds in fishes from the northeastern United States. *Environ. Sci. Technol.* **2002**, *36* (2), 146−151, DOI: 10.1021/Es010947g.

(10) Farrar, N. J.; Smith, K. E. C.; Lee, R. G. M.; Thomas, G. O.; Sweetman, A. J.; Jones, K. C. Atmospheric emissions of polybrominated diphenyl ethers and other persistent organic pollutants during a major anthropogenic combustion event. *Environ. Sci. Technol.* **2004**, *38* (6), 1681−1685, DOI: 10.1021/Es035127d.

(11) Wilford, B. H.; Shoeib, M.; Harner, T.; Zhu, J. P.; Jones, K. C. Polybrominated diphenyl ethers in indoor dust in Ottawa, Canada: Implications for sources and exposure. *Environ. Sci. Technol.* **2005**, *39* (18), 7027−7035, DOI: 10.1021/Es050759g.

(12) Tomy, G. T.; Palace, V. P.; Halldorson, T.; Braekevelt, E.; Danell, R.; Wautier, K.; Evans, B.; Brinkworth, L.; Fisk, A. T. Bioaccumulation, biotransformation, and biochemical effects of brominated diphenyl

ethers in juvenile lake trout (Salvelinus namaycush). *Environ. Sci. Technol.* **2004**, *38* (5), 1496–1504, DOI: 10.1021/Es035070v.

(13) Wan, Y.; Hu, J. Y.; Zhang, K.; An, L. H. Trophodynamics of polybrominated diphenyl ethers in the marine food web of Bohai Bay, North China. *Environ. Sci. Technol.* **2008**, *42* (4), 1078–1083, DOI: 10.1021/Es0720560.

(14) Gouin, T.; Thomas, G. O.; Cousins, I.; Barber, J.; Mackay, D.; Jones, K. C. Air-surface exchange of polybrominated diphenyl ethers and polychlorinated biphenyls. *Environ. Sci. Technol.* **2002**, *36* (7), 1426–1434, DOI: 10.1021/es011105k.

(15) Schure, A. F.; Larsson, P.; Agrell, C.; Boon, J. P. Atmospheric transport of polybrominated diphenyl ethers and polychlorinated biphenyls to the Baltic Sea. *Environ. Sci. Technol.* **2004**, *38* (5), 1282–1287, DOI: 10.1021/es0348086.

(16) Chevrier, J.; Harley, K. G.; Bradman, A.; Gharbi, M.; Sjodin, A.; Eskenazi, B. Polybrominated Diphenyl Ether (PBDE) Flame Retardants and Thyroid Hormone during Pregnancy. *Environ. Health. Persp.* **2010**, *118* (10), 1444–1449, DOI: 10.1289/ehp.1001905.

(17) Legler, J.; Brouwer, A. Are brominated flame retardants endocrine disruptors? *Environ. Int.* **2003**, *29* (6), 879–885, DOI: 10.1016/S0160-4120(03)00104-1.

(18) Mensching, D. A.; Slater, M.; Scott, J. W.; Ferguson, D. C.; Beasley, V. R. The feline thyroid gland: a model for endocrine disruption by polybrominated diphenyl ethers (PBDEs)? *J. Toxicol. Env. Heal. A* **2012**, *75* (4), 201–212, DOI: 10.1080/15287394.2012.652054.

(19) Dingemans, M. M. L.; de Groot, A.; van Kleef, R. G. D. M.; Bergman, A.; van den Berg, M.; Vijverberg, H. P. M.; Westerink, R. H. S. Hydroxylation increases the neurotoxic potential of BDE-47 to affect exocytosis and calcium homeostasis in PC12 cells. *Environ. Health Perspect.* **2008**, *116* (5), 637–643, DOI: 10.1289/Ehp.11059.

(20) Van Boxtel, A. L.; Kamstra, J. H.; Cenijn, P. H.; Pieterse, B.; Wagner, M. J.; Antink, M.; Krab, K.; Van Der Burg, B.; Marsh, G.; Brouwer, A.; Legler, J. Microarray analysis reveals a mechanism of phenolic polybrominated diphenylether toxicity in zebrafish. *Environ. Sci. Technol.* **2008**, *42* (5), 1773–1779, DOI: 10.1021/Es0720863.

(21) Wang, X. B.; Wang, Y.; Chen, J. W.; Ma, Y. Q.; Zhou, J.; Fu, Z. Q. Computational toxicological investigation on the mechanism and pathways of xenobiotics metabolized by cytochrome P450: a case of BDE-47. *Environ. Sci. Technol.* **2012**, *46* (9), 5126–5133, DOI: 10.1021/es203718u.

(22) Kierkegaard, A.; Bignert, A.; Sellstrom, U.; Olsson, M.; Asplund, L.; Jansson, B.; de Wit, C. A. Polybrominated diphenyl ethers (PBDEs) and their methoxylated derivatives in pike from Swedish waters with emphasis on temporal trends, 1967–2000. *Environ. Pollut.* **2004**, *130* (2), 187–198, DOI: 10.1016/j.envpol.2003.12.011.

(23) Sellstrom, U.; Bignert, A.; Kierkegaard, A.; Haggberg, L.; De Wit, C. A.; Olsson, M.; Jansson, B. Temporal trend studies on tetra-and pentabrominated diphenyl ethers and hexabromocyclododecane in guillemot egg from the Baltic Sea. *Environ. Sci. Technol.* **2003**, *37* (24), 5496–5501, DOI: 10.1021/es0300766.

(24) Kajiwara, N.; Noma, Y.; Takigami, H. Photolysis studies of technical decabromodiphenyl ether (DecaBDE) and ethane (DeBDethane) in plastics under natural sunlight. *Environ. Sci. Technol.* **2008**, *42* (12), 4404–4409, DOI: 10.1021/es800060j.

(25) Arnoldsson, K.; Andersson, P. L.; Haglund, P. Photochemical formation of polybrominated dibenzo-p-dioxins from environmentally abundant hydroxylated polybrominated diphenyl ethers. *Environ. Sci. Technol.* **2012**, *46* (14), 7567–7574, DOI: 10.1021/es301256x.

(26) Steen, P. O.; Grandbois, M.; McNeill, K.; Arnold, W. A. Photochemical Formation of Halogenated Dioxins from Hydroxylated Polybrominated Diphenyl Ethers (OH-PBDEs) and Chlorinated Derivatives (OH-PBCDEs). *Environ. Sci. Technol.* **2009**, *43* (12), 4405–4411, DOI: 10.1021/es9003679.

(27) Bonin, P. M.; Edwards, P.; Bejan, D.; Lo, C. C.; Bunce, N. J.; Konstantinov, A. D. Catalytic and electrocatalytic hydrogenolysis of brominated diphenyl ethers. *Chemosphere* **2005**, *58* (7), 961–967, DOI: 10.1016/j.chemosphere.2004.09.099.

(28) Konstantinov, A.; Bejan, D.; Bunce, N. J.; Chittim, B.; McCrindle, R.; Potter, D.; Tashiro, C. Electrolytic debromination of PBDEs in DE-83 technical decabromodiphenyl ether. *Chemosphere* **2008**, *72* (8), 1159–1162, DOI: 10.1016/j.chemosphere.2008.03.046.

(29) Nose, K.; Hashimoto, S.; Takahashi, S.; Noma, Y.; Sakai, S. Degradation pathways of decabromodiphenyl ether during hydrothermal treatment. *Chemosphere* **2007**, *68* (1), 120–125, DOI: 10.1016/j.chemosphere.2006.12.030.

(30) Raff, J. D.; Hites, R. A. Gas-phase reactions of brominated diphenyl ethers with OH radicals. *J. Phys. Chem. A* **2006**, *110* (37), 10783–10792, DOI: 10.1021/jp0630222.

(31) Luo, S.; Yang, S. G.; Xue, Y. G.; Liang, F.; Sun, C. Two-stage reduction/subsequent oxidation treatment of 2,2′,4,4′-tetrabromodiphenyl ether in aqueous solutions: Kinetic, pathway and toxicity. *J. Hazard. Mater.* **2011**, *192* (3), 1795–1803, DOI: 10.1016/j.jhazmat.2011.07.015.

(32) Shih, Y. H.; Tai, Y. T. Reaction of decabrominated diphenyl ether by zerovalent iron nanoparticles. *Chemosphere* **2010**, *78* (10), 1200–1206, DOI: 10.1016/j.chemosphere.2009.12.061.

(33) Xie, Q.; Chen, J. W.; Zhao, H. X.; Qiao, X. L.; Cai, X. Y.; Li, X. H. Different photolysis kinetics and photooxidation reactivities of neutral and anionic hydroxylated polybrominated diphenyl ethers. *Chemosphere* **2013**, *90* (2), 188–194, DOI: 10.1016/j.chemosphere.2012.06.033.

(34) Zhou, J.; Chen, J. W.; Liang, C. H.; Xie, Q.; Wang, Y. N.; Zhang, S.; Qiao, X.; Li, X. Quantum chemical investigation on the mechanism and kinetics of PBDE photooxidation by ·OH: a case study for BDE-15. *Environ. Sci. Technol.* **2011**, *45* (11), 4839–4845, DOI: 10.1021/es200087w.

(35) Cao, H. J.; He, M. X.; Han, D. D.; Sun, Y. H.; Zhao, S. F.; Ma, H. J.; Yao, S. D. Mechanistic and kinetic study on the reaction of 2,4-dibrominated diphenyl ether (BDE-7) with OH radicals. *Comput. Theor. Chem.* **2012**, *983*, 31–37, DOI: 10.1016/j.comptc.2011.12.017.

(36) Wang, S.; Hao, C.; Gao, Z. X.; Chen, J. W.; Qiu, J. S. Effects of excited-state structures and properties on photochemical degradation of polybrominated diphenyl ethers: a TDDFT study. *Chemosphere* **2012**, *88* (1), 33–38, DOI: 10.1016/j.chemosphere.2012.02.043.

(37) Akutsu, K.; Obana, H.; Okihashi, M.; Kitagawa, M.; Nakazawa, H.; Matsuki, Y.; Makino, T.; Oda, H.; Hori, S. GC/MS analysis of polybrominated diphenyl ethers in fish collected from the Inland Sea of Seto, Japan. *Chemosphere* **2001**, *44* (6), 1325–1333, DOI: 10.1016/S0045-6535(00)00534-8.

(38) Thomsen, C.; Lundanes, E.; Becher, G. Brominated flame retardants in plasma samples from three different occupational groups in Norway. *J. Environ. Monit.* **2001**, *3* (4), 366–370, DOI: 10.1039/B104304h.

(39) Tian, S. Y.; Zhu, L. Y.; Bian, J. N.; Fang, S. H. Bioaccumulation and Metabolism of Polybrominated Diphenyl Ethers in Carp (Cyprinus carpio) in a Water/Sediment Microcosm: Important Role of Particulate Matter Exposure. *Environ. Sci. Technol.* **2012**, *46* (5), 2951–2958, DOI: 10.1021/Es204011k.

(40) Frisch, M. J.; Trucks, G. W.; Schlegel, H. B.; Gill, P. W. M.; Johnson, B. G.; Robb, M. A.; Cheeseman, J. R.; Keith, T. A.; Petersson, G. A.; Montgomery, J. A.; Raghavachari, K.; Allaham, M. A.; Zakrzewski, V. G.; Ortiz, J. V.; Foresman, J. B.; Cioslowski, J.; Stefanov, B. B.; Nanayakkara, A.; Challacombe, M.; Peng, C. Y.; Ayala, P. Y.; Chen, W.; Wong, M. W.; Andres, J. L.; Replogle, E. S.; Gomperts, R.; Martin, R. L.; Fox, D. J.; Binkley, J. S.; Defrees, D. J.; Baker, J.; Stewart, J. P.; Head-Gordon, M.; Gonzales, C.; Pople, J. A. *Gaussian 03*; Gaussian, Inc.: Wallingford, CT, 2003.

(41) Zhao, Y.; Truhlar, D. G. Hybrid meta density functional theory methods for thermochemistry, thermochemical kinetics, and noncovalent interactions: the MPW1B95 and MPWB1K methods and comparative assessments for hydrogen bonding and van der waals interactions. *J. Phys. Chem. A* **2004**, *108* (33), 6908–6918, DOI: 10.1021/jp048147q.

(42) Qu, X. H.; Wang, H.; Zhang, Q. Z.; Shi, X. Y.; Xu, F.; Wang, W. X. Mechanistic and Kinetic Studies on the Homogeneous Gas-Phase Formation of PCDD/Fs from 2,4,5-Trichlorophenol. *Environ. Sci. Technol.* **2009**, *43* (11), 4068–4075, DOI: 10.1021/es802835e.

(43) Xu, F.; Wang, H.; Zhang, Q. Z.; Zhang, R. X.; Qu, X. H.; Wang, W. X. Kinetic Properties for the Complete Series Reactions of

Chlorophenols with OH Radicals-Relevance for Dioxin Formation. *Environ. Sci. Technol.* **2010**, *44* (4), 1399–1404, DOI: 10.1021/es9031776.

(44) Yu, W. N.; Hu, J. T.; Xu, F.; Sun, X. Y.; Gao, R.; Zhang, Q. Z.; Wang, W. X. Mechanism and Direct Kinetics Study on the Homogeneous Gas-Phase Formation of PBDD/Fs from 2-BP, 2,4-DBP, and 2,4,6-TBP as Precursors. *Environ. Sci. Technol.* **2011**, *45* (5), 1917–1925, DOI: 10.1021/es103536t.

(45) Miertus, S.; Scrocco, E.; Tomasi, J. Electrostatic interaction of a solute with a continuum. A direct utilizaion of AB initio molecular potentials for the prevision of solvent effects. *Chem. Phys.* **1981**, *55* (1), 117–129, DOI: 10.1016/0301-0104(81)85090-2.

(46) Tomasi, J.; Mennucci, B.; Cammi, R. Quantum mechanical continuum solvation models. *Chem. Rev.* **2005**, *105* (8), 2999–3093, DOI: 10.1021/cr9904009.

(47) Tomasi, J.; Persico, M. Molecular Interactions in Solution: An Overview of Methods Based on Continuous Distributions of the Solvent. *Chem. Rev.* **1994**, *94* (7), 2027–2094, DOI: 10.1021/cr00031a013.

(48) Robinson, P. J.; Holbrook, K. A. *Unimolecular Reactions*; Wiley: New York, 1972.

(49) Robertson, S. H.; Glowacki, D. R.; Liang, C.-H.; Morley, C.; Pilling, M. J. MESMER (Master Equation Solver for Multi-Energy Well Reactions), an object oriented C++ program for carrying out ME calculations and eigenvalue-eigenvector analysis on arbitrary multiple well systems. http://sourceforge.net/projects/mesmer.

(50) Gannon, K. L.; Blitz, M. A.; Liang, C. H.; Pilling, M. J.; Seakins, P. W.; Glowacki, D. R. Temperature Dependent Kinetics (195–798 K) and H Atom Yields (298–498 K) from Reactions of (CH2)-C-1 with Acetylene, Ethene, and Propene. *J. Phys. Chem. A* **2010**, *114* (35), 9413–9424, DOI: 10.1021/jp102276j.

(51) Gannon, K. L.; Blitz, M. A.; Liang, C. H.; Pilling, M. J.; Seakins, P. W.; Glowacki, D. R.; Harvey, J. N. An experimental and theoretical investigation of the competition between chemical reaction and relaxation for the reactions of (CH2)-C-1 with acetylene and ethene: implications for the chemistry of the giant planets. *Faraday Discuss.* **2010**, *147*, 173–188, DOI: 10.1039/c004131a.

(52) Roux, M. V.; Temprado, M.; Chickos, J. S.; Nagano, Y. Critically Evaluated Thermochemical Properties of Polycyclic Aromatic Hydrocarbons. *J. Phys. Chem. Ref. Data* **2008**, *37* (4), 1855–1996, DOI: 10.1063/1.2955570.

(53) Good, W. D.; Lee, S. H. The enthalpies of formation of selected naphthalenes, diphenylmethanes, and bicyclic hydrocarbons. *J. Chem. Thermodyn.* **1976**, *8*, 643–650, DOI: 10.1016/0021-9614(76)90015-x.

(54) Zhao, Y.; Gonzalez-Garcia, N.; Truhlar, D. G. Benchmark database of barrier heights for heavy atom transfer, nucleophilic substitution, association, and unimolecular reactions and its use to test theoretical methods. *J. Phys. Chem. A* **2005**, *109* (9), 2012–2018, DOI: 10.1021/jp061040d.

(55) Zeng, X.; Freeman, P. K.; Vasil'ev, Y. V.; Voinov, V. G.; Simonich, S. L.; Barofsky, D. F. Theoretical calculation of thermodynamic properties of polybrominated diphenyl ethers. *J. Chem. Eng. Data* **2005**, *50* (5), 1548–1556, DOI: 10.1021/je050018v.

(56) Cao, H. J.; He, M. X.; Han, D. D.; Sun, Y. H.; Xie, J. Theoretical study on the mechanism and kinetics of the reaction of 2,2′,4,4′-tetrabrominated diphenyl ether (BDE-47) with OH radicals. *Atmos. Environ.* **2011**, *45* (8), 1525–1531, DOI: 10.1016/j.atmosenv.2010.12.045.

(57) Cao, H. J.; He, M. X.; Sun, Y. H.; Han, D. D. Mechanical and Kinetic Studies of the Formation of Polyhalogenated Dibenzo-p-dioxins from Hydroxylated Polybrominated Diphenyl Ethers and Chlorinated Derivatives. *J. Phys. Chem. A* **2011**, *115* (46), 13489–13497, DOI: 10.1021/jp2059497.

(58) Baltaretu, C. O.; Lichtman, E. I.; Hadler, A. B.; Elrod, M. J. Primary atmospheric oxidation mechanism for toluene. *J. Phys. Chem. A* **2009**, *113* (1), 221–230, DOI: 10.1021/jp806841t.

(59) Jenkin, M. E.; Glowacki, D. R.; Rickard, A. R.; Pilling, M. J. Comment on "Primary atmospheric oxidation mechanism for toluene". *J. Phys. Chem. A* **2009**, *113*(28), 8136–8; discussion 8139–8140; DOI: 10.1021/jp903119k.

(60) Prinn, R. G.; Weiss, R. F.; Miller, B. R.; Huang, J.; Alyea, F. N.; Cunnold, D. M.; Fraser, P. J.; Hartley, D. E.; Simmonds, P. G. Atmospheric Trends and Lifetime of CH3CCI3 and Global OH Concentrations. *Science* **1995**, *269* (5221), 187–192, DOI: 10.1126/science.269.5221.187.

Chemosphere 134 (2015) 241–249

Contents lists available at ScienceDirect

Chemosphere

journal homepage: www.elsevier.com/locate/chemosphere

OH radical-initiated oxidation degradation and atmospheric lifetime of *N*-ethylperfluorobutyramide in the presence of O_2/NO_x

Yanhui Sun [a], Qingzhu Zhang [a,*], Hui Wang [b,*], Wenxing Wang [a]

[a] Environment Research Institute, Shandong University, Jinan 250100, PR China
[b] School of Environment, Tsinghua University, Beijing 100084, PR China

HIGHLIGHTS

- Oxidation degradation of EtFBA by OH radicals is investigated firstly.
- Water molecule plays an important catalytic effect during the whole degradation.
- The calculated overall rate constant is 2.50×10^{-12} cm^3 molecule^{-1} s^{-1} at 296 K.
- Primary oxidation products of the title reactions have been obtained.

ARTICLE INFO

Article history:
Received 23 February 2015
Received in revised form 20 April 2015
Accepted 21 April 2015
Available online 15 May 2015

Keywords:
N-ethylperfluorobutyramide
OH radicals
Oxidation degradation
Kinetic parameters
Atmospheric lifetime

ABSTRACT

The OH radical-initiated oxidation degradation of *N*-ethylperfluorobutyramide (EtFBA) in the presence of O_2/NO_x was investigated theoretically by using density functional theory (DFT). All possible pathways involved in the oxidation process were presented and discussed. The study shows that the H abstraction from the C^2—H^2 group in EtFBA is the most energetically favorable because of the lowest barrier and highest exothermicity. Canonical variational transition-state (CVT) theory with small curvature tunneling (SCT) contribution was used to predict the rate constants over the temperature range of 180–370 K. At 296 K, the calculated overall rate constant of EtFBA with OH radicals is 2.50×10^{-12} cm^3 molecule^{-1} s^{-1}. The atmospheric lifetime of EtFBA determined by OH radicals is short, about 4.6 days at 296 K. However, the atmospheric lifetimes of its primary oxidation products, $C_3F_7C(O)N(H)C(O)CH_3$, $C_3F_7C(O)N(H)CH_2CHO$ and $C_3F_7C(O)NH_2$, are much longer, about 30–50 days. It demonstrates the possibility that the atmospheric oxidation degradation of polyfluorinated amides (PFAMs) contributes to the burden of observed perfluorinated pollutants in the Arctic region. This study reveals for the first time that the water molecule plays an important catalytic effect on several key elementary steps and promotes the degradation potential of EtFBA.

1. Introduction

Perfluorocarboxylic acids (PFCAs, $C_nF_{2n+1}COOH$) are widespread and persistent environmental contaminants and have been detected in many biological samples including those from remote regions such as the Arctic (Stock et al., 2007; Ahrens et al., 2009; Anna et al., 2012). PFCAs are also abundant in the human population (Lee and Mabury, 2011). Some PFCAs such as perfluorooctanoic acids (PFOAs) have been shown to cause development delays and cancer and are bioaccumulative when the perfluorinated chain is more than six carbons in length (Kannan et al., 2004; Martin et al., 2003; Lau et al., 2007). The source of PFCAs in the Arctic environment is a current research priority. Long-range atmospheric transport is not expected to be important for PFCAs due to their low volatility and high water solubility. Thus, another transport mechanism must be at work. One possible mechanism is that precursor chemicals such as fluorotelomer alcohols (FTOHs), fluorotelomer acrylates (FTAs), polyfluorinated amides (PFAMs), perfluorinated alkyl sulfonamides (FOSAs) and sulfonamido ethanols (FOSEs) undergo atmospheric transport to the Arctic and subsequent oxidation degradation resulting in the formation of PFCAs (Martin et al., 2006; Butt et al., 2009; Wallington et al., 2006; D'eon et al., 2006; Jackson et al., 2013).

Polyfluorinated amides ($C_nF_{2n+1}C(O)N(H)R$, PFAMs) are a class of fluorinated compounds produced as byproducts of polyfluorinated sulfonamide synthesis by electrochemical fluorination (ECF)

* Corresponding authors.
E-mail addresses: zqz@sdu.edu.cn (Q. Zhang), wanghui@tsinghua.edu.cn (H. Wang).

http://dx.doi.org/10.1016/j.chemosphere.2015.04.059
0045-6535/© 2015 Elsevier Ltd. All rights reserved.

(Jackson and Mabury, 2013). Since 1970, up to 45000 t of perfluorooctylsulfonyl fluoride-derived compounds have been released to the environment (Paul et al., 2009). Assuming that PFAMs have an upper yield of 1% from the synthesis of sulfonamide compounds, it may emit a maximum 450 t of potential ECF PFOA precursor into the environment, which is on the same scale as PFOA deliberately produced by ECF (Jackson and Mabury, 2013). Since PFAMs are predicted to be more volatile than their sulfonamide analogs, partitioning to the atmosphere is expected to be significant. The tropospheric removal or transformation of gaseous PFAMs involves dry and wet deposition, and chemical degradation such as oxidation reactions with OH, NO_3 and atomic Cl radicals. Owing to the high volatility and sparing solubility in water, dry and wet deposition of PFAMs is of relatively minor importance as a removal pathway. Among the various oxidants, OH radicals play an essential role in the determination of the oxidizing power of the atmosphere. The reaction with OH radicals is considered to be a dominant removal pathway of PFAMs (Jackson et al., 2013; Jackson and Mabury, 2013).

Several experimental studies have been carried out to explain the source of PFCAs in the Arctic region (Ellis et al., 2003; Wallington et al., 2006; D'eon et al., 2006; Jackson et al., 2013). In 2003, Ellis et al. evaluated the atmospheric lifetime of $F(CF_2CF_2)_nCH_2CH_2OH$ ($n \geqslant 2$) determined by OH radicals, as approximately 20 days (Ellis et al., 2003). Subsequently, taking $C_8F_{17}CH_2CH_2OH$ as a model, the oxidation products were investigated (Wallington et al., 2006). In 2006, the reactions of *N*-ethyl perfluorobutanesulfonamide (NEtFBSA) with OH and atomic Cl radicals were studied at 296 K and 301 K (D'eon et al., 2006). The atmospheric lifetime of NEtFBSA was determined, 20–50 days, thus allowing long-range atmospheric transport and transformation into PFCAs. Recently, Jackson et al. proposed that the atmospheric oxidation of PFAMs was a plausible source of PFCAs in the Arctic (Jackson et al., 2013). In their smog chamber experiment, the rate coefficient for the reaction of *N*-ethylperfluorobutyramide (EtFBA) with OH radicals was measured to be $(2.65 \pm 0.50)\times10^{-12}$ cm^3 $molecule^{-1}$ s^{-1} at 296 K with an atmospheric lifetime of 4.4 days; GC–MS and LC–MS/MS techniques showed that perfluorobutyramide ($C_3F_7C(O)NH_2$) and two carbonyl compounds, $C_3F_7C(O)N(H)C(O)CH_3$ and $C_3F_7C(O)N(H)CH_2CHO$, were the primary oxidation products, which can react further to yield PFCAs. However, the detailed oxidation degradation mechanism of these fluorinated compounds in the atmosphere has not been completely elucidated, due to the fact that the transition states and radical intermediates formed in the reaction processes are very short lived and would be almost impossible for direct experimental characterization. In such a situation, theoretical calculation can be an alternative. In this work, taking EtFBA ($C_3F_7C(O)N(H)CH_2CH_3$) as an example of PFAMs, the mechanism and kinetic properties of OH radical-initiated atmospheric oxidation degradation were investigated by using the quantum chemical calculation and direct dynamic theory. The objective of the present study was to improve the understanding of the potential burden of PFCAs in the Arctic from the atmospheric oxidation degradation of PFAMs.

2. Computational method

All the electronic structure calculations were performed by using the Gaussian 09 software package (Frisch et al., 2009). Geometry optimizations of the reactants, intermediates, transition states and products were performed at the MPWB1 K (Adamo and Barone, 1998; Zhao and Truhlar, 2004) level with a standard 6-31+G(d,p) basis set. The vibrational frequencies were calculated at the same level to determine the nature of the stationary points. Each transition state was verified to connect the designated reactants and products by performing an intrinsic reaction coordinate (IRC) analysis (Fukui, 1981). Based on the optimized geometries, a more flexible basis set, 6-311+G(3df,2p), was employed to calculate the single point energies of various species. All the relative energies quoted and discussed in this work include zero-point energy (ZPE) correction with unscaled frequencies obtained at the MPWB1K/6-31+G(d,p) level. The canonical variational transition-state (CVT) theory (Baldridge et al., 1989; Gonzalez-Lafont et al., 1991; Garrett and Truhlar, 1979) with the small curvature tunneling (SCT) method (Fernandez-Ramos et al., 2007) was used to calculate the rate constants over the possible atmospheric temperature range of 180–370 K. The rotational partition functions were calculated classically, and the vibrational modes were treated as quantum–mechanical separable harmonic oscillators. The rate constant calculations were carried out with the aid of POLYRATE 9.3 program (Steckler et al., 2002; Yu et al., 2013).

3. Results and discussion

The reliability of the theoretical calculations was confirmed. Due to the absence of experimental information on the thermochemical parameters for the present reaction system, it is difficult to make a direct comparison of the calculated results with experimental data. Thus, we optimized the geometries and calculated the vibrational frequencies of $CF_3CF_2CF_3$ and $NH_2CH_2CH_3$. As shown in Fig. S1 and Table S1 of Supporting Information, the results calculated at the MPWB1K/6-31+G(d,p) level agree well with the available experimental values, and the maximum relative errors are less than 2.0% for the geometrical parameters and less than 8.0% for the vibrational frequencies (Kuchitsu, 1992; Zeroka et al., 1999; Shimanouchi, 1972; Herzberg, 1966). For the reaction of $CHF_2CF_3 + OH \rightarrow CF_2CF_3 + H_2O$, the reaction enthalpy deduced at the MPWB1K/6-311+G(3df,2p)//MPWB1K/6-31+G(d,p) level and 0 K is −11.07 kcal/mol, which agrees well with the experimental value of −12.92 kcal/mol (Atkinson et al., 2008).

3.1. Reaction with OH radicals

EtFBA is a saturated polyfluorinated amide characterized by the hydrogen atom in the amido group substituted by the ethyl group. There exist C—H, N—H and C═O bonds in the molecular structure of EtFBA. Thus, H abstraction from the C—H or N—H bonds and OH addition to the C═O bond are possible pathways for the reaction of EtFBA with OH radicals. Fig. 1 depicts the reaction scheme embedded with the potential barrier and reaction heat. For convenience of description, the C atoms and H atoms in EtFBA are numbered.

There are three different kinds of hydrogen atoms in the EtFBA molecule: two kinds in the $—CH_2CH_3$ group and one bonded with the N atom. So, three possible H abstraction pathways were identified: H abstractions from the $N—H^1$, $C^2—H^2$ and $C^3—H^4$ bonds. Thus, three transition states were located at the MPWB1K/6-31+G(d,p) level. They were confirmed with one and only one negative eigenvalue of the Hessian matrix and, therefore, one imaginary frequency. All of the H abstraction pathways are exothermic. Especially, the H abstraction from the $C^2—H^2$ bond is the most energetically favorable because of the lowest barrier and highest exothermicity. In this pathway, a pre-reactive van der Waals complex (IM2) is firstly formed before the H abstraction and releases 4.64 kcal/mol of energy. The overall reaction is strongly exothermic by 26.06 kcal/mol. In addition, the standard reaction Gibbs energies for the four pathways are given in Fig. S2 of Supplementary material. It is clear that the ΔG for the formation of IM3 is lowest (−25.85 kcal/mol), also indicating the most thermodynamic favorable of this route.

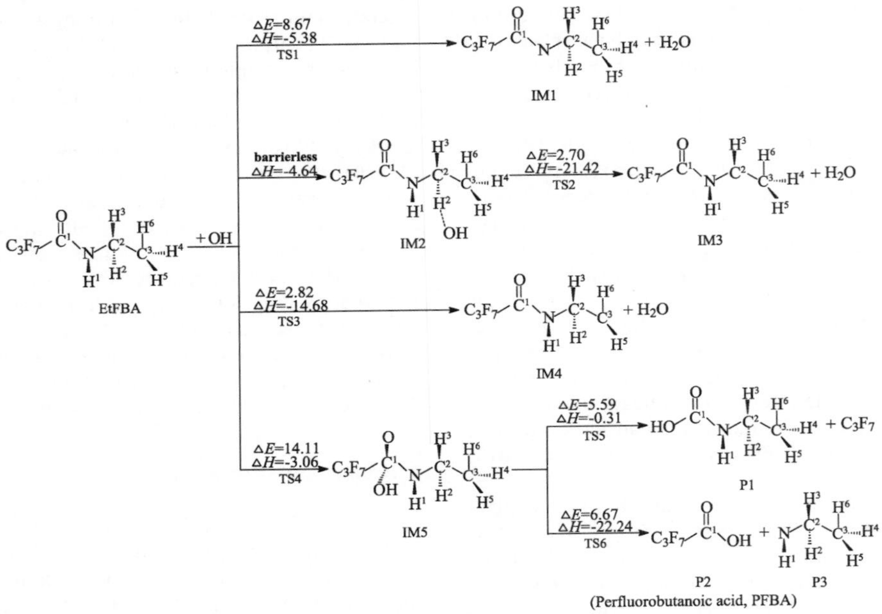

Fig. 1. OH radical-initiated reaction scheme of EtFBA embedded with the potential barriers ΔE (kcal/mol) and reaction heats ΔH (kcal/mol).

For the pathway of OH addition to the C=O bond, the barrier is 14.11 kcal/mol, which is relatively higher than other H abstraction pathways. This process is slightly exothermic by 3.06 kcal/mol. Subsequently, two possible unimolecular decomposition channels can occur from the OH-EtFBA adduct, IM5, by the cleaving of C^1—C bond or C^1—N bond. The product, perfluorobutanoic acid (C_3F_7COOH, PFBA), has been detected in the photochemical smog chamber at the Ford Motor Company (Jackson et al., 2013). C_3F_7, a perfluorinated alkyl radical, plays an important role in the formation of PFCAs. In the troposphere, the product CH_3CH_2NH (denoted as P3) will be further oxidized to yield imine, nitrosamine and nitramine (Nielsen et al., 2012).

3.2. Secondary reactions

The calculations above show that H abstractions and OH addition are energetically favorable pathways for the reaction of EtFBA with OH radicals. IM1, IM3, IM4 and C_3F_7 are important radical intermediates produced in the degradation process of EtFBA initiated by OH radicals. The conventional view is that these radical intermediates could be oxidized further by ubiquitous O_2 or O_2/NO_x and removed from the troposphere. Several important secondary pollutants can be produced from the reactions of IM1, IM3, IM4 and C_3F_7 with O_2 or O_2/NO_x.

3.2.1. Secondary reaction from IM1

IM1 is an open-shell activated intermediate. As presented in Fig. 2, five possible secondary reaction pathways are proposed from IM1. Firstly, the H^3 atom of IM1 can be directly abstracted by O_2 molecule to form an imine (denoted as IM6) and HO_2 radical. The barrier is calculated to be 16.96 kcal/mol. The process is strongly exothermic by 17.22 kcal/mol. In the area of high NO_x concentration ($\geqslant$10 ppb), IM1 can react with NO_x. Apparently, IM1 can react with NO to produce a vibrationally excited nitrosamine (denoted as P4), which will rapidly undergo photolysis in the troposphere (Tang and Nielsen, 2013). This process is barrierless and highly exothermic by 42.30 kcal/mol. There are three possible pathways for the reactions of IM1 with NO_2, yielding IM6, P6 and P7.

IM6 is an important intermediate produced for the secondary reaction of IM1. There exists an $N=C^2$ bond in the molecular structure of IM6. One subsequent pathway from IM6 is the addition reaction with OH radicals. It can be seen from Fig. 2 that OH addition to the C^2 atom is the most energetically favorable reaction leading to the formation of IM7. The energy-rich IM7 can further be oxidized by O_2/OH, or isomerized via H-migration to form $C_3F_7C(O)NH$ (IM14) and $C_3F_7C(OH)N(H)C(O)CH_3$ (P8). The reaction scheme is presented in Fig. 2. $C_3F_7C(O)NH$ was proposed by Jackson et al. (2013). A similar nitrogenous radical intermediate RC(O)NH was also suggested in the atmospheric oxidation of other amides such as formamide and acetamide (Barnes et al., 2010). Note that the isomerization from IM13 to IM14 via the direct intramolecular H-migration has an extremely high barrier, 38.68 kcal/mol. So, this process cannot occur spontaneously under the general atmospheric conditions. Water is the third most abundant species in the atmosphere after N_2 and O_2, with a typical gas-phase concentration of 7.64×10^{17} molecule cm^{-3}, corresponding to 50% of relative humidity at 298 K (DeMore et al., 1992). Water is of great chemical importance due to its ability to form hydrogen bond. It can catalyze actively the isomerization from CH_2CH_2OH to CH_3CH_2O via 1,3 H-migration or to CH_3CHOH via 1,2 H-migration (Teixeira-Dias et al., 2003). Similarly, with the aid of water, H-migration from IM13 becomes a bimolecular reaction. The barrier of the H-migration via IM13 + H_2O → IM14 + H_2O is only 13.99 kcal/mol, which is 24.69 kcal/mol lower than that of the direct H-migration via the intramolecular isomerization (red arrow in Fig. 2). Hence, the water molecule promotes the formation of IM14 from IM13 by lowering the energy barrier. The reactions of IM13 with two or three water molecules are shown in Fig. S3 of Supplementary material. The gas-phase concentration of $(H_2O)_2$ and $(H_2O)_3$ are approximately 6.00×10^{14} molecule cm^{-3} (Pfeilsticker et al., 2003) and 4.65×10^{11} molecule cm^{-3} (Ryzhkov and Ariya, 2006), which indicating smaller reaction probability than single water molecule. Further reaction from $C_3F_7C(O)NH$ (IM14) is a direct decomposition to yield C_3F_7 and isocyanic acid (HNCO). This process has a low potential barrier of 6.23 kcal/mol and is exothermic by 11.54 kcal/mol. Thus, the unimolecular decomposition of $C_3F_7C(O)NH$ would occur readily under

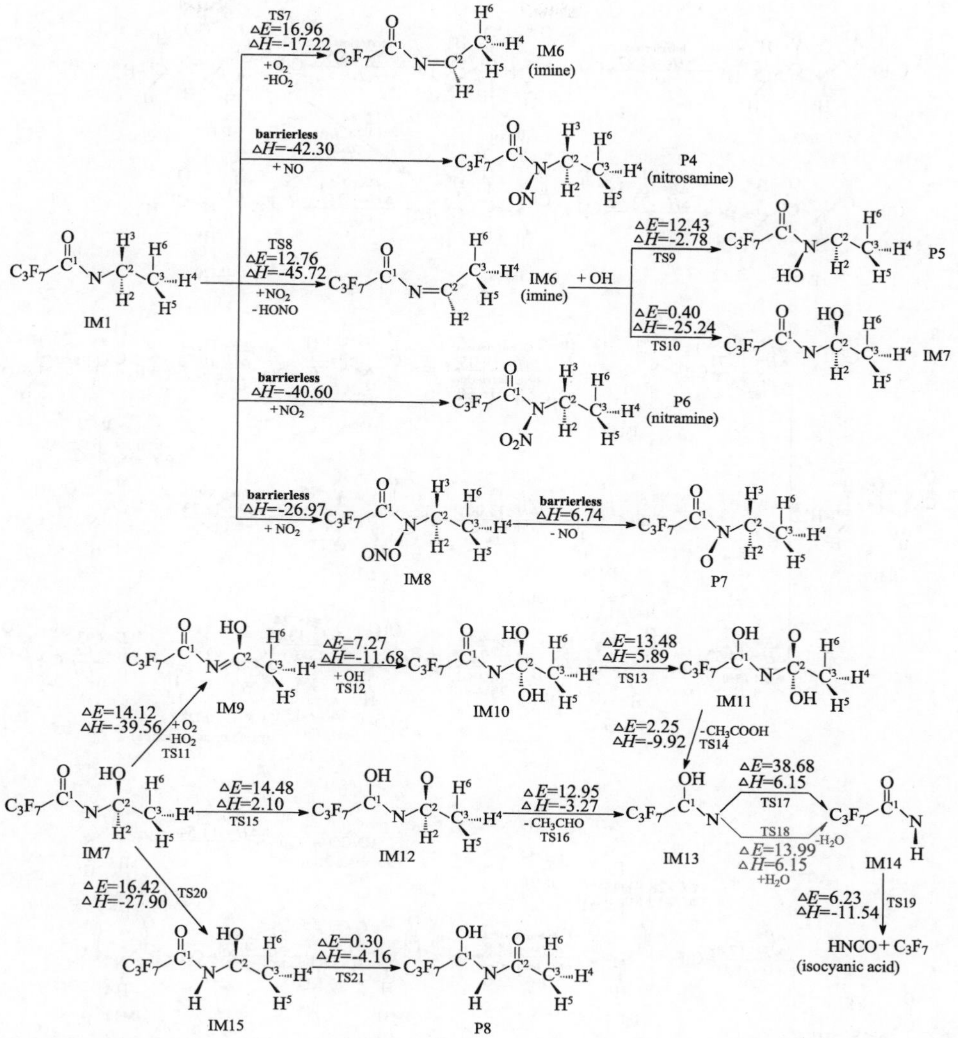

Fig. 2. Secondary reaction scheme from IM1 embedded with the potential barriers ΔE (kcal/mol) and reaction heats ΔH (kcal/mol).

atmospheric conditions. HNCO was experimentally detected by FTIR in the smog chamber (Jackson et al., 2013).

3.2.2. Secondary reaction from IM3

As described in Fig. 3, IM3 can easily react with ubiquitous O_2 in the atmosphere to form intermediate IM16. This process is barrierless and strongly exothermic by 24.27 kcal/mol. Then, IM16 will further react with NO immediately to produce a vibrationally excited IM17. The process is also a barrierless association. Subsequently, IM17 will directly decompose to yield IM18 and NO_2 via cleavage of the O—O bond with an energy barrier of 13.44 kcal/mol. There are two possible reaction pathways from IM18. In the troposphere, the active intermediates $C_3F_7C(O)N(H)C(O)CH_3$ (IM19) and $C_3F_7C(O)N(H)CHO$ (IM20) produced from the reactions of IM18 will undergo further oxidization.

Two possible reaction pathways were proposed from IM19, H abstraction and OH addition. Especially, IM19 can further react to form perfluorobutyramide $C_3F_7C(O)NH_2$ (IM28) through three elementary reaction steps. Firstly, OH adds to the C^2 atom of IM19, resulting in an adduct IM26. This elementary step is strongly exothermic by 14.62 kcal/mol with a low barrier of 5.72 kcal/mol. IM26 can further react via unimolecular decomposition and H-migration to form perfluorobutyramide $C_3F_7C(O)NH_2$ (IM28). Calculations show that the direct H-migration via intramolecular isomerization has an extremely high barrier, 35.24 kcal/mol. It inhibits the H-migration from proceeding beyond the formation of IM27. However, experimental studies showed that $C_3F_7C(O)NH_2$ (IM28) is the primary degradation product of EtFBA from the OH radical-initiated atmospheric oxidation. Besides the greenhouse effect, water plays an important role in several gas-phase chemical processes occurring in the earth's atmosphere. For example, it can participate actively the gas-phase reaction of SO_3 with H_2O to produce H_2SO_4 (Morokuma and Muguruma, 1994). An intriguing question is whether water can have an effect on the H-migration of IM27 to form perfluorobutyramide $C_3F_7C(O)NH_2$ (IM28). With the introduction of water, the H-migration process becomes a bimolecular reaction, IM27 + H_2O → IM28 + H_2O (red[1] arrow in Fig. 3). The barrier of the H-migration via the bimolecular reaction with the aid of the water molecule is 0.44 kcal/mol, which is 34.80 kcal/mol lower compared to the direct H-migration via the intramolecular isomerization.

[1] For interpretation of color in Fig. 3, the reader is referred to the web version of this article.

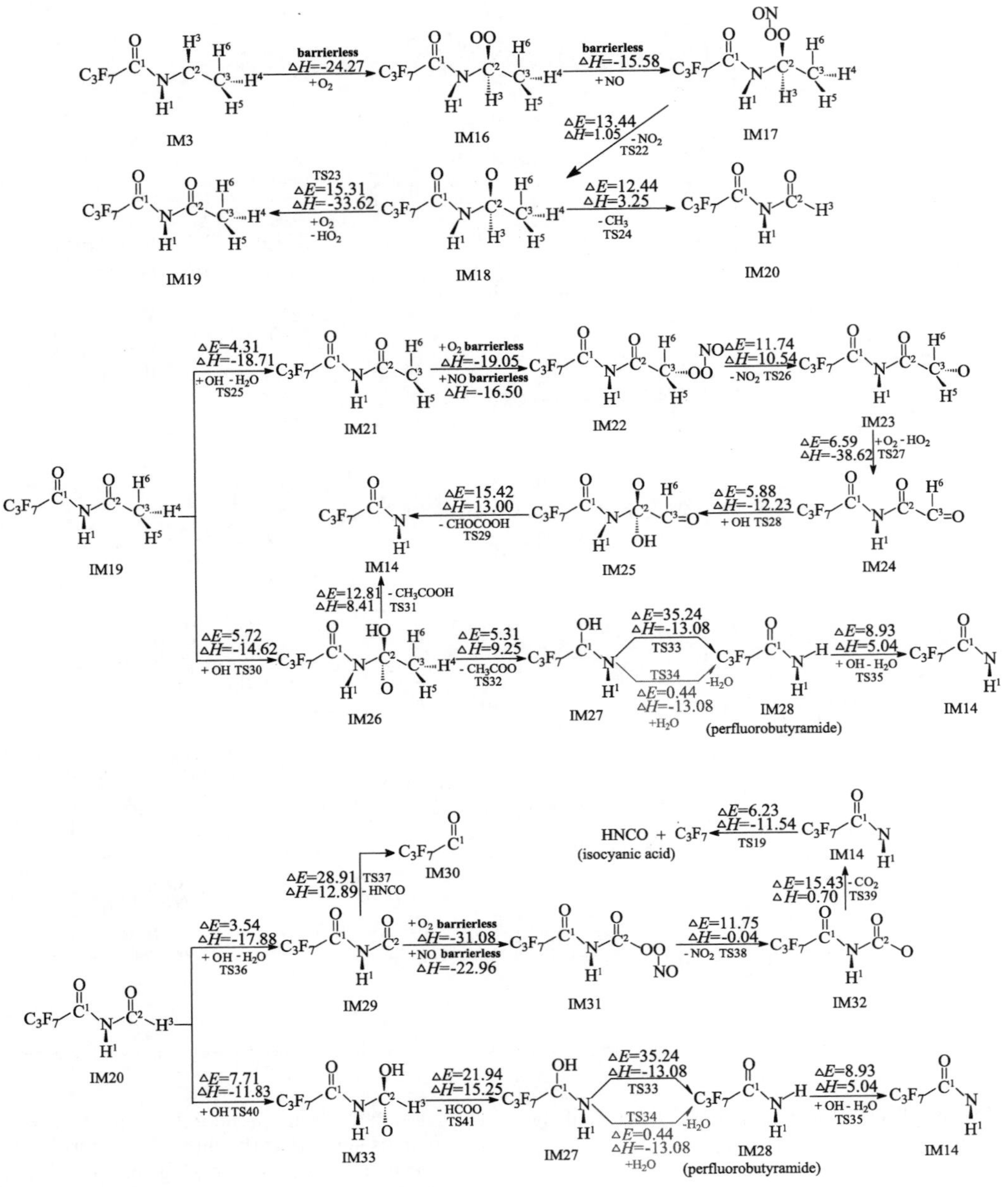

Fig. 3. Secondary reaction scheme from IM3 embedded with the potential barriers ΔE (kcal/mol) and reaction heats ΔH (kcal/mol).

As seen in Fig. S4 of Supporting Information, the direct H-migration via the intramolecular isomerization proceeds through a four-membered ring transition state (TS33), whereas the H-migration via the bimolecular reaction with the aid of water proceeds through a six-membered ring transition state (TS34). The bond angles in the four-membered ring are more bent than those in the six-membered ring (73.3°, 104.7°, 72.3° and 109.6° in TS33 vs. 103.0°, 154.6°, 88.0°, 146.7°, 103.9° and 123.7° in TS34). The ring tension imposed by the four-membered ring is larger. In the six-membered transition state, the water molecule acts as a bridge, accepting the hydrogen from the —OH group and simultaneously donating another hydrogen atom to the N atom. This reaction process breaks a bond in the original water molecule and completes the H-migration. Hence, the water molecule in the product is not the original one. The water molecule plays an important catalytic effect on the H-migration process and promotes the formation of $C_3F_7C(O)NH_2$ (IM28). Similar to IM19, IM20 can be further oxidized to yield $C_3F_7C(O)NH_2$ as well as CO_2, H_2O,NO_2, C_3F_7 et al.

3.2.3. Secondary reaction from IM4

Possible reaction routes from IM4 are depicted in Fig. 4. Similar to IM3, IM4 can also react with O_2/NO to form IM35. Then, the energy-rich IM35 decomposes to yield an alkoxy radical IM36 via the loss of NO_2. The subsequent reaction from IM36 involves two possible pathways. The first pathway is the reaction of IM36 with O_2, followed by the loss of HO_2, to yield $C_3F_7C(O)N(H)CH_2CHO$ (IM37). The reaction of IM36 with O_2 is strongly exothermic by 28.76 kcal/mol. The second pathway is the unimolecular decomposition of IM36 to yield IM38 and formaldehyde by the cleaving of C^2—C^3 bond. The decomposition process is endothermic by 7.51 kcal/mol with a barrier of 14.84 kcal/mol. Furthermore, IM37 was observed by GC–MS in the chamber (Jackson et al.,

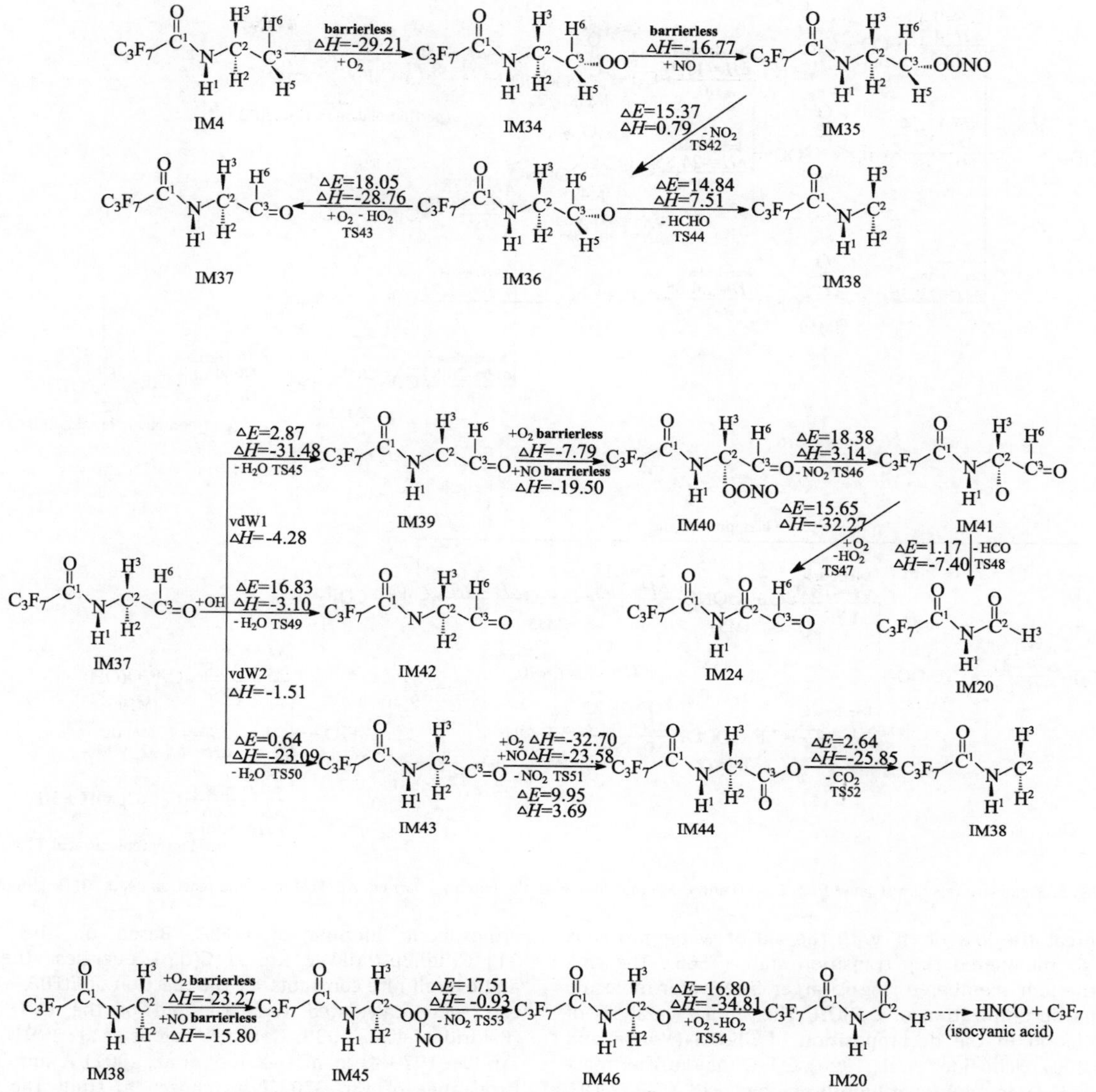

Fig. 4. Secondary reaction scheme from IM4 embedded with the potential barriers ΔE (kcal/mol) and reaction heats ΔH (kcal/mol). vdW1 and vdW2 are the van der Waals complexes between IM37 and OH radical.

2013). In the troposphere, IM37 can react with OH radicals via H abstraction mechanism. Three possible H abstraction pathways were identified: H abstractions from $N—H^1$, $C^2—H^2$ and $C^3—H^6$ bonds. Calculations show that H abstractions from $C^2—H^2$ and $C^3—H^6$ bonds are the energetically more favorable pathways. The H abstraction products, IM39 and IM43, can undergo further oxidization to yield $C_3F_7C(O)N(H)CHO$ (IM20), $C_3F_7C(O)NHC(O)CHO$ (IM24), $C_3F_7C(O)NHCH_2$ (IM38), et al. Subsequently, IM20, IM24 and IM38 can be oxidated to yield C_3F_7 and HNCO. The detailed reaction schemes were presented in Fig. 3 for IM20 and IM24 and Fig. 4 for IM38.

3.2.4. Secondary reaction of C_3F_7CO and C_3F_7

Fig. 5 exhibits the possible reaction pathways from C_3F_7CO and C_3F_7. C_3F_7CO can directly decompose to yield C_3F_7 and CO via the cleavage of $C^1—C$ bond. The barrier height is calculated to be 9.48 kcal/mol. The process is endothermic by 4.18 kcal/mol. In addition, C_3F_7CO can react with O_2 or NO_2. Calculations show that the addition of O_2 or NO_2 to C_3F_7CO is highly exothermic with a low energy barrier or without a barrier. The resulting two adducts IM48 and IM52 can subsequently react to yield perfluorobutanoic acid (C_3F_7COOH, PFBA) and O_3 etc., which have been identified experimentally (Sulbaek Andersen et al., 2003; Robert and Kerwin, 2007).

C_3F_7 is an open-shell highly activated radical. Once produced, C_3F_7 will react immediately with ubiquitous O_2. The entrance channel of the reaction is exoergic, leading to a perfluorinated peroxy radical C_3F_7OO (IM53), which promptly reacts via two possible reaction pathways. The first one is the reaction with NO to produce IM54, which will subsequently pass into unzipping cycle and release COF_2. COF_2 was identified and quantified (18% yield) by GC–MS and LC–MS/MS techniques in the chamber (Jackson et al., 2013). The second one is the reaction with HO_2 to form C_3F_7OH through the loss of ozone (Hou and Wang, 2005). In the troposphere, C_3F_7OH may decompose to yield C_3F_6O by the loss of HF. Fig. 5 shows that the barrier of the direct loss of HF via the unimolecular decomposition is extremely high, 47.24 kcal/mol. However, with the aid of the water molecule, the barrier of the loss of HF is 19.08 kcal/mol. As seen in Fig. S4 of Supporting Information, the direct loss of HF via the unimolecular decomposition proceeds through a four-membered ring transition state

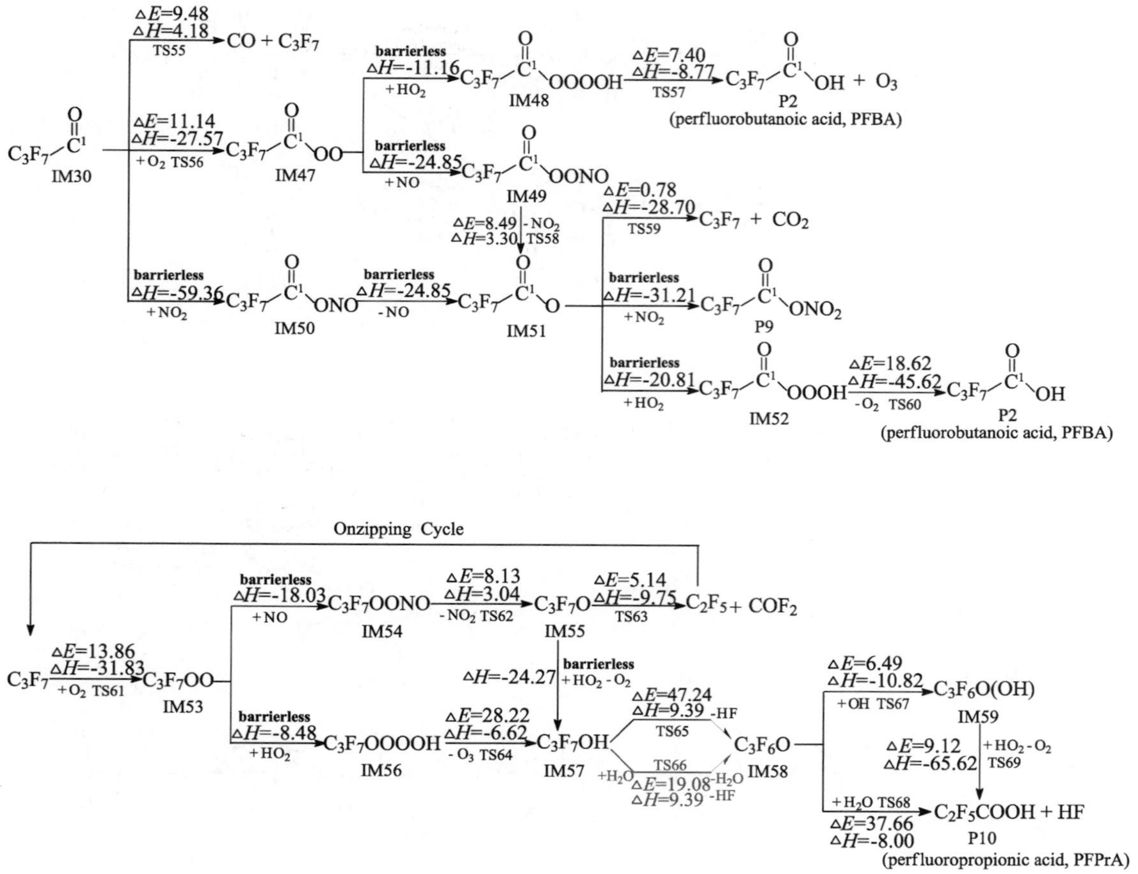

Fig. 5. Secondary reaction scheme from C_3F_7CO and C_3F_7 embedded with the potential barriers ΔE (kcal/mol) and reaction heats ΔH (kcal/mol).

(TS65), whereas the loss of HF with the aid of water proceeds through a six-membered ring transition state (TS66). The ring tension in the four-membered ring is larger. The water molecule catalyzes the decomposition of C_3F_7OH. The similar catalysis of water was found in the decomposition of CF_3OH (Young and Donaldson, 2007; Schneider et al., 1996). C_3F_6O may further react with OH or H_2O to yield perfluoropropionic acid (C_2F_5COOH, PFPrA) (Jackson et al., 2013). Similar to the formation of PFPrA, trifluoroacetic acid (CF_3COOH, TFA) can be formed from the oxidation of C_2F_5.

The theoretically predicted degradation products of EtFBA initiated by OH radicals along with the experimental findings were summarized in Table S2 of Supporting Information. The most significant pathways leading from EtFBA to PFCA, C_2F_5COOH, were described in Fig. 6. Comparison of the reaction mechanisms above show that $C_3F_7C(O)N(H)C(O)CH_3$ (IM19), $C_3F_7C(O)N(H)CH_2CHO$ (IM37) and $C_3F_7C(O)NH_2$ (IM28) are the main primary products for OH radical-initiated oxidation degradation of EtFBA, supported directly by the experimental observation (Jackson et al., 2013). The formation of IM19 and IM37 is more energetically feasible relative to IM28. Upon further oxidization, IM19 and IM37 can be transformed into IM28. This could partly explain the experimental phenomena that the yield of IM28 increases with the increasing reaction time (Jackson et al., 2013).

3.3. Rate constant calculations

To assess the transport properties and fate of EtFBA in the atmosphere, especially the contribution to the burden of PFCAs in the Arctic, it is critical to know the accurate rate constants and atmospheric lifetime of EtFBA. Based on the MPWB1K/6-311+G(3df,2p)//MPWB1K/6-31+G(d,p) energies, the individual and overall rate constants for the reaction of EtFBA with OH radicals were calculated with the aid of the CVT/SCT method (Baldridge et al., 1989; Gonzalez-Lafont et al., 1991; Garrett and Truhlar, 1979; Fernandez-Ramos et al., 2007). A suitable temperature range of 180–370 K was chosen to study the relationship between the temperature and rate constant at 1 atm. For the multichannel reaction of EtFBA with OH radicals, the individual rate constants for H abstractions from the $N—H^1$, $C^2—H^2$, $C^3—H^4$ bonds are noted as k_{abs1}, k_{abs2}, k_{abs3}, and the rate constant of OH addition to EtFBA is denoted as k_{add}. The overall rate constant for the reaction with OH radicals is noted as k_{total}, $k_{total} = k_{abs1} + 2k_{abs2} + 3k_{abs3} + k_{add}$. The calculated individual and overall rate constants for the reaction of EtFBA with OH radicals were fitted in the Arrhenius formula $k = A\exp(-E/T)$ (in cm^3 $molecule^{-1}$ s^{-1}), as given in Table 1.

The high regression coefficient ($R^2 > 0.95$) indicates the acceptability of the results. At 296 K, the calculated overall rate constant of EtFBA with OH radicals is 2.50×10^{-12} cm^3 $molecule^{-1}$ s^{-1}, which matches well with the experimental value of $(2.65 \pm 0.5)\times 10^{-12}$ cm^3 $molecule^{-1}$ s^{-1} (Jackson et al., 2013). For all the elementary reactions, H abstraction from the methylene ($—CH_2—$, namely the $C^2—H^2$ and $C^2—H^3$ bonds) group is dominant, which accounts for 49.7% of the overall rate constant.

To provide further insight into EtFBA oxidation, the atmospheric fate of the main degradation products should be examined. At 296 K, using the CVT/SCT method, the overall rate constants for the reactions of $C_3F_7C(O)N(H)C(O)CH_3$, $C_3F_7C(O)N(H)CH_2CHO$ and $C_3F_7C(O)NH_2$ with OH radicals were calculated to be

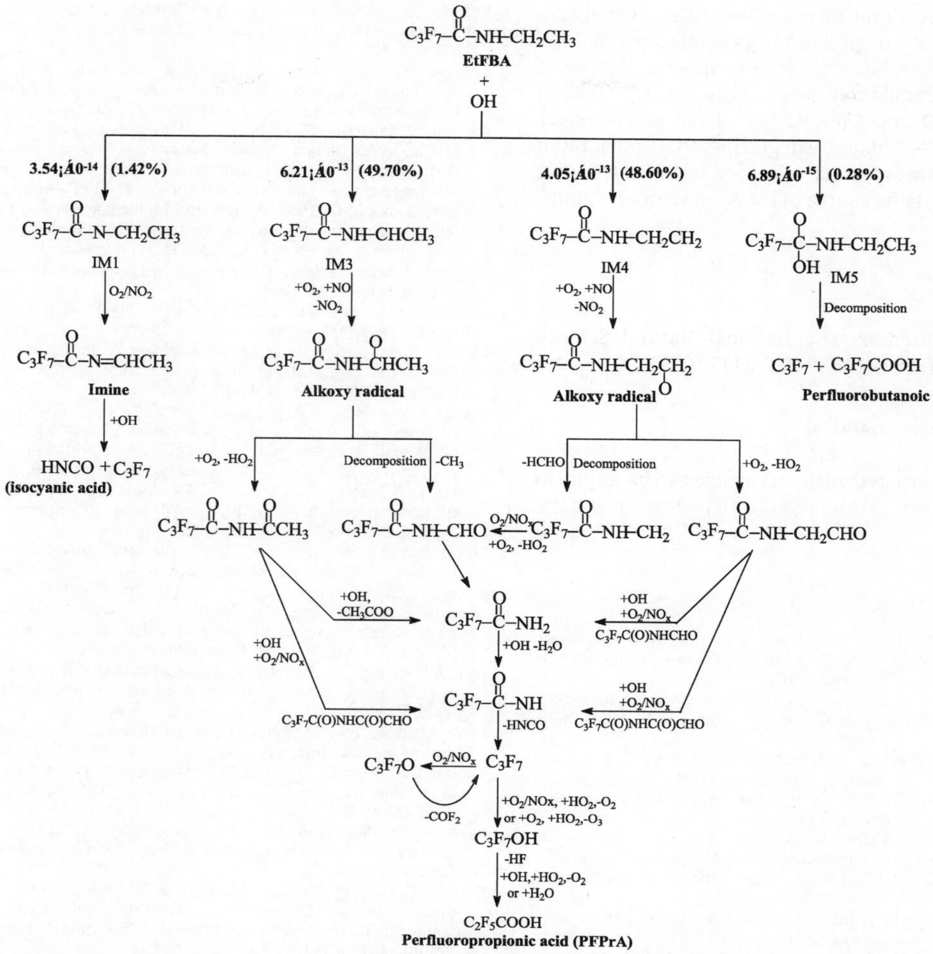

Fig. 6. The main pathways leading from EtFBA to PFCAs, C_2F_5COOH and C_3F_7COOH, embedded with individual rate constants (cm^3 $molecule^{-1}$ s^{-1}) and the percentages to the overall rate constants.

Table 1
Rate constants k (cm^3 $molecule^{-1}$ s^{-1}) at 296 K, Arrhenius formulas and the correlation coefficient (R^2) for the reaction of *N*-ethylperfluorobutyramide with OH radicals over the temperature range of 180–370 K.

Reactions	k_{296K}	Arrhenius formulas	R^2
EtFBA + OH → IM1 + H_2O	3.54×10^{-14}	$k_{abs1}(T) = (3.42 \times 10^{-13})\exp(-675.19/T)$	0.9864
EtFBA + OH → IM2 + H_2O	6.21×10^{-13}	$k_{abs2}(T) = (2.69 \times 10^{-12})\exp(-442.12/T)$	0.9963
EtFBA + OH → IM3 + H_2O	4.05×10^{-13}	$k_{abs3}(T) = (2.38 \times 10^{-12})\exp(-527.47/T)$	0.9910
EtFBA + OH → IM4	6.89×10^{-15}	$k_{add}(T) = (1.20 \times 10^{-13})\exp(-842.90/T)$	0.9923
EtFBA + OH → Products	2.50×10^{-12}	$k_{total}(T) = (1.37 \times 10^{-11})\exp(-504.21/T)$	0.9952

2.18×10^{-13}, 3.34×10^{-13} and 3.96×10^{-13} cm^3 $molecule^{-1}$ s^{-1}, respectively.

Assuming that the reaction with OH radicals is the only important sink of EtFBA, the atmospheric lifetime (τ) can be calculated by the formula: $\tau = 1/(k_{total} \cdot [OH])$. k_{total} is the overall rate constant of the initial reactions. [OH] is the atmospheric concentration of the OH radical (1.0×10^6 molecule cm^{-3}) (Lawrence et al., 2001). Thus, the atmospheric lifetimes of EtFBA as well as $C_3F_7C(O)N(H)C(O)CH_3$, $C_3F_7C(O)N(H)CH_2CHO$ and $C_3F_7C(O)NH_2$ determined by OH radicals are about 4.6, 53.1, 34.7 and 29.2 days, respectively. It takes about 1–2 months to transport a species from mid latitude to high latitude under the typical atmospheric conditions. The short atmospheric lifetime of EtFBA is not likely sufficient to account for the transport from urban or industrial sources to the high Arctic. However, its degradation products, $C_3F_7C(O)N(H)C(O)CH_3$, $C_3F_7C(O)N(H)CH_2CHO$ and $C_3F_7C(O)NH_2$, have much longer atmospheric lifetime (approximately 30–50 days), which is sufficient for long-range atmospheric transport and oxidation to yield PFCAs. The current pathways of EtFBA can be certainly extrapolated to the eight carbon PFAMs, which could have served as historical atmospheric precursors of PFOA, a contaminant of great scientific interest (Zareitalabad et al., 2013).

In summary, to evaluate the PFAM contribution to the burden of PFCA pollution observed in the Arctic, the mechanism and kinetic properties for the atmospheric oxidation degradation of EtFBA initiated by OH radicals have been investigated. Studies show that H abstraction from the C^2—H^2 bond is the dominant reaction pathway. This work reveals for the first time that the water molecule plays an important catalytic role in the several key elementary steps and promotes the degradation potential of EtFBA. Calculations show that EtFBA has a short atmospheric lifetime of 4.6 days, however, its degradation products, $C_3F_7C(O)N(H)C(O)CH_3$, $C_3F_7C(O)N(H)CH_2CHO$ and $C_3F_7C(O)NH_2$, have much longer atmospheric lifetimes (30–50 days), suggesting that substantial long-range atmospheric transport is allowed. It indicates that PFAMs can serve as atmospheric source of PFCAs in remote regions, including the Arctic.

Acknowledgment

This work was supported by the National Natural Science Foundation of China (NSFC Nos. 21337001, 21177077).

Appendix A. Supplementary material

Supplementary data associated with this article can be found, in the online version, at http://dx.doi.org/10.1016/j.chemosphere.2015.04.059.

References

Adamo, C., Barone, V., 1998. Exchange functionals with improved long-range behavior and adiabatic connection methods without adjustable parameters: the mPW and mPW1PW models. J. Chem. Phys. 108, 664–675.

Ahrens, L., Siebert, U., Ebinghaus, R., 2009. Temporal trends of polyfluoroalkyl compounds in harbor seals (*Phoca vitulina*) from the German Bight, 1999–2008. Chemosphere 76, 151–158.

Anna, R., Anna, K., Bert van, B., Anuschka, P., Frank, R., Guðjón, A.A., Gísli, V., Geir, W.G., Dorete, B., Maria, D., 2012. Increasing levels of long-chain perfluorocarboxylic acids (PFCAs) in Arctic and North Atlantic marine mammals, 1984–2009. Chemosphere 86, 278–285.

Atkinson, R., Baulch, D.L., Cox, R.A., Crowley, J.N., Hampson, R.F., Hynes, R.G., Jenkin, M.E., Rossi, M.J., Troe, J., 2008. Evaluated kinetic and photochemical data for atmospheric chemistry: volume IV-gas phase reactions of organic halogen species. Atmos. Chem. Phys. 8, 4141–4496.

Baldridge, M.S., Gordor, R., Steckler, R., Truhlar, D.G., 1989. Ab initio reaction paths and direct dynamics calculations. J. Phys. Chem. 93, 5107–5119.

Barnes, I., Solignac, G., Mellouki, A., Becker, K.H., 2010. Aspects of the atmospheric chemistry of amides. Chem. Phys. Chem. 11, 3844–3857.

Butt, C.M., Young, C.J., Mabury, S.A., 2009. Atmospheric chemistry of 4:2 fluorotelomer acrylate ($C_4F_9CH_2CH_2OC(O)CH{=}CH_2$): kinetics, mechanisms, and products of chlorine-atom- and OH-radical-initiated oxidation. J. Phys. Chem. A 113, 3155–3161.

DeMore, W.B., Sadner, S.P., Golden, D.M., Hampson, R.F., Kurylo, M.J., Howard, C.J., Ravishankara, A.R., Kolb, C.E., Molina, M.J., 1992. Chemical Kinetics and Photochemical Data for Use in Stratospheric Modeling. Jet Propulsion Laboratory.

D'eon, J.C., Hurley, M.D., Wallington, T.J., Mabury, S.A., 2006. Atmospheric chemistry of N-methylperfluorobutane sulfonamidoethanol, $C_4F_9SO_2N(CH_3)CH_2CH_2OH$: kinetics and mechanism of reaction with OH. Environ. Sci. Technol. 40, 1862–1868.

Ellis, D.A., Martin, J.W., Mabury, S.A., Hurley, M.D., Sulbaek Andersen, M.P., Wallington, T.J., 2003. Atmospheric lifetime of fluorotelomer alcohols. Environ. Sci. Technol. 37, 3816–3820.

Fernandez-Ramos, A., Ellingson, B.A., Garret, B.C., Truhlar, D.G., 2007. Variational transition state theory with multidimensional tunneling. In: Lipkowitz, K.B., Cundari, T.R. (Eds.), Reviews in Computational Chemistry. Wiley-VCH, Hoboken, NJ.

Frisch, M., Trucks, G., Schlegel, H.B., Scuseria, G., Robb, M., Cheeseman, J., Scalmani, G., Barone, V., Mennucci, B., Petersson, G., 2009. Gaussian 09, Revision A. 02; Gaussian Inc., Wallingford, CT, 270, 271.

Fukui, K., 1981. The path of chemical reactions – the IRC approach. Acc. Chem. Res. 14, 363–368.

Garrett, B.C., Truhlar, D.G.Generalized transition state theory, 1979. Classical mechanical theory and applications to collinear reactions of hydrogen molecules. J. Phys. Chem. 83, 1052–1079.

Gonzalez-Lafont, A., Truong, T.N., Truhlar, D.G., 1991. Interpolated variational transition-state theory: practical methods for estimating variational transitionstate properties and tunneling contributions to chemical reaction rates from electronic structure calculations. J. Chem. Phys. 95, 8875–8894.

Herzberg, G., 1966. Electronic Spectra and Electronic Structure of Polyatomic Molecules. Van Nostrand, New York.

Hou, H., Wang, B.S., 2005. A systematic computational study on the reactions of HO_2 with RO_2: the $HO_2 + CH_3O_2(CD_3O_2)$ and $HO_2 + CH_2FO_2$ reactions. J. Phys. Chem. A 109, 451–460.

Jackson, D.A., Mabury, S.A., 2013. Polyfluorinated amides as a historical PFCA source by electrochemical fluorination of alkyl sulfonyl fluorides. Environ. Sci. Technol. 47, 382–389.

Jackson, D.A., Wallington, T.J., Mabury, S.A., 2013. Atmospheric oxidation of polyfluorinated amides: historical source of perfluorinated carboxylic acids to the environment. Environ. Sci. Technol. 47, 4317–4324.

Kannan, K., Corsolini, S., Falandysz, J., Fillmann, G., Kumar, K.S., Loganathan, B.G., Mohd, M.A., Olivero, J., Van Wouwe, N., Yang, J.H., Aldous, K.M., 2004. Perfluorooctanesulfonate and related fluorochemicals in human blood from several countries. Environ. Sci. Technol. 38, 4489–4495.

Kuchitsu, Landolt, B., 1992. Atomic and Molecular Physics, Structure Data of Free Polyatomic Molecules, vol. 21, Springer-Verlag, Berlin.

Lau, C., Anitole, K., Hodes, C., Lai, D., Pfahles-Hutchens, A., Seed, J., 2007. Perfluoroalkyl acids: a review of monitoring and toxicological findings. Toxicol. Sci. 99, 366–394.

Lawrence, M.G., Jöckel, P., van Kuhlmann, R., 2001. What does the global mean OH concentration tell us? Atmos. Chem. Phys. 1, 37–49.

Lee, H., Mabury, S.A., 2011. A pilot survey of legacy and current commercial fluorinated chemicals in human sera from United States donors in 2009. Environ. Sci. Technol. 45, 8067–8074.

Martin, J.W., Ellis, D.A., Mabury, S.A., Hurley, M.D., Wallington, T.J., 2006. Atmospheric chemistry of perfluoroalkanesulfonamides: kinetic and product studies of the OH radical and Cl atom initiated oxidation of N-ethyl perfluorobutanesulfonamide. Environ. Sci. Technol. 40, 864–872.

Martin, J.W., Mabury, S.A., Solomon, K.R., Muir, D.C.G., 2003. Bioconcentration and tissue distribution of perfluorinated acids in rainbow trout (*Oncorhynchus mykiss*). Environ. Toxicol. Chem. 22, 196–204.

Morokuma, K., Muguruma, C., 1994. Ab initio molecular orbital study of the mechanism of the gas phase reaction $SO_3 + H_2O$: importance of the second water molecule. J. Am. Chem. Soc. 116, 10316–10317.

Nielsen, C.J., Herrmann, H., Weller, C., 2012. Atmospheric chemistry and environmental impact of the use of amines in carbon capture and storage (CCS). Chem. Soc. Rev. 41, 6684–6704.

Paul, A.G., Jones, K.C., Sweetman, A.J., 2009. A first global production, emission, and environmental inventory for perfluorooctane sulfonate. Environ. Sci. Technol. 43, 386–392.

Pfeilsticker, K., Lotter, A., Peters, C., Bösch, H., 2003. Atmospheric detection of water dimers via near-infrared absorption. Science 300, 2078–2080.

Robert, L.W., Kerwin, D.D., 2007. Atmospheric chemistry of linear perfluorinated aldehydes: dissociation kinetics of $C_nF_{2n+1}CO$ radicals. J. Phys. Chem. A 111, 2555–2562.

Ryzhkov, A.B., Ariya, P.A., 2006. The importance of water clusters $(H_2O)_n$ ($n = 2, 4$) in the reaction of Criegee intermediate with water in the atmosphere. Chem. Phys. Lett. 419, 479–485.

Schneider, W.F., Wallington, T.J., Huie, R.E., 1996. Energetics and mechanism of decomposition of CF_3OH. J. Phys. Chem. 100, 6097–6103.

Shimanouchi, T., 1972. Tables of Molecular Vibrational Frequencies, Consolidated Volume 1; NSRDS NBS-39.

Steckler, R., Chuang, Y.Y., Fast, P.L., Corchade, J.C., Coitino, E.L., Hu, W.P., Lynch, G.C., Nguyen, K., Jackells, C.F., Gu, M.Z., Rossi, I., Clayton, S., Melissas, V., Garrett, B.C., Isaacson, A.D., Truhlar, D.G., 2002. POLYRATE Version 9.3. University of Minnesota, Minneapolis.

Stock, N.L., Furdui, V.I., Muir, D.C.G., Mabury, S.A., 2007. Perfluoroalkyl contaminants in the Canadian arctic: evidence of atmospheric transport and local contamination. Environ. Sci. Technol. 41, 3529–3536.

Sulbaek Andersen, M.P., Hurley, M.D., Wallington, T.J., Ball, J.C., Martin, J.W., Ellis, D.A., Mabury, S.A., 2003. Atmospheric chemistry of C_2F_5CHO: mechanism of the $C_2F_5C(O)O_2 + HO_2$ reaction. Chem. Phys. Lett. 381, 14–21.

Tang, Y., Nielsen, C.J., 2013. Theoretical study on the formation and photolysis of nitrosamines (CH_3CH_2NHNO and $(CH_3CH_2)_2NNO$) under atmospheric conditions. J. Phys. Chem. A 117, 126–132.

Teixeira-Dias, J.J.C., Furlani, T.R., Shores, K.S., Garvey, J.F., 2003. Transition states for H atom transfer reactions in the CH_2CH_2OH radical: the effect of a water molecule. Phys. Chem. Chem. Phys. 5, 5063–5069.

Wallington, T.J., Hurley, M.D., Xia, J., Wuebbles, D.J., Sillman, S., Ito, A., Penner, J.E., Ellis, D.A., Martin, J., Mabury, S.A., Nielsen, O.J., Sulbaek Andersen, M.P., 2006. Formation of $C_7F_{15}COOH$ (PFOA) and other perfluorocarboxylic acids during the atmospheric oxidation of 8: 2 fluorotelomer alcohol. Environ. Sci. Technol. 40, 924–930.

Young, C.J., Donaldson, D.J., 2007. Overtone-induced degradation of perfluorinated alcohols in the atmosphere. J. Phys. Chem. A 111, 13466–13471.

Yu, W.N., Li, P.F., Xu, F., Hu, Zhang, Q.Z., Wang, W.X., 2013. Quantum chemical and direct dynamic study on homogeneous gas-phase formation of PBDD/Fs from 2,4,5-tribromophenol and 3,4-dibromophenol. Chemosphere 93, 512–520.

Zareitalabad, P., Siemens, J., Hamer, M., Amelung, W., 2013. Perfluorooctanoic acid (PFOA) and perfluorooctanesulfonic acid (PFOS) in surface waters, sediments, soils and wastewater – a review on concentrations and distribution coefficients. Chemosphere 91, 725–732.

Zeroka, D., Jensen, J.O., Samuels, A.C., 1999. Infrared spectra of some isotopomers of ethylamine and the ethylammonium ion: a theoretical study. J. Mol. Struct. THEOCHEM. 465, 119–139.

Zhao, Y., Truhlar, D.G., 2004. Hybrid meta density functional theory methods for thermochemistry, thermochemical kinetics, and noncovalent interactions: the MPW1B95 and MPWB1K models and comparative assessments for hydrogen bonding and van der Waals interactions. J. Phys. Chem. A 108, 6908–6918.

Environmental Science & Technology

Article

pubs.acs.org/est

Role of Water Molecule in the Gas-Phase Formation Process of Nitrated Polycyclic Aromatic Hydrocarbons in the Atmosphere: A Computational Study

Qingzhu Zhang,*,† Rui Gao,† Fei Xu,† Qin Zhou,† Guibin Jiang,‡ Tao Wang,§ Jianmin Chen,† Jingtian Hu,† Wei Jiang,† and Wenxing Wang*,†

†Environment Research Institute, Shandong University, Jinan 250100, China

‡State Key Laboratory Environmental Chemistry and Ecotoxicology, Research Center for Eco-Environmental Sciences, Chinese Academy of Sciences, Beijing 100085, China

§Department of Civil and Structural Engineering, The Hong Kong Polytechnic University, Hong Kong, China

Supporting Information

ABSTRACT: Nitro-PAHs are globally worrisome air pollutants because their high direct-acting mutagenicity and carcinogenicity. A mechanistic understanding of their formation is of crucial importance for successful prevention of their atmospheric pollution. Here, the formation of nitro-PAHs arising from the OH-initiated and NO_3-initiated atmospheric reactions of PAHs was investigated by using quantum chemical calculations. It is widely assumed that OH or NO_3 radicals attack on the C atoms of the aromatic rings in the PAH molecule, followed by the addition of NO_2 to the OH−PAH or NO_3−PAH adducts at the *ortho* position and the loss of water or nitric acid to form nitro-PAHs. However, calculations show that the direct loss of water from the OH−NO_2−PAH adducts via the unimolecular decomposition is energetically unfavorable. This study reveals for the first time that water molecule plays an important catalytic effect on the loss of water from the OH−NO_2−PAH adducts and promotes the formation of nitro-PAHs. In addition, the introduction of water unwraps new formation pathway through the addition of NO_2 to the OH−PAH or NO_3−PAH adduct at the *para* position. The individual and overall rate constants for the addition reactions of PAHs with OH and NO_3 radicals were deduced by using the Rice−Ramsperger−Kassel−Marcus (RRKM) theory.

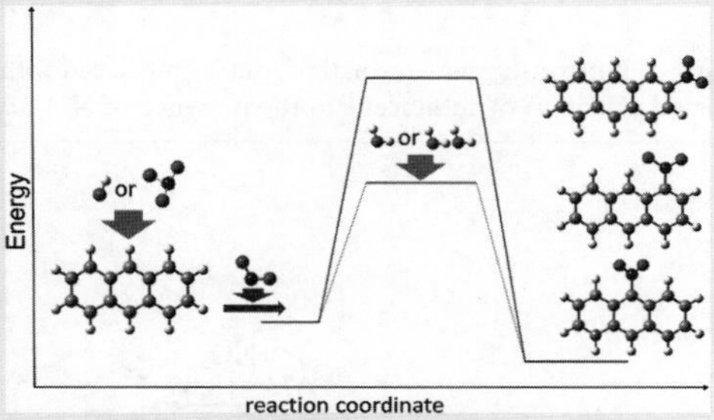

1. INTRODUCTION

Atmospheric polycyclic aromatic hydrocarbons (PAHs) and their derivatives, such as nitro-PAHs, have a strong association with human lung cancer.[1−4] In China, lung cancer has replaced liver cancer to become the predominant cause of cancer deaths (~23% of cancer deaths).[5] The atmospheric emission of 16 priority PAHs in all Asian countries accounted for 53.5% of the total global emissions (504 Gg), with the highest emission from China (106 Gg) and India (67 Gg) in 2007.[6] Consequently, the ambient air concentration of PAHs in many parts of China is orders of magnitude higher than that in developed countries.[2] Although concentration of nitro-PAHs is 1−2 orders of magnitude lower than their parent PAHs, nitro-PAHs can be 100000 times more mutagenic and 10 times more carcinogenic compared to the unsubstituted-PAHs.[7] Certain nitro-PAHs exhibit high direct-acting mutagenic potency in microbial mutagenicity bioassays and in forward mutation assays based on human cells.[1] Nitro-PAHs have been found to account for over 50% of the total direct-acting mutagenicity and the total carcinogenicity activity of ambient air.[8]

Nitro-PAHs are present in the atmosphere both as a result of primary emission from incomplete combustion, primarily by diesel engines, and as a result of the secondary formation from the atmospheric reaction of PAHs.[9] The specific congeners of nitro-PAHs observed in ambient air revealed that the secondary formation makes a significant contribution to the abundance profiles of nitro-PAHs.[10−14] For example, the main nitro-PAH, 1-nitropyrene, in diesel exhaust is generally not the predominant nitro-PAH isomer observed in ambient air. In contrast, 2-nitrofluoranthene and 2-nitropyrene, which have been shown in laboratory experiments[15,16] to be produced from the gas-phase reactions of PAHs, are usually more abundant nitro isomers in the atmosphere.[10,11,17] In addition, nitro-PAHs have lower vapor pressures than the corresponding PAHs and thus promote the formation of secondary organic aerosols.[18] Therefore, it is very meaningful to understand and predict the transformation of PAHs and the formation of nitro-PAHs in the atmosphere.

Nitro-PAHs can be formed in the atmosphere via homogeneous gas-phase reactions of PAHs with OH or NO_3 radicals in the presence of NO_2 and heterogeneous reactions of particulate PAHs with N_2O_5 or HNO_3. In the atmosphere, the fraction of

Received: February 6, 2014
Revised: March 28, 2014
Accepted: April 1, 2014
Published: April 1, 2014

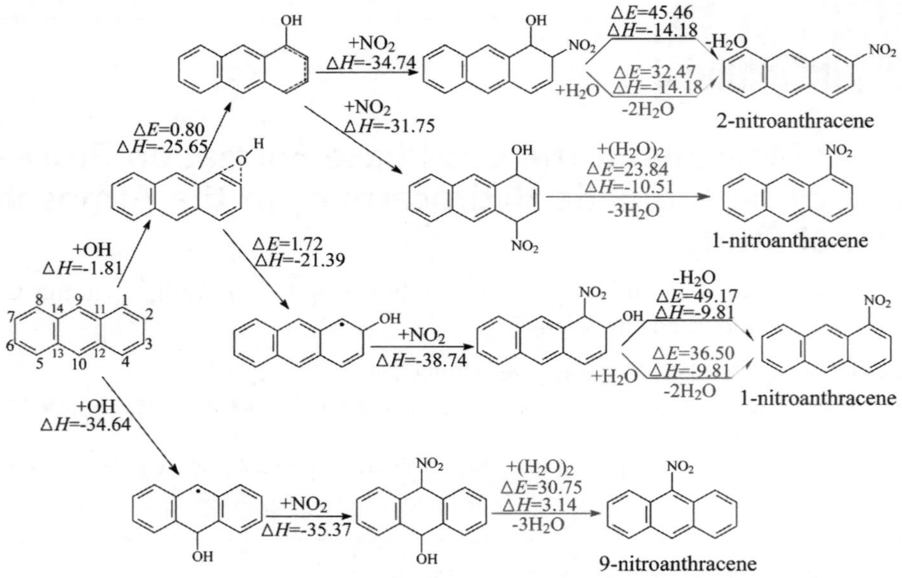

Figure 1. Nitroanthracene formation routes embedded with the potential barrier ΔE (in kcal/mol) and reaction heat ΔH (in kcal/mol) from the OH-initiated oxidation of anthracene in the presence of NO_2. ΔH is calculated at 0 K.

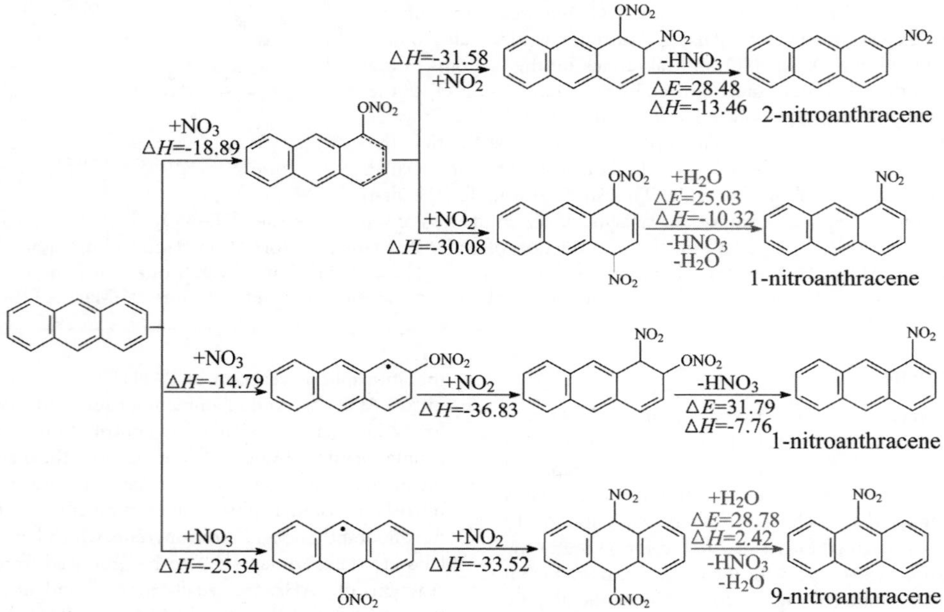

Figure 2. Nitroanthracene formation routes embedded with the potential barrier ΔE (in kcal/mol) and reaction heat ΔH (in kcal/mol) from the NO_3-initiated oxidation of anthracene in the presence of NO_2. ΔH is calculated at 0 K.

certain PAHs in the gaseous phase can reach up to 90%.[19] In addition, the gas-phase reactions of PAHs with OH and NO_3 radicals are generally faster compared to the heterogeneous reactions of particulate PAHs with N_2O_5 or HNO_3.[20] Thus, the gas-phase formation has been recognized as a significant source of nitro-PAHs in the atmosphere. It is widely assumed that OH or NO_3 radicals attack on the C atoms of the aromatic rings in the PAH molecule, followed by the addition of NO_2 to the OH-PAH or NO_3−PAH adducts at the *ortho* position and the loss of water or nitric acid to form nitro-PAHs.[11] However, the detailed formation mechanism has still not been completely elucidated, largely due to the scarcity of efficient detection schemes for radical intermediate species and commercially available standards.[11,21] In this work, we carried out a theoretical study, employing a high-accuracy quantum chemical method. The geometrical parameters and vibrational frequencies were calculated at the BB1K level[22] with the standard 6-31+G(d,p) basis set. Single-point energy calculations were carried out at the BB1*K*/6-311+G(3df,2p) level of theory. The atmospheric model places a high demand on accurate kinetic parameters.[23] Therefore, the individual and overall rate constants were evaluated by using Rice−Ramsperger−Kassel−Marcus

(RRKM) theory on the basis of the electronic structure calculations. The effect of the molecular size and steric structure of PAHs on the formation mechanism of nitro-PAHs was discussed.

2. COMPUTATIONAL METHODS

All the electronic structure calculations were carried out with the Gaussian 09 package on a supercomputer.[24] It has been shown that the hybrid density functional theory (DFT), BB1K, can give excellent transition state geometries and barrier heights, based on Becke's 1988 gradient-corrected exchange functional (Becke88 or B) and Becke's 1995 kinetic-energy-dependent dynamical correlation functional (Becke95 or B95).[22] Truhlar and co-workers pointed out that the hybrid DFT results are much less sensitive to the basis set compared to the ab initio ones, and they also emphasized the importance of diffuse functions.[25,26] Therefore, the geometries of the reactants, transition states, intermediates, and products were optimized at the BB1K level with a standard 6-31+G(d,p) basis set. For all stationary points, frequency calculations were performed at the same level to verify whether they are minima with all positive frequencies or transition states with only one imaginary frequency. The intrinsic reaction coordinate (IRC) calculations were further carried out at the BB1*K*/6-31+G(d,p) level to confirm that the transition state connects to the right minima along the reaction path. To yield more reliable energy values, a more flexible basis set, 6-311+G(3df,2p), was used for single-point energy calculations.

The rate constants were calculated by using the open source master equation program MESMER[27] (master equation solver for multienergy well reactions) based on the Rice–Ramsperger–Kassel–Marcus (RRKM) theory.[28] The reactions related in this study occur in the atmosphere. Therefore, the default pressure value was set as 1 atm. The RRKM rate constant is given by

$$k(E) = \frac{W(E)}{h\rho(E)} \tag{1}$$

where $W(E)$ is the rovibrational sum of states at the transition state, $\rho(E)$ is density of states of reactants, and h is Planck's constant. Then, canonical rate constant $k(T)$ is determined by using the usual equation

$$k(T) = \frac{1}{QT}\int k(E)\rho(E)\exp(-\beta E)\mathrm{d}E \tag{2}$$

where $Q(T)$ is the reactant partition function.

3. RESULTS AND DISCUSSION

The optimized geometries and the calculated vibrational frequencies of benzene, naphthalene, anthracene, nitrobenzene, nitronaphthalene, and 9-nitroanthracene at the BB1*K*/6-31+G-(d,p) level are in excellent agreement with the available experimental values, and the relative deviation remains within 3.0% for the bond lengths and 8.0% for the vibrational frequencies.[29−33] In order to check the reliability of the energy parameters, we calculated the reaction enthalpy for the reaction of benzene + OH + NO_2 → nitrobenzene + H_2O. The value of −76.50 kcal/mol deduced at the BB1K/6-311+G(3df,2p)//BB1K/6-31+G(d,p) level and 298.15 K shows good consistency with the corresponding experimental value of −78.28 kcal/mol obtained from the measured standard enthalpies of formation (ΔH_f^0) of benzene (19.82 kcal/mol), OH (8.93 kcal/mol), NO_2 (8.12 kcal/mol), nitrobenzene (16.38 kcal/mol), and H_2O (−57.79 kcal/mol).[29]

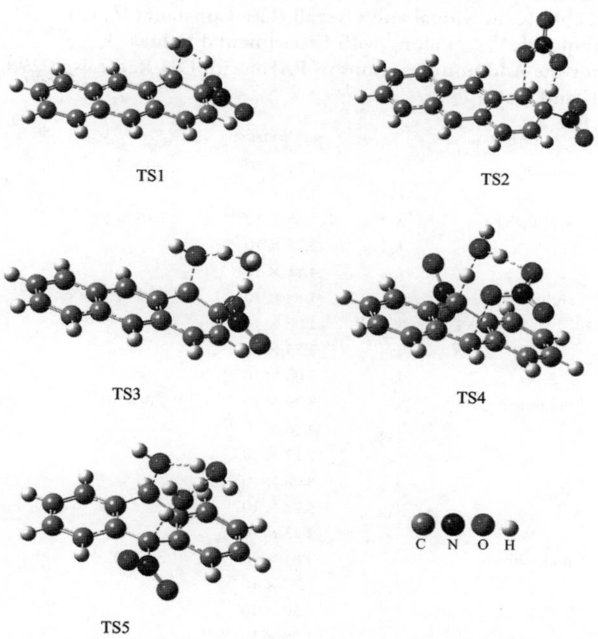

Figure 3. Configuration of the transition states for the processes of the loss of water or nitric acid. TS1: OH−NO_2−PAH → nitro-PAH + H_2O, TS2: NO_3−NO_2−-PAH → nitro-PAH + HNO_3, TS3: OH−NO_2−PAH + H_2O → nitro-PAH + 2H_2O, TS4: NO_3−NO_2−PAH + H_2O → nitro-PAH + HNO_3 + H_2O, TS5: OH−NO_2−PAH + $(H_2O)_2$ → nitro-PAH + 3H_2O.

For convenience of description, the carbon atoms in anthracene are numbered as follows:

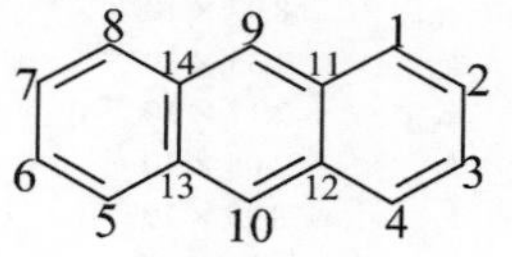

anthracene

The carbon atoms in other PAHs are numbered as well, as presented in the Supporting Information.

3.1. Formation Mechanism. Figures 1 and 2 (black arrows) depict the formation scheme of nitroanthracenes embedded with the potential barrier and reaction heat. The formation scheme of nitro-PAHs from the OH- and NO_3-initiated reactions of benzene, naphthalene, fluorene, fluoranthene, phenanthrene, acenaphthene, acenaphthylene, and pyrene is presented in the Supporting Information. As the anthracene molecule at the ground state is of D_{2h} symmetry, C atoms in anthracene fall into four groups: C1, C4, C5, and C8 atoms belong to one equivalent group; C2, C3, C6, and C7; C9 and C10; C11, C12, C13, and C14 belong to the other equivalent groups. Therefore, four different OH− or NO_3−anthracene adduct isomers can be formed through addition of OH or NO_3 to C1, C2, C9, and C11 atoms, respectively. Calculations show that the addition of OH or NO_3 to the C9 atom is the most exothermic and the most energetically favorable process compared with the addition of OH or NO_3 to C1, C2, and C11 atoms. However, the reaction pathway initiated from the addition of OH or NO_3 to the C9 atom cannot lead to the formation of nitroanthracene due to the lack of the *ortho* hydrogen. The reaction pathway through the addition of OH or NO_3 to the C11 atom followed by the addition

dx.doi.org/10.1021/es500453g | *Environ. Sci. Technol.* 2014, 48, 5051−5057

Table 1. Individual and Overall Rate Constants (k, cm^3 $molecule^{-1}$ s^{-1}) along with Experimental Values (k_{expt})[45−50] for the Addition Reactions of PAHs with OH Radicals at 298 K and 1 atm

PAHs		rate constants	k_{expt}
benzene	k	1.14×10^{-12}	1.0×10^{-12}
	k_{C1}	1.91×10^{-13}	
naphthalene	k	1.53×10^{-11}	2.16×10^{-11}
	k_{C1}	3.34×10^{-12}	
	k_{C2}	4.81×10^{-13}	
anthracene	k	1.65×10^{-10}	$(1.3-1.91) \times 10^{-10}$
	k_{C1}	1.06×10^{-11}	
	k_{C2}	3.85×10^{-13}	
	k_{C9}	6.06×10^{-11}	
fluorene	k	3.39×10^{-12}	1.3×10^{-11}
	k_{C1}	2.86×10^{-13}	
	k_{C2}	6.17×10^{-13}	
	k_{C3}	8.05×10^{-14}	
	k_{C4}	4.72×10^{-13}	
	k_{C1a}	2.23×10^{-13}	
fluoranthene	k	3.84×10^{-11}	5×10^{-11}
	k_{C1}	5.37×10^{-12}	
	k_{C2}	2.50×10^{-13}	
	k_{C3}	1.28×10^{-11}	
	k_{C6a}	4.51×10^{-16}	
	k_{C6b}	1.10×10^{-15}	
	k_{C7}	2.41×10^{-13}	
	k_{C8}	5.66×10^{-13}	
phenanthrene	k	3.98×10^{-11}	2.42×10^{-11}
	k_{C1}	2.43×10^{-12}	
	k_{C2}	1.68×10^{-13}	
	k_{C3}	3.84×10^{-13}	
	k_{C4}	7.95×10^{-12}	
	k_{C10}	8.97×10^{-12}	
acenaphthene	k	1.35×10^{-10}	1.0×10^{-10}
	k_{C2a}	2.17×10^{-11}	
	k_{C3}	2.06×10^{-11}	
	k_{C4}	8.98×10^{-13}	
	k_{C5}	2.45×10^{-11}	
acenaphthylene	k	3.64×10^{-10}	1.1×10^{-10}
	k_{C1}	1.69×10^{-10}	
	k_{C2a}	2.48×10^{-15}	
	k_{C3}	8.03×10^{-13}	
	k_{C4}	4.31×10^{-13}	
	k_{C5}	1.19×10^{-11}	
pyrene	k	6.58×10^{-11}	5×10^{-11}
	k_{C1}	1.11×10^{-11}	
	k_{C2}	2.40×10^{-14}	
	k_{C4}	5.35×10^{-12}	

of NO_2 to the C9 atom and the loss of water or nitric acid may yield 9-nitroanthracene. However, C11 is inside the "bend" of the anthracene molecule. Addition of OH or NO_3 to the C11 atom is stericly hindered due to the high barrier and/or strong endothermicity. Therefore, this formation pathway of 9-nitroanthracene is energetically unfavorable, and the possible nitroanthracene formation pathways may occur via addition of OH or NO_3 to the C1 or C2 atoms.

Calculations show that the addition of OH or NO_3 to the C1 and C2 atoms is highly exothermic with a low energy barrier or without a barrier. Addition of NO_2 to the OH− or NO_3−PAH intermediate is a barrierless and strongly exothermic process.

Table 2. Individual and Overall Rate Constants (k, cm^3 $molecule^{-1}$ s^{-1}) along with Experimental Values (k_{expt})[16,42,51−53] for the Addition Reactions of PAHs with NO_3 Radicals at 298 K and 1 atm

PAHs		rate constants	k_{expt}
benzene	k	3.66×10^{-17}	$(2.3-5.5) \times 10^{-17}$
	k_{C1}	6.10×10^{-18}	
naphthalene	k	4.83×10^{-14}	9.40×10^{-15}
	k_{C1}	6.09×10^{-15}	
	k_{C2}	5.99×10^{-15}	
anthracene	k	5.79×10^{-12}	5.45×10^{-12}
	k_{C1}	6.00×10^{-13}	
	k_{C2}	5.46×10^{-13}	
	k_{C9}	6.04×10^{-13}	
fluorene	k	8.72×10^{-14}	3.5×10^{-14}
	k_{C1}	5.73×10^{-15}	
	k_{C2}	2.02×10^{-14}	
	k_{C3}	6.03×10^{-15}	
	k_{C4}	5.66×10^{-15}	
	k_{C1a}	5.95×10^{-15}	
fluoranthene	k	5.91×10^{-13}	
	k_{C1}	5.98×10^{-14}	
	k_{C2}	5.74×10^{-14}	
	k_{C3}	6.25×10^{-14}	
	k_{C7}	5.67×10^{-14}	
	k_{C8}	5.90×10^{-14}	
phenanthrene	k	5.84×10^{-13}	1.2×10^{-13}
	k_{C1}	5.98×10^{-14}	
	k_{C2}	5.43×10^{-14}	
	k_{C3}	5.82×10^{-14}	
	k_{C4}	5.97×10^{-14}	
	k_{C10}	5.99×10^{-14}	
acenaphthene	k	4.80×10^{-13}	4.59×10^{-13}
	k_{C2a}	5.97×10^{-14}	
	k_{C3}	6.02×10^{-14}	
	k_{C4}	5.96×10^{-14}	
	k_{C5}	6.06×10^{-14}	
acenaphthylene	k	4.52×10^{-12}	5.45×10^{-12}
	k_{C1}	6.14×10^{-13}	
	k_{C2a}	2.89×10^{-14}	
	k_{C3}	5.99×10^{-13}	
	k_{C4}	4.09×10^{-13}	
	k_{C5}	6.10×10^{-13}	
pyrene	k	4.94×10^{-12}	
	k_{C1}	8.33×10^{-14}	
	k_{C2}	5.96×10^{-13}	
	k_{C4}	5.98×10^{-13}	

The OH−NO_2−PAH or NO_3−NO_2−PAH adducts may subsequently undergo unimolecular decomposition to yield nitro-PAHs through the direct loss of water or nitric acid. The process of the direct loss of water or nitric acid requires crossing a high barrier. In particular, as shown in Figure 1 and the Supporting Information, the barriers of the direct loss of water via the unimolecular decomposition are extremely high. For example, the barriers of the direct loss of water from OH−NO_2−anthracene are more than 45 kcal/mol. The transition states of the direct loss of water even lie at 5.40−20.44 kcal/mol above the energy of OH−PAH + NO_2. This inhibits the direct loss of water from proceeding beyond the formation of the adduct OH−NO_2−PAH. As seen in Figure 3, the direct loss of water proceeds via a four-membered ring transition state, whereas the loss of nitric acid proceeds via a six-membered ring transition state. The bond angles in the four-membered ring are more bent than those in the six-membered ring (for example 77.0°, 115.9°, 71.8°, and 95.2° in TS1 vs 123.7°, 117.8°, 106.8°, 168.1°, 98.9°, and 104.2° in TS2). The ring tension in the four-membered ring is also larger. However, field studies showed that the OH-initiated formation of nitro-PAHs is dominant compared with the NO_3-initiated formation. For example, the contribution of the OH-initiated formation is >90% in Denmark and >45% during summertime and >83% during wintertime in the mid-Atlantic region.[10,34] The main question is how to explain the nitro-PAH formation from the OH-initiated reactions of PAHs. There must be some other mechanism in its action.

Water is the third most abundant species in the atmosphere after N_2 and O_2, with a typical gas-phase concentration of 7.64×10^{17} molecules cm^{-3}, corresponding to 50% of relative humidity at 298 K. Besides the greenhouse effect, water plays an important role in several gas-phase chemical processes occurring in the earth's atmosphere. For example, it can participate actively the gas-phase reaction of SO_3 with H_2O to produce H_2SO_4[35] and the H-abstraction from HOCl by OH radicals[36] through its ability to form hydrogen bonds. An intriguing question is whether water can have an effect on the loss of water from the $OH-NO_2-PAH$ adducts. To answer this question, the following quantum chemical calculations were conducted. With the introduction of water, the loss of water from the $OH-NO_2-PAH$ adducts becomes a bimolecular reaction ($OH-NO_2-PAH + H_2O$). As shown in Figure 3, a six-membered ring transition state (TS3) was identified, in which water molecule acts as a bridge, accepting the hydrogen from the aromatic ring and simultaneously donating another hydrogen atom to the phenolic group. This reaction process breaks a bond in the original water molecule and leads to the formation of nitro-PAHs. Hence, the water molecule in the products is not the original one. The barriers of the loss of water via the bimolecular reaction are 9.94−14.73 kcal/mol lower compared to the direct loss of water via the unimolecular decomposition. The water molecule stabilizes the transition state and drastically lowers the reaction barrier. Thus, the water molecule plays a positive catalytic effect on the loss of water from the $OH-NO_2-PAH$ adducts and promotes the formation of nitro-PAHs.

The OH− or NO_3-PAH adduct has an unpaired electron and is a hybrid of several resonance structures, as depicted in Figure S1 of the Supporting Information. The unpaired electron can be distributed to the *ortho* position and the *para* position of the OH− or NO_3- group, and NO_2 can be added to those positions. The pathway through the addition of NO_2 to the OH−PAH adduct at the *para* position followed by the direct loss of water via unimolecular decomposition is improbable due to the large distance between oxygen and hydrogen atoms to leave. This calculation shows that the pathway through the addition of NO_2 to the NO_3-PAH adduct at the *para* position followed by the direct loss of nitric acid via unimolecular decomposition is unfeasible as well due to one of the two reasons: high barrier of the loss of nitric acid and the steric hindrance from the adjacent aromatic rings. However, with the aid of a water molecule, nitro-PAHs can be formed from the reaction pathway through the addition of NO_2 to the NO_3-PAH adduct at the *para* position followed by the loss of nitric acid via $NO_3-NO_2-PAH+H_2O \rightarrow$ nitro-PAH + HNO_3 + H_2O. Similarly, with the introduction of two water molecules, nitro-PAHs can be formed from the pathway through the addition of NO_2 to the OH−PAH adduct at the *para* position followed the loss of water via $OH-NO_2-PAH$ + H_2O + H_2O $\rightarrow$ nitro-PAH + $3H_2O$. Since the simultaneous collision of three molecules ($OH-NO_2-PAH + H_2O + H_2O$) is very improbable, in this water loss reaction, the termolecular mechanism is ruled out. Hence, the most probable mechanism is the reaction of $OH-NO_2-PAH$ with the water dimer, i.e., $OH-NO_2-PAH + (H_2O)_2 \rightarrow$ nitro-PAH + $3H_2O$. A field measurement by near-infrared spectroscopy has revealed a relatively high concentration of 6×10^{14} molecules cm^{-3} of water dimer in the atmosphere.[37,38]

9-Nitroanthracene is one of the most abundant nitro-PAHs in the atmosphere. It has been suggested that 9-nitroanthracene cannot be formed from the OH or NO_3 radical-initiated gas-phase reaction of anthracene due to the reason given above, and the presence of 9-nitroanthracene in the atmosphere is from a combination of primary emissions and heterogeneous reactions.[34,39] The present study drastically changes the traditional view[34,39] and reveals for the first time that 9-nitroanthracene can be formed from the OH or NO_3 radical-initiated gas-phase reaction of anthracene through the addition of OH or NO_3 to the C10 atom followed by the addition of NO_2 to the C9 atom (*para* position relative to the C10 atom) and the loss of water or nitric acid via the bimolecular reaction with the aid of water. The formation scheme embedded with the potential barrier and reaction heat is displayed in Figures 1 and 2 (red arrows). 9-Nitroanthracene was observed in a laboratory experiment, but it was ascribed to artifact formation during sampling.[39]

It is worth discussing the effect of the molecular size and steric structure of PAHs on the OH or NO_3 addition reaction as well as the formation mechanism of nitro-PAHs. First, we analyze three relatively simple PAHs, benzene, naphthalene, and anthracene. The ranking of the exothermicity for the addition reaction of PAHs with OH or NO_3 is as follows: benzene < naphthalene < anthracene. More aromatic rings increase the potential of the addition reaction. This is consistent with the experimental phenomena that the OH or NO_3 addition rate coefficient increases strongly from benzene to naphthalene, and to anthracene.[40−44] The potential barrier of the loss of water or nitric acid from the $OH-NO_2-PAH$ or NO_3-NO_2-PAH adducts (with or without the aid of water) increases when the PAH molecule size increases from benzene to anthracene, leading to the decrease in the nitro-PAH formation potential. Phenanthrene and anthracene have the same molecular size but different steric structure. However, the addition reaction of phenanthrene with OH or NO_3 is less exothermic than the corresponding reaction of anthracene. In addition, the loss of water or nitric acid from the $OH-NO_2$−phenanthrene or NO_3-NO_2−phenanthrene adducts has a lower barrier compared to the $OH-NO_2$−anthracene or NO_3-NO_2−anthracene adducts. Thus, the steric structure of PAHs has an effect on the formation mechanism of nitro-PAHs as well.

3.2. Rate Constant Calculations. The formation rate and yield of nitro-PAHs depend on the ambient PAH, OH and NO_3 radical concentrations, and the rate constants for the reactions of the parent PAHs with OH and NO_3 radicals. Several experimental studies are on record for the reaction rates of PAHs with OH and NO_3 radicals.[16,42,45−53] However, the reported values are the overall rate constants of the addition reaction. There is a notable absence of the individual rate constants. In this work, the individual and overall rate constants for the addition reactions of PAHs with OH and NO_3 radicals were computed with the aid of the RRKM theory at 298 K and 1 atm. The RRKM method has been successfully used to deal with several reactions.[27,54,55] The calculated values along with the available experimental data are listed in Table 1 for the addition reactions of PAHs with OH radicals and Table 2 for the addition reactions of PAHs with NO_3 radicals. Taking anthracene as an example, the individual rate constants for the OH or NO_3 addition to the C1, C2 and C9 atoms are noted as k_{C1}, k_{C2}, and k_{C9}, respectively. The overall rate constants of the addition reaction of anthracene OH or NO_3 with are noted as k, $k = 4k_{C1} + 4k_{C2} + 2k_{C9}$.

It is vital to clarify the reliability of the RRKM method. As is clear from Tables 1 and 2, the calculated overall rate constants for the addition reactions of benzene, naphthalene, anthracene, fluorene, fluoranthene, phenanthrene, acenaphthene, acenaphthylene, and pyrene with OH radicals and NO_3 radicals are in excellent agreement with the experimental values, and the discrepancy remains within 4 times. For example, the RRKM rate constant, 1.65×10^{-10} cm^3 $molecule^{-1}$ s^{-1}, agrees well with the experimental value[51] of $(1.3-1.91) \times 10^{-10}$ cm^3 $molecule^{-1}$ s^{-1} for the addition reaction of anthracene with OH radicals. From the good agreement with experimental values, it might be inferred that the RRKM individual rate constants are reasonable.

In summary, this study investigated theoretically the nitro-PAH formation arising from the OH-initiated and NO_3-initiated homogeneous reactions of benzene, naphthalene, anthracene, fluorene, fluoranthene, phenanthrene, acenaphthene, acenaphthylene, and pyrene. This work shows for the first time that water plays a positive catalytic role in the loss of water from the OH–NO_2–PAH adducts and unwraps new formation pathway through the addition of NO_2 to the OH–PAH or NO_3–PAH adduct at the *para* position. The present study reveals for the first time that 9-nitroanthracene can be formed from the OH or NO_3 radical-initiated gas-phase reaction of anthracene.

■ ASSOCIATED CONTENT

Ⓢ Supporting Information

Resonance structures of the OH–anthracene adduct; formation scheme of nitro-PAHs from the OH- and NO_3-initiated reactions of benzene, naphthalene, fluorene, fluoranthene, phenanthrene, acenaphthene, acenaphthylene, and pyrene. This material is available free of charge via the Internet at http://pubs.acs.org.

■ AUTHOR INFORMATION

Corresponding Authors

*Fax: 86-531-8836 1990. E-mail: zqz@sdu.edu.cn.
*Fax: 86-531-8836 1990. E-mail: wxwang@sdu.edu.cn.

Notes

The authors declare no competing financial interest.

■ ACKNOWLEDGMENTS

The work was financially supported by the NSFC (National Natural Science Foundation of China, Project Nos. 21337001, 21377073, and 21177076), Taishan Grand (No. ts20120522), and Independent Innovation Foundation of Shandong University (IIFSDU, Project No. 2012JC030).

■ REFERENCES

(1) Finlayson-Pitts, B. J.; Pitts, J. N., Jr. Tropospheric air pollution: ozone, airborne toxics, polycyclic aromatic hydrocarbons, and particles. *Science* **1997**, *276* (5315), 1045–1051.

(2) Zhang, Y. X.; Tao, S.; Shen, H. Z.; Ma, J. M. Inhalation exposure to ambient polycyclic aromatic hydrocarbons and lung cancer risk of Chinese population. *Proc. Natl. Acad. Sci. U.S.A.* **2009**, *106* (50), 21063–21067.

(3) Abedi-Ardekani, B.; Kamangar, F.; Hewitt, S. M.; Hainaut, P.; Sotoudeh, M.; Abnet, C. C.; Taylor, P. R.; Boffetta, P.; Malekzadeh, R.; Dawsey, S. M. Polycyclic aromatic hydrocarbon exposure in oesophageal tissue and risk of oesophageal squamous cell carcinoma in north-eastern Iran. *Gut* **2010**, *59* (9), 1178–1183.

(4) Motorykin, O.; Matzke, M. M.; Waters, K. M.; Massey Simonich, S. L. Association of carcinogenic polycyclic aromatic hydrocarbon emissions and smoking with lung cancer mortality rates on a global scale. *Environ. Sci. Technol.* **2013**, *47* (7), 3410–3416.

(5) Wang, J.; Chen, S. J.; Tian, M.; Zheng, X. B.; Gonzales, L.; Ohura, T.; Mai, B. X.; Simonich, S. L. M. Inhalation cancer risk associated with exposure to complex polycyclic aromatic hydrocarbon mixtures in an electronic waste and urban area in south China. *Environ. Sci. Technol.* **2012**, *46* (17), 9745–9752.

(6) Shen, H. Z.; Huang, Y.; Wang, R.; Zhu, D.; Li, W.; Shen, G. F.; Wang, B.; Zhang, Y. Y.; Chen, Y. C.; Lu, Y.; Chen, H.; Li, T. C.; Sun, K.; Li, B. G.; Liu, W. X.; Liu, J. F.; Tao, S. Global atmospheric emissions of polycyclic aromatic hydrocarbons from 1960 to 2008 and future predictions. *Environ. Sci. Technol.* **2013**, *47* (12), 6415–6424.

(7) Durant, J. L.; Busby, W. F., Jr.; Lafleur, A. L.; Penman, B. W.; Crespi, C. L. Human cell mutagenicity of oxygenated, nitrated and unsubstituted polycyclic aromatic hydrocarbons associated with urban aerosols. *Mutat. Res.* **1996**, *371* (3–4), 123–157.

(8) Lu, Y. M.; Ding, X. C.; Ye, S. H.; Jin, X. P. Mutagenicity of various organic fractions of diesel exhaust particles. *Environ. Health. Prev. Med.* **1999**, *4* (1), 9–12.

(9) Ringuet, J.; Leoz-Garziandia, E.; Budzinski, H.; Villenave, E.; Albinet, A. Particle size distribution of nitrated and oxygenated polycyclic aromatic hydrocarbons (NPAHs and OPAHs) on traffic and suburban sites of a European megacity: Paris (France). *Atmos. Chem. Phys.* **2012**, *12* (18), 8877–8887.

(10) Bamford, H. A.; Baker, J. E. Nitro-polycyclic aromatic hydrocarbon concentrations and sources in urban and suburban atmospheres of the Mid-Atlantic region. *Atmos. Environ.* **2003**, *37* (15), 2077–2091.

(11) Reisen, F.; Arey, J. Atmospheric reactions influence seasonal PAH and nitro-PAH concentrations in the Los Angeles basin. *Environ. Sci. Technol.* **2004**, *39* (1), 64–73.

(12) Albinet, A.; Leoz-Garziandia, E.; Budzinski, H.; Viilenave, E. Polycyclic aromatic hydrocarbons (PAHs), nitrated PAHs and oxygenated PAHs in ambient air of the Marseilles area (south of France): concentrations and sources. *Sci. Total Environ.* **2007**, *384* (2–3), 280–292.

(13) Bamford, H. A.; Bezabeh, D. W.; Schantz, M. M.; Wise, S. A.; Baker, J. E. Determination and comparison of nitrated-polycyclic aromatic hydrocarbons measured in air and diesel particulate reference materials. *Chemosphere* **2003**, *50* (5), 575–587.

(14) Arey, J.; Zielinska, B.; Atkinson, R.; Winer, A. M. Polycyclic aromatic hydrocarbon and nitroarene concentrations in ambient air during a wintertime high-NOx episode in the Los Angeles Basin. *Atmos. Environ.* **1987**, *21* (6), 1437–1444.

(15) Kamens, R. M.; Fan, Z.; Yao, Y.; Chen, D.; Chen, S.; Vartiainen, M. A methodology for modeling the formation and decay of nitro-PAH in the atmosphere. *Chemosphere* **1994**, *28* (9), 1623–1632.

(16) Fan, Z.; Chen, D.; Birla, P.; Kamens, R. M. Modeling of nitro-polycyclic aromatic hydrocarbon formation and decay in the atmosphere. *Atmos. Environ.* **1995**, *29* (10), 1171–1181.

(17) Atkinson, R.; Arey, J.; Zielinska, B.; Aschmann, S. M. Kinetics and nitro-products of the gas-phase OH and NO_3 radical-initiated reactions of naphthalene-d_8, fluoranthene-d_{10}, and pyrene. *Int. J. Chem. Kinetic.* **1990**, *22* (9), 999–1014.

(18) Lee, J. Y.; Lane, D. A. Unique products from the reaction of naphthalene with the hydroxyl radical. *Atmos. Environ.* **2009**, *43* (32), 4886–4893.

(19) Nishino, N.; Arey, J.; Atkinson, R. Formation and reactions of 2-formylcinnamaldehyde in the OH radical-initiated reaction of naphthalene. *Environ. Sci. Technol.* **2009**, *43* (5), 1349–1353.

(20) Esteve, W.; Budzinski, H.; Villenave, E. Relative rate constants for the heterogeneous reactions of NO_2 and OH radicals with polycyclic

aromatic hydrocarbons adsorbed on carbonaceous particles. Part 2: PAHs adsorbed on diesel particulate exhaust SRM 1650a. *Atmos. Environ.* **2006**, *40* (2), 201–211.

(21) Wang, L.; Atkinson, R.; Arey, J. Dicarbonyl products of the OH radical-initiated reactions of naphthalene and the C1- and C2-alkylnaphthalenes. *Environ. Sci. Technol.* **2007**, *41* (8), 2803–2810.

(22) Zhao, Y.; Lynch, B. J.; Truhlar, D. G. Development and assessment of a new hybrid density functional model for thermochemical kinetics. *J. Phys. Chem. A* **2004**, *108* (14), 2715–2719.

(23) Feilberg, A.; Kamens, R. M.; Strommen, M. R.; Nielsen, T. Modeling the formation, decay, and partitioning of semivolatile nitro-polycyclic aromatic hydrocarbons (nitronaphthalenes) in the atmosphere. *Atmos. Environ.* **1999**, *33* (8), 1231–1243.

(24) Frisch, M. J.; Trucks, G. W.; Schlegel, H. B.; Scuseria, G. E.; Robb, M. A.; Cheeseman, J. R.; Scalmani, G.; Barone, V.; Mennucci, B.; Petersson, G. A.; Nakatsuji, H.; Caricato, M.; Li, X.; Hratchian, H. P.; Izmaylov, A. F.; Bloino, J.; Zheng, G.; Sonnenberg, J. L.; Hada, M.; Ehara, M.; Toyota, K.; Fukuda, R.; Hasegawa, J.; Ishida, M.; Nakajima, T.; Honda, Y.; Kitao, O.; Nakai, H.; Vreven, T.; Montgomery, J. A., Jr.; Peralta, J. E.; Ogliaro, F.; Bearpark, M.; Heyd, J. J.; Brothers, E.; Kudin, K. N.; Staroverov, V. N.; Kobayashi, R.; Normand, J.; Raghavachari, K.; Rendell, A.; Burant, J. C.; Iyengar, S. S.; Tomasi, J.; Cossi, M.; Rega, N.; Millam, J. M.; Klene, M.; Knox, J. E.; Cross, J. B.; Bakken, V.; Adamo, C.; Jaramillo, J.; Gomperts, R.; Stratmann, R. E.; Yazyev, O.; Austin, A. J.; Cammi, R.; Pomelli, C.; Ochterski, J. W.; Martin, R. L.; Morokuma, K.; Zakrzewski, V. G.; Voth, G. A.; Salvador, P.; Dannenberg, J. J.; Dapprich, S.; Daniels, A. D.; Farkas, O.; Foresman, J. B.; Ortiz, J. V.; Cioslowski, J.; Fox, D. J. *Gaussian 09*, revision A.02; Gaussian, Inc.: Wallingford, CT, 2009.

(25) Lynch, B. J.; Truhlar, D. G. How well can hybrid density functional methods predict transition state geometries and barrier heights? *J. Phys. Chem. A* **2001**, *105* (13), 2936–2941.

(26) Lynch, B. J.; Zhao, Y.; Truhlar, D. G. Effectiveness of diffuse basis functions for calculating relative energies by density functional theory. *J. Phys. Chem. A* **2003**, *107* (9), 1384–1388.

(27) Glowacki, D. R.; Liang, C.-H.; Morley, C.; Pilling, M. J.; Robertson, S. H. MESMER: An open-source master equation solver for multi-energy well reactions. *J. Phys. Chem. A* **2012**, *116* (38), 9545–9560.

(28) Robinson, P. J.; Holbrook, K. A. *Unimolecular reactions*; John Wiley & Sons: New York, 1972.

(29) Johnson III, R. D. NIST Computational Chemistry Comparison and Benchmark Database, NIST Standard Reference Database Number 101, Release 16a, Aug 2013.

(30) Cyvin, B. N.; Cyvin, S. J. Mean amplitudes of vibration of comparatively large molecules. III. Isotopic anthracene. *J. Phys. Chem.* **1969**, *73* (5), 1430–1438.

(31) Trotter, J. Steric inhibition of resonance II. Bond lengths in meso-substituted anthracenes. *Can. J. Chem.* **1959**, *37* (4), 825–827.

(32) Wiberg, K. B. Properties of some condensed aromatic systems. *J. Org. Chem.* **1997**, *62* (17), 5720–5727.

(33) Ketkar, S. N.; Fink, M. The molecular structure of naphthalene by electron diffraction. *J. Mol. Struct.* **1981**, *77* (1–2), 139–147.

(34) Feilberg, A.; B. Poulsen, M. W.; Nielsen, T.; Henrik, S. Occurrence and sources of particulate nitro-polycyclic aromatic hydrocarbons in ambient air in Denmark. *Atmos. Environ.* **2001**, *35* (2), 353–366.

(35) Morokuma, K.; Muguruma, C. Ab initio molecular orbital study of the mechanism of the gas phase reaction SO_3 + H_2O: importance of the second water molecule. *J. Am. Chem. Soc.* **1994**, *116* (22), 10316–10317.

(36) Gonzalez, J.; Anglada, J. M.; Buszek, R. J.; Francisco, J. S. Impact of water on the OH + HOCl reaction. *J. Am. Chem. Soc.* **2011**, *133* (10), 3345–3353.

(37) Tretyakov, M. Y.; Serov, E. A.; Koshelev, M. A.; Parshin, V. V.; Krupnov, A. F. Water dimer rotationally resolved millimeter-wave spectrum observation at room temperature. *Phys. Rev. Lett.* **2013**, *110* (9), 093001.

(38) Pfeilsticker, K.; Lotter, A.; Peters, C.; Bösch, H. Atmospheric detection of water dimers via near-infrared absorption. *Science* **2003**, *300* (5628), 2078–2080.

(39) Arey, J.; Zielinska, B.; Atkinson, R.; Aschmann, S. M. Nitroarene products from the gas-phase reactions of volatile polycyclic aromatic hydrocarbons with the OH radical and N_2O_5. *Int. J. Chem. Kinetic.* **1989**, *21* (9), 775–799.

(40) Atkinson, R.; Arey, J. Atmospheric chemistry of gas-phase polycyclic aromatic hydrocarbons: formation of atmospheric mutagens. *Environ. Health. Persp.* **1994**, *102* (Suppl 4), 117.

(41) Atkinson, R.; Arey, J. Mechanisms of the gas-phase reactions of aromatic hydrocarbons and PAHs with OH and NO_3 radicals. *Polycyclic Aromat. Compd.* **2007**, *27* (1), 15–40.

(42) Atkinson, R. Kinetics and Mechanisms of the gas-phase reactions of the NO_3 radical with organic compounds. *J. Phys. Chem. Ref. Data* **1991**, *20* (3), 459–507.

(43) Brubaker, W. W., Jr.; Hites, R. A. OH reaction kinetics of polycyclic aromatic hydrocarbons and polychlorinated dibenzo-*p*-dioxins and dibenzofurans. *J. Phys. Chem. A* **1998**, *102* (6), 915–921.

(44) Berndt, T.; Boge, O. Gas-phase reaction of OH radicals with benzene: products and mechanism. *Phys. Chem. Chem. Phys.* **2001**, *3* (22), 4946–4956.

(45) Sasaki, J.; Aschmann, S. M.; Kwok, E. S. C.; Atkinson, R.; Arey, J. Products of the gas-phase OH and NO_3 radical-initiated reactions of naphthalene. *Environ. Sci. Technol.* **1997**, *31* (11), 3173–3179.

(46) Lee, W.; Stevens, P. S.; Hites, R. A. Rate constants for the gas-phase reactions of methylphenanthrenes with OH as a function of temperature. *J. Phys. Chem. A* **2003**, *107* (34), 6603–6608.

(47) Zhou, S.; Wenger, J. C. Kinetics and products of the gas-phase reactions of acenaphthene with hydroxyl radicals, nitrate radicals and ozone. *Atmos. Environ.* **2013**, *72*, 97–104.

(48) Atkinson, R. Kinetics and mechanisms of the gas-phase reactions of the hydroxyl radical with organic compounds under atmospheric conditions. *Chem. Rev.* **1986**, *86* (1), 69–201.

(49) Shannon, R. J.; Blitz, M. A.; Goddard, A.; Heard, D. E. Accelerated chemistry in the reaction between the hydroxyl radical and methanol at interstellar temperatures facilitated by tunnelling. *Nat. Chem.* **2013**, *5* (9), 745–749.

(50) Glowacki, D. R.; Pilling, M. J. Unimolecular reactions of peroxy radicals in atmospheric chemistry and combustion. *ChemPhysChem* **2010**, *11* (18), 3836–3843.

(51) Calvert, J. G.; Atkinson, R.; Becker, K. H.; Kamens, R. M.; Seinfeld, J. H.; Wallington, T. J.; Yarwood, G. *The Mechanisms of Atmospheric Oxidation of Aromatic Hydrocarbons*; Oxford University Press: New York, 2002.

(52) Atkinson, R.; Aschmann, S. M.; Pitts, J. N. Kinetics of the reactions of naphthalene and biphenyl with hydroxyl radicals and with ozone at 294 ± 1 K. *Environ. Sci. Technol.* **1984**, *18* (2), 110–113.

(53) Atkinson, R.; Aschmann, S. M.; Pitts, J. N. J. Rate constants for the gas-phase reactions of the NO_3 radical with a series of organic compounds at 296 ± 2 K. *J. Phys. Chem.* **1988**, *92* (12), 3454–3457.

(54) Kwok, E. S. C.; Atkinson, R.; Arey, J. Kinetics of the gas-phase reactions of indan, indene, fluorene, and 9,10-dihydroanthracene with OH radicals, NO_3 radicals, and O_3. *Int. J. Chem. Kinet.* **1997**, *29* (4), 299–309.

(55) Kwok, E. S. C.; Harger, W. P.; Arey, J.; Atkinson, R. Reactions of gas-phase phenanthrene under simulated atmospheric conditions. *Environ. Sci. Technol.* **1994**, *28* (3), 521–527.

THE JOURNAL OF
PHYSICAL CHEMISTRY A

Article

pubs.acs.org/JPCA

Theoretical Investigation on Mechanistic and Kinetic Transformation of 2,2′,4,4′,5-Pentabromodiphenyl Ether

Haijie Cao, Dandan Han, Mingyue Li, Xin Li, Maoxia He,* and Wenxing Wang

Environment Research Institute, Shandong University, Jinan 250100, P. R. China

Supporting Information

ABSTRACT: This study investigates the decomposition of 2,2′,4,4′,5-pentabrominated diphenyl ether (BDE99), a commonly detected pollutant in the environment. Debromination channels yielding tetrabrominated diphenyl ethers and hydrogen abstracting aromatic bromine atom formations play significant roles in the reaction of BDE99 + H, in which the former absolutely predominates bimolecular reactions. Polybrominated dibenzo-*p*-dioxins (PBDDs) and polybrominated dibenzofurans (PBDFs) can be produced during BDE99 pyrolysis, especially for PBDFs under inert conditions. The expected dominant pathways in a closed system are debromination products and PBDF formations. The bimolecular reaction with hydroxyl radical mainly leads to hydroxylated BDE99s rather than hydroxylated tetrabrominated diphenyl ethers. PBDDs are then generated from *ortho*-hydroxylated PBDEs. HO_2 radical reactions rarely proceed. The total rate constants for the BDE99 reaction with hydrogen atoms and hydroxyl radicals exhibit positive dependence on temperature with values of 1.86×10^{-14} and 5.24×10^{-14} cm^3 $molecule^{-1}$ s^{-1} at 298.15 K, respectively.

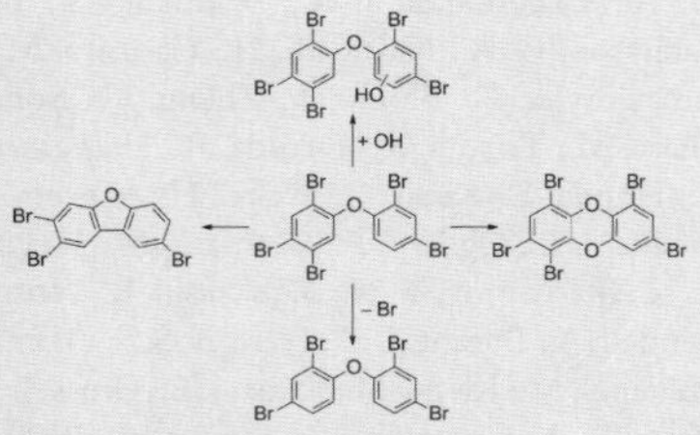

1. INTRODUCTION

Polybrominated diphenyl ethers (PBDEs) are widely used brominated flame retardants (BFRs), which are being phased out because of their potential hazards to humans. However, PBDEs remain as ubiquitous pollutants worldwide because of their stability, persistence, accumulation, and migration.[1−4] Hence, PBDEs degradation and metabolites, as well as derivatives elicit increasing attention. PBDEs are frequently detected in various samples (e.g., human tissues, fluids, sewage sludge, air, and water)[5−9] and expected to exist in the environment at extended periods.[10] Therefore, knowledge on the kinetic properties and removal pathways of PBDEs is crucial.

Furthermore, 2,2′,4,4′,5-pentabromodiphenyl ether (BDE99) was a major component of commercial pentabrominated diphenyl ethers. BDE99 is frequently detected in air dust and other medias and usually acts as a dominant congener.[11−14] Similar to other PBDEs, BDE99 exhibits an evident ichthyotoxic effect,[15] which causes learning disorders and disturbs thyroid hormones.[16] BDE99 in hepatocytes can be metabolized to form hydroxylated PBDEs (OH-PBDEs), which are more toxic than the parent compound.[17−20] Although BDE99 can be metabolized in vivo, BDE99 degradation has not been observed in soil,[21] which may be a reason behind the high BDE99 concentrations in sediment and soil samples.[22]

PBDEs debromination has been well-investigated experimentally, and the generation of sub-PBDEs are mostly observed together with different catalysts (e.g., UV irritation and zero iron).[23−29] Production, utilization, treatment, and recycling of PBDEs lead to the formation of notorious contaminant polybrominated dibenzo-*p*-dioxins (PBDDs)/polybrominated dibenzofurans (PBDFs).[30−32] PBDEs can also feasibly react with air oxidants and be converted into corresponding derivatives. The rate constants of several mono- to tetra-BDEs with hydroxyl radical have been determined experimentally[33] or theoretically.[34−36] Given the limited oxidation data of higher brominated PBDEs, the roles of hydroxyl radicals in the decomposition of higher brominated PBDEs require further investing ation.

Computational methods will compensate for the invisible reaction mechanisms. In this study, the bimolecular decomposition of BDE99 with hydrogen atoms, hydroxyls, and HO_2 radicals are investigated with regard to active radicals in combustion chemistry. This research presents the available mechanism involved in BDE99 debromination. The formation of PBDDs and PBDFs are also discussed in detail. The calculated results will provide researchers with further understanding about the removal of PBDEs from the environment. This information will facilitate the development of appropriate degradation methods under optimal conditions.

2. COMPUTATIONAL METHODS

Gaussian 09 packages[37] were used to perform all the calculations. The MPWB1K functional[38] was selected for quantum calculations. In our previous work,[39,40] the reliability of this method was proven appropriate (time-saving and accurate) in calculations involving PBDE components. The geometries of all the stable points (reactants, intermediates, and products) and transition states were optimized at the MPWB1K/6-31+G(d,p) level. Harmonic vibrational frequen-

Received: April 28, 2015
Revised: May 22, 2015
Published: May 26, 2015

DOI: 10.1021/acs.jpca.5b04022
J. Phys. Chem. A 2015, 119, 6404−6411

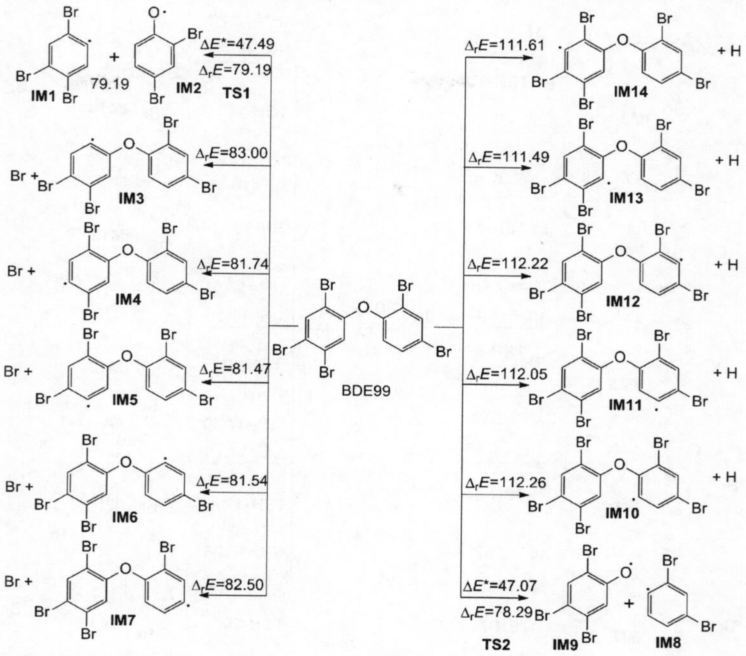

Figure 1. Schematic self-decomposition of BDE99. $\Delta_r E$ (kcal/mol) represents the reaction heat.

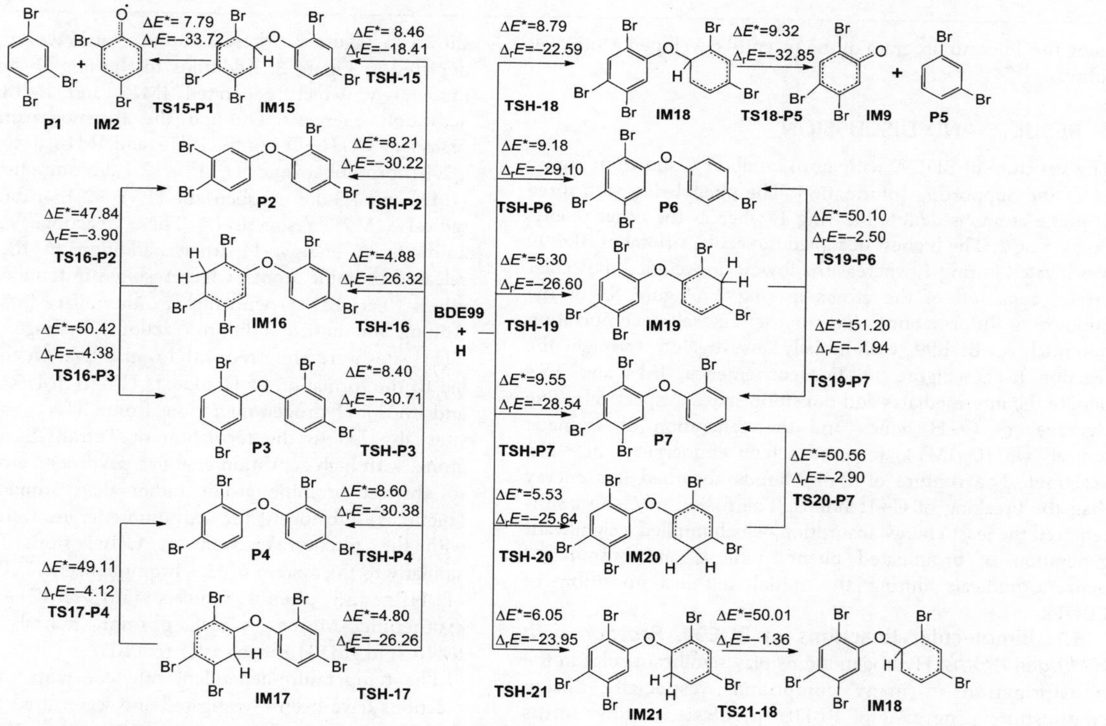

Figure 2. Schematic mechanism of addition reactions between BDE99+H. Units of reaction heats ($\Delta_r E$) and activation energies (ΔE^*) are kcal/mol.

cies and zero-point vibrational energies were calculated at the same level. Intrinsic reaction coordinates[41,42] were calculated to verify the connections between the minima and the transition states. The single-point energies of the higher level were calculated at the theoretical MPWB1*K*/6-311+G(3df,2p). Then the rovibrational frequencies and the energies were used to calculate the rate constants via canonical transition state theory (TST).[43] Quantum tunneling effect from the asymmetrical Eckart barrier was included. TST rate constants were obtained

DOI: 10.1021/acs.jpca.5b04022
J. Phys. Chem. A 2015, 119, 6404–6411

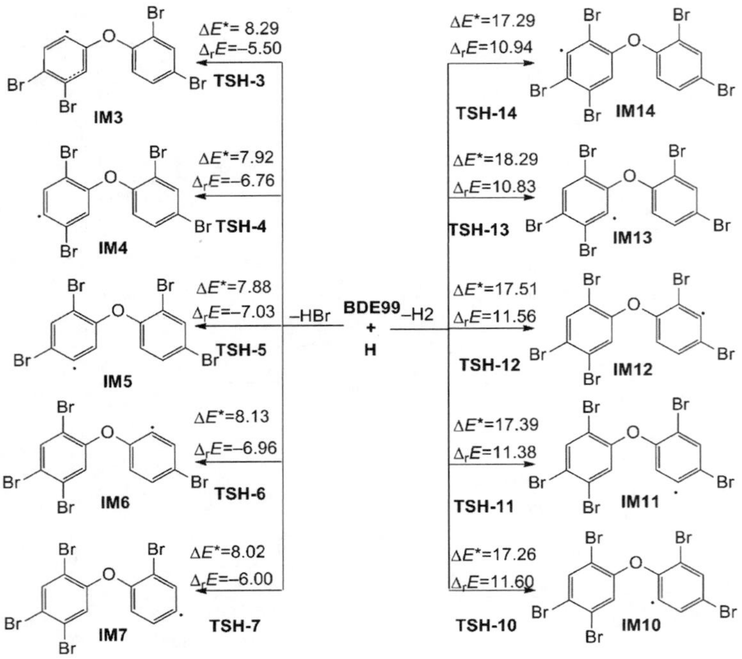

Figure 3. Schematic mechanism of abstraction reactions between BDE99 and H. Units of reaction heat ($\Delta_r E$) and activation energy (ΔE^*) are kcal/mol.

using the Thermo program of the recently developed MultiWell software.[44,45]

3. RESULTS AND DISCUSSION

The structure of BDE99 with atom numbers is shown in Figure S1 of the Supporting Information. The phenyl ring with three bromine atoms is denoted as ring 1, whereas the other phenyl ring is ring 2. The highest occupied molecular orbital of BDE99 was located in ring 1, whereas the lowest unoccupied molecular orbital consisted of the atoms in ring 2 (Figure S2 of the Supporting Information). The barrierless self-decomposition potential of BDE99 was initially investigated through the reaction heats (Figure 1). For convenience, IM's and TS's denote the intermediates and transition states, respectively. The cleavages of C−H bonds and the generation of aromatic radicals (IM10−IM14) incurred a high endoergicity of ~112 kcal/mol. The rupture of C−Br bonds absorbed less energy than the breaking of C−H bonds. The fission of C−O bonds required the least energy absorption, which implied a significant generation of brominated phenol radicals and brominated benzene radicals during the initial self-decomposition of PBDEs.

3.1. Bimolecular Reactions of BDE99.

Reaction with Hydrogen Atoms. Hydrogen atoms play significant roles in the transformation of many compounds, especially during combustion progression. PBDE pyrolysis mainly forms debromination products, indicating a replacement activity of hydrogen and bromine atoms.[23−26,29] Most environmental PBDEs result from the debromination of higher BDEs (e.g., Deca-BDE). Thus, H-reducing reactions are important to treat PBDEs.

Hydrogen reaction with the selected BDE99 reactant was investigated, which branched into 22 channels. The reaction heats and activation energies of hydrogen addition routes are shown in Figure 2, whereas those of abstraction channels are depicted in Figure 3. Additions to the *ipso*-C sites proceeded moderately, which generated IM15 and IM18 via similar activation energies. Through the thermodynamic preferred fission of C(H)−O bonds, IM15 and IM18 decomposed into 1,2,4-tribromobenzene (P1) + 2,4-dibromophenoxy radical (IM2) and 1,3-dibromobenzene (P5) + 2,4,5-tribromophenoxy radical (IM9), respectively. These reactions released large amounts of energy. Hydrogen addition to the C(Br) site released bromine atoms, which resulted in tetrabromodiphenyl ethers (TetraBDEs) with high exothermicity (~30 kcal/mol). Among the initial addition reactions, hydrogen addition to C(H) sites were preferred with lower activation energies, which led to the formation of H-adducts (IM16, IM17, IM19, IM20, and IM21). Hydrogen migrations from C(H_2) groups to C(Br) sites also led to the formation of TetraBDEs and bromine atoms with high activation energies. Hydrogen atoms preferred to abstract bromine atoms rather than aromatic hydrogen (Figure 3). Obviously, the activation energies were competitive with the addition reactions at C(Br) sites. This behaved similarly to the system of H + bromobenzene.[34] The formation of HBr and phenyl radicals (IM3−IM7) was slightly exothermic. Moreover, *ortho*-position radicals (IM3, IM6, IM10, and IM13) can transfer to PBDFs.

The temperature-dependent rate constants of bimolecular reactions have been investigated and were fitted to a modified three-parameter Arrhenius expression within 250 to 1000 K interval (Table 1). All the rate constants exhibited positive temperature dependence. The total rate constant of BDE99 + H ranged from 5.75×10^{-14} cm^3 molecule^{-1} s^{-1} at 298.15 K to 2.19×10^{-11} cm^3 molecule^{-1} s^{-1} at 1000 K. At room temperature, the orders of the rate constants generally agreed with the activation energies. Hydrogen addition to the C(H) sites dominated the whole bimolecular reactions, which

DOI: 10.1021/acs.jpca.5b04022
J. Phys. Chem. A 2015, 119, 6404−6411

Table 1. Parameters Involved in the Modified Arrhenius Expression $k(T) = AT^n \exp\{-[(E_a)/(RT)]\}$ at the Interval of 250–1000 K[a]

reaction	A	n	E_a/R
BDE99 + H → IM15	1.46×10^{-20}	2.86	2631.33
BDE99 + H → P2 + Br	3.46×10^{-22}	3.46	2230.55
BDE99 + H → IM16	1.26×10^{-18}	2.30	1380.55
BDE99 + H → P3 + Br	7.32×10^{-23}	3.64	2186.12
BDE99 + H → P4 + Br	1.03×10^{-22}	3.53	2290.78
BDE99 + H → IM17	2.28×10^{-18}	2.19	1371.47
BDE99 + H → IM18	3.02×10^{-20}	2.74	2861.06
BDE99 + H → P6 + Br	4.09×10^{-22}	3.36	2695.18
BDE99 + H → IM19	2.78×10^{-18}	2.17	1608.81
BDE99 + H → P7 + Br	2.43×10^{-22}	3.41	2817.05
BDE99 + H → IM20	1.31×10^{-18}	2.27	1670.48
BDE99 + H → IM21	3.28×10^{-19}	2.50	1800.83
BDE99 + H → IM3 + HBr	2.26×10^{-15}	1.46	3834.87
BDE99 + H → IM4 + HBr	2.58×10^{-15}	1.54	3641.45
BDE99 + H → IM5 + HBr	2.06×10^{-15}	1.51	3625.37
BDE99 + H → IM6 + HBr	7.74×10^{-16}	1.65	3698.70
BDE99 + H → IM7 + HBr	8.05×10^{-16}	1.59	3651.77
BDE99 + H → IM10 + H_2	1.69×10^{-18}	2.46	7754.96
BDE99 + H → IM11 + H_2	1.36×10^{-18}	2.47	7789.33
BDE99 + H → IM12 + H_2	1.74×10^{-18}	2.44	7892.78
BDE99 + H → IM13 + H_2	1.06×10^{-18}	2.36	8142.21
BDE99 + H → IM14 + H_2	1.90×10^{-18}	2.38	7763.59
BDE99 + ·OH → IM21	2.20×10^{-21}	2.75	−37.48
BDE99 + ·OH → P9	3.06×10^{-21}	2.71	1629.77
BDE99 + ·OH → IM22	4.79×10^{-21}	2.69	539.08
BDE99 + ·OH → P11	1.47×10^{-21}	2.72	1781.61
BDE99 + ·OH → P12	1.21×10^{-21}	2.72	2163.20
BDE99 + ·OH → IM23	3.49×10^{-21}	2.51	−460.44
BDE99 + ·OH → IM24	2.96×10^{-21}	2.53	1789.17
BDE99 + ·OH → P14	6.06×10^{-21}	2.63	2943.86
BDE99 + ·OH → IM25	1.42×10^{-20}	2.55	634.02
BDE99 + ·OH → P16	2.16×10^{-21}	2.65	2511.63
BDE99 + ·OH → IM26	8.95×10^{-21}	2.49	559.76
BDE99 + ·OH → IM27	6.88×10^{-21}	2.57	1244.67
IM3 → P19 + Br	3.52×10^{10}	0.34	6447.69
IM3 → IM28	3.79×10^{10}	0.37	3850.56

[a]Unit of A is cm^3 $molecule^{-1}$ s^{-1} for bimolecular reactions and s^{-1} for unimolecular reactions.

contributed 98% of the total rate constant, similar to the results of the H + bromobenzene system.[34] The difference between H + BDE99 and H + bromobenzene exhibited in the branching ratios of addition channel and abstraction channel. In the system of H + bromobenzene, addition channels dominated the reaction within 250–2000 K.[34] In the system of H + BDE99, the branching ratio of the addition channel decreased steadily with temperature and reduced to 45% of the whole value (Table S1 of the Supporting Information). By contrast, hydrogen abstracting bromine atoms and hydrogen addition to C(Br) sites became dominant channels at high temperatures (Table S1 of the Supporting Information). Hydrogen addition to *ipso*-C atoms evidently processed slowly, whereas the abstraction of phenyl hydrogen atoms was negligible. The rate constants for the hydrogen migration reactions of H-adducts were also calculated. However, fitting these values to the modified Arrhenius expression was inappropriate because of the huge error between the fitted data and the calculated values, especially at low temperatures. The nonfitted rate constants are shown in Table S2 of the Supporting Information. Apparently, hydrogen migrations proceeded slowly at 298.15 K but increased rapidly with temperature. At high temperatures, these processes will significantly participate in the formation of low brominated congeners. The formation of PBDDs required oxygen atom insertion in the *ortho*-position of the ether bond. Therefore, at high temperatures (e.g., >900 K), PBDFs were the dominant products rather than PBDDs under inert gas conditions.

Bimolecular Reaction with Hydroxyl Radical. OH-PBDEs are evidently important products in recycling waste printed circuit boards.[46] Hence, hydroxylation of PBDEs is a significant source of OH-PBDEs. The detailed parameters of the oxidation reactions of BDE99 and hydroxyl radicals (·OH) are shown in Figure 4. All the ·OH addition reactions to BDE99 were exothermic with moderate activation energies.

IM21 and IM24 were the intermediates formed via ·OH addition to *ipso*-C atoms. The activation energy of IM21 was extremely lower than that of IM24. Thus, hydroxyl radical addition was preferred for the *ipso*-C atom of ring 1. The activation energies of hydroxyl radical addition to C(Br) sites ranged from 4.52 to 7.14 kcal/mol. These processes were deeply exothermic (29.68 to 32.11 kcal/mol). Bromine atoms were also released simultaneously, leading to the formation of hydroxylated TetraBDEs (OH-TetraBDEs). ·OH addition to C(H) sites should overcome lower activation energies, varying from 0.26 to 3.72 kcal/mol. The corresponding intermediates (OH-adducts, IM22, IM23, IM25, IM26, and IM27) were then formed, followed by the removal of hydrogen atoms through O_2 molecules in the air. Stable hydroxylated BDE99s (OH-BDE99s) were produced once H atoms were abstracted by O_2. These pathways were barrierless, and OH-BDE99s were the preferred products. ·OH can also abstract the phenyl hydrogen of BDE99. However, these processes incurred higher activation energies of 6.27 to 7.96 kcal/mol (Figure S3 of the Supporting Information). Despite our optimal attempts, we failed to locate the transition states of the ·OH abstracting bromine atoms of BDE99. Compared with the activation energies of elementary reactions, ·OH prefers to add to the phenyl ring with three bromine atoms. This is contrary to the results of 2,4-dibromodiphenyl ether[39] and 2,4,4′-tribromodiphenyl ether,[40] indicating that higher brominated PBDEs have some different properties and should be investigated carefully.

The rate constants (Table 1) were positively dependent on temperature, and the branching ratios are summarized in Table S3 of the Supporting Information. The total rate constant of BDE99 + ·OH was 5.24×10^{-14} cm^3 $molecule^{-1}$ s^{-1} at room temperature, which was approximately 2 orders of magnitude lower than the experimental data of mono- and dibrominated diphenyl ethers.[33] Compared with the results of other PBDEs, the rate constants decreased moderately as the degree of bromine substitution of PBDEs increased (Table S4 of the Supporting Information).[36,39,40] The atmospheric lifetime of BDE99 due to tropospheric ·OH (1×10^6 molecules cm^{-3}) is 5.30×10^3 hours. The atmospheric lifetime of BDE99 is considerably longer than low brominated congeners (Table S4 of the Supporting Information). In contrast to the bimolecular reaction of BDE99 + H, the rate constants of ·OH addition to *ipso*-C atoms were more significant than the addition to C(Br) sites. Hydrogen abstraction reactions barely contributed to the total rate constant. Among the studied intervals of temperature, the hydroxyl radical addition to C(H) sites dominated the bimolecular reaction, although the branching ratio decreased

DOI: 10.1021/acs.jpca.5b04022
J. Phys. Chem. A 2015, 119, 6404–6411

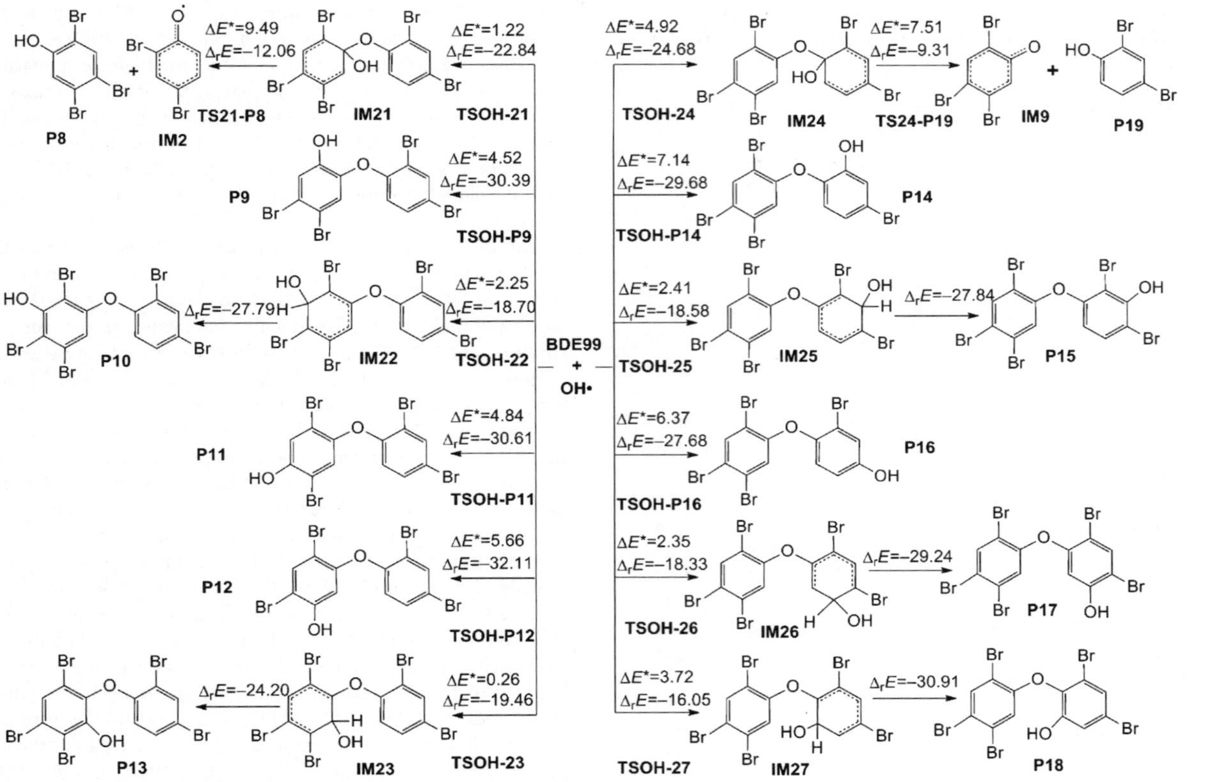

Figure 4. Schematic mechanism of bimolecular reactions between BDE99 and OH radical. Units of reaction heat ($\Delta_r E$) and activation energy (ΔE^*) are kcal/mol.

moderately with temperature. TetraBDEs (P9, P11, P12, P14, and P16), which resulted from the hydroxyl group substituting bromine atoms of BDE99, were not major products. The yield of OH-TetraBDEs was quite minimal and even negligible at room temperature. Thus, ·OH preferred to react with nonsubstituted carbon atoms of BDE99 to produce OH-adducts, which were then stabilized through the O_2 abstraction of the H atoms. Considering the rate of determining steps, OH-BDE99s were preferred products instead of OH-TetraBDEs in the presence of oxygen.

Oxidation by HO_2 Radical. Bimolecular reactions of BDE99 + HO_2· were all endothermic and incurred high activation energies (Figure S5 of the Supporting Information), indicating lower feasibility of these reactions. The steric hindrance between HO_2· and phenyl rings was seemingly enhanced, whereas the inactivation effect of Br atoms was weakened. The kinetic results (Table S4 of the Supporting Information) confirmed this assumption. The rate constant of BDE99+HO_2· showed moderately positive dependence on temperature and was determined to be 4.53×10^{-27} cm^3 $molecule^{-1}$ s^{-1} at room temperature, negligible compared with the rate constant of BDE99 + ·OH/H. Although the ambient concentration of HO_2· is higher than ·OH by ~100 times, HO_2· barely contributes to BDE99 decomposition. At 298 K, the HO_2·-determined an atmospheric lifetime of BDE99 was 6.13×10^{14} hours.

3.2. Formation of PBDDs/PBDFs. Thermal treatments of materials containing PBDEs are evidenced producing hazardous PBDDs and PBDFs.[47,48] The phenyl-type radical intermediates at the ortho-position of ether bonds (IM3, IM6, IM10, and IM13) were PBDF precursors (Figure 5).[34] Two channels leading to different PBDFs were available for IM3, that is, attacking the *ortho*-C(Br) site or *ortho*-C(H) site of the other phenyl ring. Moreover, 2,3,8-tribromodibenzofuran (2,3,8-TriBDF, P24) was generated through a concerted process, where bromine atoms were released coupled with the bond association of C−C. This process was deeply exothermic by 43.84 kcal/mol with a 13.06 kcal/mol activation energy. The formation of 2,3,6,8-tetrabromodibenzofuran (2,3,6,8-TetraBDF, P25) proceeded with a stepwise mechanism, which was initiated by the association of the C−C(H) bond. After releasing a hydrogen atom, 2,3,6,8-TetraBDF (P25) was formed. The self-decomposition of IM35 faced high activation energy and was endothermic. Hence, 2,3,8-TriBDF was the dominant product in this closed system instead of 2,3,6,8-TetraBDF. However, in the case of bromine atoms, IM35 to P25 transformation was strongly prompted via barrierless reaction. The process of IM35 + Br → P25 + HBr released substantial energy of 75.92 kcal/mol.

The association of IM3 with oxygen molecules was preferred in an oxidative atmosphere, leading to the formation of peroxyl radical intermediate (IM36). The isomerization of IM36 was divided into four endoergic channels. Terminal oxygen atom addition to the *ipso*-C atom of the same phenyl ring incurred the highest activation energy of 44.73 kcal/mol and absorbing energy of 35.39 kcal/mol. Addition of the other ring to *ortho*-C atoms was much easier, but the favored pathway was the one toward IM40 because this pathway exhibited the lowest

DOI: 10.1021/acs.jpca.5b04022
J. Phys. Chem. A 2015, 119, 6404−6411

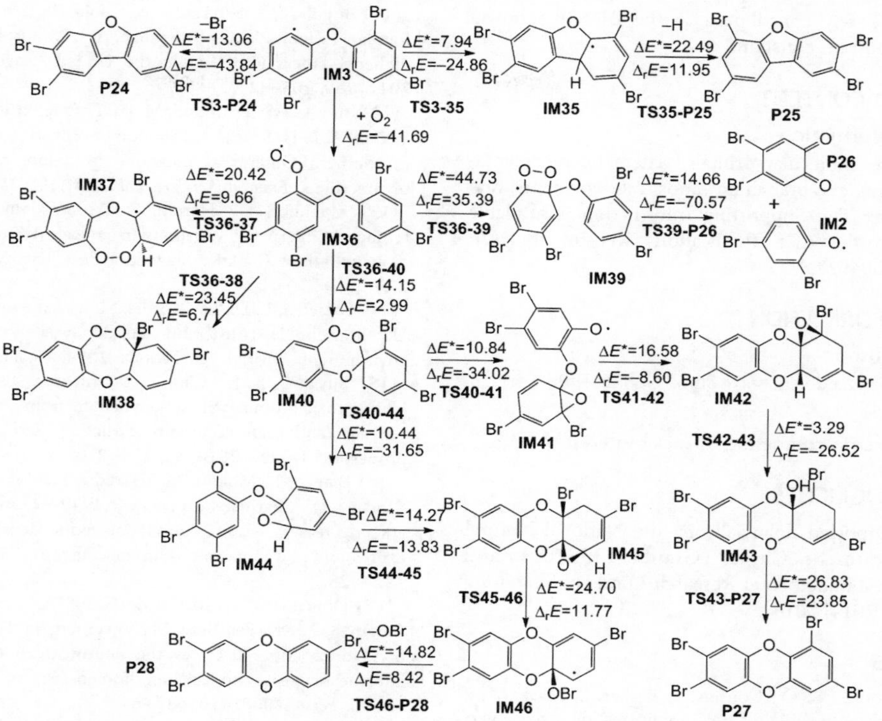

Figure 5. Reaction parameters involved in the formation of PBDD/Fs from IM3. Units of reaction heat ($\Delta_r E$) and activation energy (ΔE^*) are kcal/mol.

activation energy (14.15 kcal/mol) and a slight endoergicity (2.99 kcal/mol). The decomposition of IM40 branched into two channels. Through the fission of O−O bond and the association of O−C(Br) bond, IM41 was produced, followed by oxygen attacking the C(H) site. Hydrogen migration from aromatic rings to cyclo-oxygen atoms underwent moderate activation energy, leading to the formation of hydroxylated intermediate IM43. Subsequently, 1,3,7,8-tetrabromodibenzo-*p*-dioxin (P27) was produced through the release of hydroxyl radicals.

IM40 to IM44 transformation proceeded with the rupture of O−O bonds and association of O−C(H) bonds. This process showed similar energy barriers and reaction heats with the channel toward IM41. IM45 was formed via O−C(Br) bond association. Bromine atom migration from aromatic ring to cyclo-oxygen underwent high activation energy. After the removal of the OBr radical, 2,3,7-tribromodibenzo-*p*-dioxin (P28) was formed.

The transformations from other ortho-position radical intermediates to PBDDs/PBDFs are illustrated in Figures S5−S7 of the Supporting Information. These reactions showed similar trends to the reaction of IM3. The formation of PBDDs and PBDFs exhibited similar trends from a different phenyl-type radical intermediate. The results were in agreement with a study on the reaction of phenyl-type radical intermediate from 2,2′-dibromodiphenyl ether.[34] In a word, the number and position of bromine atoms have little impact on the formation of PBDFs and PBDDs from phenyl-type intermediates. In a word, four PBDDs (2,3,7-tribromodibenzo-*p*-dioxin, 1,3,7,8-tetrabromodibenzo-*p*-dioxin, 1,2,4,8-tetrabromodibenzo-*p*-dioxin, and 1,2,4,6,8-pentabromodibenzo-*p*-dioxin) and four PBDFs (2,3,8-tribromodibenzofuran, 2,3,6,8-tetrabromodibenzofuran, 1,2,4,8-tetrabromodibenzofuran, and 1,2,4,6,8-pentabromodibenzofuran) emerged after a series of stepwise reactions.

In addition to the aforementioned reactions, PBDDs can form through the decomposition of *ortho*-position OH-PBDEs, as confirmed both experimentally[60] and theoretically. Figure S8 of the Supporting Information depicts the mechanism of PBDDs resulting from four *ortho*-OH-PBDEs which has been verified experimentally[49] and theoretically.[40,50] Removal of phenoxyl hydrogen atoms led to essential intermediates, which were precursors of PBDDs. PBDDs formed after different ring-closing reactions. Sub-brominated PBDDs were evidently preferred products.

4. CONCLUSIONS

Hydroxyl radical- or hydrogen atom-initiated BDE99 transformations are important pathways in BDE99 decomposition. In contrast to BDE28 reactions, higher brominated phenyl rings are more reactive. In this study, hydrogen atoms preferred to attack C(H) sites and dominated the reaction at low temperatures. The removal of the bromine atom through the hydrogen atom became competitive as the branching ratio increased at high temperatures. The fission of ether bonds were insignificant at 250 to 1000 K. Hydrogen migrations of H-adducts, which formed TetraBDEs, were highly prompted by temperature. Sub-brominated PBDEs, which can be major products at low-temperature PBDFs, were important at high temperatures. PBDDs were available in the presence of oxygen molecules. For OH-initiated oxidation of BDE99, OH-BDE99s were the major components of OH-PBDEs in the presence of oxygen molecules. PBDDs can be generated through further reaction of OH-PBDEs. Thus, the reaction between BDE99

DOI: 10.1021/acs.jpca.5b04022
J. Phys. Chem. A 2015, 119, 6404−6411

and HO_2 radicals barely contributed to the BDE99 removal because of the small rate constant.

■ ASSOCIATED CONTENT

Ⓢ Supporting Information

Geometry structure, molecular orbitals, schematic representation, reaction parameters, branching ratios, rate constants, and table of parameters. The Supporting Information is available free of charge on the ACS Publications website at DOI: 10.1021/acs.jpca.5b04022.

■ AUTHOR INFORMATION

Corresponding Author

*E-mail: hemaox@sdu.edu.cn. Fax: 86-531-8836 1990.

Notes

The authors declare no competing financial interest.

■ ACKNOWLEDGMENTS

This work was supported financially by the National Natural Science Foundation of China (Grants 21337001 and 21477065) and the Fundamental Research Funds of Shandong University (Grant 2014JC014).

■ REFERENCES

(1) Gouin, T.; Thomas, G. O.; Cousins, I.; Barber, J.; Mackay, D.; Jones, K. C. Air-surface exchange of polybrominated diphenyl ethers and polychlorinated biphenyls. *Environ. Sci. Technol.* **2002**, *36*, 1426–1434.

(2) Ter Schure, A. F. H.; Larsson, P.; Agrell, C.; Boon, J. P. Atmospheric transport of polybrominated diphenyl ethers and polychlorinated biphenyls to the Baltic sea. *Environ. Sci. Technol.* **2004**, *38*, 1282–1287.

(3) Tomy, G. T.; Palace, V. P.; Halldorson, T.; Braekevelt, E.; Danell, R.; Wautier, K.; Evans, B.; Brinkworth, L.; Fisk, A. T. Bioaccumulation, biotransformation, and biochemical effects of brominated diphenyl ethers in juvenile lake trout (Salvelinus namaycush). *Environ. Sci. Technol.* **2004**, *38*, 1496–1504.

(4) Wan, Y.; Hu, J. Y.; Zhang, K.; An, L. H. Trophodynamics of polybrominated diphenyl ethers in the marine food web of Bohai Bay, North China. *Environ. Sci. Technol.* **2008**, *42*, 1078–1083.

(5) Bodin, N.; Abarnou, A.; Fraisse, D.; Defour, S.; Loizeau, V.; Le Guellec, A. M.; Philippon, X. PCB, PCDD/F and PBDE levels and profiles in crustaceans from the coastal waters of Brittany and Normandy (France). *Mar. Pollut. Bull.* **2007**, *54*, 657–668.

(6) Cincinelli, A.; Martellini, T.; Misuri, L.; Lanciotti, E.; Sweetman, A.; Laschi, S.; Palchetti, I. PBDEs in Italian sewage sludge and environmental risk of using sewage sludge for land application. *Environ. Pollut.* **2012**, *161*, 229–234.

(7) Crosse, J. D.; Shore, R. F.; Wadsworth, R. A.; Jones, K. C.; Pereira, M. G. Long-term trends in PBDEs in sparrowhawk (Accipiter nisus) eggs indicate sustained contamination of UK terrestrial ecosystems. *Environ. Sci. Technol.* **2012**, *46*, 13504–13511.

(8) Kierkegaard, A.; Bignert, A.; Sellstrom, U.; Olsson, M.; Asplund, L.; Jansson, B.; de Wit, C. A. Polybrominated diphenyl ethers (PBDEs) and their methoxylated derivatives in pike from Swedish waters with emphasis on temporal trends, 1967–2000. *Environ. Pollut.* **2004**, *130*, 187–198.

(9) Petreas, M.; Nelson, D.; Brown, F. R.; Goldberg, D.; Hurley, S.; Reynolds, P. High concentrations of polybrominated diphenylethers (PBDEs) in breast adipose tissue of California women. *Environ. Int.* **2011**, *37*, 190–197.

(10) Fu, J.; Wang, T.; Wang, P.; Qu, G.; Wang, Y.; Zhang, Q.; Zhang, A.; Jiang, G. Temporal trends (2005–2009) of PCDD/Fs, PCBs, PBDEs in rice hulls from an e-waste dismantling area after stricter environmental regulations. *Chemosphere* **2012**, *88*, 330–335.

(11) Birgul, A.; Katsoyiannis, A.; Gioia, R.; Crosse, J.; Earnshaw, M.; Ratola, N.; Jones, K. C.; Sweetman, A. J. Atmospheric polybrominated diphenyl ethers (PBDEs) in the United Kingdom. *Environ. Pollut.* **2012**, *169*, 105–111.

(12) Butt, C. M.; Diamond, M. L.; Truong, J.; Ikonomou, M. G.; Ter Schure, A. F. H. Spatial distribution of polybrominated diphenyl ethers in southern Ontario as measured in indoor and outdoor window organic films. *Environ. Sci. Technol.* **2004**, *38*, 724–731.

(13) Muenhor, D.; Harrad, S. Within-room and within-building temporal and spatial variations in concentrations of polybrominated diphenyl ethers (PBDEs) in indoor dust. *Environ. Int.* **2012**, *47*, 23–27.

(14) Schreder, E. D.; La Guardia, M. J. Flame retardant transfers from US households (dust and laundry wastewater) to the aquatic environment. *Environ. Sci. Technol.* **2014**, *48*, 11575–11583.

(15) Suyama, T. L.; Cao, Z. Y.; Murray, T. F.; Gerwick, W. H. Ichthyotoxic brominated diphenyl ethers from a mixed assemblage of a red alga and cyanobacterium: Structure clarification and biological properties. *Toxicon* **2010**, *55*, 204–210.

(16) Blanco, J.; Mulero, M.; Heredia, L.; Pujol, A.; Domingo, J. L.; Sanchez, D. J. Perinatal exposure to BDE-99 causes learning disorders and decreases serum thyroid hormone levels and BDNF gene expression in hippocampus in rat offspring. *Toxicology* **2013**, *308*, 122–128.

(17) Dingemans, M. M. L.; de Groot, A.; van Kleef, R. G. D. M.; Bergman, A.; van den Berg, M.; Vijverberg, H. P. M.; Westerink, R. H. S. Hydroxylation increases the neurotoxic potential of BDE-47 to affect exocytosis and calcium homeostasis in PC12 cells. *Environ. Health. Persp.* **2008**, *116*, 637–643.

(18) Dingemans, M. M. L.; van den Berg, M.; Westerink, R. H. Neurotoxicity of brominated flame retardants: (in)direct effects of parent and hydroxylated polybrominated diphenyl ethers on the (developing) nervous system. *Environ. Health. Persp.* **2011**, *119*, 900–907.

(19) Leijs, M. M.; Koppe, J. G.; Olie, K.; van Aalderen, W. M. C.; de Voogt, P.; ten Tusscher, G. W. Effects of dioxins, PCBs, and PBDEs on immunology and hematology in adolescents. *Environ. Sci. Technol.* **2009**, *43*, 7946–7951.

(20) Stapleton, H. M.; Kelly, S. M.; Pei, R.; Letcher, R. J.; Gunsch, C. Metabolism of polybrominated diphenyl ethers (PBDEs) by human hepatocytes in vitro. *Environ. Health. Persp.* **2009**, *117*, 197–202.

(21) Wong, F.; Kurt-Karakus, P.; Bidleman, T. F. Fate of brominated flame retardants and organochlorine pesticides in urban soil: volatility and degradation. *Environ. Sci. Technol.* **2012**, *46*, 2668–2674.

(22) Zeng, Y. H.; Luo, X. J.; Yu, L. H.; Chen, H. S.; Wu, J. P.; Chen, S. J.; Mai, B. X. Using compound-specific stable carbon isotope analysis to trace metabolism and trophic transfer of PCBs and PBDEs in fish from an e-waste site, south China. *Environ. Sci. Technol.* **2013**, *47*, 4062–4068.

(23) Fang, Z.; Qiu, X.; Chen, J.; Qiu, X. Debromination of polybrominated diphenyl ethers by Ni/Fe bimetallic nanoparticles: influencing factors, kinetics, and mechanism. *J. Hazard. Mater.* **2011**, *185*, 958–969.

(24) Li, A.; Tai, C.; Zhao, Z.; Wang, Y.; Zhang, Q.; Jiang, G.; Hu, J. Debromination of decabrominated diphenyl ether by resin-bound iron nanoparticles. *Environ. Sci. Technol.* **2007**, *41*, 6841–6846.

(25) Kim, Y. M.; Murugesan, K.; Chang, Y. Y.; Kim, E. J.; Chang, Y. S. Degradation of polybrominated diphenyl ethers by a sequential treatment with nanoscale zero valent iron and aerobic biodegradation. *J. Chem. Technol. Biotechnol.* **2012**, *87*, 216–224.

(26) Rayne, S.; Wan, P.; Ikonomou, M. Photochemistry of a major commercial polybrominated diphenyl ether flame retardant congener: 2,2′,4,4′,5,5′-hexabromodiphenyl ether (BDE153). *Environ. Int.* **2006**, *32*, 575–585.

(27) Sun, C. Y.; Zhao, D.; Chen, C. C.; Ma, W. H.; Zhao, J. C. TiO2-mediated photocatalytic debromination of decabromodiphenyl ether: kinetics and intermediates. *Environ. Sci. Technol.* **2009**, *43*, 157–162.

(28) Zhao, S.; Ma, H.; Wang, M.; Cao, C.; Xiong, J.; Xu, Y.; Yao, S. Study on the mechanism of photo-degradation of p-nitrophenol exposed to 254 nm UV light. *J. Hazard. Mater.* **2010**, *180*, 86−90.
(29) Zhuang, Y.; Ahn, S.; Luthy, R. G. Debromination of polybrominated diphenyl ethers by nanoscale zerovalent iron: pathways, kinetics, and reactivity. *Environ. Sci. Technol.* **2010**, *44*, 8236−8242.
(30) Gullett, B. K.; Wyrzykowska, B.; Grandesso, E.; Touati, A.; Tabor, D. G.; Ochoa, G. S. PCDD/F, PBDD/F, and PBDE emissions from open burning of a residential waste dump. *Environ. Sci. Technol.* **2010**, *44*, 394−399.
(31) Kajiwara, N.; Noma, Y.; Takigami, H. Photolysis studies of technical decabromodiphenyl ether (DecaBDE) and ethane (DeBD-ethane) in plastics under natural sunlight. *Environ. Sci. Technol.* **2008**, *42*, 4404−4409.
(32) Takigami, H.; Suzuki, G.; Hirai, Y.; Sakai, S. Transfer of brominated flame retardants from components into dust inside television cabinets. *Chemosphere* **2008**, *73*, 161−169.
(33) Raff, J. D.; Hites, R. A. Gas-phase reactions of brominated diphenyl ethers with OH radicals. *J. Phys. Chem. A* **2006**, *110*, 10783−10792.
(34) Altarawneh, M.; Dlugogorski, B. Z. A mechanistic and kinetic study on the formation of PBDD/Fs from PBDEs. *Environ. Sci. Technol.* **2013**, *47*, 5118−5127.
(35) Wang, S.; Hao, C.; Gao, Z.; Chen, J.; Qiu, J. Effects of excited-state structures and properties on photochemical degradation of polybrominated diphenyl ethers: a TDDFT study. *Chemosphere* **2012**, *88*, 33−38.
(36) Zhou, J.; Chen, J.; Liang, C. H.; Xie, Q.; Wang, Y. N.; Zhang, S.; Qiao, X.; Li, X. Quantum chemical investigation on the mechanism and kinetics of PBDE photooxidation by ·OH: a case study for BDE-15. *Environ. Sci. Technol.* **2011**, *45*, 4839−4845.
(37) Frisch, M.; Trucks, G.; Schlegel, H. B.; Scuseria, G.; Robb, M.; Cheeseman, J.; et al. *Gaussian 09*, revision A. 02, Gaussian. Inc.: Wallingford, CT, 2009, 200.
(38) Zhao, Y.; Truhlar, D. G. Hybrid meta density functional theory methods for thermochemistry, thermochemical kinetics, and non-covalent interactions: The MPW1B95 and MPWB1K models and comparative assessments for hydrogen bonding and van der Waals interactions. *J. Phys. Chem. A* **2004**, *108*, 6908−6918.
(39) Cao, H. J.; He, M. X.; Han, D. D.; Sun, Y. H.; Zhao, S. F.; Ma, H. J.; Yao, S. D. Mechanistic and kinetic study on the reaction of 2;4-dibrominated diphenyl ether (BDE-7) with OH radicals. *Comput. Theor. Chem.* **2012**, *983*, 31−37.
(40) Cao, H. J.; He, M. X.; Han, D. D.; Li, J.; Li, M. Y.; Wang, W. X.; Yao, S. D. OH-Initiated oxidation mechanisms and kinetics of 2,4,4′-tribrominated diphenyl ether. *Environ. Sci. Technol.* **2013**, *47*, 8238−8247.
(41) Gonzalez, C.; Schlegel, H. B. An improved algorithm for reaction path following. *J. Chem. Phys.* **1989**, *90*, 2154−2161.
(42) Gonzalez, C.; Schlegel, H. B. Reaction path following in mass-weighted internal coordinates. *J. Phys. Chem.* **1990**, *94*, 5523−5527.
(43) Pechukas, P. Transition state theory. *Annu. Rev. Phys. Chem.* **1981**, *32*, 159−177.
(44) Barker, J. R.; Ortiz, F. N.; Preses, M. J.; Lohr, L. L.; Maranzana, A.; Stimac, J. P.; Nguyen, T. L.; Kumar, T. J. D. *MultiWell-2014.1 Software*; University of Michigan, Ann Arbor: MI, 2014, http://aoss.engin.umich.edu/multiwell/.
(45) Barker, J. R. Multiple-well, multiple-path unimolecular reaction systems. I. MultiWell computer program suite. *Int. J. Chem. Kinet.* **2001**, *33*, 232−245.
(46) Ren, Z.; Bi, X.; Huang, B.; Liu, M.; Sheng, G.; Fu, J. Hydroxylated PBDEs and brominated phenolic compounds in particulate matters emitted during recycling of waste printed circuit boards in a typical e-waste workshop of South China. *Environ. Pollut.* **2013**, *177*, 71−77.
(47) Ebert, J.; Bahadir, M. Formation of PBDD/F from flame-retarded plastic materials under thermal stress. *Environ. Int.* **2003**, *29*, 711−716.
(48) Wyrzykowska-Ceradini, B.; Gullett, B. K.; Tabor, D.; Touati, A. PBDDs/Fs and PCDDs/Fs in the raw and clean flue gas during steady state and transient operation of a municipal waste combustor. *Environ. Sci. Technol.* **2011**, *45*, 5853−5860.
(49) Steen, P. O.; Grandbois, M.; McNeill, K.; Arnold, W. A. Photochemical formation of halogenated dioxins from hydroxylated polybrominated diphenyl ethers (OH-PBDEs) and chlorinated derivatives (OH-PBCDEs). *Environ. Sci. Technol.* **2009**, *43*, 4405−4411.
(50) Cao, H. J.; He, M. X.; Sun, Y. H.; Han, D. D. Mechanical and kinetic studies of the formation of polyhalogenated dibenzo-p-dioxins from hydroxylated polybrominated diphenyl ethers and chlorinated derivatives. *J. Phys. Chem. A* **2011**, *115*, 13489−13497.

DOI: 10.1021/acs.jpca.5b04022
J. Phys. Chem. A 2015, 119, 6404−6411

Science of the Total Environment 524–525 (2015) 195–200

Contents lists available at ScienceDirect

Science of the Total Environment

journal homepage: www.elsevier.com/locate/scitotenv

Adsorption and transformation mechanism of NO_2 on NaCl(100) surface: A density functional theory study

Chenxi Zhang [a,b], Xue Zhang [a], Lingyan Kang [a], Ning Wang [a], Mandi Wang [a], Xiaomin Sun [a,*], Wenxing Wang [a]

[a] *Environment Research Institute, Shandong University, Jinan 250100, PR China*
[b] *Department of Resources and Environment, Binzhou University, Binzhou 256600, PR China*

HIGHLIGHTS

- NO_2 is vertically located at the Na–Na bridge site.
- Three ways can spin pair the orbitals of two NO_2 to make a closed-shell dimmer.
- The isomerization step consists of reciprocal transformation and mutual conversion.
- The reactions of H_2O with three N_2O_4 isomers are considered.

ARTICLE INFO

Article history:
Received 28 January 2015
Received in revised form 26 March 2015
Accepted 2 April 2015
Available online 20 April 2015

Editor: Xuexi Tie

Keywords:
NO_2
NaCl(100) surface
Absorption
Dimerization
Hydrolysis

ABSTRACT

To understand the heterogeneous reactions between NO_2 and sea salt particles in the atmosphere of coastal areas, the absorption of an NO_2 molecule on the NaCl(100) surface, the dimerization of NO_2 molecules and the hydrolysis of N_2O_4 isomers at the (100) surface of NaCl are investigated by density functional theory. Calculated results show that the most favorable adsorption geometry of isolated NO_2 molecule is found to reside at the bridge site (II-1) with the adsorption energy of −14.85 kcal/mol. At the surface of NaCl(100), three closed-shell dimers can be identified as *sym*-O_2N–NO_2, *cis*-ONO–NO_2 and *trans*-ONO–NO_2. The reactions of H_2O with *sym*-O_2N–NO_2 on the (100) surface of NaCl are difficult to occur because of the high barrier (33.79 kcal/mol), whereas, the reactions of H_2O with *cis*-$ONONO_2$ and *trans*-$ONONO_2$ play the key role in the hydrolysis process. The product, HONO, is one of the main atmospheric sources of OH radicals which drive the chemistry of the troposphere.

1. Introduction

Sea salt particles containing NaCl are the largest source of tropospheric aerosol particulate matter with 10^{12} kg introduced into the atmosphere from wave action over the oceans each year (Weis and Ewing, 1999). NO_2 is one of the major pollutants in vehicle exhaust (Lawrence and Crutzen, 1999). In coastal cities with severe vehicle exhaust pollution, considerable attention has been paid to heterogeneous reactions between NaCl and NO_2, and their possible roles in atmospheric chemistry (Finlayson-Pitts and Hemminger, 2000; Karlsson and Ljungström, 1995; Rossi, 2003). The heterogeneous reaction of NO_2 on the surface of NaCl has been an intense research area of recent physical chemistry using X-ray photoelectron spectroscopy (Laux et al., 1996), infrared spectroscopy (Finlayson-Pitts, 1983; Vogt and Finlayson-Pitts, 1994; Weis and Ewing, 1999; Ye et al., 2010; Yoshitake, 2000) and Raman spectroscopy (Scolaro et al., 2009). All kinetic studies clearly show that the reaction is second order with respect to NO_2, but there remains the issue of whether the reactive molecule is NO_2 or its dimer, N_2O_4.

In many cases with NO_2 as reactant, it is not NO_2 monomer itself that directly participates in reactions, but its dimer (Finlayson-Pitts, 2003; Koda et al., 1985; Njegic et al., 2010; Raff et al., 2009; Schroeder and Urone, 1974). It has been known that dinitrogen tetroxide (N_2O_4) exists in substantial concentrations in chemical equilibrium at room temperature and atmospheric pressure. The formation process of N_2O_4 from NO_2 has been investigated by many scientists (Liu and Goddard, 2012; Pimentel et al., 2007a). The pathways of forming such dimers on the surface of NaCl, however, have no previous study.

In the marine environment, there are large amounts of water vapor in the air. Therefore, simulation of the heterogeneous reaction in the presence of water vapor much more resembles the real atmospheric environment than a dry one. Yoshitake (2000) proposed that the NO_2/H_2O/NaCl reaction is thought to keep the following mechanism:

* Corresponding author.
E-mail address: sxmwch@sdu.edu.cn (X. Sun).

http://dx.doi.org/10.1016/j.scitotenv.2015.04.010
0048-9697/

$$2NO_2 + H_2O \rightarrow HNO_3 + HONO \quad (1)$$

$$HNO_3 + NaCl \rightarrow HCl + NaNO_3. \quad (2)$$

Reaction (2) is likely fast enough to make reaction (1) rate-determining. Reaction (1) is the main source of HONO, which is one of the main atmospheric sources of OH radicals that drives the chemistry of the troposphere (Stockwell and Calvert, 1983; Stutz et al, 2004). In heavily polluted regions, HONO photolysis accounts for about 50% of the total OH production during the early morning. Finlayson-Pitts (2003) and Miller et al. (2009) indicated that the asymmetric dimer ($ONO–NO_2$) on surfaces plays a key role in NO_2 hydrolysis to form HONO and HNO_3. There are some data on the heterogeneous reaction of $NO_2/H_2O/NaCl$ (Finlayson-Pitts, 1983; Finlayson-Pitts and Hemminger, 2000; Weis and Ewing, 1999; Ye et al., 2010). Unfortunately, the chemical and physical interactions on the surface of NaCl are not well understood on a molecular scale.

Theoretical calculation can provide information for the reaction intermediates and pathways. In this work, the density functional theory (DFT) calculation (Payne et al., 1992) was carried out to investigate the possible absorption of an NO_2 molecule at the NaCl(100) surface, the dimerization of NO_2 molecules and the N_2O_4–H_2O intermolecular potential at the same surface.

2. Computational method

All calculations were performed with the program package of CASTEP in Materials Studio (version 5.5) of Accelrys Inc. (Segall et al., 2002), implementing the Perdew–Wang 1991 (PW91) version of the general gradient approximation (GGA) (Perdew et al., 1992; Perdew and Wang, 1992). The energies of three important crystallographic surfaces (100), (110) and (111) were calculated, and the stability order of this three surfaces of NaCl was found to be (100) > (110) > (111) (Khan and Ganguly, 2013). Thus, in this study, we used a (2 × 2) supercell with a slab of four layers to represent the NaCl(100) surface. Cabrera-Sanfelix et al. (2006) verified that this was enough thickness to produce consistent adsorption properties. The vacuum gap along the normal to the surface was 12 Å. This has proven to be sufficient to prevent any significant overlap of the electronic densities corresponding to different periodic images of the slab. During geometry optimization, the bottom of two NaCl layers was fixed, and the atoms in the two topmost layers as well as the NO_2 were allowed to relax. All other degrees of freedom were allowed to relax until all the components of the forces were smaller than 0.01 eV/Å. We used a plane-wave cutoff of 330 eV and a 2 × 2 × 1 Monkhorst–Pack k-point sampling, and the cutoff energy and k-point sampling were used throughout our calculations. The tolerance of self-consistent field (SCF) convergence was 1×10^{-6}.

The adsorption energy E_{ads} is defined as the energy difference between the optimized system and those of the relaxed NaCl(100) surface and the isolated NO_2 molecule.

$$E_{ads} = E_{(NO_2/slab)} - \left[E_{(NO_2)} + E_{(slab)}\right]$$

where the first term is the total energy of the slab with the adsorbed NO_2 on the surface, the second term is the total energy of free NO_2, and the third term is the total energy of the bare slab of the surface. According to the above definitions, a negative E_{ads} value corresponds to an exothermic adsorption, and the more negative is the E_{ads}, the stronger is the adsorption between NO_2 and NaCl.

In order to determine accurate activation barriers of the reaction, we chose the complete LST/QST approach to search for transition states of the reactions (Halgren and Lipscomb, 1977). In this method, the linear synchronous transit (LST) maximization was performed, followed by an energy minimization in directions conjugating to the reaction pathway to obtain an approximated transition state (TS). The approximated TS was used to perform quadratic synchronous transit (QST) maximization, and then another conjugated gradient minimization was performed. The cycle was repeated until a stationary point was located. The reaction energy (E_r) and energy barrier (E_b) of an elementary reaction on NaCl(100) surface were calculated based on the following formulas:

$$E_r = E_{Product} - E_{Reactant}$$

$$E_b = E_{TS} - E_{Reactant}.$$

3. Results and discussion

3.1. Tests of computational conditions

To confirm the reliability of the methods, we compared the bond distances (in Å) and bond angle (°) for H_2O, NO_2 and N_2O_4 with available experimental values in Table 1. We take here the energy of the H_2O, NO_2 and N_2O_4 molecules in a large unit cell of 10 Å × 10 Å × 10 Å. The calculated bond length and bond angle are in good agreement with the experimental values (Pimentel et al., 2007b). All of the calculated bond lengths are within 2% of experimental ones, and the calculated bond angles are within 1°.

3.2. NO_2 adsorption on the (100) surface of NaCl

The $^2A_1$ ground state of NO_2 has a bent structure with the unpaired electron residing on the N atom, which has three different adsorption sites on the NaCl(100) surface: the top site (I), the bridge site (II) and the 3-fold hollow site (III). Resulting NO_2/NaCl(100) structures are shown in Fig. 1.

As shown in Fig. 1 and Table 2, NO_2 could either weakly adsorb at the Na top site on NaCl(100) surface with the adsorption energy of below −10 kcal/mol, or adsorb at the Cl top site on NaCl(100) surface with the adsorption energy of above −9.5 kcal/mol. According to the definitions of E_{ads}, the more negative is the E_{ads}, the stronger is the adsorption between NO_2 and NaCl. Thus, the Na–NO_2 interaction is much stronger than the Cl–NO_2 interaction, possibly because Na is an electron donor.

For the bridge site, NO_2 is vertically located at Na–Na bridge site, Na–Cl bridge site and Cl–Cl bridge site. On the whole, the NO_2 at Na–Na bridge sites (II-1, II-4, and II-7) are the most energetically favorable among the all structure, and then the Na–Cl bridge sites (II-2, II-5, II-8 and II-9) and Cl–Cl bridge sites (II-3, II-6, and II-10).

For the 3-fold hollow sites, there are two different adsorption forms: the first one is that O atom is bound to the surface of Na atom whereas the N atom was bound to the Cl atom, and the second one is that the O atom is bound to the surface of Cl atom whereas the N atom is bound to the NaCl atom. The first adsorption form is more stable.

In all the structures, the most stable adsorption of the NO_2 on NaCl(100) surface is found to reside at the bridge site (II-1) with the adsorption energies of −14.85 kcal/mol which is taken for example in the subsequent reactions.

Table 1
The bond distances (in Å) and bond angle (°) for H_2O, NO_2 and N_2O_4.

Molecule	Parameter	Calculated value	Experimental value[a]
H_2O	O–H	0.975	0.958
	∠HOH	104.6	104.5
NO_2	N–O	1.231	1.193
	∠ONO	133.5	134.1
N_2O_4	N–N	1.807	1.782
	N–O	1.222	1.190
	∠ONO	135.4	135.4

[a] Lide, D. R. *CRC Handbook of Chemistry and Physics*; CRC Press: Boca Raton, 1996.

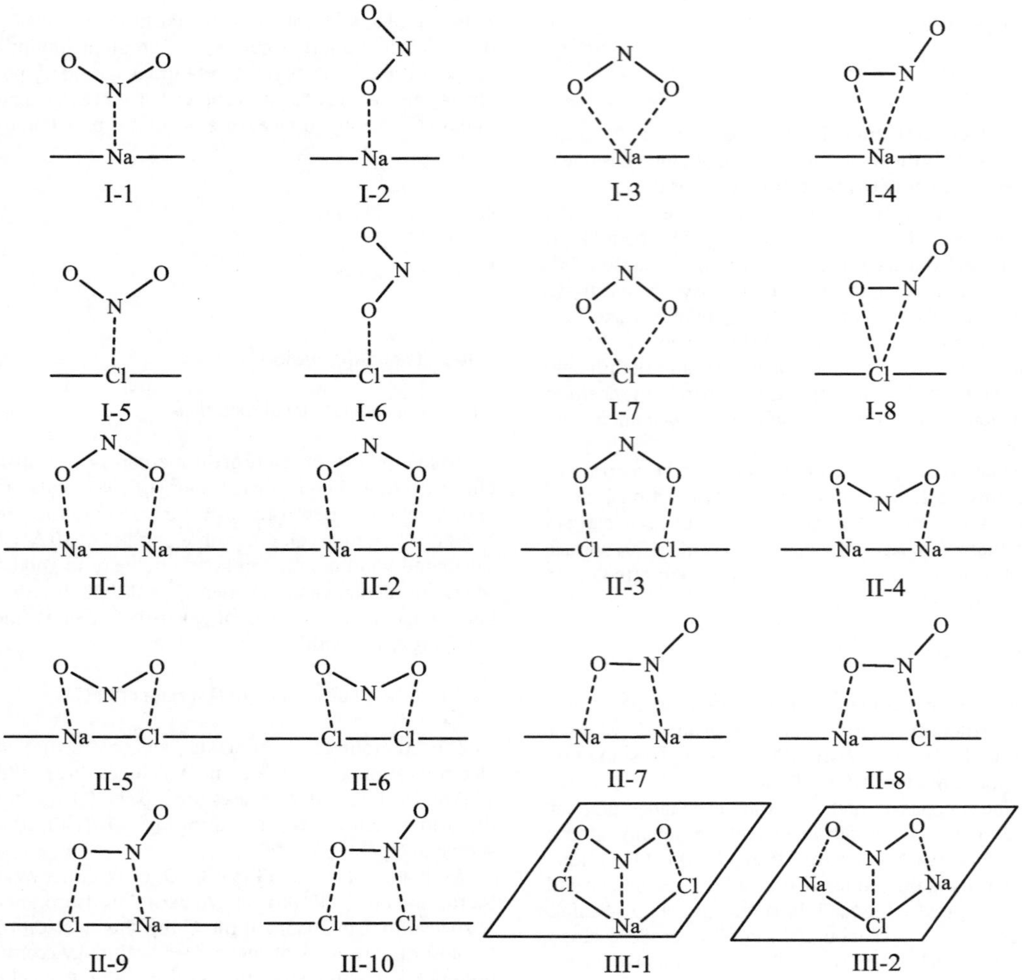

Fig. 1. Possible isomers of adsorbed NO_2 on the NaCl(100) surface.

3.3. NO_2 dimerization and isomerization on the (100) surface of NaCl

3.3.1. NO_2 dimerization

There are three ways to spin pair the orbitals of two NO_2 radicals to make a closed-shell dimer: N–N bond leading to *sym*-O_2N–NO_2(NaCl) as IM1; N–O bond leading to *cis*-ONO–NO_2(NaCl) and *trans*-ONO–NO_2(NaCl) as IM2 and IM3, respectively. The profiles of potential energy surfaces for the forming of NO_2 dimers are drawn in Fig. 2. The recombination starts from two separating NO_2 molecules with open-shell singlet configuration, and three van der Waals (vdW) complexes are formed. Then, vdW1 reaches the TS1 with a barrier height of 16.23 kcal/mol and ends up with the closed-shell O_2N–NO_2. The two NO_2 in the vdW2 and vdW3 approach each other perpendicularly to make new O–N bonds, reach the transition states with the O–N bond of 1.983 Å and 2.803 Å, and end up with the intermediate structure with an O–N bond of 1.458 Å and 1.463 Å, respectively.

Among these, the *sym*-O_2N–NO_2(NaCl) is most stable. The asymmetric dimer, ONO–NO_2, is less stable and it has been identified in the condensed phase via IR spectroscopy (Beckers et al., 2010).

Table 2
Calculated adsorption energies of adsorption isomers on the NaCl(100) surface.

Adsorption isomers	E_{ads} (kcal/mol)	Adsorption isomers	E_{ads} (kcal/mol)
I-1	−10.22	II-3	−8.64
I-2	−10.34	II-4	−13.74
I-3	−11.14	II-5	−9.81
I-4	−11.32	II-6	−10.16
I-5	−9.04	II-7	−11.23
I-6	−5.66	II-8	−10.92
I-7	−9.49	II-9	−5.70
I-8	−8.49	II-10	−6.80
II-1	−14.85	III-1	−8.33
II-2	−10.68	III-2	−11.67

3.3.2. N_2O_4 isomerization

The isomerization step consists of the reciprocal transformation between asymmetric isomer (*cis*-ONO–NO_2 or *trans*-ONO–NO_2) and the mutual conversion between the *sym*-O_2N–NO_2 and *trans*-ONO–NO_2, which was displayed in Fig. 3.

As shown in Fig. 3, the D_{2h} symmetry of the *sym*-O_2N–NO_2 is broken to form the $TS_{IM1-IM3}$, which has a three-atom interacting site replaced by an N–O–N bridge in the product, *trans*-ONO–NO_2. The barrier of this process was 51.84 kcal/mol at the surface of NaCl. Liu and Goddard (2012) study this reaction without the surface of NaCl using several ab initio methods, including MP2, CCSD(T) and GVB-RCI. At MP2 level, the barrier for *sym*-O_2N–NO_2 isomerizing to *trans*-ONO–NO_2 is 48.9 kcal/mol, whereas CCSD(T)/cc-pVDZ gives 47.2 kcal. Such a high barrier could rule out the possibility of this process playing a significant role in N_2O_4 isomerization.

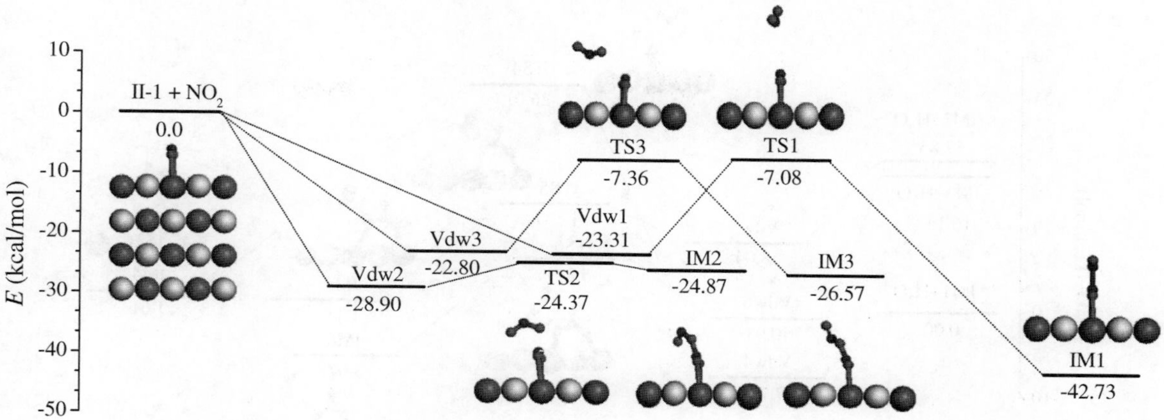

Fig. 2. The profiles of potential energy surfaces for the forming of NO_2 dimers on the NaCl(100) surface.

The *cis–trans* isomerization easily occurs, and the barrier of conversion *tran-* to *cis*-ONO–NO_2 is 7.57 kcal/mol, which can compare with the number reported by Liu and Goddard (2012), 4.4 kcal/mol. Such a low barrier could lead to a fast equilibrium between *cis-* and *trans*-ONO–NO_2 at room temperature.

3.4. Hydrolysis of N_2O_4 isomers

In the marine environment, simulation of the heterogeneous reaction in the presence of water vapor much more resembles the real atmospheric environment than a dry one. Thus, in this work, the reactions of H_2O with *sym*-O_2N–NO_2, *trans*-ONO–NO_2, and *cis*-ONO–NO_2 on the (100) surface of NaCl have been considered.

At the (100) surface of NaCl, the interaction of H_2O with *sym*-O_2N–NO_2 further forms one complex, H_2O–*sym*-O_2N–NO_2 (NaCl), with −7.78 kcal/mol association energies. The H_2O–*sym*-O_2N–NO_2 (NaCl) dissociates to HNO_3 + HONO via a five-membered ring transition state TS4 with 33.79 kcal/mol energy above H_2O + *sym*-O_2N–NO_2. As the reaction takes place, the two NO_2 units of *sym*-O_2N–NO_2 start to depart from each other. The O atom of H_2O approaches an N atom of one NO_2 unit on the surface of NaCl, while an H atom of H_2O approaches to an O atom of the other NO_2 unit. At the transition state, as shown as TS4 in Fig. 4, the two NO_2 units match the H and OH fragments of H_2O, respectively, to form the HONO and HNO_3 product molecules. As a result, the reaction is not kinetically favorable. Moreover, the reaction is endothermic by 9.39 kcal/mol, and is not thermo-dynamically favorable.

For the *cis*-ONO–NO_2 and *trans*-ONO–NO_2, two vdWs, H_2O–*cis*-ONO–NO_2(NaCl) and H_2O–*trans*-ONO–NO_2(NaCl), were formed with −9.81 kcal/mol and −17.09 kcal/mol association energies, respectively. As the two molecules get closer, an H atom of H_2O starts to break away and approach toward the O atom of NO_2 that absorbed on the surface of NaCl. And the O atom of H_2O approaches the N atom of the ONO– fragment. Then the N–O bond of the ONO– fragment breaks away, leading to the formation of HONO and HNO_3. The barriers of these two processes are 4.33 and 6.80 kcal/mol, respectively.

This indicates that the contribution of the reactions of H_2O with *sym*-O_2N–NO_2 on the (100) surface of NaCl can be ruled out due to the high barrier (33.79 kcal/mol) involved. The *cis*-ONO–NO_2 and *trans*-ONO–NO_2 were found to play the key role in the hydrolysis process. The mechanism for formation of HONO is consistent with isotope labeling experiments. Sakamaki et al. (1983) showed that NO_2 reacts in a small quartz cell at room temperature with $H_2^{18}O$ to generate exclusively $H^{18}ONO$. The formation of $H^{18}ONO$ can easily be explained by the mechanism in Fig. 4. Reaction of $H_2^{18}O$ with asymmetric ONO–NO_2 will lead to $H^{18}ONO$, rather than *sym*-O_2N–NO_2. These results are also consistent

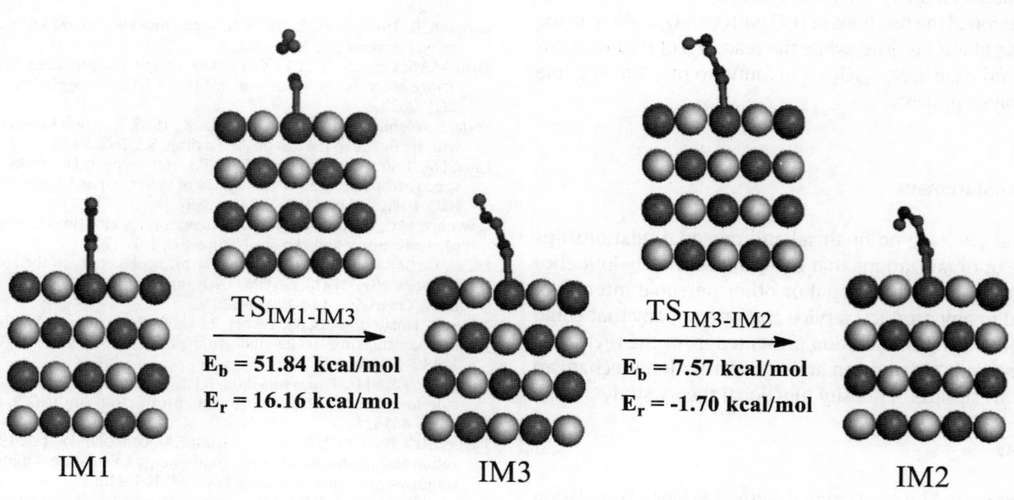

Fig. 3. The isomerization of NO_2 dimers on the NaCl(100) surface.

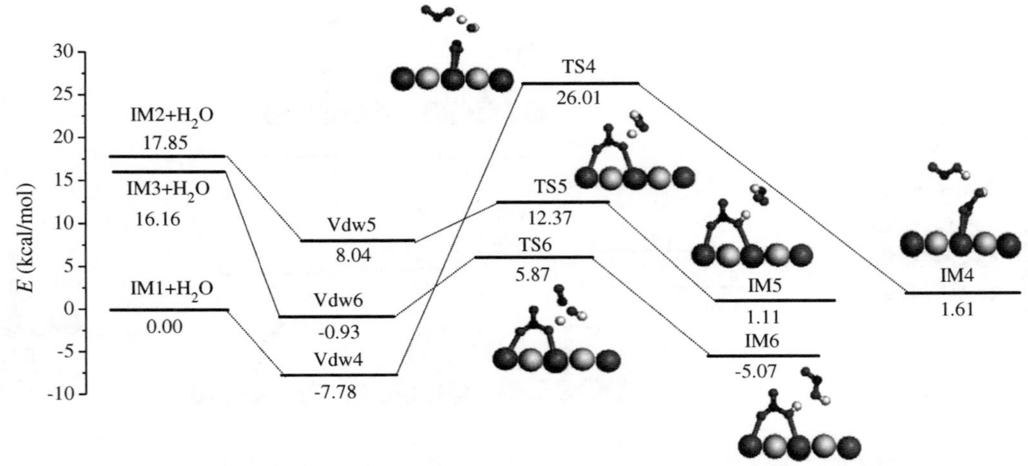

Fig. 4. The profiles of potential energy surfaces for the hydrolysis of N_2O_4 isomers on the NaCl(100) surface.

with the experimentally proposed mechanism that heterogeneous disproportionation of asymmetric ONO–NO_2 releases HONO and surface-adsorbed HNO_3 (Finlayson-Pitts et al., 2003).

4. Conclusion

The possible absorption of an NO_2 molecule at the NaCl(100) surface, the dimerization of NO_2 molecules and the hydrolysis of N_2O_4 isomers on the (100) surface of NaCl have been investigated by DFT calculation. Many of the findings may help to understand the mechanism for the absorption and transformation of NO_2 on the surface of NaCl in coastal region.

(1) The NO_2 molecule can easily adsorb on the (100) surface of NaCl. The most stable adsorption of the NO_2 on NaCl(100) surface is found to reside at the bridge site (II-1), where the NO_2 is vertically located at the Na–Na bridge site.
(2) At the (100) surface of NaCl, two NO_2 radicals can make closed-shell dimers, including *sym*-O_2N–NO_2, *cis*-ONO–NO_2 and *trans*-ONO–NO_2. Among these, the symmetric dimer is most stable, and the asymmetric dimers are less stable. All the dimers can be identified via IR spectroscopy.
(3) The isomerization of N_2O_4 dimers consists of the reciprocal transformation between asymmetric isomers ONO–NO_2 and the mutual conversion between the *sym*-O_2N–NO_2 and *trans*-ONO–NO_2.
(4) The contribution of the reactions of H_2O with *sym*-O_2N–NO_2 on the (100) surface of NaCl is little, while the reactions of H_2O with *cis*-ONO–NO_2 and *trans*-ONO–NO_2 were found to play the key role in the hydrolysis process.

Conflict of interest statement

We declare that we have no financial and personal relationships with other people or organizations that can inappropriately influence our work, and there is no professional or other personal interest of any nature or kind in any product, service and/or company that could be construed as influencing the position presented in, or the review of, the manuscript entitled, "Adsorption and transformation mechanism of NO_2 on NaCl(100) surface: A density functional theory study".

Acknowledgments

This work is supported by the National Natural Science Foundation of China (Nos. 21177076, 21277082 and 21337001), Natural Science Foundation of Shandong Province (No. ZR2014BP012), Marie Curie International Research Staff Exchange Scheme Fellowship within the 7th European Community Framework Programme (No. 295132), Program for New Century Excellent Talents in University (NCET-13-0349), Project for Science and Technology Development of Shandong Province (2014GSF117028), Promotive Research Fund for Excellent Young, Middle-aged Scientists of Shandong Province (BS2012HZ009) and Beijing National Laboratory for Molecular Science (No. 20140160).

References

Beckers, H., Zeng, X.Q., Willner, H., 2010. Intermediates involved in the oxidation of nitric oxide: chemistry of NO•O_2 and $(NO)_2$•O_2 and NO_2 dimer complexes in various cryogenic solid matrices. Chem. Eur. J. 16, 1506–1520.
Cabrera-Sanfelix, P., Arnau, A., Darling, G.R., Sanchez-Portal, D., 2006. Water adsorption and diffusion on NaCl(100). J. Phys. Chem. B 110, 24559–24564.
Finlayson-Pitts, B.J., 1983. Reaction of NO_2 with NaCl and atmospheric implications of NOCl formation. Nature 306, 676–677.
Finlayson-Pitts, B.J., 2003. The tropospheric chemistry of sea salt: a molecular-level view of the chemistry of NaCl and NaBr. Chem. Rev. 103, 4801–4822.
Finlayson-Pitts, B.J., Hemminger, J.C., 2000. Physical chemistry of airborne sea salt particles and their components. J. Phys. Chem. A 104, 11463–11477.
Finlayson-Pitts, B.J., Wingen, L.M., Sumner, A.L., Syomin, D., Ramazan, K.A., 2003. The heterogeneous hydrolysis of NO_2 in laboratory systems and in outdoor and indoor atmospheres: an integrated mechanism. Phys. Chem. Chem. Phys. 5, 223–242.
Halgren, T.A., Lipscomb, W.N., 1977. The synchronous-transit method for determining reaction pathways and locating molecular transition states. Chem. Phys. Lett. 49, 225–232.
Karlsson, R., Ljungström, E., 1995. Nitrogen dioxide and sea salt particles—a laboratory study. J. Aerosol Sci. 26, 39–50.
Khan, M.A.S., Ganguly, B., 2013. Can surface energy be a parameter to define morphological change of rock-salt crystals with additives? A first principles study. CrystEngComm 15, 2631–2639.
Koda, S., Yoshikawa, K., Okada, J., Akita, K., 1985. Reaction kinetics of nitrogen dioxide with methanol in the gas phase. Environ. Sci. Technol. 19, 262–264.
Laux, J.M., Flster, T.F., Finlayson-Pitts, B.J., Hemminger, J.C., 1996. X-ray photoelectron spectroscopy studies of the effects of water vapor on ultrathin nitrate layers on NaCl. J. Phys. Chem. 100, 19891–19897.
Lawrence, M.G., Crutzen, P.J., 1999. Influence of NO_x emissions from ships on tropospheric photochemistry and climate. Nature 402, 167–170.
Liu, W.G., Goddard, W.A.I.I.I., 2012. First-principles study of the role of interconversion between NO_2, N_2O_4, cis-ONO–NO_2, and trans-ONO–NO_2 in chemical processes. J. Am. Chem. Soc. 134, 12970–12978.
Miller, Y., Finlayson-Pitts, B.J., Gerber, R.B., 2009. Ionization of N_2O_4 in contact with water: mechanism, time scales and atmospheric implications. J. Am. Chem. Soc. 131, 12180–12185.
Njegic, B., Raff, J.D., Finlayson-Pitts, B.J., Gordon, M.S., Gerber, R.B., 2010. A catalytic role for water in the atmospheric production of ClNO. J. Phys. Chem. A 114, 4609–4618.
Payne, M.C., Teter, M.P., Allan, D.C., Arias, T.A., Joannopoulos, J.D., 1992. Iterative minimization techniques for ab initio total-energy calculations: molecular dynamics and conjugate gradients. Rev. Mod. Phys. 64, 1045–1097.
Perdew, J.P., Wang, Y., 1992. Accurate and simple analytic representation of the electron-gas correlation energy. Phys. Rev. B1 (45), 13244–13249.

Perdew, J.P., Chevary, J.A., Vosko, S.H., Jackson, A.K., Pederson, R.M., Singh, D.J., Fiolhais, C., 1992. Atoms, molecules, solids, and surfaces: applications of the generalized gradient approximation for exchange and correlation. Phys. Rev. B 46, 6671–6687.

Pimentel, A.S., Lima, F.C.A., da Silva, A.B.F., 2007a. The asymmetric dimerization of nitrogen dioxide. Chem. Phys. Lett. 436, 47–50.

Pimentel, A.S., Lima, F.C.A., da Silva, A.B.F., 2007b. The isomerization of dinitrogen tetroxide: $O_2N–NO_2 \rightarrow ONO–NO_2$. J. Phys. Chem. A 111, 2913–2920.

Raff, J.D., Njegic, B., Chang, W.L., Gordon, M.S., Dabdub, D., Gerber, R.B., Finlayson-Pitts, B.J., 2009. Chlorine activation indoors and outdoors via surface-mediated reactions of nitrogen oxides with hydrogen chloride. Proc. Natl. Acad. Sci. U. S. A. 106, 13647–13654.

Rossi, M.J., 2003. Heterogeneous reactions on salts. Chem. Rev. 103, 4823–4882.

Sakamaki, F., Hatakeyama, S., Akimoto, H., 1983. Formation of nitrous acid and nitric oxide in the heterogeneous dark reaction of nitrogen dioxide and water vapor in a smog chamber. Int. J. Chem. Kinet. 15, 1013–1029.

Schroeder, W.H., Urone, P., 1974. Formation of nitrosyl chloride from salt particles in air. Environ. Sci. Technol. 8, 756–758.

Scolaro, S., Sobanska, S., Barbillat, J., Laureyns, J., Louis, F., Petitprez, D., Brémard, C., 2009. Confocal raman imaging and atomic force microscopy of the surface reaction of NO_2 and NaCl(100) under humidity. J. Raman Spectrosc. 40, 157–163.

Segall, M., Lindan, P., Probert, M., Pickard, C., Hasnip, P., Clark, S., Payne, M., 2002. First-principles simulation: ideas, illustrations and the CASTEP code. J. Phys. Condens. Matter 14, 2717–2744.

Stockwell, W.R., Calvert, J., 1983. The mechanism of NO_3 and HONO formation in the nighttime chemistry of the urban atmosphere. J. Geophys. Res. 88, 6673–6682.

Stutz, J., Alicke, B., Ackermann, R., Geyer, A., Wang, S., White, A.B., Williams, E.J., Spicer, C.W., Fast, J.D., 2004. Relative humidity dependence of HONO chemistry in urban areas. J. Geophys. Res. 109, D03307.

Vogt, R., Finlayson-Pitts, B.J., 1994. A diffuse reflectance infrared fourier transform spectroscopic (DRIFTS) study of the surface reaction of NaCl with gaseous NO_2 and HNO_3. J. Phys. Chem. 98, 3141–3155.

Weis, D.D., Ewing, G.E., 1999. The reaction of nitrogen dioxide with sea salt aerosol. J. Phys. Chem. A 103, 4865–4873.

Ye, C.X., Li, H.J., Zhu, T., Shang, J., Zhang, Z.F., Zhao, D.F., 2010. Heterogeneous reaction of NO_2 with sea salt particles. Sci. China Chem. 53, 2652–2656.

Yoshitake, H., 2000. Effects of surface water on NO_2/NaCl reaction studied by diffuse reflectance infrared spectroscopy (DRIRS). Atmos. Environ. 34, 2571–2580.

Catalysis Science & Technology

PAPER

View Article Online
View Journal | View Issue

CrossMark
click for updates

Cite this: Catal. Sci. Technol., 2016, 6, 73

Catalytic mechanism of C–F bond cleavage: insights from QM/MM analysis of fluoroacetate dehalogenase†

Yanwei Li,[a] Ruiming Zhang,[a] Likai Du,[b] Qingzhu Zhang*[a] and Wenxing Wang[a]

Received 27th May 2015, Accepted 14th September 2015

DOI: 10.1039/c5cy00777a

www.rsc.org/catalysis

The catalytic mechanisms of fluoroacetate dehalogenase (FAcD) toward substrates fluoroacetate and chloroacetate were studied by a combined quantum mechanics/molecular mechanics (QM/MM) method. There are twenty snapshots considered for each of the three individual systems. By analyzing multiple independent snapshots, positive or negative relationships between energy barriers and structural parameters in defluorination and dechlorination processes were established. We have also shown that conformational variations may cause enzymatic preference differences toward competitive pathways. Besides residues Arg111, Arg114, His155, Trp156, and Tyr219, the importance of residues His109, Asp134, Lys181, and His280 during the defluorination process were also highlighted through electrostatic analysis. These results may provide clues for designing new biomimetic catalysts toward degradation of fluorinated compounds.

1. Introduction

Fluorinated compounds are widely used in numerous industries and presently compose up to 20% of all pharmaceuticals and 30% of all the agrochemicals.[1–3] Their large-scale applications have caused increasing environmental concerns due to their toxicity, global warming potential, environmental persistence, and bioaccumulation character.[4–6] It is thus critically important to set up strategies to minimize continued exposure of these fluorinated compounds. Environmental biotransformation, one of the most promising strategies with the lowest energy consumption, has provided some encouraging results in cleaving the highly stable C–F bond, whose dissociation energy is the highest among all the natural products (~130 kcal mol^{-1}).[1] For example, fluoroacetate dehalogenase (FAcD) discovered in bacteria *Burkholderia sp.* FA1 was found to catalyze the dehalogenation process of its natural substrate fluoroacetate (FAc).[1] FAc is very stable and toxic, and has been widely manufactured and used as a vertebrate pest control agent in many countries like the United States, Mexico, Australia, and New Zealand.[7,8] The dehalogenation process catalyzed by FAcD has attracted the most interest and currently serves as the model system for enzymatic defluorination investigations.[1,9–18]

The catalytic mechanism of FAcD has been investigated for many decades, mainly using site-directed mutagenesis and electrospray mass spectrometry.[9–11] Jitsumori *et al.* reported the first crystallization structure of FAcD (from *Burkholderia sp.* FA1), which makes the mechanical elucidation of FAcD at the molecular level possible.[13] They also found that defluorination of FAcD requires a catalytic triad Asp–His–Asp, and the aspartate acts as a nucleophile and directly ejects the fluoride anion from FAc. Because the substrate FAc was not co-crystallized in the crystal (PDB code 1Y37), Yoshizawa and coworkers predicted the binding mode of FAc with FAcD and investigated the subsequent mechanisms through quantum mechanics/molecular mechanics (QM/MM) calculations.[14,17] In these two excellent pioneering studies, the authors not only managed to determine the reaction barrier of defluorination and dechlorination, but also explored the roles of residues near the halide ion. However, the theoretically predicted binding mode of the substrate is quite different from the binding mode found in the co-crystallized FAcD–FAc complex (PDB code 3R3V) extracted in another bacterium *Rhodopseudomonas palustris* CGA009.[14,16,17] This inconformity raises interest to further investigate the catalytic mechanism of FAcD (PDB code 3R3V) and answer the question of what are the structural requirements that enable defluorination rather than dechlorination.[16] Understanding

[a] *Environment Research Institute, Shandong University, Jinan 250100, PR China. E-mail: zqz@sdu.edu.cn; Fax: +86 531 8836 1990*

[b] *Key Laboratory of Bio-based Materials, Qingdao Institute of Bio-energy and Bioprocess Technology, Chinese Academy of Sciences, Qingdao 266101, PR China*

† Electronic supplementary information (ESI) available: Hydrogen bond distances between FAcD and the substrates (Table S1), NPA charge variations (Table S2), gas phase calculations (Scheme S1); binding of FAc with FAcD (Fig. S1), root-mean-square deviation (Fig. S2), structures involved in dechlorination process of system $FAcD_{Hse155}$–ClAc (Fig. S3), correlation between potential energy barriers and dihedral $O_\omega C_\gamma C_\delta O_\varepsilon$ (Fig. S4), and correlation between potential energy barriers and bond $O_\varepsilon C_\delta$ (Fig. S5). See DOI: 10.1039/c5cy00777a

defluorination details of FAcD may be helpful in enzyme engineering or biomimetic catalysis to remove harmful fluorinated compounds from the environment. The relative locations of key active site residues and substrate FAc are illustrated in Fig. S1, ESI.†

Flexibility is one of the most intriguing characteristics of enzymes. Recent room-temperature single molecule experiments have shown that enzyme molecules exhibit large turnover rate fluctuations with a broad range of time scales (1 ms–100 s).[19,20] This leads to the proposal that each of the conformational states of an enzyme are long-lived, and correspond to a different turnover rate constant.[21,22] Thus, although it is still not common, considering multiple snapshots is highly recommended when modelling enzymatic reactions.[23–25] Multiple snapshots should be considered when theoretically exploring why FAcD prefers defluorination rather than dechlorination.

One of the main purposes of the current QM/MM analysis is to investigate what are the structural requirements for FAcD (from bacterium *Rhodopseudomonas palustris* CGA009, PDB code 3R3V) in enabling defluorination rather than dechlorination by considering twenty snapshots. This may help in designing *de novo* enzymes or biomimetic catalysts for degradation of other fluorinated compounds. The present study also tries to provide solutions on how to identify the two possible states of a neutrally charged histidine (Hsd155 or Hse155, as shown in Scheme 1) of FAcD. This is valuable because currently there are still no better solutions than visual inspection of the local hydrogen-bonding environment in distinguishing these two neutrally charged states.[26] In total, there are sixty reaction pathways studied, forty for defluorination (with Hsd155 and Hsd155) and twenty for dechlorination by FAcD.

2. Methods

2.1 MD simulation

The initial models for the present simulation were built on the basis of the X-ray crystal structure of $FAcD_{D110N}$–FAc binary complex (PDB code 3R3V, resolution 1.50 Å) obtained from the Protein Data Bank (www.rcsb.org).[16] The mutated residue (D110N) presented in the crystal structure was manually transformed back into its natural form. The missing hydrogen atoms in the crystal structure were added through a CHARMM22 force field in the HBUILD module of the CHARMM package.[27–29] The whole enzyme was dissolved in a

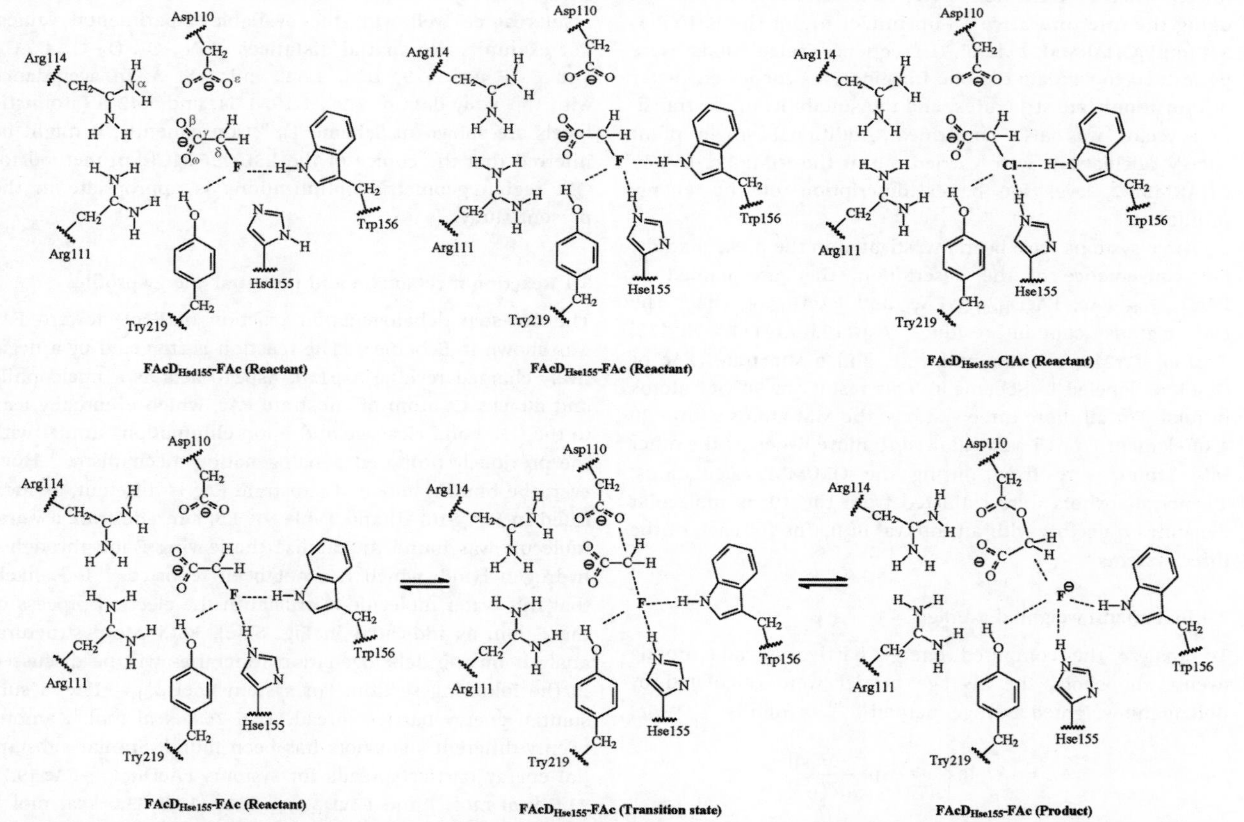

Scheme 1 The QM regions in the reactants of three studied systems ($FAcD_{Hsd155}$–FAc, $FAcD_{Hse155}$–FAc, and $FAcD_{Hse155}$–FAc) and the dehalogenation processes of system $FAcD_{Hse155}$–FAc. The boundary between the QM and MM regions are indicated by wavy lines.

water droplet (TIP3P model)[30] with a radius of 35 Å. Then, the enzyme-water system was neutralized by sodium ions *via* random substitution of solvent water molecules before being relaxed through energy minimizations. The whole system was first heated from absolute zero to 298.15 K within 50 ps (1 fs per step) and equilibrated thermally for 500 ps (1 fs per step) to reach the equilibration state. After that, a 10 ns stochastic boundary molecular dynamics (SBMD) simulation was performed at 298.15 K using a NVT ensemble for conformational sampling.[31] During the SBMD simulations, the whole system moved freely except the substrate, the coordinates of which are restrained to keep consistence with its positions in the crystal structure. The leap-frog algorithm and Langevin temperature coupling method implemented in the CHARMM program were applied during the simulations. The obtained root-mean-square deviation is provided in Fig. S2, ESI.†

2.2 QM/MM calculations

The QM/MM calculations were performed using the ChemShell[32] platform, which can integrate the programs Turbomole[33] and DL-POLY.[34] The charge shift model[35] and electrostatic embedding method[36] were used during the QM/MM calculations. The geometries of the intermediates were optimized using hybrid delocalized internal coordinates optimizer, whereas transition state searches were carried out using the microiterative TS optimizer under the B3LYP/6-31G(d,p)//CHARMM22 level.[37] Frequency calculations were performed to validate the one imaginary frequency character of transition state structures, and the suitability of the transition vector was also confirmed. Additional single point energy calculations were carried out at the RIMP2/cc-pVTZ//CHARMM22 level for better description of the energy profiles.

Three systems have been investigated in the present study. For convenience of the description, they are named as $FAcD_{Hsd155}$–FAc, $FAcD_{Hse155}$–FAc, and $FAcD_{Hse155}$–ClAc. The QM regions contain residues Asp110, Arg111, His155, Trp156, Tyr219, a water molecule, and a substrate (FAc or ClAc), as labeled in Scheme 1. This results in 90 QM atoms in total. For all these three systems, the MM atoms within 20 Å of element F or Cl were allowed to move, whereas the other MM atoms were fixed during the QM/MM calculations. Twenty snapshots were extracted from the 10 ns molecular dynamics trajectory with an interval of 0.5 ns for each of the three systems.

2.3 Boltzmann-weighted average

To analyze the computed energy barrier spreads among twenty snapshots, the average barrier were calculated by Boltzmann-weighted average method[38–40] as follows:

$$\Delta E = -RT\ \ln\left\{\frac{1}{n}\sum_{i=1}^{n}\exp\left(\frac{-\Delta E_i}{RT}\right)\right\}$$

where ΔE is the average barrier, R is the gas constant, n is the number of snapshots, ΔE_i is the energy barrier of path i, and T is the temperature. For a small n value, if the set of starting geometries include one with an anomalously low energy barrier, this will have a disproportionate effect on the Boltzmann-weighted average barrier. The disproportionate effect can be evaluated by the following equation:

$$\mathrm{DE} = \frac{\Delta E^{a-1} - \Delta E^{a}}{\Delta E^{a}} \times 100\%$$

where DE represents the disproportionate effect, ΔE^{a-1} is the Boltzmann-weighted average barrier calculated by neglecting the snapshot with the lowest energy barrier and ΔE^{a} is the Boltzmann-weighted average barrier with all the snapshots considered.

3. Results and discussion

The first step of this study is to identify the reliability of the calculation method. Due to the absence of the X-ray crystal structure of the $FAcD_{wild}$–FAc binary complex, it is difficult to make a direct comparison between the calculated results and the experimental data. To verify the reliability of the computational results, we optimized the available crystal structure of $FAcD_{D110N}$–FAc binary complex at the B3LYP/6-31G(d,p)//CHARMM22 level. The calculated results agree well with the available experimental values. For example, the spatial distances of N_α–O_β, O_β–C_γ, C_γ–C_δ, and C_δ–F are 2.79, 1.26, 1.52, and 1.44 Å, in accordance with the X-ray data of 2.98, 1.19, 1.54, and 1.42 Å (atomistic labels are shown in Scheme 1).[16] Consequently, it might be inferred that the choice of the B3LYP/6-31G(d,p) method for QM region geometric optimizations is appropriate in the present study.

3.1 Reaction mechanism and potential energy profiles

The one-step dehalogenation reaction of FAcD toward FAc was shown in Scheme 1. The reaction is triggered by a negatively charged residue Asp110. Asp110 acts as a nucleophile and attacks C_δ atom of substrate FAc, which eventually lead to the C–F bond cleavage and F ion elimination, similar with the previously proposed dehalogenation mechanism.[14] However, the binding mode of substrate FAc is different, as indicated in Fig. S1(a–d) and Table S1, ESI.† In addition, a water molecule was found to stabilize the leaving F ion through a hydrogen bond, which has not been reported.[14] It is likely that this water molecule is crucial in the ejection process of the F ion, as indicated in Fig. S1(e), ESI.† More structural analysis on the dehalogenation itineraries will be discussed in the following section. For system $FAcD_{Hsd155}$–FAc, a substantial energy barrier spread, 12.5–26.8 kcal mol^{-1}, among twenty different snapshots has been found. Similar substantial energy barrier spreads for systems $FAcD_{Hse155}$–FAc (9.7–21.5 kcal mol^{-1}) and $FAcD_{Hse155}$–ClAc (13.0–23.6 kcal mol^{-1}) were also found. By assuming that each snapshot extracted from the dynamics trajectory corresponds to a local rate constant,[41] these calculated energy barrier fluctuations may be helpful in rationalizing recent single molecule experiment findings that the reaction rate of a single enzyme molecule is not constant but exhibits large fluctuations with a broad range.[19,20,22,39]

The Boltzmann-weighted average barriers, energy barrier spreads, and the disproportionate effects for the three systems were provided in Table 1. The detailed barriers and imaginary frequencies for each reaction pathway were provided in Table 2. No anomalously low energy barriers were found for all the three systems, as indicated by the low value of disproportionate effects (2.9%, 8.8%, and 4.8%). The Boltzmann-weighted average barrier of system $FAcD_{Hsd155}$–FAc is 13.8 kcal mol^{-1}, which is 2.4 kcal mol^{-1} higher than that of system $FAcD_{Hse155}$–FAc. This implies that $FAcD_{Hse155}$ structure is slightly feasible than $FAcD_{Hsd155}$. By analyzing the energy barriers of twenty different snapshots, about 70% of the barriers in system $FAcD_{Hsd155}$–FAc were found to be higher than the barriers in corresponding snapshots in system $FAcD_{Hse155}$–FAc, as shown in Table 2. Although it is credible at a relatively high ratio (about 70%) in predicting the feasibility of competitive pathways using a single snapshot, errors may also occur. For example, if only snapshot 6 ns is used in distinguishing the competitive pathways, error occurs: $FAcD_{Hsd155}$ (ΔE = 14.6 kcal mol^{-1}) may seem more feasible than $FAcD_{Hse155}$ (ΔE = 19.3 kcal mol^{-1}).

The following dehalogenation investigations toward substrate FAc and ClAc were mainly investigated on the basis of structure $FAcD_{Hse155}$ because it is energetically feasible than structure $FAcD_{Hsd155}$. The Boltzmann-weighted average barrier of system $FAcD_{Hse155}$–FAc (11.4 kcal mol^{-1}) is 3.1 kcal mol^{-1} lower than that of system $FAcD_{Hse155}$–ClAc (14.5 kcal mol^{-1}), which indicates that defluorination is more feasible than dechlorination. Interestingly, gas phase calculations (without protein environment) performed at the RIMP2/cc-pVTZ//B3LYP/6-31G(d,p) level using Gaussian 09 program[42] showed that energy barriers for defluorination (105.8 kcal mol^{-1}) is 33.1 kcal mol^{-1} higher than that of dechlorination (72.7 kcal mol^{-1}) (Scheme S1, ESI†). By considering the contribution from side chains of Arg111 and Arg114, significant lower barriers were found for defluorination (37.7 kcal mol^{-1}) and dechlorination (18.1 kcal mol^{-1}). This highlights the importance of residues Arg111 and Arg114 in dehalogenation reactions. However, residues Arg111 and Arg114 are not responsible for the fact that FAcD prefers defluorination (11.4 kcal mol^{-1}) rather than dechlorination (14.5 kcal mol^{-1}). Discussions on this issue will be provided in detail in the following paragraphs through both structural and energetic aspects.

Table 1 Energy barrier spreads, Boltzmann weighted average barriers, and disproportionate effects calculated at RIMP2/cc-pVTZ//CHARMM22 level for the six studied systems

Systems	Barrier spreads/ (kcal mol^{-1})	Boltzmann-weighted average barriers/ (kcal mol^{-1})	Disproportionate effects
$FAcD_{Hsd155}$–FAc	12.5–26.8	13.8	2.9%
$FAcD_{Hse155}$–FAc	9.7–21.5	11.4	8.8%
$FAcD_{Hse155}$–ClAc	13.0–23.6	14.5	4.8%

3.2 Dehalogenation itineraries

Among all the twenty studied snapshots, six snapshots with lowest energy barriers in systems $FAcD_{Hse155}$–FAc (0.5 ns, 1.5

Table 2 Energy barriers and imaginary frequencies for twenty snapshots of systems $FAcD_{Hsd155}$–FAc, $FAcD_{Hse155}$–FAc, and $FAcD_{Hse155}$–FAc calculated at RIMP2/cc-pVTZ//CHARMM22 level

Systems	$FAcD_{Hsd155}$–FAc		$FAcD_{Hse155}$–FAc		$FAcD_{Hse155}$–ClAc	
Snapshots/ns	Barrier/ (kcal mol^{-1})	Imaginary frequency/(cm^{-1})	Barrier/ (kcal mol^{-1})	Imaginary frequency/(cm^{-1})	Barrier/ (kcal mol^{-1})	Imaginary frequency/(cm^{-1})
0.5	12.5	315i	14.3	294i	15.4	213i
1	18.9	312i	18.6	327i	17.8	239i
1.5	19.6	317i	15.1	298i	20.4	211i
2	22.6	329i	20.2	302i	17.6	251i
2.5	19.1	307i	10.7	330i	19.2	226i
3	21.7	297i	20.5	281i	17.4	215i
3.5	21.4	311i	16.0	309i	19.8	209i
4	20.7	321i	9.7	277i	13.6	199i
4.5	19.8	299i	15.5	289i	13.0	224i
5	22.1	329i	17.8	297i	18.5	231i
5.5	26.8	311i	21.9	301i	23.3	224i
6	14.6	300i	19.3	276i	14.8	218i
6.5	12.7	285i	15.0	296i	23.6	212i
7	22.6	333i	16.5	306i	14.6	217i
7.5	27.9	311i	15.3	292i	23.4	282i
8	13.1	352i	16.9	314i	18.0	176i
8.5	16.2	305i	16.5	309i	19.4	274i
9	18.4	308i	16.2	295i	21.0	247i
9.5	25.3	331i	21.5	308i	17.2	243i
10	25.4	328i	20.3	286i	20.0	252i

ns, 2.5 ns, 4 ns, 6.5 ns, and 7.5 ns) and FAcD$_{Hse155}$–ClAc (0.5 ns, 4 ns, 4.5 ns, 6 ns, 7 ns, and 9.5 ns) were chosen for the following dehalogenation itinerary investigations. The variations of two crucial geometry parameters, angle $O_\varepsilon C_\delta X$ and dihedral $O_\omega C_\gamma C_\delta O_\varepsilon$, along the dehalogenation processes (indicated by bond C_δ–X increase) were provided in Fig. 1. For a more direct view, the spatial locations of active site residues in the structures of reactants, transition states and products for systems FAcD$_{Hse155}$–FAc (4 ns, ΔE = 9.7 kcal mol^{-1}) and FAcD$_{Hse155}$–ClAc (4.5 ns, ΔE = 13.0 kcal mol^{-1}) were representatively displayed in Fig. 2 and Fig. S3,† respectively. Fig. 2 shows that residues Arg111 and Arg114 provide hydrogen network stabilization for the carboxy group of FAc or ClAc, whereas residues His155, Trp156, and Tyr219 provide stabilization for F or Cl. As shown in Fig. 1, the calculated C_δ–X bond distances in the reactant structures of systems FAcD$_{Hse155}$–FAc (1.42–1.44 Å) and FAcD$_{Hse155}$–ClAc (1.84–1.85 Å), and the calculated dihedral $O_\omega C_\gamma C_\delta O_\varepsilon$ in the products of system FAcD$_{Hse155}$–FAc (162.7–175.4°) are all in promising

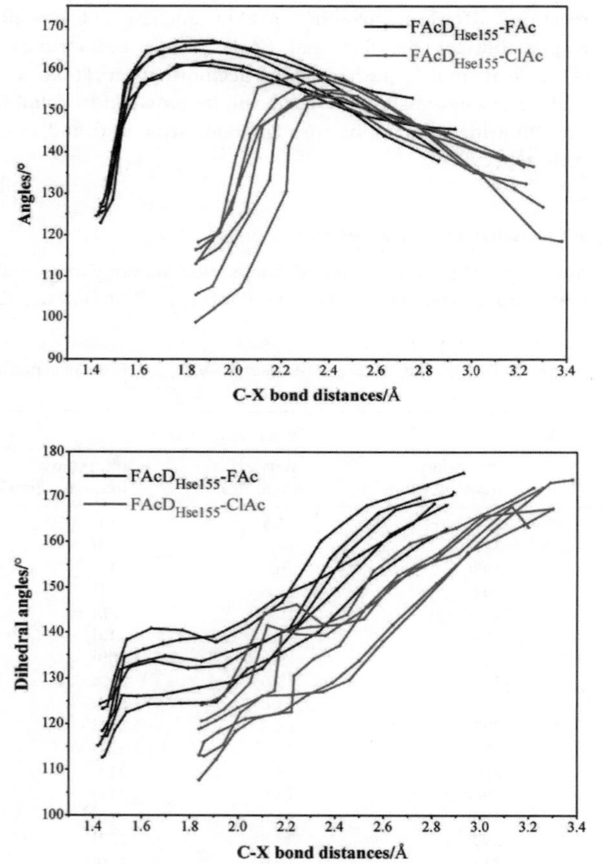

Fig. 1 Variations of angles $O_\varepsilon C_\delta F$ and $O_\varepsilon C_\delta Cl$ and dihedrals $O_\omega C_\gamma C_\delta O_\omega$ for six snapshots with the lowest energy barriers in systems FAcD$_{Hse155}$–FAc (0.5 ns, 1.5 ns, 2.5 ns, 4 ns, 6.5 ns, and 7.5 ns) and FAcD$_{Hse155}$–ClAc (0.5 ns, 4 ns, 4.5 ns, 6 ns, 7 ns, and 9.5 ns).

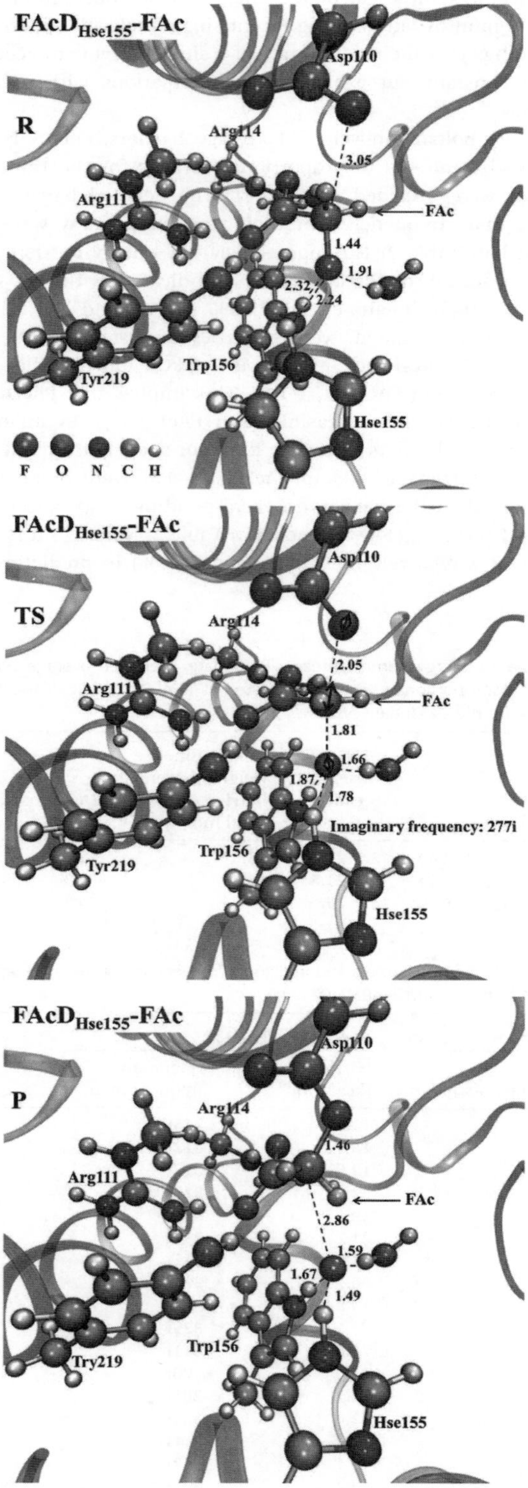

Fig. 2 Structures of reactant (R), transition state (TS), and product (P) involved in the defluorination process of system FAcD$_{Hse155}$–FAc at snapshot 4 ns. The unit for bond distances and imaginary frequency are in Å and cm^{-1}.

agreement with the available crystal data (1.42 Å, 1.79 Å, and 172.2°, respectively).[16] The angles of $O_\varepsilon C_\delta F$ and $O_\varepsilon C_\delta Cl$ in the transition states locate at the range of 161.0–166.8° and 149.3–157.9°, which are slightly deviated from the theoretical value (180°) for an S_N2 reaction. Another interesting issue is the variation of dihedral $O_\omega C_\gamma C_\delta O_\varepsilon$. Previous *ab initio* calculations in free solutions indicate an orthogonal direction (~90°) of the dihedral $O_\omega C_\gamma C_\delta O_\varepsilon$ during the dehalogenation process, whereas the crystal data of the product (3R3Y, resolution 1.15 Å) indicate a nearly coplanar dihedral $O_\omega C_\gamma C_\delta O_\varepsilon$ (172.2°).[43]

To get a more comprehensive understanding, more analysis on the dehalogenation process were performed, and a dynamic property of dihedral $O_\omega C_\gamma C_\delta O_\varepsilon$ during the dehalogenation processes was found. For example, $O_\omega C_\gamma C_\delta O_\varepsilon$ varies from 112.6–124.4° (reactants) to 125.1–138.6° (transition states) and finally to 162.7–175.4° (products) during defluorination processes by enzyme FAcD. In addition, the natural population analysis (NPA) on systems $FAcD_{Hse155}$–FAc and $FAcD_{Hse155}$–ClAc were performed and the natural charge variations are provided in Table S2, ESI.† The natural charges of the halide atoms in two systems are significantly different: natural charges of atom F changes from −0.43 ± 0.02 to −0.73 ± 0.04, whereas the natural charges of atom Cl changes from −0.18 ± 0.02 to −0.89 ± 0.02 during the dehalogenation processes. The natural charges of halide ions in the products indicate a better stabilization of FAcD toward F (−0.73 ± 0.04) than Cl (−0.89 ± 0.02).

3.3 Potential energy profiles *versus* key structural parameters

To gain a more comprehensive understanding between potential energy profiles and structural parameters, twenty energy barriers in both defluorination and dechlorination reactions as a function of the corresponding angle $O_\varepsilon C_\delta X$ (X = F or Cl) variations (from reactants to transition states), the values of angle $O_\varepsilon C_\delta X$ in the reactants, and the values of angle $O_\varepsilon C_\delta X$ in the transition states were provided, as shown in Fig. 3. Although it is not possible to establish a precise correlation, in some way the barriers tend to increase as the angle $O_\varepsilon C_\delta X$ variations becomes larger (Fig. 3a). Because distribution ranges of angle $O_\varepsilon C_\delta X$ in the transition states (within 10°) are about two or three times narrower than that in the reactants, the established barrier increasing tendency was mainly associated with the value of angle $O_\varepsilon C_\delta X$ in reactants, as shown in Fig. 3b and c. This may at least provide one suggestion for biomimetic catalyst or *de novo* enzyme designing in enhancing the C–F or C–Cl bond cleavage: try to increase the value of angle $O_\varepsilon C_\delta X$ in the reactant structures. The relatively smaller angles of $O_\varepsilon C_\delta Cl$ (91.2–118.3°) compared with $O_\varepsilon C_\delta F$ (119.5–132.5°) in the reactants were mainly due to the improper binding of ClAc in the smaller active site pocket designed for accommodating the natural substrate (FAc) of FAcD. This highlights the importance of residues Arg111, Arg114, His155, Trp156, and Tyr219 during the dehalogenation processes. For example, mutations of Arg111Lys or His155Asn may change the topology of the active site and increase the angle values. Plots displaying the potential energy barriers *versus* different structural parameters (such as dihedral $O_\omega C_\gamma C_\delta O_\varepsilon$ and bond $O_\varepsilon C_\delta$) are provided

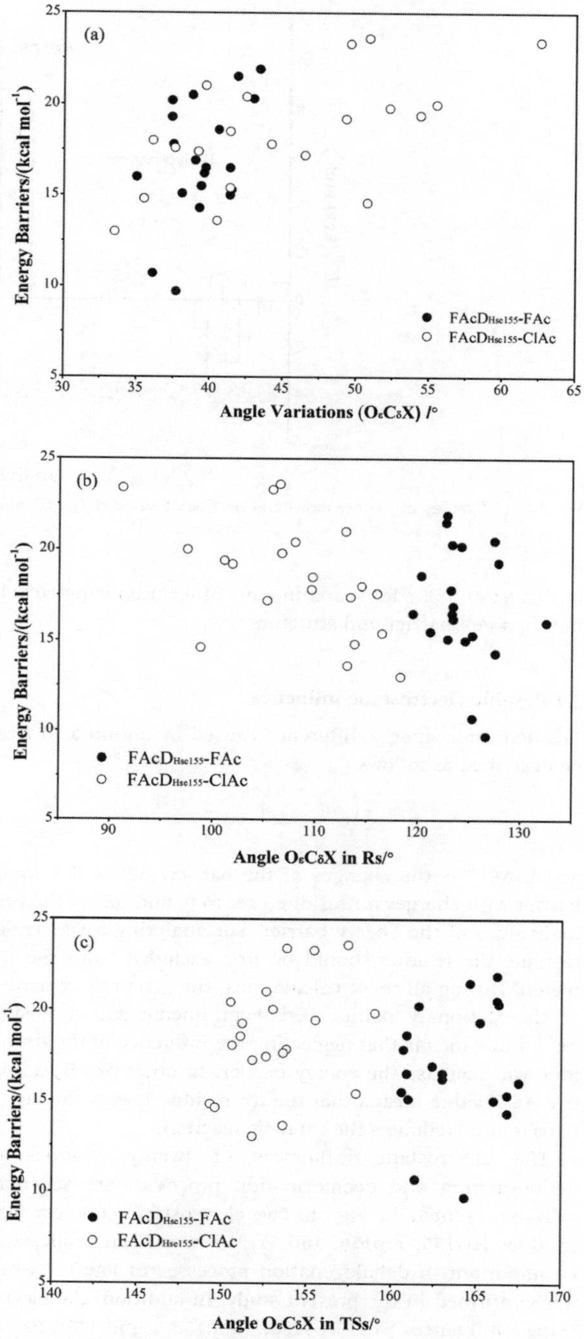

Fig. 3 a, Potential energy barriers *versus* angle $O_\varepsilon C_\delta X$ variations (X means F for $FAcD_{Hse155}$–FAc and Cl for $FAcD_{Hse155}$–ClAc); b, potential energy barriers *versus* values of angle $O_\varepsilon C_\delta X$ in reactants; c, potential energy barriers *versus* values of angle $O_\varepsilon C_\delta X$ in transition states.

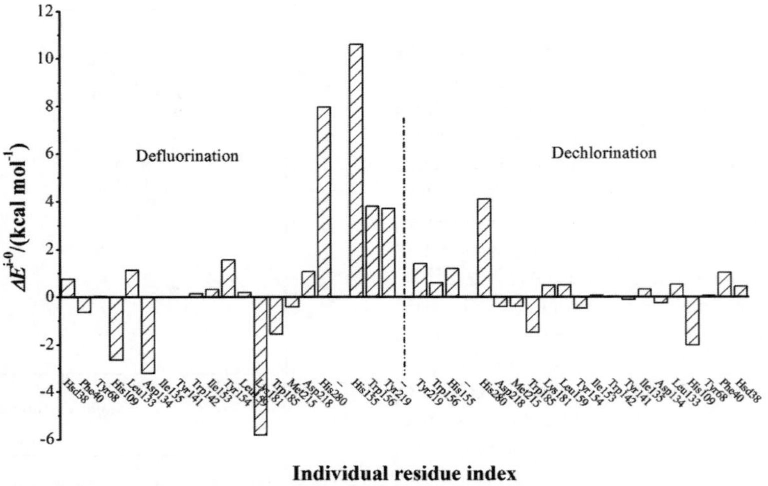

Fig. 4 ΔE^{i-0} values of twenty individual residues toward defluorination and dechlorination processes of FAcD.

in Fig. S4 and S5† for searching any other promising correlation between barrier and structure.

3.4 Residue electrostatic influence

The activation energy difference caused by amino acid i can be described as follows:

$$\Delta E^{i-0} = \Delta E^{i} - \Delta E^{0}$$

where ΔE^{i-0} is the changes of the barrier, ΔE^{i} is the energy barrier with charges on residue i set to 0, and ΔE^{0} is the original values of the energy barrier. For analyzing a QM region residue, the residue should be first excluded from the QM region. During all these calculations, the geometry structures of the stationary points were kept unchanged. A positive ΔE^{i-0} value means that neglecting the influence of the ith residue will increase the energy barrier. In other words, a positive ΔE^{i-0} value means that the ith residue lowers the energy barrier and facilitates the catalytic reaction.

The electrostatic influences of twenty residues on defluorination and dechlorination processes are schematically represented in Fig. 4. The electrostatic contacts from residues His155, Trp156, and Tyr219 have been proposed to be important in dehalogenation processes of FAcD,[14] which was confirmed in the present study. In addition, the electrostatic influences of His155, Trp156, and Ty219 on defluorination are considerably stronger than on dechlorination. Our analysis also highlights four residues (His109, Asp134, Lys181, and His280) for defluorination reactions and two residues (His109 and His280) for dechlorination reactions. These residues have a strong electrostatic influence on the reaction barrier (-2.0 kcal mol^{-1} $< \Delta E^{i-0} <$ 2.0 kcal mol^{-1}) and may serve as candidate residues for the following mutation studies. The other residues were found to have relatively weaker electrostatic influence on the reaction barrier.

4. Conclusions

By analyzing the energy barriers of twenty snapshots and comparing the Boltzmann weighted average barriers, we proved that structure $FAcD_{Hse155}$ is more energetically feasible than structure $FAcD_{Hsd155}$ for enzyme FAcD, whereas $FAcD_{Hse155}$ prefers defluorination rather than dechlorination process. A positive correlation between energy barriers and key structural parameter (angle $O_{\varepsilon}C_{\delta}X$) was found. This may help biomimetic catalyst or *de novo* enzyme designing in enhancing the C–F or C–Cl bond cleavage. Besides residues Arg111, Arg114, His155, Trp156, and Tyr219, the important role of residues His109, Asp134, Lys181, and His280 during the defluorination process were also highlighted. In addition, we found that conformational variations may cause different enzymatic preferences toward competitive pathways. Thus, studying only one snapshot in distinguishing competitive reaction pathways is not reliable.

Acknowledgements

The study was financially supported by NSFC (National Natural Science Foundation of China, project no. 21337001, 21577082, 21507073 and 21177077), SPNSFC (Shandong Provincial Natural Science Foundation, China, project no. ZR2015PB002), FRFSU (Fundamental Research Funds of Shandong University, project no. 2015GN007), and GFGCPSF (General Financial Grant from the China Postdoctoral Science Foundation, project no. 2015M570594).

References

1 P. Goldman, *Science*, 1969, **164**, 1123–1130.
2 K. Muller, C. Faeh and F. Diederich, *Science*, 2007, 317, 1881–1886.
3 B. D. Key, R. D. Howell and C. S. Criddle, *Environ. Sci. Technol.*, 1997, **31**, 2445–2454.

4 C. Douvris and O. V. Ozerov, *Science*, 2008, **321**, 1188–1190.
5 A. M. Calafat, L. Y. Wong, Z. Kuklenyik, J. A. Reidy and L. L. Needham, *Environ. Health Perspect.*, 2007, **115**, 1596–1602.
6 M. Houde, J. W. Martin, R. J. Letcher, K. R. Solomon and D. C. Muir, *Environ. Sci. Technol.*, 2006, **40**, 3463–3473.
7 X. Zhang, T. Lai and Y. Kong, *Top. Curr. Chem.*, 2012, **308**, 365–404.
8 A. T. Proudfoot, S. M. Bradberry and J. A. Vale, *Toxicol. Rev.*, 2006, 25, 213–219.
9 J. Q. Liu, T. Kurihara, S. Ichiyama, M. Miyagi, S. Tsunasawa, H. Kawasaki, K. Soda and N. Esaki, *J. Biol. Chem.*, 1998, 273, 30897–30902.
10 T. Kurihara and N. Esaki, *Chem. Rec.*, 2008, 8, 67–74.
11 T. Kurihara, T. Yamauchi, S. Ichiyama, H. Takahata and N. Esaki, *J. Mol. Catal. B: Enzym.*, 2003, 23, 347–355.
12 Y. Zhang, Z. S. Li, J. Y. Wu, M. Sun, Q. C. Zheng and C. C. Sun, *Biochem. Biophys. Res. Commun.*, 2004, 325, 414–420.
13 K. Jitsumori, R. Omi, T. Kurihara, A. Kurata, H. Mihara, I. Miyahara, K. Hirotsu and N. Esaki, *J. Bacteriol.*, 2009, **191**, 2630–2637.
14 T. Kamachi, T. Nakayama, O. Shitamichi, K. Jitsumori, T. Kurihara, N. Esaki and K. Yoshizawa, *Chem. – Eur. J.*, 2009, **15**, 7394–7403.
15 P. W. Y. Chan, M. Wong, J. Guthrie, A. V. Savchenko, A. F. Yakunin, E. F. Pai and E. A. Edwards, *Microb. Biotechnol.*, 2010, 3, 107–120.
16 P. W. Y. Chan, A. F. Yakunin, E. A. Edwards and E. F. Pai, *J. Am. Chem. Soc.*, 2011, 133, 7461–7468.
17 T. Nakayama, T. Kamachi, K. Jitsumori, R. Omi, K. Hirotsu, N. Esaki, T. Kurihara and K. Yoshizawa, *Chem. – Eur. J.*, 2012, **18**, 8392–8402.
18 C. K. Davis, R. I. Webb, L. I. Sly, S. E. Denman and C. S. McSweeney, *FEMS Microbiol. Ecol.*, 2012, **80**, 671–684.
19 H. P. Lu, L. Xun and X. S. Xie, *Science*, 1998, **282**, 1877–1882.
20 W. Min, B. P. English, G. Luo, B. J. Cherayil, S. C. Kou and X. S. Xie, *Acc. Chem. Res.*, 2005, **38**, 923–931.
21 H. Yang, G. Luo, P. Karnchanaphanurach, T. M. Louie, I. Rech, S. Cova, L. Xun and X. S. Xie, *Science*, 2003, **302**, 262–266.
22 X. S. Xie, *Science*, 2013, **342**, 1457–1459.
23 A. Warshel, P. K. Sharma, M. Kato, Y. Xiang, H. Liu and M. H. Olsson, *Chem. Rev.*, 2006, **106**, 3210–3235.
24 H. M. Senn and W. Thiel, *Angew. Chem., Int. Ed.*, 2009, **48**, 1198–1229.
25 G. G. Hammes, S. J. Benkovic and S. Hammes-Schiffer, *Biochemistry*, 2011, **50**, 10422–10430.
26 R. Lonsdale, J. N. Harvey and A. J. Mulholland, *Chem. Soc. Rev.*, 2012, **41**, 3025–3038.
27 B. R. Brooks, R. E. Bruccoleri, B. D. Olafson, D. J. States, S. Swaminathan and M. Karplus, *J. Comput. Chem.*, 1983, **4**, 187–217.
28 B. R. Brooks, C. L. Brooks III, A. D. Mackerell, L. Nilsson, R. J. Petrella, B. Roux, Y. Won, G. Archontis, C. Bartels, S. Boresch, A. Caflisch, L. Caves, Q. Cui, A. R. Dinner, M. Feig, S. Fischer, J. Gao, M. Hodoscek, W. Im, K. Kuczera, T. Lazaridis, J. Ma, V. Ovchinnikov, E. Paci, R. W. Pastor, C. B. Post, J. Z. Pu, M. Schaefer, B. Tidor, R. M. Venable, H. L. Woodcock, X. Wu, W. Yang, D. M. York and M. Karplus, *J. Comput. Chem.*, 2009, **30**, 1545–1615.
29 A. D. MacKerell, B. Brooks, C. L. Brooks, L. Nilsson, B. Roux, Y. Won and M. Karplus, *Encycl. Comput. Chem.*, John Wiley & Sons, New York, 1998, pp. 271–277.
30 W. L. Jorgensen, J. Chandrasekhar, J. D. Madura, R. W. Impey and M. L. Klein, *J. Chem. Phys.*, 1983, 79, 926–935.
31 C. L. Brooks and M. Karplus, *J. Chem. Phys.*, 1983, **79**, 6312–6325.
32 P. Sherwood, A. H. D. Vries, M. F. Guest, G. Schreckenbach, C. R. A. Catlow, S. A. French, A. A. Sokol, S. T. Bromley, W. Thiel, A. J. Turner, S. Billeter, F. Terstegen, S. Thiel, J. Kendrick, S. C. Rogers, J. Casci, M. Watson, F. King, E. Karlsen, M. Sjovoll, A. Fahmi, A. Schafer and C. Lennartz, *J. Mol. Struct.: THEOCHEM*, 2003, **632**, 1–28.
33 R. Ahlrichs, M. Bär, M. Häser, H. Horn and C. Kölmel, *Chem. Phys. Lett.*, 1989, **162**, 165–169.
34 W. Smith and T. R. Forester, *J. Mol. Graphics*, 1996, **14**, 136–141.
35 A. H. de Vries, P. Sherwood, S. J. Collins, A. M. Rigby, M. Rigutto and G. J. Kramer, *J. Phys. Chem. B*, 1999, **103**, 6133–6141.
36 D. Bakowies and W. J. Thiel, *J. Phys. Chem.*, 1996, **100**, 10580–10594.
37 S. R. Billeter, A. J. Turner and W. Thiel, *Phys. Chem. Chem. Phys.*, 2000, **2**, 2177–2186.
38 R. Lonsdale, J. N. Harvey and A. J. Mulholland, *J. Phys. Chem. B*, 2010, **114**, 1156–1162.
39 Y. Li, X. Shi, Q. Zhang, J. Hu, J. Chen and W. Wang, *Environ. Sci. Technol.*, 2014, **48**, 5008–5016.
40 Y. Li, R. Zhang, L. Du, Q. Zhang and W. Wang, *RSC Adv.*, 2015, 5, 13871–13877.
41 P. Saura, R. Suardíaz, L. Masgrau, J. M. Lluch and À. González-Lafont, *ACS Catal.*, 2014, **4**, 4351–4363.
42 M. J. Frisch, G. W. Trucks, H. B. Schlegel, G. E. Scuseria, M. A. Robb, J. R. Cheeseman, G. Scalmani, V. Barone, B. Mennucci, G. A. Petersson, H. Nakatsuji, M. Caricato, X. Li, H. P. Hratchian, A. F. Izmaylov, J. Bloino, G. Zheng, J. L. Sonnenberg, M. Hada, M. Ehara, K. Toyota, R. Fukuda, J. Hasegawa, M. Ishida, T. Nakajima, Y. Honda, O. Kitao, H. Nakai, T. Vreven, J. A. Montgomery Jr., J. E. Peralta, F. Ogliaro, M. Bearpark, J. J. Heyd, E. Brothers, K. N. Kudin, V. N. Staroverov, R. Kobayashi, J. Normand, K. Raghavachari, A. Rendell, J. C. Burant, S. S. Iyengar, J. Tomasi, M. Cossi, N. Rega, J. M. Millam, M. Klene, J. E. Knox, J. B. Cross, V. Bakken, C. Adamo, J. Jaramillo, R. Gomperts, R. E. Stratmann, O. Yazyev, A. J. Austin, R. Cammi, C. Pomelli, J. W. Ochterski, R. L. Martin, K. Morokuma, V. G. Zakrzewski, G. A. Voth, P. Salvador, J. J. Dannenberg, S. Dapprich, A. D. Daniels, O. Farkas, J. B. Foresman, J. V. Ortiz, J. Cioslowski and D. J. Fox, *Gaussian 09, revision A.02*; Gaussian, Inc., Wallingford, CT, 2009.
43 R. D. Bach, B. A. Coddens and G. J. Wolber, *J. Org. Chem.*, 1986, **51**, 1030–1033.

Environmental Science & Technology

Article

pubs.acs.org/est

Computational Evidence for the Detoxifying Mechanism of Epsilon Class Glutathione Transferase Toward the Insecticide DDT

Yanwei Li, Xiangli Shi, Qingzhu Zhang,* Jingtian Hu, Jianmin Chen, and Wenxing Wang

Environment Research Institute, Shandong University, Jinan 250100, P. R. China

Ⓢ *Supporting Information*

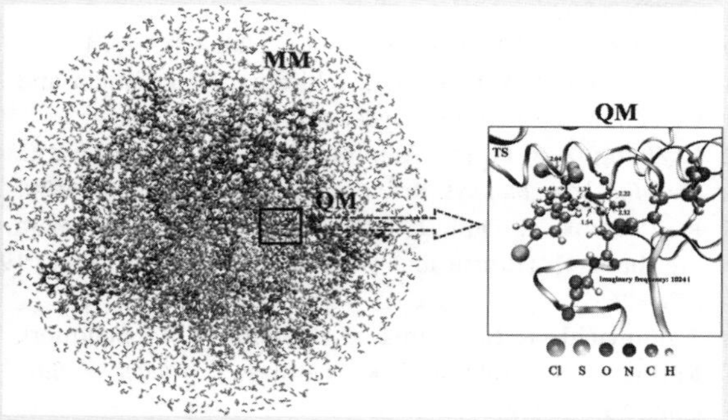

ABSTRACT: A combined quantum mechanics/molecular mechanics (QM/MM) computation of the detoxifying mechanism of an epsilon class glutathione transferases (GSTs) toward organochlorine insecticide DDT, 1,1,1-trichloro-2,2-bis(p-chlorophenyl)-ethane, has been carried out. The exponential average barrier of the proton transfer mechanism is 15.2 kcal/mol, which is 27.6 kcal/mol lower than that of the GS-DDT conjugant mechanism. It suggests that the detoxifying reaction proceeds via a proton transfer mechanism where GSH acts as a cofactor rather than a conjugate. The study reveals that the protein environment has a strong effect on the reaction barrier. The experimentally proposed residues Arg112, Glu116 and Phe120 were found to have a strong influence on the detoxifying reaction. The influence of residues Pro13, Cys15, His53, Ile55, Glu67, Ser68, Phe115, and Leu119 was detected as well. It is worth noticing that Ile55 facilitates the detoxifying reaction most. On the basis of the structure of DDT, structure **2**, $(BrC_6H_4)_2CHCCl_3$, is the best candidate among all the tested structures in resisting the detoxification of enzyme agGSTe2.

1. INTRODUCTION

Glutathione transferases (GSTs) are a superfamily of multifunctional enzymes present ubiquitously in aerobic organisms, plants, and animals.[1−13] Their primary function is to detoxify endogenous and xenobiotic electrophiles by catalyzing the conjugation of these substances to glutathione (GSH), the tripeptide γ-Glu-Cys-Gly.[11,14−16] The resultant products are more water-soluble and excretable than the non-GSH conjugated substrates.[17] In addition, GSTs are involved in a wide range of biological processes. They play an important role in protecting the cells from the harmful effects of oxidative stress and have been implicated in various biosynthetic pathways.[18−23] GSTs in insects are of particular interest because of their potential role in insecticide metabolism producing resistance.[24−27] Organochlorine insecticide DDT, 1,1,1-trichloro-2,2-bis(p-chlorophenyl)ethane, plays a prominent role in the malaria (the most severe insect transmitted disease with at least 300 million cases per year and 1.4 million deaths) eradication by controlling the population of malaria vector mosquitoes since 1950s.[28−30] However, resistance to DDT has developed in some mosquito species (e.g., *Anopheles Gambiae*), raising the threat of malaria to humans.[31,32] A mechanistic understanding of the detoxifying of GSTs toward DDT is critically warranted to understand the impact of DDT resistance and to develop more effective novel insecticides.[33]

At least six classes of insect GSTs have been identified: delta, epsilon, omega, sigma, theta, and zeta, being possibly the existence of novel GST classes.[34,35] The delta and epsilon classes are the major GSTs in insects. The GSTs from the malaria vector *Anopheles Gambiae* have been extensively studied.[30−32,36−43] Twenty eight GSTs from *Anopheles Gambiae* have been identified, in which 12 belong to the deta class and 8 to the epsilon class.[38] Although the DDT-detoxifying ability of the deta class GSTs was proved, they are not considered important in conferring DDT resistance based on the observations that none of them are overexpressed in DDT resistant strains.[39,40] However, five of the eight epsilon members have elevated expression level. Among them, agGSTe2 is overexpressed about 8-fold and represents up to 92% DDT metabolism.[41]

The cytosolic GSTs are homo or heterodimeric proteins, that is, they are formed by two subunits or polypeptide chains of approximately 25 kDa in size each. Each subunit has a *G-site* where glutathione (GSH) substrate binds and an *H-site* pocket for electrophilic substrates. When GSH binds to the *G-site*, it is activated to a negatively charged thiolate anion (GS^-) under the help of a water molecule.[11,44−49] GS^- is stabilized by a hydrogen-bond-network which distributes the negative charge to the surrounding networked atoms.[43] The nucleophilic thiolate anion is capable of attacking the electrophilic center of the lipophilic compounds. Generally, the GSTs catalyze the detoxification reactions of halogenated hydrocarbons via proton transfer and/or glutathione conjugation.[21,50,51] The proton

Received: November 25, 2013
Revised: March 2, 2014
Accepted: March 28, 2014
Published: March 28, 2014

Scheme 1. Reaction Pathways of the Proton Transfer Mechanism and the GS-DDT Conjugation Mechanism of agGSTe2 with DDT[a]

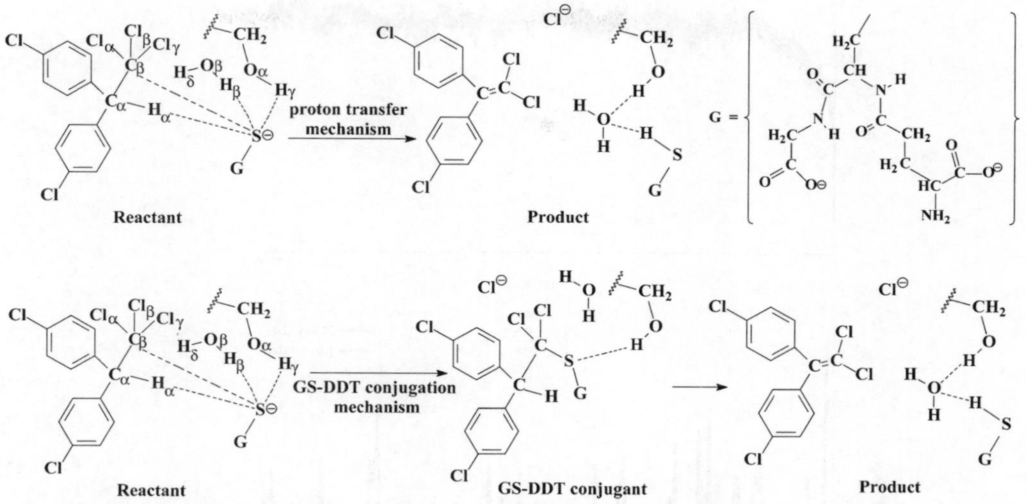

[a]The boundary between the QM and MM regions is indicated by wavy lines.

transfer is considered to be one of the possible DDT resistance mechanism in mosquitoes based on the direct detection of DDE by mixing agGSTe2 and DDT in the presence of GSH, coupled with the fact that the GS-DDT conjugant has never been identified.[51] However, there is a possibility that the GS-DDT conjugant is unstable and it may quickly decompose to produce DDE. Thus, more evidence is required to clarify the DDT detoxifying mechanism.

The crystal structures of agGSTe2 and agGSTe2−GSH binary complex were determined with a resolution up to 1.4 Å.[43] However, attempts to cocrystallize the substrate DDT with agGSTe2 as well as the metabolism product DDE with agGSTe2−GSH complex were not successful.[43] The comparison with the less active delta GST, agGST1−6, from the same mosquito vector indicates that in the structure of agGSTe2−GSH complex, residues Arg112, Glu116 and Phe120 is closer to the *G-site* resulting in a more efficient GS^--stabilizing hydrogen-bond-network and higher DDT-binding affinity.[43,52] However, the influence of residues Arg112, Glu116, and Phe120 as well as other key residues surrounding the active site on the DDT detoxifying reaction is still unknown. In the present work, the detoxifying mechanism of agGSTe2 toward DDT has been carried out with the aid of a combined quantum mechanics/molecular mechanics (QM/MM) computations. QM/MM computations of the enzyme-catalyzed reaction can provide the atomistic details of the enzyme mechanism and is therefore becoming an increasingly important tool to complement experimental enzyme chemistry. In order to compare with the detoxifying ability of agGSTe2, the detoxifying mechanism of agGST1−6 toward DDT was also investigated in this study.

2. MATERIALS AND METHODS

2.1. System Setup and Molecular Dynamics. The initial models for the present simulation were built on the basis of X-ray crystal structure of agGSTe2-GSH complex (PDB code: 2IMI) and agGST1−6−GSH complex (PDB code: 1PN9) obtained from the Protein Data Bank (www.rcsb.org). The protonation states of the ionizable residues were determined on the basis of the pK_a values obtained through the PROPKA procedure,[53] and the missing hydrogen atoms were complemented through the HBUILD facility in the CHARMM package.[54−56] DDT was built and docked with the agGSTe2−GS^- and agGST1−6−GS^- complexes through a grid-based receptor-flexible docking module (CDOCKER) by using Material Studio 4.4 and Discovery Studio 2.1 (Accelrys Software Inc.), and the best docking pose was identified by the lowest binding energy.[57,58] Then, the agGSTe2−GS^-−DDT and agGST1−6−GS^-−DDT ternary complexes were placed in a water sphere (TIP3P model[59]) with a diameter of 65 Å, which guarantees that the whole protein was completely solvated. Water molecules overlapping within 2.5 Å of the ternary complex were deleted. The whole system was neutralized with seven sodium ions. After that, the system was heated from absolute zero to 298.15 K in 50 ps, and a trajectory of 500 ps was computed to reach an equilibration state (1 fs/step). A 9 ns stochastic boundary molecular dynamics (SBMD) simulation with canonical ensemble (NVT, 298.15 K) was performed to mimic the aqueous environment.[60] During the SBMD simulations, the coordinates of GS^- were restrained to keep its positions consistent with the crystal structure. A stochastic buffer region was used and a deformable potential was imposed at the edge of the water sphere to mimic the effect of the water solvent. Leap-frog algorithm and Langevin dynamics implemented in the CHARMM package were applied during the simulation, and the friction coefficients of Langevin thermostat for protein and water molecules were set to 200.0 and 62.0 (corresponding unit is 1/ps). The snapshots of the system were saved every picosecond resulting in 9000 conformations. For the system of agGSTe2-GS^--DDT, seven snapshots were taken every 0.5 ns from 6 to 9 ns of the molecular dynamics run. They are labeled as SN-6.0, SN-6.5, SN-7.0, SN-7.5, SN-8.0, SN-8.5, and SN-9.0. For the system of agGST1−6−GS^-−DDT, three snapshots were taken from the MD run after 8.0, 8.5, and 9.0 ns (labeled as SS-8.0, SS-8.5, and SS-9.0). These structures serve as starting configurations for the following geometry optimizations and transition-state search.

2.2. QM/MM Calculations. The QM/MM calculations were performed with the aid of ChemShell[61] integrating

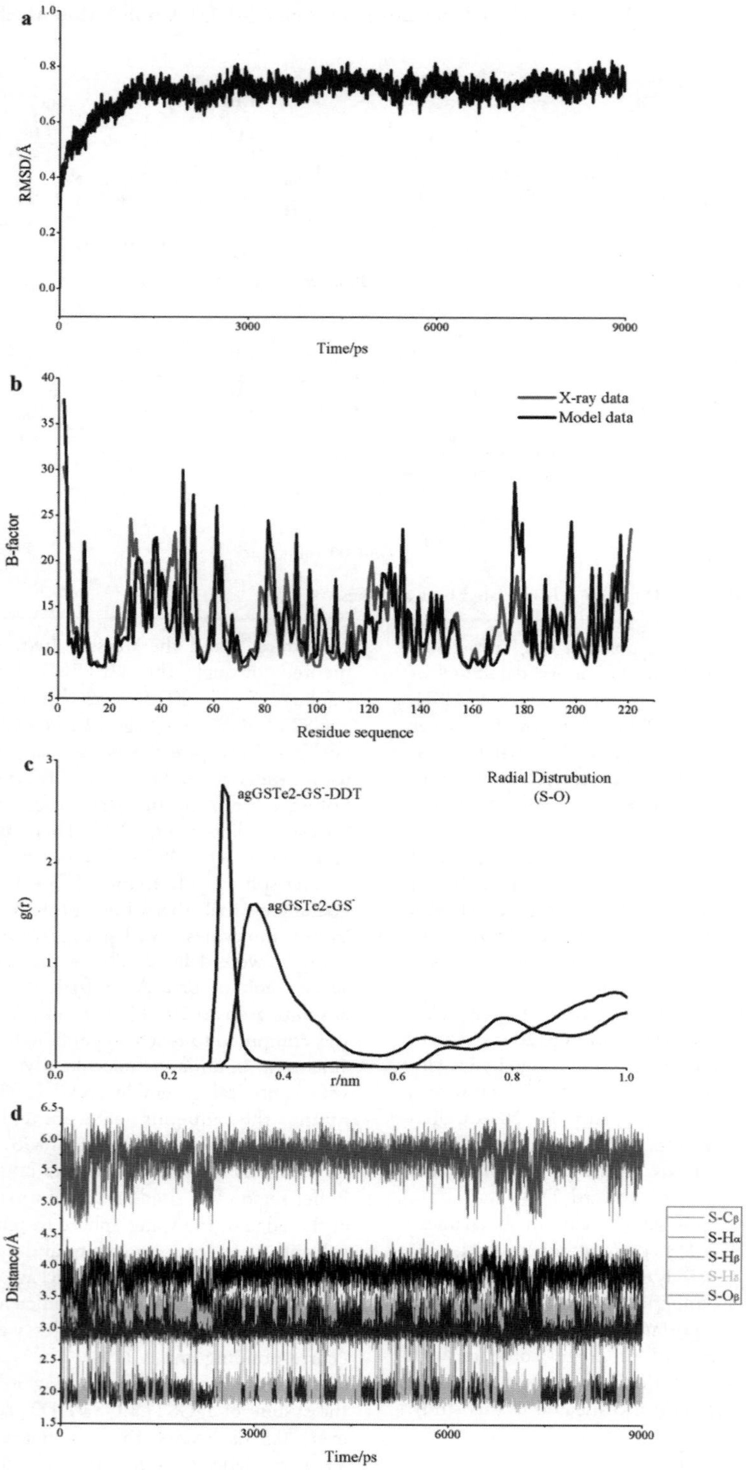

Figure 1. a, root-mean-square deviation for the system of agGSTe2-GS$^-$-DDT during 9,000 ps MD simulation. b, B-factors for the system of agGSTe2-GS$^-$–DDT (hydrogen atoms are not included). c, radial distribution of water oxygen atoms relative to sulfur atom in the systems of agGSTe2-GS$^-$ and agGSTe2-GS$^-$-DDT. d, distance variations of S–C$_\beta$, S–H$_\alpha$, S–H$_\beta$, S–H$_\delta$, and S–O$_\beta$ during 9,000 ps MD simulation.

Turbomole[62] and DL-POLY programs.[63] Hybrid delocalized internal coordinate (HDLC)[64] was adopted. The QM-region was treated by the DFT[65] method, while the CHARMM22 force field[66] was chosen for the MM-region. Covalent bonds between the QM-region and MM-region were truncated and modified by adding hydrogen link atoms to the QM side. A

Table 1. Energy Barriers and Selected QM/MM Bond Distances in the Reactant (R), Transition State (TS), and Product (P) Involved in the Proton Transfer Mechanism of agGSTe2 in Seven Pathways

		distances (Å)											
		$C_\alpha-H_\alpha$			$S-H_\alpha$			$C_\alpha-C_\beta$			$C_\beta-Cl_\alpha$		
	barriers (kcal/mol)	R	TS	P	R	TS	P	R	TS	P	R	TS	P
SN-6.0	28.2	1.09	1.76	4.08	4.23	1.48	1.34	1.56	1.42	1.34	1.79	2.17	2.90
SN-6.5	39.9	1.09	1.67	3.73	4.25	1.55	1.35	1.56	1.44	1.34	1.80	2.09	2.88
SN-7.0	20.5	1.09	1.67	4.21	4.07	1.42	1.35	1.55	1.53	1.34	1.80	2.21	2.99
SN-7.5	32.4	1.09	1.83	4.02	4.06	1.50	1.35	1.55	1.43	1.35	1.80	1.96	2.59
SN-8.0	21.3	1.09	1.68	4.27	4.28	1.53	1.35	1.55	1.46	1.35	1.80	1.91	3.52
SN-8.5	14.1	1.10	1.74	3.35	3.92	1.54	1.34	1.55	1.44	1.34	1.80	2.04	2.73
SN-9.0	21.1	1.09	1.75	3.15	4.14	1.49	1.34	1.55	1.43	1.34	1.80	2.10	2.92

charge shift model[67] was used to avoid over polarization of the QM density of the QM-region. The QM-region contains cofactor GS^-, DDT, a water molecule and part of Ser12. The truncation of Ser12 was done by cutting a C−C bond. This resulted in 71 QM atoms including a hydrogen link atom. The QM-region was optimized by the B3LYP/6-31G(d,p) method[68,69] with a charge of −3 and a spin multiplicity 1. The potential energy profile from the reactant to the product was scanned to determine the transition state structure. The corresponding structure of the highest energy point along the reaction path was extracted and further optimized by combining partitioned rational function optimizer (P-RFO) algorithm[70] and the low-memory Broyden-Fletcher-Goldfarb-Shanno (L-BFGS) algorithm.[71] Harmonic vibrational frequency calculations were performed to validate its transition state character. A larger basis set of cc-pVTZ, which is comparable to basis set 6-311++G(2df,2pd), was employed in single point energy calculations. The energies include ZPE corrections with unscaled frequencies obtained at the B3LYP/6-31G(d,p) level. The natural bond orbital (NBO) analysis was performed at the B3LYP/6-311++G(d,p)//CHARMM22 level to investigate the key atom charge variations during the reaction process.

3. RESULTS AND DISCUSSION

For convenience of description, several atoms in the QM-region are numbered, as presented in Scheme 1. The first step of this work is to identify the calculation method which not only is able to produce accurate theoretical results, but also is computationally feasible and economical for currently available hardware and software. Due to the absence of the X-ray crystal structure of the agGSTe2-GS^-−DDT ternary complex, it is difficult to make a direct comparison between the calculated results and the experimental data. Thus, we took two snapshots from the MD simulation as starting configurations of the agGSTe2−GS^- binary complex and optimized them at the B3LYP/6-31G(d,p)//CHARMM22 level. The calculated results agree well with the available experimental values. For example, the spatial distances between S and O_α (Scheme 1) are 3.24 and 3.26 Å, in agreement with the experimental value of 3.48 Å. The S−C bonds in GS^- are 1.82 and 1.82 Å, which match perfectly with the X-ray value of 1.82 Å. A larger basis set 6-311++G(d,p) was also used to evaluate the accuracy of the B3LYP/6-31G(d,p) level, and the structural parameter differences are within 2% (Supporting Information (SI) Table S1). So, it might be inferred that the B3LYP/6-31G(d,p) level is appropriate for the QM-region geometrical optimization.

3.1. Reaction Mechanism. The agGSTe2-GS^-−DDT ternary complex was extracted per picosecond during the 9,000 ps MD simulation. The corresponding root-mean-square deviation (RMSD) and B-factor analysis were provided in Figure 1a and 1b. The radial distribution function (RDF) analysis for the systems of agGSTe2−GS^- and agGSTe2-GS^-−DDT is given in Figure 1c. The water coordinate numbers for the systems of agGSTe2−GS^- and agGSTe2−GS^--DDT are about 1.6 and 5.4, which indicates that about four water molecules are excluded from the active site after DDT binding. Five distance variations ($S-C_\beta$, $S-H_\alpha$, $S-H_\beta$, $S-H_\delta$, and $S-O_\beta$) along the 9,000 ps trajectory were depicted in Figure 1d. The average $S-O_\beta$ distance (2.9 Å) and its small variation during the MD simulation indicate that a water molecule is involved in stabilizing the sulfur anion. Further analysis of $S-H_\beta$ or $S-H_\delta$ (~2 Å) suggests that the hydrogen bond is formed through either H_β or H_δ. $S-H_\alpha$ and $S-C_\beta$ distances are ~4.0 Å and ~5.5 Å, which indicates that ionized GS^- can attack either DDT α-hydrogen or β-carbon, resulting in two possible reaction mechanisms, the proton transfer mechanism and the GS-DDT conjugation mechanism, as shown in Scheme 1.

For the proton transfer mechanism, the DDT α-hydrogen transfers from DDT to GS^-. A substantial energy barrier spread from 14.1 to 39.9 kcal/mol is found among different snapshots (Table 1). This suggests that the protein environment has a strong effect on the reaction barrier. To analyze the computed results, average energy barriers of the seven snapshots were calculated by exponential average:[72,73]

$$\Delta E_{ea} = -RT\ln\left\{\frac{1}{n}\sum_{i=1}^{n}\exp\left(\frac{-\Delta E_i}{RT}\right)\right\}$$

Where, ΔE_{ea} is the average barrier, R is gas constant, n is the number of snapshots, ΔE_i is the energy barrier of path i, and T is the temperature. The calculated exponential average barrier of the proton transfer mechanism is 15.2 kcal/mol.

The feasibility of an alternative GS-DDT conjugation mechanism was also investigated. The structural details are provided in Figure S1. In this mechanism, GS^- attacks the β-carbon atom of DDT and results in a GS-DDT conjugant. The seven energy barriers calculated at the B3LYP/cc-pVTZ//CHARMM22 level vary from 41.6 to 67.9 kcal/mol with an exponential average barrier of 42.8 kcal/mol, which is 27.6 kcal/mol higher than that of the proton transfer mechanism (Figure S2). This shows that the proton transfer mechanism is more energetically feasible than the GS-DDT conjugation mechanism. This is supported indirectly by a fact that the experimentally determined specific activity of agGSTe2 toward DDT corresponds to an energy barrier ~17 kcal/mol,[41,42,74,75] which is in good agreement with the exponential average barrier of the proton transfer mechanism (15.2 kcal/mol) rather than that of the conjugation mechanism. As a result, only the proton transfer mechanism will be minutely investigated in the following study.

3.2. Catalytic Itinerary and Structural Details. For the proton transfer mechanism, lengths of bonds C_α-H_α, S−H_α, C_α-C_β and C_β-Cl_α in the reactant, transition state, and product obtained at the B3LYP/6-31G(d,p)//CHARMM22 level were provided in Table 1. Since a majority of the catalytic reactions proceed through the lowest energy barrier of 14.1 kcal/mol, the following investigations will mainly focus on the pathway SN-8.5.

Figure 2 shows the active site structures of the reactant, transition state and product for the pathway SN-8.5. In the reactant, negatively charged sulfur is stabilized by Ser12 and a water molecule (S−H_γ, 2.04 Å; S−H_β, 2.02 Å). Besides Ser12, other key residues which are spatially close to either DDT or GS^- were also depicted in SI Figure S3 and Figure S4. The effect of these residues on the reaction mechanism will be discussed below. For the transition state structure, its character is confirmed by the vibrational mode and the corresponding imaginary frequency (1024i cm^{-1}). In the product, the length of double bond C_α=C_β (1.34 Å) and the angles Cl_β-C_β-Cl_γ (114.8°), C_α-C_β-Cl_β (120.1°), and C_α-C_β-Cl_γ (123.6°) suggest the formation of DDE. It is worth mentioning that the chlorphenyl groups and the C_α=C_β bond are not in a same plane. The distance of C_β-Cl_α (2.73 Å) suggests the formation of a chloride anion. Figure 3 shows several selected bond length and atom charge variations along the pathway SN-8.5. The changes of the bond lengths of S−H_α, C_α-H_α, C_α-C_β, and C_β-Cl_α from the reactant to the product through the transition state indicate that the hydrogen transfer, the formation of the C_α=C_β double bond and the dechlorination process occur simultaneously. The anion character of Cl_α in the product was further confirmed by its negative charge (Cl_α, −0.87).

3.3. Hydrogen-Bond-Network and Individual Residue Influence. Experimental study about the hydrogen-bond-network is only focused on the interactions between GS^- and agGSTe2 in the reactant structure.[41] However, the hydrogen-bond-network in the transition state should also be investigated because of its significant influence on the reaction barrier. Here variations of six hydrogen bonds (≤2.5 Å) between GS^- and enzyme agGSTe2 along the reaction pathway of SN-8.5 are investigated (Figure 4). Variations of these hydrogen bonds fall into two groups: the hydrogen bond in the reactant is stronger than in the transition state, and the hydrogen bond in the reactant is weaker than in the transition state. For instance, the hydrogen bond variations along the reaction pathway SN-8.5 show that GS^- is more stabilized by Ile55 in the transition state than in the reactant, whereas it is less stabilized by Arg112 in the transition state than in the reactant. The result is in excellent agreement with our following electrostatic influence study where it is shown that Ile55 facilitate the reaction by decreasing the reaction barrier of about 2.3 kcal/mol ($\Delta E^{i\text{-}0}$ = 2.3 kcal/mol), whereas Arg112 suppresses the catalytic reaction by increasing the reaction barrier of about 1.9 kcal/mol ($\Delta E^{i\text{-}0}$ = −1.9 kcal/mol).

The electrostatic influence of 19 key individual residues on the energy barrier of the pathway SN-8.5 was estimated. The electrostatic influence of Ser12 is not studied because it is included in the QM-region. The activation energy difference caused by amino acid i can be described as

$$\Delta E^{i-0} = \Delta E^{i} - \Delta E^{0}$$

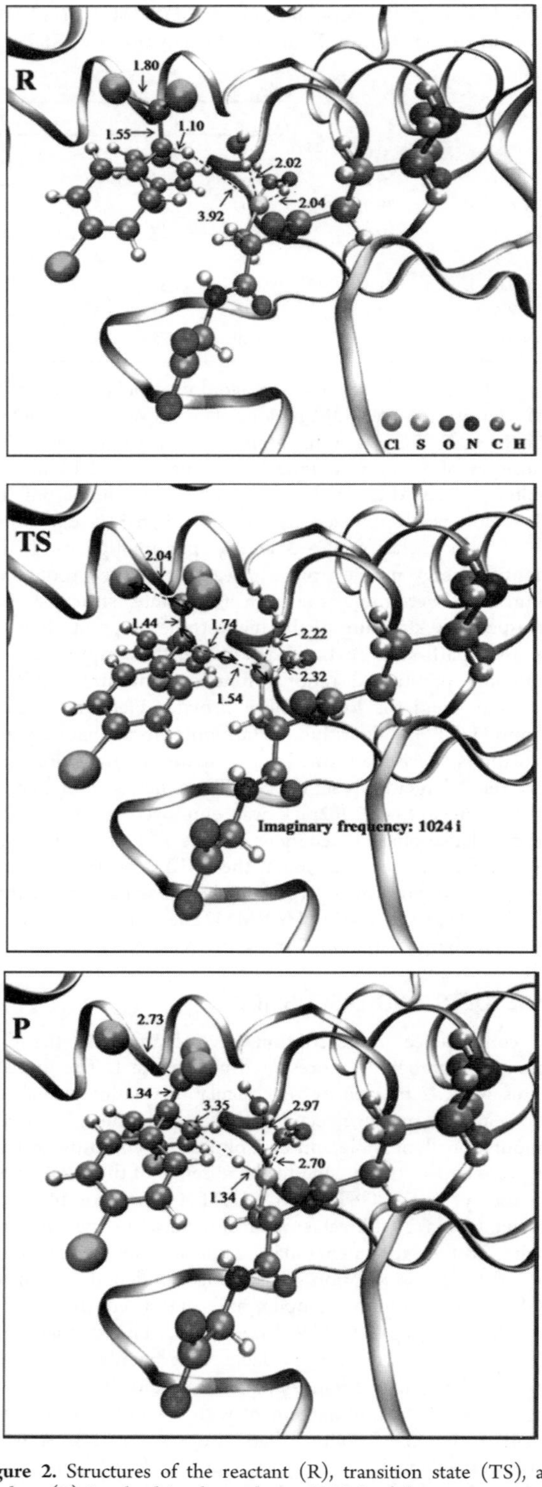

Figure 2. Structures of the reactant (R), transition state (TS), and product (P) involved in the pathway SN-8.5 of the proton transfer mechanism for agGSTe2 with DDT. QM atoms including a link hydrogen atom were shown through ball and stick representation. The unit for bond distances and imaginary frequency are in Å and cm^{-1}.

Where $\Delta E^{i\text{-}0}$ is the changes of the barrier, ΔE^{i} is the energy barrier with charges on residue i set to 0, and ΔE^{0} is the original

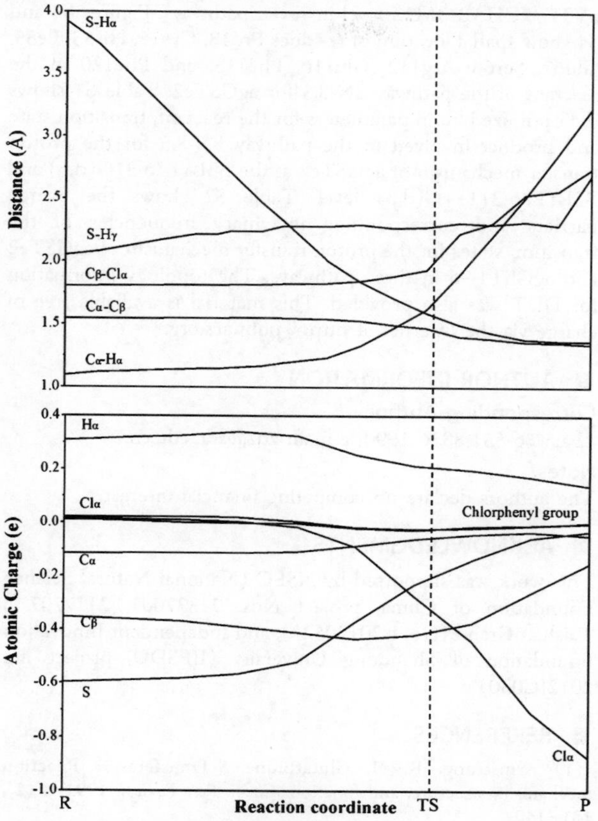

Figure 3. Variations of key bond distances and atomic charges along the pathway SN-8.5 for the proton transfer mechanism of agGSTe2 with DDT.

values of the energy barrier. During all these energy calculations, the geometry structures of the stationary points were kept unchanged. A positive $\Delta E^{i\text{-}0}$ value means that neglecting the influence of the ith residue will increase the energy barrier. In other words, the ith residue lowers the energy barrier and facilities the catalytic reaction. The $\Delta E^{i\text{-}0}$ values of 19 residues were schematically represented in Figure 5. Besides

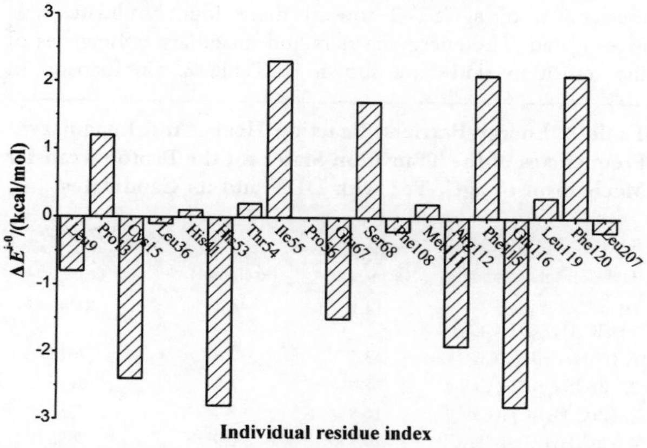

Figure 5. $\Delta E^{i\text{-}0}$ values of individual residues toward the proton transfer mechanism of agGSTe2 with DDT.

electrostatic influence of residues Ile55 and Arg112 discussed above, we have also shown that residues Cys15, His53, Glu67, and Glu116 suppress the detoxifying reaction ($\Delta E^{i\text{-}0}$ < −1 kcal/mol), whereas residues Pro13, Ser68, Phe115, and Phe120 facilitate the detoxifying reaction ($\Delta E^{i\text{-}0}$ > 1 kcal/mol). There are nine other residues that are found to have a weaker contribution (−1 kcal/mol< $\Delta E^{i\text{-}0}$ < 1 kcal/mol) on the reaction barrier. The spatial locations of the residues mentioned in the text were provided in SI Figure S3 and S4.

3.4. Structure−Activity Relationship and Comparison of AgGSTe2 with AgGST1−6.

Since resistance to DDT in some mosquito species raises the threat of malaria to humans, developing more effective novel insecticides is critically warranted. On the basis of DDT structures, four candidate structures, **1** $(C_6H_5)_2CHCCl_3$, **2** $(BrC_6H_4)_2CHCCl_3$, **3** $(BrC_6H_4)_2CHCBr_3$, and **4** $(ClC_6H_4)_2CHCH_2Cl$ were designed

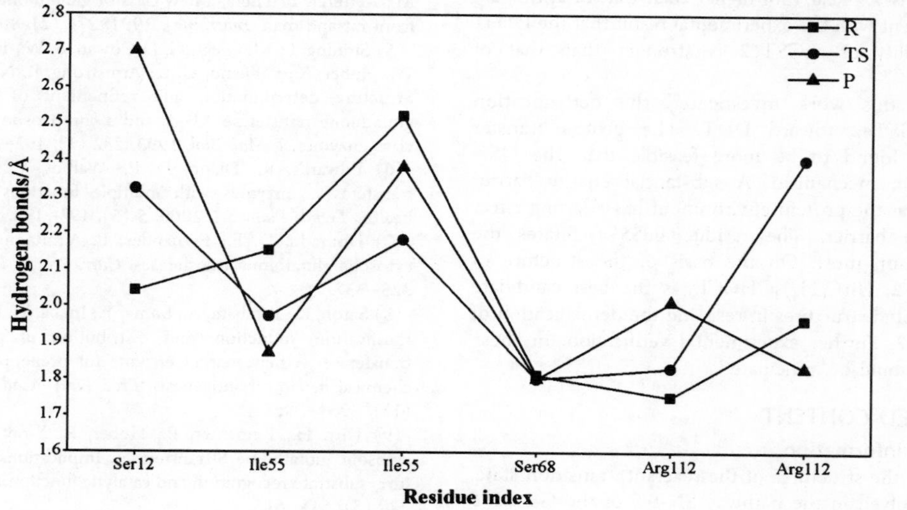

Figure 4. Hydrogen bond distances between GS⁻ and the surrounding residues in the reactant (R), transition state (TS), and product (P) structures involved in the pathway SN-8.5 for the proton transfer mechanism of agGSTe2 with DDT.

in the present study. We directly modified DDT in the agGSTe2-GS$^-$−DDT ternary complex to these four candidate structures and reoptimized the new ternary structures by using the QM/MM method. Then, the corresponding proton transfer mechanism of agGSTe2 toward these four candidates was investigated. The energy barriers and imaginary frequencies of the transition states are shown in Table 2. Comparison of

Table 2. Energy Barriers, Reaction Heats, And Imaginary Frequencies of the Transition States for the Proton Transfer Mechanism of agGSTe2 with DDT and its Candidates

DDT and its candidates	barriers (kcal/mol)	reaction heats (kcal/mol)	imaginary frequencies (cm^{-1})
DDT, $(ClC_6H_4)_2CH(CCl_3)$	14.1	−6.4	1024*i*
1, $(C_6H_5)_2CH(CCl_3)$	13.5	−7.6	761*i*
2, $(BrC_6H_4)_2CH(CCl_3)$	21.9	1.9	617*i*
3, $(BrC_6H_4)_2CH(CBr_3)$	16.8	−8.4	735*i*
4 $(ClC_6H_4)_2CH(CH_2Cl)$	20.3	−0.9	765*i*

energy barriers of DDT (14.1 kcal/mol), structure **1** (13.5 kcal/mol), and structure **2** (21.9 kcal/mol) indicates that replacement of Cl atoms in chlorphenyl group by H atoms facilitates the detoxification reaction, whereas replacement of Cl atoms by Br atoms suppresses it. Comparison of energy barriers of structures **2** (21.9 kcal/mol) and **3** (16.8 kcal/mol) shows that replacement of Cl atoms in $-CCl_3$ group by Br atoms facilitates the detoxification reaction. Comparison of energy barriers of DDT (14.1 kcal/mol) with structure **4** (20.3 kcal/mol) shows that replacement of Cl atoms in $-CCl_3$ group by H atoms suppresses the detoxification reaction. Results in Table 2 indicate that structure **2** $(BrC_6H_4)_2CHCCl_3$ is the best candidate in resisting the detoxification of agGSTe2.

The proton transfer mechanism of agGST1−6 with DDT was also studied to investigate the catalytic efficiencies of enzymes in different classes. Three pathways of agGST1−6 were investigated. The corresponding energy barriers and imaginary frequencies of the transition states were provided in SI Table S2. The exponential average energy barrier is 22.7 kcal/mol, which is 7.9 kcal/mol higher than that of agGSTe2. This is in consistent with the experimental result that the DDT-metabolizing ability of agGSTe2 is stronger than that of agGST1−6.[40−42,74,75]

In summary, this work investigated the detoxification reaction of agGSTe2 toward DDT. The proton transfer mechanism was found to be more feasible than the GS−DDT conjugation mechanism. A substantial energy barrier spread reveals that the protein environment has a strong effect on the reaction barrier. The residue Ile55 facilitates the detoxifying reaction most. On the basis of the structure of DDT, structure **2**, $(BrC_6H_4)_2CHCCl_3$, is the best candidate among all the tested structures in resisting the detoxification of enzyme agGSTe2. Further experimental verification for these new molecules would be anticipated.

■ ASSOCIATED CONTENT

Ⓢ Supporting Information

Figure S1 shows the structures of the reactant, transition state, and product involved in the pathway SN-8.5 of the GS-DDT conjugation mechanism of agGSTe2. Figure S2 shows the energy barriers of the proton transfer mechanism and the GS-DDT conjugation mechanism for agGSTe2 at the B3LYP/cc-pVTZ//CHARMM22 level in seven pathways. Figures S3 and S4 show spatial location of residues Pro13, Cys15, His53, Ile55, Glu67, Ser68, Arg112, Glu116, Phe115, and Phe120 in the reactant of the pathway SN-8.5 for agGSTe2. Table S1 shows the optimized bond parameters for the reactant, transition state and product involved in the pathway SN-8.5 for the proton transfer mechanism of agGSTe2 at the B3LYP/6-31G(d,p) and B3LYP/6-311++G(d,p) level. Table S2 shows the energy barriers and corresponding imaginary frequencies of the transition states for the proton transfer mechanism of agGSTe2 and agGST1−6 in three pathways. The topology information for DDT was also provided. This material is available free of charge via the Internet at http://pubs.acs.org.

■ AUTHOR INFORMATION

Corresponding Author

*Fax: 86-531-8836 1990; e-mail: zqz@sdu.edu.cn.

Notes

The authors declare no competing financial interest.

■ ACKNOWLEDGMENTS

This work was supported by NSFC (National Natural Science Foundation of China, project Nos. 21337001, 21177077), Taishan Grand (No. ts20120522), and Independent Innovation Foundation of Shandong University (IIFSDU, project no. 2012JC030).

■ REFERENCES

(1) Armstrong, R. N. Glutathione *S*-Transferases: Reaction mechanism, structure, and function. *Chem. Res. Toxicol.* **1991**, *4* (2), 131−140.

(2) Ali-Osman, F.; Akande, O.; Antoun, G.; Mao, J.; Buolamwini, J. Molecular cloning, characterization, and expression in Escherichia coli of full-length cDNAs of three human glutathione *S*-transferase Pi gene variants. Evidence for differential catalytic activity of the encoded proteins. *J. Biol. Chem.* **1997**, *272* (15), 10004−10012.

(3) Batist, G.; Tulpule, A.; Sinha, B. K.; Katki, A. G.; Myers, C. E.; Cowan, K. H. Overexpression of a novel anionic glutathione transferase in multidrug-resistant human breast cancer cells. *J. Biol. Chem.* **1986**, *261* (33), 15544−15549.

(4) Meyer, D. J.; Coles, B.; Pemble, S. E.; Gilmore, K. S.; Fraser, G. M.; Ketterer, B. Theta, a new class of glutathione transferases purified from rat and man. *Biochem. J.* **1991**, *274* (2), 409−414.

(5) Sinning, I.; Kleywegt, G. J.; Cowan, S. W.; Reinemer, P.; Dirr, H. W.; Huber, R.; Gilliland, G. L.; Armstrong, R. N.; Ji, X.; Board, P. G. Structure determination and refinement of human alpha class glutathione transferase A1−1, and a comparison with the Mu and Pi class enzymes. *J. Mol. Biol.* **1993**, *232* (1), 192−212.

(6) Edwards, R.; Dixon, D. P.; Walbot, V. Plant glutathione *S*-transferases: Enzymes with multiple functions in sickness and in health. *Trends Plant Sci.* **2000**, *5* (5), 193−198.

(7) Dourado, D. F.; Fernandes, P. A.; Ramos, M. J. Mammalian cytosolic glutathione transferases. *Curr. Protein Pept. Sci.* **2008**, *9* (4), 325−337.

(8) Satoh, K.; Kitahara, A.; Soma, Y.; Inaba, Y.; Hatayama, I.; Sato, K. Purification, induction, and distribution of placental glutathione transferase: A new marker enzyme for preneoplastic cells in the rat chemical hepatocarcinogenesis. *Proc. Natl. Acad. Sci. U.S.A.* **1985**, *82* (12), 3964−3968.

(9) Dirr, H.; Reinemer, P.; Huber, R. X-ray crystal structures of cytosolic glutathione S-transferases. Implications for protein architecture, substrate recognition and catalytic function. *Eur. J. Biochem.* **1994**, *220* (3), 645−61.

(10) Mannervik, B.; Alin, P.; Guthenberg, C.; Jensson, H.; Tahir, M. K.; Warholm, M.; Jornvall, H. Identification of three classes of cytosolic glutathione transferase common to several mammalian

species: Correlation between structural data and enzymatic properties. *Proc. Natl. Acad. Sci. U.S.A.* **1985**, *82* (21), 7202–7206.
(11) Widersten, M.; Björnestedt, R.; Mannervik, B. Involvement of the carboxyl groups of glutathione in the catalytic mechanism of human glutathione transferase A1–1. *Biochemistry* **1996**, *35* (24), 7731–7742.
(12) Adler, V.; Yin, Z.; Fuchs, S. Y.; Benezra, M.; Rosario, L.; Tew, K. D.; Pincus, M. R.; Sardana, M.; Henderson, C. J.; Wolf, C. R.; Davis, R. J.; Ronai, Z. Regulation of JNK signaling by GSTp. *EMBO J.* **1999**, *18* (5), 1321–1334.
(13) Koonin, E. V.; Mushegian, A. R.; Tatusov, R. L.; Altschul, S. F.; Bryant, S. H.; Bork, P.; Valencia, A. Eukaryotic translation elongation factor 1 gamma contains a glutathione transferase domain-study of a diverse, ancient protein superfamily using motif search and structural modeling. *Protein Sci.* **1994**, *3* (11), 2045–2054.
(14) Danielson, U. H.; Esterbauer, H.; Mannervik, B. Structure-activity relationships of 4-hydroxyalkenals in the conjugation catalysed by mammalian glutathione transferases. *Biochem. J.* **1987**, *247* (3), 707–713.
(15) Robertson, I. G.; Guthenberg, C.; Mannervik, B.; Jernstrom, B. Differences in stereoselectivity and catalytic efficiency of three human glutathione transferases in the conjugation of glutathione with 7 beta,8 alpha-dihydroxy-9 alpha,10 alpha-oxy-7,8,9,10-tetrahydrobenzo(a)-pyrene. *Cancer. Res.* **1986**, *46* (5), 2220–2224.
(16) Booth, J.; Boyland, E.; Sims, P. An enzyme from rat liver catalyzing conjugations with glutathione. *Biochem. J.* **1961**, *19* (3), 516–524.
(17) Habig, W. H.; Pabst, M. J.; Jakoby, W. B. Glutathione S-transferases. The first enzymatic step in mercapturic acid formation. *J. Biol. Chem.* **1974**, *249* (22), 7130–7139.
(18) Aniya, Y.; Naito, A. Oxidative stress-induced activation of microsomal glutathione *S*-transferase in isolated rat liver. *Biochem. Pharmacol.* **1993**, *45* (1), 37–42.
(19) Hayes, J. D.; Pulford, D. J. The glutathione *S*-transferase supergene family: Regulation of GST and the contribution of the isoenzymes to cancer chemoprotection and drug resistance. *Crit. Rev. Biochem. Mol. Biol.* **1995**, *30* (6), 445–600.
(20) Raza, H.; Robin, M. A.; Fang, J. K.; Avadhani, N. G. Multiple isoforms of mitochondrial glutathione *S*-transferases and their differential induction under oxidative stress. *Biochem. J.* **2002**, *366* (1), 45–55.
(21) Boyland, E.; Chasseaud, L. F. The role of glutathione and glutathione *S*-transferases in mercapturic acid biosynthesis. *Adv. Enzymol. Relat. Areas Mol. Biol.* **1969**, *32*, 173–219.
(22) Johansson, A. S.; Mannervik, B. Human glutathione transferase A3–3, a highly efficient catalyst of double-bond isomerization in the biosynthetic pathway of steroid hormones. *J. Biol. Chem.* **2001**, *276* (35), 33061–33065.
(23) Scharf, D. H.; Remme, N.; Habel, A.; Chankhamjon, P.; Scherlach, K.; Heinekamp, T.; Hortschansky, P.; Brakhage, A. A.; Hertweck, C. A dedicated glutathione *S*-transferase mediates carbon-sulfur bond formation in gliotoxin biosynthesis. *J. Am. Chem. Soc.* **2011**, *133* (32), 12322–12325.
(24) Vontas, J.; Small, G.; Hemingway, J. Glutathione *S*-transferase as antioxidant defence agents confer pyrethroid resistance in Nilaparvata lugens. *Biochem. J.* **2001**, *357* (1), 65–72.
(25) Vontas, J. G.; Small, G. J.; Nikou, D. C.; Ranson, H.; Hemingway, J. Purification, molecular cloning and heterologous expression of a glutathione *S*-transferase involved in insecticide resistance from the rice brown plant hopper, *Nilaparvata lugens*. *Biochem. J.* **2002**, *362* (2), 329–337.
(26) Enayati, A. A.; Ranson, H.; Hemingway, J. Insect glutathione transferases and insecticide resistance. *Insect Mol. Biol.* **2005**, *14* (1), 3–8.
(27) Che-Mendoza, A.; Penilla, R. P.; Rodriguez, D. A. Insecticide resistance and glutathione *S*-transferases in mosquitoes: A review. *Afr. J. Biotechnol.* **2009**, *8* (8), 1386–1397.
(28) Breman, J. G.; Alilio, M. S.; Mills, A. Conquering the intolerable burden of malaria: What's new, what's needed: A summary. *Am. J. Trop. Med. Hyg.* **2004**, *71* (Suppl 2), 1–15.
(29) Read, A. F.; Lynch, P. A.; Thomas, M. B. How to make evolution-proof insecticides for malaria control. *PLoS Bio.* **2009**, *7* (4), 1–10.
(30) Yadouleton, A. W.; Padonou, G.; Asidi, A.; Moiroux, N.; Bio-Banganna, S.; Corbel, V.; N'guessan, R.; Gbenou, D.; Yacoubou, I.; Gazard, K.; Akogbeto, M. C. Insecticide resistance status in *Anopheles gambiae* in southern Benin. *Mala. J.* **2010**, *9* (83), 1–6.
(31) David, J. P.; Strode, C.; Vontas, J.; Nikou, D.; Vaughan, A.; Pignatelli, P. M.; Louis, C.; Hemingway, J.; Ranson, H.; Beaty, B. J. The *Anopheles gambiae* detoxification chip: A highly specific microarray to study metabolic-based insecticide resistance in malaria vectors. *Proc. Natl. Acad. Sci. U.S.A.* **2005**, *102* (11), 4080–4084.
(32) Ndiath, M. O.; Sougoufara, S.; Gaye, A.; Mazenot, C.; Konate, L.; Faye, O.; Sokhna, C.; Trape, J. F. Resistance to DDT and pyrethroids and increased kdr mutation frequency in An. gambiae after the implementation of permethrin-treated nets in senegal. *PLoS One* **2012**, *7* (2), 1–6.
(33) Hemingway, J.; Field, L.; Vontas, J. An overview of insecticide resistance. *Science* **2002**, *298* (5591), 96–97.
(34) Prapanthadara, L. A.; Hemingway, J.; Ketterman, A. J. Partial purification and characterization of glutathione S-transferases involved in DDT resistance from the mosquito *Anopheles gambiae*. *Pestic. Biochem. Physiol.* **1993**, *47* (2), 119–133.
(35) Tang, A. H.; Tu, C. P. Biochemical characterization of Drosophila glutathione *S*-transferases D1 and D21. *J. Biol. Chem.* **1994**, *269* (45), 27876–27884.
(36) Huang, H. S.; Hu, N. T.; Yao, Y. E.; Wu, C. Y.; Chiang, S. W.; Sun, C. N. Molecular cloning and heterologous expression of a glutathione *S*-transferase involved in insecticide resistance from the diamondback moth *Plutella xylostella*. *Insect Biochem. Mol. Biol.* **1998**, *28* (9), 651–658.
(37) Holt, R. A.; Subramanian, G. M.; Halpern, A.; Sutton, G. G.; Charlab, R.; Nusskern, D. R.; Wincker, P.; Clark, A. G.; Ribeiro, J. M.; Wides, R.; Salzberg, S. L.; Loftus, B.; Yandell, M.; Majoros, W. H.; Rusch, D. B.; Lai, Z.; Kraft, C. L.; Abril, J. F.; Anthouard, V.; Arensburger, P.; Atkinson, P. W.; Baden, H.; de Berardinis, V.; Baldwin, D.; Benes, V.; Biedler, J.; Blass, C.; Bolanos, R.; Boscus, D.; Barnstead, M.; Cai, S.; Center, A.; Chaturverdi, K.; Christophides, G. K.; Chrystal, M. A.; Clamp, M.; Cravchik, A.; Curwen, V.; Dana, A.; Delcher, A.; Dew, I.; Evans, C. A.; Flanigan, M.; Grundschober-Freimoser, A.; Friedli, L.; Gu, Z.; Guan, P.; Guigo, R.; Hillenmeyer, M. E.; Hladun, S. L.; Hogan, J. R.; Hong, Y. S.; Hoover, J.; Jaillon, O.; Ke, Z.; Kodira, C.; Kokoza, E.; Koutsos, A.; Letunic, I.; Levitsky, A.; Liang, Y.; Lin, J. J.; Lobo, N. F.; Lopez, J. R.; Malek, J. A.; McIntosh, T. C.; Meister, S.; Miller, J.; Mobarry, C.; Mongin, E.; Murphy, S. D.; O'Brochta, D. A.; Pfannkoch, C.; Qi, R.; Regier, M. A.; Remington, K.; Shao, H.; Sharakhova, M. V.; Sitter, C. D.; Shetty, J.; Smith, T. J.; Strong, R.; Sun, J.; Thomasova, D.; Ton, L. Q.; Topalis, P.; Tu, Z.; Unger, M. F.; Walenz, B.; Wang, A.; Wang, J.; Wang, M.; Wang, X.; Woodford, K. J.; Wortman, J. R.; Wu, M.; Yao, A.; Zdobnov, E. M.; Zhang, H.; Zhao, Q.; Zhao, S.; Zhu, S. C.; Zhimulev, I.; Coluzzi, M.; della Torre, A.; Roth, C. W.; Louis, C.; Kalush, F.; Mural, R. J.; Myers, E. W.; Adams, M. D.; Smith, H. O.; Broder, S.; Gardner, M. J.; Fraser, C. M.; Birney, E.; Bork, P.; Brey, P. T.; Venter, J. C.; Weissenbach, J.; Kafatos, F. C.; Collins, F. H.; Hoffman, S. L. The genome sequence of the malaria mosquito *Anopheles gambiae*. *Science* **2002**, *298* (5591), 129–149.
(38) Ding, Y. C.; Ortelli, F.; Rossiter, L. C.; Hemingway, J.; Ranson, H. The *Anopheles gambiae* glutathione transferase supergene family: Annotation, phylogeny and expression profiles. *BMC Genomics* **2003**, *4* (35), 1–16.
(39) Ranson, H.; Paton, M. G.; Jensen, B.; McCarroll, L.; Vaughan, A.; Hogan, J. R.; Hemingway, J.; Collins, F. H. Genetic mapping of genes conferring permethrin resistance in the malaria vector *Anopheles gambiae*. *Insect Mol. Biol.* **2004**, *13* (4), 379–386.

(40) Ranson, H.; Prapanthadara, L.; Hemingway, J. Cloning and characterization of two glutathione *S*-transferases from a DDT-resistant strain of *Anopheles gambiae*. *Biochem. J.* **1997**, *324* (1), 97–102.
(41) Ranson, H.; Rossiter, L.; Ortelli, F.; Jensen, B.; Wang, X.; Roth, C. W.; Collins, F. H.; Hemingway, J. Identification of a novel class of insect glutathione *S*-transferases involved in resistance to DDT in the malaria vector *Anopheles gambiae*. *Biochem. J.* **2001**, *359* (2), 295–304.
(42) Ortelli, F.; Rossiter, L. C.; Vontas, J.; Ranson, H.; Hemingway, J. Heterologous expression of four glutathione transferase genes genetically linked to a major insecticide-resistance locus from the malaria vector *Anopheles gambiae*. *Biochem. J.* **2003**, *373* (3), 957–963.
(43) Wang, Y.; Qiu, L.; Ranson, H.; Lumjuan, N.; Hemingway, J.; Setzer, W. N.; Meehan, E. J.; Chen, L. Structure of an insect epsilon class glutathione *S*-transferase from the malaria vector *Anopheles gambiae* provides an explanation for the high DDT-detoxifying activity. *J. Struct. Bio.* **2008**, *164* (2), 228–235.
(44) Adang, A. E.; Brussee, J.; van der Gen, A.; Mulder, G. J. The glutathione-binding site in glutathione *S*-transferases. Investigation of the cysteinyl, glycyl and gamma-glutamyl domains. *Biochem. J.* **1990**, *269* (1), 47–54.
(45) Dourado, D. F. A. R.; Fernandes, P. A.; Ramos, M. J. Glutathione transferase A1–1: Catalytic role of water. *Theor. Chem. Acc.* **2009**, *124* (1–2), 71–83.
(46) Dourado, D. F. A. R.; Fernandes, P. A.; Mannervik, B.; Ramos, M. J. Glutathione transferase: New model for glutathione activation. *Chem.—Eur. J.* **2008**, *14* (31), 9591–9598.
(47) Dourado, D. F. A. R.; Fernandes, P. A.; Mannervik, B.; Ramos, M. J. Glutathione transferase A1–1: Catalytic importance of arginine 15. *J. Phys. Chem. B* **2010**, *114* (4), 1690–1697.
(48) Dourado, D. F. A. R.; Fernandes, P. A.; Ramos, M. J. Glutathione transferase classes alpha, pi, and mu: GSH activation mechanism. *J. Phys. Chem. B* **2010**, *114* (40), 12972–12980.
(49) Dourado, D. F. A. R.; Fernandes, P. A.; Ramos, M. J.; Mannervik, B. Mechanism of glutathione transferase P1–1-catalyzed activation of the prodrug canfosfamide (TLK286, TELCYTA). *Biochemistry* **2013**, *52* (45), 8069–8078.
(50) Lipke, H.; Chalkley, J. The conversion of DDT to DDE by some anophelines. *Bull. W. H. O.* **1964**, *30*, 57–64.
(51) Hemingway, J.; Hawkes, N. J.; McCarroll, L.; Ranson, H. The molecular basis of insecticide resistance in mosquitoes. *Insect Biochem. Mol. Biol.* **2004**, *34* (7), 653–665.
(52) Winayanuwattikun, P.; Ketterman, A. J. An electron-sharing network involved in the catalytic mechanism is functionally conserved in different glutathione transferase classes. *J. Biol. Chem.* **2005**, *280* (36), 31776–31782.
(53) Li, H.; Robertson, A. D.; Jensen, J. H. Very fast empirical prediction and rationalization of protein pKa values. *Proteins* **2005**, *61* (4), 704–721.
(54) Brooks, B. R.; Bruccoleri, R. E.; Olafson, B. D.; States, D. J.; Swaminathan, S.; Karplus, M. CHARMM: A program for macromolecular energy, minimization, and dynamics calculations. *J. Comput. Chem.* **1983**, *4* (2), 187–217.
(55) Brooks, B. R.; Brooks, C. L., III; Mackerell, A. D.; Nilsson, L.; Petrella, R. J.; Roux, B.; Won, Y.; Archontis, G.; Bartels, C.; Boresch, S.; Caflisch, A.; Caves, L.; Cui, Q.; Dinner, A. R.; Feig, M.; Fischer, S.; Gao, J.; Hodoscek, M.; Im, W.; Kuczera, K.; Lazaridis, T.; Ma, J.; Ovchinnikov, V.; Paci, E.; Pastor, R. W.; Post, C. B.; Pu, J. Z.; Schaefer, M.; Tidor, B.; Venable, R. M.; Woodcock, H. L.; Wu, X.; Yang, W.; York, D. M.; Karplus, M. CHARMM: The Biomolecular simulation Program. *J. Comput. Chem.* **2009**, *30* (10), 1545–1615.
(56) MacKerell, A. D.; Brooks, Jr, B.; Brooks, III, C. L.; Nilsson, L.; Roux, B.; Won, Y.; Karplus, M. CHARMM: The energy function and its parameterization with an overview of the program. *Encyclopedia of Computational Chemistry*; John Wiley & Sons: New York, 1998.
(57) Wu, G.; Robertson, D. H.; Brooks, C. L., 3rd.; Vieth, M. Detailed analysis of grid-based molecular docking: A case study of CDOCKER-A CHARMm-based MD docking algorithm. *J. Comput. Chem.* **2003**, *24* (13), 1549–1562.
(58) Vieth, M.; Hirst, J. D.; Kolinski, A.; Brooks, C. L., 3rd. Assessing energy functions for flexible docking. *J. Comput. Chem.* **1998**, *19* (14), 1612–1622.
(59) Jorgensen, W. L.; Chandrasekhar, J.; Madura, J. D.; Impey, R. W.; Klein, M. L. Comparison of simple potential functions for simulating liquid water. *J. Chem. Phys.* **1983**, *79* (2), 926–935.
(60) Brooks, C. L.; Karplus, M. Deformable stochastic boundaries in molecular dynamics. *J. Chem. Phys.* **1983**, *79* (12), 6312–6325.
(61) Sherwooda, P.; Vriesa, A. H. D.; Guesta, M. F.; Schreckenbacha, G.; Catlowb, C. R. A.; Frenchb, S. A.; Sokolb, A. A.; Bromleyb, S. T.; Thielc, W.; Turnerc, A. J.; Billeterc, S.; Terstegenc, F.; Thielc, S.; Kendrickd, J.; Rogersd, S. C.; Cascie, J.; Watsone, M.; Kinge, F.; Karlsenf, E.; Sjovollf, M.; Fahmif, A.; Schaferg, A.; Lennartzg, C. QUASI: A general purpose implementation of the QM/MM approach and its application to problems in catalysis. *J. Mol. Struct.: THEOCHEM* **2003**, *632* (1–3), 1–28.
(62) Ahlrichs, R.; Bär, M.; Häser, M.; Horn, H.; Kölmel, C. Electronic structure calculations on workstation computers: The program system turbomole. *Chem. Phys. Lett.* **1989**, *162* (3), 165–169.
(63) Smith, W.; Forester, T. R. DL_POLY_2.0: A general-purpose parallel molecular dynamics simulation package. *J. Mol. Graphics* **1996**, *14* (3), 136–141.
(64) Billeter, S. R.; Turner, A. J.; Thiel, W. Linear scaling geometry optimization and transition state search in hybrid delocalised internal coordinates. *Phys. Chem. Chem. Phys.* **2000**, *2*, 2177–2186.
(65) Burke, K.; Werschnik, J.; Gross, E. K. U. Time-dependent density functional theory: Past, present, and future. *J. Chem. Phys.* **2005**, *123* (062206), 1–9.
(66) MacKerell, A. D.; Bashford, D.; Bellott, M.; Dunbrack, R. L.; Evanseck, J. D.; Field, M. J.; Gao, J.; Guo, H.; Ha, S.; Joseph-McCarthy, D.; Kuchnir, L.; Kuczera, K.; Lau, F. T. K.; Mattos, C.; Michnick, S.; Ngo, T.; Nguyen, D. T.; Prodhom, B.; Reiher, W. E.; Roux, B., III; Schlenkrich, M.; Smith, J. C.; Stote, R.; Straub, J.; Watanabe, M.; Wiórkiewicz-Kuczera, J.; Yin, D.; Karplus, M. All-atom empirical potential for molecular modeling and dynamics studies of proteins. *J. Phys. Chem. B* **1998**, *102* (18), 3586–3616.
(67) de Vries, A. H.; Sherwood, P.; Collins, S. J.; Rigby, A. M.; Rigutto, M.; Kramer, G. J. Zeolite structure and reactivity by combined quantum-chemical-classical calculations. *J. Phys. Chem. B* **1999**, *103* (29), 6133–6141.
(68) Becke, A. D. A new mixing of Hartree-Fock and local density-functional theories. *J. Chem. Phys.* **1993**, *98* (2), 1372–1377.
(69) Lee, C.; Yang, W.; Parr, R. G. Development of the Colle-Salvetti correlation-energy formula into a functional of the electron density. *Phys. Rev. B* **1988**, *37* (2), 785–789.
(70) Baker, J. An algorithm for the location of transition states. *J. Comput. Chem.* **1986**, *7* (4), 385–395.
(71) Liu, D. C.; Nocedal, J. On the limited memory BFGS method for large scale optimization. *Math. Program* **1989**, *45* (3), 503–528.
(72) Lonsdale, R.; Harvey, J. N.; Mulholland, A. J. Compound I reactivity defines alkene oxidation selectivity in cytochrome P450cam. *J. Phys. Chem. B* **2010**, *114* (2), 1156–1162.
(73) Rommel, J. B.; Kästner, J. The fragmentation-recombination mechanism of the enzyme glutamate mutase studied by QM/MM simulations. *J. Am. Chem. Soc.* **2011**, *133* (26), 10195–10203.
(74) Prapanthadara, L.; Hemingway, J.; Ketterman, A. J. DDT-resistance in *Anopheles gambiae* (Diptera: Culicidae) from Zanzibar, Tanzania, based on increased DDT-dehydrochlorinase activity of glutathione S-transferases. *Bull. Entomol. Res.* **1995**, *85* (2), 267–274.
(75) Wongtrakul, J.; Pongjaroenkit, S.; Leelapat, P.; Nachaiwieng, W.; Prapanthadara, L. A.; Ketterman, A. J. Expression and characterization of three new glutathione transferases, an epsilon (AcGSTE2–2), omega (AcGSTO1–1), and theta (AcGSTT1–1) from *Anopheles cracens* (Diptera: Culicidae), a major Thai malaria vector. *J. Med. Entomol.* **2010**, *47* (2), 162–171.

RSC Advances

PAPER

View Article Online
View Journal | View Issue

CrossMark
click for updates

Cite this: *RSC Adv.*, 2016, 6, 4183

Dehydrochlorination mechanism of γ-hexachlorocyclohexane degraded by dehydrochlorinase LinA from *Sphingomonas paucimobilis* UT26†

Xiaowen Tang, Ruiming Zhang, Qingzhu Zhang* and Wenxing Wang

Received 15th October 2015
Accepted 1st December 2015

DOI: 10.1039/c5ra21461k

www.rsc.org/advances

This study investigated the aerobic degradation mechanism of γ-HCH to 1,3,4,6-TCDN catabolized by dehydrochlorinase LinA from *Sphingomonas paucimobilis* UT26. The enzymatic step was studied by a combined quantum mechanics/molecular mechanics (QM/MM) computation and the nonenzymatic step was investigated by the DFT method. There are three elementary steps involved in the degradation process. Two discontinuous dehydrochlorination reactions with the Boltzmann-weighted average potential barriers of 16.2 and 17.3 kcal mol^{-1} are connected by a conformational transition with a barrier of 11.1 kcal mol^{-1}. The electrostatic influence analysis of fourteen key residues surrounding the active site has been carried out. The study reveals that Phe68 facilitates the dehydrochlorination of γ-HCH, whereas Leu21 and Cys71 suppress it. Future mutation studies for improving the degradation efficiency of LinA can focus on mutating the amino acids of Leu21 and Cys71.

1. Introduction

Hexachlorocyclohexane (HCH) is an organochlorine compound with several stable isomers. Among all the isomers, only the γ isomer (γ-hexachlorocyclohexane, γ-HCH) has insecticidal properties and has been widely used as a broad-spectrum insecticide to control a wide range of agricultural, horticultural, and public health pests.[1–3] Two kinds of γ-HCH products, technical HCH (the content of γ-HCH is 10–15%) and lindane (purified γ-HCH), have been applied and about 600 000 tons were released into the environment between the 1940s and 1990s.[4,5] Due to its non-target toxicity and environmental persistence, HCH was included in the list of persistent organic pollutants (POPs) according to the Stockholm convention in 2009,[6] and has been prohibited in most countries. However, lindane is still used in some developing countries for its high efficiency and low cost.[3,7] Therefore, HCH contamination, especially γ-HCH contamination, is a serious continuous threat to the environment.

Environment Research Institute, Shandong University, Jinan 250100, P. R. China. E-mail: zqz@sdu.edu.cn; Fax: +86-531-8836-1990

† Electronic supplementary information (ESI) available: Root-mean-square deviations (RMSD) of the backbone and key distance variations along the molecular dynamic simulations (Fig. S1); the three dimensional structures of the docked structure, the MD snapshot, and the QM/MM-optimised structure in the γ-HCH and γ-PCCH-1 reaction systems (Fig. S2 and S3); additional details on the methods; the coordinates of the docked structures, MD snapshots, QM-optimized structures and QM/MM-optimized structures. See DOI: 10.1039/c5ra21461k

Microbial degradation of γ-HCH can proceed under either anaerobic or aerobic condition.[8–10] Chlorobenzene and benzene will be accumulated when γ-HCH is degraded under the anaerobic condition. The biochemical pathways for anaerobic degradation of γ-HCH are available, but unfortunately the specific genes and enzymes involved in the anaerobic degradation have not been identified yet.[4] In contrast, γ-HCH can be degraded completely into nontoxic molecules under the aerobic condition. Researches about the aerobic degradation of γ-HCH are numerous and several HCH-degrading aerobes have been described in details.[1,11–15] Most of them belong to the family of *Sphingomonadaceae*.[16] They contain a set of genes called *lin* genes, which can encode HCH degradation enzymes. The aerobic degradation pathway of γ-HCH is devoted by various enzymes, among which the HCH dehydrochlorinase (LinA) from *Sphingobium japonicum* UT26 is considered to be significant because it catalyzes the initial step of the γ-HCH aerobic degradation.[17,18]

LinA catalyzes the dehydrochlorination of γ-HCH to generate an observed intermediate γ-pentachlorocyclohexene (γ-PCCH), which is transformed to a putative product 1,3,4,6-tetrachloro-1,4-cyclohexadiene (1,3,4,6-TCDN) through another dehydrochlorination step.[19] During the LinA-catalyzed degradation process, the substrate is released from the enzyme after the first dehydrochlorination reaction and rebinds to the active site of LinA when undergoing the subsequent dehydrochlorination reaction.[17] Crystal structure studies revealed that a catalytic dyad formed by His73 and Asp25 is located in the active site of LinA.[17,18] During the enzymatic dehydrochlorination reaction,

His73 of LinA acts as a general base to grab the proton of substrate, generating a positively charged His73 residue which is stabilized though its interaction with Asp25.[20] It is generally considered that the leaving hydrogen and chlorine in the LinA-catalyzed dehydrochlorination reaction should be axial, adjacent and in antiparallel position.[21] Hence, LinA exclusively catalyzes the dehydrochlorination reaction of the substrates containing at least one adjacent *trans*-diaxial H/Cl pair. The biotransformation mechanism from γ-HCH to 1,3,4,6-TCDN is exhibited in Scheme 1A. The hydrogen atom bonded to C^1 and chlorine atom bonded to C^2, composing an adjacent *trans*-diaxial H/Cl pair (H^1/Cl^2), are removed in the dehydrochlorination reaction of γ-HCH by LinA. Enzymatic dehydrochlorination of the newly generated intermediate (γ-PCCH) must proceed through the elimination of H^4/Cl^5 pair during the transformation from γ-PCCH to 1,3,4,6-TCDN. However, neither of them in γ-PCCH is situated on axial orientation, implying that γ-PCCH is unable to transform to 1,3,4,6-TCDN by enzymatic dehydrochlorination directly. Actually, the formation of 1,3,4,6-TCDN can accomplish through the LinA-catalyzed dehydrochlorination of a PCCH conformer with an adjacent *trans*-diaxial H^4/Cl^5 pair, as γ-PCCH-1 presented in Scheme 1. It can be considered as a product of the conformational transition of γ-PCCH. Therefore, the transformation pathways from γ-PCCH to 1,3,4,6-TCDN *via* γ-PCCH-1 must be at work, in which the LinA-catalyzed dehydrochlorination reaction occurs after the conformational transition of γ-PCCH instead of eliminating the H^4/Cl^5 pair in γ-PCCH directly.

Although the LinA-catalyzed degradation process of γ-HCH have been established roughly,[4,22] the in-depth understanding of its dehydrochlorination reaction still remains indistinct. The transition states and some intermediates as well as some products formed in the catalytic process are impossible to be observed in the general experimental enzyme chemistry, for instance, 1,3,4,6-TCDN, a very short-lived metabolism product, has never been directly detected in experimental characterization.[20] Furthermore, the influence of residues Leu21, Ile109, and Thr133 as well as other key residues surrounding the active site of LinA in the γ-HCH dehydrochlorination process is still unknown. Therefore, theoretical calculation can be an alternative. In the present work, the detailed degradation mechanism from γ-HCH to 1,3,4,6-TCDN catalyzed by dehydrochlorinase LinA from *Sphingomonas paucimobilis* UT26 was investigated by theoretical calculations. The enzymatic step was studied with the aid of a combined quantum mechanics/molecular mechanics (QM/MM) method. QM/MM computations of the enzyme-catalyzed reaction can provide the atomistic details of the enzyme mechanism and is therefore becoming an increasingly important tool to supplement experimental enzyme chemistry.

2. Calculation methods

2.1 System setup and molecular dynamics

The initial enzyme model for the present simulation was built on the basis of the X-ray crystal structure of γ-hexachlorocyclohexane dehydrochlorinase LinA from *Sphingomonas paucimobilis* UT26 (PDB code: 3A76) obtained from the Protein Data Bank (http://www.rcsb.org).[17] It reveals that LinA is a homotrimer with no significant difference in backbone conformation among the three chains and the LinA-catalyzed reaction can be achieved in any chain independently.[17] Therefore, chain A of LinA was selected as the initial enzyme model for our present study. The protonation state of ionizable residues was determined on the basis of the pK_a values obtained from the PROPKA procedure.[23] Missing hydrogen atoms of the crystal structure were supplemented through CHARMM22 force field[24] in the HBUILD facility of CHARMM package.[25–27] MolProbity software was used to check the flipped Asn/Gln/His residues.[28] Substrate models (γ-HCH and γ-PCCH-1) were built by using the Material Studio 4.4 program and then docked with the dehydrochlorinase LinA through a grid-based receptor-flexible docking module (CDOCKER) installed in the Discovery Studio 2.1 program[29,30] (Accelrys Software Inc.). The binding site was defined as a sphere with a radius of 5.0 Å (coordinate: −9.806, 22.269, −5.274). Substrates were docked into the binding site with the aid of a CHARMM-based molecular dynamics (MD). Random substrate conformations were generated through high-temperature MD and translated into the binding site. Candidate poses were then created using rigid-body rotations followed by simulated annealing. A final minimization was used to refine the substrate poses. Finally, the substrate poses with interaction energy of 23.7 kcal mol^{-1} for γ-HCH and 22.8 kcal mol^{-1} for γ-PCCH-1 were select for our present work. The substrate–LinA binary complex was placed in a water sphere (TIP3P model[31]) with a diameter of 70.0 Å, which ensures that the complex was completely solvated. Water molecules overlapping within 2.5 Å of the binary complex were deleted. The whole system was neutralized with seven sodium ions at random positions. After that, the system was heated gradually from 0 K to 298.15 K within 50 ps and a trajectory of 500 ps was calculated to reach the thermal equilibration state (1 fs per step). Finally, a 6 ns stochastic boundary molecular dynamic (SBMD) simulation with canonical ensemble (NVT, 298.15 K) was performed to mimic the aqueous environment.[32] The leap-frog algorithm and Langevin dynamics attached in the CHARMM package were applied during the simulation.

2.2 QM/MM calculations

The QM/MM calculations were performed with the aid of ChemShell 3.3.01 (ref. 33) integrating Turbomole 6.2 (ref. 34) and DL-POLY[35] programs. The hybrid delocalized internal coordinate (HDLC)[36] was adopted for the calculation. The MM region was treated with the CHARMM22 force field,[24] while the QM region was calculated by the DFT[37] method. The boundary was defined by cleaving the covalent bonds between the QM and MM regions. In order to avoid over-polarization of the QM density in the QM region, hydrogen-link atoms were complemented to the QM side with the charge shift model.[38] When partitioning the QM region, some essential criteria should be considered, residues participating in bond formation or cleavage and having strong interaction with the reactive center should be classified to the QM region. Therefore, the QM region of the LinA-catalyzed dehydrochlorination reaction system in

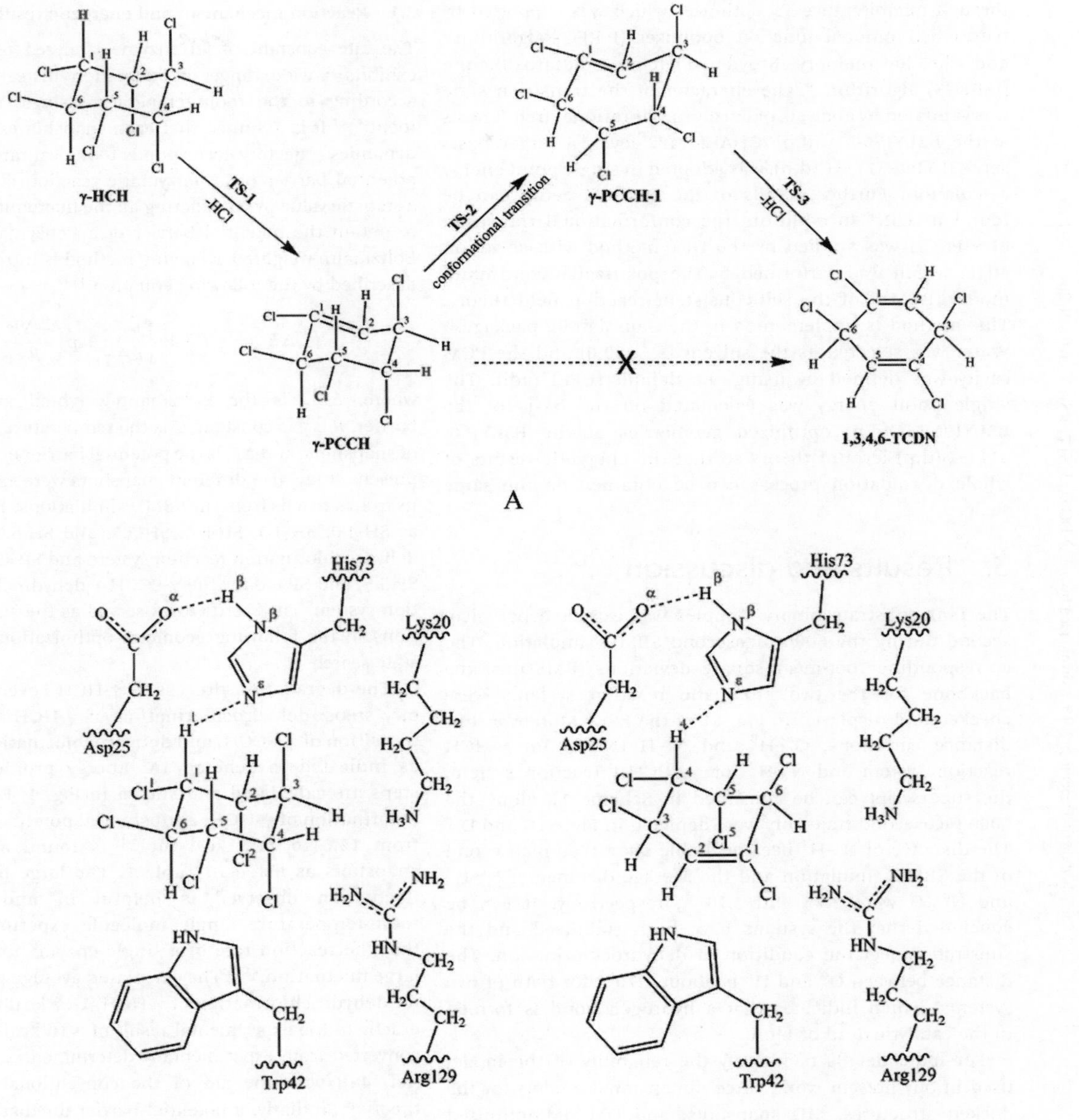

Scheme 1 (A) The degradation pathway from γ-HCH to 1,3,4,6-TCDN catabolized by dehydrochlorinase LinA. The leaving atoms are labeled in bold red. The QM region for LinA-catalyzed dehydrochlorination of γ-HCH (B) and γ-PCCH-1 (C). Several key atoms are numbered and the boundary between the QM and MM regions is indicated by wavy lines.

the present study contains residues Lys20, Asp25, Trp42, His73, Arg129 and the substrate (γ-HCH or γ-PCCH-1). Together with five hydrogen-link atoms, a total of 83 atoms were treated in the QM region. Similarly, 81 atoms were regarded as QM atoms in the γ-PCCH-1 reaction system. For both of the two systems, all the atoms within 18 Å of N^{ε} atom (Scheme 1) from His73 were selected to be the active region (about 3400 movable atoms). Atoms that lie beyond 18 Å of N^{ε} were fixed during the QM/MM calculation. The QM region was optimized by the B3LYP/6-31G(d,p) method with a charge of 1 and a spin multiplicity of 1. The transition state structure was determined by scanning the potential energy profile from the reactant to the product. The corresponding structure with the highest energy along the reaction path was selected and further optimized through microiterative TS optimizer which was supported by partitioned rational function optimizer (P-RFO) algorithm[39] and the low-memory Broyden–Fletcher–Goldfarb–Shanno (L-BFGS) algorithm.[40] The character of the transition state was validated by analysis of harmonic vibrational frequencies at the B3LYP/6-31G(d,p)//CHARMM22 level. A larger basis set, B3LYP/6-311++G(d,p), was adopted in single point energy calculation. Further details of the QM/MM setup can be found in ESI.† In addition, the conformational transition of γ-PCCH was studied by the DFT method with solvation effect which was performed by the polarizable continuum model (PCM)[41] of the self-consistent reaction field theory. This method is implemented in the Gaussian 09 package.[42] Water was selected as the solvent ($\varepsilon = 80.0$) and the PCM cavity was defined by using the default (UFF) radii. The single point energy was calculated on the basis of the B3LYP/6-31G(d,p) optimized geometries at the B3LYP/6-311++G(d,p) level of theory so that the energetic results of whole degradation process can be obtained on the same scale.

3. Results and discussion

The LinA-substrate binary complex was extracted per picosecond during the 6000 picosecond SBMD simulation. The corresponding root-mean-square deviations (RMSD) of the backbone for the two enzymatic reaction systems were checked and displayed in Fig. S1 of the ESI.† Moreover, two distance variations, O^{α}–H^{β} and N^{ε}–H (N^{ε}–H^1 for γ-HCH reaction system and N^{ε}–H^4 for γ-PCCH-1 reaction system, the superscript can be consulted in Scheme 1), along the 6000 picosecond trajectory were depicted in Fig. S1C and D.† The distance of N^{ε}–H^1 became stable after 1700 picosecond of the SBMD simulation and the average distance of N^{ε}–H^1 and N^{ε}–H^4 were 2.75 and 2.70 Å, respectively. It can be concluded that the systems have been stabilized and the substrates meet the condition of dehydrochlorination. The distance between O^{α} and H^{β} is about 1.70 Å for both of two systems, which indicates that a hydrogen bond is formed in the catalytic dyad of LinA.

For more details to identify the reliability of the model used in our present work, three dimensional models for the docked structures, MD snapshots, and QM/MM-optimised structures were exhibited in ESI.† For the γ-HCH reaction system (Fig. S2†), the substrate keeps its chair conformation with the position staying relatively stationary in the three sections. The relative position with His73 is measured through distance of N^{ε}–H^1, which is 2.63 Å in docked structure, an average of 2.75 Å in MD snapshots, and an average of 2.46 Å in QM/MM-optimised structures. Similarly, D1, D2 and D3 are also adopted to estimate the relative position with Trp42, Arg129 and Lys20, which are about 3.50 Å, 5.00 Å and 4.90 Å in the three sections. Analogously, the half-chair conformation γ-PCCH-1 (Fig. S3†) is also located in the active site with a relatively stable position. Hence, it might be inferred that the model used in our present work could be credible for the present study.

3.1 Reaction mechanism and energetic results

The rate constant of an enzyme-catalyzed reaction generally exhibits a wide range of fluctuation instead of a constant, according to the room-temperature single molecule experiment.[43,44] It is assumed that each snapshot extracted from the dynamics trajectory corresponds to a local rate constant.[45] The potential barrier of an enzymatic reaction is supposed to be a statistic value by considering all the fluctuant results. In order to obtain the potential barrier of an enzymatic reaction, the Boltzmann-weighted averaging method is introduced. It can be described by the following equation:[46,47]

$$\Delta E_{\mathrm{ea}} = -RT\ \ln\left\{\frac{1}{n}\sum_{i=1}^{n}\exp\left(\frac{-\Delta E_i}{RT}\right)\right\}$$

where, ΔE_{ea} is the Boltzmann-weighted average potential barrier, R is gas constant, T is the temperature, n is the number of snapshots, and ΔE_i is the potential barrier of pathway i. In the present study, five different snapshots were extracted every 0.5 ns from 4 to 6 ns from the SBMD simulations. They were labeled as SH-4.0, SH-4.5, SH-5.0, SH-5.5, and SH-6.0 for the γ-HCH dehydrochlorination reaction system and SP-4.0, SP-4.5, SP-5.0, SP-5.5, and SP-6.0 for the γ-PCCH-1 dehydrochlorination reaction system. These structures served as the starting configurations in the following geometry optimization and transition-state search.

The degradation process of γ-HCH covers three elementary steps: dehydrochlorination of γ-HCH, conformational transition of γ-PCCH, and dehydrochlorination of γ-PCCH-1, as indicated in Scheme 1A. Energy profiles of the three steps are calculated and shown in Fig. 1. For the dehydrochlorination of γ-HCH, a substantial potential barrier spread from 12.6 to 21.3 kcal mol^{-1} is found among different snapshots as listed in Table 1. The large potential barrier fluctuation observed is helpful in understanding the room-temperature single molecule experimental evidence that the reaction rate of a single enzyme molecule exhibits large fluctuation.[43,44] The calculated average potential barrier for dehydrochlorination of γ-HCH, 16.2 kcal mol^{-1}, conforms exactly to the experimental result of ~16 kcal mol^{-1}, which is converted from experimentally determined k_{cat} value (63.5 s^{-1} (ref. 48)) with the aid of the conventional transition-state theory.[49] Similarly, a potential barrier fluctuation spread from 13.4 to 21.5 kcal mol^{-1} listed in Table 2 is found in the second dehydrochlorination step (dehydrochlorination of γ-PCCH-1) and the calculated average potential barrier is 17.3 kcal mol^{-1}, a slightly higher than that of the dehydrochlorination of γ-HCH. For the conformational transition of γ-PCCH, the calculated potential barrier is 11.1 kcal mol^{-1}. It is worth noticing that all of the three elementary steps are exothermic, the enthalpy of reaction (ΔH, 298.15 K) is −4.7 kcal mol^{-1} for dehydrochlorination of γ-HCH, −1.0 kcal mol^{-1} for conformational transition of γ-PCCH and −19.3 kcal mol^{-1} for dehydrochlorination of γ-PCCH-1. The low potential barrier and strong exothermicity of three elementary steps indicate that they are energetically feasible. Consequently, the assumed metabolic pathway from γ-HCH to 1,3,4,6-TCDN

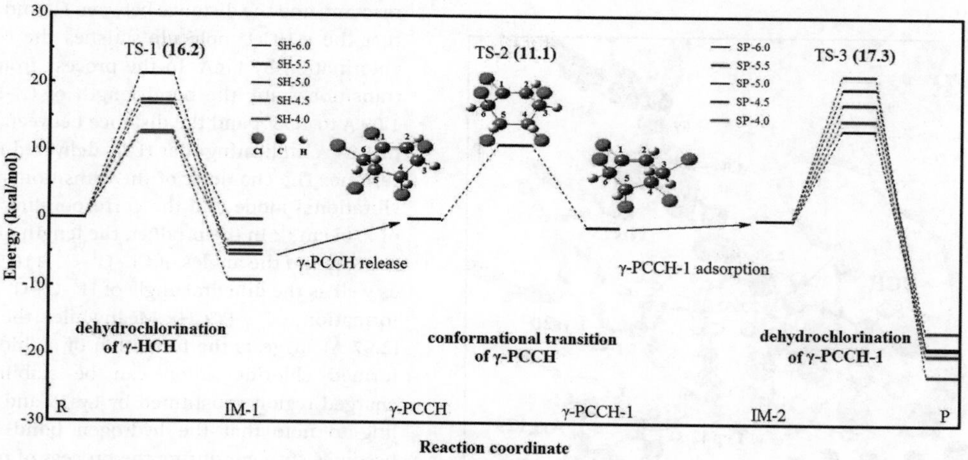

Fig. 1 Energy profiles of three elementary steps along the transformation process from γ-HCH to 1,3,4,6-TCDN. The structures of the reactant (γ-PCCH), transition state (TS-2) and product (γ-PCCH-1) involved in the conformational transition step are exhibited in ball and stick models. The potential barriers of each elementary step are provided in the braces.

Table 1 Potential barriers $\Delta E^{\ddagger}$ (in kcal mol^{-1}) and enthalpy of reaction ΔH (in kcal mol^{-1}) as well as selected internuclear distances (in Å) in the reactant (R), transition state (TS-1) and product (IM-1) involved in the LinA-catalyzed dehydrochlorination of γ-HCH in five pathways. ΔH is calculated at 298.15 K

	N^{ε}–H^1			H^1–C^1			C^1–C^2			C^2–Cl^2				
Pathway	R	TS-1	IM-1	R	TS-1	IM-1	R	TS-1	IM-1	R	TS-1	IM-1	$\Delta E^{\ddagger}$	ΔH
SH-4.0	2.42	1.23	1.02	1.09	1.57	2.60	1.53	1.48	1.34	1.82	1.95	2.91	12.6	−3.9
SH-4.5	2.41	1.23	1.01	1.09	1.55	2.97	1.53	1.48	1.33	1.83	1.96	2.97	17.4	−5.4
SH-5.0	2.37	1.21	1.01	1.09	1.57	2.60	1.53	1.48	1.33	1.82	1.93	2.92	16.8	−0.2
SH-5.5	2.58	1.24	1.01	1.09	1.53	2.61	1.52	1.48	1.33	1.82	1.93	2.97	12.8	−9.1
SH-6.0	2.53	1.22	1.01	1.09	1.56	2.78	1.53	1.48	1.33	1.82	1.94	3.16	21.3	−4.9

Table 2 Potential barriers $\Delta E^{\ddagger}$ (in kcal mol^{-1}) and enthalpy of reaction ΔH (in kcal mol^{-1}) as well as selected internuclear distances in the reactant (IM-2), transition state (TS-3) and product (P) involved in the LinA-catalyzed dehydrochlorination of γ-PCCH-1 in five pathways. ΔH is calculated at 298.15 K

	N^{ε}–H^4			H^4–C^4			C^4–C^5			C^5–Cl^5				
Pathway	IM-2	TS-3	P	IM-2	TS-3	P	IM-2	TS-3	P	IM-2	TS-3	P	$\Delta E^{\ddagger}$	ΔH
SP-4.0	2.40	1.25	1.01	1.10	1.52	2.97	1.53	1.49	1.33	1.82	1.90	4.15	14.9	−19.8
SP-4.5	2.41	1.20	1.01	1.09	1.60	3.15	1.52	1.49	1.34	1.82	1.90	4.01	21.5	−22.8
SP-5.0	2.50	1.25	1.01	1.09	1.52	3.03	1.53	1.50	1.33	1.82	1.91	3.94	19.5	−18.7
SP-5.5	2.48	1.26	1.01	1.09	1.53	2.98	1.53	1.50	1.34	1.82	1.90	4.16	17.3	−16.3
SP-6.0	2.26	1.23	1.01	1.10	1.53	2.81	1.53	1.49	1.34	1.82	1.91	3.78	13.4	−18.9

catalyzed by dehydrochlorinase LinA from *Sphingomonas paucimobilis* UT26 is reasonable.

3.2 Catalytic itinerary and structural details

For convenience of description, several key atoms in the QM region are numbered for the LinA-catalyzed dehydrochlorination of γ-HCH, as presented in Scheme 1B. Some crucial internuclear distances in the reactant, transition state, and product computed at the B3LYP/6-31G(d,p)//CHARMM22 level are provided in Table 1. Since a majority of the catalytic reactions occur through the pathway with the lowest potential barrier, the following investigation towards γ-HCH dehydrochlorination process will mainly focus on the pathway SH-5.5. For a more intuitive observation, the three dimensional structures of R, TS-1, and IM-1 involved in the γ-HCH dehydrochlorination step of the pathway SH-5.5 are displayed in Fig. 2. Obviously, an adjacent *trans*-diaxial H/Cl pair composed by H^1 and Cl^2 is situated towards the N^{ε} atom of His73 residue in the

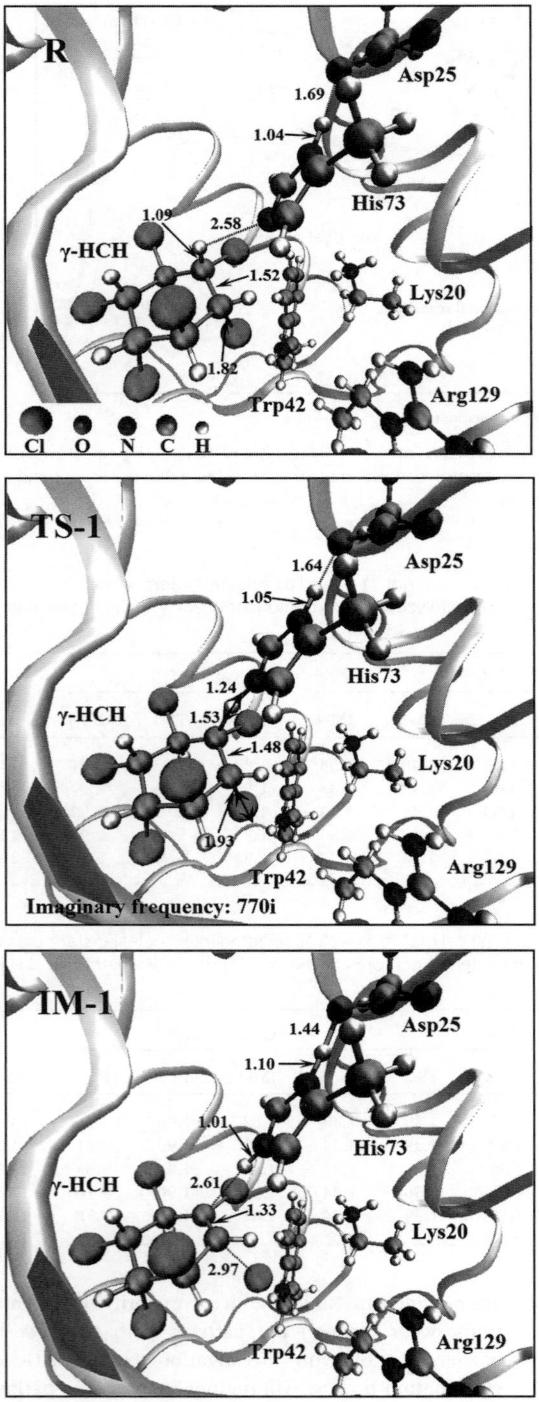

Fig. 2 The three dimensional structures of the reactant (R), transition state (TS-1), and product (IM-1) involved in the pathway SH-5.5 of the γ-HCH dehydrochlorination step. The QM atoms including link hydrogen atoms are shown in ball and stick representation. The unit of bond distances and imaginary frequency are in Å and cm^{-1}.

reactant, and the distance between H^1 and $N^ε$ is 2.58 Å. It reveals that the γ-HCH molecule satisfies the condition of dehydrochlorination by LinA. In the process from the reactant to the transition state, the bond length of C^1–H^1 is stretched from 1.09 Å to 1.53 Å and the distance between $N^ε$ and H^1 is reduced to 1.24 Å, indicating that H^1 is delivered from γ-HCH to His73 residue. The character of the transition state is verified by the vibrational mode and the corresponding imaginary frequency of 770i cm^{-1}. In the product, the length of double bond C^1=C^2 (1.33 Å) and the angles of Cl^1–C^1–C^2 (118.9°), H^2–C^2–C^1 (120.4°) as well as the dihedral angle of H^1–C^1–C^2–Cl^2 (0.8°) suggest the formation of γ-PCCH. Meanwhile, the distance of C^2–Cl^2 (2.97 Å) suggests the formation of a chloride anion. The new-formed chloride anion can be stabilized by a positively charged region constituted by Lys20 and Arg129. It is compelling to note that the hydrogen bond between $O^α$ and $H^β$ becomes stronger during the process of proton H^1 transferring from γ-HCH to His73 residue. Hence, Asp25 can distribute the positive charge in protonated imidazole ring of His73.

For a more detailed description, the internuclear distance and Mulliken population analysis charge variations are introduced. Fig. 3A shows the variations of four crucial internuclear distances along the γ-HCH dehydrochlorination process. It is evident that the dehydrogenation and dechlorination process occur simultaneously, theoretically confirming the fact that LinA catalyzes degradation of γ-HCH *via* an E2 mechanism. The atomic charge analysis of several key atoms is displayed in Fig. 3B. The negative charge of $N^ε$ has been weakened along the process, corresponding to the state of proton transfer. The anion character of Cl^2 in the product was further confirmed by its negative charge (−0.48).

The intermediate γ-PCCH will diffuse out of the enzyme when the dehydrochlorination of γ-HCH is completed.[17] As a consequence, the subsequent conformational transition of γ-PCCH is nonenzymatic. In the present work, the conformational transition step was considered by the DFT method with solvation effect. The structures of reactant, transition state and product optimized at the B3LYP/6-31G(d,p) level are exhibited in Fig. 1. During the conformational transition process, the dihedral angle of C^3–C^4–C^5–C^6 varies from −58.8° to 59.9°, indicates that the relative position of C^4 and C^5 has been inverted. The adjacent diequatorial H^4/Cl^5 pair is converted to *trans*-diaxial H^4/Cl^5 pair. The transformation from one half-chair conformer (γ-PCCH) to another half-chair conformer (γ-PCCH-1) is accomplished. It is worth noting that the dihedral angle of C^3–C^4–C^5–C^6 in transition state is approximately 0°, suggests this four carbon atoms are coplanar in the cyclohexene structure. However, all the six carbon atoms of the cyclohexene structure are not situated in the same plane. The dihedral angle of C^1–C^2–C^3–C^4 (33.7°) and C^2–C^1–C^6–C^5 (−33.2°) reveals that the transition state is a boat form structure. The character of the transition state is also verified by the vibrational mode and the corresponding imaginary frequency of 54i cm^{-1}.

For the dehydrochlorination of γ-PCCH-1, some crucial QM atoms are numbered in Scheme 1C. The degradation process was investigated at the B3LYP/6-31G(d,p)//CHARMM22 level. Four selected internuclear distances in the reactant, transition

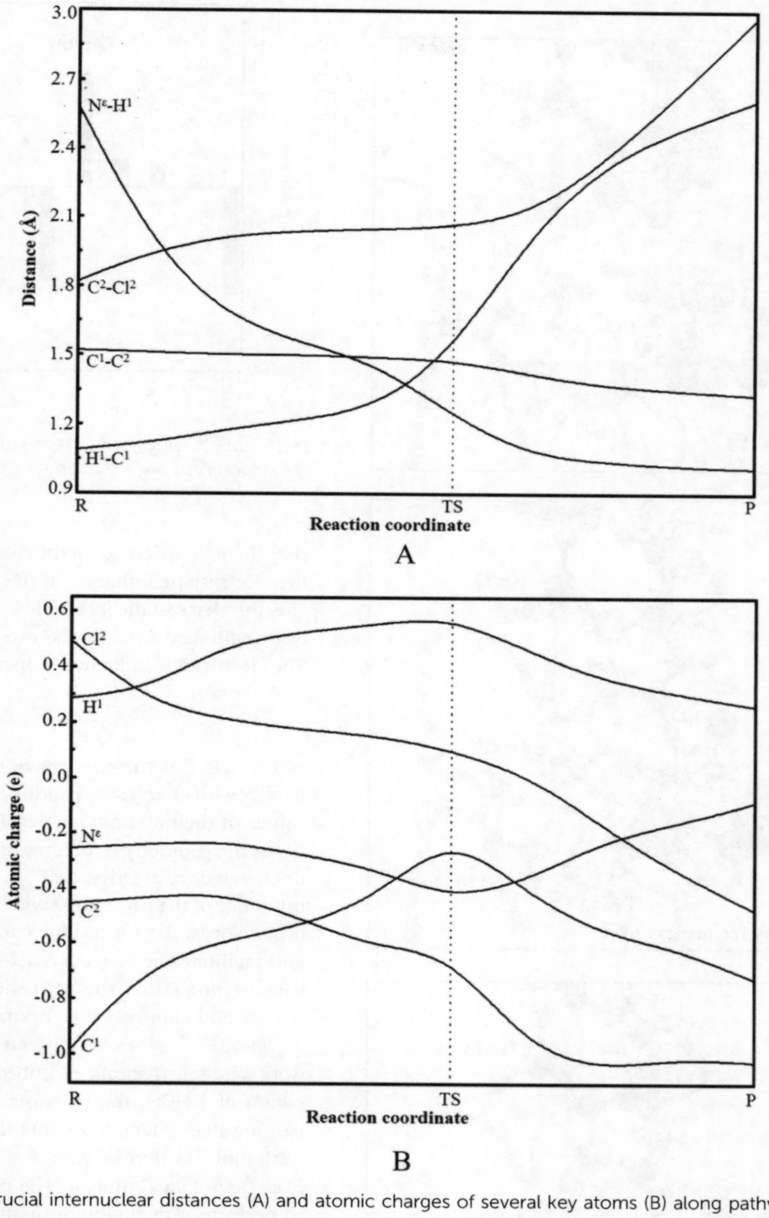

Fig. 3 Variations of four crucial internuclear distances (A) and atomic charges of several key atoms (B) along pathway SH-5.5 of the γ-HCH dehydrochlorination process.

state and product are provided in Table 2 respectively. Fig. 4 displays the active site structures of IM-2, TS-3, and P in the pathway SP-6.0 as it executes the dehydrochlorination process with the lowest potential barrier. An overall view of the reaction process indicates that the dehydrochlorination of γ-PCCH-1 is accomplished with the same mechanism as that from γ-HCH. The metabolism product 1,3,4,6-TCDN is optimized successfully, theoretically verifying the existence of the putative short-lived product. The distance between the leaving chlorine atom (Cl^5) and its interrelated carbon atom (C^5) is 3.78 Å. However, the negative charge of the leaving chlorine atom Cl^5 (−0.29) is incomprehensibly weaker than that of Cl^2 (−0.48). A reasonable explanation is that the chloride anion Cl^5 is closer to the positively charged region constituted by Lys20 and Arg129, causing a more sufficient charge dispersion.

3.3 Individual residue influence

According to previous crystal structure study, the active site of LinA is largely surrounded by fourteen residues.[17] They can make an electrostatic influence on the enzyme reaction, though

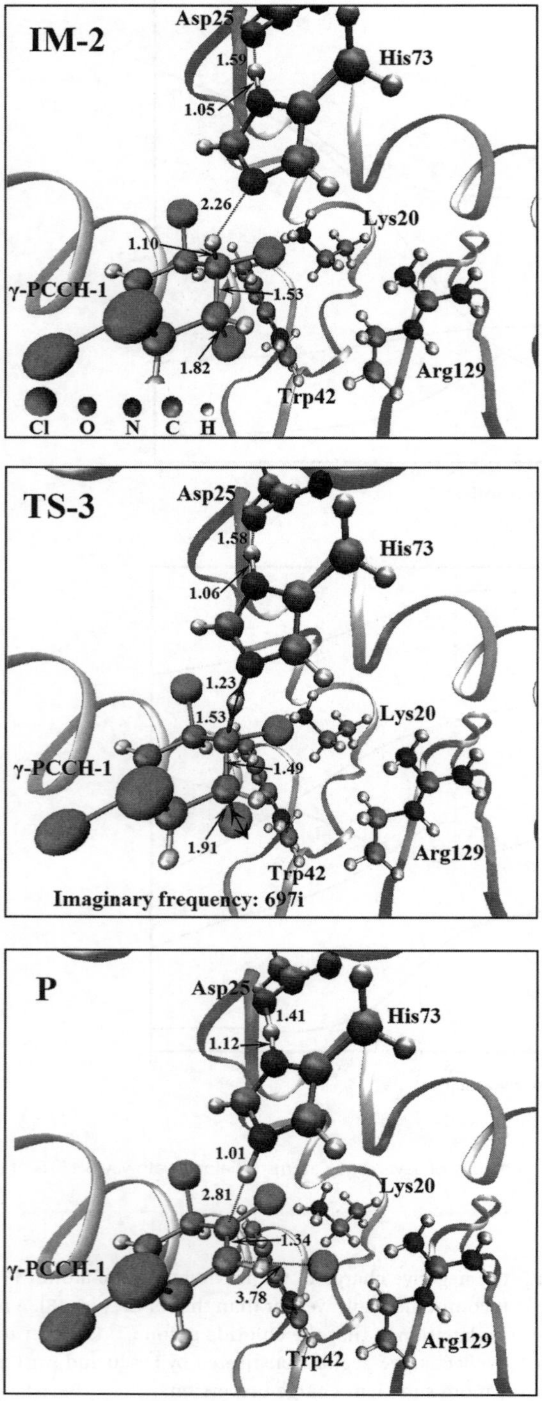

Fig. 4 The three dimensional structures of the reactant (IM-2), transition state (TS-3), and product (P) involved in the pathway SP-6.0 of the γ-PCCH-1 dehydrochlorination step. The QM atoms including link hydrogen atoms are shown in ball and stick representation. The unit of bond distances and imaginary frequency are in Å and cm^{-1}.

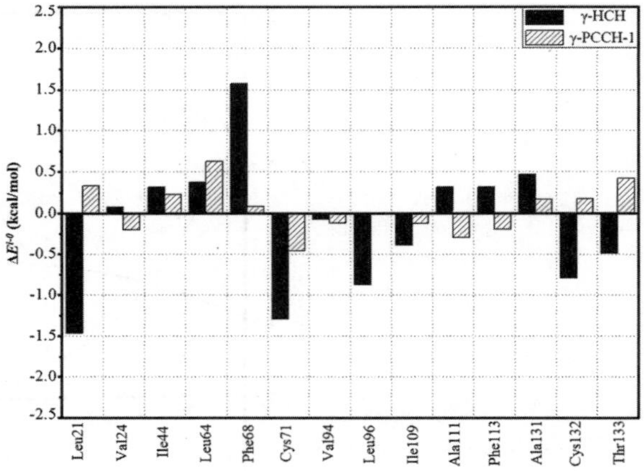

Fig. 5 ΔE^{i-0} values of fourteen individual residues toward the dehydrochlorination of γ-HCH and γ-PCCH-1.

they do not participate in the reaction directly. In order to clarify the electrostatic influence of the residues surrounding the active site, the electrostatic interaction energies of the fourteen residues were estimated towards the two dehydrochlorination processes. The electrostatic influence of an amino acid i can be described as:

$$\Delta E^{i-0} = \Delta E^{i} - \Delta E^{0}$$

where, ΔE^{i-0} is the changes of the barrier, ΔE^{i} is the potential barrier with charges on residue i set to 0, and ΔE^{0} is the original values of the potential barrier. During all these energy calculations, the geometry structures of the stationary points were kept unchanged. A positive ΔE^{i-0} value means that neglecting the influence of the ith residue will increase the potential barrier. In other words, the ith residue can diminish the potential barrier and facilitate the enzyme reaction. Contrarily, a negative ΔE^{i-0} value denotes that the ith residue can increase the potential barrier and suppress the enzyme reaction.[47]

The ΔE^{i-0} values of fourteen residues studied in the current work were schematically described in Fig. 5. For the dehydrochlorination of γ-HCH, the electrostatic influence analysis shows that residue Phe68 facilitates this degradation reaction ($\Delta E^{i-0} > 1$ kcal mol^{-1}),whereas residues Leu21 and Cys71 suppress it ($\Delta E^{i-0} < -1$ kcal mol^{-1}). The other eleven residues are found to perform a negligible influence (-1 kcal $mol^{-1} < \Delta E^{i-0} < 1$ kcal mol^{-1}) towards the dehydrochlorination of γ-HCH. This electrostatic influence analysis highlights Leu21 and Cys71 as candidate residues for future mutation studies. In addition, all the fourteen residues studied in this analysis are found to have a weaker effect (-1 kcal $mol^{-1} < \Delta E^{i-0} < 1$ kcal mol^{-1}) on the dehydrochlorination of γ-PCCH-1.

4. Conclusions

The present work investigated the biotransformation pathway from γ-HCH to 1,3,4,6-TCDN catabolized by dehydrochlorinase LinA from *Sphingomonas paucimobilis* UT26. The degradation process contains two discontinuous dehydrochlorination reactions. The product of the first dehydrochlorination step undergoes a conformational transition instead of executing the second dehydrochlorination step directly. The electrostatic influence analysis reveals that the residue Phe68 facilitates the degradation reaction most and the residues Leu21 and Cys71 suppress it. It can be a valuable base for rational design of mutants of dehydrochlorinase LinA with a more efficient activity towards the degradation of γ-HCH and further experimental verification would be anticipated.

Acknowledgements

This work was supported by NSFC (National Natural Science Foundation of China, project No. 21337001, 21177077) and the Research Fund for the Doctoral Program of Higher Education of China (project No. 20130131110058).

References

1 V. Raina, A. Hauser, H. R. Buser, D. Rentsch, P. Sharma, R. Lal, C. Holliger, T. Poiger, M. D. Muller and H. P. E. Kohler, *Environ. Sci. Technol.*, 2007, **41**, 4292–4298.
2 M. Okai, J. Ohtsuka, L. F. Imai, T. Mase, R. Moriuchi, M. Tsuda, K. Nagata, Y. Nagata and M. Tanokura, *J. Bacteriol.*, 2013, **195**, 2642–2651.
3 K. L. Willett, E. M. Ulrich and R. A. Hites, *Environ. Sci. Technol.*, 1998, **32**, 2197–2207.
4 R. Lal, G. Pandey, P. Sharma, K. Kumari, S. Malhotra, R. Pandey, V. Raina, H. P. E. Kohler, C. Holliger, C. Jackson and J. G. Oakeshott, *Microbiol. Mol. Biol. Rev.*, 2010, **74**, 58–80.
5 M. Suar, A. Hauser, T. Poiger, H. R. Buser, M. D. Müller, C. Dogra, V. Raina, C. Holliger, J. R. van der Meer and R. Lal, *Appl. Environ. Microbiol.*, 2005, **71**, 8514–8518.
6 K. Bala, B. Geueke, M. Miska, D. Rentsch, T. Poiger, M. Dadhwal, R. Lal, C. Holliger and H. P. E. Kohler, *Environ. Sci. Technol.*, 2012, **46**, 4051–4058.
7 K. Walker, D. A. Vallero and R. G. Lewis, *Environ. Sci. Technol.*, 1999, **33**, 4373–4378.
8 H. R. Buser and M. D. Muller, *Environ. Sci. Technol.*, 1995, **29**, 664–672.
9 P. J. Middeldorp, W. van Doesburg, G. Schraa and A. J. Stams, *Biodegradation*, 2005, **16**, 283–290.
10 T. M. Phillips, A. G. Seech, H. Lee and J. T. Trevors, *Biodegradation*, 2005, **16**, 363–392.
11 S. Nagasawa, R. Kikuchi, Y. Nagata, M. Takagi and M. Matsuo, *Chemosphere*, 1993, **26**, 1719–1728.
12 H. H. M. Rijnaarts, A. Bachmann, J. C. Jumelet and A. J. B. Zehnder, *Environ. Sci. Technol.*, 1990, **24**, 1349–1354.
13 S. K. Sahu, K. K. Patnaik, S. Bhuyan, B. Sreedharan, N. Kurihara, T. K. Adhya and N. Sethunathan, *J. Agric. Food Chem.*, 1995, **43**, 833–837.
14 S. K. Sahu, K. K. Patnaik and N. Sethunathan, *Bull. Environ. Contam. Toxicol.*, 1992, **48**, 265–268.
15 S. K. Sahu, K. K. Patnaik, M. Sharmila and N. Sethunathan, *Appl. Environ. Microbiol.*, 1990, **56**, 3620–3622.
16 D. Boltner, S. Moreno-Morillas and J. L. Ramos, *Environ. Microbiol.*, 2005, **7**, 1329–1338.
17 M. Okai, K. Kubota, M. Fukuda, Y. Nagata, K. Nagata and M. Tanokura, *J. Mol. Biol.*, 2010, **403**, 260–269.
18 Y. Nagata, K. Miyauchi and M. Takagi, *J. Ind. Microbiol. Biotechnol.*, 1999, **23**, 380–390.
19 R. Imai, Y. Nagata, M. Fukuda, M. Takagi and K. Yano, *J. Bacteriol.*, 1991, **173**, 6811–6819.
20 L. Trantírek, K. Hynkova, Y. Nagata, A. Murzin, A. Ansorgova, V. Sklenař and J. Damborsky, *J. Biol. Chem.*, 2001, **276**, 7734–7740.
21 P. G. Deo, N. G. Karanth and N. G. Karanth, *Crit. Rev. Microbiol.*, 1994, **20**, 57–78.
22 R. Lal, M. Dadhwal, K. Kumari, P. Sharma, A. Singh, H. Kumari, S. Jit, S. K. Gupta, D. Lal, M. Verma, J. Kaur, K. Bala and S. Jindal, *Indian J. Microbiol.*, 2008, **48**, 3–18.
23 H. Li, A. D. Robertson and J. H. Jensen, *Proteins*, 2005, **61**, 704–721.
24 A. D. MacKerell, D. Bashford, M. Bellott, R. L. Dunbrack, J. D. Evanseck, M. J. Field, J. Gao, H. Guo, S. Ha, D. Joseph-McCarthy, L. Kuchnir, K. Kuczera, F. T. Lau, C. Mattos, S. Michnick, T. Ngo, D. T. Nguyen, B. Prodhom, W. E. Reiher, B. Roux III, M. Schlenkrich, J. C. Smith, R. Stote, J. Straub, M. Watanabe, J. Wiórkiewicz-Kuczera, D. Yin and M. Karplus, *J. Phys. Chem. B*, 1998, **102**, 3586–3616.
25 B. R. Brooks, R. E. Bruccoleri, B. D. Olafson, D. J. States, S. Swaminathan and M. Karplus, *J. Comput. Chem.*, 1983, **4**, 187–217.
26 B. R. Brooks, C. L. Brooks III, A. D. Mackerell, L. Nilsson, R. J. Petrella, B. Roux, Y. Won, G. Archontis, C. Bartels, S. Boresch, A. Caflisch, L. Caves, Q. Cui, A. R. Dinner, M. Feig, S. Fischer, J. Gao, M. Hodoscek, W. Im, K. Kuczera, T. Lazaridis, J. Ma, V. Ovchinnikov, E. Paci, R. W. Pastor, C. B. Post, J. Z. Pu, M. Schaefer, B. Tidor, R. M. Venable, H. L. Woodcock, X. Wu, W. Yang, D. M. York and M. Karplus, *J. Comput. Chem.*, 2009, **30**, 1545–1615.
27 A. D. MacKerell Jr, B. Brooks III, C. L. Brooks, L. Nilsson, B. Roux, Y. Won and M. Karplus, *Encyclopedia Comput. Chem.*, 1998, **1**, 271–277.
28 V. B. Chen, W. B. Arendall III, J. J. Headd, D. A. Keedy, R. M. Immormino, G. J. Kapral, L. W. Murray, J. S. Richardson and D. C. Richardson, *Acta Crystallogr., Sect. D: Biol. Crystallogr.*, 2010, **66**, 12–21.
29 G. Wu, D. H. Robertson, C. L. Brooks III and M. Vieth, *J. Comput. Chem.*, 2003, **24**, 1549–1562.
30 M. Vieth, J. D. Hirst, A. Kolinski and C. L. Brooks III, *J. Comput. Chem.*, 1998, **19**, 1612–1622.
31 W. L. Jorgensen, J. Chandrasekhar, J. D. Madura, R. W. Impey and M. L. Klein, *J. Chem. Phys.*, 1983, **79**, 926–935.
32 C. L. Brooks and M. Karplus, *J. Chem. Phys.*, 1983, **79**, 6312–6325.
33 P. Sherwood, A. H. D. Vries, M. F. Guest, G. Schreckenbach, C. R. A. Catlow, S. A. French, A. A. Sokol, S. T. Bromley, W. Thiel, A. J. Turner, S. Billeter, F. Terstegen, S. Thiel,

J. Kendrick, S. C. Rogers, J. Casci, M. Watson, F. King, E. Karlsen, M. Sjovoll, A. Fahmi, A. Schafer and C. Lennartz, *J. Mol. Struct.: THEOCHEM*, 2003, **632**, 1–28.
34 R. Ahlrichs, M. Bär, M. Häser, H. Horn and C. Kölmel, *Chem. Phys. Lett.*, 1989, **162**, 165–169.
35 W. Smith and T. R. Forester, *J. Mol. Graphics Modell.*, 1996, **14**, 136–141.
36 S. R. Billeter, A. J. Turner and W. Thiel, *Phys. Chem. Chem. Phys.*, 2000, **2**, 2177–2186.
37 K. Burke, J. Werschnik and E. K. U. Gross, *J. Chem. Phys.*, 2005, **123**, 1–9.
38 A. H. de Vries, P. Sherwood, S. J. Collins, A. M. Rigby, M. Rigutto and G. J. Kramer, *J. Phys. Chem. B*, 1999, **103**, 6133–6141.
39 J. Baker, *J. Comput. Chem.*, 1986, **7**, 385–395.
40 D. C. Liu and J. Nocedal, *Math. Program.*, 1989, **45**, 503–528.
41 C. Amovilli, V. Barone, R. Cammi, E. Cances, M. Cossi, B. Mennucci, C. S. Pomelli and J. Tomasi, *Adv. Quantum Chem.*, 1998, **32**, 227–261.
42 M. J. Frisch, G. W. Trucks, H. B. Schlegel, P. W. M. Gill, B. G. Johnson, M. A. Robb, J. R. Cheeseman, T. A. Keith, G. A. Petersson and J. A. Montgomery, *GAUSSIAN 03*, Pittsburgh, PA, 2003.
43 H. P. Lu, L. Y. Xun and X. S. Xie, *Science*, 1998, **282**, 1877–1882.
44 W. Min, B. P. English, G. Luo, B. J. Cherayil, S. C. Kou and X. S. Xie, *Acc. Chem. Res.*, 2005, **38**, 923–931.
45 Y. W. Li, R. M. Zhang, L. K. Du, Q. Z. Zhang and W. X. Wang, *RSC Adv.*, 2015, **5**, 13871–13877.
46 R. Lonsdale, J. N. Harvey and A. J. Mulholland, *J. Phys. Chem. B*, 2010, **114**, 1156–1162.
47 Y. W. Li, X. L. Shi, Q. Z. Zhang, J. T. Hu, J. M. Chen and W. X. Wang, *Environ. Sci. Technol.*, 2014, **48**, 5008–5016.
48 D. R. B. Brittain, R. Pandey, K. Kumari, P. Sharma, G. Pandey, R. Lal, M. L. Coote, J. G. Oakeshott and C. J. Jackson, *Chem. Commun.*, 2011, **47**, 976–978.
49 J. R. Alvarez-Idaboy, A. Galano, G. Bravo-Perez and M. E. Ruiz, *J. Am. Chem. Soc.*, 2001, **123**, 8387–8395.

Front. Environ. Sci. Eng., 2016, 10(5) : 03
DOI 10.1007/s11783-016-0836-z

RESEARCH ARTICLE

Heterogeneous reaction mechanism of gaseous HNO_3 with solid NaCl: a density functional theory study

Nan ZHAO, Qingzhu ZHANG (✉), Wenxing WANG

Environment Research Institute, Shandong University, Jinan 250100, China

HIGHLIGHTS

- We studied the heterogeneous reaction mechanism of gaseous HNO_3 with solid NaCl.
- HCl is released from heterogeneous reactions between gaseous HNO_3 and solid NaCl.
- Water molecules induce surface reconstruction of NaCl to facilitate the reaction.

GRAPHIC ABSTRACT

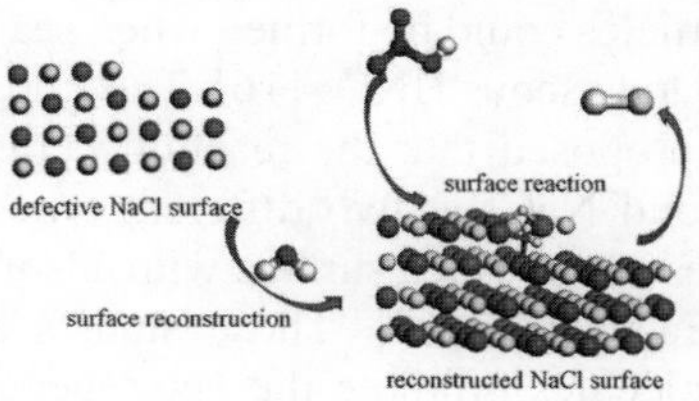

ARTICLE INFO

Article history:
Received 11 November 2015
Received in revised form 22 February 2016
Accepted 22 February 2016

Keywords:
Seasalt particles
NaCl
HNO_3
Heterogeneous reaction
Reaction mechanism
Density functional theory

ABSTRACT

Sea salt particles containing NaCl are among the most abundant particulate masses in coastal atmosphere. Reactions involving sea salt particles potentially generate Cl radicals, which are released into coastal atmosphere. Cl radicals play an important role in the nitrogen and O_3 cycles, sulfur chemistry and particle formation in the troposphere of the polluted coastal regions. This paper aimed at the heterogeneous reaction between gaseous HNO_3 and solid NaCl. The mechanism was investigated by density functional theory (DFT). The results imply that water molecules induce the surface reconstruction, which is essential for the heterogeneous reaction. The surface reconstruction on the defective (710) surface has a barrier of 10.24 $kcal \cdot mol^{-1}$ and is endothermic by 9.69 $kcal \cdot mol^{-1}$, whereas the reconstruction on the clean (100) surface has a barrier of 18.46 $kcal \cdot mol^{-1}$ and is endothermic by 12.96 $kcal \cdot mol^{-1}$. The surface reconstruction involved in water-adsorbed (710) surface is more energetically favorable. In comparison, water molecules adsorbed on NaCl (100) surface likely undergo water diffusion or desorption. Further, it reveals that the coordination number of the Cl^-_{out} is reduced after the surface reconstruction, which assists Cl^-_{out} to accept the proton from HNO_3. HCl is released from heterogeneous reactions between gaseous HNO_3 and solid NaCl and can react with OH free radicals to produce atomic Cl radicals. The results will offer further insights into the impact of gaseous HNO_3 on the air quality of the coastal areas.

1 Introduction

Chlorine (Cl) radicals are highly reactive toward a variety of volatile organic compounds (VOCs). The rate constants for reactions with atomic Cl is generally one to two orders of magnitudes larger than those of hydroxyl radical (OH), the major atmospheric oxidant [1]. Atomic Cl plays an important role in the nitrogen and O_3 cycles, sulfur chemistry and particle formation in the troposphere of the polluted coastal regions [2,3]. Field measurements explored that the maximum atomic Cl concentration can be up to 10^5 $atoms \cdot cm^{-3}$ in the early morning in coastal atmosphere [4,5].

NaCl-containing sea salt aerosols are one of the most abundant particulate masses and the largest natural sources of Cl in the troposphere of coastal areas [6–8]. Ionic composition studies revealed a depletion of chlorine in the sea salt aerosols in comparison with original seawater in the polluted marine troposphere [9,10]. The depletion of Cl was attributed by the release of gaseous chlorine-containing compounds through heterogeneous/multiphase reactions between sea salt aerosols and gaseous H_2SO_4 or nitrogen compounds such HNO_3, N_2O_5, NO_3, etc [11]. HCl is the most abundant chlorine-containing species

✉ Corresponding author
E-mail: zqz@sdu.edu.cn

released from heterogeneous/multiphase reactions of sea salt aerosols in the marine environment and reacts with OH free radicals to produce atomic Cl radicals [12]. Hence, a mechanistic understanding of these heterogeneous/multiphase reactions of NaCl is crucial for illustrating the oxidativecapacity of the atmosphere, especially in the polluted coastal regions.

HNO_3 has been proposed to react with sea salt aerosols to liberate Cl in the form of HCl [11,12]. Average concentration of HNO_3 was 16–67 nmol·m^{-3} in June and 5–21 nmol·m^{-3} in October in Osaka Prefecture [13].The heterogeneous reaction of gaseous HNO_3 with solid NaCl was investigated by several studies [14–17]. Fenter et al. characterized the titled reaction in a Teflon-coated, low-pressure flow reactor and found that HCl was the sole gas-phase product [14]. Ro et al. showed strong evidence that $NaNO_3$-containing particles could be formed when sea salt aerosols were exposed to gaseous HNO_3 [16]. The study of Finlayson-Pitts et al. proposed that the reaction between gaseous HNO_3 and solid NaCl at two different types of sites: 1) steps and edges on the NaCl surface with adsorbed water and 2) dry terrace sites [17]. These studies also indicate that water molecules promote the heterogeneous reaction between gaseous HNO_3 and solid NaCl. However, the mechanism of water adsorbed on NaCl surface affecting heterogeneous reaction still remain poorly understood.

The mechanism of the heterogeneous reaction between gaseous HNO_3 and solid NaCl has not been fully illustrated by previously experimental studies. In particular, the transition states formed in the heterogeneous reaction are very short-lived and would be almost impossible for direct experimental characterization. In this work, therefore, we carried out a theoretical study on the heterogeneous reaction of gaseous HNO_3 with solid NaCl using density functional theory (DFT). The results provide fundamental understanding of heterogeneous reaction of gaseous HNO_3 with solid NaCl, which will offer further insights into the impact of gaseous HNO_3 on the air quality of the coastal areas.

2 Computational methods

The quantum chemical calculations were performed using DMol3 package from Accelrys [18,19]. The generalized gradient approximation (GGA) with the Perdew-Burke-Ernzerhof (PBE) method was employed to describe the exchange and correlation energy [20]. The wave function was expanded in terms of numerical basis sets of double numerical quality (DNP) [18] with *d*-type polarization functions on each atom. To precisely describe the hydrogen bonding interaction, DNP includes a polarization *p*-functionon all hydrogen atoms [21]. DFT semi-empirical dispersion interaction correction module was also included to account for dispersion forces [22]. All calculations were performed with spin unrestricted. The real-space cutoff radius was 5.2 Å. The periodic system has a vacuum thickness of 10 Å, which was chosen to eliminate spurious interactions between the adsorbate and the periodic image of the bottom layer of the surface. The convergence criterion of self-consistent field (SCF), geometry optimization for energy and maximum force were 1.0×10^{-6} Ha, 1.0×10^{-5} Ha and 0.001 HaÅ^{-1}, respectively. The surfaces were modeled by slab geometries with periodic boundary conditions. Surfaces of NaCl (100), a clean ideal surface, and NaCl (710), a surface with step defect, were examined. For the NaCl (100) surface, four atomic layers of NaCl and a 3×3 super cell were adopted. Four atomic layers and a 2×2 super cell were constructed to model the NaCl (710) surface, where the layers were tilted to create a stoichiometric (100)-like step on the surface. This particular type of structure has been identified as the lowest energy step defect [23]. A $2 \times 2 \times 1$ *k*-point mesh was used for the NaCl (100) and NaCl (710) surfaces [24]. Charge transfer was calculated on the basis of the Mulliken population analysis.

The transition states were located using the synchronous method with conjugated gradient refinements [25]. This method involves linear synchronous transit (LST) maximization, followed byrepeated conjugated gradient (CG) minimizations, and then quadratic synchronous transit (QST) maximizations and repeated CG minimizations until a transition state is located. All computed stationary pointson the potential energy surface were confirmed by frequency analysis and minimum energy pathway (MEP) search based on nudged-elastic band (NEB) algorithm [26].

For the adsorption of HNO_3, H_2O or HCl on the NaCl surface, the adsorption energy (E_{ads}) was calculated as the following equation:

$$Eads = E[(\mathrm{NaCl}+\mathrm{X})] - E[\mathrm{NaCl}] - E[X\mathrm{gas}], \quad (1)$$

where, $E[(\mathrm{NaCl}+\mathrm{X})]$ is the total energy of the adsorbate-substrate system in the equilibrium state. E[NaCl] and $E[X_{gas}]$ are the energies of the NaCl substrate and adsorbate (gaseous HNO_3, H_2O or HCl), respectively.

The geometrical parameters of gaseous HNO_3, H_2O and HCl molecules and the bulk NaCl were presented to validate the reliability of our theoretical methods. As shown in Fig. S1 in Supporting Information, the values agree well with the previously reported experimental data. The calculated lattice constant of NaCl is 5.680 Å, which is consistent with the previous experimental (5.64 Å) and theoretical values (5.67 Å) [27,28]. In order to verify the reliability of the energy parameters, the adsorption energy of a H_2O molecule adsorbed on the NaCl (100) surface was calculated. The value of –13.15 kcal·mol^{-1}shows good consistency with previous experimental (–13.87 to –11.58 kcal·mol^{-1}) and theoretical values (–11.87 kcal·mol^{-1}) [29–31].

3 Results and discussion

3.1 Reaction of HNO_3 with dry NaCl

Theoretical calculations were conducted to characterize the adsorption of one HNO_3 molecule on the flat NaCl (100) as well as on the stepped NaCl (710) surfaces. Possible adsorption modes were determined and the corresponding possibilities were evaluated based on adsorption energies. The most stable configurations are shown in Fig. 1. In the case of NaCl (100), two O atoms and one H atom in the adsorbed HNO_3 molecule interact with the surface. The O-H bond length in adsorbed HNO_3 is 1.012 Å. The adsorption energy, E_{ads}, is −16.55 kcal· mol^{-1}.

The defective surfaces with steps or kinks/corners should be present inevitably in real NaCl crystals. NaCl (710) was chosen to model the surface with a step defect. As shown in Fig. 1, in the most stable adsorption mode, the adsorbed HNO_3 molecule interacts with the step through two O atoms and one H atom.The O-H bond length in adsorbed HNO_3 is 1.027 Å.The adsorption energy is −20.42 kcal·mol^{-1}. Calculations suggest that the gaseous HNO_3 molecule can be readily adsorbed on both flat and stepped NaCl surfaces, and the adsorption on the step is more energetically favorable in comparison with the adsorption on flat surface. The HNO_3 molecule adsorbed on the (710) surface can approximately maintain a planar structure as the sole gaseous HNO_3 molecule. However, the molecule adsorbed on the (100) surface is distorted, thus resulting in decreased adsorption energy.

The configuration was found to be unstable when the H of HNO_3 was transferred to the adjacent Cl^-. As illustrated

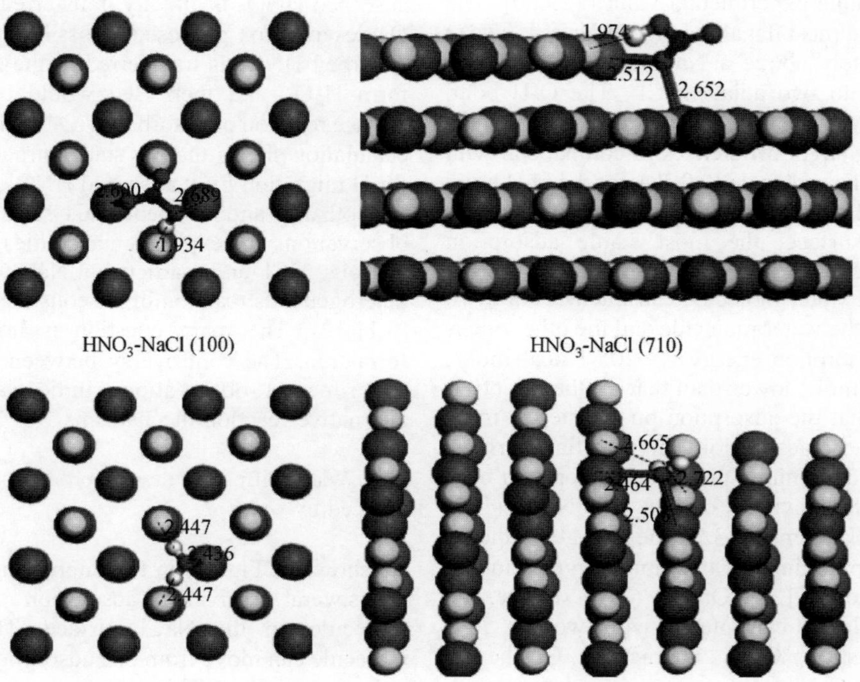

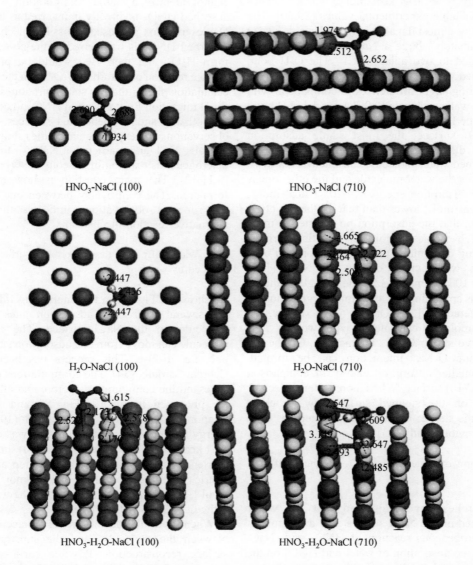

Fig. 1 The most stable configurations of HNO_3-NaCl, H_2O-NaCl and HNO_3-H_2O-NaCl adsorption systems. The distances are in angstrom. Purple ball: Na; green ball: Cl; blue ball: N; red ball: O; white ball: H

in Fig. 1, the O-H bond length in the adsorbed HNO_3 shows negligible difference in comparison with that in the sole gaseous HNO_3 molecule. It implies that the adsorbed HNO_3 molecule is inert on both flat and stepped NaCl surfaces. Thus, the reaction of HNO_3 with dry NaCl is unlikely to occur under the general atmospheric conditions.

3.2 Water adsorption on flat (100) and stepped (710) surfaces

Herein, the adsorption of water on (100) and (710) surfaces was investigated to reveal the general properties ofthe genuine sea salt particles. Different adsorption modes were analyzed and the most stable configurations are illustrated in Fig. 1. For the flat (100) surface, the calculated adsorption energy is -13.15 kcal·mol^{-1}, which matches well with the available experimental values [29,30]. The water molecule lies almost flat along the Na-Na axis.The O atom is approximately above a Na^+ and two hydrogen atoms are oriented to two adjacent Cl^-. The O-H bond length in adsorbed water is 0.978Å and the HOH angle is 103.4 Å, presenting slight differences in comparison with the sole gaseous water molecule (0.971Å for the O-H bond and 104.1° for the HOH angle).

For the (710) surface, the most stable adsorption configuration is displayed in Fig. 1. In this adsorption mode, the adsorbed water molecule can interact with two Na^+ - one is below the water molecule and the other one is in the step. The adsorption energy is –20.81 kcal·mol^{-1}, which is 7.66 kcal·mol^{-1} lower than that on the flat (100) surface, implying that the adsorption on the step is more favorable relative to adsorption on the flat surface. Ahlswede and Jug determined that water adsorption on a monatomic step was favored overon the NaCl (100) surface by about 10 kcal·mol^{-1} [32]. The strength of the O-Na^+ interaction is approximately two times larger than that of the H-Cl^- interaction [28]. On the (710) surface, the adsorbed water molecule can interact with two Na^+ ions and thus form two strong O-Na^+ interactions. On the flat surface, only one O-Na^+ interaction can be formed, resulting in a smaller adsorption energy in comparison with that on the (710) surface. When more water molecules were adsorbed on the stepped surface, awater chain structure along the step could be formed, which has been observed in a previous study [33].

3.3 Co-adsorption of water and HNO_3 on the (100) and (710) surfaces

Previous researches have suggested that the water molecule adsorbed on the NaCl surface play an important role in the heterogeneous reaction of HNO_3 with NaCl [14–17]. Herein, co-adsorption of water and HNO_3 on the flat (100) and stepped (710) surfaces was investigated. Possible adsorption modes were characterized and the most stable adsorption configurations are depicted in Fig. 1. HNO_3 tends to be close to the adsorbed water molecule on both (100) and (710) surfaces. On the flat (100) surface, the co-adsorption energy is -31.13 kcal·mol^{-1}, which is 1.43 kcal·mol^{-1} smaller than the sum (-29.70 kcal·mol^{-1}) of isolated HNO_3 and water adsorption energies; whereas, on the stepped (710) surface, the co-adsorption energy of -41.22 kcal·mol^{-1} is almost equal to the sum (-41.23 kcal·mol^{-1}) of isolated HNO_3 and water adsorption energies. On the (100) surface, a strong hydrogen bond (1.615 Å) formed between the H atom of HNO_3 and the O atom of H_2O, which stabilizes the co-adsorption system. In contrast, no hydrogen bond is formed in the co-adsorption system on the stepped (710) surface.

The mechanism of HCl formation in the HNO_3-H_2O-NaCl co-adsorption system was investigated. However, we cannot identify a stable configuration where the H of adsorbed HNO_3 is directly transferred to the adjacent Cl^-. Davies and Cox proposed a possible mechanism: the H of adsorbed HNO_3 is transferred to the co-adsorbed H_2O to form H_3O^+ and then HCl would be produced by the surface reaction of adsorbed H_3O^+ with the Cl^- ion [8]. Our calculation shows that no stable structures can be formed for H migration from adsorbed HNO_3 to co-adsorbed H_2O on both flat and stepped NaCl surfaces. However, field observations and experimental studies showed evidence of gaseous HCl and particulate $NaNO_3$ formation in the heterogeneous reaction of gaseous HNO_3 with solid NaCl [8,11,12]. The main question is how to explain their formation. The controversy between our calculation and experimental observations indicates the existence of alternative reaction mechanisms.

3.4 Water diffusion, desorption and surface reconstruction induced by water

As shown in Fig. S2 in the Supporting Information, there are several equivalent adsorption sites for the water molecule on the NaCl surface. The adsorbed water molecule can move from one adsorption site to an adjacent one, i.e. diffuse. This process has been identified by Cabrera-Sanfelix [34]. The configurations of the transition statesand the configurations before and after the diffusion are presented in Fig. 2, Fig. S3 and Fig. S4 in the Supporting Information. On the flat (100) surface, the energy barrier of the water diffusion was calculated to be 3.61 kcal·mol^{-1}. In the case of the (710) surface, the energy barriers of the diffusion along the step and on stoichiometric (100)-like plane are18.64 kcal·mol^{-1} and 19.77 kcal·mol^{-1}, respectively. It indicates that the water diffusion is difficult on the (710) surface.

When molecules adsorb on the surface, the interactions between the molecules and the substrate may induce surface reconstruction. Possible surface reconstruction processes were investigated for the NaCl surface upon adsorbing a water molecule. A Cl^- ion (defined as Cl^-_{out})

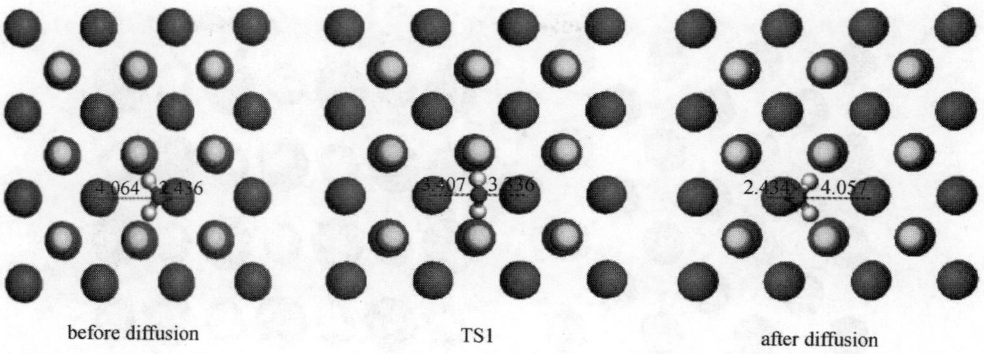

Fig. 2 Configuration of the transition state and the configurations before and after the water diffusion on the flat (100) surface. The distances are in angstrom. All color settings are the same as Fig. 1. TS1: the transition state

adjacent to the adsorbed water molecule could move out of the plane, and then the adsorbed water molecule occupies the vacancy. Our calculations show that the two processes occur simultaneously. The configurations of the transition states and the configurations before and after the surface reconstruction are described in Fig. 3 and Fig. S5 in Supporting Information. In the case of the (100) surface, the energy barrier of surface reconstruction is 18.46 kcal $\cdot mol^{-1}$. After the surface reconstruction, the water molecule embedded in the bulk NaCl interacts with NaCl through three O-Na^+ and two H-Cl^- interactions. The O-H bonds of the water molecule embedded in the bulk NaCl are 0.988 and 1.018 Å and the HOH angle is 109.1°. The Cl^-_{out} is about 1.7 Å above the plane. The coordination number of the Cl^-_{out} is reduced from 5 to 2 after the surface reconstruction. Similarly, the (710) surface can reconstruct after adsorbing a water molecule. The process requires overcoming a barrier of 10.24 kcal$\cdot mol^{-1}$ and is endothermic by 9.69 kcal$\cdot mol^{-1}$.After the surface reconstruction, the interaction between the water molecule with the bulk NaCl involves two O-Na^+ and two H-Cl^- interactions. The O-H bonds and the HOH angle in the water molecule embedded in the bulk NaCl are determined to be 0.979 Å, 1.010 Å and 108.2°. The Cl^-_{out} is about 1.4 Å above the plane. The coordination number of the Cl^-_{out} is reduced from 4 to 2 after the surface reconstruction. Thus, the reconstruction of the (710) surface is more favorable in comparison with the (100) surface, due to the lower energy barrier and less endothermicity.

For the H_2O-NaCl adsorption system, three processes likely occur: water diffusion, desorption, and surface reconstruction induced by water. The water desorption and surface reconstruction processes are endothermic, whereas the water diffusion is thermodynamically neutral. For the (100) surface, the surface reconstruction has a much higher energy barrier compared with the water diffusion. Therefore, the occurrence of water diffusion or desorption rather than the surface reconstruction is preferred on the water-adsorbed NaCl (100) surface. The energy barrier of the surface reconstruction is much lower for the (710) surface than that of the water diffusion. In addition, the surface reconstruction is much less endothermic compared with the water desorption. Hence, surface reconstruction is more feasible on the water-adsorbed (710) surface.

3.5 HCl formation from reconstructed (710) surface

The study above shows that the water-adsorbed (710) surface can readily undergo reconstruction after adsorbing a water molecule. The reaction of gaseous HNO_3 with the reconstructed (710) surface was characterized. The gaseous HNO_3 molecule can adsorb on the step of the reconstructed (710) surface and the most stable adsorption configuration is illustrated in Fig. 4 (denoted as reactant state in Fig. 4). The distance between H of adsorbed HNO_3 and Cl^-_{out} is 1.971 Å, and the O-H length of the adsorbed HNO_3 molecule is 1.031 Å. The adsorption energy is -19.99 kcal$\cdot mol^{-1}$. Then, the Cl^-_{out} above the plane moves toward H of the adsorbed HNO_3 to form a stable intermediate, IM1. This process is endothermic by 4.14 kcal$\cdot mol^{-1}$ with an energy barrier of 7.26 kcal$\cdot mol^{-1}$ through the transition state, TS6. In IM1, the distance between H of the adsorbed HNO_3 and Cl^-_{out} is 1.880 Å, and the adsorbed HNO_3 O-H length is 1.052 Å. Finally, adsorbed HCl is formed when H of HNO_3 is transferred to the Cl^-_{out} via the transition state, TS7. The process has an energy barrier of 4.86 kcal$\cdot mol^{-1}$ and is endothermic by 3.95 kcal$\cdot mol^{-1}$. In TS7, the distance between H and Cl^-_{out} is reduced to 1.482 Å, and the O-H length is further increased to 1.353 Å. In the product state, the bond length of the newly formed H-Cl_{out} is 1.375 Å, and the O-H distance is 1.568 Å, which demonstrates that H of adsorbed HNO_3 has been grabbed by Cl^-_{out}, resulting in the formation of HCl. Desorption of adsorbed HCl is endothermic by 13.66 kcal$\cdot mol^{-1}$.

The gaseous HNO_3 can also possibly adsorb on stoichiometric (100)-like plane of the reconstructed (710) surface. Hence, an alternative reaction pathway was investigated. The stable adsorption configuration is presented in Fig. 5 (denoted as reactant state in Fig. 5). The

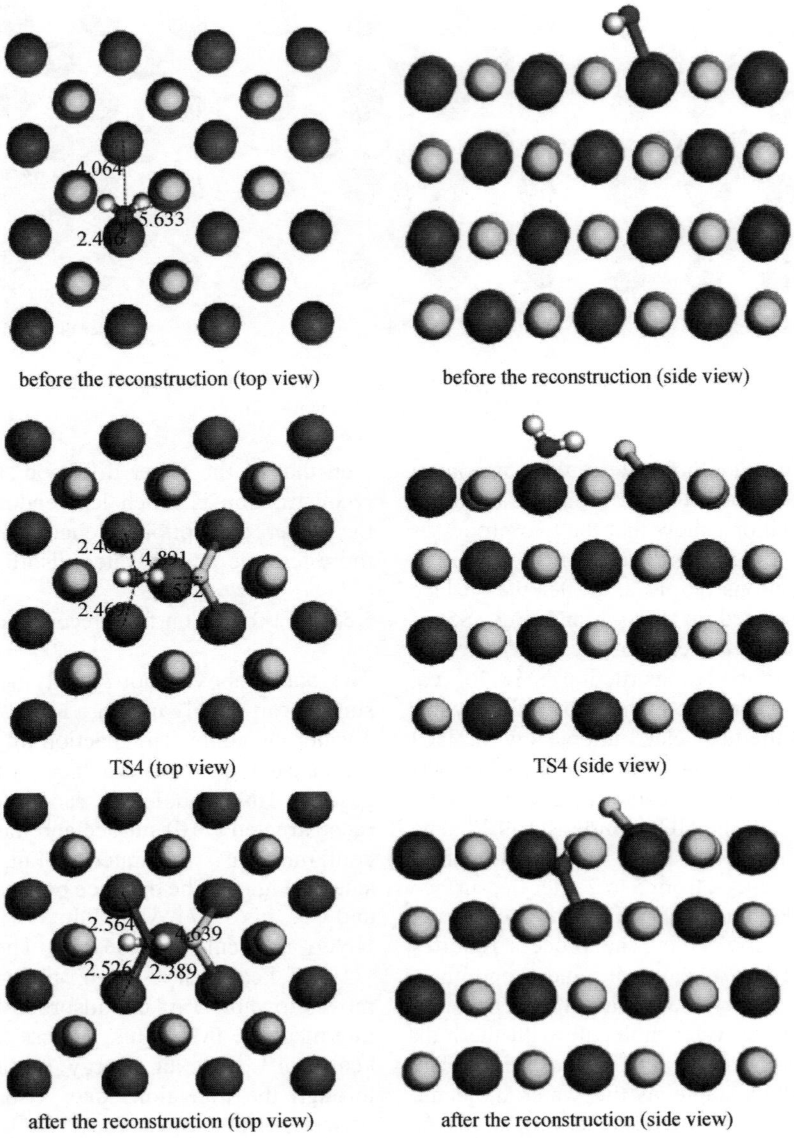

Fig. 3 Configuration of the transition state and the configurations before and after the surface reconstruction of the water-adsorbed (100) surface. The distances are in angstrom. All color settings are the same as Fig. 1. TS4: the transition state

distance between H of adsorbed HNO_3 and Cl^-_{out} is 2.000 Å, and the length of the O-H bond in the adsorbed HNO_3 molecule is 1.017 Å. Sequentially, IM2 is formed through a transition state, TS8. The formation of IM2 is slightly exothermic by 0.15 kcal·mol^{-1} and has a low energy barrier of 1.77 kcal·mol^{-1}. In the intermediate IM2, the distance between H of adsorbed HNO_3 and Cl^-_{out} is 1.876 Å, and the O-H bond in the adsorbed HNO_3 is elongated to 1.047 Å. Then, bound state HCl is formed from IM2 through a transition state, TS9. The process is endothermic by 5.53 kcal·mol^{-1} with an energy barrier of 9.63 kcal·mol^{-1}. In TS9, the distance between H of adsorbed HNO_3 and Cl^-_{out} is shortened to 1.472 Å, and the length of O-H is further elongated to 1.323 Å. Desorption of bound state HCl needs energy of 10.25 kcal·mol^{-1}.

4 Conclusions

In summary, the present theoretical study explored the mechanism of the reaction between gaseous HNO_3 and solid NaCl. The results indicate that the surface reconstruction induced by water is essential for the heterogeneous reaction. The defective surface, (710), is energetically more favorable for the surface reconstruction in comparison with the clean ideal surface, (100). The

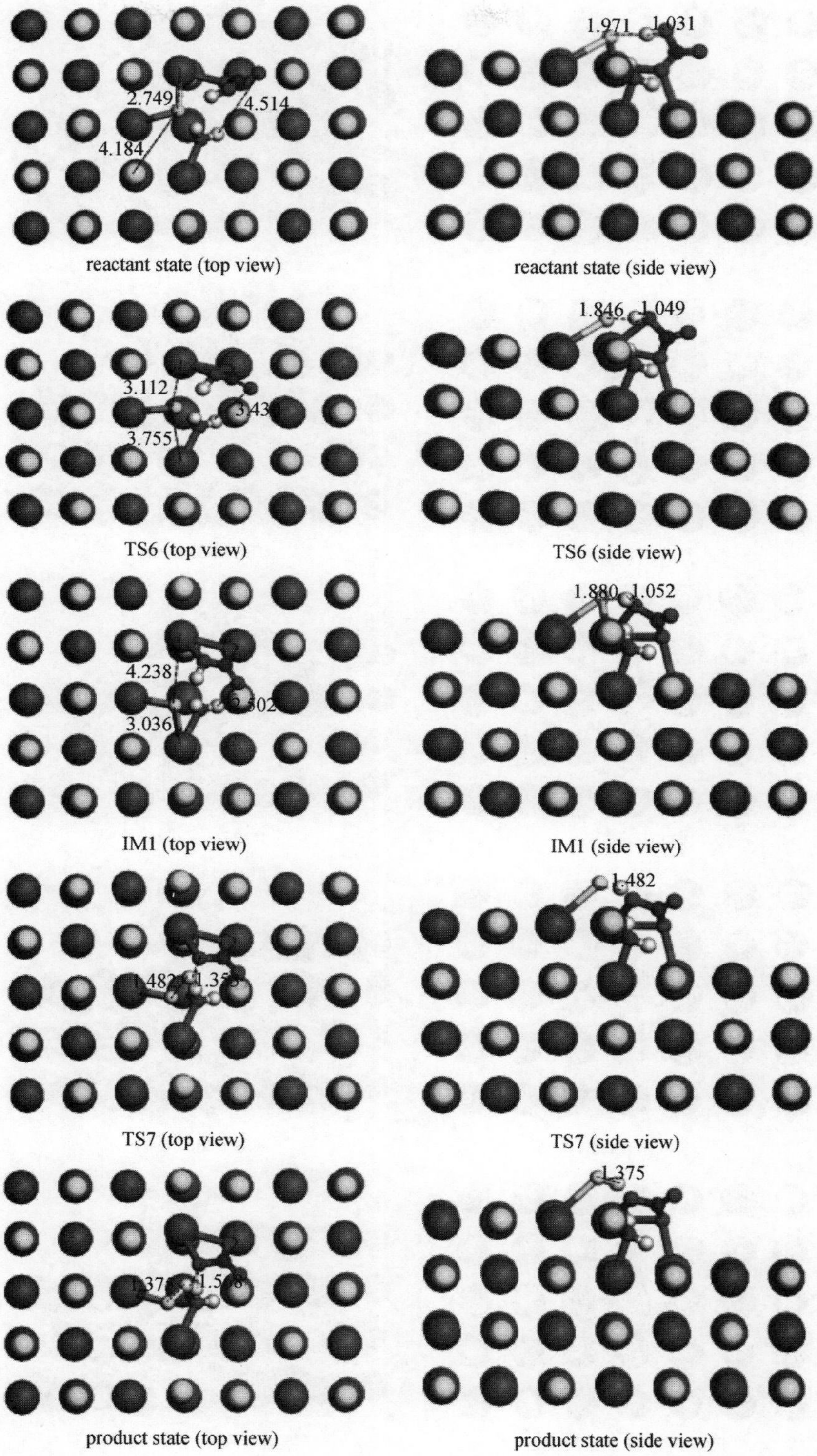

Fig. 4 Configurations of the reactant state, intermediate, transition state and the product state for the reaction of HNO_3 on the step of the reconstructed (710) surface. The distances are in angstrom. All color settings are the same as Fig. 1. TS6 and TS7: transition states. IM1: the intermediate

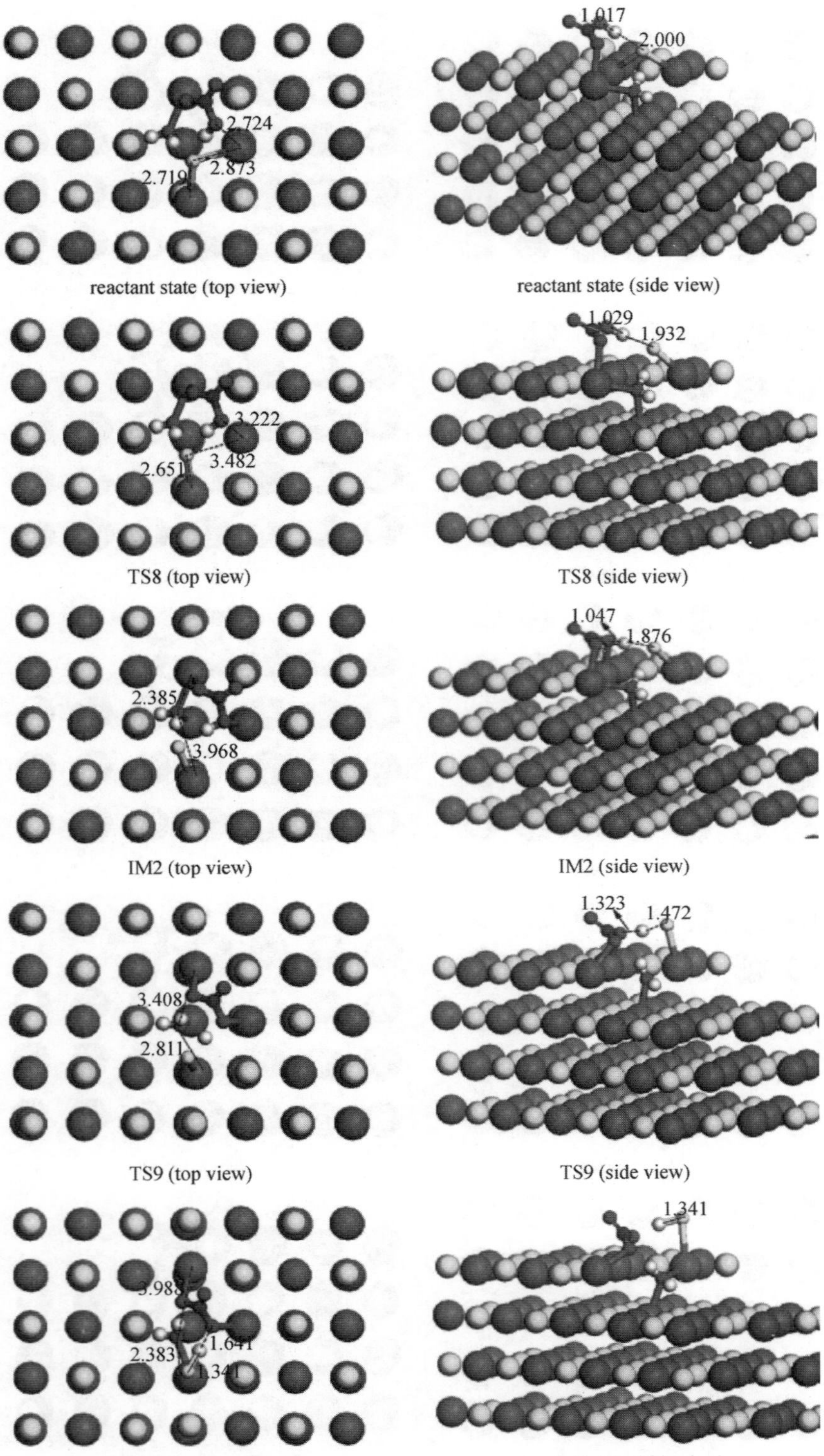

Fig. 5 Configurations of the reactant state, intermediate, transition state and the product state for the reaction of HNO_3 on the stoichiometric (100)-like plane of the reconstructed (710) surface. The distances are in angstrom. All color settings are the same as Fig. 1. TS8 and TS9: transition states. IM2: the intermediate

coordination number of the Cl^{-}_{out} is reduced after the surface reconstruction, which assists Cl^{-}_{out} to accept the proton from HNO_3. The two proposed reaction pathways are likely competitive. Gaseous HCl and particulate $NaNO_3$ can be formed in the heterogeneous reaction of gaseous HNO_3 with solid NaCl.

Acknowledgements This work was supported by the National High Technology Research & Development Program of China (No. 2012AA06A301) and the National Natural Science Foundation of China (Grant Nos. 21337001 and 21177076).

Electronic Supplementary Material Supplementary material is available in the online version of this article at http://dx.doi.org/10.1007/s11783-016-0836-zand is accessible for authorized users.

References

1. Riedel T P, Bertram T H, Crisp T A, Williams E J, Lerner B M, Vlasenko A, Li S M, Gilman J, de Gouw J, Bon D M, Wagner N L, Brown S S, Thornton J A. Nitryl chloride and molecular chlorine in the coastal marine boundary layer. Environmental Science & Technology, 2012, 46(19): 10463–10470
2. Chang S, Allen D T. Atmospheric chlorine chemistry in southeast texas: impacts on ozone formation and control. Environmental Science & Technology, 2006, 40(1): 251–262
3. Sommariva R, Glasow R. Multiphase halogen chemistry in the tropical Atlantic ocean. Environmental Science & Technology, 2012, 46(19): 10429–10437
4. Allan W, Struthers H, Lowe D C. Methane carbon isotope effects caused by atomic chlorine in the marine boundary layer: global model results compared with southern hemisphere measurements. Journal of Geophysical Research, D, Atmospheres, 2007, 112(D4): D04306
5. Keene W C, Stutz J, Pszenny A A P, Maben J R, Fischer E V, Smith A M, von Glasow R, Pechtl S, Sive B C, Varner R K. Inorganic chlorine and bromine in coastal New England air during summer. Journal of Geophysical Research, D, Atmospheres, 2007, 112(D10): D10S12
6. Philip S, Martin R V, van Donkelaar A, Lo J W, Wang Y, Chen D, Zhang L, Kasibhatla P S, Wang S, Zhang Q, Lu Z, Streets D G, Bittman S, Macdonald D J. Global chemical composition of ambient fine particulate matter for exposure assessment. Environmental Science & Technology, 2014, 48(22): 13060–13068
7. Gregory R C, Bhupesh A, Sarika K, Alessio D, Youhua T, David S, Qiang Z, Tami C B, Veerabhadran R, Aditsuda J, Pallavi M. Asian aerosols: current and year 2030 distributions and implications to human health and regional climate change. Environmental Science & Technology, 2009, 43(15): 5811–5817
8. Davies J A, Cox R A. Kinetics of the heterogeneous reaction of HNO_3 with NaCl: effect of water vapor. Journal of Physical Chemistry A, 1998, 102(39): 7631–7642
9. Evans C D, Monteith D T, Fowler D, Cape J N, Brayshaw S. Hydrochloric acid: an overlooked driver of environmental change. Environmental Science & Technology, 2011, 45(5): 1887–1894
10. Yao X, Fang M, Chan C K. Experimental study of the sampling artifact of chloride depletion from collected sea salt aerosols. Environmental Science & Technology, 2001, 35(3): 600–605
11. Ro C U, Kim H, Oh K Y, Yea S K, Lee C B, Jang M, Van Grieken R. Single-particle characterization of urban aerosol particles collected in three Korean cities using low-Z electron probe X-ray microanalysis. Environmental Science & Technology, 2002, 36(22): 4770–4776
12. Hess M, Krieger U K, Marcolli C, Peter T, Lanford W A. Uptake of nitric acid on NaCl single crystals measured by backscattering spectrometry. Nuclear Instruments & Methods in Physics Research. Section B, Beam Interactions with Materials and Atoms, 2010, 268 (11–12): 2202–2204
13. Nishikawa Y, Kannari A. Atmospheric concentration of ammonia, nitrogen dioxide, nitric acid, and sulfur dioxide by passive method within Osaka prefecture and their emission inventory. Water, Air, and Soil Pollution, 2010, 215(1–4): 229–237
14. Fenter F F, Caloz F, Rossi M J. Kinetics of nitric acid uptake by salt. Journal of Physical Chemistry, 1994, 98(39): 9801–9810
15. Finlayson-Pitts B J, Hemminger J C. Physical chemistry of airborne sea salt particles and their components. Journal of Physical Chemistry A, 2000, 104(49): 11463–11477
16. Ro C U, Oh K Y, Kim Y P, Lee C B, Kim K H, Kang C H, Osán J, de Hoog J, Worobiec A, Van Grieken R. Single-particle analysis of aerosols at Cheju Island, Korea, using low-Z electron probe X-ray microanalysis: a direct proof of nitrate formation from sea salts. Environmental Science & Technology, 2001, 35(22): 4487–4494
17. Allen H C, Laux J M, Vogt R, Finlayson-Pitts B J, Hemminger J C. Water-induced reorganization of ultrathin nitrate films on NaCl: implications for the tropospheric chemistry of sea salt particles. Journal of Physical Chemistry, 1996, 100(16): 6371–6375
18. Delley B. An all-electron numerical method for solving the local density functional for polyatomic molecules. Journal of Chemical Physics, 1990, 92(1): 508–517
19. Delley B. From molecules to solids with the DMol3 approach. Journal of Chemical Physics, 2000, 113(18): 7756–7764
20. Perdew J P, Burke K, Ernzerhof M. Generalized gradient approximation made simple. Physical Review Letters, 1996, 77 (18): 3865–3868
21. Rabuck A D, Scuseria G E. Performance of recently developed kinetic energy density functions for the calculation of hydrogen binding strengths and hydrogen-bonded structures. Theoretical Chemistry Accounts, 2000, 104(6): 439–444
22. McNellis E R, Meyer J, Reuter K. Azobenzene at coinage metal surfaces: role of dispersive van der Waals interactions. Physical Review B: Condensed Matter and Materials Physics, 2009, 80(20): 205414
23. Li B, Michaelides A, Scheffler M. Density functional theory study of flat and stepped NaCl(001). Physical Review B: Condensed Matter and Materials Physics, 2007, 76(7): 075401
24. Monkhorst H J, Pack J D. Special points for Brillouin-zone integrations. Physical Review B: Condensed Matter and Materials Physics, 1976, 13(12): 5188–5192
25. Govind N, Petersen M, Fitzgerald G, King-Smith D, Andzelm J. A generalized synchronous transit method for transition state location. Computational Materials Science, 2003, 28(2): 250–258
26. Jonsson H. Improved tangent estimate in the nudged elastic band

method for finding minimum energy paths and saddle points. Journal of Chemical Physics, 2000, 113(22): 9978–9985
27. Nickels J E, Fineman M A, Wallace W E. X-Ray diffraction studies of sodium chloride-sodium bromide solid solutions. Journal of Physical Chemistry, 1949, 53(5): 625–628
28. Yang Y, Meng S, Wang E. Water adsorption on a NaCl (001) surface: adensity functional theory study. Physical Review B: Condensed Matter and Materials Physics, 2006, 74(24): 245409
29. Bruch L W, Glebov A, Toennies J P, Weiss H. A helium atom scattering study of water adsorption on the NaCl(100) single crystal surface. Journal of Chemical Physics, 1995, 103(12): 5109–5120
30. Folsch S, Stock A, Henzler M. Two-dimensional water condensation on the NaCl(100) surface. Surface Science, 1992, 264(1–2): 65–72
31. Li B, Michaelides A, Scheffler M. How strong is the bond between water and salt? Surface Science, 2008, 602(23): L135–L138
32. Ahlswede K J. MSINDO Study of the adsorption of water molecules at defective NaCl(100) surfaces. Surface Science, 1999, 439(1): 86–94
33. Verdaguer A, Sacha G M, Luna M, Ogletree D F, Salmeron M. Initial stages of water adsorption on NaCl (100) studied by scanning polarization force microscopy. Journal of Chemical Physics, 2005, 123(12): 124703
34. Pepa C, Sanfelix A A, George R D, Daniel S. Water adsorption and diffusion on NaCl(100). Journal of Physical Chemistry B, 2006,48 (110): 24559–24564

Colloids and Surfaces A: Physicochem. Eng. Aspects 335 (2009) 194–201

Contents lists available at ScienceDirect

Colloids and Surfaces A: Physicochemical and Engineering Aspects

journal homepage: www.elsevier.com/locate/colsurfa

Adsorption and desorption of divalent mercury (Hg^{2+}) on humic acids and fulvic acids extracted from typical soils in China

Jie Zhang [a], Jiulan Dai [a], Renqing Wang [a,b,*], Fasheng Li [c], Wenxing Wang [a]

[a] Environment Research Institute, Shandong University, Ji'nan 250100, China
[b] College of Life Science, Shandong University, Ji'nan 250100, China
[c] Laboratory of Soil Pollution Control, Chinese Research Academy of Environmental Sciences, Beijing 100012, China

ARTICLE INFO

Article history:
Received 31 January 2008
Received in revised form 23 October 2008
Accepted 9 November 2008
Available online 17 November 2008

Keywords:
Adsorption
Structural characterization
Hg^{2+}
Humic acids
Fulvic acids

ABSTRACT

A series of batch equilibration experiments were conducted to assess the adsorption and desorption of divalent mercury (Hg^{2+}) by humic acids (HAs) and fulvic acids (FAs) extracted from black soil and red soil in China. The Fourier transform infrared (FTIR) spectroscopy, cross-polarization (CP) with magic-angle spinning (MAS) ^{13}C nuclear magnetic resonance (NMR) spectroscopy and scanning electron microscopy (SEM) were jointly adopted to characterize the humic substances (HSs) samples and HSs–Hg complexes. The FTIR spectra showed that the adsorption of Hg^{2+} mainly acted on O–H, C–O and C═O groups of HAs and FAs. The NMR spectra indicated that HAs are higher in Paraffin and carbonyl C content compared with corresponding FAs, while FAs exhibit higher methoxy C, O-alkyl C and carboxyl C contents. The SEM images revealed the different surface structures of HAs, FAs and HSs–Hg complexes, which explains Hg^{2+} complexation phenomenon on HSs. The study also showed that HAs have higher adsorption capacity for Hg^{2+} than those of FAs. Hg^{2+} adsorption isotherms could be well fitted with both Langmuir and Freundlich equations. The desorbed percentages of all HSs samples were less than 1%, which indicated their high binding strength for Hg^{2+}. Furthermore, HAs samples have a lower desorption ratio than FAs, HAs can play a more important role in pollution control of Hg^{2+} in environment.

1. Introduction

As one of most harmful global environmental contaminants, mercury has been recognized as a threat to human health [1,2]. Like other trace elements, the behavior and fate of Hg^{2+} in environment are mainly controlled by its adsorption and desorption processes with various adsorbents [3]. Many studies show that organic matter is the most important soil component controlling Hg^{2+} adsorption and desorption processes, affecting its retention and release in environment [2,4–7].

Humic substances (HSs) are the active part of organic matter, which in turn account for 70–80% of all soil organic matter [8]. HSs, among the most widely distributed natural organic macromolecule matters on the earth surface, contain a number of function groups such as carboxylic, carbonylic, phenolic and methoxy groups. Those function groups in HSs may combine with some heavy metals in environment. Therefore, HSs play many significant environmental roles in stabilizing soils and sediments, regulating the levels of water, metals and other components [9]. Both metals and organic contaminants can be adsorbed by HSs in a number of ways such as complexation, ion exchange and reduction [10], forming some hydrosoluble or water-insoluble complex compounds, thus significantly affect the environment. All the three main components of HSs, humic acids (HAs), fulvic acids (FAs) and humin (HU), have similar structures but different in molecular weights, elemental analysis and function group contents [11].

The objectives of this research are to give an in-depth study of the Hg^{2+} adsorption–desorption processes on HAs and FAs extracted from two typical soils (black soil and red soil) in China. The binding capacity of HSs samples was compared to reveal the related complex interactions between Hg^{2+} and HSs. Analysis using multiple methods including FTIR spectroscopy, CP/MAS ^{13}C NMR spectroscopy and SEM revealed structural characteristics and elucidated the mechanism of Hg^{2+} adsorption–desorption on typical HSs.

2. Materials and methods

2.1. Materials

Black soil collected from Gongzhuling of Jilin Province and red soil collected from Yingtan of Jiangxi Province with distinct properties were used as the sources of HAs and FAs. Both soils were

* Corresponding author at: Environment Research Institute, Shandong University, Ji'nan 250100, China. Tel.: +86 531 8836 4425; fax: +86 531 8836 9788.
E-mail address: wrq@sdu.edu.cn (R. Wang).

0927-7757/$ – see front matter. Crown Copyright © 2008 Published by Elsevier B.V. All rights reserved.
doi:10.1016/j.colsurfa.2008.11.006

Table 1
The physical and chemical characteristics of the two typical soils and humic substances extracted from them.

Samples	Soil type[a]	Location	pH (H_2O)	SOM (%)	Particle size distribution (%)[b]			Textural classification	Mercury content (mg kg^{-1})	Humic substances content (%)		Element contents (%)		
					Sand	Silt	Clay					C	H	N
Black soil	Phaeozem	Gongzhuling, Jilin Province	6.30	3.03	17.4	60.3	22.3	Silt loam	0.186	HA	2.47	18.72	2.95	1.44
										FA	10.5	6.94	0.65	0.1
Red soil	Udic Ferrisols	Yingtan, Jiangxi Province	5.03	1.33	24.9	35.4	39.7	Clay loam	0.257	HA	0.39	17.42	3.03	1.75
										FA	0.46	6.32	1.8	0.4

[a] Soil classification according to [47].
[b] Sand (0.05–2 mm), Silt (0.002–0.05 mm), Clay (<0.002 mm).

air-dried and then sieved through a 2 mm sieve for soil particle size distribution, ground to pass through a 1 mm sieve for soil pH, extraction of humic substances, and through a 0.125 mm sieve for soil organic matter and Hg in soil. The physical and chemical properties of the soils and the elemental composition of the purified HAs and FAs extracted from two soil samples were analyzed by the following procedure.

Soil pH value was determined with a pH meter (pHS-3C, Leici, China) in a 1:2.5 suspension in H_2O. Soil organic matter was determined by oxidation with potassium dichromate-titration of $FeSO_4$, which is a standard method recommended by the Chinese Society of Soil Science [12]. Soil particle size distribution was measured by the micro-pipette method [13]. The background Hg in soil was determined by Atomic Fluorescence Spectrometer (AFS 930, Beijing Jitian Instrument Co., China) after digestion in AIM600 Block Digestion System with teardrops using aqua fortis (1:1). Element contents of HSs extracted from soils were measured by Elemental Analyzer (Vario EL III, Elementar Analysensysteme GmbH., Germany). Basic properties of soils and HSs used in this study are listed in Table 1.

2.2. Extraction and purification of humic substances

The extraction and purification of HAs and FAs were performed using a method developed by the International Humic Substances Society (IHSS) [14–16]. Briefly, 100 g of soil sample was weighed and added into a beaker with 1 L 0.05 M HCl, and the solution was intermittently churned for 18 h. Then the soil was separated from supernate by centrifugation at 5000 rpm for 10 min and was washed with deionized water. Then, 1 L 0.1 M NaOH was added to the soil persistently churned for 24 h. The solution was centrifuged at 15,000 rpm for 20 min. This process was repeated at least twice to make sure the supernate was completely separated from the residue. All the supernates were collected into a clean glass beaker. The pH was adjusted to 1–2 using 6 M HCl. HAs do not dissolve in acid while FAs do. Therefore, all HAs precipitates are separated from FAs. The HAs products were obtained by centrifugal precipitation (5000 rpm, 10 min). All experiments were carried out at temperature 25 °C.

The HAs were mixed with 0.5 L HF–HCl acid mixture(0.1 M HCl + 0.3 M HF) and then shaken for 24 h at 25 °C. The silicate in HAs was dissolved in mixture acid and then removed by centrifugation at 5000 rpm for 10 min. This process was repeated three times until little ash left in HAs. Cation exchange resin was filled in dialysis bags and shaken for 24 h in HAs solution to remove ions. At last, the purified HAs were freeze-dried and kept in a refrigerator. The XAD-8 resin (purchased from Sigma Co., USA) was used in FAs purification process. The supernates separated form HAs were flowing through an ion exchange column (length: 20 cm, diameter: 1.0 cm, flow rate: 1.5–2 ml min^{-1}), left the FAs on the resin. After that, the resin was washed with deionized water till pH value at 6–7. Then, 0.1 M NaOH solution was used to elute FAs from the resin. After cation exchange was used in the elute solution, it was concentrated in a rotary evaporator at 45 °C. The purified FAs were freeze-dried and ground to fine powders for characteristic analysis and adsorption/desorption experiments.

2.3. Adsorption and desorption experiments with HAs and FAs

A batch equilibration technique was conducted in this mercury adsorption and desorption experiments. Four kinds of samples were used including two HAs and two FAs extracted from black soil and red soil. An amount of 30 mg HAs (or FAs) was precisely weighed and combined with 25 ml solution in 50 ml plastic centrifuge tubes. The adsorption solutions were prepared in 0.1 M $NaNO_3$ solution [4], with the rising Hg^{2+} concentration gradients of 0, 1.5, 2.5, 3.5

and 4.5 mg L^{-1}. All samples were duplicated. The tubes fitted with caps were then shaken for 24 h at 25 °C on an end-over-end shaker (220 rpm). The suspensions were then centrifuged at 5000 rpm for 15 min, and passed through a 0.45 μm filter. The filtrates were analyzed using AFS. Hg^{2+} adsorption amounts were calculated by the differences between the initial and equilibrium concentrations of Hg^{2+}.

After the adsorption experiment, 10 ml of 0.1 M $NaNO_3$ was added to each tube. The tubes were shaken on an end-over-end shaker (220 rpm) for 24 h at 25 °C. The suspensions were then centrifuged at 5000 rpm for 15 min, passed through a 0.45 μm filter and analyzed using AFS. Each sample was duplicated. The amount of Hg^{2+} desorbed from each sample was calculated based on the concentrations of Hg^{2+} in the desorption solutions.

The operation temperature of the AFS was at 15–30 °C. The Lamp current was 30 mA. Ultra pure argon gas was used as the carrier gas, with the pressure of 0.25 MPa and flow rate of 0.4 L min^{-1}. The KBH_4–KOH mixture (0.05% KBH_4 + 0.5% KOH) was used as reducer. The carrier flow was 2% HNO_3.

2.4. FTIR spectrum

Four kinds of HSs samples and four HSs–Hg complexes were characterized by FTIR. The FTIR spectra were recorded over the range 4000–400 cm^{-1} by a Nicolet 380 FTIR spectrophotometer (Thermo Co., USA, software Omnic 1.2) using potassium bromide pellets obtained through pressing uniformly prepared mixtures of 2 mg samples and 200 mg KBr.

2.5. Solid-state ^{13}C NMR spectrum

Cross-polarization and magic-angle spinning techniques (CP/MAS) were used to obtain the solid-state ^{13}C NMR spectra of HAs and FAs. The NMR was a Bruker AV 300 spectrometer, about 200 mg sample was filled into a 4 mm zirconium rotor with a Kel-F cap. The spectral frequency was at 75.47 MHz, spinning rate at 4 kHz, pulse width of 2.60 μs, and data acquisition time 0.3 s.

2.6. SEM measurement

The SEM images of HSs samples and HSs–Hg complexes were achieved by HITACHI S-4800 (Hitachi, Japan), and the samples were coated with gold.

2.7. Data analysis

Amounts of adsorbed Hg^{2+} were calculated by the mass balance equation as follows:

$$y = \frac{v(c_i - c)}{m} \qquad (1)$$

where y is the adsorption amount of Hg^{2+} (mg g^{-1} HSs), v is the equilibrium solution volume, c_i and c represent the initial and equilibrium concentrations of Hg^{2+} (mg L^{-1}), and m is the weighed amount of HSs.

Linear regression analysis was adopted to calculate relevant parameters for Hg^{2+} adsorption on HAs and FAs. Adsorption data were fitted with both Langmuir and Freundlich equations as follows:

Langmuir equation:

$$\frac{1}{y} = \frac{1}{y_m} + \frac{1}{(y_m k_1)}\left(\frac{1}{c}\right) \qquad (2)$$

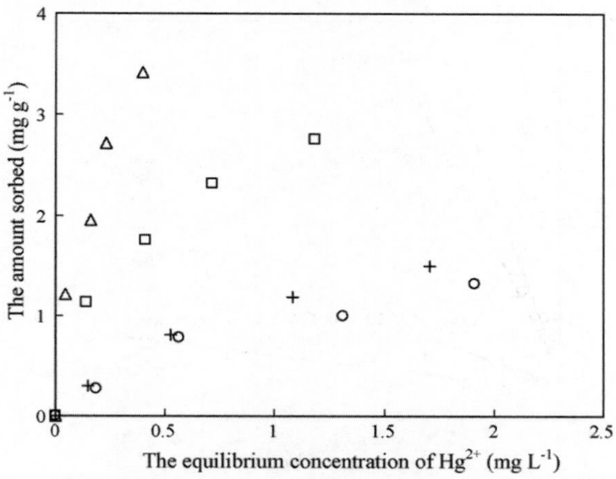

Fig. 1. Hg^{2+} adsorption isotherms of HA from black soil (□), HA from red soil (△), FA from black soil (○) and FA from red soil (+).

Freundlich equation:

$$y = k_2 c^{1/\alpha} \qquad (3)$$

where y is adsorption amount (mg g^{-1} HSs), y_m is the maximum sorption value, c is the equilibrium concentration in solution (mg L^{-1}) and k_1, k_2 are constants.

3. Results and discussion

3.1. Adsorption isotherms and desorption characteristics

To assess the effect of adsorbate concentrations in Hg^{2+} adsorption, HAs and FAs extracted from two typical soils (Table 1) were selected to conduct adsorption isotherms. Adsorption isotherms of Hg^{2+} on HAs and FAs were illustrated in Fig. 1, which shows that HA from red soil has the highest capacity to adsorb Hg^{2+} among four kinds of HSs, while FA from red soil has the lowest capacity to adsorb Hg^{2+}. Furthermore, both HAs have higher capacity to adsorb Hg^{2+} than those of two FAs. The average percentages of adsorption (data not shown) are as follows: 82.0% of Hg^{2+} in starting solution was adsorbed by HA from black soil, 94.5% of Hg^{2+} was adsorbed by HA from red soil, 50.9% of Hg^{2+} was adsorbed by FA from black soil and 55.8% of Hg^{2+} was adsorbed by FA from red soil. According to other researchers [3], most of Hg^{2+} adsorbed by HAs is considered to be chemisorptions while FAs not, which might explain the higher adsorption capacity of HAs. In order to calculate relevant parameters for Hg^{2+} adsorption by HSs, adsorption data was fitted with both Langmuir and Freundlich equations as stated above. In addition, the isotherms of Hg^{2+} on both HAs conformed better to the Freundlich model, while the isotherms of Hg^{2+} on both FAs conformed better to the Langmuir model. Linear regression analysis showed that both equations can adequately describe Hg^{2+} adsorption on four kinds of HSs (Fig. 2), and fitness and adsorption parameters were derived from these two equations (Table 2). Results from both equations indicated that the affinity of Hg^{2+} to the four kinds of HSs follows this ranking: HA of red soil > HA of black soil > FA of black soil > FA of red soil. Adsorption of Hg^{2+} on four kinds of HSs was readily because all of the values of α from Freundlich equation were above 1, and correspondingly confirms the heterogeneity of the adsorbent [17]. This result supports the theory that Hg^{2+} is adsorbed to strong sites first (at low solution Hg^{2+} concentrations), and occupies weaker sites when additional Hg^{2+} is loaded into the system [18]. The ranking of favorable adsorption capability is: HA of black soil > HA of red soil > FA of red soil = FA of

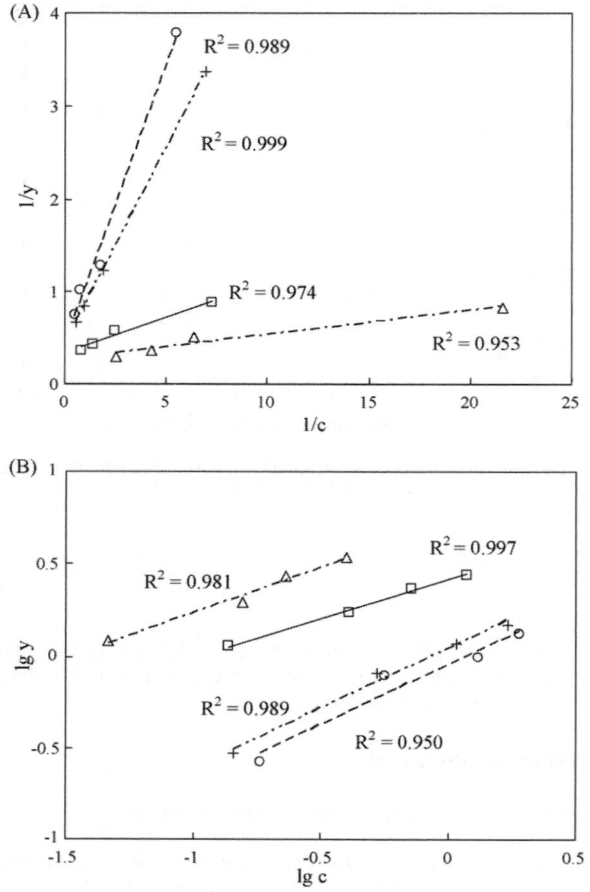

Fig. 2. Fitting curves derived from Langmuir (A) and Freundlich (B) equations of Hg^{2+} adsorption on HA from black soil (□), HA from red soil (△), FA from black soil (○) and FA from red soil (+).

black soil. This result may due to markedly higher C contents in HAs than those of FAs (Table 1).

The amounts of Hg^{2+} desorbed after adsorption ranged from 0.26% to 0.5% of the total adsorption were very little (Fig. 3), which was consistent with the findings of Jing et al. [18] about Hg^{2+} from soils and kaolin or kaolin–HA complexes. This result proves a strong affinity of Hg^{2+} to HSs and may be caused at low surface coverage, the Hg^{2+} was bound by high-energy sites, independent of the adsorbent surface [19]. These high-energy sites could either be the sites that form very stable surface complexes with Hg^{2+} [4,20], or the micro pores that trap Hg^{2+} and require high activation energy to release Hg^{2+}. Desorption percentages of both HAs were lower than those of FAs, and follows the ranking: FA of black soil > FA of red soil > HA of red soil > HA of black soil. This result indicated that HAs from different types of soils have higher binding strength for Hg^{2+} than corresponding FAs have. The investigation of distribution of humic substance-bond mercury (HS–Hg) also confirmed that the two fractions existed in soils as humic acid-bound mercury (HA-Hg) > fulvic acid-bound mercury (FA–Hg) [21]. Consequently, FAs from two typical soils have greater threat to environments than corresponding HAs, which is in agreement with other findings [22]. Experimental data of some researches also indicated that FA–Hg was an important source of Hg for lettuce [23]. The amounts of Hg^{2+} desorbed were positively correlated with those of Hg^{2+} adsorption for all HSs samples.

Table 2
Adsorption isotherm constants and characteristics derived from Langmuir and Freundlich equations for HAs and FAs.

HAs and FAs	Langmuir equation			Freundlich equation		
	$1/y = 1/y_m + 1/(y_m k_1)(1/c)$			$y = k_2 c^{1/\alpha}$		
	y_m	k_1	r	k_2	α	r
HA (black soil)	3.04	4.23	0.987**	2.60	2.38	0.998**
HA (red soil)	3.70	10.27	0.976**	5.29	2.05	0.990**
FA (black soil)	2.43	0.68	0.994**	0.90	1.52	0.975**
FA (red soil)	2.29	1.04	0.999**	1.12	1.52	0.994**

y, the amount adsorbed ($mg\,g^{-1}$ HSs); y_m the maximum sorption value ($mg\,g^{-1}$ HSs); c, the equilibrium concentration in solution ($mg\,L^{-1}$); k_1, k_2: constants.
** correlation is significant at 0.01 level.

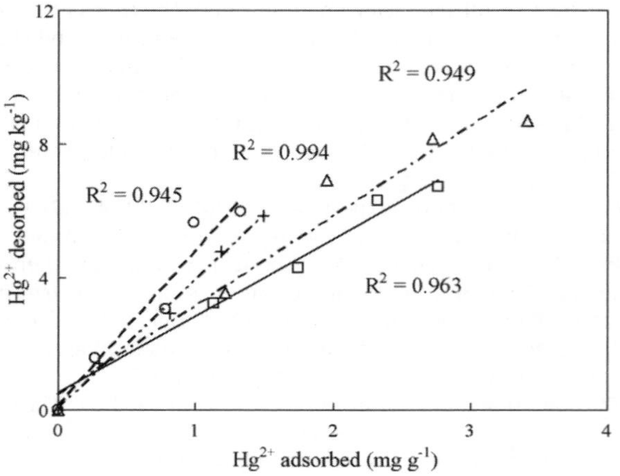

Fig. 3. Relationship between Hg^{2+} adsorption and desorption on HA from black soil (□), HA from red soil (△), FA from black soil (○) and FA from red soil (+).

3.2. FTIR spectra

Infrared (IR) spectroscopy has been widely used for the structural investigation of HSs because it can provide considerable insight into the structural arrangement of oxygen and carbon containing function groups in HSs. Different bands reveal different function groups present in complex mixtures [24]. As an important parameter for chemical characterization, IR spectra have been used to investigate the interactions of HSs with metals [25].

The FTIR spectra of HSs and HSs–Hg complexes are shown in Fig. 4. Fig. 4A is the FTIR spectra of HAs and FAs samples. The figure revealed that HSs have common peaks arise at the same wave number in the five spectra, those were 3436, 2918, 2851, 1633, 1112 and $1035\,cm^{-1}$ [11]. This indicated that all HAs and FAs contain the same function groups, which is in agreement with the findings of Xu et al. [14,26] and Mafra et al. [27]. The shoulders at $3436\,cm^{-1}$ are assigned to H-bonded O–H stretching, and to N–H stretching; the bonds in the range from 2840 to $2950\,cm^{-1}$ are attributed to aliphatic or alicyclic C–H stretching; the peaks at $1633\,cm^{-1}$ are mainly due to C=O stretching of amide, quinone and/or H-bonded conjugated ketone groups, describes a similar relative intensity for all HSs samples; the shoulders at $1112\,cm^{-1}$ and $1035\,cm^{-1}$ indicate C–O stretching of polysaccharides or polysaccharide-like substances, or Si–O of silicate impurities. The intensity of some bands appears different on spectrum of these HSs samples. For example, the FA from black soil has a strong absorption peak at $1460\,cm^{-1}$ while others not. This means that it contains more alicyclic C–H than other samples.

In order to understand the changes of FTIR spectra after Hg^{2+} adsorption on HSs and elucidate the mechanism of Hg^{2+} adsorption on HSs, the FTIR spectra of HSs–Hg^{2+} complex were characterized and shown in Fig. 4B. Comparing them with the spectra of HSs

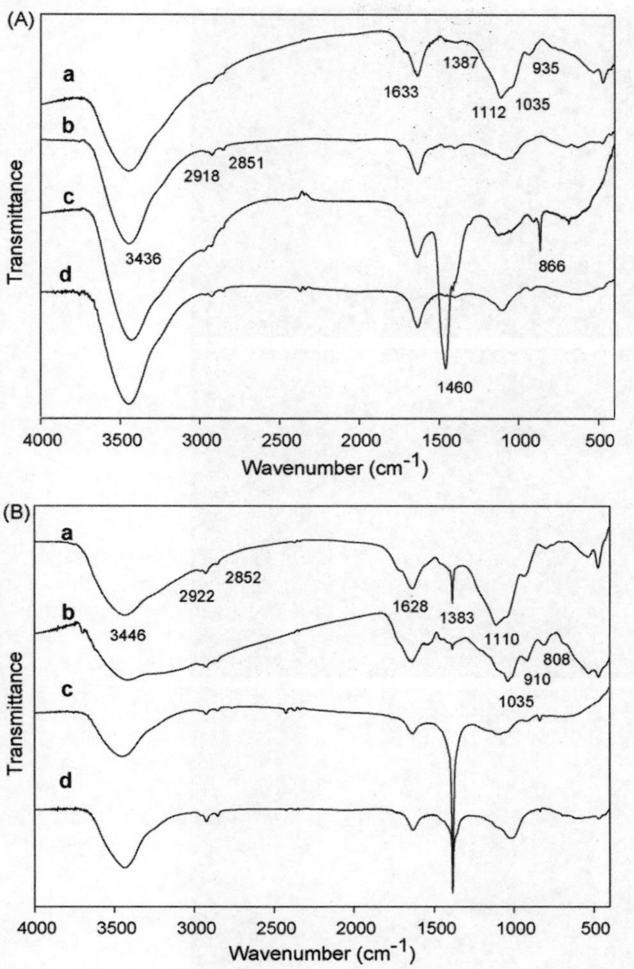

Fig. 4. The FTIR spectra of HSs samples (A) and HSs–Hg^{2+} complex (B) (a: HA–black soil, b: HA–red soil, c: FA–black soil and d: FA–red soil).

samples in Fig. 4A, some differences can be observed. It is obvious that the absorption intensity of the bonds at 3446 and 1628 cm^{-1} decreased. This suggests that the adsorption of Hg^{2+} on HSs was mainly acted on H-bonded O–H groups and C=O groups of HSs. As research [28] proved, Hg ions are able to form complex compounds with carboxylic and phenolic groups of humic substances. Xia et al. [29] also pointed out that Hg^{2+} binding to humic acid surfaces was dominated by interactions with organic thiol function groups in combination with carboxyl and phenol function groups. Compared with Fig. 4A, the narrowing of the peak at 1383 cm^{-1} may correspond to the stretching vibration of C=O and C–O groups or their O–H deformation vibration, which influences the position and width of the adsorbing peaks of these groups. The changes may cause by the formation of hydrogen bonds, and indicate that adsorption may also take place on these groups.

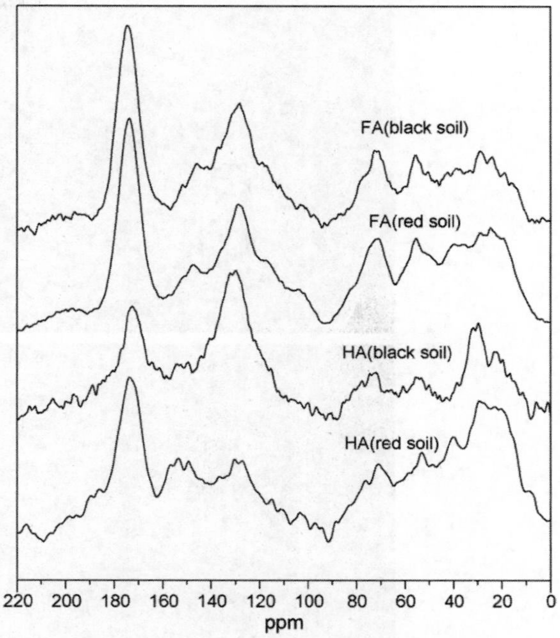

Fig. 5. Solid-state ^{13}C CP/MAS NMR spectra of HAs and FAs samples.

3.3. Solid-state CP/MAS ^{13}C NMR spectrum

The use of nuclear magnetic resonance (NMR) provides better understanding of identification of function groups, and complements to FTIR spectroscopy for investigating the structural properties of HSs [30–32]. In this study, the solid-state CP/MAS ^{13}C NMR spectra of HAs and FAs are shown in Fig. 5. Within the chemical shift range of 0–220 ppm, five regions are divided as follows: carbon atoms at 0–50 ppm are assigned to alkyl carbon, 50–110 ppm to O-alkyl carbon, 110–165 ppm to aromatic carbon, 165–190 ppm to carboxyl carbon, and 190–220 ppm to carbonyl carbon [26,33]. Relative intensity was determined by the corresponding peak areas of each region. The structural carbon distribution of the HAs and FAs was calculated and shown in Table 3.

Peak at 30 ppm is assigned to methylenes in long chains, mainly characteristic for paraffin-like structures in alkyl C (0–50 ppm) [34,35]. Spectra exhibited that FAs dominated in content in peaks at 55 ppm (methoxy C), 70 ppm (O-alkyl C) and 172 ppm (carboxyl, ester or amide C). Peaks at 130 ppm and 150 ppm pertained to C-substituted aromatic C and O-substituted aromatic C [32,35,36].

The adsorption performance and model of HSs for mercury may be affected by all these results:

(a) The general features and their intensity distributions of the spectra of four samples indicate their similar carbon functionalities [31]. Table 3 indicates that alkyl C, O-alkyl C, aromatic C and carboxyl C are the dominant C components in HAs and FAs sam-

Table 3
Structural carbon distribution (%) of the HAs and FAs.

Samples	Alkyl C Range	O-alkyl C	Aromatic C	Carboxyl C	Carbonyl C
	0–50 ppm	50–110 ppm	110–165 ppm	165–190 ppm	190–220 ppm
HA (black soil)	20.98	14.69	43.57	16.98	3.78
FA (black soil)	21.91	21.41	31.86	21.36	3.46
HA (red soil)	31.73	21.36	24.9	16.58	5.43
FA (red soil)	25.52	23.43	28.33	18.93	3.79

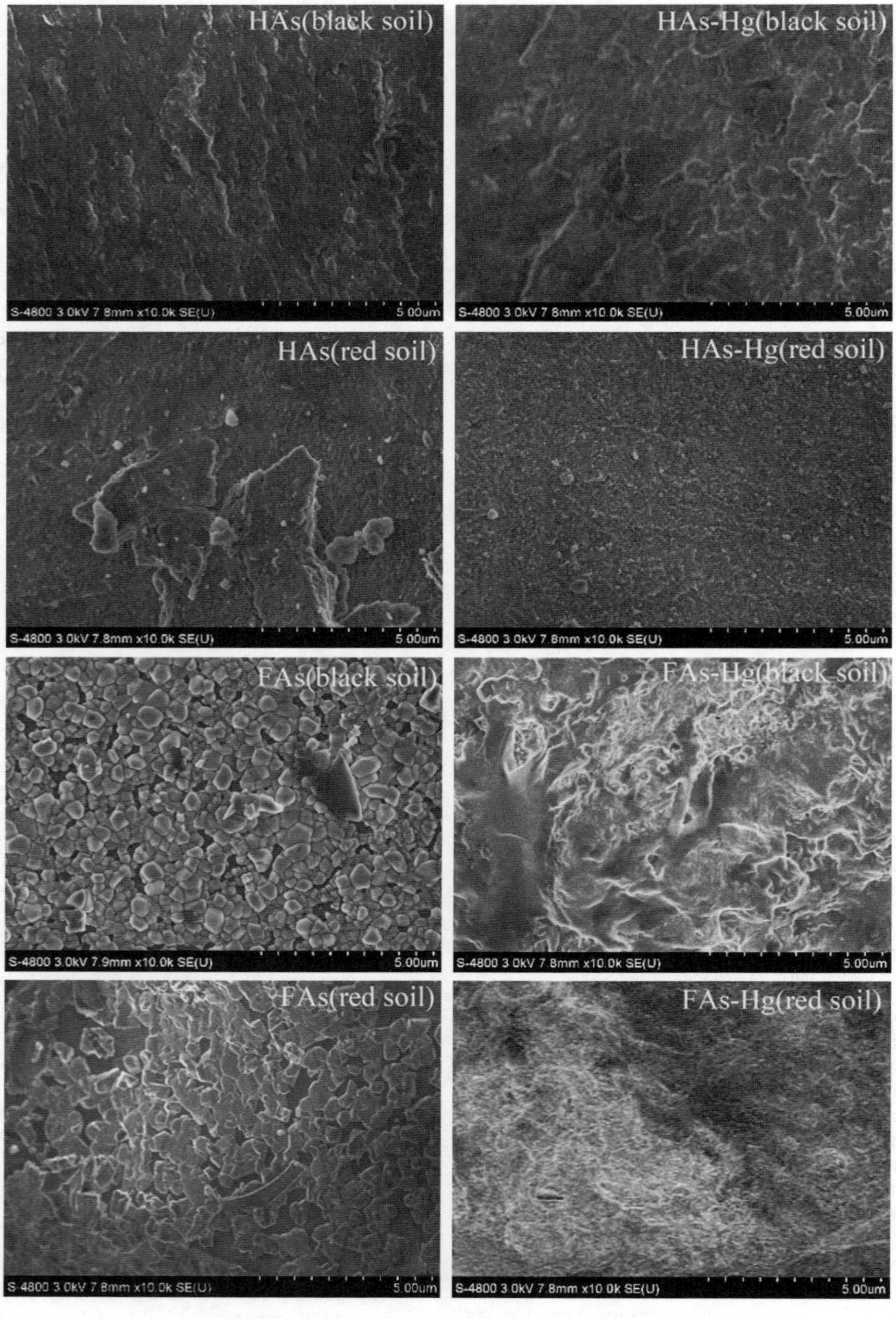

Fig. 6. The SEM images of HSs samples and HSs–Hg complexes.

ples. HAs are higher in Paraffin and carbonyl C content compared to corresponding FAs. In particular, FAs exhibit higher methoxy C, O-alkyl C and carboxyl C contents. This is consistent with their greater mobility and solubility [35], which may explain FAs' lower adsorption strength with Hg^{2+} compared with HAs.

(b) O-alkyl C, Carboxyl C and Carbonyl C are regarded as polar carbons, the content of which will positively contribute to the adsorption amount of polar contaminants [26]. The polar carbon contents of HAs from red soil and black soil are 43.37% and 35.45%, respectively. This confirms to the results of equilibrium adsorption experiments above. The adsorption amount of mercury is higher on HA from red soil than that of HA from ~~red~~ soil. The polar carbon contents of FAs from red soil and black soil are 46.23% and 46.15%, which is also consistent with their similar absorption ability.

(c) The aromaticity of the samples is calculated by expressing the content of aromatic C as the percentage of the sum of the contents of aliphatic carbon (0–110 ppm) and aromatic C (110–165 ppm) [32,37]. Values of $C_{arom}/(C_{arom}+C_{alip})$ of HA from black soil, FA from black soil, HA from red soil and FA from red soil are 55.0%, 42.4%, 31.9% and 36.7% respectively. This shows that the humification degree of HSs from black soil is heavier than that of red soil. It was reported in [38] that HSs from cultivated soils have higher concentration of aromatic C determined by CP/MAS ^{13}C NMR than HS from noncultivated soils. However, humification degree of HSs from soils did not affect the adsorption of Hg^{2+} on HSs.

3.4. SEM images

SEM is widely adopted with other methods to characterize the size, shape, structure and composition of HSs. The macromolecular structures and shapes of HSs could be directly observed under different conditions by these techniques, thus provide further information of the physicochemical reactions and to evaluate their role in natural environment [39–43].

Fig. 6 shows the SEM images of HSs samples and HSs–Hg complexes. HSs exhibited more than one type of macromolecular structure in aqueous solutions. The HAs from black soil have a sheet-like structure, while the surface of HAs from red soil seem rough and has more little particles and cracks. This suggests that adsorption capacity of the latter HAs may be higher, which coincides with other research [44] and our experiments. Previous work also found that HSs forms thin thread- and net-like structures in dilute solution and grows lager rings and sheets with increased HSs and cation concentrations [45]. Comparing the images of HSs–Hg complexes with corresponding HSs reveals some obvious changes. The surfaces of HAs from black soil becomes smoother. Homogeneous structure with barely any cracks is formed on the surface of HAs from red soil. In the images of both FAs–Hg complexes, the former distinctly separated small particles are melt and joint together on the surfaces. The structural changes may be mainly caused by metal complexation of HSs functional groups. The presence of Hg^{2+} in solution forms strong complexes with HSs and thereby makes HSs to form aggregates strongly. The formation of the aggregates is possible because of the formation of cation bridges between Hg^{2+} with carboxyl, hydrogen bonds and other functional groups of HSs [39]. In view of previous results of FTIR and NMR, C=O, C–O and O–H groups could all react with Hg^{2+} and induce macromolecular changes of HSs surface. The complexation phenomenon observed by SEM supports results of FTIR and NMR and helps explain adsorption behavior of Hg^{2+} on HSs.

The changes in HSs macromolecular structures can alter functional groups chemistry of HSs, and modify the exposed surface area of HSs in solution. Metal ions play a crucial role in HSs aggregate formation and stability. The aggregation of HSs can increase the size of the network of HSs and therefore transfers them from the dissolved phase (<0.45 μm) into the particulate one (>0.45 μm) [45,46]. The degree of complexation with metal ions strongly affect the degree of neutralization of negative surface charge, thus vary the hydrodynamic sizes of HAs. Divalent metal cation has a higher effect on the size of the HSs network than monovalent cation [46]. Therefore, differences of macromolecular structures in HSs can affect the intensity and rates of sorption and desorption with contaminants.

4. Conclusions

The present study shows that the two different components of HSs have distinct capacities for mercury adsorption. Humic acids have higher adsorption capacity of Hg^{2+} compared to fulvic acids. The affinity of Hg^{2+} to the four kinds of HSs follows this ranking: HA of red soil > HA of black soil > FA of black soil > FA of red soil. The adsorption isotherms can be adequately described by Langmuir and Freundlich equations. The desorbed amounts of Hg^{2+} after adsorption on HSs samples are rather low, which indicates HAs and FAs have a strong affinity to Hg^{2+}. Moreover, desorption capacity for HSs follows the ranking: FA of black soil > FA of red soil > HA of red soil > HA of black soil. HAs exhibit higher binding strength and lower desorption ratio for Hg^{2+} compared with FAs. Therefore, HAs have important role in pollution control of mercury in terrestrial and water system. Effective sorbent has some advantages such as widely distributed, accessible, cheap and strongest sorbed ability comparing with other sorbents.

The structural characterization of HAs and FAs is analyzed using FTIR, NMR and SEM techniques in this work. The results provide theoretical basis to control the transport of heavy metals, to modify contaminant solubility and biotransformation in water, soils and sediments.

Acknowledgements

This research is supported by the National Basic Research Program of China (973 program, no.2004CB418501), the Outstanding Young Scientists Grants of Shandong Province (no.2007BS08001) and Shandong Postdoctoral Science Foundation (no.200601005). We thank Jilin Academy of Agricultural Sciences and Jiangxi Red Soil Station of Chinese Academy of Sciences for their assistance to obtain the soil samples. We are also grateful to Prof. Xu Duanping for his suggestions.

References

[1] WHO, Environmental health criteria 1: Mercury, World Health Organization, Geneva, 1976, pp.1–132.
[2] P. Miretzky, M.C. Bisinoti, W.F. Jardim, Chemosphere 60 (2005) 1583.
[3] P. Thanabalasingam, W.F. Pickering, Environ. Pollut. B 9 (1985) 267.
[4] Y.J. Yin, H.E. Allen, C.P. Huang, P.F. Sanders, Soil Sci. 162 (1997) 35.
[5] Y.J. Yin, H.E. Allen, C.P. Huang, P.F. Sanders, Anal. Chim. Acta 341 (1997) 73.
[6] E. Schuster, Water Air Soil Pollut. 56 (1991) 667.
[7] A. Anderson, Mercury in Soil, in: J.O. Nriagu (Ed.), The Biogeochemistry of Mercury in the Environment, Elsevier, North-Holland Biomedical Press, Amsterdam, 1979, pp. 79–112.
[8] M. Schnitzer, Humic substances: chemistry and reactions, in: M. Schnitzer, S.U. Khan (Eds.), Soil Organic Matter, Elsevier Publishing, Amsterdam, 1978, pp. 1–64.
[9] E.A. Ghabbour, G. Davies, A. Fataftah, N.K. Ghali, M.E. Goodwillie, S.A. Jansen, N.A. Smith, J. Phys. Chem. B 101 (1997) 8468.
[10] S.A. Wood, Ore Geol. Rev. 11 (1996) 1.
[11] M. Schnitzer, S.U. Khan, Humic Substances in the Environment, Marcel Dekker. Inc., New York, 1972, 1–6.
[12] R.K. Lu (Ed.), The Analysis Method of Soil Agricultural Chemistry, China Agricultural Science and Technology Press, Beijing, 2000, p. p. 107 (in Chinese).
[13] W.P. Miller, D.M. Miller, Commun. Soil Sci. Plant 18 (1987) 1.
[14] D.P. Xu, Z.H. Xu, S.Q. Zhu, Y.Z. Cao, Y. Wang, X.M. Du, Q.B. Gu, F.S. Li, J. Colloid Interface Sci. 285 (2005) 27.
[15] D. Gondar, R. Lopez, S. Fiol, J.M. Antelo, F. Arce, Geoderma 126 (2005) 367.
[16] Z.M. Gu, X.R. Wang, X.Y. Gu, J. Cheng, L.S. Wang, L.M. Dai, M. Cao, Talanta 53 (2001) 1163.
[17] S.P. Mishra, D. Tiwari, R.S. Dubey, M. Mishra, Bioresour. Technol. 63 (1998) 1.
[18] Y.D. Jing, Z.L. He, X.E. Yang, Chemosphere 69 (2007) 1662.
[19] M. Arias, M.T. Barral, J. Da Silva-Carvalhal, J.C. Mejuto, D. Rubinos, Clay Miner. 39 (2004) 36.
[20] M.F. Schultz, M.M. Benjamin, J.F. Ferguson, Environ. Sci. Technol. 21 (1987) 863.
[21] G.F. Yu, H.T. Wu, X. Jiang, W.X. He, C.L. Qing, J. Environ. Sci. 18 (2006) 951.
[22] G.F. Yu, C.L. Qing, S.S. Mou, S.Q. Wei, Acta Sci. Circum. 21 (2001) 601.
[23] G.F. Yu, H.T. Wu, C.L. Qing, X. Jiang, J.Y. Zhang, Commun. Soil. Sci. Plant 35 (2004) 1123.
[24] W.M. Davis, C.L. Erickson, C.T. Johnston, J.J. Delfino, J.E. Porter, Chemosphere 38 (1999) 2913.
[25] Q.X. Wen (Ed.), Study Methods of Soil Organic Matter, Agricultural Press, Beijing, 1984, pp. 223–224 (in Chinese).
[26] D.P. Xu, S.Q. Zhu, H. Chen, F.S. Li, Colloids Surf. A: Physicochem. Eng. Aspects 276 (2006) 1.
[27] A.L. Mafra, N. Senesi, G. Brunetti, A.A.W. Miklós, A.J. Melfi, Geoderma 138 (2007) 170.
[28] H. Martyniuk, J. Wieckowska, Fuel Process. Technol. 84 (2003) 23.
[29] K. Xia, U.L. Skyllberg, W.F. Bleam, P.R. Bloom, E.A. Nater, P.A. Helmke, Environ. Sci. Technol. 33 (1999) 257.
[30] M.G. Peĭrez, L. Martin-Neto, S.C. Saab, E.H. Novotny, D.M.B.P. Milori, V.S. Bagnato, L.A. Colnago, W.J. Melo, H. Knicker, Geoderma 118 (2004) 181.
[31] L.H. Dong, J.S. Yang, H.L. Yuan, E.T. Wang, W.X. Chen, Eur. J. Soil Biol. 44 (2008) 166–171.
[32] M. Schnitzer, C.M. Preston, Plant. Soil 75 (1983) 201.
[33] B. Xing, Environ. Pollut. 111 (2001) 303.
[34] D. Fabbri, M. Mongardi, L. Montanari, G.C. Galletti, G. Chiavari, R. Scotti, Fresenius J. Anal. Chem. 362 (1998) 299.

[35] C.M. Preston, R. Hempfling, H.R. Schulten, M. Schnitzer, J.A. Trofymow, D.E. Axelson, Plant Soil 158 (1994) 69.
[36] A.P. Deshmuck, B. Chefetz, P.G. Hatcher, Chemosphere 45 (2001) 1007.
[37] N.J. Mathers, Z.H. Xu, Geoderma 114 (2003) 19.
[38] N. Mahieu, D.S. Powlson, E.W. Randall, Soil Sci. Soc. Am. 63 (1999) 307.
[39] C. Chen, X. Wang, H. Jiang, W. Hu, Colloids Surf. A: Physicochem. Eng. Aspects 302 (2007) 121.
[40] J.C. Joo, C.D. Shackelford, K.F. Reardon, J. Colloid Interface Sci. 317 (2008) 424.
[41] M. Ne'gre, P. Leone, J. Trichet, C. Deïfarge, V. Boero, M. Gennari, Geoderma 121 (2004) 1.
[42] C.A. Coles, R.N. Yong, Eng. Geol. 85 (2006) 26.
[43] N. Senesi, F.R. Rizzi, P. Dellino, P. Acquafredda, Colloids Surf. A: Physicochem. Eng. Aspects 127 (1997) 57.
[44] D. Xu, S. Zhu, H. Chen, F. Li, Colloids Surf. A: Physicochem. Eng. Aspects 276 (2006) 1.
[45] S.C.B. Myneni, J.T. Brown, G.A. Matinez, W. Meyer-Ilse, Science 286 (1999) 1335.
[46] M. Baalousha, M. Motelica-Heino, P.L. Coustumer, Colloids Surf. A: Physicochem. Eng. Aspects 272 (2006) 48.
[47] Z.T. Gong, The Classificantion System Chinese Soils, Science Publishing Company, Beijing, 1999.

Geoderma 173-174 (2012) 215-223

Contents lists available at SciVerse ScienceDirect

Geoderma

journal homepage: www.elsevier.com/locate/geoderma

Distribution and sources of petroleum-hydrocarbon in soil profiles of the Hunpu wastewater-irrigated area, China's northeast

Juan Zhang [a], Jiulan Dai [a], Xiaoming Du [c], Fasheng Li [c], Wenxing Wang [a], Renqing Wang [a,b,*]

[a] Environmental Research Institute, Shandong University, Jinan 250100, China
[b] School of Life Sciences, Shandong University, Jinan 250100, China
[c] Laboratory of Soil Pollution Control, Chinese Research Academy of Environmental Science, Beijing 100012, China

ARTICLE INFO

Article history:
Received 8 February 2011
Received in revised form 2 November 2011
Accepted 6 December 2011
Available online 4 February 2012

Keywords:
Aliphatic hydrocarbons
Geochemical indices
Multivariate statistical analysis
Wastewater irrigation

ABSTRACT

Petroleum-contamination soil is ubiquitous. The various sources of this contamination necessitate the effective evaluation of the contamination level, contaminant transport analysis and identification of the pollution sources. In the present study, soil profiles were collected from both upland and paddy fields along the strike of the irrigation canals in the Hunpu wastewater irrigation region in northeast China. The concentrations of aliphatic hydrocarbons in the soil were analyzed. The sites near the oil wells, wastewater irrigation canal and Shenyang City were found to have high concentrations of [$\sum$ n-alkanes (sum of n-alkanes): (0.8 to 32.2) $\mu g\ g^{-1}$ dry wt.; TAH (total aliphatic hydrocarbons): (5.0 to 161.2) $\mu g\ g^{-1}$ dry wt.]. The geochemical analysis results showed various degrees of petroleum pollution. The $\sum$ n-alkanes varied (0.5 to 7.5) $\mu g\ g^{-1}$ while TAH varied (1.6 to 47.2) $\mu g\ g^{-1}$ at other sites where biogenic hydrocarbons were dominant. The results from the principal components and redundancy analyses showed that the samples containing hydrocarbons from wastewater irrigation, oil wells, and atmospheric deposition were partitioned into different groups. These results further represented that the soil properties had some effects on hydrocarbon distribution especially the sand content which had significantly negative correlation with oil-related hydrocarbons. The current study showed that the early input and vertical migration of hydrocarbons resulted in a high deep-soil hydrocarbon concentration.

1. Introduction

The petroleum industry and the use of petroleum products are ubiquitous in modern society. Petroleum accidents, discharge of waste oil, reuse of wastewater or sludge containing petroleum products, and soil contamination resulting from them are frequent (Adeniyi and Afolabi, 2002; Li et al., 2005; Moreda et al., 1998; Williams et al., 2006; Xiong et al., 1997; Zhou et al., 2005). Petroleum hydrocarbons in soil pose risks to the environment and human health. They can change the physico-chemical properties of soil and affect the physioecology of flora and soil microorganisms. The toxic components migrating from soil to groundwater and crops will endanger human health. The heavy, branched-chain, and cyclic hydrocarbons which are persistent in the environment can be transformed into more toxic substances (ATSDR, 1999; CCME, 2008; Li et al., 2005; Moreda et al., 1998; Plaza et al., 2005; Rao et al., 2007; Thomas et al., 1995; Xiong et al., 1997; Ye et al., 2007).

On the basis of chemical structure and physico-chemical properties, petroleum hydrocarbons are divided into two main classes: aliphatic and aromatic. Their compositions have been widely applied to assess the petroleum contamination level and identify the different sources. Zemo (2007) reported that the chromatogram pattern, geochemical indices (such as constitute ratios and biomarkers), age-dating releases, additives, and stable isotopes can be used to recognize various hydrocarbon sources. To identify the biogenic, petrogenic and pyrolytic hydrocarbons, previous studies analyzed the concentrations and patterns of aliphatic hydrocarbons (AHs) and/or polycyclic aromatic hydrocarbons (PAHs) and some of their common indices [range of carbon number, unresolved complex mixtures (UCM), isoprenoids, low carbon number hydrocarbons/high carbon number hydrocarbons, the n-alkane with maximum concentration (C_{max}), the ratio of the sum of all n-alkanes to hexadecane (C_{16} ratio), and carbon preference index (CPI)] (Abdullah, 1997; Commendatore and Esteves, 2004; Readman et al., 2002; Tran et al., 1997). These common indices can be combined with biomarkers, weather ratios, radiocarbon age, and/or stable carbon isotope to obtain more information about sources, maturity, and depositional environment (Harji et al., 2008; Hegazi et al., 2004; Medeiros and Bicego, 2004; Scholz-Böttcher et al., 2009; Wakeham and Carpenter, 1976; Zhu et al., 2005). Multivariate statistical analysis was efficiently used to simplify the complicated data sets on AHs and/or PAHs and determine the relationship of the different groups to different sources (Aboul-Kassim and Simoneit, 1995a,b; Aboul-Kassim and Williamson,

* Corresponding author at: Environmental Research Institute, Shandong University, Jinan 250100, China. Tel.: +86 531 8836 4425; fax: +86 531 8836 1990.
E-mail address: wrq@sdu.edu.cn (R. Wang).

0016-7061/$ – see front matter © 2011 Elsevier B.V. All rights reserved.
doi:10.1016/j.geoderma.2011.12.004

2003; Kang et al., 2000; Ye et al., 2007). The relationship of hydrocarbons and samples to pollution sources are explored using R-mode and Q-mode factor analysis, respectively. The CANOCO software can simultaneously focus on both covariance relationships and inter-sample relationships and can also be used to analyze the effects of the soil properties on hydrocarbon compositions.

Hunpu region is one of the most spacious and typical wastewater irrigation areas in China and has been irrigated for more than 40 years with water from the Dahuofang Reservoir, the Hunhe River, and the Xihe River. The Xihe River is the major drainage channel of industrial effluent and municipal sewage of Shenyang City, where many petrochemical plants exist and the use of petroleum products is common (Xiao et al., 2008). The situation resulted in the excessive petroleum hydrocarbon contamination of the Xihe River and its riverbank groundwater (Song et al., 2004; Song et al., 2007), making it the most polluted river of Shenyang and a pollution source of the Hunhe River. Other than the wastewater-irrigation, there are other sources of petroleum hydrocarbons (such as oil wells and atmospheric deposition) in the region. Our objectives are to effectively evaluate petroleum contamination level of soil in the Hunpu region by analyzing the concentration, geochemical characteristics and multivariate statistical analysis of aliphatic hydrocarbons; to explain the different distributions by the possible pollution sources and the basic soil properties; and to provide the guidelines for the quantity and safety of agricultural products and soil pollution control in the Hunpu region. The sources of petroleum hydrocarbons in complex agricultural soil are rarely studied compared to sediments, and the CANOCO software is tentatively used to analyze the pollution sources in the current study.

2. Materials and methods

2.1. Soil sampling

In May 2008, soil profiles were collected at the Hunpu wastewater-irrigated area located at the southwest of Shenyang City, Liaoning Province, China. The Hunpu region has an area of 410, 000 ha with a typical temperate monsoon climate. The annual mean temperature is 7.8 °C and the annual mean precipitation is 734.4 mm. Although the north wind is dominant during winter, south is the predominant direction of the wind all year. The main soil types are sandy loam and black soil. The irrigation water was delivered by the main canal and the Xihe River canal, both of which drew water from the Hunhe River. As the dominant land-use types, four paddy fields (I-1P, I-3P, I-5P, and I-7P) and four corn fields (I-2U, I-4U, I-6U, and I-8U) along the strike of the irrigation canals were chosen as sampling sites, all of which were at least 200 m away from roads (Fig. 1). The top 100 cm soil was sampled from a 1 m × 1 m hole after harvest before the next ploughing and divided into ten layers (including 0–5, 5–10, 10–15, 15–20, 20–25, 25–30, 30–40, 40–50, 50–60 and 60–100 cm). Each sample was homogeneously mixed with five randomly selected subsamples at each site, and then a portion of each sample was filled into a pre-cleaned aluminum box and transported to the laboratory at a temperature of 4 °C. The samples were freeze-dried and ground to 1 mm for petroleum hydrocarbons analysis. The other portions were air-dried and ground to 2 mm for soil particle composition, then to 1 mm for soil pH and conductivity, and finally, to 0.125 mm for organic matter (OM) analyses, respectively.

2.2. Analytical methods

For the extraction of petroleum hydrocarbons, a soil sample (ca. 10 g) was pressurized-liquid extracted with acetone/dichloromethane (1/1, v/v) using an ASE-300 (Dionex, Beijing, China). During the extraction, the cells were pressurized to 1500 psi/1.0×10^7 Pa, and heated to 175 °C for 8 min. The static extraction was held for 5 min followed by flushing (75% of the cell volume) and purging for 60 s at 150 psi/1.0×10^6 Pa (Dionex, 2007; EPA, 2007; Richter, 2000). The aliphatic hydrocarbons were obtained through eluting with approximately 20 mL n-hexane after purification with an alumina and silica gel chromatography column (Guo et al., 2007; Richter, 2000; Van De Weghe et al., 2006) and concentrated to 1 mL for qualitative and quantitative analysis.

The aliphatic hydrocarbon fraction was analyzed by Agilent 7890A gas chromatograph with a HP-5 fused silica capillary column (30 m × 0.25 mm i.d. × 0.25 mm film thickness). A 2 mL volume was injected in the pulsed splitless mode. The initial oven temperature was 60 °C (held for 2 min), was increased to 320 °C at 8 °C min^{-1} and held for 10 min. The flame ionization detector was at 330 °C. The carrier gas was N_2. The n-alkanes (nC_{10} to nC_{40}), phytane and pristane were quantified using the external standard (Accustandard, US) method. The UCM was quantified by assuming a response factor of 1.0 based on nonadecane. The $\sum$n-alkanes value was the sum of all n-alkanes from nC_{10} to nC_{40}. The total aliphatic hydrocarbons (TAH) value was the sum of the $\sum$n-alkanes and UCM.

The analytical procedure was strictly evaluated. Calibration graphs were constructed by plotting the peak area against the reference material concentration every two days. A linear relationship with $r^2 > 0.99$ was always obtained. The recovery test was performed with a reference material-spiked soil. A control soil sample (ca. 10 g) was spiked with three kinds of reference material solutions (the concentrations of each reference material were 2, 6, and 10 μg mL^{-1}, respectively), and the spiked sample was analyzed after being left for 30 min at room temperature. The procedure was repeated twice. The recoveries for nC_{10} to nC_{40}, phytane and pristane were 60–120% with 3.4–17.0% RSD (except for nC_{10} to nC_{12} due to their volatility), as shown in Table 1.

Soil pH was measured using a pH meter (PSH-3 C, Leici, China) and the ratio of water to soil was 2.5:1. Potassium dichromate volumetric method was used to determine OM (Lu, 2000). Particle composition was determined by a micro-pipettes method (Miller and Miller, 1987). Soil conductivity was measured using a conductivity meter (DDS-12A, LIDA, China) and the ratio of water to soil was 5:1.

2.3. Statistical methods

Simple statistical analysis was done using STATISTICA 7. To describe the similarity and dissimilarity of aliphatic hydrocarbons in the eighty samples and to provide an explanation of their soil properties, the CANOCO 4.5 software bundled with CanoDraw for Windows was used for statistical analysis.

The gradient lengths of the four axes calculated from detrended correspondence analysis (DCA) were all below 3, therefore the linear model analysis methods principal components analysis (PCA), and redundancy analysis (RDA), were applied (Lepš and Šmilauer, 2003). PCA was used to investigate the hydrocarbon compositions in all samples, whereas the relationships between the hydrocarbon compositions and the measured soil properties were analyzed by RDA. The values of aliphatic hydrocarbons or soil properties were evaluated by projecting the sample points onto the arrows of aliphatic hydrocarbons or soil properties. The dissimilarities of the samples were represented by the distances between the sample points. The (linear) correlation coefficients among the aliphatic hydrocarbons or the aliphatic hydrocarbons and soil properties were indicated by the cosine values of angles between the arrows of aliphatic hydrocarbons or arrows of aliphatic hydrocarbons and soil properties. The effects of the soil properties on hydrocarbon composition were indicated by the lengths of soil properties arrows.

3. Results and discussion

3.1. The concentrations of $\sum$n-alkanes and total aliphatic hydrocarbons of soil samples

Kruskal–Wallis ANOVA results indicated that the $\sum$n-alkanes and TAH of the eight sampling sites were significantly different ($P < 0.01$).

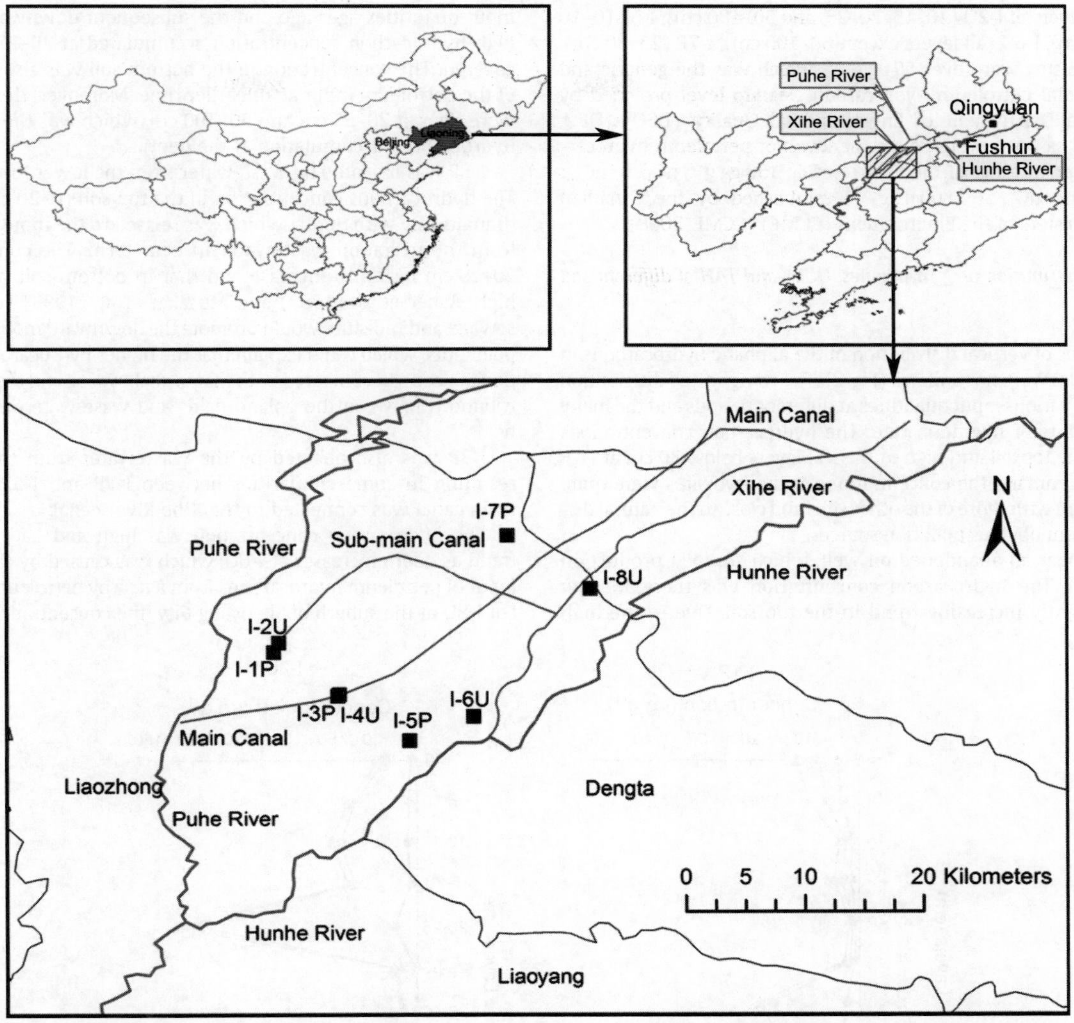

Fig. 1. Location of sampling sites in the Hunpu wastewater irrigation region. Four soil profiles from paddy fields: I-1P, I-3P, I-5P, and I-7P; four soil profiles from upland fields: I-2U, I-4U, I-6U, and I-8U.

The hydrocarbon concentrations were relatively higher at I-2U, I-6U, and I-5P, which were near the oil wells or the downstream of the wastewater irrigation canal. The $\sum$n-alkanes ranged from 1.1 to 32.2 μg g^{-1}, and the TAH ranged from 10.9 to 161.2 μg g^{-1}. The highest concentration was at I-6U. The concentrations of $\sum$n-alkanes and TAH were medial at I-7P and I-8U at the suburb of Shenyang and the middle-upper stream of the Xihe River, where the $\sum$n-alkanes and TAH were 0.8 to 13.1 μg g^{-1} and 5.0 to 65.6 μg g^{-1}, respectively. The hydrocarbon concentrations at I-1P, I-3P and I-4U were low, where the $\sum$n-alkanes value was 0.5 to 7.5 μg g^{-1} and the TAH was 1.6 to 47.2 μg g^{-1}. I-4U was the lowest concentration site. The Kruskal–Wallis test showed that there were no significant differences among the ten layers for $\sum$n-alkanes and TAH. However, the Mann–Whitney test showed that the $\sum$n-alkanes at 0–10 cm (1.7–32.2 μg g^{-1}) were significantly higher than those at the other layers (0.5–15.5 μg g^{-1}), due to biogenic hydrocarbons and fresh oil entering the topsoil. For TAH, the concentrations at 0–10 cm were highest (4.1–141.4 μg g^{-1}), the values at 20–25 cm and 50–100 cm were medial (1.8–161.2 μg g^{-1}), and those at the other layers were lower (1.6–109.1 μg g^{-1}). The higher TAH values in the deep soil were caused by both input and downward migration of difficultly-degraded petroleum hydrocarbons in the previous years.

The total aliphatic hydrocarbon concentrations were over 10 μg g^{-1} in all layers at I-2U, I-5P, I-6U, and I-7P (except 0–5 cm) and some layers at I-1P (0–5, 30–50, and 60–100 cm); I-3P (0–15 and 60–100 cm); I-4U (5–10, 30–40, and 50–100 cm); and I-8U (0–15, 20–30, 40–50, and 60–100 cm). Therefore, the cleanup was required according to the most rigorous level (10 μg g^{-1} for total petroleum hydrocarbon in soil) suggested by the Total Petroleum Hydrocarbon Criteria Working Group (TPHCWG) (Gustafson et al., 1997). Furthermore, the TAH values

Table 1
The average recoveries with RSD of n-alkanes, pristine, and phytane.

	Mean recovery ±RSD (%)		Mean recovery ±RSD (%)		Mean recovery ±RSD (%)
nC_{10}	43±8.2	nC_{19}	108±14.0	nC_{30}	83±13.9
nC_{11}	45±8.1	nC_{20}	100±12.6	nC_{31}	87±11.0
nC_{12}	50±5.7	nC_{21}	96±6.7	nC_{32}	71±8.1
nC_{13}	105±3.4	nC_{22}	79±3.1	nC_{33}	90±10.1
nC_{14}	115±13.0	nC_{23}	68±5.8	nC_{34}	92±9.3
nC_{15}	93±13.1	nC_{24}	60±7.3	nC_{35}	56±11.4
nC_{16}	74±7.8	nC_{25}	71±10.7	nC_{36}	113±9.5
nC_{17}	98±7.2	nC_{26}	60±6.5	nC_{37}	108±12.5
Pristane	73±8.1	nC_{27}	66±9.6	nC_{38}	101±7.5
nC_{18}	85±6.9	nC_{28}	69±5.8	nC_{39}	125±17.0
Phytane	83±11.0	nC_{29}	95±12.6	nC_{40}	96±4.3

in some soil layers of I-2U (10–15, 20–25, and 50–100 cm), I-5P (0–10 and 40–100 cm), I-6U (all layers except 60–100 cm), I-7P (25–30 cm), and I-8U (0–5 cm) were over 50 μg g^{-1}, which was the generic and top 2 ft soil total petroleum hydrocarbon cleanup level provided by the Oklahoma Department of Environmental Quality (DEQ) (DEQ, 2004). All soil samples were below the level for petroleum hydrocarbons (boiling point ranges from nC_{10} to nC_{16}: 150 μg g^{-1}, nC_{16} to nC_{34}: 300 μg g^{-1} and nC_{34+}: 2800 μg g^{-1}) established by the Canadian Council of Ministers of the Environment (CCME) (CCME, 2008).

3.2. Vertical distribution of $\sum$n-alkanes, UCM, and TAH of different soil profiles

The patterns of vertical distribution of the aliphatic hydrocarbons in the eight profiles were not uniform (Fig. 2). The complicated distribution was a result of various input quantities at different periods and the many influential factors of migration rate. The hydrocarbon concentrations were low in the topsoil and high in the soil layers below 30 cm at I-1P and I-4U. In summary, the concentrations at the two sites were quite lower compared with those of the other sites and close to the natural distribution without obvious pollution sources.

I-2U was near an abandoned oil well, which stopped production ten years ago. The hydrocarbon concentration was therefore low and had a slightly increasing trend in the top soil. Due to the high input quantities ages ago and the subsequent downward migration, high hydrocarbon concentration accumulated at 20–25 cm was discovered. The concentration in the bottom soil was also high because of the petroleum input at an earlier time. Moreover, the soil particles were finer at 20–25 cm and 50–100 cm, which was the reason of the hydrocarbons accumulation at the depths.

I-5P was irrigated by wastewater from the lower Xihe River canal. The hydrocarbons concentration in the top soil (0–20 cm) decreased dramatically with depth, which was related to the input of petroleum from the irrigation wastewater in near term. The concentration at 20–25 cm began to increase and that in bottom soil was especially high. Acher et al. (1989) and Muszkat et al. (1993) suggested that sewage and moisture would promote the downward mobility of organic pollutants, which could explain that the higher hydrocarbon concentration in deep soil compared with that in shallow soil and the different distributions between the upland fields and wastewater-irrigated paddy fields.

I-3P was also affected by the wastewater from the Xihe River resulting in some similarities between I-3P and I-5P because the main canal was connected to the Xihe River canal.

The hydrocarbon concentration was high and had a descending trend as depth increases at I-6U, which was caused by the continuous input of petroleum hydrocarbons from a nearby petroleum exploration. For I-8U at the suburb of Shenyang City, the concentration was medial

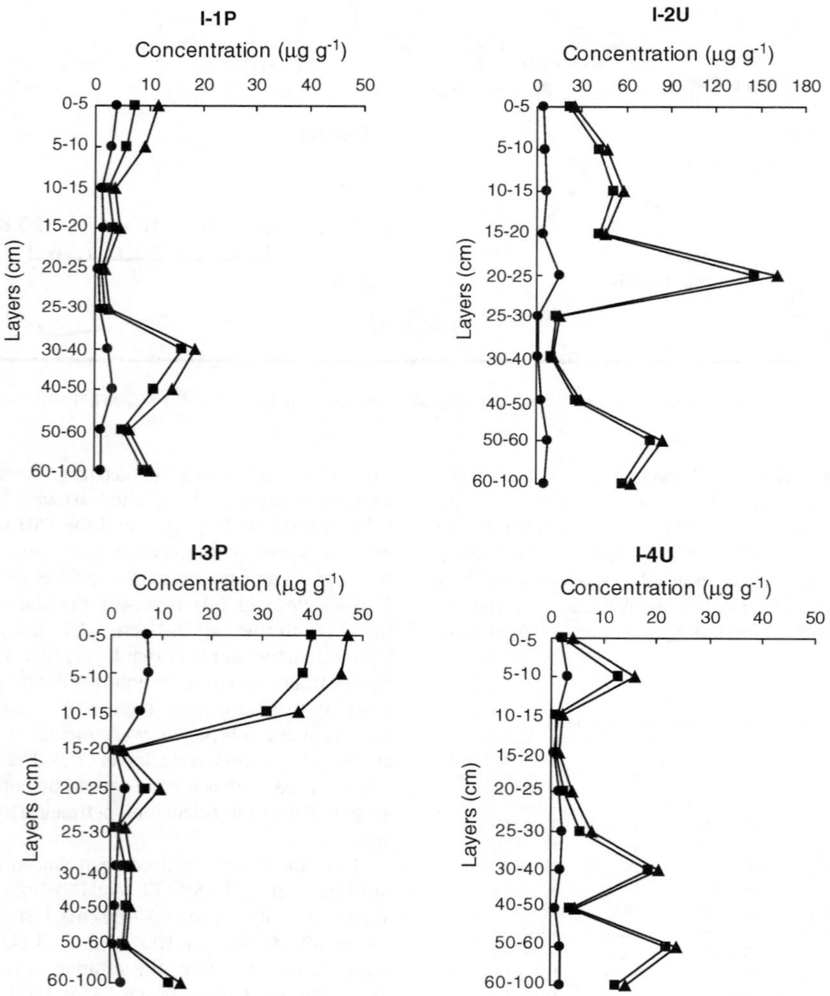

Fig. 2. The concentrations of $\sum$n-alkanes, unresolved complex mixtures (UCM), and total aliphatic hydrocarbons (TAH) at different depths of eight profiles (I-1P, I-2U, I-3P, I-4U, I-5P, I-6U, I-7P, and I-8U).

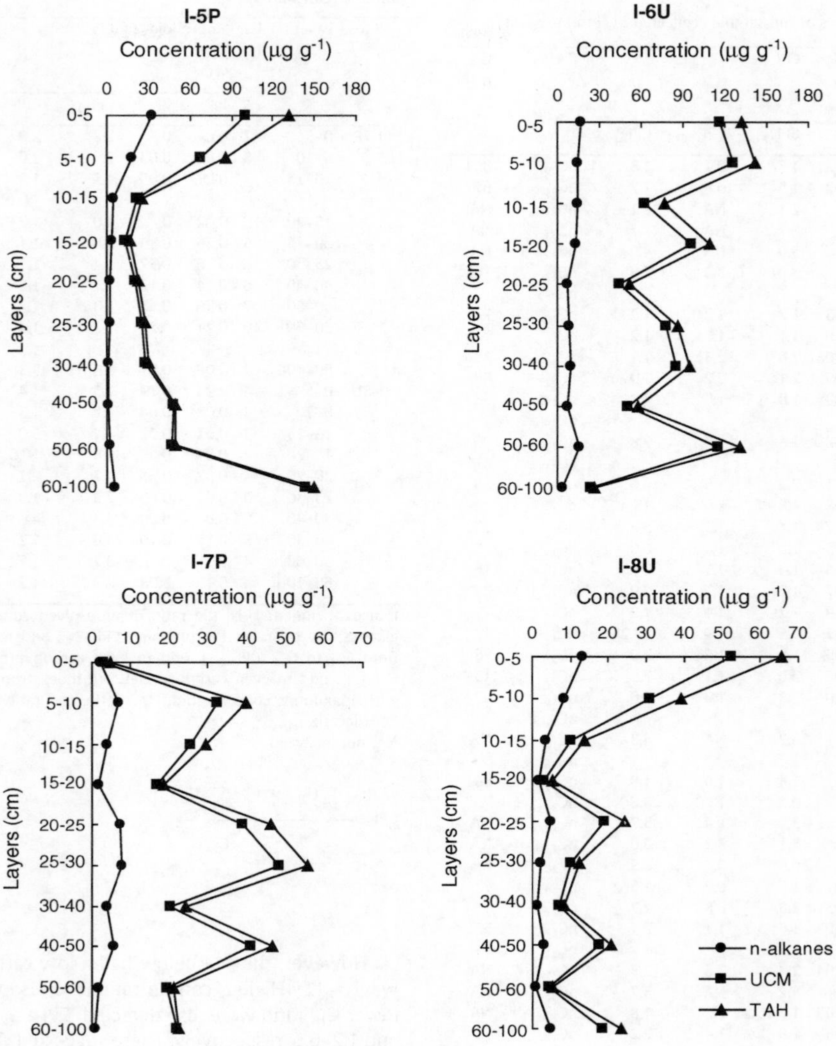

Fig. 2 (*continued*).

and decreased with the increase of depth, because the petroleum hydrocarbons continuously entered the soil through the atmospheric deposition.

The distribution in the layers below 5 cm at I-7P was similar to that at I-8U, both of which were due to long-term effect of its location relative to the Shenyang City. The quite low concentration in the layer 0–5 cm at I-7P was possibly due to washing by the treated water from the upper reach of the Xihe River approaching the 2008 Olympic Games. The differences of the water quality in the Xihe River resulted in the different concentrations and distributions between I-5P and I-7P.

The concentration of saturated hydrocarbons in Beijing was 1.5 to 54.1 $\mu g\ g^{-1}$, which decreased with depth within the top 30 cm and kept constant in the lower layers (He et al., 2008). In Tianjin, the hydrocarbon content in most samples was within 20 $\mu g\ g^{-1}$, and the maximum occurrence at the suburb area was 80 $\mu g\ g^{-1}$. The contents decreased with depth, which was identical to the distribution of PAHs in Hunpu (Bu et al., 2007; Xiao et al., 2008). The diesel range organics in soil varied from 0.2 to 4200 $\mu g\ g^{-1}$ at the three oil release sites at Big South Fork National River and Recreation Area, Tennessee and Kentucky, where the concentration of deep layer was higher compared with that of the shallow layer at some points (Williams et al., 2006).

3.3. Geochemical characteristics and source analysis of aliphatic hydrocarbons in the soil samples

The features of the determined geochemical indices were not consistent among the samples at different depths of the eight profiles (Table 2). The geochemical characteristics of I-1P and I-4U were similar. The C_{max} of soil above 30 cm at I-1P and above 25 cm at I-4U showed high carbon number, which suggested a low maturity (Scholz-Böttcher et al., 2009; Zhang et al., 2004). Low carbon number C_{max} elucidates that major hydrocarbons are from petroleum, while C_{max} higher than C_{27} means higher plants input (Aboul-Kassim and Williamson, 2003; Ye et al., 2007). Hexadecane and isoprenoid hydrocarbons were small in those samples, but they were mostly detected in petroleum hydrocarbons (Aboul-Kassim and Simoneit, 1995b; Medeiros and Bicego, 2004; Readman et al., 2002; Ye et al., 2007; Zhu et al., 2005). The ratio of UCM to the sum of n-alkanes (U/N) was also low. A U/N higher than 2 reflects a significant petroleum product contamination (Aboul-Kassim and Simoneit, 1995a; Harji et al., 2008). A separately revised CPI was selected to evaluate the contribution of biogenic hydrocarbons (Harji et al., 2008; Marzi et al., 1993; Zhu et al., 2005). High CPI was obtained and all characteristics confirmed that biogenic hydrocarbons were predominant.

Table 2
Distribution of geochemical indices of soil samples collected in Hunpu region.

Sites	Layers (cm)	U/N	Isoprenoid μg g^{-1} dry wt.		CPI			C_{max}	C_{16} ratio
			Pr	Ph	CPI_{13-35}	CPI_{13-22}	CPI_{23-35}		
I-1P	0–5	2	0	0	1.7	0.7	1.8	nC_{29}	801
	5–10	2	0.01	0.02	1.5	0.8	1.7	nC_{31}	977
	10–15	2	0.02	0	2.1	NA	2.1	nC_{31}	NA
	15–20	2	0	0	1.7	NA	1.7	nC_{31}	NA
	20–25	2	0	0	5.0	NA	4.5	nC_{27}	NA
	25–30	2	0.002	0	3.0	3.3	2.9	nC_{12}, nC_{31}	189
	30–40	7	0.20	0.15	1.2	1.2	1.2	nC_{16}	7
	40–50	3	0.15	0.16	1.1	1.1	1.2	nC_{16}	7
	50–60	5	0.003	0.008	2.0	2.3	1.7	nC_{11}	196
	60–100	7	0.07	0.05	2.1	2.2	1.9	nC_{12}	69
I-2U	0–5	5	0.09	0.08	1.8	1.5	2.1	nC_{13}, nC_{29}	58
	5–10	8	0.25	0.28	1.8	1.3	2.8	nC_{13}, nC_{31}	36
	10–15	8	0.42	0.21	2.2	1.5	5.0	nC_{12}	25
	15–20	8	0.32	0.18	1.8	1.5	15.4	nC_{13}	18
	20–25	10	1.66	1.17	1.4	1.3	5.2	nC_{13}	13
	25–30	10	0.08	0.05	1.8	1.5	6.9	nC_{11}	19
	30–40	9	0.09	0.08	1.4	1.1	27.5	nC_{11}	16
	40–50	9	0.14	0.07	1.7	1.7	6.0	nC_{12}	27
	50–60	10	1.79	1.19	1.2	1.1	1.8	nC_{15}	7
	60–100	11	0.72	0.47	1.5	1.4	27.2	nC_{13}	12
I-3P	0–5	5	0.08	0.05	1.8	1.5	2.0	nC_{31}	78
	5–10	5	0.09	0.05	2.0	1.1	2.5	nC_{33}	157
	10–15	5	0.15	0.10	2.3	1.6	3.0	nC_{19}, nC_{31}	62
	15–20	2	0	0	6.2	NA	4.3	nC_{15}, nC_{36}	NA
	20–25	2	0.01	0.07	1.4	1.0	1.8	nC_{31}	801
	25–30	1	0	0	0.7	13.9	0.6	nC_{24}	295
	30–40	2	0.03	0.12	3.0	2.4	3.7	nC_{19}, C_{31}	NA
	40–50	4	0	0	3.1	3.2	3.0	nC_{11}	NA
	50–60	4	0	0	0.9	0.3	6.9	nC_{20}	NA
	60–100	5	0.08	0.07	1.0	0.9	1.5	nC_{16}	7
I-4U	0–5	1	0.03	0.03	2.3	0.8	2.7	nC_{31}	154
	5–10	4	0.08	0.09	1.6	1.1	2.1	nC_{19}, nC_{31}	65
	10–15	1	0.07	0.004	2.4	3.5	2.3	nC_{31}	NA
	15–20	2	0	0	3.2	NA	3.2	nC_{35}	NA
	20–25	2	0.007	0.003	1.7	1.2	1.8	nC_{31}	NA
	25–30	3	0.05	0.06	1.9	1.4	2.2	nC_{19}, nC_{31}	71
	30–40	10	0.07	0.05	2.7	1.4	6.2	nC_{15}, nC_{31}	17
	40–50	5	0.03	0.11	2.1	1.5	6.6	nC_{13}, nC_{36}	31
	50–60	12	0.07	0.04	2.3	1.5	3.7	nC_{13}, nC_{31}	93
	60–100	7	0.04	0.12	1.4	1.2	3.2	nC_{13}	48
I-5P	0–5	3	0.13	0.18	1.3	0.9	1.4	nC_{31}	233
	5–10	4	0.10	0.07	1.7	0.7	2.0	nC_{31}	434
	10–15	5	0.06	0.03	2.5	1.7	2.9	nC_{31}	203
	15–20	4	0.07	0	2.2	3.4	2.0	nC_{39}	163
	20–25	7	0.09	0.05	2.3	1.6	2.9	nC_{12}, nC_{35}	27
	25–30	11	0.13	0.04	2.4	2.0	2.7	nC_{15}, nC_{33}	34
	30–40	16	0.18	0.12	2.4	2.3	2.8	nC_{21}	291
	40–50	31	0.33	0.17	3.2	2.9	4.2	nC_{15}	NA
	50–60	17	0.30	0.28	1.8	1.5	2.2	nC_{15}	NA
	60–100	23	0.67	0.29	1.8	1.7	2.4	nC_{13}	55
I-6U	0–5	7	1.06	0.76	1.6	1.4	2.6	nC_{13}	12
	5–10	9	1.71	1.14	1.7	1.4	3.5	nC_{15}	15
	10–15	4	0.81	0.78	1.3	1.1	1.8	nC_{16}	9
	15–20	7	1.22	0.88	1.5	1.3	3.5	nC_{13}	11
	20–25	6	0.71	0.67	1.4	1.2	2.5	nC_{15}	10
	25–30	9	1.21	0.87	1.4	1.3	4.3	nC_{13}	14
	30–40	8	0.89	0.67	1.6	1.5	2.3	nC_{13}	19
	40–50	8	0.47	0.35	1.5	1.3	6.2	nC_{12}	15
	50–60	7	1.64	1.82	1.2	1.0	2.6	nC_{13}	16
	60–100	8	0.51	0.43	1.3	1.1	31.8	nC_{12}	12
I-7P	0–5	2	0.04	0	2.2	2.9	2.1	nC_{31}	971
	5–10	5	0.04	0.03	2.5	2.0	2.8	nC_{33}	138
	10–15	7	0.07	0.05	1.5	1.3	1.7	nC_{19}, nC_{29}	85
	15–20	10	0.02	0	4.0	1.7	6.1	nC_{31}	19
	20–25	5	0.26	0.35	1.6	1.1	1.8	nC_{31}	24
	25–30	6	0.36	0.37	1.4	1.1	1.6	nC_{31}	21
	30–40	5	0.14	0.12	1.7	1.2	2.0	nC_{31}	20
	40–50	7	0.76	0.52	1.3	1.2	1.5	nC_{15}	13
	50–60	10	0.21	0.20	1.8	1.4	3.7	nC_{17}, nC_{31}	11
	60–100	28	0.08	0.01	5.6	3.3	NA	nC_{15}	75
I-8U	0–5	4	0.26	0.19	1.5	1.4	1.6	nC_{31}	40
	5–10	4	0.08	0.04	1.2	1.5	1.1	nC_{39}	172
	10–15	3	0.02	0.02	1.4	3.0	1.3	nC_{33}	137
	15–20	2	0.01	0	1.8	10.6	1.7	nC_{31}	433
	20–25	4	0.02	0	1.7	2.1	1.7	nC_{31}	490
	25–30	5	0.07	0.05	2.3	1.3	2.9	nC_{33}	20
	30–40	5	0.04	0.004	4.4	4.6	4.3	nC_{31}	350
	40–50	5	0.12	0.10	1.9	1.2	2.5	nC_{31}	17
	50–60	4	0	0	3.7	1.7	4.6	nC_{31}	NA
	60–100	4	0.16	0.08	2.3	1.2	2.9	nC_{31}	18

(Rows from I-7P onward are printed under the heading **Table 2 (continued)**.)

Indices included: U/N, the ratio of unresolved complex mixture to the sum of n-alkanes; Pr, pristane; Ph, phytane; CPI_{13-35}, odd to even carbon preference index from C_{13} to C_{35}; CPI_{13-22}, odd to even carbon preference index from C_{13} to C_{22}; CPI_{23-35}, odd to even carbon preference index from C_{23} to C_{35}; C_{max}, the n-alkane with maximum concentration; C_{16} ratio, the ratio of the sum of all n-alkanes to hexadecane.
NA: not analyzed

$$CPI = \frac{\left(\sum_{i=n}^{m} C_{2i+1}\right) + \left(\sum_{i=n+1}^{m+1} C_{2i+1}\right)}{2 \times \left(\sum_{i=n+1}^{m+1} C_{2i}\right)}$$

However, the soil below had a low carbon number C_{max} and U/N was 3–12. Higher concentrations of isoprenoid hydrocarbons and lower C_{16} ratio were also detected. CPI_{13-22} and CPI_{23-35} were 1.1–2.3 and 1.2–6.6, respectively. These suggested that different degrees of petroleum pollution existed. The layers 30–40 and 40–50 cm of I-1P were similar with the petroleum-polluted sediments from the Chubut River (Commendatore and Esteves, 2004).

According to the geochemical indices of I-3P, the aliphatic hydrocarbons in the top 15 cm soil showed some degree of petroleum pollution. The composition of hydrocarbons in the bottom layer was similar to petroleum, but not with the other soil layers.

I-2U and I-6U were typical petroleum-contaminated sites. They were rich in low carbon number hydrocarbons and poor in high carbon number hydrocarbons except the top two soil layers of I-2U, which was also displayed by C_{max}. Both the low carbon number C_{max} and high carbon number C_{max} existed in the top 10 cm soil of I-2U, whereas C_{max} showed low carbon number in other samples. The concentrations of UCM and isoprenoid hydrocarbons were high, and all the U/N was over 4. The C_{16} ratio was generally around 15, which indicated hydrocarbons from petroleum source; relatively, the value was up to 50 for biogenic hydrocarbons (Colombo et al., 1989; Harji et al., 2008; Tran et al., 1997). CPI_{13-22} ranged from 1.0 to 1.7, while CPI_{23-35} ranged from 1.8 to 31.8. The results indicated that the low carbon number hydrocarbons were mainly from petroleum, whereas the high carbon number hydrocarbons received more hydrocarbons from the higher plant wax.

For I-5P, the aliphatic hydrocarbons were composed of rich heavy hydrocarbons and small light hydrocarbons in the top 10 cm soil. Gas chromatogram showed a big hump in heavy hydrocarbons, a feature of heavy oil pollutants (Colombo et al., 1989; Hegazi et al., 2004;

Scholz-Böttcher et al., 2009) or weathered oil (Cripps, 1989). UCM from nC_{22} to nC_{33} and C_{max} between nC_{22} and nC_{33} were also found in sediments affected by industrial effluents in Kuwait and Daliao River watershed, China, respectively (Beg et al., 2003; Guo et al., 2007).

Thus, we deduced that I-5P was impressed by irrigation water from Xihe River which contained industrial effluents. The measured geochemical indices also showed an oil input. The concentration decreased dramatically from 10 to 20 cm and the characteristics of oil became less evident. The indices at the layers 20–30 cm revealed petroleum input. U/N of 30–60 cm was quite high, although CPI and the C_{16} ratio did not present obvious characteristics of petroleum. U/N over 5 showed significant input of degraded petroleum products (Damas et al., 2009). Therefore, contamination could have been a period of time ago. Reversely, the layer 60–100 cm contained rich light hydrocarbons and a big hump located at low carbon number part, and geochemical analysis indicated a contamination of light oil, which resembled the pattern of I-2U and I-6U.

Geochemical indices showed that the layer 0–5 cm of I-7P was dominated by biogenic hydrocarbons; however, petrogenic hydrocarbons gradually became predominant as the depth increased. Soil at 15–60 cm was obviously contaminated by petroleum, and the bottom soil was affected by degraded petroleum products. Compared to I-7P, layer 0–5 cm of I-8U was contaminated by petroleum. From 0 to 20 cm, the features of petroleum gradually disappeared with the increase of depth, but some characteristics of petroleum appeared in the soil below.

3.4. PCA and RDA analysis of n-alkanes, UCM, and isoprenoid hydrocarbons of soil samples

For PCA, the percentage variances of the aliphatic hydrocarbons explained by the first and second axes were 69.9 and 18.8, respectively (the cumulative percentage variance explained by the two axes was 88.7). The hydrocarbons from different sources were differentiated by PCA (Fig. 3). Low carbon number n-alkanes, Pr, Ph and UCM collectively defined the first axis which represented hydrocarbons mainly from petroleum pollution, while high carbon number n-alkanes assembled and defined the second axis which represented hydrocarbons dominated by higher plants or heavy oil.

The compositions of aliphatic hydrocarbons of different paddy fields as well as those of the different upland fields were not consistent (Fig. 4). The eighty samples had three main composition patterns: first, the high carbon number n-alkanes and UCM concentrations of the samples were quite high, while the low carbon number n-alkanes concentration was low (grouped in the first quadrant of Fig. 4); second, for the samples in the second or third quadrant, the concentrations of all aliphatic hydrocarbon components were low showing the pattern of I-1P and I-4U (samples presented by up-triangle symbols in Fig. 4) with minor or no contamination; third, in the fourth quadrant, the low carbon number n-alkanes, Pr, Ph, and UCM of the samples were all rich, and most samples of I-2U and I-6U (samples presented by star symbols) had this pattern because both were affected by nearby oil wells. I-3P and I-8U (samples presented by right-triangle symbols) contained two patterns: the top soil had the first pattern, while 15–100 cm of I-3P and 10–100 cm of I-8U had the second one. The samples at I-5P and I-7P (samples presented by circle symbols) were affected by wastewater-irrigation and basically belonged to the first pattern group.

The cumulative percentage variance of the aliphatic hydrocarbons explained by the measured soil basic properties in the RDA was 23.5 (four axes). The percentage variances explained by the first and second axes were 17.4 and 5.7, respectively. Unrestricted permutation under a reduced model result showed that the aliphatic hydrocarbon composition had highly significant correlation with sand content ($P=0.002$) and pH value ($P=0.006$), while the correlations between the aliphatic hydrocarbons and the other soil properties were not significant. The variances explained by sand content and pH value were 14% and 6%, respectively. In Fig. 5, it was found that the low carbon number n-alkanes, Pr, Ph, and UCM had a negative correlation with sand while having a positive correlation with fine materials (silt and clay). This meant that the soil composed of high silt and clay could potentially contain more oil or hydrocarbon than sandy soil, which was consistent with the findings of Colombo et al. (1989) and Abdullah (1997). Commendatore and Esteves (2004) did not find any correlation between fine fraction and OM or TAH. Early research also reported that n-alkanes, especially terrestrial hydrocarbons were related with coarse fraction, but UCM was related with finer

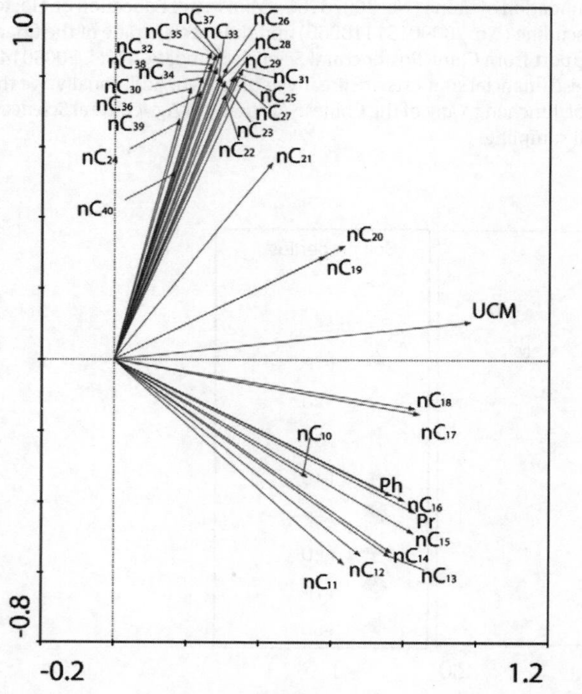

Fig. 3. Scatter plot of aliphatic hydrocarbons (AHs, included: n-alkanes (nC_{10} to nC_{40}), phytane (Ph), pristine (Pr), and UCM) from PCA analysis.

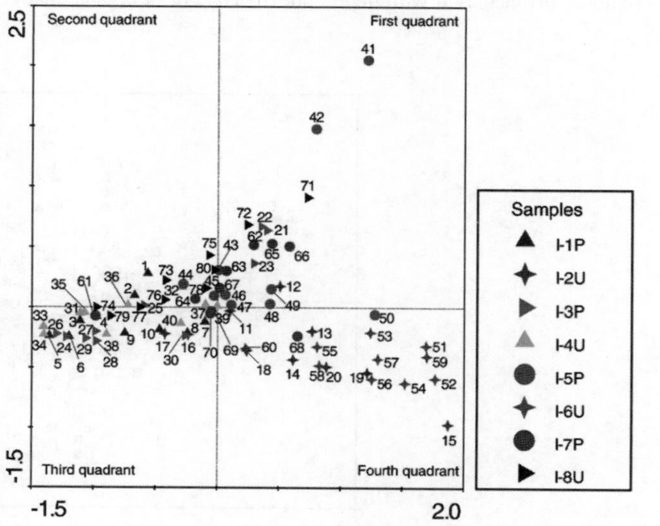

Fig. 4. Scatter plot of soil samples from PCA analysis. Samples were classified according to sampling sites (samples from top to bottom, each sites were labeled with consecutive indices).

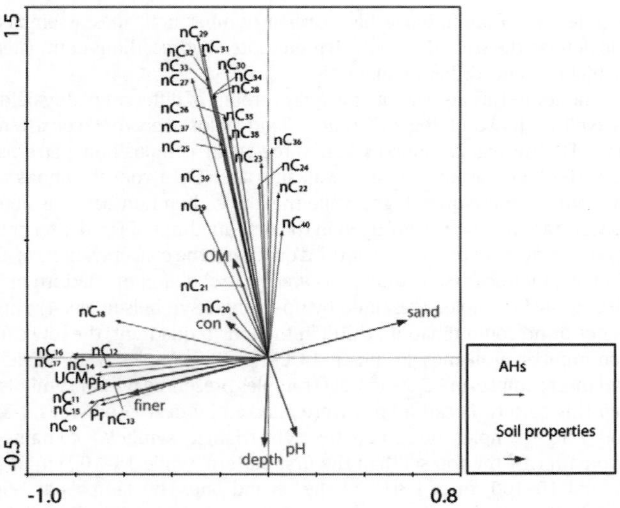

Fig. 5. Bipolt of aliphatic hydrocarbons (AHs) and soil properties from RDA analysis. AHs included: n-alkanes (nC_{10} to nC_{40}), phytane (Ph), pristine (Pr), and UCM. Soil properties included: sand content, sand; total content of silt and clay, finer; organic matters, OM; pH; conductivity, con; depth.

particles (Readman et al., 2002). High carbon number n-alkanes had a positive correlation with sand and OM. Nonpolar organic pollutants were considered to be absorb on OM when clay content of soil was low (Moreda et al., 1998).

The eight sample sites were divided into two main clusters by RDA (Fig. 6): I-5P, I-6U, I-7P, and I-8U near the Xihe River canal, deep soil at I-1P assembled on the left (the first cluster), and I-2U, I-3P, I-4U and upper layers at I-1P near the main canal clustered on the right (the second cluster). This was obviously different from the result from PCA, which meant that the pollution source rather than the soil properties determined the aliphatic hydrocarbon composition in soil. Compared to the second cluster, the soil from the Xihe River canal was finer and the differences between the upland and paddy fields were better observed. The conductivity of the upper layers of soil near the Xihe River canal was relatively higher. These differences suggested that the soil properties could be affected by pollutants. For all samples, pH increased with depth, but the content of organic matter declined. The measured soil properties and hydrocarbon compositions determined in deep soil of the eight profiles were similar, which indicated that the upper horizon soil was easily impressed by environment and human activities.

4. Conclusions

The result from the concentration analysis agreed with that from the geochemical analysis, which was successfully visualized by PCA and RDA. The soil from I-2U and I-6U (with nearby oil wells still operating or abandoned) had rich hydrocarbons ($\sum$n-alkanes: 1.1–16.4 μg g^{-1} dry wt. and TAH: 10.9–161.2 μg g^{-1} dry wt.), obvious characteristics of petroleum hydrocarbons, and were grouped together by PCA. The hydrocarbon concentrations of the samples from I-5P was also high ($\sum$n-alkanes: 1.6–32.2 μg g^{-1} and TAH: 16.7–149.7 μg g^{-1}) and displayed oil contamination led by irrigation with wastewater from the downriver of the Xihe River. The hydrocarbon concentrations of I-7P and I-8U, which were affected by the industrial city of Shenyang, were medium ($\sum$n-alkanes: 0.8–13.1 μg g^{-1} and TAH: 5.0–65.6 μg g^{-1}) and had certain characteristics of petroleum contamination. I-5P and I-7P were grouped together because of wastewater irrigation. Another group contained I-1P and I-4U ($\sum$n-alkanes: 0.5–4.0 μg g^{-1} and TAH: 1.6–23.8 μg g^{-1}), which had a small hydrocarbons. Biogenic hydrocarbons dominated in most samples and no obvious pollution source was detected. I-8U was affected by atmospheric deposition and I-3P had a similar pattern. The result from RDA indicated that the hydrocarbon compositions were also impacted by the soil properties. The differences between the results from PCA and those from RDA also suggested that the hydrocarbon composition was mainly determined by the pollution sources.

Acknowledgment

The work, which led to this paper, was sponsored by the National Basic Research Program of China (No. 2004CB418501), National Science Foundation of China (No. 30970166), Ministry of Education on Doctorial Discipline (No. 20090131110066) and the special grade of the financial support from China Postdoctoral Science Foundation (No. 200801403). These financial supports are greatly appreciated. Additionally, we thank Prof. Juncheng Yang of the Chinese Academy of Agricultural Sciences for soil sampling.

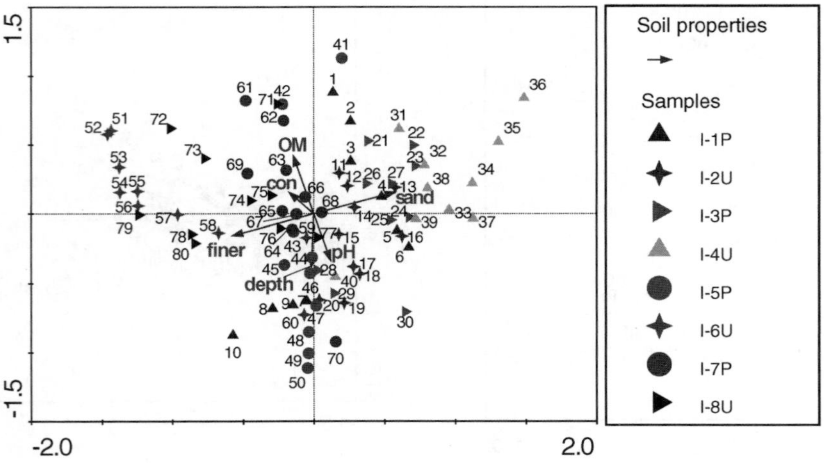

Fig. 6. Bipolt of soil properties and soil samples from RDA analysis. Soil properties included: sand content, sand; total content of silt and clay, finer; organic matters, OM; pH; conductivity, con; depth. Samples were classified according to sampling sites (samples from top to bottom, each sites were labeled with consecutive indices).

References

Abdullah, M.P., 1997. Hydrocarbon pollution in the sediment of some Malaysian coastal areas. Environmental Monitoring and Assessment 44, 443–454.

Aboul-Kassim, T.A.T., Simoneit, B.R.T., 1995a. Aliphatic and aromatic hydrocarbons in particulate fallout of Alexandria, Egypt: sources and implications. Environmental Science & Technology 29, 2473–2483.

Aboul-Kassim, T.A.T., Simoneit, B.R.T., 1995b. Petroleum hydrocarbon fingerprinting and sediment transport assessed by molecular biomarker and multivariate statistical analyses in the Eastern Harbour of Alexandria, Egypt. Marine Pollution Bulletin 30, 63–73.

Aboul-Kassim, T.A.T., Williamson, K.J., 2003. Forensic analysis and genetic source partitioning model for Portland Harbor contaminated sediments. Journal of Environmental Informatics 1, 58–75.

Acher, A.J., Boderie, P., Yaron, B., 1989. Soil pollution by petroleum products, I. Multiphase migration of kerosene components in soil columns. Journal of Contaminant Hydrology 4, 333–345.

Adeniyi, A.A., Afolabi, J.A., 2002. Determination of total petroleum hydrocarbons and heavy metals in soils within the vicinity of facilities handling refined petroleum products in Lagos metropolis. Environment International 28, 79–82.

ATSDR, 1999. Toxicological profile for total petroleum hydrocarbons (TPH). U.S Department of Health and Human Services, Public Health Service, Agency for Toxic Substances and Disease Registry (ATSDR), Atlanta, GA.

Beg, M.U., Saeed, T., Al-Muzaini, S., Beg, K.R., Al-Bahloul, M., 2003. Distribution of petroleum hydrocarbon in sediment from coastal area receiving industrial effluents in Kuwait. Ecotoxicology and Environmental Safety 54, 47–55.

Bu, Q.W., Zhang, Z.H., Wang, L., Du, M.L., 2007. Vertical distribution and environmental significance of saturated hydrocarbons in soil samples collected from Tianjin. Progress in Geography 26, 21–32.

CCME, 2008. Canada-wide standards for peteoleum hydrocarbons (PHC) in soil. April 30–May 1, 2001. Revision January 2008. Canadian Council of Ministers of the Environment (CCME), Winnipeg.

Colombo, J.C., Pelletier, E., Brochu, C., Khalil, M., Catoggio, J.A., 1989. Determination of hydrocarbon sources using n-alkane and polyaromatic hydrocarbon distribution indexes. Case study: Rio de la Plata Estuary, Argentina. Environmental Science & Technology 23, 888–894.

Commendatore, M.G., Esteves, J.L., 2004. Natural and anthropogenic hydrocarbons in sediments from the Chubut River (Patagonia, Argentina). Marine Pollution Bulletin 48, 910–918.

Cripps, G.C., 1989. Problems in the identification of anthropogenic hydrocarbons against natural background levels in the Antarctic. Antarctic Science 1, 307–312.

Damas, E.Y.C., Clemente, A.C.N., Medina, M.O.C., Bravo, L.G., Ramada, R.M., Porto, R.M.D., Rodriguez, M.R., Diaz, M.A.D., 2009. Petroleum hydrocarbon assessment in the sediments of the Northeastern Havana littoral, Cuba. Revista Internacional De Contaminacion Ambiental 25, 5–14.

DEQ, 2004. Risk-based Cleanup Levels for Total Petroleum Hydrocarbons (TPH). Oklahoma Department of Environmental Quality (DEQ). Land Protection Division, Oklahoma.

Dionex, 2007. Extraction of total petroleum hydrocarbon contamination (diesel range organics and waste oil organics) from soils using accelerated solvent extraction. Environmental Chemistry 26, 721–723.

EPA, 2007. US Environmental Protection Agency Method 3545A. Pressurized fluid extraction (PFE). EPA SW-846. US Government Printing Office, Washington DC.

Guo, W., He, M.C., Yang, Z.F., Lin, C.Y., Quan, X.C., 2007. Distribution and sources of petroleum hydrocarbons and polycyclic aromatic hydrocarbons in sediments from Daliao River watershed, China. Acta Scientiae Circumstantiae 27, 824–830.

Gustafson, J.B., Tell, J.G., Orem, D., 1997. Selection of Representative TPH Fractions Based on Fate and Transport Considerations. Amherst Scientific Publishers, Amherst, Mass.

Harji, R.R., Yvenat, A., Bhosle, N.B., 2008. Sources of hydrocarbons in sediments of the Mandovi estuary and the Marmugoa harbour, west coast of India. Environment International 34, 959–965.

He, F.P., Zhang, Z.H., Gao, D.D., 2008. Vertical distribution characteristics and composition of saturated hydrocarbon in soil profiles of Beijing. Environmental Science 29, 170–178.

Hegazi, A.H., Andersson, J.T., Abu-Elgheit, M.A., El-Gayar, M., 2004. Source diagnostic and weathering indicators of tar balls utilizing acyclic, polycyclic and S-heterocyclic components. Chemosphere 55, 1053–1065.

Kang, Y.H., Sheng, G.Y., Fu, J.M., Mai, B.X., Zhang, G., Lin, Z., Min, Y.S., 2000. The study of n-alkanes in a sedimentary core from Macao Estuary, Pearl River. Geochimica 29, 302–310.

Lepš, J., Šmilauer, P., 2003. Multivariate Analysis of Ecological Data using CANOCO. Cambridge University Press, Cambridge, UK; New York.

Li, H., Zhang, Y., Zhang, C.G., Chen, G.X., 2005. Effect of petroleum-containing wastewater irrigation on bacterial diversities and enzymatic activities in a paddy soil irrigation area. Journal of Environmental Quality 34, 1073–1080.

Lu, R.K., 2000. Soil Analysis in Agricultural Chemistry. Agricultural Science and Technology Press of China, Beijing.

Marzi, R., Torkelson, B.E., Olson, R.K., 1993. A revised carbon preference index. Organic Geochemistry 20, 1303–1306.

Medeiros, P.M., Bicego, M.C., 2004. Investigation of natural and anthropogenic hydrocarbon inputs in sediments using geochemical markers. I. Santos, SP-Brazil. Marine Pollution Bulletin 49, 761–769.

Miller, W.P., Miller, D.M., 1987. A micro-pipette method for soil mechanical analysis. Communications in Soil Science and Plant Analysis 18, 1–15.

Moreda, J.M., Arranz, A., De Betono, S.F., Cid, A., Arranz, J.F., 1998. Chromatographic determination of aliphatic hydrocarbons and polyaromatic hydrocarbons (PAHs) in a sewage sludge. Science of the Total Environment 220, 33–43.

Muszkat, L., Raucher, D., Magaritz, M., Ronen, D., Amiel, A.J., 1993. Unsaturated zone and ground-water contamination by organic pollutants in a sewage-effluent-irrigated site. Ground Water 31, 556–565.

Plaza, G., Nalecz-Jawecki, G., Ulfig, K., Brigmon, R.L., 2005. The application of bioassays as indicators of petroleum-contaminated soil remediation. Chemosphere 59, 289–296.

Rao, C.V., Afzal, M., Malallah, G., Kurian, M., Gulshan, S., 2007. Hydrocarbon uptake by roots of *Vicia faba* (*Fabaceae*). Environmental Monitoring and Assessment 132, 439–443.

Readman, J.W., Fillmann, G., Tolosa, I., Bartocci, J., Villeneuve, J.P., Catinni, C., Mee, L.D., 2002. Petroleum and PAH contamination of the Black Sea. Marine Pollution Bulletin 44, 48–62.

Richter, B.E., 2000. Extraction of hydrocarbon contamination from soils using accelerated solvent extraction. Journal of Chromatography. A 874, 217–224.

Scholz-Böttcher, B.M., Ahlf, S., Vázquez-Gutiérrez, F., Rullkötter, J., 2009. Natural vs. anthropogenic sources of hydrocarbons as revealed through biomarker analysis: a case study in the southern Gulf of Mexico. Boletín de la Sociedad Geológica Mexicana 61, 47–56.

Song, X.Y., Sun, L.N., Wang, X., Li, X.X., Sun, T.H., 2007. Organic contamination status of Xihe River surface water and its riverbank underground water. Chinese Journal of Ecology 26, 2057–2061.

Song, Y.F., Zhou, Q.X., Song, X.Y., Zhang, W., Sun, T.H., 2004. Accumulation of pollutants in sediments and their eco-toxicity in the wastewater irrigation channel of western Shenyang. Chinese Journal of Applied Ecology 15.

Thomas, K.V., Donkin, P., Rowland, S.J., 1995. Toxicity enhancement of an aliphatic petrogenic unresolved complex mixture (UCM) by chemical oxidation. Water Research 29, 379–382.

Tran, K., Yu, C.C., Zeng, E.Y., 1997. Organic pollutants in the coastal environment off San Diego, California. 2. Petrogenic and biogenic sources of aliphatic hydrocarbons. Environmental Toxicology and Chemistry 16, 189–195.

Van De Weghe, H., Vanermen, G., Gemoets, J., Lookman, R., Bertels, D., 2006. Application of comprehensive two-dimensional gas chromatography for the assessment of oil contaminated soils. Journal of Chromatography. A 1137, 91–100.

Wakeham, S.G., Carpenter, R., 1976. Aliphatic hydrocarbons in sediments of Lake Washington. Limnology and Oceanography 21, 711–723.

Williams, S.D., Ladd, D.E., Farmer, J.J., 2006. Fate and transport of petroleum hydrocarbons in soil and ground water at Big South Fork National River and Recreation Area, Tennessee and Kentucky, 2002–2003. U.S. Geological Survey, Reston, Va.

Xiao, R., Du, X.M., He, X.Z., Zhang, Y.J., Yi, Z.H., Li, F.S., 2008. Vertical distribution of polycyclic aromatic hydrocarbons (PAHs) in Hunpu wastewater-irrigated area in northeast China under different land use patterns. Environmental Monitoring and Assessment 142, 23–34.

Xiong, Z.T., Hu, H.X., Wang, Y.X., Fu, G.H., Tan, Z.Q., Yan, G.A., 1997. Comparative analyses of soil contaminant levels and plant species diversity at developing and disused oil well sites in Qianjiang oilfield, China. Bulletin of Environmental Contamination and Toxicology 58, 667–672.

Ye, B., Zhang, Z., Mao, T., 2007. Petroleum hydrocarbon in surficial sediment from rivers and canals in Tianjin, China. Chemosphere 68, 140–149.

Zemo, D., 2007. Forensic tools for petroleum hydrocarbon releases. Southwest Hydrology 6, 26–35.

Zhang, Z.H., Tao, S., Ye, B.X., Peng, Z.Q., Peng, Y.J., 2004. Pollution sources and identification of hydrocarbons in soil and sediment using molecular markers. Chinese Journal of Soil Science 35.

Zhou, Q., Sun, F., Liu, R., 2005. Joint chemical flushing of soils contaminated with petroleum hydrocarbons. Environment International 31, 835–839.

Zhu, Y.F., Liu, H., Xi, Z.Q., Cheng, H.X., Xu, X.B., 2005. Characterization of aliphatic hydrocarbons in deep subsurface soils near the outskirts of Beijing, China. Journal of Environmental Sciences 17, 360–364.

Bull Environ Contam Toxicol (2015) 94:431–436
DOI 10.1007/s00128-015-1475-y

Heavy Metal Bioaccumulation and Health Hazard Assessment for Three Fish Species from Nansi Lake, China

Pengfei Li · Jian Zhang · Huijun Xie · Cui Liu · Shuang Liang · Yangang Ren · Wenxing Wang

Received: 14 March 2014 / Accepted: 22 January 2015 / Published online: 31 January 2015

Abstract Metal accumulation in fish is a global public health concern, because the consumption of contaminated fish accounts for the primary exposure of humans to toxic metals. In this study, the concentrations of arsenic (As), cadmium (Cd), lead (Pb), and mercury (Hg) in Crucian carp (*Carassius auratus*),Yellow catfish (*Pelteobagrus fulvidraco*), and Bighead carp (*Hypophthalmichthys nobilis)* from Nansi Lake of China were evaluated, and compared with the corresponding historical values in 2001 when the government started to govern water environment effectively. Bioaccumulation of heavy metal was highest in *P.fulvidraco*, followed by *C.auratus* and *H.nobilis*. The concentrations of Pb, As, Cd were much lower than the historical values, but Hg concentration was higher, suggesting that heavy metal pollution problem in fish from Nansi Lake still exists. Health hazard assessment showed no health risk from exposure to Pb, As, Cd, and Hg by consuming fish from this lake.

Keywords Heavy metals · Fish bioaccumulation · Nansi Lake

P. Li · H. Xie · W. Wang
Environment Research Institute, Shandong University, Jinan 250100, China

J. Zhang (✉) · S. Liang · Y. Ren
School of Environmental Science and Engineering, Shandong University, Jinan 250100, China
e-mail: zhangjian00@sdu.edu.cn

C. Liu
Department of Mathematics and Statistics, Texas Tech University, Broadway and Boston, Lubbock, TX 79409-1042, USA

Fish is an important source of food for people worldwide. In 2008, 115 million tons of fish were consumed with an average of 17 kg per capita (FAO 2010). As fish contributes an important part of human diet, it is not surprising that the quality and the safety aspects of fish are of particular interest to human health. Elevated levels of heavy metals in aquatic ecosystems have raised serious public concerns around the world, due to their high potential to enter and accumulate in food chains and the correlation between heavy metals exposure and cancer in human (Zhang et al. 2011; Sirot et al. 2009).

The concentrations of heavy metals in fish have been extensively studied over the past several decades. Research has shown that the extent of accumulation of heavy metals in fish is dependent on the metal types, fish species, and the tissues respectively (Korkmaz Görür et al. 2012; Petrović et al. 2013). Water chemistry (Driscoll et al. 1994) directly affects the accumulation of heavy metal in fish. Sediment is also known to be an important factor for heavy metal accumulation in fish, as it is considered as the major source of contaminants for bottom dwelling and bottom feeding aquatic organisms (Farag et al. 1998), which in turn represents the concentrated source of metals in the diet of fish.

Nansi Lake is the largest shallow freshwater lake in North China, situated at the southwest of Shandong Province. It plays the most important role in the supply of both water and fresh water fish in Shandong Province. The total area of the lake is 1,266 km^2 with fish production reaching 2.3×10^4 t/year. However, before 2001, Nansi Lake was one of the most heavily polluted freshwater lakes in China (Pang et al. 2002). Water, sediment, and fish in the lake were moderately polluted by heavy metals (Yang et al. 2003; Shi et al. 2006). In 2001, China's South-to-North Water Transfer Project was started and Nansi Lake was one of the most important water delivery channels and storage lakes in the East Line Project (An and Li 2009). After that time, according to “Integrated Wastewater Discharge Standard in South-to-North Water Diversion Project of Shandong Province” and “Water Pollution Prevention Planning of the South-to-North Water Diversion Project (east route) of Shandong Section”, the government started to reduce the discharge standards of the industrial wastewater outfalls in the key protective zone from 1.0–0.1, 0.1–0.02, 0.05–0.005 to 0.5–0.1 mg/kg for Pb, Cd, Hg, and As respectively. Therefore, the discharges of heavy metals were reduced when the wastewater outfalls were the same. By the end of 2012, the water quality has significantly improved. However, to date, no data has been collected and analyzed regarding the current heavy metal concentrations in the lake, and particularly in fish tissues. These metals continue to be emitted in the region due to mining activities and industrial wastewater discharge leading to the potential for high toxicity in the food chain.

The main objectives of this study was to (1) summarize and compare the current and historical values of heavy metal pollution in the three representative fish species (i.e., *C. auratus, H. nobilis*, and *P. fulvidraco*) in Nansi Lake, (2) whether fish in the lake was polluted heavily by heavy metal, and (3) their risk assessment to human health. These results will help contribute towards a better understanding of the enrichment behavior of heavy metals in shallow lake ecosystems and a more systematic evaluation the outcome of heavy metal pollution control after the long-term environmental governance over the past 10 years.

Materials and Methods

The location of Nansi Lake and sampling sites (S1: 34°59′13″ N, 116°46′08″; S2, 34°56′09″ N; 116°55′28″ E; S3: 34°49′52″ N, 117°02′38″ E) are shown in Fig. 1. In this study, fish samples were caught from the Nansi Lake with the help of local fishermen, using a beach seine net (15.5 × 1.75 m) with a mesh size of 16 mm. The target species for this study were the main commercial species. In total, 136 fish were caught from the lake, all fish were 1–3 years old. The number of *C. auratus, P. fulvidraco* and *H. nobilis* were 55, 37 and 44 respectively.

The samples were held in polyethylene bags with ice, labeled and transported to the laboratory on the same day. Fish samples were kept at −20°C until analysis. Approximately 0.5 g of muscle, liver, gill or fin tissue for analysis was weighed and dried on filter paper. The moisture content of the selected tissues was determined by the loss of mass upon freeze drying. No evidence of desiccation during frozen storage was found. Throughout the preparation procedure, the samples were kept away from metallic materials to avoid contamination.

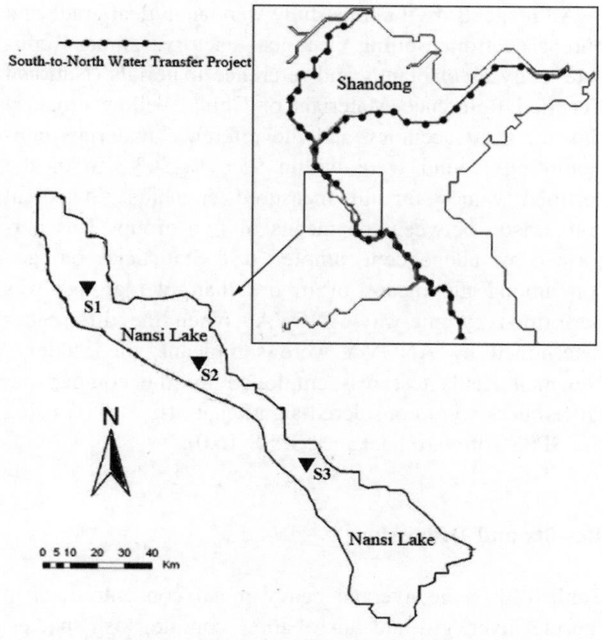

Fig. 1 The location of Nansi Lake and sample sites

Studies from field and laboratory experiments have shown that the accumulation of heavy metals in fish tissues is mainly dependent on the metal concentrations in surrounding water and sediment (Yi and Zhang 2012). Therefore, we monitored the concentrations of heavy metal in water and sediment where fishes were caught. Sediment samples were stored in polyethylene bags and placed on ice for transport to the laboratory. Coarse gravel and plant materials were removed prior to air drying and screening with a 240-mesh nylon sieve and storage in borosilicate glass bottles. Water samples were preserved by treating with nitric acid to reduce the pH < 2 in the field, placed into borosilicate glass bottles and stored in the dark.

The samples were analyzed as follows. A microwave digestion system (CEM-MDS 2000) was firstly used to prepare the fish and sediment samples for analysis, following the methods of Karadede et al. (2000). Water samples were directly analyzed. The metal analysis of Cd and Pb were carried out by using ICP-700ES (Varian, US). The absorption wavelengths were 228.8 nm for Cd and 217.0 nm for Pb, respectively. The concentration of Hg (total mercury) was analyzed by cold vapor atomic absorption spectrometry (Model F732-V, China), wave length was 253.7 nm. Concentration of As was analyzed by Atomic Fluorescence Spectrometry (AFS-930, China), wave length was 193.7 nm. The detection limits for Pb, Cd, Hg and As were 20, 2, 0.2 and 0.04 μg/L, respectively (Zhu et al. 2012; Lin and Huang 2005).

All reagents used in this study were analytical grade and purchased from Beijing Chemical Factory, China. Replicate analyses of blanks and reference materials (National Certified Reference Materials of China, yellow croaker) showed good accuracy and the reference materials concentrations found were within 95.6 %–103.3 % of the certified values for all measured elements. Statistical comparison between the means of two groups was performed by independent-sample *t* test. Statistical comparison among the means of m ore than two groups was performed by one-way ANOVA. When the differences determined by ANOVA were significant, the Student–Newman–Keuls test was employed for the comparison. Differences were considered significant at $p \leq 0.05$ using the SPSS software package (SPSS 16.0).

Results and Discussion

Table 1 lists the average heavy metal concentrations in muscle, liver, gill and fins of three common fish species. Fish species selected in this study were the same with those reported by Yang and Shi in 2001, therefore, the previous reported values and the data presented in these works were compared and discussed. The heavy metal concentrations in fish and sediment in 2001 are also shown in Table 1.

There were vast differences among the heavy metal concentrations in the muscles of different fish species. Pb, Cd and As had the highest concentration in *P. fulvidraco* muscle, *H. nobilis* accumulation most of the Hg. Significant difference in accumulation levels of Pb between *P. fulvidraco* and *H. nobilis* ($p < 0.05$). The concentrations of Cd in *P. fulvidraco* were almost 8 times of that in *C. auratus* and 2 times of that in *H. nobilis*. The concentration of As between *P. fulvidraco* and *C. auratus* also exhibited significant difference ($p < 0.05$).

It is well known that substantial bioaccumulation of heavy metals can be toxic for fish species. Studies showing that heavy metal accumulated in liver, fins and gills usually at high level and in muscles at low level (Jezierska and Witeska 2001). Results showed in Table 1 were similar to these studies.

Besides Hg, concentrations of Pb, Cd and As in 2012 in fish muscles were lower than those in 2001. The concentration of Hg in water in 2012 was 1.87 times of that in 2001($p < 0.05$). The increase of Hg level in water may result in the high concentrations of Hg in fish body in 2012. The pollution problem of Hg could be attributed to not only its discharge but also the more widespread air pollution and long-range transport of pollutants. The atmospheric deposition of Hg in watersheds are much higher (about 3–10 times higher) than the output via run-off waters (Lindqvist et al. 1991). Hg accumulation in

Table 1 Heavy metal concentrations in fishes, water and sediment in 2012 and 2001 (Yang et al. 2003; Shi et al. 2006) ($\bar{x} \pm s$)

Species (mg/kg wet weight)	Tissue	2012				2001			
		Pb	Cd	Hg	As	Pb	Cd	Hg	As
C. auratus	Muscle	0.196 ± 0.042	0.003 ± 0.002	0.231 ± 0.041	0.044 ± 0.015	0.613 ± 0.362	0.045 ± 0.023	0.158 ± 0.106	0.359 ± 0.299
	Liver	0.219 ± 0.113	0.088 ± 0.063	0.386 ± 0.053	0.114 ± 0.042	–	–	–	–
	Gill	0.163 ± 0.051	0.021 ± 0.004	0.158 ± 0.064	0.141 ± 0.053	–	–	–	–
	Fins	0.556 ± 0.211	0.042 ± 0.031	0.931 ± 0.653	0.325 ± 0.146	–	–	–	–
P. fulvidraco	Muscle	0.249 ± 0.015	0.023 ± 0.012	0.213 ± 0.064	0.083 ± 0.022	0.580 ± 0.301	0.042 ± 0.023	0.172 ± 0.151	0.292 ± 0.205
	Liver	0.321 ± 0.032	0.441 ± 0.116	0.217 ± 0.093	0.135 ± 0.043	–	–	–	–
	Gill	0.425 ± 0.143	0.307 ± 0.124	0.327 ± 0.053	0.163 ± 0.025	–	–	–	–
	Fins	0.665 ± 0.242	0.115 ± 0.034	1.037 ± 0.438	0.206 ± 0.087	–	–	–	–
H. nobilis	Muscle	0.147 ± 0.043	0.011 ± 0.004	0.243 ± 0.105	0.074 ± 0.023	0.641 ± 0.312	0.038 ± 0.029	0.132 ± 0.102	0.170 ± 0.097
	Liver	0.324 ± 0.032	0.052 ± 0.011	0.171 ± 0.063	0.142 ± 0.035	–	–	–	–
	Gill	0.435 ± 0.325	0.043 ± 0.014	0.049 ± 0.032	0.153 ± 0.064	–	–	–	–
	Fins	0.531 ± 0.215	0.032 ± 0.015	0.353 ± 0.164	0.253 ± 0.114	–	–	–	–
Water (mg/L)		0.021 ± 0.002	0.002 ± 0.001	0.001 ± 0.000	0.005 ± 0.001	0.030	0.004	0.0008	0.013
Sediment (mg/kg)		16.329 ± 4.412	0.065 ± 0.021	11.611 ± 2.452	8.632 ± 2.431	19.1	0.092	–	12.2

– No date

Table 2 Comparison of the estimated daily intake of heavy metals from fish species studied with the recommended daily dietary allowances

Year	Metal	Average concentration in fish muscle (mg/kg-wet wt.)	EDI in mg/day/person	Recommended daily dietary allowance(mg/day/person)	Maximum consumption of fish muscle(g/day)	THQ
2001	Pb	0.61	12.83×10^{-3}	0.21	343	0.28
	Cd	0.04	0.88×10^{-3}	0.06	1,428	0.07
	Hg	0.15	3.23×10^{-3}	0.03	194	0.53
	As	0.27	5.75×10^{-3}	0.13	474	0.25
2012	Pb	0.20	4.14×10^{-3}	0.21	1,065	0.15
	Cd	0.01	0.25×10^{-3}	0.06	5,000	0.04
	Hg	0.23	4.81×10^{-3}	0.03	131	0.75
	As	0.07	1.41×10^{-3}	0.13	1,940	0.06

Nansi Lake was reasonably expected at the present atmospheric deposition rate. Mercury concentrations in fish have also been negatively correlated with other water quality factors: such as alkalinity and dissolved oxygen content (ATSDR US, 1999). Although the Hg concentration in fish muscle in 2012 was higher than that in 2001, it is still below the Food and Agriculture Organization of the United Nations (FAO) limit (1 mg/kg, wet weight).This suggests that Hg contamination of Nansi Lake fishes poses minimal hazard to human health. Arsenic consumption of fish is an important source of metal accumulation in humans, there is great interest in the estimated daily intake (EDI) of heavy metals through fish consumption. The estimation of the daily intakes was recommended by Joint FAO/WHO Expert Committee on Food Additive online database (JECFA 2009). The daily intake of some selected trace metals were estimated and compared with the recommended values to assess whether or not the metal levels in the fish samples from Nansi Lake were safe for human consumption. This study was conducted only for the fish muscle as this tissue was the most important part consumed by humans. The EDI of metals for humans was determined using the following equation (Zhuang et al. 2009):

$$\mathrm{EDI} = \frac{\mathrm{C} \times \mathrm{I}}{\mathrm{W}}$$

where C is the metal concentration in fish tissue based on wet weight (mg/kg), I is the ingestion rate of fish (70 g/d), W is the body weight of adults considered to be 60 kg.

Current non-cancer risk assessment methods are typically based on the employment of the target hazard quotient (THQ), a ratio between the estimated dose of a contaminant and the reference dose below which there will not be any appreciable risk. When THQ risk is above 1, considered by the US-EPA (1989), systemic effects may occur, and it means that THQ is higher than the reference dose. As exposure is a significant environmental health concern worldwide. Lifetime cancer risk (CR) for As was obtained using the cancer slope factor (CSF), provided by EPA only for this metal. If CR risk is above the acceptable

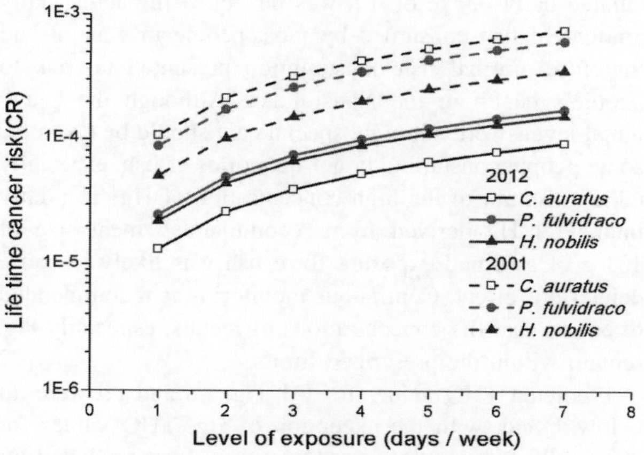

Fig. 2 Estimated lifetime cancer risk for inorganic as at different levels of consuming fishes

lifetime risk of 10^{-5}, considered by the US-EPA, and applied in this study, it indicates a probability of greater than 1 over 100,000 of an individual of developing cancer (Sirot et al. 2009).

The THQ and CR are expressed as follows (US-EPA 1989):

$$\mathrm{THQ} = \frac{\mathrm{EF} \times \mathrm{ED} \times \mathrm{FIR} \times \mathrm{C}}{\mathrm{RfD} \times \mathrm{BW} \times \mathrm{TA}} \times 10^{-3};$$

$$\mathrm{CR} = \frac{\mathrm{EF} \times \mathrm{ED} \times \mathrm{FIR} \times \mathrm{C} \times \mathrm{CSF}}{\mathrm{BW} \times \mathrm{TA}} \times 10^{-3}$$

where EF is the exposure frequency; ED is the exposure duration (70 years), equivalent to the average lifetime; FIR is the food ingestion rate (93 g/day for adults); C is the metal concentration in fish (mg/kg); BW is the average body weight (60 kg for adults), TA is the averaging exposure time (25,550 days), RfD is the oral reference dose, the RfD for Pb, Cd, Hg and As suggested by former studies and US EPA was 3.5, 1.0, 0.5 and 2.2 ug/kg/d respectively. CSF is the cancer slope factor (mg/kg/day) set by US-EPA only for inorganic As. Most arsenic is present in food in less harmful (organic) forms. In this paper, the calculation of As was conducted by assuming the inorganic As accounting for 3 % of the total concentration (Copat et al. 2013). EF, ED, FIR, BW and AT are default values provided by US-EPA (1989) for consumption limits calculation. The RfD and CSF for single contaminant were provided by EPA's Integrated Risk Information System (IRIS 2012) online database.

The EDI and THQ of heavy metals through consumption of three fish species by people are illustrated in Table 2. The dose of a toxic metal that one obtains from fish depends on not only the concentration of the specific metal in fish, but also the quantity of fish (intake) consumed. Considering normal consumption habits, the calculated daily intake of fish was far below the actual daily amount of fish consumed by most people in general and therefore, normal fish consumption presented no risk to people's health in the Nansi Lake. Although the heavy metal levels were not high, special care should be taken for some people consuming larger quantities of fish, especially taking account of the high concentration of Hg. The daily intake for Hg derived from recommended meal size of 131 g of assumed exposure form fish was likely to cause deleterious effect. Continuous monitoring is recommended to ensure that the concentrations of metals, especially Hg, remain within the prescribed limits.

Predicted THQ values for Cd, Hg, As and Pb were all below 1 and, with the exception of Hg, THQ values for 2012 collected samples were lower than those predicted for the 2001 data (Fig. 2). This means that there is no risk for developing chronic systemic effects, due to the intake of the above investigated metals. However, the pollution of Hg has become more serious. The THQ of heavy metals from fish muscle consumption follows the order Hg > Pb > As > Cd. THQ values of Pb, Cd, and As are all reduced in 2012 than in 2001.

It can be seen from Fig. 2 that CR was above the acceptable lifetime risk assumed for this study (i.e., 10^{-5}) for all levels of assumed exposure for 3 species of fish in both 2001 and 2012. The CR assessment for inorganic arsenic was critic because all of calculations were above 10^{-5} even from the minimum stage of exposure supposed (i.e. 1 meal/week in 2012). It was, therefore, assumed that a risk for cancer would be greater than the acceptable lifetime risk of 10^{-5}. The CR values predicted for the three kinds of fish were lower for 2012 collected samples relative to those collected in 2001 indicating that remediation activities have reduced the risk of cancer associated with consuming contaminated fish from Nansi Lake.

The Metal Pollution Index (MPI) was used to compare the total metal accumulation level in various tissues of different fish. The MPI values were calculated using the equation as described by Usero et al. (1997):

Table 3 MPI in different tissues and fishes in 2001 and 2012

	Year	MPI			
		Muscle	Liver	Gill	Fins
C.auratus	2012	0.059	0.127	0.093	0.289
	2001	0.198	–	–	–
P.fulvidraco	2012	0.073	0.109	0.227	0.214
	2001	0.187	–	–	–
H.nobilis	2012	0.046	0.198	0.148	0.232
	2001	0.152	–	–	–

– No date

$$\mathrm{MPI} = (\mathrm{Cf1} \times \mathrm{Cf2} \times \cdots \times \mathrm{Cfn})^{1/n}$$

where C_{fn} is the content of metal n in the sample.

The MPI values determined for 2012 fish muscle samples were significantly ($p < 0.05$) lower than those determined for the 2001 samples. This means that the tendency of heavy metal pollution in 2012 was decreased compared with 2001 (Table 3). A large number of studies (Bank et al., 2007; Mohammadnabizadeh et al., 2013) have shown that the bioaccumulation of heavy metal in fish muscle is significantly correlated with fish species. The results observed in this study were in good agreement with the above consensus. Bioaccumulation was prone to be strongest in carnivorous species (*P.fulvidraco*), followed by omnivorous (*C.auratus*) and filter-feeding (*H.nobilis*) species, and it tended to be stronger in bottom-living fish than that in pelagic fish.

In conclusion, significant differences were identified among muscle, liver, fins, and gill of the species in view of the accumulation of the selected elements. The concentrations of Pb, As, and Cd in water, sediments, and the tissues of fish species investigated in this study turned out to be lower than samples in 2001. However, regular monitoring of heavy metal concentrations still should be conducted in the future, as the level of Hg concentration was higher than the previous value. Concentrations of heavy metals such as Cd, Pb, As, and Hg do not represent a serious problem for human health. In addition, feeding habit was found to affect heavy metal concentrations in fish. The results of this study demonstrate that the remediation efforts to reduce metal contamination of Nansi Lake have reduced the bioaccumulation of heavy metals in fish species from this lake and also the potential health hazards associated with their consumption.

Acknowledgments This work was supported by the Independent Innovation Foundation of Shandong University (No. 2012JC029), Natural Science Foundation for Distinguished Young Scholars of Shandong Province (No. JQ201216), Shandong Provincical Natural Science Foundation, China (2009ZRB019Y9), Independent Innovation Foundation of Shandong University (2014JC023).

References

An WC, Li XM (2009) Phosphate adsorption characteristics at the sediment-water interface and phosphorus fractions in Nansi Lake, China, and its main inflow rivers. Environ Monit Assess 148:173–184

Bank MS, Chesney E, Shine JP, Maage A, Senn DB (2007) Mercury bioaccumulation and trophic transfer in sympatric snapper species from the Gulf of Mexico. Ecol Appl 17:2100–2110

Copat C, Arena G, Fiore M, Ledda C, Fallico R, Sciacca S, Ferrante M (2013) Heavy metals concentrations in fish and shellfish from eastern Mediterranean Sea: consumption advisories. Food Chem Toxicol 53:33–37

Driscoll CT, Yan C, Schofield CL, Munson R, Holsapple J (1994) The mercury cycle and fish in the Adirondack lakes. Environ Sci Technol 28:136A–143A

Farag AM, Woodward DF, Goldstein JN, Brumbaugh W, Meyer JS (1998) Concentrations of metals associated with mining waste in sediments, biofilm, benthic macroinvertebrates, and fish from the Coeur d'Alene River Basin, Idaho. Arch Environ Contam Toxicol 34:119–127

Food and Agriculture Organization of the United Nations (2010) Report on the Joint FAO/WHO expert consultation on the risks and benefits of fish consumption. FAO fisheries and aquaculture Report No. 978

IRIS (2012) Integrated risk information system. http://cfpub.epa.gov/ncea/iris/index.cfm?fuseaction=iris.showSubstanceList

JECFA (2009) Evaluations of the Joint FAO/WHO expert committee on food additives. Available at: <http://apps.who.int/ipsc/database/evaluations/search.aspx>

Jezierska B, Witeska M (2001) Metal toxicity to fish. Monografie. University of Podlasie, Poland

Karadede H, Ünlü E (2000) Concentrations of some heavy metals in water, sediment and fish species from the Atatürk Dam Lake (Euphrates), Turkey. Chemosphere 41:1371–1376

Korkmaz Görür F, Keser R, Akçay N, Dizman S (2012) Radioactivity and heavy metal concentrations of some commercial fish species consumed in the Black Sea Region of Turkey. Chemosphere 87:356–361

Lin j, Huang L (2005) Simultaneous determination of Arsenicand mercury in food stuffs by Microwave Digestion-Hydride. Nat Sci J of Hainan univ 23:117–121

Lindqvist O, Johansson K, Bringmark L, Timm B, Aastrup M, Andersson A, Hovsenius G, Håkanson L, Iverfeldt Å, Meili M (1991) Mercury in the Swedish environment—recent research on causes, consequences and corrective methods. Water Air Soil Poll 55:xi–261

Mohammadnabizadeh S, Afshari R, Pourkhabbaz A (2013) Metal concentrations in marine fishes collected from Hara Biosphere in Iran. Bull Environ Contam Toxicol 90:188–193

Pang Y, Long TY, Chen F, Gao x (2002) Water pollution investigation in Shandong province along the east line of south to north water transfer project. Water and Wast Eng 28(8):19–21

Petrović Z, Teodorović V, Dimitrijević M, Borozan S, Beuković M, Milićević D (2013) Environmental Cd and Zn concentrations in liver and kidney of european hare from different serbian regions: age and tissue differences. Bull Environ Contam Toxicol 90:203–207

Shi K, Liu J, Ma L (2006) Investigation and evaluation of lead, cadmium, arsenic, mercury in aquatic products of Nansihu. Food and Drug 7:020

Sirot V, Guérin T, Volatier J, Leblanc J (2009) Dietary exposure and biomarkers of arsenic in consumers of fish and shellfish from France. Sci Total Environ 407:1875–1885

US-EPA (1989) Risk assessment guidance for superfund, vol. I. Human health evaluation manual (Part A), interim final. EPA 540/1–89/002. United States Environmental Protection Agency, Washington

Usero J, González-Regalado E, Gracia I (1997) Trace metals in the bivalve molluscs *Ruditapes decussatus* and *Ruditapes philippinarum* from the atlantic coast of Southern Spain. Environ Int 23:291–298

Yang L, Shen J, Zhang Z, Zhu Y, Sun Q (2003) Multivariate analysis of heavy metal and nutrient in surface sediments of Nansihu Lake. China Environ Sci 23:206–209

Yi YJ, Zhang SH (2012) The relationships between fish heavy metal concentrations and fish size in the upper and middle reach of Yangtze River. Procedia Environ Sci 13:1699–1707

Zhang LX, Ulgiati S, Yang ZF, Chen B (2011) Emergy evaluation and economic analysis of three wetland fish farming systems in Nansi Lake area, China. J Environ Manage 92:683–694

Zhu H, Yan BX, Cao HC, Wang LX (2012) Risk assessment for methylmercury in fish from the songhua river, china: 30 years after mercury-containing wastewater outfalls were eliminated. Environ Monit Assess 184(1):77–88

Zhuang P, McBride MB, Xia H, Li N, Li Z (2009) Health risk from heavy metals via consumption of food crops in the vicinity of Dabaoshan mine, South China. Sci Total Environ 407:1551–1561

王文兴培养的研究生及其学位论文目录

一、博士毕业生及其学位论文

序号	毕业年份	姓名	单位	学位论文题目
1	2005	李英霞	北京化工大学	用于烷基化反应的绿色催化剂的制备表征和性能研究
2	2006	童　莉	北京化工大学	生态工业园区产业链设计及其系统稳定性研究——以烟台、乌鲁木齐为例
3	2007	岳钦艳	山东大学	阳离子型有机高分子水处理剂—聚环氧氯丙烷胺的研究
4	2008	范立维	北京化工大学	潜流人工湿地水力学特性及其处理废水中有机污染物的研究
5	2008	高　健	山东大学	大气颗粒物个数浓度、粒径分布及颗粒物生成——成长过程研究
6	2008	杨凌霄	山东大学	济南市大气 $PM_{2.5}$ 污染特征、来源解析及其对能见度的影响
7	2009	屈小辉	山东大学	量子化学方法研究典型有毒有机污染物的形成与降解机理
8	2010	韩晓丽	北京化工大学	BAF 强化人工湿地工艺处理生活污水试验研究
9	2010	刘晓环	山东大学	我国典型地区大气污染特征的数值模拟
10	2011	王　哲	山东大学	中国典型地区碳质气溶胶及二次有机气溶胶特征研究
11	2011	许鹏举	山东大学	济南城区大气颗粒物数浓度及粒径分布特征研究
12	2011	薛丽坤	山东大学	中国地区低对流层高层大气化学与长距离输送特征研究
13	2012	高晓梅	山东大学	我国典型地区大气 $PM_{2.5}$ 水溶性离子的理化特征及来源解析
14	2012	聂　玮	山东大学	我国典型地区大气颗粒物测量技术、粒径分布及长期变化趋势
15	2012	周　杨	山东大学	华北地区气溶胶理化特性、来源解析及实验室模拟
16	2013	孙晓艳	山东大学	典型含氧挥发性有机物和全氟磺酰胺的大气降解机理的理论研究
17	2013	王新锋	山东大学	我国典型城市与高山的大气气溶胶质量粒径分布特征及 N_2O_5 化学行为
18	2013	周东凯	北京化工大学	新型纤维挂膜填料污水处理研究
19	2014	封　红	北京化工大学	2009~2014 序批式生物膜法处理水产养殖废水研究
20	2014	高　锐	山东大学	典型氯代二噁英和硝基多环芳烃的形成机理研究
21	2014	张晨曦	山东大学	大气中典型二噁英类物质的氧化降解机理研究
22	2014	周声圳	山东大学	我国典型城市和高山地区碳质气溶胶及单颗粒混合状态研究
23	2015	徐　政	山东大学	华南典型地区大气反应性氮氧化物的污染特征、来源及大气转化过程的研究
24	2015	袁　琦	山东大学	黄河三角洲地区大气颗粒物理化特性研究
25	2016	曹海杰	山东大学	典型溴系阻燃剂降解机理的量子化学及分子模拟研究
26	2016	李鹏飞	山东大学	底栖类动物生态湿地的构建和相关运行机制研究
27	2016	姚　兰	山东大学	山东典型地区 $PM_{2.5}$ 化学成分、来源及二次生成

二、硕士毕业生及其学位论文

序号	毕业年份	姓名	单位	学位论文题目
1	1990	李金花	中国环境科学研究院	用室外烟雾箱研究燃煤飞灰上多环芳烃光解反应
2	1988	梁金友	中国环境科学研究院	酸沉降区域源解析模式的建立和应用
3	1989	束勇辉	中国环境科学研究院	燃煤烟气颗粒物上多环芳烃降解烟雾箱研究（室内烟雾箱）
4	1990	谢　英	中国环境科学研究院	甲烷在大气中光化学反应机理研究（室内烟雾箱）
5	1990	徐卫国	中国环境科学研究院	酸雨形成的大气化学过程
6	1991	石　全	中国环境科学研究院	绝对因子得分受体模式的开发和应用
7	1995	吕晓红	中国环境科学研究院	我国东部沿海地区酸沉降及其来源研究
8	2004	郭　靖	北京化工大学	酸沉降对材料破坏的经济损失估算
9	2008	王玉东	山东大学	臭氧引发的萜烯化合物大气反应机理的理论研究
10	2010	寿幼平	山东大学	济南大气 $PM_{2.5}$ 中无机离子在线研究
11	2010	王　静	山东大学	北京昌平跨奥运气溶胶无机水溶性离子的污染特征及其来源解析
12	2010	张　建	山东大学	邻苯二甲酸酯类污染物的微生物去除作用及物理吸附研究
13	2011	柏学凯	山东大学	城市河道底泥污染分析与抑制方法研究
14	2011	马　强	山东大学	新型生态浮岛设计、应用效果及微生物机理研究
15	2011	王琳琳	山东大学	北京大气污染特征研究
16	2011	于阳春	山东大学	济南市大气颗粒物中水溶性无机离子的粒径分布研究
17	2012	董　灿	山东大学	济南市公共场所室内环境中 $PM_{2.5}$ 及无机水溶性离子污染特征研究
18	2012	高明瑜	山东大学	生物岛栅对污染河水的长期净化效果及微生物机理研究
19	2012	鄢　超	山东大学	济南市室内环境大气颗粒物数浓度及粒径分布特征研究
20	2012	袁　超	山东大学	香港 $PM_{2.5}$ 水溶性离子的在线监测仪器评估及理化特征分析
21	2013	孟川平	山东大学	室内环境大气细颗粒物 $PM_{2.5}$ 中多环芳烃（PAHs）污染组成及其粒径分布特征研究
22	2014	刘　健	山东大学	胜利油田采油区土壤石油污染状况及其微生物群落结构
23	2014	芦亚玲	山东大学	区域背景采样点大气气溶胶单颗粒的理化特性研究
24	2014	杨　飞	山东大学	济南市大气颗粒物 $PM_{2.5}$ 中多环芳烃（PAHs）的污染特征及来源解析
25	2015	文　亮	山东大学	山东地区灰霾期间硝酸盐生产机制研究
26	2016	王丽玮	山东大学	华北典型地区大气亚硝酸与硝基苯酚类化合物的来源及转化
27	2016	郑龙飞	山东大学	南京地区细颗粒物污染特征及灰霾事件成因研究

三、在读研究生

序号	年级	类别	姓名	单位	序号	年级	类别	姓名	单位
1	2014	博士研究生	朱艳红	山东大学	10	2015	硕士研究生	陈天舒	山东大学
2	2015	博士研究生	杨　雪	山东大学	11	2015	硕士研究生	马晓辉	山东大学
3	2015	博士研究生	文　亮	山东大学	12	2015	硕士研究生	张亚婷	山东大学
4	2015	博士研究生	孙　雷	山东大学	13	2015	硕士研究生	姜　盼	山东大学
5	2015	博士研究生	甘延东	山东大学	14	2015	硕士研究生	于　川	山东大学
6	2016	博士研究生	赵　彤	山东大学	15	2015	硕士研究生	纵瑞涵	山东大学
7	2016	博士研究生	朱　豹	山东大学	16	2016	硕士研究生	蒋　莹	山东大学
8	2014	硕士研究生	王　浩	山东大学	17	2016	硕士研究生	雒园园	山东大学
9	2014	硕士研究生	张众志	山东大学	18	2016	硕士研究生	张　鑫	山东大学

四、指导的博士后

序号	入站年份	姓名	单位	序号	入站年份	姓名	单位
1	2004	张庆竹	山东大学	7	2008	解伏菊	山东大学
2	2005	周学华	山东大学	8	2009	李卫军	山东大学
3	2005	孙孝敏	山东大学	9	2013	徐　菲	山东大学
4	2006	孙瑞莲	山东大学	10	2014	李延伟	山东大学
5	2008	何茂霞	山东大学	11	2015	孙延慧	山东大学
6	2008	谢慧君	山东大学	12	2016	王　蕙	山东大学

后　　记

王文兴院士1952年毕业于山东大学（青岛）化学系，1955年在吉林大学化学系研究生班进修，1959年赴苏联卡尔波夫物理化学研究所进修。大学毕业后至70年代中期，先后任重工业部沈阳化工研究所工程师、化工部北京化工研究院研究室主任、天津化工研究院副院长。早期主要从事烃类催化氧化反应机理与动力学研究，编著了我国第一本《工业催化》（1978）。当时国家工业研究机构中的研究内容多与工业技术机密有关，除申请专利外，传统上不鼓励发表论文，主要看实际贡献。

1980年受邀参与筹建中国环境科学研究院，先后担任中国环境科学研究院副院长、学术委员会主任、学术顾问。连续承担国家“六五”至“九五”科技攻关项目，在大气光化学污染规律和防治、煤烟型大气污染与控制、大气环境容量、酸沉降化学与污染控制等方面，组织开展了大量的现场观测和实验室模拟工作。在中国环境研究院期间，他（含与同事合作）发表了100余篇学术论文，2008年出版的《王文兴文集》收录了从中挑选出来的代表性论文。

2002年回到母校山东大学，被聘为教授、博士生导师。2003年筹建了山东大学环境研究院，担任院长。在这里，他教书育人，指导科研。系统开展了区域性$PM_{2.5}$和雾霾形成机制研究，创建了环境量子化学计算团队和研究新领域。培养了21名博士毕业生和19名硕士毕业生，合作培养了十几名研究生，指导了十多名博士后，为我国环境学科和环境保护事业培养了优秀人才。过去十年，王文兴院士科研成果卓著，他和他的研究生发表了近200篇学术论文。本次编辑出版的《王文兴文集》（第二卷），收录了从中遴选出来的54篇代表性论文，是2008年出版的《王文兴文集》的续集。

《王文兴文集》（第二卷）编辑出版之际，欣逢王文兴院士九十华诞。在此，我们谨向这位著名的环境化学家表示衷心的祝贺，并送上温馨的祝福，祝愿他健康长寿。

《王文兴文集》（第二卷）的编辑出版，得到中华人民共和国环境保护部、中国工程院、中国环境科学研究院和山东大学等单位领导同志的大力支持和帮助。科学出版社的朱丽等同志组织开展了本书的编辑和出版工作。在此，谨向他们表示衷心的感谢。

本文集的资料收集和编辑工作是在短时间内完成的，疏漏和不妥之处在所难免，欢迎各位专家和读者朋友们批评指正。

《王文兴文集》编辑组

2016年12月